ECOTOXICOLOGY

A Comprehensive Treatment

ECOTOXICOLOGY

A Comprehensive Treatment

Michael C. Newman
William H. Clements

CRC Press is an imprint of the
Taylor & Francis Group, an **informa** business

CRC Press
Taylor & Francis Group
6000 Broken Sound Parkway NW, Suite 300
Boca Raton, FL 33487-2742

Printed in the United States of America on acid-free paper
10 9 8 7 6 5 4 3 2 1

International Standard Book Number-13: 978-0-8493-3357-6 (Hardcover)

Library of Congress Cataloging-in-Publication Data

Newman, Michael C.
Ecotoxicology : a comprehensive treatment / Michael C. Newman and William H. Clements.
p. ; cm.
"A CRC title."
Includes bibliographical references and index.
ISBN 978-0-8493-3357-6 (alk. paper)
1. Toxicology--Environmental aspects. I. Clements, William H. (William Henry), 1954- II. Title.
[DNLM: 1. Environmental Exposure--adverse effects. 2. Environmental Pollutants--toxicity. 3. Ecosystem. 4. Population Dynamics. WA 671 E196 2008]

RA1226.N48 2008
615.9'02--dc22 2007017797

Visit the Taylor & Francis Web site at
http://www.taylorandfrancis.com

and the CRC Press Web site at
http://www.crcpress.com

Dedication

To Peg, Ben, and Ian (MCN)
To Diana for her endless support over the years (WHC)

Do that which will render thee worthy of happiness

Critique of Pure Reason, I. Kant 1781

Contents

Preface

"What is the use of a book," thought Alice, "without pictures or conversations?"

Lewis Carroll (*Alice in Wonderland*)

Ecotoxicology: A Comprehensive Treatment is intended to bridge a widening gap between ecotoxicology textbooks and technical books focused on specific ecotoxicological topics. Important, narrowly focused books abound, and textbooks appear yearly but are often broad-brush treatments of the field of ecotoxicology. This treatment represents a synthesis needed to provide the student with an understanding beyond that afforded by a general textbook but, unlike that from more specialized books, remains focused on paradigms and fundamental themes. Designed to be flexible enough to meet the variety of instructional vantages, subsets of chapters may be used while de-emphasizing others. (See Table 1 for two possible chapter groupings.) Regardless of an instructor's vantage and

TABLE 1
Two Illustrations of Chapter Selections That Might Be Chosen in a 3-Credit Hour Ecotoxicology Course (The Entire Book Could Be Covered in a 4-Credit Hour Course)

Hypothetical Course 1	
Chapter 1	The Hierarchical Science of Ecotoxicology
Chapter 2	The Organismal Ecotoxicology Context
Chapter 3	Biochemistry of Toxicants
Chapter 4	Cells and Tissues
Chapter 5	Organs and Organ Systems
Chapter 6	Physiology
Chapter 7	Bioaccumulation
Chapter 8	Models of Bioaccumulation and Bioavailability
Chapter 9	Lethal Effects
Chapter 10	Sublethal Effects
Chapter 12	The Population Ecotoxicology Context
Chapter 14	Toxicants and Simple Population Models
Chapter 15	Toxicants and Population Demographics
Chapter 17	Population Genetics: Damage and Stochastic Dynamics of the Germ Line
Chapter 18	Population Genetics: Natural Selection
Chapter 20	Introduction to Community Ecotoxicology
Chapter 22	Biomonitoring and the Responses of Communities to Contaminants
Chapter 25	Disturbance Ecology and the Responses of Communities to Contaminants
Chapter 29	Introduction to Ecosystem Ecology and Ecotoxicology
Chapter 34	Fate and Transport of Contaminants in Ecosystems
Chapter 36	Conclusion

Continued

TABLE 1
Continued

Hypothetical Course 2	
Chapter 1	The Hierarchical Science of Ecotoxicology
Chapter 2	The Organismal Ecotoxicology Context
Chapter 7	Bioaccumulation
Chapter 8	Models of Bioaccumulation and Bioavailability
Chapter 9	Lethal Effects
Chapter 10	Sublethal Effects
Chapter 12	The Population Ecotoxicology Context
Chapter 14	Toxicants and Simple Population Models
Chapter 15	Toxicants and Population Demographics
Chapter 17	Population Genetics: Damage and Stochastic Dynamics of the Germ Line
Chapter 18	Population Genetics: Natural Selection
Chapter 20	Introduction to Community Ecotoxicology
Chapter 21	Biotic and Abiotic Factors That Regulate Communities
Chapter 22	Biomonitoring and the Responses of Communities to Contaminants
Chapter 23	Experimental Approaches in Community Ecology and Ecotoxicology
Chapter 25	Disturbance Ecology and the Responses of Communities to Contaminants
Chapter 29	Introduction to Ecosystem Ecology and Ecotoxicology
Chapter 31	Descriptive Approaches for Assessing Ecosystem Responses to Contaminants
Chapter 32	The Use of Microcosms, Mesocosms, and Field Experiments to Assess Ecosystem Responses to Contaminants and Other Stressors
Chapter 34	Fate and Transport of Contaminants in Ecosystems
Chapter 36	Conclusion

objectives, chapters can be selected so that students come away from any course with a respect for the importance of understanding and integrating concepts from all levels of biological organization.

Some chapters include materials that are not conventionally covered in ecotoxicology textbooks and courses (e.g., Chapters 30, 33, and 35). We disagree with any suggestion that discussion include only conventional mammalian poisons in the environment that could do harm to other biota. Any ecotoxicant,[1] even nitrogen or CO_2, is relevant if present in sufficient amounts to perturb ecological entities. Because most biology students are more familiar with principles of biochemical to organismal biology than ecological principles relevant to the second half of the book, the authors dedicated more pages in later chapters discussing relevant foundation principles.

Ecotoxicology: A Comprehensive Treatment has separate sections for the ecotoxicology of individuals, populations, communities, and ecosystems. The ecosystem section encompasses issues ranging from the conventional, discrete ecosystem (e.g., a lake) to the entire biosphere. Although these topics are treated separately, the authors' intent is to integrate ecotoxicological concepts across these hierarchical levels.

[1] Restricting discussion to the conventional context of a toxicant or poison is difficult to justify when dealing with populations, communities, and ecosystems because it is not necessary for an agent to directly interact with an individual to harm it. An agent might eliminate a pollinator species, reducing reproductive success of a plant, which depends on that pollinator. Ecotoxicant is the term applied here to ensure that agents causing indirect effects or acting as important co-stressors are also considered in addition to those causing direct effects.

The Authors

Michael C. Newman, Ph.D., is a professor at the College of William and Mary's School of Marine Science. Previously, he was a faculty member at the University of Georgia's Savannah River Ecology Laboratory. His research interests include quantitative ecotoxicology, environmental statistics, risk assessment, population effects of contaminants, metal chemistry, and bioaccumulation modeling. He has written more than 100 articles and 5 books and edited another 5 books on these topics. He has taught at universities throughout the world including the University of California–San Diego, University of South Carolina, University of Georgia, College of William and Mary, Jagiellonian University (Poland), University of Antwerp (Belgium), University of Joensuu (Finland), University of Technology—Sydney (Australia), University of Hong Kong, and Royal Holloway University of London (U.K.). He served as Dean of Graduate Studies at the College of William and Mary from 1999 to 2002.

Dr. Newman has served numerous international, national, and regional organizations including the OECD, the U.S. EPA Science Advisory Board, and the U.S. National Academy of Science NRC. In 2004, the Society of Environmental Toxicology and Chemistry awarded him its Founders Award, "the highest SETAC award, given to a person with an outstanding career who has made a clearly identifiable contribution in the environmental sciences."

William H. Clements, Ph.D., is a professor in the Department of Fish, Wildlife, and Conservation Biology and a faculty advisor in the Graduate Degree Program in Ecology at Colorado State University. He teaches graduate and undergraduate courses in ecology, experimental design, and ecotoxicology. Dr. Clements received his M.S. from Florida State University, where he studied ecology of marine seagrass communities. He received his Ph.D. from Virginia Polytechnic Institute and State University, where he examined community responses to heavy metal pollution. Dr. Clements' research interests focus primarily on community and ecosystem responses to contaminants. He is especially interested in questions that address responses to multiple perturbations and interactions between contaminants and global climate change. He has published more than 75 peer-reviewed papers and book chapters in the fields of stream ecology and aquatic ecotoxicology.

Dr. Clements is an associate editor for the *Journal of the North American Benthological Society*. He has served on numerous professional and national committees, including the National Academy of Science NRC, the Department of Interior Federal Advisory Committee, and the Board of Directors for the Society of Environmental Toxicology and Chemistry.

Part I

Hierarchical Ecotoxicology

1 The Hierarchical Science of Ecotoxicology

> I have argued that there is intrinsically only one class of explanation. It traverses the scales of space, time, and complexity to unite the disparate facts of the disciplines by consilience, the perception of a seamless web of cause and effect.
>
> **(E.O. Wilson 1998)**

1.1 AN OVERARCHING CONTEXT OF HIERARCHICAL ECOTOXICOLOGY

1.1.1 GENERAL

Ecotoxicology is the science of contaminants in the biosphere and their effects on constituents of the biosphere, including humans (Newman and Unger 2003). The scope of ecotoxicology is so necessarily encompassing that an ecotoxicologist can study fate or effect of toxicants from the molecular to the biospheric scales. Consequently, it is necessary at the onset to give the context and philosophical vantage from which ecotoxicology will be explored here. That is the goal of this brief chapter.

In the opening quote of this section, Wilson articulates a central theme of this book, that is, to explore in detail one level of biological organization at a time with a progressive linking of facts and paradigms among levels. Putting this goal in terms of conceptual systems theory, the complementary processes of differentiation and integration will be addressed. Differentiation is the pulling together of a large number of diverse facts and concepts, while maintaining discrimination among these facts and concepts (Karlins 1973). The book tries to bring the reader's capacity for differentiation beyond that provided in most introductory textbooks. It also has the sister goal of integration, the gradual interrelating of concepts and information from different levels of biological organization into a coherent whole.

These twin goals allow in-depth exploration of paradigms and important approaches at each level while emphasizing as much as possible the complementary nature of information extracted from proximate levels. Too often, studies at different levels of biological organization are perceived as producing incompatible insights, or worse, as being competing systems of explanation (Maciorowski 1988). Conclusions from one level may be judged deficient based on criteria relevant to another. Such self-inflicted nearsightedness in a field eventually assures that information produced in the context of one level cannot be translated to that of another without great difficulty and compromise. This conceptual myopia is a common, yet correctable, condition associated with the study of complex topics.

> As a question becomes more complicated and involved, and extends to a greater number of relations, disagreement of opinion will always be multiplied, not because we are irrational, but because we are finite beings, furnished with different kinds of knowledge, exerting different degrees of attention, one discovering consequences which escape another, none taking in the whole concatenation of causes and effects, and most comprehending but a very small part; each referring it to a different purpose.

> Where, then, is the wonder, that they, who see only a small part, should judge erroneously of the whole? Or that they, who see different and dissimilar parts, should judge differently from each other?
>
> **(Samuel Johnson 1753)**

Our fundamental goal is to foster understanding by presenting sound information, organizing it around explanations (paradigms) emerging at each level, and then integrating information and paradigms into a consonant whole. This goal is designed to partially satisfy that of any science—to organize knowledge around explanatory principles (Nagel 1961). But there is additional motivation to do this for our emerging science. For ecotoxicology, the knowledge to be organized emerges from the molecular to biospheric scales. We are convinced that, to be maximally useful, the organization of knowledge in ecotoxicology, as in any science, must be as congruent as possible among hierarchical levels of biological organization. While this may seem obvious, it has been a difficult goal to obtain in practice. Only in the last decade has there been perceptible movement to that end.

1.1.2 The Modified Janus Context

How can organization and integration of ecotoxicological knowledge be achieved? A straightforward approach will be applied to foster congruity in this hierarchical science. The first conceptual component of this approach was described by Koestler (1991) who defined the Janus context for hierarchical, biological systems.[1] Like Janus, any level of the biological organization presents two faces simultaneously to the scientist. It is a whole composed of parts while also being a part of a larger whole. Such a unit or subassembly within a hierarchical system was called a holon by Koestler. We assume in this book that no hierarchical level—holon—has favored status relative to causal or conceptual consonance.

Koestler's Janus context is modified slightly here to include a simple, but awkwardly phrased, "unfixed cause–effect–significance concatenation" scheme. The unfixed concatenation scheme is common in most thoughtful treatments of hierarchical topics in the natural sciences (Figure 1.1). Simply stated, mechanisms explaining an effect at a particular level are often sought at the level(s) immediately below it, while the significance of the effect is sought at the next higher level(s). For example, the observation of reproductive failure of individuals exposed to a pollutant—an observation at the organismal level—is explained by changes in endocrine function—a causal mechanism at the organ system level. The consequence or significance is sought at the population level, that is, possible risk of local population extinction in polluted areas. This concatenation of cause–effect–significance applies equally well to any level. This is the reason for describing it as unfixed. Ideally, this unfixed concatenation permits a smooth conceptual transition from one level to another without experiencing what science historian Thomas S. Kuhn describes as an abrupt conceptual gestalt switch (see Watkins 1970). In this case, the abrupt gestalt shift would involve an incongruous transition from one set to another of level-specific paradigms or explanations. As science philosopher Sir Karl R. Popper (1959) argued regarding such self-contradictory groupings of paradigms, any scientific system of causal explanations not affording a smooth transition from one level to another will be uninformative in the short term and nonviable in the long term. Without congruity, the science of ecotoxicology will eventually splinter into smaller subdisciplines. The inconsistencies would distract practitioners from formulating and establishing new and useful paradigms unique to the science. Odum (1996) addresses this concern:

> If every possible interface, and there are hundreds of possibilities, should become a new discipline with separate societies, journals, college departments, and degrees then science would become so fragmented as to not only completely bewilder the public, but also make any kind of political action even more difficult than it already is.

---

[1] Janus was a Roman deity, who was often carved at doorways with two faces looking in opposite directions.

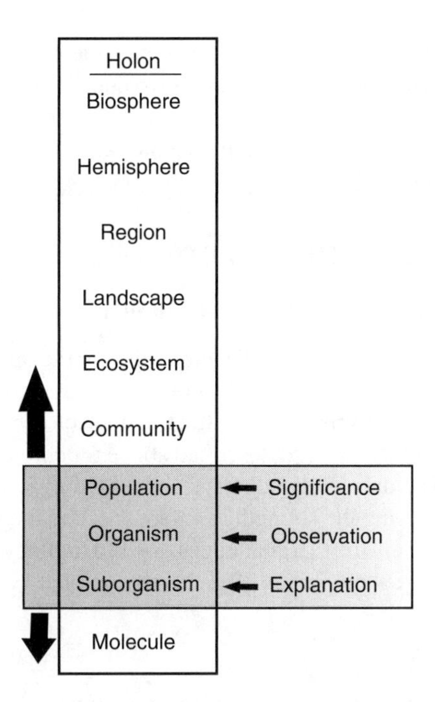

FIGURE 1.1 The "unfixed cause–effect–significance concatenation" scheme often applied to hierarchical subjects. The concatenation of explanation–observation–significance for adjacent holons can be applied anywhere within the hierarchical structure.

There are very real dangers associated with not bringing the diverse components of this emerging field together into a congruent whole that has unique emergent paradigms and theories. With this in mind, we attempt here to meld the unfixed concatenation theme with the Janus context to foster congruity of explanation in this book.

> There is a translation problem at the core of ecotoxicology: how to translate mechanisms at one level into effects of another. This problem is not unique to ecotoxicology, but arises in studies of any hierarchical system. In such systems, processes at one level take their mechanisms from the level below and find their consequences at the level above. The organismal physiology that is "mechanistic detail" to a population biologist is the pattern that the physiologist wants to explain and a higher-level integration in the eyes of a biochemist. Recognizing this principle makes it clear that there are no truly "fundamental" explanations, and makes it possible to move smoothly up and down the levels of a hierarchical system without falling into the traps of naive reductionism or pseudo-scientific holism.
>
> **(Caswell 1996)**

An essential precautionary note must follow this explanation of the unfixed concatenation scheme. The framework just described does not restrict ecotoxicologists to framing cause–effect contexts that only proceed upward through levels. Sometimes, the opposite context is more appropriate to understanding ecotoxicological phenomena. Take as an example the consequences of applying herbicide to control forbs and shrub-grass on a range land (Johnson and Johnson 1969). Herbicide application shifts plant community composition and the resulting change in habitat quality produces territorial dominance-structure shifts in an endemic rodent population, resulting in lowered population viability. Notice that this causal cascade was initiated by a change at the community level. Effects occurred at the population level with eventual consequences of diminished rodent population size. The cause–effect–significance sequence progressed down the hierarchical scheme in this case. The potential framing of such sequences is essential in ecotoxicology: the cause–effect–significance sequence should be applied from a "bottom-up" or "top-down" context as is most appropriate for a situation. The most obvious instances in this book where a top-down context would be appropriate include discussions of endocrinology and bioenergetics (Chapter 6), stress and the general adaptation syndrome (Chapter 9), and community top-down regulation (Chapter 23).

1.2 REDUCTIONISM VERSUS HOLISM DEBATE

1.2.1 Reductionism versus Holism as a False Dichotomy

Merging aspects of mammalian toxicology and ecology during the initial framing of ecotoxicology generated a protracted debate about whether reductionism or holism was the best approach. In our opinion, few topics have been more grievously distracting in ecotoxicology than this one. The associated debate is inconsistent with the context described above for deriving and organizing knowledge from hierarchical subjects. The debate also distracts students from more important topics such as methods for judging and improving the strength of inferences about ecotoxicological issues. As we will see in the following discussion, the real challenges in ecotoxicology are inference (1) without sufficient information about emergent properties, (2) without complete causal knowledge, and (3) with aggregated information.

1.2.2 Microexplanation, Holism, and Macroexplanation

Review of science philosophy provides a broad context for the reductionism/holism debate. What are usually described as qualities of reductionism are those of microexplanation, where "the properties and powers of individual things and of materials [are] due to their fine structure, that is due to the dispositions and interactions of their parts" (Harré 1972). Microexplanation emerged from the classic corpuscularian concept of structure and holds that "the global properties of individuals become functions of the properties of their parts" (Harré 1972). Within the context described above, this mode of exploring causation is profitable; however, unique properties emerge at higher levels that are very difficult or impossible to predict by microexplanation alone. Conceding the unpredictability and importance of emergent properties in ecotoxicological systems, holism aims principally at description of consistent details and predictable behaviors of the whole without becoming hopelessly entangled in a complex web of microexplanation. So, the holistic mode of generating information is also very profitable in the context just described. Finally, a third mode, macroexplanation, is used to explain "the nature and structure of the parts of an individual thing in terms of the characteristics of the whole thing" (Harré 1972). Inferences are made about the behavior of parts from the behavior of the whole, that is, from aggregate data. Macroexplanation is also valuable but has as significant challenges (i.e., the problem of ecological inference) in its application as do holism and microexplanation.

Three, not two, explanatory vantages exist: microexplanation (reductionism), holism, and macroexplanation. In the modified Janus context, they are equally appropriate vantages of explanation and inquiry applicable to the different "faces" of any holon. It is inconceivable that exclusion of any one of the three would foster growth and continuity in ecotoxicology. Each has strengths if properly used and weaknesses if misused. To facilitate experimentation, microexplanation pragmatically assumes that consequences of emergent properties are unimportant. Most thoughtful reductionists remain open to the possibility of emergent properties. Macroexplanation brings logical compromises by attempting to extract information about parts that are hidden within the measured properties of the whole. Such aggregated information makes inferences quite uncertain at times, and caution is practiced by most responsible scientists. Holistic descriptions of consistent relationships without knowledge of causal mechanisms can lead to ineffective extrapolation or prediction within and among levels. However, holistic descriptions are absolutely essential in many instances, especially those for which insufficient information exists for confidently posing explanations.

To apply only one mode of investigation is inconsistent with the modified Janus context. A cause at one level is the consistent association described at another. Identifying the significance at one level from a phenomenon observed at a lower one depends as much on the observation of a consistent association (holism) as on mechanistic underpinnings (reductionism). Holistic prediction based on consistency does not require full mechanistic knowledge. Regardless, holistic prediction is inherently limited without mechanistic knowledge. In the modified Janus context, cause is most often sought at a lower level via reductionism after observation of a consistent association at a level. The initial observation of a consistent association was probably made in what many would describe as a holistic study. Some macroexplanation may have been applied based on paradigms from the next highest level. Macroexplanation is often applied to results from studies with a strong holistic context and can be the true root for contention between reductionists and holists.

1.2.3 A Closer Look at Macroexplanation

Macroexplanation requires more comment because it is not a topic much discussed by ecotoxicologists, but it will become more important as ecotoxicologists pay increasing attention to large-scale issues. Macroexplanation has a potential problem as large as the emergent properties problem so often thrust at reductionists by holists or the limited prediction problem thrust back at holists by reductionists. Macroexplanation about global changes is becoming more common; for example, initial macroexplanation of decreases in polar ozone concentrations eventually led ecotoxicologists to chlorofluorocarbons that influence chemical reactions in the stratosphere. Although not often recognized as such, macroexplanation is also applied frequently in ecology and ecotoxicology. For example, a shift in population structure after pollution exposure may be used to infer how individual life history strategies might have changed or used to formulate testable hypotheses about changes in reproduction or survival of individuals. This may be done without specific data about how life histories of individuals actually shifted. Macroexplanation is also common when natural selection theory is invoked to suggest how individuals optimize their fitness under stressful conditions. This "process of using aggregate (i.e., 'ecological') data to infer individual-level relationships of interest when individual-level data are not available" (King 1997) is called ecological inference in sociology (King 1997) and epidemiology (Last 1983).[2] Ecological inference is used often by social scientists and epidemiologists with care and appropriate qualification.

Many relevant examples of macroexplanation can also be found in epidemiology. Radon levels measured in counties and lung cancer deaths tallied in those same counties may be used to imply the chances of an individual dying of lung cancer at a certain radon exposure concentration (King 1997). Similarly, correlations between water quality in regions of the United States and heart disease may be used to estimate risk of heart disease for an individual living in a region with water hardness of a certain level (Last 1983). With the application of ecological inference, there is a risk that a researcher's predilection will be used inappropriately to identify the "best" explanation when alternate explanations are possible. This is the ecological fallacy or problem of ecological inference.

In surprising contrast to the careful attention paid to this problem in the social sciences, it appears without much comment in ecotoxicology. When the problem of ecological inference is raised, it is usually misidentified as a flaw of holism. Perhaps our preoccupation with the reductionism/holism debate contributes to this neglect. The logical difficulty of using aggregated information derived in a holistic study to suggest a mechanism at the next lowest level makes one hesitate before designing a holistic or descriptive study.[3] The problem of ecological inference can appear while trying to extract more information from holistic studies generating "aggregate" data. This is the reductionist's

[2] "Ecological" denotes only that the data are aggregate in nature. Etymology aside, it has nothing to do with the subject of ecology.

[3] The data would only be considered aggregate if they were used to suggest behavior of parts at a level below that being measured.

counterargument to the inability to predict emergent properties with a predominantly reductionist approach. The optimal solution for ecotoxicology is to use all approaches thoughtfully and to avoid their overextension, that is, acknowledge the problem of ecological inference, limits of prediction from holistic analysis, and the possibility of emergent properties.

1.3 REQUIREMENTS IN THE SCIENCE OF ECOTOXICOLOGY

1.3.1 GENERAL

Let us adopt the modified Janus context described earlier as the correct one for planning ecotoxicological research programs and for organizing knowledge around ecotoxicological explanations. The reductionism/holism debate becomes immediately irrelevant because the modified Janus context implies that all levels are equally valid foci for exploring the cause–effect–significance concatenation. For any level, one needs to understand its parts and also how that level functions as part of a larger whole. This can only be done by simultaneous and thoughtful application of microexplanation, holism, and macroexplanation. These explanatory vantages are equally useful to all levels of organization because no level has a favored status and the cause–effect–significance structure can be shifted freely to any level. This last statement is contrary to current thinking that reductionism is more applicable at lower levels and holism is more useful at higher levels. Hopefully, the confusion is dispelled and we can move on to more useful topics in this treatment of ecotoxicology.

What is most needed in ecotoxicology is the accumulation of high-quality knowledge and the structuring of that knowledge around rigorously tested explanatory principles. Further, the present chimerical state of knowledge must be slowly transformed to foster consilience of explanatory principles (paradigms) among levels of organization. Without these changes, ecotoxicologists run the risk of scientific jerry-building.

How can one assess the relative value of research programs in accomplishing these goals if it does not matter what level of biological organization is being addressed, or whether a microexplanatory, macroexplanatory, or holistic vantage is taken? We believe that assessment of relative value for most hierarchical sciences is surprisingly straightforward if based on two qualities: the strength of associated inferences and consilience among paradigms/explanations. Consilience has already been discussed, so inferential strength is the only component needing any further explanation.

1.3.2 STRONGEST POSSIBLE INFERENCE

The concept of strong inference is described in a remarkable article by Platt (1964). At its heart is the Baconian scientific method. A working hypothesis and alternate hypothesis are formulated, and a discriminating experiment is performed. The result is used to produce additional hypotheses and the process is repeated until only one remains as a viable explanation for the observations. This process guides a researcher through a dichotomous logic tree to a final explanation for the phenomenon under study.

However, the strong inference process extends this method. The concept of multiple working hypotheses (Chamberlin 1897) is added because the Baconian method tends to favor a central hypothesis in the formulation of tests and to bias the allocation of effort toward that favored hypothesis during the falsification process. An unintentional bias can emerge with the tendency for investigators to develop a sense of ownership for a particular hypothesis or explanation. Chamberlin's solution was to advocate a multiple working hypotheses framework. All reasonable hypotheses are formulated and experimental efforts spent equitably among these hypotheses. This process diminishes the tendency to become enamored with a particular explanation and shifts the emphasis to the process of discriminating among the candidate hypotheses. The researcher owns the process, not a favored hypothesis or explanation. This process also reduces the tendency to stop testing when "the cause" is discovered, where in reality, there might be multiple causes for a phenomenon. The final and crucial aspect of strong inference is that the process must be taught and practiced consistently in the field. Using an analogy, Platt (1964) explains the benefits of this approach relative to the rate at which a scientific field can advance.

> The difference between the average scientist's informal methods and the methods of the strong-inference users is somewhat like the difference between a gasoline engine that fires occasionally and one that fires in steady sequence. If our motorboat engines were as erratic as our deliberate intellectual efforts, most of us would not get home for supper.

We suggest one essential extension to Platt's strong inference approach. This extension renders the process to what might more appropriately be called the strongest possible inference approach. The strongest possible inference approach adds more current Bayesian inferential methods to Platt's strong inference approach. Although all efforts should be made to perform tests that result in clear dichotomous outcomes, i.e., reject or do not reject the hypothesis, the discrimination afforded by many formal experiments and less structured observational studies is often less clear. The dichotomous falsification process described for the strong inference approach can be enhanced by including formal abductive inference—inference to the best explanation (Josephson and Josephson 1996). Modern statistical methods, specifically Bayesian methods, allow the conditional conclusion that a hypothesis is falsified when it becomes sufficiently improbable. Especially useful are well-established techniques such as Bayesian networks. A "reject/accept" conclusion from hypothesis tests is no longer essential (nor adequate) for rigorously evaluating an explanation.[4] This final modification produces an extremely powerful and flexible approach for making the strongest possible inferences from existing ecotoxicology evidence and, consequently, for accelerating advancement in the field. The Strongest Possible Inference approach is equally valuable for all levels of the ecotoxicological hierarchy and investigative vantages.[5] It will be described in more detail in Chapter 36.

1.4 SUMMARY

Our intent in producing this book is to bridge a gap between general textbooks and highly specialized books. Our conceptual tack is to provide more details about key paradigms and approaches, and to explore possible consilience among levels of organization. The stagnant reductionism/holism context is rejected in favor of a modified Janus context. On the basis of this context, the cause–effect–significance sequence can slide freely up and down the scales of this hierarchical science. The three vantages of investigation (microexplanation, macroexplanation, and holism) are relevant at all levels of biological organization. Strong inference augmented by modern Bayesian techniques of abductive inference; that is, the Strongest Possible Inference framework, is advocated as the best investigative mode in this new science (see Chapter 36). We believe that our collective commitment to well-founded inference and conceptual consilience will determine whether or not ecotoxicology quickly realizes its potential for becoming one of the most influential sciences of the new millennium.

1.4.1 SUMMARY OF FOUNDATION CONCEPTS AND PARADIGMS

- Ecotoxicology is the science of contaminants in the biosphere and their effects on constituents of the biosphere, including humans.
- Ecotoxicology is a hierarchical science.

---

[4] In fact, the successful codification of Popper's strict falsification principle into statistical hypothesis testing by Fisher created considerable inferential confusion. Box 10.2 provides one example relevant to ecotoxicology.

[5] Note that we will extend this theme in Chapter 29 to include also a course of inquiry called adaptive influence. With adaptive influence, researchers shift concerns between Types I and II errors with concern about Type II error being minimal during initial phases of inquiry.

- No level in the ecotoxicological hierarchy is better than another relative to identifying causation or attributing relevance.
- Essential to the growth of ecotoxicology are the application of the strongest inference possible to efficiently organize knowledge around rigorously tested paradigms (explanatory principles), and consilience of concepts and paradigms among all hierarchical levels.
- The modified Janus context allows optimal inference and organization of knowledge around paradigms at all levels of biological organization.
- Wider application of the methods of strong inference, adding modern Bayesian methods of abductive inference, would greatly accelerate progress in ecotoxicology.
- Three modes of acquiring knowledge and enhancing belief are useful: microexplanation (reductionism), macroexplanation (inference about parts from the behavior of the whole), and holism (inference from observation of a consistent pattern or behavior without the requirement of a lower-level, causal explanation).
- Inference based on holism may lead to prediction error as causal mechanisms are not necessarily known.
- Inference based on microexplanation (reductionism) may lead to prediction error as properties difficult or impossible to predict can emerge at higher levels of organization.
- Inference based on macroexplanation may lead to prediction error due to the problem of ecological inference.

REFERENCES

Caswell, H., Demography meets ecotoxicology: Untangling the population level effects of toxic substances, In *Ecotoxicology. A Hierarchical Treatment*, Newman, M.C. and Jagoe, C.H. (eds.), CRC Press/Lewis Publishers, Boca Raton, FL, 1996, pp. 255–292.

Chamberlin, T.C., The method of multiple working hypotheses, *J. Geol.*, 5, 837–848, 1897.

Harré, R., *The Philosophies of Science. An Introductory Survey*, Oxford University Press, Oxford, UK, 1972.

Johnson, S. 1753. The Adventurer, In *Samuel Johnson. Selected Writings*, Cruttwell, P. (ed.), Penguin Books, London, UK, 1968, pp. 186–204.

Johnson, D.R. and Hansen, R.M., Effects of range treatment with 2,4-D on rodent populations, *J. Wildl. Manage.*, 33, 125–132, 1969.

Josephson, J.R. and Josephson, S.G., *Abductive Inference. Computation, Philosophy, Technology*, Cambridge University Press, Cambridge, UK, 1996.

Karlins, M., Conceptual complexity and remote-associative proficiency as creativity variables in a complex problem-solving task, In *Creativity. Theory and Research*, Bloomberg, M. (ed.), College & University Press, New Haven, CT, 1973, pp. 200–228.

King, G., *A Solution to the Ecological Inference Problem*, Princeton University Press, Princeton, NJ, 1997.

Koestler, A., Holons and hierarchy theory, In *From Gaia to Selfish Gene. Selected Writings in the Life Sciences*, Barlow, C. (ed.), MIT Press, Cambridge, MA, 1991, pp. 88–100.

Last, J.M., *A Dictionary of Epidemiology*, Oxford University Press, New York, 1983.

Maciorowski, A.F., Populations and communities: Linking toxicology and ecology in a new synthesis, *Environ. Toxicol. Chem.*, 7, 677–678, 1988.

Nagel, E., *The Structure of Science. Problems in the Logic of Scientific Explanation*, Harcourt, Brace and World, Inc., New York, 1961.

Newman, M.C. and Unger, M.A., *Fundamentals of Ecotoxicology*, CRC/Lewis Press, Boca Raton, FL, 2003.

Odum, E.P., Foreward, In *Ecotoxicology. A Hierarchical Treatment*, Newman, M.C. and Jagoe, C.H. (eds.), CRC Press/Lewis Publishers, Boca Raton, FL, 1996.

Platt, J.R., Strong inference. *Science*, 146, 347–353, 1964.

Popper, K.R., *The Logic of Scientific Discovery*, T.J. Press (Padstow) Ltd., London, UK, 1959.

Watkins, J., Against "Normal Science," In *Criticism and the Growth of Knowledge*, Lakatos, I. and Musgrave, A. (eds.), Cambridge University Press, Cambridge, UK, 1970, pp. 25–37.

Wilson, E.O., *Consilience*, Alfred A. Knopf, Inc., New York, 1998.

Part II

Organismal Ecotoxicology

Conventionalities are as bad as impurities.

Uncommon Learning **(H.D. Thoreau 1851)**

Organismal ecotoxicology explores toxicant effects to individuals and, where possible, links them to effects to populations and communities. Such exploration has been at the center of ecotoxicology and its predecessors (aquatic, wildlife, and environmental toxicology) since their inceptions. Because of our proclivity toward study of toxicant effects to the soma—the body of the individual organism—much in this section should be comfortably familiar to the professional ecotoxicologist or advanced student. What might not be as familiar will be the focus on fundamental principles and linkage of these effects to those at higher levels of biological organization.

The preoccupation of ecotoxicologists with the soma emerges from the historical foundations of our new science. It is obvious during even a cursory examination of the most popular ecotoxicology textbooks (e.g., Cockerham and Shane 1994, Connell et al. 1999, Landis and Yu 1995, Newman 1998, Walker et al. 2001) that many basic concepts and techniques blended into ecotoxicology come from mammalian toxicology, a field with a justifiable emphasis on the individual. Still other concepts and techniques come from classic autecology. Used with balance and insight, this offers several advantages to the field. Ecotoxicologists can draw deeply from the mechanistic knowledge base of classic toxicology, a field focused on individuals. This knowledge is directly useful for charismatic, endangered, or threatened species that are protected by prohibiting the taking of even a single individual. It also provides a firm base at one level of biological organization from which to extend scientific insight upward to the next.

The mechanistic and technological richness of classic toxicology and autecology comes at a price. The paradigms around which phenomena are explored by ecotoxicologists are often those associated with the soma. Exploration of other important ecotoxicological phenomena are unintentionally addressed with less intensity or quietly dismissed as secondary. The rich technology associated with organismal toxicology naturally draws practitioners to these tools. The result is a rapid enrichment of the field: an enrichment that also maintains the present imbalance. Resolution of this incongruity requires application of concepts and technology in a way that does not foster any unintentional neglect of higher levels of organization and with the intent of producing predictive insight about phenomena at higher levels of organization. That is the intent of this section.

REFERENCES

Cockerham, L.G., and Shane, B.S., *Basic Environmental Toxicology*, CRC Press, Boca Raton, FL, 1994.

Connell, D., Lam, P., Richardson, B., and Wu, R., *Introduction to Ecotoxicology*, Blackwell Science Ltd., Oxford, UK, 1999.

Landis, W.G., and Yu, M.-H., *Introduction to Environmental Toxicology*, CRC Press, Boca Raton, FL, 1995.

Newman, M.C., *Fundamentals of Ecotoxicology*, CRC Press/Lewis, Boca Raton, FL, 1998.

Thorean, H.D., *Uncommon Learning*, Bickman, M. (ed.), Houghton, Mifflin, Co., Boston, 1851.

Walker, C.H., Hopkin, S.P., Sibly, R.M., and Peakall, D.B., *Principles of Ecotoxicology*, Taylor & Francis, New York, 2001.

2 The Organismal Ecotoxicology Context

2.1 OVERVIEW

The science of ecotoxicology has grown rapidly in 30 years and has brought together a vast body of facts around several explanatory systems. Explanatory systems were borrowed in necessary haste from mammalian toxicology and ecology. The immediacy of our environmental problems required this haste. Required now is coherence among the clusters of explanatory hypotheses that are rapidly coalescing at each level of biological organization. Together, these paradigms form the foundation for ecotoxicological theory.[1] If these paradigms are not made mutually supportive, the foundation of ecotoxicology will not be adequate to support further knowledge accumulation and organization. The field will break into semi-isolated scientific disciplines.

Conceptual consilience is not an intellectual nicety: it is vital to the health of any science. Without consistency among theories and facts, there is no way for the ecotoxicologist to choose from among many the explanation providing the best foundation for predicting pollutant effects.

> The requirement of consistency will be appreciated if one realizes that a self-contradictory system is uninformative. It is so because any conclusion we please can be derived from it. Thus no statement is singled out, either as incompatible or as derivative since all are derivable. A consistent system, on the other hand, divides the set of all possible statements into two: those which it contradicts and those with which it is compatible .... This is why consistency is the most general requirement for a [scientific] system ... if it is to be of any use at all.
>
> **(Popper 1959)**

As articulated by Popper, sciences lacking self-consistency are not viable. Ecotoxicological explanations need to be consistent among all levels of organization or the science of ecotoxicology will eventually fail to be useful. Beyond this, efforts spent finding consistency have another desirable effect relative to scientific logic. It can identify common causes for phenomena described at different levels of biological organization. The identification of a common cause allows the overall number of theories to be reduced. Why have two distinct theories to explain the same thing? The parsimony resulting from theory reduction—that is, intertheoretical reduction (Rosenberg 2000)—enhances any science and is particularly warranted in ecological sciences (Loehle 1988).

A final reason exists for the emphasis on integrating explanatory systems from different levels of biological organization. Not doing so allows the current condition to remain in which an ecotoxicologist trying to describe and solve a particular environmental problem may present and defend findings based on contradictory explanations. This diminishes the legal defensibility of arguments

[1] Definitions of Rosenberg (2000) are being used in this discussion. A set of explanatory principles or paradigms comprise the established scientific theory of a discipline, for example, evolutionary theory contains many explanatory principles such as genetic drift or natural selection. The paradigms have withstood rigorous testing and currently provide the best causal explanation of natural phenomena, for example, evolution theory explains genetic change in a population exposed to a toxicant.

calling for costly remediation. It also increases the risk of pathological science, science practiced with an excess loss of objectivity (Langmuir 1989, Rousseau 1992).

> It is a fault which can be observed in most disputes, that, truth being mid-way between the two opinions that are held, each side departs the further from it the greater his passion for contradiction.
>
> **(Descartes, translated by Sutcliffe 1968)**

Integration, combined with differentiation, is a major theme here because it fosters the identification of causal mechanisms that are consistent among levels of organization, is logically necessary in any healthy science, fosters resolution of environmental issues, and decreases the tendency toward pathological science.

2.2 ORGANISMAL ECOTOXICOLOGY DEFINED

2.2.1 WHAT IS ORGANISMAL ECOTOXICOLOGY?

> Every species of plant is a law unto itself, the distribution of which in space depends upon its individual peculiarities of migration and environmental requirements. ... It grows in the company with any other species of similar environmental requirements, irrespective of their normal associational affiliations.
>
> **(Gleason 1926)**

The scope of ecotoxicology is so necessarily encompassing that an ecotoxicologist can study fate or effect of toxicants from the molecular to the biospheric scales. This book attempts to discuss this wide range of topics. The focus of attention in this particular section is organismal ecotoxicology, the science of contaminants in the biosphere and their direct effects on individual organisms.

The prominence of the organismal context is so long-standing and familiar to ecologists that it has its own name, autecology. Autecology is the study of the individual organism or species, and its relationships to its physical, chemical, and biological environment. The quote above from Gleason's classic paper articulates the autecological framework.

The boundaries of autecology are often vague. Since its origins, autecology has been described as either distinct from (e.g., Emmel 1973) or synonymous with (e.g., Reid 1961) population ecology. In reality, it overlaps with population ecology but tends to characteristically emphasize species requirements, physiological tolerances, means of adaptation, and life history traits, and how these influence success or failure in certain environs. It emphasizes the soma and how it manages to survive. For example, a wildlife manager concerned with a specific game bird species might take an autecological vantage to managing that particular species. Another example of an autecological topic might be how the physiological tolerances of an estuarine crab relative to salinity and temperature influence its spatial distribution within an estuary. A study by Costlow et al. (1960) is a classic one of this sort (Box 2.1) in which tests of the physiological limits of individuals were used to suggest that salinity confines the spatial distribution of an estuarine crab. The emphasis is plainly on qualities of individuals, not complex interactions among species or even the interactions among individuals in populations of this crab species.[2]

Several fundamental laws of ecology emerge from this context. Liebig's law of the minimum (Liebig 1840), first formulated to explain how nutrients limit plant standing crop, states that the factor in the shortest supply of all required factors will limit the number or amount of individuals that a

---

[2] In contrast to autecology, the subdiscipline of ecology focused on the integrated interactions of groups of individuals within an environment is called synecology. Conventional topics of synecology are discussed principally in the last sections of this book. The population ecotoxicology section is the boundary between autecology and synecology, and covers a blend of autecology with some synecology.

Box 2.1 Autecology of a Crab: Physiological Tolerances Determine Adult Distribution

Costlow et al. (1960) reared larvae of the estuarine crab, *Sesarma cinereum*, at different combinations of salinity and temperature, hoping to gather enough information to explain the observed distribution of adult crabs in estuaries. The assumption was simple: the physiological tolerances of the larval stages, as reflected in survival rates, will determine the most likely part of the estuary in which the larvae will survive to become adults.

Salinity strongly influenced survival and development time for all larval stages. For example, the first zoea withstood higher salinities much better than lower salinities (Figure 2.1). Eggs hatched at all tested salinity and temperature combinations. However, close to 100% mortality occurred at low (<12.5‰) and high (>31.1‰) salinities for most larval stages. This suggested that those larvae of any stage that were brought into intermediate salinity (and temperature) conditions would have the highest chances of survival. Those hatching and staying in the lowest or highest salinity waters would have the poorest survival probability. The optimum salinity and temperature for each larval stage were the following:

Larval Stage	Optimal Salinity (‰)	Optimal Temperature (°C)
Zoeal Stage 1	27.9	23.5
Zoeal Stage 2	12.4	25.0
Zoeal Stage 3	24.1	26.0
Zoeal Stage 4	No maximum or minimum	
Megalops	No maximum or minimum	

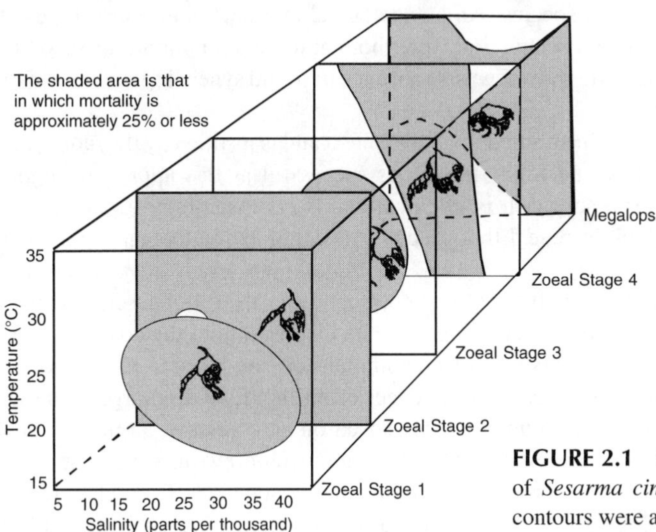

FIGURE 2.1 Salinity and temperature tolerances of *Sesarma cinereum* larvae. The 25% mortality contours were arbitrarily chosen to show the differences in tolerances among stages. (Modified from Figures 8–12 of Costlow et al. 1960 and larval drawings rendered from Figures 1–5 in Costlow and Bookhout 1960.)

In contrast to the first and third zoea, the second zoea had high tolerances of salinity and temperature. The first and second zoea had best survival at 21–31°C and 23–28‰. The last zoeal stage showed an increase in temperature tolerance (to 35°C) and a salinity tolerance down to 3‰. The megalops, the stage reached just before settling to the bottom, showed wide tolerances. The authors concluded that completion of this crab's life cycle to the adult depended primarily on the fourth zoea and that "the survival and molting to the megalops can only occur in estuarine waters." Any earlier stage larvae that were transported by water movements outside of the estuarine conditions had very low probabilities of producing megalops. Survival was less dependent on temperature or salinity once the megalops stage was reached. So, the tolerances of the larval stages determined the estuarine region in which the life cycle of this crab will be successfully completed. The weak links in the life cycle were the earlier larval stages. If the fourth zoeal larvae emerged under the appropriate salinity–temperature conditions, the relatively tolerant megalops would be produced, resulting in adults in that particular part of the estuary.

---

particular habitat can sustain. As an example, phosphorus might limit the standing crop of a nuisance blue-green algal species in a freshwater lake and, based on Liebig's law, lake managers might focus on controlling phosphorus input to the lake. Shelford's law of tolerances (the tolerance of individuals of a species over one or more environmental gradients determines the species' geographical distribution or abundance) (Shelford 1911, 1913) is another such law that is neatly illustrated by the Costlow et al. (1960) study.

The ecological niche concept was formulated originally with emphasis on individual tolerances and requirements, and only later was enlarged to include biotic factors. In fact, the niche concept theorizes that the organism occupies a realized niche in the presence of other species that is only a portion of its fundamental niche as defined by its organismal tolerances and requirements. As a classic example, the realized niche of the intertidal barnacle, *Chthamalus stellatus*, is strongly influenced by desiccation at one extreme and interspecific competition for space with *Balanus balanoides* at the other (Connell 1961).

In ecotoxicology, an autecological study might be conducted of the effects of a pollutant on individuals of a protected or threatened species. An autecological approach might also be used if synecological aspects of a species' niche occupation were thought to be unimportant or secondary. Such an approach is also taken reluctantly if there was an absence of sound synecological information available relative to contamination.

Much of ecotoxicology is done within an autecological context and is justified by the indisputable success of classic autecology (Calow and Sibly 1990). As successful as this approach might be for many situations, the autecological approach is often applied in ecotoxicology for situations in which a moment of introspection might reveal that crucial synecological factors are unjustifiably ignored.[3]

Reflecting the stage of ecology almost half a century ago, Costlow et al. (1960) described how the physicochemical tolerances of a crab species contributed to its distribution in the coastal systems. Twenty-five years later, when the understanding of contaminant effects became more and more essential, they approached the ecotoxicological consequences of drilling fluid discharge in the same way, implying whether populations would remain viable based on acute assays on the notionally most sensitive stages of a species' life cycle (Box 2.2). This approach drew on a well-established autecological approach and, in this case, produced a reasonable conclusion. They also adopted, with minimal adaptation, a technological paradigm from mammaliam toxicology—the LC50/LD50

[3] Staying with a coastal marine theme, see Harger (1972) for further discussion of the importance of species interactions in determining intertidal species distributions.

Box 2.2 Crab Autecotoxicology: Do Chromium Tolerances of Larvae Determine Adult Fate?

Twenty-four years after the study described in Box 2.1, these researchers (Bookhout et al. 1984) again described crab larval survival relative to environmental conditions. In keeping with the emerging concern about anthropogenic chemicals, they focused on a pollutant this time.[4] The intent was very similar to that of their first paper—to determine the tolerances of the larval stages to an environmental quality and, in doing so, to predict the likelihood of life cycle completion in the presence of a specified intensity of that quality.

They studied chromium used in drilling fluids to thin mud as it becomes dense. Added as ferrochrome or chrome lignosulfonate, chromium was discharged during and after the drilling. At the time of this study, there was little information on whether its use was harmful to marine species. To explore the potential hazard, Bookhout et al. (1984) exposed decapod larva to different concentrations of hexavalent chromium (as Na_2CrO_4). Results for mud crab, *Rhithropanopeus harrisii*, larvae are described here. Figure 2.2 summarizes the cumulative mortality experienced at different larval stages exposed to 0–58 μg/L of sodium chromate. Sodium chromate concentrations from 7 to 29 μg/L were considered to be sublethal concentrations because 10% or more of exposed larvae reached the first crab stage. The lethal range was above 29 μg/L. The LC50 for the complete hatch → zoea → first crab life stages was 13.7 μg/L.[5]

After integrating this information with knowledge about drilling fluid distributions around points of discharge, the authors concluded that "it is probable that Cr in drilling fluids, whether Cr^{3+} or Cr^{6+}, is not likely to reduce the population of crab larvae and other planktonic organisms in the area around oil wells except possibly in the immediate vicinity of the discharge pipes." Implied from these acute lethality tests on *individual larvae* was that the persistence of the *mud crabs population* was not jeopardized, except in the immediate vicinity of a discharge. This reasoning was adapted from that used to define the fundamental niche, that is, application of the law of tolerances to predict the habitat that a species could occupy. However, other aspects of the niche concept, such as interspecies competition, predation, and disease, were ignored. This expedient neglect seems understandable in this particular application but is not always justified.

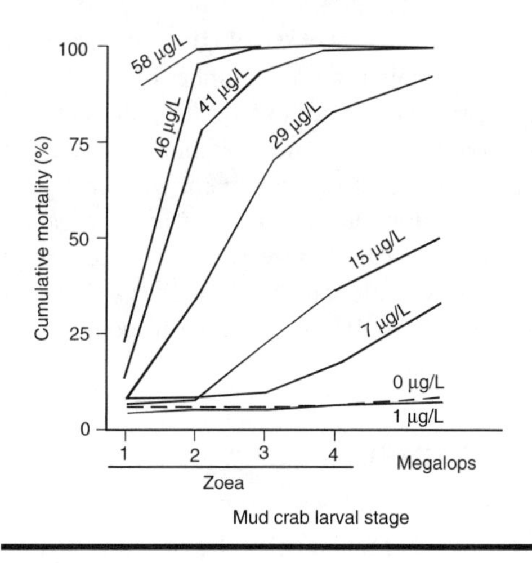

FIGURE 2.2 Cumulative mortality of mud crab larvae resulting from hexavalent chromium exposure. (Modified from Figure 4 in Bookhout et al. 1984.)

[4] The research described in Box 2.1 was published 2 years prior to Rachel Carson's watershed book, *Silent Spring* (Carson 1962). As evidenced by the shift in Costlow et al.'s research, Carson's book mandated that our research efforts become more focused on ecotoxicological questions.

[5] Note that this last metric, the LC50, was borrowed from the mammalian toxicology literature to measure toxic effects here but was absent in the temperature–salinity study described in Box 2.1.

approach. The approach taken by these authors and many others dominates ecotoxicology to this day. (See Chapter 9 for details.) It is sufficient in many cases or useful in others for quickly identifying gross problems associated with contamination. Despite its utility in such cases, this autecotoxicological approach is insufficient in many others and a synecotoxicological context is needed. The key to successful prediction of ecotoxicological consequences is being able to accurately discriminate between situations requiring autecological or synecological vantages, and being able to integrate information from both vantages into reliable predictions of exposure consequences.

2.3 THE VALUE OF ORGANISMAL ECOTOXICOLOGY VANTAGE

> If the modes of action of toxicants are better understood, we could more accurately predict their effects as pollutants; much knowledge already exists in medical sciences, and could be transferred.
>
> **(Sprague 1971)**

> Although this discussion may appear hostile to single species toxicity testing efforts, it is not intended to be. Single species tests are exceedingly useful and are presently the major and only reliable means of estimating probable damage from anthropogenic stress. Furthermore, a substantial majority, perhaps everyone in this meeting is certainly aware of the need for community and system level toxicity testing. How then does one account for the difference between awareness and performance?
>
> **(Cairns 1984)**

Just as autecology is an essential component of ecology, organismal ecotoxicology—autecotoxicology, if you will—is an essential component of ecotoxicology. Unfortunately, as exemplified in the quote by Cairns above, organismal ecotoxicology tends to overshadow equally crucial investigative vantages. In the remainder of this chapter, the many appropriate and essential applications of organismal ecotoxicology will be highlighted.

2.3.1 TRACTABILITY AND DISCRETENESS

Organismal effects are generally the most discrete and tractable of ecotoxicological effects. Few ecotoxicologists would disagree with this statement. After agreeing, a good number of ecotoxicologists would immediately identify this truism as a sad statement about the field, or point out that this condition may simply be a matter of the historical amounts of effort and thought that have gone into autecotoxicology and synecotoxicology. Here, the follow-up to this statement will simply be to demonstrate the important advantages of drawing on our comprehensive knowledge of organismal effects. The ease with which organismal effects or exposure can be assessed will be described first. Next, the relatively effective extrapolation among individuals will be detailed. Organismal information also contributes to our abilities to do reasoned extrapolations of effect to populations and communities, and to predict toxicant transfer within communities.

2.3.2 INFERRING EFFECTS TO OR EXPOSURE OF ORGANISMS WITH SUBORGANISMAL METRICS

Sprague, as quoted above, stated correctly that knowledge of suborganismal modes of action greatly improves predictions of toxicant effects to individuals. A current example is endocrine system modulation by xenobiotics. Our newfound understanding of this mode of action for diverse classes of xenobiotics, such as 17β-estradiol from oral contraceptives, 4-nonylphenols, and polychlorinated biphenyls (PCB), improves prediction of similar effects of new chemicals or from new sources of existing xenobiotics (Brown et al. 2001, Hale and La Guardia 2002, Schultz 2003). Knowledge of the suborganismal mode of action also provides a means by which diverse phenomena can be linked by a common thread. As an example, Bard (2000) opined that multixenobiotic resistance in aquatic organisms can be explored in the context of the multidrug resistance phenomena. The important insight is that aquatic toxicologists could greatly advance their understanding of multixenobiotic resistance in exposed populations by exploring the extensive literature on the role of P-glycoprotein overexpression in determining antitumor drug resistance of cancer cells. Transmembrane P-glycoproteins tend to inhibit the transport of xenobiotics into cells, reducing the concentration of xenobiotics at intracellular sites of action. The same is true whether the xenobiotic is a cancer drug or a contaminant. Environmental xenobiotic resistance and anticancer drug resistance share a common theme of adaptation by P-glycoprotein overexpression.

Suborganismal qualities also provide evidence of contaminant exposure or effects. As a surprising example, mesopelagic fish sampled at 300–1500-m depth in the open Atlantic Ocean show evidence of exposure to aryl hydrocarbon receptor antagonists, e.g., polynuclear aromatic hydrocarbons (PAHs) and coplanar-halogenated aromatic hydrocarbons (Stegemen et al. 2001). Elevated cytochrome P450 1A also suggests exposure at considerable distance from coastal sources of aryl hydrocarbon receptor antagonists.

2.3.3 Extrapolating among Individuals: Species, Size, Sex, and Other Key Qualities

Often predictions require extrapolation from data in hand to some less well-defined situation.[6] Suter (1998) describes two typical examples: extrapolation from LC50 values of Salmoniformes to those for Perciformes, and prediction of carbamate pesticide LD50 based on a test species weight. Ellersieck et al. (2003) provide a computational means for extrapolation of toxicant effects among species. Although challenging and error prone, interpolation among individuals within a species is perhaps the most credible of ecotoxicological interpolations. As an example, Newman (1995) describes interpolation among mosquitofish sexes, genotypes, and sizes relative to survival during acute mercury exposure.

2.3.4 Inferring Population Effects from Organismal Effects

Sound inference about population effects is possible based on effects on individuals as has already been discussed in our treatment of autecology. The population ecotoxicology section of this book also explores many instances of such reasoning. These instances can be rendered as the following general statements:

Population Genetics. (1) Genotypic and phenotypic qualities of individuals sampled from a study population can be used to document population consequences of toxicant exposure. (2) Toxicants can influence the germ line of a population and, in so doing, influence the phenotypes present in the population. (3) Differences in individual genotypes' fitnesses in critical life stages can be used to suggest key selection components acted on by toxicants.

Population Demographics. (1) Vital rates derived by sampling individuals from populations can be used to document current or to project future conditions of a population. These vital rates include rates of mortality, growth, natality, and migration. (2) Toxicants that lower an individual's fitness can influence the demographics of an exposed population.

Metapopulation Biology. (1) Differences in vital rates of individuals occupying habitat patches that differ in their capacity to maintain the species can produce differences in metapopulation

[6] See Suter (1998) for a general discussion of ecotoxicological extrapolation.

dynamics and persistence. (2) Spatial distributions of individuals relative to the distribution of toxicant in habitat patches will influence metapopulation dynamics and vitality.

Epidemiology. (1) Disease prevalence, incidence, and distribution in a population can be defined by measuring disease state in sampled individuals. (2) Causal knowledge derived from the suborganismal or organismal levels reinforce epidemiological inferences.

Life History. (1) In the presence of phenotypic plasticity, an organism will experience a shift in its life history traits if stressed. Ideally, such a trade-off in life history traits will optimize the individual's Darwinian fitness under the environmental conditions it finds itself. (2) Toxicants can change the life history traits of individuals in predictable ways and, in so doing, also influence the population demographics of exposed populations. (3) Phenotypic expression by individuals can involve reaction norms or polyphenisms.

It is also true that the abundance of individual-based data tempts intelligent and well-intended ecotoxicologists to make flawed inferences about population consequences from individual-based effects data. As an important example, Newman and Unger (2003) identify the weakest link incongruity: the inappropriate prediction of the most sensitive quality relative to population persistence based on the most sensitive life stage of an individual. Often, toxicity testing is done for all life stages of a species and the most sensitive life stage is identified as that stage with the lowest NOEC or LC50. The incorrect extension of such an approach is to falsely infer from this that the most sensitive life stage relative to individual fitness (i.e., survival, growth, or reproduction) is also the most crucial or sensitive relative to population persistence or vitality. Although there are cases in which this approach is adequate, it would be inconsistent with the foundation concepts of population ecology (Hopkin 1993) to assume that it is always adequate. It is demonstrably false in some cases (e.g., Kammenga et al. 1996, Petersen and Petersen 1988). Another important example is the assumption that individual-derived effect metrics are accurate, albeit conservative, predictors of concentrations that will adversely affect important population qualities. Forbes and Calow (2003) found that this was sometimes the case, but in general, individual-based metrics of adverse effect were not reliable predictors of concentrations adversely impacting populations. More information was needed to accurately infer population effects.

2.3.5 Inferring Community Effects from Organismal Effects

If done cautiously, potentially useful inferences about community consequences can be made from the effects of contaminants on individuals. These are detailed in the community ecotoxicology section. A quick review of that section reveals that many community metrics are generated with counts of individuals for the community of interest. Presence of individuals of key or indicator species is also crucial to many of the community-oriented methods. Species-specific sensitivity of individuals to toxicants can be used to develop biotic indices for implying toxicant effect to communities. Colonization or succession theory draws on individual life history qualities for its causal foundation. The autecologically oriented laws of Liebig and Shelford are used to describe the transition in community types along environmental gradients. Rapoport's rule relating species richness and latitude (or elevation) is also based on individual species' tolerances.

Ambiguously useful applications of individual-based metrics to predictions of community-level consequences are also present in ecotoxicology. A current example is the emerging species sensitivity distribution method. The LC50 (or NOEC) values are collected for all relevant species and used to produce a curve that describes the distribution of toxicant sensitivities of the tested species. The curve is used to compute the LC50 or NOEC concentration associated with only the lowest 5% (or 10%) of species. Only the most sensitive 5% of test species would have an LC50 (or NOEC) at or below that concentration. This HC_p is then used to imply a concentration below which all but a small percentage of species in a community will be protected from the adverse effects of the toxicant. Phenomena that would bring such an implication into question are discussed in detail in the population and community ecotoxicology sections. The interested reader is encouraged to browse these sections before applying the species sensitivity distribution method.

2.3.6 Inferring Potential for Trophic Transfer from Bioaccumulation

Bioaccumulation is the accumulation of a contaminant in (and occasionally also on) an individual organism. Models and associated concepts applied to toxicant bioaccumulation in individuals establish the foundation on which many community trophic transfer models are built (e.g., Mason et al. 1994, Simon and Boudou 2001). An explicit example of such a model is provided by Laskowski (1991). Consequently, knowledge of contaminant uptake, transformation, and elimination by individuals is useful in predicting transfer of contaminants within food webs.

2.4 SUMMARY

- The organismal focus in mammalian toxicology and early ecology (i.e., autecology) contributed to an organismal bias in the new science of ecotoxicology.
- Although resulting in an imbalance in ecotoxicological research, the organismal bias does have positive consequences. These include (1) transfer of new technologies rapidly into ecotoxicology, (2) providing mechanisms for effects seen at higher levels of organization, (3) providing sensitive indicators of exposure or effect, and (4) providing a highly discrete and tractable approach to any ecotoxicological questions.
- Careful interpretation of organismal data enhances our ability to predict consequences at the population and, sometimes, community levels of organization.
- Careful application of bioaccumulation data enhances our ability to predict trophic transfer of toxicants in food webs.

REFERENCES

Bookhout, C.G., Monroe, R.J., Forward, R.B., Jr. and Costlow, J.D., Jr., Effects of hexavalent chromium on development of crabs, *Rhithropanopeus harrisii* and *Callinectes sapidus*, *Water Air Soil Pollut.*, 21, 199–216, 1984.

Brown, R.P., Greer, R.D., Mihaich, E.M., and Guiney, P.D., A critical review of the scientific literature on potential endocrine-mediated effects in fish and wildlife, *Ecotoxicol. Environ. Saf.*, 49, 17–25, 2001.

Cairns, J., Jr., Are single species toxicity tests alone adequate for estimating environmental hazard? *Environ. Monitor Assess.*, 4, 259–273, 1984.

Calow, P. and Sibly, R.M., A physiological basis of population processes: Ecotoxicological implications, *Funct. Ecol.*, 4, 283–288, 1990.

Carson, R., *Silent Spring*, Houghton Mifflin, Boston, MA, 1962.

Connell, J.H., The influence of interspecific competition and other factors on the distribution of the barnacle, *Chthamalus stellatus*, *Ecology*, 42, 710–723, 1961.

Costlow, J.D., Jr. and Bookhout, C.G., The complete larval development of *Sesarma cinereum* (Bosc) reared in the laboratory, *Biol. Bull.*, 118, 203–214, 1960.

Costlow, J.D., Jr., Bookhout, C.G., and Monroe, R., The effect of salinity and temperature on larval development of *Sesarma cinereum* (Bosc) reared in the laboratory, *Biol. Bull.*, 118, 183–202, 1960.

Descartes, R., 1637, *Discourse on the Method and the Meditations*, translation by F.E. Sutcliffe, Penguin Books, London, UK, 1968.

Ellersieck, M.R., Asfaw, A., Mayer, F.L., Krause, G.F., Sun, K., and Lee, G., Acute-to-chronic estimation (ACE v 2.0) with time–concentration–effect models. Use Manual and Software, EPA/600/R-03/107, U.S. EPA, Washington, D.C., 2003.

Emmel, T.C., *An Introduction to Ecology & Population Biology*, W.W. Norton & Co., Inc., New York, 1973.

Forbes, V.E. and Calow, P., Contaminant effects on population demographics, In *Fundamentals of Ecotoxicology*, 2nd ed., Newman, M.C. and Unger, M.A. (eds.), CRC Press/Lewis Publishers, Boca Raton, FL, 2003, pp. 221–224.

Gleason, H.A., The individualistic concept of the plant association, *Bull. Torrey Bot. Club*, 53, 7–26, 1926.

Hale, R.C. and La Guardia, M.J., Emerging contaminants of concern in coastal and estuarine environments, In *Coastal and Estuarine Risk Assessment*, Newman, M.C., Roberts, M.H., Jr., and Hale, R.C. (eds.), CRC Press/Lewis Publishers, Boca Raton, FL, 2002, pp. 41–72.

Harger, J.R.E., Competitive coexistence among intertidal invertebrates, *Am. Sci.*, 60, 600–607, 1972.

Hopkin, S.P., Ecological implications of "95% protection levels" for metals in soils, *OIKOS*, 66, 137–141, 1993.

Kammenga, J.E., Busschers, M., van Straalen, N.M., Jepson, P.C., and Baker, J., Stress induced fitness is not determined by the most sensitive life-cycle trait, *Funct. Ecol.*, 10, 106–111, 1996.

Langmuir, I., Pathological science, *Phys. Today*, October, 36–48, 1989.

Laskowski, R., Are the top predators endangered by heavy metal biomagnification? *OIKOS*, 60, 387–390, 1991.

Liebig, J., *Chemistry in Its Application to Agriculture and Physiology*. Taylor and Walton, London, UK, 1840.

Loehle, C., Philosophical tools: Potential contributions to ecology, *OIKOS*, 51, 97–104, 1988.

Mason, R.P., Reinfelder, J.R., and Morel, F.M.M., Uptake, toxicity and trophic transfer of mercury in a coastal diatom, *Environ. Sci. Technol.*, 30, 1835–1845, 1994.

Newman, M.C., *Quantitative Methods in Aquatic Ecotoxicology*, CRC Press/Lewis Publishers, Boca Raton, FL, 1995.

Newman, M.C. and Unger, M.A., *Fundamentals of Ecotoxicology,* CRC Press/Lewis Publishers, Boca Raton, FL, 2003.

Petersen, R.C., Jr. and Petersen, L.B.-M., Compensatory mortality in aquatic populations: Its importance for interpretation of toxicant effects, *Ambio*, 17, 381–386, 1988.

Popper, K.R., *The Logic of Scientific Discovery*, Routledge, London, UK, 1959.

Reid, G.K., *Ecology of Inland Waters and Estuaries.* Van Nostrand Reinhold Co., New York, 1961.

Rosenberg, A., *Philosophy of Science*, Routledge, London, UK, 2000.

Rousseau, D.L., Case studies in pathological science, *Am. Sci.*, 80, 54–63, 1992.

Schultz, I., Environmental estrogens: Occurrence of ethynyestradiol and adverse effects on fish reproduction, In *Fundamentals of Ecotoxicology*, 2nd ed., Newman, M.C. and Unger, M.A. (eds.), CRC Press/Lewis Publishers, Boca Raton, FL, 2003, pp. 156–160.

Shelford, V.E., Physiological animal geography, *J. Morphol.*, 22, 551–618, 1911.

Shelford, V.E., *Animal Communities in Temperate America*, University of Chicago Press, Chicago, IL, 1913.

Simon, O. and Boudou, A., Direct and trophic contamination of herbivorous carp *Ctenopharyngodon idella* by inorganic mercury and methylmercury, *Ecotox. Environ. Saf.*, 50, 48–59, 2001.

Sprague, J.B., Measurement of pollutant toxicity to fish—III. Sublethal effects and "safe" concentrations, *Water Res.*, 5, 245–266, 1971.

Suter, G.W., Jr., Ecotoxicological effects extrapolation models, In *Risk Assessment. Logic and Measurement*, Newman, M.C. and Strojan, C.J. (eds.), CRC Press/Lewis Publishers (originally Ann Arbor Press), Boca Raton, FL, 1988, pp. 167–185.

3 Biochemistry of Toxicants

All chemical pollutants must initially act by changing structural and/or functional properties of molecules essential to cellular activities.

(Jagoe 1996)

3.1 OVERVIEW

Two themes are often explored in expositions of biochemical toxicology: the nature of the biochemical change and the mode of toxic action. Relative to the nature of the change, biochemical changes such as those associated with cytochrome P450 monooxygenases, metallothioneins, or stress proteins are considered in the context of general toxicant detoxification or sequestration phenomena. Other changes such as DNA adduct formation, enzyme inhibition, or lipid peroxidation might be viewed as evidence of a particular mode of action resulting in damage. Consequently, toxicants sharing a common mode of action are discussed together, such as the coplanar polychlorinated biphenyls (PCBs), dioxins, and furans whose common mode of action involves the aryl hydrocarbon receptor (Lucier et al. 1993). The discussion here will adopt these organizing themes because doing so facilitates integration of the chapter's content with the rich mammalian toxicology literature that is similarly organized. But, in keeping with the series emphasis on interlinking phenomena, chapter topics will also be described in an information transfer context (Figure 3.1 and also Figure 36.1 in Chapter 36).

The fields describing relevant levels of information transfer and complexity are genomics → transcriptomics → proteomics → metabolomics → bioenergetics or biochemical physiology → molecular toxicology. All these areas of study explore different, yet linked, levels of organization relative to biological information flow and complexity. Genomics explores the entire nuclear DNA complement and variations within it.[1] Toxicogenomics specifically focuses on the influence of toxicants on the nuclear DNA. The next level of the biochemical information flow emerges at transcription. Transcription initiation occurs when RNA polymerase attaches to promoter regions of DNA. Nucleotides are added according to the DNA base sequence to produce mRNA during the elongation step of translation that ends with mRNA release. Transcriptomics attempts to describe and explain the complement of mRNA transcripts and their abundances present in cells or tissues under various conditions. Through translation, pools of various proteins are created in the cytoplasm. Proteomics is the study of the full complement of these proteins, their relative abundances, changes, and interactions. Finally, metabolomics attempts to explain the metabolite complement in cells or tissues under various conditions, including toxicant exposure. Repeating an important theme in this book, the greatest insight is gained by applying combinations of these approaches to a research question.

---

[1] Despite the focus here on nuclear DNA, mitochondrial DNA can also provide valuable information about contaminant effects. Baker et al. (1999) quantified genetic damage in voles from the contaminated area surrounding the Chernobyl reactor using a portion of the mitochondrial cytochrome *b* gene. They measured heteroplasmy (DNA sequence variation within an individual) to suggest increased rates of somatic mutation in the liver of irradiated voles.

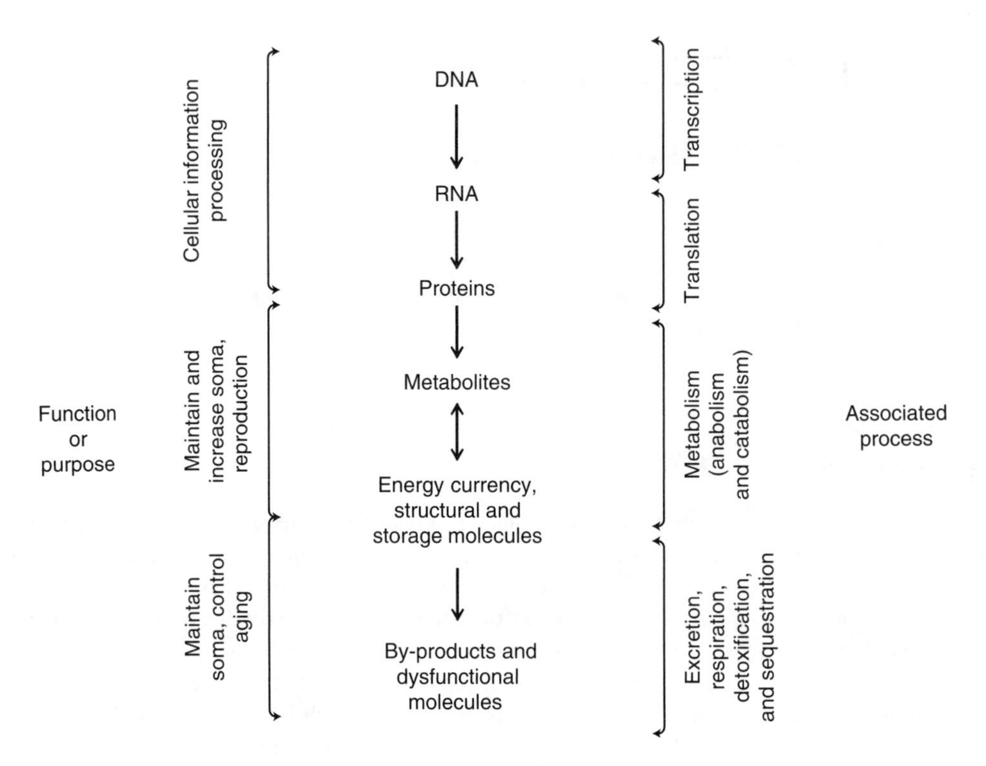

FIGURE 3.1 Hierarchical organization of biochemical effects discussed in this chapter.

The genome contains the instructions for growing and maintaining the soma. Although genomics often focuses on consequences to the germ line, somatic risks are also created by toxicant-induced changes to the genome. Carcinogenesis gives rise to the most obvious somatic risk (see Burdon (1999) for a fuller treatment of this topic). Changes in the genome will be discussed below relative to toxicant-induced modification of the DNA molecule.

Transcription and translation activities can provide evidence of response to a toxicant. As an example, El-Alfy and Schlenk (1998) discovered that up-regulation of a monooxygenase in Japanese medaka (*Oryzias latipes*) explained salinity-enhanced toxicity of aldicarb. In another study, differences in cytochrome P450 1A induction for chub (*Leuciscus cephalus*) populations with different contaminant exposure histories was taken as evidence of pollutant-induced changes in population genomics (Larno et al. 2001).

Shifts in metabolites can also suggest effects of, or responses to, toxicants. Kramer et al. (1992) measured glycolysis and Krebs cycle metabolites in mosquitofish (*Gambusia holbrooki*) exposed to mercury, finding decreased Krebs cycle flux during exposure. De Coen et al. (2001) noted increased Krebs cycle activity during *Daphnia magna* exposure to lindane, suggesting that biochemical assays be used to define the metabolic state of daphnids under stress.

Proteomics also has diverse applications in biochemical toxicology. Examples range from inducible detoxification proteins to evidence of effects at higher levels of organization. A specific example of evidence of potential effect at a higher level of biological organization is the abnormal induction of the egg protein, vitellogenin, in male fish exposed to methoxychlor (Schlenk et al. 1997) or synthetic estrogens (Schultz 2003). This induction will be discussed again in the following chapters in the context of endocrine dysfunction.

Processes ensuing at higher levels of biological organization can manifest as shifts in biochemical pools. Stressor-induced changes in bioenergetics can be detected with shifts in energy storage or pools of high-energy molecules. Biochemical by-products can also be assessed in cells, tissues, and physiological fluids. These types of biochemical shifts (e.g., shifts in heme biosynthesis) will also be discussed. The discussion of cellular, tissue, and bioenergetic effects detected with biochemical qualities will be addressed again in chapters exploring these higher levels of biological organization (i.e., Chapters 4–6).

3.2 DNA MODIFICATION

Damage to DNA occurs in several ways. It can result from strand breakage and subsequent imperfect repair. Damage can also result from chemical bonding directly to the DNA or by some similar DNA modification.

Although cancer is a paramount concern relative to somatic risk following toxicant-induced DNA modification, some DNA changes to the germ line have population consequences, and in some cases, these germ line-associated changes affect an exposed individual's Darwinian fitness. The population ecotoxicology section describes such changes and their consequences. As an example, men working in certain conditions or occupations can have elevated risks of teratogenic effects in their children or of their children developing cancer (Gardner et al. 1990, Stone 1992). In an even broader context, the mutation accumulation theory proposes that the accumulation of genetic damage determines the rate of aging for individuals (see Medvedev (1990) for details). Somatic longevity may be determined by DNA modifications accrued during an individual's life.

DNA can be damaged by contaminants or their metabolites that are free radicals or can facilitate free radical[2] generation. Free radicals can break one or both strands of the DNA molecule, or can oxidize bases in the DNA molecule. As an example of manifest breakage, Shugart (1996) noted elevated levels of double-strand breaks in DNA of sunfish from contaminated reaches of East Poplar Creek (Tennessee). As an example of base modification, Malins (1993) reported high concentrations of the guanine product, 2,6-diamino-4-hydroxy-5-formaminidopyrimidine, in tumors of English sole exposed to carcinogens in the field.

Contaminants or their metabolites can also bind covalently to DNA to form adducts. For example, Ericson and Larsson (2000) found DNA adducts in perch caught below a Kraft pulp mill. As another important example, metabolites of the carcinogen benzo[a]pyrene combined with guanine to form a guanosine adduct.

Still other modes of DNA damage are possible. Mercury cross-links DNA with proteins. Some metals bind to phosphate groups and heterocyclic bases of DNA. This changes the stability of the molecule and increases the incidence of mismatched bases.

Damage, modification, and imperfect repair of protooncogenes or tumor suppressor genes can initiate carcinogenesis (Burdon 1999). It can also accelerate the rate at which somatic mutations accumulate, and in doing so, accelerate the rate of aging. Genomic damage changes cell functioning and ultimately influences individual fitness.

3.3 DETOXIFICATION OF ORGANIC COMPOUNDS

A wide range of organic contaminants are transformed within organisms. The design behind such transformations is to render the toxic chemical more amenable to elimination; however, this is not always achieved without adverse consequences. The products of detoxification reactions can sometimes be more toxic or reactive than the original compound. Such a transformation that makes an inactive compound bioactive or an active compound more bioactive is called activation. In the case of cancer-producing agents, the original compound is a procarcinogen and the cancer-causing metabolite is called the carcinogen.

Detoxifying reactions are often classified as Phase I or II reactions. Phase I reactions produce a more reactive, and sometimes more hydrophilic, metabolite from the original compound; the product

[2] Free radicals are extremely reactive molecules possessing an unshared electron.

is more amenable to further reaction and, in some cases, elimination. The reactive groups —OH, $—NH_2$, —SH, and —COOH are added or made available by oxidation, hydrolysis, or reduction. Products of a Phase I reaction can be eliminated directly, be subject to additional Phase I transformations, or undergo Phase II transformations. Phase II reactions conjugate the compound or its Phase I metabolite(s) with some compound such as acetate, cysteine, glucuronic acid, sulfate, glycine, glutamine, or glutathione. The conjugate is more hydrophilic and readily eliminated than the compound was before conjugation.

3.3.1 Phase I Reactions

In Phase I, reactive groups are added or existing sites are made more readily available to further reactions. This can be illustrated with the metabolism of the dioxin benzo[a]pyrene (Figure 3.2). The addition of oxygen by the microsomal mixed function oxidase system (MFO, also referred to as the cytochrome P450 monooxygenase system) is the most prominent Phase I reaction. The cytochrome P450 system is present in diverse species from bacteria to vertebrates, and functions in the metabolism of endogenous (e.g., steroids and fatty acids) as well as xenobiotic compounds (Synder 2000). Associated Phase I oxidations involve two membrane-bound enzymes (cytochrome P450 isozymes and NADPH–cytochrome P450 reductase), NADPH, and molecular oxygen. The epoxidations of benzo[a]pyrene to benzo[a]-4,5-oxide, benzo[a]-7,8-oxide, and benzo[a]-9,10-oxide shown in Figure 3.2 are achieved by the MFO system. The MFO system is also responsible for the conversion of benzo[a]pyrene-7,8-dihydrodiol to benzo[a]pyrene-7,8-dihydrodiol-9,10-oxide.

Phase I enzymes also include epoxide hydrolases, esterases, and amidases that expose existing functional groups on compounds (George 1994). For example, epoxide hydrolase is responsible for

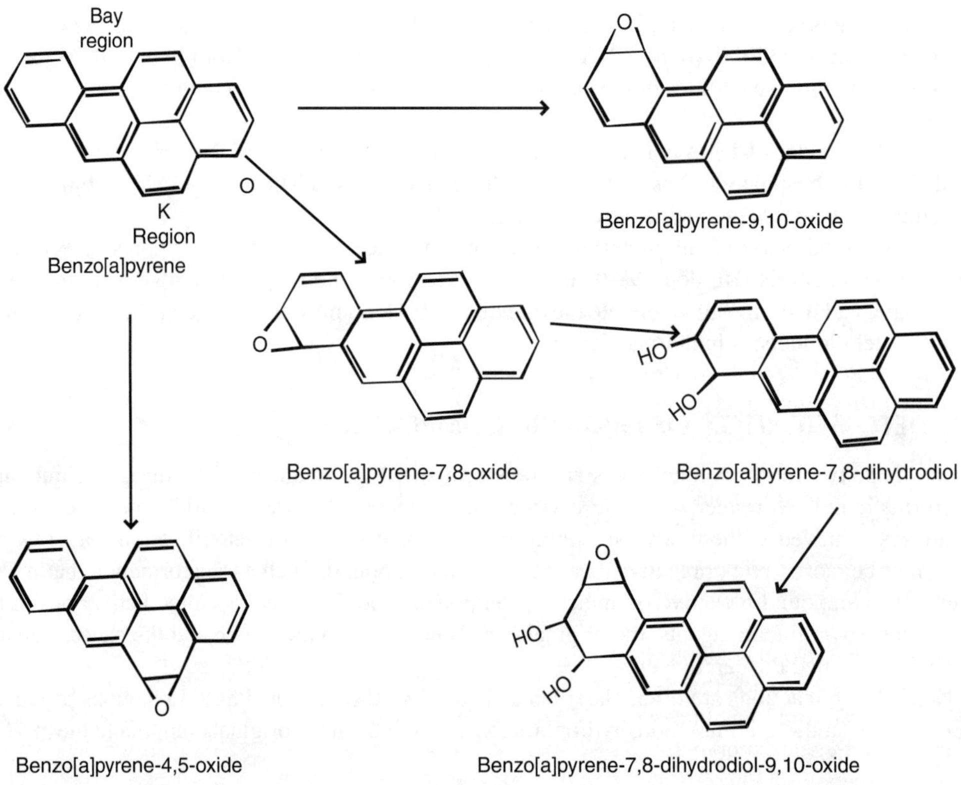

FIGURE 3.2 Phase I reactions for benzo[a]pyrene.

the Phase I conversion of benzo[a]pyrene-7,8-oxide to benzo[a]pyrene-7,8-dihydrodiol, shown in Figure 3.2. Epoxide hydrolase catalyzes the addition of water to MFO-generated epoxides. Other enzymes such as alcohol and aldehyde dehydrogenases, aldehyde oxidases, and carbonyl reductase generate products that are more rapidly eliminated than the original compound (George 1994, Parkinson 1996). As an example, ethanol is oxidized to acetaldehyde by alcohol dehydrogenase. This aldehyde is then oxidized by aldehyde dehydrogenase to acetic acid.

Type I reactions can also activate compounds to produce more poisonous or carcinogenic ones (Figure 3.2). The epoxide formed at the K region of benzo[a]pyrene (e.g., the epoxide in benzo[a]pyrene-4,5-oxide) and bay region dihydrodiols (e.g., benzo[a]pyrene-7,8-dihydrodiol) of polycyclic aromatic hydrocarbons are potent carcinogens (Timbrell 2000). These products of benzo[a]pyrene metabolism are strong electrophiles that bind to guanosine in the DNA molecule. Formation of such adducts within protooncogenes can result in cancer. Another example of Phase I activation is MFO-mediated epoxidation of the organochlorine pesticide aldrin to produce the more toxic dieldrin (Chambers and Yarbrough 1976).

3.3.2 Phase II (Conjugative) Reactions

In Phase II reactions, endogenous compounds are conjugated with contaminants or their metabolites to detoxify them or to accelerate their elimination. Phase II conjugation can occur without any Phase I reactions if the appropriate groups are already available. A compound is made more polar by binding it to some amino acid, carbohydrate derivative, glutathione, or sulfate. However, Phase II reactions can also involve methylation or acetylation that does not generally increase hydrophilicity.

Many Phase II reactions produce hydrophilic compounds readily eliminated from the individual. Conjugates are commonly organic anions that are eliminated by glomerular filtration and tubular transport in vertebrates (James 1987). Conjugation with glucuronic acid by UDP-glucuronosyltransferases involves generation of a polar, hydrophilic glucuronide by combining the compound with uridine diphosphate-glucuronic acid. As a relevant example, stimulated by concern about birth control compounds released from sewage treatment plants into waterways, Schultz (2003) studied the conjugation of the synthetic estrogen 17α-ethynylestradiol after its injection into trout. Sulfate conjugation by sulfotransferases produces hydrophilic conjugates of polyaromatic compounds, aliphatic alcohols, aromatic amines, and hydroxylamines. Xenobiotics with aromatic or aliphatic hydroxyl groups are prone to such sulfation (James 1987). Amino acids may be conjugated to carboxylic acid or aromatic hydroxylamine groups of contaminants or their metabolites. The amino acids most often involved are glycine, glutamine, and taurine (Jones 1987). Glutathione (i.e., glycine–cysteine–glutamic acid) can be conjugated by glutathione *S*-transferases with a wide array of electrophilic compounds. As examples, the benzo[a]-9,10-oxide and benzo[a]-4,5-oxides shown in Figure 3.2 can undergo further Phase I transformations and the products of these reactions conjugated with glutathione.

In contrast to the Phase II reactions just described, Phase II methylation and acetylation are reactions that do not generally produce more hydrophilic products. The reader is directed to Parkinson (1996) for more details about such reactions.

Box 3.1 There Is More to It Than Phase I and II Reactions

Our understanding of reactions associated with xenobiotic conversion and elimination has grown to include those outside the conventional Phase I and II reactions. The associated mechanisms have been referred to as Phase III reactions (Zimniak et al. 1993). The ATP-dependent glutathione *S*-conjugate export pump described by Ishikawa (1992) facilitates a Phase III reaction that removes xenobiotic Phase II metabolites from the cell. Probably the best Phase III example is the membrane-associated P-glycoprotein (P-gp) that acts as an energy-requiring efflux pump for xenobiotics and is described by Bard (2000) as the cell's first line of defense. It also eliminates metabolites from Phase I and II reactions from cells.

The P-gp mechanism for xenobiotic removal is similar to the multidrug resistance (MDR) transporter protein discovered first in cancer cells that had become resistant to chemotherapeutic agents. The cancer cell resistance results from reduced intracellular concentrations of these chemotherapeutic agents due to the overexpression of an efficient ATP-dependent membrane-bound pump, P-gp. This 170-kDa protein not only increases resistance to the original anticancer drug, but also improves resistance to unrelated chemotherapy agents. The P-gp acts as a barrier to xenobiotic absorption and accelerates their removal if they gained entry into the cell (Abou-Donia et al. 2002). The mammalian P-gp is expressed at high levels in the kidney, adrenal glands, liver, and lungs. Expression in mammalian brain capillary endothelial cells has also been shown to reduce neurotoxicity of the pesticide ivermectin (Sckinkel et al. 1994).

The multixenobiotic resistance (MXR) mechanism is similar to MDR, involving a membrane-associated transport P-gp that removes moderately hydrophobic, planar compounds (Segner and Braunbeck 1998). Bard (2000) defines its substrates as "moderately hydrophobic, amphipathic (i.e., somewhat soluble in both lipid and water), low molecular weight, planar molecules with a basic nitrogen atom, cationic or neutral but never anionic, and natural products." P-gp can be induced during exposure to xenobiotics and has regulatory genes in common with the cytochrome P450 system. It has been found in mussel (*Mytilus galloprovincialis*) cell membranes, leading Kurelec and Pivčević (1991) to speculate that this mechanism could account for the relatively high tolerance of these mussels to contaminants. The *MXR* gene was also found recently in marine fish (*Anoplarchus purpurescens*) (Bard et al. 2002), *Mytilus edulis* (Luedeking and Koehler 2004), and the Asiatic clam, *Corbicula fluminea* (Achard et al. 2004). Their levels have been correlated with elevated concentrations of a variety of toxicants ranging from crude oil (Hamdoun et al. 2002) to metals (Achard et al. 2004). Induction by metals likely reflects the fact that protein-damaging chemicals induce several systems simultaneously, including stress proteins, MXR, and cytochrome P450.

How does the P-gp work? A "flippase" model was proposed by Higgins and Gottesman (1992) in which the xenobiotic binds to the P-gp at the inner surface of the cell membrane and is "flipped" via an energy-requiring mechanism to the outside surface of the cell membrane.

The MXR's presence in many taxonomic groups and its role in detoxification of many contaminants led Smital and Kurelec (1998) to define a new group of pollutants, that is, those that modify the MXR response. In the laboratory, MXR can be readily inhibited with verapamil, so there is potential for some environmental chemicals doing the same. A water-soluble fraction of weathered crude oil, for example, appears to competitively inhibit MXR in larvae of the marine worm, *Urechis caupo* (Hamdoun et al. 2002). Bard (2000) reviewed reports of such chemosensitizers (Smital and Kurelec 1998), listing the following contaminants: pentchlorophenol, 2-acetylaminofluorene, diesel oil, and several pesticides (chlorbenside, sulfallate, and dacthal).

3.4 METAL DETOXIFICATION, REGULATION, AND SEQUESTRATION

Predicting the consequences of metal exposure is complicated because metals may be essential or nonessential. Very low concentrations of essential metals[3] can be as harmful as high concentrations (Figure 3.3, upper panel). Nonessential metals display more conventional toxicity curves, showing

[3] The essential metals are currently believed to be Co, Cr, Cu, Fe, Mn, Mo, Ni, Se, V, and Zn (Fraústo da Silva and Williams 1991, Mertz 1981).

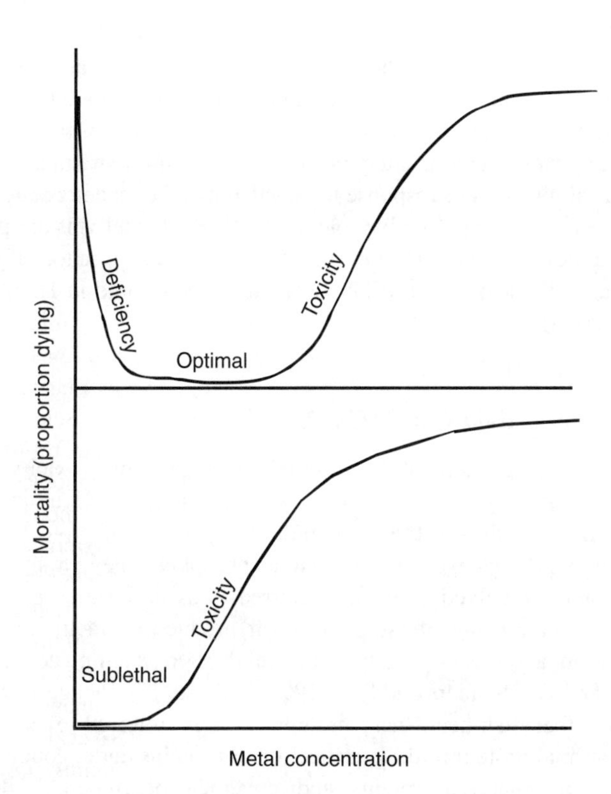

FIGURE 3.3 Mortality versus concentration for essential (upper panel) and nonessential (lower panel) metals. A deficiency occurs if an essential metal is present below a certain concentration. This is not the case for a nonessential metal. An essential metal will have an optimal range above and below which mortality begins to be expressed. Increasing concentrations of the nonessential metal will increase the level of mortality experienced in a group of exposed individuals. There might or might not be an apparent threshold concentration below which no effect is expressed.

a sigmoidal increase in proportion of exposed individuals dying with an increase in metal concentration (Figure 3.3, lower panel). Essential metal deficiencies manifest in many ways other than death. For example, insufficient intake of copper or zinc causes immunodeficiencies in mice (Beach et al. 1982, Prohaska and Lukasewycz 1981).

Understanding this dichotomy of essential and nonessential metal concentration–effect curves can still be insufficient for sound prediction of metal effects. For example, chronic exposure to the nonessential element cadmium can cause symptoms of zinc deficiency because cadmium displaces zinc in metalloenzymes. Excessive amounts of nonessential tungsten can cause an apparent deficiency of molybdenum, an essential and chemically similar element (Mertz 1981). Such an effect would appear as a shift to the left for the curve shown in the upper panel of Figure 3.3 (x-axis being the essential metal concentration). The bioactivity of some nonessential elements can also be affected by another element. For example, mercury toxicity is lowered if sufficient concentrations of selenium are also present. This would cause the curve in the lower panel of Figure 3.3 to shift to the right.

Excess metals are dealt with in two ways, elimination or sequestration. Sequestration can involve metal complexation with proteins or incorporation into granules. Sequestration in granules will be discussed in the next chapter. Biomolecules involved in lessening metal intoxication will be described here.

Metallothioneins are low-molecular-weight, cytosolic proteins that take up and facilitate transport, sequestration, and excretion of metals such as cadmium, copper, silver, mercury, and zinc. They have high cysteine content, giving them the ability to form metal–thiolate clusters. Elevated metal concentrations induce the production of metallothioneins to levels above those needed for normal metal homeostasis. Metallothioneins bind metals, lowering the concentrations of metal available to interact with sites of adverse action. Titers of metallothionein-coding mRNA or metallothionein itself are often used as biomarkers of response to elevated metal concentrations.

Phytochelatin serves a similar protective role in plants. Phytochelatins are peptides of the form (γ-glutamic acid–cysteine)$_n$-glycine where $n = 3, 5, 6$, or 7 (Grill et al. 1985). Elevated concentrations of other phytochelatin-like peptides have recently been found in zinc-tolerant green algae (Pawlik-Skowrońska 2003).

3.5 STRESS PROTEINS AND PROTEOTOXICITY

The adverse effects of some agents result from protein damage (proteotoxicity). Indeed, this mode of action is so pervasive that a general cellular stress response has evolved in most animal, plant, or microbial species. Early studies of the stress-induced synthesis of protective proteins involved the heat shock reaction—the organisms' response to an abrupt change in temperature (Craig 1985). Consequently, the proteins involved were first referred to as heat shock proteins. However, we now know that a wide range of agents stimulate their production, including metals, metalloids, ultraviolet (UV) radiation, and diverse organic compounds such as amino acid analogs, puromycin, and ethanol (Hightower 1991, Sanders and Dyer 1994, Vedel and DePledge 1995). Because of their induction by stressors other than heat, these proteins are now referred to as stress proteins. They function to facilitate normal protein folding, protection of proteins under conditions that might lead to denaturation, repair of denatured proteins, and movement of irreparably denatured protein to lysosomes (Sanders and Dyer 1994).[4] Some stress proteins are present at basal levels but others are present only after induction by some agent. Regardless of whether they were present under normal conditions or induced by proteotoxic conditions, they collectively function to maintain homeostasis by fostering essential protein levels, structure, and function.

The stress proteins are classified and named based on their molecular size. Stress70 and Stress90 are 70 and 90 kDa stress proteins, respectively. Smaller (60 kDa) stress proteins are called chaperons owing to their role in mediating proper protein folding. Chaperons are abbreviated cpn60 (Di Giulio et al. 1995). Stress70, Stress90, and cpn60 are present at basal levels that increase to reduce proteotoxicity on appropriate induction. Another group of stress proteins (20–30 kDa) are the Low Molecular Weight (LMW) stress proteins that are present only after induction.

Proteomic analysis of stress proteins is advocated by Sanders and Dyer (1994) for potentially identifying agents responsible for adverse impact on species in the field. Their argument was based on the observation that different chemicals induce different stress proteins to varying degrees. Comparison of stress protein expression in field organisms to those of organisms exposed to each candidate toxicant individually in the laboratory could provide causal insight. For example, Vedel and DePledge (1995) measured Stress70 increase in crabs (*Carcinus maenas*) after laboratory copper exposure. Currie and Tufts (1997) explored the combination of anoxia and heat stress on Stress70 induction in trout (*Oncorhychus mykiss*) red blood cells. Still other researchers focus on stress protein genomics. Hightower (1991) made the novel suggestion that we could use the change in heat shock protein genomes of various species to track the consequences of global warming. He hypothesized that, as suggested by laboratory studies and field studies of desert species, the heat shock genes will move in the direction of overexpression with adaptation to rapid warming.

---

[4] Because our focus is chemical toxicology, other stress proteins will be ignored here. However, it should be mentioned for the sake of completeness that glucose-regulated proteins (GRPs), metallothionein, hemeoxygenase, and the multidrug-resistant *p-glycoprotein* are considered by many to be stress proteins (Di Giulio et al. 1995, Hightower 1991, Sander and Dyer 1994).

3.6 OXIDATIVE STRESS

Molecular oxygen is both benign and malign. On the one hand it provides enormous advantages and on the other it imposes a universal toxicity. This toxicity is largely due to the intermediates of oxygen reduction, that is, $O_2^{\bullet -}$, H_2O_2, and $OH^\bullet$, and any organism that avails itself of the benefits of oxygen does so at the cost of maintaining an elaborate system of defenses against these intermediates.

(Fridovich 1983)

A price was levied when much of the life on Earth took on the energetic advantage of using molecular oxygen as a terminal electron acceptor for respiration. Very reactive, free oxyradicals[5] and oxyradical-producing molecules such as hydrogen peroxide are generated during aerobic metabolism. Oxyradicals oxidize lipids, proteins, and DNA, causing diverse effects ranging from membrane damage to enzyme dysfunction to cancer to accelerated aging. Consequently, organisms using aerobic respiration had to develop ways of coping with oxidative stress.

Oxidative stress is reduced in two ways. Antioxidant molecules are produced that react with oxyradicals and enzymes are synthesized that consume oxyradicals or oxyradical-generating chemicals. Antioxidants include catecholamines, glutathione, uric acid, and Vitamins A, C, and E. Enzymes include superoxide dismutase, catalase, and glutathione peroxidase that catalyze the reactions shown in Equations 3.1–3.3, respectively. (The unpaired electron in free radicals is designated as a dot by convention. GSH and GSSG in these equations are reduced and oxidized glutathione, respectively.)

$$2O_2^{\bullet -} + 2H^+ \rightarrow H_2O_2 + O_2 \quad (3.1)$$

$$2H_2O_2 \rightarrow 2H_2O + O_2 \quad (3.2)$$

$$2GSH + H_2O_2 \rightarrow GSSG + 2H_2O \quad (3.3)$$

The removal of hydrogen peroxide, which is not itself an oxyradical, is crucial because it produces the hydroxyl radical ($OH^\bullet$). This is accomplished through the Fenton reaction which, catalyzed by a transition metal ion, generates $OH^\bullet$ and OH^- from H_2O_2 (Equation 3.4). The transition metal ion can be Cu(I), Cr(V), Fe(II), Mn(II), or Ni(II) (Gregus and Klaassen 1996).

$$H_2O_2 + Fe^{2+} \rightarrow Fe^{3+} + HO^- + HO^\bullet \quad (3.4)$$

Why is this discussion relevant to environmental toxicants? Many organic chemicals become free radicals during biochemical reactions or can generate oxyradicals. For example, paraquat reacting within the MFO system becomes a charged free radical that reacts with molecular oxygen to produce the superoxide anion, $O_2^{\bullet -}$. After reacting with molecular oxygen, the paraquat becomes available again to enter the same reactions, producing more superoxide anions each time it passes through the redox cycle. Another example is carbon tetrachloride, which is converted to the trichloromethyl radical ($CCl_4 + e^- \rightarrow CCl_3^{\bullet -} + Cl^-$) during Phase I reactions (Slater 1984). As a final example, enhanced oxidative damage at high metal concentrations occurs due to hydroxyl radical formation. In such a case, more metal ion is available to catalyze the Fenton reaction and more oxyradicals are formed as a consequence.

Responses to oxidative stress are used with field and laboratory exposures as evidence for xenobiotic hazard (Livingston et al. 1990, Winston and Di Giulio 1991). As an example, glutathione and antioxidant enzymes shifted in mussels (*M. galloprovincialis*) transplanted from clean to metal-contaminated conditions (Regoli and Principato 1995). Regoli (2000) later used the total oxyradical scavenging capacity of mussels to indicate adverse effect of field exposure to metals.

[5] A free radical is a charged or uncharged molecule or molecular fragment that has an unpaired electron (Slater 1984). An oxyradical is a free radical in which the unpaired electron is associated with an oxygen atom.

3.7 ENZYME DYSFUNCTION

Metals inhibit many types of enzymes that range in function from facilitating digestion (Chen et al. 2002) to heme synthesis (Dwyer et al. 1988). Eichhorn (1975) and, more extensively, Fraústo da Silva and Williams (1991) provide details about metal binding to, and modifying the activity of, enzymes. A metal can displace another metal from an enzyme's active site or otherwise interact with the enzyme to change its secondary or tertiary structure. Metal ions can produce dysfunction by either increasing or decreasing enzyme activity (Brown 1976, Eichhorn 1975).

Organic contaminants can also modify enzyme activity and, in so doing, modify an exposed individual's fitness. For example, brain cholinesterase activity was depressed for individuals of several bird species found dying after organophosphorus or carbamate insecticide spraying (Hill and Fleming 1982). More global examples exist such as the population consequences of DDT or DDE inhibition of Ca–ATPase in the eggshell gland of birds. Its inhibition resulted in thin-shelled eggs that broke before full development and hatching (e.g., Kolaja and Hinton 1979). Inhibition of this one enzyme resulted in abrupt decreases in population size for osprey, *Pandion haliaetus* (Ambrose 2001, Spitzer et al. 1978), bald eagle, *Haliaeetus leucocephalus* (Bowerman et al. 1995), falcon, *Falco peregrinus* (Ratcliffe 1967, 1970), and brown pelican, *Pelecanus occidentalis* (Hall 1987).

3.8 HEME BIOSYNTHESIS INHIBITION

Porphyrin and heme synthesis (Figure 3.4) is central to producing hemoglobin, myoglobin, cytochromes, tryptophan pyrrolase, catalase, and peroxidase. Although all cells produce heme, mammals produce most heme in the liver and erythroid cells (Marks 1985). In the mitochondria, where

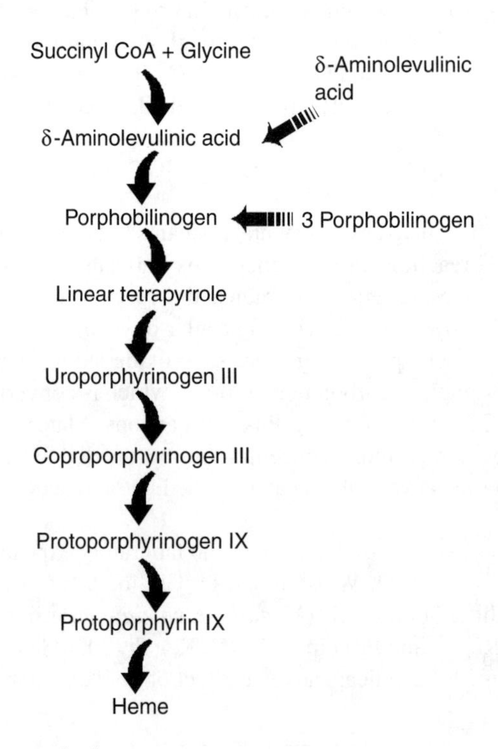

FIGURE 3.4 Steps in heme synthesis.

the tricarboxylic acid cycle generates ample succinyl CoA, succinyl CoA and glycine are converted to δ-aminolevulinic acid by δ-aminolevulinic acid synthetase. The δ-aminolevulinic acid then passes into the cytoplasm where two molecules of δ-aminolevulinic acid are then combined by δ-aminolevulinic acid dehydratase to form porphobilinogen. Four molecules of porphobilinogen are then acted on by uroporphyrinogen I synthetase to produce linear tetrapyrrole. Still in the cytoplasm, the linear tetrapyrrole is converted to uroporphyrinogen III by uroporphyrinogen III synthase and uroporphyrinogen III cosynthase. Uroporphyrinogen I synthase catalyses the initial process which, if left on its own, would eventually produce uroporphyrinogen I, a product with no known biochemical utility. However, uroporphyrinogen III cosynthase completes the process to yield uroporphyrinogen III instead. Uroporphyinogen III is converted to coproporphyrinogen III by uroporphyrinogen decarboxylase. Intermediate products are also generated by uroporphyrinogen decarboxylase including the seven, six, and five carboxy intermediates, heptacarboxyporphyrinogen, hexacarboxyporphyrinogen, and pentacarboxyporphyrinogen. Porphyrins with 4–8 carboxyl groups tend to be generated in excess during heme synthesis and are excreted in the urine (Wood et al. 1993). The remaining steps take place within the mitochondria. Coproporphyrinogen III is converted to protoporphyrin IX by coproporphyrinogen oxidase and this protoporphyrin IX is acted on by protoporphyrinogen IX oxidase and then ferrochelatase to produce heme. The heme concentration in the mitochondria completes a feedback loop that regulates further synthesis.

Porphyrin and heme synthesis can be influenced by a variety of factors, and deviations from normal synthesis can have serious effects on an individual's fitness. A human example is the genetic disorder in porphyrin synthesis called acute intermittent porphyria. Acute intermittent porphyria results from a uroporphyrinogen I synthetase deficiency that substantially diminishes this enzyme's activity. The disorder is diagnosed by urine analysis for excess porphobilinogen, the substrate for this enzyme. Because of the strong influence of hormonal changes, the disorder does not usually manifest in humans until puberty (Tschudy et al. 1975). It manifests in a range of intermittent effects including abdominal pain, constipation, hypertension, psychosis, and even death from respiratory paralysis (Becker and Kramer 1977, Goldberg 1959, Stein and Tschudy 1970). Barbiturates that induce the early steps of heme synthesis or ethanol can trigger the adverse effects of this disorder (Tschudy et al. 1975).

Inorganic toxicants also interfere with heme synthesis. Lead decreases heme production by binding to a susceptible sulfhydrl group of aminolevulinic acid dehydrase (ALAD). An excess of δ-aminolevulinic acid in urine and anemia are indicative of lead poisoning as a consequence. The inhibition of ALAD has been developed as a biomarker for effects to fish from exposure to lead, cadmium, and other metals (e.g., Dwyer et al. 1988, Johansson-Sjöbeck and Larsson 1978, 1979, Marks 1985, Schmitt et al. 1993). Mercury impairs enzymes involved in heme synthesis and also directly oxidizes reduced porphyrins (Wood et al. 1993) (Box 2.1). Prolonged imbibing of sodium arsenate in water by rats depressed δ-aminolevulinic acid synthetase activity and modified the activity of other enzymes involved in heme synthesis (Wood and Fowler 1978).

Organic chemicals can also interfere with heme synthesis. An outbreak of human porphyrias was precipitated in Turkey during the 1950s when hexachlorobenzene fungicide-treated wheat was consumed by several thousand Turks (Marks 1985). In addition to the skin lesions resulting from this unintentional consumption of treated grain, the decrease in uroporphyrinogen decarboxylase activity in afflicted individuals resulted in the accumulation of so much uroporphyrin that their urine was the color of dark red wine. The heme synthesis dysfunction appears to result from the action of a Phase I metabolite of hexachlorobenzene. A Phase I metabolite of the dioxin 2,3,7,8-tetrachlorodibenzo-p-dioxin (TCDD) also causes uroporohyrinogen decarboxylase dysfunction. Other organic chemicals affecting heme synthesis include polychlorinated biphenyls (PCBs), polybrominated biphenyls (PBBs), and pesticides (diazinon, lindane, and heptachlor) (Marks 1985).

Box 3.2 Of Mice and Men (Dentists)

Perhaps the best demonstrations of metal effects on heme synthesis are provided by Wood and his coworkers (i.e., Wood and Fowler 1978, Woods et al. 1993).

Prompted by concern about chronic sodium arsenate exposure in drinking water, Wood and Fowler (1978) exposed rats and mice to arsenate, and examined the dose–response relationships for heme synthesis. The influence of graded doses of arsenate on δ-aminolevulinic acid synthetase, uroporphyrinogen I synthase, and ferrochelatase activities are shown in Figure 3.5. Notice that uroporphyrinogen I synthase activity increased slightly, but the activities of the other two enzymes decreased. Concentrations of uroporphyrin and coproporphyrin also increased with dose. Uroprophyrin concentrations in urine for mice exposed to 20, 40, and 85 μg/L doses were 120%, 205%, and 910% of control concentrations, respectively. Similarly, coproporphyrin concentrations were 104%, 142%, and 743% of control concentrations. Heme synthesis was clearly influenced by arsenic exposure in drinking water.

Urinary porphyrins and mercury were measured in volunteer male dentists at the 1991 and 1992 American Dental Association meetings. Notionally, the dentists had been exposed to mercury while working with the silver–mercury amalgam used for dental fillings. Urinary mercury concentrations ranged from <0.5 to 556 μg/L with approximately 10% of screened dentists having concentrations exceeding 20 μg/L. (The World Health Organization had recommended an exposure limit of 25 μg/L in the urine.) Results were analyzed by splitting the dentists into those with no detected urinary mercury (<0.5 μg/L) (n = 37) and those with ≥20 μg/L of mercury in their urine (n = 56). The lower panel of Figure 3.5 shows the differences in mean concentrations of urinary pentacarboxylporphyrin, precoproporphyrin, and coproporphyrin for these two groups of dentists. All three were significantly higher in dentists with high exposures ($\alpha = 0.05$) but differences in concentrations of six- to seven-carboxyl porphyrins were not. The authors concluded that these three porphyrins were excellent biomarkers for long-term mercury exposure in humans.

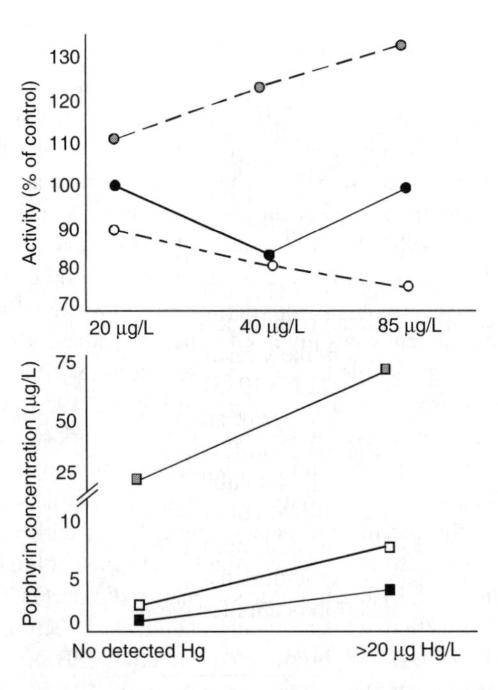

FIGURE 3.5 (Upper panel) The change in heme synthesis enzyme activities (relative to that of controls) for mice chronically exposed to sodium arsenate. (Solid circle = δ-aminolevulinic acid synthetase, shaded circle = uroporphyrinogen I synthase, and open circle = ferrochelatase activities.) (Data from Table 2 of Wood and Fowler 1978.) (Lower panel) The shift in porphyrins in dentists exposed through their occupation to elevated levels of elemental mercury. Mean concentrations of urinary pentacarboxylporphyrin (solid squares), precoproporphyrin (shaded squares), and coproporphyrin (open squares) are shown for these two groups of dentists. (Data from Table 3 of Wood et al. 1993.)

3.9 OXIDATIVE PHOSPHORYLATION INHIBITION

Some chemicals such as salicylic acid or pentachlorophenol act by uncoupling oxidative phosphorylation in the mitochondria.[6] Understandably, this leads to distinct physiological shifts such as the lowered blood carbon dioxide levels and elevated blood pH (alkalosis) seen in humans overdosed with aspirin (Timbrell 2000), or the significant increase in total oxygen consumption and gill ventilation volume for trout overdosed with pentachlorophenol (McKim et al. 1987). A number of toxicants act by this mode of action, notably substituted phenols such as 2,4-dinitrophenol, pentachlorophenol, and 2,4,5-trichlorophenol. In a study by Penttinen and Kukkonen (1998), exposure to substituted phenols predictably shifted the metabolic rate of exposed aquatic invertebrates.

Intoxications by substituted phenols are also described below as cases of narcosis with the relative toxicities of 2,4-dinitrophenol, pentachlorophenol, and 2,4,5-trichlorophenol being related to their "effects on the energy-transducing membrane by uncoupling oxidative phosphorylation" (Penttinen and Kukkonen 1998). Magnitude of effect is related to each chemical's lipophilicity (i.e., propensity to enter the membrane) and reactivity (i.e., ability to react at the appropriate receptor site). Similarly, in a study of eight phenols that were narcotics and uncouplers of oxidative phosphorylation, lipophility (log K_{ow}) and acidity (pK_a) were found to be important predictors of potency (Schüürmann et al. 1997).

3.10 NARCOSIS

Narcosis, including that brought about by many xenobiotics, results from a general and reversible disruption of cell membrane functioning. There is a general depression of biological activity due to toxicant interaction with membranes. The most familiar case of narcosis is that occurring with anesthetic administration. The exact nature of the narcotic–membrane interaction seems to be incompletely understood at the moment although changes in nerve cell membranes are clearly important in higher animals. The protein-binding theory suggests that anesthetics (narcotics) act on ion channels by directly binding to membrane proteins but the critical volume theory suggests that anesthetics enter the membrane and modify its lipid bilayer (Abernethy et al. 1988). The critical volume theory proposes that the toxicant accumulates in the lipid bilayer to such an extent that the membrane swells, causing dysfunction. The toxicant molecule's volume determines its capacity to swell the membrane. Narcosis occurs when the membrane is swollen beyond a critical volume. In contrast, the protein-binding theory suggests that the toxicant causes dysfunction by binding reversibly to critical protein sites on the membrane. Franks and Lieb (1978) applied x-ray and neutron diffraction techniques to find no change in the lipid bilayer of nerve cells, suggesting that anesthetic effect did not involve lipid bilayer swelling. Later, they subjected a model protein, luciferase, to a wide range of anesthetics and found that the anesthetics could modify the protein's activity by binding to specific receptors (Franks and Lieb 1984). The potency of an anesthetic in animal tests was also highly correlated with its ability to inhibit luciferase. Franks and Lieb argued from this evidence that anesthetic action likely results from competition with endogenous ligands for protein receptor sites. They (Franks and Lieb 1985) also explained the cutoff phenomena of many anesthetic series with this model system. A series of anesthetics appears to have increasing potency as lipophilicity increases, but only to a certain cutoff point. Potency decreases quickly beyond that point. Although the strong correlation with lipid solubility had provided support to the explanation of anesthetic cutoff point on the basis of critical volume theory, they demonstrated with luciferase exposed to n-alcohols and n-alkanes that the cutoff point was related to the anesthetic's binding to a hydrophobic protein pocket site of very specific dimensions. By analogy to the luciferase binding pocket, they suggested that a similar situation occurs for membrane-associated proteins. The importance of protein binding in

[6] Salicylic acid is produced from acetylsalicylic acid (aspirin) by a Phase I hydrolysis. It can then undergo conjugation with glucuronic acid or glycine (Timbrell 2000).

determining anesthetic potency was reinforced in another study using optical isomers of isoflurane. These optimal isomers are equally soluble in lipids but have very different potencies and binding capacities for ion channels of molluscan nerves (Franks and Lieb 1991). The remarkable work of Franks and Lieb lends strong, but not yet definitive, support for the protein-binding theory of narcosis.

Narcotics can be defined as "polar" (weak acids) and "nonpolar" (neutral or nonelectrolyte). Many of the narcotics used in the Franks and Lieb studies were nonpolar, and lipophility was adequate to predict trends in potency for them. Almost any nonelectrolyte organic compound that can become associated with the cell membrane can express a nonspecific narcosis, but chemicals commonly categorized as nonelectrolyte narcotics are ethers, alcohols, and chlorinated alkanes. Other narcotics are weak acids. The most important of these polar narcotics have already been discussed (i.e., the substituted phenols). Ionization also becomes important in predicting potency for these narcotics because the unionized form of a compound is generally believed to be the most capable of passage into lipid-rich membranes. McCarty et al. (1993) suggested that pK_a and $\log K_{ow}$ were important in predicting relative lethal effects of polar narcotics. (See Box 9.3 in Chapter 9 for related details.) The concentration of an unionized narcotic in an exposure solution can be calculated if the compound's pK_a and the medium's pH are known. The Henderson–Hasselbach relationship can be used to estimate the proportion of a weak acid that is unionized:

$$f_u = \frac{1}{1 + 10^{pH - pK_a}}. \tag{3.5}$$

The critical body residue (CBR) approach is often applied in dealing with narcotics. The concept is simply that a narcotic's action is a direct function of the whole body dose at any moment. For example, Penttinen and Kukkonen (1998) modeled effects of substituted phenols on aquatic invertebrates with a threshold model of narcotic tissue concentration versus metabolic rate.

Before leaving the topic of polar narcotics, it is important to highlight a minor inconsistency. Narcosis was described as a general phenomenon associated with cell membrane changes (Section 3.9), but several substituted phenols act specifically on oxidative phosphorylation. This specificity is inconsistent with the definition of narcosis as a nonspecific phenomenon. Although this inconsistency does not impede understanding, it does cause confusion.

3.11 SUMMARY

Many, but not all, biochemical responses and consequences of toxicant exposure were discussed in this brief chapter. Others will emerge in the next few chapters in discussions such as that addressing cellular accumulation of degradation products from oxidative damage. Others such as the important MXR transporter (Hamdoun et al. 2002) are relevant to discussions of contaminant uptake and elimination. Together, they provide strong causal insights and sensitive biomarkers of contaminant exposure or effect.

3.11.1 Summary of Foundation Concepts and Paradigms

- The fields of study describing levels in the biological information hierarchy covered in this chapter are the following: genomics → transcriptomics → proteomics → metabolomics → bioenergetics or biochemical physiology → molecular toxicology.
- Damage to DNA occurs by DNA strand breakage and subsequent imperfect repair, by chemical bonding of a toxicant or its metabolite directly with the DNA, or by some similar DNA modification such as DNA-protein cross-linking. Consequent effects to the

soma include cancer and perhaps accelerated aging (i.e., the mutation accumulation theory of aging).

- Many organic contaminants are subject to transformation within organisms that renders the toxic chemical more amenable to elimination. In some cases, the transformation products can be more toxic or reactive than the original compound. A transformation in which an inactive compound becomes bioactive or an active compound becomes more bioactive is called activation.
- A series of Phase I and II reactions can occur, which render a toxicant more amenable to elimination. Phase I reactions make compounds more reactive and sometimes more hydrophilic. Reactive groups are added or existing sites are made more readily available to further reactions. In Phase II (conjugative) reactions, endogenous compounds are conjugated with contaminants or their metabolites to accelerate their elimination. Phase II conjugation can occur without any Phase I reactions if the appropriate groups are already available.
- Toxic metals can bind with metallothioneins or phytochelatins to enhance transport, sequestration, and elimination, or they can be incorporated into granules.
- Stress proteins lessen proteotoxicity of a wide range of stressors including metals, metalloids, UV radiation, many organic compounds, and abrupt changes in temperature.
- Oxidative stress is reduced by the production of antioxidant molecules and by production of enzymes that reduce the concentrations of free radicals or free radical generating molecules.
- Organic and inorganic toxicants can also bind to enzymes, causing dysfunction.
- Heme synthesis is also sensitive to the action of organic and inorganic contaminants. Shifts in porphyrin pools in body fluids such as urine can be a sensitive biomarker as a consequence.
- Some toxicants (e.g., substituted phenols) act by uncoupling oxidative phosphorylation in mitochondria.
- Narcosis, a result of a reversible disruption of cell membrane functioning, generally depresses biological activity. Many toxicants act as narcotics. Two theories exist for narcosis but current information supports the theory emphasizing action through narcotic binding to membrane proteins and disruption of their functioning.

REFERENCES

Abernethy, S.G., Mackay, D., and McCarty, L.S., "Volume Fraction" correlation for narcosis in aquatic organisms: The key role of partitioning, *Environ. Toxicol. Chem.*, 7, 469–481, 1988.

Abou-Donia, M., Elmasry, E.M., and Abu-Qare, A.W., Metabolism and toxicokinetics of xenobiotics, In *Handbook of Toxicology*, 2nd ed., Derelanko, M.J. and Hollinger, M.A. (eds.), CRC Press, Boca Raton, FL, 2002, pp. 769–833.

Achard, M., Baudrimont, M., Boudou, A., and Bourineaud, J.P., Induction of a multixenobiotic resistance protein (MXR) in the Asiatic clam *Corbicula fluminea* after heavy metal exposure, *Aquat. Toxicol.*, 67, 347–357, 2004.

Ambrose P., Osprey revival from DDT complete in Chesapeake Bay, *Mar. Pollut. Bull.*, 42, 388, 2001.

Baker, R.J., DeWoody, J.A., Wright, A.J., and Chesser, R.K., On the utility of heteroplasmy in genotoxicity studies: An example from Chornobyl, *Ecotoxicology*, 8, 301–309, 1999.

Bard, S.M., Multixenobiotic resistance as a cellular defense mechanism in aquatic organisms, *Aquat. Toxicol.*, 48, 357–389, 2000.

Bard, S.M., Woodin, B.R., and Stegeman, J.J., Expression of P-glycoprotein and cytochrome P-450 1A in intertidal fish (*Anoplarchus purpurescens*) exposed to environmental contaminants, *Aquat. Toxicol.*, 60, 17–32, 2002.

Beach, R.S., Gershwin, M.E., and Hurley, L.S., Gestational zinc deprivation in mice: Persistence of immunodeficiency for three generations, *Science*, 218, 469–471, 1982.

Becker, D.M. and Kramer, S., The neurological manifestations of porphyria: A review, *Medicine*, 56, 411–423, 1977.

Bowerman, W.W., Giesy, J.P., Best, D.Q., and Kramer, V.J., A review of factors affecting productivity of bald eagles in the Great lakes regions: Implications for recovery, *Environ. Health Perspect.*, 103(Suppl. 4), 51–59, 1995.

Brown, G.W., Jr., Effects of polluting substances on enzymes of aquatic organisms, *J. Fish. Res. Board Can.*, 33, 2018–2022, 1976.

Burdon, R.H., *Genes and the Environment*. Taylor & Francis Ltd., Philadelphia, PA, 1999.

Craig, E.A., The heat shock response, *CRC Crit. Rev. Biochem.*, 18, 239–280, 1985.

Chambers, J.E. and Yarbrough, J.D., Xenobiotic biotransformation systems in fishes, *Comp. Biochem. Physiol. C*, 55, 77–84, 1976.

Chen, Z., Mayer, L.M., Weston, D.O., Bock, M.J., and Jumars, P.A., Inhibition of digestive enzyme activities by copper in the guts of various benthic invertebrates, *Environ. Toxicol. Chem.*, 21, 1243–1248, 2002.

Currie, S. and Tufts, B., Synthesis of stress protein 70 (Hsp70) in rainbow trout (*Oncorhynchus mykiss*) red blood cells, *J. Exp. Biol.*, 200, 607–614, 1997.

De Coen, W.M., Janssen, C.R., and Segner, H., The use of biomarkers in *Daphnia magna* toxicity testing. V. *In vivo* alterations in the carbohydrate metabolism of *Daphnia magna* exposed to sublethal concentrations of mercury and lindane, *Ecotoxicol. Environ. Saf.*, 48, 223–234, 2001.

Di Giulio, R.T., Benson, W.H., Sanders, B.M., and Van Veld, P.A., Biochemical mechanisms: Metabolism, adaptation, and toxicity, In *Fundamentals of Aquatic Toxicology*, 2nd ed., Rand, R.M. (ed.), Taylor & Francis, Washington, D.C., 1995, pp. 523–561.

Dwyer, F.J., Schmitt, C.J., Finger, S.E., and Mehrle, P.M., Biochemical changes in longear sunfish, *Lepomis megalotis*, associated with lead, cadmium and zinc from mine tailings, *J. Fish Biol.*, 33, 307–317, 1988.

Eichhorn, G.L., Active sites of biological macromolecules and their interaction with heavy metals, In *Ecological Toxicology: Effects of Heavy Metal and Organohalogen Compounds*, McIntyre, A.D. and Mills, C.F. (eds.), Plenum Press, New York, 1975, pp. 123–142.

El-Alfy, A. and Schlenk, D., Potential mechanisms of the enhancement of aldicarb toxicity to Japanese medaka, *Oryzias latipes*, at high salinity, *Toxicol. Appl. Pharmacol.*, 152, 175–183, 1998.

Ericson, G. and Larsson, A., DNA adducts in perch (*Perca fluviatilis*) living in coastal water polluted with bleached pulp mill effluents. *Ecotoxicol. Environ. Saf.*, 46, 167–173, 2000.

Franks, N.P. and Lieb, W.R., Where do general anaesthetics act? *Nature*, 274, 339–342, 1978.

Franks, N.P. and Lieb, W.R., Do general anaesthetics act by competitive binding to specific receptors? *Nature*, 310, 599–601, 1984.

Franks, N.P. and Lieb, W.R., Mapping of general anaesthetic target sites provides a molecular basis for cutoff effects, *Nature*, 316, 349–351, 1985.

Franks, N.P. and Lieb, W.R., Stereospecific effects of inhalational general anesthetic optical isomers on nerve ion channels, *Science*, 254, 427–430, 1991.

Fraústo da Silva, J.J.R. and Williams, R.J.P., *The Biological Chemistry of the Elements*, Oxford University Press, Oxford, UK, 1991.

Fridovich, I., Superoxide radical: An endogenous toxicant, *Annu. Rev. Pharmacol. Toxicol.*, 23, 239–257, 1983.

Gardner, W.S., Snee, M.P., Hall, A.J., Powell, C.A., Downes, S., and Terrell, J.D., Results of case-control study of leukaemia and lymphoma among young people near Sellafield nuclear plant in West Cumbria, *BMJ*, 300, 423–434, 1990.

George, S.G., Enzymology and molecular biology of Phase II xenobiotic-conjugating enzymes in fish, In *Aquatic Toxicology. Molecular, Biochemical and Cellular Perspectives*, Malins, D.C. and Ostrander, G.K. (eds.), CRC Press/Lewis Publishers, Boca Raton, FL, 1994, pp. 37–85.

Goldberg, A., Acute intermittent porphyria, *Q. J. Med.*, 28, 183–209, 1959.

Gregus, Z. and Klaassen, C.D., Mechanisms of toxicity, In *Casarett and Doull's Toxicology. The Basic Science of Poisons*, 5th ed., Klaassen, C.D. (ed.), McGraw-Hill, New York, 1996, pp. 35–74.

Grill, E., Winnaker, E.-L., and Zenk, M.H., Phytochelatins: The principal heavy-metal complexing peptides in higher plants, *Science*, 230, 674–676, 1985.

Hall, R.J., Impact of pesticides on bird populations, In *Silent Spring Revisited*, Marco, G.L., Hollingworth, R.M., and Durham, W. (eds.), American Chemical Society, Washington, D.C., 1987, p. 214.

Hamdoun, A.M., Griffin, F.J., and Cherr, G.N., Tolerance to biodegraded crude oil in marine invertebrate embryos and larvae is associated with expression of a multixenobiotic resistance transporter, *Aquat. Toxicol.*, 61, 127–140, 2002.

Hightower, L.E., Heat shock, stress proteins, chaperons, and proteotoxicity, *Cell*, 66, 191–197, 1991.

Hill, E.F. and Fleming, W.J., Anticholinesterase poisoning of birds: Field monitoring and diagnosis of acute poisoning, *Environ. Toxicol. Chem.*, 1, 27–38, 1982.

Ishitawa, T., The ATP-dependent glutathione S-conjugate export pump, *Trends Biochem. Sci.*, 18, 164–166, 1992.

Jagoe, C.H., Responses at the tissue level: Quantitative methods in histopathology applied to ecotoxicology, In *Ecotoxicology. A Hierarchical Treatment*, Newman, M.C. and Jagoe, C.H. (eds.), CRC Press/Lewis Publishers, Boca Raton, FL, 1996, pp. 163–196.

James, M.O., Conjugation of organic pollutants in aquatic species, *Environ. Health Perspect.*, 71, 97–103, 1987.

Johansson-Sjöbeck, M.-L. and Larsson, Å., The effect of cadmium on the hematology and on the activity of δ-aminolevulinic acid dehydratase (ALA-D) in blood and hematopoietic tissues of the flounder, *Pleuronectes flesus* L., *Environ. Res.*, 17, 191–204, 1978.

Johansson-Sjöbeck, M.-L. and Larsson, Å., Effects of inorganic lead on delta-aminolevulinic acid dehydratase activity and hematological variables in the rainbow trout, *Salmo gairdnerii*, *Arch. Environ. Contam. Toxicol.*, 8, 419–431, 1979.

Kolaja, G.L. and Hinton, D.E., DDT-induced reduction in eggshell thickness, weight, and calcium is accompanied by calcium ATPase inhibition, In *Animals as Monitors of Environmental Pollutants*, National Academy of Sciences, Washington, D.C., 1979, pp. 309–318.

Kramer, V.J., Newman, M.C., and Ultsch, G.R., Changes in concentrations of glycolysis and Krebs cycle metabolites in mosquitofish, *Gambusia holbrooki*, induced by mercuric chloride and starvation, *Environ. Biol. Fishes*, 34, 315–320, 1992.

Kurelec, B. and Pivčević, B., Evidence for a multixenobiotic resistance mechanism in the mussel *Mytilus galloprovincialis*, *Aquat. Toxicol.*, 19, 291–302, 1991.

Larno, V., LaRoche, J., Launey, S., Flammarion, P., and DeVaux, A., Responses of chub (*Leuciscus cephalus*) populations to chemical stress, assessed by genetic markers, DNA damage and cytochrome P4501A induction, *Ecotoxicology*, 10, 145–158, 2001.

Livingston, D.R., Garcia Martinez, P., Michel, X., Narbonne, J.F., O'Hara, S., Ribera, D., and Winston, G.W., Oxyradical production as a pollution-mediated mechanism of toxicity in the common mussel, *Mytilus edulis* L., and other molluscs, *Funct. Ecol.*, 4, 415–424, 1990.

Lucier, G.W., Portier, C.J., and Gallo, M.A., Receptor mechanisms and dose-response models for the effects of dioxins, *Environ. Health Perspec.*, 101, 36–44, 1993.

Luedeking, A. and Koehler, A., Regulation of expression of multixenobiotic resistance (MXR) genes by environmental factors in the bluw mussel *Mytilus edulis*, *Aquat. Toxicol.*, 69, 1–10, 2004.

Malins, D.C., Identification of hydroxyl radical-induced lesions in DNA base structure: Biomarkers with a putative link to cancer development, *J. Toxicol. Environ. Health*, 40, 247–261, 1993.

Marks, G.S., Exposure to toxic agents: The heme biosynthetic pathway and hemoproteins as indicator, *Crit. Rev. Toxicol.*, 15, 151–179, 1985.

McCarty, L.S., Mackay, D., Smith, A.D., Ozburn, G.W., and Dixon, D.G., Residue-based interpretation of toxicity and bioconcentration QSARs from aquatic bioassays: Polar narcotic organics, *Ecotoxicol. Environ. Saf.*, 25, 253–270, 1993.

McKim, J.M., Schmieder, P.K., Carlson, R.W., Hunt, E.P., and Niemi, G.J., Use of respiratory-cardiovascular responses of rainbow trout (*Salmo gairdneri*) in identifying acute toxicity syndromes in fish: Part 1. Pentachlorophenol, 2,4-dinitrophenol, tricaine methanesulfonate and 1-octanol, *Environ. Toxicol. Chem.*, 6, 295–312, 1987.

Medvedev, Z.A., An attempt at a rational classification of theories of ageing, *Biol. Rev. Camp. Philos. Soc.*, 65, 375–398, 1990.

Mertz, W., The essential trace elements, *Science*, 213, 1332–1338, 1981.

Pawlik-Skowrońska, B., When adapted to high zinc concentrations the periphytic green alga *Stigeoclonium tenue* produces high amounts of novel phytochelatin-related peptides, *Aquat. Toxicol.*, 62, 155–163, 2003.

Parkinson, A., Biotransformation of xenobiotics, In *Casarett & Doull's Toxicology. The Basic Science of Poisons*, 5th ed., Klaassen, C.D. (ed.), McGraw-Hill, New York, 1996, pp. 113–186.

Penttinen, O.-P. and Kukkonen, J., Chemical stress and metabolic rate in aquatic invertebrates: Threshold, dose–response relationships, and mode of toxic action, *Environ. Toxicol. Chem.*, 17, 883–890, 1998.

Prohaska, J.R. and Lukasewycz, O.A., Copper deficiency suppresses the immune response of mice, *Science*, 213, 599–561, 1981.

Ratcliffe, D.A., Decrease in eggshell weight in certain birds of prey, *Nature*, 215, 208–210, 1967.

Ratcliffe, D.A., Changes attributable to pesticides in egg breakage frequency and eggshell thickness in some British birds, *J. Appl. Ecol.*, 7, 67–107, 1970.

Regoli, F., Total oxyradical scavenging capacity (TOSC) in polluted and translocated mussels: A predictive biomarker of oxidative stress, *Aquat. Toxicol.*, 50, 351–361, 2000.

Regoli, F. and Principato, G., Glutathione, glutathione-dependent and antioxidant enzymes in mussel, *Mytilus galloprovincialis*, exposed to metals under field and laboratory conditions: Implications for use of biochemical biomarkers, *Aquat. Toxicol.*, 31, 143–164, 1995.

Sander, B.M. and Dyer, S.D., Cellular stress response, *Environ. Toxicol. Chem.*, 13, 1209–1210, 1994.

Schlenk, D., Stresser, D., McCants, J., Nimrod, A., and Benson, W., Influence of beta-naphthoflavone and methoxychlor pretreatment on the biotransformation and estrogenic activity of methoxychlor in channel catfish (*Ictalurus punctatus*), *Toxicol. Appl. Pharmacol.*, 145, 349–356, 1997.

Schinkel, A.D., Smit, J.J., Tellingen, O., Beijnen, J.H., Wagenaar, E., van Deemter, L., Mol, C.A., et al., Disruption of the mouse *mdr1a* P-glycoprotein gene leads to a deficiency in the blood-brain barrier and to increased sensitivity to drugs, *Cell*, 77, 491–502, 1994.

Schmitt, C.J., Wildhaber, M.L., Hunn, J.B., Nash, T., Tieger, M.N., and Steadman, B.L., Biomonitoring of lead-contaminated Missouri streams with an assay for erythrocyte δ-aminolevulinic acid dehydratase activity in fish blood, *Arch. Environ. Contam. Toxicol.*, 25, 464–475, 1993.

Schultz, I., Environmental estrogens: Occurrence of ethynylestradiol and adverse effects on fish reproduction, In *Fundamentals of Ecotoxicology*, 2nd ed., Newman, M.C. and Unger, M.A. (eds.), CRC Press/Lewis Publishers, Boca Raton, FL, 2003, pp. 156–160.

Schüürmann, G., Segner, H., and Jung, K., Multivariate mode-of-action analysis of acute toxicity of phenols, *Aquat. Toxicol.*, 38, 277–296, 1997.

Segner, H. and Braunbeck, T., Cellular response profile to chemical stress, In *Ecotoxicology*, Schüürmann, G. and Braunbeck, T. (eds.), John Wiley & Sons, New York, 1998, pp. 521–569.

Shugart, L.R., Molecular markers to toxic agents, In *Ecotoxicology. A Hierarchical Treatment*, Newman, M.C. and Jagoe, C.H. (eds.), CRC Press/Lewis Publishers, Boca Raton, FL, 1996, pp. 133–161.

Slater, T.F., Free-radical mechanisms in tissue injury, *Biochem. J.*, 222, 1–15, 1984.

Smital, T. and Kurelee, B., The chemosensitizers of multixenobiotic resistance mechanisms in aquatic invertebrates: A new class of pollutants, *Mutat. Res.*, 399, 43–53, 1998.

Spitzer, P.R., Risebrough, R.W., Walker, W., Hernandez, R., Poole, A., Pulleston, D., and Nisbet., I.C.T., Productivity of ospreys in Connecticut-Long Island increases as DDE residues decline, *Science*, 202, 333–335, 1978.

Stein, J.A. and Tschudy, D.P., Acute intermittent porphyria: A clinical and biochemical study of 46 patients, *Medicine*, 49, 1–16, 1970.

Stone, R., Can a father's exposure lead to illness in his children? *Science*, 258, 31, 1992.

Timbrell, J., *Principles of Biochemical Toxicology*, Taylor & Francis, Philadelphia, PA, 2000.

Vedel, G.R. and DePladge, M.H., Stress-70 levels in the gills of *Carcinus maenas* exposed to copper, *Mar. Pollut. Bull.*, 31, 84–86, 1995.

Winston, G.W. and Di Giulio, R.T., Prooxidant and antioxidant mechanisms in aquatic organisms, *Aquat. Toxicol.*, 19, 137–161, 1991.

Wood, J.S. and Fowler, B.A., Altered regulation of mammalian hepatic heme biosynthesis and urinary porphyrin excretion during prolonged exposure to sodium arsenate, *Toxicol. Appl. Pharm.*, 43, 361–371, 1978.

Wood, J.S., Martin, M.D., Naleway, C.A., and Echeverria, D., Urinary porphyrin profiles as a biomarker of mercury exposure: Studies on dentists with occupational exposure to mercury vapor, *J. Toxicol. Environ. Health*, 40, 235–246, 1993.

Zimniak, P., Awasthi, S., and Awasthi, Y.C., Phase III detoxification system. *Trends Biochem. Sci.*, 18, 164–166, 1993.

4 Cells and Tissues

Cells are the site[s] of primary interaction between chemicals and biological systems.

(Segner and Braunbeck 1998)

4.1 OVERVIEW

The biochemistry explored in Chapter 3 is central to all life. Also essential are the spatial differences in the distribution of biochemical activities and moieties within cells and tissues. Examples include essential differences in respiratory activities within the mitochondria versus nucleus, or glycogen synthesis differences in liver versus kidney cells. Macromolecular complexes forming membranes, organelles, cell junctions, and extracellular matrices facilitate this spatial heterogeneity.

Such differences, emerging at the levels of the membrane, organelle, cell, and tissue, also produce spatial differences in effects of, and responses to, toxicants. Cyanide inhibits mitochondrial electron transport reactions by interfering with cytochrome *a* function. Methylated forms of arsenic can damage chromosomes localized in the nucleus and, in so doing, provide a mechanism for arsenic's carcinogenicity. Differences in the biochemical processes and moieties in various cell organelles determine the site of action for poisons such as cyanide and arsenic, and spatial separation of cell types into different tissues determines which tissue is most affected by a toxicant or is most responsive to toxicant damage. High microsomal mixed function oxidase (MFO) activity in hepatocytes make liver tissue a major site of Phase I reactions. It also makes hepatocytes particularly prone to cancers initiated by strongly electrophilic metabolites of toxicants. High levels of metallothionein and lysosomal activity in vertebrate proximal tubules make renal damage an unfortunate consequence of acute cadmium poisoning. The extent of toxicant-induced cell death within a tissue and the tissue's regenerative capacity determine whether or not that tissue can support the proper functioning of the associated organ.

Histopathology is the science that focuses on cellular and tissue changes resulting from infectious and noninfectious diseases. This brief chapter explores histopathology of toxicants by building on the previous chapter. Hopefully, it also provides a bridge to the organ and organ system effects discussed next.

4.2 CYTOTOXICITY

4.2.1 Necrosis and Apoptosis

Pathological changes (lesions) in cells, tissues, or organs occur at sites of toxic action. Some lesions reflect failures to maintain a viable cellular state while others reflect only partially successful attempts to maintain optimal cellular homeostasis.

Cells die if stress is insurmountable or injury irreparable. Necrosis, cell death resulting from disease or injury, is apparent in many kinds of lesions. Necrotic cells are characteristically swollen, with swollen mitochondria and disintegrating cell membranes (Gregus and Klaassen 1996). Swollen mitochondria take in calcium with consequent and often pervasive internal precipitation of calcium phosphate. This leads to eventual breakdown of the mitochondria's inner membrane and loss of its capacity for oxidative phosphorylation (La Via and Hill 1971). Pyknosis, the condensation of the nuclear material into a dark staining mass, is also characteristic of necrotic cells. Diffuse strands of chromatin condense during cell death to form these darkly staining masses. Also karyolysis can be seen in necrotic cells. Karyolysis is the dissolution of the nucleus and its lost ability to be stained with basic stains such as hematoxylin. The nuclear membrane remains intact with karyolysis. The loss of staining qualities results from DNAase and lysosomal cathepsin destruction of the DNA (La Via and Hill 1971). Karyorrhexis might occur later with the disruption of the nuclear membrane, fragmentation of the nucleus, and breaking apart of the chromatin into small granules. Necrotic cells can be dislocated from their normal position within tissues. Often inflammation accompanies necrosis. Necrosis can occur in distinct zones (Figure 4.1, middle panel) or diffusely (Figure 4.1, lower panel) within tissues.

Necrosis would seem at first consideration to be the only kind of cell death relevant to chemical intoxication. What other kind could there be? Apoptosis, programmed cell death (PCD), can also occur by a genetically controlled series of cellular events. Cells undergoing apoptosis characteristically shrink, their nuclear material condenses, and they break into membrane-bound fragments called apoptotic bodies. Inflammation is characteristically absent. (The dead cells in the lower panel of Figure 4.1 appear like cells that have undergone apoptosis.) Remnants of cells experiencing apoptosis can be engulfed by phagocytic cells or shed from the gut lining or skin surface.

Apoptosis occurs in normal and toxicant-exposed cells. In fact, a balance between cellular mitosis and apoptosis is essential in development and maintenance of tissue homeostasis (Roberts et al. 1997). As one example involving development, some cells must die away between the developing fingers of a human fetus to facilitate normal development of the hand. Toxicant-induced imbalance between mitosis and apoptosis can produce developmental abnormalities. As another example, apoptosis is essential for developing the appropriate gaps between connecting neurons. Human neutrophils formed in and released from the bone marrow also undergo apoptosis after a brief time in circulation.[1] Apoptosis may also remove cells that become a threat to tissues, e.g., cells infected with a virus or damaged by a toxicant. For example, cadmium-induced oxidative stress in trout hepatocytes results in apoptosis of damaged cells (Risso-de Faverney et al. 2004). Similarly, apoptosis removes cells in snails (*Helix pomatia*) after exposure to cadmium-enriched food (Chabicovsky et al. 2004). As a contrasting illustration, cadmium's adverse effect on mammalian male fertility is a consequence of testicular necrosis, not apoptosis (Fowler et al. 1982). Clearly, the relative importance of necrosis and apoptosis varies with the particular toxicant and tissue.

4.2.2 Types of Necrosis

Four major categories of necrosis exist: coagulative, liquefactive, caseous, and fat. Other less general classes mentioned in the histopathology literature are Zenker's, fibrinoid, and gangrenous necrosis.

Coagulative, or coagulation, necrosis involves extensive protein coagulation throughout the dead cell. This coagulation makes the cell appear opaque, having a cloudy and weakly eosinophilic appearance. Cells might maintain their relative positions within tissues for days after coagulative necrosis occurred: cell ghosts, a term applied to the opaque dead cells, are characteristic of this type of necrosis.

Coagulative necrosis can be expected under a variety of situations, including poisonings. Ingestion of phenol or inorganic mercury by mammals produces coagulation necrosis in the intestinal lining because both toxicants rapidly denature proteins (Sparks 1972). Accordingly, this type of

[1] The term necrobiosis, coined first by Rudolf Virchow, is synonymous with apoptosis (Sparks 1972). It was used specifically for the natural aging and death of cells, such as epithelial cells, that are then replaced by new cells (Sparks 1972).

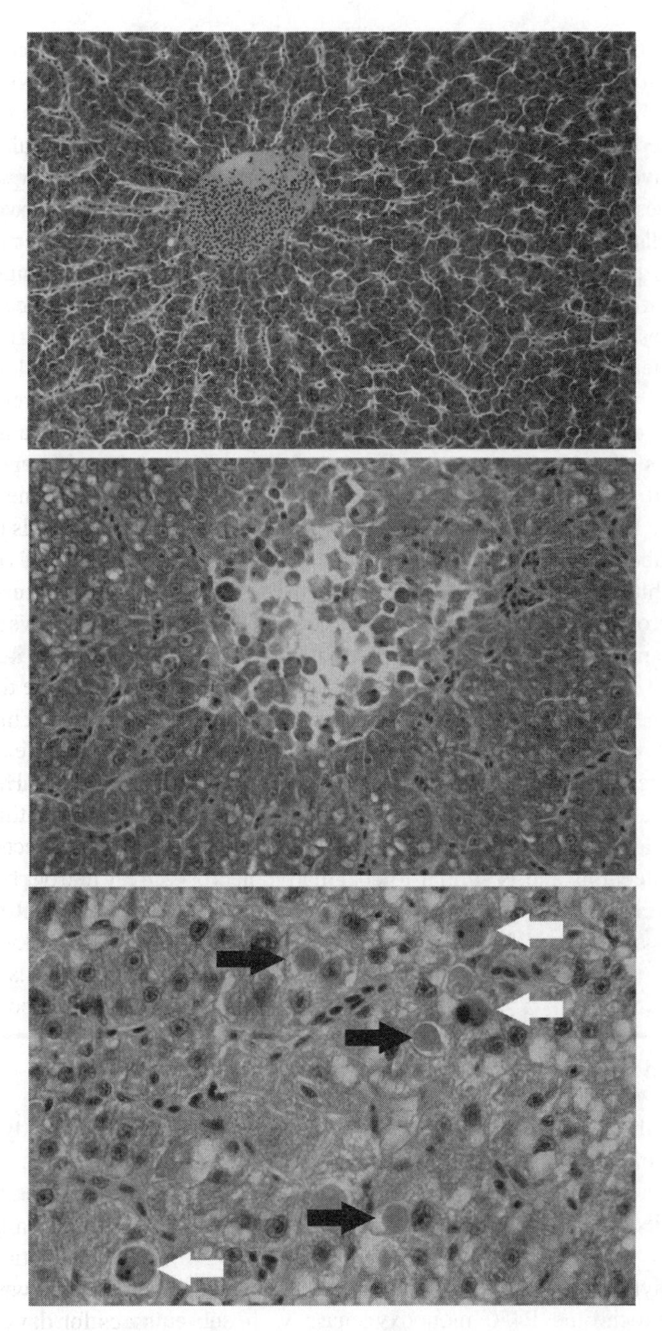

FIGURE 4.1 Liver necrosis. The upper panel is a section through a normal *F. heteroclitus* liver with branching hepatic tubules lined with hepatic sinusoids. Note that the hepatocytes are relatively uniform in size and shape. The middle panel is an example of necrosis in the liver. Notice the difference in staining between the living and dead cells. Dead cells show nuclear pyknosis and karyolysis, and loss of cell adherence. The lower panel shows necrosis of individual cells, not a localized area as seen with the necrosis shown in the middle panel. Three necrotic cells are at the tips of the dark arrows. They are round or oval remnants that stain strongly with eosin. The basophilic chromatin remnants are visible in dead cells identified by the white arrows. Such a scattering of single necrotic cells in the liver suggests the effect of a chemical toxicant (Roberts et al. 2000). (Photomicrographs and general descriptions provided by W. Vogelbein, Virginia Institute of Marine Science.)

necrosis is favored by Hinton and Laurén (1990) as a biomarker[2] of environmental toxicant exposure. Heat also produces coagulative necrosis. Ischemia, the sudden loss of oxygen supply as might occur with a myocardial infarct or a puncture wound, can also induce this type of necrosis by shifting metabolism to glycolysis and decreasing cellular pH by production of lactic acid.

With liquefactive (cytolytic or liquefaction) necrosis, the cell contents are liquefied by the cell's proteolytic enzymes, and perhaps also enzymes from leukocytes that move into the injured area. Relative to coagulative necrosis, cell liquefaction tends to be rapid and extensive. Liquefactive necrosis in tissues possessing considerable enzymatic activity can produce fluid-filled spaces in tissues. This type of necrosis is often associated with bacterial or fungal infections, and can produce cell debris-filled abscesses. It can also be associated with a brain infarct. Given these characteristics, especially its frequent association in infectious disease, this type of necrosis is a less useful indicator of toxicant effects than coagulative necrosis.

The two other common forms of necrosis, caseous and fat necrosis, are also not useful as general biomarkers of toxicant exposure. Caseous (caseation or cheesy) necrosis, named for its milk casein or soft cheese appearance, involves the complete disintegration of cells into a mass of fat and protein. It is often associated with mycobacterial infections such as the lung necrosis characteristic of tuberculosis. Fat necrosis involves deposition of calcium with released fatty acids, which imparts a white color to lesions. Fat necrosis can result from lipase and other enzyme activities (enzymatic fat necrosis) or from physical trauma to fat cells (traumatic fat necrosis). The mammalian pancreas, which can release high levels of lipases and other pertinent enzymes, is a common site of fat necrosis.

Other types of necrosis exist. Gangrenous necrosis occurs with ischemia and consequent bacterial infection. As such, gangrenous necrosis will have characteristics of liquefactive and coagulative necrosis. Fibroid necrosis is another form of necrosis that is associated with autoimmune disease (e.g., lupus erthematosis) or vessel wall necrosis with extreme hypertension. Zenker's (hyaline or waxy) necrosis is a specific condition in striated muscle that is associated with acute infections such as typhoid infections and is similar to coagulative necrosis. Although reported in goat heart muscle tissue with chronic mercury poisoning (Pathak and Bhowmik 1998), Zenker's necrosis is not generally useful as an ecotoxicological biomarker.

Box 4.1 Death by Trichloroethylene: Intentional and Otherwise

Several themes discussed to this point can be illustrated using a recent study by Lash et al. (2003). These toxicologists were interested in the effects on humans from exposure to trichloroethylene, a metal degreaser and solvent. This chemical enjoys very widespread use but has been classified by EPA as a probable carcinogen. As such, it is the subject of much justified interest.

Trichloroethylene undergoes a variety of Phase I and II reactions. It can be acted on by cytochrome P450 monooxygenase with subsequent glutathione conjugation. *S*-(1,2-dichlorovinyl)-L-cysteine (DCVC) is produced via β-lyase activity after cysteine conjugation to a cytochrome P450 monooxygenase metabolite of trichloroethylene. The β-lyase activity is primarily a result of glutamine transaminase K that is localized in the kidney's proximal tubules (Lash and Parker 2001). The DCVC causes necrosis in the human kidney. Relatively high doses of DCVC were found to be nephrotoxic to cultured proximal tubular cells of rats, inducing apoptosis.

[2] This term was used loosely in Chapter 3 but now needs to be defined more precisely. A biomarker is a cellular, tissue, body fluid, physiological, or biochemical change in living organisms used quantitatively to imply the presence of significant pollutant exposure (Newman and Unger 2003).

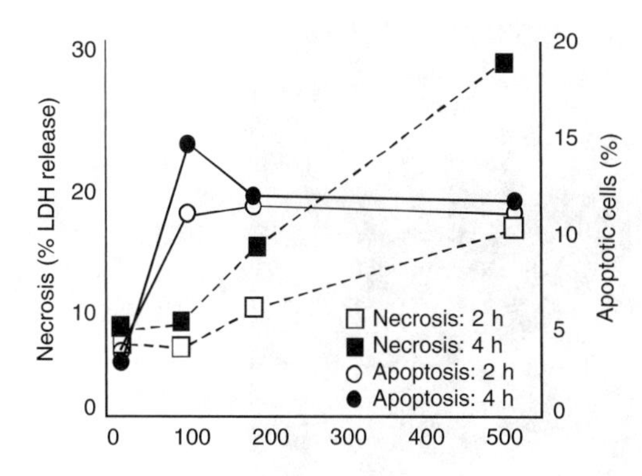

FIGURE 4.2 Necrosis and apoptosis occurring in primary cultures of human proximal tubular cells exposed to DCVCS, a nephrotoxic metabolite of trichloroethylene. Data for concentration-dependent necrosis (squares) and apoptosis (circles) are shown for DCVCS exposure durations of 2 h (open symbols) and 4 h (filled symbols). (Data extracted from Figures 2 and 5 of Lash et al. 2003.)

A flavin-containing monooxygenase can produce *S*-(1,2-dichlorovinyl)-L-cysteine sulfoxide (DCVCS) from DCVC. The potency of DCVCS was much higher than DCVC in rat proximal tubular cell cultures, leading Lash et al. (2003) to be concerned that DCVCS might also be responsible for the nephrotoxic effects of trichloroethylene exposure of humans. To assess this hypothesis, they examined injury resulting from DCVCS exposure of cultured human proximal tubular cells. Necrosis and apoptosis were measured at different DCVCS concentrations and exposure durations; however, only results for 2- and 4-h exposures are discussed here.

Necrosis was quantified in this study by measuring the amount of lactate dehydrogenase (LDH) in the cultured cells and the amount released from cells into the culture media. The more LDH measured in the media, the more necrosis. The percentage of LDH metric was simply 100 times the amount in the media divided by the sum of the LDH in the media plus the amount in the cells.

$$\text{LDH}\ (\%) = 100\frac{\text{LDH}_{\text{media}}}{\text{LDH}_{\text{cells}} + \text{LDH}_{\text{media}}}$$

The amount of necrosis present in cultures increased with DCVCS dose and exposure duration (Figure 4.2). This was also the case for results from other exposure durations (1, 8, 24, and 48 h) not shown here. In contrast, apoptosis increased at the lowest exposure concentration and remained at that elevated level at all DCVCS concentrations. This induction of apoptosis by DCVCS was consistent with apoptosis induced by DCVC. A set level of apoptosis appeared to be triggered by DCVCS but necrosis increased steadily with any increase in DCVCS. Regardless, both contributed to the net loss of cells due to DCVCS exposure.

The authors concluded that flavin-containing monooxygenase activation and subsequent sulfoxidation of DCVC play important roles in human kidney damage after exposure to trichloroethylene. Both necrosis and apoptosis contribute to kidney cell death due to trichloroethylene exposure but the pattern of response differs for necrosis and apoptosis.

Within the hierarchical framework of this book, the study illustrates that Phase I and II biochemical reactions activate xenobiotics in cells. Beyond a certain stress level, cells are unable to recuperate and death occurs due to necrosis and apoptosis. Sufficient levels of cell death within kidney tissues can result in renal failure and death of the individual.

4.2.3 Inflammation and Other Responses

Inflammation is a general response to damage or infection. It is characterized by "infiltration of leucocytes into the peripheral tissues, followed by the release of various mediators eliciting nonspecific physiological defense mechanisms" (House and Thomas 2002) (Figure 4.3). The intended

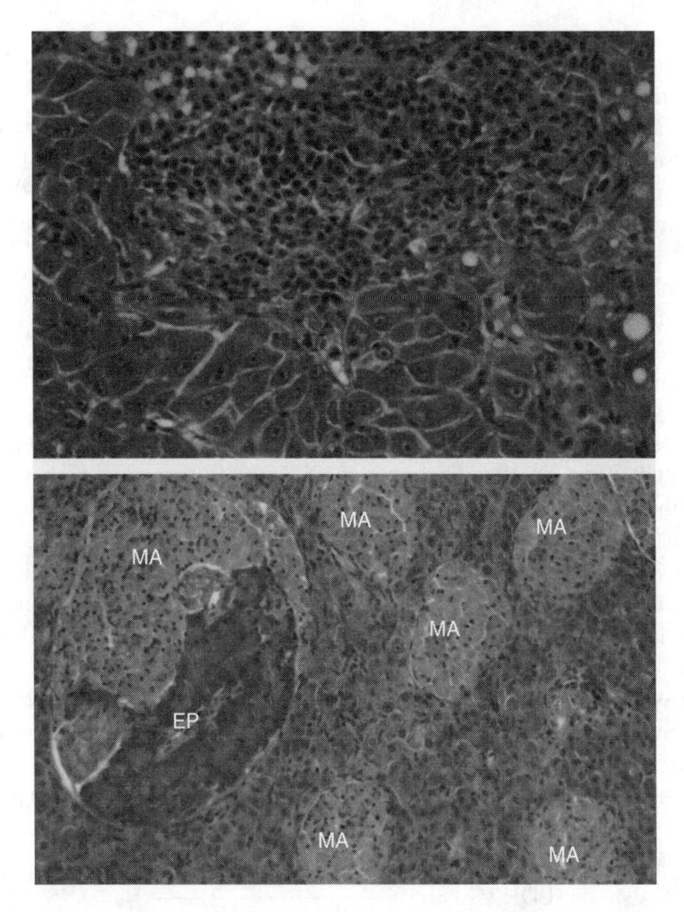

FIGURE 4.3 Inflammation in the liver of the estuarine fish, *F. heteroclitus*. At the top center of the top photomicrograph is a focus of inflammation. The bottom photomicrograph shows macrophage aggregates (MA) produced during inflammation in *Fundulus* liver. (EP is exocrinic pancreas tissue.) (Photomicrographs and general descriptions provided by W. Vogelbein, Virginia Institute of Marine Science.)

result is tissue repair with a return to a healthy state; however, chronic inflammation or inflammation after extensive damage can produce compromised tissue structure and function. With toxicant-induced injury, inflammation isolates, removes, and replaces damaged cells. Consequently, ongoing inflammation or telltale signs of past inflammation can be evidence of cell poisoning.

Classic work by Elie Metchnikoff established the scientific foundation of inflammation theory. Taking advantage of the transparency of minute invertebrates, he explored phagocytic responses in injured or infected individuals. In one set of experiments, he closely observed the cellular response of *Daphnia* to infection with *Monospora bicuspidata*. In others, he studied responses to mechanical injury. Bibel (1982) describes one of Metchnikoff's initial experiments, done while staying in a Sicilian seaport with his family. Whiling away time after resigning from the University of Odessa, Metchnikoff gazed through his microscope and hypothesized that all organisms, even the simplest, will exhibit inflammation.

> We had a few days previously organized a Christmas tree for the children on a little tangerine tree: I fetched from it a few thorns and introduced them at once under the skin of some beautiful starfish larvae as transparent as water . . . . I was so excited to sleep that night in the expectation of the result of my experiment and very early the next morning I ascertained that it had fully succeeded.
>
> **(Metchnikoff 1921)**

Although Metchnikoff's experiment and early morning anticipations were not those normally expected during a Christmas with one's family, his experiment did demonstrate phagocyte infiltration into the area of injury and, combined with similar experiments, established the universal nature of this response to injurious or infectious agents.

Much of this pioneering experimentation with invertebrates took place more than a century ago. But our understanding of symptoms of inflammation goes back still further. Most introductory discussions describe four cardinal signs of inflammation for humans: heat, redness, swelling, and pain. Cornelius Celsus identified these signs millennia ago and they were further detailed by Virchow (see footnote 1) and Metchnikoff a century ago (Plytycz and Seljelid 2003). The area of damage reddens as blood vessels dilate. Swelling of surrounding tissues with fluids (edema) occurs, imparting a feeling of heat and painful pressure.

Obviously, some of these signs are relevant only to red-blooded poikilotherms; however, the underlying processes are relevant to all animals. Typical of a tissue experiencing inflammation is leukocyte movement into the involved tissues. Diapedesis occurs when leukocytes, responding to chemotactic factors released from the damaged tissue, adhere to the vascular endothelium and then migrate through it into the involved tissues. The clumping of leukocytes at the endothelium is called margination. The cells in the area retract to facilitate leukocyte passage through interendothelial cell junctions. The leukocytes phagocytize cellular debris and remove it from the area. Starting as a mass called the granulation tissue, new vessels and connective tissue will eventually begin to grow back as the process continues. Scar tissue or collagenous connective tissue can form to cause tissue dysfunction in the case of chronic inflammation.

Diverse examples of inflammation are easy to find because inflammation is such a universal cellular response to injury. The human autoimmune disease rheumatoid arthritis involves chronic inflammation at the synovial membrane of joints. This inflammation gradually damages joint tissues. Inhalation of zinc-rich particulate matter can produce metal-fume fever, a condition arising from pulmonary inflammation and injury (Kodavanti et al. 2002). Exposure of freshwater fish to a water-soluble fraction of crude oil results in gill and liver necrosis, and consequent inflammation (Akaishi et al. 2004).

Other cellular changes such as hyperplasia and hypertrophy can indicate response to toxicants. Hyperplasia is the increase in the number of cells in a tissue. Hypertrophy is an increase in cell size (and function) that is often part of a compensatory response. Fish gill hyperplasia is evident in Figure 4.4. The upper panel of that figure shows a section through a normal gill from the estuarine fish, *Fundulus heteroclitus*. The axis of the primary lamellae is denoted with a black line and the letter "P," and one of the many secondary lamellae projecting out from the primary lamellae is denoted by the letter "S." The lower panel is a lower magnification image of a *Fundulus* gill that has undergone extensive hyperplasia. One of the primary lamellae in the image is shown with a dark line and "P," and one secondary lamella with a "S." Notice that extensive hyperplasia of epithelial cells has filled in the gaps between secondary lamellae of the labeled primary gill lamella and also of the primary lamella at the bottom right hand corner of the photomicrograph. The hyperplasia is so extensive that the primary lamellae at the center of the photograph have fused together with no discernable secondary lamellae. This can be seen easily by noting the filament cartilage (C) in the normal primary lamella (upper panel) and then locating the filament cartilage in the lower panel (C) where two of the primary lamellae have fused into one single mass of tissue. Available respiratory surface has decreased considerably because these secondary lamellae are the structures where most gas exchange occurs.

Figure 4.5 shows gills of the freshwater mosquitofish, *Gambusia holbrooki*, which exhibit chloride cell (ionocytes) hypertrophy in addition to hyperplasia as a consequence of inorganic mercury exposure. The upper panel of that figure is a gill from an unexposed fish with an arrow pointing to one of several lightly staining chloride cells on the primary lamellae. Notice in the lower panel that, in addition to chloride cell proliferation between and onto the secondary lamellae (hyperplasia),

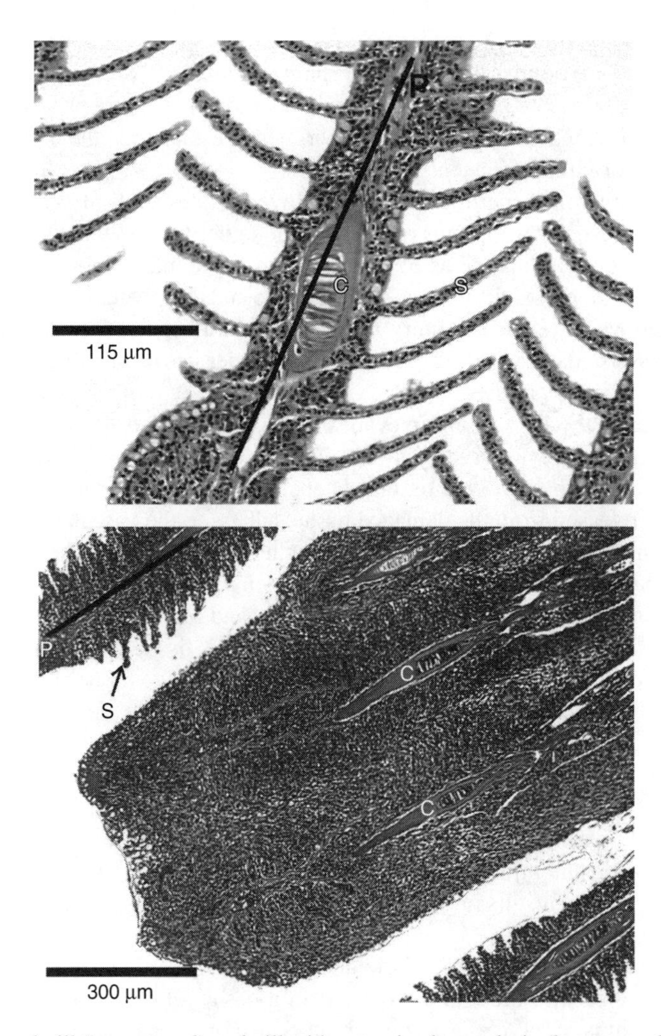

FIGURE 4.4 Normal gill (upper panel) and gill with extensive hyperplasia (lower panel) from the estuarine fish, *F. heteroclitus*. The epithelial cells have filled the gaps between secondary lamellae, causing fusion in the primary lamellae shown in the center of the bottom photomicrograph. Often such hyperplasia is accompanied by inflammation. (Photomicrographs and general descriptions provided by W. Vogelbein, Virginia Institute of Marine Science.)

the chloride cells have become enlarged (hypertrophy) (three arrows). Chloride cells function in ion transport and this hypertrophy is seen as an attempt to compensate for a loss of ion transport capabilities due to mercury damage (Jagoe et al. 1996). Other toxicants produce such compensatory hypertrophy in other tissues. The trichloroethylene metabolite DCVC, which we discussed previously, results in hypertrophy in primary cultures of rat proximal tubule epithelial cells (Kays and Schnellmann 1995). Heptocytes also display hypertrophy when zebrafish (*Danio rerio*) are injected with 2,3,7,8-tetrachlorodibenzo-*p*-dioxin (TCDD) (Zodrow et al. 2004).

4.3 GENOTOXICITY

4.3.1 Somatic and Genetic Risk

Genotoxicity is damage to the cell's genetic material by a physical or chemical agent. The individual organism is the focus of most genotoxicity studies although implications about risk factors are often

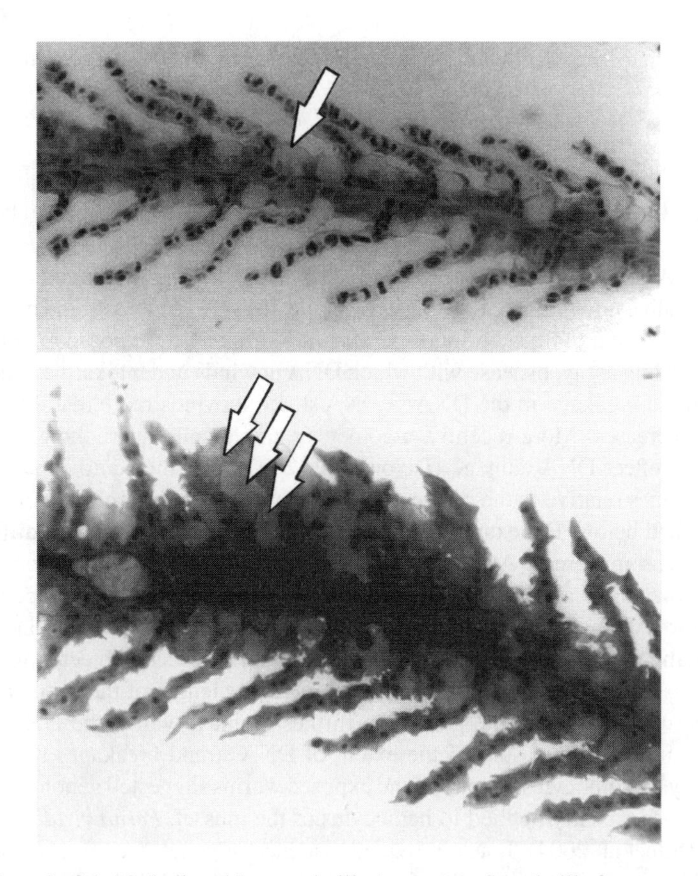

FIGURE 4.5 Mosquitofish (*G. holbrooki*) normal gills (upper panel) and gills from mercury-exposed mosquitofish (lower panel). The gill from the fish exposed to inorganic mercury shows hyperplasia and chloride cell hypertrophy. (See Jagoe et al., 1996. Courtesy C. Jagoe, Savannah River Ecology Laboratory.)

framed in a population context. By convention, genetic damage is discussed relative to somatic and genetic risk. Somatic risk is the risk to the somatic cells (soma) (e.g., genetic modifications resulting in cancer). Genetic risk involves risk to offspring of exposed individuals. Such genetic risk was mentioned briefly in Section 3.2 where examples were given of possible consequences to offspring of occupational exposure. More attention is paid to somatic than genetic risk in the field of genotoxicology primarily because of concern about cancer.

In the ecotoxicological context of this book, one could argue that population risk should be considered too. Population risk would be defined as risk of decreased population viability due to genetic damage to germ cells by a physical or chemical agent. An admittedly contrived and extreme example of such a population effect would be those intended in tsetse fly, screwfly, or medfly control programs that aim to dramatically impact population size by γ irradiation and release of large numbers of sterile males (Sterile Insect Technique, SIT) (Knipling 1955, Lindquist 1955, Lux et al. 2002). But such intense exposures are not common outside of pest control programs. Perhaps, elevated cancer incidences in small, slow-growing wildlife populations could result in population risk. Such a scenario might develop for the Beluga whales in the St. Lawrence estuary, which have high levels of cancer deaths (18% of all deaths) (Martineau et al. 2002). These whales are exposed to polycyclic aromatic hydrocarbons (PAH) and display annual cancer rates (163 in 100,000 animals) considerably higher than those of other cetacean populations. (The link between cancer and PAH genotoxicity was reinforced by Shugart (1990) who reported elevated DNA adducts in tissues of St. Lawrence Beluga whales.) Regardless, to our knowledge, few examples of immediate and significant population risk due to direct genetic damage to germ cells have emerged.

4.3.2 DNA Damage

DNA damage in cells is measured in a variety of ways. Jenner et al. (1990) applied flow cytometry to quantify differences in DNA content in individual hepatocytes of English sole (*Parophrys vetulus*), showing more DNA damage in sole from contaminated areas than those from reference sites. Shugart (1988) used an alkaline unwinding assay to get a relative measure of DNA strand breakage in bluegill (*Lepomis macrochirus*) and fathead minnow (*Pimephales promelas*) exposed to benzo[a]pyrene. In this alkaline unwinding assay, the ease with which DNA unwinds under alkaline conditions suggests the amount of strand breakage in the DNA: a DNA strand unwinds more readily as the number of breaks within it increases. More recently, a comet, or single cell electrophoresis, assay has been applied widely to reflect DNA damage (Dixon et al. 2002). For the ecotoxicologist, this method has several advantages relative to the conventional karyotyping or sister chromatid exchange (SCE) techniques described below. For example, karyotyping and SCE assays can be difficult for species with many small chromosomes. Also both methods require that cell division occur (Pastor et al. 2001). In an ecotoxicological application of the comet technique, neutrophilic coelomocytes from nickel-exposed earthworms (*Eisenia fetida*) were embedded in agarose, lysed in place with detergent, placed under alkaline conditions that unwound their DNA, and then subjected to electrophoresis. After electrophoresis and staining with ethidium bromide, the length of the "comet tails" extending from the original cell position in the gel to the furthest point to which the DNA migrated in the electric field was used as a measure of the extent of DNA strand breakage. Relative tail lengths derived from many coelomocytes of control and exposed worms suggested genotoxic effect of nickel. The comet assay was recently applied to hemocytes of the mussel, *Perna viridis*, after exposure to benzo[a]pyrene (Siu et al. 2004). It also provided evidence of genotoxic effect to white storks born near an acid and heavy metal toxic spill in Spain's Doñana National Park (Pastor et al. 2001).

4.3.3 Chromatids and Chromosomes

Section 4.3.2 describes some direct effects of toxicants on DNA including cross-linking DNA with proteins, single or double strand breaks, adduct formation, base mismatching, and point mutations. Here, the topic is addressed again but at a higher scale—that of chromatids and chromosomes. Dixon et al. (2002) use the discriminating term macrolesions for these chromatid or chromosome-level genotoxic effects in order to distinguish them from the microlesions discussed previously, which occur at the molecular DNA level. Several macrolesion assays require cells that are dividing and include SCE, chromosomal aberration, and micronuclei assays. Macrolesion-associated methods are quickly becoming valuable genotoxicity monitoring tools (Hayashi et al. 1998, Jha et al. 2000a).

Mutagenic or genotoxic effects are often correlated with rates of SCE (Dixon et al. 2002, Tucker et al. 1993). SCE involves DNA breakage followed by homologous DNA segment exchange between sister chromatids during the S phase of the cell cycle[3] (Tucker et al. 1993). To measure SCE, one chromatid in each pair comprising a chromosome is first stained with 5-bromodeoxyuridine. Cells are examined two cell cycles later under a fluorescent microscope for evidence of DNA exchange between chromatids. Each of the paired sister chromatids remains either completely stained or unstained if no exchange occurred. If exchange occurred, each chromatid will have segments that are stained and others that are not. The number of SCEs per metaphase or per chromosome is used as a metric of exchange. DNA damage is generally correlated with the level of SCE.

SCE techniques are widely applied to study human exposure to mutagens or genotoxicants, and occasionally used in ecotoxicological studies. Examples of use relative to humans include exposure

[3] S phase is the "synthesis" stage in which the DNA is replicated.

to arsenic in drinking water (Lerda 1994), pesticides in the workplace (De Ferrari et al. 1991), and phenanthrene and pyrene in coke works (Popp et al. 1997). Rates of SCE in lymphocytes are routinely used for such surveys of human exposure. Ecotoxicology applications include larvae of the mussel (*Mytilus edulis*) exposed to mutagens (Jha et al. 2000) or tributyltin (Jha et al. 2000), and adult *M. edulis* exposed to mitomycin (Dixon and Clarke 1982).

A variety of effects can manifest at the level of the chromosome. Anomalies during the cell cycle can produce chromosomal aberrations. Aberrations and spindle dysfunction can result in micronuclei, nuclear segments separated from the cell nucleus, which are not incorporated into daughter cell nuclei. Genotoxicity assays based on micronucleus formation are also well established in ecotoxicology (Dixon et al. 2002). As an example, the frequency of micronuclei in mussels (*P. viridis*) exposed to benzo[a]pyrene was used as a measure of genotoxicity (Siu et al. 2004). Micronuclei were also shown to increase in oysters (*Crassostrea gigas*) exposed to benzo[a]pyrene, copper, or paper mill effluent (Burgeot et al. 1995).

Errors in chromosome segregation can result in aneuploidy. Aneuploidy is the condition in which a cell has an atypical number of chromosomes. As a good ecotoxicological example, Lamb et al. (1991) found elevated levels of aneuploidy in red blood cells of turtles (*Trachemys scripta*) inhabiting radionuclide-contaminated waterbodies.

Chemicals causing chromosomal breaks are called clastogens. Clastogenic effects can involve the gain, loss, or rearrangement of parts of chromosomes. They do not necessarily involve direct damage to chromosomes, and can result from errors occurring during the cell cycle in which the chromosomal complement is not passed intact to the daughter cells (e.g., as would occur with spindle dysfunction).

Chromosome damage can be measured by the conventional karyological approach of visually examining metaphase chromosome preparations. Such an approach was applied by McBee et al. (1987) to study chromosomal aberrations in rodents inhabiting a petrochemical waste site. These same methods were used to prove that women exposed in the workplace to elevated lead concentrations have elevated levels of chromosomal aberrations (Forni et al. 1980) and that methylated trivalent arsenicals are clastogenic (Kligerman et al. 2003).

4.4 CANCER

4.4.1 CARCINOGENESIS

Cancer results from a hyperplasia unlike that discussed above. The hyperplasia discussed earlier relative to tissue repair is referred to as physiologic hyperplasia. There are also two pathological types of hyperplasia, compensatory and neoplastic. The former is an excessive cell proliferation in response to damage or irritation such as that shown in Figure 4.4. Neoplastic hyperplasia results when hereditary material of a cell is changed and the cell no longer responds appropriately to signals controlling cell proliferation. The cell's DNA is changed by a point mutation, deletion, addition, rearrangement, or gene insertion by a retrovirus. Neoplastic hyperplasia can produce cancer. The resulting cancer is benign if associated cells remain relatively differentiated and slow growing, or the cancer is malignant if the associated cells become undifferentiated, rapid growing, and invasive. A malignant cancer can spread to other sites when pieces separate from the original tumor and pass into the lymphatic or circulatory system.

Carcinogenesis is envisioned as a sequence of low probability, irreversible events (Figure 4.6). The initiation stage begins when some agent alters a cell's genes responsible for normal growth and differentiation: a protooncogene is changed to an oncogene. The result is inappropriate cell proliferation, differentiation, or both. Changes in suppressor genes that inhibit abnormal cell growth may be involved. Agents that start the neoplastic process are called initiators.

After initiation, a cell can pass into a promotion stage and some chemical agents act by promoting the development of cancer. As examples, cell proliferation in the area of a chemically induced

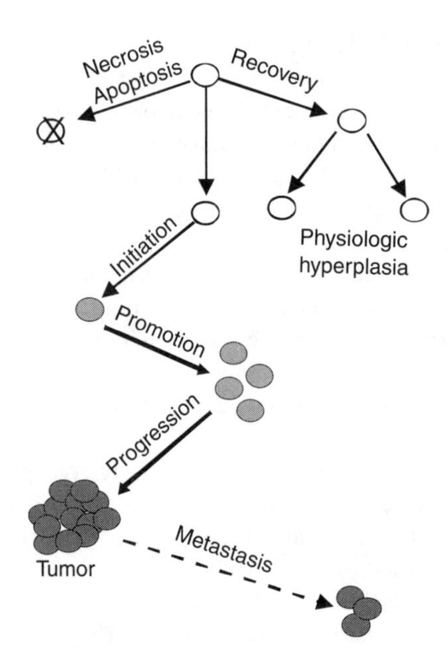

FIGURE 4.6 General stages of carcinogenesis. A damaged cell can recover and perhaps contribute to tissue recovery via physiologic hyperplasia. Return to normalcy might involve cellular exclusion, detoxification and elimination of the toxic agent, or successful repair of any DNA damage. Beyond a certain level of damage, the cell might experience necrosis or undergo apoptosis. Carcinogenesis involves initiation to produce a latent tumor cell and then promotion in which the latent tumor cell proliferates with expansion in the area of the primary tumor. However, not all latent tumor cells successfully pass through a progression stage. In the progression stage, the latent tumor cell's attributes change and a tumor manifests. Metastasis can spread the cancer from the primary tumor location to other locations within the organism.

lesion can foster development of a neoplastic lesion (tumor) after initiation. Arsenic induces liver hyperplasia in rats and this can promote tumor formation (Kotsanis and Lliopoulou-Georgudaki 1999). Hormones associated with hyperplasia can promote tumor development by a process called hormonal oncogenesis. A related consequence of cell proliferation and promotion of cancer is the accelerated growth of tumor cells remaining after surgery, as suggested by the rodent work of Rasnidi et al. (1999). Some chemical agents promote cancer by inactivating suppressor genes. Promotion might also involve the inhibition of apoptosis (Pitot and Dragan 1996, Roberts et al. 1997).

Cancer progression occurs if the qualities of the neoplastic cells change over time, leading to malignancy. Changes could involve selection among cancer cells for those best able to grow within the tissue. Pitot and Dragan (1996) emphasize that karyotype instability is important in cancer progression and add that "mechanisms [of] karyotypic instability . . . include disruption of the mitotic apparatus, alteration of telomere function . . ., DNA hypomethylation, recombination, gene amplification, and gene transposition."

4.4.2 Cancer Latency

There is a delay from cancer initiation to tumor manifestation in the afflicted individual. This should be no surprise to the reader given the multistage nature of carcinogenesis just described. This delay between exposure and detectable cancer and the added complications associated with promotion and progression make it difficult to assign causality in ecotoxicological studies of cancer. As a further confounding factor, the length of the latency period can depend on dose (Guess and Hoel 1977). Consequently, multiple lines of evidence are required such as those used in the research program addressing liver cancers in bottom-dwelling fishes (Box 13.2).

Cancer latency also makes documentation difficult for any improvement occurring after environmental remediation: reduction in cancer incidence will lag for a period after remediation. Consequently, specific computational techniques are required to estimate these cessation lags when dealing with carcinogenic effects on endemic species (Chen and Gibb 2003).

4.4.3 Threshold and Nonthreshold Models

Dose–effect models are central to describing and predicting cancer risk. Many models have the same general form as the hazard models described in Chapter 13 (Section 13.1.3.1), e.g., models of Dewanji et al. (1989), Gart et al. (1986), Hartley and Sielken (1977), Moolgavkar (1986), Moolgavkar et al. (1988), and Muirhead and Darby (1987). These models incorporate probabilities of cell transitions in schema similar to that shown in Figure 4.6. Many include estimates of minimal times to tumor presentation. Still other models include the carcinogen toxicokinetics (e.g., Yang et al. 1998).

Practical models for predicting cancer risk differ relative to inclusion of a dose threshold. Some models assume that there is risk at any dose and, consequently, work under the assumption that any lack of evidence for very low dose effects reflects our inability to detect low risk with conventional study designs (Hartley and Sielken 1977). These linear nonthreshold models arise from assumptions that only one hit is required to initiate carcinogenesis (Pitot and Dragan 1996) and that a certain, albeit extremely low, risk exists no matter how low the dose. Some irradiation-induced cancers appear to fit a nonthreshold model (Fabriant 1972). Linear nonthreshold models are often applied pragmatically during risk assessments because they provide a more conservative regulatory approach for low dose scenarios than do assessments based on threshold models (e.g., EPA 1989).

Other researchers such as Cohen (1990) and Goldman (1996) argue that threshold models seem consistent with much existing information. How a dose–cancer response threshold model might emerge can be illustrated using oxidative damage of cellular DNA (see Beckman and Ames (1997) for a recent review). The cell has a finite capacity to resist oxidative damage and accrues DNA damage above a certain dose. The cell's ability to resist damage at low dose results in a threshold relationship between dose and cancer.

4.5 SEQUESTRATION AND ACCUMULATION

4.5.1 Toxicants or Products of Toxicants

Often toxicants sequestered or accumulated in cells are taken as evidence of exposure. For example, elevated metals associated with metallothionein in kidney cells and granules of hepatopancreas cells reflect metals sequestered away from sites of action. Those associated with granules are not only sequestered to minimize toxic effect within the soma, they are also relatively unavailable for trophic transfer to grazers, predators, or parasites (Nott and Nicolaiduo 1993, Wallace and Luoma 2003, Wallace et al. 2003). Let us examine details of metal sequestration in intracellular granules as one example.

Several kinds of granules occur in invertebrate tissues (Beeby 1991). In molluscs, calcium carbonate granules can be present outside and within some cells, and are associated with maintaining calcium homeostasis (Mason and Nott 1981). Other granules found in invertebrates are iron-rich granules composed of residues of heme-containing biomolecules. Within the hepatopancreas of woodlice are other granules rich in sulfur and copper. These cuprosomes accumulate high concentrations of lead from contaminated soils (Hopkin 1989). A general class of calcium and magnesium pyrophosphate granules is found in certain invertebrate cells. These granules are also rich in metals owing to their role in metal detoxification (Howard et al. 1981). These granules have been studied extensively in molluscs but are found in many other invertebrate phyla. Toxic metals (except Period 1A metals) combine with pyrophosphate generated during anabolic activity of the cell to form very insoluble, and therefore biologically inactive, salts (Mason and Simkiss 1982). Hepatopancreatic basophil cells of gastropods contain such granules. The metal-rich granules remain in the basophilic cell as membrane-bound structures or are released in waste vacuoles into the digestive tubule lumen to eventually be voided from the gut. In laboratory experiments and field studies, Class A and a range of intermediate metals tended to be incorporated into pyrophosphate granules and Class B metals tended to be bound to sulfur-rich proteins in the cytoplasm (Mason et al. 1984, Simkiss and Taylor 1981). Mason and Simkiss (1983) interpreted the accumulation of metals in *Littorina littorea* hepatopancreas and kidney as arising from metal ligand-binding tendencies with pyrophosphate granules in cells of the hepatopancreas and with S-rich protein in cells of the kidney. Similar granules have also been found in scallop (*Argopecten irradians* and *Argopecten gibbus*) kidney epithelial cells (Carmichael et al. 1979).

Mercuric selenide granules (calculi) are found in liver connective tissue of cetaceans and are thought to sequester mercury away from potential sites of action (Martoja and Viale 1977). They are produced simultaneously with demethylation in the liver of accumulated methylmercury. They are found in high concentrations in connective tissues around the portal vascular system (Frodello et al. 2000). Examining molar Hg:Se ratios in stranded dolphin's livers, Mackey et al. (2003) speculated that selenide granules might also be involved in sequestration of other metals.

4.5.2 Cellular Materials as Evidence of Toxicant Damage

The presence of excessive lipofuscin (ceroid or age pigment) is evidence of oxidative damage. The name "age pigment" for lipofuscin indicates its gradual increase in neurons with age (La Via and Hill 1971). Oxidized lipids form pigmented, fluorescent deposits called residual bodies. Although unclear in black and white rendering, the macrophage aggregates in Figure 4.3 contain such light brown ceroid deposits. This micrographic is also incorporated in color into the front book cover which facilitates identification of the ceroid deposits.

Accumulation of lipofuscin can be indicative of free radical damage of membrane lipids, that is, a biomarker for lipid peroxidation and consequent membrane damage. For example, cytochrome P450 action on carbon tetrachloride produces a free radical that causes lipid peroxidation of hepatocyte membranes (Snyder and Andrews 1996). Hepatocytes of mullet (*Liza ramada*) exposed to the *s*-triazine herbicide atrazine display large lipofuscin granules, indicating increased lipid peroxidation (Biagianti-Risbourg and Bastide 1995). Peroxisomes, vacuoles containing peroxidative enzymes, also increased in these exposed mullet's hepatocytes. Lipid peroxidation is also involved in photo-induced toxicity[4] of anthracene as evidenced by reduced photo-induced cytotoxicity in the presence of a lipid peroxidation antagonist (Trolox) (Choi and Oris 2000). Copper exposure of squid (*Torpedo marmorata*) produced lipofuscin deposits in neurons, especially in the electric lobe that has high levels of oxidative metabolism but low levels of superoxide dismutase (Aloj Totaro et al. 1986).

Other evidence of cell or tissue damage is often sought in various body fluids. Evidence from urine provides a good example. Enzymuria, high levels of enzymes in urine, can provide insight about diverse nephrotoxicants. *N*-actetyl-β-glucoaminidase in the urine suggests general renal disease in humans (Kunin et al. 1978), and mercury damage of kidney cells is indicated in elevated urinary glutamine transaminase K (Trevisan et al. 1996). Proteinuria, abnormally high levels of protein in the urine, can indicate tubule damage due to cadmium exposure (Nogawa et al. 1978). As we saw in Chapter 3 (Section 3.8), elevated porphyrins in urine suggests exposure to several toxicants. Elevated guanine in urine of rats indicated that lead disrupted guanine aminohydrolase and crystalline concretions of guanine were found in the femor head's epiphyseal plate in these lead-exposed rats (Farkas and Stanawitz 1978). The guanine concretions were thought to be the tissue-level effect that translated into the saturnine gout characteristic of lead intoxication.

[4] Photo-induced toxicity is toxicity enhancement by sunlight that, in the case of PAHs, can involve photosensitization and photomodification. Photosensitization reactions produce singlet-state oxygen and photomodification reactions convert the parent PAH to a more toxic compound.

4.6 SUMMARY

The biochemical responses and consequences noted in the last chapter are expressed differentially in cells and tissues. Many associated consequences were described in this short chapter. Consequences not predictable solely from biochemical knowledge were also discussed, including necrosis, apoptosis, inflammation, and carcinogenesis. In the next chapter, these and other processes emerging at the organ and organ system levels will be described and linked to those described here.

4.6.1 Summary of Foundation Concepts and Paradigms

- Essential spatial differences in biochemical processes and moieties exist in cells and tissues. These differences determine which cells or tissues are most affected by, or responsive to, toxicants.
- Coagulative necrosis is the most useful type of necrosis in identifying toxicant-induced cell death.
- Cells die by a process known as necrosis if toxicant stress becomes insurmountable or injury irreparable. Cells damaged by toxicants can also die via apoptosis.
- Toxicant-induced imbalance between mitosis and apoptosis can lead to deviations in tissue homeostasis, developmental abnormalities, or cancer promotion.
- Inflammation, a general response to damage or infection, aims to isolate, remove, and replace damaged cells.
- Toxicants can also produce hyperplasia and hypertrophy in tissues.
- Genotoxicity, damage to the cell's DNA by a physical or chemical agent, can result in somatic and genetic risk. Most attention is paid to somatic risk.
- DNA damage in cells (microlesions) is measured in several ways by ecotoxicologists including measurement of DNA adducts, flow cytometry, alkaline unwinding assays, and comet (single cell) electrophoresis.
- Macrolesions occurring to the chromatids or chromosomes are detected by ecotoxicologists using a variety of approaches including SCE rates, micronucleus formation, and karyological metaphase preparations.
- Cancer, or neoplastic hyperplasia, begins with a change in a cell's hereditary materials, and involves a sequence of low probability, irreversible events. Three general stages of carcinogenesis are initiation, promotion, and progression.
- Assignment of causality in ecotoxicological studies of cancer is difficult due to the latency characteristic of carcinogenesis, the potential role of toxicants at the different stages of carcinogenesis, and ambiguity about threshold doses in cancer models.
- Evidence of toxicant effect or cellular response to toxicants can suggest realized exposure. Examples include metal-rich intracellular granules in specific cells of many invertebrates, mercury and selenium granules in cetacea, lipofuscin deposits, guanine concretions, and molecular markers in body fluids.

REFERENCES

Aloj Totaro, E., Pisanti, F., Glees, P., and Continillo, A., The effect of copper pollution on mitochondrial degeneration, *Mar. Environ. Res.*, 18, 245–253, 1986.

Akaishi, F.M., Silva de Assis, H.C., Jakobi, S.C.G., Eiras-Stofella, D.R., St-Jean, S.D., Courtenay, S.C., Lima, E.F., Wagener, A.L.R., Scofield, A.L., and Oliveira Ribeiro, C.A., Morphological and neurotoxicological findings in tropical freshwater fish (*Astyanax* sp.) after waterborne and acute exposure to water soluble fraction (WSF) of crude oil, *Arch. Environ. Contam. Toxicol.*, 46, 244–253, 2004.

Beckman, K.B. and Ames, B.N., Oxidative decay of DNA, *J. Biol. Chem.*, 272, 19633–19636, 1997.

Beeby, A., Toxic metal uptake and essential metal regulation in terrestrial invertebrates: A review, In *Metal Ecotoxicology. Concepts & Applications*, Newman, M.C. and McIntosh, A.W. (eds.), Lewis Publishers, Boca Raton, FL, pp. 65–89, 1991.

Biagianti-Risbourg, S. and Bastide, J., Hepatic perturbations induced by a herbicide (atrazine) in juvenile grey mullet *Liza ramada* (Mugilidae, Teleostei): An ultrastructural study, *Aquat. Toxicol.*, 31, 217–229, 1995.

Bibel, D.J., Centennial of the rise of cellular immunology: Metchnikoff's discovery at Messina, *ASM News*, 48, 558–560, 1982.

Burgeot, T., His, E., and Galgani, F., The micronucleus assay in *Crassostrea gigas* for the detection of seawater genotoxicity, *Mutat. Res.*, 342, 125–140, 1995.

Carmichael, N.G., Squibb, K.S., and Fowler, B.A., Metals in the molluscan kidney: A comparison of two closely related bivalve species (*Argopecten*), using X-ray microanalysis and atomic absorption spectroscopy, *J. Fish. Res. Board Can.*, 36, 1149–1155, 1979.

Chabicovsky, M., Klepal, W., and Dallinger, R., Mechanisms of cadmium toxicity in terrestrial pulmonates: Programmed cell death and metallothionein overload, *Environ. Toxicol. Chem.*, 23, 648–655, 2004.

Chen, C.W. and Gibb, H., Procedures for calculating cessation lag, *Regul. Toxicol. Pharmacol.*, 38, 157–165, 2003.

Choi, J. and Oris, J.T., Anthracene photoinduced toxicity to PLHC-1 cell line (*Poeciliopsis lucida*) and the role of lipid peroxidation in toxicity, *Environ. Toxicol. Chem.*, 19, 2699–2706, 2000.

Cohen, B.L., A test of the linear-no threshold theory of radiation carcinogenesis, *Environ. Res.*, 53, 193–220, 1990.

De Ferrari, M., Artuso, M., Bonassi, S., Cavalieri, Z., Pescatore, D., Marchini, E., Pisano, V., and Abbondandolo, A., Cytogenetic biomonitoring of an Italian population exposed to pesticides: Chromosome aberration and sister-chromatid exchange analysis in peripheral blood lymphocytes, *Mutat. Res.*, 260, 105–113, 1991.

Dewanji, A., Venzon, D.J., and Moolgavkar, S.H., A stochastic two-stage model for cancer risk assessment. II. The number and size of premalignant clones, *Risk Anal.*, 9, 179–187, 1989.

Dixon, D.R. and Clarke, K.R., Sister chromatid exchange: A sensitive method for detecting damage caused by exposure to environmental mutagens in the chromosomes of adult *Mytilus edulis*, *Mar. Biol. Lett.*, 3, 163–172, 1982.

Dixon, D.R., Pruski, A.M., Dixon, L.R.J., and Jha, A.N., Marine invertebrate eco-genotoxicology: A methodological overview, *Mutagenesis*, 17, 495–507, 2002.

EPA, *Risk Assessment Guidance for Superfund, Volume I: Human Health Evaluation Manual, EPA 540/1-89/001*, National Technical Information Service, Springfield, VA, 1989, p. 57.

Forni, A., Sciame, A., Bertazzi, P.A., and Alessio, L., Chromosome and biochemical studies in women occupationally exposed to lead, *Arch. Environ. Health*, 35, 139–146, 1980.

Fowler, A.J., Singh, D.N., and Dwivedi, C., Effect of cadmium on meiosis, *Bull. Environ. Contam. Toxicol.*, 29, 412–415, 1982.

Farkas, W.R. and Stanawitz, T., Saturnine gout: Lead-induced formation of guanine crystals, *Science*, 199, 786–787, 1978.

Frodello, J.P., Roméo, M., and Viale, D., Distribution of mercury in the organs and tissues of five toothed-whale species of the Mediterranean, *Environ. Pollut.*, 108, 447–452, 2000.

Gart, J.J., Krewski, D., Lee, P.N., Tarone, R.E., and Wahrendorf, J., *Statistical Methods in Cancer Research, Volume III—The Design and Analysis of Long-Term Animal Experiments*, International Agency for Research on Cancer, Lyon, France, 1986.

Goldman, M., Cancer risk of low-level exposure, *Science*, 271, 1821–1822, 1996.

Gregus, Z. and Klaassen, C.D., Mechanisms of toxicity, In *Casarett & Doull's Toxicology. The Basic Science of Poisons*, Klaassen, C.D. (ed.), McGraw-Hill, New York, 1996, pp. 35–74.

Guess, H.A. and Hoel, D.G., The effect of dose on cancer latency period, *J. Environ. Pathol. Toxicol.*, 1, 279–286, 1977.

Hartley, H.O. and Sielken, R.L., Jr., Estimation of "safe doses" in carcinogenic experiments, *Biometrics*, 33, 1–30, 1977.

Hayashi, M., Ueda, T., Uyeno, K., Wada, K., Kinae, N., Saotome, K., Tanaka, N., et al., Development of genotoxicity assay systems that use aquatic organisms, *Mutat. Res.*, 399, 125–133, 1998.

Hinton, D.E. and Laurén, D.J., Liver structural alterations accompanying chronic toxicity in fishes: Potential biomarkers of exposure, In *Biomarkers of Environmental Contamination*, McCarthy, J.F. and Shugart, L.R. (eds.), Lewis Publishers, Boca Raton, FL, 1990, pp. 17–57.

Hopkin, S.P., *Ecophysiology of Metals in Terrestrial Invertebrates*, Elsevier Applied Science, London, UK, 1989, p. 366.

House, R.V. and Thomas, P.T., Immunotoxicology: Fundamentals of preclinical assessment, In *Handbook of Toxicology*, 2nd ed., Derelanko, M.J. and Hollinger, M.A. (eds.), CRC Press, Boca Raton, FL, 2002, pp. 401–435.

Howard, B., Mitchell, P.C.H., Ritchie, A., Simkiss, K., and Taylor, M., The composition of intracellular granules from the metal-accumulating cells of the common garden snail (*Helix aspersa*), *Biochem. J.*, 194, 507–511, 1981.

Jagoe, C.H., Faivre, A., and Newman, M.C., Morphological and morphometric changes in the gills of mosquitofish (*Gambusia holbrooki*) after exposure to mercury (II). *Aquat. Toxicol.*, 34, 163–183, 1996.

Jenner, N.K., Ostrander, G.K., Kavanagh, T.J., Livesey, J.C., Shen, M.W., Kim, S.C., and Holmes, E.H., A flow cytometric comparison of DNA content and glutathione levels in hepatocytes of English sole (*Parophyrs vetulus*) from areas of differing water quality, *Arch. Environ. Contam. Toxicol.*, 19, 807–815, 1990.

Jha, A.N., Cheung, V.V., Foulkes, M.E., Hill, S.J., and Depledge, M.H., Detection of genotoxins in the marine environment: Adoption and evaluation of an integrated approach using the embryo-larval stages of the marine mussel, *Mytilus edulis*, *Mutat. Res.*, 464, 213–228, 2000a.

Jha, A.N., Hagger, J.A., and Hill, S.J., Tributyltin induces cytogenic damage in the early life stages of the marine mussel, *Mytilus edulis*, *Environ. Mol. Mutagen.*, 35, 343–350, 2000b.

Kays, S.E. and Schnellmann, R.G., Regeneration of renal proximal tubule cells in primary culture following toxicant injury: Response to growth factors, *Toxicol. Appl. Pharmacol.*, 132, 273–280, 1995.

Kligerman, A.D., Doerr, C.L., Tennant, A.H., Harrington-Brock, K., Allen, J.W., Winkfield, E., Poorman-Allen, P., et al., Methylated trivalent arsenicals as candidate ultimate genotoxic forms of arsenic: Induction of chromosomal mutations but not gene mutations, *Environ. Mol. Mutagen.*, 42, 192–205, 2003.

Kodavanti, U.P., Schladweiler, M.C.J., Ledbetter, A.D., Hauser, R., Christiani, D.C., Samet, J.M., McGee, J., Richards, J.H., and Costa, D.L., Pulmonary and systemic effects of zinc-containing emission particles in three rat strains: Multiple exposure scenarios, *Toxicol. Sci.*, 70, 73–85, 2002.

Kotsanis, N. and Lliopoulou-Georgudaki, J., Arsenic induced liver hyperplasia and kidney fibrosis in rainbow trout (*Oncorhynchus mykiss*) by microinjection technique: A sensitive animal bioassay for environmental metal-toxicity, *Bull. Environ. Contam. Toxicol.*, 62, 169–178, 1999.

Knipling, E.F., Possibilities of insect control or eradication through the use of sexually sterile males, *J. Econ. Entomol.*, 48, 902–904, 1955.

Martoja, R. and Viale, D., Accumulation de granules de séléniure mercurique dans le foie d'odontocétes (Mammiféres, *cetacea*): un mécanisme possible de détoxication du méthyl-mercure par le sélénium, *CR. Acad. Sci., Paris, Ser. D*, 185, 109–112, 1977.

Lamb, T.S., Bickham, J.W., Gibbons, J.W., Smolen, M.J., and McDowell, S., Genetic damage in a population of slider turtles (*Trachemys scripta*) inhabiting a radioactive reservoir, *Arch. Environ. Contam. Toxicol.*, 20, 138–142, 1991.

Lash, L.H. and Parker, J.C., Hepatic and renal toxicities associated with perchloroethylene, *Pharmacol. Rev.*, 53, 177–208, 2001.

Lash, L.H., Putt, D.A., Hueni, S.E., Krause, R.J., and Elfarra, A.A., Roles of necrosis, apoptosis, and mitochondrial dysfunction in *S*-(1,2-Dichlorovinyl)-L-cysteine sulfoxide-induced cytotoxicity in primary cultures of human renal proximal tubular cells, *J. Pharmcol. Exp. Ther.*, 305, 1163–1172, 2003.

La Via, M.F. and Hill, R.B., Jr., *Principles of Pathobiology*, Oxford University Press, London, UK, 1971.

Lerda, D., Sister-chromatid exchange (SCE) among individuals chronically exposed to arsenic in drinking water, *Mutat. Res.*, 312, 111–120, 1994.

Lindquist, A.W., The use of gamma radiation for control or eradication of the screwworm, *J. Econ. Entomol.*, 48, 467–469, 1955.

Lux, S.A., Vilardi, J.C., Liedo, P., Gaggl, K., Calcagno, G.E., Munyiri, F.N., Vera, M.T., and Manso, F., Effects of irradiation on the courtship behavior of medfly (Diptera, Tephritidae) mass reared for the sterile insect technique, *Florida Entomol.*, 85, 102–111, 2002.

Mackey, E.A., Oflaz, R.D., Epstein, M.S., Buehler, B., Porter, B.J., Rowles, T., Wise, S.A., and Becker, P.R., Elemental composition of liver and kidney tissues of rough-toothed dolphin (*Steno bredanensis*), *Arch. Environ. Contam. Toxicol.*, 44, 523–532, 2003.

Martineau, D., Lemberger, K., Dallaire, A., Labelle, P., Lipscomb, T.P., Michel, P., and Mikaelian, I., Cancer in wildlife, a case study: Beluga from the St. Lawrence estuary, Québec, Canada, *Environ. Health Perspec.*, 110, 285–292, 2002.

Mason, A.Z. and Nott, J.A., The role of intracellular biomineralized granules in the regulation and detoxification of metals in gastropods with special reference to the marine prosobranch *Littorina littorea*, *Aquat. Toxicol.*, 1, 239–256, 1981.

Mason, A.Z. and Simkiss, K., Sites of mineral deposition in metal-accumulating cells, *Exp. Cell Res.*, 139, 383–391, 1982.

Mason, A.Z. and Simkiss, K., Interactions between metals and their distribution in tissues of *Littorina littorea* (L.) collected from clean and polluted sites, *J. Mar. Biol. Ass. U.K.*, 63, 661–672, 1983.

Mason, A.Z., Simkiss, K., and Ryan, K.P., The ultrastructural localization of metals in specimens of *Littorina littorea* collected from clean and polluted sites, *J. Mar. Biol. Ass. U.K.*, 64, 699–720, 1984.

McBee, K., Bickham, J.W., Brown, K.W., and Donnelly, K.C., Chromosomal aberrations in native small mammals (*Peromyscus leucopus* and *Sigmodon hispidus*) at a petrochemical waste disposal site: I. Standard karyology, *Arch. Environ. Contam. Toxicol.*, 16, 681–688, 1987.

Metchnikoff, O., *Life of Elie Metchnikoff 1845–1916*, Houghton Mifflin, Boston, MA, 1921.

Moolgavkar, S.H., Carcinogenesis modeling: From molecular biology to epidemiology, *Ann. Rev. Public Health*, 7, 151–169, 1986.

Moolgavkar, S.H., Dewanji, A., and Venzon, D.J., A stochastic two-stage model for cancer risk assessment. I. The hazard function and the probability of tumor, *Risk Anal.*, 8, 383–392, 1988.

Muirhead, C.R. and Darby, S.C., Modeling the relative and absolute risks of radiation-induced cancers, *J. R. Statist. Soc. A*, 150, 83–118, 1987.

Newman, M.C. and Unger, M.A., *Fundamentals of Ecotoxicology*, 2nd ed., CRC/Lewis Publishers, Boca Raton, FL, 2003.

Nogawa, K., Ishizake, A., and Kawano, S., Statistical observations of the dose–response relationships of cadmium based on epidemiological studies in the Kakehashi River Basin, *Environ. Res.*, 15, 185–198, 1978.

Nott, J.A. and Nicolaiduo, A., Bioreduction of zinc and manganese along a molluscan food chain, *Comp. Biochem. Physiol.*, 104A, 235–238, 1993.

Pastor, N., López-Lázaro, M., Tella, J.L., Baos, R., Hiraldo, F., and Cortés, F., Assessment of genotoxic damage by the comet assay in white storks (*Ciconia ciconia*) after the Doñana ecological disaster, *Mutagenesis*, 16, 219–223, 2001.

Pathak, S.K. and Bhowmik, M.K., The chronic toxicity of inorganic mercury in goats: Clinical signs, toxicopathological changes and residual concentrations, *Vet. Res. Commun.*, 22, 131–138, 1998.

Pitot, H.C., III and Dragan, Y.P., Chemical carcinogenesis, In *Casarett & Doull's Toxicology. The Basic Science of Poisons*, Klaassen, C.D. (ed.), McGraw-Hill, New York, 1996, pp. 201–267.

Plytycz, B. and Seljelid, R., From inflammation to sickness: Historical perspective, *Arch. Immunol. Ther. Exp.*, 51, 105–109, 2003.

Popp, W., Vahrenholz, C., Schell, C., Grimmer, G., Dettbarn, G., Kraus, R., Brauksiepe, A., et al., DNA single strand breakage, DNA adducts, and sister chromatid exchange in lymphocytes and phenathrene and pyrene metabolites in urine of coke oven workers, *Occup. Environ. Med.*, 54, 176–183, 1997.

Risso-de Faverney, C., Orsini, N., de Sousa, G., and Rahmani, R., Cadmium-induced apoptosis through the mitrochondrial pathway in rainbow trout hepatocytes: Involvement of oxidative stress, *Aquat. Toxicol.*, 69, 247–258, 2004.

Roberts, R.A, Nebert, D.W., Hickman, J.A., Richburg, J.H., and Goldsworthy, T.L., Perturbation of the mitosis/apoptosis balance: A fundamental mechanisms in toxicology, *Fund. Appl. Toxicol.*, 38, 107–115, 1997.

Segner, H. and Braunbeck, T., Cellular response profile to chemical stress, In *Ecotoxicology*, Schüürmann, G. and Markert, B. (eds.), John Wiley & Sons, New York, 1998, pp. 521–569.

Shugart, L.R., Quantitation of chemically induced damage to DNA of aquatic organisms by alkaline unwinding assay, *Aquat. Toxicol.*, 13, 43–52, 1988.

Shugart, L.R., Detection and quantitation of benzo[a]pyrene-DNA adducts in brain and liver tissues of Beluga whales (*Delphinapterus leucas*) from the St. Lawrence and Mckenzie estuaries, In *Proceedings of the International Forum for the Future of the Beluga*, Presses de l'Université du Quebec, Quebec, 1990, pp. 219–223.

Simkiss, K. and Taylor, M., Cellular mechanisms of metal ion detoxification and some new indices of pollution, *Aquat. Toxicol.*, 1, 279–290, 1981.

Snyder, R. and Andrews, L.S., Toxic effects of solvents and vapors, In *Casarett & Doull's Toxicology. The Basic Science of Poisons*, Klaassen, C.D. (ed.), McGraw-Hill, New York, 1996, pp. 737–771.

Sparks, A.K., *Invertebrate Pathology. Noncommunicable Diseases*, Academic Press, New York, 1972.

Tucker, J.D., Auletta, A., Cimino, M.C., Dearfield, K.L., Jacobson-Kram, D., Tice, R.R., and Carrano, A.V., Sister-chromatid exchange: Second report of the Gene-Tox program, *Mutat. Res.*, 297, 101–180, 1993.

Wallace, W.G. and Luoma, S.N., Subcellular compartmentalization of Cd and Zn in two bivalves. II. The significance of trophically available metal (TAM), *Mar. Ecol. Prog. Ser.*, 257, 125–137, 2003.

Wallace, W.G., Lee, B.-G., and Luoma, S.N., Subcellular compartmentalization of Cd and Zn in two bivalves. I. Significance of metal-sensitive fractions (MSF) and biologically detoxifies metal (BDM), *Mar. Ecol. Prog. Ser.*, 249, 183–197, 2003.

Yang, R.S.H., Thomas, R.S., Gustafson, D.L., Campain, J., Benjamin, S.A., Verhaar, H.J.M., and Mumtaz, M.M., Approaches to developing alternative and predictive toxicology based on PBPK/PD and QSAR modeling, *Environ. Health Perspect.*, 106, 1385–1393, 1998.

Zodrow, J.M., Stegeman, J.J., and Tanguay, R.L., Histological analysis of acute toxicity of 2,3,7,8-tetrachlorodibenzo-*p*-dioxin (TCDD) in zebrafish, *Aquat. Toxicol.*, 66, 25–38, 2004.

5 Organs and Organ Systems

Organ systems of individuals are the highest levels of organization that are commonly studied in laboratory exposures to various toxicants, and the concept of target organ is firmly established in mammals and in aquatic organisms.

(Hinton 1994)

5.1 OVERVIEW

The cells examined in the previous chapter are the building blocks of organs. The cellular differentiation and organization that give rise to distinct organs also set the scene for differences in toxicant effects among organs. In addition to the nature of the cells incorporated into a tissue, the spatial relation of organs relative to direct environmental exposure or exposure during somatic circulation makes one organ more prone to poisoning than another. Because of these differences and the central role of organs in maintaining health, effects to target organs constitute a major theme in classic toxicology. Roughly one-third of the chapters in *Casarett & Doull's Toxicology: The Basic Science of Poisons* (Klaassen 1996) and Williams et al.'s *Principles of Toxicology* (Williams et al. 2000) are devoted to organ toxicity. Target organ toxicity is also an important theme in ecotoxicology; however, extra attention is required when reaching conclusions about organ toxicity for the myriad relevant species because many differences exist among species relative to which organ is most affected by any particular chemical agent (Heywood 1981).

Organ toxicology is discussed here using examples because any comprehensive coverage of organ toxicity for all relevant species would require several books. Discussion is also biased toward vertebrate organs because of the abundance of available information.

5.2 GENERAL INTEGUMENT

The integument, like the respiratory and digestive organs, is subject to significant toxicant exposure because of its intimate contact with the external milieu and large surface area for exchange. For these reasons, some of the classic discoveries about environmentally induced human disease involved dermal exposure: Percoval Pott's linkage in 1775 of scrotal cancer prevalence among chimney sweeps with dermal exposure to soot and pitch (polycyclic aromatic hydrocarbon, PAH) was one such watershed study.

The integument is a good, albeit imperfect, barrier to toxicants; consequently, dermal absorption constitutes one of three major routes of exposure. (The other two, inhalation and ingestion, are discussed in Sections 5.3 and 5.5.) The prominence of its role is determined by the individual organism of concern, nature of the toxicant, and relative concentrations of the toxicant in various media. For example, the terrestrial tiger salamander (*Ambystoma tigrinum*), which has moist skin that comes into close contact with soils, can have significant dermal uptake of contaminants such as 2,4,6-trinitrotoluene (TNT) (Johnson et al. 1999). Similarly, the foot of the terrestrial snail,

Helix aspersa, comes in close contact with soil-associated metals (Gomot-De Vaufleury and Pihan 2002) and can experience significant exposure. The dermal and ingestion routes can co-mingle for some species such as those that preen (birds) or groom (mammals) (Suter 1997). As an example, it is easy to imagine dermal, ingestion, and inhalation exposure all being significant for a marine mammal grooming after its coat had been soiled by an oil spill.

Some general trends exist for dermal exposure. Moist skin is generally more permeable than dry skin to hydrophilic compounds (Salminen and Roberts 2000). Human skin is permeable to low molecular weight, lipophilic, but also many hydrophilic, toxicants (Rice and Cohen 1996). Skin penetration can be predicted using toxicant K_{OW} and molecular weight (Poulin and Krishnan 2001), with penetration being fastest for small, hydrophobic compounds. A grim illustration of rapid dermal penetration of a low molecular weight lipophilic poison is the tragic death of mercury expert Dr. Karen Wetterhahn in 1997, who was fatally exposed to a few drops of dimethylmercury that rapidly seeped through her latex gloves and skin (Nierenberg et al. 1998).

The extent of toxicant metabolism in, or elimination from, the integument depends on the particular organism and toxicant. Some level of Phase I and II detoxification occurs in the mammalian integument. As an example, β-naphthoflavone exposure induces cytochrome P450 activity in sperm whale skin (Godard et al. 2004). This cytochrome P450 activity was found in endothelial cells, smooth muscle cells, and fibroblasts, and was concentration-dependent. Such induction of cytochrome P450 activity is quite relevant because cetaceans accumulate high concentrations of organic xenobiotics in skin and blubber (e.g., Valdez-Márquez et al. 2004). An amphibian example of xenobiotic metabolism within the integument is the cytochrome P450 metabolism measured in leopard frog (*Rana pipiens*) skin after exposure to 3,3′,4,4′,5-pentachlorobiphenyl (Huang et al. 2001). Unfortunately, such transformations can activate compounds, establishing a mechanism for disease manifestation in the integument (Salminen and Roberts 2000). Some PAHs within the skin (see Table 18.6 in Rice and Cohen (1996)) can also be phototoxic on exposure to ultraviolet (UV) light; free radicals form that cause lesions resembling sunburn.

The many species relevant to ecotoxicology have diverse elimination mechanisms associated with the integument. Organochlorine compounds are shed with reptile skins (Jones et al. 2005). Toothed-whales eliminate mercury in the integument by desquamation (Viale 1977). Mercury and arsenic are lost in the hair of mammals, and mercury is lost in bird feathers. Monteiro and Furness (1995) estimate that most of the mercury in Cory's shearwaters is present in the feathers at fledging time. Terrestrial isopods, commonly called woodlice or pillbugs, eliminate metals during molting (Raessler et al. 2005). Crocodiles, which possess dermal plates (osteoderms), accumulate and presumably sequester metals in these bones (Jeffree et al. 2005).

Toxicant-induced effects in the integument can manifest visibly in other ways. Chloracne is a telltale sign of human poisoning with halogenated aromatic hydrocarbons such as many polychlorinated biphenyls (PCBs), pesticides, and dioxin. Chloracne is the presence of many comedones (noninflammatory flesh, white, or dark-colored lesions that impart a rough texture to the skin) and straw-colored cysts on the face, neck, behind the ear, back, chest, and genitalia. A recent example of chloracne is apparent in December 2004 photographs of Viktor Yushckenko, a Ukrainian politician who was intentionally poisoned with dioxins. Chloracne is not the only dermal effect of poisons to humans. Chemical irritants cause a range of pathologies from allergic response to inflammation to obvious necrotic lesions. As an example, skin lesions are one of the most common features of chronic human arsenic poisoning (Yoshida et al. 2004).

Effects manifest in the integument of many nonhuman species. Mercury, at concentrations comparable to that found in integument of some whales in the St. Lawrence estuary, can produce micronuclei in Beluga whale skin fibroblasts (Gauthier et al. 1998). Orthodichlorobenzene, a compound proposed at one time as a predator barrier around shellfish beds (Loosanoff et al. 1960), causes large dermal lesions on starfish (*Asterias forbesi*) coming in physical contact with orthodichlorobenzene-coated sand (Sparks 1972). Amphibians have active dermal ion and gas exchange from water that is disrupted by contaminants. As an important example, some pyrethroid pesticides modify sodium and chloride ion transport across amphibian skin (Cassano et al. 2000, 2003).

5.3 ORGANS ASSOCIATED WITH GAS EXCHANGE

Respiratory organs have intimate contact and exchange with the external environment; therefore, it is no surprise that they are also major organs of toxicant exchange and effect. Several classic examples exist in human epidemiology. Doll et al. (1970) found high incidence of nasal and lung cancer in Welsh refinery workers who breathed air containing nickel-rich particles. Despite a long period of uncertainty (Cornfield et al. 1959, Doll and Hill 1964), linkage between lung cancer and tobacco smoke inhalation is now common knowledge. Both of these examples involved inhalation of particles that come into contact with moist surfaces of respiratory exchange.

Respiratory exposure resulting in disease is still an expanding field of study. Topics range from studies such as that linking lung cancer rates in rural China to coal burning in homes (Lan et al. 2002) to studies of exchange of toxicants across gills (e.g., Erickson and McKim 1990, Playle 2004). This section highlights issues associated with lung and gill exposure to toxicants.

5.3.1 AIR BREATHING

The chemical nature and form of a toxicant are important determinants of its ability to deliver an effective dose to air breathing animals. Its physical phase association is also extremely important.

Water solubility determines movement of a gaseous toxicant in the lungs. "Highly soluble gases such as SO_2 do not penetrate farther than the nose and are therefore relatively nontoxic . . . relatively nonsoluble gases such as ozone and NO_2 penetrate deeply . . . where they elicit toxic responses" (Witschi and Last 1996). Inhalation of volatile organic toxicants can also be a very efficient route of exposure. A human example of recent concern is inhalation of the gasoline additive methyl tertiary butyl ether (MTBE) (Buckley et al. 1997). Volatile compounds imbibed in tap water or inhaled during showering can also result in significant exposure. On the other hand, exhalation can act as a significant route of elimination for some volatile toxicants such as trichloroethene or chloroform (Pleil et al. 1998, Weisel and Jo 1996).

Realized dose for a particle-associated toxicant depends on its form and the nature of the particle with which it is associated. Generally, nonpolar toxicants in liquid aerosols are absorbed more quickly after inhalation than polar toxicants. Weak electrolytes in liquid aerosols are absorbed at pH-dependent rates (Gibaldi 1991).

Dry aerosols containing toxicants are formed in many ways. Vehicular combustion of leaded gasoline generates submicron aerosols rich in water-soluble lead halides ($PbClBr$, $PbCl_2 \cdot PbClBr$) and oxyhalides ($PbO \cdot PbClBr$, $PbO \cdot PBr_2$, $2PbO \cdot PbClBr$). Any weathering of the aerosol-associated lead produces less soluble forms (i.e., $PbSO_4$ and $PbSO_4 \cdot (NH_4)_2SO_4$) and these less soluble forms of lead also tend to be incorporated into larger particles in roadside soil (Biggins and Harrison 1980, Laxen and Harrison 1977). The small size of the initial aerosols allows deep access into the terminal bronchioles and aveoli, and the associated lead is relatively soluble. In contrast, the weathered lead associated with the larger soil particles has limited penetration down the pulmonary system and is less soluble once deposited on a moist respiratory surface.

Effects of inhaled toxicants range from fatal cancers to pulmonary edema to mild irritation. The nickel-induced cancer in Welsh smelter workers discussed above is one example of a fatal cancer. Another is the lung cancer manifested after asbestos inhalation. Some toxicants such as ozone and chlorine gas cause pulmonary edema in which fluids build up in the lungs and diminish the effectiveness of gas exchange; extensive toxicant-induced edema can kill. Combustion can produce particles rich in bioavailable zinc and inhalation of such particles by rats can produce metal-fume fever, that is, pulmonary inflammation and injury (Kodavanti et al. 2002). High densities of airborne particles cause pulmonary distress by stimulating elevated numbers of polynuclear leukocytes in airways that release reactive oxygen species (Prahalad et al. 1999). A high oxidative burden caused by these particles and other pulmonary stressors such as ozone or NO_2 can result in pulmonary damage. Oxidant toxicity resulting from free radical production by toxicants and leukocyte production of hydrogen peroxide is thought to cause lung damage such as that occurring after inhalation of paraquat (Witschi and Last 1996).

5.3.2 Water Breathing

Predicting effects of gill exposure is complicated because there are so many kinds of gills. Also many gills are not involved solely in dissolved oxygen and carbon dioxide exchange. Fish gills are also important organs for ion- and osmoregulation, and excretion of nitrogenous wastes. Chloride cells at the base of gill lamella facilitate ion exchange, and as a result, differ in freshwater and saltwater fishes. Gills of some invertebrates are feeding organs. Molluscs, such as the oyster, have gills with complex ciliary tracts for filtering and sorting of food items. In contrast to the oyster, the pulmonate gastropods have no gills, and most prosobranchs such as *Busycon* have gills with no feeding role whatsoever. Annelid gill structure can vary from gills at the base of feeding tentacles of the tube-dwelling *Amphitrite* to gill clusters along the body of the lugworm, *Arenicola*, to parapodia-associated structures of the burrowing *Glycera americana*, to the large feeding appendages of the fan worm, *Sabella*. Because of this diversity, only examples involving fish and crustaceans will be given in this short section.

Some effects to gills were discussed in the last chapter. A range of toxicants including copper (van Heerden et al. 2004), nickel (Pane et al. 2004), zinc (Matthiessen and Brafield 1973), endosulfan (Saravana Bhavan and Geraldine 2000), the anti-inflammatory drug diclofenac (Trienskorn et al. 2004), and alkyl benzene sulfonate (Scheier and Cairns 1966) induce the secondary lamellae to change in a manner similar to that shown in Figures 4.4 and 4.5. These morphological changes can persist for long periods after exposure ends (Scheier and Cairns 1966, van Heerden et al. 2004), potentially causing chronic gill dysfunction.

Other important changes occur to gills with toxicant exposure. Fish (*Perca fluviatilis* and *Rutilus rutilus*) exposed to mine drainage experienced mucus cell hyperplasia as well as chloride cell hyperplasia and hypertrophy. The epithelial thickening seen in gills is thought to decrease the rate of exchange with waters by increasing the distance between the blood and water. Changes in the amounts of phosphatidylcholine (increase) and cholesterol (decrease) in gills were associated notionally with "increased fluidity of membranes and possibly strengthen[ing of] their protective qualities" (Tkatcheva et al. 2004). Gill lipid changes were seen as an adaptive response to adverse changes in chloride cell structure; however, other lipid changes in gills reflect damage. For example, Morris et al. (1982) suggested that bioaccumulation of lipophilic organic contaminants (aromatic hydrocarbons and phthalate plasticizers) in gills of the amphipod, *Gammarus duebeni*, affects gill phospholipid composition. Also, 1,1′-dimethyl-4,4′-bipyridium dichloride (paraquat) or 2,4-dichlorophenoxyacetic acid (2,4-D) exposure causes lipid peroxidation in rainbow trout (*Oncorhynchus mykiss*) gill (Marinez-Tabche et al. 2004). So, changes in gill phospholipid content can be seen as an adverse effect or an adaptive shift upon toxicant exposure.

5.4 CIRCULATORY SYSTEM

Toxicants can have teratogenic effects on developing components of the circulatory system or direct effects on fully developed circulatory system components. Both types of effects are studied in humans and nonhuman species.

Cardiac malformation is well documented for fish development in the presence of toxicants. Zebrafish (*Danio rerio*) exposed to high PAH concentrations show such abnormalities

(Incardona et al. 2004). Ownby et al. (2002) quantified cardiovascular abnormalities in *Fundulus heteroclitus* embryos developing in the presence of creosote-contaminated sediments. Cardiovascular abnormalities have been noted for this same killifish species exposed to mercury during development (Weis et al. 1981).

Obvious effects and injury are also realized in fully developed cardiovascular systems. The tissue damage described in Chapter 4 is one example. Chronic cadmium exposure can cause human heart hypertrophy (Ramos et al. 1996). Particles in vehicular exhaust cause vasoconstriction (Tzeng et al. 2003), and long-term exposure to combustion-generated particles rich in zinc can also produce myocardial injury (Kodavanti et al. 2003). The readily bioavailable zinc in inhaled particles enters the pulmonary circulation to the heart, causing cardiac lesions, chronic inflammation, and fibrosis.

Toxicant effects on blood and blood-forming (hemapoietic) tissues occur such as the changes in heme biosynthesis described in Chapter 3. Mercury exposure of the fish, *Aphanius dispar*, elevated leukocyte numbers and blood clotting time, and decreased numbers of red blood cells, hemoatocrit, and hemoglobin titer (Hilmy et al. 1980). Relative to leukocyte changes, thrombocytes decreased but eosinophiles increased with mercury exposure. Lindane injection into *Tilapia* lowered the number of white blood cells in the pronephros, the major hematopoietic organ, and the spleen (Hart et al. 1997). Diazinon reduced hematopoiesis in the clawed frog, *Xenopus laevis* (Rollins-Smith et al. 2004). More discussion of such effects to cells associated with immune response will be provided in Section 5.8.

5.5 DIGESTIVE SYSTEM

Organs normally associated with digestion provide the third major exposure route, ingestion. However, other relevant processes occur in the digestive system. Toxicants can manifest effects in the digestive system. Detoxification can also occur in the digestive tissues.

Some digestive system features make certain species especially prone to poisoning. Many birds are prone to lead poisoning because they ingest lead shot as grit and these shot are ground together under acidic conditions in their gizzards, generating dissolved lead that is taken into the bird's system (Anonymous 1977, Kendall et al. 1996).

Adverse effects to the digestive system are easily found and have various distinguishing features. The general irritation and inflammation of the digestive system due to the action of a biological or nonbiological agent is referred to as gastroenteritis. It often manifests in an individual with digestive system poisoning and has general symptoms that most people know too well (e.g., diarrhea, weakness, fever, vomiting, and blood or mucous in feces). A severe example of poisoning after ingestion that goes beyond such irritation and inflammation is phenol poisoning (carbolism). Carbolism involves extensive protein denaturation and local necrosis of contacted tissues of the digestive system. (Phenol's rapid penetration of tissues also creates a risk of phenol poisoning via dermal exposure.)

Many effects on the digestive system are less obvious and can involve roles of digestive system components other than digestion. Cadmium interferes with chloride absorption in the intestine of the eel, *Anguilla anguilla*. It also alters the tight junctions between cells (Lionetto et al. 1998). Cadmium interferes with eel intestine carbonic anhydrase and Na^+–K^+ ATPase activities that, respectively, are involved in acid–base regulation and ion regulation (Lionetto et al. 2000). Similarly, copper changes the intestinal fluid ion composition in toadfish (*Opsanus beta*) (Grosell et al. 2004).

Detoxification in the digestive system can be significant. Cytochrome P450 enzymes have been found in the human digestive system (Ding and Kaminsky 2003), midguts of insects (Mayer et al. 1978, Stevens et al. 2000), and the crustacean stomach (James 1989, Mo 1989). Relative to Phase II detoxification, human digestive system components have sulfotransferases that act on a wide range of compounds. The human small intestine has considerable sulfotransferase activity (Chen et al. 2003). The midgut of insects produces metal-inducible intestinal mucins that modify resistance to toxicants and infective agents (Beaty et al. 2002).

5.6 LIVER AND ANALOGOUS ORGANS OF INVERTEBRATES

The liver and analogous organs of invertebrates are prone to high exposures due to reasons other than a close contact with external sources. Rather, they function in such a way that effects can manifest easily. The detoxification occurring in these organs leaves them susceptible to damage by activation products. Active hepatocyte regeneration taking place in the liver coupled with potentially activating reactions arising from its detoxification role make it particularly prone to cancer. The vertebrate liver also functions in the regulation of essential and toxic metals: a metal such as cadmium that is bound to metallothionein in the liver can remain in the liver for a very long time (Ballatori 1991). The hepatopancreas or digestive glands of invertebrates function in this way too. Squids accumulate high concentrations of metals such as cadmium in the digestive gland (Bustamante et al. 2002). Excessive accumulation of metals in molluscan digestive gland interferes with cellular activities, for example, mercury interference with calcium flux through membrane channels in mussel digestive cells (Canesi et al. 2000). Livers of vertebrates can also participate in enterohepatic circulation of toxicants, leading to higher risk of contaminant interaction with a site of action in the liver. Enterohepatic circulation (or enterohepatic cycling) occurs when, for example, conjugates of toxicants are removed from the liver via the bile, freed from their conjugated form by intestinal enzymes, reabsorbed in the intestine, and returned to the liver again. A toxicant can cycle through the liver many times in this manner. The toxicant is retained in the body longer because of this cycling and has the opportunity to cause more damage. Ballatori (1991) suggests that a similar biliary-hepatic cycle can occur for methylmercury in which the methylmercury incorporated into canalculi bile[1] can be cycled back across the biliary epithelia.

Hepatoxicity results from a variety of cellular mechanisms (Jaeschke et al. 2002). Cytochrome P450 metabolism of a toxicant such as ethanol or carbon tetrachloride in the liver can promote oxidative stress. Toxicant interactions can be explained for this oxidative stress as in the case of the class of fire retardants, polybrominated biphenyls, which induce the liver's cytochrome P450 system and, in so doing, enhance the hepatotoxicity of carbon tetrachloride (Kluwe and Hook 1978). Chronic ethanol liver damage can induce inflammation with an accumulation of macrophages and neutrophils. The phagocytic activities of these cells produce even more oxidative damage including lipid peroxidation. Relative to invertebrates, similar inflammation associated with necrosis occurred in shrimp (*Penaeus vannamei*) hepatopancreas after exposure to the fungicide benomyl (Lightner et al. 1996) and in prawns (*Macrobrachium malcolmsonii*) after exposure to the pesticide endosulfan (Saravana Bhavan and Geraldine 2000). If cadmium concentrations in the liver exceed those that can be dealt with by metallothionein or glutathione binding (Chan and Cherian 1992), hepatoxicity manifests as initial injury from the cadmium binding to sulfhydryl groups in essential biomolecules in the mitochondria followed by further damage associated with the ensuing inflammation and Kupffer cell activation (Rikans and Yamano 2000). Extensive liver necrosis resulting from oxidative damage can also give rise to an immune reaction. Chromium and cadmium cause fish hepatocyte necrosis by increasing oxidative stress (Krumschnabel and Nawaz 2004, Risso-de Faverney et al. 2004). The brominated fire retardants, hexabromocyclododecane and tetrabromobisphenol, also cause oxidative stress in the liver of fish (Ronisz et al. 2004).

These cases of cytotoxicity are not the only manifestations of hepatotoxicity. Hepatic damage can manifest at the level above the cell as in the case of cholestatis, the physical blockage of bile secretion. This blockage may or may not be associated with inflammation (Zimmerman 1993). Liver qualities reflecting its state of health or function can also change with toxicant exposure. Mink fed PCB-tainted fish had low levels of vitamin A_1 (retinol) that is essential for a variety of biological functions (Käkelä et al. 2002). Mice exposed to tralkoxydim, a component of herbicides used for cereal crops, had

[1] Bile in the liver's network of canaliculi will eventually pass into the bile ducts and then into the gallbladder.

an abnormal increase in hepatic porphyrins (i.e., porphyria) (Pauli and Kennedy 2005). Cadmium and inorganic mercury modify hepatocyte glucose metabolism (Fabbri et al. 2003). Exposure to carbofuran insecticide (Begum 2004) or the anti-inflammatory drug diclofenac (Triebskorn et al. 2004) decreases fish liver glycogen levels via stress-induced glycogenolysis. Rabbits exposed to crude oil displayed elevated bilirubin concentrations relative to control rabbits (Ovuru and Ezeasor 2004). The increase in this degradation product of hemoglobin was attributed to excessive destruction of erythrocytes and the compromised ability to remove bilirubin from the damaged liver. Histological examination revealed liver tissue necrosis, inflammation, and congestion around the central vein of the liver. Similar inflammation in livers of crude oil-exposed tropical fish was reported by Akaishi et al. (2004).

5.7 EXCRETORY ORGANS

Species vary in the form and function of their excretory organs, and in the form of nitrogen excreted. Gills remove much of the nitrogenous wastes as ammonia for most adult fish, whereas the kidneys of mammals are responsible for excretion of nitrogenous wastes. Vertebrates can excrete ammonia (fish), urea (mammals, some fish), and uric acid (birds, reptiles). Molluscs have nephridia that excrete ammonia (aquatic molluscs) or insoluble uric acid (terrestrial molluscs). The molluscan digestive gland can play a role in nitrogen waste excretion. Polychaete annelids have proto- or metanephridia as excretory organs but other tissues and cells may also be involved. Oligocheates have metanephridia that excrete ammonia or urea. Crustaceans excrete primarily ammonia via antennal glands and insects have Malpighian tubules that excrete uric acid. Both crustaceans and insects also have nephrocytes in other parts of the body that are involved in waste removal.

Goldstein and Schnellmann (1996) summarize the reasons why the kidney is particularly susceptible to toxicants despite its remarkable capacity to recover after injury. First, any toxicant in circulation will be delivered to the kidney because of the kidneys' central role in removing wastes from the blood. Second, the kidneys concentrate materials from the blood, providing an avenue for concentrating toxicants to levels damaging to kidney tissues. Metallothionein-bound cadmium concentrates in the proximal tubules and, if cadmium is present in excess of the binding capacity of kidney-associated metallothionein, it can cause damage ranging from elevated protein in the urine (proteinuria) to complete kidney failure (Goldstein and Schnellmann 1996, Faurskov and Bjerregaard 2000). Third, detoxification reactions in the kidney can produce activation products that damage the kidney or urinary tract. Cytochrome P450 monooxygenase (Omura 1999), some Phase II conjugation (Lash and Parker 2001) stress proteins (Tolson et al. 2005), and metallothioneins (Margoshes and Vallee 1957) may be associated with the kidney and can influence nephrotoxicity. As an example, chloroform's nephrotoxicity is a result of P450 monooxygenase activation to a reactive species.

Other mechanisms cause damage to the excretory system. Arsenic imbibed in drinking water is methylated and causes bladder cancer (Yoshida et al. 2004). Bismuth (Leussink et al. 2001) and cadmium (Prozialeck et al. 2003) cause epithelial cells of the proximal tubules to detach from each other and undergo necrosis or apoptosis. Rats exposed to uranium had necrosis in the kidney proximal tubules (Tolson et al. 2005). High concentrations of mercury and selenium have been correlated with fibrous nodules in rough-toothed dolphin kidneys (Mackey et al. 2003). Rainbow trout exposed to the anti-inflammatory drug diclofenac have a distinctive accumulation of protein in tubular cells, necrosis of endothelial cells, and macrophage infiltration (Triebskorn et al. 2004). Clearly, a wide range of effects are expected in the excretory organs of exposed species.

5.8 IMMUNE SYSTEM

Unquestionably, toxicants can have a strong influence on the immunocompetence of a wide range of species. A review of wildlife immunotoxicity (Luebke et al. 1997) gave examples that ranged from

PAH-reduction of fish leukocyte's ability to kill tumor cells, to reduced ability of piscivorous birds from the Great Lakes to respond properly to mitogens, to harbor seals fed contaminated fish having compromised natural killer cell function. At the other extreme, deficiency of essential elements that are also common contaminants (copper or zinc) can create humoral immunological deficiencies (Beach et al. 1982, Prohaska and Lukasewycz 1981). A human genetic disorder in copper transport, Menkes syndrome, produces a copper deficiency and an epiphenomenal high incidence of chronic infections (Kaler 1998). This brief section sketches out some kinds of immunological problems emerging from toxicant exposure of a diverse array of species.

Problems with immune system development can emerge with toxicant exposure, and researchers focused on human exposure to contaminants have begun to be particularly concerned about toxicant effects to immune system ontology (Holsapple 2003). Interest is also emerging relative to nonhuman receptors. *Xenopus laevis* tadpoles exposed to environmentally realistic diazinon concentrations have compromised stem cell abilities to populate the blood, thymus, and spleen (Rollins-Smith et al. 2004). Mixtures of agrochemicals (atrazine, metribuzine, endosulfan, lindane, aldicarb, and dieldrin) at environmentally realistic concentrations alter cellular immune processes of exposed *X. laevis* and *R. pipiens* (Christin et al. 2004).

Most ecotoxicological research focuses on either cellular or humoral immune dysfunction in adult vertebrates. Arctic breeding gulls (*Larus hyperboreus*) have elevated tissue residues of pesticides and PCBs that are correlated with white blood cell counts and antibody response to bacterial challenge (Bustnes et al. 2004). Humoral immune response of Great Tit (*Parus major*) was found to increase with increased distance from a Belgian metal smelter (Snoeijs et al. 2004). Carp (*Cyprinus carpio*) humoral and cellular immune responses were modified by vineyard-related agents, copper and chitosan (Dautremepuits et al. 2004a,b). Lindane injection into *Tilapia* decreased the white cell numbers in spleen and head kidney (pronephros that functions analogously to mammalian bone marrow) (Hart et al. 1997). The fish *Aphanius dispar* displayed higher leukocyte densities in blood (primarily due to eosinophil increase) after exposure to mercury (Hilmy et al. 1980). Metal exposure of striped bass (*Morone saxatilis*) increased (copper) or decreased (arsenic) resistance to challenge with the bacterial pathogen *Flexibacter columnaris* (MacFarlane et al. 1986).

Still other studies provide evidence that toxicants can adversely influence cellular immunological functioning of adult invertebrates. De Guise et al. (2004) suggested that malathion changed phagocytes of lobster. Similarly, tributyltin adversely influenced amoebocyte count, viability, and function of the seastar *Leptasterias polaris* (Békri and Pelletier 2004). Reactive oxygen species generation in amoebocytes of another seastar, *Asterias rubens*, was affected by PCB exposure (Danis et al. 2004). Studies of copper (Parry and Pipe 2004) and general level of environmental contamination (Fisher et al. 2000, 2003, Oliver et al. 2001, 2003) suggested that oyster hemocytes are also adversely impacted by toxicants.

An important immunotoxicity theme is emerging in ecotoxicology in response to a rapidly growing literature including the studies described above. Immunocompetence of developing and adult organisms is influenced by a variety of toxicants. These effects influence the individual's ability to cope with disease, infection, infestation, or cancer.

5.9 ENDOCRINE SYSTEM

Many contaminants influence the developing or fully developed endocrine system. Effects involve a variety of cells, glands, and functions although those associated with reproduction have gained prominence in the last decade.[2] Much interest was stimulated by observations of abnormal sexual

[2] Those associated with the General Adaptation Syndrome are also extremely relevant and are discussed in detail in Section 9.1.1 of Chapter 9.

morphology or behavior of individuals exposed to various chemical contaminants. Mosquitofish living in Kraft mill effluent displayed changes in sexual morphology and behavior (Bortone et al. 1989). Female–female pairing and nesting of western gulls (*Larus occidentalis*) were noted in Southern California nesting areas (Hunt and Hunt 1977). Blood lead concentrations as low as 3 μg/dL delayed puberty in exposed girls (Selevan et al. 2003).

Adverse effects to developing endocrine systems of diverse species are well documented, so only a few examples will be outlined here. Prenatal human exposure to natural or synthetic estrogens has been associated with increased risk of breast cancer (Birnbaum and Fenton 2003), suggesting changes in endocrine system's state in exposed women. The anabolic steriod, 17β-trenbolone, used in feedlots can enter nearby freshwaters and cause developmental abnormalities in fish (Wilson et al. 2003). Offspring of zebrafish (*D. rerio*) exposed to ethynylestradiol had reduced fecundity with males having no or poorly developed testes. This resulted in population collapse (Nash et al. 2004). Some PCBs influence sex determination in exposed slider turtle (*Trachemys scripta*) eggs (Bereron et al. 1994). This influence on slider turtle sex determination is synergistic when eggs are exposed to mixtures of endocrine disrupting compounds (Willingham 2004). As a final illustration, ammonium perchlorate, an oxidizer used by the military in rockets, changes thyroid function and mass in developing bobwhite quail (*Colinus virginianus*) (McNabb et al. 2004).

Cases of adverse endocrine effects to fully developed individuals are even easier to find in the recent literature. Most, but certainly not all, of these cases focus on reproductive consequences. This dominance of studies on reproductive effects simply reflects the current interests of scientists studying endocrine disruption. This is reasonable, given the importance of reproductive fitness in determining an organism's overall Darwinian fitness. The other effects will likely be reported more frequently in the literature in the near future.

A range of studies provide clear evidence of nonreproductive effects. Cadmium induces apoptosis of pituitary cells of rats via oxidative stress (Poliandri et al. 2003). Cadmium was also adrenotoxic to rainbow trout (*O. mykiss*) and perch (*Perca flavescens*), inhibiting cortisol secretion (Lacroix and Hontela 2004). DDD (o,p'-dichlorodiphenyldichloroethane) exposure of rainbow trout decreased corticotropic hormone-stimulated cortisol secretion by head kidney tissues (Benguira and Hontela 2000). The coplanar PCB, 3,3′,4,4′,5-pentachlorobiphenyl, affected thyroid function of lake trout (*Salvelinus namaycush*) (Brown et al. 2004).

A substantial literature is amassing relative to toxicant effects on the reproductive endocrinology of individuals. In addition to the pharmaceutical estrogens released into waterways, nonylphenol and methoxychlor are also estrogenic (Folmar et al. 2002). In juvenile summer flounder (*Paralichthys dentatus*), o,p'-DDT and p,p'-DDE are estrogenic and antiandrogenic (Mills et al. 2001). Curiously, the synthetic androgen, 17α-methyltestosterone, actually increases the egg protein, vitellogenin, in male fathead minnows, likely because it is converted to 17α-methylestradiol. It also reduces the number of eggs and fertilization rate of, and produces abnormal sexual behavior in exposed females (Pawlowski et al. 2004). Similarly, nonylphenol, methoxychlor, endosulfan, and 17β-estradiol induce vitellogenin production in male sheepshead minnows *Cyprinodon variegatus* (Hemmer et al. 2001, 2002). A field study of male walleye (*Stizostedion vitreum*) sampled near a sewage treatment plant also showed changes in serum testosterone, and elevated 17β-estradiol and vitellogenin, notionally because of the pharmaceutical estrogens discharged from the plant (Folmar et al. 2001). The synthetic steroid 17β-trenbolone changes vitellogenin and 17β-estradiol levels in exposed fathead minnows (Ankley et al. 2003).

Such reproductive effects are also to be expected in invertebrates. Albumin glands atrophied in pulmonate snails (*Lymnaea stagnalis*) exposed to β-sitosterol and, to a lesser degree, to *t*-methyltestosterone (Czech et al. 2001). Metals can also be endocrine disruptors as illustrated by the interference of cadmium on ovary growth of the crab, *Chasmagnathus granulata* (Medesani et al. 2004).

5.10 NERVOUS, SENSORY, AND MOTOR-RELATED ORGANS AND SYSTEMS

Toxicants influence nervous, sensory, and motor-related systems in numerous and complex ways. Evidence for direct effects of toxicants on nervous system tissues is abundant. There is also a large literature accumulating that documents the influence of toxicants on behavior (ethotoxicology).

A wide range of toxicants adversely impact developing or established nervous system cells and tissues. Lead causes swelling and hemorrhaging in the mammalian brain with acute exposures and cerebral vascular damage under chronic exposure (Zheng et al. 2003). The pyrethroid pesticide deltamethrin induces apotosis of cerebral cortical neurons (Wu et al. 2003). High exposure to pyrethroid insecticides also increases neurotransmitter release and activation of sodium channels, resulting in ataxia, hyperexcitation, paralysis, and possibly death. Ethanol adversely impacts cerebellar granule neurons during rat development and induces heat-shock protein production in the cerebellum (Acquaah-Mensah et al. 2001). Acrylamide, a chemical contaminant that has been found in some foods, can cause axon damage, producing loss of coordination (ataxia) and skeletal muscle weakness (LoPachin et al. 2003).

These diverse effects translate to a wide range of manifestations at the organismal level. Brook trout's (*Salvelinus fontinalis*) exposure to DDT makes lateral line nerves hypersensitive (Anderson 1968). Exposure of *Pleuroderma cinereum* tadpoles to dissolved chromium increased the interindividual variability in ventilation rate (Janssens de Bisthoven et al. 2004). Fish ventilatory behavior is also commonly affected by toxicants (e.g., Diamond et al. 1990). Honeybees exposed to insecticides displayed changes in foraging activity and performance in an olfactory discrimination behavior test (Decourtye et al. 2004). Avoidance behavior of a variety of species is adversely affected by toxicants such as cadmium and copper (McNicol and Scherer 1991, Sullivan et al. 1978, 1983). Atchison et al. (1996) provide an excellent review of behavioral changes related to toxicant exposure of aquatic species.

Even more subtle effects manifest with toxicant exposure. Exposure during development to some toxicants, such as lead, methylmercury, or PCB, can cause cognitive deficiencies (Sharbaugh et al. 2003). Selevan et al. (2003) provide another very surprising example of lead's effect on the human central nervous system at levels below the current regulatory limits. Lambs born to sheep (*Ovis aries*) that grazed on meadows treated with sewage sludge vocalized more and were less reactive at testing than were control lambs (Erhard and Rhind 2004). Male lambs, which typically display less exploratory behavior than female lambs, had the same level of exploratory behavior as their female counterparts if born by an ewe exposed to sludge from a sewage sludge treatment plant. In the context of male exploratory behavior, the sewage sludge-associated contaminants appeared to have a feminizing effect. Clearly, such effects can have significant influence on individual fitness and are beginning to get more attention.

5.11 SUMMARY

5.11.1 Summary of Foundation Concepts and Paradigms

- The cellular specialization and spatial relation of organs to direct environmental exposure or exposure via somatic circulation make one organ more prone to poisoning than another.
- The integument, like the respiratory and digestive organs, is subject to significant toxicant exposure because of its intimate contact with the external milieu and large surface area for exchange. It is one of three major exposure routes.
- Skin penetration can be predicted using toxicant K_{ow} and molecular weight. Penetration is faster for small, hydrophobic compounds.
- Respiratory organs have intimate contact and exchange with the external environment; therefore, they are major organs of toxicant exchange and effect. Respiratory exposure is the second of three major routes of exposure.

- The chemical nature and form of a toxicant are important determinants of its ability to deliver an effective dose to air breathing animals. Its physical phase association is also extremely important.
- Water solubility determines movement of a gaseous toxicant in the lungs with highly soluble gases being less able to move as deeply into the lungs as less soluble gases.
- Generally, nonpolar toxicants in liquid aerosols are absorbed quicker after inhalation than polar toxicants. Weak electrolytes in liquid aerosols are absorbed at pH-dependent rates.
- Effects of inhaled toxicants range from fatal cancers to pulmonary edema to mild irritation.
- Toxicants can have teratogenic effects on developing components of the circulatory system or direct effects on developed circulatory system components.
- Organs normally associated with digestion provide the third major route of exposure, the ingestion route.
- The liver and analogous organs of invertebrates are prone to high exposures due to reasons other than close contact with external sources. Detoxification that occurs in these organs makes them prone to damage from activation products. Active hepatocyte regeneration in the liver coupled with potentially activating reactions make the liver particularly prone to cancer.
- Enterohepatic circulation occurs when toxicants or their metabolites are removed from the liver via the bile but then are reabsorbed in the intestine and returned to the liver again.
- The kidney is susceptible to toxicants despite its remarkable capacity to recover after injury. Toxicants in circulation will be delivered to the kidney because of the kidney's central role in removing wastes from the blood. The kidneys concentrate materials from the blood, providing an avenue for concentrating toxicants to levels damaging to kidney tissues. Also, detoxification reactions in the kidney can produce activation products that damage the kidney or urinary tract.
- Toxicants have a strong influence on immunocompetence of diverse species.
- Many contaminants influence the endocrine systems of fully developed or developing individuals. A substantial literature is amassing about toxicant effects on the reproductive endocrinology of individuals.
- Toxicants influence nervous, sensory, and motor-related systems in numerous and complex ways. Evidence of direct toxicant effects on nervous system tissues is abundant. There is also a large literature accumulating that documents the influence of toxicants on behavior (ethotoxicology).

REFERENCES

Acquaah-Mensah, G.K., Leslie, S.W., and Kehrer, J.P., Acute exposure of cerebellar granule neurons to ethanol suppresses stress-activated protein kinase-1 and concomitantly induces AP-1, *Toxicol. Appl. Pharmacol.*, 175, 10–18, 2001.

Akaishi, F.M., de Assis, H.C., Jakobi, S.C., Eiras-Stofella, D.R., St.-Jean, S.D., Courtenay, S.C., Lima, E.F., Wagener, A.L., Scofield, A.L., and Ribeiro, C.A., Morphological and neurotoxicological findings in tropical freshwater fish (*Astyanax* sp.) after waterborne and acute exposure to water soluble fraction (WSF) of crude oil, *Arch. Environ. Contam. Toxicol.*, 46, 244–253, 2004.

Anderson, J.M., Effect of sublethal DDT on the lateral line of brook trout, *Salvelinusfontinalis*, *J. Fish. Res. Board. Can.*, 25, 2677–2682, 1968.

Ankley, G.T., Jensen, K.M., Makynen, E.A., Kahl, M.D., Korte, J.J., Hornung, M.W., Henry, T.R., et al., Effects of the androgenic growth promoter 17-β-trenbolone on fecundity and reproductive endocrinology of the fathead minnow, *Environ. Toxicol. Chem.*, 22, 1350–1360, 2003.

Anonymous, Waterfowl hunters must give up lead shot, *Science*, 198, 1232, 1977.

Atchison, G.J., Sandheinrich, M.B., and Bryan, M.D., Effects of environmental stressors on interspecific interactions of aquatic animals, In *Ecotoxicology. A Hierarchical Treatment*, Newman, M.C. and Jagoe, C.H. (eds.), CRC Press/Lewis Publishers, Boca Raton, FL, 1996, pp. 319–345.

Ballatori, N., Mechanisms of metal transport across liver cell plasma membranes, *Drug Metab. Rev.*, 23, 83–132, 1991.

Beach, R.S., Gershwin, M.E., and Hurley, L.S., Gestational zinc deprivation in mice: Persistence of immunodeficiency from three generations, *Science*, 218, 469–471, 1982.

Beaty, B.J., Mackie, R.S., Mattingly, K.S., Carlson, J.O., and Rayms-Keller, A., The midgut epithelium of aquatic arthropods: A critical target organ in environmental toxicology, *Environ. Health Perspect.*, 110(Suppl. 6), 911–914, 2002.

Begum, G., Carbofuran insecticide induced biochemical alterations in liver and muscle tissues of the fish *Clarias batrachus* (Linn) and recovery response, *Aquat. Toxicol.*, 66, 83–92, 2004.

Békri, K. and Pelletier, É., Trophic transfer and *in vivo* immunotoxicological effects of tributyltin (TBT) in polar seastar *Leptasterias polaris*, *Aquat. Toxicol.*, 66, 39–53, 2004.

Benguira, S. and Hontela, A., Adrenocorticotrophin- and cyclic adenosine 3′,5′-monophosphate-stimulated cortisol secretion in interrenal tissue of rainbow trout exposed *in vitro* to DDT compounds, *Environ. Toxicol. Chem.*, 19, 842–847, 2000.

Bergeron, J.M., Crews, D., and McLachlan, J.A., PCBs as environmental estrogens: Turtle sex determination as a biomarker of environmental contamination, *Environ. Health Perspect.*, 102, 780–781, 1994.

Biggins, P.D.E. and Harrison, R.M., Chemical speciation of lead compounds in street dusts, *Environ. Sci. Technol.*, 14, 336–339, 1980.

Birnbaum, L.S. and Fenton, S.E., Cancer and developmental exposure to endocrine disruptors, *Environ. Health Perspect.*, 111, 389–394, 2003.

Bortone, S.A., Davis, W.P., and Bundrick, C.M., Morphological and behavioral characteristics in mosquitofish as potential bioindication of exposure to kraft mill effluent, *Bull. Environ. Contam. Toxicol.*, 43, 370–377, 1989.

Brown, S.B., Evans, R.E., Vandenbyllardt, L., Finnson, K.W., Palace, V.P., Kane, A.S., Yarechewski, A.Y., and Muir, D.C.G., Altered thyroid status in lake trout (*Salvelinus namaycush*) exposed to co-planar 3,3′,4,4′5-pentachlorobiphenyl, *Aquat. Toxicol.*, 67, 75–85, 2004.

Buckley, T.J., Prah, J.D., Ashley, D., Zweidinger, R.A., and Wallace, L.A., Body burden measurements and models to assess inhalation exposure to methyl tertiary butyl ether (MTBE), *J. Air Waste Manag. Assoc.*, 47, 739–752, 1997.

Bustamante, P., Cosson, R.P., Gallien, I., Caurant, F., and Miramand, P., Cadmium detoxification processes in the digestive gland of cephalopods in relation to accumulated cadmium concentrations, *Mar. Environ. Res.*, 53, 227–241, 2002.

Bustnes, J.O., Hanssen, S.A., Folstad, I., Erikstad, K.E., Hasselquist, D., and Skaare, J.U., Immune function and organochlorine pollutants in Arctic breeding glaucous gulls, *Arch. Environ. Contam. Toxicol.*, 47, 530–541, 2004.

Canesi, L., Ciacci, C., and Gallo, G., Hg^{2+} and Cu^{2+} interfere with agonist-mediated Ca^{2+} signaling in isolated Mytilus digestive gland cells, *Aquat. Toxicol.*, 49, 1–11, 2000.

Cassano, G., Bellantuono, V., Quaranta, A., Ippolito, C., Ardizzone, C., and Lippe, C., Interaction of pyrethroids with ion transport pathways present in frog skin, *Environ. Toxicol. Chem.*, 19, 2720–2724, 2000.

Cassano, G., Bellantuono, V., Ardizzone, C., and Lippe, C., Pyrethroid stimulation of ion transport across frog skin, *Environ. Toxicol. Chem.*, 22, 1330–1334, 2003.

Chan, H.M. and Cherian, M.G., Protective roles of metallothionein and glutathione in hepatotoxicity of cadmium, *Toxicology*, 72, 281–290, 1992.

Chen, G., Zhang, D., Jing, N., Yin, S., Falany, C.N., and Radominska-Pandya, A., Human gastrointestinal sulfotransferases: Identification and distribution, *Toxicol. Appl. Pharmacol.*, 187, 186–197, 2003.

Christin, M.S., Ménard, L., Gendron, A.D., Ruby, S., Cyr, D., Marcogliese, D.J., Rollins-Smith, L., and Fournier, M., Effects of agricultural pesticides on the immune system of *Xenopus laevis* and *Rana pipiens*, *Aquat. Toxicol.*, 67, 33–43, 2004.

Cornfield, J., Haenszel, W., Hammond, E.C., Lilienfeld, A.M., Shimkin, M.B., and Wynder, E.L., Smoking and lung cancer: Recent evidence and a discussion of some questions, *J. Natl. Cancer Inst.*, 22, 173–203, 1959.

Czech, P., Weber, K., and Dietrich, D.R., Effects of endocrine modulating substances on reproduction in the hermaphroditic snail *Lymnaea stagnalis* L., *Aquat. Toxicol.*, 53, 103–114, 2001.

Danis, B., Goriely, S., Dubois, P., Fowler, S.W., Flamand, V., and Warnau, M., Contrasting effects of coplanar versus non-coplanar PCB congeners on immunomodulation and CYP1A levels (determined using an adapted ELISA method) in the common seastar *Asteria rubens* L., *Aquat. Toxicol.*, 69, 371–383, 2004.

Dautremepuits, C., Betoulle, S., Paris-Palacios, S., and Vernet, G., Humoral immune factors modulated by copper and chitosan in healthy or parasitised carp (*Cyprinus carpio* L.) by *Ptychobothrium* sp. (Cestoda), *Aquat. Toxicol.*, 68, 325–338, 2004a.

Dautremepuits, C., Betoulle, S., Paris-Palacois, S., and Vernet, G., Immunology-related perturbations induced by copper and chitosan in carp (*Cyprinus carpio* L.), *Arch. Environ. Contam. Toxicol.*, 47, 370–378, 2004b.

Diamond, J.M., Parson, M.J., and Gruber, D., Rapid detection of sublethal toxicity using fish ventilatory behavior, *Environ. Toxicol. Chem.*, 9, 3–11, 1990.

Decourtye, A., Devillers, J., Cluzeau, S., Charreton, M., Pham-Delégue, M.-H., Effects of imidacloprid and deltamethrin on associative learning in honeybees under semi-field and laboratory conditions, *Ecotoxicol. Environ. Saf.*, 57, 410–419, 2004.

De Guise, S., Maratea, J., and Perkins, C., Malathion immunotoxicity in the American lobster (*Homarus americanus*) upon experimental exposure, *Aquat. Toxicol.*, 66, 419–425, 2004.

Ding, X. and Kaminsky, L.S., Human extrahepatic cytochromes P450: Function in xenobiotic metabolism and tissue-selective chemical toxicity in the respiratory and gastrointestinal tracts, *Annu. Rev. Pharmacol. Toxicol.*, 43, 149–173, 2003.

Doll, R. and Hill, A.B., Mortality in relation to smoking: Ten years' observations in British doctors, *Br. Med. J.*, 248, 1399–1410, 1964.

Doll, R., Morgan, L.G., and Speizer, F.E., Cancers of the lung and nasal sinuses in nickel workers, *Br. J. Cancer*, 24, 623–632, 1970.

Erhard, H.W. and Rhind, S.M., Prenatal and postnatal exposure to environmental pollutants in sewage sludge alters emotional reactivity and exploratory behaviour in sheep, *Sci. Total Environ.*, 332, 101–108, 2004.

Erickson, R.J. and McKim, J.M., A model for exchange of organic chemicals at fish gills: Flow and diffusion limitations, *Aquat. Toxicol.*, 18, 175–198, 1990.

Fabbri, E., Caselli, F., Piano, A., Sartor, G., and Capuzzo, A., Cd^{2+} and Hg^{2+} affect glucose release and cAMP-dependent transduction pathway in isolated eel hepatocytes, *Aquat. Toxicol.*, 62, 55–65, 2003.

Faurskov, B. and Bjerregaard, H.F., Chloride secretion in kidney distal epithelial cells (A6) evoked by cadmium, *Toxicol. Appl. Pharmacol.*, 163, 267–278, 2000.

Fisher, W.S., Oliver, L.M., Winstead, J.T., and Long, E.R., A survey of oysters *Crassostrea virginica* from Tampa Bay, Florida: Associations of internal defense measurements with contaminant burdens, *Aquat. Toxicol.*, 51, 115–138, 2000.

Fisher, W.S., Oliver, L.M., Winstead, J.T., and Volety, A.K., Stimulation of defense factors for oysters deployed to contaminated sites in Pensacola Bay, Florida, *Aquat. Toxicol.*, 64, 375–391, 2003.

Folmar, L.C., Denslow, N.D., Kroll, K., Orlando, E.F., Enblom, J., Marcino, J., Metcalfe, C., and Guillette, L.J., Jr., Altered serum sex steroids and vitellogenin induction in walleye (*Stizostedion vitreum*) collected near a metropolitan sewage treatment plant, *Arch. Environ. Contam. Toxicol.*, 40, 392–398, 2001.

Folmar, L.C., Hemmer, M.J., Denslow, N.D., Kroll, K., Chen, J., Check, A., Richman, H., Meredith, H., and Grau, E.G., A comparison of the estrogenic potencies of estradiol, ethynylestradiol, diethylstilbestrol, nonylphenol, and methoxychlor *in vivo* and *in vitro*, *Aquat. Toxicol.*, 60, 101–110, 2002.

Gauthier, J.M., Dubeau, H., and Rossart, É., Mercury-induced micronuclei in skin fibroblasts of Beluga whales, *Environ. Toxicol. Chem.*, 17, 2487–2493, 1998.

Gibaldi, M., *Biopharmaceutics and Clinical Pharmacokinetics*, 4th ed., Lea & Febiger, Philadelphia, 1991, Chapter 6.

Godard, C.A.J., Smolowitz, R.M., Wilson, J.Y., Payne, R.S., and Stegeman, J.J., Induction of cetacean cytochrome P4501A1 by β-naphthoflavone exposure of skin biopsy slices, *Toxicol. Sci.*, 80, 268–275, 2004.

Goldstein, R.S. and Schnellmann, R.G., Toxic responses of the kidney, In *Casarett & Doull's Toxicology: The Basic Science of Poisons*, 5th ed., Klaassen, C.D. (ed.), McGraw-Hill, New York, 1996, pp. 417–442.

Gomot-De Vaufleury, A. and Pihan, F., Methods for toxicity assessment of contaminated soil by oral or dermal uptake in land snails: Metal bioavailavility and bioaccumulation, *Environ. Toxicol. Chem.*, 21, 820–827, 2002.

Grosell, M., McDonald, M.D., Wood, C.M., and Walsh, P.J., Effects of prolonged copper exposure in the marine gulf toadfish (*Opsanus beta*). I. Hydromineral balance and plasma nitrogenous waste products, *Aquat. Toxicol.*, 68, 249–262, 2004.

Hart, L.J., Smith, S.A., Smith, B.J., Robertson, J., and Holladay, S.D., Exposure of tilapian fish to the pesticide lindane results in hypocellularity of the primary hematopoietic organ (pronephros) and the spleen without altering activity of phagocytic cells in these organs, *Toxicology*, 118, 211–221, 1997.

Hemmer, M.J., Bowman, C.J., Hemmer, B.L., Friedman, S.D., Marcovich, D., Kroll, K.J., and Denslow, N.D., Vitellogenin mRNA regulation and plasma clearance in male sheepshead minnows (*Cyprinodon variegatus*) after cessation of exposure to 17 β-estradiol and p-nonylphenol, *Aquat. Toxicol.*, 58, 99–112, 2002.

Hemmer, M.J., Hemmer, B.L., Bowman, C.J., Kroll, K.J., Folmar, L.C., Marcovich, D., Hogland, M.D., and Denslow, N.D., Effects of *p*-nonylphenol, methoxychlor, and endosulfan on vitellogenin induction and expression in sheepshead minnow (*Cyprinodon variegatus*), *Environ. Toxicol. Chem.*, 20, 336–343, 2001.

Heywood, R., Target organ toxicity, *Toxicol. Lett.*, 8, 349–358, 1981.

Hilmy, A.M., Shabana, M.B., and Said, M.M., Haematological responses to mercury toxicity in the marine teleost, *Aphanius dispar* (Rüpp), *Comp. Biochem. Physiol.*, 67C, 147–158, 1980.

Hinton, D.E., Cells, cellular responses, and their markers in chronic toxicity of fishes, In *Aquatic Toxicology. Molecular, Biochemical and Cellular Perspectives*, Malins, D.C. and Ostrander, G.K. (eds.), Lewis Publishers, Boca Raton, FL, 1994, pp. 207–239.

Holsapple, M.P., Developmental immunotoxicity testing: A review, *Toxicology*, 185, 193–203, 2003.

Huang, Y.-W., Stegeman, J.J., Woodin, B.R., and Karasov, W.H., Immunohistochemical localization of cytochrome P4501A induced by 3,3′,4,4′,5-pentachlorobiphenyl pipiens, *Environ. Toxicol. Chem.*, 20, 181–197.

Hunt, J.L., Jr. and Hunt, M.W., Female-female pairing of western gulls (*Larus occidentalis*) in Southern California. *Science*, 196, 1466–1467, 1977.

Incardona, J.P., Collier, T.K., and Scholz, N.L., Defects in cardiac function precede morphological abnormalities in fish embryos exposed to polycyclic aromatic hydrocarbons, *Toxicol. Appl. Pharmacol.*, 196, 191–205, 2004.

Jaeschke, H., Gores, G.J., Cederbaum, A.I., Hinson, J.A., Pessayre, D., and Lemasters, J.J., Mechanisms of hepatotoxicity, *Toxicol. Sci.*, 65, 166–176, 2002.

James, M.O., Cytochrome P450 monoxygenases in crustaceans, *Xenobiotica*, 19, 1063–1076, 1989.

Janssens de Bisthoven, L., Gerhardt, A., and Maldonado, M., Behavioral bioassay with a local tadpole (*Pleuroderma cinereum*) from River Rocha, Bolivia, in river water spiked with chromium, *Bull. Environ. Contam. Toxicol.*, 72, 422–428, 2004.

Jeffree, R.A., Markich, S.J., and Tucker, A.D., Patterns of metal accumulation in osteoderms of the Australian freshwater crocodile, *Crocodylus johnstoni, Sci. Total Environ.*, 336, 71–80, 2005.

Johnson, M.S., Franke, L.S., Lee, R.B., and Holladay, S.D., Bioaccumulation of 2,4,6-trinitrotoluene and polychlorinated biphenyls through two routes of exposure in a terrestrial amphibia: Is the dermal route significant? *Environ. Toxicol. Chem.*, 18, 873–878, 1999.

Jones, D.E., Gogal, R.M., Jr., Nader, P.B., and Holladay, S.D., Organochlorine detection in the shed skins of snakes, *Ecotoxicol. Environ. Saf.*, 60, 282–287, 2005.

Käkelä, A., Käkelä, R., Hyvärinen, H., and Asikainen, J., Vitamins A_1 and A_2 in hepatic tissue and subcellular fractions in mink feeding on fish-based diets and exposed to Aroclor 1242, *Environ. Toxicol. Chem.*, 21, 397–403, 2002.

Kaler, S.G., Diagonsis and therapy of Menkes syndrome, a genetic form of copper deficiency, *Am. J. Clin. Nutr.*, 67(Suppl. 5), 1029S–1034S, 1998.

Kendall, R.J., Lacher, T.E., Bunck, C., Daniel, B., Driver, C., Grue, C.E., Leighton, F., Stansley, W., Watanabe, P.G., and Whitworth, M., An ecological risk assessment of lead shot exposure in non-waterfowl avian species: Upland game birds and raptors, *Environ. Toxicol. Chem.*, 15, 4–20, 1996.

Klaassen, C.D. (ed.), *Casarett & Doull's Toxicology: The Basic Science of Poisons*, 5th ed., McGraw-Hill, New York, 1996.

Kluwe, W.M. and Hook, J.B., Polybrominated biphenyl-induced potentiation of chloroform toxicity, *Toxicol. Appl. Pharmacol.*, 45, 861–869, 1978.

Kodavanti, U.P., Schladweiler, M.C.J., Ledbetter, A.D., Hauser, R., Christiani, D.C., Samet, J.M., McGee, J., Richards, J.H., and Costa, D.L., Pulmonary and systemic effects of zinc-containing emission particles in three rat strains: Multiple exposure scenarios, *Toxicol. Sci.*, 70, 73–85, 2002.

Kodavanti, U.P., Moyer, C.F., Ledbetter, A.D., Schladweiler, M.C., Costa, D.L., Hauser, R., Christiani, D.C., and Nyska, A., Inhaled environmental combustion particles cause myocardial injury in the Wistar Kyoto rat, *Toxicol. Sci.*, 71, 237–245, 2003.

Krumschnabel, G. and Nawaz, M., Acute toxicity of hexavalent chromium in isolated teleost hepatocytes, *Aquat. Toxicol.*, 70, 159–167, 2004.

Lacroix, A. and Hontela, A., A comparative assessment of the adrenotoxic effects of cadmium in two teleost species, rainbow trout, *Oncorhynchus mykiss*, and yellow perch, *Perca flavescens*, *Aquat. Toxicol.*, 67, 13–21, 2004.

Lan, Q., Chapman, R.S., Schreinemachers, D.M., Tian, L., and He, X., Household stove improvement and risk of lung cancer in Xuanwei, China, *J. Natl. Cancer Inst.*, 94, 826–835, 2002.

Lash, L.H. and Parker, J.C., Hepatic and renal toxicities associated with perchloroethylene, *Pharmacol. Rev.*, 53, 177–208, 2001.

Laxen, D.P.H. and Harrison, R.M., The highway as a source of water pollution: An appraisal with the heavy metal lead, *Water Res.*, 11, 1–11, 1977.

Leussink, B.T., Litvinov, S.V., de Jeer, E., Slikkerveer, A., van der Voet, G.B., Bruijn, J.A., and de Wolff, F.A., Loss of homotypic epithelial cell adhesion by selective N-cadherin displacement in bismuth nephrotoxicity, *Toxicol. Appl. Pharmacol.*, 175, 54–59, 2001.

Lightner, D.V., Hasson, K.W., White, B.L., and Redman, R.M., Chronic toxicity and histopathological studies with Benlate®, a commercial grade of benomyl, in *Penaeus vannamei* (Crustacea: Decapoda), *Aquat. Toxicol.*, 34, 105–118, 1996.

Lionetto, M.G., Giordano, M.E., Vilella, S., and Schettino, T., Inhibition of eel enzymatic activities by cadmium, *Aquat. Toxicol.*, 48, 561–571, 2000.

Lionetto, M.G., Vilella, S., Trischitta, F., Cappello, M.S., Giordano, M.E., and Schettino, T., Effects of CdCl2 on electrophysiological parameters in the intestine of the teleost fish, *Anguilla anguilla*, *Aquat. Toxicol.*, 41, 251–264, 1998.

Loosanoff, V.L., MacKenzie, C.L., Jr., and Shearer, L.W., Use of chemicals to control shellfish predators, *Science*, 131, 1522–1523, 1960.

LoPachin, R.M., Balaban, C.D., and Ross, J.F., Acrylamide axonopathy revisited, *Toxicol. Appl. Pharmacol.*, 188, 135–153, 2003.

Luebke, R.W, Hodson, P.V., Fiasal, M., Ross, P.S., Grasman, K.A., and Zelikoff, J., Aquatic pollution-induced immunotoxicity in wildlife species, *Fundam. Appl. Toxicol.*, 37, 1–15, 1997.

MacFarlane, R.D., Bullock, G.L., and McLaughlin, J.J.A., Effects of five metals on susceptibility of striped bass to *Flexibacter columnaris*, *Trans. Am. Fish. Soc.*, 115, 227–231, 1986.

Mackey, E.A., Oflaz, R.D., Epstein, M.S., Buehler, B., Porter, B.J., Rowles, T., Wise, S.A., and Becker, P.R., Elemental composition of liver and kidney tissues of rough-toothed dolphins (*Steno bredanensis*), *Arch. Environ. Contam. Toxicol.*, 44, 523–532, 2003.

Margoshes, M. and Vallee, B.L., A cadmium protein from equine kidney cortex, *J. Am. Chem. Soc.*, 79, 4813–4814, 1957.

Martinez-Tabche, L., Madrigal-Bujaidar, E., and Negrete, T., Genotoxicity and lipoperoxidation produced by paraquat and 2,4-dichlorophenoxyacetic acid in the gills of rainbow trout (*Oncorhynchus mykiss*), *Bull. Environ. Contam. Toxicol.*, 73, 146–152, 2004.

Matthiessen, P. and Brafield, A.E., The effects of dissolved zinc on the gills of the stickleback *Gasterosteus acueatus* (L.), *J. Fish Biol.*, 5, 607–613, 1973.

Mayer, R.T., Svoboda, J.A., and Weirich, G.F., Ecdysone 20-hydroxylase in midgut mitochondria of *Manduca sexta* (L.), *Hoppe Seylers Z. Physiol. Chem.*, 359, 1247–1257, 1978.

McNabb, F.M.A., Jang, D.A., and Larsen, C.T., Does thyroid function in developing birds adapt to sustained ammonium perchlorate exposure? *Toxicol. Sci.*, 82, 106–113, 2004.

McNicol, R.E. and Scherer, E., Behavioral responses of lake whitefish (*Coregonus clupeaformis*) to cadmium during preference-avoidance testing, *Environ. Toxicol. Chem.*, 10, 225–234, 1991.

Medesani, D.A., López Greco, L.S., and Rodríguez, E.M., Interference of cadmium and copper with the endocrine control of ovarian growth, in the estuarine crab *Chasmagnathus granulata*, *Aquat. Toxicol.*, 69, 165–174, 2004.

Mills, L.J., Gutjahr-Gobell, R.E., Haebler, R.A., Borsay Horowitz, D.J., Jayaraman, S., Pruell, R.J., Mckinney, R.A., Gardner, G.R., and Zaroogian, G.E., Effects of estrogenic (o,p′-DDT; octylphenol) and anti-androgenic (p,p′-DDE) chemicals on indicators of endocrine status in juvenile male summer flounder (*Paralichthys dentatus*), *Aquat. Toxicol.*, 52, 157–176, 2001.

Mo, J., Cytochrome P450 mono oxygenase in crustaceans, *Xenobiotica*, 19, 1063–1076, 1989.

Montairo, L.R. and Furness, R.W., Seabirds as monitors of mercury in the marine environment, *Water Air Soil Pollut.*, 80, 851–870, 1995.

Morris, R.J., Lockwood, A.P.M., and Dawson, M.E., Changes in the fatty acid composition of the gill phospholipids in *Gammarus duebeni* with degree of gill contamination, *Mar. Pollut. Bull.*, 13, 345–348, 1982.

Nash, J.P., Kime, D.E., Van der Ven, L.T.M., Wester, P.W., Brion, F., Maack, G., Stahlschmidt-Allner, P., and Tyler, C.R., Long-term exposure to environmental concentrations of the pharmaceutical ethynylestradiol causes reproductive failure in fish, *Environ. Health Perspect.*, 112, 1725–1733, 2004.

Nierenberg, D.W., Nordgren, R.E., Chang, M.B., Siegler, R.W., Blayney, M.B., Hochberg, F., Toribara, T.Y., Cernichiari, E., and Clarkson, T., Delayed cerebellar disease and death after accidential exposure to dimethylmercury, *N. Engl. J. Med.*, 338, 1672–1676, 1998.

Oliver, L.M., Fisher, W.S., Volety, A.K., and Malaeb, Z., Greater hemocyte bactericidal activity in oysters (*Crassostrea virginica*) from a relatively contaminated site in Pensacola Bay, Florida, *Aquat. Toxicol.*, 64, 363–373, 2003.

Oliver, L.M., Fisher, W.S., Winstead, J.T., Hemmer, B.L., and Long, E.R., Relationships between tissue contaminants and defense-related characteristics of oysters (*Crassostrea virginica*) from five Florida bays, *Aquat. Toxicol.*, 55, 203–222, 2001.

Omura, T., Forty years of cytochrome P450, *Biochem. Biophys. Res. Commun.*, 266, 690–698, 1999.

Ovuru, S.S. and Ezeasor, D.N., Morphological alterations in liver tissues from rabbits exposed to crude oil contaminated diets, *Bull. Environ. Contam. Toxicol.*, 73, 132–138, 2004.

Ownby, D.R., Newman, M.C., Mulvey, M., Vogelbein, W.K., Unger, M.A., and Arzayus, L.F., Fish (*Fundulus heteroclitus*) populations with different exposure histories differ in tolerance of cresote-contaminated sediments, *Environ. Toxicol. Chem.*, 21, 1897–1902, 2002.

Pane, E.F., Haque, A., and Wood, C.M., Mechanistic analysis of acute, Ni-induced respiratory toxicity in the rainbow trout (*Oncorhychus mykiss*): An exclusively branchial phenomenon, *Aquat. Toxicol.*, 69, 11–24, 2004.

Parry, H.E. and Pipe, R.K., Interactive effects of temperature and copper on immunocompetence and disease susceptibility in mussels (*Mytilus edulis*), *Aquat. Toxicol.*, 69, 311–325, 2004.

Pauli, B.D. and Kennedy, S.W., Hepatic porphyria induced by the herbicide tralkoxydim in small mammals is species-specific, *Environ. Toxicol. Chem.*, 24, 450–456, 2005.

Pawlowski, S., Sauer, A., Shears, J.A., Tyler, C.R., and Braunbeck, T., Androgenic and estrogenic effects of the synthetic androgen 17 α-methyltestosterone on sexual development and reproductive performance in the fathead minnow (*Pimephales promelas*) determined using the gonadal recrudescence assay, *Aquat. Toxicol.*, 68, 277–291, 2004.

Playle, R.C., Using multiple metal-gill binding models and the toxic unit concept to help reconcile multiple-metal toxicity results, *Aquat. Toxicol.*, 67, 359–370, 2004.

Pleil, J.D., Fisher, J.W., and Lindstrom, A.B., Trichloroethene levels in human blood and exhaled breath from controlled inhalation exposure, *Environ. Health. Perspect.*, 106, 573–580, 1998.

Poliandri, A.H.B., Cabilla, J.P., Velardez, M.O., Bodo, C.C.A., and Duvilanski, B.H., Cadmium induces apoptosis in anterior pituitary cells that can be reversed by treatment with antioxidants, *Toxicol. Appl. Pharmacol.*, 190, 17–24, 2003.

Poulin, P. and Krishnan, K., Molecular structure-based prediction of human abdominal skin permeability coefficients for several organic compounds, *J. Toxicol. Environ. Health A*, 62, 143–159, 2001.

Prahalad, A.K., Soukup, J.M., Immon, J., Willis, R., Ghio, A.J., Becker, S., and Gallagher, J.E., Ambient air particles: Effects on cellular oxidant radical generation in relation to particulate elemental chemistry, *Toxicol. Appl. Pharmacol.*, 158, 81–91, 1999.

Prohaska, J.R. and Lukasewycz, O.A., Copper deficiency suppresses the immune response of mice, *Science*, 213, 559–560, 1981.

Prozialeck, W.C., Lamar, P.C., and Lynch, S.M., Cadmium alters the localization of N-cadherin, E-cadherin, and β-catenin in the proximal tubule epithelium, *Toxicol. Appl. Pharmacol.*, 189, 180–195, 2003.

Raessler, M., Rothe, J., and Hilke, I., Accurate determination of Cd, Cr, Cu and Ni in woodlice and their skins—is moulting a means of detoxification? *Sci. Total Environ.*, 337, 83–90, 2005.

Ramos, K.S., Chacon, E., and Acosta, D., Jr., Toxic responses of the heart and vascular systems, In *Casarett & Doull's Toxicology: The Basic Science of Poisons*, 5th ed., Klaassen, C.D. (ed.), McGraw-Hill, New York, 1996, pp. 487–527.

Rice, R.H. and Cohen, D.E., Toxic responses of the skin, In *Casarett & Doull's Toxicology: The Basic Science of Poisons*, 5th ed., Klaassen, C.D. (ed.), McGraw-Hill, New York, 1996, pp. 529–546.

Rikans, L.E. and Yamano, T., Mechanisms of cadmium-mediated acute hepatotoxicity, *J. Biochem. Mol. Toxicol.*, 14, 110–117, 2000.

Risso-de Faverney, C., Orsini, N., de Sousa, G., and Rahmani, R., Cadmium-induced apoptosis through the mitochondrial pathway in rainbow trout hepatocytes: Involvement of oxidative stress, *Aquat. Toxicol.*, 69, 247–258, 2004.

Rollins-Smith, L.A., Hopkins, B.D., and Reinert, L.K., An amphibian model to test the effects of xenobiotic chemicals on development of the hematopoietic system, *Environ. Toxicol. Chem.*, 23, 2863–2867, 2004.

Ronisz, D., Farmen Finne, E., Karlsson, H., and Förlin, L., Effects of the brominated flame retardants hexabromocyclododecane (HBCDD), and tetrabromobisphenol A (TBBPA), on hepatic enzymes and other biomarkers in juvenile rainbow trout and feral eelpout, *Aquat. Toxicol.*, 69, 229–245, 2004.

Salminen, W.F. and Roberts, S.M., Dermal and ocular toxicology: Toxic effects of the skin and eyes, In *Principles of Toxicology. Environmental and Industrial Applications*, 2nd ed., Williams, P.L., James, R.C. and Roberts, S.M. (eds.), John Wiley & Sons, Inc., New York, 2000, pp. 157–168.

Saravana Bhavan, P. and Geraldine, P., Histopathology of the hepatopancreas and gills of the prawn *Macrobrachium malcolmsonii* exposed to endosulfan, *Aquat. Toxicol.*, 50, 331–339, 2000.

Scheier, A. and Cairns, J., Jr., Persistence of gill damage in *Lepomis gibbosus* following a brief exposure to alkyl benzene sulfonate, *Notulae Naturae Acad. Nat. Sci. Philadelphia*, 391, 1–7, 1966.

Selevan, S.G., Rice, D.C., Hogan, K.A., Euling, S.Y., Pfahles-Hutchens, A., and Bethel, J., Blood lead concentration and delayed puberty in girls, *N. Engl. J. Med.*, 348, 1527–1536, 2003.

Sharbaugh, C., Viet, S.M., Fraser, A., and McMaster, S.B., Comparable measures of cognitive function in human infants and laboratory animals to identify environmental health risks to children, *Environ. Health Perspect.*, 111, 1630–1639, 2003.

Snoeijs, T., Dauwe, T., Pinxten, R., Vandesande, F., and Eens, M., Heavy metal exposure affects the humoral immune response in a free-living small songbird, the Great Tit (*Parus major*), *Arch. Environ. Contam. Toxicol.*, 46, 399–404, 2004.

Sparks, A.K., *Invertabrate Pathology. Noncommunicable Diseases*, Academic Press, New York, 1972.

Stevens, J.L., Snyder, M.J., Koener, J.F., and Feyereisen, R., Inducible P450s of the CYP9 family from larval *Manduca sexta* midgut, *Insect. Biochem. Mol. Biol.*, 30, 559–568, 2000.

Sullivan, B.K., Buskey, E., Miller, D.C., and Ritacco, P.J., Effects of copper and cadmium on growth, swimming and predator avoidance in *Eurytemora affinis* (Copepod), *Mar. Biol.*, 77, 299–306, 1983.

Sullivan, J.F., Atchison, G.L., Kolar, D.J., and McIntosh, A.W., Changes in the predator-prey behavior of fathead minnows (*Pimephales promelas*) and large-mouth bass (*Micropterus salmoides*) caused by cadmium, *J. Fish. Res. Board Can.*, 35, 446–451, 1978.

Suter, G.W., II, Integration of human health and ecological risk assessment, *Environ. Health Perspect.*, 105, 1282–1283, 1997.

Tkatcheva, V., Hyvärinen, H., Kukkonen, J., Ryzhkov, L.P., and Holopainen, I.J., Toxic effects of mining effluents on fish gills in a subarctic lake system in NW Russia, *Ecotoxicol. Environ. Saf.*, 57, 278–289, 2004.

Tolson, J.K., Roberts, S.M., Jortner, B., Pomeroy, M., and Barber, D.S., Heat shock proteins and acquired resistance to uranium nephrotoxicity, *Toxicology*, 206, 59–73, 2005.

Triebskorn, R., Casper, H., Heyd, A., Eikemper, R., Köhler, H.-R., and Schwaiger, J., Toxic effects of the non-steroidal anti-inflamatory drug diclofenac. Part II. Cytological effects in liver, kidney, gills and intestine of rainbow trout (*Oncorhynchus mykiss*), *Aquat. Toxicol.*, 68, 151–166, 2004.

Tzeng, H.-P., Yang, R.-S., Ueng, T.-H., Lin-Shiau, S.-Y., and Liu, S.-H., Motorcycle exhaust particulates enhance vasocontriction on organ culture of rat aortas and involve reactive oxygen species, *Toxicol. Sci.*, 75, 66–73, 2003.

Valdez-Márquez, M., Lares, M.L., Camacho Ibar, V., and Gendron, D., Chlorinated hydrocarbons in skin and blubber of two blue whales (*Balaenoptera musculus*) stranded along the Baja California coast, *Bull. Environ. Contam. Toxicol.*, 72, 490–495, 2004.

Van Heerden, D., Vosloo, A., and Nikinmaa, M., Effects of short-term copper exposure on gill structure, metallothionein and hypoxia-inducible factor-1 α (HIF-1 α) levels in rainbow trout (*Oncorhynchus mykiss*), *Aquat. Toxicol.*, 69, 271–280, 2004.

Viale, D., Ecologie des cétacés en Méditerranée nord-occidentale: Leur place dans l'écosystéme, leur réaction a la pollution marine par les métaux. Thése Doct., Univ. Paris VI, Paris.

Weis, J.S., Weis, P., Heber, M., and Vaidya, S., Methylmercury tolerance of killifish (*Fundulus heteroclitus*) embryos from a polluted *vs* non-polluted environment, *Mar. Biol.*, 65, 283–287, 1981.

Weisel, C.P. and Jo, W.-K., Ingestion, inhalation, and dermal exposures to chloroform and trichloroethene from tap water, *Environ. Health Perspect.*, 104, 48–51, 1996.

Williams, P.L., James, R.C., and Roberts, S.M. (eds.), *Principles of Toxicology. Environmental and Industrial Applications*, 2nd ed., John Wiley & Sons, Inc., New York, 2000.

Willingham, E., Endocrine-disrupting compounds and mixtures: Unexpected dose-response, *Arch. Environ. Contam. Toxicol.*, 46, 265–269, 2004.

Wilson, V.S., Lambright, C., Ostby, J., and Gray, L.E., Jr., *In vitro* and *in vivo* effects of 17 β-trenbolone: A feedlot effluent contaminant, *Toxicol. Sci.*, 70, 202–211, 2002.

Witschi, H.R. and Last, J.A., Toxic responses of the respiratory system, In *Casarett & Doull's Toxicology: The Basic Science of Poisons*, 5th ed., Klaassen, C.D. (ed.), McGraw-Hill, New York, 1996, pp. 443–462.

Wu, A., Li, L., and Liu, Y., Deltamethrin induces apoptotic cell death in cultured cerebral cortical neurons, *Toxicol. Appl. Pharmacol.*, 187, 50–57, 2003.

Yoshida, T., Yamauchi, H., and Sun, G.F., Chronic health effects in people exposed to arsenic via drinking water: Dose-response relationships in review, *Toxicol. Appl. Pharmacol.*, 198, 243–252, 2004.

Zheng, W., Aschner, M., and Ghersi-Egea, J.-F., Brain barrier systems: A new frontier in metal neurotoxicological research, *Toxicol. Appl. Pharmacol.*, 192, 1–11, 2003.

Zimmerman, H.J., Hepatotoxicity, *Dis. Mon.*, 39, 675–787, 1993.

6 Physiology

> To us the intact organism is the basic unit of biological organization, but at the same time we recognize that the various component physiological systems play a vital role in maintaining the integrity of the organism.
>
> **(Vernberg and Vernberg 1972)**

6.1 OVERVIEW

Many issues and concepts relevant here are covered in other chapters. Discussions of oxidative stress (Chapter 3) or chloride cells (Chapter 4) could easily have been placed here. Similarly, there is overlap with discussions to come about physiologically based pharmacokinetic (PBPK) models (Chapter 8), stress (Chapter 9), and toxicant effects on phenotype expression (Chapter 16). Because physiological topics are blended into other chapters, this chapter might be shorter than anticipated by the reader. Physiology was developed separately here to provide sufficient focus on a few prominent physiological themes. Because physiological issues are blended into adjacent chapters, the readers may want to explore the book index if they do not find a particular physiological issue here. Regardless, this chapter's overarching theme is toxicant impact on physiology as it relates to optimal flux of materials and energy between the organism and its environment (i.e., toxicant modification of an individual's fitness).

Why commit a short chapter to physiology if most ecotoxicology books do not? As stated in *Fundamentals of Ecotoxicology* (Newman and Unger 2003):

> Deviations from homeostasis associated with sublethal exposure often reflect physiological alterations. Such physiological alterations can be used to infer a mode of action of the toxicant as well as to document a lowered capacity to maintain homeostasis or normal functioning.

Measuring physiological change helps identify mode of action and infer deviation from optimal functioning. Physiological relationships are also integral components of PBPK modeling, arguably the most effective approach to predicting movement of contaminants and their metabolites within individuals. As an important example, a common issue addressed in PBPK-based ecotoxicological models is scaling, that is, "the structural and functional consequences of changes in size or scale among otherwise similar organisms" (Schmidt-Nielsen 1984). These PBPK-based models also have very high potential for integrating processes occurring at several levels of biological organization. Finally, shifts in energy allocation and metabolism are often at the root of life history strategies. Understanding physiological processes underlying energy allocation improves understanding of how individuals within populations maintain the highest possible fitness under diverse conditions. As Stearns (1992) explains:

> Physiological trade-offs are caused by allocation decisions between two or more processes that compete directly with one another for limited resources within a single individual . . . . Microevolutionary trade-offs are defined by the response of populations to selection. Physiological trade-offs are involved in almost all microevolutionary trade-offs . . . .

So, to answer the question posed at the beginning of this paragraph, physiological discussion enhances identification and understanding of underlying mechanisms that manifest as adverse effects, are central in PBPK models, and help explain life history shifts and trade-offs.

The physiological processes sketched out below relative to ecotoxicology are those usually addressed and those that link processes within the individual to population processes. Most are involved in maintaining the soma at minimal energetic costs, and optimizing energy and material allocation to functions determining an individual's fitness.

> [The organism] is a dynamic and energetic system, consisting of many subunits finely tuned to interact harmoniously and thus insure survival of the total system. To the organism in a harsh environmental complex, mere survival is the minimal level of performance; but for the perpetuation of the species, more highly integrated processes leading to reproduction must be operational.
>
> **(Vernberg and Vernberg 1972)**

6.2 IONIC AND OSMOTIC REGULATION

All organisms must maintain an internal milieu with an acceptable ionic and osmotic state. Some conform to external osmotic conditions but none have cellular or body fluid ionic compositions identical to their environment. Physiology and behavior play roles in maintaining optimal ionic and osmotic conditions for marine, freshwater, and terrestrial organisms: both can be adversely impacted by toxicants.

Most marine organisms have internal osmotic concentrations matching that of the surrounding seawater, but the intracellular and intercellular composition of specific ions differ from seawater owing to active regulation. The ability of a marine species to tolerate different salinity ranges varies depending on the species' physiological strategy that evolved to promote optimal fitness within phylogenetic constraints (Figure 6.1). There might be no advantage for a deep-sea species to spend precious energy maintaining physiological mechanisms that allow it to tolerate wide fluctuations in salinity that never occur; however, survival of a coastal species might be utterly dependent on wide salinity tolerance. Some phylogenetic groups have such fundamental constraints

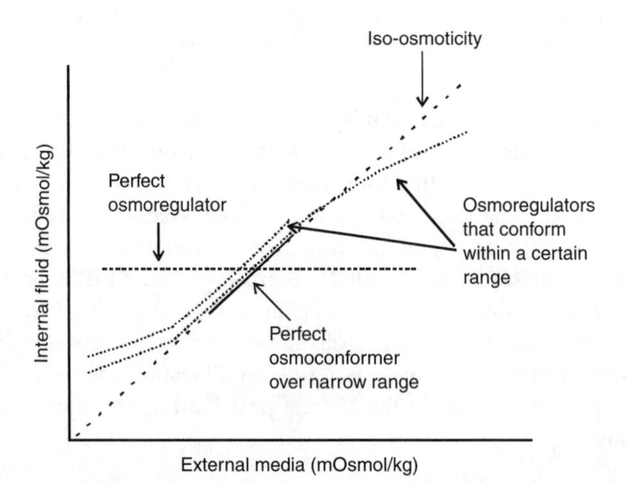

FIGURE 6.1 Within phylogenetic constraints, strategies evolve for coping with changes in external osmotic conditions. Osmoconformers have internal fluid osmotic concentrations that conform to those of external media whereas the internal fluids of osmoregulators are actively maintained constant despite changes in external concentrations. Schmidt-Nielsen (1970) (Figure 4.2) gives as examples spider crab (*Maia squinado*) as a narrow range osmoconformer, the mitten crab (*Eriocheir sinensis*) as an osmoregulator at both extremes, and the shore crab (*Carcinus maenas*) as an osmoregulator at the bottom of the osmotic range.

on their abilities to tolerate salinity changes that they are incapable of occupying certain habitats. Echinoderms have fundamental osmoregulatory constraints, so none is found in freshwaters. Some marine species actively osmoregulate, so if salinity drops below or rises above a certain level, they expend energy to maintain a higher or lower internal osmotic concentration than that of the external environment. Within cells, a pool of free amino acids (FAAs) is modified to maintain intracellular osmotic conditions. For example, teleosts maintain internal osmotic concentrations ranging from 250 to 500 mOsmol/kg while existing in aqueous environments that might have osmotic concentrations of <0.01–1000 mOsmol/kg (Karnaky 1998). Freshwater species have higher internal osmotic concentrations than surrounding waters yet, like marine species, they are pervious to water. They must restrict water intake, eliminate water and some ions via the kidneys, and actively move ions across the gills (e.g., fish) or skin (e.g., amphibians). ATPases are involved in active ion pumping in many organs such as the gills, integument, and intestine. Costs associated with ATPase ion pumping in healthy tissues range from 3% to 28% of total energy expenditures (Horn 1998), suggesting that significant costs can accrue with toxicant-induced stress or damage to ion regulatory tissues. Terrestrial animals must deal with dehydration, regulate water volume and ionic composition, and eliminate nitrogen wastes as discussed in the last chapter.

Many toxicants influence these essential processes, leading to energy-dissipating deviations from homeostasis, damage, or even death. Many such effects were described in the previous chapter. Copper interferes with chloride ion absorption across the eel gut by altering cell junctions: cadmium interferes with Na^+/K^+-ATPase function in eel intestine (see Section 5.5). Gill chloride cells that facilitate ion exchange are also influenced by toxicants (see Section 5.3.2). As an example, rainbow trout (*Oncorhynchus mykiss*) gill regulation of Na^+, Cl^-, and K^+ diminishes during copper exposure (Laurén and McDonald 1985). Other metals similarly influence ATPases (e.g., mercury and crayfish gills (Torreblanca et al. 1989), lead in rainbow trout (Rogers et al. 2003)), and plasma ions and osmolality (Grosell et al. 2004a,b).

Nonmetallic contaminants interfere with these processes in a variety of species. Transport of Na^+ and Cl^- across amphibian integument is altered by pyrethroid pesticides (Section 5.2). Low pH waters interfere with freshwater fish ion regulation (Fromm 1979). Plasma osmolality and Cl^- concentrations of seawater-acclimated Atlantic salmon (*Salmo salar*) increase with exposure to elevated concentrations of ammonia (Knoph and Olsen, 1994).

Cellular FAA pools can also reflect contaminant exposure or stress. Treatment of blue mussels (*Mytilus edulis*) with crude oil increased tissue FAA: ornithine and enzymes associated with its synthesis were particularly impacted (Soini and Rantamäki 1985). Similarly, FAA was elevated in mantle tissues of freshwater mussels (*Amblema plicata*) exposed to acid mine drainage (Gardner et al. 1981).

6.3 ACID–BASE REGULATION

Maintaining internal pH within an optimal range is critical to physiological homeostasis, and is accomplished by buffering of internal fluids and moving pH-relevant ions into and out of the organism (Claiborne 1998). Blood pH is dependent on the CO_2 concentration and associated bicarbonate buffering system equilibrium concentrations of HCO_3^-, OH^-, and H^+, and NH_4^+. The Henderson–Hasselbalch equation can be used to illustrate the influence of the bicarbonate buffering system on internal pH:

$$pH = pK + \log \frac{[HCO_3^-]}{[H_2CO_3]} \quad \text{where pK} \approx 6.1. \tag{6.1}$$

The enzyme carbonic anhydrase is important in pH regulation because it catalyzes the otherwise slow hydration of metabolically generated CO_2 to carbonic acid, H_2CO_3. Obviously from Equation 6.1, the resulting carbonic acid concentrations are pivotal in determining pH.

Organisms also exchange pH-influencing ions with their environment. Carbon dioxide is eliminated from the blood by the respiratory organs and kidneys. In fish gills, Na^+ is exchanged for H^+ or NH_4^+ (Wood 1992), illustrating the intimate linkage between ion and pH regulation. Na^+/H^+, NH_4^+-, and Cl^-/HCO_3^--dependent ATPases are also active in gill transport (Wood 1992).

Acidosis (i.e., drop in blood pH) can result for many reasons ranging from toxicant damage to physical exertion to dysfunction of acid–base regulation mechanisms. Acidosis impacts important physiological properties such as O_2-binding affinity of respiratory pigments. As an important example of toxicant effects on acid–base regulation, ATPases involved in H^+ transport or Cl^-/HCO_3^- exchange in gills can be directly impacted by toxicants or, as described in the last chapter, indirectly impacted by inducing changes to the cells, such as chloride cells, involved in these exchanges. Carbonic anhydrase activity can also be changed by inorganic toxicants (e.g., metals, organometals, and some anions). Pesticides influence catfish (*Ictalurus punctatus*) carbonic anhydrase activity (Christensen and Tucker 1976).

Many diverse instances of toxicant effects on pH regulation can be found in the literature. Exposure of rainbow trout (*O. mykiss*) to 2–18 mg/L of chloramine-T for 3.5 h caused a decrease in net acid uptake, producing acidosis (Powell and Perry 1998). Changes in ion fluxes were also reported in this study. Nitrite reduced hemolymph pCO_2, pH, and HCO_3^- concentrations of the shrimp, *Marsupenaeus japonicus* (Cheng and Chen 2002). Saponin, a plant toxin used for nuisance species control in fish ponds, also decreased shrimp (*Penaeus japonicus*) hemolymph pCO_2, pH, and HCO_3^- concentrations (Chen and Chen 1996). Associated with saponin-induced changes was a drop in oxyhemocyanin concentrations, reflecting an adverse influence of the hemolymph pH decrease on hemocyanin oxygenation.

6.4 RESPIRATION AND GENERAL METABOLISM

Given the effects explored to this point, it should be no surprise to find that respiratory and metabolic processes are influenced by toxicants. Such effects are diverse, including changes in ventilation rate, metabolic rate, tissue energy charge, and carbohydrate metabolism. Some even result from a causal cascade involving the endocrine system, the system responsible for regulating metabolism and other key processes.

Because a clear approach to studying fish respiration was established many years ago (e.g., Fry 1957), toxicant effects on fish ventilation rates were among the first sublethal effects to be quantified by aquatic toxicologists. The simple apparatus depicted in Figure 6.2 measures minute electrical signals that are easily converted to gill movement rates under various conditions. In this figure (bottom), increased ventilation rate is obvious during eight consecutive pulses of elevated zinc concentrations in water (Thompson et al. 1983).

Taking their cue from classic physiology, many early studies of sublethal physiological effects used rates of oxygen consumption to estimate metabolic rate of exposed individuals. In 1926, Cook found that respiration of the fungus *Aspergillus niger* dropped with exposure to copper, silver, or mercury. More than 30 years ago, oxygen consumption rates were measured for diverse aquatic organisms exposed to toxicants (e.g., *Daphnia pulex* exposed to dichromate) (Sherr and Armitage 1971). For reasons discussed in the next several sections, these types of studies remain useful for understanding impacts to exposed individuals, and have become even more meaningful as associated theory and computer-based signal processing methods have advanced. Rowe et al. (2001) provide one such recent study in which crayfish exposed to fly ash-contaminated sediments had elevated metabolic rates that were interpreted as reflecting the high costs of mounting a defense against the effects of chronic metal and metalloid exposure.

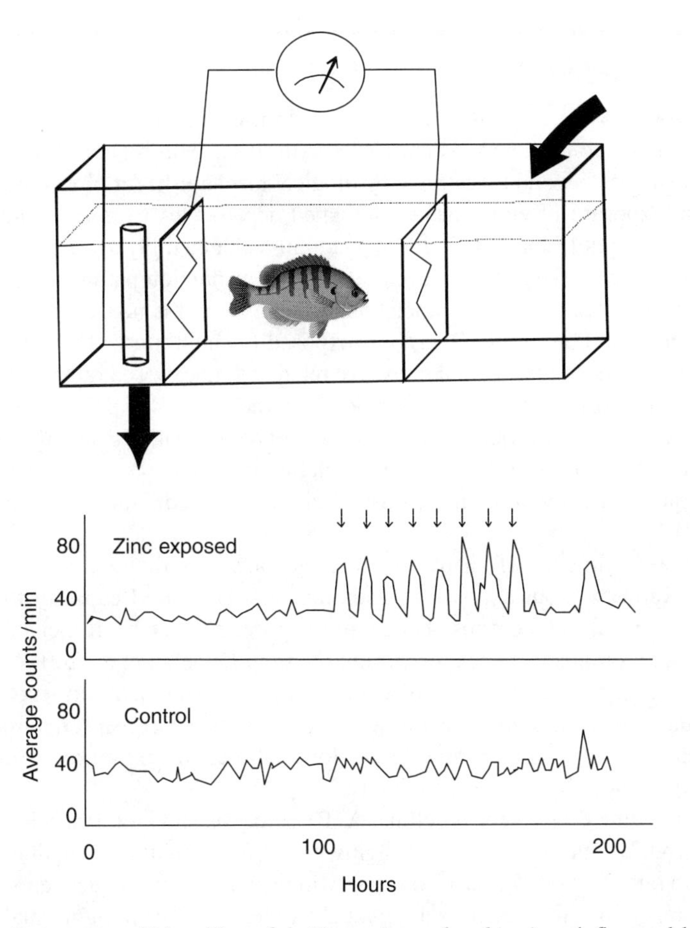

FIGURE 6.2 Ventilation rate of bluegill sunfish (*Lepomis machrochirus*) as influenced by pulses of zinc introduced to water. Juvenile bluegill sunfish are placed into a ventilatory monitoring chamber fitted with electrodes that detect minute electrical pulses as branchial muscles contract, reflecting gill ventilation movement (top). Time courses of average ventilation rates for a control bluegill and a bluegill exposed to a sequence of eight zinc concentration pulses (↓) illustrate the increase in ventilation rate during exposures. (Modified from Figures 1 and 4 of Thompson et al. (1983).)

At one point, adenylate energy charge (AEC) began to be measured in tissues of exposed individuals (e.g., Giesy and Dickson 1981, Giesy et al. 1981):

$$\mathrm{AEC} = \frac{[\mathrm{ATP}] + 0.5[\mathrm{ADP}]}{[\mathrm{ATP}] + [\mathrm{ADP}] + [\mathrm{AMP}]}, \tag{6.2}$$

where [ATP], [ADP], and [AMP] are measured concentrations of ATP, ADP, and AMP, respectively. Although useful, this metric, which was thought to reflect the net internal energy status of an individual, tends to vary widely with season (Giesy and Dickson 1981) and ambient conditions (e.g., salinity or temperature (Ivanovici 1980), and habitat (Dickson and Giesy 1981)). This makes its use as a biomarker challenging.

Direct measurement of other biochemical qualities provides insight into the effect of toxicants on metabolic processes. Arsenic dosing of rats disturbed carbohydrate metabolism (Ghafghazi et al. 1980). However, this disruption involved more than direct effects on cellular biochemistry: it also

Box 6.1 More about Cadmium and Fish Fitness

Let us look more closely at cadmium's effect on fish. Cadmium can directly inhibit oxygen uptake by fish mitochondria and enhance ATP hydrolysis (Hiltibran 1971).[1] But cadmium's influence on fish goes beyond this simple cytological mechanism for physiological disruption.

Cadmium exposure of epinephrine-stimulated hepatocytes from eel (*Anguilla anguilla*) reduced glucose output (Fabbri et al. 2003). Cadmium changed glycogen regulation and glucose release by changing the adenylyl cyclase/cAMP transduction pathway.[2] Cadmium clearly interfered with adrenergic control of metabolism; therefore, it is no surprise that Ricard et al. (1998) found that rainbow trout (*O. mykiss*) exposed to cadmium had reduced liver size and glycogen content, and smaller overall body size relative to unexposed rainbow trout. The metabolism stimulating hormones T3 (3,5,3′-triiodothyonine) and T4 (3,5,3′,5′-tetraiodothyronine or thyroxine) decreased in the plasma with exposure of adult trout, suggesting impact on thyroid function. Cadmium also disrupts endocrine regulation in rainbow trout (*O. mykiss*) and yellow perch (*Perca flavescens*) by impairing cortisol secretion from adrenocortical cells (Lacroix and Hontela 2004).

The combined influence of all of these changes lowers individual fitness with exposure to cadmium. Rainbow trout exposed to cadmium had decreased capacity to compete with nonexposed individuals, and groups of cadmium-exposed fish establish social dominance hierarchies faster than groups of nonexposed rainbow trout (Sloman et al. 2003). Further, fitness was influenced in the context of multigenerational changes in cadmium-exposed fish. After seven generations of cadmium exposure, cadmium-resistant killifish (*Heterandria formosa*) had decreased fecundity, smaller brood size, longer times to first reproduction, and shorter female life expectancy (Xie and Klerks 2004).

So, cadmium directly influences cellular ATP hydrolysis and modifies cellular metabolism. It also acts through organ systems by changing endocrine regulation. Qualities of individuals such as growth rate and body mass are adversely impacted as a consequence. At the population level, cadmium can further reduce an individual's fitness by disrupting normal social interactions. Acquisition of genetic resistance to cadmium in populations can involve trade-offs that lower fitness relative to other environmental stressors.

---

involved a change in adrenal gland regulation of glucose. Cadmium also influenced carbohydrate metabolism in rats (Chapatwala et al. 1982a,b). Cadmium-induced increases in gluconeogenic enzymes and serum glucose concentration suggested mobilization of fat deposits to produce this glucose. This conclusion was reinforced by the concomitant slower growth of dosed rats. A similar increase in gluconeogenesis was noted in crabs (*Barytelphusa guerini*) exposed to cadmium (Reddy et al. 1989).

Many studies have demonstrated the influence of toxicants on endocrine function that, in turn, changes metabolism and homeostasis. Some endocrine-modifying chemicals act via the hypothalamic-pituitary-gonadal axis (Figure 6.3). Such effects have clear impact on an individual's fitness because this axis is central to maintaining homeostasis in a changing environment. Chemicals can be hormone mimics (i.e., agonists or antagonists) or those that modify associated enzymes or signaling pathways. One very relevant example is the currently active area of study in which exposure

[1] Interestingly, metal-induced ATP hydrolysis in fish mitochondria is even further complicated in the presence of insecticides (Hiltibran 1982).

[2] Adenylyl cyclase, a membrane-associated enzyme that converts ATP to cyclic adenosine monophosphate (cAMP), is important in intracellular signaling.

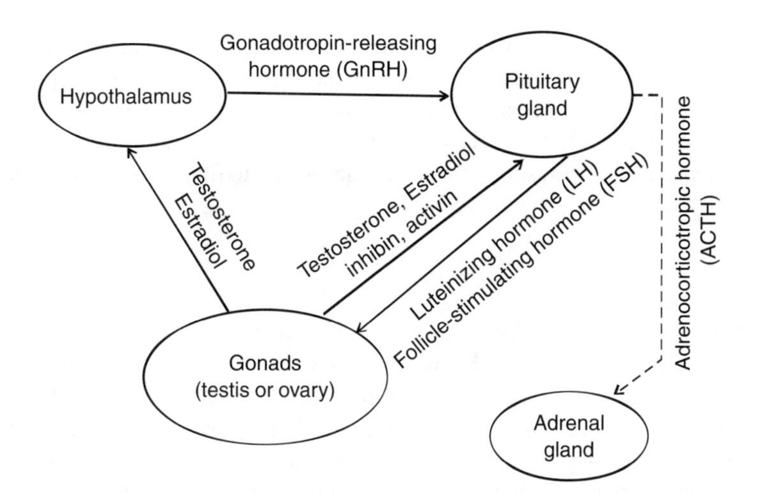

FIGURE 6.3 The hypothalamic-pituitary-gonadal axis. The hypothalamus produces gonadotropin-releasing hormone (GnRH) that stimulates LH and follicle-stimulating hormone (FSH) secretion from the pituitary gland. (The GnRH neurons receive input from other brain neurons and mediate this GnRH release.) The LH and FSH act on the testes and ovaries. In response to LH, Leydig cells in the testes and follicle-associated ovarian cells produce steroid hormones. The FSH mediates sperm production and ovarian follicle stimulation. Linked to the processes occurring in the hypothalamic-pituitary-gonadal axis is adrenocorticotrophic hormone (ACTH) release from the pituitary gland, which regulates production of cortisol (plus other glucocorticoids, androgens, and aldosterone). Cortisol, a steroid hormone, is involved in the general stress reactions discussed in Chapter 10. Testosterone and estradiol provide negative feedback to the hypothalamus and pituitary gland. Inhibin and activin from the gonads inhibit or stimulate FSH secretion by the pituitary gland.

to some xenobiotics results in the expression of an egg protein, vitellogenin, in the plasma of male teleosts (Ankley and Johnson 2004). Toxicant-induced thyroid dysfunction is another deservedly active theme in endocrine toxicity research (Brown et al. 2004, Colborn 2002). Brown et al. (2004) indicate the potential for teleost thyroid disruption with exposure to polychlorinated biphenyl (PCB) mixtures, PAH, (organochloride, carbamate, or organophosphorus) pesticides, chlorinated paraffins, compounds of cyanide, methyl bromide, trihalomethanes such as chloroform, phenols, ammonia, low pH, steroids, pharamceutical compounds, perchlorate, brominated fire retardants, and metals.

A few examples can give the reader a sense of the broad impact of thyroid disruption. Pesticides and their residues elevated T4 titers in field-exposed tree swallows (*Tachycineta bicolor*) (Mayne et al. 2005). These swallows had thyroid glands with hypotrophic epithelia and collapsed follicles. In this same study, bluebirds (*Sialia sialis*) from sprayed orchards had higher T3 titers after challenge with thyroid-stimulating hormone (TSH). The mummichog (*Fundulus heteroclitus*) from a mercury-contaminated site had higher T4 titers than mummichog from less contaminated sites: the behavior of the fish from the contaminated site was also distinct (Zhou et al. 1999). The implication of these studies is that a toxicant-stressed individual becomes metabolically compromised, resulting in diminished fitness. A study by Veldhuizen-Tsoerkan et al. (1991) clearly demonstrates such a consequence. Blue mussels (*M. edulis*) exposed to cadmium or PCB showed decreased survival times under anoxic conditions.

6.5 BIOENERGETICS

> Studies based on individuals can be useful in ecological analysis of polluted systems when based on the concepts of resource allocation-based life history analysis.
>
> **(Congdon et al. 2001)**

To maintain optimal fitness, an individual must expend its metabolic currency efficiently among essential functions including reproduction, foraging, and maintaining and growing the soma. Toxicants alter fitness by impacting associated functions. Such effects will be discussed briefly here and again in Chapter 16 (see Figure 16.2 and associated text).

Koehn and Bayne (1989) use a simple bioenergetics model to frame discussion of stressor effects to individuals

$$\mathrm{Pg} + \mathrm{Pr} = (\mathrm{C} \times \mathrm{AE}) - (R_{\mathrm{m}} + R_{\mathrm{r}}), \tag{6.3}$$

where the left-hand side of the equation is the sum of somatic (Pg) and reproductive (Pr) production, and the right-hand side is the difference between the product of energy consumption (C) and energy assimilation efficiency (AE), and the sum of the metabolic cost of maintaining the soma (R_{m}) and the cost of maintaining all other processes such as feeding, growth, and reproduction (R_{r}). The right-hand side of the equation reflects production given the energy taken in (C × AE) and the energy costs of maintaining the soma, feeding, growing, and reproducing ($R_{\mathrm{m}} + R_{\mathrm{r}}$).

> If an organism can only acquire a limited amount of materials and energy for which two processes [maintaining and growing the soma versus reproduction] compete directly, then an increase in materials allocated to one must result in a decrease to the other . . . [i.e.,] the Principle of Allocation.
>
> **(Stearns 1992)**

Components of this model are subject to modification due to toxicant influence on diverse processes such as detoxification (Janczur et al. 2000), cellular metabolism (Ghafghazi et al. 1980), organ function (van Heerden et al. 2004), tissue/organ repair (references in Chapter 4), endocrine control (Brouwer et al. 1990, Mayne et al. 2005), and behavior (Sloman et al. 2003). An individual can adjust its bioenergetics within limits to compensate, as described later in Chapter 16 (Figure 6.4). These adjustments can involve changes in foraging behavior or life history characteristics.

The various suites of bioenergetic adjustments made by the individual have life history consequences and often are subject to natural selection (Stearns 1992). Yellow perch (*P. flavescens*) exposed to metals in the field had decreased growth and distinct changes in annual glycogen and

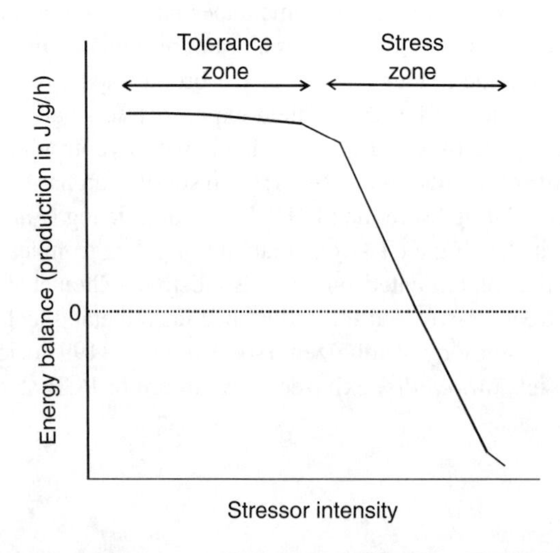

FIGURE 6.4 Zone of tolerance relative to energy balance [modified from Figure 1 of Koehn and Bayne (1989)]. The term stress used here conforms to the definition given by Koehn and Bayne, that is, "any environmental change that acts to reduce the fitness of an organism." For toxicants that are also essential nutrients, there may be a zone of stress at low concentrations.

triglyceride cycles (Levesque et al. 2002). As mentioned previously, killifish adapted to cadmium had significant changes in reproduction and life expectancy (Xie and Klerks 2004). *Daphnia magna* exposed to cadmium or copper had changes in growth rate and reproduction, but not metabolic rate (Knops et al. 2001), reflecting somatic maintenance at the cost of growth and reproduction. Life history changes were also seen for the cladoceran, *Chydorus piger*, exposed to cadmium (Dekker et al. 2002). Profenofos-exposed rainbow fish (*Melanotaenia duboulayi*) have decreased acetylcholinesterase activity and consequent changes in feeding activity and food intake (Kumar and Chapman 1998).

Toxicant effects on foraging and other behaviors can lower the fitness of individuals (e.g., Bridges 1997, Perez and Wallace 2004, Saglio and Trijasse 1998, Teather et al. 2005). In some cases, even parental exposure can adversely impact fitness-influencing behavior of offspring (Faulk et al. 1999). The mummichog mentioned above with the modified T4 titers had documented behavioral changes that the authors attempted to link to thyroid dysfunction (Zhou et al. 1999). This same group of researchers later showed the adverse impact of contamination on the mummichog's ability to capture prey (Weis et al. 2001) and speculated that this change might be associated with altered brain levels of the neurotransmitter, serotonin, in mummichog from the contaminated site (Weis et al. 2000). Predator–prey interactions of stream insects are also modified by contaminants (Clements 1999). Pollutants modify pumping activity of bivalves and geotaxis of gastropods (Salánki et al. 2003). Of course, contaminated food (e.g., Irving et al. 2003) or sediment (McCloskey and Newman 1995) can elicit avoidance behavior, reducing adverse effects of toxicant presence. But most behavior changes noted in relevant reviews (e.g., Atchison et al. 1996, Henry and Atchison 1991) reflect changes that increase energy expenditure and, in doing so, decrease Darwinian fitness.

6.6 PLANT-RELATED PROCESSES

Physiological responses of plants to toxicants are also important to understand. Qualities related to fitness are often impacted by toxicants. For example, PCB exposure of the diatom, *Thalassiosira pseudonana*, decreases photosynthesis and consequently cell growth rates (Michaels et al. 1982). Figure 6.5 provides a general example in which plant yield is adversely impacted by very low and

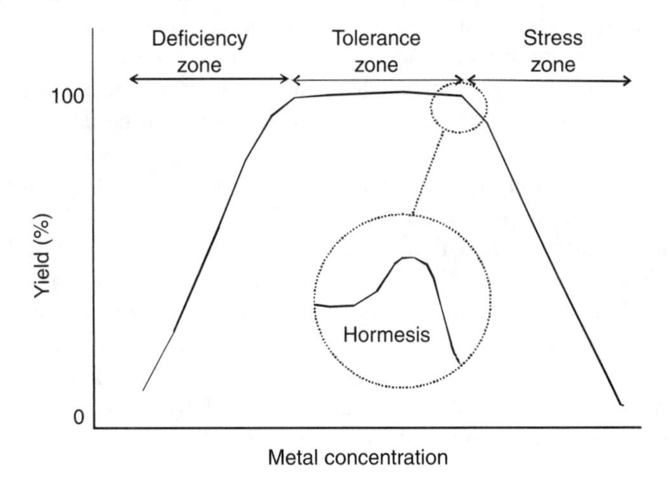

FIGURE 6.5 For some essential elements, the relationship shown in Figure 6.4 requires modification. In this modification of Figure 1 of Baker and Walker (1989), a zone of deficiency is included. Below a certain concentration, the element becomes limiting and the individual enters into a condition of deficiency (e.g., decreased plant yield). In some instances, a sublethal concentration of a toxicant such as a metal can cause an apparent beneficial or stimulatory effect (hormesis).

very high soil concentrations of an essential element such as copper or zinc. Hormesis, an apparent stimulatory effect of sublethal concentrations of a toxicant, can also occur in some cases. As a nonmetal toxicant example, low concentrations of the growth inhibitor, phosfon, actually stimulates growth of the mint, *Mentha piperita* (Calabrese et al. 1987). Hormesis can involve overcompensation, the marshaling of compensatory mechanisms in response to a subinhibitory toxicant concentration that results in an apparent increase in a fitness characteristic such as soma growth. Of course, because an individual must balance energy allocation among several functions in order to maintain optimal fitness, one could argue that such a hormetic response might not always reflect an increase in overall Darwinian fitness. The energy used in the overcompensation comes at the expense of other fitness-related functions.

A well-established literature exists for the trade-offs related to plant tolerance to heavy metals. Seedling success or height, and root elongation or respiration are different under metal contamination for metal tolerant and naïve strains (Antonovics 1975, Baker and Walker 1989, Wu et al. 1975). Metals also impact water balance and stomatal function in kidney bean plants (*Phaseolus vulgaris*) (Rauser and Dumbroff 1981). The gaseous pollutant SO_2 reduces chlorophyll concentrations, perhaps enough to cause severe chlorosis. Sulfur dioxide also influences carbohydrate concentrations in kidney beans (Koziol and Jordon 1978).

6.7 SUMMARY

Prominent physiological changes due to toxicant exposure are discussed in this chapter with additional physiological changes described elsewhere (e.g., Chapters 3, 4, 8, and 16). The most important themes are summarized in the bulleted list below.

6.7.1 Summary of Foundation Concepts and Paradigms

- Physiological impacts of toxicants determine optimal flux of materials and energy between the organism and its environment. Toxicants can diminish an individual's fitness in this way.
- Discussion of toxicant effects on physiology enhances identification and understanding of underlying mechanisms that manifest as adverse effects, are important in PBPK modeling, and explain life history shifts or trade-offs with toxicant exposure.
- Physiology and behavior have roles in maintaining optimal ionic and osmotic conditions for marine, freshwater, and terrestrial organisms: both can be adversely impacted by toxicants.
- Toxicant effects on ion- and osmoregulatory processes lead to energy-dissipating deviations from homeostasis, damage, or even death.
- Maintenance of internal pH within an optimal range is critical to physiological homeostasis, and is accomplished by buffering of internal fluids and moving pH-relevant ions into and out of the organism. Numerous toxicants disrupt pH regulation.
- Acidosis, a decrease in blood pH, can result from toxicant damage and impacts important physiological properties such as oxygen binding capacity of respiratory pigments.
- Respiratory and metabolic processes are influenced by toxicants. Such effects are diverse, including changes in ventilation rate, metabolic rate, tissue energy charge, and carbohydrate metabolism.
- The influence of certain toxicants on endocrine function can, in turn, modify metabolism and other processes influencing Darwinian fitness.
- To maintain optimal fitness, an individual must allocate its metabolic currency efficiently among various functions including reproduction, foraging, and maintaining and growing the soma. Toxicants alter an individual's fitness by impacting associated functions.

- Physiological responses of plants to toxicants are also important to understand. Qualities related to fitness are often impacted by toxicants.

REFERENCES

Ankley, G.T. and Johnson, R.D., Small fish models for identifying and assessing the effects of endocrine-disrupting chemicals, *ILAR J.*, 45, 469–483, 2004.

Antonovics, J., Metal tolerance in plants: Perfecting an evolutionary paradigm, In *International Conference on Heavy Metals in the Environment*, Vol. 2, National Research Council of Canada, Ottawa, Canada, 1975, pp. 169–186.

Atchison, G.J., Sandheinrich, M.B., and Bryan, M.D., Effects of environmental stressors on interspecific interactions of aquatic animals, In *Ecotoxicology. A Hierarchical Treatment*, Newman, M.C. and Jagoe, C.H. (eds.), CRC Press/Lewis Publishers, Boca Raton, FL, 1996, pp. 319–345.

Baker, A.J.M. and Walker, P.L., Physiological responses of plants to heavy metals and the quantification of tolerance and toxicity, *Chem. Speciation Bioavailability*, 1, 7–18, 1989.

Bridges, C.M., Tadpole swimming performance and activity affected by acute exposure to sublethal levels of carbaryl, *Environ. Toxicol. Chem.*, 16, 1935–1939, 1997.

Brouwer, A., Murk, A.J., and Koeman, J.H., Biochemical and physiological approaches in ecotoxicology, *Funct. Ecol.*, 4, 275–281, 1990.

Brown, S.B., Adams, B.A., Cyr, D.G., and Eales, J.G., Contaminant effects on the teleost fish thyroid, *Environ. Toxicol. Chem.*, 23, 1680–1701, 2004.

Calabrese, E.J., McCarthy, M.E., and Kenyon, E., The occurrence of chemically induced hormesis, *Health Phys.*, 52, 531–541, 1987.

Chapatwala, K.D., Boykin, M., Butts, A., and Rajanna, B., Effect of intraperitoneally injected cadmium on renal and hepatic gluconeogenic enzymes in rats, *Drug Chem. Toxicol.*, 5, 305–317, 1982a.

Chapatwala, K.D., Hobson, M., Desaiah, D., and Rajanna, B., Effect of cadmium on hepatic and renal gluconeogenic enzymes in female rats, *Toxicol. Lett.*, 12, 27–34, 1982.

Chen, J.-C. and Chen, K.-W., Hemolymph oxyhemocyanin, protein levels, acid–base balance, and ammonia and urea excretions of *Penaeus japonicus* exposed to saponin at different salinity levels, *Aquat. Toxicol.*, 36, 115–128, 1996.

Cheng, S.-Y. and Chen, J.-C., Study on the oxyhemocyanin, deoxyhemocyanin, oxygen affinity and acid–base balance of *Marsupenaeus japanicus* following exposure to combined elevated nitrite and nitrate, *Aquat. Toxicol.*, 61, 181–193, 2002.

Christensen, G.M. and Tucker, J.H., Effects of selected water toxicants in the *in vitro* activity of fish carbonic anhydrase, *Chem. Biol. Interact.*, 13, 181–192, 1976.

Claiborne, J.B., Acid–base regulation, In *The Physiology of Fishes*, 2nd ed., Evans, D.H. (ed.), CRC Press, Boca Raton, FL, 1988, pp. 177–198.

Clements, W.H., Metal tolerance and predator–prey interactions in benthic macroinvertebrate stream communities, *Ecol. Appl.*, 9, 1073–1084, 1999.

Colborn, T., Clues from wildlife to create an assay for thyroid disruption, *Environ. Health Perspect.*, 110, 363–367, 2002.

Congdon, J.D., Dunham, A.E., Hopkins, W.A., Rowe, C.L., and Hinton, T.G., Resource allocation-based life histories: A conceptual basis for studies of ecological toxicology, *Environ. Toxicol. Chem.*, 20, 1698–1703, 2001.

Cook, S.F., The effects of certain heavy metals on respiration, *J. Gen. Physiol.*, 9, 575–601, 1926.

Dekker, T., Krips, O.E., and Admiraal, W., Life history changes in the benthic cladoceran *Chydorus piger* induced by low concentrations of sediment-bound cadmium, *Aquat. Toxicol.*, 56, 93–101, 2002.

Dickson, G.W. and Giesy, J.P., Variation of phosphoadenylates and adenylate energy charge in crayfish (Decapoda: Astacidae) tail muscle due to habitat differences, *Comp. Biochem. Physiol.*, 70A, 421–425, 1981.

Fabbri, E., Caselli, F., Piano, A., Sartor, G., and Capuzzo, A., Cd^{2+} and Hg^{2+} affect glucose release and cAMP-dependent transduction pathway in isolated eel hepatocytes, *Aquat. Toxicol.*, 62, 55–65, 2003.

Faulk, C.K., Fuiman, L.A., and Thomas, P., Parental exposure to ortho,para-dichlorodiphenyltrichloroethane impairs survival skills of Atlantic croaker (*Micropogonias undulatus*) larvae, *Environ. Toxicol. Chem.*, 18, 254–262, 1999.

Fromm, P.O., A review of some physiological and toxicological responses of freshwater fish to acid stress, *Env. Biol. Fish.*, 5, 79–93, 1980.

Fry, F.E.J., The aquatic respiration of fish, In *The Physiology of Fishes*, Brown, M.E. (ed.), Academic Press, Inc., New York, 1957, pp. 1–63.

Gardner, W.S., Miller, W.H., III, and Imlay, M.J., Free amino acids in mantle tissues of the bivalve Amblema plicata: Possible relation to environmental stress, *Bull. Environ. Contam. Toxicol.*, 26, 157–162, 1981.

Ghafghazi, T., Ridlington, J.W., and Fowler, B.A., The effects of acute and subacute sodium arsenite administration on carbohydrate metabolism, *Toxicol. Appl. Pharmacol.*, 55, 126–130, 1980.

Giesy, J.P. and Dickson, G.W., The effect of season and location on phosphoadenylate concentrations and adenylate energy charge in two species of freshwater clams, *Oecologia* (*Berl.*), 49, 1–7, 1981.

Giesy, J.P., Denzer, S.R., Duke, C.S., and Dickson, G.W., Phosphoadenylate concentrations and energy charge in two freshwater crustaceans: Responses to physical and chemical stressors, *Verh. Internat. Verein. Limnol.*, 21, 205–220, 1981.

Grosell, M., McDonald, M.D., Wood, C.M., and Walsh, P.J., Effects of prolonged copper exposure in the marine gulf toadfish (*Opsanus beta*). I. Hydromineral balance and plasma nitrogenous waste products, *Aquat. Toxicol.*, 249–262, 2004a.

Grosell, M., McDonald, M.D., Walsh, P.J., and Wood, C.M., Effects of prolonged copper exposure in the marine gulf toadfish (*Opsanus beta*). II. Copper accumulation, drinking rate and Na^+/K^+-ATPase activity in osmoregulatory tissues, *Aquat. Toxicol.*, 68, 263–275, 2004b.

Henry, M.G. and Atchison, G.J., Metal effects on fish behavior—advances in determining the ecological significance of responses, In *Metal Ecotoxicology. Concepts and Applications*, Newman, M.C. and McIntosh, A.W. (eds.), CRC Press/Lewis Publishers, Boca Raton, FL, 1991, pp. 131–143.

Hiltibran, R.C., Effects of cadmium, zinc, manganese, and calcium on oxygen and phosphate metabolism of bluegill liver mitochondria, *J. WPCF*, 43, 818–823, 1971.

Hiltibran, R.C., Effects of insecticides on the metal-activated hydrolysis of adenosine triphosphate by bluegill liver mitochondria, *Arch. Environ. Contam. Toxicol.*, 11, 709–717, 1982.

Horn, M.H., Feeding and digestion, In *The Physiology of Fishes*, 2nd ed., Evans, D.H. (ed.), CRC Press, Boca Raton, FL, 1988, pp. 43–63.

Irving, E.C., Baird, D.J., and Culp, J.M., Ecotoxicological responses of the mayfly *Baetis tricaudatus* to dietary and waterborne cadmium: Implications for toxicity testing, *Environ. Toxicol. Chem.*, 22, 1058–1064, 2003.

Janczur, M., Kozlowski, J., and Laskowski, R., Optimal allocation, life history and heavy metal accumulation: a dynamic programming model, In, demography in Ecotoxicology, Kemmenga, J., and Laskowski, R. (eds.), John Wiley & Sons, Inc., New York, 2000, pp. 179–197.

Ivanovici, A.M., The adenylate energy charge in the estuarine mollusc, *Pyrazus ebeninus*. Laboratory studies of responses to salinity and temperature, *Comp. Biochem. Physiol.*, 66A, 43–55, 1980.

Karnaky, K.J., Jr., Osmotic and ionic regulation, In *The Physiology of Fishes*, 2nd ed., Evans, D.H. (ed.), CRC Press, Boca Raton, FL, 1988, pp. 157–176.

Knoph, M.B. and Olsen, Y.A., Subacute toxicity of ammonia to Atlantic salmon (*Salmo salar* L.) in seawater: Effects on water and salt balance, plasma cortisol and plasma ammonia levels. *Aquat. Toxicol.*, 30, 295–310, 1994.

Knops, M., Altenburger, R., and Segner, H., Alterations of physiological energetics, growth and reproduction of *Daphnia magna* under toxicant stress, *Aquat. Toxicol.*, 53, 79–90, 2001.

Koehn, R.K. and Bayne, B.L., Towards a physiological and genetical understanding of the energetics of the stress response. *Biol. J. Linnean Soc.*, 37, 157–171, 1989.

Koziol, M.J. and Jordon, C.F., Changes in carbohydrate levels in red kidney bean (*Phaseolus vulgaris* L.) exposed to sulfur dioxide. *J. Exp. Bot.*, 29, 1037–1043, 1978.

Kumar, A. and Chapman, J.C., Profenofos toxicity to the eastern rainbow fish (*Melanotaenia dubolayi*), *Environ. Toxicol. Chem.*, 17, 1799–1806, 1998.

Lacroix, A. and Hontela, A., A comparative assessment of the adrenotoxic effects of cadmium in two teleost species, rainbow trout, *Oncorhynchus mykiss*, and yellow perch, *Perca flavscens*, *Aquat. Toxicol.*, 67, 13–21, 2004.

Laurén, D.J. and McDonald, D.G., Effects of copper on branchial ionoregulation in the rainbow trout, *Salmo gairdneri* Richardson, *J. Comp. Physiol. B*, 155, 635–644, 1985.

Levesque, H.M., Moon, T.W., Campbell, P.G.C., and Hontela, A., Seasonal variation on carbohydrate and lipid metabolism of yellow perch (*Perca flavescens*) chronically exposed to metals in the field, *Aquat. Toxicol.*, 60, 257–267, 2002.

Mayne, G.J., Bishop, C.A., Martin, P.A., Boermans, H.J., and Hunter, B., Thyroid function in nestling tree swallows and Eastern bluebirds exposed to non-persistent pesticides and p,p'-DDE in apple orchards of Southern Ontario, Canada, *Ecotoxicology*, 14, 381–396, 2005.

McCloskey, J.T. and Newman, M.C., Sediment preference in the Asiatic clam (*Corbicula fluminea*) and viviparid snail (*Campeloma decisum*) as a response to low-level metal and metalloid contamination, *Arch. Environ. Contam. Toxicol.*, 28, 195–202, 1995.

Michaels, R.A., Rowland, R.G., and Wurster, C.F., Polychlorinated biphenyls (PCB) inhibit photosynthesis per cell in the marine diatom, *Thalassiosira pseudonana*, *Environ. Pollut. Ser. A*, 27, 9–14, 1982.

Newman, M.C. and Unger, M.A., *Fundamentals of Ecotoxicology*, CRC Press/Lewis, Boca Raton, FL, 2003.

Perez, M.H. and Wallace, W.G., Differences in prey capture in grass shrimp, *Palaemonetes pugio*, collected along an environmental impact gradient, *Arch. Environ. Contam. Toxicol.*, 46, 81–89, 2004.

Powell, M.D. and Perry, S.F., Acid–base and ionic fluxes in rainbow trout (*Oncorhynchus mykiss*) during exposure to chloramine-T, *Aquat. Toxicol.*, 43, 13–24, 1998.

Rauser, W.E. and Dumbroff, E.B., Effects of excess cobalt, nickel and zinc on the water relations of *Phaseolus vulgaris*, *Environ. Exp. Bot.*, 21, 249–255, 1981.

Reddy, S.L.N., Venugopal, N.B.R.K., and Ramana Rao, J.V., *In vivo* effects of cadmium chloride on certain aspects of carbohydrate metabolism in the tissues of a freshwater field crab *Barytelphusa guerini*, *Bull. Environ. Contam. Toxicol.*, 42, 847–853, 1989.

Ricard, A.C., Daniel, C., Anderson, P., and Hontela, A., Effects of subchronic exposure to cadmium chloride on endocrine and metabolic functions in rainbow trout, *Oncorhyncus mykiss*, *Arch. Environ. Contam. Toxicol.*, 34, 377–381, 1998.

Rogers, J.T., Richards, J.G., and Wood, C.M., Ionoregulatory disruption as the acute toxic mechanism for lead in the rainbow trout (*Oncorhynchus mykiss*), *Aquat. Toxicol.*, 64, 215–234, 2003.

Rowe, C.L., Hopkins, W.A., Zehnder, C., and Congdon, J.D., Metabolic costs incurred by crayfish (*Procambarus acutus*) in a trace element-polluted habitat: Further evidence of similar responses among diverse taxonomic groups, *Comp. Biochem. Physiol.*, 129C, 275–283, 2001.

Saglio, P. and Trijasse, S., Behavioral responses to atrazine and diuron in goldfish, *Arch. Environ. Contam. Toxicol.*, 35, 484–491, 1998.

Salánki, J., Farkas, A., Kamardina, T., and Rózsa, K.S., Molluscs in biological monitoring of water quality, *Toxicol. Lett.*, 140–141, 403–410, 2003.

Schmidt-Nielsen, K., *Animal Physiology*, 3rd ed., Prentice-Hall Inc., Englewood Cliffs, NJ, 1970.

Schmidt-Nielsen, K., *Scaling. Why Is Animal Size So Important?* Cambridge University Press, Cambridge, 1984.

Sherr, C.A. and Armitage, K.B., Preliminary studies of the effects of dichromate ion on survival and oxygen consumption of *Daphnia pulex* (L.), *Crustaceana*, 25, 51–69, 1971.

Sloman, K.A., Scott, G.R., Diao, Z., Rouleau, C., Wood, C.M., and McDonald, D.G., Cadmium affects the social behavior of rainbow trout, *Oncorhynchus mykiss*, *Aquat. Toxicol.*, 65, 171–185, 2003.

Stearns, S.C., *The Evolution of Life Histories*, Oxford University Press, Oxford, 1992.

Soini, J. and Rantamäki, P., Free amino acid pattern in blue mussel (*Mytilus edulis*) exposed to crude oil, *Bull. Environ. Comtam. Toxicol.*, 35, 810–815, 1985.

Teather, K., Jardine, C., and Gormley, K., Behavioral and sex ratio modification of Japanese medaka (*Oryzias latipes*) in response to environmentally relevant mixtures of three pesticides, *Environ. Toxicol.*, 20, 110–117, 2005.

Thompson, K.W., Hendricks, A.C., Nunn, G.L., and Cairns, J., Jr., Ventilatory responses of bluegill sunfish to sublethal fluctuating exposures to heavy metals (Zn^{++} and Cu^{++}), *Water Res. Bull.*, 19, 719–727, 1983.

Torreblanca, A., Del Ramo, J., and Diaz-Matans, J., Gill ATPase activity *in Procambarus clarkii* as an indicator of heavy metal pollution, *Bull. Environ. Contam. Toxicol.*, 42, 829–834, 1989.

Van Heerden, D., Vosloo, A., and Nikinmaa, M., Effects of short-term copper exposure on gill structure, metallothionein and hypoxia-inducible factor-1α (HIF-1α) levels in rainbow trout (*Oncorhynchus mykiss*), *Aquat. Toxicol.*, 69, 271–280, 2004.

Veldhuuizen-Tsoerkan, M.B., Holwerda, D.A., and Zandee, D.I., Anoxic survival and metabolic parameters as stress indices in sea mussels exposed to cadmium or polychlorinated biphenyls, *Arch. Environ. Contam. Toxicol.*, 20, 259–265, 1991.

Vernberg, W.B. and Vernberg, F.J., *Environmental Physiology of Marine Animals*, Spring-Verlag, New York, 1972.

Weis, J.S., Samson, J., Zhou, T., Skurnick, J., and Weis, P., Prey capture ability of mummichogs (*Fundulus heteroclitus*) as a behavioral biomarker for contaminants in estuarine systems, *Can. J. Fish. Aquat. Sci.*, 58, 1442–1452, 2001.

Weis, J.S., Smith, G., and Santiago-Bass, C., Predator/prey interactions: A link between the individual level and both higher and lower level effects of toxicants in aquatic ecosystems, *J. Aquat. Ecosys. Stress Recovery*, 7, 145–153, 2000.

Wood, C.M., Flux measurements as indices of H^+ and metal effects on freshwater fish, *Aquat. Toxicol.*, 22, 239–264, 1992.

Wu, B.L., Thurman, D.A., and Bradshaw, A.D., The uptake of copper and its effects upon repsiratory processes of roots of copper-tolerant and non-tolerant clones of *Agrostis stolonifera, New Phytol.*, 75. 225–229, 1975.

Xie, L. and Klerks, P.L., Fitness cost of resistance to cadmium in the least killifish (*Heterandria formosa*), *Environ. Toxicol. Chem.*, 23, 1499–1503, 2004.

Zhou, T., John-Alder, H.B., Weis, P., and Weis, J.S., Thyroidal status of mummichog (*Fundulus heteroclitus*) from a polluted versus a reference habitat, *Environ. Toxicol. Chem.*, 18, 2817–2823, 1999.

7 Bioaccumulation

Bioavailability is a widely accepted concept based on the implicit knowledge that before an organism may accumulate or show a biological response to a chemical, that element or compound must be available to the organism. While the concept of bioavailability is widely accepted, the processes that control it are poorly understood.

(Benson et al. 1994)

7.1 OVERVIEW

With the exception of radionuclides, a toxicant must first come into contact with the initial site of action before an effect can manifest. This might, in some cases, involve a straightforward contact with a biological surface where a localized effect occurs. Even in this case, the toxicant will penetrate to some extent into cells. In many other cases, the toxicant moves from the surface of first contact into the organism where it interacts elsewhere with the site(s) of action. Once within the organism, many processes modify, redistribute, or remove the toxicant. Bioaccumulation is the net result of these uptake, transformation, translocation, and elimination mechanisms. Toxicant qualities influencing bioaccumulation-associated processes will be described here.

7.2 UPTAKE

7.2.1 CELLULAR MECHANISMS

A toxicant present in an external medium or one already inside the organism can come in contact with and then be moved across a cell membrane. External media might be gaseous (e.g., inhaled air), liquid (e.g., inhaled or imbibed water), or solid (e.g., ingested food). A toxicant already in the organism, perhaps moving within the circulatory system, can be present in its original form, complexed, or transformed to some metabolite(s) or conjugate(s). Regardless, the same general mechanisms are involved in the toxicant transport into and out of cells.

Before discussing cellular transport mechanisms, it is important to mention that some substances might not pass into cells, but instead enter by moving through the tight junctions between cells. Solvent drag of a substance into a fish through gill cell junctions (see Evans et al. 1999) is one example of such movement by this paracellular pathway. Gill cell junction permeability can increase substantially under conditions that interfere with calcium's normal role of maintaining tight seals between adjacent cells (e.g., low water pH) (Booth et al. 1988, Cuthbert and Maetz 1972, Newman and Jagoe 1994). Another example of paracellular transport occurs in the lugworm, *Arenicola marina*, whose survival depends on coping with poisonous sulfide in its environment. Sulfide emanating from anoxic sediments enters the worm but is quickly oxidized to the less toxic thiosulfate. Thiosulfate then diffuses out through the body wall of the lugworm primarily via cell junctions (Hauschild et al. 1999).

Correctly so, most treatments of toxicant movement into or out of cells focus on passage across cell membranes. The cell membrane is a phospholipid bilayer with proteins interspersed between and extending into the phospholipid layers. According to Singer and Nicholson's classic fluid mosaic membrane model (Singer and Nicholson 1972), lipids float within one layer of the membrane to associate in different ways but do not often move from one phospholipid layer to the other (Simkiss 1996). Not only do consequent differences exist in lipid characteristics between the outer and inner lipid layers, patches or macrodomains of cell surface membrane lipids and proteins form within a layer (Gheber and Edidin 1999).

The lateral transport of membrane components is influenced by several factors that impart a distinctively dynamic and heterogeneous nature to membranes (Jacobson et al. 1995). A membrane component can have its movement restricted by another cluster of membrane components, or it can move about in a random or directed manner. Connection of a membrane component to the cell cytoskeleton often directs movement. As described by the raft hypothesis, resulting lateral heterogeneities in cell surface components produce functional heterogeneity on the membrane surface (Edidin 2001, Mayor and Rao 2004).

> The plasma membrane presents an intriguing mix of dynamic activities in which components may randomly diffuse, be confined transiently to small domains, or experience highly directed movements.
>
> **(Jacobson et al. 1995)**

Given the complex and dynamic nature of cell membranes, it should be no surprise that chemicals pass to and fro across cell membranes in many ways (Figure 7.1). Some nonionized, lipid-soluble chemicals diffuse passively through the lipid bilayer. This diffusion that Simkiss (1996) calls the "lipid route" forms the basis for the pH-Partition Theory discussed in Section 7.2.3. Other chemicals move by passive diffusion (filtration) through ion channels or pores. Many hydrophilic molecules smaller than 100 Da enter cells this way, although exceptions include ions with large hydration spheres that restrict movement through channels (Timbrell 2000).[1] Gated ion channels also allow passive diffusion, but how they function depends on chemical and electrical conditions. Diffusion facilitated by a carrier molecule is faster than predicted for simple diffusion, although it also does not require the expenditure of energy. Diffusion can involve two ions synchronously exchanged between

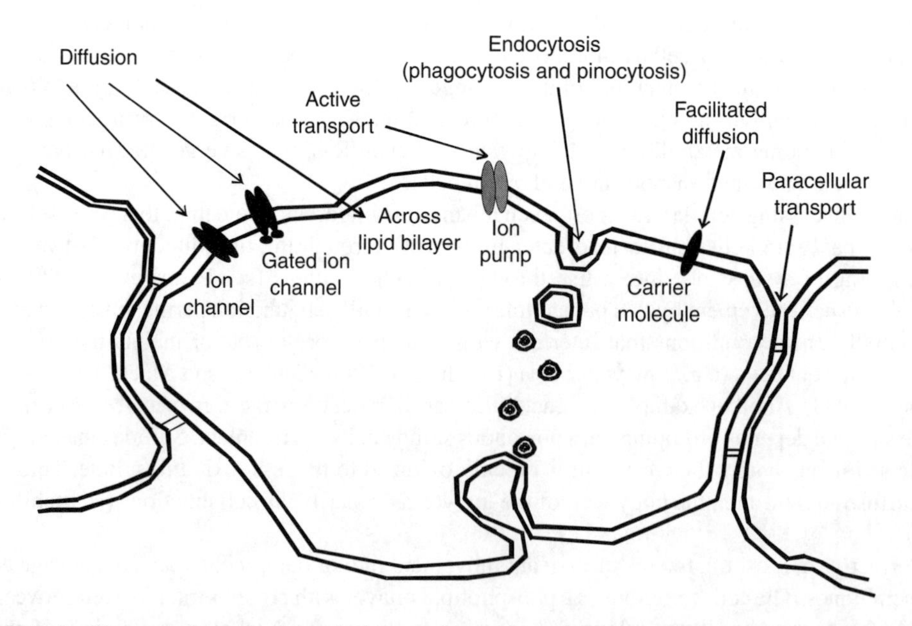

FIGURE 7.1 Diagram of routes of chemical uptake into cells and the paracellular route.

[1] For simple metal ions, hydration sphere size generally increases with increasing ion charge or decreasing ion size.

the outside and inside of the membrane. As with the ATPase-mediated exchange discussed in the last chapter, the energy-requiring active transport of a chemical can occur up an electrochemical gradient. Finally, endocytosis can be an important avenue for moving chemicals into and out of the cell. As an example, iron can be assimilated by binding to a membrane-associated transferrin protein, with subsequent movement of the iron–transferrin complex to a membrane "coated pit" region, and incorporation into a vesicle that then passes into the cell (Simkiss 1996).

At this point, some specific examples of cellular transport mechanisms might foster an appreciation for the diversity of avenues by which toxicants are taken up, transported within, and eliminated from cells. A few important ones are provided here with brief mention of detoxification mechanisms that will be described again later in this chapter. In reading these examples, it is important to understand that mechanisms often work in concert to facilitate uptake, transformation, and elimination. This point of the body's simultaneous use of several mechanisms to regulate internal chemical concentrations can be illustrated with a straightforward example peculiar to elasmobranchs that, unlike other fishes, retain urea for osmoregulatory purposes. Urea accumulates in the elasmobranch, *Squalus acanthias*, as a result of two important mechanisms (Fines et al. 2001). First, elasmobranch gill cells have a protein transport mechanism that moves urea in a direction (inward) opposite to that seen in gills of most other fishes. Second, the gill phospholipid bilayer is modified so that it is less soluble to urea than those of other fishes. A toxicological example of mechanisms working in concert is the movement and resulting toxicity of arsenic species. Low phosphate concentrations result in accelerated As(V) uptake involving a shared energy-requiring uptake mechanism. As(V) is reduced to As(III) by arsenate reductase inside the cell and then removed from the cell by an ATPase-dependent pump (Huang and Lee 1996). Removal would be very much slower if As(V) was not first converted to As(III). It should be clear from these examples that understanding of cellular movement of chemicals requires consideration of interactions among mechanisms and processes.

Active transport is important for many different toxicants or their metabolites. A relevant example is gut epithelial cell metabolism and membrane movement of benzo[a]pyrene. After entering a gut epithelial cell, benzo[a]pyrene is metabolized by Phase I reactions (CYP1A1 and CYP1B1) and then conjugated. The resulting benzo[a]pyrene-3-sulfate and benzo[a]pyrene-1-sulfate are actively transported back toward the gut lumen by ATP-binding cassette (ABC) transport proteins (Buesen et al. 2002).[2] Another example of active transport is the previously mentioned transport of metals by the gill ATPases. The basolateral membrane of gill cells, which have high Na^+/K^+-ATPase activity (Evans et al. 1999), is the site of ATPase-mediated transport of many metals such as silver in rainbow trout (*Oncorhynchus mykiss*) gills (Bury et al. 1999).

Many toxicants and natural metabolic products are organic anions. Important examples include chlorinated haloalkenes such as the solvent trichloroethylene and chlorinated phenoxyacetic acid herbicides such as 2,4-D(2,4-dichlorophenoxyacetic acid) (Sweet 2005). Some mercury complexes and many conjugated toxicants or their metabolites are also organic anions. Therefore, it should be no surprise that a family of organic anion transporters (OATs) exists to move organic anions into and out of cells. As one important example, OATs are involved in active transport across the renal proximal tubules (Sweet 2005). Ionic mercury conjugated with cysteine, *N*-acetylcysteine, and glutathione is transported by OATs of rabbit renal proximal tubule cells (Zalups and Barfuss 2002). Reduced glutathione that is involved in detoxification of some poisons and in combating oxidative stress can also be moved across cell membranes via OATs. Toxicants present as organic anions are subject to renal elimination via this mechanism, but kidney damage can occur if a toxicant was accumulated by OATs to very high concentrations in the associated cells.

---

[2] The ATP-binding ABC transporters are members of a very large family of membrane transporters that move a diversity of chemicals including phospholipids, peptides, steroids, polysaccharides, amino acids, nucleotides, organic anions, drugs, toxicants, xenobiotics, and their conjugates (Hoffmann and Kroemer 2004). Most relevant to this discussion, they pump toxicants from cells.

Box 7.1 Cadmium and Cells

Epithelial cell transport of cadmium is a good example of a toxicant movement requiring a range of mechanisms (Zalups and Ahmad 2003). Cadmium can compete for transport sites associated with movement of essential elements (calcium, iron, or zinc). For example, epithelial cells of the intestine have a zinc transporter system through which cadmium also enters cells. Cadmium also enters cultured intestinal epithelial cells via a calcium-binding protein (Pigman et al. 1997) and the crustacean gill by diffusion facilitated by a calcium-binding protein (Rainbow and Black 2005). This general type of entry route is categorized as ionic mimicry or ionic homology. Characteristic of the second general route of entry, cadmium can conjugate with thiol-containing compounds such as glutathione or cysteine, and the resulting conjugates pass through membranes by facilitated diffusion mechanisms designed for organic anion transport (Pigman et al. 1997, Zalups and Almad 2003). Aduayom et al. (2003) and Pigman et al. (1997) found evidence of such movement in cultured intestinal epithelial cells. Zalups and Ahmad (2003) refer to this second general route as molecular mimicry or molecular homology. As a third mechanism, cadmium associated with proteins such as metallothionein or albumin can enter the cell by endocytosis. Finally, Pigman et al. (1997) suggest that cadmium might also move into intestinal epithelial cells by passive diffusion.

These mechanisms of cellular transport of cadmium manifest to differing degrees throughout the body's tissues, resulting in differential uptake, distribution among the organs, localized effects, and elimination. Figure 7.2 illustrates broadly how these processes result in the complex cadmium transformations and dynamics in the mammalian body.

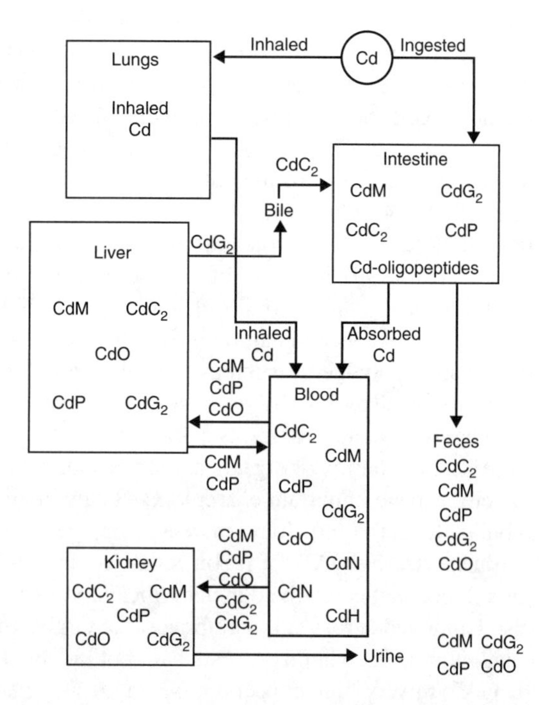

FIGURE 7.2 A general illustration of the consequences of cadmium uptake, conversion, and elimination from cells of the various organs of the mammalian body. Shown in the figure are cadmium bound to protein (CdP), metallothionein (CdM), cysteine (CdC), homocysteine (CdH), gluatathione (CdG), *N*-acetylcysteine (CdN), and other thiol-containing compounds (CdO). (Derived from Figure 1 in Zalups and Ahmad (2003) to which the reader is referred for a more comprehensive description.)

Endocytosis contributes to uptake and elimination from cells. Kupffer cells present in the mammalian liver sinusoids phagocytize some toxicants present in the blood (Timbrell 2000). Small particles of iron oxide or iron saccharate are phagocytized by cells associated with oyster gills (Galtsoff 1964). It is also well established that the cells of the molluscan digestive gland

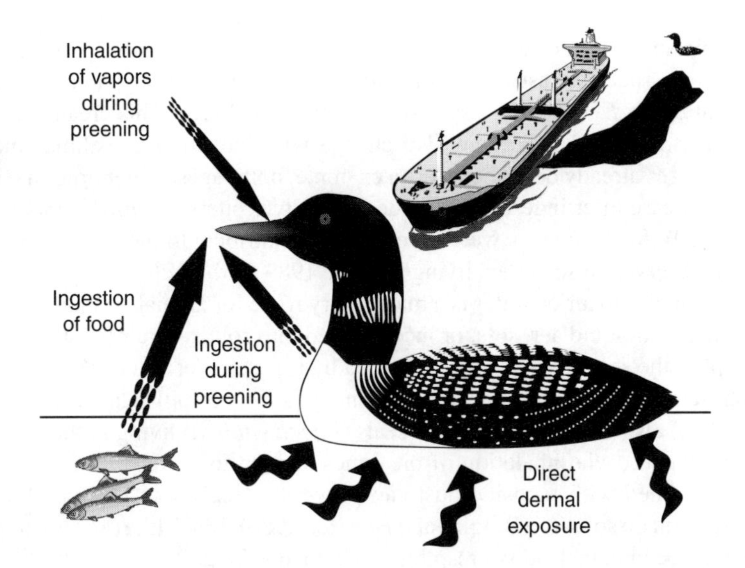

FIGURE 7.3 Three routes of exposure illustrated with a common loon coming into contact with spilt oil. The three routes include ingestion, direct dermal exposure, and inhalation. Ingestion involves consumption of tainted food and incidental ingestion during preening of oiled feathers. Dermal exposure can involve contact with dissolved oil components such as polycyclic aromatic hydrocarbons and physical contact with the floating oil. Inhalation of volatile components can occur especially during preening of oiled feathers.

(hepatopancreas) phagocytize particles (Purchon 1968) including those containing contaminants. Cells of this gland also eliminate material by expulsion from vacuoles into ducts leading to the gut. The toxicant-containing materials are then removed from the individual after incorporation into feces. Molluscs and other invertebrates sequester toxic metals in intracellular granules and can eliminate the sequestered metals by cellular release of metal-rich granules into the ducts of the digestive gland (Lowe and Moore 1979, Mason and Nott 1981, Simkiss 1981, Wallace et al. 2003).

7.2.2 Routes of Entry into Organisms

Considering contaminant movement at a higher level of organization, contaminants can enter an organism through the gut, respiratory surfaces, and dermis (Figure 7.3). These are the classic entry routes discussed thoroughly in mammalian toxicology and quantified carefully during risk assessment activities. Some aspects of toxicant movements in the associated organs have already been described briefly in Sections 5.2 (dermal), 5.3 (respiratory surfaces), and 5.5 (gut).

Oral exposure involves direct ingestion in food or imbibed water, or ingestion during grooming, preening, or pica. Some chemicals that enter initially by inhalation can also be swept back up from within the lungs, swallowed, and gain entry into the digestive tract.

The ingestion route becomes more complicated for nonhuman species. Some invertebrates possess elaborate feeding structures involved in respiration (e.g., lugworms) or locomotion (e.g., copepods). Contaminant entry to such individuals is influenced by the demands of respiration and movement in addition to feeding.

Uptake after ingestion can change if a contaminant or co-contaminant damages gastrointestinal tissues. This phenomenon of malabsorption is well studied in pharmacology. For instance, Gibaldi (1991) describes how the cancer treatment drug 5-fluorouracil damages the intestinal epithelium, allowing movement of large polar molecules that otherwise could not pass through the gut wall. Similarly, ethanol damage increases movement of toxic chemicals as large as 5000 Da across the guts of alcoholics (Bjarnason et al. 1984, Gibaldi 1991). Keshavarzian et al. (1999) recently proposed that ethanol-induced malabsorption of endotoxins is responsible for liver cirrhosis of many alcoholics. Relative to nonhuman species, gastric erosion and hemorrhaging were apparent in oiled sea otters (*Enhydra lutris*) after the *Exxon Valdez* spill (Lipscomb et al. 1996). This created a condition that would foster malabsorption. Uptake can also change with induction of cellular mechanisms or physiological processes already described. As an example, both digesta retention time in the gut and lipid absorption decrease after ingestion of crude oil by river otters (*Lontra canadensis*) (Ormseth and Ben-David 2000). Such changes were speculated to contribute to the drop in body weight and general condition observed in sea otters living near the 1989 *Exxon Valdez* spill.

Respiratory uptake from air or water is a major entry route for animals. Inhaled toxicants can be gaseous, associated with liquid aerosols, or incorporated into solids for air breathing species or life stages. As examples, the sea otters (*E. lutris*) studied by Lipscomb et al. (1996) showed pulmonary interstitial emphysema, suggesting lung contact via inhalation of oil spill-related volatile compounds. Peterson et al. (2003) also indicate that harbor seals (*Phoca vitulina*) living in the area of the *Exxon Valdez* spill were exposed via inhalation of oil fumes enough to produce brain lesions and other damage. Lead associated with roadside dust can penetrate deeply into terminal bronchioles and alveoli with subsequent dissolution (Biggins and Harrison 1980). Black kite (*Milvus migrans*) nesting near an incinerator accumulate lead via respiration (Blanco et al. 2003).

Water-breathing species can also be exposed to toxicants in gaseous, liquid (dissolved or micelles), and solid phases. The importance of the respiratory route varies for aquatic species that can have different respiratory strategies as was clearly illustrated by Buchwalter et al. (2003) with diverse aquatic insect species exposed to the organophosphate insecticide, chlorpyrifos. Most attention is focused on uptake from water and gas phases; however, as we saw earlier, relative to metal uptake from particles on oyster gills (Section 7.2.1), particulate-associated toxicants on gills can be taken up under certain conditions. For example, lead adsorbed to gibbsite can gain entry into goldfish (*Carassius auratus*) after gibbsite particles adhere to gill surfaces and the associated lead desorbs (Tao et al. 1999). Obviously, damage to respiratory surfaces can modify uptake in ways similar to that described for the ingestion route.

Considering all the species living within ecosystems, it would be a mistake to focus only on these routes of entry that emerge out of classic mammalian toxicology—even for the animal kingdom. The paradoxically high arsenic concentration in the giant clam, *Tridacna maxima*, tissues is a good illustration of this point. The Great Barrier Reef giant clams and their symbiotic zooxanthellae live in phosphorus-deficient waters. The zooxanthellae within the clam tissues actively take up and metabolize arsenate in their attempt to extract as much of the meager amount of phosphate as possible from surrounding waters. Once taken up, the arsenic is converted to various organic forms and accumulates to extremely high concentrations in various tissues of the giant clam and other invertebrates containing symbionts (Benson and Summons 1981). This symbiont exposure route does not fit neatly into the context of the three classic routes of exposure. Nor does an endoparasite's exposure to a toxicant present in a host's tissues and body fluids.

Focusing for a moment on plants, stomatal entry of gaseous air pollutants such as sulfur dioxide (Kimmerer and Kozlowski 1981), or uptake via aerial or terrestrial roots, e.g., arsenic uptake from soils (Wauchope 1983), might be important to consider. In urban areas, particulate-associated contaminants contact plants by simply settling onto their exposed surfaces. Treatment of these routes of entry would not necessarily involve minor changes to the methods applied for respiratory (stomata?), ingestion (roots?), or dermal (plant surfaces?) routes of entry for animals. Microscopic organisms can also require a different vantage for assessing toxicant entry. As an important example, treatment of toxicant entry into unicellular algae might be most effective if the context developed above for cellular movement of toxicants was adopted instead (e.g., Crist et al. 1992, Klimmek et al. 2001, Morris et al. 1984). As another, but more extreme, example involving arsenic, some microbes use arsenic oxyanions to generate energy (Oremland and Stolz 2003) and, for this reason, would require a very different vantage for discussions of exposure routes and associated uptake calculations.

7.2.3 Factors Modifying Uptake

What general rules exist regarding the uptake of contaminants by these routes? Some rules of thumb emerge despite the diversity of relevant species, toxicants, and media. Most are based on the tendency of particular toxicants to engage in specific reactions, including transport, and to preferentially associate with a certain phase. Several key themes are sketched out here.

Tendencies to partition between aqueous and lipid phases are often used to predict uptake and bioaccumulation of nonpolar organic chemicals. This point can be illustrated for fish gill uptake of nonpolar compounds differing in lipophilicity (as measured by the logarithm of the octanol:water partition coefficient, log K_{OW}) (Figure 7.4). Connell and Hawker (1988) speculate that uptake increases with log K_{OW} because membrane permeation by the chemical increases, but only to a point. At a certain point, the large molecular size of the increasingly lipophilic chemicals begins to impede their diffusion in aqueous phases of the fish: these molecules have such low diffusion coefficients that the curve plateaus above approximately log K_{OW} of 6 for many nonionic chemicals (Connell 1990). This decreasing of uptake rates results from steric hindrance as contaminant molecules attempt to pass through the cavities between membrane molecules (Connell 1990).

The influence of lipophilicity depends on the route of entry. Lipophilicity is less important for chemical uptake in lungs. Boethling and MacKay (2000) generalize, "substances with solubility in water equal to or greater than their solubility in lipids are likely to be absorbed from the lung. Polar substances are generally absorbed better from the lung than nonpolar substances due to greater

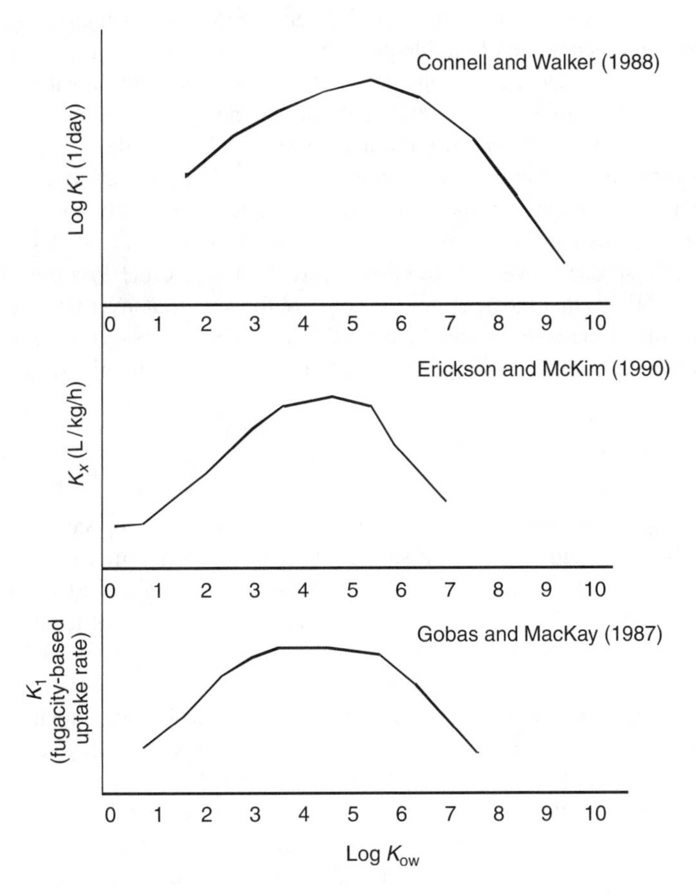

FIGURE 7.4 Gill uptake (K_1, K_x) versus log K_{OW}. (Panels derived from Figure 3 of Connell and Hawker (1988), Figure 1 of Erickson and McKim (1990), and Figure 1 of Gobas and MacKay (1987).)

water solubility." However, increasing lipophilicity of ingested chemicals does tend to increase uptake from food (Gibaldi 1991) but, similar to the limits discussed for neutral chemical uptake by gills, most chemicals with log K_{ow} values greater than roughly 5 display diminished uptake after ingestion because they are sparingly soluble in gastric juices (Boethling and MacKay 2000). For dermal exposure of mammals, compounds with higher lipid solubility tend to have greater rates of uptake than less lipid-soluble compounds.

The pH Partition Theory (Hogben et al. 1959, Shore et al. 1957) is a central one for dealing with ionizable toxicants. This theory, often invoked when dealing with absorption of ingested chemicals, is based on the simplifying (and sometimes insufficient) assumption that the gut can be envisioned as a simple lipid barrier. Nonionized forms of acidic or basic toxicants pass through this lipid barrier much more readily than ionized forms. This being the case, one can insert a compound's pK_a and the pH of the gut region where uptake is to occur into the appropriate Henderson–Hasselbalch equation (also see Equation 3.5 and associated discussion) to predict how readily the chemical might cross the lipid barrier. Henderson–Hasselbalch equations for weak (monobasic) acids (Equation 7.1) or (monoacidic) bases (Equation 7.2) are the following:

$$f_u = \frac{1}{1 + 10^{\mathrm{pH} - \mathrm{p}K_a}} \tag{7.1}$$

$$f_u = \frac{1}{1 + 10^{\mathrm{p}K_a - \mathrm{pH}}} \tag{7.2}$$

where f_u is the proportion of the toxicant remaining unionized. The calculated degree of ionization of a chemical and this theory are often adequate for approximating weak acid or base uptake by passive diffusion but, as evidenced by our previous discussions, many uptake mechanisms exist beyond simple diffusion of an uncharged chemical through the lipid route. Many of the mechanisms discussed above can facilitate uptake of an ionized compound. Regardless, based on pH Partition Theory, one can generally predict for compounds ingested by humans that weak organic bases tend to be taken up in the intestine and weak organic acids tend to be taken up in the stomach as well as the intestine (due to its high surface area and blood flow) (Abou-Donia et al. 2002).

Gibaldi (1991) discusses a very instructive elaboration of the pH Partition Theory made by Hogerle and Winne (1978) that incorporates the depth of the unstirred layer immediately adjacent to the mucosal cells, pH of the media immediately adjacent to the mucosal cell membrane, amount of surface area available for uptake, and uptake of both the ionic and nonionic forms of the chemical:

$$\text{Absorption rate} = \frac{CA}{(T/D) + \{1/P_u[f_u + f_i(P_i/P_u)]\}} \tag{7.3}$$

where C = toxicant concentration, A = area over which absorption is occurring, T = unstirred layer thickness, D = the chemical's diffusion coefficient, f_u = fraction of chemical unionized, f_i = fraction of chemical ionized, P_u = permeability of the unionized chemical, and P_i = permeability of the ionized chemical. The pH immediately adjacent to the mucosal cell membrane is used instead of the general gut lumen pH in this model. Both the unionized and ionized forms are assumed to be taken up, albeit, at distinct rates.

Another set of general theories govern our current predictions of metal uptake. A basic premise relative to metal toxicity is that the metal must first be in solution before being capable of interacting with a site of action. This is a sound premise if applied insightfully, not dogmatically. Where exactly the metal must be in solution is crucial to consider. A particulate-associated metal that is taken into the cell by phagocytosis can dissolve within the cell and cause an adverse effect. A lead halide particle inhaled deeply can release lead upon contact with moist respiratory surfaces. Finally, dissolved aluminum that encounters elevated pH conditions at the gill surface microlayer and precipitates will become associated with and cause harm due to its presence on the gill as a solid (colloid).

Beyond this point that the dissolved metal is the most bioavailable form, Mathews (1904) proposed the ionic hypothesis more than a century ago. The ionic form of any dissolved metal is the most active form relative to biological uptake or effect. Beginning in the late 1930s, this context was expanded to correlate the relative toxicities of mono-, di-, and trivalent metal ions to their respective abilities to form complexes with biomolecules (e.g., Biesinger et al. 1972, Binet 1940, Fisher 1986, Jones 1939, 1940, Jones and Vaughn 1978, Kaiser 1980, Loeb 1940, McGuigan 1954, Newman and McCloskey 1996, Newman et al. 1998, Williams and Turner 1981).

A series of ancillary theories have recently emerged around these well-established theories. The ionic hypothesis was augmented by what is now called the free ion activity model (FIAM): the free metal ion is the most important dissolved species relative to determining dissolved metal uptake and effect (Campbell and Tessier 1996). If applied with insight, the FIAM is an excellent rule of thumb; however, there are cases in which it should not be expected to apply. For example, Simkiss (1983, 1996) indicates that the neutral chloro complex of mercury, $HgCl_2^0$, is lipophilic and potentially available for uptake via the lipid route. Charged uranium complexes as well as the free uranium ion have significant bioactivity (Markich et al. 2000).

Another ancillary model goes under the name of the biotic ligand model (BLM): the bioactivity of a metal manifests if and when the amount of metal–biotic ligand complexes reaches a critical concentration (Di Toro et al. 2001, Santore et al. 2001). The BLM focuses on the activities of dissolved metal–ligand complexes and the metal–biotic ligand complexes formed at crucial sites on organism surfaces such as gill surfaces. The competition among other dissolved cations and dissolved ligands are also considered. The FIAM and BLM are often applied together to imply or predict a relationship between dissolved metal concentration (in bulk water or sediment interstitial water) and some adverse consequence. As with the FIAM, insightful use of the BLM can generate valuable explanations or predictions, but unthoughtful application can produce contradictions. As examples, the uptake rate of zinc by *Chlorella kessleri* was not directly related to the concentration of the free (aquated) ion, Zn^{2+}, leading Hassler and Wilkinson (2003) to challenge the FIAM–BLM model. The discrepancy was attributed to the synthesis of new membrane-associated zinc transporters that were involved in active transport, i.e., transport against an electrochemical gradient. Using this same algal species, Hassler et al. (2004) found that the FIAM–BLM did correctly predict lead uptake in the absence of competitors but it failed to do so in the presence of the competing ion, Ca^{2+}. Work to this point suggests that prediction from FIAM–BLM is extremely useful but must be applied with a clear understanding of important underlying processes and modifiers.

Newman et al. (1998) recently developed quantitative ion character–activity relationships (QICARs) that quantitatively predict metal activity based on Hard Soft Acid Base (HSAB) theory and FIAM–BLM. These QICARs are quantitative extensions of work by many others (e.g., Biesinger et al. 1972, Binet 1940, Fisher 1986, Jones 1939, 1940, Jones and Vaughn 1978, Kaiser 1980, Loeb 1940, McGuigan 1954, Williams and Turner 1981) showing relationships between metal bioactivity and metal–ligand binding tendencies. In some cases, such QICARs also allow bioactivity prediction for binary metal mixtures (Ownby and Newman 2003) (see Box 9.4).

Last, Di Toro et al. (1990) combined aspects of various theories (HSAB, Ionic Theory especially FIAM) to make general predictions about sediment metal availability/bioactivity. In their approach, they assumed the following: (1) the dissolved ion is the most bioactive form of a metal,[3] (2) sulfides of many metals of concern are much less soluble than iron (and manganese) sulfides found in anoxic sediments, (3) the relatively large amount of iron (and manganese) sulfides often found in anoxic sediment provides a solid-phase sink for any metal in the sediments, (4) metals associated with solid-phase sulfides are essentially unavailable relative to that dissolved in interstitial waters, and (5) metal

[3] Or, minimally, is a good indicator of bioavailable metal.

bioactivity in anoxic sediments can be predicted if one knows the amount of metal in the sediment and the amount of sulfides in the sediment available to react with and remove the metal from the interstitial waters. To this end, sediments are extracted with cold 1 N HCl, and the amounts of sulfide (acid volatile sulfides or AVS) and simultaneously extracted metal (SEM) are determined. A metal might be available to interact adversely with biota if the amount of metal in the sediment exceeds the capacity of the AVS to remove it from the interstitial waters, that is, if SEM–AVS > 0. Again, this SEM–AVS approach is very useful if used insightfully: there are cases in which considerable understanding is required to correctly interpret or predict from AVS–SEM information. For example, Lee et al. (2000) demonstrated clearly that many benthic organisms take up significant amounts of metals from solid sediment phases, contradicting a major assumption of the SEM–AVS approach.[4] Other researchers (Chen and Mayer 1999, Fan and Wang 2003) have applied biomimetic methods to provide support for the SEM–AVS approach for some metals. Chen and Mayer (1999) showed that the SEM–AVS approach gave similar predictions as their biomimetic approach except the presence of a threshold SEM–AVS suggested that other phases also contributed to availability. Fan and Wang (2003) found poor agreement between biomimetic and SEM–AVS studies of several metals.

All of these recent permutations are ancillary to the established ionic and HSAB theories. If applied thoughtfully, the FIAM, BLM, QICAR, and SEM–AVS models can provide invaluable insights. Unfortunately, their dogmatic application or rejection is the source of some confusion in the field at this time. (See Chapter 36 for further details about dogmatic rejection of emerging paradigms.)

7.3 BIOTRANSFORMATION

As described in Chapter 3, an organic toxicant can be subject to transformations that influence its retention. As a recent example, wide variation in the ability of invertebrates to metabolize PAH was found to lead to significant differences in PAH retention (Rust et al. 2004). Some might be eliminated immediately upon entry to a cell as in the case where the P-glycoprotein acts as a barrier to xenobiotic absorption (Box 3.1). If an organic toxicant gains entry into the individual, it can be eliminated in its original form by a variety of mechanisms already discussed or it can be converted to a form more amenable for elimination. It can undergo Phase I transformations and be eliminated in its new form(s). Alternatively, the Phase I metabolite(s) can be conjugated with one of a variety of endogenous compounds and then eliminated. In the example of benzo[a]pyrene metabolite conjugates given above, the metabolites are transported via the circulatory system to the gut where they are moved back into the lumen via the ABC active transport proteins. As we have already discussed, final removal of the organic chemical or its metabolites can occur via a variety of other mechanisms, for example, organic anion removal via OATs.

Inorganic toxicants can also undergo changes after entering the organism. With the simple case of sulfide in lugworms, we saw that elimination occurred with oxidation to thiosulfate and simple thiosulfate diffusion out of the worm via the paracellular route. Toxic metal transformations tend to be more involved than this. Metals can become complexed with a ligand and transported to the site of elimination in that form. Lyon et al. (1984) found that metal elimination from crayfish (*Austropotamobius pallipes*) hemolymph was linked to metal–ligand binding tendencies as predicted by HSAB theory. Metals can be taken into cell vesicles and sequestered in granules (e.g., Coombs and George 1977) (also Section 4.5.1) or cysteine-rich proteins.[5] Metal-rich granules can be emptied

[4] The complicated nature of metal uptake from sediments has stimulated the application of empirical methods for determining bioavailability. The most recent adaptation from human pharmacology is the biomimetic approach. With this approach, an aliquot of sediment, soil, or food is placed into a solution that mimics digestive juices (e.g., Leslie et al. 2002, Rodriguez and Basta 1999) or into digestive juices themselves (e.g., Mayer et al. 1996, Weston and Maruya 2002, Yan and Wang 2002). The amount of toxicant released into solution is used to predict uptake after ingestion.

[5] There is currently a regulatory movement afoot to relate toxicant effects to a critical body residue, instead of an environmental concentration. The complexity of relating a concentration in, for example, a sediment to a realized effect has prompted

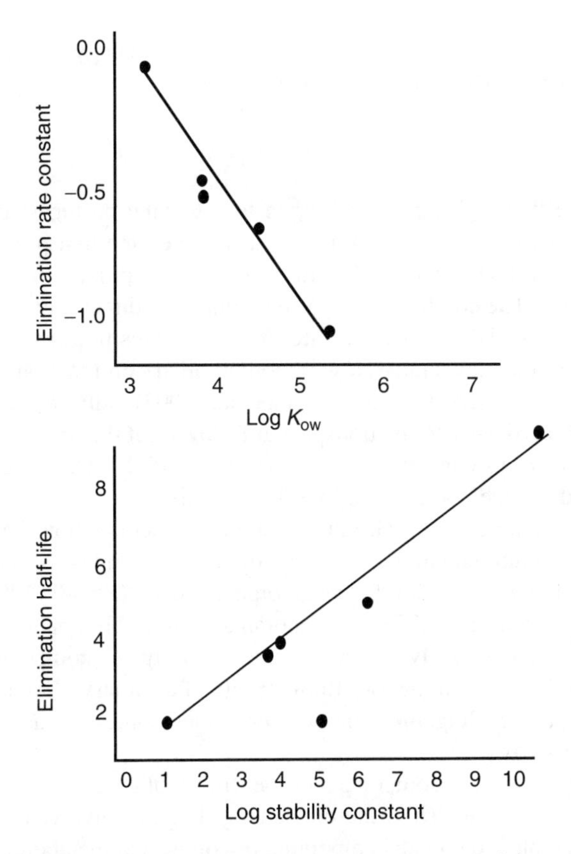

FIGURE 7.5 Rudimentary lipid solubility-based quantitative structure–activity relationships (QSARs) and ligand binding-based QICARs for toxicant elimination. The influence of chlorobenzenes lipid solubility on elimination rate constants for guppy (*Poecilia reticulata*) is clear in the upper panel. (Modified from Figure 6 of Könemann and van Leeuwen (1980).) Similar results have been derived for elimination of organophosphorous pesticides from guppies (*P. reticulata*) (De Bruijn and Hermens 1991) and chlorobiphenyls from goldfish (*Carassius auratus*) (Bruggeman et al. 1981). The change in metal half-life in crayfish hymolymph is influenced by metal–ligand binding stability as shown in the bottom panel, which uses the stability constants for metal–ethylene diamine complexes as a measure of complex stability for each metal. (Modified from Figure 3 of Lyon et al. (1984).)

into the lumen of molluscs and incorporated into feces. Protein-bound metals can be transported to organs of elimination such as the mammalian kidney.

7.4 ELIMINATION

Many of the themes discussed to this point about toxicant movement into cells and then within the whole individual pertain to elimination. As important examples, lipid solubility of many classes of nonionizing organic compounds and ligand-binding tendency of metal ions can often be correlated with rates of elimination (Figure 7.5). Instead of re-examining these molecular and cellular phenomena, the focus in this section will be elimination as facilitated by specific organs or organ systems.

---

this movement. The concept is that a specific critical body residue must be reached in order to manifest a specific adverse effect. For metals, Vijver et al. (2004) argue that metal in pools such as granules are not available to do damage and, this being the case, the total amount of metal in an organism might not be a reliable metric of the amount of metal having an effect.

Those typically identified as central to classic mammalian toxicology are discussed first and then followed by issues associated with nonmammalian species.

7.4.1 Hepatobiliary

Because the liver is the first highly perfused organ to encounter an ingested toxicant as it enters into the circulation, it can be especially prone to damage, i.e., the first pass effect (Roberts et al. 2000). Contaminants enter the liver by the hepatic artery (or lymph), becoming subject to potential hepatobiliary elimination. The compound might be eliminated directly or after metabolism in the liver. It, or its metabolite(s), is incorporated into the bile, passes into the gallbladder via the bile duct, and is released into the duodenum. How important biliary elimination is can be very species specific (Rozman and Klaassen 1996). Abou-Donia et al. (2002) indicate that passage into the bile can involve diffusion or carrier-mediated transport. It follows that the process of biliary elimination can be saturated or subject to competitive inhibition (Gibaldi 1991). For example, metals such as lead can be excreted via the mammalian liver by an active transport mechanism (Rozman and Klaassen 1996). Abou-Donia et al. (2002) also make the generalization that a molecular weight greater than 325 Da, structure containing two or more aromatic rings, or the presence of a polar group all tend to favor biliary excretion for weak organic acids. Gibaldi (1991) provides a cut-off point of molecular weight less than 300 Da for predominately renal elimination and molecular weight greater than 300 Da for predominately biliary elimination of organic compounds by rats, and about 400–500 Da for human biliary elimination. Immediately after biliary elimination or eventually in the case of the enterohepatic cycling, the compound or its metabolites are incorporated into the feces and eliminated from the body.

After release into the intestine, compounds or their metabolites can be taken up again and pass back into circulation and the liver. This enterohepatic cycling can involve the parent compound or a conjugate. A water-soluble, conjugated compound or conjugated metabolite can be broken apart, producing a less water-soluble compound that is readily taken up in the intestine again. Enterohepatic cycling can occur repeatedly, putting the liver at risk of very high exposure as the toxicant continues to come into contact with its cells during each recycling.

7.4.2 Renal

Renal elimination occurs by active transport and diffusion, and involves glomerular filtration, active tubular secretion, and passive reabsorption (Gibaldi 1991). Glomerular filtration can remove molecules as large as 60,000–70,000 Da but some are then reabsorbed (Abou-Donia et al. 2000, Rozman and Klaassen 1996). Toxicant binding with circulating plasma proteins can strongly influence their ability to participate in glomerular filtration. Tubular secretion, active transport of a compound from the blood capillary into the renal tubule, is also important (Gibaldi 1991). Anionic and cationic organic compounds are actively transported into the proximal tubules (Middendorf and Williams 2000, Rozmann and Klaassen 1996). Also occurring at this point in the process is tubular reabsorption that involves active reabsorption of chemicals such as water, salts, amino acids, and glucose. Many drugs, especially lipid-soluble compounds, are subject to passive reabsorption because of the concentration gradient created during water reabsorption (Gibaldi 1991). Passive reabsorption in the proximal tubule for weakly acid or basic compounds will be greatly influenced by pH of the fluids in the tubule (i.e., the pH Partition Theory). Elimination is favored for basic and acidic compounds under acidic and basic urine pH conditions, respectively (Abou-Donia et al. 2000).

7.4.3 Branchial

Elimination from the lungs obeys a few general rules also. Chemicals with low blood solubility and high volatility are eliminated by simple diffusion faster than those that are more soluble and less volatile (Abou-Donia et al. 2000). As an example, more than half of dosed benzene is eliminated by passive diffusion and exhalation (Trimbrell 2000). However, pulmonary elimination can be quite prolonged for some lipid-soluble compounds that are deposited in fatty tissues (Rozmann and Klaassen 1996). Toxicants associated with inhaled particles or liquid can be eliminated by the mucociliary escalator process (Dallas 2000). Mucus-entrapped toxicants are swept from the respiratory tract by the coordinated motion of a complex of cilia. The toxicant can then be ejected from the respiratory system, or swallowed and gain entry into the gastrointestinal system. Similar to malabsorption, damaged lungs can display reduced elimination efficiency. An obvious example involves damage to mucous and ciliary cells of the lungs after prolonged smoking and the consequent reduction in the lungs' ability to cleanse themselves of particulate-associated toxicants.

7.4.4 Other Elimination Mechanisms

Other elimination mechanisms can be important depending on the species and toxicant. Some toxicants can pass across the placenta and be removed in that way from the mother's body. Some contaminants become incorporated into milk and get transferred from the lactating mother to the nursing individual. Such contaminant transfer into human milk is a source of general concern (Pronczuk et al. 2002, Webster 2004). Fin whales (*Balaenoptera physalus*) appear to transfer polychlorinated biphenyl and DDT in milk (Aguilar and Borrell 1994). Egg layers such as birds and fish transfer lipophilic contaminants to eggs. Insects can eliminate metals by incorporation into the exoskeleton before molting (Lindqvist and Block 1994). Some contaminants such as metals and metalloids are lost via hair (e.g., Akagi et al. 1995) or feathers (e.g., Becker et al. 1994). Plants can eliminate contaminants by several avenues including loss in root exudates and leaf fall.

7.5 SUMMARY

Several themes emerge from this short chapter. In the next, quantitative methods for quantifying the associated processes will be described.

7.5.1 Summary of Foundation Concepts and Paradigms

- Toxicants can enter an organism via cellular uptake or by moving through cell junctions (i.e., the paracellular pathway).
- Nonionized, lipid-soluble chemicals can pass through the lipid bilayer of the cell membrane by passive diffusion (i.e., the lipid pathway). Passive diffusion of other toxicants, especially hydrophilic molecules smaller than 100 Da, can involve ion channels or pores. Some channels are gated channels.
- Toxicants can enter through cell membranes by active transport. They can also be taken in via endocytosis.
- Frequently, several mechanisms working together result in net movement of a toxicant into and out of cells.
- A family of organic anion transporters (OATs) moves organic anions across cell membranes.
- Toxicants enter the organism by three major routes: the gut, respiratory organs, and general dermis; however, other pathways can be important too.
- Damage to gastrointestinal tissues can result in malabsorption.
- Tendency to partition between aqueous and lipid phases can be used to predict bioaccumulation characteristics of nonpolar organic compounds.
- Uptake of inhaled chemicals increases with water solubility. Polar chemicals also tend to be taken up faster in the lungs than nonpolar chemicals.

- Increasing lipophilicity of ingested chemicals increases uptake from food up to a log K_{ow} of roughly 5 after which the uptake begins to decrease.
- The pH Partition Theory is based on the assumption that the gut can be envisioned as a lipid barrier. It predicts uptake based on the assumption that the nonionized form of a weakly ionizable toxicant is the form most available for movement across the lipid barrier. In some cases, the uptake of the ionized form must also be considered.
- The ionic hypothesis states that the ionic form of a metal is the most bioactive. The FIAM assumes that the most bioactive form of a dissolved metal is the free ion.
- The BLM emphasizes the crucial role of metal binding to biological ligands in order to have an effect or be taken up. It also considers the complexation competition occurring with other cations and ligands.
- On the basis of classic HSAB theory, the QICAR approach predicts metal bioactivity based on tendencies to bind to different biological ligand groups.
- The AVS–SEM method predicts metal available in sediments based on the amount of sulfides available to sequester sediment metals and reduce interstitial water metal concentrations.
- Bioaccumulation can be strongly influenced by internal transformations of the toxicant after uptake. In some cases, such as P-glycoprotein or ABC transport proteins, toxicants can be excluded upon entry into the cell membrane.
- Hepatic elimination is favored for weak organic acids with molecular weight greater than 325 Da, multiple aromatic rings, or a polar group, although the prominence of biliary elimination varies widely among species. Enterohepatic cycling occurs for some toxicants.
- Renal elimination involves active transport and diffusion. It can be influenced by urine pH.
- Branchial elimination is fastest for chemicals that are volatile but not very water soluble.

REFERENCES

Abou-Donia, M.B., Elmasry, E.M., and Abu-Qare, A.W., Metabolism and toxicokinetics of xenobiotics, In *Handbook of Toxicology*, 2nd ed., Derelanko, M.J. and Hollinger, M.A. (eds.), CRC Press, Boca Raton, FL, 2002, pp. 769–833.

Aduayom, I., Campbell, P.G.C., Denizeau, F., and Jumarie, C., Different transport mechanisms for cadmium and mercury in Caco-2 cells: Inhibition of Cd uptake by Hg without evidence for reciprocal effects, *Toxicol. Appl. Pharmacol.*, 189, 56–67, 2003.

Aguilar, A. and Borrell, A., Reproductive transfer and variation of body load of organochlorine pollutants with age in fin whales (*Balaenoptera physalus*), *Arch. Environ. Contam. Toxicol.*, 27, 546–554, 1994.

Akagi, H., Malm, O., Branches, F.J.P., Kinjo, Y., Kashima, Y., Guimaraes, J.R.D., Oliveira, R.B., et al., Human exposure to mercury due to goldmining in the Tapajos River basin, Amazon, Brazil: Speciation of mercury in human hair, blood and urine, *Water, Air, Soil Pollut.*, 80, 85–94, 1995.

Ballatori, N., Hammond, C.L., Cunningham, J.B., Krance, S.M., and Marchan, R., Molecular mechanisms of reduced glutathione transport: Role of the MRP/CFTR/ABCC and OATP/SLC21A families of membrane proteins, *Toxicol. Appl. Pharmacol.*, 204, 238–255, 2005.

Becker, P.H., Henning, D., and Furness, R.W., Differences in mercury contamination and elimination during feather development in gull and tern broods, *Arch. Environ. Contam. Toxicol.*, 27, 162–167, 1994.

Benson, A.A. and Summons, R.E., Arsenic accumulation in Great Barrier Reef invertebrates, *Science*, 211, 482–483, 1981.

Benson, W.H., Alberts, J.J., Allen, H.E., Hunt, C.D., and Newman, M.C., Bioavailability of inorganic contaminants, In *Bioavailability. Physical, Chemical, and Biological Interactions*, Hamelink, J.L., Landrum, P.F., Bergman, H.L., and Benson, W.H. (eds.), CRC Press/Lewis Publishers, Boca Raton, FL, 1994, pp. 63–71.

Biesinger, K.E. and Christensen, G.M., Effects of various metals on survival, growth, reproduction, and metabolism of *Daphnia magna*, *Can. J. Fish. Aquat. Sci.*, 29, 1691–1700, 1972.

Biggins, P.D.E. and Harrison, RM., Chemical speciation of lead compounds in street dusts, *Environ. Sci. Technol.*, 14, 336–339, 1980.

Binet, M.P., Sur la toxicité comparée des métaux alcalins et alcalino-terreux, *C.R. Acad. Paris*, 115, 251–253, 1940.

Bjarnason, I., Ward, K., and Peters, T.J., The leaky gut of alcoholism: Possible route of entry for toxic chemicals, *Lancet*, 1, 179–182, 1984.

Blanco, G., Frías, O., Jiménez, B., and Gómez, G., Factors influencing variability and potential uptake routes of heavy metals in black kites exposed to emission form a solid-waste incinerator, *Environ. Toxicol. Chem.*, 22, 2711–2718, 2003.

Boethling, R.S. and MacKay, D., *Handbook of Property Estimation Methods for Chemicals*, CRC Press/Lewis Publishers, Boca Raton, FL, 2000, p. 480.

Booth, C.E., McDonald, D.G., Simons, B.P., and Wood, C.M., Effects of aluminum and low pH on net ion fluxes and ion balance in the brook trout (*Salvelinus fontinalis*) alevins: Responses of yoke-sac and swim-up stages to water acidity, calcium, and aluminum, and recovery effects, *Can. J. Fish. Aquat. Sci.*, 45, 1563–1574, 1988.

Bruggeman, W.A., Martron, L.B.J.M., Kooiman, D., and Hutzinger, O., Accumulation and elimination kinetics of di-, tri-, and tetrachlorobiphenyls by goldfish after dietary and aqueous exposure, *Chemosphere*, 10, 811–832, 1981.

Buchwalter, D.B., Jenkins, J.J., and Curtis, L.R., Temperature influences on water permeability and chlorpyrifos uptake in aquatic insects with differing respiratory strategies, *Environ. Toxicol. Chem.*, 22, 2806–2812, 2003.

Buesen, R., Mock, M., Seidel, A., Jacob, J., and Lampen, A., Interaction between metabolism and transport of benzo[a]pyrene and its metabolites in entrocytes, *Toxicol. Appl. Pharmacol.*, 183, 168–178, 2002.

Bury, N.R., Grosell, M., Grover, A.K., and Wood, C.M., ATP-dependent silver transport across the basolateral membrane of rainbow trout gills, *Toxicol. Appl. Pharmacol.*, 159, 1–8, 1999.

Campbell, P.G.C. and Tessier, A., Ecotoxicology of metals in the aquatic environment: Geochemical aspects, In *Ecotoxicology: A Hierarchical Treatment*, Newman, M.C. and Jagoe, C.H. (eds.), CRC Press/Lewis Publishers, Boca Raton, FL, 1996, pp. 11–58.

Chen, Z. and Mayer, L.M., Assessment of sedimentary Cu availability: A comparison of biomimetic and AVS approaches, *Environ. Sci. Technol.*, 33, 650–652, 1999.

Connell, D.W., *Bioaccumulation of Xenobiotic Compounds*, CRC Press, Boca Raton, FL, 2000, p. 219.

Connell, D.W. and Hawker, D.W., Use of polynomial expressions to describe the bioconcentration of hydrophobic chemicals by fish, *Ecotoxicol. Environ. Saf.*, 16, 242–257, 1988.

Coombs, T.L. and George, S.G., Mechanisms of immobilization and detoxification of metals in marine organisms, In *Physiology and Behavior of Marine Organisms*, McLusky, D.S. and Berry, A.J. (eds.), Pergamon Press, New York, 1977, pp. 179–187.

Crist, R.H., Oberholser, K., McGarrity, J., Crist, D.R., Johnson, J.K., and Brittsan, J.M., Interaction of metals and protons with algae. 3. Marine algae, with emphasis on lead and aluminum, *Environ. Sci. Technol.*, 26, 496–502, 1992.

Cuthbert, A.W. and Maetz, J., The effects of calcium and magnesium on sodium fluxes through the gills of *Carassius auratus* L., *J. Physiol.*, 221, 633–643, 1972.

Dallas, C., Pulmonotoxicity: Toxic effects in the lung, In *Principles of Toxicology: Environmental and Industrial Applications*, 2nd ed., Williams, P.L., James, R.C., and Roberts S.M. (eds.), John Wiley & Sons, Inc., New York, 2000, pp. 169–187.

De Bruijn, J. and Hermens, J., Uptake and elimination kinetics of organophosphorous in the guppy (*Poecilia reticulata*): Correlations with the octanol/water partition coefficient, *Environ. Toxicol. Chem.*, 10, 791–804, 1991.

Di Toro, D.M., Allen, H.E., Bergman, H.L., Meyer, J.S., Paquin, P.R., and Santore, R.C., Biotic ligand model of aquatic toxicity of metals. I. Technical basis, *Environ. Toxicol. Chem.*, 20, 2383–2396, 2001.

Di Toro, D.M., Mahony, J.D., Hansen, D.J., Scott, K.J., Hicks, M.B., Mayr, S.M., and Richmond, M.S., Toxicity of cadmium in sediments: The role of acid volatile sulfide, *Environ. Toxicol. Chem.*, 9, 1487–1502, 1990.

Edidin, M., Shrinking patches and slippery rafts: Scales of domains in the plasma membrane, *Trends Cell Biol.*, 11, 492–496, 2001.

Erickson, R.J. and McKim, J.M., A simple flow-limited model for exchange of organic chemicals at fish gills, *Environ. Toxicol. Chem.*, 9, 159–165, 1990.

Evans, D.H., Piermarini, P.M., and Potts, W.T.W., Ionic transport in the fish gill epithelium, *J. Exp. Zool.*, 283, 641–652, 1999.

Fan, W. and Wang, W.-X., Extraction of spiked metals from contaminated coastal sediments: A comparison of different methods, *Environ. Toxicol. Chem.*, 22, 2659–2666, 2003.

Fines, G.A., Ballantyne, J.S., and Wright, P.A., Active urea transport and an unusual basolateral membrane composition in the gills of a marine elasmobranch, *Am. J. Physiol. Regul. Integr. Comp. Physiol.*, 280, R16–R24, 2001.

Fisher, N.S., On the reactivity of metals for marine phytoplankton, *Limnol. Oceanogr.*, 31, 443–449, 1986.

Galtsoff, P.S., The American Oyster *Crassostrea virginica* Gmelin, Fishery Bull. of Fish and Wildlife Service, Vol. 64, U.S. Government Printing Office, Washington, D.C., 1964, p. 480.

Gheber, L.A. and Edidin, M., A model for membrane patchiness: Lateral diffusion in the presence of barriers and vesicle traffic, *Biophys. J.*, 77, 3163–3175, 1999.

Gibaldi, M., *Biopharmaceutics and Clinical Pharmacokinetics*, 4th ed., Lea and Febiger, Ltd, Malvern, PA, 1991, p. 406.

Gobas, F.A.P.C. and MacKay, D., Dynamics of hydrophobic organic chemical bioconcentration in fish, *Environ. Toxicol. Chem.*, 6, 495–504, 1987.

Hassler, C.S. and Wilkinson, K.J., Failure of the biotic ligand and free-ion activity models to explain zinc bioaccumulation by *Chlorella kesslerii*, *Environ. Toxicol. Chem.*, 22, 620–626, 2003.

Hassler, C.S., Slaveykova, V.I., and Wilkinson, K.J., Some fundamental (and often overlooked) considerations underlying the free ion activity and biotic ligand models, *Environ. Toxicol. Chem.*, 23, 283–291, 2004.

Hauschild, K., Weber, W.-F., Clauss, W., and Grieshaber, M.K., Excretion of thiosulphate, the main detoxification product of sulphide, by the lugworm, *Arenicola marina* L., *J. Exp. Biol.*, 202, 855–866, 1999.

Hogben, C.A.M., Tocco, D.J., Brodie, B.B., and Schanker, L.S., On the mechanism of intestinal absorption of drugs, *J. Pharmacol. Exp. Ther.*, 125, 275–282, 1959.

Hogerle, M.L. and Winne, D., Drug absorption by the rat jejunum perfused *in situ*. Disassociation from the pH-partition theory and role of microclimate-pH and unstirred layer, *Naunyn Schmiedebergs Arch. Pharmacol.*, 322, 249, 1978.

Huang, R.-N. and Lee, T.-E., Cellular uptake of trivalent arsenite and pentavalent arsenate in KB cells cultured in phosphate-free medium, *Toxicol. Appl. Pharmacol.*, 136, 243–249, 1996.

Hoffmann, U. and Kroemer, H.K., The ABC transporters MDR1 and MDR2: Multiple functions in disposition of xenobiotics and drug resistance, *Drug Metabolism Rev.*, 36, 669–701, 2004.

Jacobson, K., Sheets, E.D., and Simson, R., Revisiting the fluid mosaic model of membranes, *Science*, 268, 1441–1442, 1995.

Jones, J.R.E., The relation between the electrolytic solution pressure of the metals and their toxicity to the stickleback (*Gasterosteus aculeatus*), *J. Exp. Biol.*, 16, 425–437, 1939.

Jones, J.R.E., A further study of the relation between toxicity and solution pressure, with *Polycelis nigra* as test animal, *J. Exp. Biol.*, 17, 408–415, 1940.

Jones, M.R. and Vaughin, W.K., HSAB theory and acute metal ion toxicity and detoxification processes, *J. Inorg. Nucl. Chem.*, 40, 2081–2088, 1978.

Kaiser, K.L.E., Correlation and prediction of metal toxicity to aquatic biota, *Can. J. Fish. Aquat.*, 37, 211–218, 1980.

Keshavarzian, A., Holmes, E.W., Patel, M., Iber, F., Fields, J.Z., and Pethkar, S., Leaky gut in alcoholic cirrhosis: A possible mechanism for alcoholic-induced liver damage, *Am. J. Gastroenterol.*, 94, 200–207, 1999.

Kimmerer, T.W. and Kozlowski, T.T., Stomatal conductance and sulfur uptake of five clones of *Populus tremuloides* exposed to sulfur dioxide, *Plant Physiol.*, 67, 990–995, 1981.

Klimmek, S., Stan, H.-J., Wilke, A., Bunke, G., and Buchholz, R., Comparative analysis of the biosorption of cadmium, lead, nickel and zinc by algae, *Environ. Sci. Technol.*, 35, 4283–4288, 2001.

Könemann, H. and van Leeuwen, K., Toxicokinetics in fish: Accumulation and elimination of six chlorobenzenes by guppies, *Chemosphere*, 9, 3–19, 1980.

Lee, B.-G., Griscom, S.B., Lee, J.-S., Choi, H.J., Koh, C.-H., Luoma, S.N., and Fisher, N.S., Influences of dietary uptake and reactive sulfides on metal bioavailability from aquatic sediments, *Science*, 287, 282–284, 2000.

Leslie, H.A., Oosthoek, A.J.P., Busser, F.J.M., Kraak, M.H.S., and Hermens, J.L.M., Biomimetic solid-phase microextraction to predict body residues and toxicity of chemicals that act by narcosis, *Environ. Toxicol. Chem.*, 21, 229–234, 2002.

Lindqvist, L. and Block, M., Excretion of cadmium and zinc during moulting in the grasshopper *Omocestus viridulus* (Orthoptera), *Environ. Toxicol. Chem.*, 13, 1669–1672, 1994.

Lipscomb, T.P., Harris, R.K., Rebar, A.H., Ballachey, B.E., and Haebler, R.J., *Exxon Valdez* oil spill state/federal natural resource damage assessment, Final report, Pathological studies of sea otters (Marine Mammal Study 6-11), June 1996, U.S. Fish and Wildlife Service, Anchorage, AK, 1996.

Loeb, J., Studies on the physiological effects of the valency and possibly the electrical charges of ions. I: The toxic and antitoxic effects of ions as a function of their valency and possibly their electric charge, *Am. J. Physiol.*, 6, 411–433, 1902.

Lowe, D.M. and Moore, M.N., The cytochemical distributions of zinc (Zn II) and iron (Fe III) in the common mussel, *Mytilus edulis*, and their relationship with lysosomes, *J. Mar. Biol. Ass. U.K.*, 59, 851–858, 1979.

Lyon, R., Taylor, M., and Simkiss, K., Ligand activity in the clearance of metals from the blood of the crayfish (*Austropotamobius pallipes*), *J. Exp. Biol.*, 113, 19–27, 1984.

Markich, S.J., Brown, P.L., Jeffree, R.A., and Lim, R.P., Valve movement responses of *Velesunio angasi* (Bivalvia: Hyriidae) to manganese and uranium: An exception to the free ion activity model, *Aquat. Toxicol.*, 51, 155–175, 2000.

Mason, A.Z. and Nott, J.A., The role of intracellular biomineralized granules in the regulation and detoxification of metals in gastropods with special reference to the marine prosobranch *Littorina littorea*, *Aquat. Toxicol.*, 1, 239–256, 1981.

Mathews, A.P., The relation between solution tension, atomic volume, and the physiological action of the elements, *Am. J. Physiol.*, 10, 290–323, 1904.

Mayer, L.M., Chen, Z., Findlay, R.H., Fang, J., Sampson, S., Self, R.F.L., Jumars, P.A., Quetel, C., and Donard, O.F.X., Bioavailability of sedimentary contaminants subject to deposit-feeder digestion, *Environ. Sci. Technol.*, 30, 2641–2645, 1996.

Mayor, S. and Rao, M., Rafts: Scale-dependent, active lipid organization at the cell surface, *Traffic* 2004, 5, 231–240, 2004.

McGuigan, H., The relation between the decomposition-tension of salts and their antifermentative properties, *Am. J. Physiol.*, 10, 444–451, 1954.

Middendorf, P.J. and Williams, P.L., Nephrotoxicity: Toxic responses of the kidney, In *Principles of Toxicology: Environmental and Industrial Applications*, 2nd ed., Williams, P.L., James, R.C., and Roberts S.M. (eds.), John Wiley & Sons, Inc., New York, 2000, pp. 129–143.

Morris, R.J., McArtney, M.J., Howard, A.G., Arbab-Zavar, M.H., and Davis, J.S., The ability of a field population of diatoms to discriminate between phosphate and arsenate, *Mar. Chem.*, 14, 259–265, 1984.

Newman, M.C. and Jagoe, C.H., Ligands and the bioavailability of metals in aquatic environments, In *Bioavailability. Physical,Chemical and Biological Interactions*, Hamelink, J.L., Landrum, P.F., Bergman, H.L., and Benson, W.H. (eds.), Lewis Publishers, Boca Raton, FL, 1994, pp. 39–61.

Newman, M.C. and McCloskey, J.T., Predicting relative toxicity and interactions of divalent metal ions: Microtox bioluminescence assay, *Environ. Toxicol. Chem.*, 15, 1730–1737, 1996.

Newman, M.C., McCloskey, J.T., and Tatara, C.P., Using metal-ligand binding characteristics to predict metal toxicity: Quantitative ion character–activity relationships (QICARs), *Environ. Health Perspect.*, 106(Suppl. 6), 1263–1270, 1998.

Oremland, R.S. and Stolz, J.F., The ecology of arsenic, *Science*, 300, 939–944, 2003.

Ormseth, O.A. and Ben-David, M., Ingestion of crude oil: Effects on digesta retention times and nutrient uptake in captive river otters, *J. Comp. Physiol. B*, 170, 419–428, 2000.

Ownby, D.R. and Newman, M.C., Advances in quantitative ion character–activity relationships (QICARs): Using metal–ligand binding characteristics to predict metal toxicity, *QSAR Comb. Sci.*, 22, 241–246, 2003.

Peterson, C.H., Rice, S.D., Short, J.W., Esler, D., Bodkin, J.L., Ballachey, B.E., and Irons, D.B., Long-term ecosystem response to the *Exxon Valdez* oil spill, *Science*, 302, 2082–2086, 2003.

Pigman, E.A., Blanchard, J., and Laird, H.E., II, A study of cadmium transport pathways using the Caco-2 cell model, *Toxicol. Appl. Pharmacol.*, 142, 243–247, 1997.

Pronczuk, J., Akre, J., Moy, G., and Vallenas, C., Global perspectives in breast milk contamination: Infectious and toxic hazards, *Environ. Health Perspect.*, 110, A349–A351, 2002.

Purchon, R.D., *The Biology of the Mollusca*, Pergamon Press, Oxford, UK, 1968, pp. 207–268.

Rainbow, P.S. and Black, W.H., Cadmium, zinc and the uptake of calcium by two crabs, *Carcinus maenus* and *Eriocheir sinesis*, *Aquat. Toxicol.*, 72, 45–65, 2005.

Roberts, S.M., James, R.C., and Franklin, M.R., Hepatotoxicity: Toxic effects on the liver, In *Principles of Toxicology: Environmental and Industrial Applications*, 2nd ed., Williams, P.L., James, R.C., and Roberts S.M. (eds.), John Wiley & Sons, Inc., New York, 2000, pp. 111–128.

Rodriguez, R.R., Basta, N.T., Castell, S.W., and Pace, L.W., An *in vitro* gastrointestinal method to estimate bioavailable arsenic in contaminated soils and solid media, *Environ. Sci. Technol.*, 33, 642–649, 1999.

Rozman, K.K. and Klaassen, C.D., Absorption, distribution, and excretion of toxicants, In *Casarett & Doull's Toxicology. The Basic Science of Poisons*, Klaassen, C.D. (ed.), McGraw-Hill, New York, 1996, pp. 91–112.

Rust, A.J., Burgess, R.M., Brownawell, B.J., and McElroy, A.E., Relationship between metabolism and bioaccumulation of benzo[a]pyrene in benthic invertebrates, *Environ. Toxicol. Chem.*, 23, 2587–2593, 2004.

Santore, R.C., Di Toro, D.M., Paquin, P.R., Allen, H.E., and Meyer, J.S., Biotic ligand model of the acute toxicity of metals. 2. Application to acute copper toxicity in freshwater fish and *Daphnia*, *Environ. Toxicol. Chem.*, 20, 2397–2402, 2001.

Shore, P.A., Brodie, B.B., and Hogben, C.A.M., The gastric secretion of drugs: A pH partition hypothesis, *J. Pharmacol. Exp. Ther.*, 119, 361–369, 1957.

Simkiss, K., Cellular discrimination processes in metal accumulating cells, *J. Exp. Biol.*, 94, 317–327, 1981.

Simkiss, K., Lipid solubility of heavy metals in saline solutions, *J. Mar. Biol. Ass. U.K.*, 63, 1–7, 1983.

Simkiss, K., Ecotoxicants at the cell-membrane barrier, In *Ecotoxicology. A Hierarchical Treatment*, Newman, M.C. and Jagoe, C.H. (eds.), CRC Press/Lewis Publishers, Boca Raton, FL, 1996, pp. 59–83.

Singer, S.J. and Nicolson, G.L., The fluid mosaic model of the structure of cell membranes, *Science*, 175, 720–731, 1972.

Sweet, D.H., Organic anion transporter (Slc22a) family members as mediators of toxicity, *Toxicol. Appl. Pharmacol.*, 204, 198–215, 2005.

Tao, S., Liu, C., Dawson, R., Cao, J., and Li, B., Uptake of particulate lead via the gills of fish (*Carassius auratus*), *Arch. Environ. Contam. Toxicol.*, 37, 352–357, 1999.

Timbrell, J., *Principles of Biochemical Toxicology*, 3rd ed., Taylor & Francis, Philadelphia, PA, 2000, p. 394.

Vijver, M.G., Van Gestel, C.A.M., Lanno, R.P., Van Straalen, N.M., and Peijnenburg, W.J.G.M., Internal metal sequestration and its ecotoxicological relevance: A review, *Environ. Sci. Technol.*, 38, 4705–4712, 2004.

Wallace, W.G., Lee, B.-G., and Luoma, S.N., Subcellular compartmentalization of Cd and Zn in two bivalves. I. Significance of metal-sensitive fractions (MSF) and biologically detoxified metal (BDM), *Mar. Ecol. Prog. Ser.*, 249, 183–197, 2003.

Wauchope, R.D., Uptake, translocation and phytotoxicity of arsenic in plants, In *Arsenic: Industrial, Biomedical, Environmental Perspectives*, Lederer, W.H. and Fensterheim, R.J. (eds.), Van Nostrand Reinhold Co., New York, 1983, pp. 348–377.

Webster, B., Exposure to fire retardants on the rise, *Science*, 304, 1730, 2004.

Weston, D.P. and Maruya, K.A., Predicting bioavailability and bioaccumulation with *in vitro* digestive fluid extraction, *Environ. Toxicol. Chem.*, 21, 962–971, 2002.

Williams, M.W. and Turner, J.E., Comments on softness parameters and metal ion toxicity, *J. Inorg. Nucl. Chem.*, 43, 1689–1691, 1981.

Yan, Q.-L. and Wang, W.-X., Metal exposure and bioavailability to a marine deposit-feeding Sipuncula, *Sipunculus nudus*, *Environ. Sci. Technol.*, 36, 40–47, 2002.

Zalups, R.K. and Ahmad, S., Molecular handling of Cadmium in transporting epithelia, *Toxicol. Appl. Pharmocol.*, 186, 163–188, 2003.

Zalups, R.K. and Barfuss, D.W., Renal organic anion transport system: A mechanism for the basolateral uptake of mercury–thiol conjugates along the oars recta of the proximal tubule, *Toxicol. Appl. Pharmacol.*, 182, 234–243, 2002.

8 Models of Bioaccumulation and Bioavailability

8.1 OVERVIEW

> The object of this work will consist in the derivation of general mathematical relations from which it is possible, at least for practical purposes, to describe the kinetics of distribution of substances in the body.
>
> **(Teorell 1937a)**

There are well-established ways to quantify bioaccumulation. Some mathematical models give parsimonious description or restricted prediction for a particular exposure scenario. As reflected in the above quote from Teorell's classic paper in which these methods were first introduced, such models are focused on practicality. More complicated models describe or predict bioaccumulation in a more general way based on physiological, biochemical, and anatomical features. Most simple models employ mathematical compartments while complicated models describe exchange among several interconnected physical or biochemical compartments. Many combine features of both modeling extremes. Although initially appearing as a jumble of competing approaches, this blend of approaches makes sense in ecotoxicology: some scenarios require a simple, pragmatic model but more involved models might be needed to capture the essential features of the situation under study. This is consistent with the general tenet that a model should be no more complicated than needed to answer the question being asked. For cases in which accurate description or prediction is paramount for many diverse species, contaminants, or conditions, a more complicated model incorporating physiological, biochemical, and anatomical features likely will be the best alternative. Otherwise, many simple models each of which only describes one relevant species/toxicant/condition combination would have to be employed.

Similarly, estimations of contaminant bioavailability from relevant sources can be produced for a particular situation using simple empirical models or more generally by applying predictive models based on in-depth mechanistic knowledge. Like bioaccumulation models, which bioavailability approach is the most appropriate depends on the study goals and the desired generality of the results.

8.2 BIOACCUMULATION

Most bioaccumulation models translate an external concentration to an internal concentration that is then related to an effect. The form of the model depends on the media containing the contaminant, qualities of the environment in which the exposure takes place, and the qualities of the organism itself. Models for nonionic organic compound uptake via gills might incorporate general equations based on lipid solubility relationships. Models for weakly acidic or basic compounds ingested in food might incorporate pH Partition Theory using the Henderson–Hasselbalch equations. Models for dissolved metal uptake across gill surfaces might be formulated using free ion activity model (FIAM) or biotic ligand model (BLM) based relationships. For example, a FIAM relationship might be developed for a dissolved metal being taken up under different pH, temperature, or water quality conditions that influence chemical speciation. Allometric[1] power equations might be incorporated if the organism grows substantially during the period of bioaccumulation. Still other models need to accommodate the interactions between physical and biological qualities. For example, temperature's influence on chlorpyrifos uptake is different for insect groups differing in respiratory strategy (Buchwalter et al. 2003). There are basic similarities among most models although a model might take slightly different forms depending on the exposure scenario and objectives of the modeling effort. These fundamental similarities will be highlighted in the next few pages.

8.2.1 Underlying Mechanisms

The exact formulation of a bioaccumulation model depends on the underlying mechanisms of uptake, internal redistribution, and elimination. As detailed in Chapter 7, some compounds are taken up or eliminated by mechanisms that can be saturated or modified by competition with other compounds. Bioaccumulation of others is influenced by factors such as urine or gut pH, and inclusion of these factors in the associated model might be required. Still other processes can be modified by acclimation or damage. Some compounds are subject to internal breakdown but others are not.

8.2.2 Assumptions of Models and Methods of Fitting Data

> Models are developed based on three different formulations: rate-constant-based, clearance-volume-based, and fugacity-based formulations. All are equivalent in their basic forms, but each formulation has its own advantages and disadvantages.
>
> **(Newman and Unger 2003)**

Models are based on mathematical expediency (descriptive models), processes and structures described in earlier chapters (mechanistic models), or a blending of both. The most common assumptions revolve around reaction order for the relevant processes so it is worth taking a moment to review the fundamentals of zero, first, and mixed order reactions. Recollect that order for a reaction or process involving one reactant refers to the power to which concentration is raised in the differential equation describing the associated kinetics, for example,

$$\text{Zero order:} \quad \frac{\mathrm{d}C}{\mathrm{d}t} = -kC^0 = -k,$$

$$\text{First order:} \quad \frac{\mathrm{d}C}{\mathrm{d}t} = -kC^1 = -kC.$$

The equations above are expressed for elimination so the change in internal concentration is denoted by $-k$ or $-kC$: concentrations are decreasing with time. For uptake, the sign would be positive. The generic k denoted here is a simple proportionality or rate constant. Continuing the example with elimination, a plot of concentration (C_t) versus time (t) will produce a straight line for zero order processes: the absolute value of the slope of that line is an estimate of k. The zero order k has units of C/t. For first order processes, ln C_t is plotted against t to produce a straight line and the absolute value of the slope is k. Alternatively, one could plot the $\ln(C_t/C_{t=0})$ against time for processes following first order kinetics. The slope would then be k (Piszkiewicz 1977). The units for the first order k are $1/t$; however, we will see that some slightly more involved bioaccumulation models apply a "first order k" for some processes that have different units.

Saturation kinetics used to describe carrier-mediated membrane transport or enzyme-mediated breakdown of a compound are slightly more complicated. Let S be the compound being acted on, by

[1] Allometry is the study of organism size and its consequences.

either enzymatic conversion or transport by a carrier molecule, E be the enzyme or carrier molecule, and P be the product of enzyme conversion or the compound successfully transported across the membrane via the carrier:

$$\mathrm{S} + \mathrm{E} \underset{k_{+1}}{\overset{k_{-1}}{\leftrightarrows}} \mathrm{ES} \underset{k_{+2}}{\overset{k_{-2}}{\leftrightarrows}} \mathrm{E} + \mathrm{P}.$$

The differential equation describing the change in substrate through time would be

$$\frac{\mathrm{d}C_\mathrm{S}}{\mathrm{d}t} = -k_{+1}C_\mathrm{E}C_\mathrm{S} + k_{-1}C_\mathrm{ES}, \tag{8.1}$$

where C_S, C_E, and C_ES are the concentrations of S, E, and ES, respectively. If one were to plot the curve of C_S versus time, the apparent order would depend on the initial C_S and might change as C_S changes through time. Above a certain C_S, the enzyme or membrane transport system would be saturated and incapable of converting/transporting S any faster than a characteristic maximum velocity (V_max): the C_S versus time curve would appear to conform to zero order kinetics above the saturation concentration. If one started with very low C_S relative to the saturation concentration, the curve would appear to describe first order kinetics. The Michaelis–Menten equation predicting the rate (v) at which conversion occurs for such a process is the following:

$$v = \frac{V_\mathrm{max}C_\mathrm{S}}{k_\mathrm{m} + C_\mathrm{S}}, \tag{8.2}$$

where k_m is the C_S at which v is half of V_max. Like concentration–time curves for zero and first order processes, there are several ways to estimate parameters for saturation kinetics, including the most commonly applied double-reciprocal (Lineweaver–Burk) plot and three less common plots (Eadie–Hofstee, Scatchard, and Woolf plots). Traditionally, a series of C_S are established in mixture with the same concentration of enzyme and the rate of disappearance of S (or appearance of P) is measured for each concentration. The C_S and conversion rates (v) are then plotted or fit by linear regression to estimate V_max and k_s. Raaijmaker (1987) discusses the statistical concerns associated with these transformations, concluding that the Woolf plot functions best.

$$\text{Lineweaver–Burk:} \quad \frac{1}{v} = \frac{k_\mathrm{m} + C_\mathrm{S}}{V_\mathrm{max}C_\mathrm{S}} = \frac{1}{V_\mathrm{max}} + \frac{k_\mathrm{m}}{V_\mathrm{max}C_\mathrm{S}}, \tag{8.3}$$

$$\text{Eadie–Hofstee:} \quad v = V_\mathrm{max} - \frac{k_\mathrm{m}v}{C_\mathrm{S}}, \tag{8.4}$$

$$\text{Scatchard:} \quad \frac{v}{C_\mathrm{S}} = \frac{V_\mathrm{max}}{k_\mathrm{m}} - \frac{v}{k_\mathrm{m}}, \tag{8.5}$$

$$\text{Woolf:} \quad \frac{C_\mathrm{S}}{v} = \frac{k_\mathrm{m}}{V_\mathrm{max}} + \frac{C_\mathrm{S}}{V_\mathrm{max}}. \tag{8.6}$$

Piszkiewicz (1977) provides details for deriving these transformations and relating one to the other.

Of course, a nonlinear model can also be fit directly to the v versus C_S data to parameterize the Michaelis–Menten model. If such fitting involves an iterative maximum likelihood approach, the estimates from one of the above linearizing plots might be used as initial values. It might be more convenient during some modeling efforts to assume zero order kinetics above a certain concentration and then first order when concentrations drop below saturation during a time course. Because the time it will take for the concentration to fall below saturation depends on the initial concentration, it would be convenient to be able to estimate when one should shift from zero to first order computations. Wagner (1979) provides a convenient relationship for estimating the time to transition from zero to first order (t^*) as a function of the initial concentration (C_{S0}):

$$t^* = \left[\frac{1-(1/e)}{V_{max}}\right] C_{S0} + \frac{k_m}{V_{max}}. \tag{8.7}$$

Box 8.1 Silver Transport across Membranes Exhibits Saturation Kinetics

Because the silver ion, Ag^+, is highly toxic to freshwater fishes, its transport across and effects on gills is a very active area of research. As one example, Bury et al. (1999) characterized silver transport by a Na^+/K^+-ATPase located on the basolateral membranes of gill cells using a conventional saturation kinetics model. This study is ideal for illustrating here the relevance of saturation kinetics for one feature of contaminant bioaccumulation.

Bury et al. (1999) describe studies published preceding theirs in which the Ag^+ ion was shown to be transported into gill cells by a Na^+ channel subject to saturation kinetics. Because P-type ATPases[2] had been documented for other metals, Bury et al. hypothesized that Ag^+ might also be transported by ATPases in rainbow trout (*Oncorhynchus mykiss*) gill membrane vesicles. They removed gills from trout and produced membrane vesicles by a sequence of homogenization, agitation, and centrifugation steps. Radioactive silver (^{110m}Ag) was then used to measure movement of Ag^+ into isolated gill cell membrane vesicles.

Supporting the hypothesis of Bury and coworkers, the Ag^+ transport into the vesicles was found to be ATP dependent. Competitive inhibition was also apparent from experiments showing that Ag^+ transport slowed if Na^+ or K^+ concentrations were increased in the media surrounding the vesicles. Michaelis–Menten parameters were estimated by fitting the nonlinear Michaelis–Menten model (Equation 8.2) to vesicle uptake (nmol of Ag^+/mg protein/min) versus Ag concentration (μmol). The V_{max} was 14.3 nmol/mg membrane protein/min and the K_m was 62.6 μmol. An Eadie–Hofstee plot produced straight lines and also was used to fit these data by linear regression methods. Bury et al. concluded that there was a P-type ATPase transport mechanism for silver in trout gills and defined the characteristics of the saturation kinetics in vesicle preparations.

8.2.3 Rate Constant-Based Models

> [The assumption of a single compartment] may not be applied to all drugs. For most drugs, concentrations in plasma measured shortly after iv injection reveal a distinct distributive phase. This means that a measurable fraction of the dose is eliminated before attainment of distribution equilibrium. These drugs impart the characteristics of a multicompartment system upon the body. No more than two compartments are usually needed to describe the time course of drug in the plasma. These are often called the rapidly equilibrating or central compartment and the slowly equilibrating or peripheral compartment.
>
> **(Gibaldi 1991)**

Without a doubt, the most commonly applied bioaccumulation model in ecotoxicology is a one-compartment model with one first order uptake and one first order elimination term. After gaining entry, the toxicant is assumed to instantly distribute itself uniformly within that compartment. There

[2] P-type ATPases are one of three categories of ATPases. They are ubiquitous in living systems, facilitating cation transport for a variety of functions. The formation of a phosphorylated intermediate during transport leads to their designation as P-type.

is no hysteresis, that is, the likelihood of a molecule of the toxicant leaving the compartment is independent of how long it has been in the compartment. The relevant model can be constructed easily with the information covered already. The elimination from the compartment would simply be the following:

$$\frac{dC_i}{dt} = -k_e C_i, \tag{8.8}$$

where C_i is the internal concentration (or amount) of the compartment and k_e is the first order rate constant $(1/t)$. As already discussed, the k_e for such elimination can be estimated by fitting a linear regression line to $\ln C_i$ at different times ($\ln C_{i,t}$) versus time (t). The antilog of the y-intercept of the regression line can also be used to estimate the initial concentration in the compartment, $C_{i,0}$. Because this estimate can be biased, Newman (1995) provides a method of removing any bias from an estimated $C_{i,0}$. This differential equation (8.8) can be integrated to predict the concentration remaining in the compartment through time, perhaps after a source has been removed[3] or the organism has been dosed once and then allowed to eliminate the toxicant:

$$C_{i,t} = C_{i,0}\, e^{-k_e t}. \tag{8.9}$$

Useful metrics associated with this simple elimination model include the biological half-life of the toxicant in the compartment ($t_{1/2}$) (Equation 8.10) and the mean residence (or turnover) time of a toxicant molecule in the compartment (τ) (Equation 8.11):

$$t_{1/2} = \frac{\ln 2}{k_e} \tag{8.10}$$

$$\tau = \frac{1}{k_e}. \tag{8.11}$$

Equation 8.12 can be used to model elimination if there are two components of elimination that remove the toxicant from one compartment:

$$C_{i,t} = C_{i,0}\, e^{-(k_{e,1}+k_{e,2})t}. \tag{8.12}$$

If a compartment were composed of two subcompartments with no exchange between them, the change in the total concentration or amount in the two combined subcompartments (1 and 2) could be estimated by

$$C_{i,t} = C_{0,1}\, e^{-k_{e,1}t} + C_{0,2}\, e^{-k_{e,2}t}. \tag{8.13}$$

Figure 8.1 depicts the change in total concentration (or amount) in such a situation. In that figure, the two subcompartments are designated "fast" and "slow." A plot of $\ln C$ for the compartment that appears to be composed of the two subcompartments versus time of elimination will result in a curve composed of two linear segments. The linear segment for the later portion of the total curve will reflect the change in concentration (or amount) in the slow subcompartment because the compound in the fast compartment will have been eliminated by that time in the course of depuration. The linear segment at the beginning of the depuration period will reflect the combined concentration in both the slow and fast subcompartments. The two elimination components can be modeled by nonlinear regression or a

[3] An experimental design or action in which an organism containing a toxicant is removed to a clean environment where it can eliminate the toxicant is called a depuration design. The elimination of toxicant after movement to a clean environment is called depuration.

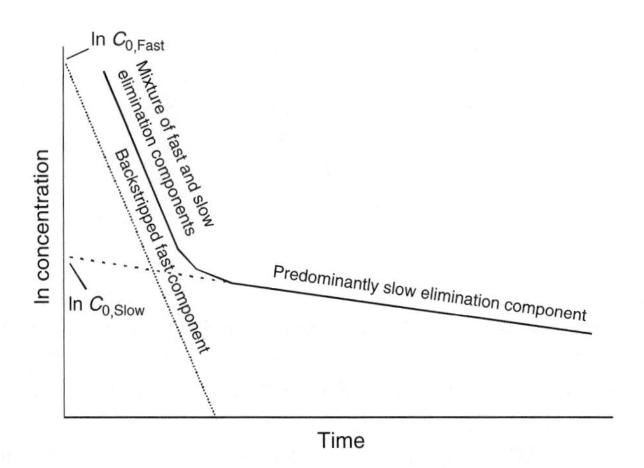

FIGURE 8.1 The elimination of compound from a compartment composed of two subcompartments with no exchange of compound between compartments. Both subcompartments have one, first order elimination component.

conventional backstripping method. To begin the backstripping approach, a line is fit to the portion of the curve that is predominantly associated with "slow" elimination. The y-intercept of the regression line is the $\ln C_{0,\text{Slow}}$ and the absolute value of the slope is the $k_{\text{e,Slow}}$. The $C_{0,\text{Fast}}$ and $k_{\text{e,Fast}}$ are then estimated using the data for the line segment associated with the combined concentrations in the two compartments through time and the regression model. The regression model for the slow component is used to predict how much of the concentration of compound measured during the initial linear phase of depuration was associated with the slow component. These predictions are subtracted from the observed concentrations to estimate the amount associated only with the fast elimination component. In essence, the concentration associated with the slow component is stripped away, leaving only that associated with the fast component. These "fast elimination" predictions for each data point would be distributed about the "backstripped" line depicted in Figure 8.1. A linear regression model is fit to these predictions, and the y-intercept and slope used as just described for the slow elimination component to predict $C_{0,\text{Fast}}$ and $k_{\text{e,Fast}}$. This same general procedure can also be used if more than two components are present.

The uptake from an external source with a constant concentration (C_{X}) can be defined as the following:

$$\frac{\mathrm{d}C_{\text{i}}}{\mathrm{d}t} = k_{\text{u}} C_{\text{X}}, \tag{8.14}$$

where k_{u} is the first order rate constant for uptake ($1/t$). As we will see, the units of k_{u} will change when Equations 8.8 and 8.14 are combined and applied to model bioaccumulation in many cases. Combining these two equations, bioaccumulation for a single compartment model can be described with the following:

$$\frac{\mathrm{d}C_{\text{i}}}{\mathrm{d}t} = k_{\text{u}} C_{\text{X}} - k_{\text{e}} C_{\text{i}}. \tag{8.15}$$

The above equation can be integrated to yield the conventional single compartment model with one uptake and one elimination term, both of which conform to first order kinetics:

$$C_{\text{i,t}} = C_{\text{X}} \frac{k_{\text{u}}}{k_{\text{e}}} [1 - \mathrm{e}^{-k_{\text{e}} t}]. \tag{8.16}$$

A simple adjustment to Equation 8.16 provides some insight about conditions as concentrations inside the organism approach steady state conditions with the external source. As t gets very large, the bracketed term in Equation 8.16 approaches 1; the bracketed term falls out of the equation as concentrations approach steady state. If both sides of the equation are then divided by C_S, it becomes apparent that the quotient of the concentration in the organism at steady state and external concentration (C_∞/C_X) is equal to k_u/k_e. If this model described uptake from water, the k_u/k_e would then be an estimate of the bioconcentration factor (BCF).

It is important to note that the constants are conditional if this model described the kinetics for compartments of different sizes. The size of the biological compartment and that of the external source compartment influence the k_u value. If two individuals of identical volume were placed into sources of different volumes but identical concentrations, the estimated values of k_u would be different for each. Similarly, the k_u values would be different for two different sized individuals exposed to the same concentration of contaminant contained in the same volume of media. In reality, the k_u is the flux or clearance rate of the source by an individual of a specific size; therefore, the units of k_u are flow/mass, for example, (mL/h)/g, for the equation to balance properly. (See Appendix 5 in Newman and Unger (2003) for a detailed dimensional analysis.) We will return to this important point of considering compartment volumes after a brief elaboration on single-compartment bioaccumulation models.

The rudimentary model defined by Equation 8.15 or 8.16 can be changed to meet the needs of the modeler. Equation 8.16 can accommodate the confounding factor of the organism represented by the compartment having an initial concentration (C_0) before the exposure being modeled

$$C_{i,t} = C_X \frac{k_u}{k_e}[1 - e^{-k_e t}] + C_{i,0}\, e^{-k_e t}. \tag{8.17}$$

Two elimination (Equation 8.18) or uptake (Equation 8.19 for uptake from water and food) terms can be included in the model also:

$$C_{i,t} = C_X \frac{k_u}{k_{e1} + k_{e2}}[1 - e^{-(k_{e1}+k_{e2})t}], \tag{8.18}$$

where k_{e1} and k_{e2} are the elimination rate constants for processes 1 and 2:

$$C_{i,t} = \frac{C_W k_{uw} + \alpha R C_f}{k_e}[1 - e^{-k_e t}], \tag{8.19}$$

where C_W and C_f are concentrations in water and food, respectively, α is amount of compound absorbed per amount of compound ingested, and R = the weight-specific ration. Obviously, the exact form of any model will change to the most convenient one as sources change but the general framework remains the same.

Returning to our discussion of volumes, compartment volumes also become relevant if the organism was modeled as having two or more compartments that exchange compound. But how are these volumes measured? As a simple illustration, a dose of the compound (D) is applied in a one-compartment model and allowed enough time to evenly distribute in the compartment. The compartment is sampled to determine the concentration (C), and then the compartment volume (V) is estimated as $D/C = V$. The estimation of volumes becomes complicated if more than one compartment is involved. How volumes are handled in such a situation can be illustrated with the conventional, two-compartment models used often in pharmacology and ecotoxicology (Figure 8.2). With such multiple compartments, volumes are expressed as apparent or effective volumes of distribution (V_d) and treated as mathematically defined compartments that might or might not be easily related to a physical compartment. The most common situation in which such volumes are employed

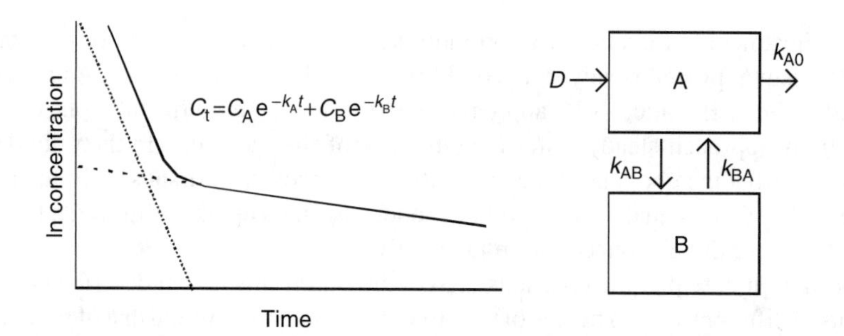

FIGURE 8.2 The estimation of compartment volumes for an organism modeled as two compartments with exchange between compartments and a single dose, D. First order microconstants are also derived from the macroconstants for concentration–time curve of the reference compartment ($C_{0,A}$, $C_{0,B}$, k_A, k_B). In many cases, the reference compartment (A) is the blood (or plasma) and the peripheral compartment includes the tissues with which the compound exchanges with the blood. More complex models are often warranted and require more detailed computations.

would be one in which a compound is introduced into the blood and the concentrations are then followed through time in the blood. The compound in the blood is envisioned as exchanging with some other compartment. The V_d for the nonblood (peripheral) compartment is expressed in units of volume of the blood (reference) compartment. For the two-compartment model depicted in Figure 8.2, the microconstants and volumes of distribution for the compartments can be estimated with the following relationships:

$$k_{AB} = \frac{C_A k_B + C_B k_A}{C_A + C_B} \tag{8.20}$$

$$k_{A0} = \frac{k_A k_B}{k_{AB}} \tag{8.21}$$

$$k_{AB} = k_A + k_B - k_{AB} - k_{A0} \tag{8.22}$$

$$V_{dA} = \frac{D}{C_A + C_B} \tag{8.23}$$

$$V_{dB} = V_{dA}\left[\frac{k_{AB}}{k_{BA}}\right]. \tag{8.24}$$

The steady-state V_d for the entire organism consisting of the two compartments is the sum of V_{dA} and V_{dB}. It is important to note that the steady state V_d reflects the amount of compound in a unit volume of the organism expressed in terms of the equivalent volume of the reference (source) compartment that would contain that same amount of compound. So, the volume of the peripheral compartment is expressed in units of the reference (blood) compartment. This holds true also if the source (reference) compartment was the water surrounding the organism; the steady-state V_d for the organism compartment would be a measure of the BCF because it expresses the volume of organism compartment that holds an equivalent amount of the chemical as a unit volume of the water source compartment.

8.2.4 Clearance Volume-Based Models

Often, especially in the fields of pharmaco- or toxicokinetics, bioaccumulation models are formulated in the context of clearance. Clearance is the volume of a compartment that is cleared of a substance per unit time. In these kinds of models, one compartment, such as the blood or plasma, is selected as the reference compartment and clearances are calculated relative to that compartment. Clearances can be expressed as volume/time or as mass-normalized clearances [(volume/time)/mass]. A quick glance back to our previous discussions of k_u in bioaccumulation models will show that the k_u was actually a clearance. Clearances (Cl) can be calculated for simple bioaccumulation models from terms already discussed above (i.e., $Cl = k_e V_d$). By substitution, the rate constant-based model shown in Equation 8.16 can be converted to a clearance-based model:

$$C_{i,t} = C_x V_d [1 - e^{-(Cl/V_d)t}]. \tag{8.25}$$

Following the reasoning used already for V_d at steady state, it is easy to see again from Equation 8.25 that V_d at steady state is equal to the quotient of the steady-state concentration in the organism (compartment) to the source.

8.2.5 Fugacity-Based Models

Bioaccumulation models sometimes are formulated in terms of fugacity, the escaping tendency for a substance from a medium or phase. Formulations using fugacity have the distinct advantage that states and rates for all components in complex models can be expressed in the same units. This makes mass balance equations much more tractable than with other approaches (Mackay 1979). Fugacity (f) is expressed as a pressure (e.g., units of Pascal or Pa) and is derived from concentrations (C in units of mol/m^3), i.e., $f = Z/C$ where Z is the fugacity capacity of the phase [mol/(m^{3}Pa)]. The fugacity capacity is "a kind of solubility or capacity of a phase to absorb the chemical" (Mackay 1979) so a modeled compound tends to accumulate in phases or compartments with high fugacity capacities. Because of this direct relationship between f and C, the steady state quotient of the concentration in one phase or compartment to that in another (e.g., the BCF) is also equal to the quotient of the fugacity capacities of the two compartments.

Only one definition and two identities are needed to convert the simple bioaccumulation models shown in Equation 8.16 to a fugacity-based model. In fugacity-based models, the movement of a substance between compartments is expressed as a transport rate (N, mol/h) and is calculated with a transport constant (D, mol/h × Pa) and the difference in fugacities for the two compartments ($f_1 - f_2$):

$$N = D[f_1 - f_2]. \tag{8.26}$$

The rate constant, D, can be used for a diverse range of relevant processes including chemical reactions, diffusion, or advection (Mackay 2004), making it a very convenient term in complex environmental models. The conversion of k_u and k_e to terms used in fugacity modeling is straightforward (Gobas and Mackay 1987):

$$k_e = \frac{D_O}{V_O Z_O}, \tag{8.27}$$

$$k_u = D_O V_O Z_S, \tag{8.28}$$

where D_O = transport constant for the organism being modeled, V_O = the volume of the organism being modeled, Z_O = a proportionality constant called the fugacity capacity for the organism, and Z_S = the fugacity capacity for the source of the substance being accumulated. With these definitions, a simple fugacity-based model can be produced.

$$C_{O,t} = C_x \frac{Z_O}{Z_S} [1 - e^{-[D_O/(V_O Z_O)]t}] \tag{8.29}$$

This model can be expressed in terms of fugacities instead of concentrations also (Gobas and Mackay 1987):

$$f_O = f_S[1 - e^{-[D_O/(V_O Z_O)]t}]. \tag{8.30}$$

Box 8.2 Unit and Real World Renderings with Fugacity Models

Fugacity simplifies and clarifies the relationship between equilibrium concentrations in various fluids and solids. Rather than relate two concentrations using a partition coefficient . . . , each concentration is independently related to fugacity, and the two fugacities equated.

(Mackay 2004)

It seemed to me that by reformulating equations in terms of fugacity, environmental mass balances could be done more easily, especially for systems involving disparate phases . . . .

(Mackay 2004)

With a toolbox containing f, Z, and D, the modeler has the key tools to quantify chemical fate in a vast variety of situations from transport across a cell membrane to estimating global distribution of chemical of commerce.

(Mackay 2004)

A brief look at the remarkable work of Don Mackay and his colleagues seems an appropriate way to illustrate the value and universal applicability of fugacity-based contaminant modeling. Fugacity-based models have been so influential that an entire issue of the journal *Environmental Toxicology and Chemistry* (vol. 23, no. 10) was recently dedicated to them. Beyond MacKay's first paper introducing the fugacity concept for contaminant modeling (Mackay 1979), particularly useful or exemplary publications applying this approach include Cahill et al. (2003), Czub and McLachlan (2004), Gobas and Mackay (1987), Hickie et al. (1999), Mackay (1979, 2001), Mackay and Wania (1995), and Wania and Mackay (1995). As described above in Mackay's own words, the great advantage of the fugacity approach is the ability to simplify models by simplifying units. Equations for different processes such as diffusion, advection, or chemical reaction could be made more consistent in this manner.

Mackay also facilitated environmental modeling by developing a series of fugacity-based models of increasing complexity. These conceptual microcosms or "unit worlds" provided the starting point for addressing numerous real-world questions. The simpler unit worlds included a few compartments such as air, sediment, soil, and water. Among the more complicated unit worlds is a Level III fugacity model including air, soil, water, settled and suspended sediments, and fish (e.g., Mackay et al. 1985). Elaboration upon unit-world fugacity models proved an effective way to expedite application to diverse, real-world situations.

The reader might also have begun to realize that the approach developed by Mackay fosters consilience and translation among levels of biological organization, a central theme of this book. Extensions of models based on the classic approach begun by Teorell (1937a,b) to include many abiotic phases and processes are possible but much more difficult than elaboration of Mackay's fugacity models. Some representative studies can be used to illustrate this point.

As one example, the general fugacity-based bioaccumulation model of hydrophobic organic compounds by fish incorporates relationships between key rates and qualities, and the K_{ow} (Gobas and Mackay 1987). The model assumed uptake from water via the gills of compounds that do not degrade. It estimated BCFs and displayed good agreement with experimental data.

As another example, Hickie et al. (1999) produced a physiologically based pharmacokinetic (PBPK) fugacity model (see Section 8.2.6) for bioaccumulation in Beluga whale (*Delphinapterus leucas*) exposed to polychlorinated biphenyls (PCB) over long periods of time. As a still more complicated example involving human exposure via fish consumption, a fugacity model including a five environmental phase unit world was used to predict bioaccumulation in edible fish (Mackay et al. 1985). A fugacity model called ACC-HUMAN was applied in 2004 (Czub and McLachlan 2004) to predictions of PCB congener trophic transfer to human milk. At a subcontinental scale, Mackay and Wania modeled the movement of organochlorine contaminants in the Arctic. General (Wania and Mackay 1995) and chemical-specific (Wania et al. 1999) fugacity-based models have been applied successfully to model the global movement of organic contaminants. Clearly, the fugacity-based approach pioneered by Mackay and co-investigators allows contaminant issues to be addressed at remarkably diverse temporal and spatial scales.

8.2.6 Physiologically Based Pharmacokinetic Models

Another approach to modeling bioaccumulation is called either physiologically based pharmacokinetic (PBPK) or physiologically based toxicokinetic (PBTK) modeling. Such models can be formulated in rate constant-, clearance-, or fugacity-based terms, but all PBPK have in common the creation of compartments and expressions of transfers among compartments based on real physical, biochemical, physiological, or anatomical compartments and processes. As a common example, a PBPK compartment might be the liver or fatty tissues. The transfer of compound among compartments might be calculated using blood flow rates, partitioning constants between blood and the tissue of interest, or enzymatic conversion within a tissue. Allometric equations might be employed to adjust compartments and processes among individuals differing in size. As an example, size effects on gill surface area might be incorporated into a fish PBPK model. These types of models have even been applied to general issues such as the accumulation and effects of contaminant mixtures (Haddad and Krishnan 1998).

Examples are easily to find of PBPK models applied to environmental contaminant bioaccumulation. Relevant PBPK models of gill exchange include those for hydrophobic chemicals (Erickson and McKim 1990, Nichols et al. 1990, 1991, Yang et al. 2000) and metals (Anderson and Spear 1980). The lifetime bioaccumulation of hydrophobic, persistent organic contaminants in Beluga whales (*D. leucas*) was modeled with this type of model (Hickie et al. 1999). Whole body concentrations of arsenic in tilapia (*Oreochromis mossambicus*) were estimated with a PBPK model in a recent study of human risk due to fish consumption (Ling et al. 2005) (Figure 8.3). These types of models, because they originated in the medical sciences, are found most often in that part of the ecotoxicology literature focused on human exposure and risk from contaminants. As additional examples, Clewell and Hearhart (2002) used PBPK models to estimate infant exposure to chemicals in breast milk. Yang et al. (1998) included lipophilicity-based quantitative structure–activity relationships (QSARs) to enrich current PBPK models of organic contaminant exposure.

8.2.7 Statistical Moments Formulations

The PBPK models just described are among the most complicated bioaccumulation models applied by ecotoxicologists. At the other extreme are statistical methods that avoid explicit models yet provide very useful information.

The classic statistical moments approach of Yamaoka et al. (1978) begins with a simple curve of concentration (C_t) in a compartment such as the blood or urine through time (t) after introduction of

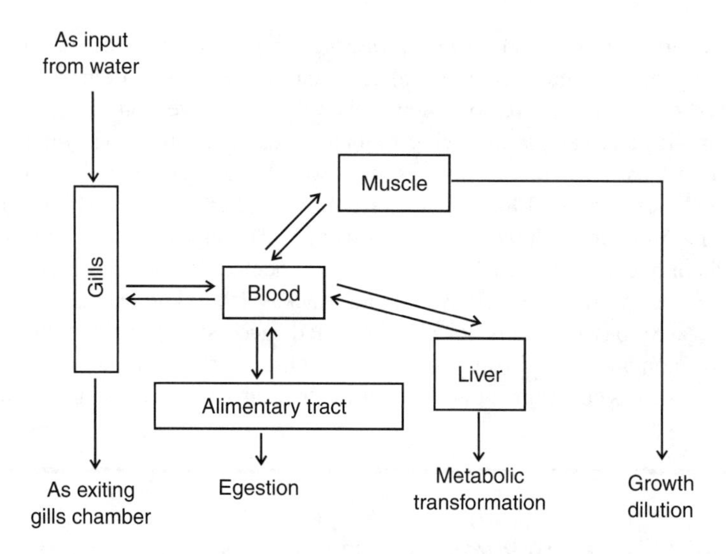

FIGURE 8.3 An example of a PBPK model. This figure depicts the compartment model used by Ling et al. (2005) to predict arsenic accumulation in tilapia consumed by people inhabiting an area of Taiwan with high incidence of arsenic poisoning (blackfoot disease). Dissolved arsenic passing over the gills can be taken up and enter the circulation. Arsenic in the circulation can then accumulate in several tissues. It can also be diluted as the fish grows (growth dilution), resulting in a decrease in concentration. It can be eliminated via the gut or metabolism in the liver.

a substance, and estimates the zero to second moments of the curve:

$$\mathrm{AUC} = \int_0^\infty C_\mathrm{t}\,\mathrm{d}t, \tag{8.31}$$

$$\mathrm{MRT} = \frac{\int_0^\infty tC\,\mathrm{d}t}{\int_0^\infty C\,\mathrm{d}t}, \tag{8.32}$$

$$\mathrm{VRT} = \frac{\int_0^\infty (t - \mathrm{MRT})^2 C\,\mathrm{d}t}{\int_0^\infty C\,\mathrm{d}t}, \tag{8.33}$$

where AUC (zero moment) = the area under the concentration–time curve, MRT = mean residence time of the contaminant in the body, and VRT = the variance of the residence time. The MRT, first moment of the concentration–time curve, is derived with the area under the curve of $C \times t$ versus t, that is, the area under the moments curve (AUMC). The MRT is AUMC/AUC. It is the time needed to eliminate 63.2% of the dose under the assumption of first order elimination (Gibaldi 1991). The AUC, MRT, and VRT are easily estimated from the C versus t data by methods described by Yamaoka et al. (1978) and again by Newman (1995). Of course, if one specifies a model as done in the methods described earlier, these moments can still be calculated. The unique aspect of this approach is that a model is not required to produce useful information.

Several very useful metrics can be obtained with these methods. As an example, clearance after intravenous (iv) injection can be estimated to be simply D_iv/AUC (Medinsky and Klaassen 1996). Also, if a dose was administered and the MRT calculated, the first order rate constant k_e can be estimated

$$k_\mathrm{e} = \frac{1}{\mathrm{MRT}}. \tag{8.34}$$

The first order rate constant for absorption (k_a) via the gut can be estimated from MRT values estimated if one administers a dose separately by iv (MRT_{iv}) and then by ingestion (MRT_{oral}) (Gibaldi 1991)

$$MAT = MRT_{oral} - MRT_{iv}, \tag{8.35}$$

where MAT is the mean absorption time of the chemical:

$$k_a = \frac{1}{MAT} \tag{8.36}$$

More will be made of statistical moments techniques in the next section.

8.3 BIOAVAILABILITY

> Bioavailability is the extent to which a contaminant in a source is free for uptake. In many definitions, especially those associated with pharmacology or mammalian toxicology, bioavailability of a contaminant implies the degree to which the contaminant is free to be taken up *and to cause an effect at the site of action.*
>
> **(Newman and Unger 2003)**

8.3.1 Conceptual Foundation: Concentration→Exposure→Realized Dose→Effect

The foundation for the bioavailability concept is plain to see. The presence of a specified contaminant concentration may or may not present a danger to an individual. Whether the organism is exposed to the contaminant concentration will depend on whether it comes within close proximity of the media containing the contaminant. The organism must be in appropriate contact in order to absorb, ingest, imbibe, or inhale the contaminated material. Whether or not this contact results in a realized dose in the organism will depend on a wide range of factors that collectively determine bioavailability. We have already discussed most of the important factors in previous chapters. If one defines bioavailability in the restricted sense of availability to a site of action then toxicokinetics within the organism must be considered. In such a case, assessment of bioavailability is concerned with predicting an effective dose, i.e., the dose delivered to the site of action. To summarize, bioavailability is a metric for translating contaminant exposure to a realized or effective dose. Its context may differ depending on the researcher's methods and intentions.

Box 8.3 What Is Involved in Oral Bioavailability?

Slightly differing terms and methods are used by ecotoxicologists and environmental risk assessors to describe and study bioavailability of substances in food. Penry (1998) pointed out that the processes of digestion, absorption, and assimilation all contribute to oral bioavailability, but their contributions are often misunderstood and variable. This leads to confused discussions of bioavailability and its measurement. Penry emphasizes how important it is to understand the context in which "bioavailability" is discussed, providing the following distinctions.

Digestion is the process by which ingested materials are broken down so that they are capable of being absorbed across the gut lining. Digestion efficiency can be estimated by measuring the amount of the substance in the ingested food (F_0) and the amount of that same substance remaining in egested feces (F_e):

$$\text{Digestive efficiency} = 1 - \frac{F_e}{F_0}. \quad (8.37)$$

The next step, absorption, is the uptake of the products of digestion into cells of the gut lining. Absorption efficiency reflects the fraction of the compound in the gut (F_g) that passes through the membranes of cells lining the gut lumen (F_c):

$$\text{Absorption efficiency} = 1 - \frac{F_c}{F_g}. \quad (8.38)$$

Assimilation involves the actual incorporation of the substance into the organisms' tissues. Assimilation efficiency is a measure of the fraction of total amount absorbed (F_a) that is actually assimilated into the tissues of the organism (F_t):

$$\text{Assimilation efficiency} = 1 - \frac{F_t}{F_a}. \quad (8.39)$$

According to Penry (1998), digestive and absorption efficiencies may be inappropriately used as analogous to assimilation efficiency in many bioavailability studies. Obviously, digestion, absorption, and assimilation all contribute to the realized bioavailability but each is influenced by different processes that alone do not suffice to quantify bioavailability. For example, selective sorting of particles by deposit feeders complicates estimation of digestion efficiency, because some contaminant-containing particles will not be subject to digestive processes, but instead, will simply pass out of the gut. And absorption does not necessarily lead to assimilation. As discussed in the previous chapter, the uptake of benzo[a]pyrene by cells of the gut, its conversion to benzo[a]pyrene-3-sulfate and benzo[a]pyrene-1-sulfate, and consequent transport of these conjugates back into the gut lumen by ABC transport proteins is one example of absorption into gut cells that does not result in direct delivery to the organism's tissues for incorporation. Some substances can be metabolized in gut cells and fail to move into circulation. Gibaldi (1991) provides several examples of drugs that undergo such presystemic metabolism. Similarly, binding of contaminant entering the circulatory system and rapid incorporation into bile can make it unavailable for assimilation into tissues. Finally, in the context of delivery to the site of action, enterohepatic cycling of a substance would complicate any simplistic use of digestion, absorption, or assimilation efficiencies alone to infer bioavailability.

8.3.2 Types and Estimation of Bioavailability

Bioavailability is important to understand for dermal, inhalation, and ingestion routes although most focus is on the ingestion route. Our previous discussions about dermal exposure and uptake from water and air provide detail about the less commonly discussed routes. Often, bioavailability is expressed as the fraction of the amount of the potentially available chemical that makes it into the circulation or as the amount in a medium available for uptake based on some criterion. The first expression will be used here.

Bioavailabilities can be either absolute or relative. The AUC approach for ingested chemicals can be used to illustrate absolute and relative bioavailability. The fraction (F) of an ingested dose (D) that enters the circulation can be estimated as the AUC in the blood (or plasma) measured after ingestion (AUC_{oral}) divided by the AUC measured after iv injection of the same dose (AUC_{iv}):

$$F = \frac{AUC_{oral}}{AUC_{iv}}. \quad (8.40)$$

In the above equation, the AUC estimated after iv administration is the reference AUC because the availability of the iv administered dose is, by definition, 100%. Therefore, the F estimates the fraction of the ingested dose that is available to reach the circulation. If the AUC increases linearly with dose within the range being administered, an F can be calculated when the doses administered intravenously and orally differ:

$$F = \frac{AUC_{oral} D_{iv}}{AUC_{iv} D_{oral}}. \quad (8.41)$$

Similarly, relative bioavailabilities can be estimated with AUCs produced by administering doses from two different sources. In this case, the AUC for one of the sources becomes the reference AUC. For example, AUCs can be estimated if a dose of a chemical was delivered in one type of food and then in another:

$$F = \frac{AUC_{Food\ A}}{AUC_{Food\ B}}. \quad (8.42)$$

In this example, the relative bioavailability indicates how much more or less a dose is available to enter the circulation from the two different sources. Of course, other estimated parameters such as the rate constants for absorption (Equation 8.36) from AUC analysis also suggest relative availabilities for the ingested materials.

Box 8.4 Taking a Few Moments to Estimate Methylmercury Bioavailability

The accumulation of methylmercury in fish has become a major issue relative to human exposure. Consequently, methylmercury bioaccumulation dynamics and bioavailability have become important research themes. In methylmercury bioavailability studies using goldfish (*Carassius auratus*) (Sharpe et al. 1977) and rainbow trout (*O. mykiss*) (Giblin and Massaro 1973), the conventional mass balance approach was used to estimate bioavailability. Briefly, this involves introducing a known amount of methylmercury in food, and after ample time passes to allow unassimilated methylmercury to be eliminated from the body, measuring the amount of methylmercury retained in the fish. This method can be compromised if the time needed to remove unassimilated methylmercury is underestimated or if the elimination of the compound of interest is very fast and a fraction of the assimilated methylmercury is eliminated before it can be measured in the fish tissue.

McCloskey et al. (1998) applied AUC methods just described as a convenient means of estimating oral methylmercury bioavailability to channel catfish (*Ictalurus punctatus*). Each catfish was fitted with an aortic cannula that allowed blood samples to be taken through time after dose administration. Each fish first received an oral methylmercury dose and the time course of methylmercury in the blood was followed through time. The same catfish were then given an intra-arterial dose and the time course of methylmercury concentration in the blood followed again. The blood methylmercury concentrations were normalized to an average hematocrit (17%) during computations because most of the methylmercury in the blood was associated with the red blood cells and the hematocrit varied with time and among catfish.

The dose-adjusted AUC approach (Equation 8.41) was applied to estimate oral bioavailability (F) for each catfish, using both noncompartment and compartment model–based methods. The noncompartment-based method simply used the linear trapezoidal method to measure AUC. The compartment model involved fitting of a triexponential model for the intra-arterial injected dose

$$C_{\text{blood,t}} = \text{A}\,\text{e}^{-\pi t} + \text{B}\,\text{e}^{-\alpha t} + \text{C}\,\text{e}^{-\beta t}$$

and estimating the AUC as

$$\text{AUC}_{0\rightarrow\infty} = \frac{\text{A}}{\pi} + \frac{\text{B}}{\alpha} + \frac{\text{C}}{\beta}.$$

A more involved model with a lag time before gut absorption of the dose (D) and a biexponential decline in blood concentrations was used for the orally administered dose. The blood methylmercury concentration was modeled under the assumption that it exchanged by first order kinetics between central and peripheral compartments (k_{21}), and gut absorption could be defined with a first order rate constant (k_{a}) after an initial lag (lag) of approximately 4 h before absorption:

$$C_{\text{blood,t}} = k_{\text{a}} D \left(\frac{F}{V_{\text{central}}}\right)\left[\frac{(k_{21}-\alpha)\text{e}^{-\alpha(t-\text{lag})}}{(k_{\text{a}}-\alpha)(\beta-\alpha)} + \frac{(k_{21}-\beta)\text{e}^{-\beta(t-\text{lag})}}{(k_{\text{a}}-\beta)(\alpha-\beta)} + \frac{(k_{21}-k_{\text{a}})\text{e}^{-k_{\text{a}}(t-\text{lag})}}{(\alpha-k_{\text{a}})(\beta-k_{\text{a}})}\right].$$

The F and V_{central} in this model are the bioavailability and volume of the central compartment, respectively.

Figure 8.4 shows the bioavailabilities (F) estimated with both approaches. The first point to make from this plot is that bioavailability is strongly influenced by feeding rates. Bioavailability was highest for fish receiving the least amount of food. Also, the compartment model method gave similar, but slightly lower, estimates of bioavailability as the noncompartmental method.

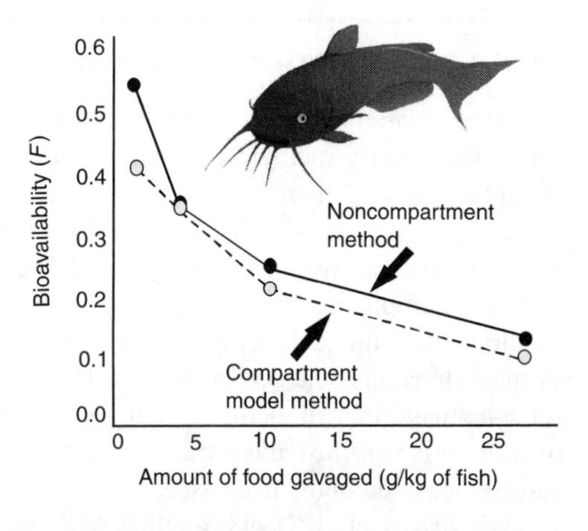

FIGURE 8.4 Bioavailability (F) estimated for methylmercury fed to channel catfish in varying amounts of food. The AUC estimates were generated with two methods (model free and a model-based method) (McCloskey et al. 1998).

Bioavailability can be estimated in other ways. As mentioned above, the amount of a delivered dose that becomes incorporated in the tissues or not egested has been used to estimate bioavailability. Bias can emerge in such an estimated bioavailability if the compound is eliminated very quickly so that a significant amount of the compound making it to the circulation is lost during the time that one is waiting for the unassimilated contaminant in the gut to clear. An opposite bias can emerge if insufficient time is allowed for gut clearance and unassimilated contaminant in the gut is incorrectly included in estimates of that incorporated into the organism's tissues. A dual-isotope approach can be used to minimize the risk of having unassimilated chemical in the organism's system at the time chosen to estimate F. Reinfelder and Fisher (1991) used such a method to estimate F for a series of metals in food fed to marine copepods. An inert (notionally unavailable) radionuclide tracer such as ^{51}Cr is incorporated into the food along with a radioactive form of the chemical of interest (e.g., ^{65}Zn). The assumption is made that both radionuclides are similarly incorporated into the food. The dual-labeled food is given to the organism and the radioactivity associated with the inert tracer measured until it has been completely eliminated, indicating that unassimilated material has passed from the organism by that time. The amount of the other chemical of interest (e.g., ^{65}Zn) that remains in the organism at that point is used to estimate F. This general method has been extremely useful and can be modified even if a completely inert tracer is not available. The dual-labeled isotope approach can be biased if the inert tracer is not incorporated into the food identically to the element/chemical for which F is being estimated. Lydy and Landrum (1993) provide one example in which ^{51}Cr was not incorporated into sediment similarly to the toxicant of interest (benzo[a]pyrene). In combination with the selective feeding by *Diporeia* spp., heterogeneous labeling produced unacceptable estimates of bioavailability. Instead, estimation for sediment-associated [^{14}C]benzo[a]pyrene required a mass-balance approach with carbon normalization. Total organic carbon and [^{14}C]benzo[a]pyrene were determined in fecal pellets and sediment from the exposure systems. A selectivity index (SI) was estimated in order to calculate F in the presence of selective feeding on sediment particles.

$$\mathrm{SI} = \frac{[\mathrm{TOC_{feces}}/(1 - \mathrm{RC})]}{\mathrm{TOC_{sediment}}}, \tag{8.43}$$

where $\mathrm{TOC_{feces}}$ and $\mathrm{TOC_{sediment}}$ are TOC fraction of the feces or sediment, respectively, and RC is the fractional loss of carbon during passage through the gut as obtained from the literature in this case. The assimilation of benzo[a]pyrene could then be calculated using the following equation:

$$F = \frac{\mathrm{SI} \times C_{\mathrm{sediment}} - C_{\mathrm{feces}}}{\mathrm{SI} \times C_{\mathrm{sediment}}}. \tag{8.44}$$

Another means of estimating F in such a situation is to use SI and feeding rates, as done by Kukkonen and Landrum (1995) in their assessment of assimilation of sediment-associated PAH and PCB congeners.

8.4 SUMMARY

In this chapter, the basic quantitative approaches applied to bioaccumulation were described as well as the basic approaches to estimating bioavailability. Rate constant-, clearance-, and fugacity-based formulations were described. Many of these methods can be used to generate simple or complex PBPK models. Fugacity-based models have a distinct advantage relative to the major goal of this book of fostering consilience among levels of complexity. The concept of bioavailability and methods for its estimation were also explored.

8.4.1 Summary of Foundation Concepts and Paradigms

- Bioaccumulation can be modeled with rate constant-, clearance-, and fugacity-based formulations.
- The kinetics associated with the bioaccumulation process are most often described using first or mixed order (saturation) formulations although other reaction orders are possible.

- Bioaccumulation models can describe mathematical compartments/processes or physical compartments/processes as in the case of PBPK (or PBTK) models.
- Noncompartment statistical moments methods can also generate extremely useful information about bioaccumulation processes and bioavailability.
- Procedures used to estimate bioavailability vary. Each method has advantages and slight differences, making it important to understand fully what each actually indicates.
- Bioavailabilities can be described as absolute or relative depending on the purpose of the associated study.

REFERENCES

Anderson, P.D. and Spear, P.A., Copper pharmacokinetics in fish gills—I. Kinetics in pumpkinseed sunfish, *Lepomis gibbosus*, of different body size, *Water Res.*, 14, 1101–1105, 1980.

Buchwalter, D.B., Jenkins, J.J., and Curtis, L.R., Temperature influences on water permeability and chlorpyrifos uptake in aquatic insects with differing respiratory strategies, *Environ. Toxicol. Chem.*, 22, 2806–2812, 2003.

Bury, N.R., Grosell. M., Grover, A.K., and Wood, C.M., ATP-dependent silver transport across the basolateral membrane of rainbow trout gills, *Toxicol. Appl. Pharmacol.*, 159, 1–8, 1999.

Clewell, R.A. and Gearhart, J.M., Pharmacokinetics of toxic chemicals in breast milk: Use of PBPK models to predict infant exposure, *Environ. Health Perspect.*, 110, A333–A337, 2002.

Czub, G. and McLachlan, M.S., A food chain model to predict the levels of lipophilic organic contaminants in humans, *Environ. Toxicol. Chem.*, 23, 2356–2366, 2004.

Erickson, R.J. and McKim, J.M., A model of exchange of organic chemicals at fish gills: Flow and diffusion limitations, *Aquat. Toxicol.*, 18, 175–198, 1990.

Gibaldi, M. *Biopharmaceutics and Clinical Pharmacokinetics*, 4th ed., Lea and Febiger, Ltd., Malvern, PA, 1991, p. 406.

Giblin, F.J. and Massaro, E.J., Pharmacodynamics of methyl mercury in the rainbow trout (*Salmo gairdneri*): Tissue uptake, distribution and excretion, *Toxicol. Appl. Pharmacol.*, 24, 81–91, 1973.

Gobas, F.A.P.C. and Mackay, D., Dynamics of hydrophobic organic chemical bioconcentration in fish, *Environ. Toxicol. Chem.*, 6, 493–504, 1987.

Haddad, S. and Krishnan, K., Physiological modeling of toxicokinetic interactions: Implications for mixture risk assessment, *Environ. Health Perspect.*, 106, 1377–1384, 1998.

Hickie, B.E., Mackay, D., and de Koning, J., Lifetime pharmacokinetic model for hydrophobic contaminants in marine mammals, *Environ. Toxicol. Chem.*, 18, 2622–2633, 1999.

Kukkonen, J. and Landrum, P.F., Measuring assimilation efficiencies for sediment-bound PAH and PCB congeners by benthic organisms, *Aquat. Toxicol.*, 32, 75–92, 1995.

Ling, M.-P., Liao, C.-M., Tsai, J.-W., and Chen, B.-C., A PBTK/TD modeling-based approach can assess arsenic bioaccumulation in farmed Tilapia (*Oreochromis massambicus*) and human health risks, *Int. Environ. Assess. Man.*, 1, 40–54, 2005.

Lydy, M.J. and Landrum, P.F., Assimilation efficiency for sediment-sorbed benzo[a]pyrene by *Diporeia* spp., *Aquat. Toxicol.*, 26, 209–224, 1993.

Mackay, D., Finding fugacity feasible, *Environ. Toxicol. Chem.*, 13, 1218–1223, 1979.

Mackay, D., *Multimedia Environmental Models: The Fugacity Approach*, 2nd ed., CRC Press/Lewis Publishers, Boca Raton, FL, 2001, p. 272.

Mackay, D., Peterson, S., Cheung, B., and Neely, W.B., Evaluating the environmental behavior of chemicals with a Level III fugacity model, *Chemosphere*, 14, 335–374, 1985.

Mackay, D. and Wania, F., Transport of contaminants to the Arctic: Partitioning, processes and models, *Sci. Total Environ.*, 160/161, 25–38, 1995.

McCloskey, J.T., Schultz, I.R., and Newman, M.C., Estimating the oral bioavailability of methylmercury in channel catfish (*Ictalurus punctatus*), *Environ. Toxicol. Chem.*, 17, 1524–1529, 1998.

Medinsky, M.A. and Klaassen, C.D., Toxicokinetics, In *Casarett & Doull's Toxicology. The Basic Science of Poisons*, Klaassen, C.D. (ed.), McGraw-Hill, New York, 1996, pp. 187–198.

Newman, M.C., *Quantitative Methods in Aquatic Ecotoxicology*, CRC Press/Lewis Publishers, Boca Raton, FL, 1995, p. 426.

Newman, M.C. and Unger, M.A., *Fundamentals of Ecotoxicology*, CRC Press/Lewis Publishers, Boca Raton, FL, 2003.

Nichols, J.W., McKim, J.M., Andersen, M.E., Gargas, M.L., Clewell, H.J., III, and Erickson, R.J., A physiologically based toxicokinetic model for the uptake and disposition of waterborne organic chemicals in fish, *Toxicol. Appl. Pharmacol.*, 106, 433–447, 1990.

Nichols, J.W., McKim, J.M., Lien, G.J., Hoffman, A.D., and Bertelsen, S.L., Physiologically based toxicokinetic modeling of three waterborne chloroethanes in rainbow trout (*Oncorhynchus mykiss*), *Toxicol. Appl. Pharmacol.*, 110, 374–389, 1991.

Penry, D.L., Applications of efficiency measurements in bioaccumulation studies: Definitions, clarifications, and a critique of methods, *Environ. Toxicol. Chem.*, 17, 1633–1639, 1998.

Piszkiewicz, D., *Kinetics of Chemical and Enzyme-Catalyzed Reactions*, Oxford University Press, New York, 1977, p. 235.

Raaijmakers, J.G.W., Statistical analysis of the Michaelis–Menten equation, *Biometrics*, 43, 793–803, 1987.

Reinfelder, J.R. and Fisher, N.S., The assimilation of elements ingested by marine copepods, *Science*, 251, 794–796, 1991.

Sharpe, M.S., DeFreitas, A.S.W., and McKinnon, A.E., The effect of body size on methylmercury clearance by goldfish (*Carassius auratus*), *Environ. Biol. Fish.*, 2, 177–183, 1977.

Teorell, T., Kinetics of distribution of substances administered to the body. I. The extravascular modes of administration, *Arch. Int. Pharmacodyn.*, 57, 205–225, 1937a.

Teorell, T., Kinetics of distribution of substances administered to the body. I. The intravascular modes of administration, *Arch. Int. Pharmacodyn.*, 57, 226–240, 1937b.

Wagner, J.G., *Fundamentals of Clinical Pharmacokinetics*, Drug Intelligence Publications, Hamilton, IL, 1979, p. 461.

Wania, F. and Mackay, D., A global distribution model for persistent organic chemicals, *Sci. Total Environ.*, 160/161, 211–232, 1995.

Wania, F., Mackay, D., Li, Y.-F., Bidleman, T.F., and Strand, A., Global chemical fate of α-hexachlorocyclohexane. I. Evaluation of a global distribution model, *Environ. Toxicol. Chem.*, 18, 1390–1399, 1999.

Yamaoka, Y., Nakagawa, T., and Uno, T., Statistical moments in pharmacokineties, *J. Pharmaco. Biopharmac.*, 6-6, 547–558, 1978.

Yang, R.S.H., Thomas, R.S., Gustafson, D.L., Campain, J., Benjamin, S.A., Verhaar, H.J.M., and Mumtaz, M.M., Approaches to developing alternative and predictive toxicology based on PBPK/PD and QSAR modeling, *Environ. Health Perspect.*, 106, 1385–1393, 1998.

Yang, R., Thurston, V., Neuman, J., and Randall, D.J., A physiological model to predict xenobiotic concentration in fish, *Aquat. Toxicol.*, 48, 109–117, 2000.

9 Lethal Effects

9.1 OVERVIEW

When toxicologists added the prefix eco to the field of toxicology, so that the word became ecotoxicology, they continued primarily to make the same measurements they made before the name changed.

(Cairns 1992)

Death can result from acute or chronic exposures to toxicants contained in many diverse sources. The distinction between acute and chronic exposure duration, adopted from human toxicology, is based as much on pragmatism as sound toxicology. A lethal exposure is customarily categorized as acute if it is a relatively brief and intense one to a poison. Standard durations are espoused for conducting acute lethality tests. For example, Sprague (1969) argued for 96 h after observing that "For 211 of 375 toxicity tests reviewed, acute lethal action apparently ceased with 4 days, although this tabulation may have been biased ...." This kind of correlative analysis and the convenience of fitting a test within the workweek motivated the initial codification of a 96-h test.

It is important to note that Sprague stated in his 1969 monograph that his intentions were to describe "profitable bioassay methods" about which there was ample "room for healthy disagreement." Along the vein of healthy disagreement, one could conclude from these same data that a 96-h duration was insufficient for characterizing acute lethality in more than 4 out of 10 tests (Figure 9.1). Further, Sprague notes that the tests considered in making his recommendation included many static tests[1] in which toxicant concentrations probably decreased substantially during the exposures and that those results from continuous flow tests that had much less chance of substantial toxicant concentration decrease during the tests generally indicated a longer duration was needed than did the static tests. Given the urgency in the 1960s for standard tools for dealing with pervasive pollution, the assumption that mortality by 96 h accurately reflected that occurring during any acute exposure duration is an understandable regulatory stance. However, it is scientifically indefensible and insufficient for today's needs. Consequently, many thoughtful ecotoxicologists now generate lethal effect metrics several times during acute toxicity tests.[2] And, as we will see, alternative approaches exist that avoid this issue altogether.

A similar blend of science and pragmatism contributed to the current selection of test durations for chronic exposures. By recent convention, chronic exposure occurs if exposure duration exceeds 10% of an organism's lifetime (Suter 1993); however, this has not always been the convention and 10% is an arbitrary cut-off point. Consequently, other durations are specified in some standard chronic test protocols and associated results are reported throughout the peer-reviewed literature.

Test protocols have emerged for exposures differing relative to the medium containing the toxicant(s) as well as exposure duration. For example, test protocols for acute (e.g., EPA 2002a) and chronic (e.g., EPA 2002b) water exposures quantify lethality under these two general categories of exposure duration. Exposures occur by oral, dermal, and respiratory routes, and accordingly, testing techniques have emerged that accommodate these routes (e.g., EPA (2002) for sediments).

[1] Generally, the toxicant is introduced into the test tanks at the beginning of a static aquatic toxicity test and not renewed for the test duration. Such tests are often characterized by substantial decreases in toxicant concentrations as the toxicant degrades, volatilizes, adsorbs to solids, or otherwise leaves solution. Such dosing problems in early, static tests have been reduced in current techniques by either periodic renewal of toxicant solutions or supplying a continuous flow of toxicant solution into exposure tanks (see for more detail Buikema et al. 1982).

[2] Sprague (1969) recommended this strategy to increase the information drawn from acute lethality tests.

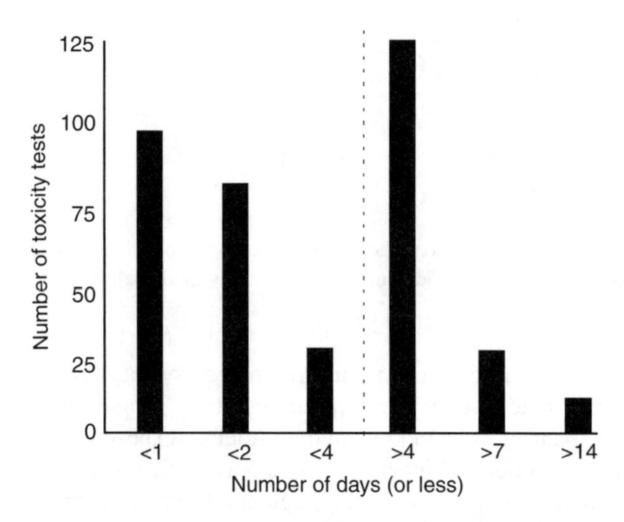

FIGURE 9.1 The number of early toxicity tests tabulated by Sprague (1969) in which acute mortality appeared to be completely expressed in exposed individuals by the specified exposure duration. Sprague noted that this data set included results from many static exposure tests in which the toxicant solutions were not changed and, as a consequence, the toxicant concentrations likely decreased substantially during testing. The tests are categorized here based on the time interval thought to be adequate for full expression of acute mortality, for example, "<1" = complete acute lethality expressed in 1 day or shorter.

Unfortunately, standard methods incorporating predictions of mortality from pulsed exposures are yet to be codified, but methods for dealing with these exposure scenarios are becoming increasingly seen as necessary to consider by ecological risk assessors. Those accommodating simultaneous exposure to several sources are also less common than warranted.

Approaches for characterizing or predicting lethal effects of single toxicant exposures are well established although some potentially useful approaches have yet to be explored sufficiently. This being the case, conventional and emerging approaches will be described in this chapter after discussion of some examples of lethality as manifested at the whole organism level of biological organization.

9.1.1 Distinct Dynamics Arising from Underlaying Mechanisms and Modes of Action

Molecular, cellular, anatomical, and physiological alterations that contribute to somatic death were sketched out in preceding chapters. Here, organismal consequences of such processes as narcosis, uncoupling of oxidative phosphorylation, and general stress will be explored. Hopefully, these examples demonstrate that all lethal responses to poisonings are not identical and that understanding the suborganismal processes resulting from exposure is extremely helpful for predicting consequences to individuals and populations.

Narcosis is often described as a reversible, chemically induced decrease in general nervous system functioning. The decrease in nervous system function results from disruption of nerve cell membrane functioning in higher animals as explained earlier (Chapter 3, Section 3.10); however, narcotic effects due to pervasive membrane dysfunction also manifest as a general depression of biological activity in organisms lacking nervous systems. Narcosis of sufficient intensity and duration lowers biological activities of any organism below those essential to maintaining the soma, resulting in death. But, because narcosis is reversible, postexposure mortality may be low relative to that resulting from damage which requires more time to repair. For example, grass shrimp (*Palaemonetes pugio*) acutely exposed for 48 or 60 h to polycyclic aromatic hydrocarbons (1-ethylnaphthalene,

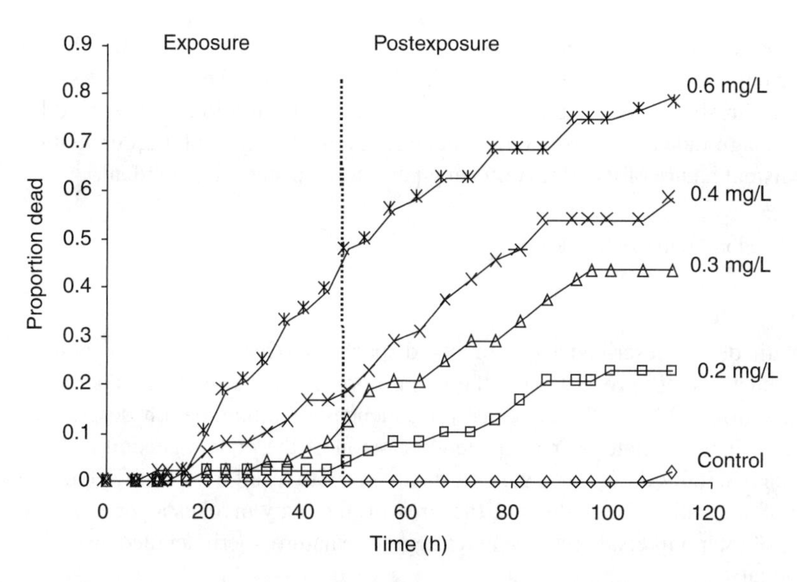

FIGURE 9.2 Cumulative mortality, including postexposure mortality, of amphipods (*Hyalella azteca*) exposed to four concentrations of dissolved copper. (Modified panel from Figure 1 of Zhao and Newman (2004).) Note that substantial mortality occurred after copper exposure ended.

2,6-dimethylnaphthalene, and phenanthrene) showed minimal postexposure mortality (Unger et al. 2007). In contrast, mortality experienced by amphipods (*Hyalella azteca*) after exposure to dissolved copper was quite high (Figure 9.2) because, as discussed in previous chapters, metals cause extensive biochemical, cellular, and tissue damage that takes considerable time to repair (Zhao and Newman 2004).

Another specific mechanism that can produce mortality is oxidative phosphorylation uncoupling. Such disruption of this essential mitochondrial process is typical of many substituted phenols (see Chapter 3, Section 3.9). At the organismal level, consequences range from elevated blood pH to disruption of normal respiratory processes to somatic death. Like the narcosis-related mortality just described, there can be minimal postexposure death in an exposed population. For example, amphipods acutely exposed to sodium pentachlorophenol showed minimal postexposure mortality (Zhao and Newman 2004). The pentachlorophenol is quickly eliminated from this amphipod and effects are reversible (Nuutinen et al. 2003). Mosquitofish (*Gambusia holbrooki*) acutely exposed to pentachlorophenol showed similar minimal postexposure mortality for the same reasons (Newman and McCloskey 2000).

In contrast to the lethal dynamics of such poisons, some toxicants cause pervasive changes or damage that requires considerable time to recover. The copper damage that resulted in the post-exposure mortality shown in Figure 9.2 is one example. The tissue damage resulting from metal exposure took considerable time to repair and, consequently, mortality continued well beyond termination of exposure. Similarly, mosquitofish (*G. holbrooki*) acutely exposed to high concentrations of sodium chloride showed prolonged and high postexposure mortality (Newman and McCloskey 2000). The cellular and tissue damage caused by the associated isomotic and ionic conditions takes time to repair. Fish, succumbing after exposure ends, did not have enough time or energy reserves to recuperate.

The nature of the lethal response can vary in other important ways. Some toxicants will display a concentration or dose threshold below which no lethal consequences are apparent. Mosquitofish exposure to high concentrations of sodium chloride is one obvious example in which death will not occur as long as the individual is able to osmo- and ionoregulate sufficiently at the particular sodium chloride concentration. However, the energetic burden imposed on the individual might result in decreased fitness in other aspects of the individual's life cycle. In addition, some, but not all, toxicants are characterized by a minimum time to die: the individual simply cannot die faster than this threshold time regardless of the exposure concentration or dose (Gaddum 1953). The presence and magnitude of a threshold time depends on the toxicant's bioaccumulation kinetics and the suborganismal nature of its effect upon any particular species or individual.

Complete freedom from stress is death.

(Selye 1973)

The somatic deaths described above involved specific modes of action but some somatic deaths involve the general stress process. Like the inappropriate toxicant-induced apoptosis described in Chapter 4 (Section 4.2.1) or the adverse consequences of inflammation described in Chapter 4 (Section 4.2.3), inappropriate or inadequate expression of the body's general reaction to stressors can lead to death of individuals. Such somatic death is said to result from what Selye (1984) described as a disease of adaptation. Regardless of the stressor, the body invokes a general suite of reactions that, because of their universal presence and integrative nature, merit detailed discussion at the level of the individual.[3]

The endocrinologist Hans Selye was the first to describe biological stress (Selye 1936). He defined stress as all nonspecific responses induced by intense demands placed on the organism. He named the associated syndrome, the general adaptation syndrome (GAS) (Selye 1950). The GAS has three phases: the alarm reaction, resistance, and exhaustion phases. The alarm phase is easily recognized as the immediate one in which the soma's resources are mustered suddenly to cope with a stressor. Rapid hormonal changes cause an organism's pulse and blood pressure to increase, putting it into a "flight or fight" state that takes considerable energy to maintain. Other immediate changes include those to breathing, blood flow to muscles, the immune system, behavior, and even, memory. At the cellular level, secretory granules discharge from cells of the adrenal cortex (Selye 1950). Characteristics emerging later in the resistance phase that Selye first identified in stressed rats are adrenal cortex enlargement with reappearance of normal levels of secretory granules, thymus and lymph node atrophy, and appearance of gastric ulcers. In mammals, such changes are brought about by the hypothalamic-pituitary-adrenal system's response to a stressor (see Tsigos and Chrousos (2002) for details). Analogous systems are involved in other vertebrates (i.e., the hypothalamo-pituitary-interrenal system of fishes and amphibians). The glucocorticoid cortisol and the catecholamines dopamine, norepinephrine, and epinephrine are prominent facilitators of the stress response. The resistance or adaptation phase is reached only if stress is sufficiently prolonged, resulting in organ and physiological changes such as those mentioned above. These shifts are intended to resist changes associated with a stressor by using less energy than changes associated with the alarm phase, and also, to maintain homeostasis. Examples of changes are adrenal gland enlargement to produce glucocorticoids that modify metabolism and also shifts in the immune system so that the body generally has reduced ability to express an inflammation response.[4] Selye refers to this state of artificially increased homeostasis as heterostasis. If stress continues and eventually exceeds the individual's finite adaptive energy, the exhaustion phase is entered in which the individual gradually loses its ability to maintain any semblance of essential stasis in the presence of the stressor.

---

[3] A reasonable argument could be made that this issue, because of the essential role played by hormones, should have been discussed in Chapter 6. However, the associated processes involve the integration of many biochemicals, organs, tissues, and organ systems within the individual, so it is more appropriate to discuss it here. The fact that it could be covered in either chapter attests to the soundness of the central theme of this book that making linkages among levels of biological organization is important and possible in ecotoxicology.

[4] The body's response to a local stressor is called a Local Adaptation Syndrome (LAS) and will be coordinated within the GAS. An example of such coordination is the influence of the GAS on the degree to which the body expresses inflammation locally in a damaged tissue.

Box 9.1 The Pharmacologist of Dirt

As a University of Prague medical student in 1925, Selye noticed a consistent syndrome with patients suffering from different, but intense, demands on their bodies (Selye 1973, 1984). A decade later as a young researcher studying sex hormones, he saw the same syndrome manifest in laboratory rats after injection with ovarian extracts. Rats showed a distinct syndrome in which the adrenal cortex enlarged, lymphatic structures (thymus, spleen, and lymph nodes) shrunk, and stomach ulcers appeared. He later found that injection of extracts from other tissues and even formalin elicited this same syndrome.

Because his original intent had been to identify novel sex hormones by injecting ovarian extracts into rats, his findings were extremely disheartening. That tissues other than ovaries elicited the same response might be an acceptable finding because tissues other than gonads were known at that time to produce sex hormones. But the appearance of the syndrome after formalin injection was inexplicable by any mechanism involving a sex hormone. After performing several more permutations of his experiments, he reluctantly came to the conclusion that the syndrome was not a specific one to an extracted hormone, but a general defense response to demands placed on the soma by a stressor. But his mood gradually changed from despair to fascination. He had found a general adaptive response, yet medical convention at that time focused solely on telltale effects produced by specific disease agents. Contrary to convention, he had discovered a nonspecific, defensive response. He shared his excitement about this novel vantage with a valued mentor who, after failing to dissuade him from further work along this theme, exclaimed, "But, Selye, try to realize what you are doing before it is too late. You have now decided to spend your entire life studying the pharmacology of dirt." After recovering from the sting of this comment, Selye spent his career studying what later became known as the theory of stress. Along the way, he published 1500 articles and 30 books that established a completely new discipline. Fortunately, the label "dirt pharmacology" never caught on.

What is the point? To use Selye's own thoughts about his experience, "My advice to a novice scientist is to look for the mere outlines of the big things with his fresh, untrained, but still unprejudiced mind" (Selye 1984). Respect, but do not be confined by, the current thinking in your field (see also Chapter 36).

Many changes that appeared during the alarm stage and abated during the resistance phase can reappear during the exhaustion phase (Selye 1950). Death occurs at the end of the exhaustion phase.

What is the significance of the GAS-associated shifts relative to coping with an infectious or noninfectious stressor? Selye breaks these changes down into responses facilitated by syntoxic and catatoxic hormones. Syntoxic hormones facilitate an individual's ability to coexist with the stressor during the period of challenge (e.g., those modulating the inflammation response during a general infection). Specific examples include cortisone and cortisol inhibition of inflammation as well as their altering of glucose metabolism. The catatoxic hormones are designed to enhance stressor destruction, "mostly through the induction of poison-metabolizing enzymes in the liver" (Selye 1984). Dysfunctions of these responses are called diseases of adaptation because they reflect health-enhancing processes gone awry. Human diseases of this sort include hypertension, some heart and kidney diseases, and rheumatoid arthritis. The activation of chemicals by liver enzymes discussed in previous chapters fit into this category of diseases also. Regardless, the reader will probably recognize at this point that the syntoxic and catatoxic hormones are pivotal to integrating the diverse defense mechanisms described in earlier chapters at the organismal level.

Not only can stress cause direct mortality of exposed individuals but can also, as suggested by the immunological changes described above, modify an individual's risk of death from toxicants or infectious agents. Friedman and Lawrence (2002) describe such exacerbation by stress of environmentally induced human maladies. Contaminants can also modify the stress response of exposed species. Hontela (1998) reported that low, chronic toxicant field exposures of fish appeared to reduce plasma corticosteroid levels, suggesting a compromised ability to respond to other stressors. Amphibians (*Necturus maculosus*) exposed in the field to polychlorinated biphenyls and organochlorine pesticides also demonstrated reduced ability to produce corticosterone when stressed (Gendron et al. 1997). As a final example, Benguira and Hontela (2000) documented reduced ability of rainbow trout (*Oncorhychus mykiss*) interrenal tissue to secrete cortisol with adrenocorticotropic hormone stimulation after *in vitro* exposure to *o,p′*-dichlorodiphenyldichloroethane (DDD).

So, toxicant-induced death can result from specific and nonspecific effects to or responses of individuals. This conclusion should create in the reader an anticipation that a diversity of mortality dynamics exist within groups of exposed individuals. In the next section, the focus will shift to the nature of these differences among lethally exposed individuals.

9.1.2 Lethality Differences among Individuals

> It has been recognized that in bioassays, the least and most resistant individuals in a group show much greater variability in response than individuals near the median. A good deal of accuracy may therefore be gained by measuring some average response rather than a minimum or maximum response ....
>
> **(Sprague 1969)**

Not surprisingly, toxicologists see variability in the resistance of individuals to lethal agents. Several factors contribute to this variability including allometric scaling, sex, age, genetics, and random chance. Even in the earliest publications quantifying lethal effects (e.g., Gaddum 1933), the influences of these factors were known. Except for random chance, which will be discussed in Sections 9.1.2.1 and 9.1.2.2, these factors will be described briefly here.

Scaling is simply the influence of organism size on structural and functional characteristics (Schmidt-Nielsen 1986). Many relevant processes such as those determining bioaccumulation (Anderson and Spear 1980), structures such as gill exchange surface area (Hughes 1966), and states such as metal body burden (Newman and Heagler 1991) are subject to scaling, so it is no surprise that the risk of death can be influenced by organism size. In fact, allometry, the science of scaling, is used to quantitatively predict differences in mortality for individuals differing in size (see Newman (1995) for details). Bliss (1936) developed a general power model that, in its various forms, currently enjoys widespread use for scaling lethal effects. As an important example, Anderson and Weber (1975) extended Bliss's approach to predict the mortality expected in a toxicity test if tested fish differed in size:

$$\text{Probit}(P) = a - b\log(M/W^h), \quad (9.1)$$

where Probit(P) = the probit transform[5] of the proportion of exposed fish dying, M = the toxicant concentration, W = the weight of the exposed fish for which prediction was being made, and h = an exponent adjusting mortality predictions for fish weight. Hedtke et al. (1982) used Equation 9.1 successfully to quantify the influence of Coho salmon (*Oncorhynchus kisutch*) size on the lethal effects of copper, zinc, nickel, and pentachlorophenol. Anderson and Weber (1975) advocated that this relationship be applied generally; however, some studies such as Lamanna and Hart (1968) show that not all data sets fit this relationship. As will be discussed later in this chapter, scaling effects on mortality can also be easily accommodated using survival time modeling, as implemented by many statistical programs.

[5] See Section 9.2.2 for details about the probit transformation.

Sex and age can influence the risk of dying during toxicant exposure. Several studies have shown differences in sensitivity between the sexes including Kostial et al. (1974) and Newman et al. (1989). Age is commonly an important factor determining sensitivity of toxicants (e.g., Hogan et al. 1987) although its influence is often confounded by its positive correlation with size. A cursory review of the previous chapters should reveal important biochemical, physiological, and anatomical differences that could give rise to sex- and age-dependent sensitivities. Some of these differences can produce unexpected results in combination. As an example, Williamson (1979) found that age and size of the land snail (*Cepaea hortensis*) had opposite effects on cadmium accumulation and probably the adverse effects of this toxic metal.

As a quick glance ahead to Chapters 16 through 18 will confirm, many opportunities exist for genetic qualities to contribute to tolerance differences.[6] There is no need to discuss genetic tolerance further at this point except to point out that one example described in Box 18.1 can be linked to the GAS. In that example, mosquitofish differed in the genetically determined form of a glycolytic enzyme (glucosephosphate isomerase) that is pivotal in the processing of glucose through metabolic pathways. Glucosephosphate isomerase-2 genotypes differed in their survival probabilities under stress and these differences were correlated with those in changes in glycolytic flux under general stress. Downward in the biological hierarchy, explanation for these response differences could notionally be linked to syntoxic hormone (glucocorticoid) responses in which blood glucose increases under stress. As done in Chapter 16, the glucosephosphate isomerase genotype differences during stress can also be projected upward in the biological hierarchy as one mechanism contributing to phenotypic plasticity and associated changes in life history strategies.

9.1.2.1 Individual Effective Dose Hypothesis

> On this theory, the dosage-mortality curve is primarily descriptive of the variation in susceptibility between individuals of a population ... the susceptibility of each individual may be represented by a smallest dose which is sufficient to kill it, the individual lethal dose.
>
> **(Bliss 1935)**

> The distributions of the individual effective doses and the results of the tests are in most cases "lognormal" ....
>
> **(Gaddum 1953)**

In modeling lethal effects, the variation in response among tested individuals is most often explained in the context of the individual effective dose or lethal tolerance hypothesis. The two quotes above present the essential features of this hypothesis. There is a minimum dose (or concentration) that is characteristic of each individual in a population at or above which it will die, and below which it will survive under the specified exposure conditions. For most populations, the distribution of such tolerances is believed to be described best by a log normal distribution with some individuals being very tolerant (Figure 9.3). Early toxicologists conjectured mechanisms for differences based on the then-popular Weber–Fechner Law[7] or conventional adsorption laws such as the Langmuir isotherm model. The context from which these conjectures emerged was conventional laboratory toxicity testing in which most variables such as animal age, sex, and size were controlled, so the tolerance differences being explained were inherent—perhaps genetic—qualities. However, because conventional ecotoxicity test data are generated for diverse inbred laboratory lines or field-collected

[6] See Mulvey and Diamond (1991) for a general review.

[7] A field called psycho-physics emerged during the first half of the nineteenth century in an attempt to quantify the intensity of human sensation resulting from a stimulus of a specified magnitude. The Weber–Fechner Law of psycho-physics states that the magnitude of the sensation (expressed on an arithmetic scale) increases in proportion to the logarithm of the stimulation. Extending this law, early toxicologists related the magnitude of toxic response to the logarithm of the dose or exposure concentration.

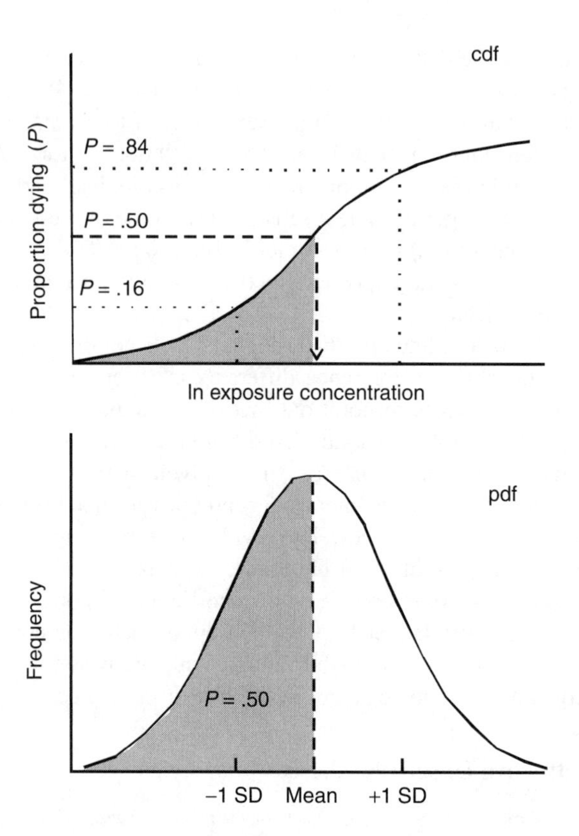

FIGURE 9.3 The upper panel shows the typical sigmoid concentration- (or dose-) mortality curve. The logarithm of the exposure concentration is plotted on the x-axis against the proportion of individuals dying during the exposure (P). This sigmoid curve can be described as a cumulative density function (cdf, upper panel) in which $P = .16$, .50, and .84 correspond to approximately –1 standard deviation below the mean, the mean, and +1 standard deviation above the mean. The antilogarithm of the x-value associated with $P = .50$ is an estimate of the median lethal concentration (LC50) or dose (LD50). The bottom panel shows the same data expressed as a probability density function, that is, as the conventional normal "bell curve." The cumulative area to the left of the mean is .50, corresponding to $P = .50$ in the cdf above.

individuals, it is difficult to imagine a genetic mechanism that consistently produced a log normal distribution of tolerances for most populations and toxicants. Mono- and multigenetic differences in tolerance (see Chapters 17 and 18) could produce a variety of distributions from ecotoxicity testing. Moreover, some conventional tests use metazoan clones (e.g., *Daphnia magna* or *Lemna minor*) or unicellular algal or bacterial cultures. It is difficult to invoke a genetic mechanism that produces a log normal distribution of tolerances for these diverse clones, laboratory strains, and field-caught individuals. It is more plausible that phenotypic plasticity (see Chapter 16) might generate variability in many of these cases but there does not seem to be a clear mechanism associated with phenotypic plasticity that would consistently produce a log normal distribution of tolerances. Regardless, this concept of a log normal distribution of inherent tolerance differences in all test populations was the first, and remains the dominant, explanation presented in the current ecotoxicology literature.

9.1.2.2 Probabilistic Hypothesis

> If it is seriously believed that there is some physical property more or less stably characterizing each organism, which determines whether or not it succumbs, then it is justifiable to advance the hypothesis of tolerances. In that case one should be prepared to suggest the nature of this characteristic so that

> the hypothesis may be capable of corroboration by independent experiments. If on the other hand the [log normal] formulation is only that of a "mathematical model" then it would be [better] . . . not to create any hypothetical tolerances . . . .
>
> **(Berkson 1951)**

This quote by Berkson precedes his counterargument that it is better to apply a log logistic model than a log normal one to toxicity data. But, more generally, it is an eminently reasonable point that remains inadequately addressed more than half a century later (see Box 12.2 in Chapter 12). Disinterest with the underlying mechanism by the founders of modern toxicology arises from pragmatism as is evident in the following quote from Finney's seminal book (1947):

> The validity and appropriateness of the logarithmic transformation in the analysis of experimental data are not dependent on the truth or falsity of any hypotheses relating to adsorption; use of the log concentration . . . requires no more justification than it introduces a simplification into the analyses.

In his arguments, Berkson (1951) related one experiment involving human tolerances to high altitude conditions that did not support the individual tolerance hypothesis, suggesting instead that differences in individual tolerances during testing were mostly random. Such a conclusion gives rise to an alternate explanation (probabilistic or stochasticity hypothesis) that most of the variation among similar individual's results from a random process (or processes) that is best modeled with a log normal or a similar skewed distribution. Which specific individual dies within a treatment is a matter of chance. Nearly half a century later, Newman and McCloskey (2000) tested these two hypotheses, rejecting the customary assumption that the individual tolerance hypothesis was the sole explanation for observed differences in response of lethally exposed individuals. The stochasticity hypothesis was supported in two cases and the individual tolerance hypothesis in another. Neither hypothesis alone was adequate to explain the observed differences. Similar conclusions were recently made by Zhao and Newman (2007) for amphipods (*H. azteca*) exposed to copper or sodium pentachlorophenol.

Two questions may have occurred to the critical reader at this point. First, why was the underlying mechanism for a foundation approach in classic toxicology left undefined for so long? Second, why is an understanding of the underlying mechanism important to the practicing ecotoxicologist? An inkling of an answer to the first question emerges from statements of prominent toxicologists of the time such as that of Finney above. Originally, the log normal model was applied to quantify relative poison toxicity or drug potency so it did not matter what the underlying mechanism was. Within the context of the laboratory bioassay, one chemical was or was not more potent than another. Classic toxicology could progress just fine without knowing the reason that data seemed to fit a skewed distribution. Precipitate explanation was presented without much scrutiny and the methods were broadly applied in studies of poisons and drugs. Unfortunately, because many ecotoxicologists tend to feel that anything good for mammalian toxicologists is good enough for them, it has been erroneously supposed that the underlying mechanism is also an esoteric issue in ecotoxicology, the science concerned with effects ranging from those to individuals to those to the biosphere. The error in this supposition can be shown in several ways but we will illustrate it here using only population consequences under repeated toxicant exposures. Suppose that a population was exposed for exactly 96 h to a toxicant concentration that kills half of the exposed individuals. Only the most tolerant individuals remain alive according to the individual tolerance theory but the stochasticity hypothesis would predict that, after recovery, the tolerances of the survivors will be the same as those of the original population. During a second exposure, the concentration-response curve could be very different (individual tolerance theory) or the same (stochasticity theory) as that for the original population during the first exposure. Indeed, during a sequence of such exposures, the survivors would drop in numbers by 50% during the first exposure and then remain at that number under the individual tolerance hypothesis but would drop

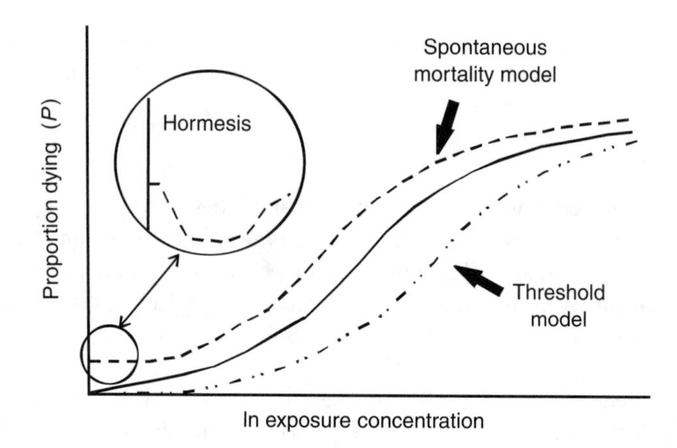

FIGURE 9.4 Conventional sigmoid and sigmoid models with spontaneous (natural) mortality or a dose/concentration threshold. The inset illustrates hormesis at sublethal concentrations.

down 50% with each exposure under the stochasticity hypothesis. The likelihood of local population extinction is quite different depending on which hypothesis is most appropriate or if both manifest in combination. Knowing which hypothesis is correct should be important to the ecotoxicologist attempting to predict population and associated community changes resulting from multiple exposures.

9.1.3 Spontaneous and Threshold Responses

The model shown in Figure 9.3 can have an additional feature in some cases. If the test involves a prolonged exposure relative to the longevity of the test organism or tested life stage of the organism, there can be a certain level of spontaneous (natural) mortality. Unfortunately, in still other cases in which the husbandry of the test species is imperfect, there may be background mortality associated with the general stress placed on the test organisms. In these cases, the mortality curve will take on an additional feature as shown in Figure 9.4.

Another change in Figure 9.3 is required if a threshold concentration or dose is characteristic of a chemical agent (Cox 1987). Like the minimum time-to-death described in Section 9.1.1, some toxicity test data appear to have a minimum concentration or dose that must be exceeded before any deaths occur in the test treatments (Figure 9.4).

9.1.4 Hormesis

> The nature of toxicologically-based dose-response relationships has a long history that is rooted in the development and interpretation of the bioassay. While the general features of the bioassay were clearly established in the 19th century, the application of statistical principles and techniques to the bioassay is credited to Trevan and the subsequent contributions of Bliss and Gaddum . . . [which] described the nature of the S-shaped dose-response relationship and the distribution of susceptibility within the context of the normal curve . . . . Despite this long history . . . of the S-shaped dose-response relationship, a substantial number of toxicologically-based publications from the 1880s to the present indicate that biologically relevant activity may occur below the NOAEL . . . .[8]
>
> **(Calabrese and Baldwin 1998)**

[8] The NOAEL (no observed adverse effect level) is a statistically derived measure often used to imply a threshold concentration or dose below which no effect will be observed. See Chapter 10, Section 10.3 for more detail.

As described in the above quote, the sigmoid model that emerged out of a long history of bioassay research has gained a well-deserved place in the mammalian and ecological toxicology literatures. However, its prominence comes at the expense of some important features. One such example has already been discussed (i.e., the weak foundation for the oft-assumed individual tolerance theory). Another is associated with the lower end of the dose/concentration–(lethal or sublethal) effect model. Hormesis is the apparent stimulatory effect of a toxicant at subinhibitory concentrations or doses. With hormesis, the sigmoid curve is not monotonic and, instead, drips down at very low doses or concentrations (Figure 9.4). Superficially, hormesis might seem counterintuitive. How can a small amount of a poison be "good" for an exposed individual? However, as we have seen, a stressor can evoke the GAS or some other process, creating the potential for overcompensation at low levels. To use Selye's terms, it can produce a state of heterostasis in which one aspect of fitness is conditionally enhanced. In Chapter 16, related shifts in phenotypes such as those associated with life history strategies under harsh environmental conditions, also provide a rationale for such "stimulation" under subinhibitory doses or concentrations.

Hormesis has been recognized for some time, being established at various periods under the labels of Arndt-Schultz law or Hueppe's rule; however, it is only recently being discussed as a general phenomenon, rather than a surprising oddity. Further discussion of hormesis and associated models can be found in Calabrese et al. (1987), Calabrese and Baldwin (1998, 2001), and Sagan (1987).

9.1.5 Toxicant Interactions

To this point, the lethal effects of single toxicants have been emphasized, but many exposures involve simultaneous exposure to several toxicants that can interact. There are two traditional vantages for discussing the joint action of toxicants: mode of action and additivity based.

Relative to mode of action, toxicants are said to have similar joint action if they act through the same mechanism. The joint lethal effects of two similarly acting toxicants can be predicted by knowing the dose or concentration of each toxicant and adjusting these concentrations for the relative potencies of each (Finney 1947). If toxicants have independent joint action, they have different modes of action and prediction of mixture effects is not as straightforward. In instances of potentiation, one chemical that is not toxic under the exposure conditions being considered can worsen—potentiate—the effect of another chemical. Synergistic action is the final joint action mode for which prediction is possible only after one has a sound understanding of the means by which one toxicant synergizes (increases) or antagonizes (decreases) the action of the other. Antagonism between chemical agents can result from a variety of mechanisms. A functional antagonism occurs if the two chemicals counterbalance one another by affecting the same process in opposite directions. Two chemicals combine to form a less potent product with chemical antagonism. Dispositional antagonism involves chemicals that influence the uptake, movement or deposition within the body, or elimination of each other in a way that lessens their joint effect. Finally, receptor antagonism occurs if one chemical blocks the other from a receptor involved in its action and, in doing so, lowers its ability to adversely affect the exposed organism.

Mixture treatment in terms of additivity is based on deviations from simple addition of two or more toxicant effects. Two or more chemicals are said to be (effect) additive if their combined effect in mixture is simply the sum of the effects expected for each if each were administered separately. If their effects together are less than additive, they are said to be acting antagonistically. If their effects together are greater than additive, they are said to be acting synergistically. This approach will not be described in further detail because it provides less potential for linkage between suborganismal and organismal population-level effects than the vantage based on mode of action.

9.2 QUANTIFYING LETHALITY

9.2.1 GENERAL

> In 1927 Trevan drew attention to the fact that the threshold dose [of a drug] varies enormously even when the animals are as uniform as possible, and proposed that toxicity testing should be based on the median lethal dose, which kills 50 per cent of the animals.
>
> **(Gaddum 1953)**

The toxicity test methods employed by modern ecotoxicologists have their roots in mammalian toxicology, where the aim was to determine the relative toxicity of poisons or relative potencies of drugs (e.g., Bliss and Cattell 1943). In contrast, results of toxicity testing are used by the ecotoxicologist to infer consequences to valued individuals, populations, and ecological communities. Borrowing methods from classical toxicology accelerated the establishment of ecotoxicity tests when they were sorely needed. Unfortunately, the differences in goals in applying these methods led to the amassing of interpretive incongruities, lethality metrics that are less useful than other metrics, and consequently, a habit of being reluctantly satisfied with weak scientific inferences about lethal consequences. The strong and weak points of the conventional approach will be described in Section 9.2.2.

9.2.2 DOSE or CONCENTRATION–RESPONSE MODELS QUANTIFYING LETHALITY[9]

A well-established approach exists for quantifying lethal effects from data sets of concentration versus proportion of exposed individuals dying. The most common is the log normal model discussed above which involves log transformation of the concentration and then fitting of the data to the following model (Finney 1947):

$$P = \frac{1}{\sigma\sqrt{2\Pi}} \int_{-\infty}^{x_0} e^{-[(1/2\sigma^2)(x-\mu)^2]}\, dx, \qquad (9.2)$$

where P = the proportion expected to die, x_0 = the concentration for which predictions are being made, μ = the mean, and σ = the standard deviation.

Early in the formulation of quantitative methods for dealing with concentration-effect data, there was a need to transform data into terms that could easily be dealt with using simple logarithm tables and mechanical adding machines. The model above was transformed accordingly by expressing the proportions responding in units of standard deviations from the mean of the normal distribution.[10] The name given to this transformed proportion was the normal equivalent deviation (NED). This transformation still resulted in some computational inconvenience at that time because NED values for proportions below 0.5 were negative numbers. Simply to avoid negative numbers in computations, five was added to the NED to produce the probit transformation: $\text{Probit}(P) = \text{NED}(P) + 5$. A plot of NED or probit versus log concentration should produce a straight line if the log normal model was appropriate for a data set. Now, using some method such as maximum likelihood estimation, these types of data could be fit to a model such as the following:

$$\text{Probit}(P) = a + b(\log C) + \varepsilon \qquad (9.3)$$

---

[9] For convenience, concentration will be used in discussions in this section although both dose and concentrations can be applied in the described data analysis methods.

[10] That is to say the data are normally distributed when the logarithm of the concentration is used instead of concentration.

where C = the exposure concentration, a = an estimated regression intercept, and b = an estimated regression parameter accounting for the influence of exposure concentration. Because no advantage exists for using the probit transform after the advent of modern computers, models also are formulated using the NED instead of the probit. Nonlinear fitting can also be done with standard software without any computational difficulty.

The simple generalized model can be specified based on the cumulative normal function ($\Phi()$),[11]

$$P = \Phi[a + b(\log C)]. \tag{9.4}$$

Spontaneous mortality (P_S = the proportion of unexposed individuals dying) can be included in this model. If $P \geq P_S$,

$$P = P_S + (1 - P_S)(\Phi(a + b(\log C))) \tag{9.5}$$

and $P = P_S$ at $C = 0$. A lethal threshold can also be included in Equation 9.4 for concentrations (C) greater than the threshold concentration (C_T),

$$P = \Phi\{a + b[\log(C - C_T)]\}. \tag{9.6}$$

The P approaches 0 if $C \leq C_T$ for this model. This model can be modified further to include natural mortality (e.g., Equation 9.5) in which case $P = P_S$ if $C \leq C_T$. Including hormesis in these models is more involved but can be done as demonstrated by Bailer and Oris (1994).

Several other functions are commonly fit to these kinds of data. Those associated with the log logistic (or logit) model are the most common alternatives to the log normal functions just described. Conventionally, the log odds or logit transformation is applied:

$$\text{Logit}(P) = \ln\left[\frac{P}{1-P}\right]. \tag{9.7}$$

For historical reasons of convenience such as those just described for the probit transform, this logit is often transformed further to avoid negative numbers and to produce values similar to probit values:

$$\text{Transformed logit} = \frac{\text{Logit}(P)}{2} + 5. \tag{9.8}$$

A less common, but very useful, transformation is the Wiebull transformation:

$$\text{Wiebit}(P) = \ln[-\ln(1 - P)]. \tag{9.9}$$

Christensen (Christensen 1984, Christensen and Nyholm 1984) used the Wiebull function very effectively in modeling ecotoxicity effects. The next most commonly applied is the Gompertz or extreme value function (Gompit transformation). Newman (1995) provides an example of applying a Gompertz model to ecotoxicity data. All of these models can be applied to concentration-lethal response data after appropriate substitutions into Equations 9.4 through 9.6.

---

[11] To illustrate the ease to which these calculations can now be done, invoking the Excel™ function NORMINV(Probability, Mean, Standard Deviation) where Probability = the proportion for which the calculation is to be done, Mean = the distribution mean, and Standard Deviation = the distribution standard deviation, calculates the NED if mean = 0 and standard deviation = 1, that is, for the unit normal curve, $N(0,1)$. As an example, NORMINV(.84134474,0,1) will return 1. The following function would return the probit, NORMINV() + 5.

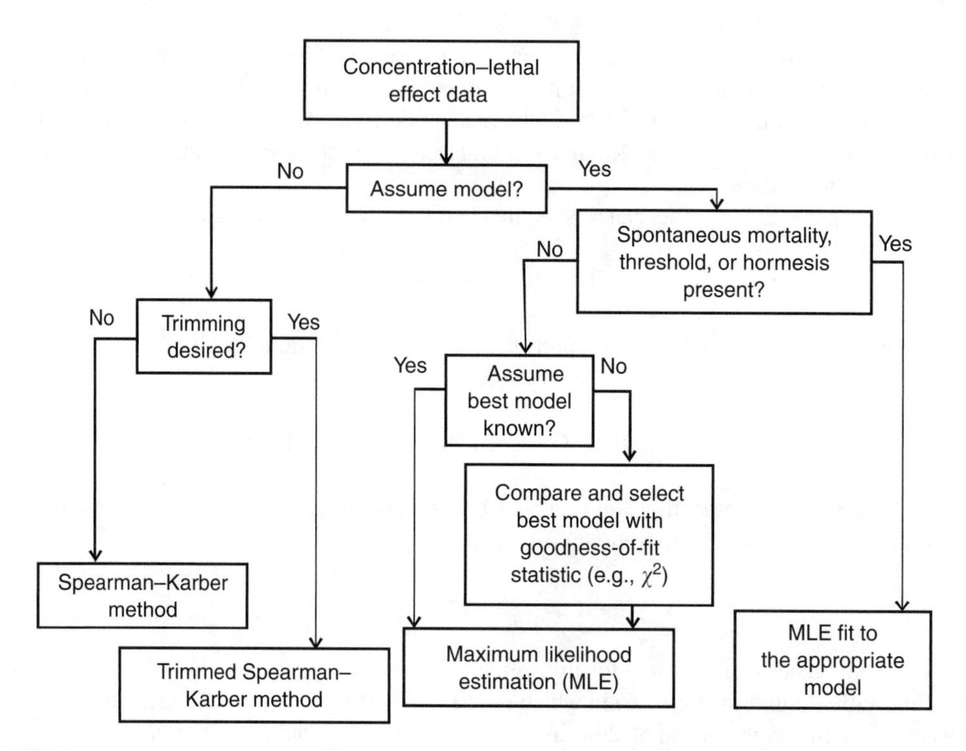

FIGURE 9.5 Methods for estimating LC50 and associated confidence limits from dose/concentration versus proportion dying data sets. Although not shown in the diagram, these summary statistics can be eked out of data sets in which all of the treatments had either complete or no mortality at all. A binomial method can provide an estimate for such data sets with no partial kills.

Parametric and nonparametric methods exist for analyzing data from concentration-lethal response tests. Many can also be applied for nonlethal effects. Each (Figure 9.5) carries advantages and disadvantages. The best methods can be applied if one assumes an explicit model. The presence of spontaneous or threshold mortality, or hormesis requires a model incorporating these features. Such more complicated models are available that use maximum likelihood methods to estimate the associated model parameters and lethality metrics such as the LC50. Most concentration-lethal effect data are analyzed using simpler models that assume a specific model. If there is no a priori reason to select one model over another, for example, log normal over the log logistic, Gompertz, or Weibull, the data can be fit to all of the candidate models and then the results compared. Comparison usually involves plotting the actual data and model predictions, and also calculating a goodness-of-fit statistic such as the χ^2-statistic. The model providing the best fit is selected for estimating model parameters and predicting metrics such as the LC50 and its 95% fiducial (confidence) limits. Nonparametric methods can be used to estimate the LC50 and 95% fiducial limits if an acceptable model was not apparent. The Spearman–Karber method, with or without trimming, is the most commonly applied nonparametric approach. Most applications of the Spearman–Karber approach conform to recommendations of Hamilton et al. (1977), especially those about trimming rules. In some applications of toxicity testing, there are no partial kills: each treatment in the test has either no mortality or complete mortality. Stephan (1977) suggested that an LC50 and associated confidence limit could be estimated from such data using a binomial method. Essentially, the LC50 can be estimated from the highest concentration treatment with no mortality (C_{NO}) and the lowest concentration treatment with complete mortality (C_{ALL}),

$$\mathrm{LC50} = \sqrt{C_{\mathrm{NO}} C_{\mathrm{ALL}}}. \tag{9.10}$$

The interval from C_{NO} to C_{ALL} is at least the 95% confidence interval for the LC50 if the number of individuals exposed in each treatment was five or more. He suggests that the exact percentage for the estimated confidence interval is the following if the same numbers of individuals were exposed in the C_{NO} and C_{ALL} treatments:

$$\text{Coefficient} = 100[1 - 2(0.5)^n], \tag{9.11}$$

where n is the number of individuals in the treatment. If the numbers of individuals were different for the C_{NO} (n_{NO}) and C_{ALL} (n_{ALL}), the following equation is used:

$$\text{Coefficient} = 100[1 - (0.5)^{n_{\mathrm{NO}}} - (0.5)^{n_{\mathrm{ALL}}}]. \tag{9.12}$$

Models for joint action of toxicants in mixture build upon these models. If two independently acting chemicals, A and B, were combined at specific concentrations in an exposure solution, the proportion dying (or probability of dying) of individuals exposed to the mixture ($P_{\mathrm{A+B}}$) can be predicted from the proportion/probability of death if the individuals were exposed to A alone at the specified concentration (P_{A}) and the proportion/probability of death if the individuals were exposed to B alone at the specified concentration (P_{B}).

$$P_{\mathrm{A+B}} = P_{\mathrm{A}} + P_{\mathrm{B}}(1 - P_{\mathrm{A}}) = P_{\mathrm{A}} + P_{\mathrm{B}} - P_{\mathrm{A}}P_{\mathrm{B}}. \tag{9.13}$$

The reason $P_{\mathrm{A+B}}$ is not simply the sum of P_{A} and P_{B} is easily understood in terms of probabilities.[12] If an outcome can result from two independent processes with associated probabilities of P_{A} and P_{B}, the probability of the event occurring is defined by Equation 9.13. The term, $-P_{\mathrm{A}}P_{\mathrm{B}}$, is needed to adjust for the fact that, if an organism dies from A, it is not available to die from B. This model can be expanded to include many toxicants:

$$P_{\mathrm{A+B+C+\cdots}} = 1 - (1 - P_{\mathrm{A}})(1 - P_{\mathrm{B}})(1 - P_{\mathrm{C}}) \cdots . \tag{9.14}$$

A slightly different model is required if the two toxicants in mixture display similar action. To implement this approach, Finney (1947) noted that toxicity curves are parallel for toxicants with similar action. The influence of each toxicant alone could be modeled with a conventional probit model and then the two models combined as shown below to predict the joint effect. Let Equations 9.15 and 9.16 be the probit (i.e., log normal) models for each toxicant alone:

$$\text{Probit}(P_{\mathrm{A}}) = \text{Intercept}_{\mathrm{A}} + \text{Slope}(\log C_{\mathrm{A}}), \tag{9.15}$$

$$\text{Probit}(P_{\mathrm{B}}) = \text{Intercept}_{\mathrm{B}} + \text{Slope}(\log C_{\mathrm{B}}). \tag{9.16}$$

The log of the relative potency of A and B ($\log \rho$) can be estimated from these two models:

$$\log \rho_{\mathrm{B}} = \frac{\text{Intercept}_{\mathrm{B}} - \text{Intercept}_{\mathrm{A}}}{\text{Slope}}. \tag{9.17}$$

This relative potency measure can now be used to combine both toxicants into one model:

$$\text{Probit}(P_{\mathrm{A+B}}) = \text{Intercept}_{\mathrm{A}} + \text{Slope} \cdot \log(C_{\mathrm{A}} + \rho_{\mathrm{B}} C_{\mathrm{B}}). \tag{9.18}$$

More similarly acting toxicants can be included in the model using the appropriate relative potencies.

---

[12] This is a simple case of the general probability law of independence.

9.2.3 Time–Response Models Quantifying Lethality

Uncertainty enters predictions of exposure consequences because all field exposure durations are not identical to those set in the concentration-effect tests described in the last section. As discussed at the beginning of this chapter, an argument of convenience was made that most acute mortality manifests within a specific time and the LC50 value for that duration is sufficient for predicting effects at all other acute exposures. After frequent repetition, but weak scrutiny, such arguments have become generally accepted.

> Concerning acute toxicity to fish, there seems to be a working consensus that it occurs within the first 100 hr of exposure .... Of 375 cases [examined], 211 or 56 per cent showed a lethal threshold in 4 days or less. Only 42 cases are clearly longer than 4 days .... The overall distribution tends to substantiate that 4 days is a reasonable limit of occurrence of acutely lethal toxicity of most pollutant ... [but] Caution in generalizing too much from these results is particularly necessary since such a tabulation may apparently be easily biased.
>
> **(Sprague 1969)**

What was really being advocated in this and similar statements that emerged during a period when environmental issues required immediate, pragmatic solutions? A critical reading of Sprague's argument indicates that 4 days was not sufficient for 4 of 10 tests. Sprague also indicates that considerable data used in his tabulations came from static toxicity tests. The exposure solutions were not changed in these static tests and, therefore, it is very plausible that the toxicant concentrations dropped substantially during the test. Although Sprague's inference was weak, he attempted responsibly to address the expediency of establishing a way to approach very real and immediate problems: it was a pragmatic stance that early ecotoxicologists accepted in order to move forward. Unfortunately, the position is still taken uncritically in the current literature and applied to problems requiring much more certitude than it can afford. Important studies that try to relate LC50 values derived for one duration to consequences of importance to ecotoxicologists and risk assessors at other durations are still done (e.g., Stark 2005).

Slight changes to the conventional concentration/dose-lethal effect framework allow the ecotoxicologist to obtain the increasingly essential information about exposure duration effects. Proportions dying in each treatment might be noted at several times during a test and LC50 values estimated for each; however, the estimates might be suboptimal for many durations because the test treatment concentrations are normally selected to give the best distribution of responses at one duration. Classic toxicological approaches include estimation of LC50 values for a set of durations and then producing plots such as a logarithm of LC50 versus logarithm of duration to extrapolate from one duration to another. This involves many tests and organisms if treatment concentrations are optimized at each duration (see Gaddum (1953), Newman (1995), Sprague (1969) for details). Methods even exist for extrapolating from the abundant acute lethality metrics to chronic lethal effects using a variety of approximations (Mayer et al. 1994). Shareware is available to facilitate the associated calculations (Ellersieck et al. 2003). Recently, Duboudin et al. (2004) used species distributions to do such extrapolations. However, each of these approaches can fall short of predicting with the necessary certainty the lethality expected for different exposure durations or the associated mortality rates needed for population modeling.

> Although adequate to address questions posed when they were first established, current methods for generating and summarizing mortality data are inadequate for answering the complex questions associated with ecological risk assessment. Time to event methods have the potential to improve this situation. The two critical components of exposure intensity (concentration or dose) and duration, can be included in the associated predictions.
>
> **(Newman and McCloskey 2002)**

Originally defined as reaction time assays, a wide range of survival time or time-to-event methods that steer clear of many of the shortcomings just described exist. The associated experimental design is similar to that used in conventional concentration-effect tests except that the time-to-death for each individual is noted during testing. Time-to-death might be noted as an exact time or as occurring within an interval such as "between 4 and 8 h of exposure." Qualities of individuals (e.g., sex or size) can also be included in the data set because the response variable (time-to-death) is associated with individuals instead of the tank or cage of individuals (e.g., proportion dying). The substantial increase in information afforded for lethality tests by measuring time-to-death instead of proportion of exposed individuals responding at a specified time has been known since the advent of classic methods (e.g., Bliss and Cattell 1943). Time is also incorporated directly into the time-to-death models. Although these methods were underutilized for decades by ecotoxicologists, they are now being applied increasingly to relevant problems (Crane et al. 2002, Newman 1995, Newman and Dixon 1996).

Survival analysis can be conducted in a variety of ways that share common characteristics. Most important, discrete events are noted through time in this approach. Although in the case of death, the event can only occur once for an individual, other events such as time-to-partition, time-to-stupification, or time-to-flower can occur several times and can be accommodated in time-to-event models. Also common is the presence of survivors (nonresponders) at the end of the test. Such individuals are identified as right censored (i.e., having times-to-death longer than the test duration) and incorporated into the models accordingly. Because it is common to have censored individuals, most survival time models are fit by a computationally intense method such as maximum likelihood estimation. If times-to-death were noted within wide intervals for logistical reasons, the associated times-to-death are recorded as having occurred within the interval instead of at a particular moment. Maximum likelihood methods can accommodate such interval censoring.

Nonparametric, semiparametric and fully parametric survival methods exist for analyzing time-to-death data sets (Figure 9.6). All of these methods are described in detail elsewhere and are only described here enough for the reader to understand the general advantages and disadvantages of

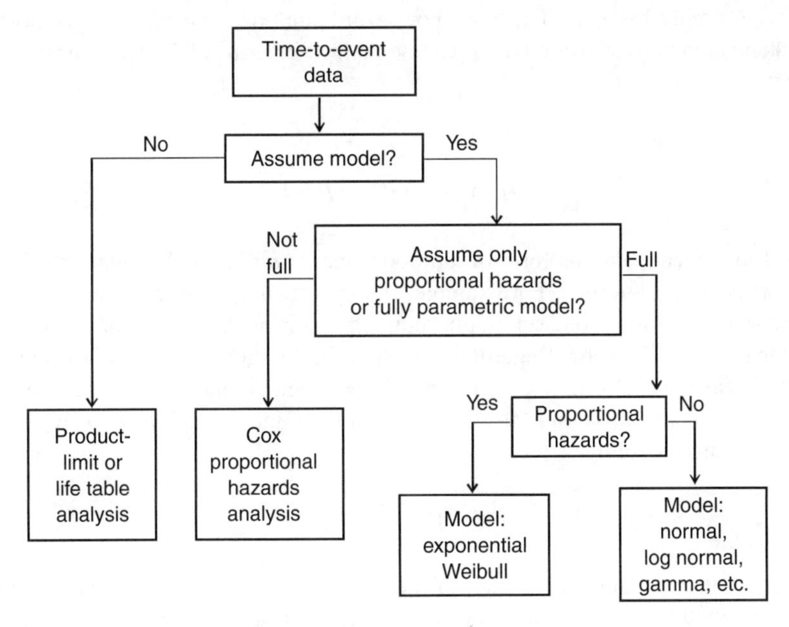

FIGURE 9.6 Methods for analyzing time-to-event data including nonparametric, semiparametric and fully parametric methods.

each approach. The reader is directed to Miller (1981), Cox and Oakes (1984), Marubini and Grazia Valsecchi (1995), Newman (1995), and Crane et al. (2002) for more detail.

Nonparametric methods include the Product Limit (also called Kaplan–Meier) and life table methods. Life table methods will not be described here because they are described in Chapter 15. The Product Limit approach allows estimation of survival through a time course and the associated variance for each estimate. The estimated cumulative survival for a group of individuals can be calculated with the following equation:

$$\hat{S}(t_i) = \prod_{j=1}^{i} \left(1 - \frac{d_j}{n_j}\right), \tag{9.19}$$

where t_j = a specific failure time, n_j = the number of individuals alive before and available to die at t_i, and d_j = the number dying at t_j. Obviously, cumulative mortality ($\hat{F}()$) is estimated as $1-\hat{S}()$. The $\hat{S}()$ is appropriate for all times up to the end of the exposure experiment and is undefined thereafter. Greenwood's formula can be used to estimate the variance associated with each estimated $\hat{S}()$ (or ($\hat{F}()$):

$$\hat{\sigma}^2 = \hat{S}^2(t_j) \sum_{j=1}^{i} \frac{d_j}{n_j(n_j - d_j)}. \tag{9.20}$$

Nonparametric methods including several rank sum tests are available if differences in survival curves are to be tested statistically.

Fully parametric methods fit a specific model to the survival time data, allowing data description, testing for significant effects of covariates, and communication of lethal risk. A few statistics must be defined before the different models can be understood. Cumulative mortality at any time ($F(t)$) is the number of individuals dead at t divided by the total number of individuals exposed. The cumulative survival ($S(t)$) is simply $1 - F(t)$. The hazard or hazard rate ($h(t)$ = instantaneous mortality rate at a moment, t) and cumulative hazard ($H(t)$) can be defined in terms of $F(t)$ and $S(t)$:

$$h(t) = -\frac{1}{S(t)} \frac{\mathrm{d}S(t)}{\mathrm{d}t}, \tag{9.21}$$

$$H(t) = -\ln[1 - F(t)]. \tag{9.22}$$

A survival model can take the form of a proportional hazard model (Equation 9.23) if hazards remain proportional through time. For example, if survival curves for males and females are generated, the proneness to die of one sex might remain proportional by the same amount to the other throughout the exposure duration. Regardless of exposure duration, one simply uses one proportion to predict the hazard rate of one sex from that of the other. Similarly, the influence of the logarithm of exposure concentration on $h(t)$ would remain the same through time. The general form of a proportional hazard model is

$$h(t, x_i) = \mathrm{e}^{f(x_i)} h_0(t), \tag{9.23}$$

where $f(x_i)$ is some conventional function of the covariate x, and $h_0()$ is the baseline hazard rate that is being modified by the covariate. As an example, a linear function of logarithm of exposure concentration ($a + b(\log \text{Concentration}_i)$) might modify the baseline proneness to die of the exposed individuals: the rate of mortality is changed by the covariate to the same degree regardless of duration of exposure.

Box 9.2 Survival of Salted Fish

Newman and Aplin (1992) exposed female mosquitofish (*G. holbrooki*) to a series of sodium chloride concentration treatments to illustrate survival modeling of ecotoxicologically relevant data. Briefly, subsets of female mosquitofish differing in size were exposed to one of six concentrations of sodium chloride ranging from 10.3 to 20.1 g/L, and time-to-death for fish recorded at 4-h intervals for 96 h. In addition to the time-to-death for each fish, duplicate tank to which the fish had been assigned within an exposure treatment, exposure concentration (g/L), and individual fish wet weight (g) were included in the data set. After nonparametric rank sum methods testing of survival curves for duplicate tanks detected no significant difference ($\alpha = 0.05$), data for duplicate tanks were combined for each treatment concentration and a parametric survival model generated using a variety of underlying distributions (Newman and Dixon 1996).

The accelerated failure time model assuming a log logistic model will be used here for purposes of illustration.

$$\text{Time-to-death} = \mathrm{e}^{15.2860}\,\mathrm{e}^{-4.2129(\ln[\mathrm{NaCl}])}\,\mathrm{e}^{0.2545(\ln \mathrm{Wgt})}\,\mathrm{e}^{0.2081L_p},$$

where L_p = a value for the log logistic function corresponding with the desired proportion dead (p) for which the prediction is being made. The units of [NaCl] and Wgt are g/L and g wet weight, respectively. The L_p is simply the value obtained if the p was inserted into the logistic function $\{\ln[p/(1-p)]\}$. With this model, the proportion of exposed fish dying can be predicted for combinations of exposure concentration and duration. Given a particular L_p predicted for some combination of concentration and duration, the corresponding p dying is calculated from this relationship or extracted from a table such as Appendix 7 in Newman (1995). In fact, an entire p response surface for all combinations of concentration and duration (within the tested ranges) can be generated.

Other useful lethality metrics can be produced. The LC50 can be calculated for any exposure duration within the tested range by rearranging the model

$$\mathrm{LC50} = \frac{\ln t - 15.2860 - 0.2545(\ln \mathrm{Wgt}) - 0.2081L_{0.5}}{-4.2129},$$

where LC50 = the LC50 for the duration of interest (t), $L_{0.5} = 0$, and Wgt = the wet weight of the fish for which predictions are being made.

Often, hazard rates do not remain proportional through time and an accelerated failure time model is more appropriate

$$\ln t_i = f(x_i) + \varepsilon_i, \tag{9.24}$$

where ε = the error term. In this formulation, the time-to-death instead of the hazard rate is being modified directly by the covariates.

In other cases, it may not be possible or desirable to generate a fully parametric survival model. For instance, the underlying distribution might be uninteresting but the relative hazards of tested groups might be important. Specific examples might be a study interested in determining the relative risk of two populations under different exposure scenarios or a study of the success of a remediation activity. In both cases, the proportional hazards are the focus, not the nature of the underlying distribution (h_0 in Equation 9.23). The Cox proportional hazard method is a semiparametric method that allows one to estimate the proportional hazards without obligating the user to fully define h_0.

9.3 LETHALITY PREDICTION

> The fundamental premise is that the structure of a chemical implicitly determines its physical and chemical properties and reactivities, which, in interaction with a biological system, determine its biological/toxicological properties.
>
> **(McKinney et al. 2000)**

Often the relative potencies of similar chemicals can be predicted based on their chemical properties. Which qualities are most useful depends on the chemical class and effects for which predictions are being made. Examples of such predictions are provided below for lipophilic organic compounds, ionizable organic compounds, and metal cations. Some of the basic concepts associated with these examples were provided already in Chapters 3 and 7, and as a consequence, will get only brief mention here.

9.3.1 ORGANIC COMPOUNDS AND THE QSAR APPROACH

One of the earliest forms of what would later become known as quantitative structure–activity relationships (QSARs) (or physicochemical property–activity relationship) is the Meyer–Overton rule. This rule states that the potency of candidate anesthetics can be quantitatively predicted within a class of compounds from their oil:water or oil:air partition coefficients. The general theme that bioactivity or bioavailability of nonpolar organic compounds can be related to lipophilicity has expanded vastly during the last century to include diverse classes of compounds and species, including description of deviations from the rule (e.g., Cantor 2001). Even exceptions can be quantitatively predicted with QSAR that incorporate other molecular qualities. Rich medical, pharmacological, and ecotoxicological literatures now exist for the QSAR approach (McKinney et al. 2000). Depending on the class of compounds of interest, these QSAR may require good information about nucleophilicity (i.e., how readily the compound donates electrons to form a covalent bond), electrophilicity (i.e., how readily the compound accepts electrons), topology (e.g., molecular connectivity), or steric qualities (e.g., steric hindrance or total molecular surface area). They use such qualities to predict effects of specific compounds from a group that shares a common mechanism, such as narcosis, acetylcholinesterase inhibition, membrane irritation, or respiratory uncoupling (McKim et al. 1987, Ren 2003, Schultz and Cronin 2003). Therefore, some knowledge of mode of action is also extremely helpful for effective QSAR generation. In addition to data about compound qualities and modes of action, QSAR development requires statistical or mechanistic models to construct and then validate predictions (Schultz and Cronin 2003). Models vary greatly in complexity. Some modern QSAR use complex computational models (e.g., 3D-QSAR) to predict potential effects based on availability and configuration of reactive regions on molecules (Chen et al. 2004, McKinney et al. 2000, Tong et al. 2003). QSAR are even applied to assess potential interactions as illustrated in the work of Altenburger et al. (2005) with algae exposed to mixtures of nitrobenzenes. The predictive utility of QSAR continues to improve as our knowledge of modes of action grows and computational tools become widespread. A wide range of computer programs that facilitate the implementation of QSAR in ecotoxicological studies are now available (Moore et al. 2003). Simple and complex QSAR have become essential tools in many regulatory activities because testing is impossible for all new organic compounds introduced annually (Zeeman et al. 1995).

Box 9.3 Narcosis by the Numbers

McKim et al. (1987) inferred common modes of action for organic compounds, including narcotics, using distinct physiological and biochemical shifts in exposed organisms. They identified what they called the fish acute toxicity syndrome by multivariate statistical analysis of these changes. Narcosis was characterized in their fish acute toxicity syndrome scheme by a dramatic drop in respiratory and cardiovascular functions, and a range of hematological adjustments to consequent hypoxia. Despite the similarity in symptoms for polar and nonpolar narcotics, differences in lethality existed in the QSAR developed for fish exposed to these two classes of narcotics. Adjustment for ionization (see Section 3.10 in Chapter 3) did not always resolve differences among polar and nonpolar narcotic compounds relative to their lethal potency. However, one team (Vaes et al. 1998) was able to account for these differences and their approach is outlined briefly here.

Utrecht University workers generated a nice illustration of the QSAR approach for predicting lethal effects of organic compounds with a narcosis mode of action (Vaes et al. 1998). This was done by focusing carefully on the differences in partitioning between water and the phospholipids of the cell membrane. Lethality data (LC50) were collected for guppies (*Poecilia reticulata*) and fathead minnows (*Pimephales promelas*) exposed to polar or nonpolar narcotic compounds. Consideration of polar and nonpolar narcotics together can require inclusion of the pK_a for the polar narcotics but, in this study, conditions where used in which polar compounds remained unionized. This allowed polar and nonpolar narcotics to be combined into the same QSAR that focused only on lipid solubilities.

First, as done with many conventional QSAR, the logarithm of each LC50 value for the polar and nonpolar narcotics was plotted against the logarithm of the corresponding octanol–water partition coefficient (K_{OW}). Two distinct linear relationships became apparent (Figure 9.7): the unionized polar narcotic compounds were more toxic than nonpolar narcotics with similar lipophilicities. Reasoning that the K_{OW} was not an ideal metric for partitioning of a compound between actual membrane phospholipids and water, Vaes's group built another QSAR using a more realistic partition coefficient for the cell membrane. When a partition coefficient for L-α-dimyristoylphosphatidylcholine (DMPC) and water was used, data for all narcotics (except quinoline) converged into a single line ($r^2 = .98$). The unionized polar and nonpolar compounds fit to the same QSAR when a more realistic partition coefficient was used.

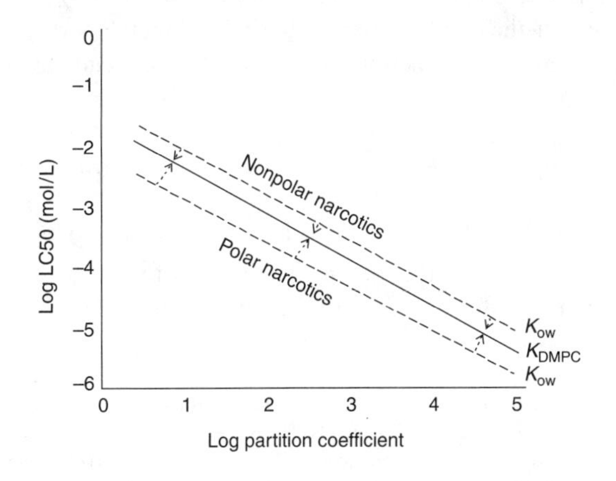

FIGURE 9.7 Partition coefficient-based QSAR models for polar and nonpolar narcotics based on K_{OW} (separate dashed lines) and K_{DMPC} (single solid line). (This figure is a composite of both panels of Figure 1 of Vaes et al. (1998).)

9.3.2 Metals and the QICAR Approach

> Quantitative structure–activity relationships (QSARs) are applied widely to predict bioactivity (e.g., toxicity or bioavailability) of organic compounds. In contrast, models relating metal ion characteristics to their bioactivity remain underexploited.
>
> **(Newman et al. 1998)**

As described briefly in Chapters 7 (Section 7.2.3) and 8 (Section 8.2), qualitative rules exist for predicting trends in metal bioactivity. More than a century ago, Matthews (1904) found that the ionic form of a metal is generally its most toxic one. This concept later was referred to as the Ionic Hypothesis. He correlated metal bioactivity with metrics of metal binding to oxygen, nitrogen, or sulfur donor atoms of biomolecules. A series of researchers (Babich et al. 1986a,b, Biesinger et al. 1972, Binet 1940, Jones 1939, 1940, Jones and Vaughn 1978, Kaier 1980, Loeb 1940, McGuigen 1954, Turner et al. 1985, Willams and Turner 1981) expanded this approach, incorporating a range of metals, species, and metal qualities into their qualitative models. Most notably, Jones (1939), Jones and Vaughn (1978), Williams and Turner (1981), and Turner et al. (1985) squarely framed these trends within the context of fundamental Hard Soft Acid Base (HSAB) theory.[13] All that remained to be done to produce QSAR-like models for metals was to extract metrics of metal binding tendencies from the literature and then apply statistical methods to the trends noted by these early workers. Newman and coworkers (McCloskey and Newman 1996, Newman and McCloskey 1996, Newman et al. 1998, Ownby and Newman 2003, Tatara et al. 1998) did this, using what they called a quantitative ion character–activity relationship (QICAR) approach. Newman et al. (1998) and Ownby and Newman (2003) provide a general description of this QSAR-like approach, discussing the ion characters most useful in producing predictive QICAR models.

Box 9.4 Metal Interactions by the Numbers

As mentioned above, Newman and coworkers described the QICAR approach to quantitatively predict relative metal bioactivities. Newman and McCloskey (1996) briefly assessed whether the QICAR approach could be applied to binary metal mixtures. Using a Microtox® bioassay, Ownby and Newman (2003) conducted a more extensive study, applying the concepts associated with independent action (Equation 9.13). The similarities in binding tendencies between the paired metals were quantified using the softness index (σ_p), a measure of the metal ions tendency to share electrons with ligand donor atoms. Here, these same data are analyzed again using the following modification of Equation 9.13 in which an interaction coefficient (β) is added. The interaction coefficient would be 1 if the joint actions of the mixed metals were perfectly independent. It would deviate from 1 as the joint actions deviated from the assumed independence of action.

$$P_{\mathrm{A+B}} = P_{\mathrm{A}} + P_{\mathrm{B}} - \beta P_{\mathrm{A}} P_{\mathrm{B}}. \tag{9.25}$$

The value of the interaction coefficient for paired metals was estimated by fitting Equation 9.25 to data from a series of metal mixture tests (upper panel of Figure 9.8). The x-axis of Figure 9.8 is the absolute value of the difference between the softness indices of the paired metals. A very

---

[13] HSAB theory provides a general scheme for quantifying metal binding tendencies. The hard–soft label has to do with the propensity for the metal's outer electron shell to deform during interactions with ligands. The A class metals in the periodic table (IA, IIA, IIIA) tend to be hard relative to the softer b class metals (in periods from Mo to W, and Sb to Bi) (Fraústo da Silva and Williams 1991, Jones and Vaughn 1978). Hard metals are not as polarizable as soft metals. Other metals are intermediate to the soft and hard metals. Acid–base refers to the Lewis acid (accepting an electron pair) or base (donating an electron pair) context for predicting the nature and stability of the metal interaction with ligands.

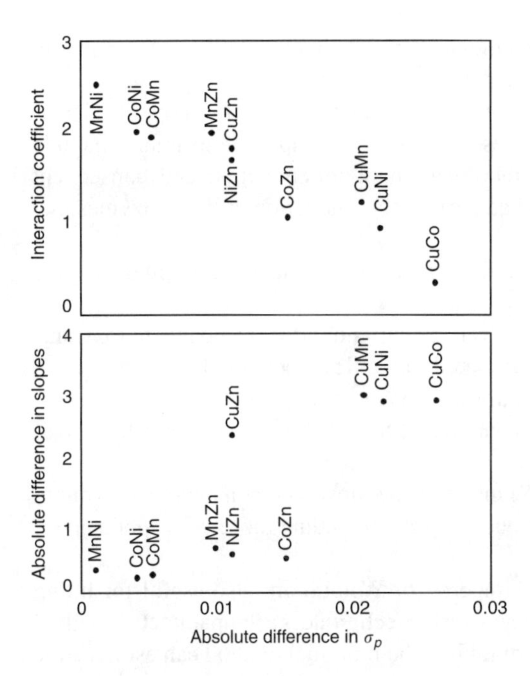

FIGURE 9.8 A σ_p-derived QICAR predicting binary metal effects based on joint action models for independent (upper panel) and similar (lower panel) acting toxicants. (Recalculated using data generated by Ownby and Newman (2003).)

small absolute difference would indicate very similar binding tendencies for the paired metals and a very large absolute difference would indicate very dissimilar binding tendencies. On the basis of the hypothesis that similar binding tendencies result in similar modes of action, Ownby and Newman reasoned that deviations from perfect independence of the paired metals could be predicted from the absolute difference in their binding tendencies as quantified with σ_p. The data clearly supported this assumption.

The bottom panel was generated from the vantage of similar joint action model. A review of our discussions about Equations 9.15 through 9.18 should make it clear that paired metals with similar joint action should have identical slopes. Therefore, the absolute difference in slopes for the probit models for the paired metals (Equations 9.15 and 9.16) should be very small for metals with similar joint action but become increasingly larger as the paired metals deviate from similar joint action. It is also clear from this plot that deviation from similar action was related to differences in binding tendencies. Although not as clear a relationship as that based on joint independent action (Equation 9.25), this approach did not require a full mixture experiment, only the slopes from single metal concentration-response models. So, in general, metal interactions could be predicted based on binding tendencies using either independent or similar joint action models.

9.4 SUMMARY

The manner in which lethal effects are addressed was detailed in this chapter. Topics included underlying mechanisms and dynamics, factors contributing to individual differences in tolerance, types of models used to quantify mortality, models for predicting lethal effects of toxicant mixtures, QSAR, and QICAR. Below are some of the major points covered in this chapter.

9.4.1 Summary of Foundation Concepts and Paradigms

- Lethal effects are characterized as acute or chronic although the distinction is often blurry.

- It is important to keep in mind that the dynamics by which lethality manifests differ among modes of action, exposure route, and species. Some exposures display thresholds.
- In addition to death due to conventional toxicant modes of action, diseases of adaptation can result from health-enhancing processes gone awry. Important examples include processes controlled by syntoxic (e.g., modification of inflammation and consequent immunocompetence of an individual) and catatoxic (e.g., induction of liver enzymes that activate toxicants) hormones.
- Size, sex, age, genetic qualities of the individual, and chance can influence the consequence of lethal exposure.
- The widespread explanation of a log normal variation in individual responses to toxicants (individual effective dose) should not be assumed true unless shown to be so.
- Hormesis can appear at sublethal or subinhibitory concentrations or doses.
- Antagonism between chemicals in mixture can result from functional, chemical, dispositional, and receptor antagonism.
- The assumption of a log normal ("Probit") model is the most common approach to quantifying lethal effects for concentration-effect data sets, including those with spontaneous mortality, a lethal threshold, or hormesis.
- Other models such as the log logistic, Gompertz, or Weibull are also useful for fitting concentration-effect data sets. The Spearman–Karber technique is a nonparametric method for point estimation of LC50. With no partial kills, the binomial method can estimate the LC50.
- Mixture lethal effects can be estimated using models based on independent (Equation 9.13) or similar (Equation 9.18) action.
- Survival time methods allow optimal inclusion of exposure duration into predictions of lethal effects. Relative to conventional test methods, they also have higher statistical power and capacity to include characteristics associated with individual organisms. They include nonparametric, semiparametric, and fully parametric techniques.
- Molecular structural and physiochemical qualities of organic compounds determine their partitioning and reactivities. QSAR for lethal potency within classes of organic poisons can be formulated based on these qualities.
- QICAR can be used to predict relative lethal potency of metals based on metal–ligand binding theory (e.g., HSAB theory).

REFERENCES

Altenburger, R., Schmitt, H., and Schürmann, G., Algal toxicity of nitrobenzenes: Combined effect analysis as a pharmacological probe from similar modes of interaction, *Environ. Toxicol. Chem.*, 24, 324–333, 2005.

Anderson, P.D. and Spear, P.A., Copper pharmacokinetics in fish gills—I. Kinetics in pumpkinseed sunfish, *Lepomis gibbosus*, of different body sizes, *Water Res.*, 14, 1101–1105, 1980.

Anderson, P.D. and Weber, L.J., Toxic response as a quantitative function of body size, *Toxicol. Appl. Pharmacol.*, 33, 471–483, 1975.

Babich, H., Puerner, J.A., and Borenfreund, E., *In vitro* cytotoxicity of metals to bluegill (BF-2) cells, *Arch. Environ. Contam. Toxicol.*, 15, 31–37, 1986a.

Babich, H., Shopsis, C., and Borenfreund, E., *In vitro* cytotoxicity testing of aquatic pollutants (cadmium, copper, zinc, nickel) using established fish cell lines, *Ecotoxicol. Environ. Saf.*, 11, 91–99, 1986b.

Bailer, A.J. and Oris, J.T., Assessing toxicity of pollutants in aquatic systems, In *Case Studies in Biometry*, Lange, N., Ryan, L., Billard, L., Brillinger, D., Conquest, L., and Greenhouse, J. (eds.), John Wiley & Sons, New York, 1994, pp. 25–40.

Benguira, S. and Hontela, A., Adrenocorticotropin- and cyclic adenosine 3′,5′-monophosphate-stimulated cortisol secretion in interrenal tissue of rainbow trout exposed *in vitro* to DDT compounds, *Environ. Toxicol. Chem.*, 19, 374–381, 2001.

Berkson, J., Why I prefer logits to probits, *Biometrics*, 7, 327–339, 1951.

Biesinger, K.E. and Christensen, G.M., Effects of various metals on survival, growth, reproduction, and metabolism of *Daphnia magna*, *Can. J. Fish. Aquat. Sci.*, 29, 1691–1700, 1972.

Binet, M.P., Sur la toxicité comparée des métaux alcalins et alcalino-terreux, *C.R. Acad. Sci. Paris*, 115, 225–253, 1940.

Bliss, C.I., The calculation of the dosage-mortality curve, *Ann. Appl. Biol.*, 22, 134–307, 1935.

Bliss, C.I., The size factor in the action of arsenic upon silkworm larvae, *Exp. Biol.*, 13, 95–110, 1936.

Bliss, C.I. and Cattell, M., Biological assay, *Ann. Rev. Physiol.*, 5, 479–539, 1943.

Bradbury, S.P., Russom, C.L., Ankley, G.T., Schultz, T.W., and Walker, J.D., Overview of data and conceptual approaches for derivation of quantitative structure–activity relationships for ecotoxicological effects of organic chemicals, *Environ. Toxicol. Chem.*, 22, 1789–1798, 2003.

Buikema, A.L., Jr., Niederlehner, B.R., and Cairns, J., Jr., Biological monitoring. Part IV—Toxicity testing, *Water Res.*, 16, 239–262, 1982.

Cairns, J., Jr., The threshold problem in ecotoxicology, *Ecotoxicology*, 1, 3–16, 1992.

Calabrese, E. and Baldwin, L.A., A general classification of U-shaped dose-response relationships in toxicology and their mechanistic foundations, *Hum. Exp. Toxicol.*, 17, 353–364, 1998.

Calabrese, E.J. and Baldwin, L.A., U-shaped dose-responses in biology, toxicology, and public health, *Annu. Rev. Public Health*, 22, 15–33, 2001.

Calabrese, E.J., McCarthy, M.E., and Kenyon, E., The occurrence of chemically induced hormesis, *Health Physics*, 52, 531–541, 1987.

Cantor, R.S., Breaking the Meyer–Overton rule: Predicted effects of varying stiffness and interfacial activity on the intrinsic potency of anesthetics, *Biophys. J.*, 80, 2284–2297, 2001.

Chen, D., Yin, C., Wang, X., and Wang, L., Holographic QSAR of selected esters, *Chemosphere*, 57, 1739–1745, 2004.

Christensen, E.R., Dose–response functions in aquatic toxicity testing and the Weibull model, *Water Res.*, 18, 213–221, 1984.

Christensen, E.R. and Nyholm, N., Ecotoxicological assays with algae: Weibull dose–response curves, *Environ. Sci. Technol.*, 18, 713–718, 1984.

Cox, C., Threshold dose-response models in toxicology, *Biometrics*, 43, 511–523, 1987.

Cox, D.R. and Oakes, D., *Analysis of Survival Data*, Chapman & Hall, London, 1984.

Crane, M., Newman, M.C., Chapman, P.F., and Fenlon, J., *Risk Assessment with Time to Event Models*, CRC Press/Lewis Publishers, Boca Raton, FL, 2002.

Duboudin, C., Ciffroy, P., and Magaud, H., Acute-to-chronic species sensitivity distribution extrapolation, *Environ. Toxicol. Chem.*, 23, 1774–1785, 2004.

Ellersieck, M.R., Asfaw, A., Mayer, F.L., Krause, G.F., Sun, K., and Lee, G., *Acute-to-Chronic Estimation (ACE v 2.0) with Time-Concentration-Effect Models*, EPA/600/R-03/107, December 2003. US EPA Office of Research Development, Washington, D.C., 2003.

EPA, *Methods for Measuring the Toxicity and Bioaccumulation of Sediment-associated Contaminants with Freshwater Invertebrates*, 2nd ed., EPA/600/R-99/064, March 2000. NTIS, Washington, D.C., 2000.

EPA, *Short-Term Methods for Estimating the Acute Toxicity of Effluents and Receiving Water to Freshwater and Marine Organisms*, 5th ed., EPA/821/R-02/012, October 2002. NTIS, Washington, D.C., 2002a.

EPA, *Short-Term Methods for Estimating the Chronic Toxicity of Effluents and Receiving Waters to Freshwater Organisms*, 4th ed., EPA821/R-02/013, October 2002. NTIS, Washington, D.C., 2002b.

Finney, D.J., *Probit Analysis. A Statistical Treatment of the Sigmoid Response Curve*, Cambridge University Press, Cambridge, UK, 1947, p. 256.

Fraústo da Silva, J.J.R. and Williams, R.J.P., *The Biological Chemistry of the Elements*, Oxford University Press, Oxford, UK, 1991.

Friedman, E.M. and Lawrence, D.A., Environmental stress mediates changes in neuroimmunological interactions, *Toxicol. Sci.*, 67, 4–10, 2002.

Gaddum, J.H., Reports on biological standards. III. Methods of biological assay depending on a quantal response, *Br. Med. Res. Council Special Report Series*, 183, A2–85, 1933.

Gaddum, J.H., Bioassays and mathematics, *Pharmacol. Rev.*, 5, 87–134, 1953.

Gendron, A.D., Bishop, C.A., Fortin, R., and Hontela, A., *In vivo* testing of the functional integrity of the corticosterone-producing axis in mudpuppy (Amphibia) exposed to chlorinated hydrocarbons in the wild, *Environ. Toxicol. Chem.*, 16, 1694–1706, 1997.

Hamilton, M.A., Russo, R.C., and Thurston, R.V., Trimmed Spearman–Karber method for estimating median lethal concentrations in toxicity bioassays, *Environ. Sci. Technol.*, 11, 714–719, 1977.

Hedtke, J.L., Robinson-Wilson, E., and Weber, L.J., Influence of body size and developmental stage of Coho salmon (*Oncorhynchus kisutch*) on lethality of several toxicants, *Fund. Appl. Toxicol.*, 2, 67–72, 1982.

Hogan, G.R., Cole, B.S., and Lovelace, J.M., Sex and age mortality responses in zinc acetate-treated mice, *Bull. Environ. Contam. Toxicol.*, 39, 156–161, 1987.

Hontela, A., Interrenal dysfuntion in fish from contaminated sites: *In vivo* and *in vitro* assessment, *Environ. Toxicol. Chem.*, 17, 44–48, 1998.

Hughes, G.M., The dimensions of fish gills relative to their function, *J. Exp. Biol.*, 45, 177–195, 1966.

Jones, J.R.E., The relation between the electrolytic solution pressures of the metals and their toxicity to the stickleback (*Gasterosteus aculeatus* L.), *J. Exp. Biol.*, 16, 425–437, 1939.

Jones, J.R.E., A further study of the relation between toxicity and solution pressure, with *Polycelis nigra* as test animal, *J. Exp. Biol.*, 17, 408–415, 1940.

Jones, M.R. and Vaughn, W.K., HSAB theory and acute metal ion toxicity and detoxification processes, *J. Inorg. Nucl. Chem.*, 40, 2081–2088, 1978.

Kaiser, K.L.E., Correlation and prediction of metal toxicity to aquatic biota, *Can. J. Fish. Aquat. Sci.*, 37, 211–218, 1980.

Kostial, K., Maljkovic, T., and Jugo, S., Lead acetate toxicity in rates in relation to age and sex, *Arch. Toxicol.*, 31, 265–269, 1974.

Lamanna, C. and Hart, E.R., Relationship of lethal toxic dose to body weight of the mouse, *Toxicol. Appl. Pharmacol.*, 13, 307–315, 1968.

Loeb, J., Studies on the physiological effects of the valency and possibly the electrical charges of ions. I. The toxic and antitoxic effects of ions as a function of their valency and possibly their electrical charge, *Am. J. Physiol.*, 6, 411–433, 1940.

Marubini, E. and Grazia Valsecchi, M., *Analysing Survival Data from Clinical Trials and Observational Studies*, John Wiley & Sons, Chichester, UK, 1995.

Matthews, A.P., The relation between solution tension, atomic volume, and the physiological action of the elements, *Am. J. Physiol.*, 10, 290–323, 1904.

Mayer, F.L., Krause, G.F., Buckler, D.R., Ellersieck, M.R., and Lee, G., Predicting chronic lethality of chemicals to fishes from acute toxicity test data: Concepts and linear regression analysis, *Environ. Toxicol. Chem.*, 13, 671–678, 1994.

McCloskey, J.T., Newman, M.C., and Clark, S.B., Predicting the relative toxicity of metal ions using ion characteristics: Microtox® bioluminescence assay, *Environ. Toxicol. Chem.*, 15, 1730–1737, 1996.

McGuigan, H., The relation between the decomposition-tension of salts and their antifermentative properties, *Am. J. Physiol.*, 10, 444–451, 1954.

McKim, J.M., Bradbury, S.P., and Niemi, G.J., Fish acute toxicity syndromes and their use in the QSAR approach to hazard assessment, *Environ. Health Perspec.*, 71, 171–186, 1987.

McKinney, J.D., Richard, A., Waller, C., Newman, M.C., and Gerberick, F., The practice of structure activity relationships (SAR) in toxicology, *Toxicol. Sci.*, 56, 8–17, 2000.

Miller, R.G., Jr., *Survival Analysis*, John Wiley & Sons, New York, 1981.

Moore, D.R.J., Breton, R.L., and MacDonald, D.R., A comparison of model performance for six quantitative structure–activity relationship packages that predict acute toxicity to fish, *Environ. Toxicol. Chem.*, 22, 1799–1809, 2003.

Mulvey, M. and Diamond, S.A., Genetic factors and tolerance acquisition in populations exposed to metals and metalloids, In *Metal Ecotoxicology, Concepts and Applications*, Newman, M.C. and McIntosh, A.W. (eds.), CRC Press/Lewis Publishers, Boca Raton, FL, 1991, pp. 301–321.

Newman, M.C., *Quantitative Methods in Aquatic Ecotoxicology*, CRC Press/Lewis Publishers, Boca Raton, FL, 1995, p. 426.

Newman, M.C. and Aplin, M., Enhancing toxicity data interpretation and prediction of ecological risk with survival time modeling: An illustration using sodium chloride toxicity to mosquitofish (*Gambusia holbrooki*), *Aquatic Toxicol.*, 23, 85–96, 1992.

Newman, M.C., Diamond, S.E., Mulvey, M.E., and Dixon, P.M., Allozyme genotype and time to death of mosquitofish, *Gambusia affinis* (Baird and Girard), during acute toxicant exposure: A comparison of arsenate and inorganic mercury, *Aquatic Toxicol.*, 15, 141–159, 1989.

Newman, M.C. and Dixon, P.H. Ecologically meaningful estimates of lethal effect on individuals, in *Ecotoxicology: A Hierarchical Treatment*, Newman, M.C. and Jagoe C.H. (eds.), CRC Press/Lewis Publishers, Boca Raton, FL, 1996, pp. 225–253.

Newman, M.C. and Heagler, M.G., Allometry of metal bioaccumulation and toxicity, in *Metal Ecotoxicology. Concepts and Applications*, Newman, M.C. and McIntosh, A.W. (eds.), CRC Press/Lewis Publishers, Boca Raton, FL, 1991, pp. 91–130.

Newman, M.C. and McCloskey, J.T., Predicting relative toxicity and interactions of divalent metal ions: Microtox® bioluminescence assay, *Environ. Toxicol. Chem.*, 15, 275–281, 1996.

Newman, M.C. and McCloskey, J.T., The individual tolerance concept is not the sole explanation for the probit dose-effect model, *Environ. Toxicol. Chem.*, 19, 520–526, 2000.

Newman, M.C. and McCloskey, J.T., Applying time to event methods to assess pollutant effects on populations, In *Risk Assessment with Time to Event Models*, Crane, M., Newman, M.C., Chapman, P.F. and Fenlon, J. (eds.), CRC Press/Lewis Publishers, Boca Raton, FL, 2002, pp. 23–38.

Newman, M.C., McCloskey, J.T., and Tatara, C.P., Using metal-ligand binding characteristics to predict metal toxicity: Quantitative Ion Character-Activity Relationships (QICARs), *Environ. Health Perspec.*, 106 (Suppl. 6), 1419–1425, 1998.

Nuutinen, S., Landrum, P.F, Schuler, L.J., Kukkonen, J.V.K., and Lydy, M.J., Toxicokinetics of organic contaminants in *Hyalella azteca*, *Arch. Environ. Contam. Toxicol.*, 44, 467–476, 2003.

Ownby, D.R. and Newman, M.C., Advances in Quantitative Ion Character-Activity Relationships (QICARs): Using metal-ligand binding characteristics to predict metal toxicity, *Quant. Struct.-Act. Relat.*, 22, 1–6, 2003.

Ren, S., Ecotoxicity prediction using mechanism- and non-mechanism-based QSARs: A preliminary study, *Chemosphere*, 53, 1053–1065, 2003.

Sagan, L.A., What is homresis and why haven't we heard about it before? *Health Phys.*, 52, 521–525, 1987.

Schmidt-Nielsen, K., *Scaling. Why Is Animal Size So Important?* Cambridge University Press, Cambridge, UK, 1986, p. 241.

Schultz, T.W. and Cronin, M.T.D., Essential and desirable characteristics of ecotoxicity quantitative structure–activity relationships, *Environ. Toxicol. Chem.*, 22, 599–607, 2003.

Selye, H., A syndrome produced by diverse nocuous agents, *Nature*, 138, 32, 1936.

Selye, H., Stress and the general adaptation syndrome, *Br. Med. J.*, 4667, 1383–1392, 1950.

Selye, H., The evolution of the stress concept, *Amer. Sci.*, 61, 692–699, 1973.

Selye, H., *The Stress of Life*, McGraw-Hill Companies, New York, 1984, p. 515.

Sprague, J.B., Measurement of pollutant toxicity to fish. I. Bioassay methods for acute toxicity. *Water Res.*, 3, 793–821, 1969.

Stark, J.D., How closely do acute lethal concentration estimates predict effects of toxicants on populations? *Int. Environ. Assess. Manag.*, 1, 109–113, 2005.

Stephan, C.E., Methods for calculating an LC50, In *Aquatic Toxicology and Hazard Evaluation, ASTM STP 634*, Mayer, F.L. and Hamelink, J.L. (eds.), American Society for Testing and Materials, Philadelphia, PA, 1977, pp. 65–84.

Suter, G.W., II, *Ecological Risk Assessment*, CRC Press/Lewis Publishers, Boca Raton, FL, 1993, p. 538.

Tatara, C.P., Newman, M.C., McCloskey, J.T., and Williams, P.L., Use of ion characteristics to predict relative toxicity of mono-, di-, and trivalent metal ions: *Caenorhabditis elegans* LC50, *Aquatic Toxicol.*, 42, 255–269, 1998.

Tong, W., Welsh, W.J., Shi, L., Fang, H., and Perkins, R., Structure–activity relationship approaches and applications, *Environ. Toxicol. Chem.*, 22, 1680–1695, 2003.

Tsigos, C. and Chrousos, G.P., Hypothalamic-pituitary-adrenal axis, neuroendocrine factors and stress, *J. Psychosom. Res.*, 53, 865–871, 2002.

Turner, J.E., Williams, M.W., Jacobson, K.B., and Hingerty, B.E., Correlations of acute toxicity of metal ions and the covalent/ionic character of their bonds, In *QSAR in Toxicology and Xenobiochemistry*, Tichy, M. (ed.), Elsevier, Amsterdam, 1985, pp. 171–177.

Unger, M.A., Newman, M.C., and Vadas, G.G., Predicting survival of grass shrimp (*P. pugio*) during ethylnaphthalene, dimethylnaphthalene, and phenanthrene exposures differing in concentration and duration. *Environ. Toxicol. Chem.* 26, 528–534, 2007.

Vaes, W.H.J., Ramos, E.U., Verhaar, H.J.M., and Hermens, J.L.M., Acute toxicity of nonpolar versus polar narcosis: Is there a difference?, *Environ. Toxicol. Chem.*, 17, 1380–1384, 1998.

Williams, M.W. and Turner, J.E., Comments on softness parameters and metal ion toxicity, *J. Inorg. Nucl. Chem.*, 43, 1689–1891, 1981.

Williamson, P., Opposite effects of age and weight on cadmium concentrations of a gastropod mollusc, *Ambio*, 8, 30–31, 1979.

Zeeman, M., Auer, C.M., Clements, R.G., Nabholz, J.V., and Boethling, R.S., U.S. EPA regulatory perspectives on the use of QSAR for new and existing chemical evaluation, *SAR QSAR Environ. Res.*, 3, 179–201, 1995.

Zhao, Y. and Newman, M.C., Shortcomings of the laboratory-derived median lethal concentration for predicting mortality in field populations: Exposure duration and latent mortality, *Environ. Toxicol. Chem.*, 23, 2147–2153, 2004.

Zhao, Y. and Newman, M.C., The theory underlying dose-response models influences predictions for intermittent exposures. *Environ. Toxicol. Chem.*, 26, 543–547, 2007.

10 Sublethal Effects

10.1 OVERVIEW

> Although sublethal effects are often more subtle than those inducing direct mortality, they can equally impact overall community structure.
>
> **(Bridges 1997)**

The above statement could not have been made in the 1960s and 1970s without requiring immediate qualification. Then instances of acutely lethal exposures to human products and byproducts were blatant and preoccupied most wildlife and aquatic toxicologists. Acutely lethal exposures still occur (e.g., accidental bird poisonings with pesticides (Stansley and Roscoe 1999)) but our attention is being drawn increasingly to effects that do not produce outright death. This is not to imply that sublethal effects were completely ignored in the past: published contributions to the sublethal effects literature began well before the 1960s. Indeed, sublethal effects featured prominently in the opening chapter of Rachel Carson's *Silent Spring* (1962). Sublethal effects were simply treated as secondary in importance in the 1960s and 1970s. However, interest in and effort spent on sublethal effects has appropriately come into balance during the past two decades with those focused on lethal effects. Most sublethal studies explore one or more of five fitness-related features of individuals: reproduction, growth, development, behavior, and physiology. Very often, growth and reproduction are examined simultaneously.

> Significance tests are for situations where we do not understand, in any theoretical sense, what is happening.
>
> **(Hacking 2001)**

A few key points can be made at the onset about the current sublethal effects literature based on a quick survey of 114 randomly selected studies published between 1968 and 2006. Eighty-two percent of these papers applied experimental designs appropriate for analysis of variance (ANOVA) or analysis of covariance (ANCOVA), and did not draw on available ecological models to quantify or interpret results. Most were designed to detect change in response to increasing exposure level and interpreted results in that context. Basic hypothesis testing dominated analyses though descriptive regression models were common. Theory-based experimental designs were applied in fewer than 10% of the surveyed publications. This is surprising given the observation that 47% of these papers attempted to link results to fitness, demographic, or bioenergetics consequences in their discussions. This observation and the above quote suggest that the pervasive use of significance testing results from a lack of trust in or knowledge of available theory that could be applied to design better experiments.

Perhaps one cause of this incongruity is the longstanding regulatory stance that sublethal effects are best addressed with hypothesis testing. Pragmatic coping with past problems gives justification for the emergence of this position but its continued maintenance becomes less justifiable with each passing year. A clear tendency away from the dominance of hypothesis testing of sublethal effects now seems to be emerging in Environmental Protecting Agency (EPA) and other agency documents. Because the field is currently shifting relative to how we deal with sublethal effects, the reader will find considerable discussion of hypothesis testing in this chapter but will also find very relevant theory and studies in the chapters that follow. The theory contained in those chapters bridge

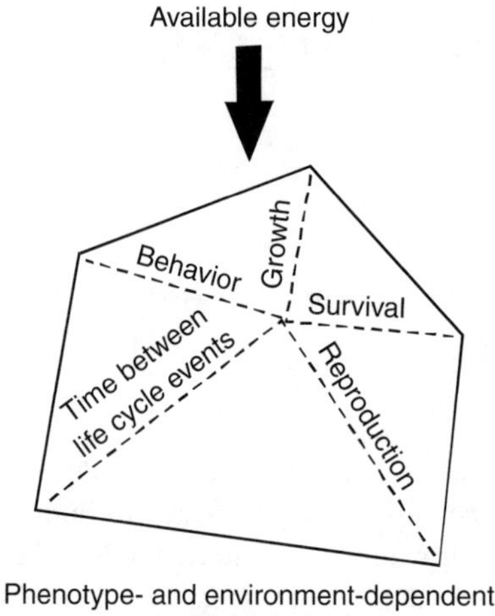

FIGURE 10.1 Sublethal and lethal effects are interlinked and patterns of effects should be interpreted within this context. Strategies evolve within the phylogenetic constraints of an organism in order to optimize Darwinian fitness in a range of environments. Coordinated reallocation occurs of energy resources to maintain the soma (survival), increase the soma (grow), make transitions among life stages, and reproduce. Behavior also can be modified during the expression of phenotype. A strategy designed by natural selection to maximize fitness will be taken given a particular environment and phenotypic plasticity.

the conceptual gulf over which sublethal effects information cannot currently pass without taking on considerable—likely unacceptable—uncertainty and ambiguity.

In addition to methodological changes taking place about how we deal with sublethal effects, a less obvious evolution is emerging relative to how we go about interpreting the rapidly growing body of sublethal effect data. Is it best to interpret changes based solely on the mechanisms described in earlier chapters or are there "emergent properties" (i.e., higher-order phenomena) that must be considered too? As we will discuss again in Chapter 16, trade-offs and energy allocations are made in complex ways by individuals faced with variable environments (Figure 10.1). Some of the most important involve allocation of energy among growth, somatic maintenance (survival), and reproduction. Other equally important shifts are associated with the timing of life-cycle events and foraging behavior. Ideally, these trade-offs integrate in a manner that maximizes an individual's fitness within the confines of the particular environment in which the individual finds itself (Figure 10.2). It is unreasonable to assume that evolved life history strategies and trade-offs associated with optimizing fitness in changing environments do not also manifest within sublethal effect data sets. In many cases, mechanisms and paradigms associated with such strategies might be equally or more relevant than those associated with lower levels of organization. For example, Brown Sullivan and Spence (2003) conclude from a conventional, lower $\rightarrow$ higher-level interpretative vantage that some sublethal effects of atrazine and nitrate appear inexplicably to be antagonistic while others are synergistic. Perhaps these seemingly contradictory effects from the vantage of suborganismal mechanisms could be interpreted successfully using life history strategy theory. Similarly, Heinz and Hoffman noted from their studies of selenium and mercury effects on mallard ducks (*Anas platyrhynchos*) that "mercury and selenium may be antagonistic to each other for adults and synergistic to young, even in the same experiment." Such differences are not easily explained based on suborganismal

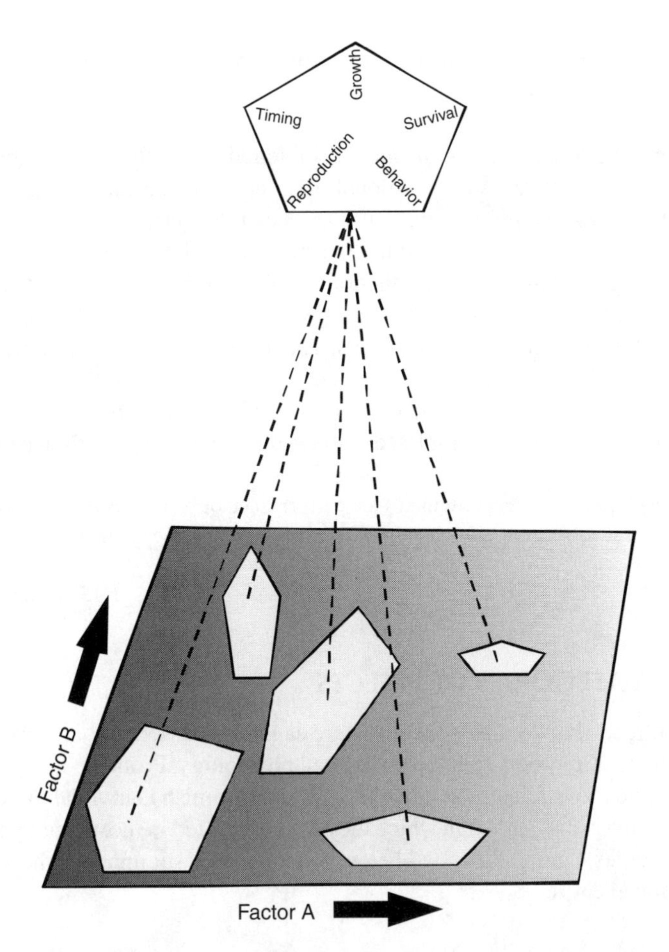

FIGURE 10.2 In differing environments, including those containing toxicant stressors, Darwinian fitness is optimized for an individual of a particular species by shifting energy resources allocated to growth, survival, and reproduction, and changing behavior and the timing associated with achieving life history events such as metamorphosis to a sexual adult. The hypothetical factors A and B shown here can be physical (e.g., habitat), chemical (e.g., salinity or chemical toxicant), or biological (e.g., competitors) factors.

mechanisms alone but could be explained and further scrutinized with life history strategy theory. Higher-level controls are exerted on many suborganismal processes and predicted by optimality or life history strategy theories.

Regardless, is it possible to effectively address these higher-order phenomena in studies of sublethal effects? The unambiguous answer is yes. Handy et al. (1999) explored copper effects on rainbow trout (*Oncorhynchus mykiss*) locomotion and detoxification in terms of trade-off theory. Knops et al. (2001) examined the balance between metabolic costs of resisting toxicant effects versus those of reproduction and growth. Sibly and Calow (1989) and Atchison et al. (1996) write convincingly about linking ecotoxicology to life-cycle theory and optimization theory. Books such as Stearns' *The Evolution of Life Histories* (1992) provide ample detail about relevant processes and concepts. Higher-order phenomena can be integrated into ecotoxicity studies if one seeks out the appropriate methods and theories instead of remaining fixed in the conventional mode of interpretation that emerged to address the immediate issues confronting ecotoxicologists in the 1960s and 1970s.

Complementary interpretative paradigms are now blending in an exciting way in the area of sublethal toxicant effects. An emerging consensus from associated studies is that higher-order processes can be as important as those suborganismal mechanisms discussed in earlier chapters. Associated higher-level theories also provide an essential way of translating organismal effects into those to populations.

The current blending of concepts in the area of sublethal effects leads to difficulty in presenting the associated information here. A conventional approach will be taken that hopefully will not leave the reader with the false impression that all issues relevant to sublethal effects are covered in this chapter. Previous chapters discussed the mechanistic underpinnings of sublethal effects and chapters that follow will discuss important theories explaining higher-order properties that modify suborganismal processes. This short chapter will generally discuss fitness consequences and then conventional ways of quantifying sublethal effects. The reader is urged to explore earlier chapters for detailed discussion of suborganismal mechanisms giving rise to these effects and later chapters for relevant higher-order mechanisms influencing sublethal effects and responses. Concepts essential to understanding sublethal effects will be missed if those materials were overlooked by the reader.

> Toxicant exposure often completely eliminates the performance of behaviors that are essential to fitness and survival in natural ecosystems, frequently after exposures of lesser magnitude than those causing significant mortality.
>
> **(Scott and Sloman 2004)**

10.2 GENERAL CATEGORIES OF EFFECTS

Sublethal stressor effects that dominate the literature can be discussed in four convenient groupings: growth and development, reproduction, behavior, and physiology. From the first such publication to the most recent, implications remain on how these effects diminish Darwinian fitness. Although not often stated as such, this is the context in which they are interpreted in most regulatory processes. The diversity of these studies can be illustrated below with the brief summary of the surveyed research publications mentioned earlier.

10.2.1 DEVELOPMENT AND GROWTH

Slightly less than a third of the 114 research publications that we surveyed described stressor effects on growth and roughly the same fraction described effects on development. Many addressed them together.

Developmental studies included those examining conventional teratogenic effects, those emphasizing sexual development, and those examining more subtle developmental effects (e.g., changes in an individual's behavior due to exposure during development). Early examples of the first type are Pechkurenkov et al. (1966), D'Agostino and Finney (1974), Ward et al. (1981), Bookhout et al. (1984), and Conrad (1988) who noted morphological changes in fish, crustaceans, or molluscan eggs or larvae during exposure to a range of chemicals. More recent studies apply similar approaches but tend to address more relevant than conventional test species (e.g., DeWitt et al. 2006) and stressor (e.g., Moreels et al. 2006, Williams et al. 2003, Wollenberger et al. 2005) combinations. Many recent studies such as Carr et al. (2003), Degitz et al. (2000), Fordham et al. (2001), Rowe et al. (1996), Brown Sullivan and Spence (2003), and Tietge et al. (2005) explore the important issue of stressor effects on amphibian development. A few emphasize developmental stability (e.g., Green and Lochman 2006), which is described in Chapter 16. These more recent publications also have a greater tendency to frame studies in a mechanistic context; for example, endocrine or reproductive system modification by contaminants (e.g., Boudreau et al. 2004). Those publications that focused on reproductive effects are currently dominated by fish studies (e.g., Bortone et al. 1989, Parrott et al. 2003, Penske et al. 2005, Teather et al. 2005, Toft et al. 2004) but developmental effects to important bird (Fry and Toone 1981), insect (Delpuech and Meyet 2003), and molluscan (Bryan and Gibbs 1991) species are also common. Many emphasize either reduced reproductive fitness consequences or imposex (the development of male characteristics such as a penis in females). Still other studies examined nonmorphological effects. As an example, Samson et al. (2001) explored delayed functional effects to zebra fish (*Danio rerio*) after exposure to methylmercury, including behavioral effects. Weis and Weis (1987) provide a good, but somewhat dated, review of such developmental effects.

Studies of toxicant effects on growth were similarly diverse and slightly more plentiful because growth is relatively easy to measure. As a very welcome exception to their rarity in most other areas of organismal ecotoxicology, plant studies were common. Plant growth studies ranged widely, including those involving terrestrial macrophytes (e.g., Baker and Walker 1989, Boutin et al. 1995, 2000, Fletcher et al. 1988, 1996, Wilson 1988), aquatic macrophytes (Lewis and Wang 1999, Lytle and Lytle 2005, Marwood et al. 2003, Stewart et al. 1999), and microscopic plants (Mayer et al. 2000, McGarth et al. 2004). Reports of hormesis (see Chapter 9) appeared in several plant studies (e.g., Calabrese et al. 1987, Stebbing 1982). Growth as affected by toxicants was also commonly studied for terrestrial (e.g., Coeurdassier et al. 2001, Inouye et al. 2006, Spurgeon et al. 1994) and aquatic (e.g., De Schamphelaere and Janssen 2004, Gray et al. 1998, Ingersoll et al. 1998, Knops et al. 2001) invertebrates. Growth studies of vertebrates included such diverse study groups as birds (Bishop et al. 2000, Hanowski et al. 1997, Woodford et al. 1998), fish (Al-Yakoob et al. 1996, Cook et al. 2005, Hansen et al. 2004, Munkittrick and Dixon 1988, Schmidt et al. 2005), and amphibians (e.g., Diana et al. 2000, Relyea 2004, Schöpf Rehage et al. 2002, Wojtaszek et al. 2004, 2005).

10.2.2 Reproduction

Thirty percent of the surveyed sublethal effects publications examined reproduction alone or in combination with another effect. Studies of reproductive effects are so common that standard assays have been established such as those for the fathead minnow (Ankley et al. 2001, Bringolf et al. 2004). In contrast to the older publications such as Arnold (1971) or Bodar et al. (1988), the more recent reproductive effects publications tended to focus less on conventional test species such as *Daphnia magna* and more on species and exposure scenarios relevant to the particular situation of concern. As examples using aquatic crustacea, Wirth et al. (2002) examined grass shrimp (*Palaemonetes pugio*) chronically exposed to endosulfan, and Cold and Forbes (2004) studied growth of *Gammarus pulex* experiencing pulsed exposures to esfenvalerate. Additional examples are easily found for terrestrial species including soil-associated species (e.g., Collembola exposed to arsenic) (Crouau and Cazes 2005), and annelids exposed to chemicals from explosives (Dodard et al. 2005) or metals (Kuperman et al. 2006).

Birds are common subjects of reproduction studies as a consequence of historical events involving avian reproduction such as dichlorodichloroethylene (DDE)-linked eggshell thinning (Hickey and Anderson 1968) and current problems such as selenium's effect on Kesterson Wildlife Refuge (California) waterfowl and wading birds (Ohlendorf 2002). Typical of well-done field studies of avian reproduction are those of Bishop et al. (2000) of the possible impact of apple orchard-associated pesticides on Tree swallows (*Tachyecineta bicolor*) and Eastern bluebirds (*Sialia sialis*), and Ohlendorf (2002) who surveyed birds associated with a selenium-contaminated region of San Joaquin Valley. Also typical are field studies of birds particularly prone to chronic exposure such as piscivorous birds exposed to dietary mercury (e.g., Elbert and Anderson (1998) and Meyer et al. (1998)). Similarly, laboratory studies of toxicant effects on avian reproduction are accumulating in the literature (e.g., Heinz and Hoffman (1998)). Standard methods have been established for quantifying avian reproductive effects although Mineau et al. (1994) suggest important shortcomings in these tests relative to predicting effects in the field.

10.2.3 Behavior

Surprisingly, nearly 50% of the surveyed publications contained descriptions of behavioral changes of one sort or another. Some were straightforward reports of changes in locomotor behavior.

Zebrafish (*Brachydanio rerio*) general swimming (Grillitsch et al. 1999, Vogl et al. 1999), amphipod (*Gammarus lawrencianus*) swimming direction (Wallace and Estephan 2004), annelid (*Lumbriculus variegates*) helical swimming (O'Gara et al. 2004), and woodlouse (*Oniscus asellus*) movement (Bayley et al. 1997) are a few of the locomotion changes related to toxicant exposure in studies. Still other straightforward studies assessed an individual's ability to simply avoid high concentrations of toxicants (e.g., Kynard 1974, McCloskey et al. 1995, Roast et al. 2001, Sprague 1968).

Other behavioral studies focus on endpoints with implicit connection to an individual's general fitness. Examples include the influence of mercury on the foraging behavior of fish (Weis and Khan 1990) and the sediment reworking activities of a benthic oligochaete (Landrum et al. 2004). Social (intraspecies) interactions are another important set of behaviors that can be changed by toxicant exposure. These include aggression (e.g., Janssens et al. 2003), social structuring (e.g., Sloman et al. 2003), schooling (e.g., Nakayama et al. 2005), and mating (e.g., Hunt and Warner Hunt 1977) behaviors occurring within groups of individuals of the same species. Finally and as detailed in later chapters, interspecies interactions are also important, including balancing activities associated with foraging and predator avoidance (e.g., Hui 2002, Perez and Wallace 2004, Preston et al. 1999, Riddell et al. 2005, Schulz and Dabrowski 2001, Sullivan et al. 1978, Tagatz 1976, Webber and Haines 2003). All of these behaviors influence the overall fitness of an individual with or without the influence of toxicant exposure.

10.2.4 Physiology

Physiological effects can decrease fitness directly or indirectly. Even the behavioral changes just described create the potential for physiological shifts. For example, Sloman et al. (2003) noted that subordinate rainbow trout (*O. mykiss*) accumulated more copper than dominant rainbow trout, notionally creating a difference in potential for physiological effects in different trout within a social hierarchy. Shifts in physiology, including shifts associated with energy expenditure, reduce an individual's options relative to optimizing energy allocation. Such energetic costs have been measured directly in carp (*Cyprinus carpio*) exposed to copper (De Boeck et al. 1997); bivalve molluscs (*Pisidium amnicum*) and salmon (*Salmo salar*) exposed to pentachlorophenol (Penttinen and Kukkonen 2006); and bivalve molluscs (*P. amnicum*), Chironomid larvae (*Chironomus riparius*), and oligochaetes (*L. variegates*) exposed to 2,4,5-trichlorophenol (Penttinen et al. 1996). Consequent to all of the issues described above, an increasingly common interpretation of sublethal effects is based on energy allocation (e.g., Handy et al. 1999).

10.3 QUANTIFYING SUBLETHAL EFFECTS

> Results of almost all life-cycle, partial life-cycle, and early life-stage toxicity tests have been calculated using hypothesis tests in conjunction with analysis of variance to detect statistically significant differences from the control treatment, whereas results of almost all acute tests have been calculated using regression analysis. Because the experimental designs for these two types of toxicity tests usually are very similar, both hypothesis testing and regression analysis can be used to calculate results of both acute and chronic toxicity tests.
>
> **(Stephan and Rogers 1985)**

Conventional tests for sublethal effect consist of replicate groups of organisms exposed to a series of concentrations or doses[1] for a specified duration. The treatment levels include one or more types of

[1] Although sometimes mistaken as synonyms, it is important to remember that concentration and dose are not the same. Concentration is the mass of the chemical per unit of mass or volume of the relevant media to which the organism is exposed. Dose is an amount administered to or entering an individual such as the amount ingested by an individual. The related term, dosage is simply a body mass normalized dose (e.g., 5 mg/kg of body weight).

reference (no toxicant) treatment and a series of increasing concentrations or doses. Some protocols specify that effects should be noted at a few durations, not a single one. Sometimes, a contaminated medium such as an effluent, soil, or sediment is mixed with uncontaminated media to produce the graded series of test exposure treatments. A variety of manuals provide details for conducting such tests and analyzing results; for example, the well-established *EPA Short-Term Methods for Estimating the Chronic Toxicity of Effluents and Receiving Waters to Freshwater Organisms* (EPA 2002) and the recent OECD *Current Approaches in the Statistical Analysis of Ecotoxicology Data: A Guidance to Application* (OECD 2006). In these manuals, recommended effect metrics are calculated using either hypothesis testing or point estimation methods. Which is the best approach has been vigorously debated for at least two decades (e.g., Stephan and Rogers 1985, Chapman et al. 1996, Crane and Newman 2000); therefore, the salient points of the debate are summarized below.

As described in the above quote from Stephan and Rogers (1985), the same data set can be analyzed with these two methods. However, optimization of experimental design is not the same for both methods. On the basis of well-accepted design principles, optimization of hypothesis testing might involve more replicates per treatment but optimizing point estimation by regression might involve spreading the experimental units among more concentrations and selecting concentrations closer to the level of effect for which estimation is being done (Figure 10.3). Or optimization for regression analysis could involve another distribution of experimental units depending on the model being applied. For example, the best distribution of experimental units for an exponential model would be different from that for the simple model depicted in Figure 10.3. More replicates in the control or reference treatments improve power of many hypothesis tests (see pages 17–31 in Cochran and Cox (1957)), but the best distribution of experimental units to produce good regression estimates is dependent on the applied model and the point being predicted (see pages 86–89 in Draper and Smith (1998)).

10.3.1 Hypothesis Testing and Point Estimation

The hypothesis testing approach attempts to identify the highest concentration or dose that has no effect (i.e., either a biological or proof-of-hazard threshold).[2] This renders in practice to statistically comparing the level of effect measured in a series of experimental treatments to that of the reference or control treatment(s). This is done with conventional hypothesis tests that compare means, medians, or other distributional qualities for the experimental units of the treatments (see Newman (1995) for detailed descriptions). Statistical tests identify the lowest concentration that is significantly different from the control or reference treatment(s) (i.e., the lowest observed effect concentration or level (LOEC or LOEL)). The highest concentration that is not statistically significant different from the reference or control treatment is the no observed effect concentration or level (NOEC or NOEL). In common practice, the NOEC and LOEC calculated in a full life-cycle or partial life-cycle test[3] are conditionally used to establish "Safe Concentrations." Appropriately, results from this hypothesis testing approach are increasingly being judged insufficient to estimate safe concentrations unless they are associated with enough statistical power to detect a relevant change in a key ecotoxicological effect and, equally important, their interpretation incorporates appropriate biological theory.

---

[2] The reader should know that, although commonly done, it is not strictly valid to make a judgment about biological thresholds with the usual hypothesis testing methods. Cautious users of hypothesis testing interpret results in a proof-of-hazard context instead: no hazard is assumed if no evidence exists at a particular dose or concentration level. The evidence is a significant deviation from the null hypothesis. The threshold becomes not a strict biological threshold but what could be referred to as a threshold of toxicological concern (DeWolf et al. 2005). The problem is that it is often treated incorrectly as a proof of safety threshold. The reader is invited to read Hauschke (1997) or OECD (2006) for more details.

[3] A life-cycle test is one in which key components of individual's life cycle are assessed for contaminant adverse effects. For example, reproduction, development, growth, and survival of a test species might be examined in a battery of tests and effect metrics determined for each. Because of the expense associated with a full life-cycle test, partial life cycle tests have been developed that focus only at the notionally "most sensitive" stages of an organism's life cycle, usually the early stages.

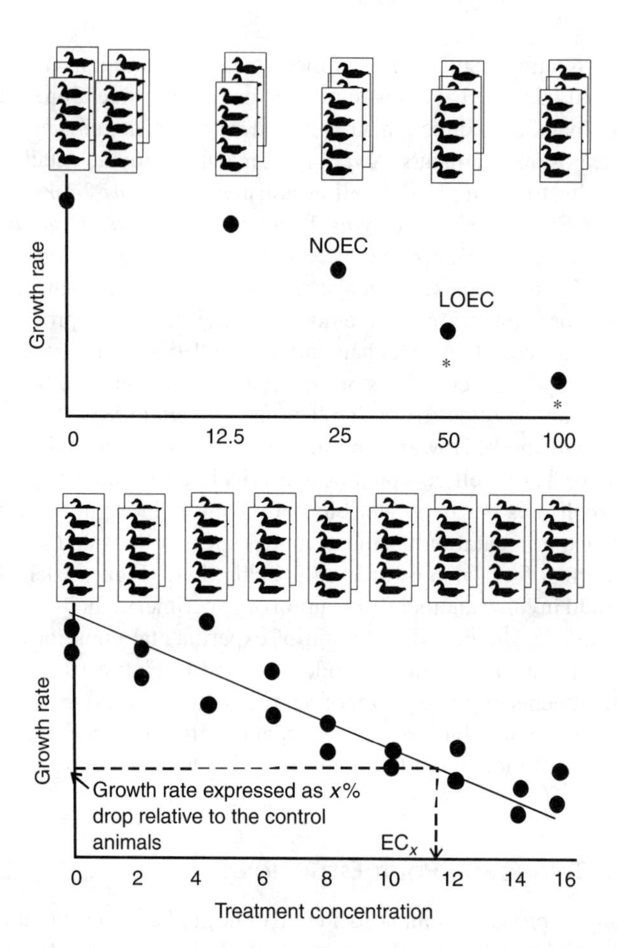

FIGURE 10.3 A hypothetical case illustrating hypothesis testing and point estimation, and changes to design necessary to optimize effect metric generation. The top panel shows an experimental design with a control treatment and four concentration treatments (12.5, 25, 50, and 100). Each noncontrol treatment has triplicate cages containing five waterfowl each. The mean growth rate for each cage is used as the effect variable. To optimize power of the hypothesis testing, the number of control replicates (cages) is set at the number of replicates in each noncontrol treatment times the square root of the number of comparisons or $\sqrt{4} \times 3 = 2 \times 3 = 6$. (See Newman (1995) or OECD (2006) for details and references.) The NOEC and LOEC are obtained on the basis of the treatments with mean growth rates significantly different (*) from the mean control growth rate. If the experiment was intended to produce an EC_x, the design would be optimized in a slightly different manner (bottom panel) because the emphasis would be on producing a good point estimate instead of optimizing power. More treatment concentrations near the suspected EC_x might be chosen with fewer replicates within each treatment. Depending on the selected model, the treatment concentrations might also be spaced to optimize estimation.

Point estimation methods begin by assigning a level of effect that is deemed unacceptable and then estimating the corresponding concentration or dose. Several approaches can be used, ranging from fully parametric regression modeling to nonparametric estimation. Models range from simple dose–response to very complex models. They might or might not include thresholds, natural baseline response levels, or hormesis. The OECD (2006) report mentioned above has an especially good discussion of such models although discussion of biologically-based models is restricted to only one of several potential mechanistic models.

Point estimation methods predict the concentration or dose associated with a given level of effect. Often, this involves prediction from a regression model fit to the ecotoxicity data set. Predictions are referred to in such terms as the EC_x, the effective concentration calculated to have an $x\%$ change in the response.[4] Models fit to effects data sets are often those described in Chapter 9.

As already mentioned, the virtues and shortcomings of these two approaches have been and still are debated. This debate extends back to the origins of these approaches, that is, back to human health assessment where No Observed and Lowest Observed Adverse Effect Concentrations (NOEAC and LOEAC), and Benchmark Doses are still used in a similar fashion.

There are situations in which point estimation is a compromised tool. The first major drawback of point estimation is highlighted in the above Hacking quote. Hypothesis testing is preferable to modeling if one has no understanding of the relationship between the effect and the toxicant concentration (or dose). Second, hypothesis testing might be the only option if the variability in the data set does not allow one to identify and fit a model. Several ecotoxicologists have argued that a third shortcoming is that an x must be defined a priori with point estimation and insufficient insight often exists with which this has to be done with acceptable certainty. However, the hypothesis testing approach does not avoid the crucial issue of determining what is a biologically unacceptable level of effect except during its misapplication (i.e., when statistical significance is mistakenly equated with biological relevance). Hypothesis testing simply puts the question off. Is it not better to confront these uncertainties at the onset of an investigation? Fourth, some of the model fitting techniques are necessarily iterative and situations exist in which they fail to converge on an acceptable solution. The associated parameter estimates and predictions are unacceptable in that case and the hypothesis testing might be the only tool available.

The shortcomings of hypothesis testing are more commonly discussed than those for point estimation. Point estimation is often presented as a superior technique that will eventually replace hypothesis testing as the preferred means of producing sublethal effects metrics. Most of the arguments to abolish the hypothesis test-associated metrics (e.g., Crane and Newman 2000, Kooijman 1996, Laskowski 1995) emerge from the pervasive misapplication of hypothesis tests and misinterpretation of test results, not any fundamental disagreement with Neyman–Pearson theory. The first shortcoming is that statistical significance is not a reliable indicator of biological significance or relevance of an effect. The ecotoxicological literature is replete with instances in which the two are confused. Newman and Unger (2003) refer to this pervasive confusion as the maulstick incongruity. As an example, the estimation of a hazardous concentration (HC_p) for p percent of species in a community from a collection of NOEC values of those species (e.g., Van Straalen and Denneman 1989) requires one to assume that the NOEC is the concentration at or below which there is no *biologically* significant effect. This is clearly an overextension of the concept because the NOEC/LOEC values are extremely dependent on the experimental design, variation in the data, and the particular significance test applied. The subtle extension of the NOEC/LOEC values to vaguely infer hazard or risk can be found in many sources such as the following quote:

> The parameter p in HC_p was considered equivalent to risks estimated for industrial accidents, cancer risks from radiation, etc. Consequently, HC_5 could be considered as a concentration with an ecological risk of 5%, with risk in this case being the probability of finding a species exposed to a concentration higher than its NOEC . . . .
>
> **(Van Straalen and van Leeuwen 2002)**

Although such statements are motivated by well-intended pragmatism, it is difficult to separate inferences of statistical significance and biological relevance with such effect metrics. Does exceedance of an NOEC constitute an ecological risk as inferred? Kooijman (1987) and Newman (1995) articulated concern about making such inferences in hazardous concentration estimations.

[4] The EC_x is similar to the human toxicologist's Benchmark Dose (BMD) approach that uses regression model predictions for a specified effect level (benchmark response) instead of hypothesis testing. Often the lower 95% confidence limit for such an estimated BMD (BMDL) is used to set exposure limits in human risk assessment (Crump 1984, Falk Filipsson et al. 2003).

Further, as a second shortcoming, the NOEC/LOEC metrics are relatively static metrics whereas a regression model can estimate a range of effect concentrations as our knowledge of the level having a relevant biological effect changes with time.

Several related concerns have been added to the two already mentioned. Third, the values of the NOEC/LOEC metrics are dependent on design and statistical features of the process, not simply the biological properties being studied. A fourth criticism is that the values of the NOEC/LOEC can change depending on the power of the specific hypothesis test applied to the data set. A related fifth criticism is that the conventional methods used to estimate these metrics generally have the power to detect effects at the level of roughly a 20% change, yet some effects will have biological relevance with much smaller changes. However, this criticism could be addressed to a degree by changing the conventional design. The sixth and seventh criticisms are related to the metrics derived from the hypothesis tests. The manner in which the NOEC/LOEC metrics are derived from the results dictates that they can only take on the value of a treatment. The spacing and selection of the treatments have a strong influence on the NOEC/LOEC values (criticism 6). Next, because they can only take on the value of an experimental treatment, a standard error cannot be produced for the metric estimate (criticism 7). The last several criticisms combine in practice in such an undesirable way that poor design and/or wide within-treatment variability are rewarded with higher NOEC/LOEC metrics (criticism 8). Relative to the conservative application of the effect metrics during environmental decision making, it would be preferable if the opposite were true. The ninth criticism is that the conventional value of .05, or perhaps .01, for the Type I error (α) is an arbitrary one. A critical biological effect with an associated p of .06 might be ignored while a trivial biological effect with a p of .01 might be used to generate the NOEC/LOEC metrics. Obviously, this last criticism is invalid if proper biological insight and judgment were integrated into the procedure including appropriate changes to error rates. Unfortunately, application of such insight is only now becoming obligatory in reports and publications.

> Statisticians are often stunned by the over-zealous use of some particular statistical tool or methodology on the part of an experimenter, and we offer the following caveat. Experimenters, when you are doing "statistics" do not forget what you know about your subject-matter field! Statistical techniques are most effective when combined with appropriate subject-matter knowledge. The methods are an important adjunct to, not a replacement for, the natural skill of the experimenter.
>
> **(Box et al. 1978)**

A tenth criticism relates to treatment assignment and associated error. As generally described by Montgomery (1997), the levels of treatment can be inaccurate in many experiments (e.g., all nominal 100 mg/L treatment replicates are not actually 100 mg/L when measured such that non-trivial differences occur in "replicate" concentrations) and regression methods are more applicable in such cases. An eleventh criticism from Stephan and Rogers (1985) relates to the manner in which sublethal effects testing is done. Often the same experiment is used to test for significant effects of growth, reproduction, and other sublethal effects. The question should be answered in such a case of whether or not the separate hypothesis tests for each effect are independent. A very strong argument could be made that the experimentwise Type I error rate should be adjusted because the tests are not independent. Stephan and Rogers (1985) suggest a twelfth shortcoming of the hypothesis testing approach. The models used for hypothesis tests are rudimentary ones that do not provide ecotoxicologists with an avenue for extending explorations to other models more directly relevant to the biological mechanism or specific context for which inferences are to be made. This criticism will likely become less serious as ecotoxicologists slowly come to appropriately balance the use of hypothesis testing and point estimation from biologically well-founded models. Hypothesis testing would then be an invaluable first step in a progression of studies, ending in ecotoxicologically meaningful point estimates.

A quick review of the materials just presented results in many more shortcomings for the hypothesis test approach than for the point estimation approach. This does not mean that point estimation should completely displace hypothesis testing in quantifying sublethal effects metrics. As stated by Hacking (2001) above, hypothesis tests are useful for exploring poorly understood situations to which a theory-based model cannot be selected and applied with enough certainty (but see Box 10.2 for a major concern about current misuses of hypothesis tests). The abundant criticisms do suggest that point estimation should be favored over hypothesis test methods more than is currently the general practice. Hypothesis testing is valuable at initial stages of the investigative process and does not result in the best metric of effect or the most insight. The methods just described as "point estimation" approaches should be applied once hypothesis testing has created enough information in the investigative process to effectively generate sound quantitative models.

Box 10.1 Hypothesis Testing from the OECD Vantage

The most commonly used methods for determining the NOEC are not necessarily the best.

(OECD 2006)

As described earlier, the approaches for understanding and quantifying sublethal effects are rapidly evolving, and these changes are reflected in the related regulatory documents. The recent exceptionally clear and thorough OECD (2006) description of hypothesis tests is the best example with which to illustrate this evolution. The same basic structure remains but incremental improvements are apparent. It will be summarized here relative to the analysis of sublethal effects data but it contains useful insight about analysis of lethal effects data too.

The OECD committee begins by making the distinction between best methods for single-step and step-down approaches. Single-step approaches compare the mean (or some other statistic such as the median) of the control treatment with those of each of the other dose/concentration treatments. A step-down test first tests for a significant difference between the control mean (or some other statistic) and that of the highest treatment. If there is a significant difference, the test is then repeated for the control and the next highest treatment. This process is repeated, stepping down the concentrations in the series and testing for a significant difference between each noncontrol treatment and the control treatment. Testing stops when a test results in a nonsignificant difference. In this way, step-down approaches limit the number of comparisons being done for the data set and gain some power relative to the single-step techniques. In contrast to step-down tests, there is no order to the multiple tests done in a single-step test. All tests are done between the control and each of the noncontrol treatments. Examples from the EPA flow chart of single-step and step-down methods are Dunnett's and Williams' tests, respectively. The original schema in EPA (2002) does not show step-down methods; however, step-down testing has been presented in the text of several such EPA documents. More so than the EPA, the OECD committee recommends a blend of single-step and step-down tests. The OECD highlights some different tests based on their enhanced power, robustness to assumption violations, or appropriateness for continuous or quantal data.

One difference between the EPA and OECD documents involves error rate adjustments if multiple comparisons are made. The EPA suggests the conventional Bonferroni adjustment that tends to be slightly conservative.[5] Other, slightly less conservative adjustments can be made including the Holm modification of the Bonferroni recommended by the OECD committee.

[5] A method is more conservative than another if it is less prone to reject the null hypothesis of no significant difference when a difference exists. A Bonferroni-adjusted error rate tends to be conservative because it does not set the α. Instead it sets an upper limit for the α: the Bonferroni-adjusted α associated with the test might vary slightly from 0.05. Less conservative adjustments include the Dunn-Šidák adjustment. Ury (1976), Day and Quinn (1989), and Wright (1992) provide good discussions of this issue.

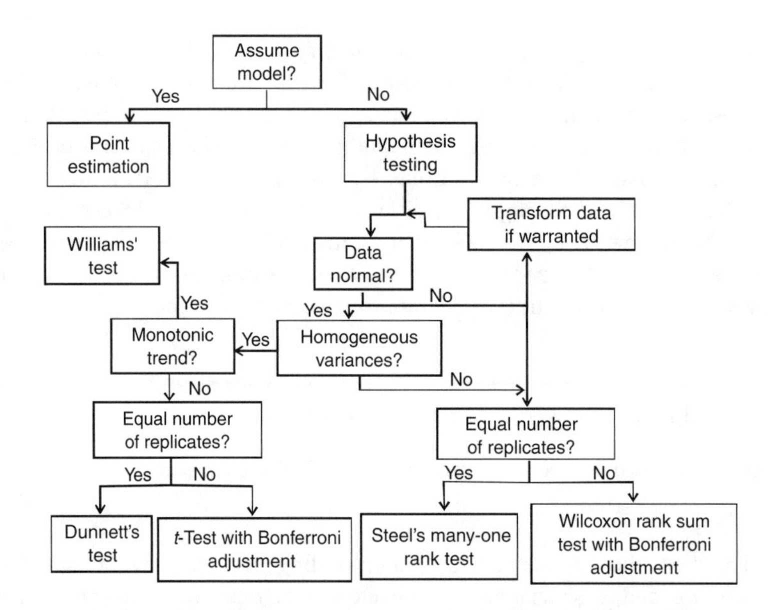

FIGURE 10.4 Typical flow chart for hypothesis testing of sublethal effects data. (Modified from Figure 2 and associated text of EPA (2002).) The EPA manual indicates that the data can be discrete or continuous. Data might be transformed to meet the assumptions of monotonicity, normality, homogeneity of treatment variances, or the distribution of observations among treatments of the associated tests. The most powerful tests tend to have the most assumptions; for example, Williams' test, which assumes a monotonic change in response with increasing treatment intensity in addition to the formal requirements of normal data with the same variance for all treatments. The least powerful tests assume the least, for example, the Wilcoxon rank sum test with a Bonferroni adjustment of Type I error rates.

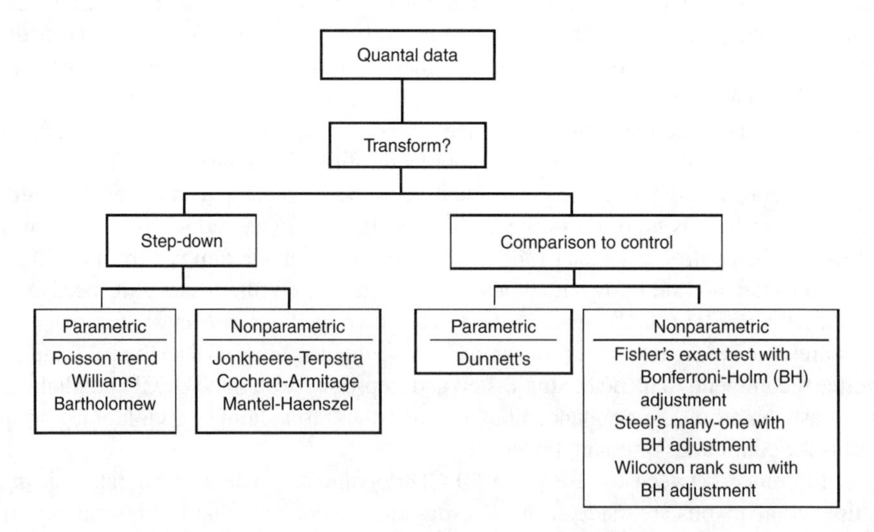

FIGURE 10.5 Alternative flow chart of methods available for testing the statistical significance of discrete sublethal effects data sets. (Modified from Figure 5.1 of OECD (2006).) Not all recommended tests are shown.

In contrast to the EPA flow chart shown in Figure 10.4, a distinction between best methods for quantal (e.g., not responding versus responding, Figure 10.5) and continuous (e.g., growth, Figure 10.6) data is also made by the OECD committee. The methods recommended for quantal data are similar for the tests that compare the control to the other treatments in a single step; however, those for step-down tests include several new tests including some that

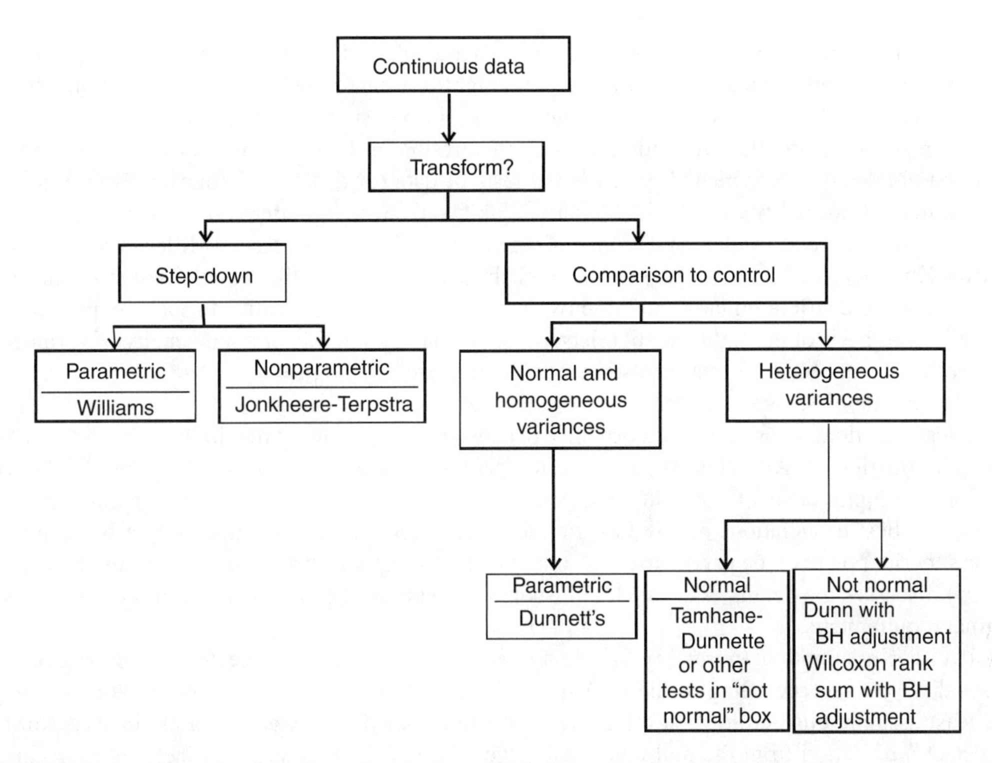

FIGURE 10.6 Alternative flow chart of methods available for testing the statistical significance of continuous sublethal effects data sets. (Modified from Figures 5.2 and 5.3 of OECD (2006).) Not all recommended tests are shown.

are nonparametric. These differences are generally the same for the continuous data but the Tamhane–Dunnette test is recommended if the data are normal yet variances are not equivalent among treatments.

10.3.1.1 Basic Concepts and Assumptions of Hypothesis Tests

Several EPA documents establish convention for assessing sublethal effects data sets using hypothesis testing (e.g., EPA 2002). Figure 10.4 is typical of the schema described in these documents. The emphasis in these schema is assuring that professionals with very different statistical backgrounds can perform tests with a specified Type I error rate (e.g., .05) without violating the strict assumptions of the various tests. Although not often emphasized in such documents, the conventional tests tend to be robust to violations of many assumptions and more power can be gained with insightful deviations from this flow chart (Newman 1995). Specifically, some deviations from normality and heterogeneity of variance for tests such as Dunnett's test have little influence on the outcome. The number of replicates does not have to be the same for the Dunnett's test: in fact, there are very good reasons to have more replicates in the control treatment than in the noncontrol treatment. Also, some deviations from monotonicity in the data can be accommodated with Williams' test. Finally, these documents often do not highlight the enhanced power available if one can use a one-sided instead of a two-sided test (Cohen 1988).

The first step in analyzing sublethal effect data with the EPA scheme (Figure 10.4) is deciding whether the data should be fit to a specific model. If not, many options are available depending on the nature of the data. They include the hypothesis tests shown in the chart and also bioequivalence tests described at the end of this section.

If hypothesis testing is the chosen approach, data are assessed relative to formal requirements of the candidate hypothesis tests.[6] The original data might be transformed in some manner to facilitate conformity to such requirements. For example, proportions such as proportion of exposed individuals responding are often transformed using the arcsine of the square root of the proportion as described in detail in Newman (1995). Formal tests of data (or data transforms) conformity to the requirement of normality include the Shapiro–Wilk's and several goodness-of-fit tests such as the Kolmogorov–Smirnov, Anderson–Darling, Cramer–von Mises, or χ^2 tests. (Miller (1986) argued that the Kolmogorov–Smirnov and χ^2 tests should be avoided because the deviations of most concern are those at the distribution tails and these two tests are relatively insensitive to such deviations.) If the null hypothesis of normality is not rejected, the formal assumption of homogeneity of variances for the treatments (i.e., all treatments have the same variances, $\sigma_0^2 = \sigma_1^2 = \sigma_2^2 = \cdots = \sigma_i^2$) is tested with one of several tests including the Hartley, Bartlett, Cochran, or Levene tests. Usually these tests are done after those for normality because several formally require that the data set be normally distributed. Although ANOVA and the parametric methods shown in the flow chart have the formal requirements of normality and homogeneity of variances, Miller (1986) points out that they are robust to violations of these assumptions. He suggests that violations might be accepted if the superior power of the parametric tests relative to the nonparametric tests is important. Newman (1995) discusses the details associated with such a decision to ignore minor deviations from these formal requirements.

If either assumption (normality and homogeneity of variance) is rejected, a nonparametric method is selected according to the EPA-derived flow chart (Figure 10.4). These methods carry the least assumptions but do so at the expense of some statistical power, i.e., a diminished ability to detect a deviation from the null hypothesis when the null hypothesis is, in fact, false. According to the EPA approach, Steel's many-one rank test can be used if there are equal numbers of observations per treatment or the Wilcoxon rank sum (= Mann–Whitney U) test can be used if there are not.

If neither of the assumptions was rejected, several more powerful parametric methods can be applied. The most powerful Williams' test can be applied if the data set also conforms to the assumption of monotonicity (i.e., if the effect is a consistent decrease or increase with concentration/dose). In practice, Williams' test can be applied even with small deviations from monotonicity in the data set. Several alternatives are available if the assumption of monotonic trends is inappropriate. According to the EPA flow chart, Dunnett's test can be used if all treatments have equal numbers of observations or the slightly less powerful t-test with Bonferroni adjustment of error rates can be used if the number of observations is not equal for all treatments. (Newman (1995) provides discussion of deviations from this scheme to enhance statistical power.)

Box 10.2 But What Does a Positive Test Result Really Mean?

Confusion about hypothesis testing is pervasive in science today (Ioannidis 2005, Sterne and Smith 2001, Wacholder et al. 2004), and it influences how ecotoxicologists interpret effect concentrations such as the NOEC or LOEC. A treatment concentration might be designated the LOEC in a toxicity test if its associated test statistic has P smaller than a specified Type I

---

[6] Another fundamental requirement of these tests is random assignment of experimental units to treatments: observations are independent within a concentration/dose treatment. For example, catching batches of ten fish from a holding tank and placing each batch into replicate tanks starting at the control tanks and ending at the highest concentration tanks could easily insert an extraneous factor into the test. Perhaps the most easily caught fish were the smallest. The fish in the experimental tanks would tend to have the smallest fish in the control tanks and the largest in the highest concentration tanks. This would violate a basic assumption of independence. Similarly, nonrandom placement of treatment replicates into an incubator such that some treatments were closer to a light or temperature source could invalidate the experiment.

error rate (α). The ecotoxicologist identifying the LOEC reasons that the chance of getting the test statistic for that treatment concentration due to chance alone is so low (1 in 20 or less) and, conversely, the chance of there being an effect is so high—perhaps 19 in 20 or better. Therefore, it is not reasonable to maintain the position that there was no effect: there is an effect. This common conclusion that there is an effect at that concentration is wrong. The false positive result probability (FPRP) described by Wacholder et al. (2004) or its converse (positive predictive value or PPV) (Ioannidis 2005) are good tools with which to show the reason.

> [The] α level is the probability of a statistically significant finding given the null hypothesis is true, whereas FPRP is the probability that the null hypothesis is true, given that the statistical test is statistically significant.
>
> **(Wacholder et al. 2004)**

The FPRP is the probability that the null hypothesis of no effect was true given that the statistical test was significant. Its converse, the PPV, is the probability that the null hypothesis was false (i.e., an effect exists), given a significant statistical test.

As in the example given above, α is being confused with the PPV in most studies to identify the LOEC and its associated NOEC. As assumed in the example, it is true that the probability of an effect actually being present given a positive test does depend on α. However, it also depends on the power of the test $(1 - \beta)$ and the prior probability of an association between a concentration treatment and an effect (π). This can be seen in the formulae used to estimate the FPRP and PPV.

$$\text{FPRP} = \frac{\alpha(1-\pi)}{\alpha(1-\pi)+\pi(1-\beta)}$$

$$\text{PPV} = \frac{R(1-\beta)}{R-\beta R+\alpha}$$

where R = the prior quotient of "true relationships" to "no relationships" in the field. The π in the equation for FPRP is estimated as $R/(R+1)$. The PPV is the post-study probability that a positive test correctly identified a true effect.

The P associated with the Type I error is often misused to suggest the magnitude of PPV. The difference can be shown easily using a conventional sublethal effects test with five treatments (including the reference—"0" concentration—treatment) for which the results are assessed using Dunnett's test. Let's assume in our illustration that selecting treatments carefully in a conventional test generally results in two of the four pairwise tests being positive. The NOEC would be the second concentration treatment and the LOEC the third treatment concentration. This suggests a prior probability of .5 for an LOEC/NOEC experiment for which the associated data might be analyzed with Dunnett's test. The conventional α would be set at .05 but the test power $(1 - \beta)$ would vary.

For such a test, how would power and the prior probability influence the FPRP or PPV? To answer the question of power's effect, let's use a range of powers from .5 to .9. This range includes anticipated powers of conventional toxicity tests. For example, Alldredge (1987), Cohen (1988), Oris and Bailer (1993), and Thursby et al. (1997) included .8 as a realistic power for good tests. The influence of π can be explored by comparing the π of .5 for the above test with that of a test with one more treatment than the above test and assuming that π was .4, i.e., 2 of 5 treatment concentrations usually result in a positive test. (The corresponding R values for PPV calculations are 1 and .666.) The resulting FPRP and PPV estimates are provided in the table below.

$1 - \beta$	FPRP ($\pi = .5$)	PPV ($\pi = .5$)	FPRP ($\pi = .4$)	PPV ($\pi = .4$)
.5	.09	.91	.13	.87
.6	.08	.92	.11	.89
.7	.07	.93	.10	.90
.8	.06	.94	.09	.91
.9	.05	.95	.08	.92

The PPV—the probability of correctly identifying a treatment as having an effect given a positive test—is not as high as commonly thought. A relatively high power (.9) and a prior of .5 were needed in this example to get a PPV of .95 for a test with $\alpha = .05$. With a moderate power (.70) and design with a prior of .40, the PPV was estimated to be 9 in 10, not 19 in 20. The certainty in the LOEC being a true effect concentration is not .95 or better as often thought with a test α of .05.

Some generalizations can be made about hypothesis test FPRP and PPV. The FPRP "is always high when α is much greater than π, and even more so when $1 - \beta$ is low" (Wacholder et al. 2004). Also, a positive result from a hypothesis study is more likely to be true if $(1 - \beta)\pi > \alpha$ (Ioannidis 2005).

Although applied narrowly here in the context of sublethal toxicity testing, this error in judging the plausibility of an effect is pervasive in science. In many instances, the problem is worse than described for toxicity tests. For example, Sterne et al. (2001) estimated that human epidemiology studies generally have π and powers of approximately .1 and .5, respectively. In epidemiology studies examined by Sterne et al. (2001), they found that "Of the 95 studies that result in a significant ($P < .05$) result, 45 (47%) are true null hypotheses and so are 'false alarms' ..." Clearly, traditional misinterpretations of hypothesis test results foster muddled judgments about adverse effects. In an ideal world, the PPV or FPRP would be used instead of the α to judge the plausibility of research conclusions from hypothesis testing.

Obvious changes would reduce the magnitude of this problem. As stated above, widespread application of the FPRP or PPV would greatly enhance interpretation. Requiring that power estimates be reported would allow calculation of the FPRP with which the quality of any inferences could be judged. (Power, like α, must be established a priori while following this recommendation (Cohen 1988)). Also helpful would be the abandonment of arbitrary divisions (i.e., $\alpha = .05$) in interpreting hypothesis test results. Applying the general trends described by Wacholder et al. (2004) and Ioannidis (2005) during design, experimentation, and hypothesis testing would greatly enhance the value tests. As a final and more encompassing recommendation, instead of making inferences from results of single studies, research programs composed of a series of studies might be recognized as a more profitable approach to assessing the plausibility of an effect. In a research program, the prior probability (π) can be adjusted continually as more experiments and observations contribute to the research program.

Considering statistical power provides a direct way to design an experiment to be precise enough to detect effects of toxicological importance, but it is an indirect way to reach the conclusion that a treatment has no effect.

(Hauschke 1997)

The classic null hypothesis (no difference between treatments) is inappropriate if the intent is to prove that some treatment is "safe" or has no effect.

(Dixon 1998)

Practitioners of drug testing have developed alternate methods, called bioequivalence or equivalence testing, that avoid several of the major drawbacks of hypothesis testing. In contrast to the conventional testing of the null hypothesis of no significant difference, bioequivalence techniques test if an actual difference is outside some prespecified region of equivalence. Bioequivalence testing can be done using single-step or step-down approaches (Bofinger and Bofinger 1995). The treatments are essentially equivalent if this hypothesis is rejected. One advantage of this approach is that it requires one to decide a priori what an ecotoxicologically significant difference is. Another is that, in contrast to the conventional null hypothesis tests described above, high variability or poor power does not result in the associated NOEC increasing. Bioequivalence procedures are so well established that textbooks such as Chow and Liu (2000) are available to guide their application in medical sciences and the pharmaceutical industry. Unfortunately, they have been underutilized in ecotoxicology despite very clear presentations of their utility.

There are several excellent discussions and examples of applying bioequivalence tests in ecotoxicology. Dixon (Dixon 1998, Dixon and Garrett 1994) contrasts conventional null hypothesis tests with equivalence tests relative to ecotoxicity testing, describing parametric and nonparametric approaches. Erickson and McDonald (1995) provide a similar discussion. Shukla et al. (2000) apply bioequivalence testing in the whole effluent toxicity (WET) testing context, emphasizing that the adoption of bioequivalence testing will resolve power issues present during conventional WET testing.

10.3.1.2 Basic Concepts and Assumptions of Point Estimation Methods

Unlike the hypothesis testing methods, most of the relevant issues associated with point estimation were covered in the previous chapter. Only a few additional issues require discussion. The distinction should be made between approaches most appropriate for discrete and continuous data. The existence of a convenient and common interpolation method also requires discussion.

Many of the methods described in Chapter 9 are relevant to quantal data for sublethal effects. For example, an EC50 might be estimated by maximum likelihood fitting of those data to a probit (log normal) model. Also, the logistic regression models described in Chapter 13 are relevant. For continuous data, many conventional regression methods can be applied including those with thresholds, hormetic responses, or toxicant interactions.

Fitting to a parametric model is not necessary in some cases as illustrated with the EPA linear interpolation method (Appendix M of EPA (2002)). With this approach, a concentration associated with a specified level of effect is estimated by linear interpolation between treatments. Bootstrap confidence intervals are then generated with the treatment replicate data. The result is an estimated IC_p (inhibition concentration associated with a specified percent (p) reduction relative to the control's smoothed mean for some effect metric) and its bootstrap confidence intervals. The only specific assumptions made are that the response is monotonically decreasing with concentration/dose; is predictable using a piecewise linear response function between treatments; and the observations are random, independent, and representative (EPA 2000). Obviously, an increasing response can be converted to a decreasing one by simple transformation and some deviation from monotonic trend can be accommodated by smoothing.

10.4 SUMMARY

The context for sublethal effects studies and statistical means of estimating sublethal effects are discussed in this short chapter. Most sublethal effect studies have as their explicit or implicit goal assessment of changes in an individual's fitness under toxicant exposure. A wide range of hypothesis testing and point estimation techniques are available for quantifying sublethal effects. Used appropriately, both approaches have a role to play and provide useful insight. Unfortunately, studies often stop after producing an effect metric with hypothesis tests, contributing to a large data base that is not easily linked to theory-based, predictive models.

10.4.1 Summary of Foundation Concepts and Paradigms

- Most studies of sublethal effects apply experimental designs appropriate for ANOVA or ANCOVA and do not take full advantage of available ecological models to quantify or interpret results.
- Higher-order phenomena can be integrated into ecotoxicity studies if one seeks out the appropriate methods and theories instead of using solely the conventional mode of analysis and interpretation that emerged in the 1960s and 1970s.
- Sublethal stressor effects that dominate the literature are those to growth and development, reproduction, behavior, and physiology. The goal of these studies is assessing how exposure diminishes Darwinian fitness.
- Conventional tests for sublethal effects consist of replicate groups of organisms exposed to a series of concentrations or doses for a single duration.
- Sublethal effect metrics are calculated using either hypothesis testing or point estimation approaches. The same data set can be analyzed with these methods; however, optimization of experimental design is not the same for these two approaches.
- The hypothesis testing approach attempts to identify the highest concentration or dose that has no effect and the lowest concentration or dose that has an effect.
- The point estimation approach begins by assigning a level of effect that is deemed unacceptable and then estimating the corresponding concentration or dose.
- Bioequivalence testing avoids some major shortcomings of the conventional null hypothesis testing approach but remains underexploited in sublethal effects studies. Bioequivalence methods test if the "true" difference is outside some prespecified region of equivalence.

REFERENCES

Alldredge, J.R., Sample size for monitoring of toxic chemical sites, *Environ. Monit. Assess.*, 9, 143–154, 1987.

Al-Yakoob, S.N., Gundersen, D., and Curtis, L., Effects of the water-soluble fraction of partially combusted crude oil from Kuwait's oil fires (from Desert Storm) on survival and growth of the marine fish *Menidia Beryllina*. *Ecotox. Environ. Safety*, 35, 142–149, 1996.

Ankley, G.T., Jensen, K.M., Kahl, M.D., Korte, J.J., and Makynen, E.A., Description and evaluation of a short-term reproduction test with the fathead minnow (*Pimephales promelas*), *Environ. Toxicol. Chem.*, 20, 1276–1290, 2001.

Arnold, D.E., Ingestion, assimilation, survival, and reproduction in *Daphnia pulex* fed seven species of blue-green algae, *Limnol. Oceanogr.*, 16, 906–920, 1971.

Atchison, G.J., Sandheinrich, M.B., and Bryan, M.D., Effects of environmental stressors on interspecific interactions of aquatic animals, In *Ecotoxicology. A Hierarchical Treatment*, Newman, M.C. and Jagoe, C.H. (eds.), CRC Press/Lewis Publishers, Boca Raton, FL, 1996, pp. 319–345.

Baker, A.J.M. and Walker, P.L., Physiological responses of plants to heavy metals and the quantification of tolerance and toxicity, *Chem. Spec. Bioavail.*, 1, 7–18, 1989.

Bayley, M., Baatrup, E., and Bjerregaard, P., Woodlouse locomotor behavior in the assessment of clean and contaminated field sites, *Environ. Toxicol. Chem.*, 16, 2309–2314, 1997.

Bishop, C.A., Ng, P., Mineau, P., Quinn, J.S., and Struger, J., Effects of pesticide spraying on chick growth, behavior, and parental care in tree swallows (*Tachycineta bicolor*) nesting in an apple orchard in Ontario, Canada, *Environ. Toxicol. Chem.*, 19, 2286–2297, 2000.

Bodar, C.W.M., Van der Sluis, I., Voogt, P.A., and Zandee, D.I., Effects of cadmium on consumption, assimilation and biochemical parameters of *Daphnia magna*: Possible implications for reproduction, *Comp. Biochem. Physiol.*, 90C, 341–346, 1988.

Bofinger, E. and Bofinger M., Equivalence with respect to a control: Stepwise tests, *J. Roy. Statis. Soc.*, 57, 721–733, 1995.

Bookhout, C.G., Monroe, R.J., Forward, Jr., R.B., and Costlow, Jr., J.D., Effects of hexavalent chromium on development of crabs, *Rhithropanopeus harrisii* and *Callinectes sapidus*, *Water Air Soil Pollut.* 21, 199–216, 1984.

Bortone, S.A., Davis, W.P., and Bundrick, C.M., Morphological and behavioral characters in mosquito fish as potential bioindication of exposure to Kraft mill effluent, *Bull. Environ. Contam. Toxicol.* 43, 370–377, 1989.

Boudreau, M., Courtenay, S.C., MacLatchy, D.L., Bérubé, C.H., Parrott, J.L., and Van der Kraak, G.J., Utility of morphological abnormalities during early-life development of the estuarine mummichog, *Fundulus heteroclitus*, as an indicator of estrogenic and antiestrogenic endocrine disruption, *Environ. Toxicol. Chem.*, 23, 415–425, 2004.

Boutin, C., Freemark, K.E., and Keddy, C.J., Overview and rationale for developing regulatory guidelines for nontarget plant testing with chemical pesticides, *Environ. Toxicol. Chem.*, 14, 1465–1475, 1995.

Boutin, C., Lee, H.-B., Peart, E.T., Batchelor, P.S., and Maguire, R.J., Effects of the sulfonylurea herbicide metasulfuron methyl on growth and reproduction of five wetland and terrestrial plant species, *Environ. Toxicol. Chem.*, 19, 2532–2541, 2000.

Bouton, S.N., Frederick, P.C., Spalding, M.G., and McGill, H., Effects of chronic, low concentrations of dietary methylmercury on the behavior of juvenile Great Egrets, *Environ. Toxicol. Chem.*, 18, 1934–1939, 1999.

Box, G.E.P., Hunter, W.G., and Hunter, J.S., *Statistics for Experimenters. An Introduction to Design, Data Analysis, and Model Building*, John Wiley & Sons, New York, 1978, p. 14.

Bridges, C.M., Tadpole swimming performance and activity affected by acute exposure to sublethal levels of carbaryl, *Environ. Toxicol. Chem.*, 16, 1935–1939, 1997.

Bringolf, R.B., Belden, J.B., and Summerfelt, R.C., Effects of atrazine on fathead minnow in a short-term reproduction assay, *Environ. Toxicol. Chem.*, 23, 1019–1025, 2004.

Brown Sullivan, K. and Spence, K.M., Effects of sublehtal concentrations of atrazine and nitrate on metamorphosis of the African clawed frog, *Environ. Toxicol. Chem.*, 22, 627–635, 2003.

Bryan, G.W. and Gibbs, P.E., Impact of low concentrations of tributyltin (TBT) on marine organisms: A review, In *Metal Ecotoxicology. Concepts & Applications*, Newman, M.C. and McIntosh, A.W. (eds.), CRC Press/Lewis Publishers, Boca Raton, FL, 1991, pp. 323–361.

Calabrese, E.J., McCarthy, M.E., and Kenyon, E., The occurrence of chemically induced hormesis, *Health Phys.*, 52, 531–541, 1987.

Carr, J.A., Gentiles, A., Smith, E.E., Goleman, W.L., Urquidi, L.J., Thett, K., Kendall, R.J., et al., Response of larval *Xenopus laevis* to atrazine: Assessment of growth, metamorphosis, and gonadal and laryngeal morphology, *Environ. Toxicol. Chem.*, 22, 396–405, 2003.

Carson, R.L., *Silent Spring*, Houghton Mifflin Co., New York, 1962.

Chapman, P.F., Crane, M., Wiles, J., Noppert, F., and McIndoe, E., Improving the quality of statistics in regulatory ecotoxicity tests, *Ecotoxicology*, 5, 169–186, 1996.

Chow, S.-C. and Liu, J.-P., *Design and Analysis of Bioavailability and Bioequivalence Studies*, 2nd ed., Marcel Dekker, Inc., New York, 2000.

Cochran, W.G. and Cox, G.M., *Experimental Designs*, 2nd ed., John Wiley & Sons, New York, 1957.

Coeurdassier, M., Saint-Denis, M., Gomot-de Vaufleury, A., Ribera, D., and Badot, P.-M., The garden snail (*Helix aspersa*) as a bioinidcator of organophosphorus exposure: Effects of dimethoate on survival, growth, and acteylcholinesterase, *Environ. Toxicol. Chem.*, 20, 1951–1957, 2001.

Cohen, J., *Statistical Power Analysis for the Behavioral Sciences*, 2nd ed., Lawrence Erlbaum Assoc., Hillsdale, NJ, 1988.

Cold, A. and Forbes, V.E., Consequences of a short pulse of pesticide exposure for survival and reproduction of *Gammarus pulex*, *Aquat. Toxicol.*, 67, 287–299, 2004.

Conrad, G.W., Heavy metal effects on cellular shape changes, cleavage, and larval development of the marine gastropod mollusk, (*Ilyanassa obsoleta Say*), *Bull. Environ. Contam. Toxicol.*, 41, 79–95, 1988.

Cook, L.W., Paradise, C.J., and Lom, B., The pesticide malathion reduces survival and growth in developing zebrafish, *Environ. Toxicol. Chem.*, 24, 1745–1750, 2005.

Crane, M. and Newman, M.C., What level of effect is a No Observed Effects? *Environ. Toxicol. Chem.*, 19, 516–519, 2000.

Crouau, Y. and Cazes, L., Unexpected reduction in reproduction of Collembola exposed to an arsenic-contaminated soil, *Environ. Toxicol. Chem.*, 24, 1716–1720, 2005.

Crump, K.S., A new method for determining allowable daily intakes, *Fun. Appl. Toxicol.*, 4, 854–871, 1984.

D'Agostino, A., Finney, C., The effect of copper and cadmium on the development of *Tigriopus japonicus*, In *Pollution and Physiology of Marine Organisms*, Vernberg, F.J., and Vernberg, W.B. (eds.), Academic Press, New York, 1974, pp. 445–463.

Day, R.W. and Quinn, G.P., Comparisons of treatments after an analysis of variance in ecology, *Ecol. Monogr.*, 59, 433–463, 1989.

De Boeck, G., Borger, R., Van der Linden, A., and Blust, R., Effects of sublethal copper exposure on muscle energy metabolism of common carp. Measured by ^{31}P-nuclear magnetic resonance spectroscopy, *Environ. Toxicol. Chem.*, 16, 676–684, 1997.

Degitz, S.J., Kosian, P.A., Makynen, E.A., Jensen, K.M., and Ankley, G.T., Stage- and species-specific developmental toxicity of all-trans retinoic acid in four native North American Ranids and *Xenopus laevis. Toxicol. Sci.* 57, 264–274, 2000.

De Schamphelaere, K.A.C. and Janssen, C.R., Effects of chronic dietary exposure on growth and reproduction of *Daphnia magna*, *Environ. Toxicol. Chem.*, 23, 2038–2047, 2004.

DeWitt, J.C., Millsap, D.S., Yeager, R.L., Heise, S.S., Sparks, D.W., and Henshel, D.S., External heart deformities in passerine birds exposed to environmental mixtures of polychlorinated biphenyls during development, *Environ. Toxicol. Chem.*, 25, 541–551, 2006.

De Wolf, W., Siebel-Sauer, A., Lecloux, A., Koch, V., Holt, M., Feijtel, T., Comber, M., and Boeije, G., Mode of action and aquatic exposure thresholds of no concern, *Environ. Toxicol. Chem.*, 24, 479–485, 2005.

Diana, S.G., Resetarits, W.J., Jr., Schaeffer, D.J., Beckmen, K.B., and Beasley, V.R., Effects of atrazine on amphibian growth and survival in artificial aquatic communities, *Environ. Toxicol. Chem.*, 19, 2961–2967, 2000.

Dixon, P.M. Assessing effect and no effect with equivalence tests, In *Risk Assessment. Logic and Measurement*, Newman, M.C. and Strojan, C.L. (eds.), CRC Press/Lewis Publishers, Boca Raton, FL, 1998, pp. 275–301.

Dixon, P.M. and Garrett, K.A., Statistical issues for field experimenters, In *Wildlife Toxicology and Population Modeling. Integrated Studies of Agroecosytsems*, Kendall, R.L. and Lacher, T.E., Jr. (eds.), CRC Press/Lewis Publishers, Boca Raton, FL, 1994, pp. 439–449.

Doddard, S.G., Sunahara, G.I., Kuperman, R.G., Sarrazin, M., Gong, P., Ampleman, G., Thiboutot, S., and Hawari, J., Survival and reproduction of enchytraeid worms, Oligochaete, in different soil types amended with energetic nitramines, *Environ. Toxicol. Chem.*, 24, 2579–2587, 2005.

Draper, N.R. and Smith, H., *Applied Regression Analysis*, 3rd ed., John Wiley & Sons, New York, 1998.

Elbert, R.A. and Anderson, D.W., Mercury levels, reproduction, and hematology in Western grebes from three California lakes, USA, *Environ. Toxicol. Chem.*, 17, 210–213, 1998.

EPA, *Short-Term Methods for Estimating the Chronic Toxicity of Effluents and Receiving Waters to Freshwater Organisms,* 4th ed, EPA821/R-02/013, October 2002. NTIS, Washington, D.C., 2002.

Erickson, W.E. and McDonald, L.L., Tests of bioequivalence of control media and test media in studies of toxicity, *Environ. Toxicol. Chem.*, 14, 1247–1256, 1995.

Falk Filipsson, A., Sand, S., Nilsson, J., and Victorin, K., The benchmark dose method—review of available models, and recommendations for application in health risk assessment, *Critical Rev. Toxicol.*, 33, 505–542, 2003.

Fletcher, J.S., Johnson, F.L., and McFarlane, J.C., Database assessment of phytotoxicity data published on terrestrial vascular plants, *Environ. Toxicol. Chem.*, 615–622, 1988.

Fletcher, J.S., Pfleeger, T.G., Ratsch, H.C., and Hayes, R., Potential impact of low levels of chlorsulfuron and other herbicides on growth and yield of nontarget plants, *Environ. Toxicol. Chem.*, 15, 1189–1196, 1996.

Fordham, C.L., Tessari, J.D., Ramsdell, H.S., and Keefe, T.J., Effects of malathion on survival, growth, development, and equilibrium posture of bullfrog tadpoles (*Rana catesbeiana*), *Environ. Toxicol. Chem.*, 20, 179–184, 2001.

Fry D.M. and Toone C.K., DDT-induced feminization of gull embryos, *Science*, 213, 922–924, 1981.

Gray, B.R., Emery, V.L., Jr., Brandon, D.L., Wright, R.B., Duke, B.M., Farrar, J.D., and Moore, D.W., Selection of optimal measures of growth and reproduction for the sublethal *Leptocheirus plumulosus* sediment bioassay, *Environ. Toxicol. Chem.*, 17, 2288–2297, 1998.

Green, C.C. and Lochmann, S.E., Fluctuating asymmetry and condition in golden shiner (*Notemigonus crysoleucas*) and channel catfish (*Ictalurus punctatus*) reared in sublethal concentrations of isopropyl methylphosphonic acid, *Environ. Toxicol. Chem.*, 25, 58–64, 2006.

Grillitsch, B., Vogl, C., and Wytek, R., Qualification of spontaneous undirected locomotor behavior of fish for sublethal toxicity testing. Part II. Variability of measurement parameters under toxicant-induced stress, *Environ. Toxicol. Chem.*, 18, 2743–2750, 1999.

Hacking, I., *An Introduction to Probability and Inductive Logic*, Cambridge University Press, Cambridge, UK, 2001.

Handy, R.D., Sims, D.W., Giles, A., Campbell, H.A., and Musonda, M.M., Metabolic trade-off between locomotion and detoxification for maintenance of blood chemistry and growth parameters by rainbow trout (*Oncorynchus mykiss*) during chronic dietary exposure to copper, *Aquat. Toxicol.*, 47, 23–41, 1999.

Hanowski, J.M., Niemi, G.J., Lima, A.R., and Regal, R.R., Do mosquito control treatments of wetlands affect Red-winged Blackbird (*Agelaius pheoniceus*) growth, reproduction, or behavior? *Environ. Toxicol. Chem.*, 16, 1014–1019, 1997.

Hansen, J.A., Lipton, J., Welsh, P.G., Cacela, D., and MacConnell, B., Reduced growth of rainbow trout (*Oncorhychus mykiss*) fed a live invertebrate diet pre-exposed to metal-contaminated sediments, *Environ. Toxicol. Chem.*, 23, 1902–1911, 2004.

Hauschke, D., Statistical proof of safety in toxicological studies, *Drug Inform. J.*, 31, 357–361, 1997.

Heinz, G.H. and Hoffman, D.J., Methylmercury chloride and selenomethionine interactions on health and reproduction in mallards, *Environ. Toxicol. Chem.*, 17, 139–145, 1998.

Henry, M.G. and Atchison, G.J., Metal effects on fish behavior—Advances in determining the ecological significance of responses, In *Metal Ecotoxicology. Concepts & Applications*, Newman, M.C. and McIntosh, A.W. (eds.), CRC Press/Lewis Publishers, Boca Raton, FL, 1991, pp. 131–143.

Hickey, J.J. and Anderson, D.W., Chlorinated hydrocarbons and eggshell changes in raptorial and fish-eating birds, *Science*, 162, 271–273, 1968.

Hui, C.A., Lead burdens and behavioral impairments of the lined shore crab, *Pachygrapsus crassipes*, *Ecotoxicol.*, 11, 417–421, 2002.

Hunt, G.L., Jr. and Warner H.M., Female-female pairing in Western gulls (*Larus occidentalis*) in Southern California, *Science*, 196, 1466–1467,1977.

Ingersoll, C.G., Brunson, E.L., Dwyer, F.J., Hardesty, D.K., and Kemble, N.E., Use of sublethal endpoints in sediment toxicity tests with the amphipod *Hyalella azteca*, *Environ. Toxicol. Chem.*, 17, 1508–1523, 1998.

Inouye, L.S., Jones, R.P., and Bednar, A.J., Tungsten effects on survival, growth, and reproduction in the earthworm, *Eisenia fetida*, *Environ. Toxicol. Chem.*, 25, 763–768, 2006.

Ioannidis, J.P.A., Why most published research findings are false, *PLoS Med.*, 2 (e124) 696–701, 2005.

Janssens, E., Dauwe, T., Van Duyse, E., Beernaert, J., Pinxten, R., and Eens, M., Effects of heavy metal exposure on aggressive behavior in a small territorial songbird, *Arch. Environ. Contam. Toxicol.*, 45, 121–127, 2003.

Knops, M., Altenburger, R., and Segner, H., Alterations of physiological energetics, growth, and reproduction of *Daphnia magna* under toxicant stress, *Aquat. Toxicol.*, 53, 79–90, 2001.

Kooijman, S.A.L.M., A safety factor for LC_{50} values allowing for differences in sensitivity among species, *Wat. Res.*, 21, 269–276, 1987.

Kooijman, S.A.L.M., An alterative for NOEC exists, but the standard model has to be abandoned first, *OIKOS*, 75, 310–316, 1996.

Kuperman, R.G., Checkai, R.T., Simini, M., Phillips, C.T., Speicher, J.A.,and Barclift, D.J., Toxicity benchmarks for antimony, barium, and beryllium determined using reproduction endpoints for *Folsomia candida*, *Eisenia fetida*, and *Encytraeus crypticus*, *Environ. Toxicol. Chem.*, 25, 754–762, 2006.

Kynard, B., Avoidance behavior of insecticide susceptible and resistant populations of mosquitofish to four insecticides, *Trans. Amer. Fish. Soc.*, 3, 557–561, 1974.

Landrum, P.F., Leppänen, M., Robinson, S.D., Gossiaux, D.C., Burton, G.A., Greenberg, M., Kukkonen, J.V.K., Eadie, B.J., and Lansing, M.B., Effect of 3,4,3′,4′-tetrachlorobiphenyl on the reworking behavior of *Lumbriculus variegates* exposed to contaminated sediment, *Environ. Toxicol. Chem.*, 23, 178–186, 2004.

Laskowski, R., Some good reasons to ban the use of NOEC, LOEC and related concepts in ecotoxicology, *OIKOS*, 73, 140–144, 1995.

Lewis, M.A. and Wang, W., Biomonitoring using aquatic vegetation, *Environ. Sci. Forum*, 96, 243–274, 1999.

Lytle, T.F. and Lytle, J.S., Growth inhibition as indicator of stress because of atrazine following multiple toxicant exposure of the freshwater macrophyte, *Juncus effusus* L., *Environ. Toxicol. Chem.*, 24, 1198–1203, 2005.

Marwood, C.A., Bestari, K.T.J., Gensemer, R.W., Solomon, K.R., and Greenberg, B.M., Cresote toxicity to photosynthesis and plant growth in aquatic microcosms, *Environ. Toxicol. Chem.*, 22, 1075–1985, 2003.

Mayer, P., Nyholm, N., Verbruggen, E.M.J., Hermens, J.L.M., and Tolls, J., Algal growth inhibition test in filled, closed bottles of volatile and sorptive materials, *Environ. Toxicol. Chem.*, 19, 2551–2556, 2000.

McCloskey, J.T. and Newman, M.C., Sediment preference in the Asiatic clamm (*Corbicula fluminea*) and viviparid snail (*Campeloma decisum*) as a response to low-level metal and metalloid contamination, *Arch. Environ. Contam. Toxicol.*, 28, 195–202, 1995.

McGrath, J.A., Parkerton, T.F., and Di Toro, D.M., Application of the narcosis target lipid model to algal toxicity and deriving predicted-no-effect concentrations, *Environ. Toxicol. Chem.*, 23, 2503–2517, 2004.

Medesani, D.A., López Greco, L.S., Rodriguez, E.M., Interference of cadmium and copper with the endocrine control of ovarian growth, in the estuarine crab, *Chasmagnathus granulata*, *Aquat. Toxicol.*, 69, 165–174, 2004.

Meyer, M.W., Evers, D.C., Hartigan, J.J., and Rasmussen, P.S., Patterns of Common Loon (*Gavia immer*) mercury exposure, reproduction, and survival in Wisconsin, USA, *Environ. Toxicol. Chem.*, 17, 184–190, 1998.

Miller, R.J., Jr., *Beyond ANOVA. Basics of Applied Statistics*, John Wiley & Sons, New York, 1986.

Mineau, P., Boersma, D.C., and Collins, B., An analysis of avian reproduction studies submitted for pesticide registration, *Ecotox. Environ. Safety*, 29, 304–329, 1994.

Montgomery, D.C., *Design and Analysis of Experiments*, 4th ed., John Wiley & Sons, New York, 1997, pp. 552–554.

Moreels, D., Lodewijks, P., Zegers, H., Rurangwa, E., Vromant, N., Bastiaens, L., Diels, L., Springael, D., Merckx, R., and Ollevier, F., Effect of short-term exposure to methyl-*tert*-butyl ether and *tert*-butyl alcohol on the hatch rate and development of the African catfish, *Clarias gariepinus*, *Environ. Toxicol. Chem.*, 25, 514–519, 2006.

Munkittrick, K.R. and Dixon, D.G. Growth, fecundity, and energy stores of white sucker (*Catostomus commersoni*) from lakes containing elevated levels of copper and zinc, *Can. J. Fish. Aquat. Sci.*, 45, 1355–1365, 1988.

Nakayama, K., Oshima, Y., Hiramatsu, K., Shimasaki, Y., and Honjo, T., Effects of polychlorinated biphenyls on the schooling behavior of Japanese medaka (*Oryzias latipes*), *Environ. Toxicol. Chem.*, 24, 2588–2593, 2005.

Newman, M.C., *Quantitative Methods in Aquatic Ecotoxicology*, CRC Press/Lewis Publishers, Boca Raton, FL, 1995, p. 426.

Newman, M.C. and Ungar, M.A., *Fundamentals of Ecotoxicology*, CRC Press/Lewis Publishers, Boca Raton, FL, 2003.

OECD, *Current Approaches in the Statistical Analysis of Ecotoxicology Data: A Guidance to Application*, OECD Environmental Health and Safety Publications Series on Testing and Assessment, No. 54, Environment Directorate, Paris, 2006, p. 147. Available in pdf format from the OECD website, www.oecd.org/document/30/0,2340,en_2649_34377_1916638_1_1_1_1,00.html

O'Gara, B.A., Bohannon, V.K., Teague, M.W., and Smeaton, M.B., Copper-induced changes in locomotor behaviors and neuronal physiology of the freshwater oligochaete, *Lumbriculus variegates*, *Aquat. Toxicol.*, 69, 51–66, 2004.

Ohlendorf, H.M., The birds of Kesterson Reservoir: A historical perspective, *Aquat. Toxicol.*, 57, 1–10, 2002.

Oris, J.T. and Bailer, A.J., Statistical analysis of the *Ceriodaphnia* toxicity test: Sample size determination for reproductive effects, *Environ. Toxicol. Chem.*, 12, 83–90, 1993.

Parrott, J.L., Wood, C.S., Boutot, P., and Dunn, S., Changes in growth and secondary sex characteristics of fathead minnows exposed to bleached sulfite mill effluent, *Environ. Toxicol. Chem.*, 22, 2908–2915, 2003.

Pechkurenkov, V.L., Shekhanovc, I.A., and Telysheva, I.G., The effect of chronic small dose irradiation on the embryonic development of fishes and the validity of various assessment methods, In *New Research on the Ecology and Propagation of Phytophagous Fish*, Sukhanova, A.I. (ed.), Nauka Press, Moscow, 1966, pp. 71–79.

Penttinen, O.P. and Kukkonen, J., Body residues as dose for sublethal responses in alevins of landlocked salmon (*Salmo salar* m. *sebago*): A direct calorimetry study, *Environ. Toxicol. Chem.*, 25, 1088–1093, 2006.

Penttinen, O.P., Kukkonen, J., and Pellinen, J., Preliminary study to compare body residues and sublethal energetic responses in benthic invertebrates exposed to sediment-bound 2,4,5-trichlorophenol, *Environ. Toxicol. Chem.*, 15, 160–166, 1996.

Perez, M.H. and Wallace, W.G., Differences in prey capture in grass shrimp, *Palaemonetes pugio*, collected along an environmental impact gradient, *Arch. Environ. Contam. Toxicol.*, 46, 81–89, 2004.

Preston, B.L., Cecchine, G., and Snell, T.W., Effects of pentachlorophenol on predator avoidance behavior of the rotifer *Brachionus calyciflorus*, *Aquat. Toxicol.*, 201–212, 1999.

Relyea, R.A., Growth and survival of five amphibian species exposed to combinations of pesticides, *Environ. Toxicol. Chem.*, 23, 1737–1742, 2004.

Riddell, D.J., Culp, J.M., and Baird, D.J., Behavioral responses to sublethal cadmium exposure within an experimental aquatic food web, *Environ. Toxicol. Chem.*, 24, 431–441, 2005.

Roast, S.D., Widdows, J., and Jones, M.B., Impairment of mysid (*Neomysis integer*) swimming ability: an environmentally realistic assessment of the impact of cadmium exposure, *Aquat. Toxicol.*, 52, 217–227, 2001.

Rowe, C.L., Kinney, O.M., Fiori, A.P., and Congdon, J.D., Oral deformities in tadpoles (*Rana catesbeiana*) associated with coal ash deposition: Effects on grazing ability and growth, *Freshwater Biol.*, 36, 723–730, 1996.

Samson, J.C., Goodridge, R., Olobatuyi, F., and Weis, J.S., Delayed effects of embryonic exposure of zebrafish (*Danio rerio*) to methylmercury (MeHg), *Aquat. Toxicol.*, 51, 369–376, 2001.

Schmidt, K., Staaks, G.B.O., Pflugmacher, S., and Steinberg, C.E.W., Impact of PCB mixture (Aroclor 1254) and TBT and a mixture of both on swimming behavior, growth and enzymatic biotransformation activities (GST) of young carp (*Cyprinus carpio*), *Aquat. Toxicol.*, 71, 49–59, 2005.

Schöpf Rehage, J., Lynn, S.G., Hammond, J.I., Palmer, B.D., and Sih, A., Effects of larval exposure to triphenyltin on the survival, growth, and behavior of larval and juvenile *Ambystoma barbouri* salamanders, *Environ. Toxicol. Chem.*, 21, 807–815, 2002.

Schulz, R. and Dabrowski, J.M., Combined effects of predatory fish and sublethal pesticide contamination on the behavior and mortality of mayfly nymphs, *Environ. Toxicol. Chem.*, 20, 2537–2543, 2001.

Scott, G.R. and Sloman, K.A., The effects of environmental pollutants on complex fish behavior: Integrating behavioral and physiological indicators of toxicity, *Aquat. Toxicol.*, 369–392, 2004.

Shukla, R., Wang, Q., Fulk, F., Deng, C., and Denton, D., Bioequivalence approach for whole effluent toxicity testing, *Environ. Toxicol. Chem.*, 19, 169–174, 2000.

Sibly, R.M. and Calow, P., A life-cycle theory of responses to stress, *Biol. J. Linnean Soc.*, 37, 101–116, 1989.

Sloman, K.A., Scott, G.R., Diao, Z., Rouleau, C., Wood, C.M., and McDonald, D.G., Cadmium affects the social behavior of rainbow trout, *Oncorhynchus mykiss*, *Aquat. Toxicol.*, 65, 171–185, 2003.

Sprague, J.B., Avoidance reactions of rainbow trout to zinc sulphate solutions, *Water Res.*, 2, 367–372, 1968.

Spurgeon, D.J., Hopkin, S.P., and Jones, D.T., Effects of cadmium, copper, lead and zinc on growth, reproduction and survival of the earthworm *Eisenia fetida* (Savigny): assessing the environmental impact of point-source metal contamination in terrestrial ecosystems, *Environ. Pollut.*, 84, 123–130, 1994.

Stansley, W. and Roscoe, D.E., Chordane poisoning of birds in New Jersey, USA, *Environ. Toxicol. Chem.*, 18, 2095–2099, 1999.

Stearns, S.C., *The Evolution of Life Histories*, Oxford University Press, Oxford, UK, 1992.

Stebbing, A.R.D., Hormesis—The stimulation of growth by low levels of inhibitors, *The Sci. Total Environ.*, 22, 213–234, 1982.

Stephan, C.E. and Rogers, J.W., Advantages of using regression analysis to calculate results of chronic toxicity tests, In *Proceedings of the Aquatic Toxicology and Hazard Assessment, Eight Symposium, ASTM STP891*, Bahner, R.C. and Hansen, D.J. (eds.), American Society for Testing and Materials, Philadelphia, PA, 1985, pp. 328–338.

Sterne, J.A.C. and Davey Smith G., Sifting the evidence—What's wrong with significance tests? *BMJ*, 322, 226–230, 2001.

Stewart, P.M., Scrbailo, R.W., and Simon, T.P., The use of aquatic macrophytes in monitoring and in assessment of biological integrity, *Environ. Sci. Forum*, 96, 275–302, 1999.

Sullivan, J.F., Atchison, G.J., Kolar, D.J., and Mcintosh, A.W., Changes in the predator-prey behavior of fathead minnows (*Pimephales promelas*) and largemouth bass (*Micropterus salmoides*) caused by cadmium, *J. Fish. Res. Board Can.*, 35, 446–451, 1978.

Tagatz, M.E., Effect of Mirex on predator-prey interaction in an experimental estuarine ecosystem, *Trans. Am. Fish. Soc.*, 4, 546–549, 1976.

Teather, K., Jardine, C., and Gormley, K., Behavioural and sex ratio modification of Japanese medaka (*Oryzias latipes*) in response to environmentally relevant mixtures of three pesticides, *Environ. Toxicol.*, 20, 110–117, 2005.

Thetge, J.E., Holcombe, G.W., Flynn, K.M., Kosian, P.A., Korte, J.J., Anderson, L.E., Wolf, D.C., and Degitz, S.J., Metamorphic inhibition of *Xenopus laevis* by sodium perchlorate: Effects on development and thyroid histology, *Environ. Toxicol. Chem.*, 24, 926–933, 2005.

Thursby, G.B., Heltshe, J., and Scott, K.J., Revised approach to toxicity test acceptability criteria using a statistical performance assessment, *Environ. Toxicol. Chem.*, 16, 132–1329, 1997.

Tietge, J.E., Holcombe, G.W., Flynn, K.M., Kosian, P.A., Korte, J.J., Anderson, L.E., Wolfe, D., and Degitz, S.J., Metamorphic inhibition of *Xenopus laevis* by sodium perchlorate: Effects on development and thyroid histology, *Environ. Toxicol. Chem.*, 24, 926–933, 2005.

Toft, G., Baatrup, E., and Guillette, L.J., Jr., Altered social behavior and sexual characteristics in mosquitofish (*Gambusia holbrooki*) living downstream of a paper mill, *Aquat. Toxicol.*, 70, 213–222, 2004.

Ury, H.K., A comparison of four procedures for multiple comparisons among means (pairwise contrasts) for arbitrary sample sizes, *Technometrics*, 18, 89–97, 1976.

Van Straalen, N.M. and Denneman, C.A.J., Ecotoxicological evaluation of soil quality criteria, *Ecotoxicol. Environ. Saf.*, 18, 241–251, 1989.

Van Straalen, N.M., and Van Leeuwen, C.J., European history of species sensitivity distribution, In *Species Sensitivity, Distributions in Ecotoxicology*, Posthuma, L., Suter, G.W., II, and Traas, T.F. (eds.), CRC Press/Lewis Publishers, Boca Raton, FL, 2002, pp. 19–34.

Vogl, C., Grillitsch, B., Wytek, R., Spieser, O.H., and Scholz, W., Qualification of spontaneous undirected locomotor behavior of fish for sublethal toxicity testing. Part I. Variability of measurement parameters under general test conditions, *Environ. Toxicol. Chem.*, 18, 2736–2742, 1999.

Wacholder, S., Chanock, S., Garcia-Closas, M., El ghormli, L., and Rothman, N., Assessing the probability that a positive report is false: An approach for molecular epidemiology studies, *J. Natl. Cancer. Inst.*, 96, 434–442, 2004.

Wallace, W.G. and Estephan, A., Differential susceptibility of horizontal and vertical swimming activity to cadmium exposure in a gammaridean amphipod (*Gammarus lawrencianus*), *Aquat. Toxicol.*, 69, 289–297, 2004.

Ward, G.S., Parrish, P.R., and Rigby, R.A., Early life stage toxicity tests with a saltwater fish: Effects of eight chemicals on survival, growth, and development of sheepshead minnows (*Cyprinedon variegatus*), *J. Toxicol. Environ. Health*, 8, 225–240, 1981.

Webber, H.M. and Haines, T.A., Mercury effects on predator-prey behavior of the forage fish, golden shiner (*Notemigonus crysoleucas*), *Environ. Toxicol. Chem.*, 22, 1556–1561, 2003.

Weis, J.S. and Khan, A.A., Effects of mercury on the feeding behavior of the mummichog, *Fundulus heteroclitus* from a polluted habitat, *Mar. Environ. Res.*, 30, 243–249, 1990.

Weis, J.S. and Weis, P., Pollutants as developmental toxicants in aquatic organisms. *Environ. Health Perspect.* 71, 77--85, 1987.

Williams, J., Roderick, C., and Alexander, R., Sublethal effects of Orimulsion-400 on eggs and larvae of Atlantic herring (*Clupea harengus* L.), *Environ. Toxicol. Chem.*, 22, 3044–3048, 2003.

Wilson, J.B., The cost of heavy-metal tolerance: An example, *Evolution*, 42, 408–413, 1988.

Wirth, E.F., Lund, S.A., Fulton, M.H., and Scott, G.I., Reproductive alterations in adult grass shrimp, *Palaemonetes pugio*, following sublethal, chronic endosulfan exposure, *Aquat. Toxicol.*, 59, 93–99, 2002.

Wojtaszek, B.F., Buscarini, T.M., Chartrand, D.T., Stephenson, G.R., and Thompson, D.G., Effect of Release® herbicide on mortality, avoidance response, and growth of amphibian larvae in two forest wetlands, *Environ. Toxicol. Chem.*, 24, 2533–2544, 2005.

Wojtaszek, B.F., Staznik, B., Chartrand, D.T., Stephenson, G.R., and Thompson, D.G., Effects of Vision® herbicide on mortality, avoidance response, and growth of amphibian larvae in two forest wetlands, *Environ. Toxicol. Chem.*, 23, 832–842, 2004.

Wollenberger, L., Dinan, L., and Breitholtz, M., Brominated flame retardants: Activities in a crustacean development test and in an ecdysteroid screening assay, *Environ. Toxicol. Chem.*, 24, 400–407, 2005.

Woodford, J.E., Karasov, W.H., Meyer, M.W., and Chambers, L., Impact of 2,3,7,8-TCDD on survival, growth, and behavior of ospreys breeding in Wisconsin, USA, *Environ. Toxicol. Chem.*, 17, 1323–1331, 1998.

Wright, S.P., Adjusted P-values for simultaneous inferences. *Biometrics* 48, 1005–1013, 1992.

11 Conclusion

11.1 GENERAL

This chapter summarizes what has been discussed to this point. Chapter 1 presented an overview of the science of ecotoxicology. Chapters 2 through 10 discussed a range of autecotoxicological issues, including biomolecular to whole organism issues. Particularly relevant points are summarized in the paragraphs below.

11.2 SOME PARTICULARLY KEY CONCEPTS

Ecotoxicology is the science of contaminants in the biosphere and their effects on constituents of the biosphere, including humans. It is a hierarchical science in which no level in the biological hierarchy is superior to another in the context of identifying causation or attributing relevance. Accordingly, the modified Janus context was advocated for optimal inference and organization of knowledge around paradigms at all levels of biological organization. Two factors were described as most needed in ecotoxicology at the moment: the application of the strongest possible inferences for the purpose of efficiently organizing knowledge around rigorously tested paradigms, and the consilience of concepts and paradigms among all hierarchical levels.

The advantages of an autecological exploration of ecotoxicology were outlined in Chapter 2. They include the rapid transfer of new technologies from mammalian toxicology into ecotoxicology, provision of mechanistic understanding for effects seen at higher levels of organization, provision of sensitive indicators of exposure or effect, and creation of a highly discrete and tractable approach to any ecotoxicological questions. Careful interpretation of organismal data improves our ability to predict consequences at the population, and commonly, community levels of organization. Also, models of bioaccumulation in individuals set the stage for models predicting trophic transfer of toxicants in food webs.

Biochemical issues relevant to ecotoxicology were outlined in Chapter 3. The fields of study describing relevant levels of information transfer as influenced by toxicants were genomics, transcriptomics, proteomics, metabolomics, and bioenergetics. Alterations to an individual's genome and consequences were detailed with emphasis on cancer. Transformations of organic and inorganic toxicants were described at the molecular level, including those resulting in activation, accelerated elimination, sequestration, and tissue localization. Finally, a series of general phenomena were described including oxidative stress, proteotoxicity, enzyme dysfunction, uncoupling of oxidative phosphorylation, and narcosis.

Chapter 4 considered spatial differences in biochemical processes and moieties that exist in cells and tissues, and determine the cells or tissues most affected by or responsive to toxicants. Coagulative necrosis and inflammation were identified as useful cell or tissue markers for toxicant-induced harm. Hyperplasia and hypertrophy in tissues were similarly described as telltale responses to toxicant exposure. Toxicant-induced necrosis and apoptosis were described in detail and linked to possible effects at higher levels of organization (e.g., developmental abnormalities or malignant cancer promotion in exposed individuals). Genotoxicity was explored with emphasis placed on carcinogenesis.

Discussions moved to the level of organs and organ systems in Chapter 5 where the cellular specialization in organs, and spatial relation of organs to direct environmental exposure or exposure via somatic circulation were described as critical to making some organs more prone to poisoning than others. Qualities influencing entry into, effects within, and potential transformations of toxicants were described for the integument, respiratory organs, liver and analogous organs, digestive organs, and excretory organs. Similar discussions were developed for the immune, circulatory, nervous and endocrine systems, and the associated effects were related to consequences at higher levels of organization.

Chapter 6 was a relatively short one in which selected physiological issues not addressed sufficiently elsewhere were discussed. A strong theme was the impact of toxicants on the optimal movement of materials and energy between the organism and its environment (i.e., Darwinian fitness). Accordingly, most changes were also discussed in the contexts of life history shifts or trade-offs with toxicant exposure. Homeostasis-related issues included pH, ion, and osmotic regulation.

The emphasis shifted in Chapter 7 to accumulation of toxicants in individuals. Cellular and paracellular movements were described for different classes of toxicants including highly lipid-soluble organic, ionizable organic, and ionic inorganic toxicants. Discussion began at the cellular level and moved to the organismal level in discussions of movement associated with ingestion, respiration, and dermal routes. General principles surrounding toxicant movement into and out of cells were presented first. Next, such principles were outlined for the ingestion, respiration, and dermal routes. Applicable theory included the lipid solubility theory for nonionic organic compounds; pH Partition Theory for weakly ionizable compounds; and the ionic hypothesis, free ion activity model, biological ligand model, Hard Soft Acid Base (HSAB) theory, and quantitative ion character–activity relationships for metals. The bioaccumulation and bioavailability themes continued into Chapter 8, which provided a quantitative treatment of related issues. Rate constant-, clearance-, and fugacity-based formulations of bioaccumulation models were presented after general discussion of issues such as reaction order. These formulations were extended to the PBPK context. Noncompartment statistical moments methods were illustrated and used to explore issues of bioavailability.

By Chapters 9 and 10, enough information had been explored so that discussion could turn to lethal and sublethal effects on individuals. How ecotoxicologists treat lethal and sublethal effects is strongly influenced by regulatory history and perceived needs. Lethal effects were characterized as acute or chronic, but the reader was reminded that the distinction is blurry. The reader was also shown that the dynamics by which lethality manifests differs among modes of action, exposure route, and species. The often overlooked issue of diseases of adaptation was discussed including shifts in syntoxic and catatoxic hormones. The foundation and various forms of toxicity curves were explored, suggesting the need for more research in this area. Mixture effects were discussed and quantified on the basis of mode of action. The survival time approach was discussed as an alternative to conventional methods for quantifying lethal effects. The influences of previously discussed molecular and ionic qualities were addressed again in the context of quantitative models of lethality.

Sublethal effects discussed in Chapter 10 were framed initially in the context of Darwinian fitness and then in the context of regulatory methods for their quantification. Sublethal stressor effects that dominate the literature are those to growth and development, reproduction, physiology, and behavior. Although most studies of sublethal effects apply rudimentary conceptual and experimental designs that do not take full advantage of available ecological models, there is a clear movement away from this condition. Sublethal effect metrics are currently estimated using either hypothesis testing or point estimation approaches. Bioequivalence testing, a promising approach that avoids many of the shortcomings of the conventional hypothesis tests, was discussed briefly as an underexploited approach to assessing sublethal effects studies.

11.3 CONCLUDING REMARKS

In the next section, population ecotoxicology will be discussed in detail. This theme has characteristics of both autecology and synecology. Consequently, the associated materials provide many exciting concepts and techniques with the blending of organismal ecotoxicology and the ecotoxicology of communities and ecosystem.

Part III

Population Ecotoxicology

> The emergence of ecological toxicology as a coherent discipline is perhaps unique in that it combines aspects of toxicology and ecology, both of which are in and of themselves synthetic sciences .... Chemicals may affect every level of biological organization (molecules, cells, tissues, organs, organ systems, organisms, populations, communities) contained in ecosystems. Any one of these levels is a potential unit of study for the field, as are the interdependent structures and relationships within and between levels.
>
> **(Maciorowski 1988)**

A central concern of ecotoxicologists is toxicant impact on populations. Population concerns were highlighted in the first ecotoxicology textbook (Moriarty 1983). Population consequences are implicitly at the core of regulatory concerns about toxicant impact.

Traditionally, information generated for assessing ecological risk was extracted with an autecological emphasis despite the acknowledged need of prediction of population effects (Barnthouse et al. 1987). During the late 1970s and into the early 1980s, this incongruity between information that was required to assess population consequences and available information began to be addressed effectively by more and more ecotoxicologists. Today, population ecotoxicology is emerging as a central research theme and is more commonly applied in assessments of exposure consequences. Excellent books are emerging on this topic (e.g., Kammenga and Laskowski 2000). This being the case, it is important that the practicing regulator and advanced student understand the essentials of population ecotoxicology. Fostering such an understanding is the goal of this section.

REFERENCES

Barnthouse, L.W., Suter, G.W., II, Rosen, A.E., and Beauchamp, J.J., Estimating responses of fish populations to toxic contaminants, *Environ. Toxicol. Chem.*, 6, 811–824, 1987.

Kammenga, J. and Laskowski, R. (eds.) *Demography in Ecotoxicology*, John Wiley & Sons, Chichester, UK, 2000.

Maciorowski, A.F., Populations and communities: Linking toxicology and ecology in a new synthesis, *Environ. Toxicol. Chem.*, 7, 677–678, 1988.

Moriarty, F., *Ecotoxicology. The Study of Pollutants in Ecosystems*, Academic Press, Inc., London, UK, 1983.

12 The Population Ecotoxicology Context

12.1 POPULATION ECOTOXICOLOGY DEFINED

12.1.1 WHAT IS A POPULATION?

Intent influences one's definition of a population. An ecologist might envision a population as a collection of individuals of the same species that occupy the same space at the same time. Suggested in this definition is a boundary defining some space although no distinct boundary might exist. So the spatial context for a population can be strict or operational depending on how clear spatial boundaries are. The temporal context for a population may be blurry too. Groups of individuals of the same species may come together and disperse through time, making it difficult to distinguish populations.

A more realistic conceptualization of many populations emerges if one considers the dynamics of a group of contemporaneous individuals of the same species occupying a habitat with patches that differ markedly in their capacity to foster survival, growth, and reproduction. Differences among patches produce differences in fitnesses among individuals. Good habitat in the mosaic is a source of individuals because excess production of young is possible, while less favorable habitat might be a sink for these excess individuals. A population living within such a habitat mosaic is called a metapopulation. Metapopulation dynamics in source–sink habitats have unique features that should be understood by ecotoxicologists. For example, a sink habitat created by contamination may still possess high numbers of individuals, a condition inexplicable based on conventional ecotoxicity test results but easily explained in a metapopulation context. Also, the loss of a small amount of habitat to contamination can have dire consequences if the lost habitat was a source habitat sustaining the metapopulation components in adjacent, clean habitats. Such keystone habitats are crucial for maintaining the population in adjacent areas and some species are particularly sensitive to keystone habitat loss (O'Connor 1996).

The aforementioned concept of a population requires one more quality to be complete. A population may be defined as a collection of individuals of the same species occupying the same space at the same time and within which individuals may exchange genetic information (Odum 1971). Gene flow would now be included in the identification of population boundaries. Population boundaries can be clear (e.g., a pupfish species population in an isolated desert spring) or necessarily operational (e.g., mosquitofish in a stream branch). Spatial clines in gene flow become common because individuals in populations are more likely to mate with nearby neighbors than with more distant neighbors. Temporal changes in population boundaries should also be considered. As an extreme example, if females store sperm and a toxicant kills all males after the breeding season, the dead males are still part of the effective population contributing genes to the next generation.

Mitton (1997) provides an additional context for populations that is relevant to population ecotoxicology. A species population can be studied in the context of all existing individuals throughout the species' range. The influence of some contaminant, alone or in combination with factors such as habitat loss or fragmentation, might be suspected as the cause of a species' decline or imminent extinction over its entire range. Such a broad biogeographic perspective is at the heart of one explanation for the current rapid decline in many amphibian populations throughout the biosphere. Sarokin and Schulkin (1992) describe several other instances of large-scale population changes and suggest potential linkage to widespread contaminants. In these instances, the population of concern is the entire collection of individuals comprising the species, and not a local population. Assuming that toxicant-linked extinctions are undesirable, there is obvious value to studying contaminant influence on the biogeographic distribution and character of a species population.

12.1.2 Definition of Population Ecotoxicology

Population ecotoxicology is the science of contaminants in the biosphere and their effects on populations. In this section, a population is defined as a collection of contemporaneous individuals of the same species occupying the same space and within which genetic information may be exchanged. Population ecotoxicology explores contaminant effects in the context of epidemiology, basic demography, metapopulation biology, life-history theory, and population genetics. Accordingly, the chapters of this section are organized around these topics.

12.2 THE NEED FOR POPULATION ECOTOXICOLOGY

12.2.1 General

Why commit eight chapters to population ecotoxicology? Is there sufficient merit to develop a population context to this science and to imposing this context on our present methods of environmental stewardship? The answers to these questions are easily formulated on the basis of the scientific and practical advantages of doing so.

Although not often envisioned as such, landmark studies in population biology (e.g., population dynamics of agricultural pests) and evolutionary genetics (e.g., industrial melanism) involved pollutants. These ecotoxicological topics are currently associated with other disciplines such as population ecology and genetics, because ecotoxicology is only now emerging as a distinct science and the researchers who conducted those studies were affiliated with other disciplines. Toxicants served as useful probes for teasing meaning from wild populations. Just as individuals with metabolic disorders are studied by medical biochemists to better understand the metabolic processes taking place within healthy individuals, populations exposed to toxicants help scientists to understand the behavior of healthy populations. Often, they provide an accelerated look at processes such as natural selection, adaptation, and evolution that usually occur over time periods too long to study directly.

Equally clear are the practical advantages of better understanding toxicant effects on populations. Early problems involving pollutants centered on consequences to populations. Widespread applications of dichlorodiphenyltrichloroethane (DDT) (2,2-*bis*-[*p*-chlorophenyl]-1,1,1-trichloroethane) and DDD (1,1-dichloro-2,2-*bis*-[*p*-chlorophenyl]-ethane) had unacceptable consequences to populations of predatory birds. Within 15 years of Paul Müeller receiving the 1948 Nobel Prize in medicine for discovering the insecticidal qualities of DDT, convincing evidence had emerged worldwide about population declines of raptors and fish-eating birds induced by DDT and its degradation product, DDE (1,1-dichloro-2,2-*bis*-[*p*-chlorophenyl]-ethene) (Carson 1962, Dolphin 1959, Hickey and Anderson 1968, Ratcliffe 1967, 1970, Woodwell et al. 1967).

Our current environmental concerns remain focused on population viability. Important examples include the presently unexplained drop in amphibian populations throughout the world (Wake 1991), the decline in British bird populations putatively due to widespread pesticide use (Beaumont 1997, Newman et al. 2006), and the population consequences of estrogenic contaminants (Fry and Toone 1981, Luoma 1992, McLachlan 1993). These concerns are predictable manifestations of the general impingement on species populations by human populations that have expanded to "use 20–40% of the solar energy that is captured in organic materials by land plants" (Brown and Maurer 1989). This level of consumption by humans and the manner in which it is practiced could not but come at the expense of other species populations.

More and more authors are expressing the importance of population-level information in making environmental decisions, for example, "the effects of concern to ecologists performing assessments are those of long-term exposures on the persistence, abundance, and/or production of populations" (Barnthouse et al. 1987) and "Environmental policy decision makers have shifted emphasis from physiological, individual-level to population-level impacts of human activities" (Emlen 1989). The phrasing of many federal laws and regulations likewise reflects this central concern for populations.

> During the past two decades, toxicological endpoints (e.g., acute and chronic toxicity) for individual organisms have been the benchmarks for regulations and assessments of adverse ecological effects.... The question most often asked regarding these data and their use in ecological risk assessment is, "What is the significance of these ecotoxicity data to the integrity of the population?" More important, can we project or predict what happens to a pollutant-stressed population when biotic and abiotic factors are operating simultaneously in the environment?
>
> Protecting populations is an explicitly stated goal of several Congressional and [Environmental Protection] Agency mandates and regulations. Thus, it is important that ecological risk assessment guidelines focus upon the protection and management at the population, community, and ecosystem levels. . . .
>
> **(EPA 1991)**

The practical value of using population-level tools in ecotoxicology is also clear in risk assessment. Both human and ecological risk assessments draw methods from epidemiology, the science of disease in populations. Epidemiological methods were applied in the Minamata Bay area to ferret out the cause for a mysterious disease in the local population. Since this early outbreak of pollutant-induced disease in a human population, epidemiology has become crucial in fostering human health in an environment containing complex mixtures of contaminants. Although used much less than warranted, epidemiological methods could be equally helpful in studying nonhuman populations.

12.2.2 Scientific Merit

So many examples come immediately to mind in considering the scientific merit of population ecotoxicology that the issue becomes selecting the best, not finding a convincing one. Natural selection in wild populations seems the most general illustration. Industrial melanism, a topic mentioned in nearly all biology textbooks, is a population-level consequence of air pollution (Box 12.1). "Industrial melanism in the peppered moth (*Biston betularia*) is the classic example of observable evolution by natural selection" (Grant et al. 1998). Further, the evolution of metal tolerance in plant species growing on mining wastes is a clear example of natural selection in plants (Antonovics et al. 1971). Numerous additional examples of toxicant-driven microevolution include rodenticide resistance (Bishop and Hartley 1976, Bishop et al. 1977, Webb and Horsfall 1967), insecticide resistance in target species (Comins 1977, McKenzie and Batterham 1994, Whitton et al. 1980), and nontarget species resistance to toxicants (Boyd and Ferguson 1964, Klerks and Weis 1987, Weis and Weis 1989). It appears that, with the important exception of sickle cell anemia in human populations, the clearest and best-known examples of microevolution are those associated with anthropogenic toxicants.

Box 12.1 Industrial Melanism: There and Back Again (Almost)

Industrial melanism is universally acknowledged as one of the harbingers of our initial failure to create an industrial society compatible with ecological systems. Less well known, but perhaps equally important, it is also one of the clearest indicators of a widespread improvement in air quality (Figure 12.1). Recent shifts in the occurrence of the color morphs of the peppered moth (*B. betularia*) (Figure 12.2) suggest that the money and effort put into controlling air pollutants in several industrialized countries are having positive effects.

Before roughly 1848, melanistic (dark-colored) morphs of the peppered moth were extremely rare. The conventional, and still sound, explanation for this observation is that (1) while quiescent during the day, this moth depends on its coloration to blend into its background, (2) this crypsis is focused on avoiding notice by visual predators, especially birds, (3) light coloration favors the moth if it rests on natural vegetation including light-colored lichens, (4) dark morphs appear rarely due to mutation, (5) dark morphs are less cryptic than light morphs relative to evading visual predators, (6) rare dark morphs are quickly taken by visual predators, and, consequently (7) light morphs predominate as rare dark morphs quickly disappear from natural populations (Kettlewell 1973).

British industrialization of the nineteenth century changed this situation by producing air pollutants that darkened surfaces and reduced the surface coverage by light-colored lichens. Crypsis began to favor the dark or *carbonaria* morph as birds took increasing numbers of light morphs. The shift from a preponderance of light to dark moths was quite rapid because of large fitness differences among color morphs relative to avoiding notice of predators and the genetic dominance of the *carbonaria* allele over those for light morphs. [The light phenotypes are controlled by four recessive genes producing various pale to intermediate phenotypes (Berry 1990, Lees and Creed 1977).] Whereas one dark moth was observed around Manchester in 1848, moths of that area were composed almost entirely of dark morphs by 1895 (Clarke et al. 1985).

FIGURE 12.1 Normal and melanistic color morphs of the peppered moth, *Biston betularia*. (Photograph courtesy of Bruce S. Grant, College of William & Mary.)

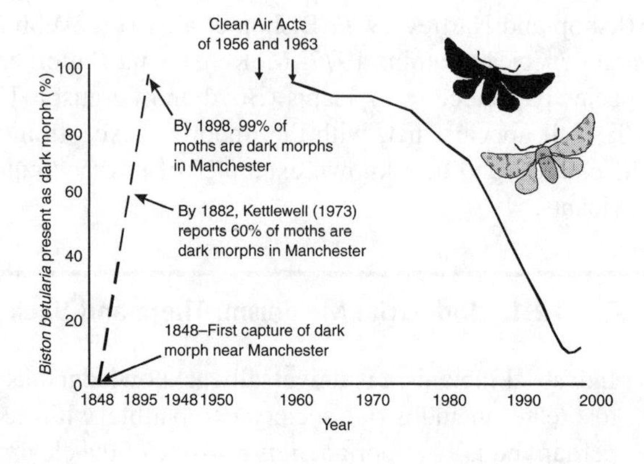

FIGURE 12.2 Rise and fall in the proportion of *B. betularia* of the melanistic morph caught near Liverpool, UK. Information for the decline in the dark morph comes from Clarke and Grant (Clarke, C.A., et al., 1994, Grant, B.S. and Clarke, C.A., 1999) who monitored a moth population outside of Liverpool from 1959 to the present.

A subsequent series of events resulted in a second shift in the balance between light and dark morphs. Unacceptable consequences of poor air quality (including outright human and livestock illness and death) in the United Kingdom resulted in the passage and implementation of the 1956 and 1963 British Clean Air Acts (Grant et al. 1995). Air quality improved and dark morphs began to rapidly decline in numbers. In a comprehensive documentation of this change, Clarke and Grant (Clarke et al. 1994, Grant and Clarke 1999) report the clear decline in dark morphs from 1959 to present at Caldy Common, a location about 18 km outside of Liverpool (Figure 12.2). The frequency of the dark morph dropped quickly until 1996 but fluctuated thereafter in the range of 7.1–11.5% (Grant and Clarke 1999). A similar leveling off at a low frequency occurred at a Nottingham location (Grant and Clarke 1999). Thus, although the population appears to be shifting back to its original condition, the moths have not returned to their preindustrial state where the frequency of the *carbonaria* morph was extremely low. Perhaps there is a final part to this story yet to be written on the basis of this new state with a low proportion of the once-rare, dark morph persisting in moth populations.

B. betularia is an extremely widespread species and similar declines in pollution-related melanism have been documented in other countries [e.g., the United States (Grant et al. 1995, 1998, West 1977) and the Netherlands (Brakefield 1990)] after enactment of air quality legislation. The frequency of the *carbonaria* morph declined when air quality improved. Peppered moth populations in Japan provide the exception that proves the rule. In Japan, unlike European industrial areas, the distribution of moths and industry was distinct. Thus, the conditions leading to the industrial melanism in other countries were not present (Asami and Grant 1994). Japanese studies serve as persuasive, negative controls for assessment of the relationship between pollution and melanism in *B. betularia* populations.

The industrial melanism story continues. A significant proportion of all *B. betularia* in relevant U.K. and U.S. populations is still the *carbonaria* morph. Perhaps the dark morph will again become rare with further improvements in air quality. Recent studies (Grant and Howlett 1988) indicate that Kettlewell's explanation based primarily on differential predation on adults by birds (Kettlewell 1955, 1973) may not be the complete story. The preadult stage has differences in viability (i.e., survival) fitness among color morphs (Mani 1990). Genetic shifts may be at least partially due to processes affecting preadults (i.e., nonvisual selection) (Mani 1990). Further, multivariate statistical studies suggest that the best correlation between *B. betularia carbonaria* frequency in moth populations and air quality is with sulfur dioxide (Grant et al. 1998, Mani 1990). Although there is considerable opportunity for the problem of ecological inference to emerge here, it is possible that other mechanisms of selection associated with sulfur dioxide's effect on plants and animals might be important (e.g., acid precipitation-related direct changes to larval fitness or indirect effects by influencing vegetation quality). This classic example of population response to pollutants will likely yield more valuable insights as studies continue.

12.2.3 Practical Merit

> Extrapolations from laboratory bioassays to response in natural systems at the population level are effective if the environmental realism of the bioassay is sufficiently high. When laboratory systems are poor simulations of natural systems, gross extrapolation errors may result. The problem of extrapolating among levels of biological organization has not been given the serious attention it deserves.
>
> **(Cairns and Pratt 1989)**

Examples of the practical application of population ecotoxicology are also easily found. Examples range from demographic analyses of toxicity test data (Caswell 1996, Green and Chandler 1996, Karås et al. 1991, Mulvey et al. 1995, Pesch et al. 1991, Postma et al. 1995) to surveys of field population qualities (Ginzburg et al. 1984, Sierszen and Frost 1993) to epidemiological analysis of populations in polluted areas (Hickey and Anderson 1968, Osowski et al. 1995, Spitzer et al. 1978) to using enhanced tolerance as an indicator of pollutant effect (Beardmore 1980, Guttman 1994, Klerks and Weis 1987, Mulvey and Diamond 1991). What follows is an illustration of the consequences of *not* considering population-level metrics of effect in practical ecotoxicology. The example illustrates the logical flaws incurred during predictions of effects to populations based on conventional toxicity test results.

Current ecotoxicity test methods have their roots in mammalian toxicology. Methods developed to infer the mammalian toxicity of various chemicals focused initially on lethal thresholds (Gaddum 1953). A dose or concentration was estimated below which no mortality would be expected. Because the statistical error associated with such a metric was quite high, effort shifted toward identification of a dose or concentration killing a certain percentage of exposed individuals (e.g., the LD50 or LC50) (Trevan 1927). A metric of toxicity was generated with a relatively narrow confidence interval. This proved suitable for measuring relative toxicity among chemicals or for the same chemical under different exposure situations. Ecotoxicologists adopted this approach in the mid-1940s to 1950s (Cairns and Pratt 1989) as a measure of toxicant effect (Cairns and Pratt 1989, Maciorowski 1988). To improve the metric, details such as different exposure durations (i.e., acute and chronic LC50), exposure pathways (e.g., oral LC50 and dissolved LC50), and life stages (i.e., larval LC50, juvenile LC50, and adult LC50) were added. By the 1960s, these were the metrics of effect on organisms exposed to environmental toxicants that were "generally accepted as a conservative estimate of the potential effects of test materials in the field" (Parrish 1985). These tests were extended further to predict field consequences of toxicant release by focusing testing on the most sensitive stage of a species' life cycle (e.g., early life stage tests).

Can tests that use such responses of individuals provide sufficiently accurate predictions of consequences to populations? Does the application of a metric that is not focused on population qualities compromise our ability to predict consequences to field populations? Four potential problems of using these metrics to predict population consequences come immediately to mind.

First, toxicity test interpretation is often based on the most sensitive life stage paradigm: if the most sensitive stage of an individual is protected, the species population will be protected. However, the most sensitive stage of an individual's life history might not be the most crucial for maintaining a viable population (Hopkin 1993, Petersen and Petersen 1988). Newman (1998) uses the phrase "weakest link incongruity" for this false assumption that the most sensitive stage of an individual's life history is the most crucial to population viability. For many species, there is an overproduction of individuals at the sensitive early life stage. Loss of sexually mature individuals might be more damaging to population persistence than a much higher loss of sensitive neonates. The loss of 10% of oyster larvae from a spawn may be trivial to the maintenance of a viable oyster population because oyster populations can accommodate wide fluctuations in annual spawning success. At the other extreme, sparrow hawk (kestrel) populations remain viable despite a loss of 60% of breeding females each year (Hopkin 1993). As a more ecotoxicological example, the most sensitive stage of the nematode, *Plectus acuminatus*, was not the most crucial stage in determining population effects of cadmium exposure (Kammenga et al. 1996). Inattention to population parameters can create a practical problem in prediction from ecotoxicity test results.

Second, metrics such as the 96-h LC50 cannot be fit into ordinary demographic analyses without introducing gross imprecision. Life tables require mortality information over the lifetime of a typical individual but LC*x* [or no observed effect concentration (NOEC)] metrics derived from one or a few observation times during the test are inadequate for filling in a life table. This problem would be greatly reduced if survival time models were produced from toxicity tests of the appropriate duration instead of a LC50 calculated for some set time (Dixon and Newman 1991, Newman and Aplin 1992, Newman and Dixon 1996, Newman and McCloskey 1996). Appropriate methods exist but are used infrequently because of our preoccupation with metrics of toxicity to individuals without enough concern for translation to the next hierarchical level, that is, the population. This preoccupation with a traditional, statistically reliable metric of toxicity to individuals confounds appropriate analysis of mortality data and accurate prediction of population-level effects. Fortunately, there is now clear, albeit slow, movement away from such a preoccupation.

Third, although of less import when applying LC50-like metrics to determine toxicity in mammalian studies, postexposure mortality of individuals exposed to a toxicant can make predictions of population-level effects grossly inaccurate on the basis of a LC50-like metric. Considerable mortality can occur for many toxicants after exposure ends. As an example, 12% of mosquitofish (*Gambusia holbrooki*) exposed to 13 g/L of NaCl died by 144 h of exposure but another 44% died in the weeks immediately following termination of exposure (Newman and McCloskey 2000). More recently, Zhao and Newman (2004) estimated that the 48-h LC50 for amphipods (*Hyalella azteca*) actually killed 65–85% of exposed individuals if postexposure mortality was considered. This postexposure mortality is irrelevant in the use of the LC50-like metrics in mammalian toxicology to measure relative toxicity but is extremely important in ecotoxicology where the population consequences of exposure are to be predicted. Postexposure mortality in a population cannot continue to be treated as irrelevant in ecotoxicology.

Finally, as described in Box 12.2, the preoccupation with toxicity metrics borrowed from mammalian toxicology has distracted ecotoxicologists from important ambiguities about the underpinnings of the models used to predict effect. Ecotoxicology textbooks (e.g., Connell and Miller 1984, Landis and Yu 1995) and technical books (e.g., Finney 1947, Forbes and Forbes 1994, Suter 1993) explain the most widely used model (log normal or probit model) for concentration (or dose) effect data with the individual tolerance or individual effective dose concept. The development of this model assumes that each individual has an innate dose at or above which it will die. The distribution of individual effective doses in a population is thought to be a log normal one. However, another explanation for observed log normal distributions is that the same stochastic processes are occurring in all individuals. The probability of dying is the same for all individuals and is best described by a log normal distribution. These two alternative hypotheses remain poorly tested, but, in the context of population consequences of toxicant exposure, result in very different predictions.

Practical problems emerge due to our preoccupation with measuring effects in a way more appropriate for predicting fate of exposed individuals than of exposed populations. Current tests to predict population-level consequences are no less peculiar than one described in the poem *Science* by Alison Hawthorne Deming (1994) in which the mass of the soul is estimated by weighing mice before and after they were chloroformed to death. The incongruity of the test is more fascinating than its predictive power. Fortunately, ecotoxicology is moving toward more effective approaches to predicting population effects.

Box 12.2 Probit Concentration (or Dose)–Effect Models: Measuring Precisely the Wrong Thing?[1]

The first application of what eventually became the probit method was in the field of psychophysics. Soon thereafter, it was applied in mammalian toxicology to model quantal response data (e.g., dead or alive) generated from toxicity assays. Gaddum (as ascribed by Bliss and Cattell (1943)) hypothesized an explanation for its application called alternately, the individual effective dose or individual tolerance hypothesis. Which name was used seemed to depend on whether the toxicant was administered as a dose or in some other way, such as an exposure concentration. The concept was the same regardless of the exact name. Each individual was assumed to have an innate tolerance often expressed as an effective dose. The individual

[1] See Sections 9.1.3.1 and 9.1.3.1 in Chapter 9 for further discussion of this issue.

would survive if it received a dose below its effective dose but would die if its effective dose were reached or exceeded. Studies of drug or poison potencies conducted on individuals suggested that individual effective doses were log normally distributed in populations. This provided justification for fitting quantal data to a log normal (probit) model (Bliss 1935, Finney 1947, Gaddum 1953). For example, a common assay to determine the potency of a digitalis preparation was to slowly infuse an increasing dose of the preparation into individual cats until each one's heart just stopped beating. If enough cats were so treated, the distribution of effective doses would appear log normal.

Surprisingly, this central hypothesis has not been rigorously tested. The reason seems more historical than scientific. First, in the context of the early toxicity assays, the theory was presented primarily to support the application of a log normal model. Second, it was easy to find genetic evidence of differences in tolerance among individuals. However, no studies defined the general magnitude of these differences among individuals in populations or the rationale for why these differences should always be log normally distributed in populations. Third, the correctness of the theory was not as important in this context as in the one into which ecotoxicologists have thrust it.

Another explanation, already mentioned, exists and will be labeled the stochasticity hypothesis (Newman 1998, Newman and McCloskey 2000). Instead of a lethal dose being an innate characteristic of each individual, the risk of dying is the same for all individuals because the same stochastic processes are occurring in all individuals. Whether one or another individual dies at a particular dose is random with the resulting distribution of doses killing individuals described best by a log normal distribution. Gaddum (1953) described a random process involving several "hits" at the site of action to cause death that resulted in a log normal distribution of deaths. Berkson (1951) describes an experiment supporting the stochastic theory. The experiment was done when he was hired as a consultant to analyze tolerances to high altitude conditions of candidate aviators. Candidates were screened by being placed into a barometric chamber and then noting whether they fainted at high altitude conditions. The premise was that those men with an inherently low tolerance to high altitude conditions would be poor pilots. Berkson broke from the screening routine to challenge this individual tolerance concept. He asked that a group of pilots be retested to see if individuals retained their relative rankings between trials. They did not, indicating that the test and the individual effective dose concept were not valid in this case. In contrast, zebrafish (*Brachydanio rerio*) tolerance to the anesthetic, benzocaine, did more recently provide some limited support for the individual effective dose concept (Newman and McCloskey 2000).

The crucial difference between these two models is whether the dose that actually kills or otherwise affects a particular individual is determined by an innate quality of the individual or by a random process taking place in all individuals. Determining under what conditions, which one, or combination of these hypotheses is correct is important in determining the population consequences of exposure.

The importance of discerning between these two hypotheses can be illustrated with a simple thought experiment (Newman 1998). Assume that a concentration of exactly one LC50 results from a discharge into a stream for exactly 96 h. During the release, a population of similar individuals is exposed for 96 h to one LC50 and then to no toxicant for enough time to recover. For simplicity, we assume no postexposure mortality. After ample time for recovery, the survivors in the population are exposed again. This process is repeated several times. Under the individual effective dose or individual tolerance hypothesis, 50% of the individuals would die during the first exposure. During any exposure thereafter, there would be no, or minimal, mortality because all survivors of the first exposure would have individual tolerances greater than the LC50. In contrast, the stochasticity hypothesis predicts a 50% loss of exposed individuals during each 96-h exposure. The population consequences are very different with these

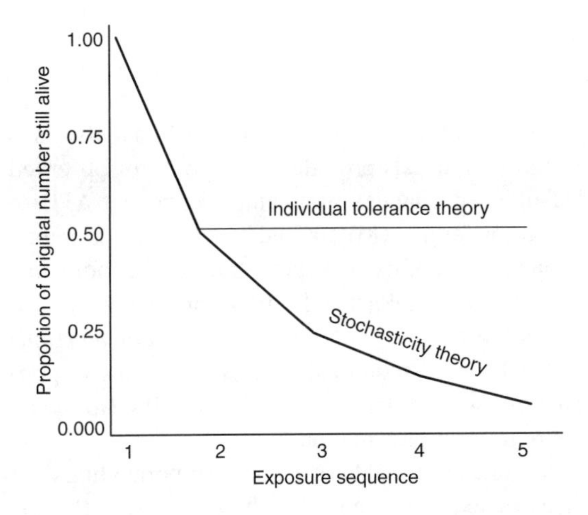

FIGURE 12.3 The predicted decrease in size for a population receiving repeated exposures to one LC50 for exactly 96 h with ample time between exposures for recovery. Highly divergent outcomes are predicted on the basis of the individual tolerance (individual effective dose) or stochasticity hypotheses. A blending of the two hypotheses (both processes are important in determining risk of death) would produce curves in the area between those for the individual tolerance and stochasticity theories.

two hypotheses. In this thought experiment, the population remains extant (individual effective dose hypothesis) or eventually goes locally extinct (stochasticity hypothesis) (Figure 12.3). With some deliberation, the reader can likely find other situations in which it would be crucial to determine the appropriate theory in order to predict population fate under toxicant exposure.

It would be surprising if the individual effective dose hypothesis were applicable to all or most ecotoxicity data to which the probit model is now applied. The probit method is applied to data for different effects under a variety of conditions to many species. It is applied to both clonal (e.g., *Daphnia magna*, *Lemna minor*, and *Vibrio fisheri*) and nonclonal collections of individuals. Nonclonal groups of individuals might be inbred, laboratory bred, or collected from the wild. It would be remarkable if the same explanation fit all diverse effects to such diverse collections of individuals. Indeed, recent work with sodium chloride toxicity to mosquitofish (*G. holbrooki*) suggests that the individual effective dose concept is an inadequate explanation for all applications of the probit (log normal) model (Newman and McCloskey 2000).

Again, why has this ambiguity remained unresolved for so long? Because, following the lead of mammalian toxicologists, ecotoxicologists have focused on the effects of toxicants on individuals and paid less attention than warranted to translating effect metrics to population consequences.

12.3 INFERENCES WITHIN AND BETWEEN BIOLOGICAL LEVELS

In Chapter 1, several avenues for inference within and between biological levels were discussed. Microexplanation (reductionism) might be possible for population behavior based on the qualities of individuals. Acknowledging the unpredictable influence of emergent properties, a description (explanation without a strict knowledge of the underlying mechanism) might be made in a holistic study of a consistent response at the population level. Careful speculation from the population level to the level of the individual (macroexplanation) might be possible as long as the problem of ecological inference is acknowledged. Finally, one could project from the response of populations to plausible consequences to communities. Here, again, emergent properties might compromise predictions.

12.3.1 Inferring Population Effects from Qualities of Individuals

Unquestionably, predicting population consequences from organismal and suborganismal effects is a major pursuit in ecotoxicology. Toxicant effects in individuals provide explanation for observed changes in field populations; for example, DDT-induced changes in calcium-dependent ATPase in the eggshell gland with consequent bird population failure (Kolaja and Hinton 1979). Subtle changes such as fluctuating asymmetry or developmental stability have potential as field indicators of contaminant influence on populations (Graham et al. 1993, Zakharov 1990). Changes to individuals can imply changes in vital rates such as described for white sucker (*Catostomus commersoni*) in a metal-contaminated lake (McFarlane and Frazin 1978). Theoretical models for disease in populations (Moolgavkar 1986) and population impact of toxicants (Callow and Sibly 1990, Holloway et al. 1990) are also based on organismal and suborganismal information.

More and more frequently, vital rates derived from individuals in laboratory populations are applied to projections of population consequences of exposure (e.g., Pesch et al. 1991, Postma et al. 1995). In some studies, problems associated with ecological inference can arise. For example, Pesch et al. (1991) were required by the characteristics of their experimental species (the polychaete, *Neanthes arenaceodentata*) to derive fecundity and survival schedules from aggregated data instead of data for individuals.

12.3.2 Inferring Individual Effects from Qualities of Populations

Inferences about individuals from population measurements (e.g., ecological inference) are sometimes desirable and yield meaningful knowledge if the problem of ecological inference is addressed carefully. This is the major challenge during macroexplanation.

Surveys of genetic markers in toxicant-exposed populations exemplify inference about individuals based on population qualities. Allele frequencies determined for subsamples of populations inhabiting sites that differ in toxicant concentrations might be used to determine a relationship between allele frequency and exposure intensity (e.g., Sloss et al. 1998). The observation of such a statistically significant correlation often leads the investigator to argue that selection occurred against certain genotypes. Aggregate data (allele frequencies) are used to imply behavior of genotypes (individuals) as a consequence of exposure. However, alternate explanations exist such as toxicant-induced genetic bottlenecks or accelerated drift (Newman 1995). Genotype frequencies that are found to deviate significantly from Hardy–Weinberg expectations might be used to conclude that certain genotypes have lower fitness during exposure than others. However, other factors can produce departures from Hardy–Weinberg expectations including low population size, high migration rates, nonrandom breeding, and population substructuring; that is, the Wahlund effect seen by Woodward et al. (1996). Although inferences from these types of data are very useful, without supporting experimentation on individuals, the problem of ecological inference precludes definitive conclusions about individuals. Chapters 17 and 18 will cover these topics in population genetics in more detail.

Epidemiological studies often make inferences about individuals on the basis of aggregated, population-level data. King (1997) gives the example of incorrectly predicting an individual's risk of dying of radon-induced lung cancer based on correlations between regional radon levels and fatal lung cancer rates. Without monitoring an individual's radon exposure for many years, ecological inference is the only recourse for the epidemiologist. Inferences about acceptable radon levels must be made by regulators without such extremely expensive information for radon exposure to individuals. As practical, assessment of human risk from air pollutants has slowly moved toward monitoring of individual exposures for this and other reasons (e.g., see Ryan's (1998) treatment of this topic). Because of constraints imposed by limited time and money, deriving plausible inferences about individual risk based on aggregated data will continue to be a major challenge in epidemiology, the subject of the next chapter.

12.3.3 Inferring Community Effects from Qualities of Populations

Inferences about communities from population or individual measurements are commonplace but definitive only if the question of emergent properties is addressed carefully. This is the general challenge of any attempt at microexplanation and can result in inaccurate prediction. For example, community structure can be indirectly changed by the removal of one keystone species, a species that influences the community by its activity or role, and not its numerical dominance. Removal of sea urchins from a rocky intertidal pool results in a very dramatic shift in the composition of the epilithic algal species and biomass (Paine and Vadas 1969). A toxicant that kills urchins could produce a fundamental change in algal communities without having a direct effect on algal species. Field (Post et al. 1995) and laboratory (Taylor et al. 1995) studies with freshwater invertebrates also suggest changes in community interactions (predator–prey relationships) as a consequence of toxicant exposure. Whether microexplanation would have predicted the importance of these interactions among populations based on community status is uncertain.

Box 12.3 Protection of Communities Based on Species NOEC Values

Most knowledge applied to setting effluent discharge limits or to ecological risk assessment comes from assays of toxicant effects on individuals (e.g., 96-h LC50, IC50, or NOEC). With some exceptions, the typical test involves exposure of groups of individuals to a series of toxicant concentrations for a predetermined time. A metric is derived on the basis of such effects as the number of individuals dying, change in growth of exposed individuals, or change in reproductive output of exposed individuals. This preponderance of individual-based knowledge and the immediate need to protect ecological communities have led to the application of individual-level test results to infer community consequences. As an important example, U.S. numerical water quality criteria have been based on "an estimate of a concentration that will protect most but not all aquatic life; this concentration is defined as that which affects no more than 5% of taxa, with a proviso that important single species will also be protected" (Niederlehner et al. 1986). Although emergent properties can invalidate inferences of this sort, there are documented cases in which chronic effects to individuals did seem adequate for predicting community-level consequences (e.g., cadmium (Niederlehner et al. 1986) and diazinon (Giddings et al. 1996) in freshwater systems). Risk assessments are appearing on the basis of this assumption that the distribution of effects metrics from single species tests allows adequate prediction of a concentration below which only an acceptable proportion of a community will be impacted. Provided keystone and dominant species are protected, redundancy in ecological communities may allow for a certain level of species loss before an adverse impact to the community occurs. Some recent examples of risk assessments based on this concept include estimations of risk associated with atrazine (a herbicide) use in North America (Solomon et al. 1996), and cadmium and copper in the Chesapeake Bay watershed (Hall et al. 1998).

This single species test approach, growing out of regulatory necessity, was proposed in the mid-1980s. The uncertainty associated with extrapolation from acute, individual-based data for many species to ecosystem-level effects was studied by Slooff et al. (1986). They produced a regression model predicting the log of the NOEC for a community from the log of the lowest LC50 or EC50 value from toxicity test data. The relationship had an r of .77, so roughly 59% (or r^2) of the variation in the log of NOEC for a community could be explained with the model.

Considering that there was clearly scatter in these data and that log transformations of NOECs were being predicted, this suggests that prediction would be only grossly accurate. The authors recommended that a large uncertainty factor be employed to compensate in a conservative manner for this uncertainty. With this relationship and uncertainty factor, the authors concluded that acute toxicity information was useful for predicting "ecosystem" effects.

It would have been useful to apply cross-validation methods to these data as this is the best way to actually estimate the predictive value of such a regression model. Our own cross-validation analysis of their data using prediction sum of squares suggests that the overall predictive value of the regression model was acceptable. [See cross-validation procedures in statistical books such as Neter et al. (1990), pp. 465–470, for more detail.]

Several changes have occurred in this general approach. Kooijman (1987) developed a log logistic model for predicting the concentration affecting 5% of tested species (Figure 12.4, top drawing) and giving conditions under which this could be used for limited prediction of community protection. Kooijman (1987) also provided estimates of the optimum number of species needed to make predictions, basing these estimates on parametric methods. Kooijman was aware of the potential applications of his analyses and began his manuscript by stating

> The purpose of this paper is to define the concept of a hazardous concentration for sensitive species. ... [it] can be regarded as a lower bound for concentrations that can be expected to be harmful for a given community .... All honest scientific research workers will feel rather uncomfortable with such a task; and the author is no exception.

Van Straalen and Denneman (1989) extended this approach in risk assessment activities to protect soil dwelling organisms from contaminants. They modified Kooijman's mathematics to allow prediction of concentrations protecting all but a specified, low proportion of all species (p). Wagner and Løkke (1991) modified the approach further by fitting data to a log normal instead of a log logistic model. More importantly, they suggested that the lower 95% tolerance limit of

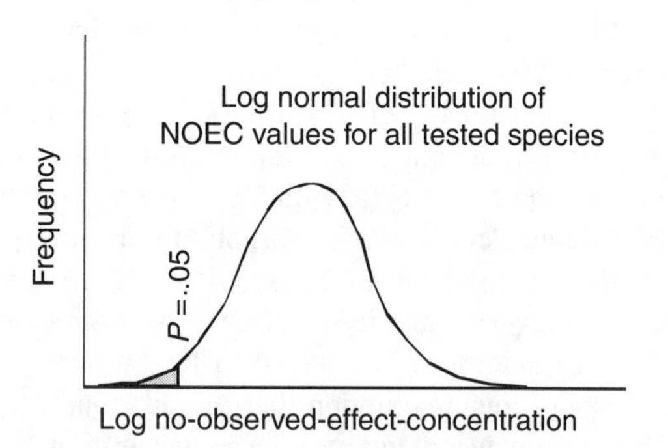

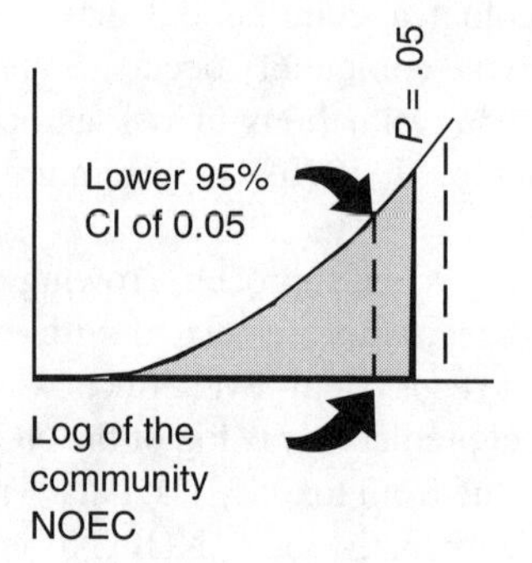

FIGURE 12.4 Prediction of community NOEC based on NOEC values for individual species notionally reflecting those in the community to be protected. The upper panel is the log normal (or log logistic) distribution of NOEC values for all tested species. The species are assumed to adequately represent those in an ecological community. The lower panel depicts the derivation of a community NOEC or HCp from this distribution. One could select the NOEC value at $p = 5\%$ of species as the community NOEC. The assumption is that all but 5% of species in the community would be protected at this NOEC. A more conservative community NOEC can be established at the lower 95% tolerance limit of the 5% value.

the predicted concentration affecting 5% of species be used as a conservative estimate of the hazardous concentration (HC*p*) (Figure 12.4, bottom drawing). Consequently, the hazardous concentration was fixed according to a specific level of confidence (expressed as a probability) in knowing that no more than a specified proportion of all species would be adversely affected. Wagner and Løkke (1991) specified that 95% of the time that such an analysis was done, no more than 5% of the species would be adversely affected at the calculated concentration. This compensates in a conservative direction for prediction uncertainty.

Considerable thought was put into the limitations of these methods as they began to be applied in regulations and assessments. Kooijman's original work (Kooijman, 1987) and later comment by Hopkin (1993), Jagoe and Newman (1997), and Newman et al. (2000) focused on ecological and statistical limits to inference from this approach. The major concerns were summarized by Newman et al. (2000) as the following:

- LC50, EC50, NOEC, and MATC data have significant deficiencies as measures of effect on natural populations. Any method based on extrapolation from these metrics shares their shortcomings.
- The assumption that redundancy in communities permits a certain proportion of species to be lost is based on the redundant species hypothesis. The alternate hypothesis, the rivet popper hypothesis, suggests the opposite would be true and that any loss of species weakens a community (see Pratt and Cairns (1996) for more details). Work with grassland communities lends some support to the rivet popper hypothesis (Naeem et al. 1994, Tilman 1996, Tilman and Downing 1994, Tilman et al. 1996). Until ecologists determine which is the correct hypothesis it would be best to be conservative and to adopt the rivet popper context for regulatory action.
- The species sensitivity distribution approach requires substantial knowledge of dominant and keystone species, and the importance of species interactions. Adequate knowledge is often not available to the ecotoxicologist. Also, even when important species and species interactions are considered, consideration is sometimes done short shrift.
- *In situ* exposure differs among species because of differences in their life histories, feeding habits, life stages, microhabitat, and other qualities. These differences are poorly reflected in the species-sensitivity distribution method because of the toxicity test design.
- There is a bias toward mortality data despite the likelihood of sublethal effects playing a crucial role in local extinction of exposed populations.
- A specific distribution, for example, the log normal (probit) model, is assumed more out of tradition than careful consideration of alternatives. Newman et al. (2000) and Jagoe and Newman (1997) demonstrated the inaccuracy of assuming such a model and advocated a bootstrap method for estimating the HC*p*. Grist et al. (2002) then published an improved bootstrap approach. The bootstrap method does not require a distribution to provide the HC*p* and associated confidence limits. Therefore, this particular shortcoming is easily resolvable.
- Determination of the optimum number of species required to produce a sound estimate of the HC*p* and the sample representativeness of the community to be protected are often given inadequate attention. However, the suboptimal sample size shortcoming can be resolved with more consistent application of parametric (Kooijman 1987) and nonparametric (Newman et al. 2000) methods for estimating optimum sample size.

These problems include technically resolvable issues and problems that can be characterized as ambiguities about emergent properties. Key to the use of this species-sensitivity distribution method to predict community-level consequences are thoughtfulness and caution based on an in-depth knowledge of the population qualities. Otherwise, inference across two or more levels of organization is of questionable value and, at worst, is a source of false and distracting information. Gross approximations, for example, implying species disappearance from a community based on a 96-h LC50, are forced on the ecotoxicologist as a regulatory necessity. Associated uncertainty must be dealt with by making estimates as conservative as reasonable.

12.4 SUMMARY

This chapter introduced the reader to population ecotoxicology, the science of contaminants in the biosphere and their effects on populations. Topics relevant to population ecotoxicology are epidemiology of toxicant-linked disease, basic demography, metapopulation biology, life-history theory, and population genetics.

The need for exploring population ecotoxicology was defined in terms of its scientific and practical values. The emerging regulatory emphasis on predicting population consequences was contrasted with the preoccupation with generating information directly relevant to individuals. Happily, more population-based work is being done every year so the present bias is likely to become less of an issue in coming years.

Inferences within and between biological levels were explored from the perspective of the population. Application of individual and subindividual data to predict population consequences was described, emphasizing the importance of thoughtful awareness of emergent properties. Brief discussion of inferring individual behaviors based on population-level information demonstrated the problem of ecological inference, as well as the need for such inference. Epidemiology was identified as one relevant field in which such inferences occur. Finally, we discussed inferences based on individual- or population-level information to predict community fate upon toxicant exposure. A species-sensitivity distribution approach was used as an example. The problem of unforeseen emergent properties was again discussed as inhibiting clear inference about community effects from toxicity test results.

12.4.1 Summary of Foundation Concepts and Paradigms

- Laws, regulations, and standards aim to protect the viability of natural populations in ecological communities. Consequently, population ecotoxicology has practical value.
- Effects to individuals measured in standard toxicity tests are intended for use in predicting consequences to exposed populations. They do this with ambiguous precision and accuracy.
- Classic studies in population biology and genetics involve response to toxicants. Consequently, population ecotoxicology has already demonstrated its scientific value.
- Inferences about individuals from population studies, for example, some epidemiological studies, are valuable but prone to the problem of ecological inference.
- Inferences about communities from population studies are valuable but prone to error due to the emergence of unique properties at higher levels of biological organization.

REFERENCES

Antonovics, J., Bradshaw, A.D., and Turner, R.G., Heavy metal tolerance in plants, In *Advances in Ecological Research*, Vol. 7., Cragg, J.B. (ed.), Academic Press, London, UK, 1971, pp. 1–83.

Asami, T. and Grant, B., Melanism has not evolved in Japanese *Biston betularia* (Geometridae), *J. Lepid. Soc.*, 49, 88–91, 1994.

Barnthouse, L.W., Suter, G.W., II, Rosen, A.E., and Beauchamp, J.J., Estimating responses of fish populations to toxic contaminants, *Environ. Toxicol. Chem.*, 6, 811–824, 1987.

Beardmore, J.A., Genetical considerations in monitoring effects of pollution, *Rapp. P-v. Réun. Cons. Int. Explor. Mer.*, 179, 258–266, 1980.

Beaumont, P., Where have all the birds gone? *Pesticides News*, 30, 3, 1997.

Berkson, J., Why I prefer logits to probits, *Biometrics*, 7, 327–339, 1951.

Berry, R.J., Industrial melanism and peppered moths (*Biston betularia* (L.)), *Biol. J. Linn. Soc.* 39, 301–322, 1990.

Bishop, J.A. and Hartley, D.J., The size and structure of rural populations of *Rattus norvegicus* containing individuals resistant to the anticoagulant poison Warfarin, *J. Anim. Ecol.*, 45, 623–646, 1976.

Bishop, J.A., Hartley, D.J., and Partridge, G.G., The population dynamics of genetically determined resistance to warfarin in *Rattus norvegicus* from mid-Wales, *Heredity*, 39, 389–398, 1977.

Bliss, C.I., The calculation of the dosage-mortality curve, *Ann. Appl. Biol.*, 22, 134–307, 1935.

Bliss, C.I. and Cattell, M., Biological assay, *Ann. Rev. Physiol.*, 5, 479–539, 1943.

Boyd, C.E. and Ferguson, D.E., Susceptibility and resistance of mosquito fish to several insecticides, *J. Economic Entomol.*, 57, 430–431, 1964.

Brakefield, P.M., A decline of melanism in the peppered moth *Biston betularia* in the Netherlands, *Biol. J. Linn. Soc.*, 39, 327–334, 1990.

Brown, J.H. and Maurer, B.A., Macroecology: The division of food and space among species on continents, *Science*, 243, 1145–1150, 1989.

Cairns, J., Jr. and Pratt, J.R., The scientific basis of bioassays, *Hydrobiologia*, 188/189, 5–20, 1989.

Calow, P. and Sibly, R.M., A physiological basis of population processes: Ecotoxicological implications, *Funct. Ecol.*, 4, 283–288, 1990.

Carson, R., *Silent Spring*, Houghton-Mifflin Co., Boston, MA, 1962.

Caswell, H., Demography meets ecotoxicology: Untangling the population level effects of toxic substances, In *Ecotoxicology. A Hierarchical Treatment*, Newman, M.C. and Jagoe, C.H. (eds.), CRC Press/Lewis Publishers, Boca Raton, FL, 1996, pp. 255–292.

Clarke, C.A., Grant, B., Clarke, F.M.M., and Asami, T., A long term assessment of *Biston betularis* (L.) in one U.K. locality (Caldy Common near Kirby, Wirral), 1959–1993, and glimpses elsewhere, *The Linnean*, 10, 18–26, 1994.

Clarke, C.A., Mani, G.S., and Wynne, G., Evolution in reverse: Clean air and the peppered moth, *Biol. J. Linn. Soc.*, 26, 189–199, 1985.

Comins, H.N., The development of insecticide resistance in the presence of migration, *J. Theor. Biol.*, 64, 177–197, 1997.

Connell, D.W. and Miller, D.J., *Chemistry and Ecotoxicology of Pollution*, John Wiley & Sons, New York, 1984.

Deming, A.H., *Science and Other Poems*, Louisiana State University Press, Baton Rouge, LA, 1994.

Dixon, P.M. and Newman, M.C., Analyzing toxicity data using statistical models of time-to-death: An introduction, In *Metal Ecotoxicology. Concepts & Application*, Newman, M.C. and McIntosh, A.W. (eds.), CRC Press/Lewis Publishers, Chelsea, MI, 1991, pp. 207–242.

Dolphin, R., Lake County mosquito abatement district gnat research program. Clear Lake Gnat (*Chaoborus astictopus*), In *Proceedings of 27th Annual Conference of the California Mosquito Control Association*, 1959, Sacramento, CA, pp. 47–48.

Emlen, J.M., Terrestrial population models for ecological risk assessment: A state-of-the-art review, *Environ. Toxicol. Chem.*, 8, 831–842, 1989.

EPA, Summary Report on Issues in Ecological Risk Assessment, EPA/625/3-91/018 February 1991, NTIS, Springfield, VA, 1991.

Finney, D.J., *Probit Analysis. A Statistical Treatment of the Sigmoid Response Curve*, Cambridge University Press, Cambridge, UK, 1947.

Forbes, V.E. and Forbes, T.L., *Ecotoxicology in Theory and Practice*, Chapman & Hall, London, 1994.

Fry, D.M. and Toone, C.K., DDT-induced feminization of gull embryos, *Science*, 213, 922–924, 1981.

Gaddum, J.H., Bioassays and mathematics, *Pharmacol. Rev.*, 5, 87–134, 1953.

Giddings, J.M., Biever, R.C., Annunziato, M.F., and Hosmer, A.J., Effects of diazinion on large outdoor pond microcosms, *Environ. Toxicol. Chem.*, 15, 618–629, 1996.

Ginzburg, L.R., Johnson, K., Pugliese, A., and Gladden, J., Ecological risk assessment based on stochastic age-structure models of population growth, In *Statistics in the Environmental Sciences, ASTM STP 845*, Gertz, S.M. and London, M.D. (eds.), American Society for Testing and Materials, Philadelphia, PA, 1984, pp.31–45.

Graham, J.H., Freeman, D.C., and Emlen, J.M., Developmental stability: A sensitive indicator of populations under stress, In *Environmental Toxicology and Risk Assessment, ASTM STP 1179*, Landis, W.G., Hughes, J.S., and Lewis, M.A. (eds.), American Society for Testing and Materials, Philadelphia, PA, 1993, pp. 136–158.

Grant, B.S. and Clarke, C.A., An examination of intrraseasonal variation in the incidence of melanism in peppered moths, *Biston betularia*, *J. Lepid. Soc.*, 53, 99–103, 1999.

Grant, B.S., Cook, A.D., Clarke, C.A., and Owen, D.F., Geographic and temporal variation in the incidence of melanism in peppered moth populations in America and Britain, *J. Hered.*, 89, 465–471, 1998.

Grant, B.S. and Howlett, R.J., Background selection by the peppered moth (*Biston betularia* Linn.): Individual differences, *Biol. J. Linn. Soc.*, 33, 217–232, 1988.

Grant, B.S., Owen, D.F., and Clarke, C.A., Decline of melanic moths, *Nature*, 373, 565, 1995.

Grant, B.S., Owen, D.F., and Clare, C.A., Parallel rise and fall of melanic peppered moths in America and Britain, *J. Hered.*, 87, 351–357, 1996.

Green, A.S. and Chandler, G.T., Life-table evaluation of sediment-associated chlorpyrifos chronic toxicity to the benthic copepod, *Amphiascus tenuiremis*, *Arch. Environ. Contam. Toxicol.*, 31, 77–83, 1996.

Grist, E.P.M., Leung, K.M.Y., Wheeler, J.R., and Crane, M., Better bootstrap estimation of hazardous concentration thresholds for aquatic assemblages, *Environ. Toxicol. Chem.*, 21, 1515–1524, 2002.

Guttman, S.I., Population genetic structure and ecotoxicology, *Environ. Health Perspect.*, 102(Suppl. 12), 97–100, 1994.

Hall, L.W., Jr., Scott, M.C., and Killen, W.D., Ecological risk assessment of copper and cadmium in surface waters of Chesapeake Bay watershed, *Environ. Toxicol. Chem.*, 17, 1172–1189, 1998.

Hickey, J.J. and Anderson, D.W., Chlorinated hydrocarbons and eggshell changes in raptorial and fish-eating birds, *Science*, 162, 271–273, 1968.

Holloway, G.J., Sibly, R.M., and Povey, S.R., Evolution in toxin-stressed environments, *Functional Ecol.*, 4, 289–294, 1990.

Hopkin, S.P., Ecological implications of '95% protection levels' for metals in soil, *OIKOS*, 66, 137–141, 1993.

Jagoe, R.H. and Newman, M.C., Bootstrap estimation of community NOEC values, *Ecotoxicology*, 6, 293–306, 1997.

Kammenga, J.E., Busschers, M., Van Straalen, N.M., Jepson, P.C., and Baker, J., Stress induced fitness is not determined by the most sensitive life-cycle trait, *Funct. Ecol.*, 10, 106–111, 1996.

Karås, P., Neuman, E., and Sandström, O., Effects of a pulp mill effluent on the population dynamics of perch, *Perca fluviatilis*, *Can. J. Fish. Aquat. Sci.*, 48, 28–34, 1991.

Kettlewell, H.B.D., Selection experiments on industrial melanism in the Lepidoptera, *Heredity*, 9, 323–342, 1955.

Klerks, P.L. and Weis, J.S., Genetic adaptation to heavy metals in aquatic organisms: A review, *Environ. Pollut.*, 45, 173–205, 1987.

Kolaja, G.J. and Hinton, D.E., *Animals as Monitors of Environmental Pollutants*, National Academy of Sciences, Washington, D.C., 1979, pp. 309–318.

Kooijman, S.A.L.M., A safety factor for LC_{50} values allowing for differences in sensitivity among species, *Wat. Res.*, 21, 269–276, 1987.

Landis, W.G. and Yu, M.-H., *Introduction to Environmental Toxicology: Impacts of Chemicals upon Ecological Systems*, CRC Press, Boca Raton, FL, 1995.

Lee, D.R. and Cred, E.R., The genetics of the *insularia* forms of the peppered moth, *Biston betularia*, *J. Anim. Ecol.*, 39, 67–73, 1977.

Luoma, J.R., New effect of pollutants: Hormone mayhem. *New York Times*, May 24, 1992.

Maciorowski, A.F., Populations and communities: Linking toxicology and ecology in a new synthesis, *Environ. Toxicol. Chem.*, 7, 677–678, 1988.

Mani, G.S., Theoretical models of melanism in *Biston betularia*—A review, *Biol. J. Linn. Soc.*, 39, 355–371, 1990.

McFarlane, G.A. and Franzin, W.G., Elevated heavy metals: A stress on a population of white suckers, *Catostomus commersoni*, in Hamell Lake, Saskatchewan, *J. Fish. Res. Board Can.*, 35, 963–970, 1978.

McKenzie, J.A. and Batterham, P., The genetic, molecular and phenotypic consequences of selection for insecticide resistance, *TREE*, 9, 166–169, 1994.

McLachlan, J.A., Functional toxicology: A new approach to detect biologically active xenobiotics, *Environ. Health Perspect.*, 101, 386–387, 1993.

Mitton, J.B., *Selection in Natural Populations*, Oxford University Press, Inc., Oxford, UK, 1997.

Moolgavkar, S.H., Carcinogenesis modeling: From molecular biology to epidemiology, *Ann. Rev. Public Health*, 7, 151–169, 1986.

Mulvey, M. and Diamond, S.A. Genetic factors and tolerance acquisition in populations exposed to metals and metalloids, In *Metal Ecotoxicology. Concepts & Applications*, Newman, M.C. and McIntosh, A.W. (eds.), CRC Press/Lewis Publishers, Chelsea, MI, 1991, pp. 301–321.

Mulvey, M., Newman, M.C., Chazal, A., Keklak, M.M., Heagler, H.G., and Hales, S., Genetic and demographic changes in mosquitofish (*Gambusia holbrooki*) populations exposed to mercury, *Environ. Toxicol. Chem.*, 14, 1411–1418, 1995.

Naeem, S., Thompson, L.J., Lawler, S.P., Lawton, J.H., and Woodfin, R.M., Declining biodiversity can alter performance of ecosystems, *Nature*, 368, 734–737, 1994.

Neter, J., Wasserman, W., and Kutner, W.H., *Applied Linear Statistical Models. Regression, Analysis of Variance, and Experimental Design.*, Richard D. Irwin, Inc., Homewood, IL, 1990.

Newman, M.C., *Fundamentals of Ecotoxicology*, CRC Press/Ann Arbor Press, Boca Raton, FL, 1998.

Newman, M.C., *Quantitative Methods in Aquatic Ecotoxicology*, CRC Press/Lewis Publishers, Boca Raton, FL, 1995.

Newman, M.C. and Aplin, M., Enhancing toxicity data interpretation and prediction of ecological risk with survival time modeling: An illustration using sodium chloride toxicity to mosquitofish (*Gambusia holbrooki*), *Aquat. Toxicol.*, 23, 85–96, 1992.

Newman, M.C., Crane, M., and Holloway, G., Does pesticide risk assessment in the European Union assess long-term effects? *Rev. Environ. Contam. Toxicol.*, 198, 1–65, 2006.

Newman, M.C. and Dixon, P.M., Ecologically meaningful estimates of lethal effect on individuals, In *Ecotoxicology: A Hierarchical Treatment*, Newman, M.C. and Jagoe, C.H. (eds.), CRC/Lewis Press, Boca Raton, FL, 1996, pp. 225–253.

Newman, M.C. and McCloskey, J.T., The individual tolerance concept is not the sole explanation for the probit dose-effect model, *Environ. Toxicol. Chem.*, 19, 520–526, 2000.

Newman, M.C., Ownby, D.R., Mézin, L.C.A., Powell, D.C., Christensen, T.R.L., Lerberg, S.B., and Anderson, B.-A., Applying species sensitivity distributions in ecological risk assessment: Assumptions of distribution type and sufficient numbers of species, *Environ. Toxicol. Chem.*, 19, 508–515, 2000.

Niederlehner, B.R., Pratt, J.R., Buikema, A.L., Jr., and Cairns, J., Jr., Comparison of estimates of hazard derived at three levels of complexity, In *Community Toxicity Testing*, *ASTM STP 920*, Cairns, J., Jr. (ed.), American Society for Testing and Materials, Philadelphia, PA, 1986, 30–48.

O'Connor, R.J., Toward the incorporation of spatiotemporal dynamics in ecotoxicology, In *Population Dynamics in Ecological Space and Time*, Rhodes, O.E., Jr., Chesser, R.K., and Smith, M.H. (eds.), University of Chicago Press, Chicago, IL, 1996, pp. 281–317.

Odum, E.P., *Fundamentals of Ecology*, 3rd ed., W.B. Saunders Co., Philadelphia, PA, 1971.

Osowski, S.L., Brewer, L.W., Baker, O.E., and Cobb, G.P., The decline of mink in Georgia, North Carolina, and South Carolina: The role of contaminants, *Arch. Environ. Contam. Toxicol.*, 29, 418–423, 1995.

Paine, R.T. and Vadas, R.L., The effects of grazing by sea urchins, *Strongylocentrotus spp.*, on benthic algal populations, *Limnol. Oceanogr.*, 14, 710–719, 1969.

Parrish, P.R. Acute toxicity tests, In *Fundamentals of Aquatic Toxicology*, Rand, G.M. and Petrocelli, S.R. (eds.), Hemisphere Publishing Corp., Washington, D.C., 1985, pp. 31–57.

Pesch, C.E., Munns, W.R., Jr., and Gutjahr-Gobell, R., Effects of a contaminated sediment on life history traits and population growth rate of *Neanthes arenaceodentata* (Polychaeta: Nereidae) in the laboratory, *Environ. Toxicol. Chem.*, 10, 805–815, 1991.

Petersen, R.C., Jr. and Petersen, L.B.-M., Compensatory mortality in aquatic populations: Its importance for interpretation of toxicant effects, *Ambio*, 17, 381–386, 1988.

Post, D.M., Frost, T.M., and Kitchell, J.F., Morphological responses by *Bosmina longirostris* and *Eubosmina tubicen* to changes in copepod predator populations during a whole-lake acidification experiment, *J. Plankton Res.*, 17, 1621–1632, 1995.

Postma, J.F., van Kleunen, A., and Admiraal, W., Alterations in life-history traits of *Chironomus riparius* (Diptera) obtained from metal contaminated rivers, *Arch. Environ. Contam. Toxicol.*, 29, 469–475, 1995.

Pratt, J.R. and Cairns, J., Jr., Ecotoxicology and the redundancy problem: Understanding effects on community structure and function, In *Ecotoxicology. A Hierarchical Treatment*, Newman, M.C. and Jagoe, C.H. (eds.), CRC Press/Lewis Publishers, Boca Raton, FL, 1996, pp. 347–370.

Ratcliffe, D.A., Decrease in eggshell weight in certain birds of prey, *Nature*, 215, 208–210, 1967.

Ratcliffe, D.A., Changes attributable to pesticides in egg breakage frequency and eggshell thickness in some British birds, *J. Appl. Ecol.*, 7, 67–107, 1970.

Ryan, P.B., Historical perspective on the role of exposure assessment in human risk assessment, In *Risk Assessment: Logic and Measurement*, Newman, M.C. and Strojan, C.L. (eds.), Ann Arbor Press, Chelsea, MI, 1998, pp. 23–43.

Sarokin, D. and Schulkin, J., The role of pollution in large-scale population disturbances. Part 1: Aquatic populations, *Environ. Sci. Technol.*, 26, 1476–1484, 1992.

Sierszen, M.E. and Frost, T.M., Response of predatory zooplankton populations to the experimental acidification of Little Rock Lake, Wisconsin, *J. Plankton Res.*, 15, 553–562, 1993.

Slooff, W., Van Oers, J.A.M., and De Zwart, D., Margins of uncertainty in ecotoxicological hazard assessment, *Environ. Toxicol. Chem.*, 5, 841–852, 1986.

Sloss, B.L., Romano, M.A., and Anderson, R.V., Pollution-tolerant allele in fingernail clams (*Musculium transversum*), *Arch. Environ. Contam. Toxicol.*, 35, 302–308, 1998.

Solomon, K.R., Baker, D.B., Richards, R.P., Dixon, K.R., Klaine, S.J., La Point, T.W., Kendall, R.J., et al., Ecological risk assessment of atrazine in North American surface waters, *Environ. Toxicol. Chem.*, 15, 31–76, 1996.

Spitzer, P.R., Risebrough, R.W., Walker, W., II, Hernandez, R., Poole, A., Puleston, D., and Nisbet, I.C.T., Productivity of ospreys in Connecticut-Long Island increase as DDE residues decline, *Science*, 202, 333–335, 1978.

Suter, II, G.W., *Ecological Risk Assessment*, Lewis Publishers, Chelsea, MI, 1993.

Taylor, E.J., Morrison, J.E., Blockwell, S.J., Tarr, A., and Pascoe, D., Effects of lindane on the predator-prey interaction between *Hydra oligactis* Pallas and *Daphnia magna* Strauss, *Arch. Environ. Contam. Toxicol.*, 29, 291–296, 1995.

Tilman, D., Biodiversity: Population versus ecosystem stability, *Ecology*, 77, 350–363, 1996.

Tilman, D. and Downing, J.A., Biodiversity and stability on grasslands, *Nature*, 367, 363–365, 1994.

Tilman, D., Wedlin, D., and Knops, J., Productivity and sustainability influenced by biodiversity in grassland ecosystems, *Nature*, 379, 718–720, 1996.

Trevan, J.W., The error of determinations of toxicity, *Proc. R. Soc. Lond. B. Biol. Sci.*, 101, 483–514, 1927.

Van Straalen, N.M. and Denneman, C.A.J., Ecotoxicological evaluation of soil quality criteria, *Ecotox. Environ. Safety*, 18, 241–251, 1989.

Wagner, C. and Lokke, H., Estimation of ecotoxicological protection levels from NOEC toxicity data, *Water Res.*, 25, 1237–1242, 1991.

Wake, D.B., Declining amphibian populations, *Science*, 253, 860, 1991.

Webb, R.E. and Horsfall, Jr., F, Endrin resistance in the pine mouse, *Science*, 156, 1762, 1967.

Weis, J.S. and Weis, P., Tolerance and stress in a polluted environment, *BioScience*, 39, 89–95, 1989.

West, D.A., Melanism in *Biston* (Lepidoptera: Geometridae) in the rural Central Appalachians, *Heredity*, 39, 75–81, 1977.

Whitton, M.J., Dearn, J.M., and McKenzie, J.A., Field studies on insecticide resistance in the Australian sheep blowfly, *Lucilia cuprina*, *Aust. J. Biol.*, 33, 725–735, 1980.

Woodward, L.A., Mulvey, M., and Newman, M.C., Mercury contamination and population-level responses in chironomids: Can allozyme polymorphisms indicate exposure? *Environ. Toxicol. Chem.*, 15, 1309–1316, 1996.

Woodwell, G.M., Wurster, C.F., Jr., and Isaacson, P.A., DDT residues in an East Coast estuary: A case of biological concentration of a persistent insecticide, *Science*, 156, 821–823, 1967.

Zakharov, V.M., Analysis of fluctuating asymmetry as a method of biomonitoring at the population level, In *Bioindicators of Chemical and Radioactive Pollution*, Krivolutsky, D.A. (ed.), CRC Press, Boca Raton, FL, 1990, pp. 187–198.

Zhao, Y. and Newman, M.C., Shortcomings of the laboratory-derived median lethal concentration for predicting mortality in field populations: Exposure duration and latent mortality, *Environ. Toxicol. Chem.*, 23, 2147–2153, 2004.

13 Epidemiology: The Study of Disease in Populations

All scientific work is incomplete—whether it be observational or experimental. All scientific work is liable to be upset or modified by advancing knowledge. That does not confer upon us a freedom to ignore the knowledge we already have, or to postpone the action that it appears to demand at a given time.

(Sir Austin Hill 1965)

13.1 FOUNDATION CONCEPTS AND METRICS IN EPIDEMIOLOGY

In environmental toxicology, methods may be applied to populations with two different purposes. The goal might be either protection of individuals or an entire population. This distinction is often confused in ecotoxicology, a science that must consider many levels of biological organization in its deliberations.

When dealing with contamination-associated disease in human populations, information is collected to protect individuals with certain characteristics such as high exposure or hypersensitivity. The emphasis is on identifying causal and etiological factors that put one individual at higher risk than another, and quantifying the likelihood of the disease afflicting an individual characterized relative to risk factors. In contrast, in the study of nonhuman species, the focus shifts more toward maintaining viable populations than toward minimizing risk to specific individuals. Important exceptions involve the protection of endangered, threatened, or particularly charismatic species. In such cases, individuals may be the protected entities. Another situation is the natural resource damage assessment context in which lost individuals might be estimated and compensation for resource injury estimated on the basis of lost individuals.

The focus in this chapter will be on epidemiology, the science concerned with the cause, incidence, prevalence, and distribution of disease in populations. More specifically, we will focus on ecological epidemiology, that is, epidemiology applied to assess risk to nonhuman species inhabiting contaminated sites (Suter 1993). Methods described will provide insights of direct use for protecting individuals and describing disease presence in populations, and of indirect use for implying population consequences.

13.1.1 FOUNDATION CONCEPTS

In the above paragraph describing epidemiology, mention was made without explanation of causal and etiological factors. Let us take a moment to explain these terms and some associated concepts.

A causal agent is one that causes something to occur directly or indirectly through a chain of events. Although seemingly obvious, this definition carries many philosophical and practical complications.

Causation, a change in state or condition of one thing due to interaction with another, is surprisingly difficult to identify. One can identify a cause by applying the push-mechanism context of Descartes (Popper 1965) or Kant's (1934) concept of action. In this context, some cause has an innate power to produce an effect and is connected with that effect (Harré 1972). As an example, one body might pull (via gravity) or push (via magnetism) another by existing relative to that other. The result is motion. The presence and nature of the object cause a consequence and the effect diminishes with distance between the objects.

Alternatively, a cause may be defined in the context of succession theory as something preceding a specific event or change in state (Harré 1972). Kant (1934) refers to this as the Law of Succession in Time. The consistent sequence of one event (e.g., high exposure to a toxicant) followed by another (e.g., death) establishes an expectation. On the basis of past observations or observations reported by others, one comes to expect death after exposure to high concentrations of the toxicant.

Building from the thoughts of Popper (1959) regarding qualities of scientific inquiry, other qualities associated with the concept of causation emerge. Often there is an experimental design within which an effect is measured after a single thing is varied (i.e., the potential cause). The design of the experiment in which one thing is selected to be changed determines directly the context in which the term, cause, is applied. That which was varied causes the effect; for example, increasing temperature caused an increase in bacterial growth rate. If another factor (e.g., an essential nutrient) had been varied in the experiment, it could have also caused the effect (e.g., increased growth rate). The following quote by Simkiss (1996) illustrates the importance of context and training in formulating causal structures.

> Thus, the problem took the form of habitat pollution → DDE accumulated in prey species → DDE in predators → decline in brood size → potential extermination. The same phenomenon can, however, be written in a different form. Lipid soluble toxicant → bioaccumulation in organisms with poor detoxification systems (birds metabolize DDE very poorly when compared with mammals) → vulnerable target organs (i.e., the shell gland has a high Ca flux) → inhibition of membrane-bound ATPases at crucial periods → potential extermination. Ecologists would claim a decline in population recruitment, biochemists—an inhibition of membrane enzymes.

Clearly the context of observations and experiments, and measured parameters determined the causal structure for the ecologist (i.e., DDE spraying causes bird population extinctions) and biochemist (i.e., DDE bioaccumulation causes shell gland ATPase inhibition) studying the same phenomenon.

Controlled laboratory experiments remain invaluable tools for assigning causation as long as one understands the conditional nature of associated results. A coexistence of potential cause and effect is imposed unambiguously by the experimental design (Kant 1934), for example, death occurred after 24-h exposure to 2 μg/L of dissolved toxicant in surrounding water. With this unambiguous co-occurrence and simplicity (low dimensionality), a high degree of consistency is expected from structured experiments. Also, one is capable of easily falsifying the hypothesized cause–effect relationship during structured experimentation. Inferences about causation are strengthened by these qualities of experiments. Information on causal linkage emerging from such a context is invaluable in ecological epidemiology but it is not the only type of useful information. Valuable information is obtained from less structured, observational "experiments" possessing a lower ability to identify causal structure. Epidemiology relies heavily on such observational information.

Other factors complicate the process by which we effectively identify a cause–effect relationship in a world filled with interactions and change. According to Kant (1934), our minds are designed to create or impose useful structures of expectation that are not necessarily as grounded in objective reality as we might want to believe. We survive by developing webs of expectations based on unstructured observations of the world and by then, pragmatically assigning causation within this complex. With incomplete knowledge and increasing complexity (high dimensionality), we often are compelled to build causal hypotheses from correlations (a probabilistic expectation based on past experience that depends heavily on the Law of Succession) and presumed mechanisms (linked cause–effect relationships leaning heavily on the concept of action). This is called pseudoreasoning in cognitive studies and is a wobbly foundation of everyday "common sense" and the expert opinion approach in ecological risk assessment. Unfortunately, habits applied in our informal reasoning are remarkably bad at determining the likelihood of one factor being a cause of a consequence if several candidate causes exist. Piattelli-Palmarini (1994) concluded that, when we use our natural mental economy, "we are *instinctively* very poor evaluators of probability and equally poor at choosing between alternative possibilities." It follows from this sobering conclusion that accurate assignment of causation in ecotoxicology can more reliably be made by formal methods, for example, Bayesian logic or belief networks (Jensen 2001, Pearl 2000), than by informal expert opinions and weight-of-evidence methods. This is especially important to keep in mind in ecological epidemiology.

These aspects of causation can be summarized in the points below. They provide context for judging the strength of inferences about causal agents from epidemiological studies.

- Causation is most commonly framed within the concept of action and the Law of Succession.
- Causation emerges as much from our "neither rational nor capricious" (Tversky and Kahneman 1992) cognitive psychology as from objective reality.
- Causal structure emerges from the framework of the experiment or "question" as well as objective reality.
- Accurate identification of causation is enhanced by (1) clear co-occurrence in appropriate proximity of cause and effect, (2) simplicity (low dimensionality) of the system being assessed, (3) high degree of consistency from the system under scrutiny, and (4) formalization of the process for identifying causation.

Many of the conditions required to best identify causation are often absent in epidemiological studies. Therefore, when assessing effects of environmental contaminants, we resort to a blend of correlative and mechanistic (cause–effect) information. Uncertainty about cause–effect linkages tempers terminology and forces logical qualifiers on conclusions. For example, a contaminant might be defined as an etiological agent, that is, something causing, initiating, or promoting disease. Notice that an etiological agent need not be proven to be the causal agent. Indeed, with the multiple causation structures present in the real world and the human compulsion to construct subjective cause–effect relationships, the context of etiological agent seems more reasonable at times than that of causal agent.

Often, epidemiology focuses on qualities of individuals that predispose them to some adverse consequence. In the context of cause–effect, such a factor is seen more as contributing to risk than as the direct cause of the effect. Such risk factors for human disease include genetic makeup of individuals, behaviors, diet, and exercise habits. The presence of a benthic stage in the life cycle of an aquatic species might be viewed as a predisposing risk factor for the effects of a sediment-bound contaminant. Possession of a gizzard in which swallowed "stones" are ground together under acidic conditions could be considered a risk factor for lead poisoning of ducks dabbling in marshes spattered with lead shot from a nearby skeet range. Dabbling ducks tend to include lead shot among the hard objects retained in their gizzards and, as a consequence, are at high risk of lead poisoning.

The exact meanings of two terms that will be used throughout our remaining discussion, risk and hazard, need to be clarified at this point. They are not synonymous terms in ecological epidemiology. The general meaning of risk is a danger or hazard, or the chance of something adverse happening. This is close to the definition that we will use. Hazard is defined here as simply the presence of a potential danger. For example, the hazard associated with a chemical may be grossly assessed by dividing its measured concentration in the environment by a concentration shown in the laboratory to cause an adverse effect. A hazard quotient exceeding one implies a potentially hazardous concentration.[1] The concept of risk implies more than the presence of a potential danger. Risk is the probability of

[1] Hazard will be defined differently when survival time modeling is discussed later in this chapter.

a particular adverse consequence occurring because of the presence of a causal agent, etiological agent, or risk factor. The concept of risk involves not only the presence of a danger but also the probability of the adverse effect being realized in the population when the agent is present (Suter 1993). For example, the risk of a fatal cancer is 1 in 10,000 for a lifetime exposure to 0.5 mg/day/kg of body mass of chemical X.

Although defined as a probability, the concept of risk may be conveyed in other ways such as loss in life expectancy, for example, a loss of 870 days from the average life span due to chronic exposure to a toxicant in the work environment. In the context of comparing populations or groups, it could be expressed as a relative risk, for example, the risk of death at a 1 mg dose versus the risk of death at a 5 mg dose. It can also be expressed as an odds ratio (OR) or an incidence rate. These metrics are described in more detail below.

13.1.2 Foundation Metrics

There are several straightforward metrics used in epidemiological analyses. Here they will be discussed primarily with human examples but they are readily applied to other species. In fact, because of ethical limits on human experimentation, some metrics such as those generated from case–control or dose–effect studies are much more easily derived for nonhuman species than for humans.

Disease incidence rate for a nonfatal condition is measured as the number of individuals with the disease (N) divided by the total time that the population has been exposed (T). Incidence rate (I) is often expressed in units of individuals or cases per unit of exposure time being considered in the study, e.g., 10 new cases per 1000 person-years (Ahlbom 1993). The T is expressed as the total number of time units that individuals were at risk (e.g., per 1000 person-years of exposure):

$$\hat{I} = \frac{N}{T}. \tag{13.1}$$

The number of individuals with the disease (N) is assumed to fit a Poisson distribution because a binomial error process is involved—an individual either does or does not have the disease. Consequently, the estimated mean of N is also an estimate of its variance. Knowing the variance of N, its 95% confidence limits can be estimated. Then, the 95% confidence limits of I can be estimated by dividing the upper and lower limits for N by T.

There are several ways of estimating the 95% confidence limits of N. Approximation under the assumption of a normal distribution instead of a Poisson distribution produces the following estimate (Ahlbom 1993):

$$\text{Number of cases} \approx \hat{N} \pm 1.96\sqrt{\hat{N}}. \tag{13.2}$$

To get the 95% confidence limits for I, those for N are divided by T. This and the other normal approximations described below can be poor estimators if the number of disease cases is small. The reader is referred to Ahlbom (1993) and Sahai and Khurshid (1996) for necessary details for such cases.

Estimated disease prevalence ($\hat{p}$) is the incidence rate (I) times the length of time (t) that individuals were at risk:

$$\hat{p} = \hat{I} \times t. \tag{13.3}$$

For example, if there were 27 cases per 1,000 person-years, the prevalence in a population of 10,000 people exposed for 10 years (i.e., 100,000 person-years) would be (27 cases/1,000 person-years) (100,000 person-years) or 2,700 cases. Prevalence also emerges from a binomial error process, and its variance and confidence limits can be approximated as described above for incidence rate (Ahlbom 1993).

Sometimes it is advantageous to express the occurrence of disease in a population relative to that in another: often one population is a reference population. Differences in incidence rates can be used in such a comparison. For example, there may be 227 more cases per year in population A than in population B. Differences are often normalized to a specific population size (e.g., 227 more cases per year in a population of 10,000 individuals) because populations differ in size.

Let us demonstrate the estimation of incidence rate difference and its confidence limits by considering two populations with person-exposure times of T_1 and T_2, and case numbers of N_1 and N_2 during those person-year intervals. The incidence rate difference (IRD) is estimated by the simple relationship

$$\hat{\text{IRD}} = \frac{N_1}{T_1} - \frac{N_2}{T_2}. \tag{13.4}$$

The variance and confidence limits for the incidence rate difference are approximated by Equations 13.5 and 13.6, respectively (Sahai and Khurshid 1996):

$$\text{Variance of } \hat{\text{IRD}} = \frac{N_1}{T_1^2} + \frac{N_2}{T_2^2} \tag{13.5}$$

$$\text{IRD} \pm Z_{\alpha/2}\sqrt{\frac{N_1}{T_1^2} + \frac{N_2}{T_2^2}}. \tag{13.6}$$

These equations can be applied during surveys of populations or to case–control studies. The N_1 and T_1 could be associated with one population and N_2 and T_2 with another. Or N_1 and T_1 could reflect the disease incidence rate for N_1 individuals who have been exposed to an etiological agent, and N_2 and T_2 could reflect the effect incidence rate for N_2 individuals with no known exposure. Individuals designated as a control or noncase group are compared to a group of individuals who have been exposed in such retrospective case–control studies. The magnitude of the IRD suggests the influence of the etiological factor on the disease incidence.

The relative occurrence of disease in two populations can be expressed as the ratio of incidence rates (rate ratio [RR]). The following equation provides an estimate of the rate ratio for two populations:

$$\hat{\text{RR}} = \frac{\hat{I}_1}{\hat{I}_0} \tag{13.7}$$

where I_1 = incidence rate in population 1, and I_0 = incidence rate in the reference or control population. For example, twenty diseased fish found during an annual sampling of a standard sample size of 10,000 individuals taken from a bay near a heavily industrialized city may be compared to an annual incidence rate of 5 fish per 10,000 individuals from a bay adjacent to a small town. The relative risk in these populations would be estimated with a rate ratio of 4. Implied by this ratio is an influence of heavy industry on the risk of disease in populations. Obviously, an estimate of the variation about this ratio would contribute to a more definitive statement.

The variance and confidence limits for incidence rate ratios are usually derived in the context of the ln of rate ratios. The approximate variance and 95% confidence limits for the ln of rate ratio are defined by Equations 13.8 and 13.9. The antilogarithm of the confidence limits approximates those for the rate ratio (Sahai and Khurshid 1996).

$$\text{Variance of } \ln(\text{RR}) \approx \frac{1}{N_1} + \frac{1}{N_0} \quad (13.8)$$

$$\ln \text{RR} \pm Z_{\alpha/2}\sqrt{\frac{1}{N_1} + \frac{1}{N_0}}. \quad (13.9)$$

Box 13.1 Differences and Ratios as Measures of Risk

Cancer Incidence Rate Differences at Love Canal

The building of the Love Canal housing tract around an abandoned waste burial site in New York resulted in one of the most public and controversial of human risk assessments. Approximately 21,800 tons of chemical waste were buried there, starting in the 1920s and ending in 1953. Then the number of housing units in the area increased rapidly, with 4,897 people living on the tract by 1970. Public concern about the waste became acute in 1978. Enormous amounts of emotion and resources were justifiably expended trying to determine the risk to residents due to their close proximity to the buried waste. On the basis of chromosomal aberration data, the 1980 Picciano pilot study suggested that residents might be at risk of cancer but the results were not definitive. Ambiguity arose because of a lack of controls and disagreement about extrapolation from chromosomal aberrations to cancer and birth defects (Culliton 1980). Benzene and chlorinated solvents that were known or suspected to be carcinogens were present in the waste. However, extensive chemical monitoring by the Environmental Protection Agency (EPA) suggested that the general area was safe for habitation and only a narrow region near the buried waste was significantly contaminated (Smith 1982a,b).

TABLE 13.1
Cancer Incidences for Residents of Love Canal as Compared to Expected Incidences

	Males			Females		
Cancer	Observed	Expected	95% CI	Observed	Expected	95% CI
(A) 1955–1965						
Liver	0	0.4	0–2	2	0.3	0–1[a]
Lymphomas	3	2.5	0–5	2	1.8	0–4
Leukemias	2	2.3	0–5	3	1.7	0–4
(B) 1966–1977						
Liver	2	0.6	0–2	0	0.4	0–2
Lymphomas	0	3.2	0–6	4	2.5	0–5
Leukemias	1	2.5	0–5	2	1.8	0–4
(C) 1955–1977						
Liver	2	1.0	0–3	2	0.7	0–2
Lymphomas	3	5.6	2–11	6	4.3	1–8
Leukemias	3	4.8	1–9	5	3.5	0–7

[a] Although seemingly significant, the linkage of the waste chemicals and liver cancer is unlikely as the two liver cancer victims lived in a Love Canal tract away from the waste location.

Because of their mode of action and toxicokinetics, benzene and chlorinated solvents would most likely cause liver cancer, lymphoma, or leukemia (Janeich et al. 1981). Although these contaminants were present in high concentrations at some locations, it was uncertain whether this resulted in significant exposure to Love Canal residents. A study of cancer rates at the site was conducted. Archived data were split into pre- and post-1966 census information because the quality of data from the New York Cancer Registry improved considerably in 1966. Data were then adjusted for age differences and tabulated separately for the sexes.

Table 13.1 provides documented cancer incidences for residents compared to expected incidences based on those for New York State (excluding New York City) for the same period (Janeich et al. 1981). Despite the perceived risks by residents and the Picciano report of elevated numbers of chromosomal aberrations, no statistically significant increases in cancer risk were detected for people living at Love Canal (Figure 13.1). The perceived risk was inconsistent with the actual risk of cancer from the wastes. (Actual risk being estimated as the difference in expected and observed cancer incidence rates.) Nevertheless, considerable amounts of money were spent moving many families away from the area.

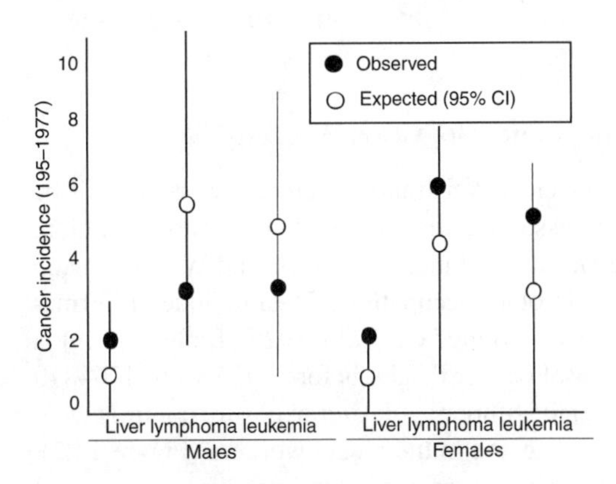

FIGURE 13.1 Cancer incidence rates (1955–1977) associated with the Love Canal community (●) compared to those expected for New York State (exclusive of New York City) (○). Vertical lines around the expected rates are 95% confidence intervals.

TABLE 13.2
Lung and Nasal Cancer in Nickel Industry Workers versus English & Welsh Workers in Other Occupations

Year of First Employment	Number of Men	Number of Person-Years[a]	Nasal Cancer Cases		Ratio of Rates	Lung Cancer Cases		Ratio of Rates
			Observed	Expected		Observed	Expected	
Before 1910	96	955.5	8	0.026	308	20	2.11	9.5
1910–1914	130	1060.5	20	0.023	870	29	2.75	10.5
1915–1919	87	915.0	6	0.015	400	13	2.29	5.7
1920–1924	250	1923.0	5	0.043	116	43	6.79	6.3
1925–1929	77	1136.0	0	0.014	—[b]	4	2.27	1.8
1930–1944	205	2945.0	0	0.022	—[b]	4	3.79	1.1

[a] Number of person-years at risk (1939–1966).

[b] Ratio of rate cannot be calculated because observed rate is 0.

Source: Modified from Tables I and II of Doll et al. (1970).

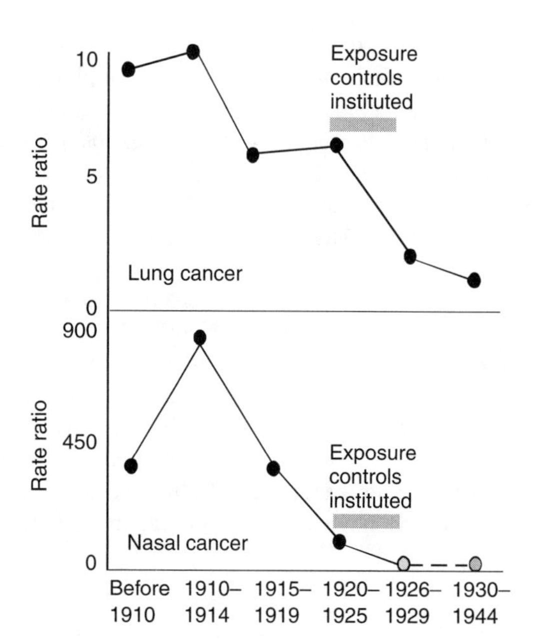

FIGURE 13.2 Rate ratios for lung and nasal cancers in nickel workers compared to English and Welsh workers in other occupations. The rate ratios for both cancers dropped for nickel workers as measures to reduce exposure via particulates were instituted beginning in approximately 1920.

Cancer Incidence Rate Ratio: Nasal and Lung Cancer in Nickel Workers

A classic study of job-related nasal and lung cancer in Welsh nickel refinery workers will be used to illustrate the application of rate ratios in assessing disease in a human subpopulation. Doll et al. (1970) documented the cancer incidence ratio of nickel workers, and Welshmen and Englishmen of similar ages who were employed in other occupations. Data included information gathered after exposure control measures were instituted ca. 1920–1925 (Table 13.2). It is immediately obvious from the rate ratios that nasal cancer deaths before 1925 were 116–870 times higher for nickel workers than for other men of similar age. After exposure controls were implemented, deaths from nasal cancer were not detected in the nickel workers (Figure 13.2). Similarly, lung cancer deaths were much higher in nickel workers before installation of control measures but dropped to levels similar to men in other occupations after exposure control. The risk ratios clearly demonstrated a heightened risk to nickel processing workers and a tremendous drop in this risk after exposure control measures were established.

Relative risk can be expressed as an odds ratio (OR) in case–control studies. Case–control studies identify individuals with the disease and then define an appropriate control group. The status of individuals in each group relative to some risk factor (e.g., exposure to a chemical) is then established and possible linkage assessed between the risk factor and disease.

Odds are simply the probability of having (p) the disease divided by the probability of not having ($1-p$) the disease. The number of disease cases (individuals) that were (a) or were not (b) exposed, and the number of control individuals free of the disease that were (c) or were not (d) exposed to the risk factor are used to estimate the OR (Ahlbom 1993, Sahai and Khurshid 1996):

$$\mathrm{OR} = \frac{a/b}{c/d} = \frac{ab}{bc}. \tag{13.10}$$

For illustration, let us assume that a disease was documented in 50 individuals: 40 cases were associated with individuals previously exposed to a toxicant (a) and 10 of them (b) were associated with people never exposed to the chemical. In a control or reference sample of 75 people with no signs of the disease, 20 had been exposed (c) and 55 (d) had no known exposure. The OR in this study would be (40)(55)/(10)(20) or 11. The OR suggests that exposure to this chemical influences proneness to the disease: an individual's odds of getting the disease are eleven times higher if they had been exposed to the chemical.

Approximate variance and confidence intervals for the OR can be generated from those for the natural logarithm of the OR (Ahlbom 1993, Sahai and Khurshid 1996),

$$\ln \text{OR} = \ln \frac{a}{N_1 - a} - \ln \frac{c}{N_0 - c} \quad (13.11)$$

where N_1 and N_0 are the number of cases (individuals) in the exposed and control groups, respectively:

$$\text{Variance of } \ln \text{OR} \approx \frac{1}{a} + \frac{1}{b} + \frac{1}{c} + \frac{1}{d}. \quad (13.12)$$

The confidence limits for ln OR can be approximated with the following equation:

$$\ln \text{of OR} \pm Z_{\alpha/2}\sqrt{\frac{1}{a} + \frac{1}{b} + \frac{1}{c} + \frac{1}{d}}. \quad (13.13)$$

As useful as these tools are for analyzing observational data, it is important to keep in mind the inherently compromised ability to infer causal association with the context from which the observations are derived. Although the difficulties in inferring causation from observational data may be obvious, we will continue to emphasize them as epidemiological studies may be particularly vulnerable to this flaw. As an example of the caution required in applying observational information to inferring linkage between a potential risk factor and disease, Taubes (1995) provides a thorough explanation of the difficulties of taking any action, including communicating risk to the public, based on such studies. He describes several cancer risk factors arising from valid and highly publicized, but inferentially weak, studies (Table 13.3).

TABLE 13.3
Examples of Weak Risk Factors for Human Cancer

Risk Factor	Relative Risk	Cancer Type
High cholesterol diet	1.65	Rectal cancer in men
Eating yogurt more than once/month	2	Ovarian cancer
Smoking more than 100 cigarettes/lifetime	1.2	Breast cancer
High fat diet	2	Breast cancer
Regular use of high alcohol mouthwash	1.5	Mouth cancer
Vasectomy	1.6	Prostate cancer
Drinking >3.3 L of (chlorinated?) fluid/day	2–4	Bladder cancer
Psychological stress at work	5.5	Colorectal cancer
Eating red meat five or more times/week	2.5	Colon cancer
On-job exposure to electromagnetic fields	1.38	Breast cancer
Smoking two packs of cigarettes daily	1.74	Fatal breast cancer

13.1.3 Foundation Models Describing Disease in Populations

Numerous models exist for describing disease in populations and potential relationships with etiological agents such as toxicants. Easily accessible textbooks such as those written by Ahlbom (1993), Marubini and Valsecchi (1995), and Sahai and Khurshid (1996) describe statistical models applicable to epidemiological data. Most models focus on human epidemiology and clinical studies but there are no inherent obstacles to their wider application in ecological epidemiology. Although most remain underutilized in ecotoxicology, they are applied more frequently in ecotoxicology each year. The most important are described below.

13.1.3.1 Accelerated Failure Time and Proportional Hazard Models

Accelerated failure time and proportional hazard models are used to estimate the magnitude of effects, test for the statistical significance of risk factors including contaminant exposure concentration, and to express these effects as probabilities or relative risks. This is done by modeling discrete events that occur through time such as time-to-death, time-to-develop cancer, time-to-disease onset, or time-to-symptom presentation (Figure 13.3).[2]

An explanation of the terms, survival, mortality, and hazard functions is needed before specific methods can be described. Let us begin by assuming an exposure time course with individuals dying during a period, T. The mortality of individuals within the population or cohort can be expressed by a probability density function, $f(t)$, or a cumulative distribution function, $F(t)$. The straightforward estimate of the cumulative mortality, $F(t)$, is the total number of individuals dead at time, t, divided by the total number of exposed individuals,

$$\hat{F}(t) = \frac{\text{Number dead}_t}{\text{Total number exposed}}. \tag{13.14}$$

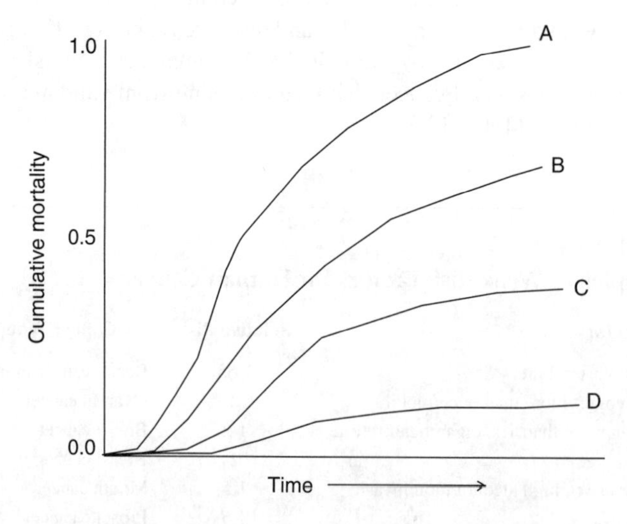

FIGURE 13.3 Data resulting from a time-to-event analysis. Several treatments (A–D) are studied relative to time-to-death. Cumulative mortality of individuals in each treatment is plotted against duration of exposure (time).

[2] See Section 9.2.3 of Chapter 9 for a similar discussion of survival time methods.

Equally intuitive, the cumulative survival function, $S(t)$, is the number of individuals surviving to t divided by the total number of individuals exposed to the toxicant or, expressed in terms of $F(t)$,

$$\hat{S}(t) = 1 - F(t). \tag{13.15}$$

The hazard rate or function, $h(t)$, is the rate of deaths occurring during a time interval for all individuals who had survived to the beginning of that interval. The hazard rate has also been called the force of mortality, instantaneous failure or mortality rate, or proneness to fail. It is definable in terms of $f(t)$ and $F(t)$, or $S(t)$:

$$\hat{h}(t) = \frac{f(t)}{1 - F(t)} = \frac{-1}{S(t)}\frac{\mathrm{d}S(t)}{\mathrm{d}t}. \tag{13.16}$$

The cumulative hazard function, $H(t)$, can be estimated from cumulative mortality, $F(t)$,

$$H(t) = \int_{-\infty}^{t} h(t)\mathrm{d}t = -\ln(1 - F(t)). \tag{13.17}$$

By simple rearrangement of Equation 13.17, cumulative mortality can be expressed in terms of $H(t)$,

$$F(t) = 1 - \mathrm{e}^{-H(t)}. \tag{13.18}$$

Please note that, although death is being used in this description of terms, other events may be analyzed with these methods. Events may be any "qualitative change that can be situated in time" (Allison 1995). The only restriction is that a discrete event occurs. Often the assumption is made that the event occurs only once (e.g., death). However, modifications to these methods allow accommodation for deviations from this condition (e.g., events such at giving birth that can occur more than once for an individual).

Life (actuarial) table and product-limit (Kaplan–Meier) methods are the two most commonly used nonparametric approaches for time-to-death analysis. Bootstrapping methods can also be applied (Manly 2002) but will not be discussed. None of these methods requires a specific form for the underlying survival distribution. Actuarial tables produce estimates of $S(t)$ for a fixed sequence of intervals (e.g., yearly age classes). Miller (1981) provides a basic discussion of computations for applying life tables in epidemiology. Life tables are discussed in more detail in Chapter 15. With the product-limit approach, the time intervals can vary in length. General details for this method are given below with additional information available from Cox and Oakes (1984), Marubini and Valsecchi (1995), and Miller (1981).

The product-limit estimate of $S(t)$ was originally described by Kaplan and Meier (1958) and an associated maximum likelihood method by Kalbfleisch and Prentice (1980). The notation here is that applied in widely used manuals of the SAS Institute (SAS 1989):

$$\hat{S}(t_i) = \prod_{j=1}^{i}\left(1 - \frac{d_j}{n_j}\right), \tag{13.19}$$

where $i =$ there are i failure times, t_i, $n_j =$ the number of individuals alive just before t_j, and $d_j =$ the number of individuals dying at time, t_j.

Although this product-limit estimate of $S(t)$ is appropriate for all times up to the end of the exposure (T), it must remain undefined for times after T if there were survivors. (The Π function in

Equation 13.19 is similar to the Σ function except the product is taken over i observations instead of the sum.)

The variance for the product–limit estimate is generated using Greenwood's formula,

$$\hat{\sigma}^2 = \hat{S}(t_i)^2 \sum_{j=1}^{i} \frac{d_j}{n_j s_j}, \tag{13.20}$$

where $s_j = n_j - d_j$. (Note that this equation is incorrect in Newman (1995) and Newman and Dixon (1996) because $\hat{S}(t_i)^2$ was unintentionally omitted from the formula.) Greenwood's estimate of variance reduces to Equation 13.21 for all times before T if there was no censoring before termination of the experiment, that is, survival times are known for all individuals dying before T (Dixon and Newman 1991):

$$\hat{\sigma}^2(t_j) = \frac{\hat{S}(t_j)[1 - \hat{S}(t_j)]}{N}, \tag{13.21}$$

where N = the total number of individuals exposed.

The confidence interval for these estimates can be generated using the square root of the variance estimated in Equation 13.20 or 13.21 in the following equation:

$$\text{CI} = \hat{S}(t_j) \pm Z_{\alpha/2}\hat{\sigma}_j. \tag{13.22}$$

These methods allow estimation of $S(t)$ for a group of individuals. Resulting survival curves for different classes (e.g., toxicant exposed versus unexposed) can be tested for equivalence with nonparametric methods. The log-rank and Wilcoxon rank tests check for evidence that the observed times-to-death for the various classes did not come from the same population.

Time-to-event data can also be analyzed with semiparametric and parametric methods. These semiparametric and fully parametric models are expressed either as proportional hazard or as accelerated failure time models. With proportional hazard models, the hazard of a reference group or type is used as a baseline hazard and the hazard of another group is scaled (made proportional) to that baseline hazard. For example, the hazard of contracting a liver cancer for fish living in a creosote-contaminated site might be made proportional to the baseline hazard for fish living in an uncontaminated site. A statement might be made that the hazard is ten times higher than that of the reference population. The hazards remain proportional by the same amount among classes regardless of the duration of exposure. Spurgeon et al. (2003) quantified survival using such a proportional hazard during their analysis of copper and cadmium exposure on earthworm demography. In contrast, accelerated failure models use functions that describe the change in ln time-to-death resulting from some change in covariates. As is true with proportional hazard models, covariates can be class variables such as site or continuous variables such as animal weight. Hazards do not necessarily remain proportional by the same amount through time with accelerated failure time models. Continuing the fish liver cancer example, the effect of creosote contamination on ln time-to-fatal cancer might be estimated with an accelerated failure model. The median time-to-fatal cancer appearance might be 230 days earlier than that of the reference population. Both forms of survival models are described below.

The general expression of a proportional hazard model is the following:

$$h(t, x_i) = e^{f(x_i)} h_0(t), \tag{13.23}$$

where $h(t, x_i)$ = the hazard at time, t, for a group or individual characterized by value x_i for the covariate x, $h_0(t)$ = the baseline hazard, and $e^{f(x_i)}$ = a function relating $h(t, x_i)$ to the baseline hazard.

The $f(x_i)$ is a function fitting a continuous variable such as animal weight or a class variable such as exposure status. A vector of coefficients and a matrix of covariates can be included if more than one covariate is required.

The proportional hazard models described above assume that a specific distribution fits the baseline hazard, $h_0(t)$ and that hazards among classes remain proportional regardless of time (t). But a specified distribution for the baseline hazard is not an essential feature of proportional hazard models. A semiparametric Cox proportional hazard model can be applied if the distribution was not apparent or was irrelevant to the needs of the study. This semiparametric model retains the assumption of proportional hazard but empirically applies a (Lehmann) set of functions to the baseline hazard. No specific model is needed to describe the baseline hazard. Cox proportional hazard models are commonly applied in epidemiology because, in many cases, the underlying distribution is unimportant and the relative hazards for the classes are more important to understand.

As mentioned, another form of survival model is the accelerated failure time model. In this case, the ln time-to-death is modified by $f(x_i)$:

$$\ln t_i = f(x_i) + \varepsilon_i, \tag{13.24}$$

where t_i = the time-to-death, $f(x_i)$ = a function that relates $\ln t_i$ to the covariate(s), and ε_i = the error term.

13.1.3.2 Binary Logistic Regression Model

Logistic regression of a binary response variable (e.g., disease present or not, or individual dead or alive) can be used for analyzing epidemiological data associated with contamination. It is one of the most common approaches for analyzing epidemiological data of human disease (SAS 1995). The resulting statistical model predicts the probability of a disease occurrence on the basis of values for risk factors:

$$\mathrm{Prob}(Y = 1 \mid X) = [1 + e^{-XB}]^{-1}. \tag{13.25}$$

The probability of a disease, i.e., a cancer ($Y = 1$) given by a vector of risk factors (X), is predicted with the logistic function ($P = [1 + e^{-XB}]^{-1}$) where XB is $B_0 + B_1X_1 + B_2X_2 + B_3X_3 + \cdots B_kX_k$. The B values are the regression coefficients for the effects of the potential risk factors or etiological agents (X values).

One can also express the logistic model directly in terms of the logarithm of the OR (Ahlbom 1993). In the following equation, the $\ln[P/(1-P)]$ transformation is the logit or "log odds" of the disease occurring:

$$\ln \frac{P}{1-P} = \alpha + XB. \tag{13.26}$$

Like results of the time-to-event models, the results of the logistic regression allow informed judgment about (1) potential agents that contribute to disease occurrence, (2) the probability of disease occurring, given the presence of some agent or risk factor, and (3) the contribution of the agent or risk factor to the chance of disease occurrence relative to those of other agents or risk factors.

13.2 DISEASE ASSOCIATION AND CAUSATION

13.2.1 HILL'S NINE ASPECTS OF DISEASE ASSOCIATION[3]

Emerging from the logic of causation (Section 13.1.1) are specific rules for enhancing belief in the association of noninfectious disease with chemical exposure. Sir Austin Hill (1965) provides one of the clearest and most relevant set. They are meant to be used together to enhance belief but they are not rigid hypotheses that, if rejected, lead to only one conclusion. Hill's nine aspects of disease association are the following:

- A strong association enhances belief.
- Consistency of an observed association enhances belief.
- Specificity of an association can enhance belief.
- Consistent temporal sequence (cause present as a precondition to seeing the disease or high incidence of the disease) can enhance belief.
- A biological gradient (higher amounts of an agent produce higher levels or chance of an effect) can enhance belief.
- Existence of a plausible biological mechanism can enhance belief.
- Coherence with our general knowledge base can enhance belief.
- Presence of experimental evidence can greatly enhance belief.
- Analogy drawn from another disease-causing agent can enhance belief.

Strength of association is very important in Hill's opinion. For instance, belief in an association between smoking and lung cancer is greatly increased if one sees an incidence of lung cancer-related deaths that is 30 times higher for heavy smokers than nonsmokers. Similarly, the very high incidence of imposex (imposition of male characters like a penis and vas deferens on female individuals) in populations of the snail, *Nucella lapillus*, from regions of coastal England that have high concentrations of the antifouling agent tributyltin (TBT) greatly reinforces belief that TBT causes imposex (Bryan and Gibbs 1991). However, it alone does not prove TBT is the causative agent.

Consistency of the association is also very important. Is there a higher incidence of lung cancer-related deaths in smokers versus nonsmokers regardless of ethnicity, sex, or cigarette brand? Here and elsewhere in this approach, it is important to be mindful of possible correlations with other factors. For example, in a study of correlations between smoking and cardiovascular disease, it would be useful to know if smokers tend to exercise less than nonsmokers. Lack of exercise, and not smoking per se, could be the reason for increased cardiovascular disease in smokers. For the TBT–imposex example, documentation of imposex in more than 40 species of neogastropods from TBT-contaminated locations around the world (Bryan and Gibbs 1991, Poloczanka and Ansell 1999) reinforces belief that TBT causes imposex in *N. lapillus* populations. TBT was a major component of marine paints used on boat and ship hulls. Therefore, TBT concentrations rise with increasing levels of boating and shipping activities in harbors, estuaries, and bays. Certainly, other possible etiological agents also increase with increasing boating and shipping traffic activities. However, TBT is also used in plastics production and belief would be fostered by the presence of imposex in neogastropod populations inhabiting aquatic systems influenced by the plastics industry.

Belief is fostered if the disease emerges from very specific conditions; for example, a specific toxicant is present in the air of a particular working environment in which a disease is seen in high incidence. A good example of disease emergence from specific conditions might be the extreme susceptibility of raptors and piscivorous birds to effects of dichlorodiphenyltrichloroethane (DDT),

[3] Hill's rules are applicable to other levels of biological organization as evidenced by their application again to communities in Chapter 22 (Box 22.3).

a chemical with a high capacity to biomagnify through trophic levels (Woodwell 1967). Reproductive failure of top avian predators, but not large numbers of bird species that feed lower in the trophic web, reinforced belief that high concentration of DDT due to biomagnification was the cause of avian reproductive damage.

Temporal sequence is obviously important. It is logically mandatory that an effect not appear before the cause is present (Last 1983). But the exact time of exposure is difficult to define with some contaminant-induced diseases. Exposure over a long period may be required before a disease is manifest: there may be a long latency period between exposure to a carcinogen and the appearance of cancer. Regardless, temporal succession is central to causation as discussed above and is extremely helpful if confirmed. Continuing the DDT example, belief was greatly enhanced by the close correspondence between the rapid decline in certain bird populations throughout the world and the onset of widespread use of DDT. Further, the recovery of certain bird populations such as osprey (*Pandion haliaetus*) (Spitzer et al. 1978) corresponded closely with the general ban on DDT use.

A clear biological gradient, such as an increased concentration of a toxicant correlated with an increase in effect, fosters belief; although, it might not be essential in order to assign an association between an etiological agent and a disease. For example, the observation that prevalence of *N. lapillus* imposex increases with proximity to TBT-contaminated harbors (Bryan and Gibbs 1991) reinforces the suggestion that TBT is the cause of imposex in this neogastropod. Other biological gradients can be more difficult to document. Some concentration–effect relationships have threshold concentrations below which there might not be a discernable effect. Also, accurate measurement or estimation of the right exposure concentration can be difficult and preclude establishment of a biological gradient.

Existence of a plausible biological mechanism can enhance belief as already discussed in our explorations of microexplanation. The discovery that DDT and DDE interfered with ATPase enzymes critical to calcium deposition in the shell gland and resulted in excessively fragile eggs for birds greatly enhanced our belief that DDT was the cause of reproductive failure in populations of birds (Simkiss 1996). As discussed in Chapter 12, the mechanism of differential bird predation on color morphs of peppered moths greatly fostered belief in the phenomenon of industrial melanism.

But biological plausibility is not always essential at initial phases of enhancing belief in disease association with some noninfectious agents. Limited knowledge may preclude ready assignment of biological plausibility in some cases.

Coherence with known facts is important regardless of our ability to identify a plausible mechanism. Baker et al. (1996) studied genetic damage to voles living around the Chernobyl reactor. They found extraordinary base-pair substitution rates for the mitochondrial cytochrome *b* gene: rates were orders of magnitude higher than expected. This very high mutation rate and the apparent viability of the vole populations seemed inconsistent with prevailing knowledge of mutation and cancer rates associated with radiation (see Hinton (1998) for details). High substitution rates in other genes are often associated with dysfunction. In fact, Baker et al. (1997) retracted their findings after realizing that an error had been made while reading associated DNA sequences. Belief was correctly hindered by a lack of coherence with existing biological knowledge. It caused the authors to go back and more carefully review their data.

Experimental results have high inferential strength if produced and interpreted competently. Such results can be very useful in assigning association or enhancing belief. Experimental information is rare or its generation is often unethical in human epidemiology. In those instances in which some "natural experiment" has occurred, the associated information can be applied in a very powerful way. This is much less of an obstacle with nonhuman species because experimental data supporting the accumulation of knowledge about disease association is much more abundant.

Analogy can increase confidence in an association. Our present knowledge of the developmental effects of thalidomide to humans makes us more likely to believe in the potential effects of other chemicals on embryonic development. The early discovery of DDT biomagnification to harmful levels allowed more rapid acceptance of similar biomagnification and effects of other contaminants such as polychlorinated biphenyls (PCBs) (Evans et al. 1991), toxophene (Evans et al. 1991), and dibenzo-*p*-dioxins and dibenzofurans (Broman et al. 1992, Rolff et al. 1993).

> Here then are nine different viewpoints from all of which we should study association before we cry causation. What I do not believe—and this has been suggested—is that we can usefully lay down some hard-and-fast rules of evidence that must be obeyed before we accept cause and effect. None of my nine viewpoints can bring indisputable evidence for or against the cause-and-effect hypothesis and none are required as *sine qua non*. What they do, with greater or less strength, is to help us to make up our minds on the fundamental question—is there any other way of explaining the set of facts before us, is there any other answer equally, or more, likely than cause and effect?
>
> **(Sir Austin Hill 1965)**

Hill's nine aspects of disease association are simply specific points consistent with the general characteristics of causation. Other sets of rules exist in addition to Hill's, including Evans's postulates (Evans 1976) and Fox's ecoepidemiology criteria (1991). Both Hill and Evans attempt to compensate a bit for the high dimensionality and less structured experimental context, which slows "the movement of thought toward belief" (Josephson and Josephson 1996). Regardless, only a subjective measurement is available for expression of confidence in the final assessment.

Box 13.2 Hockey Sticks, Mud, and Fish Livers: Hill's Nine Aspects of Disease Association

Let us illustrate the application of Hill's nine aspects of disease association with information gathered for hepatic cancer prevalence in English sole (*Pleuronectes vetulus*) populations of Puget Sound (Washington). During the 1970s, surveys began of bottom dwelling fish in bays, estuaries, and inlets of Puget Sound. Biological qualities (tissue lesions, demographic qualities, and biomarkers of effect and exposure) and chemical qualities (sediment and fish tissue PAH and other pollutant concentrations, and fluorescent aromatic compounds in bile) were measured at a series of contaminated and clean sites (Myers et al. 1990). Some supportive laboratory experiments were also conducted. This information was compared to the published literature, primarily the mammalian literature, and is evaluated here according to Hill's nine aspects.

1. *Strength of association enhances belief*: Horness et al. (1998) analyzed data for English sole inhabiting areas with sediment concentrations of polycyclic aromatic hydrocarbons ranging from 0 to 6300 ng/g dry weight of sediment. At the lowest concentrations, the prevalence of all types of liver lesions was extremely low but increased to approximately 60% at the most contaminated sites. Similarly, prevalence of neoplastic lesions was very high (ca. 10%) at the most contaminated sites relative to the clean sites.
2. *Consistency of an observed association enhances belief*: A consistent increase in lesion prevalence in English sole is seen at contaminated sites (Horness et al. 1998, Myers et al. 1990, 1994). This statement is valid for various lesion types reflecting a progression toward hepatic neoplasia including necrotic → proliferative → preneoplastic → neoplastic lesions. Necrotic lesions were thought to reflect cytotoxic effects of polycyclic aromatic hydrocarbons and their metabolites. Proliferative lesions reflect cell proliferation in compensation for this cytotoxicity. Neoplasia can eventually arise from the cells involved in this proliferation. Preneoplastic ("foci of cellular alteration") and neoplastic lesions eventually appear as altered cells increase in numbers and tumors become apparent in the liver. Close examination

of lesions by transmission electron microscopy revealed no evidence of viral infection (Myers et al. 1990), that is, the cancers did not seem to be caused by a viral agent.

3. *Specificity of an association can enhance belief*: Using logistic regression, hepatic lesion prevalences in English sole from a series of Pacific Coast sites were compared to a variety of contaminants in sediments (i.e., low molecular weight polycyclic aromatic hydrocarbons, high molecular weight polycyclic aromatic hydrocarbons, PCBs, DDT and its derivatives, chlordanes, dieldrin, hexachlorobenzene, and several metals) (Myer et al. 1994). The polycyclic aromatic hydrocarbons, PCBs, DDT and its derivatives, chlordane, and dieldrin were all found to be significant risk factors. This suggests that specificity may not be high for this association.
4. *Consistent temporal sequence can enhance belief*: The field studies could not address this aspect directly. This reflects the common challenge with chemical carcinogenesis because the ability to make a temporal linkage is hampered by the characteristically long period of cancer latency. Regardless, an appropriate temporal sequence was suggested by the observation that neoplastic lesions were not often seen in field-collected young sole but lesions thought to occur early in the progression toward neoplasia were found in these young sole (Myers et al. 1990, 1998). Laboratory experiments suggested that the exposure to polycyclic aromatic hydrocarbons resulted in lesions characteristic of early stages of a progression toward liver neoplasia (Myers et al. 1990).
5. *A biological gradient can enhance belief*: Figure 13.4 shows the consistent threshold ("hockey stick") exposure–effect curve for polycyclic aromatic hydrocarbons versus preneoplastic ("foci of cellular alteration") and neoplastic lesions in sole liver. A biological gradient with a threshold is suggested in the reports of Horness et al. (1998) and Myers et al. (1998).
6. *Existence of a plausible biological mechanism can enhance belief*: A clear mechanism for liver neoplasia appearance exists on the basis of P450-mediated production of free radicals, which form DNA adducts. Such adduct formation was clearly documented in English sole from contaminated sites and was correlated with lesions thought to lead to neoplasia (Myers et al. 1998).
7. *Coherence with general knowledge can enhance belief*: The results described for English sole are consistent with a wide literature concerning chemical carcinogenesis including specifics of polycyclic aromatic hydrocarbon induction of cancers in diverse animal models (Moore and Myers 1994). The lesion progression described for English sole closely parallels that described for rodents (Myers et al. 1990).

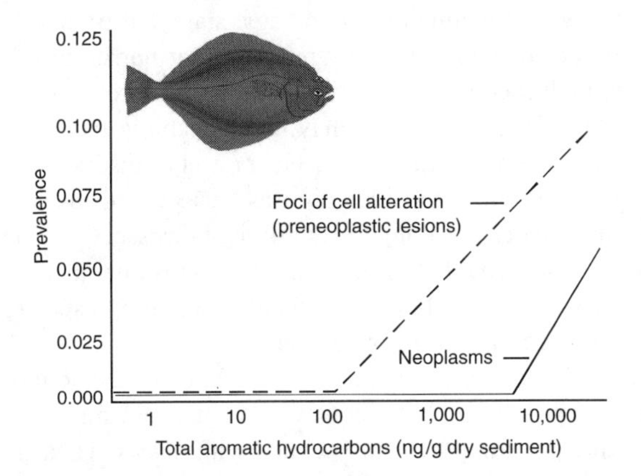

FIGURE 13.4 Prevalence of foci of cellular alteration (preneoplastic lesions) and neoplastic lesions in livers of English sole (*Pleuronectes vetulus*) sampled from areas with widely differing sediment concentrations of aromatic hydrocarbons. Note the clear threshold or "hockey stick" dose–response curve for both types of lesions. (This figure combines information from Figures 1A, B of Horness, B.H., et al., 1998.)

8. *Presence of experimental evidence can greatly enhance belief*: As mentioned above, laboratory exposure to polycyclic aromatic hydrocarbons resulted in lesions indicative of a progression toward hepatic neoplasia (Myers et al. 1998).
9. *Analogy drawn from another disease-causing agent can enhance belief*: The specific findings regarding hepatic tumor appearance in English sole following exposure to polycyclic aromatic hydrocarbons are consistent with a diverse literature on chemical carcinogenesis.

In summary, the information available for hepatic cancer in English sole from the Puget Sound area strongly supports its association with polycyclic aromatic hydrocarbon contamination. The only aspects that do not strongly support this association are the nonspecificity of hepatic cancer and the lack of a clearly documented temporal sequence because the time required between exposure and full expression of a neoplastic lesion is so long. However, laboratory studies with sole show a progression after exposure to lesions leading to cancer, and studies with other fishes more amenable to laboratory manipulation have documented a consistent temporal sequence from polycyclic aromatic hydrocarbon exposure and liver cancer (Moore and Myers 1994). It would be very difficult to defend a statement that the other carcinogens noted above were not contributing risk factors also. Regardless, the preponderance of evidence suggests that polycyclic aromatic hydrocarbons play a dominant role in determining the prevalence of hepatic cancer in English sole.

Clearly, some methods are better than others for extracting epidemiological information from populations. Also, some exploration methods are better than others for gathering evidence of disease association in populations. The strength of evidence hierarchy described below focuses on the inferential value of evidence emerging from different types of studies.

13.2.2 Strength of Evidence Hierarchy

All epidemiological evidences are not equally valuable for determining the true state of a cause–effect relationship. As we have already discussed, causal relationships are often defined as much by context as objective reality. For instance, factors leading to disease can be categorized on the basis of context (Last 1983) as either, predisposing, enabling, precipitating, or reinforcing factors. Predisposing factors create a situation conducive to disease appearance. For example, a chemical that causes immune suppression could be envisioned as a predisposing factor for the development of infectious disease. Enabling factors are those fostering or diminishing the expression of disease. They contribute by making the individual more or less inclined to be in some state that positively or negatively influences the chance of disease. For example, poverty-related, poor nutrition may allow disease to be manifested or, conversely, high income-related use of health services may be correlated with more rapid recovery from disease. Economic status may be an enabling factor for some human diseases. Precipitating factors are those associated with the clear onset of disease such as high exposure to a toxicant. Precipitating factors are often identified as "causes" of disease. Reinforcing factors tend to encourage the appearance of or prolong the duration of disease. Creation of a marginal habitat with multiple "stressors" during remediation in addition to a residual level of toxicant may reinforce the manifestation of disease in a population. Frequent foraging of a species in a contaminated environment may also be a reinforcing factor for disease.

On careful review of the characteristics of causation, it is clear that these overlapping distinctions are based partially on experimental context and partially on how closely a factor conforms to the qualities of a causative agent. For example, a precipitating factor associated with disease is easily identified as the cause if it was necessary—must be present—for the disease to occur. (In the context of disease causation, the terms, necessary and sufficient are given specific meanings. If something is necessary, it must always be present for the disease to be manifested. If something is sufficient, it will initiate or produce the disease. Its presence is sufficient for the disease to be expressed (Last 1983).) However, other factors may be equally necessary for the expression of disease and context could determine which of several necessary precipitating factors caused the disease. At the other end of the spectrum, an enabling factor would be difficult to identify as the cause because disease can occur in its absence or may not occur in its presence. It is neither necessary nor sufficient.

Strength of evidence can be categorized on the basis of the approach used to produce it (Green and Byar 1984). The weakest information emerges from anecdotal reports of disease. For example, contaminants were implicated in a die-off of seals (Dickson 1988). Dickson bolstered this implication by quoting Lies Vedder, a Dutch veterinary surgeon:

> We have never seen so many problems with bacterial infections with seal pups as this year; for example, I have not seen a single healthy umbilical scar. There seems to be no way of treating them, and the immune system does not seem able to cope; we have no proof that there is immune suppression, but there are certainly signs.

Although the implications could very easily be correct, this is fairly weak evidence for assigning causation. Better evidence is produced from case series without controls and even better evidence from case series with literature controls. The next best evidence comes from computer analyses of disease cases with consideration of disease expectations for unexposed individuals. Quality of evidence improves dramatically for the four approaches described next because they include formal control or reference cases. Case–control observational studies have already been discussed relative to ORs (e.g., Equation 13.10). In such studies, information is collected for disease cases and appropriate controls ("references" or "noncases") in the population for a specified interval. Even stronger inferences are derived from a series of studies based on historical control groups. Finally, more powerful evidence can be produced in clinical or experimental trials. These highly structured experiments are superior to any discussed to this point because of the ability to randomly assign observations to treatments and to control confounding factors. Opportunity for generation of this type of information is higher for nonhuman species than for humans. A single, controlled laboratory study produces more powerful information than methods discussed above and the associated information is only inferior to that emerging from a series of confirming, controlled, and randomized laboratory studies. Again, careful examination will show that this hierarchy of strength of evidence emerges directly from the guidelines for determining the strength of inferences about cause–effect relationships (Section 13.1.1).

Box 13.3 Belief Quantified

> Our assent ought to be regulated by the grounds of probability.
>
> **(John Locke 1690)**

The EPA recently applied qualitative tools like those described in this chapter to identify stressors in ecological risk assessments (e.g., EPA 2000). This guidance attempts to establish a standard approach but is incomplete because they do not provide a quantitative measure of belief on the basis of available evidence. Fortunately, Bayesian statistics provides a way to meet Locke's aspiration of assigning belief based on probability of a plausible explanation or outcome being true. A simple, fictitious example of Bayesian Belief Networks (BBN) demonstrates this point.

A population of an endangered fish species disappears within the mixing zone of a discharge. A legal action ensues and the pivotal question becomes, "Did the discharge cause the local extinction?" The defendant counters that a protozoan disease caused the local extinction.

Evidence must prove culpability beyond a reasonable doubt for this case so the plaintiff develops a "balance of probabilities" strategy, which based on judicial history (Cohen 1977), requires an evidence threshold of roughly .70–.90, that is, the evidence indicates a 70–90% chance that the discharge was the cause.

A simple BBN is developed from available evidence (Figure 13.5). Relative to the protozoan disease hypothesis, a particular reservoir host density above a certain threshold greatly increases the likelihood of the disease agent being present. The probability of protozoan presence jumps from a low .02 to .98 if the reservoir host population is present above the threshold. If the protozoan is present, the likelihood of a disease outbreak is .10. Relative to the discharge causal hypothesis, the probability of exceeding a concentration threshold (1 ppm) for longer than 3 days is estimated to be .75 based on a frequency distribution of concentrations measured during past discharge monitoring. This exposure level and duration is predicted to kill 80% or more of exposed individuals. According to demographic studies of this species, the probability of extinction is extremely high (.95) at that level of mortality.

The probabilities associated with the possible outcomes can be calculated using this information and the chain rule. According to the chain rule, the joint probability for an outcome distribution is the product of all specified potential states (Figure 13.6). For example,

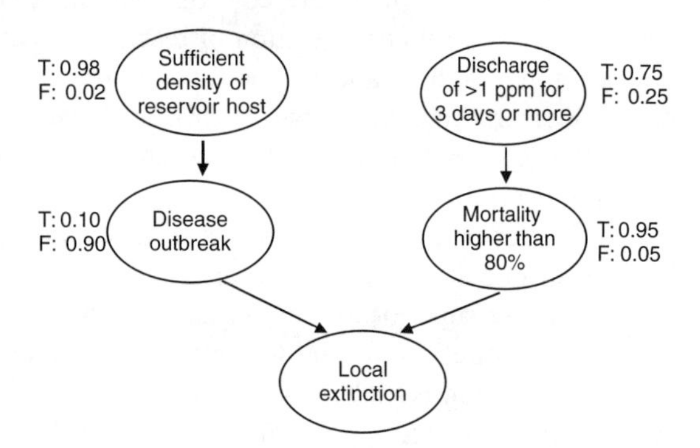

FIGURE 13.5 The causal network for the local extinction of the endangered fish species population below a discharge.

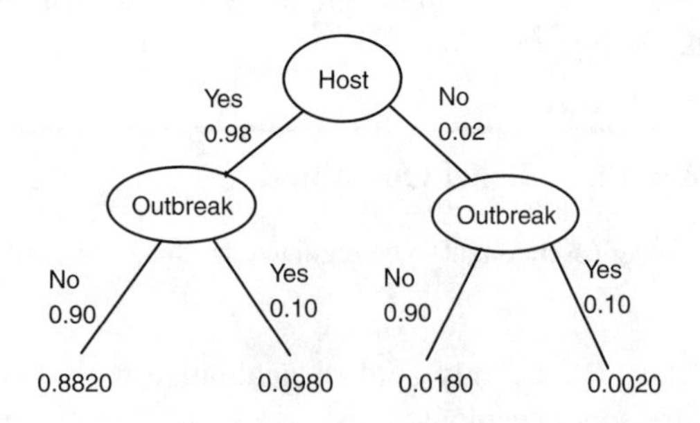

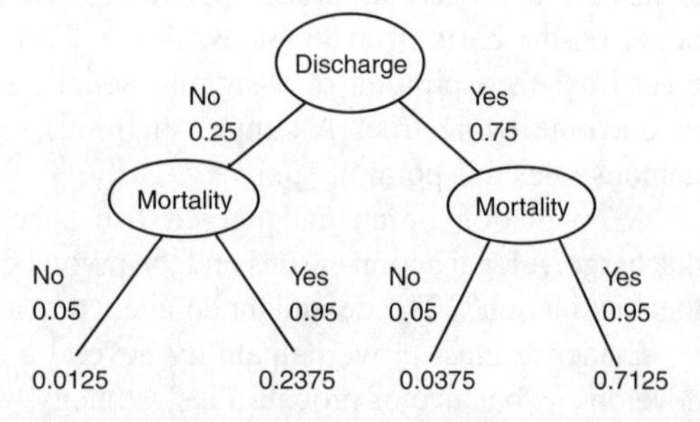

FIGURE 13.6 The decision diagram for the causal network drawn in Figure 13.5.

the probability of the state "Yes" for high host densities multiplied by the probability of the state "Yes" for outbreak is the joint probability of "Yes," i.e., the probability of "Yes" for the disease causal hypothesis. One can apply this chain rule to all possible state combinations. The results of doing these calculations for the protozoan disease and discharge hypotheses are given at the top and bottom of Figure 13.6, respectively. Notice that the outcome probabilities for each hypothesis must sum to 1, that is, $.8820 + .0980 + .0180 + .0020 = 1$.

What is the probability that the local extinction was due to the protozoan? Two of the four outcomes for this causal hypothesis involve local extinction ("Yes," "Yes," and "No," "Yes") so the probability of local extinction due to the protozoan is $.0980 + .0020$ or .1000.

What is the probability that the local extinction was due to the company's discharge? Again, the sum of the probabilities for the two outcomes resulting in local extinction is .95.

The probabilities associated with each hypothesis quantify plausibility, allowing one's level of belief to be quantified on the basis of available evidence. Given a local extinction, the odds of the discharge being the cause versus the protozoan disease is 9.5:1. This exceeds the usual tipping point for winning a preponderance of evidence-based trial.

13.3 INFECTIOUS DISEASE AND TOXICANT-EXPOSED POPULATIONS

Odum (1971) described the principle of instant pathogen in which a sudden outbreak of disease is induced by either, an introduction of rapidly reproducing species into a system lacking the ability to counterbalance its progress, or a rapid change of the environment, which tips the balance to favor the pathogen. Later, Odum (1985) listed a series of ecosystem alterations anticipated with chemical stress, including an increase in "negative interactions" (i.e., parasitism and disease). This theme of increased infectious disease with increased stress or pollution is repeated many times throughout the ecotoxicology literature.

The paradigm of the infectious disease triad provides a more comprehensive view of the influence of pollution on changes in infectious disease. As Figure 13.7 suggests, any environmental change can influence the outcome of the disease process. The final balance between health and disease is a result of environmental influences on both partners (host and infectious agent) in the disease process. As an example, summer oyster (*Crassostrea virginica*) mortality in Delaware Bay due to the sporozoan parasite *Haplosporidium nelsoni* depends on temperatures experienced during the preceding winter (Ford and Haskin 1982).

The triad paradigm for disease extends to environmental factors such as anthropogenic agents. TBT oxide decreased oyster (*C. virginica*) resistance to the protozoan *Perkinus marinus* (Fisher et al. 1999). Copper decreased catfish (*Ictalurus punctatus*) resistance to the protozoan *Ichthyophthirius multifiliis* (Ewing et al. 1982). Stretching the example of temperature's effects on disease outcome to the extreme case of thermal pollution, alligators (*Alligator mississippiensis*) inhabiting thermal effluents from nuclear reactors had diminished resistance to infection by the bacterium *Aeromonas hydrophilia* (Glassman and Bennett 1978). Obviously, chemicals or physical agents compromising immunological competence such as some pesticides (Bennett and Wolke 1987, Grant and Merhle 1973) will influence the disease process (Anderson 1990).

Although the focus in the above discussion was the increase in likelihood of disease due to pollution, the infectious disease triad paradigm implies that changes in the environment can also tip the balance in favor of the host (Figure 13.7). For example, although high concentrations of some metals increased bacterial infection in striped bass (*Morone saxatilis*) by *Flexibacter columnaris*,

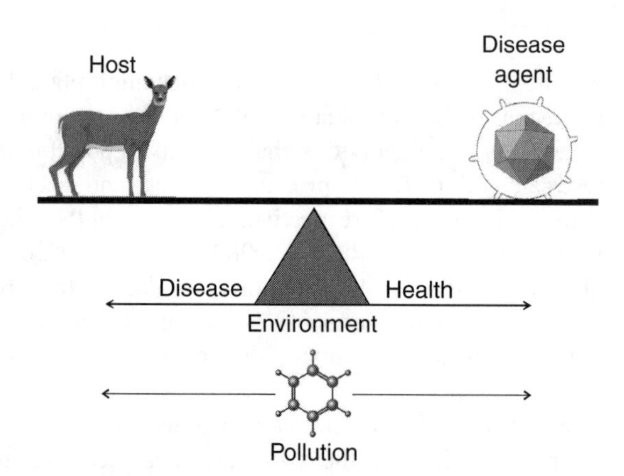

FIGURE 13.7 The infectious disease triad paradigm in which environmental factors (the fulcrum here) can shift the balance of health–disease processes to favor the host or infectious agent. Pollutants, as components of the environment can also shift the fulcrum to favor the host or infectious agent. Together, natural and anthropogenic components in the environment determine the final balance in the health–disease process.

elevated levels of other metals decreased infection (MacFarlane et al. 1986). Copper and zinc reduced the longevity and infectivity of cercariae of the trematode, *Echinoparypium recurvatum*, suggesting a potential impact on parasite transmission (Evans 1982a). These metals also reduced shedding of *Notocotylus attenuatus* cercariae from the snail host (*Lymnaea stagnalis*) (Evans 1982b). Mortality of trout (*Salmo gairdneri*) from *Aeromonas hydrophila* infection was lower in individuals injected with low doses of PCBs than in controls, perhaps due to increased blood leucocrits (Snarski 1982). In contrast to immunosuppression by contaminants, the trout response to infection seems to be a case of heightened immunological competence due to PCB injection before disease challenge. In these same studies, preexposure to copper did not influence disease outcome.

To summarize, pollutants are components of the complex milieu in which hosts and infectious agents interact. As such, they can influence the infectious disease process to favor the host or disease agent. This conclusion is consistent with the infectious disease triad paradigm and is more inclusive than Odum's (1971) principle of the instant pathogen. Regardless, such environmental influence on infectious disease is invoked primarily in speculations about infectious disease outbreaks (e.g., Dickson 1988, Sarokin and Schulkin 1992). More extensive ecotoxicological research into contaminant influences on infectious disease is needed.

13.4 DIFFERENCES IN SENSITIVITY WITHIN AND AMONG POPULATIONS

Differences in risk from contaminants exist among individuals in a population. As we will discuss in Chapter 18, some of these differences are related to the genetic qualities of individuals. Others are associated with changes occurring in an individual's life cycle; for example, younger or older individuals may be more sensitive to a particular contaminant. Still others are associated with interactions between genetic and environmental qualities (Chapter 18). These differences can influence population characteristics and fate as described in the next few chapters. Because populations often occupy heterogeneous landscapes, differences in risk to contaminants may also occur in a spatial context. In such cases, keystone habitats may play a critical role in determining the nature and persistence of the population. Ignoring these differences in populations can result in poor prediction of contaminant effects. The goal of the next several chapters is to provide ample understanding so that predictions of population effects can be made in such situations. Many of the methods described in this chapter can be combined with this understanding to describe and predict population effects of contaminants.

13.5 SUMMARY

In this chapter, we explored concepts and metrics applied in epidemiology, the science of disease in populations. A brief review of the topics discussed should reveal a tendency to focus on individuals within populations and to emphasize risk to individuals. In the next few chapters, we will explore methods designed specifically to project qualities of populations.

Logical and mathematical constructs were described for teasing out causation or strong association from epidemiological data. Respectively, Hill's aspects of disease association and the strength of evidence hierarchy were examined as useful means to infer disease association and to judge the inferential strength of evidence from diverse types of studies. The disease triad paradigm was selected as the most inclusive for understanding the influence of pollutants on the manifestation and outcome of infectious disease. Finally, we briefly described the factors that make individuals within a population more or less prone to the effects of pollutants.

13.5.1 Summary of Foundation Concepts and Paradigms

- The descriptive nature of much epidemiological information results in relatively weak inferences.
- Strength of inferences can be enhanced by logical rules such as Hill's nine aspects of disease association and the value of evidence supporting inferences judged by the strength of evidence hierarchy.
- Mechanistic knowledge is not required for inferences about disease association but its existence greatly enhances inferential strength.
- Disease prevalence and incidence are sound metrics of disease dynamics in populations. Differences in these metrics within and among populations are useful in describing disease in populations.
- Some dose–effect relationships have thresholds while others do not.
- Binary logistic, accelerated failure time, and proportional hazard models are adequate to describe most toxicant exposure–disease associations.
- The likelihood of disease is a function of the interaction among the host, the disease agent, and the environment. Pollutants, as part of the environmental milieu in which the host and disease agent interact, can modify the likelihood of disease.
- Toxicants can weaken individuals (e.g., immunosuppression), resulting in an increase in infectious disease in the population. However, enhanced immunological competence is also possible due to pollutant exposure, resulting in a decrease in infection.
- Toxicants can increase or decrease parasite load as a complex function of relative toxicant effects to the host, parasite, or another host population.
- Individuals differ genetically relative to their risk of an adverse effect after toxicant exposure.
- Nongenetic risk factors also vary among individuals, resulting in differences in risk upon exposure.
- Populations can differ in their responses to toxicant exposure due to the differences in nongenetic and genetic risk factors of individuals in each population.
- Population differences in risk factors can lead to a keystone habitat or "keystone population" context of effect in a landscape with a nonrandom distribution of contamination.
- Individuals can vary in risk of disease at different stages of their lives.
- Correlations of factors with disease within and among populations can result in incorrect inferences about association or causation.

REFERENCES

Ahlbom, A., *Biostatistics for Epidemiologists*, CRC Press/Lewis Publishers, Boca Raton, FL, 1993.

Allison, P.D., *Survival Analysis Using the SAS System: A Practical Guide*, SAS Institute, Inc., Cary, NC, 1995.

Anderson, D.P., Immunological indicators: Effects of environmental stress on immune protection and disease outbreaks, *Amer. Fish. Soc. Sympos.*, 8, 38–50, 1990.

Baker, R.J., Van den Bussche, R.A., Wright, A.J., Wiggins, L.E., Hamilton, M.J., Reat, E.P., Smith, M.H., et al., High levels of genetic change in rodents of Chernobyl, *Nature*, 380, 707–708, 1996.

Baker, R.J., Van den Bussche, R.A., Wright, A.J., Wiggins, L.E., Hamilton, M.J., Reat, E.P., Smith, M.H., et al., High levels of genetic change in rodents of Chernobyl—Retraction, *Nature*, 390, 100, 1997.

Bennet, R.O. and Wolke, R.E., The effect of sublethal endrin exposure on rainbow trout, *Salmo gairdneri* Richardson. I. Evaluation of serum cortisol concentrations and immune responsiveness, *J. Fish Biol.*, 31, 375–385, 1987.

Broman, D., Näf, C., Rolff, C., Zebühr, Y., Fry, B., and Hobbie, J., Using ratios of stable nitrogen to estimate bioaccumulation and flux of polychlorinated dibenzo-*p*-dioxins (PCDDs) and dibenzofurans (PCDFs) in two food chains from the Northern Baltic, *Environ. Toxicol. Chem.*, 11, 331–345, 1992.

Bryan, G.W. and Gibbs, P.E., Impact of low concentrations of tributyltin (TBT) on marine organisms: A review, In *Metal Ecotoxicology. Concepts & Applications*, Newman, M.C. and McIntosh, A.W. (eds.), CRC Press/Lewis Publishers, Chelsea, MI, 1991, pp. 323–361.

Cohen, L.J., *The Probable and the Provable*, Clarendon Press, Oxford, UK, 1977.

Cox, D.R. and Oakes, D., *Analysis of Survival Data*, Chapman & Hall, London, UK, 1984.

Culliton, B.J., Continuing confusion over Love Canal, *Science*, 209, 1002–1003, 1980.

Dickson, D., Mystery disease strikes Europe's seals, *Science*, 241, 893–895, 1988.

Dixon, P.M. and Newman, M.C., Analyzing toxicity data using statistical models for time-to-death: An introduction, In *Metal Ecotoxicology. Concepts & Applications*, Newman, M.C. and McIntosh, A.W. (eds.), CRC Press/Lewis Publishers, Chelsea, MI, 1991, pp. 207–242.

Doll, R., Morgan, L.G., and Speizer, F.E., Cancers of the lung and nasal sinuses in nickel workers, *Br. J. Cancer*, 24, 624–632, 1970.

EPA, *Stressor Identification Guidance Document*, EPA/822/B-00/025, December 2000. NTIS, Washington, D.C., 2000.

Evans, A.S., Causation and disease: The Henle–Koch postulates revisited, *Yale J. Biol. Med.*, 49, 175–195, 1976.

Evans, M.S., Noguchi, G.E., and Rice, C.P., The biomagnification of polychlorinated biphenyls, toxaphene, and DDT compounds in a Lake Michigan offshore food web, *Arch. Environ. Contam. Toxicol.*, 20, 87–93, 1991.

Evans, N.A., Effect of copper and zinc upon the survival and infectivity of *Echinoparyphium recurvatum* cercariae, *Parasitology*, 85, 295–303, 1982a.

Evans, N.A., Effects of copper and zinc on the life cycle of *Notocotylus attenuatus* (Digenea: Notocotylidae), *Int. J. Parasit.*, 12, 363–369, 1982b.

Ewing, M.S., Ewing, S.A., and Zimmer, M.A., Sublethal copper stress and susceptibility of channel catfish to experimental infections with *Ichthyophthirius multifiliis*, *Bull. Environm. Contam. Toxicol.*, 28, 676–681, 1982.

Fisher, W.S., Oliver, L.M., Walker, W.W., Manning, C.S., and Lytle, T.F., Decreased resistance of eastern oysters (*Crassostrea virginica*) to a protozoan pathogen (*Perkinsus marinus*) after sublethal exposure to tributyltin oxide, *Mar. Environ. Res.*, 47, 185–201, 1999.

Ford, S.E. and Haskin, H.H., History and epizootiology of *Haplosporidium nelsoni* (MSX), an oyster pathogen in Delaware Bay, 1957–1980, *J. Invert. Pathol.*, 40, 118–141, 1982.

Fox, G.A., Practical causal inference for ecoepidemiologists, *J. Toxicol. Environ. Health*, 33, 359–373, 1991.

Glassman, A.B. and Bennett, C.E., Responses of the alligator to infection and thermal stress, In *Energy and Environmental Stress in Aquatic Systems*, Thorp, J.H. and Gibbons, J.W. (eds.), NTIS, Springfield, VA, pp. 691–702, 1978.

Grant, B.F. and Mehrle, P.M., Endrin toxicosis in rainbow trout (*Salmo gairdneri*), *J. Fish. Res. Bd. Can.*, 30, 31–40, 1973.

Green, S.B. and Byar, D.P., Using observational data from registries to compare treatments: The fallacy of omnimetrics, *Statist. Med.*, 3, 361–370, 1984.

Harré, R., *The Philosophies of Science. In Introductory Survey*, Oxford University Press, Oxford, UK, 1972.

Hill, A.B., The environment and disease: Association or causation? *Proc. R. Soc. Med.*, 58, 295–300, 1965.

Hinton, T.G., Estimating human and ecological risks from exposure to radiation, In *Risk Assessment. Logic and Measurement*, Newman, M.C. and Strojan, C.L. (eds.), CRC Press/Ann Arbor Press, Chelsea, MI, 1998, pp. 143–166.

Horness, B.H., Lomax, D.P., Johnson, L.L., Myers, M.S., Pierce, S.M., and Collier, T.K., Sediment quality thresholds: Estimates from hockey stick regression of liver lesion prevalence in English sole (*Pleuronectes vetulus*), *Environ. Toxicol. Chem.*, 17, 872–882, 1998.

Janeich, D.T., Burnett, W.S., Feck, G., Hoff, M., Nasca, P., Polednak, A.P., Greenwald, P., and Vianna, N., Cancer incidence in the Love Canal area, *Science*, 212, 1404–1407, 1981.

Jensen, F.V., *Bayesian Networks and Decision Graphs*, Springer-Verlag, New York, 2001.

Josephson, J.R. and Josephson, S.G., *Abductive Inference. Computation, Philosophy, Technology*, Cambridge University Press, Cambridge, UK, 1996.

Kalbfleish, J.D. and Prentice, R.L., *The Statistical Analysis of Failure Time Data*, John Wiley & Sons, New York, 1980.

Kant, I., *Critique of Pure Reason*, J.M. Dent, London, UK, 1934.

Kaplan, E.L. and Meier, P., Nonparametric estimation from incomplete observations, *J. Am. Statist. Assoc.*, 53, 457–481, 1958.

Last, J.M., *A Dictionary of Epidemiology*, Oxford University Press, Oxford, UK, 1983.

Locke, J., *An Essay Concerning Human Understanding*, Dover Publications, New York, 1690.

MacFarlane, R.D., Bullock, G.L., and McLaughlin, J.J.A., Effects of five metals on susceptibility of striped bass to *Flexibacter columnaris*. *Trans. Amer. Fish. Soc.*, 115, 227–231, 1986.

Manly, B.F.J., Time-to-event analyses in ecology, In *Risk Assessment with Time-to-Event Models*, CRC Press, Boca Raton, FL, 2002, pp. 121–140.

Marubini, E. and Valsecchi, M.G., *Analyzing Survival Data from Clinical Trials and Observational Studies*, John Wiley & Sons Ltd., Chichester, UK, 1995.

Miller, R.G., Jr., *Survival Analysis*, John Wiley & Sons Ltd., Chichester, UK, 1981.

Moore, M.J. and Myers, M.S., Pathobiology of chemical-associated neoplasia in fish, In *Aquatic Toxicology. Molecular, Biochemical and Cellular Perspectives*, Malins, D.C. and Ostrander, G.K. (eds.), CRC Press/Lewis Publishers, Boca Raton, FL, 1994, pp. 327–386.

Myers, M.S., Johnson, L.L., Hom, T., Collier, T.K., Stein, J.E., and Varanasi, U., Toxicopathic hepatic lesions in subadult English sole (*Pleuronectes vetulus*) from Puget Sound, Washington, USA: Relationships with other biomarkers of contaminant exposure, *Mar. Environ. Res.*, 45, 47–67, 1998.

Myers, M.S., Landahl, J.T., Krahn, M.M., Johnson, L.L., and McCain, B.B., Overview of studies on liver carcinogenesis in English sole from Puget Sound; Evidence for a xenobiotic chemical etiology. I: Pathology and epizootiology, *Sci. Total Environ.*, 94, 33–50, 1990.

Myers, M.S., Stehr, C., Olson, O.P., Johnson, L.L., McCain, B.B., Chan, S.-L., and Varanasi, U., Relationships between toxicopathic hepatic lesions and exposure to chemical contaminants in English sole (*Pleuronectes vetulus*), starry flounder (*Platichthys stellatus*), and white croaker (*Genyonemus lineatus*) from selected marine sites on the Pacific Coast, USA, *Environ. Health Perspect.*, 102, 200–215, 1994.

Newman, M.C., *Quantitative Methods in Aquatic Ecotoxicology*, CRC Press/Lewis Publishers, Boca Raton, FL, 1995.

Newman, M.C. and Dixon, P.M. 1996. Ecologically meaningful estimates of lethal effect in individuals, In *Ecotoxicology. A Hierarchical Treatment*, Newman, M.C. and Jagoe, C.H. (eds.), CRC Press/Lewis Publishers, Boca Raton, FL, 1996, pp. 225–253.

Odum, E.P., *Fundamentals of Ecology*, W.B. Saunders Co., Philadelphia, PA, 1971.

Odum, E.P., Trends expected in stressed ecosystems, *Bioscience*, 35, 419–422, 1985.

Pearl, J., *Causality*, Cambridge University Press, Cambridge, UK, 2000.

Piattelli-Palmarini, M., *Inevitable Illusions*, John Wiley & Sons, New York, 1994.

Poloczanska, E.S. and Ansell, A.D., Imposex in the whelks *Buccinum undatum* and *Neptunea antiqua* from the west coast of Scotland. *Mar. Environ. Res.*, 47, 203–212, 1999.

Popper, K.A., *The Logic of Scientific Discovery*. Routledge, New York, 1959.

Popper, K.A., *Conjectures and Refutations: The Growth of Scientific Knowledge.* Harper & Row Publishers, London, UK, 1965.

Rench, J.D., Environmental epidemiology, In *Basic Environmental Toxicology*, Cockerham, L.G. and Shane, B.S. (eds.), CRC Press/Lewis Publishers, Boca Raton, FL, 1994, pp. 477–499.

Rolff, C., Broman, D., Näf, C., and Zebühr, Y., Potential biomagnification of PCDD/Fs—New possibilities for quantitative assessment using stable isotope trophic position, *Chemosphere*, 27, 461–468, 1993.

Sahai, H. and Khurshid, A., *Statistics in Epidemiology. Methods, Techniques and Application*, CRC Press, Boca Raton, FL, 1996.

Sarokin, D. and Schulkin, J., The role of pollution in large-scale population disturbances. Part 1: Aquatic populations, *Environ. Sci. Technol.*, 26, 1476–1484, 1992.

SAS Institute, Inc., *SAS/STAT User's Guide, Version 6*, 4th ed., Vol. 2, SAS Institute, Inc., Cary, NC, 1989.

SAS Institute, Inc., *Logistic Regression Examples Using the SAS System*, Version 6, 1st ed., SAS Institute Inc., Cary, NC, 1995.

Simkiss, K., Ecotoxicants at the cell-membrane barrier, In *Ecotoxicology. A Hierarchical Treatment*, Newman, M.C. and Jagoe, C.H. (eds.), CRC Press/Lewis Publishers, Boca Raton, FL, 1996, pp. 59–83.

Smith, R.J., Love Canal study attracts criticism, *Science*, 217, 714–715, 1982a.

Smith, R.J., The risks of living near Love Canal, *Science*, 217, 808–811, 1982b.

Snarski, V.M., The response of rainbow trout *Salmo gairdneri* to *Aeromonas hydrophila* after sublethal exposures to PCB and copper, *Environ. Pollut. Ser. A*, 28, 219–232, 1982.

Spitzer, P.R., Risebrough, R.W., Walker, W., II, Hernandez, R., Poole, A., Puleston, D., and Nisbet, I.C.T., Productivity of ospreys in Connecticut-Long Island increase as DDE residues decline, *Science*, 202, 333–335, 1978.

Spurgeon, D.J., Svendsen, C., Weeks, J.M., Hankard, P.K., Stubberud, H.E., and Kammenga, J.E., Quantifying copper and cadmium impacts on intrinsic rate of population increase in the terrestrial oligochaete *Lumbricus rubellus*, *Environ. Toxicol. Chem.*, 22, 1465–1472, 2003.

Suter, G.W., II, *Ecological Risk Assessment*, CRC Press/Lewis Publishers, Boca Raton, FL, 1993.

Taubes, G., Epidemiology faces its limits, *Science*, 269, 164–169, 1995.

Tversky, A. and Kahneman, D., Advances in prospect theory: Cumulative representation of uncertainty, *J. Risk and Uncertainty*, 5, 297–323, 1992.

Woodwell, G.M., Toxic substances and ecological cycles, *Sci. Am.*, 216, 24–31, 1967.

14 Toxicants and Simple Population Models

14.1 TOXICANTS EFFECTS ON POPULATION SIZE AND DYNAMICS

14.1.1 THE POPULATION-BASED PARADIGM FOR ECOLOGICAL RISK

> The greatest scandal of philosophy is that, while around us the world of nature perishes . . . philosophers continue to talk, sometimes cleverly and sometimes not, about the question of whether this world exists.
>
> **(Popper 1972)**

Every scientific discipline is built around a collection of conceptual and methodological paradigms that are "revealed in its textbooks, lectures, and laboratory exercises" (Kuhn 1962). These paradigms define what the discipline encompasses—and what it does not. During professional training, a scientist also learns the rules by which business within his or her discipline is to be conducted. A scientist understands that there is a "hard core" of irrefutable beliefs that are not to be questioned and a "protective belt of auxiliary hypotheses" that are actively tested and enriched (Lakatos 1970). To venture outside the accepted borders of a discipline or to question a core paradigm is courting professional censure. Yet, when a core paradigm fails too obviously and another is available to take its place, significant shifts do occur in a discipline. Oddly enough, a clearly inadequate paradigm will remain central in a discipline if a better one is not available to replace it (Braithwaite 1983). Because scientists are human, the shift from one core paradigm to another is characterized by as much discomfort and bickering as excitement.

Although originating from illogical roots, the dogmatic tendency to cling to a paradigm does have a positive consequence (Popper 1972). Any group of scientists who tends to drop a central paradigm too quickly will experience many disappointments and false starts. A key character of any scientific discipline is a healthy, not pathological, tenacity of central paradigms.

In writing this and several of the remaining chapters, we are caught between the risk of being censured for discussing topics out of balance with their perceived importance in ecotoxicology and the conviction that, until recently, ecotoxicologists have been dawdling in accepting a useful, new core paradigm for evaluating ecological effects. Much like the negligent philosophers described in the quote above, ecotoxicologists were enjoying the exploration of the innumerable details of their protective belt of auxiliary paradigms and hypotheses while important questions remained poorly addressed by an individual-based paradigm. Fortunately, ecotoxicologists are now focusing much more on population-based metrics of effect. It is obvious that prediction of population consequences cannot be adequately done by simply modifying the present individual-based metrics. Instead of adding to the protective belt around this collapsing paradigm, ecotoxicologists are now producing more population-level metrics of effects.

What is needed is even more effort to clearly articulate a new population-based paradigm. Also, nontraditional methods must be explored carefully in order to generate a belt of auxiliary hypotheses around this new population-based paradigm. Since the early 1980s (e.g., Moriarty 1983), the argument for population-based methods taking precedence over individual-based metrics of effect has been voiced with increasing frequency in scientific publications, regulatory documents, and federal legislation. Recently, Forbes and Calow (1999) reiterated this theme and provided more evidence to support it by comparing individual- and population-based metrics for ecological impact assessment. Also, individual-based models for populations (e.g., DeAngelis and Gross 1992) have emerged to bridge the gap between individual- and population-based metrics for judging ecological risk.

14.1.2 Evidence of the Need for the Population-Based Paradigm for Risk

The quotes below are chosen to reflect the transition taking place in our thinking about ecotoxicological risk assessment. Early quotes point to the underutilization of population-based metrics of toxicant effect. The need for more population-based predictions is then expressed in a series of regulation-oriented publications. Finally, statements made during the past few years show that methods are now available and are being applied with increasing frequency to address population-level questions.

> Ecologists have used the life table since its introduction by Birch (1948) to assess survival, fecundity, and growth rate of populations under various environmental conditions. While it has proved a useful tool in analyzing the dynamics of natural populations, the life table approach has not, with few exceptions . . . , been used as a toxicity bioassay.
>
> **(Daniels and Allan 1981)**

> There is an enormous disparity between the types of data available for assessment and the types of responses of ultimate interest. The toxicological data usually have been obtained from short-term toxicity tests performed using standard protocols and test species. In contrast, the effects of concern to ecologists performing assessments are those of long-term exposures on the persistence, abundance, and/or production of populations.
>
> **(Barnthouse 1987)**

> Environmental policy decision makers have shifted emphasis from physiological, individual-level to population-level impacts of human activities. This shift has, in turn, spawned the need for models of population-level responses to such insults as contamination by xenobiotic chemicals.
>
> **(Emlen 1989)**

> Protecting populations is an explicitly stated goal of several Congressional and Agency mandates and regulations. Thus it is important that ecological risk assessment guidelines focus upon protection and management at the population, community, and ecosystem levels . . .
>
> **(EPA 1991)**

> The Office of Water is required by the Clean Water Act to restore and maintain the biological integrity of the nation's waters and, specifically, to ensure the protection and propagation of a balanced population of fish, shellfish, and wildlife.
>
> **(Norton et al. 1992)**

> The translation from a pollutant's effects on individuals to its effects on the population can be accomplished using life-history analysis to calculate the effect on the population's growth rate.
>
> **(Sibly 1996)**

> In this chapter, I am concerned with the translation from individuals to populations using demographic models as a link. Individual organisms are born, grow, reproduce and die, and exposure to toxicants alters the risks of these occurrences. The dynamics of populations are determined by the rates of birth, growth,

> fertility, and mortality that are produced by these individual events .... By incorporating individual rates into population models, the population effects of toxicant-induced changes in those rates can be calculated.
>
> **(Caswell 1996)**

> Fortunately, traditional population and demographic analyses can be used to predict the possible outcomes of exposure and their probabilities of occurring. Although most toxicity testing methods do not produce information directly amenable to demographic analysis, some ecotoxicologists have begun to design tests and interpret results in this context.
>
> **(Newman 1998)**

> Our conclusion is that r [the population growth rate] is a better measure of responses to toxicants than are individual-level effects, because it integrates potentially complex interactions among life-history traits and provides a more relevant measure of ecological impact.
>
> **(Forbes and Calow 1999)**

What is needed is a complete understanding of these approaches and their merits, and the resolve to move further to this new context. As suggested in the above quote by Caswell (1996), individual-based information can be used to assess population-level effects if individual-based metrics are produced with translation to the population level in mind. Valuable time and effort are wasted if we are not mindful of the need for hierarchical consilience. Sufficient understanding will foster the generation of more population-based data and its eventual application in routine ecological risk assessments. It will also foster the infusion of methods from disciplines such as conservation biology, fisheries and wildlife management, and agriculture that have similar goals and relevant technologies. Toward these ends, this and the next chapter will build a fundamental understanding of population processes. Some supporting detail including methods for fitting data to these models can be found in Newman (1995).

14.2 FUNDAMENTALS OF POPULATION DYNAMICS

14.2.1 GENERAL

Initially, we assume that a population is composed of similar individuals occupying a spatially uniform habitat. Because the qualities of individuals are lost in models with such assumptions, they are often called phenomenological models—models focused on describing a phenomenon but not linked intimately to causal mechanics (i.e., not mechanistic models). Events occurring in individuals such as birth, growth, reproduction, and death are aggregated into summary statistics such as population rate of increase. Exploration of these models creates an understanding of population behaviors possible under different conditions. However, without details for individuals and inclusion of interactions with other species populations, insights derived from these models should not be confused with certain knowledge. The problem of ecological inference may appear if results are used to imply behavior of individuals. Alternatively, if results were applied to predicting population fate *in a contaminated ecosystem*, problems may arise because an important emergent property might have been overlooked (e.g., see Box 16.1).

Modeled populations can display continuous or discrete growth dynamics depending on the species and habitat characteristics in question. Continuous growth dynamics are anticipated for a species with overlapping generations and discrete growth dynamics are anticipated for species with nonoverlapping generations. Nonoverlapping generations are common for many annual plant or insect populations. Continuous and discrete growth dynamics are described below with differential and difference equations, respectively. Some of the differential models will also be integrated to allow prediction of population size through time.

14.2.2 Projection Based on Phenomenological Models: Continuous Growth

The change in size (N) of a population experiencing unrestrained, continuous growth is described by the differential equation

$$\frac{\mathrm{d}N}{\mathrm{d}t} = bN - dN = (b - d)N = rN, \tag{14.1}$$

where r = the intrinsic rate of increase or per capita growth rate. The r parameter is the difference between the overall birth (b) and death (d) rates (Birch 1948). Obviously, population numbers decline if $b < d$ (i.e., $r < 0$) or increase if $b > d$ (i.e., $r > 0$). Integration of Equation 14.1 yields Equation 14.2 and allows estimation of population size at any time based on r and the initial population size, N_0,

$$N_t = N_0\, \mathrm{e}^{rt}. \tag{14.2}$$

The amount of time required for the population to double (population doubling time, t_d) is $(\ln 2)/r$.

This model may be applicable for some situations such as the early growth dynamics of a population introduced into a new habitat or a *Daphnia magna* population maintained in a laboratory culture with frequent media replacement. However, most habitats have a finite capacity to sustain the population. This finite capacity slowly comes to have a more and more important role in the growth dynamics as the population size increases. The change in number of individuals through time slows as the population size approaches the maximum size sustainable by the habitat (the carrying capacity or K). This occurs because $b - d$ is not constant through time. Birth and death rates change as population size increases. More than 150 years ago, Verhulst (1838) accommodated this density dependence with the term $1 - (N/K)$ producing the logistic model for population density-dependent growth in the following equation:

$$\frac{\mathrm{d}N}{\mathrm{d}t} = rN\left(1 - \frac{N}{K}\right). \tag{14.3}$$

The per capita growth rate ($r_\mathrm{dd} = r[1 - (N/K)]$) is now dependent on the population density. As population size increases, birth rates decrease and death rates increase. These rates are $b = b_0 - k_\mathrm{b}N$ and $d = d_0 + k_\mathrm{d}N$ where b_0 and d_0 are the nearly density-independent birth and death rates experienced at very low population densities. The terms k_b and k_d are slopes for the change in birth and death rates with change in population density. The logistic model can be expressed in these terms (Wilson and Bossert 1971),

$$\frac{\mathrm{d}N}{\mathrm{d}t} = [(b_0 - k_\mathrm{b}N) - (d_0 + k_\mathrm{d}N)]N. \tag{14.4}$$

The carrying capacity (K) can also be expressed in these terms, $K = [(b_0 - d_0)/(k_\mathrm{b} + k_\mathrm{d})]$ (Wilson and Bossert 1971).

The model described by Equation 14.3 carries the assumption that there is no delay in population response, that is, there is an instantaneous change in r_dd due to any change in density. A delay (T) can be added to Equation 14.3:

$$\frac{\mathrm{d}N}{\mathrm{d}t} = rN\left(1 - \frac{N_{t-T}}{K}\right). \tag{14.5}$$

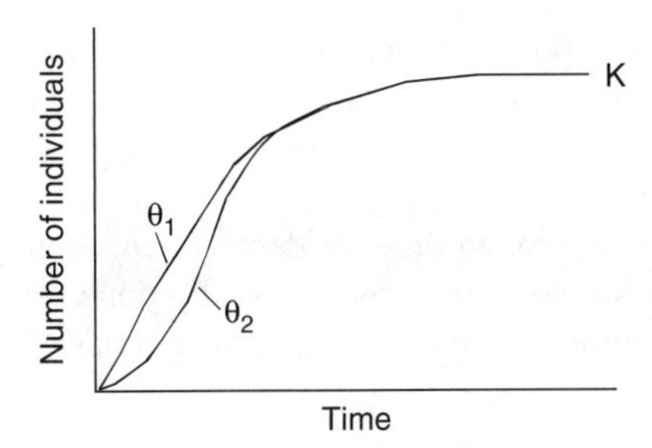

FIGURE 14.1 Logistic increase with population growth symmetry being influenced by the θ parameter in the θ-Ricker model (Equation 14.7).

A time lag (g) before the population responds favorably to a decrease in density can also be included. Such a lag might be applied for species populations in which individuals must reach a certain age before they are sexually mature:

$$\frac{dN}{dt} = rN_{t-g}\left(1 - \frac{N_{t-T}}{K}\right). \tag{14.6}$$

Gilpin and Ayala (1973) found that the shape of the logistic model was not always the same for different populations and added a term (θ) to Equation 14.3 to make the logistic model more flexible. This flexible model (Equation 14.7) is called the θ-logistic model (Figure 14.1):

$$\frac{dN}{dt} = rN\left[1 - \left(\frac{N}{K}\right)^{\theta}\right]. \tag{14.7}$$

Obviously, delays could be placed into Equation 14.7, if necessary, to produce a model of density-dependent growth for a population with time lags in responding to density changes, continuous growth, and growth symmetry defined by θ.

A density-independent effect (I) on population growth such as that of a toxicant can be added to Equation 14.3:

$$\frac{dN}{dt} = rN\left(1 - \frac{N}{K}\right) - I. \tag{14.8}$$

The I can also be expressed as some toxicant-related "loss," "take," or "yield" from the population at any moment ($E_{\mathrm{Toxicant}}N$), where E_{Toxicant} is the proportion of the existing number of individuals (N) taken owing to toxicant exposure

$$\frac{dN}{dt} = rN\left(1 - \frac{N}{K}\right) - E_{\mathrm{Toxicant}}N. \tag{14.9}$$

In words, this model predicts the change in number of individuals per unit of time as a function of the intrinsic rate of increase, density-dependent growth dynamics, and a density-independent decrease in numbers of individuals as a result of toxicant exposure. In this form, it is identical to a rudimentary harvesting model for natural populations (e.g., commercial fish harvesting) and is amenable to analysis of population sustainability and recovery (see Everhart et al. 1953, Gulland 1977, Hadon 2001, Murray 1993). The difference is that harvesting involves toxicant action instead of fishing. We will discuss this point later, but it is important now to realize that toxicant-induced changes in K, r, T, g, and θ are possible and none of these parameters should be ignored in the analysis of population fate with toxicant exposure. If necessary to produce a realistic model, time delays and θ values can be added to Equation 14.8. The underlying processes resulting in these delays could be influenced by toxicant exposure. For example, a toxicant may influence the time required for an individual to reach sexual maturity.

Prediction of population size through time for density-dependent growth of a population with continuous growth dynamics and no time delays is usually done using Equations 14.10 or 14.11. These equations are different forms of the sigmoidal growth model. May and Oster (1976) provide other useful forms:

$$N_t = \frac{N_0 K\, e^{rt}}{K + N_0(e^{rt} - 1)} \tag{14.10}$$

$$N_t = \frac{K}{1 + [(K - N_0)/N_0]e^{-rt}} \tag{14.11}$$

Newman (1995) describes methods for fitting data to these differential and integrated equations and relates them to ecotoxicology.

14.2.3 Projection Based on Phenomenological Models: Discrete Growth

Unrestrained growth of populations displaying discrete growth (nonoverlapping generations) is described with the difference equation,

$$N_{t+1} = \lambda N_t, \tag{14.12}$$

where N_t and N_{t+1} are the population sizes at times t and $t+1$ respectively, and λ is the finite rate of increase, which can be related to r (intrinsic or infinitesimal rate of increase) with Equation 14.13. It is the number of times that the population multiplies in a time unit or step (Birch 1948). The time step may be arbitrary (e.g., time between census episodes) or associated with some aspect of reproduction (e.g., time between annual calvings):

$$\lambda = \frac{N_{t+1}}{N_t} = e^r. \tag{14.13}$$

The characteristic return time (T_r) can be estimated from r or λ. It is the estimated time required for a population changing in size through time to return toward its carrying capacity or, more generally, toward its steady state number of individuals (May et al. 1974). It is the inverse of the instantaneous growth rate, r (i.e., $T_r = 1/r$). The T_r gets shorter as the growth rate, r, increases: faster growth results in a faster approach toward steady state. In Section 4.3, the influence of T_r on population stability will be described.

This difference equation (Equation 14.12) can be expanded to include density-dependence using several models (see Newman 1995). Equations 14.14 and 14.15 are the classic Ricker and a modification of it that includes Gilpin's θ parameter (the θ-Ricker model), respectively:

$$N_{t+1} = N_t\, e^{r(1-(N_t/K))} \tag{14.14}$$

$$N_{t+1} = N_t\, e^{r[1-(N_t/K)^{\theta}]}. \tag{14.15}$$

As done with the differential models, we have accommodated differences in growth curve symmetry by including a θ term in Equation 14.15. But what about adding lag terms? Because the form of the difference equations implies an inherent lag from t to $t+1$, these models may not need additional terms to accommodate lags. If a lag time different from the time step (t to $t + 1$) is required, it can be added by using N_{t-1}, N_{t-2}, or some other past population size instead of N_t where appropriate in these models. We can add an effect of a density-independent factor such as toxicant exposure to the logistic model. The difference models above are modified by inserting an I term as done in Equations 14.8 and 14.9. The modification made by Newman and Jagoe (1998) to the simplest difference model (Equation 4.16) is provided as follows (Equations 4.17 and 4.18).

$$N_{t+1} = N_t \left[1 + r \left(1 - \frac{N_t}{K} \right) \right] \tag{14.16}$$

$$N_{t+1} = N_t \left[1 + r \left(1 - \frac{N_t}{K} \right) \right] - I \tag{14.17}$$

$$N_{t+1} = N_t \left[1 + r \left(1 - \frac{N_t}{K} \right) \right] - E_{\text{Toxicant}} N_t \tag{14.18}$$

I is the number of individuals "taken" from the parental population by the toxicant at each time step. Again, these models of toxicant effect are comparable to those used to manage harvested, renewable resources such as a fishery [e.g., Equations 2.13 and 2.15 in Haddon (2001)]. Alternately, Gard (1992) expresses the influence of a toxicant directly in terms of the instantaneous growth rate (r) at time, t,

$$r_t = r_0 - r_1 C_{T(t)}, \tag{14.19}$$

where r_0 is the intrinsic rate of increase in the absence of toxicant, $C_{T(t)}$ is a time-dependent effect of the toxicant on the population, and r_1 is a units conversion parameter. Gard's model is composed of three differential equations that link temporal changes in environmental concentrations of a toxicant, concentrations in the organism, and population growth (Gard 1990). At this point, it is only necessary to note that Gard's equations reduce r directly as a consequence of toxicant exposure. Any change in r can influence population stability, as we will see in Section 14.3.

14.2.4 Sustainable Harvest and Time to Recovery

The expressions of toxicant-impacted population growth described to this point are equivalent to those general models explored by Murray (1993) for population harvesting. Therefore, his expansion of associated mathematics and explanations are translated directly in this section into terms of toxicant effects on populations. Let us assume that natality is not affected but the loss of individuals from the population is affected by toxicant exposure. For the differential model (Equations 14.8 and 14.9), Murray defines a harvest or yield that is analogous to I in Equation 14.8 and a corresponding new steady-state population size of N_{h}. This harvest is equivalent to $E \cdot N$ where E is a measure of the harvesting intensity and N is the size of the population being harvested. The E is identical by intent to E_{Toxicant} in Equation 14.9. With "harvesting" or loss upon toxicant exposure, the population will not have a steady-state size of K. Instead, it will have the following steady-state size if r is larger than E_{Toxicant},

$$N_{\text{L}} = K \left[1 - \frac{E_{\text{Toxicant}}}{r} \right], \tag{14.20}$$

where N_{L} is equivalent to Murray's N_{h} except that loss is now due to toxicant exposure. From Equation 14.20, it is clear that the population at steady state will drop to zero if the intensity of the toxicant effect (E_{Toxicant}) is equal to or greater than r.

Let us extend Murray's expression of yield from a harvested population in order to gain further insight into the loss that a population can sustain from toxicant exposure without being irreparably damaged. The yield in Murray's Equation 1.43 is modified to Equation 14.21 in order to define the loss of individuals (L) expected at a certain intensity of effect (E_{Toxicant}),

$$L_{E_{\text{Toxicant}}} = E_{\text{Toxicant}} K \left[1 - \frac{E_{\text{Toxicant}}}{r} \right]. \tag{14.21}$$

In words, the population loss or "yield" due to toxicant exposure ($L_{E_{\text{Toxicant}}}$) is the new carrying capacity (N_{L}) multiplied by the E_{Toxicant}: the yield is the number of individuals available to be taken times the toxicant-induced fraction "taken." Applying Equation 14.21, $rK/4$ is the maximum sustainable loss to toxicant exposure (analogous to the maximum sustainable yield where $E_{\text{Toxicant}} = r/2$). The new steady-state population size (N_{L}, equivalent to N_{h}) will be $K/2$ at this point of maximum sustainable loss or "yield." The population is growing maximally under these conditions. Population growth becomes suboptimal if E_{Toxicant} increases further and may even become negative if E_{Toxicant} exceeds r. Figure 14.2 illustrates this general estimation of toxicant take or loss for a hypothetical population that is growing according to the logistic model.

Moriarty (1983) makes several important points regarding this approach to analyzing toxicant effects on populations. First, growth measured as a change in number between times t and $t+1$ will not necessarily decrease with increasing loss from the population due to toxicant exposure (Figure 14.2). It might increase. Surplus young produced in populations allows a certain level of mortality without an adverse affect on population viability. Different populations have characteristic ranges of loss that can be accommodated. Low losses potentially increase the rate at which new individuals appear in a population and high losses push the population toward local extinction. Second, the carrying capacity of the population will decrease as losses due to toxicant exposure increase. Third, there can be two population sizes that produce a particular yield on either side of the N_t for maximum yield. Increases

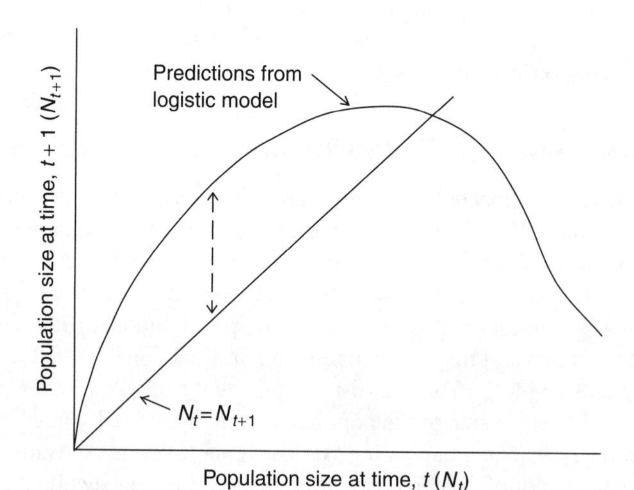

FIGURE 14.2 The maximum sustainable yield can be visualized by comparing the curve for the population size at time steps N_t and N_{t+1} to the line for $N_t = N_{t+1}$. The population is not changing from one time to another along the line for $N_t = N_{t+1}$, i.e., the population is at steady state. The vertical distance between the curve of population size at time steps N_t and N_{t+1} and the line for $N_t = N_{t+1}$ defines the sustainable yield resulting from surplus production in the population each time step. The vertical dashed line shows the yield that is maximal for this population. The reader is encouraged to review Waller et al. (1971) as an example of using this type of curve with zinc-exposed fathead minnow (*Pimephales promelas*) populations. (Modified from Figure 2.11 of Moriarty (1983) and Figure 2.2a of Murray (1993).)

or decreases in toxicant exposure can produce the same results in the context of population change. Failure to recognize this possibility could lead to muddled interpretation of results from monitoring of populations in contaminated habitats. An important advantage of the sustainable yield context just developed is a more complete understanding of population consequences at various intensities of loss due to toxicant exposure.

There is another advantage to ecotoxicologists taking an approach used by renewable resource managers. Often, ecological risk assessments focus on recreational or commercial species, for example, consequences of toxicant exposure to a salmon or blue crab fishery. Expressing toxicant effects to populations with the same equations used by fishery or wildlife managers attempting to regulate annual harvest allows simultaneous consideration of losses from fishing and pollution.

Another characteristic of harvested populations that is useful to the ecotoxicologist is the time to recovery. The time to recover (return to an original population size) after harvest can be estimated in terms of loss due to toxicant exposure. The time to recover (T_R) will increase as $E_{Toxicant}$ increases. This follows from our discussion that characteristic return time increases as r increases and that $E_{Toxicant}$ has the opposite effect on population growth rate as r. Figure 14.3 shows the general shape of this relationship for a logistic growth model.

The phenomenological models described to this point might have to be modified if interest shifts to smaller and smaller population sizes. Just as a population has a maximum population size (e.g., K) it can also have a minimum population size. The population fails below this minimum number, e.g., the smallest number of individuals in a dispersed population needed to have a chance of sufficient mating and reproductive success, or the minimum number of a social species needed to maintain a viable group. This minimum population size (M) can be placed into the logistic model (Equation 14.3) (Wilson and Bossert 1971),

$$\frac{dN}{dt} = rN\left(1 - \frac{N}{K}\right)\left(1 - \frac{M}{N}\right). \tag{14.22}$$

The population will go locally extinct if N falls below M.

More discussion of population loss, recovery time, and minimum population size in the context of fishery management can be found in books by Gulland (1977) and Everhart et al. (1953), and

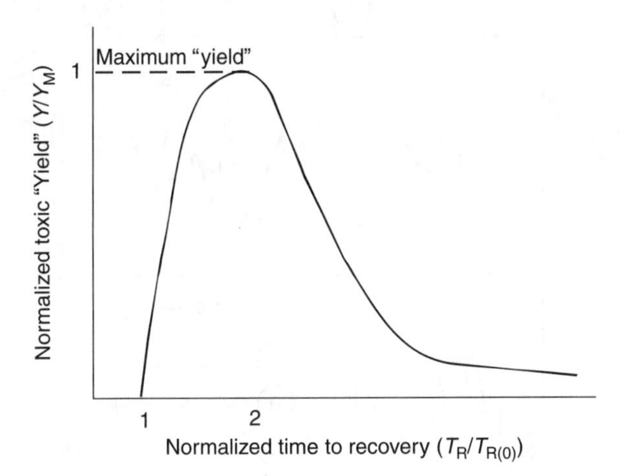

FIGURE 14.3 The time to recover (T_R) will increase as the "yield" or "take" due to toxicant exposure ($E_{Toxicant}$) increases. This modification of Figure 1.16a in Murray (1993) shows the general shape of this relationship for a logistic growth model. The $T_{R(0)}$ is the theoretical recovery time in the absence of toxicant exposure ($T_{R(0)} = 1/r$) and T_R is the recovery time at a particular yield, Y, for the steady-state population. Yield is normalized in this figure by dividing it by the maximum possible yield (Y_M).

formulations relative to discrete growth models are provided in Murray (1993) and Haddon (2001). Because some fisheries models based on commercial yield consider monetary costs, the application of a common model also provides an opportunity in risk management decisions to integrate monetary gain from fishing with monetary loss due to toxicant exposure. A management failure of a fishery would certainly occur if, by ignoring toxicant effects, one optimized solely on the basis of commercial fishery harvest. Perhaps the additional loss due to toxicant exposure would put the combined consequences to the population beyond the optimal yield and the fishery would slowly begin an inexplicable decline.

14.3 POPULATION STABILITY

Until approximately 25 years ago, the dynamics in population size described by models such as Equations 14.3, 14.14, and 14.16 were thought to consist of an increase to some steady-state size (e.g., K) as depicted in Figure 14.1. Deviations from this monotonic increase toward K were attributed to random processes. In 1974, Robert May published a remarkably straightforward paper in *Science* demonstrating that this was not the complete story. Even the simple models described in this chapter can display complex oscillations in population size and, at an extreme, chaotic dynamics. Some populations do monotonically increase to a steady-state size (i.e., Stable Point in Figure 14.4). Others tend to overshoot the carrying capacity, turn to oscillate back and forth around the carrying capacity, and eventually settle down to the carrying capacity (i.e., Damped Oscillation in Figure 14.4). Sizes of other populations oscillate indefinitely around the carrying capacity (i.e., Stable Cycles in Figure 14.4). These oscillations may be between 2, 4, 8, 16, or more points. Beyond population conditions resulting in stable oscillations, the number of individuals in a population at any time may be

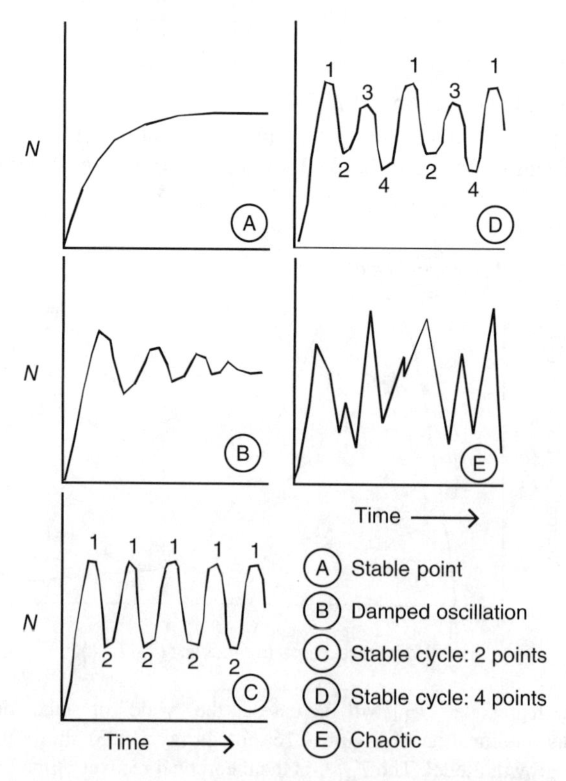

FIGURE 14.4 Temporal dynamics that might arise from the differential and difference models of population growth.

best defined as chaotic (Chaotic in Figure 14.4). Population size changes in an unpredictable fashion through time. The exact size at any time is very dependent on the initial conditions: on average, trajectories for populations with slightly differing initial sizes separate exponentially through time (Schaffer and Kot 1986). Although chaotic dynamics have been noted for only a few species populations (e.g., flour beetles, *Tribolium castaneum* (Costantino et al. 1997)), these complex population dynamics are fostered by high r, long time lags, and periodic forcing functions like those used to model impacts of weather extremes or insecticide spraying of target populations. There is no reason to reject the notion that nontarget species with high rates of increase and/or significant time lags that live near agricultural fields might exhibit chaotic dynamics.

What specific qualities determine a population's growth dynamics? The dynamics tend to move from stable point to damped oscillations to stable cycles to chaotic dynamics as the intrinsic rate of increase (r), time lag (T) and/or θ increase. The rate at which population size approaches K increases as the r increases: at a certain r, the population tends to overshoot K and move the temporal dynamics into more complicated oscillations. Similarly, if the time lag (T) increases, the populations size begins to oscillate more and more as the population tends to overshoot and then undershoot the carrying capacity. The exact conditions producing these different dynamics (i.e., the stability regions) have been determined for several of the simple models described in this chapter. Table 14.1 provides those for Equations 14.5 and 14.14 through 14.16. The derivation of these stability criteria is detailed in May et al. (1974). May (1976a) extends the graphical approach used in Figure 14.2 to show the conditions leading to different population dynamics.

Several points relevant to population ecotoxicology emerge from these considerations of population dynamics. During an ecological risk assessment, the population size at a contaminated site might be compared to that of a reference site. The observation of a smaller population at the contaminated site relative to the uncontaminated site often leads to the conclusion of an adverse effect on the population. As demonstrated by the models above, some populations will characteristically have wide variations in size through time. Others will be more stable. To compare sizes of populations from reference and contaminated sites using data from one or a few field samplings can lead to invalid conclusions if populations at both sites normally displayed wide oscillations. In addition, because toxicants can affect r, T, and θ in these simple models, there is no reason to believe that toxicants will not impact population dynamics. So, it may be important to consider pollutant effects on population

TABLE 14.1
Stability Regions for Differential (Equation 14.5) and Difference (Equations 14.15, 14.14, and 14.16) Models of Population Growth

Stability Region	Differential Equation 14.5	Difference Equations: Equation 14.15	Difference Equations: Equation 14.14	Difference Equations: Equation 14.16
Stable point	$0 < rT < e^{-1}$	$0 < r\theta < 1$	$2 > r > 0$	$2 > r > 0$
Damped oscillation	$e^{-1} < rT < 0.5\pi$	$1 < r\theta < 2$		
Stable cycles	$0.5\pi < rT$	$2 < r\theta < 2.69$		
Between 2 points			$2.526 > r > 2.000$	$2.449 > r > 2.000$
Between 4 points			$2.656 > r > 2.526$	$2.544 > r > 2.449$
Between 8 points			$2.685 > r > 2.656$	$2.564 > r > 2.544$
Between 16 or more points			$2.692 > r > 2.685$	$2.570 > r > 2.564$
Chaotic dynamics		$2.69 < r\theta$	$r > 2.692$	$r > 2.570$

Note: Stability region information for Equations 14.5, 14.15, 14.14, and 14.16 was obtained from May (1976), Thomas et al. (1980), May (1974), and May (1974), respectively.

dynamics in addition to population size. The likelihood of a local population extinction is greatly increased by a toxicant exposure that produces wide oscillations in addition to lowering the population carrying capacity. The lowering of the carrying capacity brings the population numbers closer to the minimal population number (M) and the oscillation troughs periodically bring the population numbers even closer to M.

Higher order interactions are also possible on population dynamics. Simkiss et al. (1993) examined the growth of blowfly (*Lucilia sericata*) under different combinations of food and cadmium. Food deprivation and cadmium concentration were additive in their effects on key growth components (maximum larval size, development period, pupal weight, adult weight at emergence, and fecundity). Many of these effects change r and time lags. So, the combination of cadmium exposure and limited food availability can influence population dynamics. Nicholson (1954) had previously shown that limited food alone produced population oscillations with *Lucilia cuprina*. Simkiss et al. (1993) predicted from their studies of food and cadmium effects on blowfly populations that "sublethal levels of cadmium might therefore lead to smaller-amplitude fluctuations without affecting the mean population level."

Box 14.1 Extinction Probabilities for Fruit Fly Populations under Nutritional Stress

The influence of environmental carrying capacity on the likelihood of *Drosophila* sp. population extinction was quantified by Philippi et al. (1987) by manipulating the amount of food available to cultures of different fruit fly species. In one set of experiments, food was varied from a very restrictive 3 mL per 120 mL bottle to an excessive amount of 40 mL per 120 mL bottle. Flies were periodically transferred to new bottles of media and the results were fit to a difference logistic model that included Gilpin's θ parameter:

$$\Delta N = N_{t+1} - N_t = rN_t - \frac{r}{K^\theta} N_t^{(\theta+1)}. \tag{14.23}$$

The premise was that, as they had seen with species populations competing with one another for limited resources, isolated fly populations provided with limited resources would exhibit very wide fluctuations in size. These fluctuations would increase the chance of population extinction. In a second set of experiments, they varied the density of flies and measured survival and reproduction of individuals at these different densities. The resulting data were used to assess the relative contributions of chaotic dynamics, carrying capacity reduction, and environmental stochasticity to population persistence.

Let us assume in interpreting their results for nutritional stress that, according to Equations 14.9, 14.18, and 14.20 and the work of Simkiss et al. (1993), toxicant exposure will similarly impact fruit fly populations by decreasing carrying capacity. Under this assumption, this study of food limitation has direct relevance to populations exposed to toxicants.

Much to their surprise, the food-deprived populations showed lower variability than those with unlimited food: the stressed populations had reduced variability in their numbers. Higher observed rates of extinction in food-deprived populations were a result of a reduced carrying capacity and variance in growth dynamics due to environmental variability, not a shift toward more complicated dynamics. (The stability regions for this θ-logistic model are those given for Equation 14.15 in Table 14.1.) The environmental variability involved differences in the amount of food placed into each bottle, the humidity, and level of bacterial contamination introduced into cultures during handling. They conclude that the minimum viable population size (M) is determined by deterministic and stochastic population processes but, in this case, a deterministic shift in population dynamics toward wider fluctuations under stressful conditions was not responsible for the observed accelerated rates of extinction.

Sensitivity analysis for the model provided further insight into the relative importance of changes in r and θ on the probability of population extinction. Simulations demonstrated that extinction probabilities over a wide range of r and θ values were determined by chaotic dynamics regardless of the level of environmental noise only if $r \cdot \theta$ was greater than 3. At the other extreme, systems with low r and θ values (i.e., those that we are assuming would be characteristic of pollutant-stressed populations), recover very slowly from minor perturbations. In this situation, extinction probability increases with environmental variability. The lowest probabilities of extinction were in regions with $0.5 < r \cdot \theta < 2$.

14.4 SPATIAL DISTRIBUTIONS OF INDIVIDUALS IN POPULATIONS

In Section 14.2, the convenient assumption was made that a population is composed of similar individuals occupying a spatially uniform habitat. However, individuals are often distributed heterogeneously within a habitat that may not be homogenous itself. Some consequences of this heterogeneity will be outlined in this section.

14.4.1 DESCRIBING DISTRIBUTIONS: CLUMPED, RANDOM, AND UNIFORM

Individuals may be distributed uniformly, randomly, or in clusters within a habitat (Figure 14.5). Uniform distributions are rare. Random and clumped or aggregated distributions are much more common. The driving force for the clumping may be innate to the species (e.g., the social gathering of individuals to enhance foraging or reproductive success), result from extrinsic factors (e.g., a landscape that is a mosaic of habitats widely differing in quality), or emerge from an interaction of intrinsic and extrinsic factors.

Conformity of individuals within a population to these patterns can be formally tested by methods described in various sources (i.e., Krebs 1989, Ludwig and Reynolds 1988, and Newman 1995) by assuming that positive binomial, Poisson, and negative binomial models describe uniform, random,

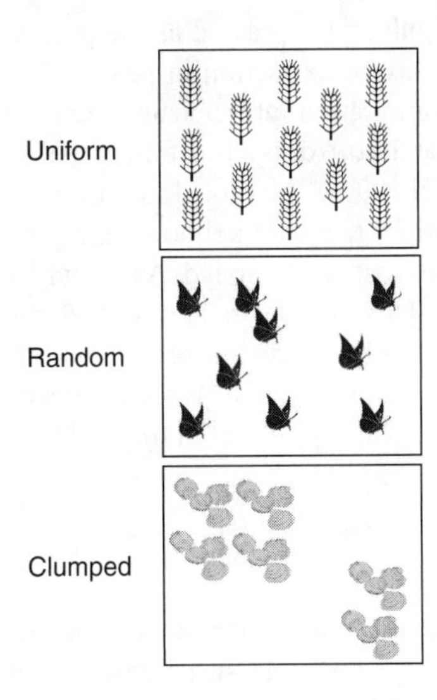

FIGURE 14.5 General distributional patterns of individuals within populations including uniform (upper panel), random (middle panel), and clumped (lower panel) distributions.

and clumped patterns, respectively. These models can also be fit to distributional data for individuals using methods described in the cited sources.

14.4.2 Metapopulations

What are the consequences of an uneven distribution of individuals within a habitat? Do things simply average out over the entire habitat and, as a consequence, conform to the simple population dynamics described so far? The simple answer is no. Unique and important qualities emerge in the dynamics and persistence of a metapopulation. Hanski (1996) described a metapopulation as "a set of local populations which interact via dispersing individuals among local populations; though not all local populations in a metapopulation interact directly with every other local population." Unique qualities of metapopulations must be understood to appreciate the influence of toxicant exposure on populations.

14.4.2.1 Metapopulation Dynamics

Subpopulations or local populations occupy patches of the available habitat that differ slightly or greatly relative to the ability to foster individual survival, growth, and reproduction. Consequently, individual fitness differs among landscape patches. High quality patches may produce so many individuals that they act as sources to other patches. Low quality patches may be so inferior that they do not produce surplus individuals. To remain occupied, inferior sink subpopulations rely on an influx of individuals from source patches. Some patches may be so superior relative to other marginal patches that they are keystone habitats (O'Connor 1996) without which the metapopulation might disappear. Which patches are sources and which are sinks may change through time depending on factors like weather, disease, competition, or predation pressures. In other situations in which patches are physical islands, the source–sink structure might remain stable through time. A source–sink dynamic emerges as essential in understanding metapopulation size and persistence on the landscape scale (Figure 14.6).

Levins (1969, 1970; cited in Hanski (1996)) explored metapopulation dynamics with a simple model:

$$\frac{\mathrm{d}p}{\mathrm{d}t} = mp(1-p) - ep, \tag{14.24}$$

where p is the size of the metapopulation expressed as the proportion of available patches that are occupied, e is the rate or probability of extinction in patches, and m is the rate or probability of population reappearance in (or immigration into) vacant patches of the landscape mosaic. In more general terms, this model states that the $\mathrm{d}p/\mathrm{d}t$ is a function of the immigration and extinction rates for patches in a habitat mosaic (Gotelli 1991). The probability of patch extinction is independent of the regional occurrence of subpopulations (p): the likelihood of a patch being vacated is not influenced by the proportion of nearby patches that are occupied. As we will see in the following text, this may or may not be a good assumption. This model can be modified into a form analogous to the logistic model for population size "with $m - e$ being the intrinsic rate of metapopulation increase for a small metapopulation (when p is small), while $1 - (e/m)$ is the equivalent of the local 'carrying capacity,' the stable equilibrium point toward which p moves in time" (Hanski 1996):

$$\frac{\mathrm{d}p}{\mathrm{d}t} = (m-e)p\left(1 - \frac{p}{1-(e/m)}\right). \tag{14.25}$$

Gotelli (1991) provides more detail for this and related metapopulation models. He also highlights several themes in metapopulation dynamics. First, he describes the rescue effect as a decrease in probability of patch extinction because of the influx of individuals from nearby subpopulations, i.e., the assumption in Equation 14.25 is avoided that the probability of patch extinction is independent

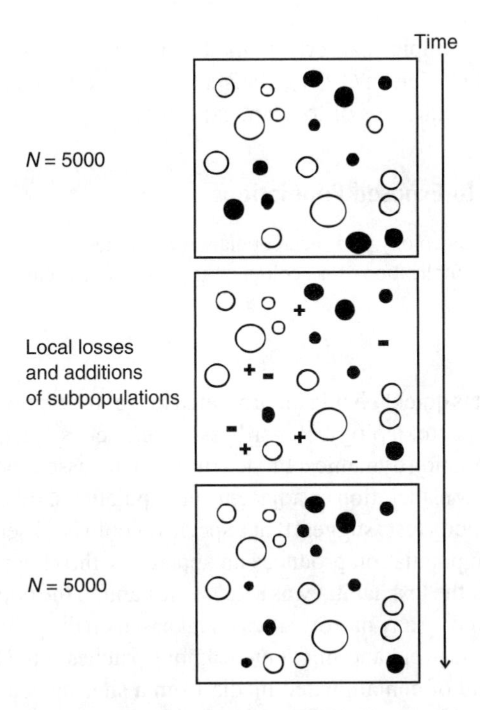

FIGURE 14.6 Stable metapopulation dynamics within a habitat mosaic despite periodic local extinctions. Subpopulations vary in their ability to act as sources (size of open circles) and sinks (size of filled circle), resulting in different probabilities of local extinction through time (upper panel). Because of the exchange among patches, the metapopulation persists through a dynamic steady state of extinctions (− in middle panel) and reestablishments via influx from nearby sources. New populations are also established through time via influx from sources (+ in middle panel). The net result is metapopulation persistence through time (lower panel).

of the regional occurrence of populations (p). The model of Hanski (described in Gotelli (1991)) includes the rescue effect by slight modification of Equations 14.24 to 14.26. In this model, the probability of local extinction decreases as p (proportion of patches occupied) increases: emigration from neighboring subpopulations reduces the likelihood of local extinction:

$$\frac{\mathrm{d}p}{\mathrm{d}t} = mp(1-p) - ep(1-p). \tag{14.26}$$

In Equation 14.24, the extinction rate increases as p increases. In Equation 14.26, this is true up to a certain p. The extinction rate then begins to decline as p continues to increase.

A source of propagules such as a seed bank or dormant stage can produce a “propagule rain” that bolsters a waning subpopulation and can influence metapopulation dynamics. In such a case, the regional occurrence of subpopulations (p) does not impact the rate of reappearance of individuals (m) and the rain of propagules increases the m by $m(1-p)^2$ (Gotelli 1991). Under this condition and an e that is independent of regional occurrence, a better description of metapopulation dynamics is Equation 14.27 (Gotelli 1991),

$$\frac{\mathrm{d}p}{\mathrm{d}t} = m(1-p) - ep. \tag{14.27}$$

Equation 14.28 combines the propagule rain and rescue effects:

$$\frac{\mathrm{d}p}{\mathrm{d}t} = m(1-p) - ep(1-p). \tag{14.28}$$

Additional details and examples can be obtained in Lewin (1989), Gotelli (1991), Gilpin and Hanski (1991), Pulliam and Danielson (1991), Hanski (1996), and Pulliam (1996). O'Connor (1996) reviews metapopulation consequences of toxicant exposure.

14.4.2.2 Consequences to Exposed Populations

> The paradigms of landscape ecology and metapopulation dynamics ... have introduced new concepts of spatial dynamics whose implications for ecological risk assessment have only just begun to receive attention.
>
> **(O'Connor 1996)**

The metapopulation context is quickly being incorporated into conservation biology efforts but is only slowly being considered in ecotoxicology. Regardless, several consequences become obvious from this brief sketch of metapopulation dynamics. First, a rudimentary assessment of population status in a contaminated area requires consideration of adjacent subpopulations; otherwise, observations might be inexplicable. Perhaps toxicity tests suggest that a species should be absent from a contaminated site but the presence of a source population produces an apparently thriving population on the site (i.e., the rescue effect). Second, if the lost habitat was a keystone habitat, the population consequences will be much worse than suggested by any narrow assessment based on the percentage of total habitat lost. Third, the creation of corridors to enhance movement among patches could be more beneficial in some cases than complete removal of contaminated media from a site. Indeed, remediation often causes considerable disruption of habitat: a thoughtful balance of removal of polluted media from patches, creation of corridors among patches, and building of barriers around other highly contaminated patches could result in optimal remediation. Fourth, among migrating individuals within the mosaic of habitats, some will have spent time in heavily contaminated patches. The result might be that individuals exposed in one patch will have their population-level consequences manifested in a subpopulation removed from that contaminated site. Spromberg et al. (1998) call this the effect at a distance hypothesis because the action of a toxicant exposure occurs at a place spatially distant from the contamination. Fifth, a sublethal effect that reduces migration-related behavior could decrease the stability or persistence of a metapopulation by affecting the rate at which vacant habitat is refilled from adjacent areas.

Box 14.2 Computer Projections of Metapopulation Risk in a Contaminated Habitat

Spromberg et al. (1998) developed phenomenological models for subpopulations in a habitat with patchy distributions of individuals and toxicants. Their intent was to explore consequences of such a metapopulation configuration and to relate the results to risk assessment activities and possible remedial actions. They described their conceptual framework with Equations 14.24 and 14.26, and added the diffusion reaction model of Wu et al. (1993),

$$\frac{\mathrm{d}N_i}{\mathrm{d}t} = N_i f(N_i) + \sum_{j=i} [d_{ij}(N_j - N_i)], \tag{14.29}$$

where d_{ij} is the migration rate from patch i to patch j, N_i and N_j are the number of individuals in patches i and j, respectively, and $f()$ is a function of N that defines the population growth rate. Note that, unlike previous models, the numbers of individuals in the patches is being modeled in Equation 14.29, not the proportion of all patches that are occupied (p).

Equation 14.29 is used to simulate the dynamics of a metapopulation under different contamination scenarios involving three patches (Figure 14.7). Simulations included a contaminant that quickly disappeared from a patch (e.g., a quickly degraded pesticide) and a persistent contaminant. The model assumed the following: (1) density-dependent growth, (2) density-dependent patch immigration and emigration, and (3) distance-dependent movement of individuals between patches. As an example of distance-dependent movement among patches, the distance between all three patches were similar for the model in the upper panel of Figure 14.7, but the distance between the two outer patches in the model at the bottom of this figure was twice as far as the distance from any one of these outer patches to the center patch. For the model in the lower panel, the distance-dependent movement between an outer patch and the central one was much higher than the movement between the two outside patches. Computation of migration rate from patch i to j was done with the simple equation, $d_{ij} = (N_iH_j)/D_{ij}$, where H_j is the habitat available to be occupied in patch j, and D_{ij} is the distance between the two patches.

Other model assumptions included a constant carrying capacity and minimum population size for a patch, no avoidance of contaminated habitat, no compensatory reproduction due to the presence of toxicant, a Poisson distribution to define the probability of an individual being exposed to a toxicant in the contaminated patch, constant bioavailability of toxicant, and the occurrence of no other stochastic disturbances (e.g., no weather-related mortalities).

The results of the simulations are easily summarized. A toxic effect can be seen in subpopulations of nearby, uncontaminated patches due to the movement of individuals into those patches from a contaminated patch. Again, the authors refer to this as the hypothesis of effect at distance. Toxicant-induced reduction in subpopulation size in one patch results in a higher rate of movement of individuals into that patch. In the model in the lower panel of Figure 14.7, even 100% mortality in the contaminated patch does not produce a local extinction. Individuals move into the vacated patch from the other patches. One noteworthy conclusion derived from the simulations was that a reference population picked from near a contaminated site might not produce useful information in an ecological risk assessment. Although not contaminated, the subpopulation occupying that clean patch may still be impacted by the toxicant due to migration of individuals from the contaminated patch. Conversely, remediation of a site can result in improvements in population viability in patches outside that containing the toxicant.

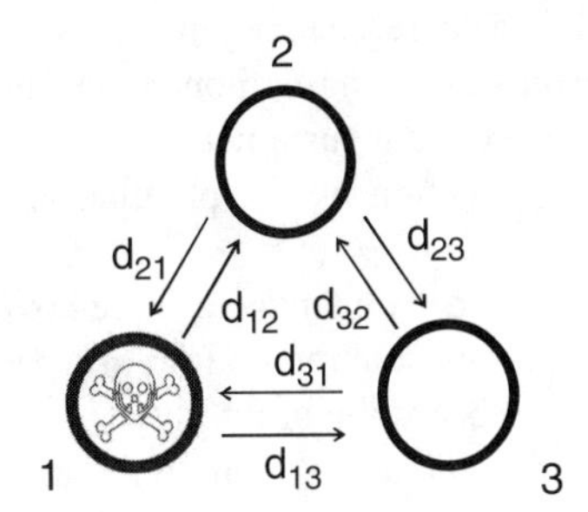

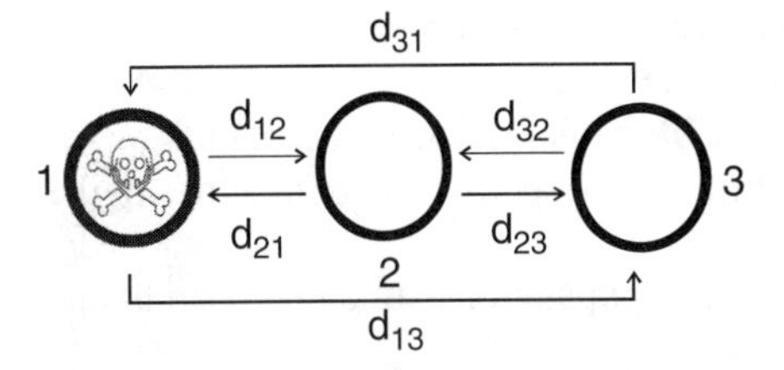

FIGURE 14.7 The three patch metapopulations simulated by Spromberg et al. (1998). Patches with skull and cross bones indicate contaminated patches and the open symbol represents uncontaminated patches. As described in the text, d_{ij} indicates the migration rate from patch i to patch j. (Modified from Figure 1 of Spromberg et al. (1998).)

14.5 SUMMARY

Population-based assessment endpoints are beginning to occupy a more prominent role in ecological risk assessments. To accelerate this change, relevant concepts and methods need to be described so that ecotoxicologists feel comfortable with their application. Population-based metrics of effect will need to accumulate in databases such as the AQUIRE database maintained by the U.S. Environmental Protection Agency. Obviously, consilience among levels is essential as the new, population-centered paradigm emerges. Linkage to individual-based effects can be made with individual-based population models such as some of the demographic analyses described in the next chapter. However, data must be collected so that these links can be made. Presently, they often are not collected in an appropriate manner. Linkage to community-level effects is also possible with the context developed in this chapter. For example, the metapopulation formulations (e.g., Equation 14.25) are closely related to the community succession models (MacArthur–Wilson model) used by Cairns et al. (1986) to predict toxicant effects to protozoan community processes.

14.5.1 SUMMARY OF FOUNDATION CONCEPTS AND PARADIGMS

- A population-based paradigm for assessing ecological risk is slowly beginning to share importance with the currently dominant paradigm based on metrics for effect to individuals.
- Phenomenological models of population dynamics provide an understanding of possible behaviors of populations, including those experiencing toxicant exposure.
- Toxicant exposure can change the population qualities reflected in the model parameters, r, λ, K, T, g, T_R, M, and θ.
- Effects of toxicant exposure can be included in conventional growth models as a density-independent population loss.
- Toxicant exposure can reduce population size.
- Toxicant exposures can change population dynamics.
- Toxicant exposure can result in an increased probability of population extinction.
- Population sustainability and recovery with toxicant exposure can be modeled with modifications to methods used to manage harvested, renewable resources such as a commercial or recreational fishery.
- Increased toxicant exposure may not necessarily result in a decrease in population production rate. Populations can loose a number of their surplus offspring without a significant change in population viability. On either side of the maximum sustainable loss to toxicant exposure ("yield") are equal points of excess production by the population (see Section 14.2.4).
- Expression of loss due to toxicant exposure in terms used by managers of renewable resources, for example, commercial fishery managers, allows the integration of toxicant and fishing/harvesting activities in assessments of resource sustainability.
- Temporal dynamics of populations can take several forms including monotonic increase to carrying capacity, damped oscillations to carrying capacity, stable oscillations about carrying capacity, or chaotic dynamics. These dynamics are determined by combinations of r, T, and θ (see Table 14.1). Because r, T, and θ can be changed by toxicant exposure, population dynamics can be changed by toxicant exposure.
- Individuals can be nonrandomly distributed in available habitat. Distribution of individuals in a population can be influenced by innate qualities of the species and/or qualities of the environment, including the presence of toxicants.
- Metapopulation dynamics influence the probability of population extinction in landscape mosaics contaminated with toxicants.

- Habitats have finite, but fluctuating, capacities to support a species population and toxicants can lower (e.g., reduction of amount of food or suitable habitat) or increase (e.g., remove a competitor or predator) the carrying capacity of a habitat.
- Assessment of toxicant exposure consequences to a metapopulation must consider source–sink dynamics, keystone habitats, and the possibilities of the rescue effect and a significant propagule rain effect.
- Creation of corridors between patches or isolation of the contaminated patch may greatly influence metapopulation viability. These actions should be considered in addition to conventional removal of contaminated media in remediation plans.

REFERENCES

Barnthouse, L.W., Suter, G.W., II, Rosen, A.E., and Beauchamp, J.J., Estimating responses of fish populations to toxic contaminants, *Environ. Toxicol. Chem.*, 6, 811–824, 1987.

Birch, L.C., The intrinsic rate of natural increase of an insect population, *J. Anim. Ecol.*, 17, 15–26, 1948.

Braithwaite, R.B., The structure of a scientific system, In *The Concept of Evidence*, Achinstein, P. (ed.), Oxford University Press, Oxford, UK, 1983, pp. 44–62.

Cairns, J., Jr., Pratt, J.R., Niederlehner, B.R., and McCormick, P.V., A simple cost-effective multispecies toxicity test using organisms with a cosmopolitan distribution, *Environ. Monit. Assess.*, 6, 207–220, 1986.

Caswell, H., Demography meets ecotoxicology: Untangling the population level effects of toxic substances, In *Ecotoxicology. A Hierarchical Treatment*, Newman, M.C. and Jagoe, C.H. (eds.), CRC Press/Lewis Publishers, Boca Raton, FL, 1996, pp. 255–292.

Costantino, R.F., Desharnais, R.A., Cushing, J.M., and Dennis, B., Chaotic dynamics in an insect population, *Science*, 275, 389–391, 1997.

Daniels, R.E. and Allan, J.D., Life table evaluation of chronic exposure to a pesticide, *Can. J. Fish. Aquat. Sci.*, 38, 485–494, 1981.

DeAngelis, D.L. and Gross, L.J. (eds.), *Individual-based Models and Approaches in Ecology*, Chapman & Hall, New York, 1992.

Emlen, J.M., Terrestrial population models for ecological risk assessment: A state-of-the-art review, *Environ. Toxicol. Chem.*, 8, 831–842, 1989.

EPA, *Summary Report on Issues in Ecological Risk Assessment, EPA/625/3-91/018* February 1991, NTIS, Springfield, 1991.

Everhart, W.H., Eipper, A.W., and Youngs, W.D., *Principles of Fishery Science*, Cornell University Press, Ithaca, NY, 1953.

Forbes, V.E. and Calow, P., Is the per capita rate of increase a good measure of population-level effects in ecotoxicology? *Environ. Toxicol. Chem.*, 18, 1544–1556, 1999.

Gard, T.C., A stochastic model for the effects of toxicants on populations, *Ecol. Modell.*, 51, 273–280, 1990.

Gard, T.C., Stochastic models for toxicant-stressed populations, *Bull. Math. Biol.*, 54, 827–837, 1992.

Gilpin, M.E. and Ayala, F.J., Global models of growth and competition, *Proc. Natl. Acad. Sci. USA*, 70, 3590–3593, 1973.

Gilpin, M.E. and Hanski, I. (eds.), *Metapopulation Dynamics: Empirical and Theoretical Investigations*, Harcourt Brace Jovanovich, London, UK, 1991.

Gotelli, N.J., Metapopulation models: The rescue effect, the propagule rain, and the core-satellite hypothesis, *Am. Nat.*, 138, 768–776, 1991.

Gulland, J.A., *Fish Population Dynamics*, John Wiley & Sons, London, UK, 1977.

Haddon, M., *Modelling and Quantitative Methods in Fisheries*, Chapman and Hall/CRC Press, Boca Raton, FL, 2001.

Hanski, I., Metapopulation ecology, In *Population Dynamics in Ecological Space and Time*, Rhodes, O.E., Jr., Chesser, R.K., and Smith, M.H. (eds.), University of Chicago Press, Chicago, IL, 1996, pp. 13–43.

Krebs, C.J., *Ecological Methodology*, Harper Collins Publishers, Inc., New York, 1989.

Kuhn, T.S., *The Structure of Scientific Revolutions*, University of Chicago Press, Chicago, IL, 1962.

Lakatos, I., Falsification and the methodology of scientific research programmes, In *Criticism and the Growth of Knowledge*, Lakatos, I. and Musgrave, A. (eds.), Cambridge University Press, Cambridge, UK, 1970, pp. 91–196.

Levins, R., Some demographic and genetic consequences of environmental heterogeneity for biological control, *Bull. Entomol. Soc. Am.*, 15, 237–240, 1969.

Levins, R., Extinction, *Lect. Math. Life Sci.*, 2, 75–107, 1970.

Lewin, R., Sources and sinks complicate ecology, *Science*, 243, 477–478, 1989.

Ludwig, J.A. and Reynolds, J.F., *Statistical Ecology. A Primer on Methods and Computing*, John Wiley & Sons, New York, 1988.

May, R.M., Biological populations with nonoverlapping generations: Stable points, stable cycles, and chaos, *Science*, 186, 645–647, 1974.

May, R.M., *Theoretical Ecology. Principles and Applications*, W.B. Saunders Co., Philadelphia, PA, 1976a.

May, R.M., Simple mathematical models with very complicated dynamics, *Nature*, 261, 459–467, 1976b.

May, R.M., Conway, G.R., Hassell, M.P., and Southwood, T.R.E., Time delays, density-dependence and single-species oscillations, *J. Anim. Ecol.*, 43, 747–770, 1974.

May, R.M. and Oster, G.F., Bifurcation and dynamic complexity in simple ecological models, *Am. Nat.*, 110, 573–599, 1976.

Moriarty, F., *Ecotoxicology. The Study of Pollutants in Ecosystems*, Academic Press, Inc., London, UK, 1983.

Murray, J.D., *Mathematical Biology*, Springer-Verlag, Berlin, 1993.

Newman, M.C., *Quantitative Methods in Aquatic Ecotoxicology*, CRC Press/Lewis Publishers, Boca Raton, FL, 1995.

Newman, M.C., *Fundamentals of Ecotoxicology*, CRC Press/Lewis Publishers, Boca Raton, FL, 1998.

Newman, M.C. and Jagoe, R.H., Allozymes reflect the population-level effect of mercury: Simulations of the mosquitofish (*Gambusia holbrooki* Girard) GPI-2 response, *Ecotoxiology*, 7, 141–150, 1998.

Nicholson, A.J., An outline of the dynamics of animal populations, *Aust. J. Zool.*, 2, 9–65, 1954.

Norton, S.B., Rodier, D.J., Gentile, J.H., Van der Schalie, W.H., Wood, W.P., and Slimak, M.W., A framework for ecological risk assessment at the EPA, *Environ. Toxicol. Chem.*, 11, 1663–1672, 1992.

O'Connor, R.J., Toward the incorporation of spatiotemporal dynamics into ecotoxicology, In *Population Dynamics in Ecological Space and Time*. Rhodes, O.E., Jr., Chesser, R.K. and Smith, M.H. (eds.), University of Chicago Press, Chicago, IL, 1996, pp. 281–317.

Philippi, T.E., Carpenter, M.P., Case, T.J., and Gilpin, M.E., *Drosophila* population dynamics: Chaos and extinction, *Ecology*, 68, 154–159, 1987.

Popper, K.R., *Objective Knowledge. An Evolutionary Approach*, Clarendon Press, Oxford, UK, 1972.

Pulliam, H.R., Sources and sinks: Empirical evidence and population consequences, In *Population Dynamics in Ecological Space and Time*, Rhodes, O.E., Jr., Chesser, R.K., and Smith, M.H. (eds.), University of Chicago Press, Chicago, IL, 1996, pp. 45–69.

Pulliam, H.R. and Danielson, B.J., Sources, sinks, and habitat selection: A landscape perspective on population dynamics, *Am. Nat.*, 137, S50–S66, 1991.

Schaffer, W.M. and Kot, M., Chaos in ecological systems: The coals that Newcastle forgot. *TREE*, 1, 58–63, 1986.

Sibly, R.M., Effects of pollutants on individual life histories and population growth rates, In *Ecotoxicology. A Hierarchical Treatment*, Newman, M.C. and Jagoe, C.H. (eds.), CRC Press/Lewis Publishers, Boca Raton, FL, 1996, pp. 197–223.

Simkiss, K., Daniels, S., and Smith, R.H., Effects of population density and cadmium toxicity on growth and survival of blowflies, *Environ. Pollut.*, 81, 41–45, 1993.

Spromberg, J.A., John, B.M., and Mandis, W.G., Metapopulation dynamics: Indirect effects and multiple distinct outcomes in ecological risk assessment, *Environ. Toxicol. Chem.*, 17, 1640–1649, 1998.

Thomas, W.R., Pomerantz, M.J., and Gilpin, M.E., Chaos, asymmetric growth and group selection for dynamical stability, *Ecology*, 6, 1312–1320, 1980.

Verhulst, P.F., Notice sur la loi que la population suit dans son accroissement, *Corr. Math. et Phys.*, 10, 113–121, 1838.

Waller, W.T., Dahlberg, M.L., Sparks, R.E., and Cairns, J., Jr., A computer simulation of the effects of superimposed mortality due to pollutants of fathead minnows (*Pimephales promelas*), *J. Fish. Res. Board Can.*, 28, 1107–1112, 1971.

Wilson, E.O. and Bossert, W.H., *A Primer of Population Biology*, Sinauer Associates, Inc., Sunderland, MA, 1971.

Wu, J., Vankat, J.L., and Barlas, Y., Effects of patch connectivity and arrangement on animal metapopulation dynamics: A simulation study, *Ecol. Modell.*, 65, 221–254, 1993.

15 Toxicants and Population Demographics

There's a special providence in the fall of a sparrow. If it be now, 'tis not to come if it be not to come, it will be now; if it be not now, yet it will come: the readiness is all . . . .

(*Hamlet* Act V, SC II)

15.1 DEMOGRAPHY: ADDING INDIVIDUAL HETEROGENEITY TO POPULATION MODELS

Discussion so far grew from phenomenological models involving identical and uniformly distributed individuals to metapopulation models incorporating spatial heterogeneity. Now, demography, the quantitative study of death, birth, age, migration, and sex in populations, will be explored. Differences among individuals produce distinct vital rates, that is, rates of death, birth, transition to the next life stage, and migration. Combined, vital rates determine a population's overall characteristics. In fact, population vital rates were aggregated earlier into summary statistics such as the intrinsic rate of increase, resulting in hidden information and incomplete insight. Finally, metapopulation models including demographic vital rates can be discussed briefly to get the fullest description of and most realistic predictions of population consequences of toxicant exposure. Variation in vital rates can be added also to render a stochastic model. Such a model could be applied to estimate the probability of local extinction for a metapopulation based on contaminant-induced changes in vital rates.

Demographic analysis allows the qualities and fate of toxicant-exposed populations to be determined. In the recent past, conventional ecotoxicological precepts suggested that a species population will remain viable if the most sensitive life stage of the species is "protected," e.g., toxicant concentrations do not exceed the no observed effect level (NOEC) or MATC concentration for that life stage. Early life stage testing results were applied under the premise that the *population* will remain viable if the weakest link in an *individual's* various life stages was protected. But this is not always true. Newman (1998) refers to this false paradigm as the weakest link incongruity. The most sensitive stage of an individual's life cycle might not be the most crucial relative to population vitality or viability (Kammenga et al. 1996, Petersen and Petersen 1988). This will become obvious as we discuss reproductive value, elasticity, and related topics below. Fortunately, ecotoxicology is rapidly moving toward a more balanced inclusion of demographic analysis (e.g., Bechmann 1994, Chaumot et al. 2003, Daniels and Allan 1981, Koivisto and Ketola 1995, Martinez-Jerónimo et al. 1993, Münzinger and Guarducci 1988, Pesch et al. 1991, Spurgeon et al. 2003). Required now is a sustained and insightful integration of demography into assessments of ecological risk. The intent of this chapter is to contribute to this integration by describing foundation demographic concepts and methods. Straightforward algebraic (e.g., Marshall 1962) and matrix (e.g., Caswell 1996, Lefkovitch 1965, Leslie 1945, 1948) formulations will be described because both are applied in population ecotoxicology.

15.1.1 STRUCTURED POPULATIONS

Age-, stage-, and sex-dependent vital rates will be considered in this section. Age data may be applied when available or, alternatively, analyses might focus on vital rates at different life stages such as larval → juvenile → and adult stages. For example, the effects of dioxin and polychlorinated biphenyls (PCBs) on *Fundulus heteroclitus* populations were modeled by considering the following life stages: embryos → larvae → 28-day larvae → 1-year-old adults → 2-year-old adults → 3-year-old adults (Munns et al. 1997). Sex-dependent vital rates can be important too but our focus here will remain primarily on females of differing ages or stages.

15.1.2 Basic Life Tables

Life tables or schedules are constructed either for mortality alone, both mortality and birth (natality), or mortality, natality, and migration combined. Obviously, analysis of a metapopulation requires the inclusion of movement among subpopulations. In this chapter, we will only show calculations that are relevant to populations with no migration; however, inclusion of these methods in metapopulation models would be possible using concepts described in the last chapter.

Data for life tables are gathered in three ways. To produce a cohort life table, a cohort of individuals is followed through time with tabulation of mortality alone, or mortality and natality. As an example, a group of 1000 young-of-the-year (YOY0+) may be tagged during the calving season and survival of these calves followed through the years of their lives. Other cohorts present in the population are ignored. In contrast, a horizontal life table includes measurements about all individuals in the population at a particular time and several cohorts are included. All individuals within the various age classes are counted and the associated data summarized in a horizontal table. An important point to note here is that the results of cohort and horizontal life tables will not always be identical for the same population. They would be identical only if environmental conditions were sufficiently stable so that vital rates remain fairly independent of time, that is, independent of the specific cohort(s) from which they were derived. In a composite life table, data are collected for several cohorts and combined. For example, a team of game managers might tag newborns during four consecutive calving seasons, follow the four cohorts through time, and then combine the final results in one table.

15.1.2.1 Survival Schedules

Oh, Death, why canst thou sometimes be timely?

(Melville, *Moby Dick* 1851)

Sometimes life schedules quantify death only. Life insurance companies or some ecological risk assessors might correctly pay most attention to the likelihood of dying and consider natality as irrelevant. The associated tabulations are called l_x schedules because, by convention, the symbol l_x designates the number or proportion of survivors in the age class, x. Often, l_x is expressed as a proportion of the original number of newborns surviving to age x. In that form, it also estimates the probability of survival to age x.

From l_x schedules, simple estimates are made of the number of deaths ($d_x = l_x - l_{x+1}$), rate of mortality (q_x), and expected lifetime for an individual surviving to age x (e_x). Like l_x, if d_x is expressed as a proportion dying instead of actual number dying, d_x estimates the probability of dying in the interval x to $x+1$. These estimates may be expressed as a simple quotient (e.g., $q_x = d_x/l_x$) or normalized to a specific number of individuals in the age class such as deaths per 1000 individuals (e.g., $1000q_x = 1000(d_x/l_x)$) (Deevey 1947).

The mean expected length of life beyond age x for an individual who survived to age x (e_x) can be estimated for any age class (x) by dividing the area under the survival curve after x by the number of individuals surviving to age x (Deevey 1947),

$$e_x = \frac{\int_x^\infty l_x \, \mathrm{d}x}{l_x}. \tag{15.1}$$

With a basic l_x table, the e_x in the above equation can be approximated with the l_x and L_x (number of living individuals between x and $x + 1$ in age):

$$L_x = \int_x^{x+1} l_x \, dx. \tag{15.2}$$

A simple linear approximation of L_x in the above equation is $L_x = (l_x + l_{x+1})/2$. Obviously, the ∞ in the summations here and elsewhere become the age at the bottommost row of the completed life table. These L_x approximations are summed in the life table from the bottommost row up to and including the age of interest (x). The e_x value for an age class is then estimated by dividing this sum (T_x) by l_x (i.e., $e_x = T_x/l_x$). (The T_x is the total years lived by all individuals in the x age class.) The e_0 or expected life span for an individual at the beginning of the life table (i.e., a neonate), and its associated variance are estimated by Leslie et al. (1955) and described in detail by Krebs (1989).

A quick check of Section 13.1.3.1, Accelerated Failure Time and Proportional Hazard Models, will show a striking similarity between those epidemiological methods for modeling mortality and these simple life table methods. In fact, the method just described is simply one method for summarizing survival information. Methods, models, and hypothesis tests described in Section 13.1.3.1 or 9.2.3 can be, and often are, applied in demography. As an example, Spurgeon et al. (2003) applied a Weibull model to survival data for metal-exposed earthworms.

Box 15.1 Death, Decline, and Gamma Rays

As the possibility of nuclear war emerged in the 1950s and 1960s, researchers began to explore the ecological effects of intense irradiation. Ecological entities from individuals (e.g., Casarett 1968) to populations (e.g., Marshall 1962) to entire ecosystems (e.g., Woodwell 1962, 1963) were irradiated in numerous studies to determine the consequences. One study placed cultures of *Daphnia pulex* (50 individuals per culture) at a series of distances from a 5000 Curie cobalt (^{60}Co) source. The *Daphnia* experienced continuous gamma irradiation at dose rates of 0, 22.8, 47.9, 52.2, 67.5, and 75.9 R/h. Survival was monitored for 35 days and life schedules constructed for each irradiated population (Table 15.1). Instead of estimating a simple LD50 at a set time, Marshall (1962) used demographic methods to summarize the population consequences of irradiation. This allowed estimation of the change in average life expectancy as a consequence

TABLE 15.1
Survival Rates (l_x as a Proportion of the Original Population) for *Daphnia pulex* Continuously Irradiated with Radiocobalt

	Dose Rate (R/h)					
Days (x)	**0**	**22.8**	**47.9**	**52.2**	**67.5**	**75.9**
0	1.00	1.00	1.00	1.00	1.00	1.00
7	0.98	0.98	0.98	0.98	0.98	0.96
14	0.98	0.96	0.98	0.94	0.96	0.94
21	0.98	0.88	0.48	0.16	0.12	0.02
28	0.19	0.53	0.00	0.00	0.00	0.00
35	0.00	0.00	0.00	0.00	0.00	0.00

Source: Modified from Table I in Marshall (1962).

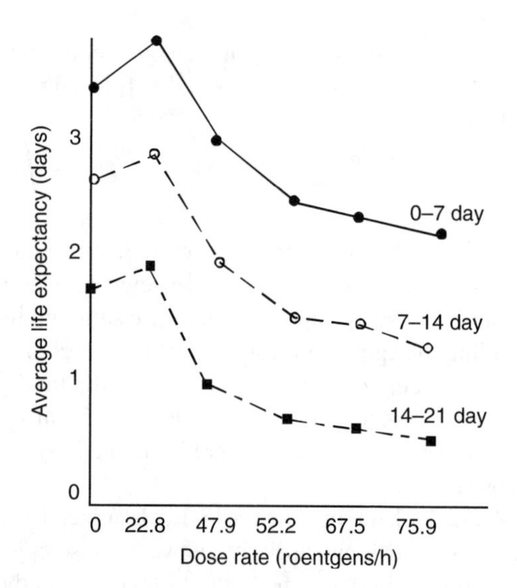

FIGURE 15.1 Calculated life expectancies for three age classes of *Daphnia pulex* as a function of gamma irradiation dose rate.

of dose rate (Figure 15.1). For the sake of brevity, calculations were done here by using weekly age classes, not daily age classes as done in the original publication. Even with this simplified analysis, the decrease in average life expectancy for the different age classes was obvious. Note that in Figure 15.1 there is a suggestion of a hormetic effect at 22.8 R/h (see Sections 9.1.4 and 16.2 for more discussion of hormesis).

Obviously, survival functions and life expectancies provide valuable insights into population consequences and, when combined later with natality data (Box 15.2), of population fate under different intensities of irradiation.

15.1.2.2 Mortality–Natality Tables

> There is an appointed time for everything, and a time for every affair under the heavens. A time to be born, a time to die . . . .
>
> **(Ecclesiastes 3)**

The inclusion of information on births (natality, m_x) in addition to mortality (l_x) allows expansion of this approach. The resulting schedules are called $l_x m_x$ tables. Often, $l_x m_x$ tables quantify information for females alone because the reproductive contribution of males to the next generation is much more difficult to estimate than that of females. An m_x is estimated for females as the average number of female offspring produced per female of age x. Several useful population qualities can be estimated after the age-specific birth rates (m_x) and l_x values are known. The expected number of female offspring produced in the lifetime of a female or net reproductive rate (R_0) is defined by the following equation (Birch 1948):

$$R_0 = \int_0^{\infty} l_x m_x \, dx. \qquad (15.3)$$

This ratio of female births in two successive generations is estimated as the sum of the products $l_x m_x$ for all age classes: $R_0 = \Sigma l_x m_x$. Knowing R_0, a mean generation time (T_c) can be calculated by dividing the sum of all the $x l_x m_x$ values by R_0. (The midpoint of interval x to $x + 1$ is used as "x" in generating the product, $x l_x m_x$. For example, $(0 + 1)/2$ or 0.5 would be used for x of the interval 0 to 1-year-old.) It can also be estimated with the following equation; however, an estimate of the intrinsic rate of increase (r) would be needed:

$$T_c = \frac{\ln R_0}{r}. \tag{15.4}$$

The intrinsic rate of increase (r) could be grossly estimated with Equation 15.5, which is a simple rearrangement of Equation 15.4:

$$r = \frac{\ln R_0}{T_c}. \tag{15.5}$$

This rough estimate of r can then be used as an initial estimate in the Euler–Lotka equation (Equation 15.6) (Euler 1760, Lotka 1907), which becomes Equation 15.7 for the approximate method applied to simple life tables (Birch 1948):

$$\int_0^\infty e^{-rx} l_x m_x \, dx = 1, \tag{15.6}$$

$$\sum_{x=0}^{\omega} l_x m_x e^{-rx} = 1, \tag{15.7}$$

where ω indicates the result for the bottommost row of the life table. The x, l_x, and m_x values, and the initial estimate of r from Equation 15.5 are placed into Equation 15.7, and the equation solved. Next, the value of r is changed slightly and the equation is solved again. This process is repeated with different estimates of r until an r is found for which the equality is "close enough." This final value of r is the best estimate from the life table. The assumptions here are that the population is increasing exponentially and the population is stable; however, Stearns (1992) states that this approach is robust to violations of the assumption of a stable age structure.

A stable population is one in which the distribution of individuals among the various age (or stage) classes remains constant through time. The structure of such a population is called its stable age structure. Any population with a constant r or λ will eventually take on a stable age structure: the eventual distribution of individuals among the age classes will be a consequence of age-specific birth and death rates. The proportion of all individuals in age class x for a stable population (C_x) is defined by Equation 15.8 (see Birch (1948), Caswell (1996), Newman (1995), or Stearns (1992) for more details):

$$C_x = \frac{\lambda^{-x} l_x}{\sum_{i=0}^{\omega} \lambda^{-i} l_i}. \tag{15.8}$$

Remember from the last chapter that $\lambda = e^r$.

Reproductive value (V_A) is a measure of the number of females that will be produced by a female of age A under the assumption of a stationary population. A stationary population is one in which simple replacement is occurring (i.e., $R_0 = 1$ or $r = 0$). Therefore, by definition, neonates will have a V_A ($=V_0$) of 1 because each will just replace herself in a stationary population. Postreproductive females will have V_A values of 0. It follows that the V_A can be envisioned as the reproductive value for a specific class, x, divided by that of a neonate (i.e., $V_A = V_x/V_0$).

Age- or stage-specific reproductive values for a population are a valuable set of measures of the contribution of offspring to be expected from each age class to the next generation. The relative sizes of V_A values for the different age classes suggest the value of each age class in contributing new individuals to the next generation. It takes simultaneously into account the facts that a female has survived to age x and that she has an age-specific capacity to produce young. (See Stearns (1992) or Wilson and Bossert (1971) for a detailed description of V_A and stepwise derivation of equations associated with V_A. Newman (1995) provides a detailed example of applying V_A to ecotoxicology.)

$$V_A = \sum_{x=A}^{\omega} \frac{l_x}{l_A} m_x. \tag{15.9}$$

Goodman (1982) (detailed in Stearns 1992) provides Equation 15.10, a modification of the Euler–Lotka equation, to describe V_A in an exponentially growing population. The lower contribution of offspring born later relative to the contribution of those born earlier is included in this equation (Stearns 1992, Wilson and Bossert 1971),

$$V_A = \frac{e^{r(A-1)}}{l_A} \sum_{x=A}^{\omega} e^{-rx} l_x m_x. \tag{15.10}$$

This demographic metric provides valuable insights relevant to the weakest link incongruity. The reproductive value (V_A) suggests the loss of individuals that would otherwise come into the next generation if one individual of a certain age class were removed from the population. The most valuable individuals in this context are not always the young stages that are most sensitive to toxicant action. In general, one could argue that individuals just entering their reproductive stage might be more valuable as they usually have very high reproductive values (Wilson and Bossert 1971). Regardless, conventional generalizations are insufficient that protection of the most sensitive stage based on life stage testing will ensure a viable population. This point will be reinforced later in discussions of sensitivity and elasticity. A demographic analysis should be done in order to make any judgments about the population consequences of toxicant exposure.

There is also a definite linkage between this demographic concept of reproductive value and those described earlier for sustainable harvest. Owing to aggregation of information, stimulation of harvest based solely on total numbers would be less effective than estimation based on a fuller knowledge of age- or size-specific harvests and reproductive values. Stock assessment models including size-specific harvesting gear have direct relevance to age-specific mortality in populations due to toxicant exposure.

Box 15.2 Death, Decline, Gamma Rays, and Birth

Marshall (1962) measured natality in addition to mortality for *D. pulex* exposed to gamma radiation. Let us add these natality data (Table 15.2) to that already analyzed for mortality (Table 15.1). Again, data are pooled here into weekly age classes.

The Euler–Lotka equation (Equation 15.7) was used to estimate the intrinsic rates of increase for the irradiated populations (Figure 15.2). Notice the general decrease in r until it drops below 0 at approximately 67.5 R/h. At that point, the population would slowly drop in size until extinction occurred. The stable population structures (Figure 15.3) show a trend from a control population with many young to highly dosed populations with proportionally fewer young and many more old individuals. Given this shift, it is interesting to note that Aubone (2004) found decreased population stability with fishery practices that skewed the stable age structure toward juveniles. From the lowest to the highest dose, the generation times dropped rapidly from 13.6 to only 4.8–6.0 days.

TABLE 15.2
Natality (m_x) for *D. pulex* Continuously Irradiated with Radiocobalt

	Dose Rate (R/h)					
Day (x to $x+1$)	**0**	**22.8**	**47.9**	**52.2**	**67.5**	**75.9**
1–6	2.63	2.29	1.94	1.88	0.94	0.39
7–13	14.64	10.84	1.60	0.45	0.18	0.22
14–20	3.29	1.06	0.02			
21–27	0.35					
28–35	0.31					

Source: Modified from Table II in Marshall (1962).

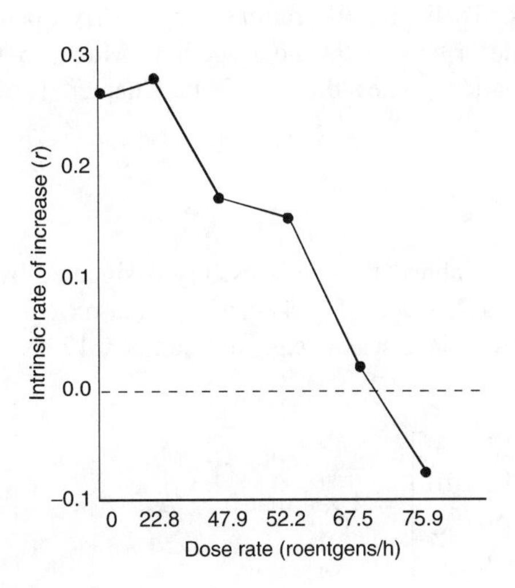

FIGURE 15.2 Drop in intrinsic rate of increase (r) with dose rate for *D. pulex* cultures.

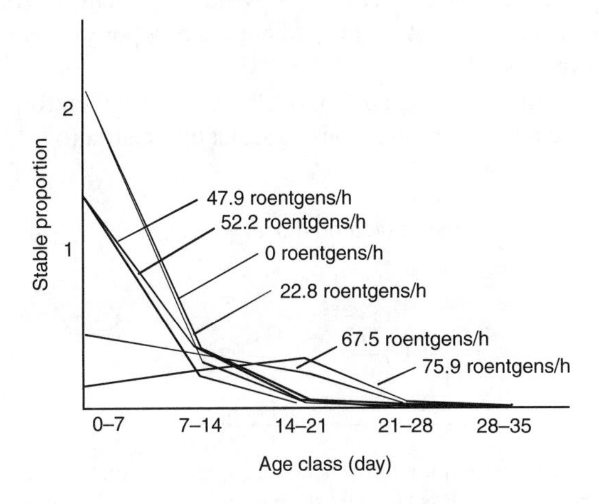

FIGURE 15.3 Shift in stable population structure for *D. pulex* cultures exposed to different dose rates of gamma radiation.

Clearly, meaningful information relative to population changes and consequences were obtained from this simple demographic analysis. Irradiation reduced average life expectancy and generation time. Population growth rate decreased with dose until it fell below simple replacement at approximately 67.5 R/h. Populations receiving such doses would disappear after a few generations. The age structure of the populations shifted to a preponderance of older individuals. In our opinion, these insights are much more meaningful than those provided by LD50 and NOEC data.

15.2 MATRIX FORMS OF DEMOGRAPHIC MODELS

To this point, discussion was simplified by avoiding matrix algebra. However, the approach becomes much more effective with matrix formulations for demographic qualities (Figure 15.4). Matrix formulations have existed for some time: Leslie (1945, 1948) articulated the foundation matrix approach to age-structured demographics. To begin, the rudimentary matrix operations needed to apply a matrix approach will be described in the next section. Much, but not all, of the description of basic matrix mathematics comes directly from Chapter 1 of Emlen (1984).

15.2.1 Basics of Matrix Calculations

A matrix is simply a rectangular array of numbers or variables. Its size is usually designated by the number of rows (i) and columns (j), for example, a 4×1 or 4×4 matrix. A matrix composed of only one row is called a vector. A 1×1 matrix is a scalar, e.g., the number, 12, is a scalar:

$$\begin{bmatrix} 2 \\ 5 \\ 6 \\ 3 \end{bmatrix} = 4 \times 1 \text{ matrix} = \mathbf{a}, \qquad \begin{bmatrix} 12 & 5 & 17 & 3 \\ 23 & 0 & 12 & 10 \\ 5 & 5 & 5 & 3 \\ 12 & 7 & 13 & 5 \end{bmatrix} = 4 \times 4 \text{ matrix} = \mathbf{A}.$$

Matrices are conventionally designated with boldfaced, capital letters (e.g., **A**), except vectors that are designated as boldface, small letters (e.g., **a** above). Scalars are written as small letters without boldfacing. A matrix can be designated generally as $\mathbf{A} = \{a_{ij}\}$ where i is row position and j is column position. For example, element $\mathbf{a}_{13}$ in **A** is 17.

We will need to do simple matrix multiplication in the demographic models that follow; therefore, a quick review of matrix multiplication is presented here. Multiplication of a scalar by a matrix ($\mathrm{b} \times \mathbf{A}$)

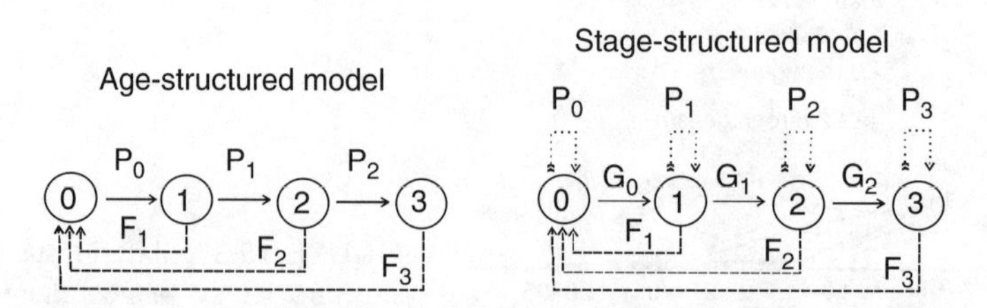

FIGURE 15.4 An illustration of age- and stage-structured population models. The age-structured model specifies natality (F) and probability of moving to the next age class (P). The stage-structured model specifies the natality (F), probability of moving to the next stage class (G), and probability of remaining in the stage class (P).

is very straightforward. Each individual element of the matrix is simply multiplied by the scalar. Let b be a scalar with value 12 and **A** be a 2 × 2 matrix:

$$12 \times \mathbf{A} = 12 \begin{bmatrix} a_{11} & a_{12} \\ a_{21} & a_{22} \end{bmatrix} = \begin{bmatrix} 12a_{11} & 12a_{12} \\ 12a_{21} & 12a_{22} \end{bmatrix}.$$

Multiplication of a matrix **A** by another matrix **B** is more tedious but no more difficult to grasp. The cross products of the rows of **A** and columns of **B** are generated and summed. Let us use the **A** matrix (2 × 2) described immediately above and multiply it by another 2 × 2 matrix, **B**.

$$\mathbf{A} \times \mathbf{B} = \begin{bmatrix} a_{11} & a_{12} \\ a_{21} & a_{22} \end{bmatrix} \times \begin{bmatrix} b_{11} & b_{12} \\ b_{21} & b_{22} \end{bmatrix} = \begin{bmatrix} a_{11}b_{11} + a_{12}b_{21} & a_{11}b_{12} + a_{12}b_{22} \\ a_{21}b_{11} + a_{22}b_{21} & a_{21}b_{12} + a_{22}b_{22} \end{bmatrix}.$$

For example,

$$\mathbf{A} \times \mathbf{B} = \begin{bmatrix} 1 & 3 \\ 5 & 2 \end{bmatrix} \times \begin{bmatrix} 4 & 1 \\ 5 & 6 \end{bmatrix} = \begin{bmatrix} 4 + 15 & 1 + 18 \\ 20 + 10 & 5 + 12 \end{bmatrix} = \begin{bmatrix} 19 & 19 \\ 30 & 17 \end{bmatrix}.$$

Multiplication of a matrix (**A**) and a vector (**b**) is done in the same way,

$$\mathbf{A} \times \mathbf{b} = \begin{bmatrix} a_{11} & a_{12} \\ a_{21} & a_{22} \end{bmatrix} \times \begin{bmatrix} b_{11} \\ b_{21} \end{bmatrix} = \begin{bmatrix} a_{11}b_{11} + a_{12}b_{21} \\ a_{21}b_{11} + a_{22}b_{21} \end{bmatrix}.$$

We can demonstrate this multiplication by modifying the above example,

$$\mathbf{A} \times \mathbf{b} = \begin{bmatrix} 1 & 3 \\ 5 & 2 \end{bmatrix} \times \begin{bmatrix} 4 \\ 5 \end{bmatrix} = \begin{bmatrix} 4 + 15 \\ 20 + 10 \end{bmatrix} = \begin{bmatrix} 19 \\ 30 \end{bmatrix}.$$

We will also need to transpose a matrix in one of the following calculations. In this simple procedure, one simply makes the rows of the original matrix (**A**) into the columns of the matrix transpose ($\mathbf{A}^T$)

$$\mathbf{A}^T = \begin{bmatrix} 1 & 3 \\ 5 & 2 \end{bmatrix}^T = \begin{bmatrix} 1 & 5 \\ 3 & 2 \end{bmatrix}.$$

In the preceding text, we noted that a matrix multiplied by a vector results in a column vector:

$$\begin{bmatrix} 2 \\ 5 \\ 6 \\ 3 \end{bmatrix}.$$

Please note that, in the following application, applying multiplication of a matrix transpose and a vector, the result will be a row vector,

$$[2 \quad 5 \quad 6 \quad 3].$$

With these simple matrix operations, the matrix formulations of demographic models can now be explored.

15.2.2 The Leslie Age-Structured Matrix Approach

More than half century ago, Leslie (1945, 1948) took natality and mortality rates from life tables and arranged them into simple matrices. He placed the probability (P_x) of a female alive in age class x being alive to enter age class $x+1$ in the subdiagonal of a matrix. This probability can be approximated as the number of individuals alive in age class $x+1$ divided by the number alive in age class x. The numbers of daughters (F_x) born in the time interval t to $t+1$ per female in this age class were placed in the top row of a square ($\omega \times \omega$) matrix (**L**). The remaining matrix elements were zeros. The conditions for the Leslie matrix being valid are $0 < P_x < 1$ and $F_x \geq 0$,

$$\mathbf{L} = \begin{bmatrix} 0 & F_1 & F_2 & F_3 & \cdots & F_\omega \\ P_0 & 0 & 0 & 0 & \cdots & 0 \\ 0 & P_1 & 0 & 0 & \cdots & 0 \\ 0 & 0 & P_2 & 0 & 0 & 0 \\ \cdots & \cdots & \cdots & \cdots & \cdots & \cdots \\ 0 & 0 & 0 & 0 & P_{\omega-1} & 0 \end{bmatrix}.$$

As an example of the use of such a matrix approach in ecotoxicology, Laskowski and Hopkin (1996) generated the following Leslie matrix for common garden snails (*Helix aspersa*) exposed to a mixture of metals in food. (See Box 15.3 and Laskowski (2000) for additional discussion.)

$$\begin{bmatrix} 0 & 0 & 54 & 54 & 54 & 54 \\ .05 & 0 & 0 & 0 & 0 & 0 \\ 0 & .20 & 0 & 0 & 0 & 0 \\ 0 & 0 & .25 & 0 & 0 & 0 \\ 0 & 0 & 0 & .25 & 0 & 0 \\ 0 & 0 & 0 & 0 & .15 & 0 \end{bmatrix}$$

Among the many convenient aspects of this matrix formulation of demographic vital rates, this matrix (**L**) can be multiplied by a vector ($\mathbf{n}_t$) of the number of individuals at the various x ages to predict the number of individuals in each age class at some time in the future (e.g., the *Daphnia* populations described in Tables 15.1 and 15.2).

$$\mathbf{L} \times \mathbf{n} = \begin{bmatrix} F_0 & F_1 & F_2 & F_3 & \cdots & F_\omega \\ P_0 & 0 & 0 & 0 & \cdots & 0 \\ 0 & P_1 & 0 & 0 & \cdots & 0 \\ 0 & 0 & P_2 & 0 & 0 & 0 \\ \cdots & \cdots & \cdots & \cdots & \cdots & \cdots \\ 0 & 0 & 0 & 0 & P_{\omega-1} & 0 \end{bmatrix} \times \begin{bmatrix} n_{0,t} \\ n_{1,t} \\ n_{2,t} \\ n_{3,t} \\ \cdots \\ n_{\omega,t} \end{bmatrix} = \begin{bmatrix} n_{0,t+1} \\ n_{1,t+1} \\ n_{2,t+1} \\ n_{3,t+1} \\ \cdots \\ n_{\omega,t+1} \end{bmatrix} \tag{15.11}$$

The Leslie matrix can then be multiplied by this new vector of age class sizes for $t+1$ to project the age class sizes at time $t+2$. The process can be repeated for $t+3$, and so on, through many time steps. Emlen (1984) provides the following simple example of this process. Let the initial population be composed of 200 neonates with the population demographics summarized by the Leslie matrix, **L**,

$$\mathbf{n}_0 = \begin{bmatrix} 200 \\ 0 \\ 0 \end{bmatrix}.$$

Box 15.3 Quick Calculations for Snails Ingesting Contaminated Food

Let us quickly illustrate some calculations using the garden snails exposed to zinc in their food (Laskowski and Hopkin 1996). The Leslie matrix for these snails after exposure to approximately 3000 mg of zinc per kg of food was the following:

$$\begin{bmatrix} 0 & 0 & 54 & 54 & 54 & 54 \\ .05 & 0 & 0 & 0 & 0 & 0 \\ 0 & .20 & 0 & 0 & 0 & 0 \\ 0 & 0 & .25 & 0 & 0 & 0 \\ 0 & 0 & 0 & .25 & 0 & 0 \\ 0 & 0 & 0 & 0 & .15 & 0 \end{bmatrix}$$

According to Laskowski (2000), this particular metal exposure had no discernible affect on survival, but there was an approximately 28% reduction in reproduction relative to reference snail populations.

The tedious calculations required to estimate the population growth rate, stable, and structure and reproductive value can be rendered easy by applying one of several software packages. Here, let us use the shareware, PopTools (Hood 2004). The estimated eigenvalue (λ) is 0.904 and the R_0 is 0.714. Both metrics project that the population will decline through time. The mean generation time is estimated to be 3.3 years. The stable age structure (right eigenvector) and reproductive values (left eigenvector) are estimated to be the following using the matrix calculation to be described in the next paragraphs.

Age	Age Structure (%)	Reproductive Value (%)
0	93.3	0.3
1	5.2	5.8
2	1.1	26.4
3	0.3	25.6
4	0.1	22.5
5	0.0	19.3

From these numbers, we can project that, as it declines, the population will be composed primarily of young but the age 2–5 adults contribute the most to the population reproduction.

The vector of age-class sizes after one time step, $\mathbf{n}_1$ is equal to $\mathbf{L} \times \mathbf{n}_0$,

$$\mathbf{n}_1 = \mathbf{L} \times \mathbf{n}_0 = \begin{bmatrix} 0 & 1 & 4 \\ 0.5 & 0 & 0 \\ 0 & 0.25 & 0 \end{bmatrix} \times \begin{bmatrix} 200 \\ 0 \\ 0 \end{bmatrix} = \begin{bmatrix} 0 \\ 100 \\ 0 \end{bmatrix}.$$

Obviously, from the F_0 element of $\mathbf{L}$, the neonates do not reproduce during their first x to $x + 1$ period of life so the number of newborns at time step 1 is 0. Half of the yearlings die in x to $x + 1$;

so the size of this cohort drops from the original 200 to 100. And with a second time step,

$$\mathbf{n}_2 = \mathbf{L} \times \mathbf{n}_1 = \begin{bmatrix} 0 & 1 & 4 \\ 0.5 & 0 & 0 \\ 0 & 0.25 & 0 \end{bmatrix} \times \begin{bmatrix} 0 \\ 100 \\ 0 \end{bmatrix} = \begin{bmatrix} 100 \\ 0 \\ 25 \end{bmatrix}.$$

Now, the 100 individuals have moved into a reproductive stage of their lives, resulting in 100 ($=100 \times 1$) newborns. Because the survival of the original cohort was only expected to be 0.25 for the next step, only 100×0.25 or 25 remain. Additional iterations could be carried out to track the population further through time but the method has been demonstrated sufficiently with these few steps.

Other calculations can be performed with this approach and only a few are presented here. As examples, the right and left eigenvectors of the Leslie matrix define the stable age-structured and age-specific reproductive values for the population, respectively. The matrix can also be used to estimate λ. Let us take a moment to show a few of these calculations.

The dominant eigenvalue (specific growth rate or λ) is straightforward to compute. An estimate of λ at time, t (i.e., λ_t) can be produced by dividing the total number of individuals in the population at time $t+1$ by the total number of individuals at time t. An estimate (λ_n) of the λ when the population reaches a stable age structure can be produced several ways with the Leslie matrix. Perhaps the most straightforward estimate of λ can be produced by using the population projections produced by multiplying the Leslie matrix by the population size vector until the age structure becomes constant with time (Donovan and Welden 2002). The asymptotic estimate of λ can be applied in equations such as Equations 14.13 and 14.14. The right and left eigenvectors are also extremely useful for drawing insight from the matrix approach to population demographics as will be described below during our discussions of stage-structured matrix models.

Migration into the population at each time step can be included as $\mathbf{n}_{t+1} = \mathbf{Ln}_t + \mathbf{m}_t$, where $\mathbf{m}_t$ is a vector containing the number of migrants of the various age classes appearing during the time step. Growth can be included in the matrix. The reader is directed to Leslie (1945, 1948) or Caswell (1989, 1996) for further details. Poptools, a free Excel™ add-in program that does these and other related computations, can be downloaded from www.cse.csiro.au/CDG/poptools. Donovan and Welden (2002) provide simple Excel™ programs and explanations for doing many of these calculations.

15.2.3 The Lefkovitch Stage-Structured Matrix Approach

Demographic analysis of populations can, as described above, take the form of an age-structured population. Models based on life stage also can be generated and are extremely informative for many species populations (Caswell 2001, Donovan and Welden 2002, Vandermeer and Goldberg 2003). Nacci et al. (2002) provide one of an ever-increasing number of ecotoxicologically oriented studies using stage-structured demographic models. As described for age-structured populations, the matrix approach is applied to stage-structured populations but the Leslie matrix is replaced by a Lefkovitch matrix (Lefkovitch 1965),

$$\begin{bmatrix} P_0 & F_1 & F_2 & F_3 \\ G_0 & P_1 & 0 & 0 \\ 0 & G_1 & P_2 & 0 \\ 0 & 0 & G_2 & P_3 \end{bmatrix}. \tag{15.12}$$

Now, population projections are done using the fertility for each stage (F), survival probability from one stage to the next (G), and probability of a surviving individual remaining at a particular stage (P) during the interval being considered. In the Lefkovitch approach, survival information includes both the probability of remaining at a stage and the probability of moving into the next stage.

Population projections can be made by multiplying the Lefkovitch matrix by the population vector as described for the Leslie matrix approach. The λ and other population metrics can be estimated with this Lefkovitch matrix as done with the Leslie matrix.

Computation of valuable population metrics will be illustrated with this stage-structured matrix approach. As discussed for age-structured models, a time-specific λ can be estimated from projected population sizes for time steps t and $t+1$ (i.e., $\lambda_t = N_{t+1}/N_t$).[1] Repeated projections through many time steps should eventually produce a population vector in which the proportions of the total population present in the different stages remains stable, that is, the stable stage distribution is achieved. The associated estimates of λ_t should have converged on the matrix eigenvalue, λ. This stable stage structure can be expressed conveniently as a vector of proportions of individuals present in each stage by dividing the number of individuals present for each stage by the total number of individuals in all stages.

Often a practicing ecotoxicologist does a complete or partial life cycle test to determine the "critical" stage of an organism at which it is most sensitive to the toxicant of interest and, as we discussed, then incorrectly suggests that this is also the most at-risk stage relative to population viability (e.g., the weakest link incongruity). In reality, to make such a judgment about population viability, an ecotoxicologist needs to understand which vital rate associated with the particular stages of a life cycle influences population growth rate the most. A matrix approach to sensitivity and elasticity analyses as implemented by Caswell (2001) allows this to be done. To begin these analyses, the stable age structure is estimated using methods just described. The right eigenvector (**w**) is estimated by expressing the asymptotic number of individuals at each stage as a column vector of proportions of the total number of individuals. Next, we need the left eigenvector (**v**) of the matrix that reflects the reproductive value for each stage in the matrix. According to Donovan and Welden (2002) and Vandermeer and Goldberg (2003), the easiest way to produce this row vector **v** is by transposing the Lefkovitch matrix (**L**). The transposed matrix ($\mathbf{L}^T$) is then multiplied by the population size vector repeatedly as done with the **L** until the population reaches a stable stage structure. The final population numbers for each stage at reaching stable age structure are then expressed as a row vector (**v**) of proportions that approximate the reproductive values for the specified stages. This row vector of proportions is the left eigenvector of **L**.

As an aside, note that it is often more convenient to express reproductive value relative to a first stage value of 1. This can be done easily by dividing all values in **v** by the value for the first stage

$$[\mathbf{v}_1 \quad \mathbf{v}_2 \quad \mathbf{v}_3 \quad \mathbf{v}_4]$$

becomes

$$\begin{bmatrix} \dfrac{\mathbf{v}_1}{\mathbf{v}_1} & \dfrac{\mathbf{v}_2}{\mathbf{v}_1} & \dfrac{\mathbf{v}_3}{\mathbf{v}_1} & \dfrac{\mathbf{v}_4}{\mathbf{v}_1} \end{bmatrix}.$$

Returning to the topic, sensitivity of the λ to changes in life stage vital rates can be assessed with the right and left eigenvectors, **w** and **v**. If we let the elements in **w** and **v** be designed as v_i (reproductive value for stage i) and w_j (stable age proportion for stage j), then sensitivity can be calculated as the following (Caswell 2001):

$$s_{ij} = \frac{w_j v_i}{\langle \mathbf{v} \times \mathbf{w} \rangle}, \tag{15.13}$$

where the bottom term on the left side of this equation is the product of the two vectors, **v** and **w**.

[1] The time interval in a stage-specific model must be specified, for example, numbers of individuals in each life cycle stage during successive spring mating periods.

These sensitivities for the different elements of the matrix are expressed in different units because they can be associated with either probabilities or number of births. As a consequence, they are not very useful for directly comparing the contribution of the various elements to λ. So discussion of sensitivities will end now and we will focus on a transformation of the sensitivities that permits easier interpretation. The elasticity ("rate of change in the log of λ with respect to the log of an element of [**L**])" (Vandermeer and Goldberg 2003) can be defined as the following:

$$e_{ij} = \frac{p_{ij}}{\lambda} v_i w_j, \tag{15.14}$$

where p_{ij} = the relevant element of interest such as neonate survival. Because the sum of all of the elasticities for the entire matrix is 1, "e_{ij} is the proportional sensitivity of λ to changes in p_{ij}" (Vandermeer and Goldberg 2003).

Let us create an ecotoxicologically relevant example using a stage-structured matrix model similar to, but having one more stage than Equation 15.12. Perhaps this fabricated population might be exposed to a toxicant and we are concerned about which stages of its life cycle and associated vital rates are most vulnerable with respect to changing λ. Assume that we calculated the following elasticities for the elements:[2]

$$\begin{bmatrix} 0 & 0.0079 & 0.0169 & 0.0157 & 0.0860 \\ 0.1265 & 0 & 0 & 0 & 0 \\ 0 & 0.1186 & 0 & 0 & 0 \\ 0 & 0 & 0.1017 & 0 & 0 \\ 0 & 0 & 0 & 0.0860 & 0.4408 \end{bmatrix}.$$

Speculating from this elasticity matrix, one could say that a toxicant effect on fertility ($e = 0$ to 0.0860 for F_0 to F_4) would have much less of an impact on the value of λ than any change in survival. Survival in this population, not reproduction, is the most critical quality and deserves the most attention. Note that the elasticity for P_4 was 0.4408: roughly 44% of the value of λ would be determined by survival at that stage. Focusing remediation actions on reproduction or neonate survival of this species population would not be the best strategy relative to fostering population persistence.

Ecotoxicological applications of elasticity and related methods are beginning to be published. As one example, elasticity analysis of the freshwater snail, *Biomphalaria glabrata*, exposed chronically to cadmium suggested that juvenile survival had the greatest effect on population growth (Salice and Miller 2003). Jensen et al. (2001) describe an equally informative elasticity analysis for the gastropod, *Potamopyrus antipodarum*, exposed to cadmium. Forbes and Calow (2003) recently published a general discussion including elasticity analysis of contaminant-exposed populations.

15.2.4 Stochastic Models

The certainty of death is attended with uncertainties in time, manner, places.

(Thomas Browne, cited in Deevey 1947)

If vital rates were defined as distributions of possible values, the deterministic matrix approaches just described could be rendered to stochastic ones. For example, the replicate *Daphnia* cultures for the six gamma irradiation treatments could have been used to define the variance to be anticipated in vital rates. At each time step, the vital rates are drawn randomly from distributions and applied

[2] Data taken from Example 18.9 in Caswell (2001).

as described above. The population size and structure would then be characterized by a stochastic trajectory through time. If this projection process was repeated many times, as might be done with Monte Carlo simulation, a family of possible outcomes could be generated. The probability of local extinction or of dropping below a certain minimum population size (M) could be estimated from the outcomes of such simulations. For example, 234 of 1000 simulations of a toxicant-exposed population might have produced populations that fell to size 0, suggesting that nearly one-quarter of populations are predicted to go locally extinct under those exposure conditions. Because the Allele effect suggests that some populations might have minimal sizes (M) above 0 that must be maintained in order to remain viable, some other threshold population size might be used instead of 0. The RAMAS program (Ferson and Akçakaya 1990) performs the calculations described here for deterministic and stochastic models. This affords the expression of population change due to toxicant exposure as a true risk. (A statement of risk specifies the probability of an adverse effect and the magnitude of the effect.) For example, a specific exposure may result in a 1 in 10 chance of the population size dropping by 50% during the 10 years that the toxicant remains above a certain threshold concentration in the species' habitat. Such models may also be developed in a metapopulation framework.

15.3 SUMMARY

This chapter describes the basics of demography and their utility in population ecotoxicology. For example, the analysis of *D. pulex* population response to gamma irradiation described here is much more meaningful than the conventional ecotoxicology approach in which a LD50 for lethality and NOEC for reproductive effects are generated. With the demographic methods, a clear consequence is indicated by the r falling below 0 at a dose rate of approximately 67.5 R/h. Even more useful information would be obtained with the inclusion of stochastic considerations. In contrast, the gross metrics of LC50 or NOEC would force the application of large uncertainty factors in order to accommodate the associated inaccuracies of these metrics of effect. Another example includes the application of elasticity analysis instead of the dubious assumption that the most sensitive stage of an individual's life cycle is the one most critical relative to population viability. Fortunately, more and more demographic analyses are being done for the effects of pollutants. Sibly (1996) provides a literature search of such studies, indicating the value of the approach. Hopefully, the trend toward such population methods will continue during the next decade.

15.3.1 SUMMARY OF FOUNDATION CONCEPTS AND PARADIGMS

- Populations have structure relative to age (or stage) and sex, and this structure can be influenced by toxicant exposure.
- Toxicant exposure can modify vital rates and, consequently, population qualities and viability.
- Conventional life table and matrix methods allow description and quantitative prediction of population qualities.
- Results of life table analyses complement those described in Chapters 9 and 13 for survival analysis.
- Life table analysis is possible for groups of individuals exposed in laboratory toxicity tests.
- Metrics from demographic analysis are useful for defining population status under the influence of toxicant exposure.
- Demographic qualities of some species make them more or less susceptible to toxicant effects and, consequently, metrics derived for effects to individuals only are poor predictors of population effects to some species.

- Demographic metrics are compatible with wildlife management, fisheries stock management, and conservation biology metrics of population status.
- Potential measures of effect include r, λ, V_A, stable population structure, and probability of local extinction.
- Toxicants can influence migration into and out of populations by modifying mechanisms such as avoidance, drift, and territoriality.

REFERENCES

Aubone, A., Loss of stability owing to a stable age structure skewed toward juveniles. *Ecol. Modell.*, 175, 55–64, 2004.

Bechmann, R.K., Use of life tables and LC50 tests to evaluate chronic and acute toxicity effects of copper on the marine copepod *Tisbe furcata* (Baird), *Environ. Toxicol. Chem.*, 13, 1509–1517, 1994.

Casarett, A., *Radiation Biology*. Prentice-Hall, Inc., Englewood Cliffs, NJ, 1968.

Caswell, H., *Matrix Population Models: Construction, Analysis, and Interpretation*, Sinauer Associates, Inc., Sunderland, MA, 1989.

Caswell, H., Demography meets ecotoxicology: Untangling the population level effects of toxic substances, In *Ecotoxicology. A Hierarchical Treatment*, Newman, M.C. and Jagoe, C.H. (eds.), CRC Press/Lewis Publishers, Boca Raton, FL, 1996, pp. 255–292.

Chaumot, A., Charles, S., Flammarion, P., and Auger, P., Ecotoxicology and spatial modeling in population dynamics: An illustration with brown trout, *Environ. Toxicol. Chem.*, 22, 959–969, 2003.

Daniels, R.E. and Allan, J.D., Life table evaluation of chronic exposure to a pesticide, *Can. J. Fish. Aquat. Sci.*, 38, 485–494, 1981.

Deevey, E.S., Jr., Life tables for natural populations of animals, *Q. Rev. Biol.*, 22, 283–314, 1947.

Donovan, T.M. and Welden, C.W., *Spreadsheet Exercises in Conservation Biology and Landscape Ecology*, Sinauer Assoc. Inc., Sunderland, MA, 2002.

Emlen, J.M., *Population Biology. The Coevolution of Population Dynamics and Behavior.* MacMillan Publishing Company, New York, 1984.

Euler, L., Recherches générales sur la mortalité: La multiplication du benre humain, *Mem. Acad. Sci., Berlin*, 16, 144–164, 1760.

Ferson, S. and Akçakaya, H.R., *Modeling Fluctuations in Age-structured Populations. RAMAS/age User Manual.* Applied Biomathematics, Setauket, 1990.

Forbes, V.E. and Calow, P., Contaminant effects on population demographics, In *Fundamentals of Ecotoxicology*, Newman, M.C. and Unger, M.A. (eds.), CRC Press/Lewis Publishers, Boca Raton, FL, 2003, pp. 221–224.

Forbes, V.E., Calow, P., and Sibly, R.M., Are current species extrapolation models a good basis for ecological risk assessment? *Environ. Toxicol. Chem.*, 20, 442–447, 2001.

Goodman, D., Optimal life histories, optimal notation, and the value of reproductive value, *Am. Nat.*, 119, 803–823, 1982.

Hood, G.M., PopTools version 2.6.4. Available at: http://www.cse.csiro.au/poptools, 2004.

Jensen, A., Forbes, V.E., and Parker, E.D., Jr., Variation in cadmium uptake, feeding rate, and life-history effects in the gastropod *Potamopyrgus antipodarum*: Linking toxicant effects on individuals to the population level, *Environ. Toxicol. Chem.*, 20, 2503–2513, 2001.

Kammenga, J.E., Busschers, M., Van Straalen, N.M., Jepson, P.C., and Baker, J., Stress induced fitness is not determined by the most sensitive life-cycle trait, *Funct. Ecol.*, 10, 106–111, 1996.

Koivisto, S. and Ketola, M., Effects of copper on life-history traits of *Daphnia pulex* and *Bosmina longirostris, Aquat. Toxicol.*, 32, 255–269, 1995.

Krebs, C.J., *Ecological Methodology*, Harper Collins Publishers, New York, 1989.

Laskowski, R., Stochastic and density-dependent models in ecotoxicology, In *Demography in Ecotoxicology*, Kammenga, J. and Laskowski, R. (eds.), John Wiley & Sons, Chichester, UK, 2000, pp. 57–71.

Laskowski, R. and Hopkin, S.P., Effect of Zn, Cu, Pb, and Cd on fitness in snails (*Helix aspersa*), *Ecotoxicol. Environ. Saf.*, 34, 59–69, 1996.

Lefkovitch, L.P., The study of population growth in organisms grouped by stages, *Biometrics*, 21, 1–18, 1965.

Leslie, P.H., On the use of matrices in certain population mathematics, *Biometrika*, 33, 183–212, 1945.
Leslie, P.H., Some further notes on the use of matrices in population mathematics, *Biometrika*, 35, 213–245, 1948.
Leslie, P.H., Tener, J.S., Vizoso, M., and Chitty, H., The longevity and fertility of the Orkney vole, *Microtus orcadensis*, as observed in the laboratory, *Proc. Zool. Soc. Lond.*, 125, 115–125, 1955.
Lotka, A.J., Studies on the mode of growth of material aggregates, *Am. J. Sci.*, 24, 199–216, 1907.
Marshall, J.S., The effects of continuous gamma radiation on the intrinsic rate of natural increase of *Daphnia pulex*, *Ecology*, 43, 598–607, 1962.
Martinez-Jerónimo, F., Villaseñor, R., Espinosa, F., and Rios, G., Use of life-tables and application factors for evaluating chronic toxicity of Kraft mill wastes on *Daphnia magna*, *Bull. Environ. Contam. Toxicol.*, 50, 377–384, 1993.
Melville, H., *Moby Dick or the white whale*, Armont Publishing Co., New York, 1851.
Munn, W.R., Jr., Black, D.E., Gleason, T.R., Salomon, K., Bengtson, D., and Gutjanr-Gobell, R., Evaluation of the effects of dioxin and PCBs on *Fundulus heteroclitus* populations using a modeling approach, *Environ. Toxicol. Chem.*, 16, 1074–1081, 1997.
Münzinger, A. and Guarducci, M.-L., The effect of low zinc concentrations on some demographic parameters of *Biomphalaria glabrata* (Say), mollusca: Gastropoda, *Aquat. Toxicol.*, 12, 51–61, 1988.
Nacci, D.E., Gleason, T.R., Gutjahr-Gobell, R., Huber, M., and Munns, W.R., Jr., Effects of chronic stress on wildlife populations: A population modeling approach and case study, In *Coastal and Estuarine Risk Assessment*, Newman, M.C., Roberts, M.H., Jr., and Hale, R.C. (eds.), CRC Press/Lewis Publishers, Boca Raton, FL, 2002, pp. 247–272.
Newman, M.C., *Quantitative Methods in Aquatic Ecotoxicology*, CRC Press/Lewis Publishers, Boca Raton, FL, 1995.
Newman, M.C., Fundamentals of Ecotoxicology, Ann Arbor/Lewis/CRC Press, Boca Raton, FL, 1998.
Pesch, C.E., Munns, W.R. Jr., and Gutjahr-Gobell, R., Effects of a contaminated sediment on life history traits and population growth rate of *Neanthes arenaceodentata* (Polychaeta: Nereidae) in the laboratory, *Environ. Toxicol. Chem.*, 10, 805–815, 1991.
Petersen, R.C., Jr. and Petersen, L.B.-M., Compensatory mortality in aquatic populations: Its importance for interpretation of toxicant effects, *Ambio*, 17, 381–386, 1988.
Salice, C.J. and Miller, T.J., Population-level responses to long-term cadmium exposure in two strains of the freshwater gastropod *Biomphalaria glabrata*: Results from a life-table response experiment, *Environ. Toxicol. Chem.*, 22 678–688, 2003.
Sibly, R.M., Effects of pollutants on individual life histories and population growth rates, In *Ecotoxicology. A Hierarchical Treatment*, Newman, M.C. and Jagoe, C.H. (eds.), CRC Press/Lewis Publishers, Boca Raton, FL, 1996, pp. 197–223.
Spurgeon, D.J., Svendsen, C., Weeks, J.M., Hankard, P.K., Stubberud, H.E., and Kammenga, J.E., Quantifying copper and cadmium impacts on intrinsic rate of population increase in the terrestrial oligochaete *Lumbricus rubellus*, *Environ. Toxicol. Chem.*, 22, 1465–1472, 2003.
Stearns, S.C., *The Evolution of Life Histories*, Oxford University Press, Oxford, UK, 1992.
Vadermeer, J.H. and Goldberg, D.E., *Population Ecology. First Principles*, Princeton University Press, Princeton, NJ, 2003.
Wilson, E.O. and Bossert, W.H., *A Primer of Population Biology*, Sinauer Associates, Inc., Sunderland, MA, 1971.
Woodwell, G.M., Effects of ionizing radiation on terrestrial ecosystems, *Science*, 138, 572–577, 1962.
Woodwell, G.M., The ecological effects of radiation, *Sci. Am.*, 208, 2–11, 1963.

16 Phenogenetics[1] of Exposed Populations

Physiological response to toxicants at an individual level can be related rigorously to predict population dynamics responses. The link can be achieved straightforwardly provided functional relationships between various physiological responses and survivorship, fecundity and developmental rates [i.e., vital rates] can be established.

(Calow and Sibly 1990)

16.1 OVERVIEW

16.1.1 The Phenotype Vantage

The translation from a pollutant's effects on individuals to its effects on the population can be accomplished using life-history analysis to calculate the effect on the population's growth rate.

(Sibly 1996)

Most descriptions of contaminant effects to populations do not give sufficient discussion of phenotypic changes to life history traits. With notable exceptions (i.e., Kammenga et al. 1996, Maltby 1991, McFarlane and Franzin 1978), coordinated changes in life history traits that are predictable from studies in other ecological fields are not carefully explored in ecotoxicology. Phenotype is often considered solely in the context of biochemical, physiological, toxicokinetic, and toxicodynamic characteristics of individuals. The result can be misleading conclusions or inexplicable responses. This chapter tries to develop a vantage of population effects emerging from toxicant-induced changes to phenotypes. Phenotypic variations in rates of growth and development, onset and rate of reproduction, and rate of mortality will be described and linked to emergent population consequences. The overriding themes are the following: (1) translation of phenotypic differences in life history traits of individuals to population consequences, (2) changes in Darwinian fitness associated with these life history phenotypes, and (3) potential for adaptation and microevolution. These themes are based on the simple causal sequence that phenotypic variation in life history traits of exposed individuals produce changes in population vital rates. These shifts in population vital rates can then give rise to changes in population viability, demographic shifts, or life history adaptation and microevolution. Often selection acts on suites of these traits, resulting in coordinated shifts in vital rates to achieve maximum fitness under changing conditions.

The influence of toxicants on the capacity within a population to produce a consistent phenotype (i.e., developmental stability) will also be described at the end of the chapter. Deviations from perfect developmental stability will be discussed as population-level indicators of contaminant effect.

16.1.2 An Extreme Case Example

Species populations differ in how suites of life history traits are expressed under various conditions. This can be illustrated by reviewing the classic *r*- (opportunistic) and *K*- (equilibrial) selection context

[1] "The study of genetic and environmental influences on the development of the phenotype" within populations (Zakharov and Graham 1992).

TABLE 16.1
Themes in Populations Dominated by *r*- or *K*-Selection Strategies

Characteristic	*r*-Strategy	*K*-Strategy
Strategy	Opportunistic, emphasis on producing large numbers of offspring quickly	Efficient interspecific competitor for resources; emphasis on producing few, high-quality offspring
Habitat	Temporary or variable, unpredictable, often "ecological vacuums" not requiring efficient competition (May 1976)	Stable and predictable
Mortality	High; migration may be important, Deevey Type I mortality curve (Deevey 1947)	Returns to *K* after perturbation; Deevey Type II or III mortality curve (Deevey 1947)
Strategy favors	Rapid development to sexual maturity	Slow development to sexual maturity
	High *r*, "boom-bust" population dynamics	Efficient competitor in community
	Low investment in individual offspring	High investment in individual offspring
	Small body size	Large body size
	Short lifespan and generation time	Long lifespan and generation time
	Semelparity (single reproductive event/individual)	Iteroparity (several reproductive events/individual)

Source: Modified from Neuhold (1987) and May (1976).

for population strategy (Table 16.1). Species at the two extremes of this scheme respond differently to stressors, and as a consequence, *r*- and *K*-selected species populations can have different fates under the same exposure scenarios. Acute exposure that eliminates nearly all individuals from a habitat might have a trivial consequence to the persistence of an *r*-selected species that will quickly repopulate the habitat via migration and rapid reproduction. The opportunistic *r*-strategist has life history traits that, in combination, allow it to quickly reestablish itself after an unpredictable stress occurs. Because *r*-selected species tend toward semelparity (i.e., an individual reproduces in a single pulse during its lifetime), the rate of recolonization for an *r*-strategist might be quite stochastic and dependent on season. In contrast, a *K*-selected species population might take much longer to recover owing to its lower rate of reproduction. The general community-level consequence is that toxicant-exposed community composition can shift away from *K*-strategists in favor of *r*-strategists (May 1976, Odum 1985). Relationships have also been suggested between tolerance to toxicants and the *r*- and *K*-strategies of fish (Neuhold 1987) and nematodes (Bongers and Ferris 1999). Indeed, a soil nematode community metric of toxicant effect is based on the shift in community composition to favor *r*-strategists (Bongers 1990, Bongers and Ferris 1999). For both types of strategists, the range of phenotypes and coordinated changes in life history-related phenotypes with exposure dictate short- and long-term consequences to populations and communities.

The *r*- and *K*-strategist scheme is an oversimplification; still, its exploration here provides a clear illustration of how life history traits can influence population consequences. In reality, most species fit along a continuum of life history strategies ranging from pure *r*- to pure *K*-strategists. Some species tend toward one strategy at one stage of their life cycle (e.g., larval amphibians living in temporary ponds) and a second strategy at another stage (e.g., terrestrial adult amphibians). The suite of life history traits that combine to create strategies optimizing fitness under different conditions is subject to natural selection.

Many changes in phenotype in response to toxicant exposure can be associated readily with enhanced individual fitness, for example, survival is enhanced by induction of a detoxification mechanism. But all crucial changes are not so obvious (e.g., Box 16.1) such as shifts in suites of individual life history characteristics. Our discussion of such changes, beginning here with the extremes of *r*- and *K*-strategists under toxicant stress, will now be expanded to include these life history shifts.

Box 16.1 Compensatory Mortality in Caddisfly Populations

In a study of pulp mill contaminants, Petersen and Petersen (1988) exposed different larval stages of the net-spinning hydropsychid caddisfly, *Hydrophsyche siltalai*, to 4,5,6-trichloroguaiacol. Because significant density-dependent mortality occurs at early stages (≤9 days) of this insect's life, Petersen and Petersen hypothesized that a clear concentration–response relationship might be difficult to produce for these putatively most sensitive stages. Reduced larval densities in the highest concentration tanks due to toxicity-induced mortality would result in less natural, density-dependent mortality in the ≤9-day larvae tanks: A combination of natural, density-dependent mortality and density-independent, toxicant-induced mortality would occur in the test chambers. Older stages of this caddisfly exhibit much lower levels of natural, density-dependent mortality than younger stages, so toxicant-induced mortality might be clearer for older than younger stages.

The results of Petersen and Petersen's experiments (Figure 16.1) showed the confounding effect of density-dependent, natural mortality for early stages (≤9 days of age), but not for older larvae. The density dependence of early life stage mortality for this species must be considered in making accurate predictions of toxicant effects *to populations*. And compensatory mortality is relevant to other species, including many fish species (Petersen and Petersen 1988).

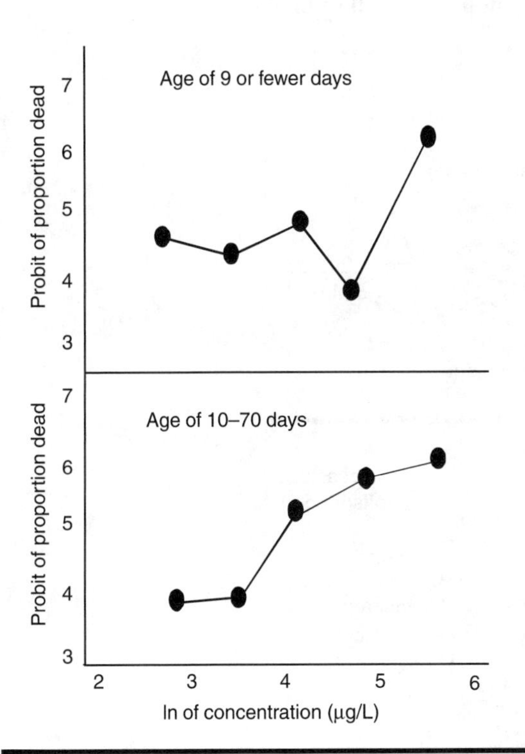

FIGURE 16.1 Density-dependent, natural mortality can obscure the concentration–effect relationship for caddisfly larvae exposed to 4,5,6-trichloroguaiacol. There was no discernible relationship for larvae <9 days old, an age class with high levels of natural, density-dependent death. Note the high mortality in all treatments. (Probit values of 4 and 5 correspond to 16% and 50% mortality, respectively.) There was a clear relationship between mortality and toxicant concentration of older larvae (>9–70 days old). (Modified from Figure 4A and B of Petersen and Petersen (1988).)

Evolutionary consequences of life history trait shifts are also important to understand. The potential exists for rapid evolution of life history traits in response to toxicant exposure. The rate of evolution of life history traits in response to nontoxicant environmental factors can be rapid (Svensson 1997) and similar rates might be expected for toxicant-related changes.

16.2 TOXICANTS AND THE PRINCIPLE OF ALLOCATION (CONCEPT OF STRATEGY)

Organisms have a finite amount of energy available to fulfill their diverse life functions, and as a consequence, they will follow basic rules when using it. This can be seen in a variety of contexts. As an example explored previously (Section 9.1.1 of Chapter 9), Selye (1956, 1973) described the general adaptation syndrome (GAS) for any individual's response to stress. The central theme of the GAS is that mechanisms are induced to resist change to normal homeostasis of the individual. Short-term, but energetically demanding, responses occur immediately after stress occurs. Lower cost, more long-term mechanisms emerge if stress persists. Although valuable alone, the Selyean stress theory describes energy-demand compensating changes within individuals without consideration of functions other than survival.

Expanding the GAS to a life cycle context considering functions other than survival, the κ-rule ("Kappa-rule") specifies that energy is meted out in a particular way among somatic maintenance, growth, development, and reproduction (Figure 16.2, top). A portion (κ) is first used for essential maintenance and growth with any remaining being allocated to reproduction and/or development (Kooijman 1993). Maintenance is given top priority with somatic growth being the next priority. Any energy for development is first spent to maintain a basal rate of maturation with an increase from this basal rate only if additional energy is available. Such straightforward rules for energy allocation have manifestations at the level of population vital rates. In the remainder of this section and Section 16.2.3, more involved rules of energy utilization will be explored.

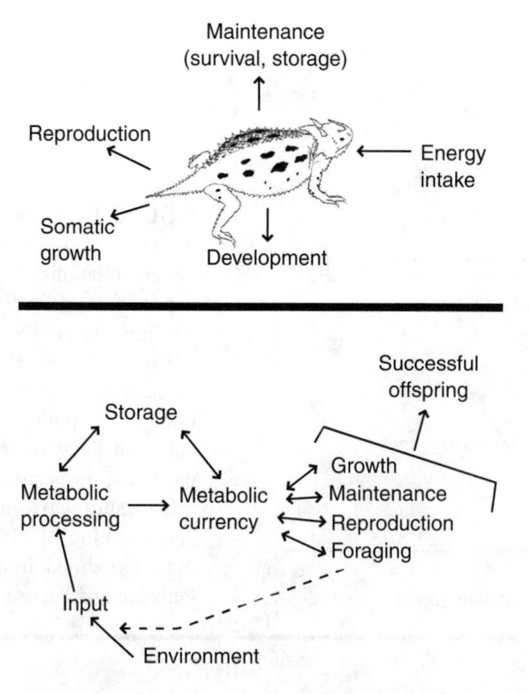

FIGURE 16.2 The allocation of energy resources among essential functions of an individual. Organisms must optimize the environment-specific trade-off in resource allocation into life history components in order to achieve the highest Darwinian fitness. Energy spent on maintenance can include that associated with detoxification mechanisms, toxicant-induced changes in structures, active elimination of toxicants and metabolites, repair of damage, and other mechanisms associated with tolerance enhancement. (Modified from Figure 1 in Watt (1986).)

Expanding the κ-rule to the general principle of allocation (concept of strategy) (Levins 1968), an individual with a specific genetic make-up and living in a particular environment must allocate energy resources to maximize its Darwinian fitness (Figure 16.2, bottom). The principle of allocation stipulates that trade-offs exist with each allocation. Energy spent on one function, process, or structure cannot be used for another. If *Daphnia pulex* puts resources into growing spines that reduce predation risk (Kreuger and Dodson 1981), these resources are unavailable for egg production. Similarly, energy spent by an exposed individual to produce a detoxification enzyme, to maintain homeostasis under toxic stress, or to replace toxicant-damaged proteins is not available for somatic growth or reproduction.[2] As a simple example of such a trade-off, Adams et al. (1992) noted that redbreast sunfish (*Lepomis auritus*) exposed to a mixture of contaminants in East Fork Poplar Creek (Oak Ridge, Tennessee) displayed increased detoxification enzyme levels but low lipid levels, decreased growth rate, and depressed reproduction. As another example, *Daphnia magna* exposed to elevated metals had reduced growth and reproduction, but not metabolic rate (Knops et al. 2001).

Strategies exist for doling out resources so as to maintain optimal fitness and these strategies are subject to natural selection. Obviously, phylogenetic differences dictate limits to the possible range of traits a species can display and, bounded by these constraints, phenotypic variations exist within a species. As examples of phenotypic limitations, phylogenetically correlated differences in species groupings such as endotherm versus ectotherm bring about differences and limits for basic allocation strategies. Traits most often considered in optimality studies include metabolic maintenance, somatic growth, reproduction, survival (including longevity), and foraging (Atchison et al. 1996, Stearns 1992).

The context of optimizing strategies can help explain some processes noted in earlier chapters. Importantly, the principle of allocation is relevant to the enigmatic phenomenon of hormesis. Hormesis, the stimulatory effect exhibited with low, subinhibitory exposures to some toxicants, is often observed for life history traits such as somatic growth, reproduction, and survival (see Figure 15.1). For example, application of low doses of the growth retardant phosfon to the mint, *Mentha piperita*, actually stimulates growth (Calabrese et al. 1987).

Hormesis is often presented in ecotoxicology as an exceptional and counterintuitive phenomenon—a peculiarity. Enhanced DNA repair due to radiation at low levels was recently described by Sagan (1989) as a "surprising possibility." Earlier, Sagan (1987) explained that "—frequent reports in the literature of 'anomalies' at low doses ... have been referred to as 'hormetic' effects." Oddly, a century before these statements were made, the principle of hormesis was sufficiently well known to be identified as Arndt–Schulz Law, Hueppe's Rule, Sufficient Challenge, and hormoligosis (Calabrese et al. 1987, Stebbing 1982).

Although difficult to explain with core ecotoxicological paradigms, hormesis is one obvious manifestation of the principle of allocation. Reallocation of energy among somatic growth, reproduction, and maintenance (survival) will vary among some, but not necessarily all, species–toxicant combinations. An increase in growth at low concentrations might be expected if energy is shifted toward somatic maintenance and away from reproduction, so hormesis is a predictable outcome of the reallocation of energy resources as environmental conditions shift. It is not surprising that effect magnitude does not always conform to a monotonic trend with toxicant concentration. Compensatory reallocation of resources combines in a complex way with toxic effects at the various exposure levels. Depending on the capacity of the individual to reallocate resources and its particular strategy, hormesis could be a predictable outcome of low-level exposure. Hormesis appears enigmatic only when one life history trait is studied in isolation from others. A more complete explanation of response to exposure would result if tests were conducted to include shifts in all relevant vital rates and allocation theory used to interpret the data.

---

[2] Protein turn-over and synthesis costs are in excess of 15% of basal metabolism in mammals, so the additional cost of toxicant-related protein replacement or production of detoxification proteins can be high (Sibly and Calow 1989).

16.2.1 Phenotypic Plasticity and Norms of Reaction

> Reaction norms transform environmental variation into phenotypic variation .... Reaction norms can be either inflexible, in which a characteristic once determined is never changed later in the organism's life, or they can be flexible, in which a characteristic can be altered more than once in the development of the same individual.
>
> **(Stearns 1989)**

The phenotype is a consequence of genetic (G) and environmental (E) factors, and the interaction between these factors (G × E). It follows that an individual has the potential to express a range of phenotypes depending on its environmental setting and genetic make-up. The reaction norm concept (Woltereck 1909) is used to describe phenotypic plasticity relative to life history traits and is often applied to study optimization of Darwinian fitness.[3] Typically, reaction norms are represented as shown in Figure 16.3. Under a range of intensities of some environmental quality (e.g., temperature, moisture, or toxicant concentration), the genetic characteristics of the individual can translate into one of a range of phenotypes (Figure 16.3, top). The reaction norm might be linear (Figure 16.3, top) or curvilinear (Figure 16.3, bottom), and genotypes within a population can have distinct reaction norms (Genotypes A, B, and C in Figure 16.3, bottom). A G × E interaction is indicated when reaction norms for different genotypes (or genetic lineages) cross as shown for Genotypes B and C in the lower panel of Figure 16.3. The reaction norm and conceptually similar approaches have been applied to relative fitness of metal tolerant and metal intolerant plants under a range of metal

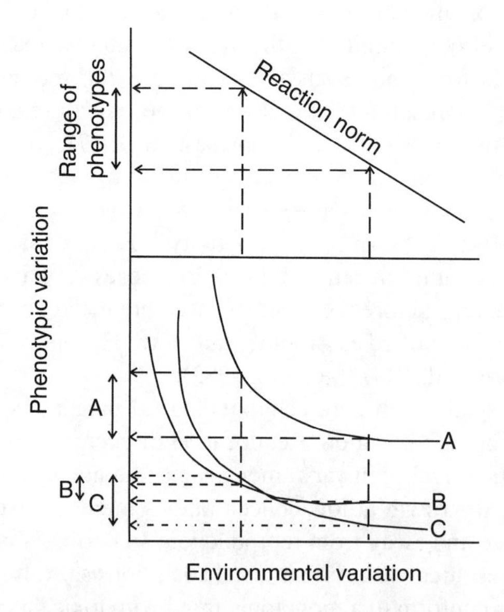

FIGURE 16.3 The reaction norm shows the translation of genotype into a range of phenotypes as a consequence of environmental variation. In the upper panel (modified from Stearns (1989)), variation in the environment (x-axis) results in variation in expressed phenotype (y-axis) for a particular genotype. The relationship is linear in this example. In the lower panel, reaction norms for several genotypes are shown as curvilinear. A genetic × environmental interaction exists if reaction norms for different genotypes cross one another. Note the distinct ranges of potential phenotypes for the three genotypes (double-headed arrows on y-axis of lower panel).

[3] The opposite of phenotypic plasticity is environmental canalization in which a consistent phenotype is produced regardless of environmental conditions (Stearns 1989).

exposures (Hickey and McNeilly 1975), populations of metal-exposed midges (Posthuma et al. 1995), *Drosophila melanogaster* experiencing diverse environmental stressors (Clark 1997), and responses of zooplankton populations to metal contamination (Forbes et al. 1999).

Box 16.2 Applying Reaction Norms to Mercury-Exposed Mosquitofish

Mulvey et al. (1995) applied the reaction norm approach to assess the influence of mosquitofish (*Gambusia holbrooki*) genotype on mercury sensitivity. Four mesocosms containing 985 mosquitofish each were split into two groups; a pair of untreated mesocosms and a pair of mesocosms spiked weekly to a water concentration of 18 μg/L of mercury. After 111 days, fish were harvested and measurements made of a variety of genetic and life history traits. Female life history traits measured for the different genotypes included the number of late stage embryos carried by a gravid female, percentage of adult females that were gravid, and fish size (standard length). The fish populations in the control mesocosms doubled in number but those in the 18 μg/L treatment dropped approximately 25% by the end of the experiment.

Putative genotypes for the glycolytic enzyme, glucosephosphate isomerase-2 (Gpi-2), were determined by starch gel electrophoresis. Three alleles (38, 66, and the 100 common allele) were scored but, because of the low frequency of some genotypes in some mesocosms, the uncommon genotypes (38 and 66) were pooled and designated as "+" to produce three possible genotypes of 100/100, 100/+, and +/+. Responses of females of these genotypic designations were plotted for the two treatments (Figure 16.4).

Female standard length remained relatively unaffected by mercury treatment for the Gpi-$2^{100/100}$ homozygotes but was greater for the Gpi-$2^{100/+}$ and Gpi-$2^{+/+}$ genotypes. Note that crossing of the reaction norms for the different genotypes indicates a genetic x environment (G × E) interaction: the different genotypes responded differently to the presence of mercury in the environment. Similarly, the number of late stage embryos carried by the females of the Gpi-$2^{+/+}$ genotype changed little with mercury additions. In contrast, the other genotypes (Gpi-$2^{100/100}$ and Gpi-$2^{100/+}$) had significantly fewer late stage embryos. Again, there was a clear G × E interaction. The genotypes responded differently to mercury exposure relative to important life history traits, suggesting the potential for selection at the Gpi-2 locus (or a closely linked locus) during multiple generations of mercury exposure.[4]

Although not shown in Figure 16.4, there were other differences in reproduction among exposed genotypes. The percentage of adult females that were gravid was similar (c. 70%) for the three genotypes in the control mesocosms. The percentages of gravid females of the Gpi-$2^{100/+}$ and Gpi-$2^{+/+}$ genotypes from the mercury-treated mesocosm were also approximately 70%, but only 43% of the Gpi-$2^{100/100}$ females in the mercury-treated mesocosm were gravid. If plotted as a reaction norm, a G × E interaction would be clear for the Gpi-$2^{100/100}$ genotype; again, suggesting a shift away from resource allocation to reproduction by the Gpi-$2^{100/100}$ homozygote during mercury exposure.

These differences in fish size (standard length), fecundity (number of late stage embryos), and percentage gravid females can be viewed from the vantage of trade-offs with toxicant exposure. For all genotypes, fish size increased and investment in young (late stage embryos) decreased with mercury exposure. Trends in size were strongest for the females of the Gpi-$2^{100/+}$ and Gpi-$2^{+/+}$ genotypes and those for fecundity were strongest for the Gpi-$2^{100/100}$ and Gpi-$2^{100/+}$ genotypes. The clearest differences from reproduction to growth were seen for the Gpi-$2^{100/+}$ heterozygote.

---

[4] See Tatara et al. (1999) for a description of the multiple generation consequences to mosquitofish in these mesocosms.

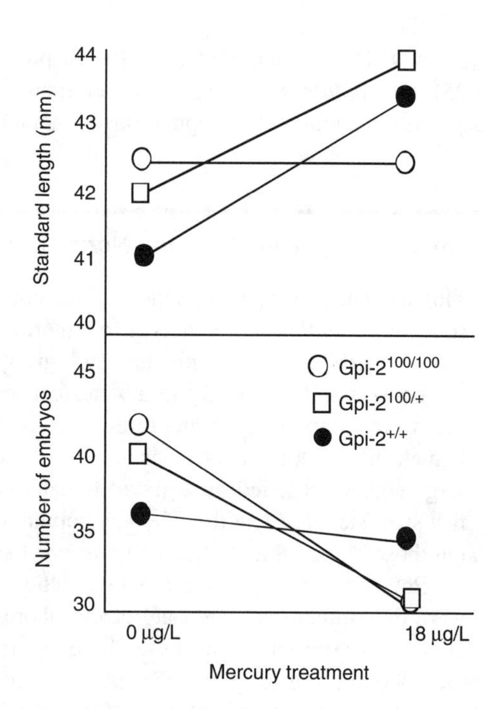

FIGURE 16.4 The influence of mercury treatment on size (standard length) and fecundity (number of late stage embryos/gravid female) of female mosquitofish. (Modified from Figure 3 in Mulvey et al. (1995).)

The trends seen here conform to the notion of a trade-off of somatic growth versus reproduction under stress but an alternate hypothesis confounds interpretation based solely on energy allocation. Size-dependent mortality during mercury exposure could have contributed to the differences seen in standard lengths among genotypes. However, actual growth information (growth increments of sagittal otoliths) for the mesocosms suggested little difference with mercury treatment, although genotype-specific growth rates could not be measured. The differences in fish length seen in Figure 16.4 could include consequences of size- and genotype-dependent mortality under mercury exposure. This size- and genotype-dependent mortality would be consistent with other work by this group (i.e., Diamond et al. 1989, Heagler et al. 1993, Newman et al. 1989).

Dismissing the size differences from further discussions of trade-offs, genotype-dependent decreases in reproductive allocation remain clear with mercury exposure. An inadequate, alternative explanation for differences in fecundity between treatments could be size-dependent fecundity, and not a trade-off strategy. Fish were generally bigger in the exposed tanks and reproductive allocation in bigger fish may be different from that in smaller fish; however, Mulvey et al. (1995) applied analysis of covariance and found no evidence to support this alternative hypothesis.

The genotype-dependent shifts in reproduction were life history responses to mercury exposure. Increased demands on energy resources to individual maintenance during mercury exposure resulted in diminished allocation of resources to reproduction. Genotypes at the Gpi-2 locus differed in how resource allocation occurred.[5]

Continuous variation in phenotype with environmental change is not the only type of phenotypic plasticity. Polyphenism occurs if an environmental cue triggers expression of one phenotype or another with no intermediate phenotypes being expressed. Two or more distinct phenotypes from

[5] This example describes only one component of a larger study of mercury and mosquitofish genetics. Other aspects will be explored in Chapter 18.

one genotype may be produced as a consequence of developmental switching. An apropos example of polyphenism is the influence of xenobiotic estrogens on sex determination of slider turtles (Bergeron et al. 1994). For the red-eared slider (*Trachemys scripta*), polyphenism takes place with nest incubation temperature being the usual cue determining hatchling sex. At a temperature that would normally produce a preponderance of male hatchlings, spotting the shell of developing red-eared slider turtle eggs with a polychlorinated biphenyl (2′,4′,6′-trichloro-4-biphenylol) solution dramatically increased the proportion of hatchlings emerging as females. The polychlorinated biphenyl (PCB) triggered an estrogen-mediated switch responsible for sex determination.

16.2.2 Toxicants and Aging

The [evolutionary] cost of division into the germ line and soma was death.

(Stearns 1992)

Binary fission in microbes or clonal reproduction in some metazoans produces essentially immortal lines. For other species, such as humans, specialization of cells into somatic and germ cell lines results in individuals (soma) with finite life spans.

The aging and eventual death of the soma are subject to selection in complex ways. Differences in fitness of postreproductive individuals may be trivial relative to continuance of the germ line; therefore, natural selection will not act strongly on such fitness differences. More than a century ago, Weismann (1882, cited in Medvedev (1990)) extended this premise to suggest that postreproductive individuals drain the limited resources available to a population because of competition with individuals that are still contributing to the germ line. Age-related death removes this impediment to germ line success.

What factors determine how long an individual lives? People have long been fascinated with how genetic and environmental factors, alone or in combination, dictate longevity. Many theories and paradigms surround this aspect of life history as a consequence. Medvedev (1990) argued that no single cause for aging exists and noted that many of the current theories are complementary. It follows that several disparate theories will probably coalesce into a more inclusive whole in the near future. For example, the limited life span theory, Gompertz Law, and selection-based theory could easily be combined. The limited life span theory holds that each individual has a genetically predetermined maximum life span (Brooks et al. 1994, Curtsinger et al. 1992). Such a powerful, genetic influence on longevity is clear in the work by Kenyon et al. (1993), who produced a long-lived nematode (*Caenorhabditis elegans*) strain that differed by only one gene (*daf*-2) from the wild type. Gompertz Law, debated recently by Barinaga (1992) and Carey et al. (1992), holds that death rate increases inexorably and exponentially with age. Despite any genetic predisposition for longevity, death rate increases in this manner with age. Selection-based theory holds that genetic factors favored in young individuals can become disadvantageous as individuals age. An antagonistic pleiotropy exists because there is no selective advantage to sparing postreproductive individuals from harm emerging later from genetic factors: overall fitness of an individual is determined more by advantageous traits of young (prereproductive or reproductive) individuals than disadvantageous traits of old individuals (Parsons 1995, Rose and Charlesworth 1980). A quick review of these three theories should make it obvious that they are easily combined. Selection produces differences in longevity among individuals; however, mortality rate generally increases with age. Genetic factors tend to favor survival of prereproductive and reproductive individuals over survival of postreproductive individuals. In any such blending of theory, it should be kept in mind that different strategies will emerge if species populations differ in their environmental challenges and genetic composition.

Other theories focus on environmental factors and their influence on longevity. They are directly relevant to understanding phenotypic changes in populations under toxicant stress and contribute to understanding the mechanisms for toxicant effects on aging. The brief discussion of aging models described below is not inclusive and the reader is directed to Medvedev (1990), Stearns (1992), and Parsons (1995) for more comprehensive reviews.

16.2.2.1 Stress-Based Theories of Aging

Parson (1995) extends the limited life span theory to the rate of living theory of aging. A particular genotype has a fixed total metabolic allocation: longevity is influenced by how quickly or slowly this allocation is spent. Supporting this theory is the consistent observation that reduced metabolic rate associated with a calorie-restricted diet increases longevity of rodents (Pieri et al. 1992) and some nonmammalian species (Sohal and Weindruch 1996). It follows that toxicant stress can influence longevity because metabolic rate commonly increases with stress. Stress could accelerate the expenditure of the organism's fixed amount of energy, decreasing its life span. As a supporting example, Kramer et al. (1992) found differences in glycolytic flux in mercury-exposed mosquitofish differing in Gpi-2 allozyme genotype and correlated them to differential survival during acute exposure. Glycolytic flux increased for the mercury-sensitive genotype but not for the other genotypes.

Parsons (1995) describes the stress theory of aging in the following manner. Natural selection occurs for stress resistance, and individuals most resistant to stress will be those with extreme longevity. Longevity is a consequence of selection for stress resistance. Evidence supportive of this theory can be found in *Drosophila* populations in which longevity was correlated with low metabolic rate and elevated antioxidant activity.

The correlation of antioxidant activity with longevity noted by Parsons is also consistent with the oxidative stress hypothesis for aging. Further support is provided by increased free radical production and consequent increased oxidative damage in rodents fed an unrestricted diet in the Pieri et al. (1992) study cited above. The oxidative stress hypothesis of aging holds that aerobic (and photosynthetic) organisms produce reactive oxygen metabolites (e.g., hydrogen peroxide, H_2O_2, the superoxide radical $O_2^{\bullet-}$, and the hydroxyl radical, $^{\bullet}OH$), which cause oxidative damage and the accumulation of such oxidative damage is a major factor determining life span. Damage to membranes (via lipid peroxidation), DNA, and proteins causes dysfunction and, in the case of DNA, potential for mutation. "The basic tenet of the oxidative stress hypothesis is that senescence-related loss of function is due to the progressive and irreversible accrual of molecular oxidative damage" (Sohal and Weindruch 1996). Consequently, chronological age is correlated with cumulative oxidative damage: individuals with different rates of oxidative damage accumulation will have different longevities.

If the oxidative stress hypothesis was valid, individuals experiencing elevated levels of oxidative stress should age at an accelerated rate. Chemical (e.g., the herbicide, paraquat, polycyclic aromatic hydrocarbons, and several heavy metals) and physical (e.g., radiation, asbestos fibers, or cigarette smoke) agents enhance production of reactive oxidants, accelerating the accumulation of oxidative damage. Oxyradicals (e.g., alkoxyradicals $RO^{\bullet}$, and peroxyradicals, $ROO^{\bullet}$ where R indicates some organic compound) produced from xenobiotics during detoxification transformations can cause oxidative damage. Other toxicants, including some metals, are involved in production of reactive oxygen metabolites (i.e., H_2O_2, $O_2^{\bullet-}$, and $^{\bullet}OH$) via the catalyzed Haber–Weiss reaction. Damage can be reduced by production of antioxidants that react with oxyradicals and by synthesis of enzymes that reduce levels of reactive oxygen metabolites (Burdon 1999). The oxidative stress response exhibits phylogenetic variation and is subject to natural selection. So, adaptation is possible for moderating the oxidative stress-related aging of exposed individuals.

16.2.2.2 Disposable Soma and Related Theories of Aging

The force of selection decreases with age because an individual's contribution to continuance of the germ line decreases. The soma—the individual—gradually becomes increasingly disposable as it contributes less and less to the germ line. According to the antagonistic pleiotropy hypothesis of aging, pleiotropic genes that have a positive effect on a young individual but a negative effect on an old one are favored by selection, and those with negative effect on young but positive on old are selected against (Hughes and Charlesworth 1994). Consequently, genes positively affecting young and negatively affecting old individuals accumulate through time in species.

According to the mutation accumulation theory of aging, the intrinsic qualities of the soma (e.g., slow accumulation of somatic mutations) combine with the decreasing force of selection, as the individual grows older. The accumulation of mutations leads to diminished functioning of individuals, e.g., lowered physiological functioning or enhanced risk of cancer. According to this theory, aging results from (1) specialization of individual's cells into soma or germ line, (2) antagonistic pleiotropy, and (3) accumulation of somatic mutations (Stearns 1992). Toxicants that change the duration of various life cycle stages can influence the timing of these adverse manifestations. This combination of toxicant-modified life cycle timing, antagonistic pleiotropy, and accumulation of mutations can diminish the fitness of exposed individuals.

16.2.3 Optimizing Fitness: Balancing Somatic Growth, Longevity, and Reproduction

> Life history evolution makes the simplifying claim that the phenotype consists of demographic traits—birth, age, and size at maturity, number and size of offspring, growth and reproductive investment, length of life, death—connected by constraining relationships, trade-offs. These traits interact to determine individual fitness.
>
> **(Stearns 1992)**

Given a change in environmental conditions such as the introduction of a toxicant, individuals will reallocate energy resources among essential functions. Perhaps, as suggested by κ-rule triage, more resources will be used for maintaining (survival) and less for increasing (growth) the soma. Or a more complex allocation could be set into motion to optimize fitness (e.g., Table 16.2).

Each individual has a certain plasticity in its life history traits that can be invoked, but each also has limits to this plasticity that it cannot exceed. As an example, an individual undergoing chronic stress might have decreased its size and age at sexual maturity so that it can minimize any reduction in fitness (Stearns and Crandall 1984). It might also increase the time between litters (Reznick and Yang 1993).[6] Exactly how individual size and age at maturity change is determined by the constraints on life history plasticity of the species and the genetic constitution of the particular individual. Described below are some theoretical schemes for the optimal allocation of energy resources under toxicant exposure.

Several general predictions are made on the basis of life history theory (Table 16.3). An obvious one occurring with increased stress and insufficient phenotypic plasticity would be high mortality and reproductive failure. As likely as this scenario is, it is a not a life history response to be explored here: it is a failure to respond. The first two general predictions of life history responses are derived directly from Reznick's work with guppies (*Poecilia reticulata*) (Reznick 1990, Reznick et al. 1990) and a third is a more involved scheme emerging from work by Sibly and coworkers (Holloway et al. 1990, Sibly 1996, Sibly and Calow 1989). First, if environmental conditions tend to increase adult mortality, individuals will shift toward early maturation and increased reproductive

[6] In this context, the fitness of an individual might be reflected in its reproductive value, that is, V_A in Equation 15.9. Reproductive value incorporates both the cumulative production of young and the timing of parturition, which affects when young begin contributing to the germ line. The r may also be used cautiously to measure the "fitness" of a lineage or population.

TABLE 16.2
Selected Examples of Life History Shifts during Exposure to Diverse Stressors

Stressor	Species	Life History Trait Shifts	Citation
Plants			
Low moisture, salt, low nutrients	*Spartina patens*	Slow growth rates	Silander and Antonovics (1979)
Heavy metals	*Agrostis capillaris* (predominantly clonal growth strategy)	Increased vegetative growth of tolerant strains versus sensitive strains grown on contaminated soil	Wilson (1988)
Heavy metals	*Plantago lanceolata*	Slow growth rates	Antonovics and Primack (1982)
Invertebrates			
Thermal	*Asellus aquaticus*	Decreased longevity Bred at younger age and smaller size	Aston and Milner (1980)
Coal mine effluent	*A. aquaticus*	Lower investment in reproduction Fewer but larger offspring	Maltby (1991)
Vertebrates			
Heavy metals	*Catostomus commersoni*	Increased growth (length and weight) Increased fecundity (but smaller eggs) Decreased overall reproductive success Decreased age of maturity Decreased longevity	McFarlane and Franzin (1978)
Food limitation	*Poecilia reticulata*	Age of maturity increases Size at maturity and growth rate decrease	Reznick (1990)
		Larger (heavier) offspring Longer period between broods	Reznick and Yang (1993)
High predation on adults	*P. reticulata*	Early maturation Higher reproductive effort More but smaller young	Reznick et al. (1990) Reznick et al. (1996)
High overall predation	*P. reticulata*	Early maturation at smaller size More but smaller young Higher reproductive effort	Reznick and Bryga (1996)

effort (Reznick et al. 1990). Second, if environmental conditions increase juvenile mortality, individuals will shift toward late maturation and decreased reproductive effort (Reznick et al. 1990). Third and more complex, a juvenile strikes a balance between growth and survival in any stressful environment. If a toxicant presents a high risk of killing the individual outright, the individual will allocate resources to maintain the highest fitness (survival) possible under these conditions. Alternately, the toxicant can impact the individual's somatic production, which can be measured as a decrease in its scope of growth. (Scope of growth is the amount of all energy taken in by the individual minus its metabolic losses.) Toxicants can decrease productivity by increasing

TABLE 16.3
Predictions for Life History Shifts with Mortality and Growth Stress[a]

Stress Type	Response
Mortality	1. Increased growth with increasing stress (Sibly and Calow 1989)
	2. Less investment in defense and repair with increasing stress (Sibly and Calow 1989)
	3. Short generation time favored, reducing risk of death between birth and reproduction (Sibly and Calow 1989)
	4. (Adult mortality) Early maturation and increased reproductive effort (Reznick et al. 1990)
	5. (Juvenile mortality) Late maturation and decreased reproductive effort (Reznick et al. 1990)
Growth	1. Increased growth with increasing stress (Sibly and Calow 1989)
	2. Less investment in defense and repair with increasing stress (Sibly and Calow 1989)

[a] Stress is defined here as anything that reduces the Darwinian fitness of an individual.

metabolic costs, e.g., losses due to protein synthesis or turnover, or diminished osmo- and ionoregulatory capacity. Sibly and coworkers noted that numerous strategies occur for balancing juvenile growth against survival. A trade-off curve exists and, if in possession of sufficient phenotypic plasticity, individuals will move along that curve to a point of highest fitness for any particular situation.

Sibly and coworkers (Holloway et al. 1990, Sibly 1996, Sibly and Calow 1989) visualize life history responses to stress with trade-off curves as shown in Figure 16.5. To use this approach for prediction, an understanding is needed of the species' phenotypic plasticity relative to growth and survival and the general mode of impact for the toxicant on the species. Does the toxicant influence the species by primarily decreasing survival or by decreasing productivity? The trade-off curve will shift upward if mortality stress is present, or to the left if growth stress is present (Figure 16.5, middle panel). Sibly and Calow (1989) give the example of a predator as a mortality stressor and lowered ambient temperature for an ectotherm as a simple growth stressor, and suggest that many toxicants are likely mixed stressors affecting both survival and growth. Sometimes the trade-off curve can change shape with toxicant exposure (Figure 16.5, lower panel) and confound prediction. If stress does not change the shape of the trade-off curve, Sibly suggests that adaptation under mortality or growth stress (middle panel of Figure 16.5) should favor less investment in defense and faster growing individuals (Sibly and Calow 1989). (See Stearns (1992), Table 4.1 for a more detailed matrix of possible life history trade-offs.)

Microevolution is suggested by a change through successive generations toward optimum fitness on these trade-off curves. This would produce a slow shift of phenotypes into a tight cluster around the highest possible fitness value. The result can be microevolution for life history traits, a remarkably fast process in studies of predation (Reznick et al. 1990), habitat stability (Stearns 1983), and toxicant exposure (Maltby 1991).

Several conditions can produce deviations from these predictions. As mentioned, toxicant exposure can change the shape of the trade-off curve, resulting in predictions other than those highlighted in Table 16.3 (Walker et al. 1996). Several tempering points made by Stearns (1992) provide details about other complicating factors. Energy resources may not be limiting in a particular environment so predictions based on the premise of optimal energy allocation might be irrelevant. Some trade-offs involve switches, not continuous functions as shown above. Finally, some functions are relatively insensitive to the differences of energy resources in the environment.

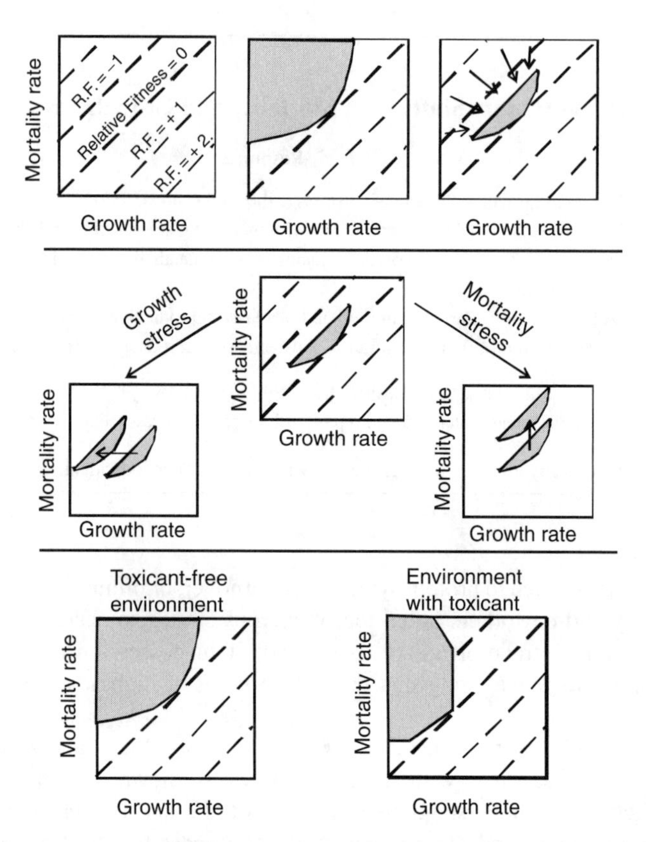

FIGURE 16.5 Trade-off curves for life history traits under toxicant stress. The upper panel shows lines of identical fitness for various combinations of growth and mortality rates (left box, modified from Figure 1 in Sibly and Calow (1989)), the lines of equal fitness plus the range of phenotypes actually possible for individuals in the population (middle box), and the hypothetical shift in phenotypes to optimize fitness (right box) in an unchanging environment. The middle panel shows the shifts in phenotypes (growth and mortality rates) relative to toxicant-induced changes in growth or mortality. (Modified from Figures 2 and 10 in Sibly and Calow (1989) and Sibly (1996), respectively.) Notice for mortality stress that individuals at the bottom of the shifted crescent have lower mortality, perhaps due to higher investment in detoxification, but slower growth than those at the top of the crescent. The lower panel shows that the shape of the trade-off curve can change in the presence of the toxicant. (Modified from Figure 11 in Sibly (1996).)

16.3 DEVELOPMENTAL STABILITY IN POPULATIONS

> Developmental homeostasis refers to the stabilized flow of the developmental trajectory. This stabilized flow has two components: canalization and developmental stability. Canalization is the stability of development in different environments .... Developmental stability, in contrast, is the stability of development in the same environment.
>
> **(Graham et al. 1993)**

> ... a genotype with a low maintenance requirement can support growth over a wide range of environmental conditions. Furthermore, such individuals should show substantial homeostasis in response to stressful environments and should live longest. Since fluctuating morphological asymmetry, FA, is a measure of homeostasis ... in various environments ..., the least asymmetric individuals therefore should live longest.
>
> **(Parsons 1995)**

Box 16.3 Metals and Suckers: Effects on Life History Traits

Life history characteristics of individuals in two white sucker (*Catostomus commersoni*) populations near a base-metal smelter (Flin Flon, Manitoba) were studied relative to their metal exposures. Aerial deposition of metals from the Flin Flon smelter to nearby lakes began in 1930 and continued at a high rate until 1974 when installation of a superstack decreased deposition in nearby lakes. Hamell Lake (higher metal loadings) and Thompson Lake (lower metal loadings) were selected on the basis of their contrasting Cd and Zn burdens. Gill nets were set to survey suckers from both lakes, including surveys during the spring spawning. Age was estimated from growth rings of pectoral fin rays with tags being placed on some spawning fish during the first survey to allow verification of this technique. Egg counts were made for subsets of females at the appropriate seasons.

Fish length was similar for both lakes' populations until age 2. At age 2, Hamell Lake fish began to grow faster than those of the less contaminated Thompson Lake. Adult fish from the more contaminated Hamell Lake were longer than adults from Thompson Lake and generally weighed more than Thompson Lake fish after normalizing weight to fish length. A reduction in life expectancy for the Hamell Lake suckers was suggested as mean ages were 4.3 and 6.2 years for Hamell and Thompson Lake suckers, respectively. Mortality in Hamell Lake occurred earlier in life (prenatal or early larval stage) than in Thompson Lake. Older fish had similar mortality rates (c. 0.8/year) in both lakes. So, in general, suckers from the more contaminated Hamell Lake grew faster (after year 2), were more rotund, but died earlier than those from Thompson Lake.

Reproductive effort was also distinct for fish from the two lakes. Hamell Lake suckers matured at an earlier age. Mean egg diameter for Hamell Lake fish (1.74 mm) was significantly smaller than for Thompson Lake fish (1.82 mm). Curiously, Hamell Lake females surveyed in June–July of 1976 had not spawned, whereas all of Thompson Lake fish had completed spawning. Egg reabsorption for Hamell Lake female suckers was evident. McFarlane and Franzin estimated that Hamell Lake suckers experienced a 50% failure in reproduction in the summer of 1976.

The white sucker population of Hamell Lake might have appeared after a typical survey to be unaffected as it had a fast growth rate, heavy individuals, early sexual maturation, and increased egg production. A closer examination revealed shortened longevity, reduced reproductive success and egg size, and low population density as reflected in catch-per-unit-effort data. Increased juvenile mortality was evident. Only by looking at a suite of traits together could the extent of effect be completely evaluated.

Population phenogenetics will be explored in this section relative to developmental homeostasis. Developmental homeostasis, or more specifically developmental stability, has been applied for decades as a metric of population condition but has only recently caught the attention of ecotoxicologists. Simple, but meaningful, measurements possible for assessing the impact of anthropogenic stressors on populations are described by Zakharov (1990) and Graham et al. (1993a,b). A population phenogenetics approach has several advantages for assessment of toxicant effects. (1) In a discipline preoccupied with complex measurement instrumentation and methodologies (i.e., Medawar's *idola quantitatis* (1982)), a refreshing feature of this approach is that a simple ruler and, if the anatomical structures of interest are minute, a dissecting microscope might be the only instruments required to take all necessary measurements. Measured anatomical traits may be meristic (i.e., phenotypes are expressed in discrete, integral terms such as number of pectoral fin rays) or continuous terms (i.e., phenotypes are expressed along a continuum such as length of the fifth pectoral fin ray). Computer-based, image analysis systems can be applied to questions requiring many measurements. (2) Equally valuable in a field biased toward producing effects data for individuals, information useful for inferring toxicant effects to populations is acquired with these simple tools. (3) The approach is so general that it can be used for extremely diverse species, for example, plants (Graham et al. 1993a), invertebrates (Graham et al. 1993b), and vertebrates (Ferguson 1986, Leary et al. 1987). (4) Metrics can potentially be used to detect population degradation before extensive or irreversible damage occurs (Zakharov 1990). As implied by Graham and Parsons above, these kinds of studies quantify the diminished capacity to produce a population composed of optimally fit individuals owing to some environmental condition such as toxicant exposure.

In contrast to the first part of this chapter where emphasis was optimizing fitness by allocation of energy resources among essential functions, the focus here is fitness consequences of imperfect anatomical development. Of course, assuming that development follows the most energetically favorable path in any given situation (Graham et al. 1993c), developmental stability could be fit into the above energy allocation framework. Regardless, the focus here is on changes in fitness occurring because of changes in anatomical development, and not underlying physiological or biochemical processes.

Developmental stability is the capacity to produce similar anatomical phenotypes ("minimize random accidents of development within a trajectory" (Graham et al. 1993a)) in a particular environment. Developmental stability is important because studies show that deviation outside a certain optimal range of phenotypes results in an individual with lowered fitness. Environmental stressors, including chemical toxicants, can decrease developmental stability.

Developmental stability can be estimated in several ways but fluctuating asymmetry (FA) is the most commonly applied metric for bilaterally symmetrical individuals or structures. FA is simply the random (nondirectional) difference (d) in features measured from the left and right sides of individuals (Equation 16.1) (Figure 16.6). These paired measurements are taken from n individuals in the population. If the distribution of d_i values ($i = 1$ to n) is normal with a mean (Equation 16.2) of zero, the variance of the population (Equation 16.3) is a measure of the level of FA in the population

$$d_i = \text{Measurement}_{\text{right } i} - \text{Measurement}_{\text{left } i}, \tag{16.1}$$

$$M_d = \frac{\sum d_i}{n}, \tag{16.2}$$

$$\hat{\sigma}^2 = \frac{\sum (d - M_d)^2}{n - 1}. \tag{16.3}$$

Differences measured for a sample of individuals are used to estimate deviations from perfect bilateral symmetry in the population. An increasing FA implies a diminished ability to maintain developmental stability—a decline in the fitness of individuals making up the population. As a particularly unsettling example, FA for teeth of children born to alcoholic mothers was found to be higher than FA for teeth of children born to nonalcoholic mothers (Kieser 1992). The implication is that *in utero* alcohol exposure caused fetal stress. More directly relevant to ecotoxicology, Valentine and Soulé (1973) demonstrated increased FA using pectoral fin ray numbers in grunion (*Leuresthes tenuis*) exposed to dichlorodiphenyltrichloroethane (DDT), suggesting that FA can be a valuable ecotoxicological tool (Valentine et al. 1973). Several studies have correlated FA with measures of fitness (Mitton 1997, Naugler and Leech 1994, Parsons 1992). Increases in FA were associated in these studies with a decrease in Darwinian fitness for individuals in the population. Further, shifts back over several generations from high FA

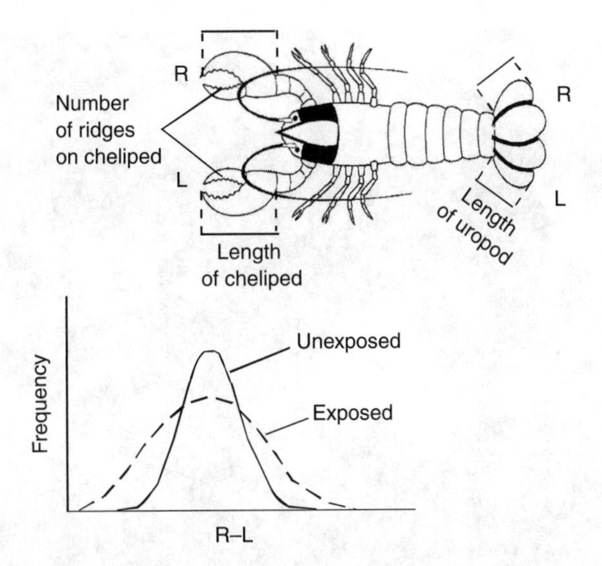

FIGURE 16.6 Developmental stability is often measured as fluctuating asymmetry. Morphological features (e.g., length of cheliped or number of ridges on the inside surface of the cheliped) are measured on the right (R) and left (L) sides of individuals with bilateral symmetry. Notice that cheliped or uropod lengths are continuous traits but number of cheliped ridges is a meristic trait. Measurements are made for a sample of individuals from the population and the distribution of differences between R and L measurements used to suggest the level of developmental stability in the population. The larger the variance, the less stable is the development. The bottom set of normal curves illustrates the example of more stability in the unexposed population than in the exposed population.

for exposed populations can provide evidence of microevolution. Hoffmann and Parsons (1991, p. 107) used laboratory studies showing a drop in FA with stress adaptation of diptera (Mather 1953, McKenzie and Clarke 1988, Reeve 1960) to suggest that such microevolution could occur in the wild.

FA is only one of several asymmetries that can be expressed by bilateral organisms. Directional asymmetry occurs if the mean of the d_i distribution is nonzero. There is a tendency for the asymmetry to be biased to one side. Antisymmetry can occur if there is a bimodal distribution for d_i. The familiar example of antisymmetry is the large claw of adult male fiddler crabs. Males will have a very large claw on the right or left (e.g., Figure 16.7). The result is a bimodal distribution in d_i values for the population. Graham et al. (1993c) provide a detailed review of these developmental conditions for bilateral organisms or structures.

16.4 SUMMARY

> Collusive lying occurs when two parties, knowing full well that what they are saying or doing is false, collude in ignoring the falsity.
>
> **(Bailey 1991)**

Most honest and informed ecotoxicologists, including the authors, would reluctantly admit to participating in a certain level of collusive lying as they assess ecological risk. By applying 96-h LC50, no observed effect concentration (NOEC), MATC, or other values derived from individual-based data, risk assessors reluctantly ignore knowledge about population-level processes such as those described here and adhere to the belief that individual-based data are all that are needed to successfully regulate

FIGURE 16.7 As an example of antisymmetry, male red-jointed fiddlercrabs (*Uca minax*) can have either an enlarged right or left claw.

toxicants. Our attempt in this chapter to translate phenotypic variation to population consequences is intended to alleviate some of this dependence on such behavior. Such adherence to clearly false paradigms is as corrosive to the framework of the science as it is expeditious for environmental regulation.

Box 16.4 Lead, Benzene, and Asymmetric Flies

In an uncommon ecotoxicological study, Graham et al. (1993b) explored the change in FA of *D. melanogaster* populations exposed to increasing concentrations of lead chloride or benzene. In this laboratory study, these toxicants were added separately to the fly's culture media. Neither lead nor benzene at any concentration affected the number of adult flies produced; however, emergence time was earlier for flies held at the highest benzene concentration (10,000 mg/kg of media) than for flies held at other concentrations.

The numbers of sternopleural bristles on the right and left sides of adults emerging in these treatments were measured. The frequency distribution of differences in the number of bristles was generated for each concentration of each toxicant. The null hypothesis of a mean of zero for d_i values was assessed with t-tests. Conventional tests of normality such as the Shapiro–Wilk's test were not applied. Instead, estimates of kurtosis and skewness were used. An F_{max} test was used to test for significant differences in FA among treatments.

With one exception, all concentration treatments for both toxicants had mean d (M_d) values that were not significantly different from zero. Male fly data for the 10,000 mg/kg benzene treatment were omitted from this analysis because data examination for this treatment suggested directional asymmetry. Although males had fewer bristles than females, the FA for both sexes was similar, so FA data were pooled for the sexes. The FA increased significantly ($\alpha = .05$) with increasing concentrations of lead or benzene. Figure 16.8 shows the results for both sexes, including the males exposed to 10,000 mg/kg of benzene.

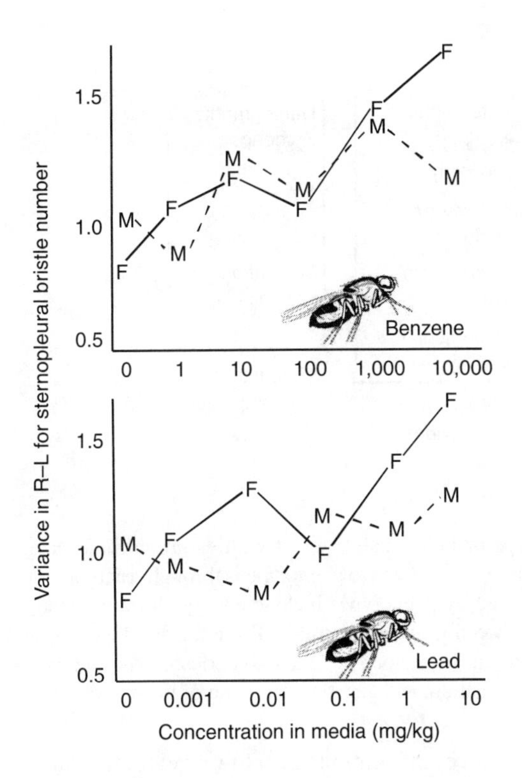

FIGURE 16.8 The influence of lead and benzene concentrations in media on the developmental stability of *D. melanogaster* (data from Table 3 in Graham et al. 1993b). Sternopleural bristle number was counted on the right and left sides of each individual. M = male and F = female.

This laboratory study clearly indicates the value of FA as a measure of population consequence of toxicant exposure. Its straightforward design and analysis also shows the ease with which laboratory toxicity tests could be developed for FA, a population-level metric of effect.

The genotype exposed to toxicant can produce a range of phenotypes as described by reaction norms or distinct phenotypes as described by polyphenisms. A spectrum of phenotypes is possible (Figure 16.9) but the range is restricted by inherent limits of each species. Associated changes include those affecting energy allocation, sex determination, aging and longevity, trade-offs between growth and survival, development time, and developmental stability. These traits are subject to natural selection. They translate into changes in demographic qualities of populations, likelihood of local population extinction, and potentially, adaptation and microevolution.

16.4.1 Summary of Foundation Concepts and Paradigms

- The phenotype is a consequence of genetic, environmental, and genetic × environmental factors.
- Changes to individuals can be translated into population consequences by considering processes controlling effects to phenotypes within populations. Theory linkage can occur from physiology→individual life history traits→population phenomena.
- Life history traits of *r*- and *K*-strategists under toxicant exposure result in shifts in community composition favoring *r*-strategists.
- The principle of allocation holds that a finite amount of energy is available to fulfill the diverse life functions of individuals. Rules emerge for the use of energy resources as individuals modify their phenotypes to optimize Darwinian fitness.

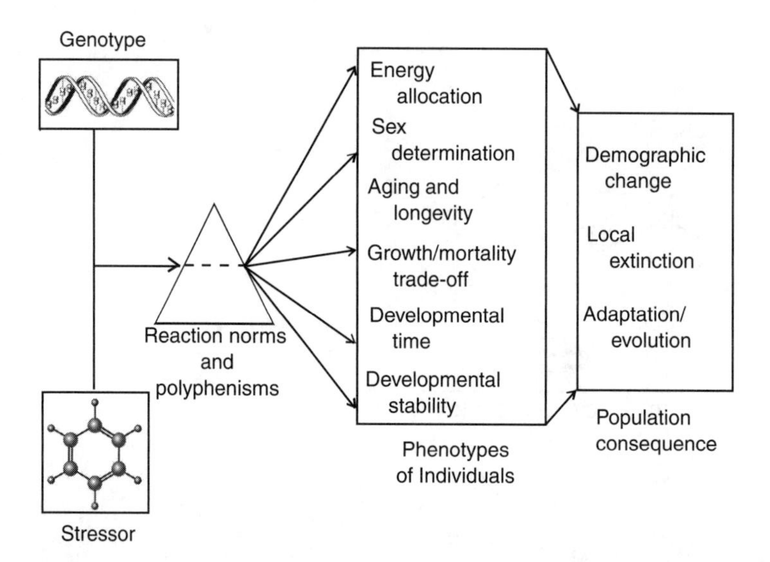

FIGURE 16.9 Summarizing the chapter, the genotype of individuals interact with environmental factors, including toxicants, to generate phenotypes. Translation from genotype to phenotype is through reaction norms and polyphenisms. Important life history traits manifested by phenotypes include energy allocation tendencies, sex determination, aging, longevity, trade-offs between growth and survival, rate of development, and developmental stability. The net consequence of changes in life history traits due to toxicant exposure can be demographic change, local extinction, physiological acclimation, and genetic adaptation or microevolution.

- Hormesis can be a consequence of energy resource reallocation under low levels of toxicant exposure.
- The reaction norm concept describes phenotypic plasticity relative to life history traits of individuals. Reaction norms transform the variation in an environment to variation in the phenotype.
- Variation in phenotype as a consequence of environmental variation may also involve polyphenism, the production of two or more distinct phenotypes from one genotype.
- Toxicants can influence aging in a manner predictable from several theories of aging.
- Many theories of aging are mutually inclusive and will likely be merged into a larger explanatory framework in the future. Some current and toxicant-relevant theories of aging include the limited life span, rate of living, stress-based including oxidative stress, and selection-based theories including the disposable soma theory.
- Trade-offs for life history traits are predicted for individuals under stress as described in Table 16.3 and Stearns's Table 4.1. Trade-offs are complex functions of species characteristics and genetic variation present in the relevant population.
- Developmental stability can be influenced by toxicants. A decrease in developmental stability implies a failure to produce a consistent phenotype in the environment and suggests a reduction in fitness for individuals in the population. Such instability is often measured as FA in populations although directional asymmetry and antisymmetry can also be useful.

REFERENCES

Adam, S.M., Crumby, W.D., Greeley, M.S., Jr., Ryon, M.G., and Schilling, E.M., Relationships between physiological and fish population responses in a contaminated stream, *Environ. Toxicol. Chem.*, 11, 1549–1557, 1992.

Antonovics, J. and Primack, R.B., Experimental ecological genetics in *Plantago*. VI. The demography of seedling transplants of *P. lanceolata*, *J. Ecol.*, 70, 55–75, 1982.

Aston, R.J. and Milner, G.P., A comparison of populations of the isopod *Asellus aquaticus* above and below power stations in organically polluted reaches of the River Trent, *Freshwater Biol.*, 10, 1–14, 1980.

Atchison, G.J., Sandheinrich, M.B. and Bryan, M.D., Effects of environmental stressors on interspecific interactions of aquatic animals, In *Ecotoxicology. A Hierarchical Treatment*, Newman, M.C. and Jagoe, C.H. (eds.), CRC Press/Lewis Publishers, Boca Raton, FL, 1996, pp. 319–345.

Bailey, F.G., *The Prevalence of Deceit*. Cornell University Press, Ithaca, NY, 1991.

Barinaga, M., Mortality: Overturning received wisdom. *Science*, 258, 398–399, 1992.

Bergeron, J.M., Crews, D., and McLachlan, J.A., PCBs as environmental estrogens: Turtle sex determination as a biomarker of environmental contamination, *Environ. Health Perspect.*, 102, 780–781, 1994.

Bongers, T., The maturity index: An ecological measure of environmental disturbance based on nematode species composition, *Oecologia*, 83, 14–19, 1990.

Bongers, T. and Ferris, H., Nematode community structure as a bioindicator in environmental monitoring, *TREE*, 14, 224–228, 1999.

Brooks, A., Lithgow, G.J., and Johnson, T.E., Mortality rates in a genetically heterogeneous populations of *Caenorhabditis elegans*, *Science*, 263, 668–671, 1994.

Burdon, R.H., *Genes in the Environment*, Taylor & Francis Inc., Philadelphia, PA, 1999.

Calabrese, E.J., McCarthy, M.E., and Kenyon, E., The occurrence of chemically induced hormesis, *Health Phys.*, 52, 531–541, 1987.

Calow, P. and Sibly, R.M., A physiological basis of population processes: Ecotoxicological implications, *Funct. Ecol.*, 4, 283–288, 1990.

Carey, J.R., Liedo, P., Orozco, D., and Vaupel, J.W., Slowing of mortality at older ages in large Medfly cohorts, *Science*, 258, 457–461, 1992.

Clark, A.G., Stress and metabolic regulation in *Drosophila*, In *Environmental Stress, Adaptation and Evolution*, Bijlsma, R. and Loeschcke, V. (eds.), Birkhäuser Verlag, Basel, Switzerland, 1997, pp. 117–132.

Curtsinger, J.W., Fukui, H.H., Townsend, D.R., and Vaupel, J.W., Demography of genotypes: Failure of the limited life-span paradigm in *Drosophila melanogaster*, *Science*, 258, 461–463, 1992.

Deevey, E.S., Life tables for natural populations, *Q. Rev. Biol.*, 22, 283–314, 1947.

Diamond, S.A., Newman, M.C., Mulvey, M., Dixon, P.M., and Martinson, D., Allozyme genotype and time to death of mosquitofish, *Gambusia affinis* (Baird and Girard), during acute exposure to inorganic mercury, *Environ. Toxicol. Chem.*, 8, 613–622, 1989.

Ferguson, M.M., Developmental stability of rainbow trout hybrids: Genomic coadaptation or heterozygosity? *Evolution*, 40, 323–330, 1986.

Forbes, V.E., Møller, V., Browne, R.A., and Depledge, M.H., The influence of reproductive mode and its genetic consequences on the responses of populations to toxicants: A case study, In *Genetics and Ecotoxicology*, Forbes, V.E. (ed.), Taylor & Francis, Inc., Philadelphia, PA, 1999, pp. 187–206.

Freeman, D.C., Graham, J.H., and Emlen, J.M., Developmental stability in plants: Symmetries, stress, and epigenetic effects, In *Developmental Stability: Origins and Evolutionary Significance*, Markow, T. (ed.), Kluwer, Dordrecht, the Netherlands, 1993, pp. 97–119.

Graham, J.H., Freeman, D.C., and Emlen, J.M., Developmental stability: A sensitive indicator of populations under stress, In *Environmental Toxicology and Risk Assessment, ASTM STP 1179*, Landis, W.G., Hughes, J.S., and Lewis, M.A. (eds.), American Society for Testing and Materials, Philadelphia, PA, 1993a, pp. 136–158.

Graham, J.H., Roe, K.E., and West, T.B., Effects of lead and benzene on the developmental stability of *Drosophila melanogaster*, *Ecotoxicology*, 2, 185–195, 1993b.

Graham, J.H., Freeman, D.C., and Emlen, J.M., Antisymmetry, directional asymmetry, and dynamic morphogenesis, *Genetica*, 89, 121–137, 1993c.

Heagler, M.G., Newman, M.C., Mulvey, M., and Dixon, P.M., Allozyme genotype in mosquitofish, *Gambusia holbrooki*, during mercury exposure: Temporal stability, concentration effects and field verification, *Environ. Toxicol. Chem.*, 12, 385–395, 1993.

Hickey, D.A. and McNeilly, T., Competition between metal tolerant and normal plant populations: A field experiment on normal soil, *Evolution*, 29, 458–464, 1975.

Hoffmann, A.A. and Parsons, P.A., *Evolutionary Genetics and Environmental Stress*, Oxford University Press, Oxford, UK, 1991.

Holloway, G.J., Sibly, R.M., and Povey, S.R., Evolution in toxin-stressed environments, *Funct. Ecol.*, 4, 289–294, 1990.

Hughes, K.A. and Charlesworth, B., A genetic analysis of senescence in *Drosophila*, *Nature*, 367, 64–66, 1994.

Kammenga, J.E., Van Koert, P.H.G., Riksen, J.A.G., Korthals, G.W., and Bakker, J., A toxicity test in artificial soil based on the life-history strategy of the nematode *Plectus acuminatus*, *Environ. Toxicol. Chem.*, 15, 722–727, 1996.

Kenyon, C., Chang, J., Gensch, E., Rudner, A., and Tabtiang, R., A *C. elegans* mutant that lives twice as long as wild type, *Nature*, 366, 461–464, 1993.

Kieser, J.A., Fluctuating asymmetry and maternal alcohol consumption, *Ann. Hum. Biol.*, 19, 513–520, 1992.

Knops, M., Altenburger, R., and Segner, H., Alterations of physiological energetics, growth and reproduction of *Daphnia magna* under toxicant stress, *Aquat. Toxicol.*, 53, 79–90, 2001.

Kooijman, S.A.L.M., *Dynamic Energy Budgets in Biological Systems*, Cambridge University Press, Cambridge, UK, 1993.

Kramer, V.J., Newman, M.C., Mulvey, M., and Ultsch, G.R., Glycolysis and Krebs cycle metabolites in mosquitofish, *Gambusia holbrooki*: Girard 1859, exposed to mercuric chloride: allozyme genotype effects, *Environ. Toxicol. Chem.*, 11, 357–364, 1992.

Kreuger, D.A. and Dodson, S.I., Embryological induction and predation ecology in *Daphnia pulex*, *Limnol. Oceanogr.*, 26, 219–223, 1981.

Leary, R.F., Allendorf, F.W., and Knudsen, K.L., Differences in inbreeding coefficients do not explain the association between heterozygosity at allozyme loci and developmental stability in rainbow trout, *Evolution*, 41, 1413–1415, 1987.

Levins, R., *Evolution in Changing Environments*. Princeton Press, Princeton, NJ, 1968.

Maltby, L., Pollution as a probe of life-history adaptation in *Asellus aquaticus* (Isopoda), *Oikos* 61, 11–18, 1991.

Mather, K., Genetic control of stability in development, *Heredity*, 7, 297–336, 1953.

May, R.M., *Theoretical Ecology. Principles and Applications.* W.B. Saunders Co., Philadelphia, PA, 1976.

McFarlane, G.A. and Franzin, W.G., Elevated heavy metals: A stress on a population of white suckers, *Catostomus commersoni*, in Hamell Lake, Saskatchewan., *J. Fish. Res. Board Can.*, 35, 963–970, 1978.

McKenzie, J.A. and Clarke, G.M., Diazinon resistance, fluctuating asymmetry and fitness in the Australian sheep blowfly, *Lucilia cuprina*, *Genetics*, 120, 213–220, 1988.

Medawar, P.B., *Pluto's Republic*. Oxford University Press, Oxford, UK, 1982.

Medvedev, Z.A., An attempt at a rational classification of theories of ageing, *Biol. Rev.*, 65, 375–398, 1990.

Mitton, J.B., *Selection in Natural Populations*, Oxford University Press, Oxford, UK, 1997.

Mulvey, M., Newman, M.C., Chazal, A., Keklak, M.M., Heagler, M.G., and Hales, L.S., Jr., Genetic and demographic responses of mosquitofish (*Gambusia holbrooki* Girard 1859) populations stressed by mercury, *Environ. Toxicol. Chem.*, 14, 1411–1418, 1995.

Naugler, C.T. and Leech, C.M., Fluctuating asymmetry and survival ability in the forest tent caterpillar moth *Malacosoma disstria*: Implications for pest management, *Entomologia Exp. Appl.*, 70, 295–298, 1994.

Neuhold, J.M., The relationship of life history attributes to toxicant tolerance in fishes, *Environ. Toxicol. Chem.*, 6, 709–716, 1987.

Newman, M.C., Diamond, S.A., Mulvey, M., and Dixon, P.M., Allozyme genotype and time to death of mosquitofish, *Gambusia affinis* (Baird and Girard), during acute toxicant exposure: A comparison of arsenate and inorganic mercury, *Aquat. Toxicol.*, 15, 141–159, 1989.

Odum, E.P., Trends expected in stressed ecosystems, *Bioscience*, 35, 419–422, 1985.

Parsons, P.A., Fluctuating asymmetry: A biological monitor of environmental and genomic stress, *Heredity*, 68, 361–364, 1992.

Parsons, P.A., Inherited stress resistance and longevity: A stress theory of ageing, *Heredity*, 75, 216–221, 1995.

Petersen, R.C., Jr. and Petersen, L.B.-M., Compensatory mortality in aquatic populations: Its importance for interpretation of toxicant effects, *Ambio*, 17, 381–386, 1988.

Pieri, C., Falasca, M., Recchioni, R., Moroni, F., and Marcheselli, F., Diet restriction: A tool to prolong lifespan of experimental animals. Models and current hypotheses of action, *Comp. Biochem. Physiol.*, 103A, 551–554, 1992.

Posthuma, J.F., van Kleunen, A., and Admiraal, W., Alterations in life-history traits of *Chironomus riparius* (Diptera) obtained from metal contaminated rivers, *Arch. Environ. Contam. Toxicol.*, 29, 469–475, 1995.

Reeve, E.C.R., Some genetic tests on asymmetry of sternopleural chaeta number in *Drosophila*, *Genet. Res.*, 1, 151–172, 1960.

Reznick, D.N., Plasticity in age and size at maturity in male guppies (*Poecilia reticulata*): An experimental evaluation of alternate models of development, *J. Evol. Biol.*, 3, 185–203, 1990.

Reznick, D.N., Life-history evolution in guppies (*Poecilia reticulata*: Poeciliidae). V. Genetic basis of parallelism in life histories, *Am. Nat.*, 147, 339–359, 1996.

Reznick, D.N., Bryga, H., and Endler, J.A., Experimentally induced life-history evolution in a natural population, *Nature*, 346, 357–359, 1990.

Reznick, D.N., Rodd, F.H., and Carenas, M., Life-history evolution in guppies (*Poecilia reticulata*: Poeciliidae). IV. Parallelism in life-history phenotypes, *Am. Nat.*, 147, 319–338, 1996.

Reznick, D.N. and Yang, A.P., The influence of fluctuating resources on life history: Patterns of allocation and plasticity in female guppies, *Ecology*, 74, 2011–2019, 1993.

Rose, M. and Charlesworth, B., A test of evolutionary theories of senescence, *Nature* 297, 141–142, 1980.

Sagan, L.A., What is hormesis and why haven't we heard about it before? *Health Phys.*, 52, 521–525, 1987.

Sagan, L.A., On radiation, paradigms, and hormesis, *Science*, 245, 574–575, 1989.

Selye, H., *The Stress of Life*, McGraw-Hill Book Co., New York, 1956.

Selye, H., The evolution of the stress concept, *Am. Sci.*, 61, 692–699, 1973.

Sibly, R.M., Effects of pollutants on individual life histories and population growth rates, In *Ecotoxicology. A Hierarchical Treatment*, Newman, M.C. and Jagoe, C.H. (eds.), CRC Press/Lewis Publishers, Boca Raton, FL, 1996, pp. 197–223.

Sibly, R.M. and Calow, P., A life-history theory of responses to stress, *Biol. J. Linn. Soc.*, 37, 101–116, 1989.

Silander, J.A. and Antonovics, J., The genetic basis of the ecological amplitude of *Spartina patens*. I. Morphometric and physiological traits, *Evolution*, 33, 1114–1127, 1979.

Sohal, R.S. and Weindruch, R., Oxidative stress, caloric restriction, and aging, *Science*, 273, 59–63, 1996.

Stearns, S.C., A natural experiment in life-history evolution: Field data on the introduction of mosquitofish (*Gambusia affinis*) to Hawaii, *Evolution*, 37, 601–617, 1983.

Stearns, S.C., The evolutionary significance of phenotypic plasticity, *BioScience*, 39, 436–445, 1989.

Stearns, S.C., *The Evolution of Life Histories*, Oxford University Press, Oxford, UK, 1992.

Stearns, S.C. and Crandall, R.E., Plasticity of age and size at sexual maturity: A life history response to unavoidable stress, In *Fish Reproduction*, Potts, G. and Wootton, R. (eds.), Academic Press, London, UK, 1984, pp. 13–34.

Stebbing, A.R.D., Hormesis—The stimulation of growth by low levels of inhibitors, *The Sci. Tot. Environ.*, 22, 213–234, 1982.

Svensson, E., The speed of life-history evolution, *TREE*, 12, 380–381, 1997.

Tatara, C., Mulvey, M., and Newman, M.C., Genetic and demographic responses of mosquitofish (*Gambusia holbrooki*) populations exposed to mercury for multiple generations, *Environ. Toxicol. Chem.*, 18, 2840–2845, 1999.

Valentine, D.W. and Soulé, M.E., Effect of p,p'-DDT on developmental stability of pectoral fin rays in the grunion, *Leuresthes tenuis*, *Fish. Bull.*, 71, 921–926, 1973.

Valentine, D.W., Soulé, M.E., and Samollow, P., Asymmetry analysis in fishes: A possible statistical indicator of environmental stress, *Fish. Bull.*, 71, 357–370, 1973.

Walker, C.H., Hopkin, S.P., Sibly, R.M., and Peakall, D.B., *Principles of Ecotoxicology*, Taylor & Francis, Ltd., London, UK, 1996.

Watt, W.B., Power and efficiency as indexes of fitness in metabolic organization, *Am. Nat.*, 127, 629–653, 1986.

Wilson, J.B., The cost of heavy-metal tolerance: An example, *Evolution*, 42, 408–413, 1988.

Woltereck, R., Weitere experimentelle Untersuchungen über Artveränderung, speziell über das Wesen quantitativer Artunterschiede bei Daphniden, *Verh. D. Tsch. Zool. Ges.*, 1909, 110–172, 1909.

Zakharov, V.M., Analysis of fluctuating asymmetry as a method of biomonitoring at the population level, In *Bioindications of Chemical and Radioactive Pollution*, Krivolutsky, D.A. (ed.), CRC Press/Mir Publishers, Boca Raton, FL, 1990, pp. 187–198.

Zakharov, V.M., Introduction, *Acta Zool. Fennica*, 191, 4–5, 1992.

Zakharov, V.M. and Grahan, J.H., Developmental stability in natural population, *Acta Zoologica Fennica*, 191, 1–15, 1992.

17 Population Genetics: Damage and Stochastic Dynamics of the Germ Line

Because they offer neither advantage nor liability, neutral mutations are either lost or fixed by stochastic changes in allele frequency from generation to generation. Thus the evolutionary dynamics of neutral mutations are adequately described by equations employing population size, N, effective population size, N_e, neutral mutation rate, u, and migration rate, m. Neutral theory has had a tremendous impact on population genetics, and many empirical patterns are consistent with predictions arising from neutral theory.

(Mitton 1997)

17.1 OVERVIEW

This chapter describes key processes in population genetics other than adaptation and natural selection. Initial discussion outlines briefly how toxicants can damage DNA and then stochastic dynamics of population genetics are described. Understanding toxicant effects on stochastic processes is as important as understanding toxicant-driven natural selection.

Qualities of toxicant-exposed populations can be directly influenced by stochastic or neutral processes. "Neutral" is used here only to indicate genetic processes or phenomena not involving natural selection. Ecotoxicologists often focus on adaptation via natural selection and pay less attention than warranted to neutral processes. At best, neutral processes are invoked as null hypotheses during testing for selection. Current applications of such hypothesis tests by ecotoxicologists are prone to neglect experimentwise Type I errors, that is, prone to inappropriately favor the "statistical detection" of selection and to reject the neutral theory-based null hypothesis. In the lead chapter of *Genetics and Ecotoxicology* (Forbes 1999), Forbes states, "The ten contributions to this volume address a number of key issues that, taken together, summarize our current understanding of the relationship between genetics and ecotoxicology." Despite the clear value of Forbes's book, this statement is dismaying. Aside from one chapter discussing genotoxic effects, no chapter focuses primarily on neutral processes. Several chapters (e.g., Chapter 4) do present discussion of neutral processes but most retain a predominant theme of selection. In contrast, basic textbooks of population genetics (e.g., Ayala 1982, Crow and Kimura 1970, Hartl and Clark 1989) contain nearly as much discussion of neutral processes as adaptation and selection.

This preoccupation of ecotoxicologists biases the early literature by frequent neglect of obvious alternate explanations for observed changes in exposed populations. To counter this bias and appropriately balance discussion of neutral and selection-based processes, discussion of adaptation and selection will be put off until Chapter 18. Processes leading to a change in the genome, including genotoxicity, will be discussed and then followed by anticipated changes in allele and genotype composition in populations owing to genetic drift, population size, isolation, and population structure. Finally, genetic diversity and the potential influence of toxicants are discussed in the context of long-term population viability. Genetic diversity and heterozygosity discussions create a conceptual bridge to selection-based topics in Chapter 18.

17.2 DIRECT DAMAGE TO THE GERM LINE

Spontaneous and toxicant-induced changes in DNA (mutations) have diverse consequences (see also Section 4.3 in Chapter 4). Consequences of mutation range from innocuous to minimal to catastrophic relative to individual fitness. Temporal scales of impact on the species population can be immediate (e.g., nonviable offspring from afflicted individuals) or long term (e.g., evolutionary). Effects may be primarily to the soma, as in the case of carcinogenesis, or to the germ line. In this chapter, effects to the soma will be ignored and discussions will focus on those to the germ line.

17.2.1 GENOTOXICITY

Genotoxicity, damage to genetic materials by a physical or chemical agent, occurs by several mechanisms, but at the heart of most genotoxic events is a chemical alteration of the DNA. This alteration may be associated with free radical formation near the DNA molecule (e.g., radiation damage) or direct reaction of a chemical agent with the DNA. The result is a modified DNA molecule that might not be repaired with absolute fidelity (e.g., base pair changes). DNA damage could result in a single- or double-strand break. Some instances of chromosome damage can even lead to chromosomal aberrations, aneuploidy, or polyploidy. The consequence to the germ line is often an adverse genetic change.

Genotoxicants modify DNA by several mechanisms (Burdon 1999). Some toxicants alkylate the DNA molecule (Figure 17.1). The locations most prone to react with electrophilic alkylating groups are position 2, 3, and 7 nitrogens and position 6 oxygen of guanine; position 1, 3, 6, and 7 nitrogens of adenine; position 3 and 4 nitrogens and position 2 oxygen of cytosine; and position 3 nitrogen and positions 2 and 4 oxygens of thymine (Burdon 1999). Monofunctional alkylating agents (e.g., ethyl methane sulfonate in Figure 17.1 or ethylnitrosourea) bind covalently to only one site. Bifunctional alkylating agents (e.g., sulfur mustards) or the antitumor agent, *cis*-$[PtCl_2(NH_3)_2]$ bind to two sites, potentially crosslinking the two DNA strands. Metabolites of other xenobiotics can also bind to DNA to form adducts, covalently bound chemical additions to the DNA (Figure 17.2). For example, benzo[a]pyrene is rendered more water soluble by a series of Phase I detoxification transformations, but some products of Phase I detoxification (e.g., diol epoxide) readily bind with the nitrogenous bases of the DNA molecule.

Chemicals and ionizing radiation that produce free radicals (Figure 17.3) can modify both the bases and deoxyribose of the DNA molecule. Depending on the nature of the compound or radiation, the result might be a single- or double-strand break in the DNA. As illustrated in Figure 17.3, the reaction with deoxyribose results in a DNA single-strand break. Some forms of radiation can release large amounts of energy in short ionization tracks as they pass through tissue and interact with water molecules. This results in high local concentrations of free radicals and consequent high levels of breakage in a local region. This increases the chances of a double-strand break. Class b metals such as bismuth, cadmium, gold, lead, mercury, and platinum also bind covalently to N groups in the DNA molecule (Fraústo da Silva and Williams 1993). This binding and associated DNA damage enables the medical use of bismuth, gold, and platinum as antitumor agents. The $Pt(NH_3)_2^{2+}$ of the antitumor agent, *cis*-$[PtCl_2(NH_3)_2]$ avidly binds to DNA by forming two covalent bonds with bases within and between the DNA strands (Fraústo da Silva and Williams 1993). Metals also influence the hydrogen bonding between DNA strands (Figure 17.4) and, because this hydrogen bonding is crucial to proper pairing of complementary bases, can either enhance or reduce the accuracy of base pairings. Metals can also generate free radicals from molecular oxygen via redox cycling and

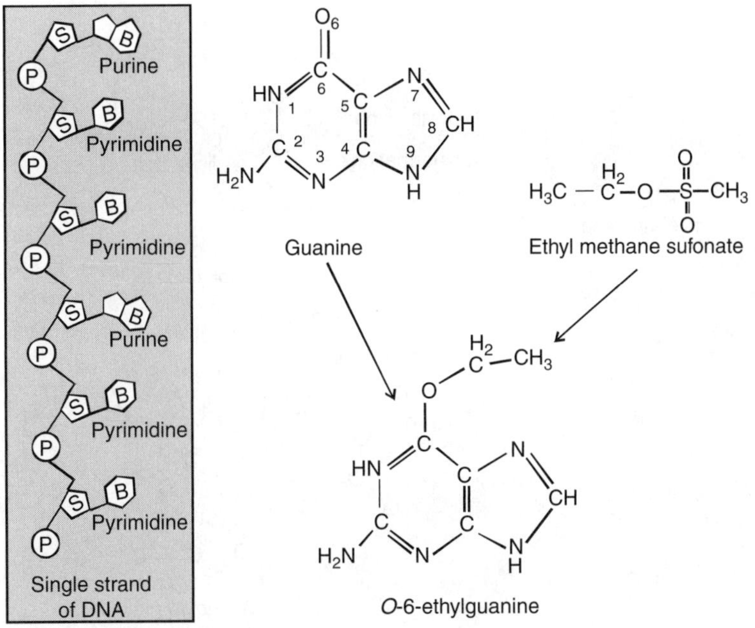

FIGURE 17.1 The modification of the purine base, guanine, by the alkylating agent, ethyl methane sulfonate. The DNA molecule (left shaded box: P = phosphate, S = deoxyribose sugar, B = purine or pyrimidine base) is modified at the nitrogenous base by such alkylating agents. Here guanine is covalently linked to an alkylating compound with only one site for potential binding. Guanine alkylated at the position 6 oxygen as shown here often mispairs with thymine and leads to a G:T→A:T transition sequence (Hoffman 1996). (With a transition, one purine is replaced by another or one pyrimidine is replaced by another.) DNA alkylation can also lead to base loss. For example, an alkyl adduct at position 7 nitrogen of guanine weakens the bond between the base and deoxyribose, and promotes base loss.

can interfere with transcription of DNA to RNA by binding to associated molecules. All of these mechanisms result in varying degrees and types of DNA damage. Although cells have several DNA repair mechanisms, some damage is more readily repaired than others. Mutations not repaired are perpetuated via the DNA replication process. The result is a wide range of potential modifications to the germ line.

17.2.2 Repair of Genotoxic Damage

Several mechanisms for DNA repair and damage tolerance have been described. For example, pyrimidine dimers formed during exposure to ultraviolet (UV) light may be enzymatically repaired. Photolyase cleaves these dimers and returns the DNA to its original state. A damage tolerance mechanism for these dimers allows the replication process to skip over the dimer and proceed normally in its presence. A gap is created in the new DNA strand that is filled later by repair mechanisms. This process also allows replication and subsequent repair in the presence of damage in the presence of DNA adducts.

Alkyltransferases are capable of removing alkyl groups from modified bases (e.g., the ethyl group attached to guanine at position 6 oxygen in Figure 17.1). Burdon (1999) indicates that, because alkyltransferase is inactivated by binding of the alkyl group to cysteine, cells have finite repair capacities. Repair is overwhelmed beyond a certain level of exposure and alkylation damage accumulates. Examples of repair by excision (Bootma and Hoeijmakers 1994) have been described for coping with larger adducts: damaged bases are removed and proper bases are inserted back into the DNA.

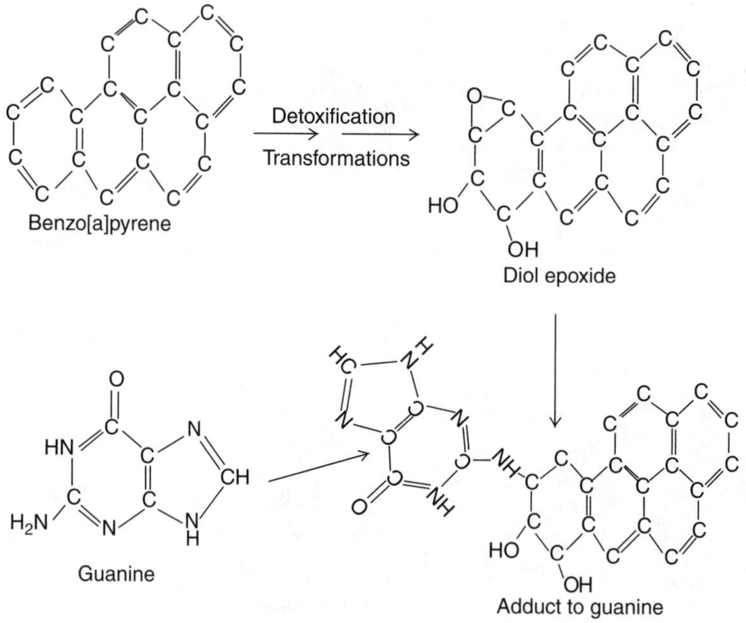

FIGURE 17.2 Cytochrome P450 monooxygenase-mediated conversion of the polynuclear aromatic hydrocarbon, benzo[a]pyrene, to a diol epoxide (7b,8a-diol-9a,10a-epoxy-7,8,9,10-tetrahydrobenzo[a]pyrene) that forms an adduct by covalently binding to the purine base, guanine. (Modified from Figure 2.5 in Burdon (1999).)

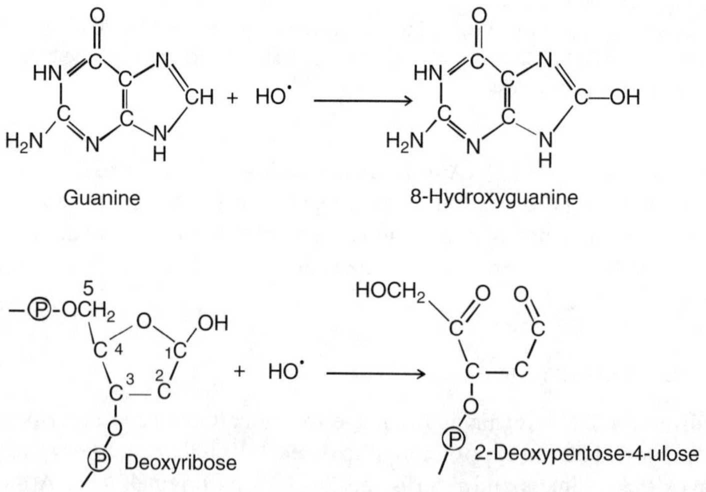

FIGURE 17.3 Interaction of the hydroxyl radical with base (guanine) and sugar (deoxyribose) components of the DNA molecule. Notice that the reaction shown with the deoxyribose results in a break in the DNA strand. (Modified from Figures 2.8 and 2.10 in Burdon (1999).)

Also, DNA ligase can insert bases into breaks in strands. Mismatched bases can be corrected via a mismatch repair process. Hoffman (1996) gives an example of mismatch repair that occurs with deamination of 5-methylcytosine.

These examples should illustrate that diverse types of DNA damage occur and that a variety of mechanisms exist for coping with the damage. Differences in types of damage and repair fidelities produce differences in genotoxicity among chemicals. For example, DNA damage due to chromium

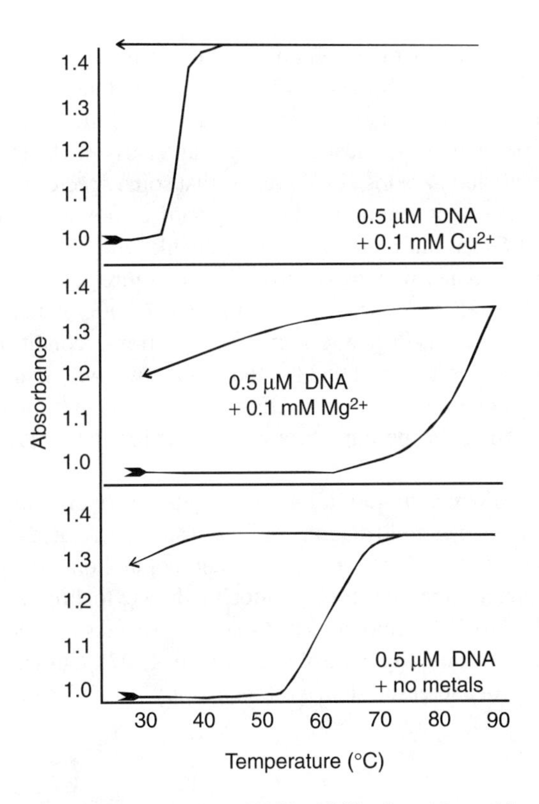

FIGURE 17.4 The influence of divalent metals on DNA stability is evidenced by changes in double-/single-stranded DNA composition of DNA solutions that are slowly heated and then cooled. Optical absorbance is low when most of the DNA is present in the double-stranded state and slowly increases as more and more DNA becomes single stranded. DNA begins to convert to predominantly single-stranded DNA (unwinding) as it is heated without metals to temperatures above circa 50°C. It remains as single-stranded DNA as it cools to temperatures below 40°C (bottom panel). The DNA double-stranded structure is stabilized by Mg^{2+}. In the presence of Mg^{2+}, the DNA unwinding occurs at a higher temperature and more DNA reverts to the double-stranded state during cooling. In contrast, the presence of Cu^{2+} results in unwinding at lower temperatures and reversion to double-stranded DNA during cooling is inhibited. The Cu^{2+} clearly interferes with proper base pairing between the strands of the DNA molecule. (Modified from Figure 6.10 in Eichhorn (1974).)

(as chromate) has lower repair fidelity than that from mercury. Mercury tends to produce single-strand breaks whereas chromate produces more protein–DNA crosslinking. Chromium is more carcinogenic of the two metals because single-strand breaks are repaired with higher fidelity than protein–DNA crosslink (Robison et al. 1984). Similarly, DNA single-strand breaks caused by thallium are repaired less effectively than those from mercury (Zasukhina et al. 1983). Imperfect repair can result in mutations within the germ line as well as cancers of the soma. Chronic exposure of male rats to thallium resulted in elevated prevalence of dominant lethal mutations among the embryos they sired (Zasukhina et al. 1983). In contrast, epidemiological studies have found male-mediated genotoxicity associated with Hiroshima atomic bomb survivors to be insignificant (Stone 1992). Indeed, mutation risk is believed to be minor relative to cancer risk in assessing radiation effects to humans (NCRP 1993).

17.2.3 Mutation Rates and Accumulation

The natural rate at which mutations appear varies among genes and species. Rates for bacteriophage, bacteria, and vertebrate species range from 4×10^{-10} to 1×10^{-4} mutations per gene per generation

(Table 1.4 in Ayala (1982)). Mutation rates for humans range from 4.7×10^{-6} to 1×10^{-4} mutations per gene per generation (Table 13.2 in Spiess (1977)). Microbes that have no distinct somatic and germ cell lines have mutation rates generally lower than those of metazoans, that is, approximately 10^{-9} to 10^{-6} mutations per cell per replication (Wilson and Bossert 1971).

Interestingly, Hoffmann and Parsons (1997) report that some species respond to increased stress by increasing mutation rates. For example, abrupt upward or downward changes in temperature increase mutation rates of *Drosophila melanogaster*. Jablonka and Lamb (1995) suggest that stress-induced increases in mutation rates may be adaptive because more genetically variable offspring are produced: The likelihood increases for producing an individual better fit to the extreme environment. However, this is envisioned as a desperate response to extreme conditions since the likelihood of an adverse mutation increases very quickly, too. Here, we will ignore such a response and focus only on increased mutation rate due to DNA damage. Such damage might involve direct genotoxic action or indirect damage, perhaps through increased oxidative stress caused by toxicants or stressors.

Stressors can clearly influence mutation rate in the laboratory and this influence is often dose dependent (Figure 17.5). However, field demonstrations of stressor-related increases in mutation rates are much less common. On the basis of sampling of field populations, Baker et al. (1996) reported extraordinary base-pair substitution rates for the mitochondrial cytochrome *b* gene (2.3 to 2.7×10^{-4} versus the anticipated 10^{-6} to 10^{-8} mutations per year) in a species of vole, but later retracted their conclusions based on a lapse in quality control (Baker et al. 1997). Convincing evidence from field studies has been reported for increased damage (aneuploidy) in slider turtles (*Trachemys scripta*)

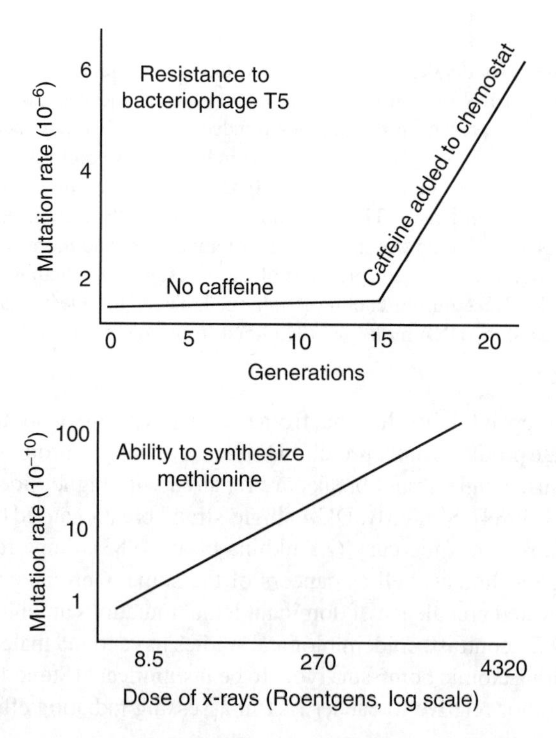

FIGURE 17.5 Genotoxic action of caffeine and x-ray irradiation on bacterial mutation rate. Bacteria maintained in a chemostat displayed an abrupt shift in their resistance to bacteriophage T5 after the addition of caffeine to the media (upper panel, modified from Figure 7 in Hartl and Clark (1989)). Such shifts in mutation rates are often concentration-dependent as evidenced by mutation rates for *E. coli* exposed to increasing doses of x-ray irradiation (lower panel, modified from Figure 2 in Wilson and Bossert (1971)).

exposed to radioactive contaminants (Lamb et al. 1991) and DNA strand breakage for mosquitofish (*Gambusia affinis*) inhabiting radionuclide-contaminated ponds (Theodorakis and Shugart 1999).

17.3 INDIRECT CHANGE TO THE GERM LINE

17.3.1 STOCHASTIC PROCESSES

Stochastic processes can have a strong influence on the genetic composition of a species population. Key stochastic determinants are effective population size, the spatial distribution of individuals within the population, mutation rate, and migration rate. Population size, specifically effective population size (N_e), determines how many individuals are available to carry a particular allele into the next generation. Small populations carry the increased risk of a random loss of an allele if too few individuals are contributing to allele transfer into future generations. Mutation rates, although very low, can influence the long-term genetic diversity of populations. Migration among subpopulations can dramatically influence the risk of allele loss or fixation. These population genetic parameters are explored below in a quantitative manner. However, before doing this, protein and DNA methods applied in the following studies are described briefly in Box 17.1.

Box 17.1 Methods Applied in Ecotoxicology to Define Genetic Qualities of Individuals

Advances in molecular genetic techniques have made the collection of genetic data for toxicological studies relatively easy and cost effective. A variety of molecular genetic markers (protein and DNA) provide powerful tools to investigate population demographic patterns, genetic variability in natural populations, gene flow, and ecological and evolutionary processes.

Environmental toxicologists are often interested in physiological or biochemical phenotypes, e.g., susceptibility, resistance, or tolerance to toxicants that are not readily assessed at the population level because they may be under the complex control of many genes and may be subject to environmental perturbation. Molecular genetic markers reflect simple genetic underpinnings. Markers may be chosen that behave as neutral markers of population processes or markers thought to be targets for selection can be examined in detail or monitored in populations. Numerous methods for acquisition of molecular genetic markers are available. Investigators must select from among them the technique that provides the requisite genetic information or variation to address each question (Table 17.1).

TABLE 17.1
A Summary of Molecular Genetic Markers and Data Provided for Uses in Ecotoxicology

Method	Number of Loci	Number of Individuals
Protein electrophoresis	Many	Many
RFLP	Few	Many
RAPD	Many	Many
Microsatellites	Few to many	Few to many
DNA sequencing	Few	Few

Protein Electrophoresis

Protein electrophoresis has been used to evaluate population genetic processes in field studies of toxicant impact and in laboratory toxicity studies. Proteins are separated on or in a supporting medium (e.g., starch, polyacrylamide, or cellulose acetate) using an electric field. Specific enzymes or proteins are visualized using histochemical stains. Differences in mobility are associated with charge differences among the proteins. A basic assumption of this method is that these charge differences reflect changes in the DNA sequence encoding the amino acids of the proteins. The bands of activity seen on gels following staining may be isozymes (functionally similar products of different gene loci, e.g., *Gpi*-1 and *Gpi*-2) or allozymes (allelic variants of specific loci, e.g., *Gpi*-2^{100} and *Gpi*-2^{165}). Banding patterns are interpreted to be genetically based, heritable, and co-dominant. Interpretation of banding patterns is well established and follows Mendelian inheritance rules.

Protein electrophoresis is a convenient and cost effective method to obtain information for many loci for many individuals or populations. Detailed descriptions of electrophoretic methods can be found in Richardson et al. (1986) and Hillis et al. (1996).

DNA Analysis

Nuclear, mitochondrial, or chloroplast genomes may be studied using DNA methods. DNA may be extracted from fresh, frozen, ethanol-preserved, or dried specimens. Gene sequences are routinely obtained by taking advantage of the polymerase chain reaction (PCR). Thermally stable DNA polymerases amplify DNA sequences from small quantities of template DNA. PCR requires short-DNA fragment primers to initiate DNA synthesis. Primers can be random or gene specific.

Restriction fragment length polymorphisms (RFLP) are determined when whole organelle genomes or amplified DNA products are digested with restriction enzymes. Restriction enzymes recognize and cleave double-stranded DNA at specific sites. These sites usually consist of four to six DNA base pairs. Following digestion of DNA with a series of restriction enzymes, the sample is subjected to electrophoresis on agarose gels. The DNA fragments are separated based on their size (number of base pairs). Data consist of the number and size of the resulting fragments. Variation arises from base pair substitutions, insertions, deletions, sequence rearrangements (which may result in the gain or loss of a restriction enzyme cutting site), or differences in overall size of the DNA fragment.

Williams et al. (1990) described a method to amplify random, anonymous DNA sequences using PCR. Random amplification of polymorphic DNA (RAPD) uses a single, short primer (approximately 10 bp) for the PCR. PCR products are DNA fragments flanked by sequences complementary to the primer. PCR products are separated by size on agarose or polyacrylamide gels. Data consist of scores of present or absent for the size-separated fragments and, therefore, display a dominant-recessive genetic pattern. Commercially available primer kits make screening for informative markers relatively easy. The RAPD approach is most useful for intraspecific studies.

Microsatellite DNA analysis can provide highly polymorphic multilocus genotype data comparable with that obtained with protein electrophoresis. Microsatellite loci behave as codominant Mendelian markers and are useful to evaluate genetic variation within and among conspecific populations. Microsatellite loci are identified by tandem repeats of short (2–4 bp) DNA sequences (e.g., CA_n or CTG_n, where n = number of tandem repeats). Changes in the number of repeat units give rise to the scored polymorphism. The PCR technique is used to obtain microsatellites. Microsatellite products are separated by size on agarose or polyacrylamide gels. Difficulties encountered with this technique include the need to screen for polymorphic loci and to develop highly specific primer pairs for the PCRs.

Each of the molecular genetic approaches discussed above provides indirect (protein electrophoresis) or incomplete (RFLP) assessment of genetic characteristics. Direct assessment of genetic traits may be obtained with DNA sequencing. The widespread availability of PCR methods and automated DNA sequencers has made this technique increasingly cost effective. DNA sequencing usually involves larger (20–30 bp) specific primers to amplify target sequences. DNA fragments of different lengths are generated using ddNTPs in the PCR for chain termination. Polyacrylamide gels are used to separate the fragments and the base sequence of DNA is determined.

17.3.2 Hardy–Weinberg Expectations

The Hardy–Weinberg principle states that the frequencies of genotypes within populations remain stable through time if (1) the population is a large (effectively infinite) one of a randomly mating, diploid species with overlapping generations, (2) no natural selection is occurring, (3) mutation rates are negligible, and (4) migration rates are negligible. For a locus with two alleles (e.g., alleles designated as 100 and 165) with allele frequencies of p for 100 and q for 165, the genotype frequencies will be p^2 for 100/100, $2pq$ for 165/100, and q^2 for 165/165. For a three allele locus (e.g., 66, 100, and 165), the genotype frequencies will be r^2 for 66/66, $2rp$ for 66/100, $2rq$ for 66/165, p^2 for 100/100, $2pq$ for 100/165, and q^2 for 165/165. Such a polynomial relationship can be visualized with a De Finetti diagram (De Finetti 1926) (Figure 17.6).

A χ^2 test can be used to test for significant deviation from Hardy–Weinberg expectations,

$$\chi^2 = \sum_{i=1}^{n} \frac{(\text{Observed}_i - \text{Expected}_i)^2}{\text{Expected}_i}, \tag{17.1}$$

where n = the number of possible genotypes (e.g., 3 for a two allele locus or 6 for a three allele locus), Observed_i = observed number of individuals of the ith genotype, and Expected_i = number of individuals of the ith genotype and expected based on the allele frequencies and the Hardy–Weinberg model. The degrees of freedom for the test is the number of possible genotypes minus the number of alleles (e.g., 3 − 2 = 1 for a two allele locus).

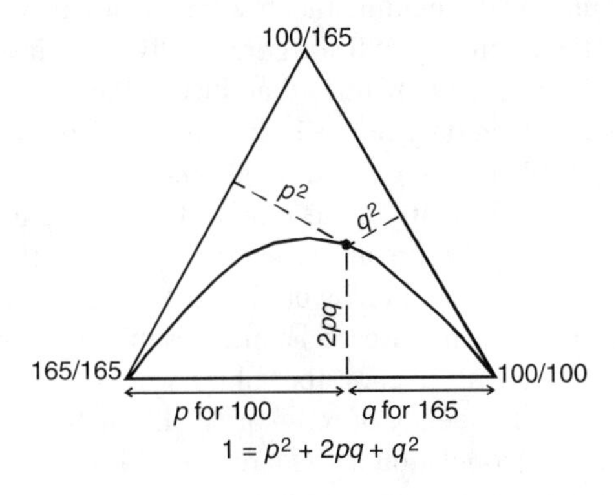

FIGURE 17.6 De Finetti diagram illustrating the Hardy–Weinberg principle. Conformity to Hardy–Weinberg expectations for any combination of allele frequencies (e.g., for alleles designated 100 and 165) are indicated by genotype combinations laying on the arc within the 100/100, 165/165, and 100/165 triangle. Points off this arc reflect deviations from expectations. The statistical significance of such a deviation can be tested with a χ^2 test.

If the χ^2 test with adequate statistical power failed to reject the null hypothesis, the conclusion is made that there is no evidence that the conditions for Hardy–Weinberg equilibrium were not met. If the null hypothesis was rejected, one or more of the assumptions was violated. As a word of warning, too often ecotoxicologists assume that rejection of the null hypothesis indicates that selection is occurring and ignore the other assumptions on which the Hardy–Weinberg relationship is based. Such studies must be read with caution.

17.3.3 Genetic Drift

Genotype frequencies do change in populations because of finite population size, population structure, migration, and nonrandom mating. An oft-observed consequence of toxicant exposure is a decrease in population size. Population migration rates or direction of migration can be influenced by toxicant avoidance increasing emigration or increased immigration after the toxicant removes a portion of the endemic population and presents vacant habitat to migrating individuals. Population structure can be influenced as toxicants create barriers, impediments, or disincentives to movement; e.g., patches of highly contaminated sediment or a large contaminant plume in a river or stream.

17.3.3.1 Effective Population Size

Genetic drift occurs in all finite populations. Drift can be continuous if the population is always small or intermittent if the population size fluctuates widely. Intermittent drift can produce genetic bottlenecks during times of small population sizes. Due to sampling error, a small population producing future generations will likely carry only a subset of the total genetic variability present in the large parent population.

Genetic drift will accelerate as the number of individuals contributing genes to the next generation (effective population size, N_e) decreases. This fact can be illustrated with a simple, random sampling experiment. Assume that a bowl is filled with 5000 red and 5000 blue marbles. We take 5000 marbles randomly from the bowl to produce the "next generation." We do this random sampling experiment 1000 times and get an average red:blue ratio each time. With these large numbers, a frequency of red marbles of 0.50 is expected with a modest amount of variation among the 1000 trials. Our sample size is so large that sampling error will be minimal. However, if we sampled only 10 marbles each time, the variation around 0.50 would be much wider than when we sampled 5000 marbles. In fact, in many more cases, the frequency will shift drastically to produce a "next generation" with a very different frequency of red or blue marbles than that of the parent generation. Indeed, there would be many more cases in which only red or blue marbles were available to produce the next generation. Drift in frequency of marble color through generations could be simulated by using the new "generational" frequency from 10 marbles to fill the bowl again with 10,000 red and blue marbles, and repeating the experiment for many generations. Clearly, the sampling error associated with taking only 10 marbles each "generation" would result in a drift in frequency away from that for the original bowl of marbles. In some cases, blue marbles might be lost completely with fixation occurring for "red." The opposite with fixation for "blue" would occur in other cases. Further, as the frequency of one allele (e.g., frequency of red marbles in the bowl) decreases, the risk of that allele (color) being lost from the population also increases. With intermittent drift and associated bottlenecks, populations can experience founder effects (a population started by a small number of individuals will differ genetically from the parent population due to high sampling error). Small populations bring to future generations a subset of the alleles present in a parent population and allele frequencies vary stochastically from those of the parent population.

The effective population size (N_e) is often smaller than the actual or census population size because all individuals do not contribute to the next generation. How many contribute to the next generation is a complex function of demographic and life history qualities. In general, N_e for a population with nonoverlapping generations is estimated as the harmonic mean of population sizes measured at a series of times (N_i) and the number of generations over which the population measurements were made (t) (Hartl and Clark 1989).

$$\frac{1}{N_e} = \left[\frac{1}{t}\right]\left[\frac{1}{N_1} + \frac{1}{N_2} + \cdots + \frac{1}{N_t}\right]. \tag{17.2}$$

The advantage of this estimate of N_e is that it weights generations with small population sizes more heavily than those with larger populations sizes. Genetic drift accelerates in a nonlinear manner as population size decreases so this heavy weighting of smaller population sizes is appropriate.

Effective population size is also influenced by sex ratio. As is evident from the use of the harmonic mean again in Equation 17.3, the sex present in the lowest number has the most influence on the estimated N_e. If the number of females and males were not equal in the population, the effective population size can be estimated with Equations 17.3 or 17.4 which is a rearrangement of Equation 17.3 (Crow and Kimura 1970).

$$\frac{1}{N_e} = \frac{1}{4N_{\text{Males}}} + \frac{1}{4N_{\text{Females}}}, \tag{17.3}$$

$$N_e = \frac{4N_{\text{Males}}N_{\text{Females}}}{N_{\text{Males}} + N_{\text{Females}}}. \tag{17.4}$$

The $\frac{1}{4}$ values in Equation 17.3 come from the fact that "the probability that two genes in different individuals in generation t are both from a male [or female] in generation $t - 1$ is $\frac{1}{4}$; and that they come from the same male [or female] is $1/4N_{\text{male}}$ [or $1/4N_{\text{female}}$]" (Crow and Kimura 1970).

If generations are overlapping in time, the assumption $N_e \approx N/2$ can be made or the following equation can be applied:

$$N_e = \frac{4N_a L}{\sigma_n^2 + 2}, \tag{17.5}$$

where N_a = the natality over a period of time, L = the mean generation time, and σ_n^2 = the brood size variance.

Genetic drift would eventually lead to loss or fixation of an allele in the absence of an effectively infinite population. How quickly or slowly this occurs is a function of N_e and the initial frequency of the allele in question. Equations 17.6 and 17.7 estimate the average number of generations needed to reach allele fixation ($p \rightarrow 1$) or loss ($p \rightarrow 0$), respectively. Wilson and Bossert (1971) grossly estimate that alleles are lost at a rate of 0.1 to 0.01 per locus per generation if N_e is 10 to 100, 0.0001 per locus per generation if N_e is approximately 10,000, and that loss is trivial if N_e is greater than 100,000. Ayala (1982) suggests that random drift is unlikely to determine allele frequencies if $4Nx$ is very much smaller than 1 (x = rate of mutation (u), rate of migration (m), or the selection coefficient (s)). (The m is estimated as the number of individuals migrating/total number of individuals that potentially could migrate; the rate of mutation is defined as the number of mutations expected per gamete per generation; the selection coefficient will be defined in Chapter 18.) Values of $4Nx > 1$ implied that mutation, migration, and/or selection will dominate changes in allele frequencies. Regardless, excluding times in which the allele is lost, the average number of generations to fixation ($p \rightarrow 1$) for an allele is the following:

$$\bar{t}_1 = -\frac{1}{p}[4N_e(1-p)\ln(1-p)]. \tag{17.6}$$

Alternatively, excluding the times when the allele becomes fixed, the average number of generations to allele loss ($p \rightarrow 0$) is the following:

$$\bar{t}_0 = -4N_e[p/(1-p)]\ln p. \tag{17.7}$$

Crow and Kimura (1970) extend these equations to consider the case of a (neutral) mutation that appears in an individual within a population. (The allele frequency, p, is set to $1/(2N)$ to derive these relationships.) Equations 17.6 and 17.7 become Equations 17.8 and 17.9, respectively. The probability of a neutral allele becoming established in the population increases as N_e decreases. Excluding cases in which it is lost from the population, a neutral mutant takes about $4N_e$ generations to reach fixation:

$$\bar{t}_1 \approx 4N_e, \tag{17.8}$$

$$\bar{t}_0 \approx 2(N_e/N)\ln(2N). \tag{17.9}$$

Why are the above details important to population ecotoxicology? First, the genetic composition of a population can be strongly impacted by a toxicant's influence on the effective population size. The toxicant can influence N_e by decreasing the total population size (Equation 17.2) through time, affecting the numbers of each sex present at any time (Equations 17.3 and 17.4), or modifying generation time or variance in brood size (Equation 17.5). Accelerated drift, genetic bottlenecks, and founder effects can result in loss of genetic information and produce strong shifts in genetic composition of populations (Equations 17.6 and 17.7). If a mutation appears in an individual in a population, its chance of fixation increases as N_e decreases. It might be helpful to re-emphasize at this point in our discussions that natural selection has nothing to do with these potential changes in the germ line. Nevertheless, toxicant exposure can lead to microevolution because allele frequencies have changed.

17.3.3.2 Genetic Bottlenecks

Drastically reduced population or subpopulation size due to toxicant exposure can result in a genetic bottleneck and consequent founder effect (Gillespie and Guttman 1999, Newman 1995, 1998). An acute toxic exposure, such as that associated with pesticide spraying and subsequent very high mortality, is the most straightforward example of an ecotoxicological event that could result in a bottleneck. Low levels of genetic variation among cheetah (O'Brien et al. 1987), Florida panther (Facemire et al. 1995), Lake Erie yellow perch (Strittholt et al. 1988), and Great Lakes brown bullhead (Murdoch and Hebert 1994) have been attributed to genetic bottlenecks. The last three examples putatively involved toxicant exposures. The underlying concern associated with bottlenecks is the potential loss of genetic information. Genetic variation in the short term may be associated with physiological or biochemical flexibility and, in the long term, with evolutionary potential and persistence in a changing environment. As an example, conservation biologists are concerned about the ability of the remaining wild cheetahs to cope with feline distemper, a serious infectious disease.

There is a lower, but finite, chance that a population experiencing a bottleneck will emerge with more genetic variation than the parent population because the variation among bottlenecked populations increases as N_e decreases. Whether the genetic variation increases or decreases simply depends on which individuals happen to make it through the bottleneck. However, the chances of a decrease are greater than those of an increase, especially with repeated or periodic bottlenecks, as might be associated with occasional or accidental release of toxicants. Gillespie and Guttman (1999) discussed this possibility of an increase in genetic variation following toxicant exposure but cautioned that maladaptive combinations of rare alleles have a higher chance of occurring in such cases.

17.3.3.3 Balancing Drift and Mutation

From our discussions to this point, the question might arise why genetic drift does not result in a gradual trend toward genetic uniformity. That would be the eventual fate of populations in the absence of mutation. Let us examine the balance between drift and mutation rates by assuming that the relevant genes are neutral. In Chapter 18, we will add details associated with differences in fitness among genotypes.

As mentioned above, the rate of change in a population of N diploid individuals owing to a mutation is $2Nu$ and that associated with drift is defined by Equations 17.6 through 17.9 and the associated text. The number of novel mutant alleles (M) that appear during each generation, eventually to become fixed, is defined by Spiess (1977),

$$M = (2N\bar{u}/2N) = \bar{u}, \tag{17.10}$$

where $\bar{u}$ = the average of the mutation rates for all alleles. Mutation rate (u) balanced against loss owing to genetic drift ($1/(2N)$) results in a steady-state level of genetic variation. Again, this explanation for the maintenance of genetic variation is conditional on neutrality of alleles. Crow and Kimura (1970) and Mitton (1997) indicate that effective population size (N_e) and mutation rate (u) determine the average heterozygosity of a population at equilibrium relative to the influences of genetic drift and mutation rate: $\bar{H} \approx (4N_e u)/(4N_e u + 1)$. Here, $\bar{H}$ is the average of the $2pq$ proportions for all scored loci where p and q are the allele frequencies for two allele loci. Obviously, the calculation is modified to include loci with more than two alleles. Populations should be expected to differ in their levels of heterozygosity. Some differences could reflect the influence of toxicant exposure on N_e, and perhaps, u.

17.3.4 Population Structure

What are the genetic consequences of population structure? Generally, an uneven distribution of individuals suggests nonrandom mating; therefore, N_e will be influenced by population structure. Hartl and Clark (1989) indicate that the density of breeding individuals in an area (δ) and the amount of dispersion between an individual's location of birth and that of the birth of its progeny (σ^2) influence N_e,

$$N_e = 4\pi\delta\sigma^2. \tag{17.11}$$

Clearly, a quality as basic as N_e is strongly influenced by population structure. Other important qualities are discussed in detail below as they often are neglected in ecotoxicological studies.

17.3.4.1 The Wahlund Effect

The Wahlund effect occurs after mixing of populations, each with distinct allele frequencies and in Hardy–Weinberg equilibrium. Mixing may occur during sampling if population structure was cryptic, i.e., individuals were unintentionally taken from two subpopulations and then pooled for analysis. Mixing may occur naturally if migration were taking place between subpopulations previously isolated by a barrier to movement. The frequency of the heterozygote in the mixed sample will be lower than predicted under the assumption that the sample came from a single, randomly mating population. For example, assume that equal numbers of individuals are mixed together from two populations with allele (100, 165) frequencies of $p_1 = 0.9$, $q_1 = 0.1$ and $p_2 = 0.1$, $q_2 = 0.9$. In Hardy–Weinberg equilibrium, the frequencies of the 100/100, 100/165, and 165/165 genotypes in these two populations would be the following:

Population 1:	$p_1^2 = 0.81$	$2p_1q_1 = 0.18$	$q_1^2 = 0.01$
Population 2:	$p_2^2 = 0.01$	$2p_2q_2 = 0.18$	$q_2^2 = 0.81$

Let us assume that 100 individuals from each population were mixed into a pooled sample. From population 1, there would be eighty-one 100/100 individuals, eighteen 100/165 individuals, and one 165/165 individual. From population 2, there would be one 100/100 individual, eighteen 100/165 individuals, and eighty-one 165/165 individuals. Therefore, the number of individuals of each genotype in the pooled sample would be the following: $100/100 = 82$ individuals, $100/165 = 36$ individuals, and $165/165 = 82$ individuals. In the pooled sample, $p(\bar{p})$ and $q(\bar{q})$ values are 0.5 each. The expected number of each genotype predicted from the Hardy–Weinberg principle $(1 = p^2 + 2pq + q^2)$ would be the following: $100/100 = 50$ individuals, $100/165 = 100$ individuals, and $165/165 = 50$ individuals. There is an apparent excess of both homozygotes or, stated another way, an apparent deficit of heterozygous genotypes. Figure 17.7 is a modified De Finetti diagram that visually illustrates this principle.

These same consequences arise if more than two populations were involved. Under conditions giving rise to the Wahlund effect (mixing of individuals from several populations and sampling before reproduction), the average frequency of heterozygotes can be generally described based on the $\bar{p}$ and $\bar{q}$ for a mixed sample involving k populations (Cavalli-Sforza and Bodmer 1971). (Assume equal numbers of individuals being contributed by each of the k populations to the sample.)

$$\bar{H} = 2\bar{p}\bar{q}[1 - (\sigma^2/\bar{p}\bar{q})], \quad (17.12)$$

where σ^2 is the variance in gene frequencies among k populations:

$$\sigma^2 = \left(\sum p_i^2/k\right) - \bar{p}^2. \quad (17.13)$$

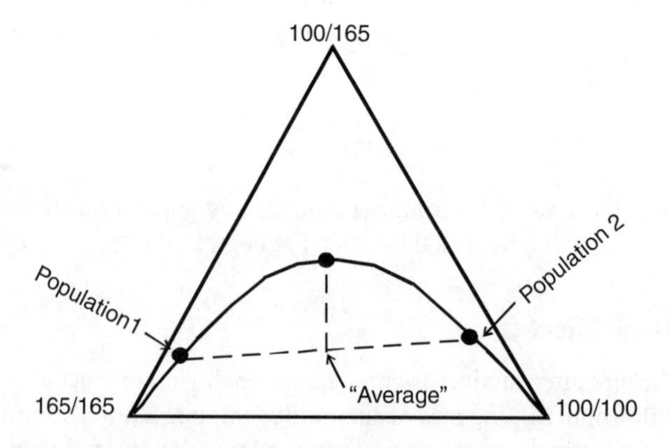

FIGURE 17.7 De Finetti diagram illustrating the Wahlund principle. In this example, equal numbers of individuals from two populations are mixed, resulting in an "average" for the genotype frequencies, 100/100, 100/165, and 165/165. The position defined by these genotype frequencies is off the arc representing all possible solutions to the Hardy–Weinberg polynomial, $1 = p^2 + pq + q^2$. The point on the arc immediately above the average reflects the expected frequencies of the three genotypes. On the basis of these expectations, there is an apparent deficiency of heterozygotes. (Modified from Hartl and Clark (1989).)

The deficiency of heterozygotes will be roughly twice the σ^2 (Cavalli-Sforza and Bodmer 1971).

Selander's *D* (1981) is a straightforward measure of the deviation from expectations. Selander's *D* is equal to the $H_{obs} - H_{exp}$ where H_{obs} is the observed proportion of heterozygotes in the sample and H_{exp} is the expected proportion of heterozygotes in the sample based on Hardy–Weinberg expectations. A negative D indicates a deficit of heterozygotes.

Samples produced by pooling individuals from several groups of a cryptically structured population might lead the unwary ecotoxicologist to conclude that the heterozygote was less fit than the two homozygotes and underrepresented in the sampled population due to selection associated with toxicant exposure. Such a conclusion must be considered conditional until the possibility of a Wahlund effect was explored carefully.

Box 17.2 Midges, Mercury, and Too Many Missing Heterozygotes: Evidence of a Wahlund Effect

Woodward et al. (1996) examined allozyme frequencies in midge (*Chironomus plumosus*) larvae from Clear Lake (California). Midges of this species emerge as adults to form mating swarms over the lake. Masses containing hundreds of eggs each are deposited on the lake surface by females and the hatched larvae drop to the bottom to become deposit feeders.

Samples of larvae were taken along a transect beginning at the Sulfur Bank Mercury Mine where mine tailings had been deposited in the lake for many decades. Six sites on the transect were sampled by boat using an Eckman dredge. Dredge samples were taken at each site until ample numbers of larvae were collected. Forty midges were deemed an adequate sample for an allozyme survey. On average, chironomids from approximately 10 dredge hauls were pooled to obtain the sample size of 40 midges per site.

Twelve polymorphic loci were examined by starch gel electrophoresis (see Figure 17.8 for an illustration) at the six sites; therefore, 72 χ^2 tests for deviation from Hardy–Weinberg

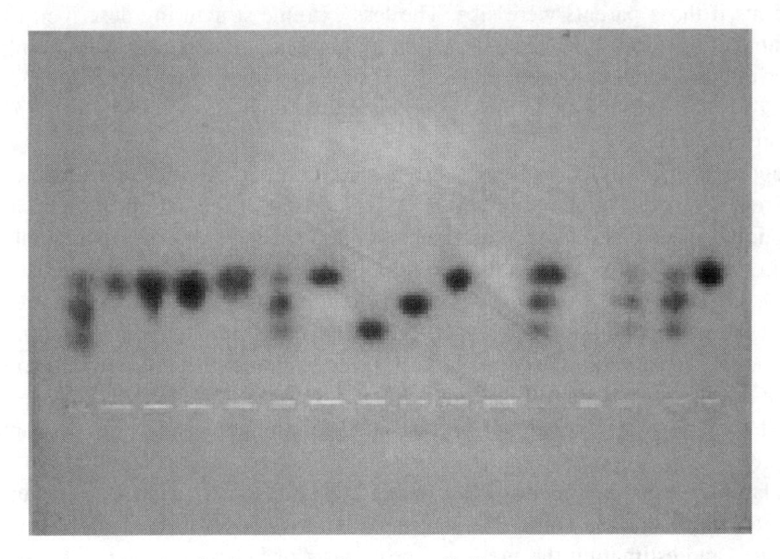

FIGURE 17.8 A starch gel stained to score allozymes, allelic variants of enzymes. Supernatants from tissue homogenates are loaded into slots in the gel. Many lanes can be loaded in each gel so that several individuals can be scored on a single gel. After protein separation in an electric field, gels are stained for specific enzyme activities and presumptive genotypes scored based on the pattern of spots in each lane. This particular gel is stained for the enzyme isocitrate dehydrogenase (*Icd-2* and *Icd-2*) from the tissue of 14 mosquitofish (*Gambusia holbrooki*).

expectations were performed. On the basis of an α of 0.05, only three or four "false" rejections of the null hypothesis (i.e., Type I errors) would have been expected to occur by chance alone. A surprisingly high proportion of the tests (18 of 72) resulted in the rejection of the null hypothesis. One or more assumptions of Hardy–Weinberg equilibrium were being violated. In 16 cases, Selander's D values were negative, indicating that rejection was associated with a lower than expected proportion of heterozygotes. A review of the sampling methods and egg depositing behavior of the midge suggested a Wahlund effect. Approximately 10 dredge samples were taken as the anchored boat drifted over the site. Perhaps individuals pooled from these dredge samples reflected cryptic population structure at each site. Small scale population structure could result from nonuniform settling of larvae from egg masses, aggregation of siblings, differences in settling behavior relative to sediment characteristics, or some other factor. The alternate explanation of selection at several loci was judged to be less likely than a Wahlund effect based on Ockham's razor (i.e., all else being equal, the explanation requiring the fewest assumptions is the most likely) because it requires that selection is occurring against heterozygotes at 9 of 12 loci. Explanation based on the Wahlund effect requires only one assumption: the population is structured. In further support of this conclusion, deviations from Hardy–Weinberg expectations that might suggest the presence of selection was not correlated with the level of mercury contamination in site sediments.

Woodward et al.'s study will be explored further (Box 17.3) after discussion of ways to quantify structuring of populations.

17.3.4.2 Isolated and Semi-Isolated Subpopulations

Violations of the assumption for the Hardy–Weinberg model due to nonrandom mating can lead to a deficit of heterozygotes, e.g., the Wahlund effect that appears during sampling of structured populations. Inbreeding can also result in deficits of heterozygotes: an individual's heterozygosity measured as the proportion of all scored loci for which the individual is heterozygous will be lower than its parents if those parents were sibs. The less extreme structuring described to this point is similar to inbreeding because individuals within the total population are not randomly mating due to the degree of their isolation by geographic distance. Population structuring will lead to a decrease in heterozygosity for individuals within a subpopulation relative to that anticipated in the absence of population structure.

The proportion of all individuals that are heterozygotes for a particular locus can be quantified at different levels of "pooling" to get an understanding of the nature of population structure. The assumption is made that, like increased inbreeding, increased structure in populations results in a decrease in the proportion of individuals that are heterozygotes. Wright's F statistics (Nei 1973, Wright 1943, 1951) are based on consideration of heterozygosity at the individual (I), subpopulation (S), and the total population (T) levels. Individuals are sampled from subpopulations of the total population to generate the associated metrics; for example, fish are sampled from creeks and tributaries of a large river system. The heterozygosity estimated at the individual level (H_{I}) is the observed heterozygosity averaged over all sites (i.e., all subpopulations of the population). The heterozygosity for the subpopulations (H_{S}) is estimated under the assumption that individuals in the subpopulations are mating randomly (i.e., using $2pq$ estimated for each subpopulation). The total heterozygosity (H_{T}) is that measured after pooling all individuals from all samples, calculating total p and q values, and estimating the predicted proportion of heterozygotes (i.e., $2pq$ for the entire population). Hartl and Clark (1989) give the following formulae to calculate H_{I}, H_{S}, and H_{T}:

$$H_{\mathrm{I}} = \sum_{i=1}^{k} \frac{H_i}{k}, \tag{17.14}$$

where H_i = the heterozygosity for subpopulation i and k = the number of subpopulations (e.g., sites or sampling locations) from which heterozygosity was estimated.

$$H_S = 1 - \sum_{i=1}^{k} p_{i,s}^2, \tag{17.15}$$

where $p_{i,s}$ = the proportion for the ith alleles in the s subpopulation. The p^2 values are summed for the k alleles. The H_S estimates for all subpopulations can be averaged to get the $\bar{H}_S$ used in Equations 7.18 and 17.19.

$$H_T = 1 - \sum_{i=1}^{k} \bar{p}_i^2, \tag{17.16}$$

where p_i = the frequency of allele i averaged over all of the subpopulations.

Three hierarchical F statistics can be generated from H_I, H_S, and H_T to evaluate the influence of population structure on the genotype frequencies at a locus. The F statistics scale the estimated heterozygosity at the levels of individual, subpopulation, and population. Recalling our discussions linking inbreeding effects and population structuring effects on heterozygosity, it becomes obvious that several of these metrics are comparable to inbreeding coefficients. The overall inbreeding coefficient, F_{IT}, is defined by Equation 17.17 using H_T and H_I from Equations 17.16 and 17.14. It is the "reduction in heterozygosity of an individual relative to the total population" (Hartl and Clark 1989). It quantifies heterozygosity of the individual (H_I) relative to that of the entire population (H_T),

$$F_{IT} = 1 - \frac{H_I}{H_T} = \frac{H_T - H_I}{H_T}. \tag{17.17}$$

Similarly, the reduction in heterozygosity of individual due to nonrandom mating within its subpopulation (Equation 17.18) or due to genetic drift (Equation 17.19) can be derived. The F_{IS} statistic quantifies the heterozygosity of the individual (H_I) relative to that of the (average) subpopulation ($\bar{H}_S$) whereas F_{ST} quantifies the heterozygosity of the (average) subpopulation ($\bar{H}_S$) relative to that of the total population (H_T).

$$F_{IS} = 1 - \frac{H_I}{\bar{H}_S} = \frac{\bar{H}_S - H_I}{\bar{H}_S}, \tag{17.18}$$

$$F_{ST} = 1 - \frac{\bar{H}_S}{H_T} = \frac{H_T - \bar{H}_S}{H_T}, \tag{17.19}$$

where $\bar{H}_S$ = the average of the H_S values from all subpopulations or sites. The $F_{ST} = 0$ if all subpopulations (e.g., sites) were in Hardy–Weinberg equilibrium. In cases where migration among subpopulations produces deviations from these expectations, F_{ST} is approximately equal to $1/(1 + 4Nm)$ (Rousset and Raymond 1997). Therefore, the effective number of migrants per generation (Nm) can be estimated if F_{ST} is known (Bossart and Prowell 1998). (Note that Ouborg et al. (1999) described Markov Chain Monte Carlo (Beerli 1998), Bayesian (Rannala and Mountain 1997), maximum likelihood (Beerli and Felsenstein 1999), and pseudomaximum likelihood (Rannala and Hartigan 1996) methods that are more effective than this F_{ST}-based methods for estimating effective migration and, in some cases, population structure from molecular genetics data.)

Box 17.3 More on Midges, Mercury, and Missing Heterozygotes

In Box 17.2, the potential for a Wahlund effect was identified for a survey of midge allozymes sampled along a gradient of sediment mercury contamination (Woodward et al. 1996). Differences were tentatively assigned to a Wahlund effect for reasons described in Box 17.2. Wright's F_{IS}, F_{IT}, and F_{ST} statistics were calculated for 12 loci sampled along the mercury gradient in order to understand the observed differences in heterozygosity. A summary for Wright's F statistics for the following 12 isozymes is provided in Table 17.2: aspartate aminotransferase (Aat), adenosine deaminase (Ada), esterase (Est), glycylleucine peptidase (gl), hexokinase (Hk), isocitrate dehydrogenase (Icd-I and Icd-2), leucylglycylglycine peptidase (lgg), malate dehydrogenase (Mdh), malic enzyme (Me), mannose-6-phosphate isomerase (Mpi), and phosphoglucomutase (Pgm). A quick glance at this table shows that the deficiencies in heterozygotes for many loci were associated with the site (subpopulation, F_{IS}) level. This supports the explanation that sampling of cryptically structured populations produced the apparent deficiency in heterozygotes, that is, a Wahlund effect. Selection remained an unlikely explanation for reasons discussed in Box 17.2. Inbreeding was another potential mechanism but the lack of any obvious barriers to adult mating as they swarm above the water surface does not support this explanation.

To assess this population structure-based hypothesis further, fine scaled sampling was done at one lake site. Forty larvae were sampled from each of fifteen adjoining, 1 × 1 m quadrats and scored for nine isozymes (Aat, Ada, Est, gl, Hk, Icd-1, Icd-2, lgg, and Pgm). This transect of 15 quadrats was constructed in a shallow (5 m) region of the lake to enhance the accuracy of dredge placement. The length of the transect was chosen to approximate the length of the average site sampled in the original study (Table 17.2). The results from this fine scaled sampling

TABLE 17.2
F_{IS}, F_{IT}, and F_{ST} Statistics for Chironomid Larvae Collected at Six Sites along a Sediment-Associated Mercury Gradient in Clear Lake (California)

Allozyme Locus	F Statistic		
	F_{IS}	F_{IT}	F_{ST}
Aat	0.165	0.181	0.019
Ada	0.107	0.114	0.007
Est	0.231	0.248	0.022
Gl	0.078	0.087	0.010
Hk	−0.116	−0.099	0.015
Icd-1	0.219	0.225	0.009
Icd-2	0.259	0.275	0.022
Lgg	0.128	0.130	0.003
Mdh	−0.059	−0.026	0.031
Me	0.618	0.627	0.023
Mpi	0.142	0.151	0.011
Pgm	0.107	0.116	0.010
Mean	0.125	0.137	0.014

Source: Modified from Table 2 of Woodward et al. (1996).

TABLE 17.3
F_{IS}, F_{IT}, and F_{ST} Statistics for Chironomid Larvae Collected from 15 Quadrats of a Transect in Clear Lake (California)

Allozyme Locus	F Statistic		
	F_{IS}	F_{IT}	F_{ST}
Aat	0.202	0.210	0.010
Ada	0.116	0.138	0.025
Est	0.024	0.247	0.019
Gl	0.250	0.264	0.019
Hk	0.024	0.030	0.006
Icd-1	0.107	0.115	0.009
Icd-2	0.021	0.034	0.014
Lgg	0.102	0.124	0.024
Pgm	0.010	0.020	0.011
Mean	0.134	0.150	0.018

Source: Modified from Table 4 of Woodward et al. (1996).

are given in Table 17.3. Wright's F_{IT} and F_{IS} statistics indicated a deficiency in heterozygotes within the transect and quadrats. This clearly indicated small scale population structure of the chironomid larvae.

Woodward et al. (1996) concluded that a Wahlund effect, not mercury-related selection or inbreeding, was the most likely explanation for the deficiencies of heterozygous genotypes. Their conclusion was based on the following observations and rules of logic:

1. Departures from Hardy–Weinberg expectations involved a deficiency of heterozygotes (Hardy–Weinberg expectations and Selander's D values).
2. There was no correlation between mercury contamination and genotype frequencies, i.e., no evidence of a cause–effect relationship or a concentration–effect gradient.
3. Ockham's razor (principle of parsimony) favors explanations with the fewest assumptions.
4. Mating swarm and egg mass deposition patterns provide a mechanism for clustering of genetically distinct larvae as they settle nonrandomly onto the sediments.
5. Wright's F statistics suggest considerable structure along the mercury gradient and within the smaller scale transect.
6. There is no obvious obstacle to adult mating that would lead to inbreeding.

The potential for a Wahlund effect should always be kept in mind when interpreting population genetics data for populations exposed to toxicants. Woodward et al. (1996) provide only one example of the importance of such thoughtfulness, but other examples exist. Lavie and Nevo (1986) suggested from laboratory testing that one could examine a suite of species and focus on the proportion of heterozygotes relative to homozygotes in populations. This conclusion was based on results from five gastropod species lethally exposed to cadmium and homozygote:heterozygote ratios for the enzyme, glucosephosphate isomerase, in survivors. They state that there seems to be a relationship for this proportion with pollution intensity and "[t]his pattern seems to have been established by natural selection." They attribute the difference in survival to the higher stability of the homodimer than the heterodimer of this dimeric enzyme: homozygotes had an enzyme form that was more resistant to inactivation by cadmium. Clearly, such an approach would be valid only in the demonstrated absence of a Wahlund effect.

Computer intensive methods are now widely available to augment or to eventually replace the metrics just described for estimating gene flow and population structure. For example, a personal computer can now quickly produce bootstrap confidence intervals for F_{ST} (Rousset and Raymond 1997). Q statistics (Nei 1973) can also be generated for structured populations in a manner analogous to performing statistical variance component analysis (Bossart and Prowell 1998, Rousset and Raymond 1997). More involved computer models incorporating geographical distances in algorithms allow more specific analysis of genetic data. For more information, the interested reader is directed to Bossart and Prowell (1998), who recently reviewed conventional and new means of assessing gene flow in structured populations.

17.3.5 Multiple-Locus Heterozygosity and Individual Fitness

At this point, concepts related to neutral theory have been explored with the aim of demonstrating how toxicants can influence population genetics in the absence of differences in individual fitness. In the remainder of this chapter, two bridging topics will be mentioned between neutral theory and selection-based theory. The general consequences of different levels of heterozygosity of individuals will be discussed relative to overall fitness. Then long-term evolutionary consequences for species having decreased genetic diversity will be explored briefly.

Numerous studies have demonstrated that an individual's overall heterozygosity can influence its fitness. However, a number of publications have shown that it might not (e.g., Koehn et al. 1988). Multiple-locus heterozygosity refers here to the number of scored loci for which the individual is heterozygous. Relevant measures of fitness for which heterozygosity did influence fitness vary widely and include survival (Pemberton et al. 1988, Samallow and Soulé 1983), developmental rate (Danzmann et al. 1985, 1986, 1988), developmental stability (Ferguson 1986, Mulvey et al. 1994), metabolic rate (Danzmann et al. 1987, Mitton et al. 1986), metabolic cost (Garton et al. 1984), and growth rate (Bush et al. 1987, Garton et al. 1984, Koehn and Gaffney 1984, McAndrew et al. 1986). Mitton's book *Selection in Natural Populations* (1997) provides extensive discussion of such correlations between fitness metrics and heterozygosity. Some studies suggest that heterozygosity can also influence susceptibility to toxicants (e.g., Nevo et al. 1986). However, careful analysis of one study of mercury toxicity (Diamond et al. 1989) and a similar study of arsenic toxicity (Newman et al. 1989) demonstrated that the observed relationships were artifacts reflecting the sum of individual locus effects, not an effect of heterozygosity per se (Newman et al. 1989). (Unfortunately, Table 6 of Gillespie and Guttman (1999) incorrectly lists the results of Diamond et al. (1989) and Newman et al. (1989) as supporting evidence for a relationship between heterozygosity and fitness.)

Why should there be a relationship between fitness and heterozygosity? There are three common explanations: inbreeding depression, multiple-locus heterosis, and optimal metabolic efficiency.

The inbreeding depression explanation can be understood from our discussions of inbreeding and heterozygosity. A decrease in heterozygosity can indicate increased inbreeding. Heterozygosity might simply reflect the degree of inbreeding experienced by individuals. Inbreeding can lead to lowered fitness (via inbreeding depression) as the probability of an individual being homozygous for deleterious genes increases. Low heterozygosity for the entire genome was approximated with scored loci and was correlated with a general inbreeding depression (Smouse 1986). Although inbreeding is a plausible mechanism, Leary et al. (1987) described one case in which inbreeding was not the explanation for the relationship between heterozygosity and a measure of individual fitness, developmental stability.

The multiple heterosis explanation extends the phenomenon of heterosis to include many loci. Single-locus heterosis is the superior performance of heterozygotes relative to homozygotes. It can also be defined in terms of hybrids (Zouros and Foltz 1987). The performance of hybrids produced from two lines is often superior to that of the two parent lines. Multiple heterosis is the sum of heterotic effects at several loci. Heterozygosity will be positively correlated with hybrid vigor and can be envisioned as representing the opposite of the negative relationship just described between heterozygosity-correlated inbreeding and fitness.

Optimal metabolic efficiency may be linked to high fitness (Dykhuizen et al. 1987, Zouros and Foltz 1987). Individuals heterozygous for major glycolytic and Krebs cycle-related loci (enzymes typically used in these studies) may be more metabolically efficient or flexible than homozygous genotypes. Allozymes, allelic variants of enzymes, can differ in their properties, including kinetic properties (Hines et al. 1983) and resistance to toxicant inactivation (Kramer and Newman 1994). A homozygote at a particular locus will have only one form of the enzyme available, but the heterozygote will have two (or more for multimeric enzymes) forms available. A homozygote (e.g., 100/100) for a dimeric enzyme will produce only one protein subunit and only one dimer will be produced. Heterozygotes (e.g., 100/165) synthesize two proteins (100 and 165) and three functional dimeric enzymes 100/100, 100/165, and 165/165 will be produced. Individuals with more heterozygous loci may have more metabolic options available to cope with changing environmental demands. Relative to their homozygous counterparts, highly heterozygous individuals might be more efficient over a wider range of conditions. If some allozymes are inactivated more readily than others by metals (Eichhorn 1974, Kramer and Newman 1994, Lavie and Nevo 1982) or are less tolerant to high temperatures (Zimmerman and Richmond 1981), an individual's fitness may be enhanced by having several forms present to catalyze essential reactions. Parsons (1997) provides a general review of stress and genetic variation.

Box 17.4 Tolerance of Fish to Stressors Increases with Heterozygosity . . . Sometimes

Studies of allozyme variation by Guttman and coworkers (Kopp et al. 1992, Schlueter et al. 1995) assessed the relationship between effects of stressor exposure and individual heterozygosity. Two of their studies will be used to demonstrate the variation possible in results from such studies.

Heterozygosity Does Influence Sensitivity

Responses to high metal and low pH conditions were studied in populations of the central mud minnow (*Umbra limi*) (Kopp et al. 1992). The concern addressed by this study was the consequence of acid precipitation on populations of fish endemic to the Adirondack Mountains (New York, USA). These researchers noted that individuals from water bodies with high aluminum and low pH had higher levels of heterozygosity at enzyme-determining loci than those from reference sites. Slightly more than 200 mud minnows from impacted and reference sites were exposed in the laboratory to assess whether mud minnows from contrasting sites responded similarly during acute exposure (96 h) to high aluminum (7.5 mg/L) and low pH (4.5) conditions. Pooling data for both sexes, Kopp et al. (1992) found that the distribution of fish among three sensitivity classes (sensitive, intermediate, and tolerant) was positively correlated with heterozygosity. These data supported the concept that fitness measures tend to increase as individual heterozygosity increases.

Heterozygosity Does Not Influence Sensitivity

A statistically more robust experimental design was applied by Schlueter et al. (1995) to address the relationship between genetic variation at enzyme-determining loci and differential survival of more than a thousand fathead minnow (*Pimephales promelas*) exposed to copper. Survival time analyses were applied so the gross assignment of fish to sensitive, intermediate, and tolerant classes could be avoided. A model predicting survival time based on fish weight and number of heterozygous loci was produced and null hypotheses of no significant effect of these two covariates tested with a χ^2 test. Although fish size was significantly ($\alpha = .05$) and positively related to survivorship, there was no apparent effect of number of heterozygous loci on survival of fathead minnows during copper exposure. These data clearly did not support the premise that measures of fitness tend to increase as individual heterozygosity increases. In another case, this failure to observe an effect could have been attributed to a lack of statistical power; however, the experiment involved large numbers of individuals and a powerful analysis technique.

17.4 GENETIC DIVERSITY AND EVOLUTIONARY POTENTIAL

Although discussed to this point as an indicator of population state or change, genetic diversity itself is crucial to the long-term viability of species populations. Mutation rates are extremely low relative to the rates of toxicant-accelerated drift, and the balance between drift and mutation is complex as evidenced by our above discussions. Regardless, an emerging concern of many population ecotoxicologists (e.g., Kopp et al. 1992, Mulvey and Diamond 1991) is the ratcheting downward of genetic diversity due to neutral theory-related consequences of pollution. Because genetic variation is the raw material on which natural selection works, the evolutionary potential of species might be lowered by toxicant exposure. Such long-term impacts of toxicant exposure are very difficult to quantify and are rarely addressed in the development of regulations.

17.5 SUMMARY

Several kinds of mechanisms for DNA damage are described in this chapter, suggesting ample opportunity for direct mutagenic effects of toxicants on species germ lines. Indirect effects on the germ line are discussed based on neutral population genetic theory. In the absence of natural selection, modified stochastic processes as a consequence of toxicant exposure can have a significant impact on population genetic characteristics. The dynamics of such changes and tools for identifying them are described. On the basis of neutral theory, the potential for loss of genetic diversity is discussed. Ecotoxicology will benefit from incorporation of these concepts into the assessment of toxicant effects on populations. Natural selection as described in the next chapter could greatly modify these predictions of toxicant-driven reductions in genetic diversity (Mulvey and Diamond 1991).

17.5.1 SUMMARY OF FOUNDATION CONCEPTS AND PARADIGMS

- Toxicants that are mutagens can influence the germ line directly.
- Stochastic processes can influence the germ line.
- Hardy–Weinberg equilibrium in a population of diploid species is based on the assumptions of an effectively infinite population, no natural selection, and negligible mutation rates and migration rates. Violations of any of these assumptions can result in deviations from Hardy–Weinberg expectations.

- Genetic drift can be accelerated by a toxicant-related decrease in the effective population size. The decrease in effective population can result from toxicant effects on total population size, sex ratio, natality, mean generation time, population structure, and brood size variance.
- In addition to accelerated drift, abrupt decreases in effective population size can lead to genetic bottlenecks and consequent founder effects.
- Genetic diversity is maintained by a balance between mutation rate and drift. In a structured population, migration also influences genetic diversity.
- Effective population size is influenced by the distribution of individuals within a population.
- Sampling a cryptically structured population or the presence of migration into a population can result in an apparent deficit of heterozygotes (i.e., the Wahlund effect).
- Wright's F statistics can be used to describe population genetic structure.
- Increases in multiple-locus heterozygosity are often, but not always, correlated with increases in individual fitness, including that associated with toxicant stress.
- Loss of genetic diversity due to toxicant exposure can reduce the evolutionary potential of a species population.

REFERENCES

Ayala, F.J., *Population and Evolutionary Genetics*, The Benjamin/Cummings Publishing Co., Menlo Park, CA, 1982.

Baker, R.J., Van den Bussche, R.A., Wright, A.J., Wiggins, L.E., Hamilton, M.J., Reat, E.P., Smith, M.H., Lomakin, M.D., and Chesser, R.K., High levels of genetic change in rodents of Chernobyl, *Nature*, 380, 707–708, 1996.

Baker, R.J., Van den Bussche, R.A., Wright, A.J., Wiggins, L.E., Hamilton, M.J., Reat, E.P., Smith, M.H., Lomakin, M.D., and Chesser, R.K., High levels of genetic change in rodents of Chernobyl, *Nature*, 390, 100, 1997.

Beerli, P., 1998. MIGRATE. http://evolution.genetics.washington.edu/lamarc/migrate.html.

Berrli, P. and Felsenstein, J., Maximum-likelihood estimation of migration rates and effective population numbers in two populations using a coalescent approach, *Genetics*, 152, 763–773, 1999.

Bootma, D. and Hoeijmakers, J.H.J., The molecular basis of nucleotide excision repair syndromes, *Mutat. Res.*, 307, 15–23, 1994.

Bossart, J.L. and Prowell, D.P., Genetic estimates of population structure and gene flow: Limitations, lessons and new directions, *TREE*, 13, 202–206, 1998.

Burdon, R.H., *Genes and The Environment*, Taylor & Francis, Inc., Philadelphia, PA, 1999.

Bush, R.M., Smouse, P.E., and Ledig, F.T., The fitness consequences of multiple-locus heterozygosity: The relationship between heterozygosity and growth rate in pitch pine (*Pinus rigida* Mill.), *Evolution*, 41, 787–798, 1987.

Cavalli-Sforza, L.L. and Bodmer, W.F., *The Genetics of Human Populations*, W.H. Freeman and Company, San Francisco, CA, 1971.

Crow, J.F. and Kimura, M., *An Introduction to Population Genetics Theory*, Harper & Row Publishers, New York, 1970.

Danzmann, R.G., Ferguson, M.M., and Allendorf, F.W., Does enzyme heterozygosity influence developmental rate in rainbow trout, *Heredity*, 56, 417–425, 1985.

Danzmann, R.G., Ferguson, M.M., and Allendorf, F.W., Heterozygosity and oxygen-consumption rate as predictors of growth and developmental rate in rainbow trout, *Physiol. Zool.*, 60, 211–220, 1987.

Danzmann, R.G., Ferguson, M.M., and Allendorf, F.W., Heterozygosity and components of fitness in a strain of rainbow trout, *Biol. J. Linn. Soc.*, 33, 219–235, 1988.

Danzmann, R.G., Ferguson, M.M., Allendorf, F.W., and Knudsen, K.L., Heterozygosity and developmental rate in a strain of rainbow trout (*Salmo gairdneri*), *Evolution*, 40, 86–93, 1986.

De Finetti, B., Considerazioni matematiche sul l'ereditarieta mendeliana, *Metron*, 6, 1–41, 1926.

Diamond, S.A., Newman, M.C., Mulvey, M., Dixon, P.M., and Martinson, D., Allozyme genotype and time to death of mosquitofish, *Gambusia affinis* (Baird and Girard), during acute exposure to inorganic mercury, *Environ. Toxicol. Chem.*, 8, 613–622, 1989.

Dykhuizen, D.E., Dean, A.M., and Hartl, D.L., Metabolic flux and fitness, *Genetics*, 115, 25–31, 1987.

Eichhorn, G.L., Active sites of biological macromolecules and their interaction with heavy metals, In *Ecological Toxicology Research. Effects of Heavy Metals and Organohalogen Compounds*, McIntyre, A.D. and Mills, C.F. (eds.), Plenum Press, New York, 1974, pp. 123–214.

Facemire, C.F., Gross, T.S., and Guillette, L.J., Jr., Reproductive impairment in the Florida panther: Nature or nuture? *Environ. Health Perspect.*, 103(Suppl. 4), 79–86, 1995.

Ferguson, M.M., Developmental stability of rainbow trout hybrids: Genomic coadaptation or heterozygosity? *Evolution*, 40, 323–330, 1986.

Forbes, V.E., *Genetics and Ecotoxicology*, Taylor & Francis, Inc., Philadelphia, PA, 1999.

Fraústo da Silva, J.J.R. and Williams, R.J.P., *The Biological Chemistry of the Elements. The Inorganic Chemistry of Life*, Clarendon Press, Oxford, UK, 1993.

Garton, D.W., Koehn, R.K., and Scott, T.M., Multiple-locus heterozygosity and the physiological energetics of growth in the coot clam, *Mulinia lateralis*, from a natural population, *Genetics*, 108, 445–455, 1984.

Gillespie, R.B. and Gutmann, S.I., Chemical-induced changes in the genetic structure of populations: Effects on allozymes, In *Genetics and Ecotoxicology*, Forbes, V.E. (ed.), Taylor & Francis, Inc., Philadelphia, PA, 1999, pp. 55–77.

Hartl, D.L. and Clark, A.G., *Principles of Population Genetics*, Sinauer Associates, Inc., Sunderland, MA, 1989.

Hillis, D.M., Mortiz, G., and Mable, B.K., *Molecular Systematics*, Sinauer Associates, Inc., Sunderland, MA, 1996.

Hines, B.A., Philipp, D.P., Childers, W.F., and Whitt, G.S., Thermal kinetic differences between allelic isozymes of malate dehydrogenase (Mdh-B locus) of largemouth bass, *Micropterus salmoides, Biochem. Genet.*, 21, 1143–1151, 1983.

Hoffman, G.R., Genetic toxicology, In *Casarett and Doull's Toxicology. The Basic Science of Poisons*, 5th ed., Klaassen, C.D. (ed.), McGraw-Hill Health Professions Division, New York, 1996, pp. 269–300.

Hoffmann, A.A. and Parsons, P.A., *Extreme Environmental Change and Evolution*, Cambridge University Press, Cambridge, UK, 1997.

Jablonka, E. and Lamb, M., *Epigenetic Inheritance and Evolution*, Oxford University Press, Oxford, UK, 1995.

Koehn, R.K., Diehl, W.J., and Scott, T.M., The differential contribution by individual enzymes of glycolysis and protein catabolism to the relationship between heterozygosity and growth rate in the coot clam, *Mulinia lateralis, Genetics*, 118, 121–130, 1988.

Koehn, R.K. and Gaffney, P.M., Genetic heterozygosity and growth rate in *Mytilus edulis*, *Mar. Biol.*, 82, 1–7, 1984.

Kopp, R.L., Guttman, S.I., and Wissing, T.E., Genetic indicators of environmental stress in central mudminnow (*Umbra limi*) populations exposed to acid deposition in the Adirondack Mountains, *Environ. Toxicol. Chem.*, 11, 665–676, 1992.

Kramer, V.J. and Newman, M.C., Inhibition of glucosephosphate isomerase allozymes of the mosquitofish, *Gambusia holbrooki*, by mercury, *Environ. Toxicol. Chem.*, 13, 9–14, 1994.

Lamb, T., Bickham, J.W., Gibbons, J.W., Smolen, M.J., and McDowell, S., Genetic damage in a population of slider turtles (*Trachemys scripta*) inhabiting a radioactive reservoir, *Arch. Environ. Contam. Toxicol.*, 20, 138–142, 1991.

Lavie, B. and Nevo, E., Heavy metal selection of phosphoglucose isomerase allozymes in marine gastropods, *Mar. Biol.*, 71, 17–22, 1982.

Lavie, B. and Nevo, E., Genetic selection of homozygote allozyme genotpyes in marine gastropods exposed to cadmium pollution, *Sci. Total Environ.*, 57, 91–98, 1986.

Leary, R.B., Allendorf, F.W., and Knudsen, K.L., Differences in inbreeding coefficients do not explain the association between heterozygosity at allozyme loci and developmental stability in rainbow trout, *Evolution*, 41, 1413–1415, 1987.

McAndrew, B.J., Ward, R.D., and Beardmore, J.A., Growth rate and heterozygosity in the plaice, *Pleuronectes platessa, Heredity*, 57, 171–180, 1986.

Mitton, J.B., *Selection in Natural Populations*, Oxford University Press, Oxford, UK, 1997.

Mitton, J.B., Carey, C., and Kocher, T.D., The relation of enzyme heterozygosity to standard and active oxygen consumption and body size of tiger salamanders, *Ambysoma tigrinum*, *Physiol. Zool.*, 59, 574–582, 1986.

Mulvey, M. and Diamond, S.A., Genetic factors and tolerance acquisition in populations exposed to metals and metalloids, In *Metal Ecotoxicology. Concepts and Applications*, Newman, M.C. and McIntosh, A.W. (eds.), Lewis Publishers, Chelsea, MI, 1991, pp. 301–321.

Mulvey, M., Keller, G.P., and Meffe, G.K., Single- and multiple-locus genotypes and life-history responses of *Gambusia holbrooki* reared at two temperatures, *Evolution*, 48, 1810–1819, 1994.

Murdoch, M.H. and Hebert, P.D.N., Mitochondrial DNA diversity of brown bullhead from contaminated and relatively pristine sites in the Great Lakes, *Environ. Toxicol. Chem.*, 8, 1281–1289, 1994.

National Council on Radiation Protection and Measurements, *Risk Estimates for Radiation Protection*, NCRP Report 115, NCRP, Bethesda, MD, 1993.

Nei, M., Analysis of gene diversity in subdivided populations, *Proc. Natl. Acad. Sci. USA*, 70, 3321–3323, 1973.

Nevo, E., Noy, R., Lavie, B., Beiles, A., and Muchtar, S., Genetic diversity and resistance to marine pollution, *Biol. J. Linn. Soc.*, 29, 139–144, 1986.

Newman, M.C., *Quantitative Methods in Aquatic Ecotoxicology*, CRC Press/Lewis Publishers, Boca Raton, FL, 1995.

Newman, M.C., *Fundamentals of Ecotoxicology*, CRC/Ann Arbor Press, Boca Raton, FL, 1998.

Newman, M.C., Diamond, S.A., Mulvey, M., and Dixon, P., Allozyme genotype and time to death of mosquitofish, *Gambusia affinis* (Baird and Girard) during acute toxicant exposure: A comparison of arsenate and inorganic mercury, *Aquat. Toxicol.*, 15, 141–156, 1989.

O'Brien, S.J., Roelke, M.E., Marker, L., Newman, A., Winkler, C.E., Meltzer, D., Colly, L., Evermann, J.F., Bush, M., and Wildt, D.E., Genetic basis for species vulnerability in the cheetah, *Science*, 227, 1428–1434, 1987.

Ouborg, N.J., Piquot, Y., and van Groenendael, J.M., Population genetics, molecular markers and the study of dispersal in plants, *J. Ecol.*, 87, 551–568, 1999.

Parsons, P.A., Stress-resistance genotypes, metabolic efficiency and interpreting evolutionary change, In *Environmental Stress, Adaptation and Evolution*, Bijlsma, R. and Loeschcke, V. (eds.), Birkhäuser, Basel, Switzerland, 1997, pp. 291–305.

Pemberton, J.M., Albon, S.D., Guinness, F.E., Clutton-Brock, T.H., and Berry, R.J., Genetic variation and juvenile survival in red deer, *Evolution*, 42, 921–934, 1988.

Rannala, B. and Hartigan, J., Estimating gene flow in island populations, *Genet. Res.*, 67, 147–158, 1996.

Rannala, B. and Mountain, J., Detecting immigration by using multilocus genotypes, *Proc. Natl. Acad. Sci. USA*, 94, 9197–9201, 1997.

Richardson, B.J., Baverstock, P.R., and Adams, M., *Allozyme Electrophoresis: A Handbook for Animal Systematics and Population Studies*, Academic Press, Sydney, Australia, 1986.

Robison, S.H., Cantoni, O., and Costa, M., Analysis of metal-induced DNA lesions and DNA-repair replication in mammalian cells, *Mutat. Res.*, 131, 173–181, 1984.

Rousset, F. and Raymond, M., Statistical analyses of population genetic data: New tools, old concepts, *TREE*, 12, 313–317, 1997.

Samollow, P.B. and Soulé, M.E., A case of stress related heterozygote superiority in nature, *Evolution*, 37, 646–649, 1983.

Schlueter, M.A., Guttman, S.I., Oris, J.T., and Bailer, A.J., Survival of copper-exposed juvenile fathead minnows (*Pimephales promelas*) differs among allozyme genotypes, *Environ. Toxicol. Chem.*, 14, 1727–1734, 1995.

Selander, R.K., Behavior and genetic variation in natural populations, *Am. Zool.*, 10, 53–66, 1970.

Smouse, P.E., The fitness consequences of multiple-locus heterozygosity under the multiplicative overdominance and inbreeding models, *Evolution*, 40, 946–957, 1986.

Spiess, E.B., *Genes in Populations*, John Wiley & Sons, New York, 1977.

Stone, R., Can a father's exposure lead to illness in his children? *Science*, 258, 31, 1992.

Strittholt, J.R., Guttmann, S.I., and Wissing, T.E., Low levels of genetic variability of yellow perch (*Perca flavescens*) in Lake Erie and selected impoundments, In *The Biogeography of the Island Region of Western Lake Erie*, Downhower, J.F. (ed.), Ohio State University Press, Columbus, OH, 1988, pp. 246–257.

Theodorakis, C.W. and Shugart, L.R. 1999. Natural selection in contaminated environments: A case study using RAPD genotypes, In *Genetics and Ecotoxicology*, Forbes, V.E. (ed.), Taylor & Francis, Philadelphia, PA, 1999, pp. 123–149.

Williams, J.G., Kubelik, A.R., Livak, K.J., Rafalski, J.A., and Tingey, S.V., DNA polymorphisms amplified by arbitrary primers are useful as genetic markers, *Nucleic Acids Res.*, 18, 6531–6535, 1990.

Wilson, E.O. and Bossert, W.H., *A Primer of Population Biology*, Sinauer Associates, Inc., Sunderland, MA, 1971.

Woodward, L.A., Mulvey, M., and Newman, M.C., Mercury contamination and population-level responses in chironomids: Can allozyme polymorphism indicate exposure? *Environ. Toxicol. Chem.*, 15, 1309–1316, 1996.

Wright, S., Isolation by distance, *Genetics*, 28, 114–138, 1943.

Wright, S., The genetical structure of populations, *Ann. Eugen.*, 15, 323–354, 1951.

Zasukhina, G.D., Vasilyeva, I.M., Sdirkova, N.I., Krasovsky, G.N., Vasyukovich, L.Ya., Kenesariev, U.I., and Butenko, P.G., Mutagenic effect of thallium and mercury salts on rodent cells with different repair activities, *Mutat. Res.*, 124, 163–173, 1983.

Zimmerman, E.G. and Richmond, M.C., Increased heterozygosity at the Mdh-B locus in fish inhabiting a rapidly fluctuating thermal environment, *Am. Fish. Soc.*, 110, 410–416, 1981.

Zouros, E. and Foltz, D.W., The use of allelic isozyme variation for the study of heterosis, In *Isozymes: Current Topics in Biological and Medical Research* 13, Rattazzi, M.C., Scandalios, J.G. and Whitt, G.S. (eds.), Alan R. Liss, Inc., New York, 1987, pp. 1–59.

18 Population Genetics: Natural Selection

18.1 OVERVIEW OF NATURAL SELECTION

> Natural Selection acts exclusively by the preservation and accumulation of variations, which are beneficial under the organic and inorganic conditions to which each creature is exposed at all periods of life.
>
> **(Darwin 1872)**

18.1.1 GENERAL

Natural selection will now be described in order to complete our discussion of pollutant-influenced evolution. More specifically, natural selection resulting in microevolution will be explored. Microevolution is evolution within a species in contrast to macroevolution that focuses on evolutionary processes and trends encompassing many species. Emphasis will be placed on microevolution leading to enhanced resistance.

The terms resistance and tolerance will be used interchangeably as done elsewhere (Forbes and Forbes 1994, Newman 1991, 1998, Weis and Weis 1989). Some authors object to this synonymy, reserving resistance to mean the enhanced ability to cope with toxicants because of genetic adaptation and tolerance to mean the enhanced ability to cope with toxicants because of physiological, biochemical, or some other acclimation.

Natural selection is the change in relative genotype frequencies through generations resulting from differential fitnesses of the associated phenotypes. Pertinent differences in phenotype fitness can involve viability (survival) or reproductive aspects of an individual's life. Natural selection has the same basic qualities regardless of the life cycle component(s) in which it manifests. It has three required conditions and two consequences (Figure 18.1) as summarized by Endler (1986). The first requisite condition is the existence of variation among individuals relative to some trait. The second is fitness differences associated with differences in that trait, i.e., differences in survival or reproductive success among phenotypes. The third condition is inheritance: the trait must be heritable. Of course, another implied requisite is Thomas Malthus's that individuals in populations are capable of producing offspring in numbers exceeding those needed to simply replace themselves. Excess production of individuals in each generation combined with heritable differences in fitness among individuals have predictable consequences.

As the first consequence of these conditions, the frequency of a heritable trait will differ among age or life stage classes of a population. As detailed in Chapter 15, differences in survival and reproduction among individuals in demographic classes result in differences in the reproductive value (V_A) of individuals. This leads to the second consequence. The frequency of the trait from adult to offspring, i.e., across generations, will change due to trait-related differences in fitness. This change will be larger than expected due to random drift alone (i.e., due to stochastic processes alone). The net result is natural selection.

Differences in fitness can manifest in two ways. Differences may be controlled by one locus with the appearance of distinct fitness classes. In such cases of "Mendelian genetics," one genotype may be intolerant, another tolerant, and a third intermediate between the two. For example, Yarbrough et al. (1986) studied cyclodiene pesticide resistance in a population of mosquitofish (*Gambusia affinis*) endemic to an agricultural region of Mississippi and found resistance to be determined by a single, autosomal gene. Three distinct phenotypes were present for resistance. During acute cyclodiene

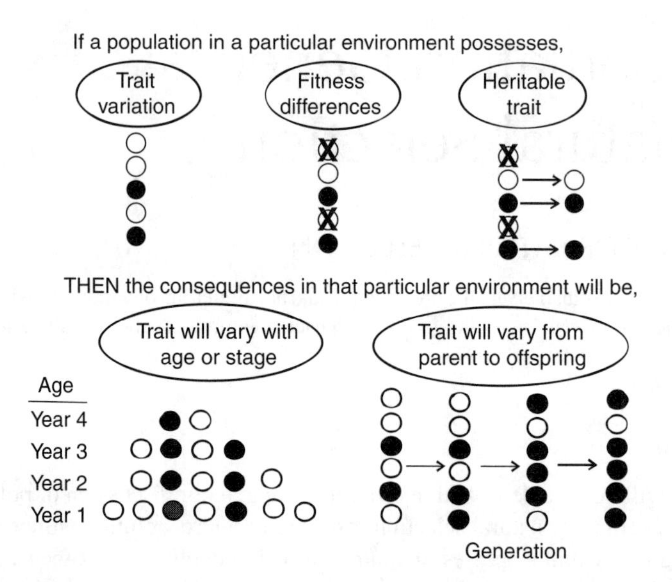

FIGURE 18.1 Syllogism of natural selection (Endler 1986). If the three conditions of trait variation, trait-related fitness differences, and trait heritability exist, then the trait frequency will vary in a predictable manner among age/stage classes and generations of a population.

exposure, resistance of heterozygotes (R/S) was intermediate to that of the sensitive (S/S) or resistant (R/R) homozygotes. Alternately, phenotype can be determined by several or many genes, resulting in a continuum of fitness states in a population. Such instances of "quantitative traits" are treated differently from instances of Mendelian genetics, and the rate of adaptation is different from that expected for a trait controlled by a single gene, e.g., selection is more rapid for traits under monogenic control versus those under polygenic control (Mulvey and Diamond 1991). Quantitative genetics methods for measuring toxicant-induced effects will be applied in Section 18.2.2.

Selection can be directional, stabilizing, or disruptive (Figure 18.2). Directional selection involves the tendency toward higher fitness at one side of the distribution of phenotypes (quantitative trait) or for a particular homozygous phenotype (Mendelian trait). The cyclodiene insecticide resistance in mosquitofish reported by Yarbrough et al. (1986) would result in directional selection. Stabilizing selection tends to favor intermediate phenotypes. Disruptive selection would favor the extreme phenotypes. Changes in the frequency of allozymes in pollution stressed gastropod species mentioned in Chapter 17 (Lavie and Nevo 1986a) suggested higher fitnesses of homozygotes than heterozygotes. In such a case, disruptive selection might be anticipated.

Several concepts associated with this overview of natural selection require comment at this point. (1) Differences in fitness are specific to a particular environment and the relative fitnesses of genotypes can change if the environment changes sufficiently. Natural selection and fitness are specific to the environmental conditions under which individuals in the population exist, e.g., a species population that has adapted successfully to an environmental toxicant will not necessarily be optimally adapted for a clean habitat. (2) Natural selection leading to successful adaptation relative to one environment or environmental condition does not necessarily result in optimal adaptation for another environment or environmental condition. For example, adaptation to cope with a particular pollutant may not necessarily result in a population of individuals well adapted to another or to a natural stressor. (3) Consistent, environment-specific differences in fitness are needed for natural selection to occur. Natural selection would not be possible if relative fitnesses of genotypes shifted randomly in direction and magnitude among generations. Natural selection can involve consistent relative fitnesses of genotypes or average relative fitness differences among genotypes in a fluctuating environment. The

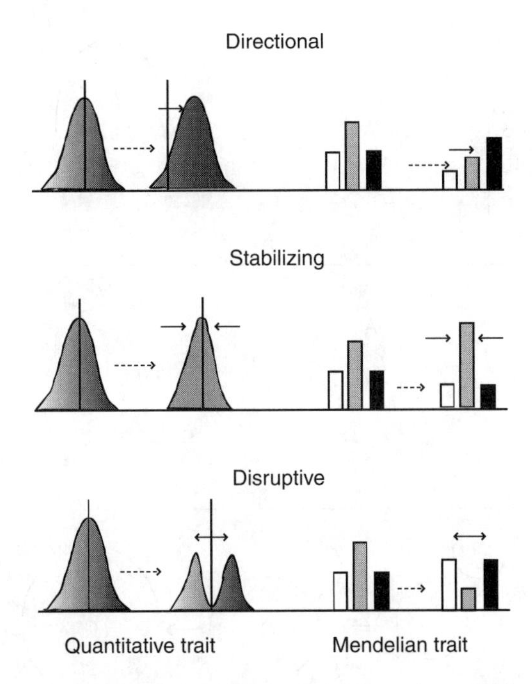

FIGURE 18.2 An illustration of directional, stabilizing, and disruptive selection for quantitative (left-hand side) and Mendelian (right-hand side) traits. (Modified from Figure 1.3 of Endler (1986) and Figure 2 of Mulvey and Diamond (1991).)

magnitude of the fitness differences may change somewhat, but the relative fitness of one genotype to another cannot abruptly and randomly change from one generation to another. (4) Without sufficient genetic variability, a species population may fail to adapt and will become locally extinct. (5) Because most environments are temporally and spatially variable, microevolution by natural selection can involve a population genome that shifts from one "best obtainable" state to another.

Natural selection for traits or trait complexes within genetic subpopulations (demes) can impart to individuals within demes temporally and spatially defined optimal fitness, i.e., Wright's shifting balance theory (Wright 1932, 1982) (Figure 18.3). A species population occupying a landscape through time might be composed of many demes shifting continually to obtain the highest fitness of associated individuals. Demes continually climb toward the highest obtainable fitness peak in a changing "adaptive landscape." Random genetic drift allows the deme to explore the adaptive landscape and natural selection then moves the deme to the nearest optimal fitness peak. This process is repeated, resulting in demes that continually explore the adaptive landscape and establish themselves on obtainable adaptive peaks. According to Wright's shifting balance theory, there may be interdemic selection within a shifting landscape of environmental factors (Hoffmann and Parsons 1997). However, caution should be used when applying this last concept of interdemic selection, i.e., group selection working on competing demes within an adaptive landscape (Coyne et al. 1997, 2000, Hartl and Clark 1989). Although some studies suggest a certain amount of support (e.g., Ingvarsson 1999 and references therein), Sewall Wright's theory of interdemic selection has not been generally supported by observational or experimental data. Regardless, important and relevant components of the shifting balance theory are demonstrably accurate (Coyne et al. 1997, 2000). The theory is mentioned here only to indicate that, *through genetic drift and natural selection on individuals*, demes tend to shift continually within an adaptive landscape to occupy local peaks of optimal fitness. These peaks shift through time as the environment changes and natural selection *working on individuals* moves the deme toward a new optimal fitness peak. Genetic drift allows exploration of nearby regions from

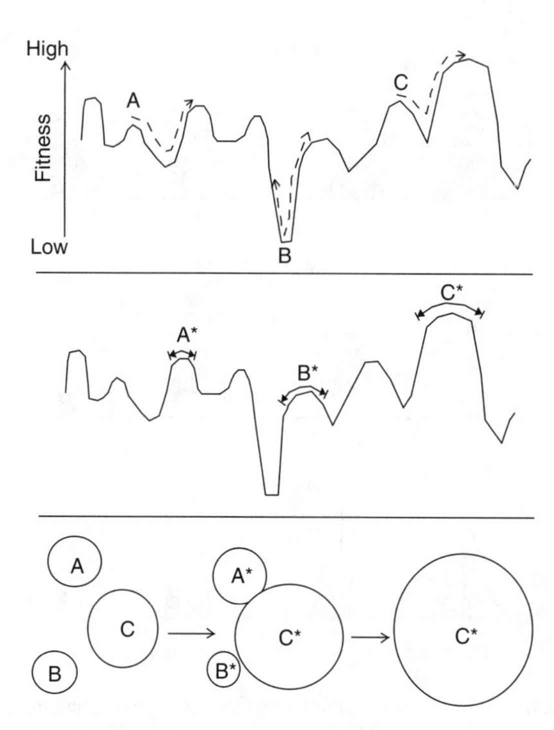

FIGURE 18.3 Shifting balance theory (Wright 1932). The three phases of this theory combine genetic drift and natural selection to produce interdeme selection (i.e., group selection). Demes undergo genetic drift (upper panel), which allows them to move from one adaptive peak through an adaptive valley to another peak (Phase I). Then, selection within demes maintains each at an adaptive peak (middle panel, Phase II). The best adapted deme will increase in size (number of individuals) and displace less well adapted demes (lower panel, Phase III), i.e., selection of a group (deme). Although Phase III is not supported by observational or experimental studies, Phases I and II are and can be important processes in natural populations (Coyne 1997).

a currently occupied fitness peak. Local populations under continual environmental pressure survive or even grow larger because of the increase in frequency of the fittest genotypes. It follows that demes will fail to adjust to changing environmental conditions unless they possess a certain level of genetic variation.

18.1.2 Viability Selection

Perhaps the most conspicuous and commonly studied type of selection by toxicants is that associated with differential survival, i.e., somatic viability of individuals. Viability selection can occur throughout the lifetime of an individual and includes fitness differences relative to development of the zygote, growth after birth, and survival to a sexual adult. For example, winter survival of juvenile red deer (*Cervus elaphus*) was correlated with allozyme genotype at several enzyme loci. Selection was implied from the observed fitness differences (Pemberton et al. 1988). A well-studied example involving pollution is the industrial melanism described in Chapter 12.

Differential survival is the most habitually studied quality in studies of pollutant-related viability. Many early studies involved the acquisition of tolerance to poisons in target and some nontarget species populations. Much of this work demonstrated rapid change of pest populations to chemicals applied to control them. Carson's *Silent Spring* (1962) includes many pages that discuss the rapid increase in survival of individuals in insect populations due to natural selection. Webb and Horsfall (1967) described the rapid decrease in pine mouse (*Pitymys pinetorum*) mortality after several years of control with endrin. Whitten et al. (1980) studied survival of insecticide-adapted sheep blowflies and Partridge (1980) described rodenticide (Warfarin) resistance in rats. Given this initial focus on the lose of pesticide efficacy, it is not surprising that survival came to dominate studies of adaptation to toxicants.

Much recent work with differential survival applied allozyme methods to identify tolerant or sensitive genotypes. Beardmore, Battaglia, and coworkers (Battaglia et al. 1980, Beardmore 1980, Beardmore et al. 1980) and Nevo, Lavie, and coworkers (Lavie and Nevo 1982, 1986a,b, Nevo et al. 1981) were among the first to apply these methods for exploring the genetic consequences of toxicant exposure for natural populations. In typical studies, field surveys were done to correlate allozyme genotype frequencies with degree of toxicant contamination. To augment these observations, individuals differing in allozyme genotypes were subjected to acutely toxic concentrations of toxicants in laboratory tests. The distribution of genotypes among survivors and dead individuals after exposure was used to imply differential fitness for the putative genotypes. The results would then be used to speculate about potential consequences to field populations exposed to much lower concentrations for longer periods of time. Speculation was normally based on the assumption that viability selection was the sole or dominant component of selection and that differences noted at high concentrations reflect differences at low concentrations.

These allozyme-based experiments continue because allozyme genotypes are relatively easy to determine and provide genetic markers for population processes. In North America, Chagnon and Guttman (1989) and Gillespie and Guttman (1989) used this approach and suggested differential survival of mosquitofish (*Gambusia holbrooki*) and central stonerollers (*Campostoma anomalum*) of specific allozyme genotypes during acute exposure to metals. Results were compared with or used to imply a mechanism for changes in field populations. Similar studies also indicated differential fitness among acutely exposed, allozyme genotypes (e.g., Keklak et al. 1994, Morga and Tanguy 2000, Schlueter et al. 1995, 2000); however, the more powerful survival analysis methods introduced by Diamond et al. (1989) and Newman et al. (1989) were applied (see Chapter 13). Newman (1995) and Newman and Dixon (1996) provide details for analyzing such allozyme-survival time data. Mulvey and Diamond (1991) and Gillespie and Guttman (1999) provide reviews of studies relating allozyme genotype and toxicant exposure.

Box 18.1 Mercury, Mosquitofish, Metabolic Allozyme Genotype, and Survival

Chagnon and Guttman (1989) suggested a relationship between survival of acute metal exposure and allozyme genotype, but many crucial facets of this relationship remained unexplored. Studies of mosquitofish and mercury were undertaken to provide an in-depth study of allozyme genotype-related fitness effects during metal exposure and to examine the major qualities of such a relationship. In the first study (Diamond et al. 1989), nearly a thousand mosquitofish (*G. holbrooki*) were exposed to 0 mg/L or 1 mg/L inorganic mercury, and times-to-death were noted at 3- to 4-h intervals for 10 days. The sex and wet weight of each fish were noted at death and individual fish were frozen for later allozyme analysis. In contrast to the negligible mortality in the reference tanks, 548 of 711 (77%) fish died in the mercury exposure tanks. Survival time methods were used to fit data to multivariate models (ln of time-to-death (TTD) $= f$(fish wet weight, sex, genotypes at 8 isozyme loci)) and to test for significant effect ($\alpha = 0.05$) of the covariates on time-to-death. Not surprisingly, fish sex and size had significant influences on time-to-death: survival time was shorter for males than females and shortened as fish weight decreased. But a remarkable three of the eight isozyme loci (isocitrate dehydrogenase-1, *Icd*-1; malate dehydrogenase-1, *Mdh*-1; and glucosephosphate isomerase-2, *Gpi*-2 = *Pgi*-2) had statistically significant effects on time-to-death. The first two of these enzymes were Krebs cycle enzymes and the last was a glycolytic enzyme.

A common explanation for relationships between allozyme genotypes and survival is that different genetically determined forms of the enzymes (e.g., different allozymes of GPI-2) differ in their capacity to bind metals and, consequently, to have their catalytic activities affected by metals. However, other studies (e.g., Watt et al. 1985) suggest that, due to the crucial roles of these enzymes in metabolism, it was equally plausible that the different allozymes produced differences in metabolic efficiencies for stressed mosquitofish. Some genotypes might be metabolically more fit under stress than others. To assess these competing hypotheses, the experiment was repeated with a different toxicant (arsenate) that had a distinct mode-of-action (i.e., interference with oxidative phosphorylation). The binding of the oxyanion, arsenate, to enzymes would be quite different than that of the mercury cation. If binding with consequent enzyme dysfunction were the mechanism for the differential effect of mercury on *Gpi*-2 genotypes, the trend noted for mercury-exposed fish would not be predicted for arsenate-exposed fish.

In addition, it was possible that sampling of the fish from the source population unintentionally resulted in subsampling a structured population with lineages differing in tolerance and having more or less of one particular genotype by chance alone. Allozyme genotypes could merely be correlated with lineages that differed in their tolerances for one or more reasons (see Section 18.2.2). This possibility was reinforced by mosquitofish reproductive and ecological characteristics combined with the highly structured pond from which the fish were taken. Another exposure study was done several months after the first during another annual reproductive pulse, allowing the source population time to grow and change structuring via lineages.

The results (Newman et al. 1989) indicated that the *Gpi*-2 effect on TTD was present for arsenate as well as mercury exposure. The most sensitive genotype ($Gpi^{38/38}$) was the same for both toxicants. This suggested that the enzyme inactivation hypothesis was incorrect for the *Gpi*-2 effect on survival. The relationships involving the other two loci were not seen again, suggesting that sampling artifacts from a structured source population likely produced these last two relationships. (See Lee et al. (1992) below (Box 18.4) for supporting justification for this conclusion.)

Heagler et al. (1993) found this *Gpi*-2 effect on TTD during mercury exposure to be consistent through time. Similar results were obtained when the mercury exposure was repeated several years after the Diamond et al. (1989) and Newman et al. (1989) studies. Her work further supported the premise that the *Gpi*-2 effect was not an artifact associated with ephemeral population structuring. During the 1993 testing, groups of fish from the same source population were exposed to several mercury concentrations. Although GPI-2 did influence TTD at most concentrations, differences in allozyme fitness were obscured above a certain mercury concentration.

Kramer and Newman (1994) further tested the assumption that differential fitness of allozyme genotypes resulted from metal inactivation of the enzymes. Mosquitofish *GPI*-2 allozymes were partially purified and subjected *in vitro* to a series of mercury concentrations. The degree of inactivation of these *Gpi*-2 allozymes was not correlated with the differential survival of the *Gpi*-2 genotypes, suggesting again that inactivation was not the mechanism for the observed differential survival. Kramer et al. (1992a,b) also examined glycolysis and Krebs cycle metabolites in fish with different *Gpi*-2 genotypes and found that the sensitive genotype ($Gpi\text{-}2^{38/38}$) displayed shifts in metabolism during exposure to mercury that were distinct from the other *Gpi*-2 genotypes. These differences in allozyme genotype sensitivity were a function of metabolic differences under toxicant stress, not differences in metal binding to and inactivation of allozymes.

The results suggested that *Gpi*-2 genotype frequencies might be useful as a marker of population level response to stressors. However, potential effects of population structure, toxicant concentration, and intensity of other stressors must also be understood and controlled in any such exercise. As will be discussed in the next section, the potential for selection also occurring for reproductive traits could complicate prediction based solely on differences in survival.

18.1.3 Selection Components Associated with Reproduction

Selection can occur at other equally important components of an organism's life cycle (Figure 18.4). This was evident from the very first elucidation of the concept of natural selection as evidenced by Charles Darwin's phrase "at all periods of life" in the opening quote of this chapter. The first selection component (viability selection, SC1) involves survival differences and other fitness differences from zygote formation to sexual maturity. Viability selection could be measured for different age classes (e.g., Christiansen et al. 1974). There might be differences in development from zygote to a mature adult. These differences might involve survival or growth rates as discussed briefly in Chapter 16. Obviously, any increase in the probability of an individual reaching sexual maturity and surviving for a long period as a sexually active adult will also enhance reproductive success.

Selection component SC2 (sexual selection) in Figure 18.4 involves differential success of adults in finding, attracting, or retaining mates. For example, Watt et al. (1985) found differential mating success in *Colias* butterflies that differed in genotype at a phosphoglucose isomerase (*Gpi*) locus. Like Kramer et al. (1992a,b) above, they attributed these differences in fitness to metabolic differences among *Gpi* genotypes. Sexual selection can occur for males (male sexual selection) or females (female sexual selection). Sexual selection might also involve differential success of mating pairs. Some genotype pairs may have a higher probability than others of being successful mates.

Three additional selection components involve the processes of gamete production and successful zygote formation. Meiotic drive (SC3) involves the differential production of the possible gamete types by heterozygotes. Sperm or ova may be produced with unequal allele representation by heterozygous individuals, leading to a higher probability of production of certain offspring genotypes. Gametic selection (SC4) can occur if certain gametes produced by heterozygotes have a higher probability of being involved in fertilization than others. Fecundity selection (SC5) can occur if pairs of certain genotypes have more offspring than others.

Endler (1986) makes the important observation that several selection components often co-occur and it is essential to understand the balance between fitnesses at these different components. A careful

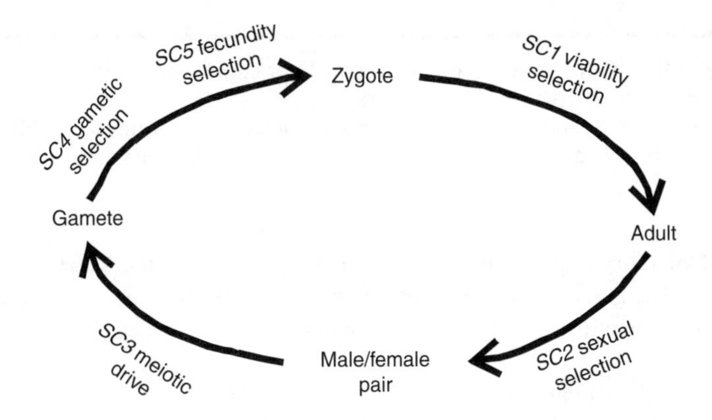

FIGURE 18.4 Selection components in the life cycle of individuals (see text for details).

re-examination of Box 12.1 will show that a preoccupation at one point (adult predation by visual predators) distracted researchers for some time from selection at other life cycle stages (pre-adult survival). Prediction from one component (e.g., viability during acute toxicant exposure) can lead to inaccurate conclusions regarding selection consequences. In fact, there are indications that selection for reproductive components may be much more common than viability selection (Clegg et al. 1978, Nadeau and Baccus 1981).

Selection components analysis is possible for many species (e.g., Bungaard and Christiansen 1972, Christiansen and Frydenberg 1973, Christiansen et al. 1973, Nadeau et al. 1981, Siegismund and Christiansen 1985, Williams et al. 1990). The analysis requires known parent–offspring combinations and scoring of genotypes for a series of demographic classes (e.g., mother–offspring pairs), adult females (gravid or nongravid), and adult males. The sequence of hypotheses (Table 18.1) are tested for these data with χ^2 statistics. The hypotheses in selection component analysis are tested sequentially and testing stops after a hypothesis is rejected. Each hypothesis test in the sequence is based on the assumption that the previously tested hypotheses were not "false," i.e., not rejected in a statistical test.

TABLE 18.1
Sequential Hypotheses Tested in Selection Component Analysis

Sequence	Hypothesis
First	Half of the offspring of heterozygous females are heterozygous (implying that there is no selection among female's gametes). Rejection implies gametic selection.
Second	The frequency of transmitted male's gametes is independent of the genotype of a female. Rejection of this hypothesis implies nonrandom mating with female sexual selection.
Third	The frequency of transmitted male gametes is equal to the frequency in adult males. Rejection implies differential male mating success and gametic selection in males.
Fourth	The genotype frequencies are equal among gravid and nongravid adult females. Rejection implies differential female mating success.
Fifth	Genotype frequencies are equal for male and female adults. Rejection implies that zygotic (viability) selection is not the same for males and females.
Sixth	The adult genotype frequency is the same as that estimated for the zygotic population. Rejection implies zygotic (viability) selection.

Source: From Table IV of Christiansen and Frydenberg's (1973) as modified by Newman (1995).

Box 18.2 Selection Components for Mercury-Exposed Mosquitofish

Most studies of natural selection contain three major faults: (1) no estimates of lifetime fitness; (2) consideration of only a few traits; and (3) unknown or poorly known trait function.

(Endler 1986)

Our studies of mercury-exposed mosquitofish attempt to avoid the shortcomings described above by Endler. The glycolytic differences noted in mercury-exposed mosquitofish genotypes define a *Gpi*-2 trait function potentially resulting in fitness differences. Points 1 and 2 will now be addressed.

The work of Mulvey et al. (1995) (Box 16.1) was used to illustrate the concepts of reaction norms and energy allocation trade-offs. Unsatisfied with predictions from the viability differences described in Box 18.1, Mulvey et al. (1995) used selection component analysis to explore the possibility of reproduction-related fitness differences in populations chronically exposed to

TABLE 18.2
Results of Selection Component Analysis for the *Gpi*-2 Locus of Mercury-Exposed Mosquitofish

Hypothesis	*P* Values from χ^2 Test for Each Replicate Mesocosm			
	Control Mesocosms		Mercury-Spiked Mesocosms	
Female gametic selection?	0.54	0.64	0.07	0.71
Random mating?	0.76	0.96	0.51	0.91
Male reproductive selection?	0.70	0.88	0.07	0.73
Female sexual selection?	0.55	0.52	**0.01**	**0.09**
Zygotic selection equal in sexes?	0.54	0.18	0.009	0.26
Zygotic selection?	0.68	0.19	0.42	0.58

Note: Boldfaced *P* values are judged to indicate selection.
Source: Modified from Table 4 in Mulvey et al. (1995).

mercury. This was possible because the mosquitofish is a prolific, live-bearing species amenable to mesocosm study and selection components analysis. Two mesocosm populations were grown with weekly additions of 18 μg/L of inorganic mercury and two mesocosm populations were grown in untreated water. After 111 days, all fish were collected and their sex, size, reproductive status (gravid/nongravid), and number of late stage embryos per gravid female determined. Selection components analysis as just described was performed for several allozyme loci; however, only *Gpi*-2 results are relevant here. The methods of Christiansen et al. (1973) as implemented with the FORTRAN program listed in Appendix 29 of Newman (1995) were used to test a series of hypotheses like those in Table 18.2. As described in Box 16.1, rare *Gpi*-2 alleles were combined in the analyses. An analysis of covariance (ANCOVA) was then applied to the number of late stage embryos carried by each gravid female to assess whether fecundity selection was occurring.

Female sexual selection was suggested from the results of the selection component analysis (Table 18.2). For the two control mesocosms, *P* values from the hypothesis testing (SC2) were .55 and .52. This suggested no female sexual selection was occurring under control conditions. However, the *P* values for the mercury-spiked mesocosms were .01 and .09. These low *P* values were taken to indicate female sexual selection and no further hypotheses were evaluated.

Whether a mature female was gravid or not was dependent on its *Gpi*-2 genotype. Approximately 68–71% of females were gravid for all genotypes and treatments, with one important exception. Only 43% of *Gpi*-$2^{100/100}$ homozygous females were gravid in the mercury-spiked mesocosms. ANCOVA also indicated ($P = .01$) that, if gravid, a *Gpi*-$2^{100/100}$ female carried fewer developing embryos than the other genotypes.

These results indicating a reproductive disadvantage for *Gpi*-$2^{100/100}$ genotypes are particularly important because the genotype least likely to survive acute mercury exposure was the *Gpi*-$2^{38/38}$ homozygote. The potential exists for balancing selection components, that is, viability selection balanced against female sexual and fecundity selection. Under some conditions, one component might outweigh another in determining the selection-driven changes in allele frequencies of a population. The results allowed a complete description of fitness differentials for several selection components, avoiding the second shortcoming listed above by Endler for studies of natural selection.

Aware that balancing selection was possible and that wild populations of mosquitofish experience wide variation in effective population size and migration, Newman and Jagoe (1998) conducted simulations of *Gpi*-2 allele frequency changes in mosquitofish populations exposed acutely and chronically to mercury for many generations. In this way, overall fitness consequences (Endler's fault 1 above) could be defined more fully under different conditions. Results indicated that *Gpi*-2 allele frequencies did change in predictable ways despite the potentially confounding effects of balancing selection, accelerated genetic drift, and migration. In general, viability selection seemed to overshadow reproductive selection components and toxicity-related acceleration of genetic drift. These results supported field studies by Heagler et al. (1993) suggesting that cautious use of *Gpi*-2 as a marker of population-level effects was possible.

18.2 ESTIMATING DIFFERENTIAL FITNESS AND NATURAL SELECTION

To understand natural selection, and for predictive purposes, it is not sufficient merely to demonstrate that selection occurs; we need to know its rate, at least in the populations under study. Rates are estimated and predicted for selection coefficients and differentials.

(Endler 1986)

18.2.1 FITNESS, RELATIVE FITNESS, AND SELECTION COEFFICIENTS

How are differences in fitness quantified? The conventional presentation of methods (Ayala 1982, Gillespie 1998) begins with a trait determined by one locus with two alleles (i.e., A_1 and A_2). Under the assumptions of the Hardy–Weinberg relationship, the A_1A_1, A_1A_2, and A_2A_2 genotype frequencies are predicted by $1 = q^2 + 2pq + p^2$ where $q =$ the A_1 allele frequency and $p =$ the A_2 allele frequency. However, Equation 18.1 depicts the expected genotype frequencies if there are relative fitnesses to be considered for the three genotypes, w_{11}, w_{12}, w_{22}. Assume, for example, that fitness differences in viability are determined using the frequencies of A_1A_1, A_1A_2, and A_2A_2 genotype for neonates and then again for adults. The relationship among the genotypes for the neonates would be $1 = q^2 + 2pq + p^2$. However, prediction of genotype frequencies for adults would involve an additional factor—differential fitnesses.

$$\overline{w} = p^2 w_{11} + 2pq w_{12} + q^2 w_{22}, \tag{18.1}$$

where $\overline{w} =$ the average fitness for all genotypes. Equation 18.1 can be rearranged to normalize fitness to the average fitness:

$$1 = p^2 \frac{w_{11}}{\overline{w}} + 2pq \frac{w_{12}}{\overline{w}} + q^2 \frac{w_{22}}{\overline{w}}. \tag{18.2}$$

Now, the frequencies of the three genotypes are predicted as a function of Hardy–Weinberg expectations (e.g., p^2) adjusted for the normalized fitness values (e.g., $(w_{11}/\overline{w})$) of each genotype. Predicted frequencies of alleles A_1 and A_2 after such selection are defined by the following equations (Ayala 1982, Gillespie 1998):

$$p_1 = p^2 \frac{w_{11}}{\overline{w}} + pq \frac{w_{12}}{\overline{w}}, \tag{18.3}$$

$$q_1 = pq \frac{w_{12}}{\overline{w}} + q^2 \frac{w_{22}}{\overline{w}}. \tag{18.4}$$

The change in p ($\Delta_p = p_1 - p$) and q ($\Delta_q = q_1 - q$) frequencies per generation are predicted from the following equations (Ayala 1982, Gillespie 1998, Spiess 1977):

$$\Delta_p = \frac{pq\,[p(w_{11} - w_{12}) + q(w_{12} - w_{22})]}{\overline{w}}, \tag{18.5}$$

$$\Delta_q = \frac{pq\,[p(w_{12} - w_{11}) + q(w_{22} - w_{12})]}{\overline{w}}. \tag{18.6}$$

The change in the average fitness can be estimated iteratively over many generations to visualize the changes due to selection in the population. According to Fisher's theorem of natural selection (Fisher 1930), higher levels of variation in fitness in populations will result in higher rates of change in average fitness. Conversely, low variation results in slow or minimal selection. (See Hartl and Clark (1989) for details and equations for applying the above approach to multiple allele genes.)

Relative fitnesses can also be estimated for genotypes. The quotient of w for each genotype can be used with the w in the denominator being that for the most fit genotype, for example, w_{11}/w_{11}, w_{12}/w_{11}, and w_{22}/w_{11} where A_1A_1 is assumed to have the highest fitness. These relative fitness values can also be expressed in terms of a selection coefficient ($s = 1 - w$ where w is the relative fitness value):

Genotype	Fitness (w)	Relative Fitness	Selection Coefficient (s)
A_1A_1	w_{11}	$w_{11}/w_{11} = 1$	$1 - (w_{11}/w_{11}) = 0$
A_1A_2	w_{12}	w_{12}/w_{11}	$1 - (w_{12}/w_{11})$
A_2A_2	w_{22}	w_{22}/w_{11}	$1 - (w_{22}/w_{11})$

As an example, selection against a recessive gene with a $w_{22} = 0.5$ would produce the following values:

Genotype	Fitness (w)	Relative Fitness	Selection Coefficient (s)
A_1A_1	$w_{11} = 1$	$w_{11}/w_{11} = 1$	$1 - (w_{11}/w_{11}) = 1 - 1 = 0$
A_1A_2	$w_{12} = 1$	$w_{12}/w_{11} = 1$	$1 - (w_{12}/w_{11}) = 1 - 1 = 0$
A_2A_2	$w_{22} = 0.5$	$w_{22}/w_{11} = 0.5/1 = 0.5$	$1 - (w_{22}/w_{11}) = 1 - 0.5 = 0.5$

Gillespie (1998) expresses selection by including a heterozyous effect (h). The relative fitness values for A_1A_1, A_1A_2, and A_2A_2 become $w_{11}/w_{11} = 1$, $w_{12}/w_{11} = 1 - hs$, and $w_{22}/w_{11} = 1 - s$. The h is 0 if A_1 is completely dominant, 1 if A_1 is completely recessive, or between 0 and 1 if A_1 is partially recessive. Relationships described in Equations 18.5 and 18.6 can be defined in terms of selection coefficients (i.e., Endler 1986, Spiess 1977). For example, if A_2 is recessive, $h = 0$, and selection is occurring for A_2, Equation 18.7 (Endler 1986) is relevant:

$$\Delta_p = \frac{spq^2}{1 - sq^2}. \tag{18.7}$$

Box 18.3 Relative Fitness of Mosquitofish Genotypes Exposed to Mercury

After completing the studies described in Boxes 18.1 and 18.2, Newman and Jagoe (1998) developed computer models to assess the potential use of *Gpi*-2 allele frequency changes as markers of population level effects. They explored shifts in allele frequencies under different exposure scenarios. Many of the scenarios involved differential survival during acute mercury exposure and estimates of relative fitness for exposed fish were needed for modeling. The approach of Newman (1995) was used to convert survival time model results to relative fitness values. The original TTD data were fit to a proportional hazard model (see Section 13.1.3.1 of Chapter 13) that included effects of fish sex, wet weight, and *Gpi*-2 genotype. The following equation describes the resulting model in terms of median TTD:

$$\text{MTTD} = e^{4.134}e^{0.358^*\text{Sex}}e^{3.157^*\text{Weight}}e^{\beta^*\text{Genotype}}e^{-0.188},$$

where sex is 1 for females and 0 for males, weight is expressed in grams of wet weight, and genotype = 0 for *Gpi*-$2^{38/38}$ and 1 for all other genotypes. The last term ($e^{-0.188}$) was estimated from the model's estimated scale factor (σ) of 0.514 and the parameter for the median of a Weibull distribution (−0.36651), that is, $e^{(0.514^*-0.36651)}$. This allowed prediction of the median TTD. The β values (Table 18.3) were estimated for each genotype and reflect the sensitivity of each genotype relative to an arbitrary reference genotype (*Gpi*-$2^{38/38}$).

The above model is a proportional hazard model: the hazard to each genotype was constant through time and the relative hazards of one genotype to any other remain constant. This allows the expression of the β coefficients as relative risks: relative risk = $e^{\beta/\sigma}$. These relative risks are estimates of genotype fitness that can be transformed to relative fitness values by simply dividing the relative risk of the most tolerant genotype by the relative risk of each genotype. The *Gpi*-$2^{66/100}$ was the most tolerant genotype as its relative risk was the smallest of all genotypes, so 0.402 is divided by all genotypes' relative risks to estimate relative fitness values (w) for each genotype. These differences in w and s values were quite large. Hartl and Clark (1989) indicate that selection coefficients with as small a difference as 1% can have very significant influence on allele frequencies in a population.

These calculations allowed the results of the toxicity trial to be included in conventional genetic models to predict changes in allele frequencies under selection. The further inclusion of reproductive differences in fitness during chronic exposures (i.e., Box 18.2) allowed predictions based on the combination of viability selection during acute exposure and female sexual and fecundity selection during chronic exposure to mercury.

TABLE 18.3
Estimation of Relative Fitness Values for Mosquitofish *Gpi*-2 Genotypes

Genotype	β	Relative Risk	Relative Fitness (w)	Selection Coefficient (s)
100/100	0.370	$e^{-0.3700/0.514} = 0.487$	0.402/0.487 = 0.82	1 − 0.82 = 0.18
100/66	0.468	$e^{-0.468/0.514} = 0.402$	0.402/0.402 = 1.00	1 − 1 = 0.00
100/38	0.362	$e^{-0.362/0.514} = 0.494$	0.402/0.494 = 0.81	1 − 0.81 = 0.19
66/66	0.389	$e^{-0.389/0.514} = 0.469$	0.402/0.469 = 0.86	1 − 0.86 = 0.14
66/38	0.339	$e^{-0.339/0.514} = 0.517$	0.402/0.517 = 0.78	1 − 0.78 = 0.22
38/38	0	$e^{-0/0.514} = 1.000$	0.402/1.000 = 0.40	1 − 0.40 = 0.40

Source: Modified from Example 18 in Newman (1995).

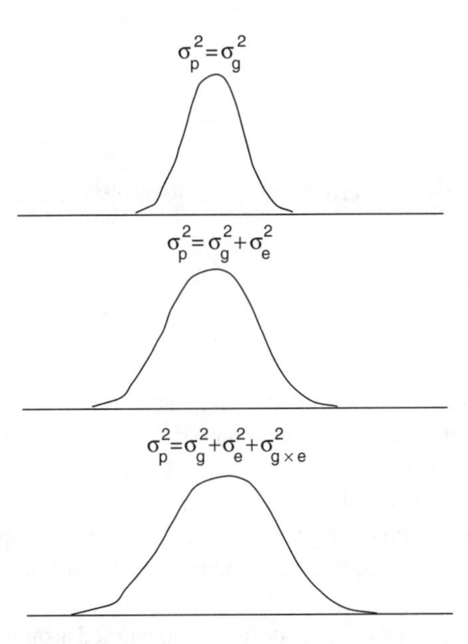

FIGURE 18.5 The components contributing the variance in a phenotypic trait (see text for details).

18.2.2 HERITABILITY

Heritability of polygenetic traits can be quantified under the assumption that variation in the phenotypes expressed among individuals results from a combination of genetic variation, environmental variation, and perhaps the interaction or covariation of genetic and environmental factors (Figure 18.5).

$$\sigma_p^2 = \sigma_g^2 + \sigma_e^2 + \sigma_{g\times e}^2, \tag{18.8}$$

where σ_p^2 = the phenotypic variance, σ_g^2 = the genetic variance, σ_e^2 = the environment-related variance, and $\sigma_{g\times e}^2$ = the variance due to the genetic × environment interaction. This last component is the covariance between genetic and environmental factors. The environmental component (σ_e^2) includes differences among phenotypes due to diverse environmental factors such as nutrition, microhabitat, and developmental conditions. The genetic component (σ_g^2) of the phenotypic variance can be separated into the additive variance arising from genetic factors (σ_a^2), a dominance component accounting for the influence of differences in gene dominances (σ_d^2), and epistatic variance (σ_i^2). The additive component is simply the sum of the effects of each of the individual genes contributing to the phenotype. The dominance component includes the influence of gene dominance on phenotypic expression. Potential epistatic interactions among relevant genes are included in σ_i^2,

$$\sigma_p^2 = \sigma_a^2 + \sigma_d^2 + \sigma_i^2 + \sigma_e^2 + \sigma_{g\times e}^2. \tag{18.9}$$

Three simplifying assumptions are applied initially in many studies of heritability: (1) quantitative traits are often normally distributed (Ayala 1982, Gillespie 1998), (2) gene × environment covariance can be ignored initially, and (3) only additive effects need be considered (i.e., Equation 18.10). Notice that the first assumption of a normally distributed, quantitative trait is inconsistent with the individual effective dose or individual tolerance hypothesis described in Box 12.2, which assumes a lognormal distribution for toxicant tolerance. Notice also that, because the dominance component (σ_d^2) of the total phenotype variance is not subject to individual selection, it is not necessary to include it below (Equation 18.11) in estimating heritability.

$$\sigma_{\mathrm{p}}^2 = \sigma_{\mathrm{a}}^2 + \sigma_{\mathrm{e}}^2. \tag{18.10}$$

In such a case, the narrow sense heritability (h^2) is simply the additive variance divided by the phenotypic variance:

$$h^2 = \frac{\sigma_{\mathrm{a}}^2}{\sigma_{\mathrm{p}}^2}. \tag{18.11}$$

Narrow sense heritability can be estimated using linear regression of the measured trait of offspring (dependent variable) versus that of a parent who shares half of its genes (independent variable). For example, the quantitative trait of interest (e.g., TTD in an acute toxicity test) is measured for the mother and also for her offspring. This is done for many parent–offspring pairs and a regression slope (b) calculated using least squares regression. The regression slope (b) is used to estimate h^2 because $b = h^2/2$. Division by 2 is required here because only half of the genes in the offspring are shared with its mother.

Heritability can also be estimated using experiments designed around twins and sibs (Ayala 1982, Spiess 1977). Alternately, the mean of a trait for both parents ("midparent trait") can be compared to that of offspring (Spiess 1977). Although these methods are infrequently applied in ecotoxicology, Klerks and Levinton (1989) estimated the heritability of metal tolerance for the oligochaete (*Limnodrilus hoffmeisteri*) using regression of the log of the midparent survival time versus the log of the mean offspring survival time. Heritability estimates indicated high levels of heritable variation for metal tolerance. McNeilly and Bradshaw (1968) provide a similar example of estimating heritability for plants species exposed to heavy metals. Posthuma et al. (1993) estimated heritability of metal tolerance for a soil springtail (*Orchesella cincta*) by employing a variety of these methods.

The qualifier "narrow sense" is applied above to heritability without any explanation. This phrase is added to distinguish it from the more inclusive, broad sense heritability. Broad sense heritability (H^2) includes the other components defined in Equation 18.9:

$$H^2 = \frac{\sigma_{\mathrm{g}}^2}{\sigma_{\mathrm{p}}^2} = \frac{\sigma_{\mathrm{a}}^2 + \sigma_{\mathrm{d}}^2 + \sigma_{\mathrm{i}}^2}{\sigma_{\mathrm{p}}^2}. \tag{18.12}$$

Being the proportion of the total phenotypic variation attributable to all genetic components, broad sense heritability simply indicates the amount of genetic influence on the trait (Spiess 1977).

Box 18.4 Phenotype Variance and Heritability of Mosquitofish Exposed to Mercury

In studying allozyme effects on TTD of mosquitofish (*G. holbrooki*) acutely exposed to mercury, Newman, Mulvey, and coworkers became concerned about the amount of phenotypic variation (TTD) due to environmental, genetic, and genetic × environment interactions (see Boxes 18.1 through 18.3 for details). The wild caught fish used for these experiments were collected from an abandoned farm pond with dense masses of submerged vegetation. Because adult mosquitofish readily consume newborn mosquitofish, clutches of newly born mosquitofish tend to remain hidden in nearby vegetation until they are large enough not to be readily eaten. This clustering of broods in vegetation clumps could create discrete lineages that are captured in a biased manner with dip netting. If high frequencies of some allozyme genotypes are correlated with specific broods, spurious correlations may occur between allozyme genotypes and TTD.

Perhaps environmental or genetic × environmental interactions produce lineages that differ in phenotype, not allozyme genotype. Earlier correlations between time-to-death and genotypes at two allozyme loci (Diamond et al. 1989, Newman et al. 1989) were found to be temporally inconsistent (Heagler et al. 1993). To address this possibility of spurious correlations due to temporally variable lineages, Lee et al. (1992) did the following experiment with fish from the reference population. Central to the analysis of the results are the concepts of sources of phenotypic variation and heritability.

Gravid females were dip netted from the reference pond and placed into 40 L tanks. They were allowed to give birth to young that were transferred to 120 L plastic pools. Each pool contained one brood (sibship) and all pools were situated adjacent to one another outdoors. This isolation of broods in adjacent pools simulated the spatial isolation occurring in the reference pond.

The mothers were placed into separate, perforated containers suspended in an exposure tank 3 weeks after producing their broods. Each female's TTD was then recorded during exposure to 1 μg/L of mercury. These TTDs were later compared to those of their offspring.

Offspring were reared as described above until they reached sexual maturity. Females from each brood were exposed as just described for their mothers and the average brood TTD calculated for each sibship. Males were omitted from exposures and later comparison to their mothers because previous studies (Diamond et al. 1989, Newman et al. 1989) showed that TTD differed between males and females.

Narrow sense heritability was estimated by linear regression of mother versus average daughter TTD (i.e., $b = h^2/2$). The slope of the resulting regression line was not significantly different from 0, i.e., no narrow sense heritability was detected.

The environmental effects associated with sibship isolation during development was also assessed by applying ANCOVA to TTD data for daughters. Included in the model were sibship ("Brood") and mean wet weight of exposed daughters in each sibship. There was a highly significant effect of sibship ($P = .004$) on TTD of the daughters. (The F test for significant effect of wet weight on TTD resulted in a $P = .070$.)

These analyses indicate that sharing of a mother and a common habitat during maturation influenced phenotype (TTD), but there was no evidence of an additive genetic component to this variation. The concerns of Lee et al. (1992) regarding the potential influence of lineage on correlations between TTD and allozyme genotypes were quite justified.

18.3 ECOTOXICOLOGY'S TRADITION OF TOLERANCE

> ... it was the advent of DDT and all its many relatives that ushered in the true Age of Resistance.
>
> **(Rachel Carson 1962)**

Populations exposed to toxicants and not eliminated outright may eventually display increased tolerance given sufficient fitness differences, selection pressure, and time. Tolerance may be gauged with one or more phenotypic traits such as enhanced survival, development, growth, or reproduction. However, enhanced tolerance will not emerge and local population will become extinct in the absence of sufficient genetic variation relative to tolerance, consistent selection pressure for tolerance, and sufficient time.

Key factors influencing the process of tolerance acquisition are summarized in Table 18.4 (modified from Table 1 in Mulvey and Diamond (1991)). Genetic qualities influencing tolerance acquisition include initial allele frequencies, dominance, number of genes involved in determining the tolerance phenotype, and the magnitude of the fitness differences among genotypes. Tolerance acquisition will be slower if it involves a rare allele instead of a more common allele. In addition, there is a higher

TABLE 18.4
Factors Influencing the Rate at Which Tolerance Is Acquired in Populations

Quality	Description of Influence
Genetic qualities	
Allele frequency	Tolerance will increase more rapidly if the tolerance allele is common, not rare, in the population before selection begins.
Gene dominance	Tolerance increases more rapidly in early generations if the tolerance allele is dominant.
Monogenic or polygenic	Tolerance will increase more rapidly if controlled by one gene versus many genes.
Fitness differences	Tolerance will increase more rapidly if differences in fitness among genotypes are large (versus small).
Selection components	Selection at different components of a life cycle can negate directional selection for one tolerance trait by counterbalancing selection for another trait, or accelerate the rate of selection for tolerance traits by reinforcing the advantage of a particular genotype.
Cotolerance	Preadaptation to a toxicant can result in elevated tolerance in a population if adaptation took place in the past for a related toxicant, e.g., a plant population adapted to one herbicide is later exposed to a similar herbicide.
Reproductive qualities	
Generation time	Tolerance will increase more rapidly for species with short generation times versus long generation species.
Intrinsic rate of increase	Tolerance will increase more rapidly for populations with high intrinsic rates of increase versus populations with low intrinsic rates of increase.
Size of population (N_e)	Genetic variation decreases with a decrease in N_e. Consequently, large populations will have more variation in tolerance and the possibility of tolerance increasing is higher for large populations versus small populations.
Ecological qualities	
Migration	Influx of nontolerant genotypes due to immigration can slow the rate of tolerance acquisition: increased levels of migration will slow tolerance acquisition.
Refugia	The presence of refugia, i.e., uncontaminated habitat, allows intolerant individuals to remain in the population and can slow the rate of tolerance acquisition.
Life stage sensitivity	Selection effectiveness can be influenced by the most sensitive life stage.

risk of allele loss if the allele is initially rare in a population experiencing high levels of mortality as can be the case in toxicant-exposed populations. If the tolerance gene is dominant, tolerance will emerge rapidly relative to the case in which it is associated with a recessive gene. As an example, Yarbrough and colleagues (Chambers and Yarbrough 1979, Wise et al. 1986, Yarbrough et al. 1986) demonstrated enhanced pesticide tolerance of mosquitofish (*G. affinis*) in agricultural areas. Pesticide resistance in mosquitofish was defined by one dominant gene. In another example, Martínez and Levinton (1996) found rapid metal tolerance acquisition controlled by a single gene in an oligochaete population.

Consistent with Fisher's theorem of natural selection (see Section 18.2.1), the rate of change in the average tolerance of the population increases as the genetic variability of the tolerance trait in the initial population increases (e.g., Equations 18.5 and 18.6). The larger the difference in fitness among the genotypes, the faster the increase in average population tolerance will occur. The industrial melanism of the peppered moth (Box 12.1) is a clear example of rapid microevolution due to the large fitness differences among color morphs and dominance of dark over light color morphs. As described in Sections 18.1.2 and 18.1.3, selection can occur at different stages of an organism's life cycle. Selection at one component of an organism's life cycle can counterbalance selection for a tolerance trait associated with another component of the organism's life cycle. Previous discussion of longevity (i.e., antagonistic pleiotropy in Chapter 16) illustrates the importance of life cycle stage on favoring or disfavoring a particular trait. On the other hand, if tolerance traits at different selection components disfavor the same genotypes, tolerance acquisition can be accelerated due to the reinforcing of the genotype disadvantage by several tolerance traits. Therefore, it is critical to understand all selection components before making predictions about tolerance acquisition. Finally, a previous adaptation to one toxicant can result in cross-resistance or co-tolerance. This co-tolerance can result in very rapid accommodation to a novel toxicant. For example, plants tolerant to a particular *s*-triazine herbicide produce a herbicide-binding protein that can impart an elevated level of tolerance if these adapted plants are later exposed to a novel *s*-triazine herbicide (Erickson et al. 1985). Isopod tolerance to copper imparts a cotolerance to lead (Brown 1978).

Factors other than genetics influence tolerance acquisition. Short generation time and rapid population growth rates increase the rate of tolerance acquisition. Effective population size and the associated changes in population genetic variability influence selection for tolerance. All else being equal, larger populations generally possess more genetic variability than smaller ones and, consequently, the chance of enhanced tolerance emerging and the rate of tolerance change will be higher for larger than for smaller populations. Relative to metapopulation considerations, an increase in migration into an exposed population will slow the rate at which the average population tolerance increases (e.g., Newman and Jagoe 1998). The presence of refugia or source demes of nontolerant genotypes will also slow the rate at which average tolerance level increases in the exposed population.

A rich literature describes toxicant-related tolerance in natural populations. Those focused on plants are particularly thorough (e.g., Antonovics et al. 1971, Baker and Walker 1989, Nacnair 1997, Pitelka 1988, Wilson 1988). Good reviews exist for the topic relative to insect resistance to pesticides (e.g., Mallet 1989) and animal tolerance to pollutants (e.g., Forbes and Calow 1997, Klerks 1990, Klerks and Weis 1987, Mulvey and Diamond 1991).

18.4 SUMMARY

In this brief chapter, the characteristics and dynamics of microevolution were described relative to natural selection and tolerance acquisition. The conditions giving rise to and the consequences of natural selection were defined. The genetic dynamics of demes are explored using parts of Wright's shifting balance theory, especially the complementary exploration of the adaptive landscape afforded by genetic drift and the shift onto adaptive peaks due to natural selection. Selection components were described and the importance of exploring all relevant selection components emphasized. Metrics of differential fitness and selection coefficients were explored as where those for narrow and broad sense heritability. Finally, tolerance acquisition was described relative to fundamental factors influencing rates of increase for average population tolerance.

18.4.1 Summary of Foundation Concepts and Paradigms

- Genetic drift and natural selection are complementary processes giving rise to evolutionary change.
- Natural selection is the change in relative genotype frequencies through generations due to differential fitnesses of the associated phenotypes. Pertinent differences in phenotype fitness can involve viability or reproductive aspects of an individual's life.
- Four required conditions for natural selection are the following: (1) the existence of variation among individuals relative to some trait, (2) fitness differences associated with differences in that trait, (3) heritability of that trait, and (4) the Malthusian premise that individuals in populations can produce excess offspring.
- There are two consequences of the four conditions for natural selection. The frequency of a heritable trait will differ among age or life stage classes. In addition, the frequency of the trait from adult to offspring (i.e., across generations) will differ due to trait-related differences in fitness.

- Fitness differences may be controlled by one locus with the appearance of distinct fitness classes (Mendelian trait) or by several or many genes, resulting in a continuum of fitness phenotypes in a population (quantitative trait).
- Natural selection can be directional, disruptive, or normalizing.
- Differences in fitness are specific to a particular environment and the relative fitness of genotypes can change if the environment changes.
- Consistent, environment-specific differences in fitness are needed for natural selection to occur.
- Lacking sufficient, appropriate genetic variability, a species population may fail to adapt and will become locally extinct.
- Wright argued that, through genetic drift and natural selection on individuals, demes tend to shift continually within an adaptive landscape to occupy local fitness peaks. These peaks shift through time as the environment changes and natural selection working on individuals move the deme toward a new optimal fitness peak.
- Selection can occur at several stages of an organism's life cycle. Selection components include viability selection, sexual (male and female) selection, meiotic drive, gametic selection, and fecundity selection.
- Viability selection includes fitness differences in development of the zygote, growth after birth, and survival.
- Four other selection components involve reproduction. Sexual selection involves differences in adult success in finding, attracting, or retaining a mate. Meiotic drive involves the differential production of the possible gamete types by heterozygotes. The gametic selection occurs if certain gametes produced by heterozygotes have a higher probability of being involved in fertilization than others. Fecundity selection occurs if pairs of certain genotypes have more young than others.
- Selection for several selection components often occurs, making it essential to understand the net balance between fitnesses at these different components. Prediction from one component (e.g., viability during acute toxicant exposure) can lead to inaccurate predictions of selection consequences.
- Differences in fitness can be quantified as fitness (w), relative fitness, or selection coefficients (s).
- Heritability can be quantified by assuming that phenotypic variation among individuals results from a combination of genetic variation (including additive-, epistatic- and dominance-associated variance), environmental variation and, perhaps, the interaction between genetic and environmental factors.
- Narrow sense heritability (h^2) is the additive genetic variance divided by the total phenotypic variance and broad sense heritability (H^2) includes the other variance components as defined in Equation 18.9.
- The rate of tolerance acquisition in an exposed population is a function of genetic, ecological, and reproductive factors as summarized in Table 18.4.

REFERENCES

Antonovics, J., Metal tolerance in plants, In *Advances in Ecological Research*, Vol. 7, Cragg, J.B. (ed.), Academic Press, New York, 1971, pp. 1–85.

Ayala, F.J., *Population and Evolutionary Genetics*, Benjamin Cummings, Menlo Park, CA, 1982.

Baker, A.J.M. and Walker, P.L., Physiological responses of plants to heavy metals and the quantification of tolerance and toxicity, *Chem. Spec. Bioavail.*, 1, 7–18, 1989.

Battaglia, J.A., Bisol, P.M., Fossato, V.U., and Rodino, E., Studies of the genetic effect of pollution in the sea, *Rapp. P-v Reun., Cons. Int. Explor. Mer*, 179, 267–274, 1980.

Beardmore, J.A., Genetical considerations in monitoring effects of pollution, *Rapp. P-v Reun., Cons. Int. Explor. Mer*, 179, 258–266, 1980.

Beardmore, J.A., Barker, C.J., Battaglia, B., Payne, J.F., and Rosenfeld, A., The use of genetical approaches to monitoring biological effects of pollution, *Rapp. P-v Reun., Cons. Int. Explor. Mer*, 179, 299–305, 1980.

Brown, B.E., Lead detoxification by a copper-tolerant isopod, *Nature*, 276, 388–390, 1978.

Bungaard, J. and Christiansen, F.B., Dynamics of polymorphism, I: Selection components in an experimental population of *Drosophila melanogaster*, *Genetics*, 71, 439–460, 1972.

Carson, R., *Silent Spring*, Houghton Mifflin Co., New York, 1962.

Chagnon, N.L. and Guttman, S.I., Differential survivorship of allozyme genotypes in mosquitofish populations exposed to copper or cadmium, *Environ. Toxicol. Chem.*, 8, 319–326, 1989.

Chambers, J.E. and Yarbrough, J.D., A seasonal study of microsomal mixed-function oxidase components in insecticide-resistant and susceptible mosquitofish, *Gambusia affinis*, *Toxicol. Appl. Pharmacol.*, 48, 497–507, 1979.

Christiansen, F.B. and Frydenberg, O., Selection component analysis of natural polymorphisms using population samples including mother–offspring combinations, *Theor. Popul. Biol.*, 4, 425–445, 1973.

Christiansen, F.B., Frydenberg, O., Gyldenholm, A.O., and Simonsen, V., Genetics of *Zoarces* populations, VI: Further evidence based on age group samples, of a heterozygote deficit in the EST III polymorphism, *Hereditas*, 77, 225–236, 1974.

Christiansen, F.B., Frydenberg, O., and Simonsen, V., Genetics of *Zoarces* populations: IV. Selection component analysis of an esterase polymorphism using population samples including mother–offspring combinations, *Hereditas*, 73, 291–304, 1973.

Clegg, M.T., Kahler, A.L., and Allard, R.W., Estimation of life cycle components of selection in an experimental plant population, *Genetics*, 89, 765–792, 1978.

Coyne, J.A., Barton, N.H., and Turelli, M., Perspective: A critique of Sewall Wright's shifting balance theory of evolution, *Evolution*, 51, 643–671, 1997.

Coyne, J.A., Barton, N.H., and Turelli, M., Is Wright's shifting balance process important in evolution? *Evolution*, 54, 306–317, 2000.

Darwin, C., *The Origin of Species*, Penguin Putnam Inc., New York, 1872.

Diamond, S.A., Newman, M.C., Mulvey, M., Dixon, P.M., and Martinson, D., Allozyme genotype and time to death of mosquitofish, *Gambusia affinis* (Baird and Girard), during acute exposure to inorganic mercury, *Environ. Toxicol. Chem.*, 8, 613–622, 1989.

Endler, J.A., *Natural Selection in the Wild*, Princeton University Press, Princeton, NJ, 1986.

Erickson, J.M., Rahire, M., and Rochaix, J.-D., Herbicide resistance and cross-resistance: Changes at three distinct sites in the herbicide-binding protein, *Science*, 228, 204–207, 1985.

Fisher, R.A., *The Genetical Theory of Natural Selection*, Clarendon Press, Oxford, UK, 1930.

Forbes, V.E. and Calow, P., Responses of aquatic organisms to pollutant stress: Theoretical and practical implications, In *Environmental Stress, Adaptation and Evolution*, Bijlsma, R. and Loeschcke, V. (eds.), Birkhäuser Verlag, Basel, Switzerland, 1997, pp. 25–41.

Forbes, V.E. and Forbes, T.L., *Ecotoxicology in Theory and Practice*, Chapman & Hall, London, UK, 1994.

Gillespie, J.H., *Population Genetics. A Concise Guide*, The John Hopkins University Press, Baltimore, MD, 1998.

Gillespie, R.B. and Guttman, S.I., Effects of contaminants on the frequencies of allozymes in populations of the central stoneroller, *Environ. Toxicol. Chem.*, 8, 309–317, 1989.

Gillespie, R.B. and Gutmann, S.I., Chemical-induced changes in the genetic structure of populations: Effects on allozymes. In Forbes, V.E. (ed.), *Genetics and Ecotoxicology*, Taylor & Francis, Philadelphia, PA, 1999.

Hartl, D.L. and Clark, A.G., *Principles of Population Genetics*, Sinauer Associates, Inc., Sunderland, MA, 1999, pp. 55–77.

Heagler, M.G., Newman, M.C., Mulvey, M., and Dixon, P.M., Allozyme genotype in mosquitofish, *Gambusia holbrooki*, during mercury exposure: Temporal stability, concentration effects and field verification, *Environ. Toxicol. Chem.*, 12, 385–395, 1993.

Hoffmann, A.A. and Parsons, P.A., *Extreme Environmental Change and Evolution*, Cambridge University Press, Cambridge, UK, 1997.

Ingvarsson, P.K., Differential migration from high fitness demes in the shining fungus beetle, *Phalacrus substriatus*, *Evolution*, 54, 297–301, 2000.

Keklak, M.M., Newman, M.C., and Mulvey, M., Enhanced uranium tolerance of an exposed population of the eastern mosquitofish (*Gambusia holbrooki* Girard 1859), *Arch. Environ. Contam. Toxicol.*, 27, 20–24, 1994.

Klerks, P.L., Adaptation to metals in animals, In *Heavy Metals Tolerance in Plants: Evolutionary Aspects*, Shaw, A.J. (ed.), CRC Press, Boca Raton, FL, 1990, pp. 311–321.

Klerks, P.L. and Levinton, J.S., Rapid evolution of metal resistance in a benthic oligochaete inhabiting a metal-polluted site, *Biol. Bull.*, 176, 135–141, 1989.

Klerks, P.L. and Weis, J.S., Genetic adaptation to heavy metals in aquatic organisms: A review, *Environ. Pollut.*, 45, 173–205, 1987.

Kramer, V.J. and Newman, M.C., Inhibition of glucosephosphate isomerase allozymes of the mosquitofish, *Gambusia holbrooki*, by mercury, *Environ. Toxicol. Chem.*, 13, 9–14, 1994.

Kramer, V.J., Newman, M.C., Mulvey, M., and Ultsch, G.R., Glycolysis and Krebs cycle metabolites in mosquitofish, *Gambusia holbrooki*, Girard 1859, exposed to mercuric chloride: Allozyme genotype effects, *Environ. Toxicol. Chem.*, 11, 357–364, 1992a.

Kramer, V.J., Newman, M.C., and Ultsch, G.R., Changes in concentrations of glycolysis and Krebs cycle metabolites in mosquitofish, *Gambusia holbrooki*, induced by mercuric chloride and starvation, *Environ. Biol. Fish*, 34, 315–320, 1992b.

Lavie, B. and Nevo, E., Heavy metal selection of phosphoglucose isomerase allozymes in marine gastropods, *Mar. Biol.*, 71, 17–22, 1982.

Lavie, B. and Nevo, E., Genetic selection of homozygote allozyme genotypes in marine gastropods exposed to cadmium pollution, *Sci. Total Environ.*, 57, 91–98, 1986a.

Lavie, B. and Nevo, E., The interactive effects of cadmium and mercury pollution on allozyme polymorphisms in the marine gastropod, *Cerithium scabridum*, *Mar. Pollut. Bull.*, 17, 21–23, 1986b.

Lee, C.J., Newman, M.C., and Mulvey, M., Time to death of mosquitofish (*Gambusia holbrooki*) during acute inorganic mercury exposure: Population structure effects, *Arch. Environ. Contam. Toxicol.*, 22, 284–287, 1992.

Macnair, M.R., The evolution of plants in metal-contaminated environments, In *Environmental Stress, Adaptation and Evolution*, Bijlsma, R. and Loeschcke, V. (eds.), Birkhäuser Verlag, Basel, Switzerland, 1997, pp. 3–24.

Mallet, J., The evolution of insecticide resistance: Have the insects won? *TREE*, 4, 336–340, 1989.

Martínez, D.E. and Levinton, J., Adaptation to heavy metals in the aquatic oligochaete *Limnodrilus hoffmeisteri*: Evidence for control by one gene, *Evolution*, 50, 1339–1343, 1996.

McNeilly, T. and Bradshaw, A.D., Evolutionary processes in populations of copper-tolerant *Argostis tenuis* Sibth, *Evolution*, 22, 108–118, 1968.

Moraga, D. and Tanguy, A., Genetic indicators of herbicide stress in the Pacific oyster *Crassostrea gigas* under experimental conditions, *Environ. Toxicol. Chem.*, 19, 706–711, 2000.

Mulvey, M. and Diamond, S.A., Genetic factors and tolerance acquisition in populations exposed to metals and metalloids, In *Metal Ecotoxicology. Concepts & Applications*, Newman, M.C. and McIntosh, A.W. (eds.), Lewis Publishers, Chelsea, MI, 1991, pp. 301–321.

Mulvey, M., Newman, M.C., Chazal, A., Keklak, M.M., Heagler, M.G., and Hales, L.S., Jr., Genetic and demographic responses of mosquitofish (*Gambusia holbrooki* Girard 1859) populations stressed by mercury, *Environ. Toxicol. Chem.*, 14, 1411–1418, 1995.

Nadeau, J.H. and Baccus, R., Selection components of four allozymes in natural populations of *Peromyscus maniculatus*, *Evolution*, 35, 11–20, 1981.

Nadeau, J.H., Dietz, K., and Tamarin, R.H., Gametic selection and the selection component analysis, *Genet. Res.*, 37, 275–284, 1981.

Nevo, E., Perl, T., Beiles, A., and Wool, D., Mercury selection of allozyme genotypes in shrimps, *Experientia*, 37, 1152–1154, 1981.

Newman, M.C., *Quantitative Methods in Aquatic Ecotoxicology*, CRC Press/Lewis Publishers, Boca Raton, FL, 1995.

Newman, M.C., *Fundamentals of Ecotoxicology*, CRC Press/Lewis Publishers, Boca Raton, FL, 1998.

Newman, M.C., Diamond, S.A., Mulvey, M., and Dixon, P., Allozyme genotype and time to death of mosquitofish, *Gambusia affinis* (Baird and Girard) during acute toxicant exposure: A comparison of arsenate and inorganic mercury, *Aquatic Toxicol.*, 15, 141–156, 1989.

Newman, M.C. and Dixon, P.M., Ecologically meaningful estimates of lethal effect in individuals, In *Ecotoxicology. A Hierarchical Treatment*, Newman, M.C. and Jagoe, C.H. (eds.), CRC Press/Lewis Publishers, Boca Raton, FL, 1996, pp. 225–253.

Newman, M.C. and Jagoe, R.H., Allozymes reflect the population-level effect of mercury: Simulations of mosquitofish (*Gambusia holbrooki* Girard) GPI-2 response, *Ecotoxicology*, 7, 141–150, 1998.

Partridge, G.G., Relative fitness of genotypes in a population of *Rattus norvegicus* polymorphic for Warfarin resistance, *Heredity*, 43, 239–246, 1979.

Pemberton, J.M., Albon, S.D., Guinness, F.E., Clutton-Brock, T.H., and Berry, R.J., Genetic variation and juvenile survival in red deer, *Evolution*, 42, 921–934, 1988.

Pitelka, L.F., Evolutionary responses of plants to anthropogenic pollutants, *TREE*, 3, 233–236, 1988.

Posthuma, L., Hogervorst, R.F., Joose, N.G., and van Straalen, N.M., Genetic variation and covariation for characteristics associated with cadmium tolerance in natural populations of the springtail *Orchesella cincta* (L.), *Evolution*, 47, 619–631, 1993.

Schlueter, M.A., Guttman, S.I., Duan, Y., Oris, J.T., Huang, X., and Burton, G.A., Effects of acute exposure to fluoranthene-contaminated sediment on the survival and genetic variability of fathead minnows (*Pimephales promelas*), *Environ. Toxicol. Chem.*, 19, 1011–1018, 2000.

Schlueter, M.A., Guttman, S.I., Oris, J.T., and Bailer, A.J., Survival of copper-exposed juvenile fathead minnows (*Pimephales promelas*) differs among allozyme genotypes, *Environ. Toxicol. Chem.*, 14, 1727–1734, 1995.

Siegismund, H.R. and Christiansen, F.B., Selection component analysis of natural polymorphisms using population samples including mother–offspring combinations, III, *Theor. Popul. Biol.*, 27, 268–297, 1985.

Spiess, E.B., *Genes in Populations*, John Wiley & Sons, New York, 1997.

Walker, C.H., Hopkin, S.P., Sibly, R.M., and Peakall, D.B., *Principles of Ecotoxicology*, Taylor & Francis, London, UK, 1996.

Watt, W.B., Carter, P.A., and Blower, S.M., Adaptation at specific loci. IV. Differential mating success among glycolytic allozyme genotypes of *Colias* butterflies, *Genetics*, 109, 157–175, 1985.

Webb, R.E. and Horsfall, F., Jr., Endrin resistance in the pine mouse, *Science*, 156, 1762, 1967.

Weis, J.S. and Weis, P., Tolerance and stress in a polluted environment: The case of the mummichog, *BioScience*, 39, 89–95, 1989.

Whitten, M.J., Dearn, J.M., and McKenzie, J.A., Field studies on insecticide resistance in the Australian sheep blowfly, *Lucilia cuprina*, *Aust. J. Biol. Sci.*, 33, 725–735, 1980.

Williams, C.J., Anderson, W.W., and Arnold, J., Generalized linear modeling methods for selection component experiments, *Theor. Popul. Biol.*, 37, 389–423, 1990.

Wilson, J.B., The cost of heavy-metal tolerance: An example, *Evolution*, 42, 408–413, 1988.

Wise, D., Yarbrough, J.D., and Roush, T.R., Chromosomal analysis of insecticide resistant and susceptible mosquitofish, *J. Hered.*, 77, 345–348, 1986.

Wright, S., The roles of mutation, inbreeding, crossbreeding and selection in evolution, *Proc. 6th Int. Cong. Genet.*, 1, 356–366, 1932.

Wright, S., The shifting balance theory and macroevolution, *Ann. Rev. Genet.*, 16, 1–19, 1982.

Yarbrough, J.D., Roush, R.T., Bonner, J.C., and Wise, D.A., Monogenic inheritance of cyclodiene resistance in mosquitofish, *Gambusia affinis*, *Experientia*, 42, 851–853, 1986.

19 Conclusion

To conceive of it with a total apprehension I must for the thousandth time approach it as something totally strange.

(Thoreau 1859, cited in Bickman (1999))

19.1 OVERVIEW

This section explored ecotoxicology from the vantage of the population. Detail relative to populations was provided to enhance the reader's differentiation and integration of population-related information. This volume also tries to bridge concepts and techniques in the two sections on organismal and community ecotoxicology. Hopefully, by this initial effort to translate concepts and metrics among hierarchical levels, consilience might gradually emerge as a more central strategic goal of ecotoxicology during the next decades.

Let us review for the moment what has been presented in this section. The vantage of population ecotoxicology, the science of contaminants in the biosphere and their effects on populations, was argued to be crucial for predicting extinction risk for populations under contaminant exposure. Such prediction is a central objective of much environmental legislation. With the exception of federal acts focused on human health or endangered species, the intent of key U.S. environmental laws is an assurance of species population viability in environments containing toxicants. This can be done more directly with population-based concepts and data than with individual-based concepts, models, and information alone. Potential contributions to the potential effectiveness of prediction can be found in the subdisciplines of epidemiology, population dynamics, demography, metapopulation biology, life history theory, and population genetics. Related concepts and techniques afford effective description and prediction of population consequences.

19.2 SOME PARTICULARLY KEY CONCEPTS

19.2.1 EPIDEMIOLOGY

Epidemiology provided a mode of describing toxicant-related disease in populations and quantitatively comparing disease in different populations or study groups. Models identifying risk factors for individuals within populations were described, including proportional hazard, accelerated failure, and binary logistic regression models. Methods were demonstrated with examples of human disease; however, they are easily applied to other species.

Results from epidemiological studies also contribute to predicting genetic consequences of exposure (i.e., population consequences) as described for mercury-exposed mosquitofish (Chapter 18). Epidemiological studies also allow convenient estimation of mortality rates applied in simple population growth models, demographic life tables, and metapopulation models of exposed populations.

Interpretation of epidemiological results is susceptible to logical errors, so evaluation of results has to be done thoughtfully. The foundations of causality were quickly reviewed. The intent was to describe common errors so that they might be avoided and, to borrow a phrase from Alan Watts (1968), to cultivate a "wisdom of insecurity" about cause–effect relationships. Hill's aspects of disease association were explored as a specific set of rules commonly used to improve the process of identifying disease associations. Hill's rules are not the only ones relevant to ecotoxicology. The reader may also want to review those of Fox (1991) and Evans (Evans 1976). Because of the difficulty in assigning causality, formal Bayesian methods of enhancing belief would be extremely valuable in epidemiological surveys by ecotoxicologists. General references for Bayesian methods include Howson and Urbach (1989), Box and Tiao (1992), Retherford and Choe (1993), and Josephson and Josephson (1996). These methods are useful at all levels of the ecological hierarchy.

The possibility of contaminants influencing the infectious disease process in populations was explored briefly. The paradigm that toxicants increase the risk of infectious disease by weakening hosts was judged to be less inclusive than the disease triad paradigm (Figure 13.5). Toxicants, as components of the environment in which the host and parasite/pathogen are interacting, can favor either the host or parasite/pathogen. Infectious disease may be fostered or discouraged by exposure to toxicants.

19.2.2 Simple Models of Population Dynamics

Phenomenological models of population dynamics were explored in Chapter 14, assuming a homogeneous distribution of identical individuals. They provided important insights despite simplification and the aggregation of information into basic parameters. Models allowed a clearer understanding of contaminant influence on the temporal dynamics of populations than afforded by conventional, individual-based methods alone. Some population effects noted during ecological risk assessments would be inexplicable or only vaguely explicable without such an understanding. Density-independent mortality due to toxicant exposure was added to classic population growth models. The possibility of enhanced population productivity ("yield") as well as reduced productivity was demonstrated with the incorporation of toxicant exposure into models used to predict yield for harvested fish and wildlife populations. Methods for estimating population consequences and time for recovering were described based on these basic models.

19.2.3 Metapopulation Dynamics

The consequences of uneven distribution of individuals in a contaminated environment were explored with metapopulation models. The risk of local population extinction or lowered carrying capacity was assessed most accurately with this metapopulation context, a context only now being introduced into ecotoxicology (O'Connor 1996).

It is crucial to understand the source–sink dynamics of the habitat mosaic populated by a species. Some poor habitats can contain a number of individuals only if a source habitat is nearby and individuals move among habitats. Keystone habitats and corridors for migration among segments of the population become crucial to predicting population consequences of contaminant exposure. Accurate prediction also depends on knowledge of other important population qualities within a landscape mosaic such as potential propagule rain and rescue effects. The metapopulation context also provides explanation for toxicant effects to individuals outside of the contaminated area.

19.2.4 The Demographic Approach

Applying basic demographic techniques, discussion moved beyond phenomenological models to include heterogeneity among individuals. Lamentably, much of the lethality and reproductive information currently generated for regulatory purposes—for protecting populations in contaminated habitats—is not gathered in a manner directly useful in demographic methods. Despite the slow evolution of standard methods relative to effectively generating and applying ecotoxicology data to prediction of population consequences, demographic methods are being used with increasing frequency in ecotoxicology. Techniques consistent with demographic methods exist for analyzing toxicological data, for example, the survival time and LTRE (Caswell 1996) methods. Simple and matrix-based demographic methods were described and means of including stochastic aspects of population projections were discussed.

Although applied widely by ecotoxicologists today, the most sensitive stage paradigm was identified as a false paradigm (weakest link incongruity). Predictions relying on demographic metrics such as elasticity or reproductive value should replace those based on the most sensitive stage paradigm.

19.2.5 Phenogenetics Theory

Although emergent properties may confound predictions, life history theory can link contaminant-related changes in phenotype to population vital rates (Calow and Sibly 1990, Kooijman et al. 1989, Sibly 1996, Sibly and Calow 1989). The principle of allocation suggests that an individual with a specific genetic make-up and living in a particular environment must allocate energy resources so as to maximize Darwinian fitness. Therefore, predictable rules for energy allocation should be identifiable, albeit expressed slightly differently, for individuals within populations. Shifts in energy allocation under different environmental conditions produce differences in population vital rates. For example, a contaminant may require increased energy expenditure for detoxification and repair of soma in order for an individual to survive to reproductive age. Once arriving at sexual maturity, that individual might have less energy reserve available for reproduction. Adjustment in the rate at which an individual becomes reproductively viable might also occur. There could be other life history changes. Such effects taken together for all individuals in a population result in changes in vital rates that could result in a change in population vitality or risk of local extinction.

Reaction norms define environment-dependent shifts in phenotype (i.e., phenotypic plasticity). Reaction norms can be inflexible in which case phenotype does not change once it is expressed. Some reaction norms can change during the life of an individual. Reaction norms for life history characteristics allow exploration of toxicant-induced shifts affecting population vital rates (e.g., Box 16.2).

Polyphenism occurs if an environmental cue triggers expression of one phenotype or another with no intermediate phenotypes being expressed. Polyphenisms are directly relevant to assessing effects of endocrine-modifying contaminants on population consequences. As an example, exposure to an endocrine-modifying contaminant could determine the sex of hatchlings that will make up the next generation of a turtle population.

Developmental stability is a valuable population-level metric quantifying the ability of individuals in a population to develop into a narrow range of phenotypes within a particular environment. Beyond a certain level of variation, deviations in phenotype expression implies a decrease in fitness of associated individuals. Metrics such as fluctuating asymmetry allow easy detection of changes in developmental stability due to contaminant exposure.

19.2.6 Population Genetics: Stochastic Processes

Population genetics can be affected by toxicant exposure. Direct changes to DNA can occur and, unless repaired, these changes lead to the appearance of mutations. Stochastic processes determining genetic qualities of populations can also be influenced by contaminants. Contaminants can modify the spatial distribution of individuals within the population, effective population size, mutation rate, and migration rate.

Several quantitative tools allow assessment of stochastic consequences to population genetics. The Hardy–Weinberg principle predicts genotype frequencies if (1) the population is a large, one of randomly mating individuals, (2) no natural selection is occurring, (3) mutation rates are negligible, and (4) migration rates are negligible. Deviations from Hardy–Weinberg expectations indicate violation of one or more of these conditions. Models of genetic drift as a function of effective population size allow prediction of genetic change with toxicant-induced reduction in population size.

Selander's D statistics quantify the deficiency of heterozygotes. Those deficiencies could be a function of selection, inbreeding, or population structure (e.g., a Wahlund effect). Wright's F statistics can provide understanding of the genetic structure of a population potentially comprised of many demes or having genetic clines. Insight about normal genetic structure is necessary for properly interpreting genetic trends seen in field populations. Finally, genetic diversity itself is crucial to the long-term viability of species populations. Without variation, a population lacks the raw material with which to adapt to changes in its environment and will eventually disappear when the environment changes.

19.2.7 Population Genetics: Natural Selection

Natural selection can be another important process occurring in populations exposed to contaminants. Natural selection can result in enhanced tolerance, the enhanced ability to cope with toxicants owing to physiological, biochemical, anatomical, or some other genetically based change in phenotype. However, natural selection resulting in enhanced tolerance requires genetic variation in the tolerance trait and populations lacking adequate variability are at higher risk of extinction than those with adequate variability.

Viability selection is often the focus of tolerance studies; however, other important selection components can be involved. They include male and female sexual selection, meiotic drive in heterozygotes, gametic selection in heterozygotes, and fecundity selection. Several selection components can occur simultaneously, perhaps resulting in balancing one component against another. It is important in predicting consequences of toxicant-driven selection that all selection potential components be assessed carefully.

19.3 CONCLUDING REMARKS

Hopefully, this short treatment of population ecotoxicology has been, simultaneously, informative and convincing. Several of the key concepts or relationships described here broaden one's understanding of toxicant effects in ecological systems. Hopefully, the reader is convinced of the importance of the population context in scientific and practical ecotoxicology.

REFERENCES

Box, G.E.P. and Tiao, G.C., *Bayesian Inference in Statistical Analysis*, John Wiley & Sons, New York, 1992.

Calow, P. and Sibly, R.M., A physiological basis of population processes: Ecotoxicological implications, *Funct. Ecol.*, 4, 283–288, 1990.

Caswell, H., Demography meets ecotoxicology: Untangling the population level effects of toxic substances, In *Ecotoxicology. A Hierarchical Treatment*, Newman, M.C. and Jagoe, C.H. (eds.), CRC Press/Lewis Publishers, Boca Raton, FL, 1996, pp. 255–292.

Evans, A.S., Causation and disease: The Henle–Koch postulates revisited, *Yale J. Biol. Med.*, 49, 175–195, 1976.

Fox, G.A., Practical causal inference for ecoepidemiologists, *J. Toxicol. Environ. Health*, 33, 359–373, 1991.

Howson, C. and Urbach, P., *Scientific Reasoning. The Bayesian Approach*, Open Court, La Salle, IL, 1989.

Josephson, J.R. and Josephson, S.G., *Abductive Inference. Computation, Philosophy, Technology*, Cambridge University Press, Cambridge, UK, 1996.

Kooijman, S.A.L.M., Van der Hoeven, N., and Van der Werf, D.C., Population consequences of a physiological model for individuals, *Funct. Ecol.*, 3, 325–336, 1989.

O'Connor, R.J., Toward the incorporation of spatiotemporal dynamics into ecotoxicology, In *Population Dynamics in Ecological Space and Time*, Rhodes, O.E., Jr., Chesser, R.K., and Smith, M.H. (eds.), The University of Chicago Press, Chicago, IL, 1996, pp. 281–317.

Retherford, R.D. and Choe, M.K., *Statistical Models for Causal Analysis*, John Wiley & Sons, New York, 1993.

Sibly, R.M., Effects of pollutants on individual life histories and population growth rates, In *Ecotoxicology. A Hierarchical Treatment*, Newman, M.C. and Jagoe, C.H. (eds.), CRC Press/Lewis Publishers, Boca Raton, FL, 1996, pp. 197–223.

Sibly, R.M. and Calow, P., A life-history theory of responses to stress, *Biol. J. Linn. Soc.*, 37, 101–116, 1989.

Thoreau, H.D., 1859. Quote from *Henry Thoreau on Education*, Bickman, M. (ed.), Houghton Mifflin Co., Boston, MA, 1999.

Watts, A.W., *The Wisdom of Insecurity*, Random House, New York, 1968.

Part IV

Community Ecotoxicology

> Chemicals are pre-tested against a few individuals, but not against living communities.
>
> **(Rachel Carson 1962)**

As described in the previous section, understanding direct effects of chemical stressors on populations is a fundamental concern of ecotoxicologists and regulators. However, interactions among species often transcend population-level responses and may play a significant role in structuring communities in nature. Occupying an intermediate level of complexity in the hierarchy of biological organization, communities are distinct from, but have intimate linkages to, populations and ecosystems. Contemporary questions in basic community ecology address the strength, ubiquity, and transience of species interactions (Strong et al. 1984), as well as the environmental factors that regulate species diversity. Understanding effects of contaminants on species interactions is considered a primary justification for testing effects at higher levels of biological organization (Cairns 1983). Furthermore, most monitoring programs developed to measure effects of contaminants on aquatic ecosystems rely heavily on community-level assessments. Because of its rich history of investigating indirect effects, the theoretical models and empirical studies in basic community ecology can be used as a framework for predicting contaminant effects (Rohr et al. 2006). In addition to improving our understanding of life history characteristics and other autecological features that determine susceptibility of organisms to chemicals, we believe that ecotoxicologists should also consider species interactions and how contaminants affect these interactions. The goal of this section is to apply basic principles developed in community ecology to improve our understanding of how groups of interacting species respond to contaminants and other anthropogenic stressors.

REFERENCES

Cairns, J., Jr., Are single species toxicity tests alone adequate for estimating environmental hazard? *Hydrobiologia*, 100, 47–57, 1983.

Carson, R., *Silent Spring*, Houghton Mifflin, Boston, MA, 1962.

Rohr, J.R., Kerby, J.L., and Sih, A., Community ecology as a framework for predicting contaminant effects, *Trends Ecol. Evol.*, 21, 606–613, 2006.

Strong, D.R., Simberloff, D., Abele, L.G., and Thistle, A.B., *Ecological Communities: Conceptual Issues and Evidence*, Princeton University Press, Princeton, NJ, 1984.

20 Introduction to Community Ecotoxicology

20.1 DEFINITIONS—COMMUNITY ECOLOGY AND ECOTOXICOLOGY

> Ecology is the science of communities.
>
> **(Shelford 1913)**

> There is little agreement among ecologists about what a community is and how its structure is regulated.
>
> **(Ricklefs 1990)**

> Doing science at the community level presents daunting problems because the database may be enormous and complex.
>
> **(Begon et al. 1990)**

20.1.1 Community Ecology

A community is defined as a group of interacting populations that overlap in time and space. However, the study of communities transcends simple descriptions of demographic and life history characteristics of individual populations. Instead of describing birth rates, death rates, and other autecological features of isolated populations, community ecologists focus on the interactions among these populations in nature. Rather than measuring fluctuations in abundance of a particular species over time or quantifying differences in population density between locations, the community ecologist considers changes in species diversity and composition of dominant taxa. The primary goal of community ecology is to describe patterns in the organization of communities and to explain the underlying processes that regulate these patterns (Wiens 1984). In particular, the community ecologist seeks to quantify the relative importance of biotic and abiotic factors that influence temporal and spatial variation in community structure. Key issues in contemporary community ecology include questions such as "Why are more species found in some habitats than in others?" or "How important are species interactions relative to abiotic factors in regulating community composition?"

The boundaries of communities have been defined based on spatial overlap of populations, trophic structure, strength of species interactions, and taxonomic relationships. In our coverage of community ecology and ecotoxicology, we will not restrict our definition of a community to any arbitrarily selected taxonomic group, although this is a common practice in terrestrial ecology (e.g., a subalpine forest bird community). We feel that interactions among different taxonomic groups (e.g., between fish and zooplankton or between birds and terrestrial insects) are at least as relevant to ecotoxicology as interactions within these groups. Similarly, instead of limiting our definition of a community to populations within a single trophic level, we will adopt a "vertical" definition of communities that includes populations within several trophic levels. Our reasoning is that the potential interactions between predators and their prey are among the most interesting, best studied, and most relevant to the field of ecotoxicology. Resource–consumer interactions form the basis for the transfer of energy and contaminants across trophic levels. Finally, we distinguish between the terms "community" and "assemblage" based on spatial scale and the potential for interactions among populations. Both terms refer to groups of populations; however, a community consists of populations that have the potential to interact, whereas an assemblage generally consists of populations at a larger spatial scale with no implied interactions.

20.1.2 Community Ecotoxicology

> It is time to use ecological theory more extensively to understand contaminant effects, and for ecologists to examine their own systems more thoroughly in light of chemical contamination.
>
> **(Rohr et al. 2006)**

Community ecotoxicology is the study of the effects of chemicals on species abundance, diversity, and interactions. Community ecotoxicologists are also interested in describing patterns in community structure (e.g., number of species or trophic organization) and explaining mechanisms responsible for these patterns. However, unlike research in basic ecology, community ecotoxicologists are especially concerned with separating effects of anthropogenic disturbance, such as chemical stressors, from natural variability. Community ecotoxicology is distinct from population and ecosystem ecotoxicology. While an understanding of the life history characteristics, habitat requirements, and other autecological features of a particular species is important for predicting consequences of exposure to chemical stressors, the endpoints investigated in community ecotoxicology typically integrate responses of numerous species. Finally, community ecotoxicology is unique from ecosystem ecotoxicology in its focus on structural measures such as species diversity and trophic organization instead of ecosystem processes such as energy flow, detritus processing, and nutrient cycling.

Community ecotoxicology has adopted many of the approaches and modified many of the questions derived from basic community ecology to predict effects of chemical stressors. For example, just as community ecologists quantify patterns of species diversity along natural habitat gradients (e.g., elevation, vegetation type), similar study designs allow community ecotoxicologists to measure changes in community composition along pollution gradients. Some researchers have advocated better integration of basic ecological theory into ecotoxicology and noted the benefits of using an ecological framework to improve our understanding of the underlying mechanisms of contaminant effects (Relyea and Hoverman 2006, Rohr et al. 2006). Empirical studies in basic community ecology have provided important insight into how communities respond to contaminants and other anthropogenic disturbances. In particular, the study of community responses to natural disturbance has been a productive area of research in ecology for the past 40 years. Community ecotoxicologists have used these results to help understand ecological responses to chemical stressors. Many of the characteristics of successional change in community composition over time are analogous to patterns of recovery from anthropogenic disturbance. Finally, basic research on food webs and trophic interactions in community ecology has greatly improved our ability to predict contaminant transport among trophic levels and their effects on trophic structure.

20.2 Historical Perspective of Community Ecology and Ecotoxicology

Although the basic definition of a community seems obvious in light of the hierarchical nature of biological organization (e.g., individuals → populations → communities → ecosystems), it underscores several of the more controversial aspects of community ecology. Since the early 1900s, ecologists have struggled to delineate communities and their spatiotemporal boundaries. The early history of ecology reveals considerable disagreement over use of terms such as community, association, assemblage, and guild. A review of major ecology textbooks reveals considerable variation in the definitions of these terms (Fauth et al. 1996). Our definition of community ecotoxicology

TABLE 20.1
Historical Developments in Community Ecology and Their Influence on Community Ecotoxicology

Historical Development in Community Ecology	Reference	Implications for Community Ecotoxicology
Debate between proponents of holism and reductionism	Clements (1936), Gleason (1926)	Limitations of single species toxicity tests for predicting ecological effects on communities and ecosystems
Importance of food webs and energy flow	Elton (1927), Lindeman (1942)	Food chain transfer of contaminants; importance of trophic structure on contaminant levels in top predators
Rise of experimental ecology	Connell (1961), Paine (1966)	Use of microcosms, mesocosms, and ecosystem manipulations for measuring ecological effects

emphasizes the spatial and temporal overlap of populations and the potential for interspecific interactions. We recognize that species interactions in some communities are relatively weak; therefore, the patterns observed are best explained by autecological processes affecting individual populations. However, in communities where interspecific interactions do play an important structuring role, the relative strength of these interactions will influence how communities respond to anthropogenic disturbance. The potential interactions among species represent emergent properties of communities (Odum 1984; Box 20.1) that define this level of biological organization.

Our treatment of community ecology also highlights three significant developments in the history of ecology that greatly influenced the way ecotoxicologists study the fate and effects of contaminants (Table 20.1). First, the deep-rooted philosophical differences between proponents of holism and reductionism in ecology are at least partially responsible for the emergence of ecotoxicology as a distinct discipline. More importantly, because the fields of ecology and toxicology developed in relative isolation, there was little opportunity to infuse ecological concepts and theories into the field of toxicology. Criticism of the underlying assumption that protection of individual species will protect communities and ecosystem processes motivated researchers to question traditional approaches in toxicology (Cairns 1983, 1986). Second, recognition of the importance of trophic interactions by early researchers influenced a generation of ecologists and significantly contributed to the development of contaminant transport models employed by ecotoxicologists. Finally, the experimental approaches developed by field ecologists who recognized the shortcomings of purely descriptive studies are slowly being integrated into ecotoxicological research. We will show that these historical developments had a profound influence on community ecology and continue to influence the current generation of ecotoxicologists.

20.2.1 Holism and Reductionism in Community Ecology and Ecotoxicology

> The relationship between classical ecologists and environmental toxicologists has never been a strong one, and an uncharitable person might well describe it as tenuous.
>
> **(Cairns and Niederlehner 1995)**

While few ecologists disagree with the definition of communities as groups of interacting populations, the relative importance of these interactions in structuring communities has been the focus of intense debate throughout the history of ecology. Some ecologists argue that species interactions are a basic property of all communities, whereas others describe communities as a random collection of populations that coincidentally occupy the same habitat because of their similar environmental requirements. Thus, since its inception the field of community ecology has struggled to define itself within the broader context of ecology (Box 20.1).

Box 20.1 Historical Perspective of Holism and Reductionism in Community Ecology

As in other sciences, the philosophical division between proponents of holism and reductionism is prevalent in ecology. Adherents of holistic approaches argue that complex systems have certain emergent properties that cannot be understood by studying component parts in isolation. Supporters of reductionism counter that there are no emergent properties of systems and that the most efficient way to describe the functioning of a system is by a detailed study of the component parts. There are few examples in the history of ecology where the debates between holism and reductionism have been more contentious than in the field of community ecology.

One of the most significant developments in the history of ecology was the recognition that different geographic locations supported unique and often predictable associations of plants and animals. As nineteenth-century naturalists began their intercontinental travels to collect field observations on the distribution and abundance of organisms, they were intrigued by the similarity of plant associations that occurred in similar climates. In the early 1900s, Frederick E. Clements, a plant ecologist studying grasslands in Nebraska, proposed that in the absence of disturbance, plant communities progressed in an orderly fashion to a final climax community. This predictable sequence of changes in vegetation, termed succession, was determined primarily by competitive interactions among species and resulted in predictable and discrete boundaries between plant communities. Clements's "superorganism" concept, which likened the functioning of a community to that of an individual organism, was undoubtedly one of the more extreme holistic interpretations of community ecology. His viewpoints were rigorously challenged by other plant ecologists, particularly Henry A. Gleason, who argued that plant communities lacked definite boundaries and consisted only of fortuitous associations of species. To Gleason, communities were nothing more than stochastic collections of independent species. Because species interactions are the emergent properties that define communities, these ideas challenged Clements's view not only on succession but also on the very existence of communities. If species interactions are relatively weak or unimportant, then communities may simply represent ecologists' futile attempts to force random associations of species into nonexistent organizational units. The ultimate demise of Clements's superorganism hypothesis was in part a result of the shift from the study of whole systems to individual populations that began in the 1940s (Simberloff 1980).

Debate over the relative importance of species interactions and the existence of emergent properties of communities is ongoing among contemporary ecologists and ecotoxicologists. At the very least, the concept that communities are organized into functional units has a "long and troubled history" (Wilson 1997). Strong et al. (1984) note that ecology has historically been dominated by the neo-Malthusian perspective that interspecific competition is the major force structuring communities. Some ecologists take the extreme position that communities lack any predictable patterns and have questioned the validity of community ecology as a legitimate science (Schrader-Frechette and McCoy 1993).

Although most contemporary ecologists readily dismiss Clements's superorganism concept (but see papers on the Gaia hypothesis (Lovelock 1979)), there is much support for the hypothesis that communities are more than the sum of their component populations. Predictable patterns in species associations exist, and these patterns are often determined by species interactions. Experimental research on multilevel selection theory (Goodnight 1990a,b) suggests that communities are shaped by natural selection and possess functional organization (Wilson 1997).

Indeed, some researchers have noted a resurgence of the holistic paradigm in ecology and argue that Clements's superorganism concept provided a foundation for the study of systems ecology (Simberloff 1980). There is also evidence that species interactions can play a major role in structuring communities (Diamond 1978, Schoener 1974, 1983). However, this evidence emerged slowly because of the historical focus on descriptive approaches and the late development of experimental procedures in community ecology. The credibility of hypotheses concerning the relative importance of species interactions was further undermined when researchers invoked untestable explanations, such as the "ghost of competition past" (Connell 1980), to explain the negative results of competition experiments. Because conducting meaningful experiments on communities is challenging, most community ecologists have relied on anecdotal accounts, observations, and mathematical formulations to argue for the importance of species interactions. As described below, the transition of community ecology from a descriptive to an experimental science has greatly increased the credibility of this discipline. We argue that a similar transition is slowly occurring in community ecotoxicology.

The debate between proponents of holistic and reductionist approaches has been especially acrimonious in the field of community ecotoxicology. Because of the need to make definitive regulatory decisions, often without an ecological perspective, there has been a historical focus on reductionist approaches in toxicology (Cairns 1983, 1986). The implicit but often untested assumption that results of single-species laboratory toxicity tests can predict the effects of contamination on more complex systems in nature is a classic example of pragmatic reductionism in ecotoxicology. Many field assessments of natural systems, especially in terrestrial habitats, also emphasize population-level analyses and dismiss community-level approaches. However, the focus of ecotoxicological research on populations can lead to misleading conclusions regarding the broader impacts of environmental pollutants on higher levels of biological organizations. There is an inherent bias that results from the emphasis on economically important or charismatic species, which often receive special attention under the natural resource damage assessment laws of the United States. For example, some ecologists argue that failure to account for responses of all taxa, including those resilient to oil, provided an incomplete picture of the responses of seabird communities following the 1989 *Exxon Valdez* oil spill in Prince William Sound, Alaska (Wiens et al. 1996).

Because of the opportunity to evaluate the responses of numerous species simultaneously, we suggest that community ecotoxicology can provide a much broader context for the assessment of environmental contamination than the study of individual species. Owing to differences in life history characteristics and tolerance, different species in a community respond differentially to contaminants and other stressors. Thus, the composition of communities at different locations or at two points in time provides useful information about these environmental conditions. Communities also provide the "ecological and evolutionary context for populations" (Angermeier and Winston 1999). Variation in responses among taxa due to differences in physiology, feeding habits, habitat use, and reproductive characteristics can provide insight into the direct mechanisms of toxic effects on species.

As illustrated by the quotes at the beginning of this chapter, there is an opinion that results of community and ecosystem studies are complex, highly variable, and difficult to interpret. For example, Luoma and Carter (1991) state that "at no level of biological organization is it more difficult to adequately understand the dose of a metal to the system than at the level of community." The primary difficulty in studying higher levels of biological organization is the need to understand both direct and indirect effects of contaminants. Direct effects of contaminants may result in reduction or elimination of local populations and are generally easier to interpret than indirect effects. In contrast, indirect effects of contaminants, such as increased susceptibility to predation or the elimination of an important prey species in the diet of a predator, are much more difficult to detect and interpret.

These indirect effects generally occur when a contaminant exerts a disproportionate effect on one species, thereby altering its interactions with another species. We suggest that a better appreciation for the importance of indirect effects is fundamental to predicting how communities respond to anthropogenic disturbances.

20.2.2 Trophic Interactions in Community Ecology and Ecotoxicology

The study of trophic interactions in communities represents the second major development in the history of ecology that has greatly influenced ecotoxicology. Since Lindeman's thermodynamic formalization of Elton's trophic pyramids in the mid-1900s (Lindeman 1942), ecologists have used feeding relationships to characterize the structure of communities. This development triggered a long-standing controversy among ecologists who argued that, systems with high diversity and trophic complexity are more stable than less complex systems. Hutchinson's "Homage to Santa Rosalia" (1959) and the classic paper published by Hairston et al. (1960) stimulated a flurry of research attempting to relate population abundance and community structure to trophic complexity.

Information on feeding habits and trophic relationships is of fundamental importance for predicting the transfer of contaminants through communities. It is well established that trophic position greatly influences levels of some contaminants in organisms. The mechanistic explanation for elevated concentrations of organochlorines and other persistent contaminants observed in top predators represented one of the first attempts to integrate basic ecological principles (e.g., trophic ecology) into toxicology. In aquatic ecosystems, an understanding of the relative importance of dietary and aqueous exposure to contaminants is required to predict bioaccumulation (Dallinger et al. 1987). Recent studies have shown that, in addition to trophic position, the number of trophic levels determines the levels of certain contaminants in top predators.

20.2.3 Importance of Experiments in Community Ecology and Ecotoxicology

The final, and perhaps most significant, development in basic ecology that influenced the field of community ecotoxicology was the recognition that experimental studies are necessary to demonstrate cause-and-effect relationships. The historical focus in ecology was almost entirely on descriptive studies. Early ecologists characterized natural history and habitat requirements, described patterns of plant and animal associations, and relied exclusively on observational studies to determine which biotic and abiotic factors limited the distribution and abundance of organisms. Reliance on these descriptive approaches is at least partially responsible for the relatively slow progress in ecology from the early 1920s until the 1960s. Ecology emerged as a rigorous science only after ecologists began to employ manipulative experiments to test explicit hypotheses. In particular, the pioneering experiments by researchers assessing species interactions in the marine rocky intertidal zone (Connell 1961, Paine 1966) revolutionized the way a generation of community ecologists investigated nature. The profusion of field experiments that followed these classic studies has greatly increased our understanding of the importance of species interactions and our appreciation of the complexity of ecological systems.

The field of community ecotoxicology has experienced a similar transformation from purely observational approaches to the use of experimental procedures in the past 20 years. Before 1980, most research in community ecotoxicology was limited to descriptive studies that related species richness, diversity, and community composition to measured levels of chemical stressors. Comparative studies of reference and polluted sites can provide support for the hypothesis that a chemical stressor is responsible for observed differences in community composition. Descriptive studies contribute significantly to our understanding of how communities respond to specific chemicals and remain the primary focus of state and federal monitoring programs in the United States. However, as in basic community ecology, the major shortcoming of descriptive approaches is the inability to show cause-and-effect relationships between stressors and community responses. Manipulative approaches, such as mesocosms, ecosystem experiments, and natural experiments, have played an increasingly important role in ecotoxicological research over the past 20 years.

20.3 ARE COMMUNITIES MORE THAN THE SUM OF INDIVIDUAL POPULATIONS?

Although general ecology textbooks devote significant coverage to the topic of communities, the focus in most ecotoxicological investigations has been on individuals and populations. Moriarty (1988) questioned the need to study effects of contaminants on communities and concluded that, for ecotoxicology, populations are the most appropriate level of organization. Interestingly, Suter's (1993) excellent treatment of ecological risk assessment includes separate chapters on organism, population, and ecosystem-level effects, but there is no corresponding chapter describing community-level responses. Dickson (1995) suggests that the historical emphasis on individuals and populations in ecotoxicological research is unlikely to change because water resource managers and the general public do not appreciate the significance of responses at higher levels of organization. It is much easier to argue for the protection of an economically important or charismatic species than for the need to maintain ecosystem functional characteristics such as detritus processing or nutrient cycling. However, the study of communities will likely uncover patterns not readily observable through population analyses. We agree with the statement of Sir Robert May (1973) that "if we concentrate on any one particular species our impression will be one of flux and hazard, but if we concentrate on total community properties (such as biomass in a given trophic level) our impression will be one of pattern and steadiness."

20.3.1 THE NEED TO UNDERSTAND INDIRECT EFFECTS OF CONTAMINANTS

If communities were abstractions and only represented a tidy way to organize populations into manageable units, then predicting the effects of contaminants at higher levels of organization would be greatly simplified. For example, suppose we knew the direct toxicological effects (e.g., LC50 or EC50 values) of a particular chemical on all species in a community. If species interactions and indirect effects were unimportant, predicting responses of communities would simply be a matter of bookkeeping. With a matrix showing the species names, abundances, and LC50 values for all species we could predict the community-level effects at a particular concentration. We know, however, that in many situations species interactions are important and indirect effects complicate ecological assessments. Just as laboratory toxicologists recognize the influences of certain abiotic factors (e.g., temperature, water hardness, dissolved organic carbon) on chemical effects, community ecotoxicologists understand that responses of individual populations cannot be measured in isolation and that understanding indirect effects is of critical importance. In some instances, these indirect effects of contaminants may be equally important or even greater than direct effects (Fleeger et al. 2003).

One of the more revealing examples demonstrating the importance of indirect effects occurred when the World Health Organization (WHO), in an attempt to eliminate malaria-bearing mosquitoes, sprayed the pesticides DDT and dieldrin on numerous villages in Borneo. In addition to controlling mosquito populations, the pesticides contaminated cockroaches, which formed the base of an unnatural food chain in the villages. The cockroaches were consumed by geckos, which were ultimately ingested by cats. Biomagnification of DDT and dieldrin by cats resulted in significant mortality and a subsequent increase in rat populations. The somewhat artificial food chain was eventually restored by parachuting large numbers of cats into the villages, a program referred to as "Operation Cat Drop" by the WHO and Royal Air Force.

Numerous examples of indirect effects of contaminants on populations have been reported in the literature (see the comprehensive review by Fleeger et al. 2003); however, separating direct and indirect effects is difficult and often requires field experimentation. Ecosystem manipulation experiments conducted by Schindler (1987) demonstrated that reductions in lake trout abundance resulted from loss of forage fish and not from direct toxicological effects of lower pH. Similar whole-lake manipulations have demonstrated the importance of predator–prey interactions in regulating aquatic communities (Box 20.2). Indirect effects have long been recognized as important causes of reduced abundance of bird populations exposed to pesticides (Powell 1984). Pesticide spray programs are designed to eliminate large numbers of insects, and it should not be surprising that reductions in insect prey may negatively affect bird populations. In addition, spray programs often coincide with critical periods of nestling growth and development because many species have adapted to take advantage of large numbers of prey during periods of insect outbreaks. Reduced prey abundance has been associated with reduced nestling growth and increased risk of predation, presumably because parents are spending more time away from nests searching for prey.

Box 20.2 Trophic Cascades in Aquatic and Terrestrial Communities

The most convincing examples demonstrating tight linkages among species and the relative importance of trophic interactions are from a series of studies investigating trophic cascades in aquatic and terrestrial communities (Chapter 27). Whole-lake manipulations conducted by Carpenter and Kitchell (1993) have investigated the relative importance of nutrients and top predators on lake productivity. Much of the limnological research conducted in the 1970s focused on the role of nutrients, especially phosphorus, in controlling productivity of lakes. According to the "bottom-up" hypothesis, discharge of nutrients increased phytoplankton biomass, providing greater resources for higher trophic levels. Although there was anecdotal support for the bottom-up hypothesis, it could not explain all of the variation in productivity of the world's lakes. More recent studies have tested the hypothesis that while nutrients determine the potential range of productivity, predation regulated actual productivity measured in lakes. In a simple three-level food chain, planktivorous fish reduce abundance of algal-grazing zooplankton and allow phytoplankton populations to expand (Figure 20.1). On the basis of the trophic cascade hypothesis, it is expected that algal biomass and primary productivity are generally greater in systems with three trophic levels. In a four-level food chain typical of many lakes, piscivorus fish (e.g., lake trout, bass) control abundance of planktivorous fish, thereby allowing densities of algal-grazing zooplankton to increase. Thus, increased abundance of top predators releases grazing zooplankton from predation and ultimately limits primary productivity. This "top-down" hypothesis has been tested in a number of biomanipulation experiments where top predators are added or planktivorous fishes are removed (Carpenter and Kitchell 1993). These manipulations have been employed as management tools to control moderate eutrophication in lentic systems (see Box 27.1 in Chapter 27).

Analogous cascading trophic relationships between producers and consumers have been observed in terrestrial communities with three trophic levels. Long-term investigations on Isle Royale National Park (Michigan, USA) have shown that density of moose populations is largely determined by wolf predation. The studies also provided strong evidence for top-down control by demonstrating close linkages between balsam fir, the winter forage of moose, and moose density (McLaren and Peterson 1994).

These examples show that indirect effects and species interactions can play a major role in regulating communities. Because of the importance of species interactions and the difficulty in predicting these indirect effects, community responses to chemical stressors cannot

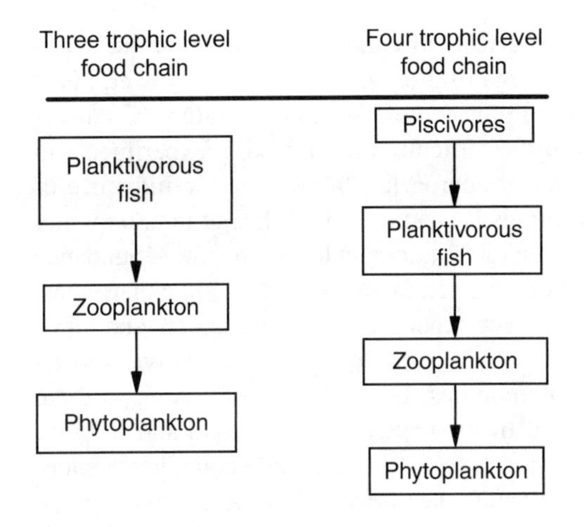

FIGURE 20.1 Trophic cascades in three- and four-level food chains. The relative size of each compartment reflects the biomass of the trophic level. In the three trophic level food chain planktivorous fish control zooplankton, thus allowing phytoplankton to increase in abundance. The addition of piscivores in a four trophic level food chain reduces abundance of planktivorous fish and allows zooplankton to regulate phytoplankton populations.

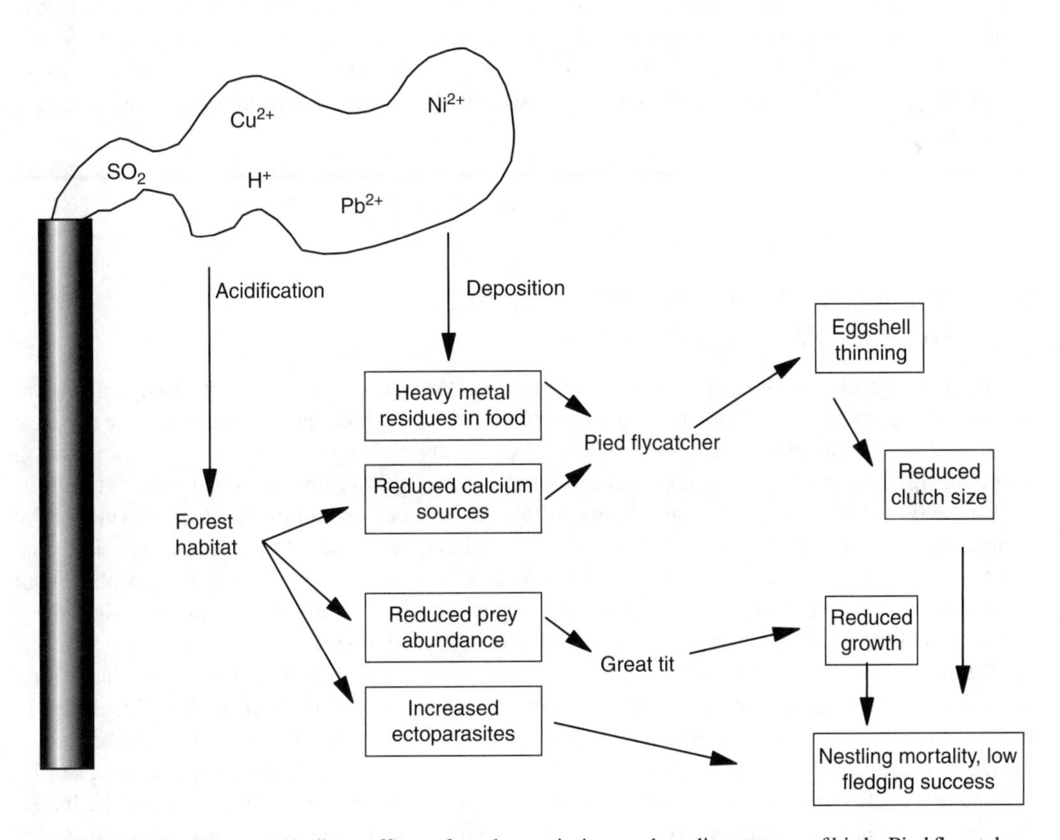

FIGURE 20.2 Direct and indirect effects of smelter emissions on breeding success of birds. Pied flycatchers are directly affected by exposure to heavy metals in food, resulting in eggshell thinning and reduced clutch size. In contrast, Great Tits suffer reduced growth due to lower food abundance. (Modified from Figure 12 in Eeva et al. (1997).)

be understood by merely studying individual populations in isolation. To measure effects of contaminants on these systems, it will be necessary to account for interactions within and among trophic levels. For example, without information on the role of trophic cascades and top-down control, it would be impossible to predict that the loss of top predators from a chemical stressor could actually increase primary productivity.

In summary, understanding potential indirect effects of contaminants is often cited as a primary justification for testing at higher levels of organization (Cairns 1983). However, indirect effects have received relatively little attention in the ecotoxicological literature (Clements 1999). This paucity of information results from the difficulty of conducting experiments to isolate direct and indirect effects. Developing novel approaches to estimate the influence of contaminants on species interactions and associations is a significant challenge in community ecotoxicology. Although experimental manipulation of contaminant levels and prey abundance is the best way to distinguish direct and indirect effects, conducting planned field experiments in ecotoxicology is difficult. In the absence of direct experimentation, an understanding of natural history requirements of individual species can be used to assess the relative importance of direct and indirect effects on natural communities. Eeva et al. (1997) compared the breeding success of the pied flycatcher and great tit to smelter emissions (Cu and SO_4) in southwestern Finland. Because of known differences in feeding habits and habitat preferences between species, these researchers were able to separate the direct toxicological effects of Cu exposure from the indirect effects of reduced prey abundance (Figure 20.2). In the absence of experimental evidence, comparative studies among species can provide a reasonable way to estimate the relative influence of direct and indirect effects. In Chapter 21, we will describe experimental and descriptive approaches for assessing the influence of contaminants on species interactions.

20.4 COMMUNITIES WITHIN THE HIERARCHY OF BIOLOGICAL ORGANIZATION

Communities represent an intermediate level of complexity in the hierarchy of biological organization. They are distinct from populations and ecosystems, but have close linkages to these lower and higher levels of organization (Figure 20.3). The changes in community composition observed at polluted sites are a result of differences in sensitivity among populations as well as the interactions between populations. Many of the measures of contaminant effects developed by community ecotoxicologists exploit these known differences in sensitivity among species. The most frequent observation at polluted sites is the loss of sensitive species and their replacement by tolerant species. The presence or absence of known pollution-tolerant and pollution-sensitive species enables community ecotoxicologists to estimate the relative degree of contamination in the field. For example, since the early 1900s, community-level measures of contamination have been employed to assess organic enrichment. The Saprobien system of classification, first developed in Europe, was used to characterize streams as either clean or polluted based on the abundance of sensitive and tolerant species (Kolkwitz and Marsson 1909). Contemporary approaches used by community ecotoxicologists to quantify pollution are generally more sophisticated and include a diverse assortment of biotic, comparative, and diversity indices (Johnson et al. 1993). However, most measures are still based on the simple assumptions that the absence of pollution-sensitive taxa and the presence of pollution-tolerant taxa are indicative of degradation. Pollution indices, such as Hilsenhoff's (1987) biotic index, integrate estimates of species-specific sensitivity to pollutants with measures of relative abundance to assess the levels of degradation in aquatic ecosystems. The application of these approaches for assessing contaminant effects in the field is described in Chapter 22.

There are also close connections between community-level properties and higher levels of biological organization. Recent studies have shown that structural characteristics of communities (e.g., species diversity, community composition) influence the functioning of ecosystems (Chapin et al. 1998). In a series of field experiments, Tilman and coworkers have shown that species diversity

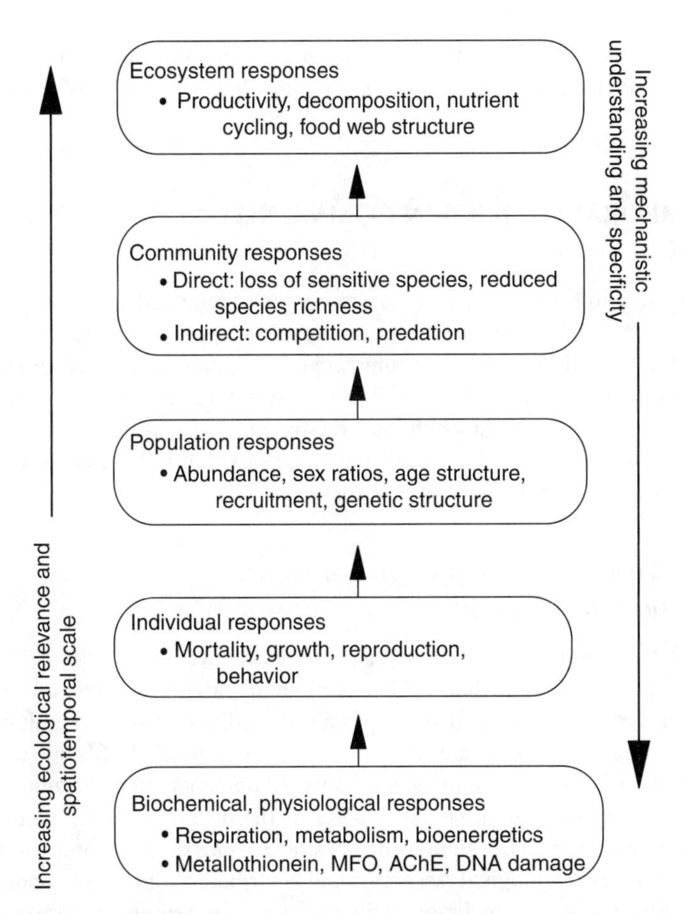

FIGURE 20.3 Effects of contaminants across levels of biological organization. Responses at lower levels of biological organization (biochemical, physiological) are generally more specific and are better understood in terms of mechanisms. Consequently, cause-and-effect relationships are more obvious with subindividual responses. Responses at higher levels of biological organization (communities and ecosystems) occur at broader spatiotemporal scales and have greater ecological relevance but often lack mechanistic explanations. (Modified from Figure 1 in Clements (2000).)

significantly influences plant productivity and nitrogen dynamics, with more diverse plots having greater productivity (Tilman et al. 1997). Similar results from model ecosystem experiments showed that depauperate communities had lower productivity and reduced CO_2 uptake compared to species-rich communities (Naeem et al. 1994). In addition to demonstrating a linkage between community composition and ecosystem function, these results have important implications for the study of anthropogenic disturbance, especially contamination. These experiments suggest that loss of functionally important species due to chemical stressors is likely to impact ecosystem processes.

The study of community-level processes also provides insight into the potential effects of natural and anthropogenic disturbance on community stability. The enduring controversy in basic ecology concerning the relationship between species diversity and stability has been tested in grassland plots subjected to severe drought (Tilman and Downing 1994). As predicted, diverse communities were better able to withstand disturbance than species-poor communities. If the relationship between resistance and species diversity illustrated in these experiments is applicable to anthropogenic disturbances, we would expect that naturally depauperate communities would be especially susceptible to chemical stressors. Furthermore, we speculate that communities that have lost species because of exposure to chemical stressors would be more susceptible to other anthropogenic disturbance (e.g., global climate change). The relationship between community structure and functional characteristics of ecosystems and its relevance to ecotoxicology is discussed in Chapter 33.

20.5 CONTEMPORARY TOPICS IN COMMUNITY ECOTOXICOLOGY

The contemporary research topics in community ecotoxicology fall into five general categories: (1) better understanding of the basic ecological factors that regulate communities, (2) development and application of improved biomonitoring approaches, (3) integration of experimental approaches into community ecotoxicology, (4) influence of trophic structure on food chain transport of contaminants, and (5) influence of global atmospheric stressors on community responses to contaminants. Each of these topics will be covered in detail in the chapters that follow. Here, we briefly describe some of the key issues that will be presented in these chapters.

20.5.1 THE NEED FOR AN IMPROVED UNDERSTANDING OF BASIC COMMUNITY ECOLOGY

For community ecotoxicology to thrive as a discipline, researchers must acquire a better understanding of the biotic and abiotic factors that regulate community structure. As noted above, the absence of a sensitive species at a contaminated site is often assumed to be a direct result of contamination. Alternatively, the absence of this species may be a result of a myriad of biotic and abiotic factors unrelated to the stressor. In order to understand how contaminants affect community structure, it is critical that ecotoxicologists develop better tools for distinguishing between natural and anthropogenic variation in communities. The relationship between species diversity and ecosystem function is a good example where basic ecological research could contribute to our understanding of community responses to contaminants. An appreciation of the role that community structure plays in controlling ecosystem function will improve our ability to predict the consequences of reduced biodiversity on higher levels of organization. Similarly, current interest in the relationship between species diversity and stability has significant implications for community ecotoxicology. An understanding of the quantitative relationship between species diversity and natural disturbance may allow community ecotoxicologists to predict which communities will be most sensitive to anthropogenic disturbance. We will examine the influence of biotic and abiotic factors on species associations and interactions within the context of community ecotoxicology in Chapter 21.

Continued research on successional changes in plant and animal communities following natural disturbance will allow ecotoxicologists to predict trajectories in systems recovering from anthropogenic disturbance. Our failure to establish well-defined goals for measuring the success of contaminant remediation impedes our ability to fully characterize ecological recovery. The rapidly emerging field of restoration ecology relies extensively on concepts developed by early plant ecologists studying community succession. Finally, basic community ecologists continue to address issues of spatiotemporal scale in their investigations. Because the spatial and temporal scales of anthropogenic stressors do not necessarily coincide with the endpoints being measured in an ecological assessment (Suter 1993), community ecotoxicologists must also consider scale when investigating effects of contaminants.

20.5.2 DEVELOPMENT AND APPLICATION OF IMPROVED BIOMONITORING TECHNIQUES

The second general area of research in contemporary community ecotoxicology is the development and application of improved biomonitoring approaches. One major goal of these improvements is to streamline biological assessments and reduce costs so that biological monitoring programs can be more efficiently implemented by state and federal regulatory agencies. The cost of quantitative assessments of community structure, especially those that require species-level identification of taxonomically difficult groups, is considered a major impediment to these programs. Research questions related to the appropriate number of samples and the necessary level of taxonomic resolution required to characterize disturbed sites are receiving considerable attention. Chapter 22 will focus on descriptive approaches in community ecotoxicology and highlight recent attempts to streamline biological monitoring programs. Other improvements in biological monitoring involve application of more sophisticated statistical procedures, especially multivariate techniques and the development of regional multimetric indices. Because responses of communities to contaminants and environmental factors are inherently multivariate, the statistical approaches employed to analyze these data should reflect this complexity. Although the relative merits of multivariate and multimetric approaches have been debated in the literature (Fausch et al. 1990, Fore et al. 1996), the recent attempt to integrate these approaches is a promising development in biomonitoring research. We will describe the application of multimetric and multivariate approaches in community ecotoxicology in Chapter 24. The traditional focus on benthic community metrics based exclusively on taxonomic groupings is currently being reevaluated by some researchers. The use of functional characteristics such as species traits to investigate relationships between environmental variables and species distributions has received considerable attention in stream benthic ecology (Poff et al. 2006). Interest in the application of species traits in biological monitoring is at least partially driven by the emergence of large-scale, spatially extensive sampling programs in the United States and Europe. Perhaps the most significant application of this approach in community ecotoxicology is the identification of unique combinations of species traits that respond to specific environmental stressors.

Finally, biomonitoring studies designed explicitly to distinguish the effects of natural and anthropogenic variation are a significant improvement in community-level assessments. Natural variation in community composition is a serious problem in most biomonitoring studies and often confounds interpretation of field results. Situations where natural variation in abiotic characteristics can be quantified and used as covariates provide the best opportunity to assess the relative importance of natural and anthropogenic effects. Similarly, study designs that allow community ecotoxicologists to separate effects of multiple and potentially interacting stressors are also necessary. The practical but simplistic emphasis of toxicology on single stressors provides a very unrealistic perspective of nature. Natural communities are often exposed to several anthropogenic stressors simultaneously, and identifying the relative importance of each stressor is necessary to understand observed community responses.

20.5.3 Application of Contemporary Food Web Theory to Ecotoxicology

The third area of significant research in community ecotoxicology is the integration of contemporary food web theory into fate and transport models to predict the movement of contaminants through communities. It is well established that levels of contaminants in predators are influenced by physiological (e.g., lipid content or metabolism) and life history (e.g., age, sex or feeding habits) features. However, recent studies have shown that ecological characteristics, such as the length and type of food chain, also explain a significant amount of variation. Research on the influence of predator–prey interactions and trophic structure on energy flow will make significant contributions to our understanding of contaminant transport through food chains. The application of stable isotope analyses to characterize feeding relationships and to quantify contaminant transport among trophic levels is a significant development in community ecotoxicology (Kiriluk et al. 1995). Although most research describing the relationship between food chain structure and contaminant levels has been conducted in aquatic ecosystems (Rasmussen et al. 1990), it is likely that similar patterns will be observed in terrestrial habitats. Studies showing trophic linkages between ecosystem types (e.g., terrestrial and aquatic) demonstrate the importance of energy input and potential for contaminant movement between communities. For example, field experiments conducted by Nakano et al. (1999) illustrated that limiting input of terrestrial arthropods in small headwater streams had dramatic indirect effects on a benthic community food web. We will examine the direct application of food web theory and new methodological approaches to the study of contaminant transport in Chapter 27 and 34.

20.5.4 The Need for Improved Experimental Approaches

The fourth general area of contemporary research in community ecotoxicology is the application of experimental procedures, both laboratory and field, to assess effects of contaminants. Motivated by the realization that observational studies alone cannot show causal relationships and the need for better mechanistic understanding of contaminant effects, ecotoxicologists are beginning to employ more complex experimental procedures in community-level assessments. Experimental approaches that include microcosms, mesocosms, and field manipulations have been used to validate traditional single species toxicity tests (Pontasch et al. 1989) and to support results of descriptive studies (Clements et al. 1988). Recently, more complex factorial designs have been employed to assess interactions of multiple stressors (Genter 1995) and quantify the influence of trophic structure (Pratt and Barreiro 1998), location (Kiffney and Clements 1996), and previous exposure to stress on community-level responses to contaminants (Courtney and Clements 2000). The most serious limitation of microcosm and mesocosm experiments is the loss of ecological realism that occurs when studies are conducted at smaller spatial scales. Some investigators are especially critical of small-scale experiments and have suggested that microcosm studies have little relevance in ecology (Carpenter 1996). Understanding the influence of spatial and temporal scale on responses to contaminants is critical for predicting how communities will respond in natural systems. In Chapter 23, we will highlight the transition of community ecotoxicology from a descriptive to an experimental science and discuss the important trade-offs between spatial scale, replication, and ecological realism.

20.5.5 Influence of Global Atmospheric Stressors on Community Responses to Contaminants

The final area of research in contemporary community ecotoxicology relates to the interactions between chemical and global atmospheric stressors. Although responses to global atmospheric stressors are generally not considered in most ecotoxicological investigations, increased CO_2, ultraviolet radiation (UVR), and acidification are major environmental issues that will significantly affect natural communities. In addition to their well-documented direct effects, these stressors will likely influence the way communities respond to contaminants. In fact, some researchers speculate that indirect effects of global warming, acidification, and UVR on communities will be greater than direct effects (Field et al. 1992). Increased temperatures resulting from global climate change will likely influence contaminant bioavailability, uptake, and depuration in complex and often unpredictable ways. The photoactivation of certain contaminants after exposure to UV-B radiation, most notably the polycyclic aromatic hydrocarbons, is well documented in the toxicological literature (Oris and Giesy 1986). Finally, decreases in pH of soils and in aquatic ecosystems as a result of acid deposition will increase concentrations and bioavailability of certain metals.

In addition to the direct and indirect influence of global atmospheric stressors on community responses to contaminants, interactions among global warming, UV-B radiation, and acidification are also possible. For example, acidic deposition and climate-induced changes in hydrologic characteristics of watersheds will likely alter the quality and quantity of dissolved organic material (DOM) in aquatic ecosystems. Because DOM plays an important role in reducing light penetration and controlling contaminant bioavailability, these changes will influence exposure of aquatic communities to

UV-B radiation and chemical stressors. The effects of global atmospheric stressors on communities and the interactions among these stressors are discussed in Chapter 26.

20.6 SUMMARY

Describing patterns in the distribution and abundance of organisms and understanding the biotic and abiotic factors that determine these patterns are fundamental goals of community ecology. Community ecotoxicologists focus on one particular set of these abiotic factors: contaminants and other anthropogenic stressors; however, they are also concerned with separating effects of anthropogenic stressors from natural variability. The transformation in community ecotoxicology from purely observational approaches to the use of laboratory and field experimental techniques to quantify natural variability relative to effects of contaminants is considered a major development in this field.

Communities are an intermediate level of complexity in the hierarchy of biological organization, with intimate linkages to lower and higher levels; however, communities are distinct from populations and ecosystems. Community ecotoxicologists recognize that life history characteristics and population dynamics influence responses to contaminants, but the focus is generally on assemblages of interacting species. Similarly, while the close connection between structural characteristics and ecosystem processes is well established, community ecotoxicologists are primarily interested in how contaminants and other stressors affect patterns in species diversity and community composition.

Considerable research effort in basic community ecology has been devoted to understanding the relative importance of species interactions. This topic, which is at the heart of the debate between proponents of holistic and reductionist approaches, deserves a similar level of attention in community ecotoxicology. Assuming that communities are not simply a random collection of species that occupy the same habitat because of similar environmental requirements, an understanding of how contaminants affect species interactions is fundamental. For example, integration of contemporary food web theory into fate and transport models will significantly improve our ability to predict the movement of contaminants through communities. It is surprising that effects of contaminants on species interactions and trophic structure have received relatively little attention in the toxicological literature, particularly since the potential for indirect effects is a primary justification for testing at higher levels of organization. Development and application of more sophisticated experimental techniques will be necessary to quantify indirect effects of contaminants on communities.

20.6.1 SUMMARY OF FOUNDATION CONCEPTS AND PARADIGMS

- Communities represent an intermediate level of complexity in the hierarchy of biological organization; they are distinct from populations and ecosystems, but have close linkages to these lower and higher levels of organization.
- Community ecotoxicology is the study of the effects of chemicals on species abundance, diversity, and interactions.
- The study of communities transcends simple descriptions of demographic and life history characteristics of individual populations.
- Debate over the relative importance of species interactions and the existence of emergent properties of communities is ongoing among contemporary ecologists and ecotoxicologists.
- The boundaries of communities have been defined based on spatial overlap of populations, trophic structure, strength of species interactions, and taxonomic relationships.
- The debate between proponents of holistic and reductionist approaches has been especially acrimonious in the field of community ecotoxicology.

- The primary difficulty in studying higher levels of biological organization is the necessity to understand both direct and indirect effects of contaminants.
- One of the most significant developments in basic ecology that influenced the field of community ecotoxicology was the recognition that experimental studies are necessary to demonstrate cause-and-effect relationships.
- The field of community ecotoxicology has experienced an important transformation from purely observational approaches to the use of experimental procedures in the past 20 years.
- Developing novel approaches to estimate the influence of contaminants on species interactions and associations is a significant challenge in community ecotoxicology.
- In order to understand how contaminants affect community structure, it is critical that ecotoxicologists develop better tools for distinguishing between natural and anthropogenic variation.
- For community ecotoxicology to thrive as a discipline, researchers must acquire a better understanding of the biotic and abiotic factors that regulate community structure.
- Because responses of communities to contaminants and environmental factors are inherently multivariate, the statistical approaches employed to analyze these data should reflect this complexity.
- Although responses to global atmospheric stressors are generally not considered in most ecotoxicological investigations, increased CO_2, UVR, and acidification are major environmental issues that will significantly affect natural communities.

REFERENCES

Angermeier, P.L. and Winston, M.R., Characterizing fish community diversity across Virginia landscapes: Prerequisite for conservation, *Ecol. Appl.*, 9, 335–349, 1999.

Begon, M., Harper, J.L., and Townsend, C.R., *Ecology: Individuals, Populations and Communities*, Blackwell Scientific Publications, Cambridge, MA, 1990.

Cairns, J., Jr., Are single species toxicity tests alone adequate for estimating environmental hazard? *Hydrobiologia*, 100, 47–57, 1983.

Cairns, J., Jr., The myth of the most sensitive species, *Bioscience*, 36, 670–672, 1986.

Cairns, J., Jr. and Niederlehner, B.R., *Ecological Toxicity Testing*, Lewis Publishers, Boca Raton, FL, 1995.

Carpenter, S.R., Microcosm experiments have limited relevance for community and ecosystem ecology, *Ecology*, 77, 677–680, 1996.

Carpenter, S.R. and Kitchell, J.F., *The Trophic Cascade in Lakes*, Cambridge University Press, New York, 1993.

Chapin, F.S., Sala, O.E., Burke, I.C., Grime, J.P., Hooper, D.U., Lauenroth, W.K., Lombard, A., et al., Ecosystem consequences of changing biodiversity—Experimental evidence and a research agenda for the future, *Bioscience*, 48, 45–52, 1998.

Clements, F.E., Nature and structure of the climax, *J. Ecol.*, 24, 252–284, 1936.

Clements, W.H., Metal tolerance and predator–prey interactions in benthic macroinvertebrate stream communities, *Ecol. Appl.*, 9, 1073–1084, 1999.

Clements, W.H., Integrating effects of contaminants across levels of biological organization: An overview, *J. Aquat. Eco. Stress Recov.* 7, 113–116, 2000.

Clements, W.H., Cherry, D.S., and Cairns, J., Jr., The impact of heavy metals on macroinvertebrate communities: A comparison of observational and experimental results, *Can. J. Fish. Aquat. Sci.*, 45, 2017–2025, 1988.

Connell, J.H., The influence of interspecific competition and other factors on the distribution of the barnacle *Chthamalus stellatus*, *Ecology*, 42, 710–723, 1961.

Connell, J.H., Diversity and the coevolution of competitors, or the ghost of competition past, *Oikos*, 35, 131–138, 1980.

Courtney, L.A. and Clements, W.H., Sensitivity to acidic pH in benthic invertebrate assemblages with different histories of exposure to metals, *J. N. Am. Benthol. Soc.*, 19, 112–127, 2000.

Dallinger, R., Prosi, F., Segner, H., and Back, H., Contaminated food and uptake of heavy metals by fish: A review and a proposal for further research, *Oecologia*, 73, 91–98, 1987.
Diamond, J.M., Niche shifts and the rediscovery of competition: Why did field biologists so long overlook the widespread evidence for interspecific competition that had already impressed Darwin? *Am. Sci.*, 66, 322–331, 1978.
Dickson, K.L., Progress in toxicity testing—An academic's viewpoint, In *Ecological Toxicity Testing: Scale, Complexity, and Relevance*, Cairns, J., Jr., and Niederlehner, B.R. (eds.), CRC Press, Inc., Boca Raton, FL, 1995, pp. 209–216.
Eeva, T., Lehikoinen, E., and Pohjalainen, T., Pollution-related variation in food supply and breeding success in two hole-nesting passerines, *Ecology*, 78, 1120–1131, 1997.
Elton, C., *Animal Ecology*, Macmillan, New York, 1927.
Fausch, K.D., Lyons, J., Karr, J.R., and Angermeier, P.L., Fish communities as indicators of environmental degradation, *Am. Fish. Soc. Symp.*, 8, 123–144, 1990.
Fauth, J.E., Bernardo, J., Camara, M., Resetarits, W.J., VanBuskirk, J., and McCollum, S.A., Simplifying the jargon of community ecology: A conceptual approach, *Am. Nat.*, 147, 282–286, 1996.
Field, C.B., Chapin, I.F.S., Matson, P.A., and Mooney, H.A., Responses of terrestrial ecosystems to the changing atmosphere: A resource-based approach, *Ann. Rev. Ecol. Syst.*, 23, 201–235, 1992.
Fleeger, J.W., Carman, K.R., and Nisbet, R.M., Indirect effects of contaminants in aquatic ecosystems, *Sci. Total. Environ.*, 317, 207–233, 2003.
Fore, L.S., Karr, J.R., and Wisseman, R.W., Assessing invertebrate responses to human activities: Evaluating alternative approaches, *J. N. Am. Benthol. Soc.*, 15, 212–231, 1996.
Genter, R.B., Benthic algal populations respond to aluminum, acid, and aluminum–acid mixtures in artificial streams, *Hydrobiologia*, 306, 7–19, 1995.
Gleason, H.A., The individualistic concept of the plant association, *Bul. Torrey Bot. Club*, 53, 1–20, 1926.
Goodnight, C.J., Experimental studies of community evolution. I. The response to selection at the community level, *Evolution*, 44, 1614–1624, 1990a.
Goodnight, C.J., Experimental studies of community evolution. II. The ecological basis of the response to community selection, *Evolution*, 44, 1625–1636, 1990b.
Hairston, N.G., Sr., Smith, F.E., and Slobodkin, L.B., Community structure, population control, and interspecific competition, *Am. Nat.*, 94, 421–425, 1960.
Hilsenhoff, W.L., An improved biotic index of organic pollution, *Great Lakes Entomol.*, 20, 31–39, 1987.
Hutchinson, G.E., Homage to Santa Rosalia or why are there so many kinds of animals? *Am. Nat.*, 93, 145–159, 1959.
Johnson, R.K., Wiederholm, T., and Rosenberg, D.M., Freshwater biomonitoring using individual organisms, populations, and species assemblages of benthic macroinvertebrates, In *Freshwater Biomonitoring and Benthic Macroinvertebrates*, Rosenberg, D.M. and Resh, V.H. (eds.), Chapman & Hall, New York, 1993, pp. 40–158.
Kiffney, P.M. and Clements, W.H., Effects of metals on stream macroinvertebrate assemblages from different altitudes, *Ecol. Appl.*, 6, 472–481, 1996.
Kiriluk, R.M., Servos, M.R., Whittle, D.M., Cabana, G., and Rasmussen, J.B., Using ratios of stable nitrogen and carbon isotopes to characterize the biomagnification of DDE, mirex, and PCB in a Lake Ontario pelagic food web, *Can. J. Fish. Aquat. Sci.*, 52, 2660–2674, 1995.
Kolkwitz, R. and Marsson, M., Okologie der tierischen Saprobien, *Int. Rev. Gesaten.*, 2, 126–152, 1909.
Lindeman, R.L., The trophic-dynamic aspect of ecology, *Ecology*, 23, 399–418, 1942.
Lovelock, J.E., *Gaia: A New Look at Life on Earth*, Oxford University Press, Oxford, UK, 1979.
Luoma, S.N. and Carter, J.L., Effects of trace metals on aquatic benthos, In *Ecotoxicology of Metals: Current Concepts and Applications*, Newman, M.C. and McIntosh, A.W. (eds.), Lewis, Chelsea, MI, 1991, pp. 261–300.
May, R.M., *Stability and Complexity in Model Ecosystems*, Princeton University Press, Princeton, NJ, 1973.
McLaren, B.E. and Peterson, R.O., Wolves, moose, and tree rings on Isle Royale, *Science*, 266, 1555–1558, 1994.
Moriarty, F., *Ecotoxicology: The Study of Pollutants in Ecosystems*, Academic Press, New York, 1988.
Naeem, S., Thompson, L.J., Lawler, S.P., Lawton, J.H., and Woodfin, R.M., Declining biodiversity may alter the performance of ecosystems, *Nature*, 368, 734–737, 1994.

Nakano, S., Miyasaka, H., and Kuhara, N., Terrestrial-aquatic linkages: Riparian arthropod inputs alter trophic cascades in a stream food web, *Ecology*, 80, 2435–2441, 1999.

Odum, E.P., The mesocosm, *Bioscience*, 34, 558–562, 1984.

Oris, J.T. and Giesy, J.P., Jr., Photoinduced toxicity of anthracene to juvenile bluegill sunfish (*Lepomis macrochirus* rafinesque): Photoperiod effects and predictive hazard evaluation, *Environ. Toxicol. Chem.*, 5, 761–768, 1986.

Paine, R.T., Food web complexity and species diversity, *Am. Nat.*, 100, 65–75, 1966.

Poff, N.L., Olden, J.D., Vieira, N.K.M., Finn, D.S., Simmons, M.P., and Kondratieff, B.C., Functional trait niches of North American lotic insects: Trait-based ecological applications in light of phylogenetic relationships, *J. N. Am. Benthol. Soc.*, 25, 730–755, 2006.

Pontasch, K.W., Niederlehner, B.R., and Cairns, J., Jr., Comparisons of single-species, microcosm and field responses to a complex effluent, *Environ. Toxicol. Chem.*, 8, 521–532, 1989.

Powell, G.V.N., Reproduction by an altricial songbird, the red-winged blackbird, in fields treated with the organophosphate insecticide fenthion, *J. Appl. Ecol.*, 21, 83–95, 1984.

Pratt, J.R. and Barreiro, R., Influence of trophic status on the toxic effects of a herbicide: A microcosm study, *Arch. Environ. Contam. Toxicol.*, 35, 404–411, 1998.

Rasmussen, J.B., Rowan, D.J., Lean, D.R.S., and Carey, J.H., Food chain structure in Ontario lakes determines PCB levels in lake trout (*Salvelinus namaycush*) and other pelagic fish, *Can. J. Fish. Aquat. Sci.*, 47, 2030–2038, 1990.

Relyea, R. and Hoverman, J., Assessing the ecology in ecotoxicology: A review and synthesis in freshwater systems, *Ecol. Lett.*, 9, 1157–1171, 2006.

Ricklefs, R.E., *Ecology*, 3rd ed., W.H. Freeman and Company, New York, 1990.

Rohr, J.R., Kerby, J.L., and Sih, A., Community ecology as a framework for predicting contaminant effects, *Trends Ecol. Evol.*, 21, 606–613, 2006.

Schindler, D.W., Detecting ecosystem responses to anthropogenic stress. *Can J. Fish. Aquat. Sci.*, Suppl., 6–25, 1987.

Schoener, T.W., The species–area relationship within archipelagos: Models and evidence from island land birds, In *Proceedings of the 16th International Ornithological Congress*, Firth, H.J. and Calby, J.H. (eds.), Australian Academy of Science, Canberra, 1974, pp. 629–642.

Schoener, T.W., Field experiments on interspecific competition, *Am. Sci.*, 122, 240–285, 1983.

Schrader-Frechette, K.S. and McCoy, E.D., *Method in Ecology: Strategies for Conservation*, Cambridge Press, Cambridge, UK, 1993.

Shelford, V.E., *Animal Communities in Temperate America*, University of Chicago Press, Chicago, IL, 1913.

Simberloff, D.S., A succession of paradigms in ecology: Essentialism to materialism and probabilism, *Synthese*, 43, 3–39, 1980.

Strong, D.R., Simberloff, D., Abele, L.G., and Thistle, A.B., *Ecological Communities: Conceptual Issues and Evidence*, Princeton University Press, Princeton, NJ, 1984.

Suter, G.W., Jr., A critique of ecosystem health concepts and indexes, *Environ. Toxicol. Chem.*, 12, 1533–1539, 1993.

Tilman, D. and Downing, J.A., Biodiversity and stability in grasslands, *Nature*, 367, 363–365, 1994.

Tilman, D., Knops, J., Wedin, D., Reich, P., Ritchie, M., and Siemann, E., The influence of functional diversity and composition on ecosystem processes, *Science*, 277, 1300–1302, 1997.

Wiens, J.A., On understanding a non-equilibrium world: Myth and reality in community patterns and processes, In *Ecological Communities: Conceptual Issues and the Evidence*, Strong, D.R., Simberloff, D., Abele, L.G., and Thistle, A.B. (eds.), Princeton University Press, Princeton, NJ, 1984, pp. 439–457.

Wiens, J.A., Crist, T.O., Day, R.H., Murphy, S.M., and Hayward, G.D., Effects of the *Exxon Valdez* oil spill on marine bird communities in Prince William Sound, *Ecol. Appl.*, 6, 828–841, 1996.

Wilson, D.S., Biological communities as functionally organized units, *Ecol. Appl.*, 78, 2018–2024, 1997.

21 Biotic and Abiotic Factors That Regulate Communities

21.1 CHARACTERIZING COMMUNITY STRUCTURE AND ORGANIZATION

> The organization of a community results from the outcome of interspecific competition for the available resources, and is expressed both in the relative abundance and the spatial distribution of constituent species.
>
> **(Hairston 1959)**

> Despite recent advances, both in the acquisition of data and in its analysis, I doubt that any multispecies community is sufficiently well understood for us to make confident predictions about its response to particular disturbances, especially those caused by man.
>
> **(May 1984)**

As with most scientific endeavors, the field of ecology is concerned with identifying patterns in the natural world and then explaining the underlying processes responsible for these patterns. Community ecologists specifically focus on characterizing variation in the numbers and types of species found at different locations and understanding the role of biotic and abiotic processes responsible for these differences (Bellwood and Hughes 2001). Changes in species diversity across broad environmental gradients or between habitats have occupied the interest of community ecologists for several decades. Variation in the distribution and abundance of species may be a result of broad geographical patterns (e.g., "Why are there so many species in the tropics compared to temperate regions?") or small-scale, local phenomena (e.g., "Why is community composition different between headwater streams and mid-order streams?"). An appreciation of factors that determine natural spatial and temporal variation in community composition is essential for ecotoxicologists. In order to characterize community responses to contaminants and other anthropogenic disturbances, we must first understand the influence of natural spatiotemporal variation on species diversity and composition. This natural variation in community structure is of practical importance because it complicates assessments of anthropogenic disturbances. Similarly, temporal changes in species diversity and community composition provide the context for understanding how communities will recover from anthropogenic disturbance.

In their attempt to quantify predictable features of communities, ecologists have identified numerous ways to categorize communities. Taxonomic groupings, trophic organization, morphological features, and life history traits are a few of the characteristics that ecologists have employed to classify community structure. As evidenced by Hairston's quote, for many ecologists, community structure was synonymous with species interactions—specifically competition. Other ecologists felt that definitions of a community should include both biotic and abiotic characteristics. Recognizing that community structure was influenced by factors other than competition, Roughgarden and

Diamond (1986) proposed the idea of "limited membership" as a unifying theme for defining community structure. Basically, their approach focuses on a single question: "Why does the unique combination of species found in a particular location or region represent only a subset of what *could* occur?" Roughgarden and Diamond argue that membership of any species in a community is a result of three primary factors: the physical environment, dispersal ability, and species interactions. The relative importance of these three factors will vary among community types and across habitats.

Another way to characterize community structure is to consider factors that limit membership in a community as a series of filters operating at different spatial and temporal scales. This idea was proposed by Poff (1997) to describe associations of species traits across spatial scales from microhabitats to entire watersheds. Using this model, Roughgarden and Diamond's (1986) concept of limited membership could be extended to include factors at regional and global scales (Figure 21.1). While species interactions, physical characteristics, dispersal ability, and anthropogenic factors play a prominent role at local scales, evolutionary and biogeographical factors determine species composition at global and regional scales. As we proceed from global to local filters, the characteristics that limit community membership become increasingly fine. The concept of limited community membership is attractive because it requires that we consider factors operating at the local level as well as historical and biogeographical characteristics. Using this model, species-specific sensitivity to contaminants is simply another filter that restricts community membership. If we are to make significant progress in predicting how communities respond to chemical stressors, an understanding of factors that limit community membership at these different spatial and temporal scales is required.

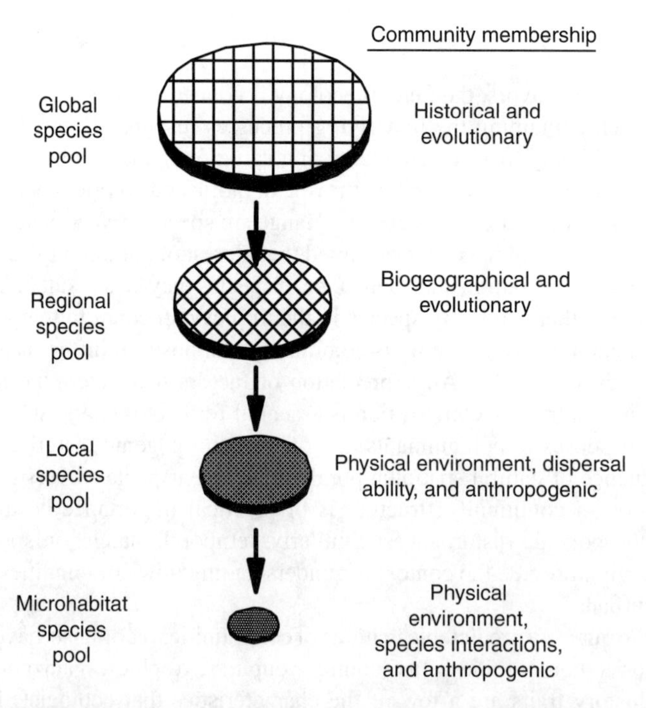

FIGURE 21.1 Historical, biogeographical, and environmental factors that determine membership of species in a community. Each factor is represented as a filter that operates at different spatial and temporal scales to determine regional, local, and microhabitat species pools. The pore size of each filter reflects its relative influence on species pools. Using this model, contaminants and other anthropogenic stressors are simply additional filters that determine community composition. (Modified from Figure 1 in Poff (1997).)

21.1.1 Colonization and Community Structure

Ecologists recognize that historical factors and regional-scale processes often interact to regulate local community composition. Colonization studies of newly created habitats provide opportunities to assess the influence of historical factors and species' dispersal abilities on community composition. If communities were regulated entirely by local deterministic factors, we would expect that communities established in similar habitats would have similar composition. Jenkins and Buikema (1998) tested this hypothesis by measuring structural and functional characteristics of zooplankton communities in 12 newly established ponds. Samples collected over a 1-year period showed that physical and chemical characteristics of these ponds were essentially identical. However, communities established in each of the ponds were distinct, reflecting the unique colonization abilities of dominant zooplankton species. Dispersal ability regulated composition among ponds because species that arrived first had a lasting effect on community structure. These results have important implications for how we view the establishment and regulation of communities. Failure to account for regional processes may explain the apparent stochastic behavior observed in some communities. The results also demonstrate that historical factors can have lasting, subtle impacts on communities, thus complicating our ability to locate reference sites and assess the importance of anthropogenic stressors (Landis et al. 1996, Matthews et al. 1996).

21.1.2 Definitions of Species Diversity

A variety of approaches have been developed by community ecologists to define and quantify species diversity. Species richness is a simple count of the number of different species within a local habitat or a region. Some ecologists are uncomfortable with measures of species richness because rare and common species are treated equally. Assuming that abundance of a species is related to its ecological importance, estimating relative abundance of different species may be a more effective way to characterize community structure. Diversity indices that account for both species richness and distribution of individuals among species are commonly used in biological assessments. These measures are described in Chapter 22. Here, our discussion of spatial and temporal patterns in diversity will focus on the number of species within a sample or within a region. To characterize spatial variation in community structure, ecologists distinguish among three different measures of species diversity. Alpha diversity refers to the species richness within a local area. Because assessments of anthropogenic disturbance are generally site specific, alpha diversity is the measure most relevant to ecotoxicologists. Beta diversity is the change in number of species and is an expression of species turnover between two adjacent habitats. Gamma diversity is the total number of species within a relatively large geographic area and represents the species pool available to colonize local habitats. Gamma diversity is a product of alpha and beta diversity and therefore will be greatest in regions with high local diversity and high species turnover.

Although concern about the global loss of species has increased awareness of the importance of biodiversity, this is a relatively recent phenomenon. Ecology textbooks published in the 1940s and 1950s made little mention of species diversity, attributing differences in community structure among locations primarily to historical and evolutionary events (Schluter and Rickleffs 1993). In contrast, experimental studies conducted in the 1960s and 1970s emphasized local regulation of diversity by species interactions and environmental heterogeneity, almost to the exclusion of historical features. Today, we know that spatial and temporal variation in diversity results from a complex interplay of historical, evolutionary, climatic, energetic, environmental, and anthropogenic phenomena. The challenge in community ecology is to understand the relative influence of these different factors on species diversity. The challenge in ecotoxicology is to interpret anthropogenic effects on species diversity within the context of these local and historical features. Some progress has been made with the recognition that natural variation and historical factors can influence community responses to contaminants (Clements 1999, Landis et al. 1996, Matthews et al. 1996).

21.2 CHANGES IN SPECIES DIVERSITY AND COMPOSITION ALONG ENVIRONMENTAL GRADIENTS

Natural changes in community composition and species diversity across environmental gradients have fascinated ecologists for many years. Early explorers frequently reported broad scale changes in species diversity and community composition with latitude and elevation. Because these changes were often predictable, ecologists developed confidence that underlying biotic and abiotic mechanisms could be identified by analysis of spatial patterns. In some instances, the observed transition from one community type to another was relatively abrupt, whereas in others it was much more gradual. These differences were most often related to species-specific tolerance for a particular environmental factor such as temperature, moisture, or soil type. Some species within a community are more tolerant of environmental variation and will be distributed across a broader range of habitats than others. Whittaker (1975) noted that patterns of species replacement along environmental gradients fall into several categories (Figure 21.2). The forest communities studied by Whittaker and coworkers showed that species replacement was gradual and that species behaved independently of each other. In contrast, marine invertebrate communities in the rocky intertidal zone show relatively abrupt transitions resulting from strong environmental gradients and intense species interactions. Finally, longitudinal changes in stream communities described in the River Continuum Concept (Vannote et al. 1980) and geographic changes in community composition across broad latitudinal gradients are relatively gradual, but often show distinct community types.

An understanding of how species respond to natural environmental gradients has direct relevance to community ecotoxicology. First, because contaminants are often distributed along a concentration gradient, the same analytical techniques employed to study natural patterns (e.g., gradient analysis or ordination) can be used to investigate community responses to chemical stressors.

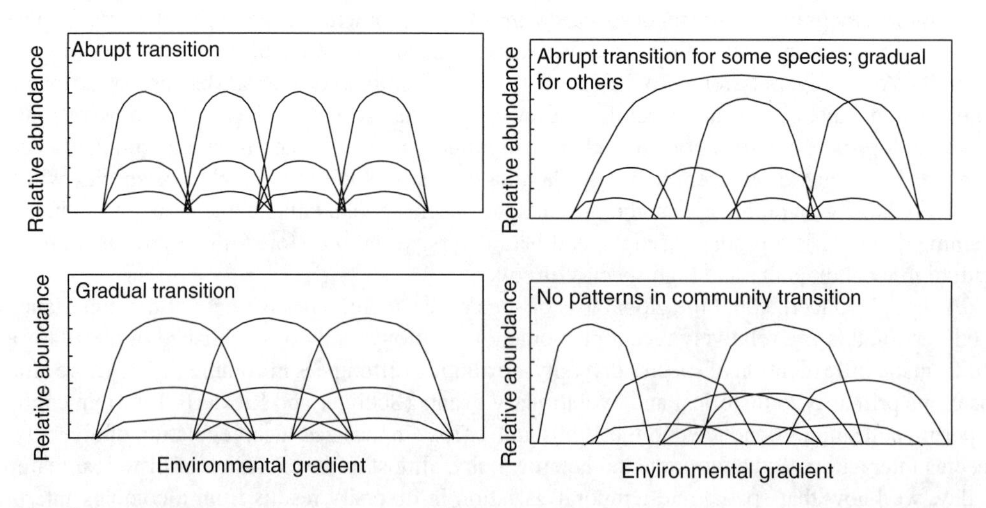

FIGURE 21.2 Hypothetical changes in relative abundance of species along an environmental gradient. Some communities show relatively abrupt transition in abundance of dominant species, while others are characterized by gradual changes. Abrupt transitions in community composition are often a result of interspecific interactions (competition, predation). Changes in community composition along a contaminant gradient are likely to be abrupt for some species and gradual for others depending on relative sensitivity to the stressor. (Modified from Whittaker (1975).)

Second, understanding the processes responsible for species replacement along natural gradients will allow ecotoxicologists to develop improved models for assessing contaminant-induced variation. We expect that changes in communities in response to contaminants will be relatively abrupt, but that recovery along a contaminant gradient may be more gradual. Finally, natural environmental gradients are often superimposed on contaminant gradients and will complicate biological assessments of community structure. In order to predict community responses to chemical stressors, ecotoxicologists require information on how these natural changes will modify and interact with contaminants.

21.2.1 Global Patterns of Species Diversity

The most consistent response to an environmental gradient reported by community ecologists at a large spatial scale is the increased species diversity from the arctic to tropical ecosystems. This pattern has been observed for most groups of organisms, and a variety of hypotheses have been proposed to explain the greater diversity in tropical communities (Table 21.1). Tropical ecosystems are more productive, predictable, structurally complex, and are less influenced by extreme climatic events compared to arctic and temperate ecosystems. It is important to note that these four hypotheses are not mutually exclusive, and it is likely that each will play a role in accounting for changes in diversity across latitudinal gradients. For example, Connell and Orias (1964) dismissed environmental harshness per se as an explanation for the paucity of species in extreme habitats. Their conceptual model predicts that greater species diversity will be observed in productive habitats with high stability. In his classic paper "Homage to Santa Rosalia or Why are there so many kinds of animals?," G.E. Hutchinson (1959) speculated that the earth's rich biodiversity was a result of an interplay among energetics, evolution, species interactions, and habitat complexity.

In an assessment of progress over the past 20 years since the publication of Hutchinson's paper, Brown (1981) noted that the inability of contemporary ecology to answer the question "Why are there so many kinds of animals?" resulted from the failure to focus on energetics. He noted that soon after publication of Hutchinson's seminal paper, ecologists were divided between two camps. The "ecosystem processes camp" considered energetics, but the research questions were not directed toward community ecology. The "species interactions camp" focused on community dynamics, but largely ignored the importance of energetics. Brown (1981) proposed a general theory of biodiversity based on the availability of energy, the apportionment of energy among species, and environmental harshness. More recently, Brown and Lomolino (1998) presented a more synthetic explanation for patterns of species diversity that included elements of productivity, abiotic stress, and species interactions, all within a broad historical context of time and space. According to this model, abiotic stress in extreme environments limits community composition to a few widely distributed, stress-tolerant

TABLE 21.1
Four Hypotheses to Explain the Increased Biological Diversity from Arctic to Tropical Ecosystems

Hypothesis	Explanation
Productivity	Tropical ecosystems have greater primary productivity, thus providing more food resources and greater food web complexity.
Heterogeneity	Tropical ecosystems are physically more complex and heterogeneous, thus providing more habitats and opportunities for specialization.
Stability	Tropical ecosystems are more stable and predictable, thus allowing species to specialize on a particular resource.
Evolutionary time	Tropical ecosystems are "older" in the sense that they have not been subjected to recent glaciation, thus providing more time for speciation.

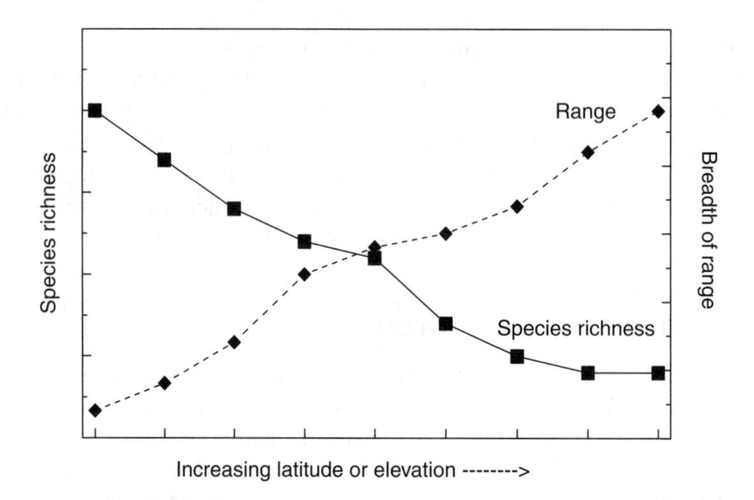

FIGURE 21.3 Hypothetical example of Rapoport's rule showing the relationship between species richness and breadth of distribution along an environmental gradient. Although the total number of species is reduced at higher elevations and at higher latitudes, the tolerance of individual species for environmental conditions is greater. These results suggest that species living in stable environments are less able to tolerate extreme conditions. Variation in tolerance may have important implications for understanding how species from different environments respond to anthropogenic stressors.

species capable of dividing up the limited resources. Biotic interactions in this harsh environment play a relatively minor role. In contrast, abiotic factors are less important in benign environments where predators and competitors limit densities of most species, allowing a large number of relatively uncommon species to partition the abundant resources.

Brown and Lomolino's (1998) synthetic explanation of community organization is intellectually satisfying for several reasons. First, it recognizes the importance of several key factors in controlling species diversity across broad environmental gradients. It is also consistent with the observation that species found in more variable habitats have a greater tolerance for environmental conditions compared to species occupying benign environments (Figure 21.3). The positive relationship between the range of latitudes occupied by a species and the latitude of its center of distribution is called Rapoport's rule (Rapoport 1982). A similar phenomenon has also been observed in communities across elevation gradients. The implication is that species found in stable habitats are less able to tolerate variation in environmental conditions than species occupying harsh conditions of higher latitudes or higher elevations. The inverse relationship between species diversity and elevation is probably a result of lower productivity and greater stress of high elevation habitats. This pattern, which has been observed for molluscs, birds, mammals, and trees, may provide important insights into variation in sensitivity to contaminants among locations. Similarly, lower diversity of some plant communities that has been observed along gradients of increased aridity and salt stress is most likely associated with the increased physical harshness of these environments. This explanation is consistent with Menge and Sutherland's (1987) hypothesis of environmental stress gradients, which has been used to account for local patterns of species diversity in benign and stressful environments (see Section 21.5.1). Factors influencing local patterns of species diversity are of particular interest to ecotoxicologists because they may help us understand how communities respond to contaminant gradients. Assuming this pattern is consistent across communities, it suggests that species occupying more predictable environments may be more sensitive to anthropogenic disturbances than species from harsh environments. This hypothesis could be tested by comparing responses of communities from different locations to the same anthropogenic stressor.

Another consistent pattern across broad geographical regions relates to changes in abundance distributions from temperate and tropical habitats. In general, tropical communities are characterized

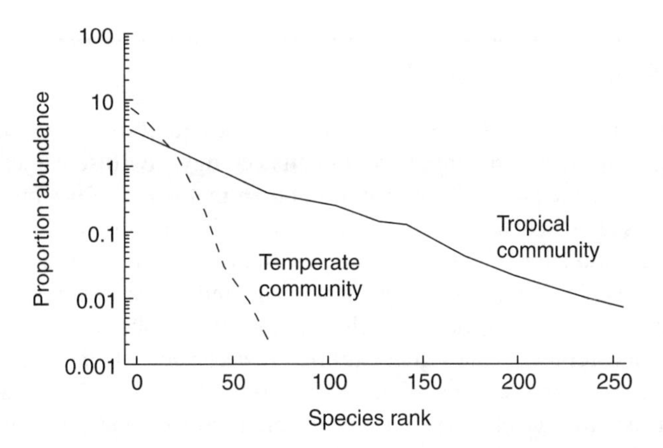

FIGURE 21.4 Variation in species abundance curves between temperate and tropical communities. In contrast to temperate systems, tropical communities are characterized by greater species richness and a more even distribution of individuals among species. The shape of species abundance curves is considered a result of species interactions and environmental conditions and has been used to characterize effects of anthropogenic disturbance (see details in Chapter 22).

by a more even distribution of individuals among species (Figure 21.4). In other words, tropical communities not only contain many more species than temperate communities, but also the most common species account for a relatively small portion of the total community. In contrast, temperate communities are often dominated by a relatively few species that account for most of the individuals and biomass. Similar patterns have been observed across elevation gradients, suggesting that this may be a general phenomenon (Brown and Lomolino 1998). Because dominance of some species increases in response to stressors, the distribution of individuals among species is a sensitive indicator of anthropogenic disturbance and has been used in biomonitoring studies. These concepts will be further developed in Chapter 22.

21.2.2 Species–Area Relationships

One of the most predictable relationships in community ecology is the increase in number of species with area. The species–area relationship, described as one of the few laws in ecology (Schoener 1974), has been reported across most taxonomic groups and a variety of habitats. In addition to explaining differences in species richness on islands with different area and varying distances from a source of colonists (Box 21.1), the species–area relationship has been applied to conservation biology and the design of wildlife refuges. Contemporary research questions regarding the size, shape, and degree of isolation of wildlife refuges and other natural areas have been addressed using this relationship.

The species–area relationship takes the form:

$$S = cA^z, \tag{21.1}$$

where S = the number of species, c is a constant, A = area, and z represents the slope of the relationship between S and A when both are plotted on a logarithmic scale. Although the constant c varies among taxonomic groups, various field studies have reported that the exponent z is approximately 0.25. The consistency of z among taxonomic groups suggests that some universal principle may be operating (May and Stumpf 2000); however, recent attempts to estimate the slope of the species–area relationship across a range of habitats have reported greater variation than previously believed. Crawley and Harral (2001) measured species richness of plant communities across a wide

Box 21.1 The Special Case of Islands

MacArthur and Wilson's (1963) theoretical treatment of the equilibrium theory of island biogeography was a major conceptual advance in community ecology. Few discoveries in ecology have had greater impact, and the practical applications of their mathematically simple, but conceptually elegant, models are still being realized decades later. The equilibrium theory was developed to explain the observation that island flora and fauna often represent a subset of species available from the mainland species pool. Distance from the mainland source of colonists and island area were primarily responsible for variation in the equilibrium number of species among islands (Figure 21.5). Small, remote islands generally had fewer species than larger islands close to a mainland source of colonists. MacArthur and Wilson (1963) also recognized that while the actual number of species was relatively consistent, community composition varied significantly due to species replacement and turnover. The importance of species turnover was evidenced by studies of the recolonization of Krakatau Islands following a massive volcanic eruption in 1883. Surveys of these islands several decades later showed a relatively constant numbers of species, supporting the equilibrium perspective; however, community composition changed significantly over time.

Experimental support for the equilibrium theory of island biogeography was provided by a large-scale manipulation of insect communities in the Florida Keys. Daniel Simberloff, a graduate student working with Wilson, fumigated mangrove islands with the pesticide methyl bromide and followed subsequent recolonization (Simberloff and Wilson 1969, 1970). Results generally supported the equilibrium theory and showed that isolated islands had lower rates of colonization and a lower equilibrium number of species compared to islands located near a mainland species pool.

While much of the research on island size has focused on structural measures (e.g., community composition and species richness), there is evidence that ecosystem function may also be related to area. The theoretical motivation for this concept is based on the observation that individual species in a community are important regulators of ecosystem processes. Wardle et al. (1997) tested this hypothesis in an island archipelago of a Swedish boreal forest. Several ecosystem processes, including respiration, decomposition, and nitrogen loss, varied with island area because of differences in community composition. Variation in community composition among islands resulted from the greater frequency of fires due to lightning strikes on larger islands. These results show that historical events (e.g., frequency of fire) play an important

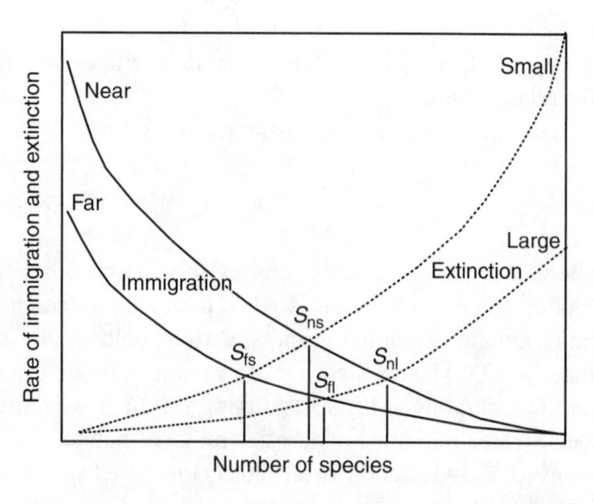

FIGURE 21.5 The relationship between number of species and rates of immigration and extinction on islands. Immigration and extinction rates are influenced by island size (large vs. small) and distance from a mainland source of colonists (near vs. far), resulting in a unique equilibrium number of species (S) for each island type. Because recovery of communities from disturbance is largely determined by immigration rate and the proximity of local colonists, these theoretical relationships have important implications for community responses to anthropogenic stressors. (Modified from MacArthur and Wilson (1967).)

role in determining both community composition and ecosystem function. By developing a better appreciation for the role of historical events, we can begin to understand how natural communities will respond to anthropogenic disturbance.

---

range of habitat scales (0.01–10^8 m^2). They reported that z-values were lowest at small spatial scales (due to interactions among species) and at very large spatial scales (due to low species turnover with distance). The greatest rate of species accrual was observed at intermediate scales, where increases in area resulted in increases in habitat diversity. These findings indicate that while species accrual rates may be similar within a small range of spatial scales, different processes operate to determine species diversity across geographic regions.

Despite its intuitive appeal and broad explanatory power in community ecology, the species–area relationship has not received much attention in ecotoxicology. The basic principles of island biogeography have important applications to the study of contaminant effects and recovery. The rate of recovery and the composition of communities during the recovery process are greatly influenced by distance from the source of colonists and colonization abilities of species. These ideas will be considered in Chapter 25.

21.2.3 Assumptions about Equilibrium Communities

MacArthur and Wilson's (1963) equilibrium theory of island biogeography was consistent with the predominant view of ecology at the time. Many ecologists believed that natural communities are orderly, balanced, and maintain a natural equilibrium unless subjected to extrinsic disturbance. Although ecologists recognize the dynamic nature of this equilibrium, the underlying assumption that communities are regulated primarily by biotic interactions remains prevalent in ecology. The emergence of equilibrium theories in ecology was supported by our deep-seated belief that attributes of natural communities are predictable and that historical factors, stochastic events, and small-scale environmental perturbations are relatively unimportant. Much of the controversy surrounding the relative importance of species interactions results from this uncritical acceptance that communities are at equilibrium (see Section 21.4).

Ecologists now recognize that few communities are regulated exclusively by predictable, deterministic processes. Long-term data collected from a variety of systems reveal temporal changes in abundance of dominant species that do not appear to be regulated by equilibrium processes. For example, detailed studies of grassland bird communities have shown few consistent patterns and little indication that biotic interactions are important (Wiens 1984). The most likely explanation for the observed nonequilibrium characteristics of these communities relate to the stochastic environmental conditions of prairie and shrub-steppe habitats.

Studies conducted in streams suggest that communities may shift from equilibrium to nonequilibrium conditions seasonally or among locations along a river continuum (Minshall et al. 1985). In his classic paper "The paradox of plankton," Hutchinson (1961) observed that the high diversity of phytoplankton in simple, homogenous environments was contrary to deterministic predictions of the competitive exclusion principle. The proposed explanation for this paradox was that planktonic communities did not achieve equilibrium conditions. Interestingly, recent studies conducted in lakes suggest that resource competition can structure communities even in environments where equilibrium conditions are rarely observed. Interlandi and Kilham (2001) reported a strong relationship between the number of limiting resources (nitrogen, phosphorus, silicon, and/or light) and diversity of phytoplankton in lakes (Figure 21.6). Clearly, the dichotomy between equilibrium and nonequilibrium communities is somewhat artificial. Instead of defining communities as either equilibrium or nonequilibrium, Wiens (1984) proposes that communities should be arrayed along a gradient based on a suite of characteristics. This model is analogous to the continuum between r-selected and K-selected species described in population ecology.

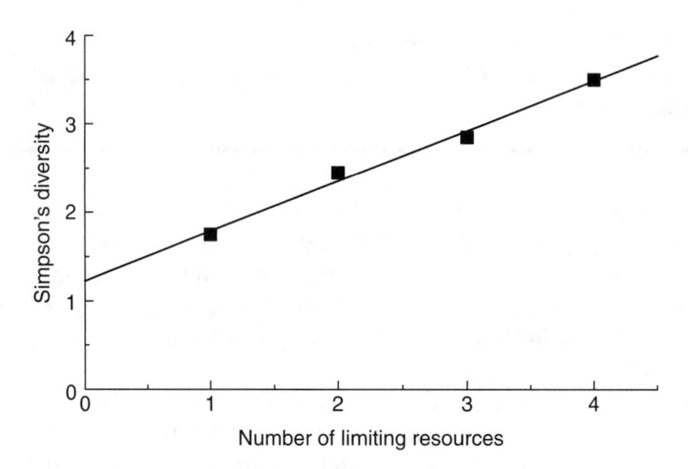

FIGURE 21.6 The relationship between the number of limiting resources and species diversity of plankton communities. (Modified from Figure 6 in Interlandi and Kilham (2001).)

21.3 THE ROLE OF KEYSTONE SPECIES IN COMMUNITY REGULATION

It is generally accepted that some species have disproportionate effects on community composition and ecosystem function (Power et al. 1996). These "keystone" species are often large, highly mobile consumers, which are especially susceptible to habitat loss and chemical stressors. Because of their impact on communities, loss of keystone species is expected to influence other species in the community. Identifying species that play a significant role in structuring communities is necessary for predicting ecological consequences of contaminants and other anthropogenic stressors.

Determining the relative importance of a species in a community will often require experimental manipulations. Experiments conducted in the marine rocky intertidal zone demonstrated that removal of the predatory starfish *Pisaster ochraceus* had significant effects on other species in the community (Paine 1966). Selective predation of *Pisaster* on mussels, the competitively dominant species in the community, maintained a diverse assemblage of subordinate species. Paine (1969) introduced the keystone species concept to describe a species that has significantly greater effects on a community than expected based on its abundance or biomass. Since the publication of Paine's conceptual paper, investigators have identified keystone species in a variety of ecosystems (Power et al. 1996), and the keystone species concept has been referred to as a "central organizing principle" in community ecology (Menge et al. 1994). Currently, we know that keystone species are widely distributed among many ecosystem types and that their effects on structure and function are often far-reaching (Table 21.2).

Paine's initial experiments described effects of a keystone predator, and most subsequent studies of keystone species have focused on similar resource–consumer interactions. However, a broad definition of a keystone species should also include effects such as physical restructuring of the environment (ecosystem engineers such as beavers in the Pacific Northwest) and mutualistic interactions (plant–pollinator systems). Similarly, we know that the effects of keystone species extend well beyond regulation of species diversity and include effects on community structure, productivity, nutrient cycling, and energy flow (Erenst and Brown 2001). In fact, an operational definition of keystones species should include any species that has a disproportionate impact on a community, regardless of the mechanism (Power et al. 1996). Figure 21.7 shows the relationship between total community impact and relative abundance or biomass in a community. Species that fall on the diagonal line influence the community in proportion to their abundance. Species to the right of the diagonal are dominant in the community but their impact is less than expected based on abundance

TABLE 21.2
Examples of Suspected or Likely Keystone Species and Their Target Groups in a Variety of Aquatic and Terrestrial Habitats

Habitat	Keystone Species	Target Group	Reference
Rocky intertidal	Predatory starfish	Mussels	Paine (1966)
Coral reefs	Sea urchins	Algal communities	Carpenter (1990)
Lakes and ponds	Planktivorous fish	Zooplankton	Brooks and Dodson (1965)
Rivers and streams	Predatory steelhead and omnivorous minnow	Invertebrates and fish fry	Power (1990)
Grasslands	Rabbits	Herbs and grasses	Tansley and Adamson (1925)
Woodlands	Wolves	Moose	McLaren and Peterson (1994)
Desert	Kangaroo rats	Seeds	Brown and Heske (1990)

Source: Modified from Power et al. (1996).

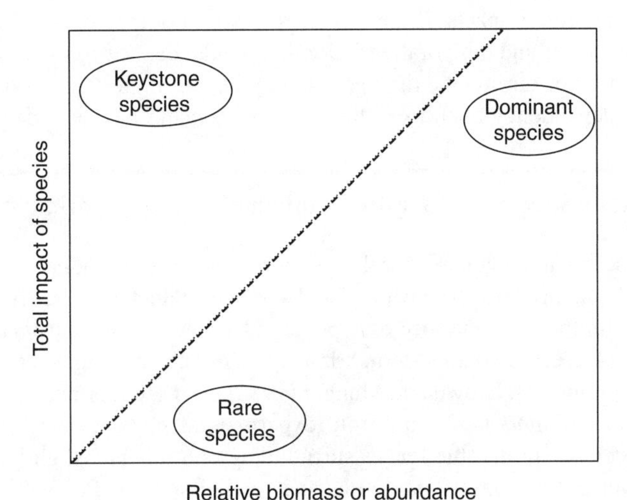

FIGURE 21.7 Relationship between total community impact and relative abundance or biomass of a species. The effects of a species on community structure and function are not necessarily related to its abundance or biomass. Species in the upper right hand quadrant are dominant in the community, but their impact is less than expected based on their abundance or biomass. Species to the left of the diagonal have greater impact than expected and are considered true keystone species. (Modified from Figure 3 in Power et al. (1996).)

or biomass. Species to the left of the diagonal are defined as true keystone species because they have a disproportionate influence on community structure or function.

21.3.1 Identifying Keystone Species

Identifying keystone species and quantifying their effects on community structure are not trivial issues, and criticism of the keystone species concept is at least partially a result of these difficulties. While manipulating density of an individual species remains the most direct approach for assessing its role in a community, conducting experiments at appropriate spatial and temporal scales is logistically challenging. Conclusions about the importance of species interactions relative to abiotic processes are clearly scale dependent. Because the size of study plots and the duration of experiments may influence patterns that we observe, it follows that more attention should be given to spatial and temporal scale when interpreting results of manipulative experiments (see Chapter 23). The tremendous success of rocky intertidal ecologists is at least partially a result of the relative ease with which these communities can be manipulated. The reluctance of some ecologists to embrace the keystone species concept is likely due to difficulty obtaining experimental evidence in systems that are less amenable to manipulation (Box 21.2). Comparative approaches and natural experiments, in which community structure and function are measured in areas with and without a particular species, are practical alternatives to actual manipulation and will play an important role in identifying keystone species. Well-designed studies that take advantage of species reintroductions could also contribute to our understanding of keystone species. For example, recovery of otter populations off the California coast and beaver populations in the Pacific Northwest provide excellent opportunities to track ecosystem changes due to increased abundance of keystone species. Community viability analysis (CVA), an approach analogous to population viability analysis, has been used to quantify the effects of species loss on a community (Ebenman et al. 2004). Deterministic CVA models used to predict effects of species loss have the potential to help researchers identify keystone species and fragile communities that are especially susceptible to species loss.

Quantifying the relative impacts of keystone species on community structure and function is also complicated by spatial and temporal variation in abundance. Species that regulate community structure in one location or during one time period may be less important in other areas or at other times. This context dependency of the keystone species concept has been demonstrated in rocky

Box 21.2 Keystone Species in Terrestrial Communities: An Experimental Demonstration

Because most experimental evidence for the keystone species concept has been obtained from aquatic ecosystems, many terrestrial ecologists have been reluctant to accept this hypothesis. However, support for the keystone species hypothesis has been obtained from a long-term study in the Chihuahuan Desert of southeastern Arizona (USA). To investigate competitive interactions in a rodent community, Brown and Munger (1985) used semipermeable fences to exclude larger kangaroo rats (*Dipodomys* spp.) from experimental plots. Control and experimental plots were identical except that the fences surrounding controls had slightly larger holes that allowed free movement of both kangaroo rats and smaller species. Populations were sampled monthly for over 20 years, providing one of the few long-term assessments of the effects of species removal. Results of this study showed that the competitively superior kangaroo rats suppressed abundance and altered foraging behavior and habitat use of the smaller species. More importantly, kangaroo rats were shown to have larger than expected impacts on energy flow and community composition of plants, thus satisfying the definition of a keystone species.

Between 1977 and 1996, six species of seed-eating rodents colonized experimental plots (where kangaroo rats were removed) at densities approximately twice as high as controls. Nonetheless, these species only consumed about 14% of the available energy consumed by kangaroo rats on the control plots. Thus, for almost 20 years, there was no evidence of compensation by subordinate species following removal of the keystone species. Remarkably, this changed in 1996 when a species of pocket mouse (*Chaetodipus baileyi*), never previously observed in the study site, colonized the experimental plots at densities 20 times greater than controls (Erenst and Brown 2001). Within 2 years, this species consumed most of the resources and was able to compensate for the excluded kangaroo rats. This experiment demonstrates that previously rare species are capable of restoring community structure and function. It also demonstrates the difficulty of identifying keystone species without experimental and/or long-term data.

intertidal communities (Menge et al. 1994). Power et al. (1996) also present several scenarios in which the structuring role of a species could be modified under different environmental circumstances. However, our understanding of the physical, chemical, and biological factors that influence the impact of a particular species is incomplete. Research in rocky intertidal habitats has focused primarily on the role of physical disturbance, and conceptual models have been developed to quantify the relative importance of species interactions under varying levels of physical stress (Menge and Sutherland 1987). These ideas are quite relevant for community ecotoxicology because chemical stressors may directly influence abundance of keystone species as well as modify their structuring role in communities.

21.4 THE ROLE OF SPECIES INTERACTIONS IN COMMUNITY ECOLOGY AND ECOTOXICOLOGY

No living thing is so independent that its abundance and distribution are unaffected by other species.

(Brown and Lomolino 1998)

Considerable research in community ecology is devoted to assessing the relative importance of biotic interactions on distribution and abundance of organisms. As discussed in Chapter 20, while many ecologists define communities by the strength of species interactions, determining the role of competition, predation, mutualism, and so forth in community regulation is challenging. Although the evolutionary consequences of competition can be studied at broad scales by measuring character displacement and resource partitioning, these studies cannot demonstrate that competition regulates communities (Schluter and Ricklefs 1993). Despite acrimonious debate among community ecologists over the importance of species interactions, most would agree that positive and negative interactions are common in nature. For example, all heterotrophic organisms necessarily interact with their food resources. However, we do not always know if these interactions significantly influence the abundance or distribution of prey species relative to other factors. In the following sections, we will show that contaminants have the potential to change the outcome of species interactions and therefore influence community structure.

Although there is empirical support for the hypothesis that species interactions are common and can play a pervasive role in structuring communities (Diamond 1978, Menge and Sutherland 1987, Schoener 1983), the effects of contaminants on species interactions have largely been ignored by ecotoxicologists. This is somewhat surprising given the prominent role that research on predation, competition, mutualism, and so forth has played in basic community ecology. Previous reviews that focused on aquatic ecosystems (Clements 1997, Sandheinrich and Atchison 1990) showed that chemical stressors frequently alter the outcome of species interactions. We suggest that the failure to consider indirect effects of contaminants on species interactions is a major limitation of single species toxicity tests.

21.4.1 DEFINITIONS

Species interactions in natural systems are generally defined by the direction and magnitude of effects (Table 21.3). Although most basic research on species interactions has focused on predation and competition, other types of interactions occur in communities and may be affected by exposure to contaminants. In particular, strong mutualistic interactions, such as those observed in obligate plant–pollinator systems, may be especially sensitive to chemical stressors. One potential limitation to understanding the importance of species interactions in nature has been the emphasis on simple pair-wise interactions. A review of 1253 papers published in *Ecology* between 1981 and 1990 showed that >60% considered only one or two species (Kareiva 1994). The emerging view from contemporary

TABLE 21.3
Types of Species Interactions Considered in Community Ecology

Type of Interaction	Effects on Species A	Effects on Species B
Competition	−	−
Predation	−	+
Parasitism	−	+
Mutualism	+	+
Commensalism	+	0

Note: Interactions between two species may be positive (+), negative (−), or neutral (0).

ecologists is that higher-order interactions or interaction modifications are probably common in natural systems. Trophic cascades are probably the best examples of higher-order interactions that have been studied in aquatic communities. Another noteworthy example involves defensive strategies that some plants employ to reduce herbivory. Plants produce a variety of chemicals designed to reduce the incidence of attack by herbivores. Some plants have significantly improved this defensive strategy and "enlisted" the support of other species. Kessler and Baldwin (2001) showed that emissions of volatile organic compounds by plants during attack by herbivores actually attracted insect predators, thus reducing the number of herbivores by 90%. Identifying higher-order interactions in nature will require that ecologists significantly expand the scope of their investigations beyond typical pair-wise studies. The greater challenge will be to determine when these complex interactions play an important role in organizing community structure.

21.4.2 Experimental Designs for Studying Species Interactions

The long history of theoretical research on species interactions has provided an important conceptual framework for designing laboratory and field studies. Relatively simple mathematical models for predation and competition predict how changes in abundance of one species will influence abundance of another species. While verifying these models with experimental studies has proved challenging, manipulations that involved removal or addition of species have provided the most convincing evidence for the importance of competition and predation. A variety of enclosure and exclosure experiments have been conducted in aquatic and terrestrial habitats to quantify species interactions in the field. Key strengths of manipulative experiments are the potential for replication (thus allowing the appropriate use of inferential statistics) and the ability to control confounding variables. One of the more basic questions in studies of competition involves assessing the relative importance of interspecific and intraspecific interactions. Different experimental designs have been used by ecologists to measure the strength of interactions within and between species (Table 21.4). An additive experimental design allows researchers to determine if the presence of a competitor has any effect on a second species, all other factors being equal. This design would be especially useful for studying impacts of an exotic species on a native species (Fausch 1998). A substitutive experimental design holds total density constant and allows researchers to quantify the importance of interspecific competition relative to intraspecific competition. This design would be most appropriate for assessing the effects of contaminants on species interactions.

Because field experiments are often limited in spatial and temporal scale, some researchers advocate the use of natural experiments for assessing the role of species interactions (Diamond 1986).

TABLE 21.4
Two Types of Experimental Designs Used to Assess the Importance of Interspecific and Intraspecific Competition

Design	Treatment 1	Treatment 2	Treatment 3
Additive			
Density of species A	5	5	
Density of species B		5	5
Substitutive			
Density of species A	10	5	
Density of species B		5	10

The table shows the number of individuals that would be included in different treatments for additive and substitutive experimental designs. Additive designs allow for assessment of the presence of a competitive effect. Substitutive designs allow for the quantification of the magnitude of this effect relative to intraspecific competition. After Fausch (1998).

The merits of field experiments and natural experiments are discussed in Chapter 23. Natural experiments often involve comparison of abundances, morphological features, and habitat use of species in sympatric and allopatric populations. For example, if competition played an important role in community organization, we would expect that morphological features related to resource use (e.g., beak size in the Galapagos finches is related to feeding habits) would show greater dissimilarity in sympatric populations compared to allopatric populations. This comparative approach has been especially effective for assessing the long-term evolutionary consequences of species interactions over broad spatial scales.

Experimental designs for assessing the influence of contaminants on species interactions add another layer of complexity because they require manipulation of both contaminant levels and predator/competitor abundances. Accomplishing this in the field will be difficult in many types of communities. Previous studies have compared the importance of species interactions in different habitats or under different levels of environmental stress (Menge and Sutherland 1987, Peckarsky et al. 1990). Conducting enclosure or exclosure experiments at sites with and without contaminants would allow researchers to determine if stressors modified the outcome of species interactions (Clements 1999).

21.4.3 The Influence of Contaminants on Predator–Prey Interactions

Research measuring effects of contaminants on predator–prey interactions fall into two general categories. Some studies consider the ecological consequences associated with alterations in predation intensity. For example, contaminant-induced changes in predation in communities regulated by top-down effects may alter the structure of lower trophic levels. Others studies are primarily concerned with developing a mechanistic understanding of how contaminants influence predator–prey interactions. Many of these laboratory studies predominately have attempted to relate changes in prey capture efficiency or predatory avoidance to individual bioenergetics.

Much of the laboratory and field research on predator–prey interactions has considered alterations in prey abundance due to direct mortality. However, a more subtle influence on prey populations, which may be more common in some systems, is predator-induced alterations in prey behavior. Indeed, these nonlethal influences of consumers on prey resources may often be as important as direct lethal effects. Meta-analysis that compared effects of density- and trait-mediated interactions found

TABLE 21.5
Behavioral Characteristics of Predators and Prey Known to Be Sensitive to Contaminant Exposure

Predator Behavior	Prey Behavior
Prey selection	Predator detection
Searching ability for prey	Predator avoidance and escape responses
Capture and handling time of prey	Defense mechanisms against predators

that foraging and other costs associated with prey intimidation was as strong as direct consumption (Preisser et al. 2005). Werner and Peacor (2006) reported that lethal effects of odonates feeding on herbivores were very strong relative to nonlethal effects and that the outcome of these interactions was dependent on system productivity. Changes in prey foraging rates due to predator avoidance may have important consequences for prey fitness (Ball and Baker 1995, 1996). Studies in aquatic systems have shown that prey organisms will alter their behavior in response to biochemical cues emitted by predators (Stirling 1995). Peckarsky and McIntosh (1998) reported that mayflies responded to fish odors by reducing the time spent on grazing, resulting in lower size at emergence and reduced fecundity. Increased algal biomass in experimental streams where mayflies were subjected to these chemical cues resulted in a "behavioral trophic cascade." These subtle responses of prey to predators will be much more difficult to detect than direct prey mortality.

Contaminants may influence various aspects of predation, and mechanistic studies of predator–prey interactions generally focus on behavioral changes in either predators or their prey (Table 21.5). In order for a predator to feed, it must locate, select, capture, and handle its prey. Any one of these behaviors may be influenced by exposure to contaminants. Predators generally rely on visual, olfactory, and/or auditory cues to locate prey species. Prey selection is often an immediate behavioral response to prey abundance and availability; however, items included in the diet of a predator may be ultimately determined by costs and benefits. Finally, prey capture is a function of predator efficiency (the number of captures per attack) and handling time. Assuming that diet is influenced by natural selection, we expect that prey are selected to maximize caloric gains and minimize expenditures and risks associated with foraging (Werner and Hall 1974). These basic predictions of the optimal foraging theory have been demonstrated in a variety of organisms including fish, birds, and mammals. Because optimal foraging theory integrates several important aspects of prey selection, capture, and handling, it provides a useful conceptual framework from which to evaluate stressor-induced changes in diet.

Not surprisingly, most field studies of contaminant effects on predator–prey interactions tend to focus on the consequences of reduced foraging success, but are unable to demonstrate clear mechanistic explanations. Field studies of birds have shown reduced foraging success in areas contaminated by organophosphate pesticides compared to uncontaminated habitats (Grue et al. 1982). These reductions in feeding may cause lower growth rates of adults or poor survival of dependent fledglings. However, because pesticides have both direct and indirect effects, it is often difficult to determine if these changes are a result of poor performance by the birds or reduced prey abundance.

Some researchers have attempted to distinguish the direct toxicological effects of contaminants from the indirect effects due to reduced prey abundance. The best examples of this research are from large-scale studies of bird populations exposed to pesticides. Aerial application of insecticides to control grasshoppers in grasslands of the United States often exceeds 1 million ha/year (USDA 1987). Because grasshoppers and other nontarget species are important prey items for many grassland birds, indirect effects are expected. Furthermore, the breeding season of many birds coincides with peak abundance of grasshoppers, the period when sprays are most likely to occur. Fair et al. (1995) measured the direct and indirect effects of the insecticide carbaryl on killdeer (*Charadrius vociferous*) in a large-scale experimental study in North Dakota. Despite dramatic reductions in abundance of grasshoppers and other prey species, killdeer foraging rate was actually greater in sprayed plots compared to controls. Increased foraging rate was attributed to greater numbers of available prey resulting from prey immobilization after pesticide exposure. The increased availability of exposed prey species creates the intriguing possibility that killdeer could receive a large dose of carbaryl while foraging on intoxicated prey.

Mechanistic-based studies of contaminant effects on predator–prey interactions have generally been restricted to the laboratory (Table 21.6). In order to distinguish effects on predator foraging from prey avoidance and escape responses, many experiments focus on behavior of either predators *or* their prey. These experiments have been criticized because they fail to consider the ecological consequences of alterations in predator–prey interactions and because they do not pose hypotheses that can be tested in the field. Sandheinrich and colleagues (Bryan et al. 1995, Sandheinrich and Atchison 1989, 1990) have conducted some of the most comprehensive analyses of contaminant effects on predator–prey interactions. Their focus on behavioral ecology has provided a solid mechanistic understanding of how contaminants affect various aspects of foraging success.

Equally important in determining the outcome of predator–prey interactions are changes in vulnerability of prey species to predation. The ability of an organism to detect, avoid, escape from or defend itself from predators is likely to be influenced by contaminant exposure. The majority of studies that have exposed both predators and prey species to chemical stressors have shown that prey vulnerability is increased (Beitinger 1990). Similarly, much of the research in terrestrial and wildlife populations has reported that alterations in the behavior of prey species, such as increased activity, will increase susceptibility to predation (Martin et al. 1998). Buerger et al. (1991) observed increased predation on birds exposed to pesticides compared to unexposed individuals. It was unclear if greater susceptibility to predation resulted from inability to detect or avoid predators. Lefcort et al. (1998) reported that exposure of Columbia spotted frog tadpoles (*Rana luteiventris*) to metals decreased predator avoidance response. In a subsequent study, Lefcort et al. (1999) showed that predation-induced shifts in habitat use by *R. luteiventris* decreased ingestion of metal-rich sediments and increased ingestion by competing snails. Schulz and Dabrowski (2001) reported a synergistic interaction between sublethal exposure to pesticides and fish predators resulted in greater mortality for mayflies. In a community-level assessment of predator impacts, Clements (1999) reported that several macroinvertebrate species collected from a metal-polluted habitat were more sensitive to stonefly predation than those collected from an unpolluted stream. These results suggest that alterations in predator–prey interactions may occur as a result of previous exposure to stressors.

In a novel experiment that investigated the influence of cadmium on foraging success, Wallace et al. (2000) exposed grass shrimp (*Palaeomonetes pugio*) to prey organisms collected from contaminated sites in the Hudson River (Foundry Cove, New York). Experiments showed that prey capture was significantly reduced in predators exposed to cadmium compared to unexposed organisms. These researchers also showed that capture success decreased with increased body burdens of cadmium and with the fraction of metals bound to high molecular weight proteins (Figure 21.8). The significance of this study is that environmentally realistic levels of a contaminant in the field significantly altered the outcome of predator–prey interactions.

In summary, the majority of studies attempting to measure the influence of contaminants on predator–prey interactions have shown significant effects (Fleeger et al. 2003). In some instances, effects were observed at concentrations below those considered toxic based on single species toxicity tests (Clements et al. 1989, Ham et al. 1995, Kiffney 1996, Sandheinrich and Atchison 1990, Sullivan et al. 1978). These findings highlight not only the sensitivity of behavioral endpoints to contaminants, but also the inadequacy of testing procedures based exclusively on single species. This does not imply that results of single-species toxicity tests are totally ineffective for predicting indirect effects. The most consistent pattern that emerges from an analysis of these data is that the outcome of predator–prey interactions is dependent on the relative susceptibility of predators and prey to a particular stressor. Thus, information on species-specific differences in sensitivity derived from single species tests may provide some insight into the direction of effects (e.g., increased or

TABLE 21.6
Examples of Experiments Conducted with Fish and Invertebrates Investigating the Effects of Chemical Stressors on Predator–Prey Interactions

Predator	Prey	Stressor	Result	Proposed Mechanism	Reference
Brook trout	Atlantic salmon	Organophosphate	Increased predation	Impaired learning ability of prey	Hatfield and Anderson (1972)
Largemouth bass	Mosquitofish	Gamma radiation	Increased predation	Abnormal behavior of prey	Goodyear (1972)
Largemouth bass	Mosquitofish	Mercury	Increased predation	Impaired escape behavior of prey	Kania and O'Hara (1974)
Largemouth bass	Fathead minnows	Cadmium	Increased predation	Greater prey vulnerability	Sullivan et al. (1978)
Largemouth bass	Mosquitofish	Ammonia	Decreased predation	Lower prey consumption	Woltering et al. (1978)
Largemouth bass	*Daphnia*	Pentachlorophenol	Decreased predation	Lower prey capture rate	Brown et al. (1987)
Smallmouth bass	*Daphnia* and tubificids	Acidification	Decreased predation	Lower visual acuity and reduced capture success	Hill (1989)
Bluegill	*Daphnia*, *Hyalella*, and damselflies	Copper	Decreased predation	Lower capture success and increased handling time	Sandheinrich and Atchison (1989)
Atlantic salmon	Brine shrimp	Fenithrothion	No significant effects	No change in capture success; increased reactive distance	Morgan and Kiceniuk (1990)
Rockfish	Chinook salmon	Fungicide	Increased predation	Greater prey susceptibility	Kruzynski and Birtwell (1994)
Bluegill	*Daphnia*	Cadmium	Lower predator growth	Lower attack rates	
Stonefly	Macroinvertebrate communities	Copper	Increased predation	Greater vulnerability of prey	Clements et al. (1989)
Stonefly	Macroinvertebrate communities	Mixture of heavy metals	Increased predation	Greater vulnerability of prey	Kiffney (1996)
Hydra	*Daphnia*	Lindane	Variable results	Differential effects on prey recruitment	Taylor et al. (1995)
Turbellarian	Isopod	Cadmium	Reduced predation	Lower predator capture or reduced hunger	Ham et al. (1995)
Rotifer	Rotifer	Pentachlorophenol	Increased risk of predation	Greater prey swimming speeds increased encounter rates with predators	Preston et al. (1999)
Grass shrimp	Brine shrimp	Cadmium	Lower predation	Reduced capture success	Wallace et al. (2000)
Cape galaxias	*Baetis* mayflies	Fenvalerate	Increased predation	Pesticide-induced drift behavior	Schulz and Dabrowski (2001)

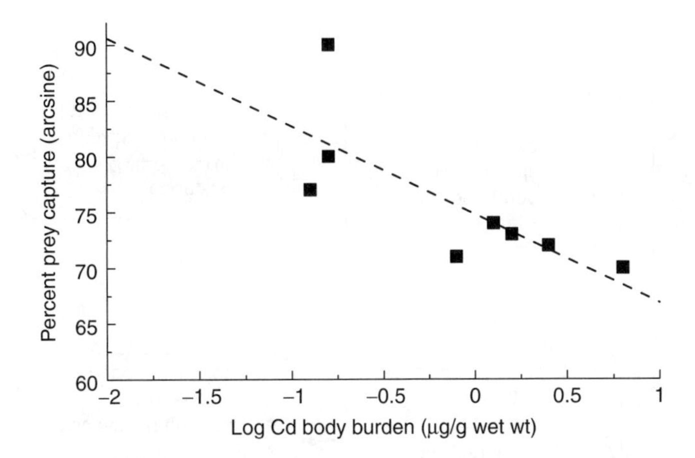

FIGURE 21.8 The influence of cadmium levels in grass shrimp (*Palaeomonetes pugio*) on prey capture ability. Predators fed cadmium-contaminated prey showed reduced capture success compared with unexposed predators. (Modified from Figure 5 in Wallace et al. (2000).)

decreased predation); however, understanding the magnitude and ecological consequences of these effects relative to direct toxicity will require integration of field experiments with mechanistic-based laboratory research.

21.4.4 The Influence of Contaminants on Competitive Interactions

While the evidence that predation is an important organizing force in communities is generally unequivocal, the role of competition in nature has been the subject of intense debate. In contrast to the direct and readily observable effects of predation, competition is generally much more subtle and difficult to quantify. While predation almost invariably involves the removal of individuals from a population, effects of competition may include habitat shifts, changes in feeding habitats, reduced growth, and delayed reproduction. Ecologists recognize that these subtle changes have important consequences for fitness, but there is serious disagreement over their importance relative to abiotic factors.

In general, relatively few studies have measured the influence of contaminants on competitive interactions. Early research on competition and chemical stressors was initiated by Antonovics et al. (1971) and their classic studies of metal tolerance in plants (see Chapter 18 for a detailed description of these experiments). Observations that metal-tolerant species performed poorly when grown on uncontaminated soils suggested these species were at a competitive disadvantage. Hickey and McNeilly (1975) measured competitive interactions in four species of metal-tolerant plants. Results showed that fitness and competitive ability of tolerant species was significantly lower than for intolerant species. Taylor et al. (1994) report that alterations in forest communities due to air pollution may result from both direct phytotoxic effects and changes in competitive ability. They suggest that phytotoxicity can reduce growth and ability to acquire resources, thus changing competitive relationships among dominant species.

Several studies have tested the hypothesis that acidification can alter competitive interactions among species. Hunter et al. (1986) measured growth rates of black ducks (*Anas rubripes*) in acidic and nonacidic ponds. They noted significant overlap in the diets of ducklings and fish and speculated that the higher growth rates of ducks in acidic ponds resulted from the elimination of fish competitors. Observations of treefrog (*Hyla andersonii*) populations showed that the distribution of this species was primarily limited to acidic ponds (Pehek 1995). Competition experiments between *H. andersonii* and two other anuran species tested the hypothesis that acidity created a refuge from predation for

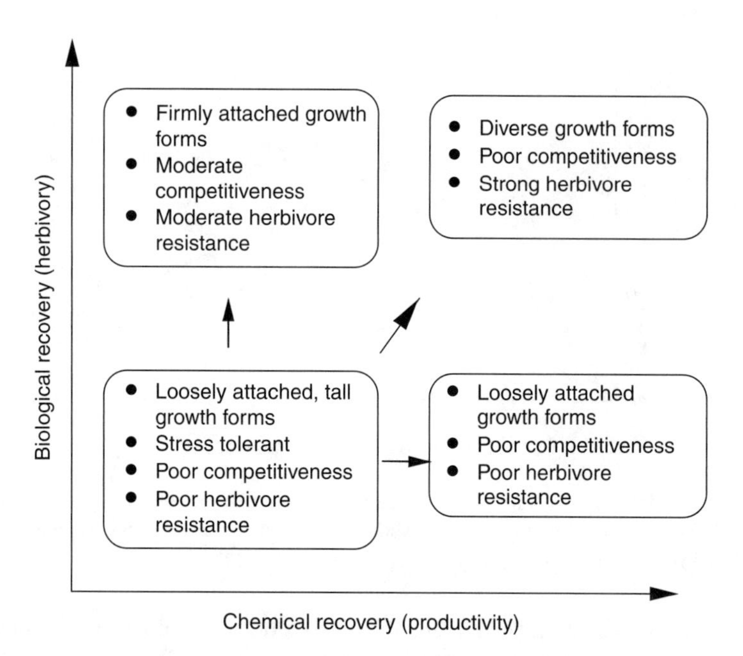

FIGURE 21.9 Conceptual model contrasting how biological and chemical recovery will result in different growth forms of algae in an acidified lake. (Modified from Figure 6 in Graham and Vinebrooke (1998).)

this acid-tolerant species. Despite strong competitive interactions among the three species, there were no differences in breeding success between low (pH = 3.9) or ambient (pH = 6.2) treatments. Graham and Vinebrooke (1998) conducted exclosure experiments to investigate trade-offs between resistance to grazing and competitive ability in periphyton communities from acidified lakes. Certain algal growth forms are known to be highly sensitive to grazing and competition. While filamentous growth forms generally outcompete closely attached, adnate species for light and nutrients, these species are generally more sensitive to herbivory. Graham and Vinebrooke developed a conceptual model to contrast how recovery of grazers and improvements in water quality in acidified lakes differentially affected these growth forms (Figure 21.9). They suggest their model can be used to understand the relationship between chemical and biological recovery in acidified lakes.

Most studies of the influence of chemical stressors on competition have either measured changes in population abundance in the field or focused on mechanisms in the laboratory. For example, Blockwell et al. (1998) developed a laboratory bioassay to measure the effects of several contaminants on competition between amphipods (*Gammarus pulex*) and isopods (*Asellus aquaticus*). Results showed that effects of the pesticide lindane on amphipod feeding rates were greater in the presence of the competitor (Figure 21.10). While these laboratory investigations are useful for demonstrating that species interactions are sensitive to chemical stressors, they provide little context for understanding the significance of these changes in natural systems. If ecotoxicologists are to develop an understanding of the role of species interactions, integration of laboratory and field experiments is essential. Lefcort et al. (1999) used field and laboratory experiments to develop a mechanistic understanding of heavy metal effects on competition between snails (*Lymnaea palustris*) and spotted frogs (*R. luteiventris*). Results showed that in the absence of heavy metals, tadpoles were able to reduce snail recruitment. However, because tadpoles were more sensitive to metals than snails, the presence of metals eliminated this competitive advantage and had a net positive effect on snails. This research was especially significant because it not only described ecological changes associated with altered competitive interactions but also identified the mechanisms responsible for these interactions. Contaminant-induced changes in competition have also been observed in terrestrial communities. Sheffield and Lochmiller (2001) exposed a small mammal community to diazinon in replicate 0.1 ha

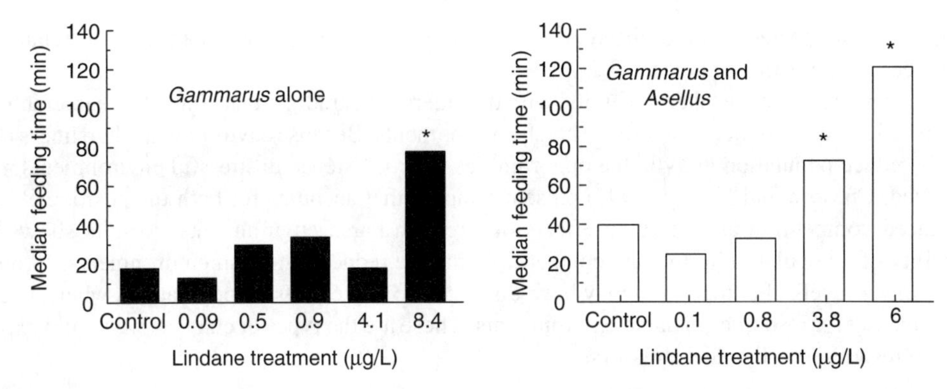

FIGURE 21.10 Effects of lindane on competitive interactions between amphipods (*Gammarus*) and isopods (*Asellus*). Feeding time of *Gammarus* is significantly increased only at the highest lindane concentration in the absence of competitors. When *Asellus* is present, effects of lindane on feeding time are increased due to competitive interactions. (Data from Tables 1 and 3 in Blockwell, S.J., et al., *Arch. Environ. Contam. Toxicol.*, 34, 41–47, 1998.)

(32 × 32 m) enclosures. Results showed that the normally strong competitive interactions between hispid cotton rats (*Sigmodon hispidus*) and prairie voles (*Microtus ochrogaster*) were altered by insecticide exposure that favored the competitively inferior species.

21.5 ENVIRONMENTAL FACTORS AND SPECIES INTERACTIONS

After several decades of attempting to identify individual factors that organize communities, ecologists now accept that multiple and often interacting factors are most likely responsible for the patterns observed in nature. There is also general agreement that the importance of biotic and abiotic processes varies with location, trophic level, and spatial scale. Simple theoretical treatments of the relative importance of disturbance, environmental variability, or species interactions have been replaced by more sophisticated models that integrate each of these processes. In a 10-year analysis of factors that organize stream fish communities, Grossman et al. (1998) determined that environmental variation was much more important than predation or competition. There is also increased awareness that environmental factors can interact with biotic processes in complex and often unpredictable ways. Peckarsky et al. (1990) reported that the role of predation in community regulation decreased with environmental harshness. Although ecologists have long recognized the direct influence of abiotic factors on populations, there have been few attempts to determine how these environmental characteristics influence species interactions. Dunson and Travis (1991) attribute this shortcoming to a cultural gap between community ecologists and physiologists. A similar cultural gap between ecologists and toxicologists may account for our poor understanding of how contaminants influence species interactions.

Recognition that the "winners" and "losers" in resource competition depend on environmental conditions is nothing new. Indeed, early laboratory experiments investigating species interactions showed that the outcome of competition was influenced by abiotic conditions (Park 1954). Dunson and Travis (1991) provide a conceptual framework for fish communities, suggesting that the ability of an organism to tolerate physiological stress is inversely related to its competitive ability. They argue that, in addition to limiting the pool of species in a specific area, abiotic factors may also determine the outcome of species interactions. By exposing closely related species to a variety of stressors, they show that differences in physiological tolerance can strongly influence resource competition. This finding has significant implications for ecotoxicological investigations because it suggests that species-specific differences in tolerance to chemical stressors may be related to differences in competitive ability.

Some researchers have taken exception with the emerging paradigm that competitive interactions are reduced and coexistence is favored in harsh environments. Because environmental harshness may directly reduce population growth, the opportunities for coexistence in stressful environments may be limited. Chesson and Huntly (1997) present a model that accounts for both the positive effects of reduced competition and negative effects of stress in harsh environments. Results show that the ability of a population to tolerate competition may be reduced in harsh environments. In other words, lower levels of competition may have disproportionate effects on populations when species are competing for resources in harsh environments. These are the types of changes we would expect to see in response to chemical stressors.

21.5.1 Environmental Stress Gradients

Menge and Sutherland's (1987) model of community regulation is a promising development in the field of ecology with direct applications in ecotoxicology. The Menge and Sutherland (hereafter, MS) model presents a conceptual framework of community organization that recognizes the importance of disturbance, competition, and predation along gradients of environmental stress. The model integrates several previous attempts to synthesize factors that determine community organization, including Hairston, Smith, and Slobodkin's (1960) trophic model and the intermediate disturbance hypothesis. Although developed in marine rocky intertidal systems, MS suggest that their model could be applied to a variety of terrestrial and aquatic habitats. More importantly, because environmental stress gradients may include physical and chemical stressors, the model is relevant to the study of contaminants.

One major goal of the MS model is to provide a framework for testing the hypothesis that communities respond predictably to variation in disturbance, competition, and predation. The model also examines how these processes vary along a gradient of environmental stress (Figure 21.11). The stress gradient may be physical (e.g., waves crashing into organisms on the rocky intertidal shore), chemical (e.g., exposure to contaminants), or physiological (e.g., temperature and desiccation

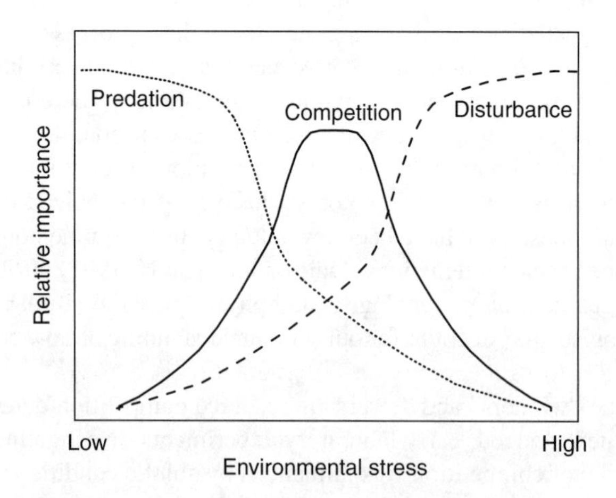

FIGURE 21.11 Conceptual model showing the influence of environmental stress on the relative importance of species interactions. At low levels of environmental stress, predation is considered the major factor regulating community structure. As environmental stress increases, predators become less effective and interspecific competition regulates the community. The role of species interactions is generally reduced under high stress conditions. On the basis of this model, we expect that contaminants would reduce the importance of species interactions in a community. (Modified from Figure 4 in Menge and Sutherland (1987).)

stress of organisms exposed during low tide). Community responses to changes in the relative importance of disturbance, competition, and predation include alterations in species diversity, food chain length, and trophic complexity. The MS model predicts that predation will be a major regulator of communities under conditions of low environmental stress, whereas competition plays an increasingly important role at intermediate levels of stress. Under harsh environmental conditions, the importance of these biotic interactions will diminish and communities will be controlled largely by disturbance. The model also considers the influence of trophic level and recruitment on these processes.

The strongest support for the MS model is from rocky intertidal communities where the primary limiting resource is space (Menge and Sutherland 1987). Attempts to verify this model in more complex systems have met with mixed success. Locke (1992) analyzed zooplankton communities from acidified lakes and found that only 3 of 10 studies showed the expected increase in species richness at intermediate levels of pH stress. In a subsequent study, Locke and Sprules (1994) analyzed zooplankton communities from 46 lakes (pH range = 3.8–7.2) sampled in the 1970s and again in 1990. The results supported two of the four predictions of the MS model (increased food web complexity and food chain length with stress). The presence of tolerant fish predators in some acidic lakes was cited as a potential explanation for the poor performance of the model. The relationship between physical disturbance and chemical stressors will be described in Chapter 25.

Despite relatively weak support in lentic communities, the MS model should be tested in other aquatic and terrestrial systems. The MS model is of particular relevance to ecotoxicology because it can be applied directly to chemical stressors. One key requirement is to locate systems with well-defined stressor gradients, a task familiar to researchers conducting environmental assessments of contaminants. Clearly one of the critical questions that must be addressed before the MS model can be applied to the study of contaminants is how effects of physical disturbances will compare to those of chemical stressors.

A refinement of the MS model that may have greater applicability to ecotoxicology was proposed by Menge and Olson (1990). They distinguish between two types of environmental stress models: consumer stress models (CSMs) and prey-stress models (PSMs) (Figure 21.12). They hypothesize that the influence of environmental stress on the outcome of consumer–resource interactions is a result of differences in species-specific sensitivity. If consumers are more sensitive to the stressor than their prey, as predicted by CSMs, consumer effects should be reduced in stressful habitats. Conversely, if prey are more sensitive to the stressor, as predicted by PSMs, consumers should have greater effects on prey populations in stressed habitats. Results shown in Table 21.6 indicate that there is support for both PSMs and CSMs in the literature. Similar models could be developed to predict the outcome of competitive interactions based on species-specific sensitivity to other environmental stressors (Dunson and Travis 1991).

21.6 SUMMARY

One of the greatest challenges in ecotoxicology is to develop an understanding of the potential indirect effects of species loss on communities. By definition, the loss of a keystone species due to an anthropogenic stressor will have disproportionate impacts on a community. In keystone-dominated communities, other species have relatively minor effects and are often considered redundant in terms of structure and function. Long-term consequences of the loss of keystone species may be influenced by the ability of these redundant species to compensate and assume similar roles as the keystone species (Ernest and Brown 2001, Navarrete and Menge 1996).

As the previous 30 years of experiments in ecology has shown, demonstrating that species interactions such as predation and competition are important organizing forces in communities has been difficult. Quantifying the influence of chemical stressors on species interactions will be especially challenging and may not be possible in many systems because of difficulties conducting experiments

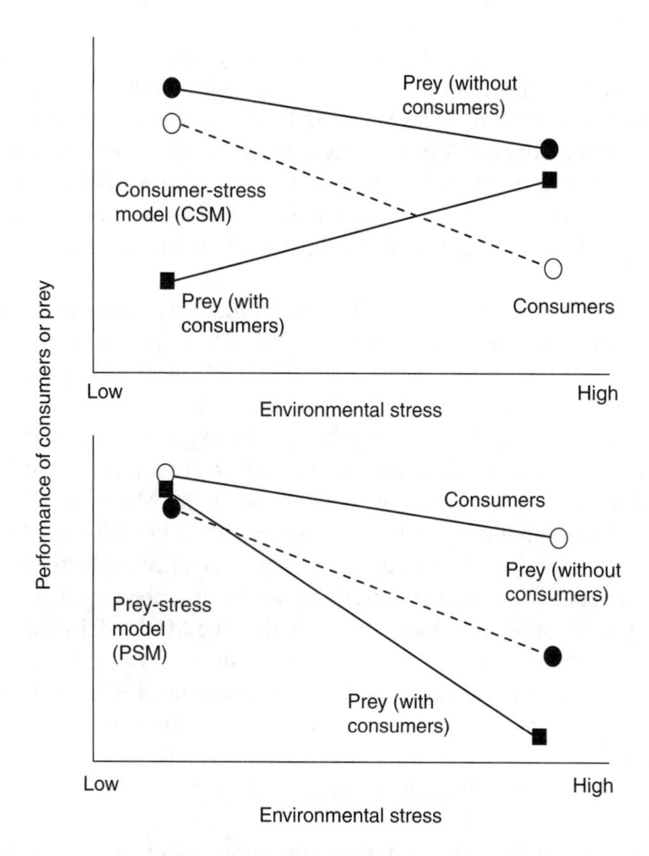

FIGURE 21.12 Consumer- and prey-stress models showing expected differences in performance of consumers and prey based on differences in sensitivity to stress. If consumers are more sensitive to stress than prey species, the CSM predicts that abundance of prey species will increase and abundance of consumers will decrease along a stress gradient. If prey species are more sensitive, the PSM predicts that abundance of prey will decrease more in the presence of predators. (Modified from Figure 2 in Menge and Olson (1990).)

at appropriate spatial and temporal scales. However, there is strong evidence that species interactions are context-dependent and that environmental factors will determine the intensity and outcome of these interactions. If community ecotoxicologists accept that species interactions are important, then some focused research should be directed at understanding the influence of contaminants on these interactions. Adopting the conceptual framework described by Rohr et al. (2006) in which contaminant-induced mortality is considered analogous to effects of a selective predator could significantly improve our understanding of these indirect effects. Because of the sensitivity of behavioral endpoints to contaminants, it may be possible to use behavioral responses as an assay to measure species interactions. Clearly, behavioral avoidance of predators is adaptive, and alteration of this response would be detrimental to populations in the field. Thus, one relatively simple test would be to measure behavioral avoidance in the presence or absence of chemical stressors. Stirling (1995) developed a "behavioral bioassay" with *Daphnia* to detect the presence of predatory fish. Similar experiments investigating alterations in behavioral responses in other communities could provide an efficient way to assess the indirect effects of chemical stressors.

21.6.1 Summary of Foundation Concepts and Paradigms

- In order to characterize community responses to contaminants and other anthropogenic disturbances, we must first understand the influence of natural spatiotemporal variation on species diversity and composition.

- Because dominance of some species increases in response to stressors, the distribution of individuals among species is a sensitive indicator of anthropogenic disturbance and has been used in biomonitoring studies.
- The emergence of equilibrium theories in ecology was supported by our deep-seated belief that attributes of natural communities are predictable and that historical factors, stochastic events, and small-scale environmental perturbations are relatively unimportant.
- Identifying species that play a significant role in structuring communities is necessary for predicting ecological consequences of contaminants and other anthropogenic stressors.
- Although there is empirical support for the hypothesis that species interactions are common and can play a pervasive role in structuring communities, the effects of contaminants on species interactions have been largely ignored by ecotoxicologists.
- Experimental designs for assessing the influence of contaminants on species interactions are complex because they require manipulation of both contaminant levels and predator/competitor abundances.
- Mechanistic studies of the effects of contaminants on predator–prey interactions generally been restricted to the laboratory and focus on behavioral changes in either predators or their prey.
- Assessing the impacts of pesticides on bird populations in the field is challenging because they require that we separate direct toxicological effects from the indirect effects due to reduced prey abundance.
- The majority of studies attempting to measure the influence of contaminants on predator–prey interactions have shown significant effects, and these effects are often observed at concentrations below those considered toxic based on single species tests.
- Relatively, few studies have measured the influence of contaminants on competitive interactions.
- Quantifying the influence of chemical stressors on species interactions will be challenging; however, there is strong evidence that species interactions are context-dependent and that contaminants will determine the outcome of these interactions.

REFERENCES

Antonovics, J., Bradshaw, A.D., and Turner, R.G., Heavy metal tolerance in plants, *Adv. Ecol. Res.*, 7, 1–85, 1971.

Ball, S.L. and Baker, R.L., The non-lethal effects of predators and the influence of food availability on life history of adult *Chironomus tentans* (Diptera: Chironomidae), *Freshw. Biol.*, 34, 1–12, 1995.

Ball, S.L. and Baker, R.L., Predator-induced life history changes: Antipredator behavior costs or facultative life history shifts? *Ecology*, 77, 1116–1124, 1996.

Beitinger, T.L., Behavioral reactions for the assessment of stress in fishes, *J. Great Lakes Res.*, 16, 495–528, 1990.

Bellwood, D.R. and Hughes, T.P., Regional-scale assembly rules and biodiversity of coral reefs, *Science*, 292, 1532–1534, 2001.

Blockwell, S.J., Taylor, E.J., Jones, I., and Pascoe, D., The influence of fresh water pollutants and interaction with *Asellus aquaticus* (L.) on the feeding activity of *Gammarus pulex* (L.), *Arch. Environ. Contam. Toxicol.*, 34, 41–47, 1998.

Brooks, J.L. and Dodson, S.I., Predation, body size and composition of plankton, *Science*, 150, 28–35, 1965.

Brown, J.A., Johansen, P.H., Colgan, P.W., and Mathers, R.A., Impairment of early feeding behavior of largemouth bass by pentachlorophenol exposure: A preliminary assessment, *Trans. Am. Fish. Soc.*, 116, 71–78, 1987.

Brown, J.H., Two decades of homage to Santa Rosalia: Toward a general theory of diversity, *Am. Zool.*, 21, 877–888, 1981.

Brown, J.H. and Heske, E.J., Control of a desert-grassland transition by a keystone rodent guild, *Science*, 250, 1705–1707, 1990.

Brown, J.H. and Lomolino, M.V., *Biogeography*, 2nd ed., Sinauer Associates, Inc., Sunderland, MA, 1998.

Brown, J.H. and Munger, J.C., Experimental manipulation of a desert rodent community: Food addition and species removal, *Ecology*, 66, 1545–1563, 1985.

Bryan, M.D., Atchison, G.J., and Sandheinrich, M.B., Effects of cadmium on the foraging behavior and growth of juvenile bluegill, *Lepomis macrochirus*, *Can. J. Fish. Aquat. Sci.*, 52, 1630–1638, 1995.

Buerger, T.T., Kendall, R.J., Mueller, B.S., De Vos, T., and Williams, B.A., Effects of methyl parathion on northern bobwhite survivability, *Environ. Toxicol. Chem.*, 10, 527–532, 1991.

Carpenter, R.C., Mass mortality of *Diadema antillarium*. I. Long-term effects on sea urchin population-dynamics and coral reef algal communities, *Mar. Biol.*, 104, 67–77, 1990.

Chesson, P. and Huntly, N., The roles of harsh and fluctuating conditions in the dynamics of ecological communities, *Am. Nat.*, 150, 519–553, 1997.

Clements, W.H., Effects of contaminants at higher levels of biological organization in aquatic systems, *Rev. Toxicol.*, 1, 269–308, 1997.

Clements, W.H., Metal tolerance and predator–prey interactions in benthic macroinvertebrate stream communities, *Ecol. Appl.*, 9, 1073–1084, 1999.

Clements, W.H., Cherry, D.S., and Cairns, J., Jr., The influence of copper exposure on predator–prey interactions in aquatic insect communities, *Freshw. Biol.*, 21, 483–488, 1989.

Connell, J.H. and Orias, E., The ecological regulation of species diversity, *Am. Nat.*, 98, 399–414, 1964.

Crawley, M.J. and Harral, J.E., Scale dependency in plant biodiversity, *Science*, 291, 864–868, 2001.

Diamond, J., Overview: Laboratory experiments, field experiments, and natural experiments, In *Community Ecology*, Diamond, J. and Case, T.J. (eds.), Harper & Row, New York, 1986, pp. 3–22.

Diamond, J.M., Niche shifts and the rediscovery of competition: Why did field biologists so long overlook the widespread evidence for interspecific competition that had already impressed Darwin? *Am. Sci.*, 66, 322–331, 1978.

Dunson, W.A. and Travis, J., The role of abiotic factors in community organization, *Am. Nat.*, 138, 1067–1091, 1991.

Ebenman, B., Law, R., and Borrvall, C., Community viability analysis: The response of ecological communities to species loss, *Ecology*, 85, 2591–2600, 2004.

Ernest, S.K. and Brown, J.H., Delayed compensation for missing keystone species by colonization, *Science*, 292, 101–104, 2001.

Fair, J.M., Kennedy, P.L., and McEwen, L.C., Effects of carbaryl grasshopper control on nesting killdeer in North Dakota, *Environ. Toxicol. Chem.*, 14, 881–890, 1995.

Fausch, K.D., Interspecific competition and juvenile Atlantic salmon (*Salmo salar*): On testing effects and evaluating evidence across scales, *Can. J. Fish. Aquat. Sci.*, 55(Suppl. 1), 218–231, 1998.

Fleeger, J.W., Carman, K.R., and Nisbet, R.M., Indirect effects of contaminants in aquatic ecosystems, *Sci. Total Environ.*, 317, 207–233, 2003.

Goodyear, C.P., A simple technique for detecting effects of toxicants or other stresses on a predator–prey interaction, *Trans. Am. Fish. Soc.*, 101, 367–370, 1972.

Graham, M.D. and Vinebrooke, R.D., Trade-offs between herbivore resistance and competitiveness in periphyton of acidified lakes, *Can. J. Fish. Aquat. Sci.*, 55, 806–814, 1998.

Grossman, G.D., Ratajczak, R.E., Crawford, M., and Freeman, M.C., Assemblage organization in stream fishes: Effects of environmental variation and interspecific interactions, *Ecol. Monogr.*, 68, 395–420, 1998.

Grue, C.E., Powell, G.V.N., and McChesney, M.J., Care of nestlings by wild female starlings exposed to an organophosphate pesticide, *J. Appl. Ecol.*, 19, 327–335, 1982.

Hairston, N.G., Species abundance and community organization, *Ecology*, 40, 404–416, 1959.

Hairston, N.G., Sr., Smith, F.E., and Slobodkin, L.B., Community structure, population control, and interspecific competition, *Am. Nat.*, 94, 421–425, 1960.

Ham, L., Quinn, R., and Pascoe, D., Effects of cadmium on the predator–prey interaction between the turbellarian *Dendrocoelum lacteum* (Muller, 1774) and the isopod crustacean *Asellus aquaticus* (L.), *Arch. Environ. Contam. Toxicol.*, 29, 358–365, 1995.

Hatfield, C.T. and Anderson, J.M., Effects of two insecticides on the vulnerability of Atlantic salmon (*Salmo salar*) parr to brook trout (*Salvelinus fontinalis*) predation, *J. Fish. Res. Bd. Can.*, 29, 27–29, 1972.

Hickey, D.A. and McNeilly, T., Competition between metal tolerant and normal plant populations: A field experiment on normal soil, *Evolution*, 29, 458–464, 1975.

Hill, J., Analysis of six foraging behaviors as toxicity indicators, using juvenile smallmouth bass exposed to low environmental pH, *Arch. Environ. Contam. Toxicol.*, 18, 895–899, 1989.

Hunter, M.L., Jr., Jones, J.J., Gibbs, K.E., and Moring, J.R., Duckling responses to lake acidification: Do black ducks and fish compete? *Oikos*, 47, 26–32, 1986.

Hutchinson, G.E., Homage to Santa Rosalia or Why are there so many kinds of animals? *Am. Nat.*, 93, 145–159, 1959.

Hutchinson, G.E., The paradox of plankton, *Am. Nat.*, 95, 137–145, 1961.

Interlandi, S.J. and Kilham, S.S., Limiting resources and the regulation of diversity in phytoplankton communities, *Ecology*, 82, 1270–1282, 2001.

Jenkins, D.G. and Buikema, A.L., Jr., Do similar communities develop in similar sites? A test with zooplankton structure and function, *Ecol. Monogr.*, 68, 421–443, 1998.

Kania, H.J. and O'Hara, J., Behavioral alterations in a simple predator–prey system due to a sublethal exposure to mercury, *Trans. Am. Fish. Soc.*, 103, 134–136, 1974.

Kareiva, P., Higher order interactions as a foil to reductionist ecology, *Ecology*, 75, 1527–1528, 1994.

Kessler, A. and Baldwin, I.T., Defensive function of herbivore-induced plant volatile emissions in nature, *Science*, 291, 2141–2144, 2001.

Kiffney, P.M., Main and interactive effects of invertebrate density, predation, and metals on a Rocky Mountain stream macroinvertebrate community, *Can. J. Fish. Aquat. Sci.*, 53, 1595–1601, 1996.

Kruzynski, G.M. and Birtwell, I.K., A predation bioassay to quantify the ecological significance of sublethal responses of juvenile chinook salmon (*Oncorhynchus tshawytscha*) to the antisapstain fungicide TCMTB, *Can. J. Fish. Aquat. Sci.*, 51, 1780–1790, 1994.

Landis, W.G., Matthews, R.A., and Matthews, G.B., The layered and historical nature of ecological systems and the risk assessment of pesticides, *Environ. Toxicol. Chem.*, 15, 432–440, 1996.

Lefcort, H., Meguire, R.A., Wilson, L.H., and Ettinger, W.F., Heavy metals alter the survival, growth, metamorphosis, and antipredatory behavior of Columbia spotted frog (*Rana luteiventris*) tadpoles, *Arch. Environ. Contam. Toxicol.*, 35, 447–456, 1998.

Lefcort, H., Thomson, S.M., Cowles, E.E., Harowicz, H.L., Livaudais, B.M., Roberts, W.E., and Ettinger, W.F., Ramifications of predator avoidance: Predator and heavy-metal-mediated competition between tadpoles and snails, *Ecol. Appl.*, 9, 1477–1489, 1999.

Locke, A., Factors influencing community structure along stress gradients: Zooplankton responses to acidification, *Ecology*, 73, 903–906, 1992.

Locke, A. and Sprules, W.G., Effects of lake acidification and recovery on the stability of zooplankton food webs, *Ecology*, 75, 498–506, 1994.

MacArthur, R.H. and Wilson. E.O., An equilibrium theory of insular zoogeography, *Evolution*, 17, 373–387, 1963.

MacArthur, R.H. and Wilson, E.O., *The Theory of Island Biogeography*, Princeton University Press, Princeton, NJ, 1967.

Martin, P.A., Johnson, D.L., Forsyth, D.J., and Hill, B.D., Indirect effects of the pyrethroid insecticide deltamethrin on reproductive success of chestnut-collared longspurs, *Ecotoxicology*, 7, 89–97, 1998.

Matthews, R.A., Landis, W.G., and Matthews, G.B., The community conditioning hypothesis and its application to environmental toxicology, *Environ. Toxicol. Chem.*, 15, 597–603, 1996.

May, R.M. An overview: Real and apparent patterns in community structure, In *Ecological Communities: Conceptual Issues and the Evidence*, Strong, D.R., Simberloff, D., Abele, L.G., and Thistle, A.B. (eds.), Princeton University Press, Princeton, NJ, 1984, pp. 3–16.

May, R.M. and Stumpf, M.P.H., Species–area relations in tropical forests, *Science*, 290, 2084–2086, 2000.

McLaren, B.E. and Peterson, R.O., Wolves, moose, and tree rings on Isle Royale, *Science*, 266, 1555–1558, 1994.

Menge, B.A., Berlow, E.L., Blanchette, C.A., Navarrete, S.A., and Yamada, S.Y., The keystone species concept: Variation in interaction strength in a rocky intertidal habitat, *Ecol. Monogr.*, 64, 249–286, 1994.

Menge, B.A. and Olson, A.M., Role of scale and environmental factors in regulation of community structure, *Trends Ecol. Evol.*, 5, 52–57, 1990.

Menge, B.A. and Sutherland, J.P., Community regulation: Variation in disturbance, competition, and predation in relation to environmental stress and recruitment, *Am. Nat.*, 130, 730–757, 1987.

Minshall, G.W., Petersen, R.C., Jr., and Nimz, C.F., Species richness in streams of different size from the same drainage basin, *Am. Nat.*, 125, 16–38, 1985.

Morgan, M.J. and Kiceniuk, J.W., Effect of fenitrothion on the foraging behavior of juvenile atlantic salmon, *Environ. Toxicol. Chem.*, 9, 489–495, 1990.

Navarrete, S.A. and Menge, B.A., Keystone predation and interaction strength: Interactive effects of predators on their prey, *Ecol. Monogr.*, 66, 409–429, 1996.

Paine, R.T., Food web complexity and species diversity, *Am. Nat.*, 100, 65–75, 1966.

Paine, R.T., A note on trophic complexity and community stability, *Am. Nat.*, 103, 91–93, 1969.

Park, T., Experimental studies of interspecies competition. II. Temperature, humidity, and competition in two species of *Tribolium*, *Physiol. Zool.*, 27, 177–229, 1954.

Peckarsky, B.L., Horn, S.C., and Statzner, B., Stonefly predation along a hydraulic gradient: A field test of the harsh-benign hypothesis, *Freshw. Biol.*, 24, 181–191, 1990.

Peckarsky, B.L. and McIntosh, A.R., Fitness and community consequences of avoiding multiple predators, *Oecologia*, 113, 565–576, 1998.

Pehek, E.L., Competition, pH, and the ecology of larval *Hyla andersonii*, *Ecology*, 76, 1786–1793, 1995.

Poff, N.L., Landscape filters and species traits: Towards mechanistic understanding and prediction in stream ecology, *J. N. Am. Benthol. Soc.*, 16, 391–409, 1997.

Power, M.E., Effects of fish in river food webs, *Science*, 250, 811–814, 1990.

Power, M.E., Tilman, D., Estes, J.A., Menge, B.A., Bond, W.J., Mills, L.S., Daily, G., Castilla, J.C., Lubchenco, J., and Paine, R.T., Challenges in the quest for keystone species, *Bioscience*, 46, 606–620, 1996.

Preisser, E.L., Bolnick, D.I., and Benard, M.F., Scared to death? The effects of intimidation and consumption in predator–prey interactions, *Ecology*, 86, 501–509, 2005.

Preston, B.L., Cecchine, G., and Snell, T.W., Effects of pentachlorophenol on predator avoidance behavior of the rotifer *Brachionus calyciflorus*, *Aquat. Toxicol.*, 44, 201–212, 1999.

Rapoport, E.H., *Aerography: Geographical strategies of species*, Pergamon Press, New York, 1982.

Rohr, J.R., Kerby, J.L., and Sih, A., Community ecology as a framework for predicting contaminant effects, *Trends Ecol. Evol.*, 21, 606–613, 2006.

Roughgarden, J. and Diamond, J., Overview: The role of species interactions in community ecology, In *Community Ecology*, Diamond, J.M. and Case, T.J. (eds.), Harper & Row, New York, 1986, pp. 333–343.

Sandheinrich, M.B. and Atchison, G.J., Sublethal copper effects on bluegill, *Lepomis macrochirus*, foraging behavior, *Can. J. Fish. Aquat. Sci.*, 46, 1977–1985, 1989.

Sandheinrich, M.B. and Atchison, G.J., Sublethal toxicant effects on fish foraging behavior: Empirical vs. mechanistic approaches, *Environ. Toxicol. Chem.*, 9, 107–119, 1990.

Schluter, D. and Ricklefs, R.E., Species diversity—an introduction to the problem, In *Species Diversity in Ecological Communities: Historical and Geographical Perspectives*, Ricklefs, R.E. and Schluter, D. (eds.), University of Chicago Press, Chicago, IL, 1993, pp. 1–10.

Schoener, T.W., The species-area relationship within archipelagos: Models and evidence from island land birds, In *Proceedings of the 16th International Ornithological Congress*, Firth, H.J. and Calby, J.H. (eds.), Australian Academy of Science, Canberra, Australia, 1974, pp. 629–642.

Schoener, T.W., Field experiments on interspecific competition, *Am. Sci.*, 122, 240–285, 1983.

Schulz, R. and Dabrowski, J.M., Combined effects of predatory fish and sublethal pesticide contamination on the behavior and mortality of mayfly nymphs, *Environ. Toxicol. Chem.*, 20, 2537–2543, 2001.

Sheffield, S.R. and Lochmiller, R.L., Effects of field exposure to diazinon on small mammals inhabiting a semienclosed prairie grassland ecosystem. I. Ecological and reproductive effects. *Environ. Toxicol. Chem.*, 20, 284–296, 2001.

Simberloff, D.S. and Wilson, E.O., Experimental zoogeography of islands: The colonization of empty islands, *Ecology*, 50, 278–296, 1969.

Simberloff, D.S. and Wilson, E.O., Experimental zoogeography of islands: A two-year record of colonization, *Ecology*, 51, 934–937, 1970.

Stirling, G., Daphnia behaviour as a bioassay of fish presence or predation, *Funct. Ecol.*, 9, 778–784, 1995.

Sullivan J.F., Atchison G.J., Kolar D.J., and MacIntosh A.W., Changes in the predator–prey behavior of fathead minnows (*Pimephales promelas*) and largemouth bass (*Micropterus salmoides*) caused by cadmium, *J. Fish. Res. Bd. Can.*, 35, 446–451, 1978.

Tansley, A.G. and Adamson, R.S., Studies of the vegetation of the English chalk. III. The chalk grasslands of the Hampshire–Sussex border, *J. Ecol.*, 13, 177–223, 1925.

Taylor, E.J., Morrison, J.E., Blockwell, S.J., Tarr, A., and Pascoe, D., Effects of lindane on the predator–prey interaction between *Hydra oligactis* pallas and *Daphnia magna* Strauss, *Arch. Environ. Contam. Toxicol.*, 29, 291–296, 1995.

Taylor, G.E., Jr., Johnson, D.W., and Anderson, C.P., Air pollution and forest ecosystems: A regional to global perspective, *Ecol. Appl.*, 4, 662–689, 1994.

United States Department of Agriculture, *Rangeland Grasshopper Cooperative Management Program, Final Environmental Impact Statement*, FEIS 87-1, Animal and Plant Health Inspection Service, Washington, D.C., 1987.

Vannote, R.L., Minshall, G.W., Cummins, K.W., Sedell, J.R., and Cushing, C.E., The river continuum concept, *Can. J. Fish. Aquat. Sci.*, 37, 130–137, 1980.

Wallace, W.G., Brouwer, T.M.H., Brouwer, M., and Lopez, G.R., Alterations in prey capture and induction of metallothioneins in grass shrimp fed cadmium-contaminated prey, *Environ. Toxicol. Chem.*, 19, 962–971, 2000.

Wardle, D.A., Zackrisson, O., Hornberg, G., and Gallet, C., The influence of island area on ecosystem properties, *Science*, 277, 1296–1299, 1997.

Werner, E.E. and Hall, D.J., Optimal foraging and the size selection of prey by the bluegill sunfish *Lepomis macrochirus*, *Ecology*, 55, 1042–1052, 1974.

Werner, E.E. and Peacor, S.D., Lethal and nonlethal predator effects on an herbivore guild mediated by system productivity, *Ecology*, 87, 347–361, 2006.

Whittaker, R.H., *Communities and Ecosystems* 2nd ed., Macmillan, New York, 1975.

Wiens, J.A., On understanding a non-equilibrium world: Myth and reality in community patterns and processes, In *Ecological Communities: Conceptual Issues and the Evidence*, Strong, D.R., Simberloff, D., Abele, L.G. and Thistle, A.B. (eds.), Princeton University Press, Princeton, NJ, 1984, pp. 439–457.

Woltering, D.M., Hedtke, J.L., and Weber, L.J., Predator–prey interactions of fishes under the influence of ammonia, *Trans. Am. Fish. Soc.*, 107, 500–504, 1978.

22 Biomonitoring and the Responses of Communities to Contaminants

22.1 BIOMONITORING AND BIOLOGICAL INTEGRITY

Biomonitoring is defined as the use of biological systems to assess the structural and functional integrity of aquatic and terrestrial ecosystems. Karr and Dudley (1981) define biological integrity as the ability of an ecosystem "to support and maintain a balanced, integrated, adaptive community of organisms having a species composition, diversity, and functional organization comparable to natural habitats in the region." Measurements (endpoints) used to assess biological integrity may be selected from any level of biological organization; however, the historical focus has been on populations, communities, and ecosystems. Community-level biological monitoring, which is the focus of this chapter, is based on the assumption that composition and organization of communities reflect local environmental conditions and respond to anthropogenic alteration of those conditions. A second important assumption of community-level biomonitoring is that species differ in their sensitivity to anthropogenic stressors, resulting in structural and functional changes at polluted sites.

Karr and Dudley's definition of biological integrity underscores the two most significant challenges to the development and implementation of community-level monitoring: the selection of endpoints and the identification of reference conditions. Although Karr and Dudley provide some suggestions for endpoints (e.g., species diversity and composition), there is little consensus among ecologists as to what key features of communities are the most appropriate indicators of biological integrity. There is, however, widespread agreement that no single measure will be effective and that approaches integrating several endpoints are often necessary to assess effects of contaminants.

The selection of appropriate reference sites and the determination of what exactly constitutes "natural habitats in the region" have been equally troublesome to natural resource managers. Identifying reference conditions and separating natural variation from contaminant-induced changes are currently major areas of research interest. Community ecotoxicologists have utilized a variety of study designs to distinguish the effects of contaminants from natural variation. If natural changes in community composition are predictable and occur along well-defined gradients (e.g., the longitudinal changes in stream communities along a river continuum), then this variation can be explained using an appropriate study design and statistical analyses. In situations where natural variation is more stochastic, it may be difficult to quantify all but the most extreme examples of perturbation. Regardless, an understanding of the natural spatial and temporal variation of community structure is essential for any biomonitoring program.

Although biomonitoring studies have been conducted in almost every type of aquatic and terrestrial ecosystem, community-level assessments of contaminant effects are largely restricted to aquatic habitats. Excellent historical descriptions of the early development of biological monitoring in aquatic habitats have been published (Cairns and Pratt 1993, Davis 1995). Biological monitoring of community attributes in aquatic systems has occurred since the early 1900s. More recently, conservation biologists have begun to employ community-level monitoring techniques to estimate biodiversity and to prioritize sites for preservation. However, assessments of contaminant effects at the level of communities are much less common in terrestrial systems. We consider the lack of information on responses of terrestrial communities to contaminants to be a significant research limitation in ecotoxicology.

22.2 CONVENTIONAL APPROACHES

Conventional approaches in biological monitoring begin with a species list (or some other taxonomic category) for the study site or sampling unit. The species list consists of species names and the numbers of individuals present for each. Depending on the taxonomic group, other units besides individuals might be used, such as species biomass or groundcover. Some lists may indicate simple presence or absence from the sample instead of the actual numbers of individuals. None of the methods retain information on the spatial relationship among individuals in the community other than the implicit understanding that all organisms came from the same sampling unit. An associated sampling site is defined operationally based on tractability and the assumption of homogeneity within the site (Pielou 1969). The species being enumerated might all be associated with a particular part of the habitat or microhabitat (e.g., a benthic community) or with a specific taxonomic group (e.g., tree canopy insects). Interpretation of the resulting indices must be done thoughtfully because the data will never reflect the entire ecological community.

Species diversity or heterogeneity indices include both evenness and richness. This blending may be seen as convenient or confounding depending on one's ultimate goal. Due to the computational ease for calculating these indices, tandem computation of species richness, evenness, and diversity seems the best way of extracting the most meaningful information. A few of the more common community indices are described below, with alpha diversity (see Chapter 21) being considered the most relevant for ecotoxicological investigations. The reader is referred to Pielou (1969), May (1976), Ludwig and Reynolds (1988), Magurran (1988), Newman (1995), and Matthews et al. (1998) for more detail and theory associated with these metrics.

22.2.1 INDICATOR SPECIES CONCEPT

The impacts of degraded water quality on biological communities were first noted in the early 1900s by German biologists describing effects of organic enrichment on benthic fauna. The Saprobien system of classification (Kolwitz and Marsson 1909) distinguished three categories of streams (polysaprobic, mesosaprobic, and oligosaprobic) based on the abundance of pollution-tolerant and pollution-sensitive species. The partially subjective index was based on well-established lists of species and their observed tolerances of conditions at various distances from a waste source. Primary among the factors considered is oxygen tolerance as it strongly influences the ability of a species to flourish in the different zones below the discharge. These early attempts to characterize water quality based on presence or absence of indicator species launched a significant but highly controversial period in biological monitoring. The use of indicator species, which are defined as species known to be sensitive or tolerant to a specific class of environmental conditions, has received considerable attention in the literature (Cairns and Pratt 1993).

Although their specific life history characteristics will vary, pollution-tolerant species generally include organisms with high intrinsic rates of increase, rapid colonization ability, and/or morphological and physiological adaptations that allow them to withstand exposure to toxic chemicals or habitat alteration (see Chapter 25). In contrast, pollution-sensitive species are defined as those species that are consistently absent from systems with known physical or chemical disturbances. The classic example of indicator organisms in aquatic systems, which figured prominently in development of the original Saprobien system, are the large numbers of pollution-tolerant chironomids (*Diptera:*

Chironomidae) and oligochaete worms that commonly replace sensitive mayflies (Ephemeroptera) and stoneflies (Plecoptera) at sites with high levels of organic enrichment.

While the notion that presence or absence of a particular species could indicate the degree of environmental degradation has intuitive appeal, there are obvious limitations with this approach. The indicator species concept has received rather unfavorable reviews in the United States (Cairns 1974). One of the most obvious shortcomings of this approach is the difficulty in defining pollution tolerance for species without resorting to inherently tautological arguments (e.g., species are defined as pollution-sensitive because they are absent from polluted habitats). The second limitation, which is considerably more serious, is the need to distinguish the relative importance of chemical stressors from the multitude of other biotic and abiotic factors that influence the presence or absence of a species. This is especially problematic in aquatic systems because many of the species that are sensitive to chemical stressors are also sensitive to other natural or anthropogenic disturbances. The absence of a pollution-sensitive species from a contaminated site provides only weak support for the hypothesis that its absence is due to contamination. Similarly, the presence of pollution-tolerant species (e.g., chironomids and oligochaetes in aquatic systems) does not necessarily imply that a site is degraded. Roback (1974) summarized his opinion of the indicator species concept, which is probably shared by many stream ecologists, stating that, "the presence or absence of any species in a stream indicates no more or less than the bald fact of its presence or absence."

Before dismissing the indicator species concept, we should recognize its general contributions to biological monitoring and its applications outside of water quality assessments. Although the absence of a particular species tells little about environmental conditions, its presence may be much more informative. For example, in the Pacific Northwest, the endangered spotted owl (*Strix occidentalis*) is a habitat specialist known to be highly dependent on old growth forests. Because factors other than the availability of old growth forests can influence its distribution, the *absence* of spotted owls from an area is not especially informative. However, the presence of this old growth specialist provides useful information on habitat suitability. Similarly, the presence of a species known to be sensitive to a particular type of pollutant provides strong evidence that the chemical is either not present or not bioavailable. With careful application, the indicator species concept could be employed to locate potential reference sites or to document recovery following pollution abatement. Because of the ability of some species to either acclimate or adapt to chemical stressors (Mulvey and Diamond 1991, Newman 2001, Wilson 1988), it is important to consider that tolerance developed during exposure may allow sensitive organisms to persist in polluted habitats.

The hasty abandonment of the Saprobien system and the indicator species concept is at least partially responsible for the relatively slow progress in the field of biological monitoring. Cairns and Pratt (1993) note that the unwillingness of stream ecologists to accept the indicator species concept supported the dominant viewpoint that water quality monitoring programs could focus exclusively on physical and chemical measures. Despite the poor initial support, the indicator species concept and Saprobien system are credited with initiating interest in the development of numerical criteria (Davis 1995). Furthermore, the modern approach of using indicator communities to assess environmental perturbation was at least partially inspired by this early work.

22.3 BIOMONITORING AND COMMUNITY-LEVEL ASSESSMENTS

22.3.1 Species Abundance Models

During the early history of ecology, field biologists were satisfied to characterize communities based on extensive species lists showing the presence or absence of individual taxa. There were few attempts to quantify species abundance distributions or to propose ecological explanations for these patterns. Frank Preston's (1948) seminal paper on the "Commonness and rarity of species" was considered a significant turning point in the maturation of community ecology. Ecologists had long observed that some species in nature are quite rare and represented by relatively few individuals whereas other species are very abundant. Preston's contribution provided one of the first opportunities to quantify this relationship.

Species abundance models are a useful way to summarize data from community surveys. Models are fit to tabulated species abundances, and model parameters become the summary statistics for the data set. However, more useful information can be extracted from these models (Pielou 1975), such as estimates of the total number of species in the community. Some variables, such as the parameter of the log series model, are commonly employed diversity indices. The steepness of species abundance curves (Figure 22.1, upper panel) suggests the evenness with which individuals are distributed among species (Tokeshi 1993). As will be shown shortly, evenness increases in the following model sequence: geometric series < log series < discrete log normal < broken stick.

Although many models exist (Tokeshi 1993), abundance data are commonly fit to only four models: logarithmic series, geometric series, discrete log normal, and broken stick. All have been

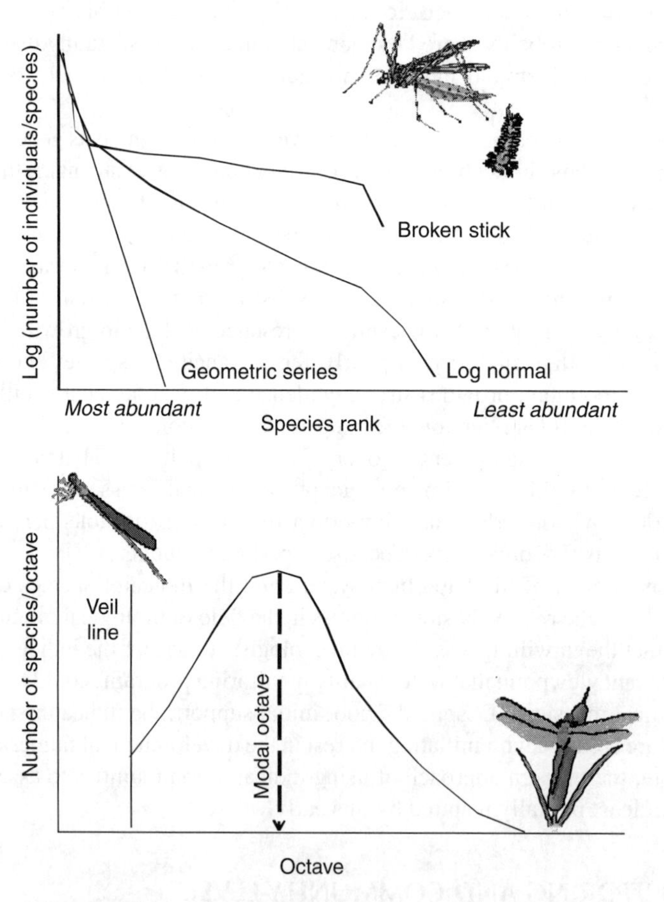

FIGURE 22.1 Species abundance curves for summarizing community data. The top panel depicts three conventional models including the extremes (geometric series and broken stick) and most commonly used (log normal) models. The bottom panel illustrates Preston's (1948) approach to analyzing species abundance data with a log normal model. Notice that there is a veil line on the x-axis. For most such curves, there is some minimal count (e.g., one individual/species), below which abundance cannot be quantified. Much of the mathematics associated with Preston's analysis of the log normal model is associated with estimating distributional parameters with such a left-truncated curve.

interpreted in the context of resource competition, with the relative species abundance being used to imply the portion of resources or niche volume secured by a species. Whether competition is a reasonable foundation for such a model depends very much on the community, species assemblage, or taxonomic group being studied. It may be very appropriate for studying an ecological guild but quite inadequate for a collection of functionally divergent species. Although the explanations based on realized niche and resource allocation "are useful in suggesting possibilities underlying community organization" (Tokeshi 1993), interpretation based on competition theory should be done cautiously (Hughes 1986). Some researchers prefer to view species abundance models as statistical models because of this loose theoretical foundation. However, the cost of such freedom from theory is a severely restricted ability to assign ecological meaning to results.

The simplest and earliest model, the geometric series (Motomura 1932), is based on the niche preemption concept (Figure 22.1). According to this model, one species takes kth of the available niche space, leaving only $1 - k$ for the remaining species to share. A second species then takes kth of the remaining $1 - k$ niche space. This niche preemption sequence continues until all species have secured their portion of the available niche space. Any variation from k among species is attributed to stochasticity.

There will be a few very abundant species in such a community, as might be expected during early stages of succession in which r-selected strategies dominate or for a community associated with a severe environment in which one or a few factors determine species success (May 1976). The associated model is given in the following equation (Magurran 1988):

$$N_i = kN\left[\frac{1}{1-(1-k)^s}\right][1-k]^{i-1}. \tag{22.1}$$

A log series model is similar to the geometric series except that species arrive and occupy niche space randomly, not in the regular intervals as described for the geometric series. The result is a community with a few dominants and more rare species than the geometric model would predict. The curve for the log series would be intermediate between the geometric series and log normal models in Figure 22.1. The expected number of species with n individuals is $\alpha x^n/n$, with x being a sample size-dependent constant less than 1 and α being a community-dependent constant. The log series model is often described as the model most useful for "samples from small, stressed, or pioneer communities" (Hughes 1986).

The discrete log normal model fits most communities (Magurran 1988) and is often advocated as universally acceptable for species abundance modeling (May 1976). The competition theory behind it is that a species' success in occupying niche space is determined by many factors. The result is more intermediate abundance species and fewer rare species than for the geometric series model (Figure 22.1). In contrast to the geometric series model in which r-selection strategists often dominate, this model might be more suggestive of equilibrium or K-selection strategies such as those occurring in climax or unstressed communities.

The log normal model cannot be fit by simply calculating the central tendency and dispersion parameters, because values for some observations to the left of the veil point are not known (Figure 22.1, lower panel). Preston (1948) speculated that log normal distributions were truncated because of the difficulty sampling all rare species in a community and that the distribution would shift to the right with larger sample sizes. Preston developed the classic method for analyzing the truncated log normal species abundance curve by first separating all species into abundance classes. The most convenient abundance categories were octaves, grouped by doubling in numbers such as 1 to 2, 2 to 4, 4 to 8, 8 to 16, 16 to 32, and so forth. The number of species in each octave was plotted to produce a graph similar to the lower panel of Figure 22.1. The octaves are often labeled relative to the modal octave (e.g., $R = 0$ denotes the modal octave, $R = -1$ denotes one octave to the left of the mode, and $R = 2$ denotes two octaves to the right of the mode). In samples containing large numbers of species, a normal distribution is obtained when the log abundance of species is plotted against the number of species in each category.

The original method of Preston (1948) or the more simplified approach of Newman (1995) can be used to estimate the distribution parameters and subsidiary information such as the estimated number of species in the community. The predicted number of species in octave R (S_R) is estimated from the number of species in the modal octave (S_0) and the variance of the log normal distribution, σ^2.

Preston's log normal distribution was found to be widely applicable for explaining the rank abundance of many taxonomic groups. Although Preston did not provide an ecological explanation for the generality of log normal distributions in nature, other ecologists discussed the evolutionary implications. Using the broken stick model, MacArthur (1960) proposed that species abundance distributions resulted from interspecific competition and allocation of resources among species. According to this model, the niche space available to any species is allocated much as a length of stick would be if a stick were randomly snapped along its length to produce S pieces. In more formal terms, $S - 1$ points are randomly identified along the length of the stick and the stick is broken at these points. The length of each segment reflects the amount of niche space (inferred from species abundance) allocated to each species. In such a model, the niche space would be randomly distributed among the S species to produce a community with many moderately abundant species but relatively few rare or extremely abundant species (Figure 22.1 bottom panel). As such, this model is most likely to describe an equilibrium assemblage of very similar species (e.g., a specific guild in a climax community).

Magurran (1988) provides estimators of the expected number of individuals (N_i) for the ith most abundant species (Equation 22.2) and the expected number of species (S_n) for the nth abundance class (Equation 22.3) based on the broken stick model:

$$S_n = S_0\, \mathrm{e}^{-(1/\sqrt{2\sigma^2})^2 R^2}, \tag{22.2}$$

$$N_i = \frac{N}{S} \sum_{n=i}^{S} \frac{1}{n}. \tag{22.3}$$

Which specific model best fits the data statistically can be determined by deferring to the advice of experts (e.g., May's preference for the log normal model), or by applying conventional goodness-of-fit methods. Magurran (1988), Ludwig and Reynolds (1988), and Newman (1995) provide the details for formally assessing relative model goodness-of-fit. Regardless of how relative model goodness-of-fit is examined, one is ultimately faced with the difficult task of deciding which model best fits the ecological reality of the species assemblage being studied.

In general, attempts to seek underlying biological processes for log normal distributions were unsuccessful. Recent analyses of log normal distributions and MacArthur's broken stick model have revealed their statistical inevitability (Gotelli and Graves 1996). Despite the lack of an evolutionary explanation, comparisons of the distribution of individuals among species are a powerful tool in community ecology and ecotoxicology. Because of differences in sensitivity among species, shifts in the relative abundance of tolerant and sensitive species at polluted sites should be reflected in the shape of species abundance curves (Figure 22.2). As the classic example, Patrick (1971) used the shapes of such curves to interpret shifts in diatom communities impacted by pollution. Because the shape of the log normal distribution also reflects whether the contaminant is toxic or has a stimulatory influence (e.g., nutrient enrichment), the curves could be employed to distinguish between stressors. Thus, species abundance models extract more information than simple species lists, but are applied much less frequently than diversity, evenness, and richness metrics.

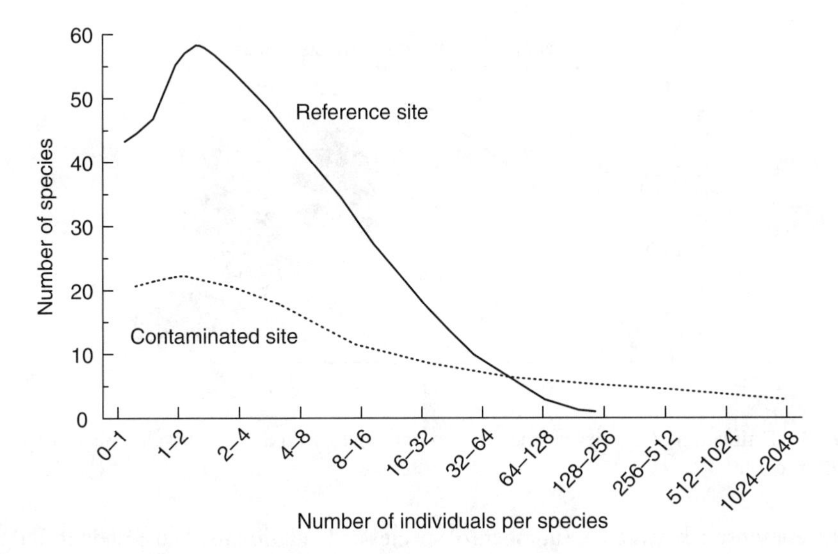

FIGURE 22.2 The predicted rank abundance distribution of species collected from reference and polluted communities (Preston 1948). The figure shows the number of species within each abundance class. The community from the reference site approximates a log normal distribution, whereas the community from the contaminated site is characterized by lower richness and increased abundance of tolerant species. This is a typical response of algal and benthic macroinvertebrate communities to organic pollution.

22.3.2 The Use of Species Richness and Diversity to Characterize Communities

22.3.2.1 Species Richness

As noted in Chapter 21, patterns of species richness across local, regional, and global scales have intrigued community ecologists for several decades. Community ecotoxicologists have routinely employed species richness as an indicator of ecological integrity. Rapport et al. (1985) include reduced species richness as one of five general indicators of the "ecosystem distress syndrome" (Chapter 25). Among the scores of measures used by community ecotoxicologists to assess effects of contaminants, reduced species richness is probably the most consistent (and least controversial) response. Because of the perceived value of biodiversity to the lay public, measures of species richness also have high societal relevance.

Species richness is defined as the number of species present in a prescribed sampling unit. Richness (R) can be determined by sampling more and more individuals from a site and keeping a running tally of the number of species that appear (Equations 22.4 and 22.5). The results can be used to estimate the total number of species in the community. Plots of the cumulative number of species versus sampling effort (e.g., number of dredge hauls, km^2 searched, biomass sampled, or number of individuals captured) will show an initial rapid increase in the number of species followed by a more gradual increase until becoming asymptotic (Figure 22.3). In most situations, this measure of species richness can be quite difficult to determine. In others, it might be undesirable to do such exhaustive sampling of a community if sampling was destructive or disruptive.

The number of species in a community can also be approximated with specific models (e.g., a log normal model) or indices that assume specific models linking sample size (number of individuals in the sample or N) and species richness (Equations 22.4 and 22.5) (Ludwig and Reynolds 1988, Magurran 1988, Matthews et al. 1998). All of these methods rely on the law of frequencies (Fisher et al. 1943), which holds that a relationship exists between the number of species and number of individuals in any ecological community. However, the law of frequencies does not dictate

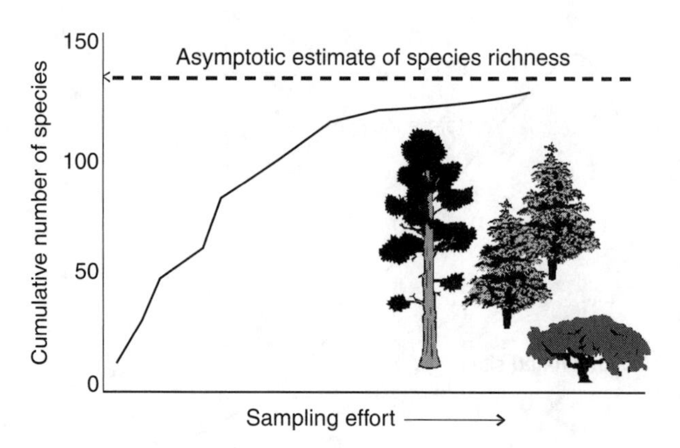

FIGURE 22.3 Estimation of species richness for a community with a cumulative number of species versus sampling effort curve.

a particular relationship between the numbers of species and individuals. Thus, Ludwig and Reynolds (1988) argue that, unless shown to be true, the assumption of a specific relationship between S and N in these models or metrics should be handled cautiously:

$$R_{\text{Margalef}} = \frac{S-1}{\ln N}, \tag{22.4}$$

$$R_{\text{Menhinick}} = \frac{S}{\sqrt{N}}. \tag{22.5}$$

Despite broad support for the use of species richness to assess biological integrity, estimating the number of species in the field is often problematic. Except in a few examples where all species in a habitat can be completely sampled (e.g., bird communities on small islands), we rarely know the total number of species in a community. Furthermore, species richness is highly dependent on area (Chapter 21) and increases asymptotically with sample size and the number of individuals collected (May 1973). Consequently, comparisons of the number of species among sites should be standardized for area and number of individuals (Vinson and Hawkins 1996). This is not a serious limitation in most biomonitoring studies because the same sampling effort will presumably be employed in both reference and impacted sites; however, it does complicate making comparisons with historic data or comparing results from different studies. One proposed solution to this problem is the use of a procedure known as rarefaction (Simberloff 1972), in which samples are selected randomly from the entire dataset to derive a quantitative relationship between number of species and total abundance. Rarefaction procedures estimate the expected number of species based on samples with standard sample sizes. The advantage of the rarefaction estimate is that samples of different sizes can be compared. The disadvantage is that information is lost when the actual sample size taken at a site is larger than the sample size for which the number of species is being estimated. The equation for estimating species richness by rarefaction is:

$$\hat{S}_n = \sum_{i=1}^{S} 1 - \frac{\binom{N-N_i}{n}}{\binom{N}{n}}, \tag{22.6}$$

where N = the number of individuals in the sample, N_i = the number of individuals of species i in the sample, S = the number of species in the sample, and n = the sample size (number of individuals) to which normalization is being done.

A second more pervasive problem is that measures of species richness do not account for differences in abundance among species. Theoretically, two locations could have very different total abundances and a very different distribution of individuals among species and still have the same species richness. Measures of species diversity, which account for both richness and the distribution of individuals among species, have been developed to resolve this problem. Although used routinely to compare communities in different locations, most diversity measures have received intense criticism from ecologists and ecotoxicologists. Diversity indices have been attacked based on theoretical, statistical, and conceptual arguments (Fausch et al. 1990, Green 1979, Hurlbert 1971). Despite the criticism, diversity measures continue to be widely used in biomonitoring studies and have appeared to multiply in the literature.

22.3.2.2 Species Diversity

Many ecologists, including ecotoxicologists, condense large species abundance data sets into diversity indices. There are two general types of diversity indices, those based on dominance and those derived from information theory. Both types include a species richness component and an evenness component of diversity; however, the relative importance of rare species differs between the two approaches. Simpson's index (1949), the most widely used measure of dominance, is given as

$$\hat{\lambda} = \sum_{i=1}^{S} \frac{1}{p_i^2}, \tag{22.7}$$

where λ is the measure of diversity and p_i is the proportion of the ith species in the sample. The value of λ ranges from 1 to S (where S = species richness), with larger values representing greater diversity. Community evenness reflects the distribution of individuals among species. If all species in a community have the same relative abundance, the value of λ is maximized and equals species richness. In practice, Equation 22.8 is often used to avoid bias associated with estimating p_i with N_i/N and from diversity estimation for the entire community based on a sample:

$$\lambda = \sum_{i=1}^{S} \frac{N_i(N_i - 1)}{N(N - 1)}. \tag{22.8}$$

Simpson's modified index as given in Equation 22.8 is converted in practice to $1 - \lambda$ so that any increase in the index reflects an increase in diversity. This weighted mean of the species proportions is very sensitive to dominant species and relatively insensitive to rare species. Thus, the main criticism of Simpson's index is that rare species contribute relatively little to the index value.

Two common diversity indices based on information theory, the Shannon–Wiener and Brillouin indices, are more sensitive to rare species (Qinghong 1995) and, in our opinion, are more relevant to ecotoxicology. The distinction between the two indices is simply that the Shannon–Wiener index (Equation 22.9) estimates diversity for the community from which the sample was taken, whereas Brillouin's index (Equation 22.10) estimates diversity for the sample itself. The Shannon–Wiener index can be described as the uncertainty of predicting the species of a randomly selected individual from the community. This uncertainty increases as more species are present in the community and as the individuals are more evenly distributed among those species (Ludwig and Reynolds 1988). Although calculated here using natural logarithms, both diversity indices can be calculated with base 10 or 2. Therefore, it is important to note units in published diversity (and related evenness) indices before using them together.

$$H' = -\sum_{i=1}^{S} p_i \ln p_i \cong -\sum_{i=1}^{S} \frac{N_i}{N} \ln\left(\frac{N_i}{N}\right) \tag{22.9}$$

$$H = \frac{1}{N} \ln \frac{N!}{\prod_{i=1}^{S} N_i!} \tag{22.10}$$

In Equations 22.9 and 22.10, the units of diversity are units of information per individual. If $\log_{10}$ or $\log_2$ were applied, the units would have been decits/individual or bits/individual, respectively. Like Simpson's index, Shannon–Wiener diversity is maximized (H_{MAX}) when all species are equally abundant in a sample.

22.3.2.3 Species Evenness

How equally the individuals in a community are distributed among the species can be measured with a variety of indices. The first two to be illustrated (Pielou 1969) are based on H' and H. They are simply H' or H divided by their estimated maxima, and consequently, the resulting evenness indices are those for the entire community (J') or for the sample itself (J). The maxima are used because they would be the values for H' and H if individuals were uniformly distributed among the available species:

$$J' = \frac{H'}{\ln S} \tag{22.11}$$

$$J = \frac{H}{H_{\text{MAX}}}. \tag{22.12}$$

H_{MAX} is defined by the following formula:

$$H_{\text{MAX}} = \frac{1}{N} \ln\left[\frac{N!}{([N/S]!)^{S-r}\{[(N/S)+1]!\}^r}\right],$$

where $[N/S]$ = the integer part of the quotient, N/S, and $r = N - S[N/S]$ (Magurran 1988).

The third evenness index (Alatalo 1981) is insensitive to species richness and combines both Hill's and Shannon–Wiener's indices (Equation 22.13). It is a modification of Hill's index ($[1/\lambda]/[e^{H'}]$), a measure that quantifies the proportion of common species in the sample. In the modified Hill's index, $e^{H'}$ reflects the number of abundant species and $1/\lambda$ reflects the number of very abundant species. The modification consists only of subtracting the maxima (i.e., 1) from each of the estimates, $1/\lambda$ and $e^{H'}$:

$$E = \frac{(1/\hat{\lambda}) - 1}{e^{H'} - 1}. \tag{22.13}$$

22.3.2.4 Limitations of Species Richness and Diversity Measures

The Simpson, Shannon–Wiener, and Brillouin indices are three examples from a long list of diversity measures that have been employed by community ecotoxicologists to assess effects of contaminants. Studies comparing performance and sensitivity of diversity measures have shown that each has

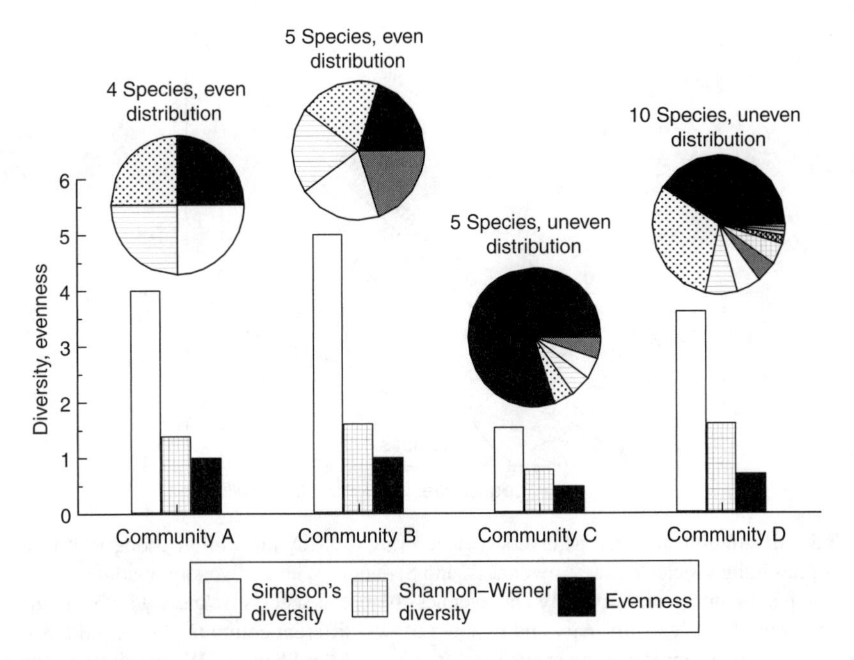

FIGURE 22.4 The influence of species richness and evenness on Shannon–Wiener and Simpson's diversity in four communities. The pie diagrams show the relative abundance of each species in the community. Note that both measures of diversity increase as species richness and evenness increase.

specific limitations (Boyle et al. 1990). Thus, it is not possible to recommend an index that will be useful in all situations. Indices that are sensitive to dominant species will be more appropriate when stressors, such as organic enrichment, favor a particular group. In contrast, because rare species are often the first to be eliminated from polluted sites, it may be more appropriate to employ an index sensitive to rare species when assessing effects of toxic chemicals.

The dependence of the Shannon–Wiener diversity index on both species richness and evenness is considered a serious shortcoming by some researchers (Qinghong 1995). Because decreases in species richness can be offset by increases in evenness (or vice versa), a single value of H' can be derived from numerous combinations of richness and evenness values. For example, in Figure 22.4, diversity (H') is the same (1.61) in two hypothetical communities (B and D), despite large differences in species richness and evenness. In practical terms, this means that changes in species diversity may go undetected even though large shifts in community composition have occurred. To address this problem, Qinghong (1995) proposed a simple model of species diversity that expresses changes in richness and evenness graphically (Figure 22.5). Using this approach, differences between any two points (e.g., two sampling locations or two points in time) on a plot of diversity versus richness can be attributed to a change in diversity, richness, or evenness.

The most serious criticism of simple community-level endpoints such as species richness and diversity is the loss of information that occurs when details of community composition are aggregated in a single number. While species abundance plots such as those developed by Preston (1948) describe how individuals are distributed among species (Figure 22.1), they do not provide information on community composition. Because sensitive species may be replaced by tolerant species at contaminated sites, it is conceivable that two communities could have a strikingly different composition but still have similar richness and diversity. An alternative approach that retains important information about community composition relevant to contaminants is the use of biotic indices. These indices (Section 22.3.3) are designed to integrate estimates of relative abundance with measures of species-specific sensitivity, thus capturing in a single index the fraction of a community consisting of tolerant and sensitive organisms.

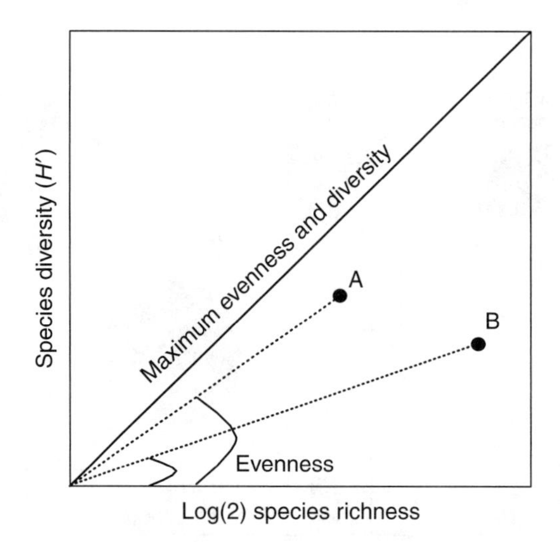

FIGURE 22.5 Illustration showing the diversity monitoring (DIMO) model (Qinghong 1995), an alternative approach for presenting species richness, evenness, and Shannon–Wiener diversity in communities. The diagonal line is the maximum species diversity and evenness based on species richness within a community. The two points represent the species diversity and richness of two different communities (A and B). The angle of the vector for each point represents the evenness component of the Shannon–Wiener diversity index. In this hypothetical example, community B has greater species richness but lower species diversity than community A because of the lower evenness.

Community ecologists have recently begun to appreciate the importance of rare species (e.g., those species that occur at low densities or are infrequently encountered in a community), especially in terms of preservation of biological diversity. However, the importance of rare species in ecotoxicology and bioassessment has received little attention (Cao et al. 1998, Fore et al. 1996). Barbour and Gerritsen (1996) argue that it is unnecessary and would be fiscally prohibitive to include rare species in biological monitoring programs. For practical reasons and because of the assumption that rare species contribute relatively little to ecosystem function, a common practice in biological assessments is to remove rare species from data analyses. However, because rare species may account for a disproportionate number of the total species at undisturbed sites (Gotelli and Graves 1996), removing them from the analysis may decrease our ability to detect differences among locations. In addition, rare species are more prone to local extinction because of low population densities. Finally, recent studies conducted in aquatic systems indicate that censoring data to eliminate rare species may underestimate effects of anthropogenic perturbations. Cao et al. (1998) showed that differences between reference and impacted sites were reduced if rare species were removed from the analyses (Figure 22.6). These researchers also showed that the small sample sizes typical of most biomonitoring studies often miss rare species, resulting in greater underestimation of species richness at reference sites compared to polluted sites.

22.3.3 Biotic Indices

Measures of total abundance, diversity, and species richness may not respond to some types of anthropogenic perturbations if sensitive species are simply replaced by tolerant species. Because sensitivity to contaminants often varies among species, the relative abundance of sensitive and tolerant taxa in a community could be employed to assess the degree of contamination. Biotic indices were developed early in the history of ecotoxicology with the intent of assessing the state of a community based on abundance of sensitive and tolerant species. Although Matthews et al. (1998) note the subjective nature of many tolerance rankings and the existence of different rankings

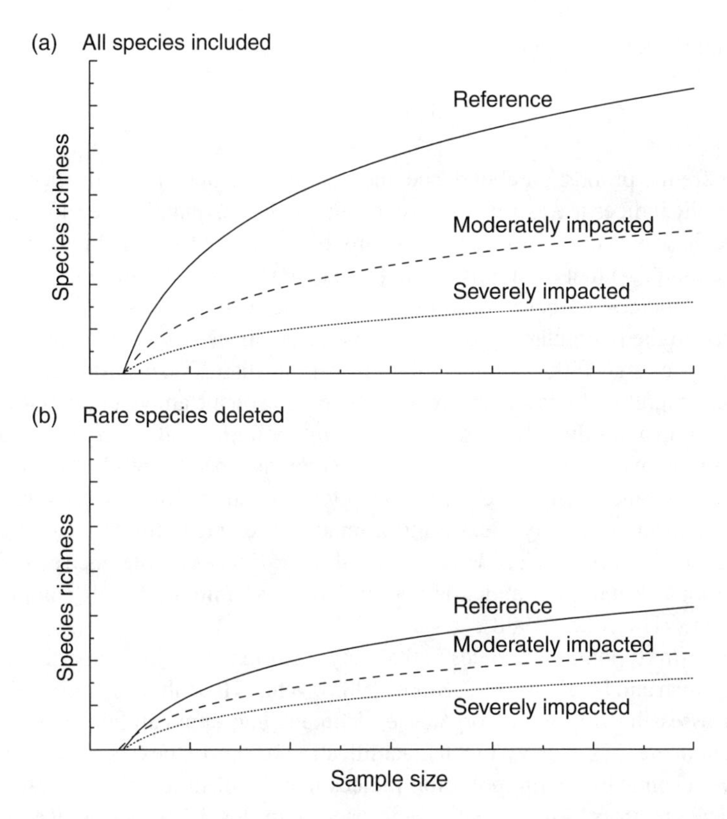

FIGURE 22.6 The relationship between sample size and species richness at reference, moderately impacted and severely impacted sites when all species are included (a) and when rare species are deleted (b). Because rare species often comprise a greater portion of communities at reference sites, the difference between reference and impacted sites diminishes when rare species are deleted. (Modified from Figure 2 in Cao et al. (1998).)

for the same species used in different regions, they conclude that biotic indices are used effectively throughout Europe today.

Innumerable biotic indices exist (Matthews et al. 1998), and all have similar features. Biotic indices assign values to individual taxa based on their relative sensitivity or tolerance to a specific type of pollution. These values are often generated based on expert opinion of ecologists with knowledge of the communities being impacted. This approach allows more information and the most relevant information to be combined in comparison to the simple diversity, evenness, and richness indices discussed earlier. However, it also makes subjective the selection of particular community qualities and the assignment of scores or weights to these qualities. In addition, the indices are relative. A score from a site suspected of being impacted is meaningful only relative to the score expected for an unimpacted site. Finally, and as a consequence of the previous points, the indices tend to be useful in a limited context, and must be modified thoughtfully to be applied elsewhere.

Because biotic indices account for both species-specific sensitivity and relative abundance, they are strongly influenced by pollution-induced changes in community composition. For example, it is well established that mayflies (Ephemeroptera), caddis flies (Trichoptera), and stoneflies (Plecoptera) are relatively sensitive to organic enrichment, whereas chironomids (Diptera) are generally tolerant. Indices such as Hilsenhoff's Biotic Index (Hilsenhoff 1987) take advantage of these differences in sensitivity and categorize sites based on the relative abundance of sensitive and tolerant species.

Hilsenhoff's biotic index is given as

$$\text{Biotic Index} = \Sigma p_i / t_i \tag{22.14}$$

where p_i and t_i are the proportion abundance and tolerance values of the ith species, respectively. Because most biotic indices use estimates of relative abundance, quantitative sampling is not necessary to calculate these measures. This feature is particularly useful for rapid bioassessment protocols (RBPs) (see Section 22.4) that often rely upon qualitative measures of community composition.

Because biotic indices are based on differences in species-specific sensitivity, their usefulness is often restricted to the particular region where tolerance values (t_i) were developed. Hilsenhoff's biotic index uses species-specific tolerance values from more than 2000 macroinvertebrate collections from polluted and unpolluted Wisconsin streams. Depending on the amount of variation in sensitivity among species within a family or higher taxonomic unit, pollution indices based on coarse levels of taxonomic resolution may be an effective solution to regional specificity. Chessman (1995) showed that family-level tolerance values were necessary for Australian streams because of the lack of taxonomic keys and the difficulty identifying immature life stages for some groups. A modified version of Hilsenhoff's biotic index based on family-level estimates of tolerance provided reasonable estimates of biological condition and was appropriate as an initial screening approach for water quality assessments (Hilsenhoff 1988).

Another criticism of pollution indices is that they are often specific to a particular class of contaminants (Chessman and McEvoy 1998, Slooff 1983). While Hilsenhoff's biotic index is especially well suited for assessing impacts of organic enrichment, the applicability of this index to other classes of contaminants (e.g., heavy metals, acidification, or pesticides) is uncertain. From a practical perspective, chemical-specific pollution indices may be of little value in systems affected by multiple chemical stressors. An alternative approach is to develop biotic indices that respond to more general classes of perturbations. Lenat (1993) published an extensive list of tolerance values for benthic macroinvertebrates in North Carolina (USA) streams. Unlike other pollution indices, Lenat's North Carolina Biotic Index (NCBI) is intended to provide a more general assessment of water quality, regardless of pollution type. Comparisons of species-specific tolerance values from Hilsenhoff's biotic index and the NCBI revealed many differences; however, mean tolerance values for major taxonomic groups were similar (Lenat 1993). These results are encouraging and suggest that sensitivity of some groups may be independent of the type of perturbation.

A key advantage of developing chemical-specific biotic indices is the potential to identify stressors based on biological measures. Chessman and McEvoy (1998) proposed a suite of biotic indices, each responding to a particular type of perturbation. A diagnostic index, based on family-level responses, was developed for several types of physical and chemical perturbations. Chessman and McEvoy (1998) concluded that, while diagnostic indices had promise, differences in sensitivity among species within a family hindered their performance. If chemical-specific biotic indices can be developed, these indices may be useful for quantifying the importance of individual chemicals in systems receiving multiple stressors (Box 22.1).

Box 22.1 Experimental Determination of Species-Specific Sensitivity

Perhaps the most serious criticism of biotic indices concerns the subjective assignment of tolerance values to individual species (Clements et al. 1988, 1992, Herricks and Cairns 1982, Matthews et al. 1998). While best professional judgment applied to survey data can provide legitimate estimates of species-specific sensitivity, these data should be supported by experimental evidence. In a review of biomonitoring approaches, Johnson et al. (1993) recognized the need to integrate laboratory-derived tolerance values with field data. The subjectivity and tautological reasoning inherent in biotic indices could be avoided by validating tolerance values

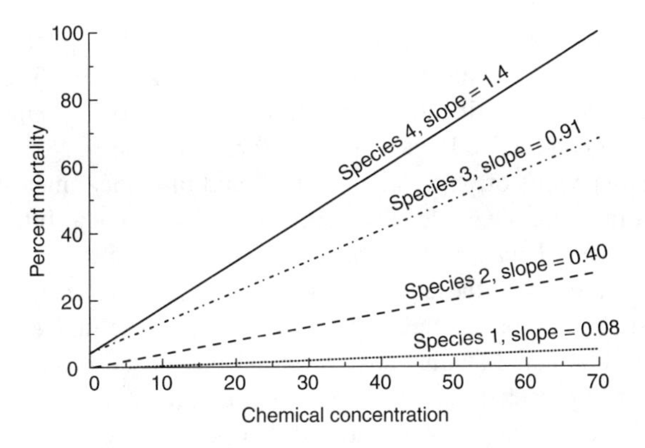

FIGURE 22.7 Results of community-level toxicity tests comparing the hypothetical responses of four species to a contaminant. The slope of the relationship between percent mortality and concentration is an indicator of relative sensitivity to the chemical and can be used in the development of biotic indices. In this example, species 1 is relatively tolerant to the chemical whereas species 4 is highly sensitive.

experimentally. Because of the opportunity to test responses of numerous species to the same chemical or mixture of chemicals simultaneously, community-level toxicity tests conducted in microcosms or mesocosms are an efficient way to obtain species-specific estimates of sensitivity. Standard toxicological endpoints (e.g., LC50, EC50) could be used to estimate relative sensitivity among species in a mesocosm experiment. Alternatively, experimental designs that use regression analyses to establish concentration–response relationships can provide objective estimates of species-specific sensitivity for numerous taxa (Figure 22.7). Estimates of relative sensitivity to chemicals derived experimentally could be integrated with field measures of relative abundance to produce pollution indices for different classes of contaminants. Clements et al. (1992) used this approach to develop an index of community sensitivity for benthic macroinvertebrates in metal-polluted streams. Benthic macroinvertebrate communities collected from a reference site were exposed to heavy metals in stream microcosms. Experimentally derived estimates of relative sensitivity were integrated into a biotic index (the index of metals impact), which was used to evaluate the degree of metal pollution downstream from the input of metals in a natural system.

In summary, while biotic indices have been employed extensively in European and other countries, they have received considerably less attention in the United States. These indices have been most successful when limited to a single class of stressors, especially organic enrichment. It should not be surprising when indices based on sensitivity to one chemical stressor fail to distinguish other types of perturbation. Bruns et al. (1992) rated several biological indicators based on their ecosystem conceptual basis, variability, uncertainty, ease of use, and cost-effectiveness. Litter decomposition and taxonomic richness received the highest ratings, whereas a biotic index received the lowest rating, primarily because it lacked information on responses of taxa to specific chemical toxicants. Finally, it is important to remember that the presence of tolerant taxa or the absence of sensitive taxa may result from numerous factors other than contaminants (Cairns and Pratt 1993). Biotic indices in isolation cannot demonstrate effects of pollution, only that a site is dominated by pollution-tolerant or pollution-sensitive organisms. However, biotic indices could be employed to evaluate potential reference sites in biomonitoring studies. A community dominated by species that are sensitive to a particular chemical provides reasonable evidence for the absence of that chemical.

22.4 DEVELOPMENT AND APPLICATION OF RAPID BIOASSESSMENT PROTOCOLS

One frequent criticism of community-level biomonitoring studies is the high cost of these approaches compared to physicochemical measures or single species toxicity tests. Because of the patchy spatial distribution of natural populations and the resulting high variability, large numbers of replicate samples are often necessary to detect differences between reference and contaminated sites. The time required for sample processing and species-level identification of taxonomically difficult groups may also be prohibitive, particularly for agencies conducting large-scale monitoring programs. Niemi et al. (1993) compared the cost and explanatory value of physical, chemical, and biological measures of recovery rates in streams. Biological measures (e.g., density, primary production, leaf litter decomposition) were considerably more expensive because of the greater variability and the need to collect large numbers of replicate samples. However, these authors acknowledged that because of their greater explanatory power, high cost should not preclude the use of biological variables in ecological assessments. We should also note that some studies have reported that costs of biological monitoring were competitive with other approaches for assessing water quality. An analysis conducted by the Ohio Environmental Protection Agency (EPA) showed that per sample costs of invertebrate and fish surveys were actually less than physical and chemical analyses of water quality (Karr 1993).

While there is evidence that biological assessments can be conducted cost-effectively, it is likely that the expense and logistical difficulties of conducting these surveys has limited our ability to assess the status of communities at larger spatial scales. Resolving the often conflicting goals of large-scale, spatially extensive monitoring with the need for intensive, long-term biological assessments requires innovative techniques that will improve efficiency but not sacrifice data quality. Rapid assessment programs (RAPs) and their aquatic counterparts, RBPs, were developed independently in the fields of conservation biology and biomonitoring to address these concerns. Both approaches attempt to streamline biological assessments by employing a variety of cost-saving but somewhat controversial procedures. Rapid assessment programs have been used extensively in conservation biology, especially in tropical ecosystems, where researchers must quickly estimate biodiversity and prioritize sites for preservation without the luxury of exhaustive biological surveys. The validity of many of these programs is based on the assumption that diversity of one group of organisms can be used as an indicator of total biological diversity within a region. For example, conservation biologists have used surveys of well-known flora and fauna (flowering plants, birds, and mammals) to estimate diversity of more difficult taxonomic groups (invertebrates). Using species diversity of one group to predict diversity of other groups has intuitive appeal and could significantly reduce costs of biological surveys (Blair 1999); however, the underlying assumption that diversity across broad taxonomic groups is regulated by the same ecological processes remains to be tested.

Innovations in rapid bioassessment procedures that streamline biological monitoring programs and reduce costs have accelerated the development of several large-scale monitoring programs in the United States, including the U.S. EPA's Environmental Monitoring and Assessment Program (EMAP) and the U.S. Geological Survey's National Water-Quality Assessment (NAWQA) program (Resh et al. 1995). The long-term goals of these programs are to assess the status and trends of terrestrial and aquatic ecosystems using a combination of probabilistic sampling designs and large-scale (regional) analyses. Given the limited funds available for routine monitoring in the United States, it is unlikely that these programs could accomplish their objectives without the cost savings provided by RBPs. More importantly, the reduced collection and processing costs allow researchers to sample a larger number of sites or increase the frequency of sampling.

In aquatic ecosystems, RBPs reduce sample collection and processing costs by (1) using qualitative sampling techniques, (2) subsampling and fixed-count processing, (3) eliminating replication and pooling samples collected from individual sites, and (4) relaxing the level of taxonomic resolution (Plafkin et al. 1989, Resh and Jackson 1993). Each of these four cost-saving measures involves important trade-offs that must be considered when implementing biomonitoring programs, regardless of whether sampling is conducted within a single stream or at a regional level. Resh et al. (1995) acknowledged the widespread acceptance of these cost-saving measures, noting that in our haste to expand biomonitoring programs, the consequences of reduced data quality have not been critically evaluated. In a review of RPBs, Hannaford and Resh (1995) reported that, while RBPs may be appropriate for prioritizing sites, their ability to produce legally defensible data or for routine impact assessments remains questionable. Later, we consider the limitations of each of the cost-saving measures used in RBPs.

22.4.1 Application of Qualitative Sampling Techniques

The abandonment of quantitative sampling techniques in many RBPs is an issue that requires serious consideration. Because of the time required to process quantitative samples, especially those collected from aquatic habitats, qualitative surveys of community composition have become increasingly common in biological assessments. Qualitative sampling techniques generally limit our ability to express data in terms of numbers of organisms per unit area or volume. Because interactions that structure communities are determined largely by absolute numbers of organisms and not their relative abundance, qualitative assessments do not provide insight into factors that regulate community composition. Furthermore, statistical analyses of biomonitoring results based on qualitative or quantitative data may lead to important differences. Figure 22.8 shows responses of several benthic macroinvertebrate metrics to heavy metals and compares statistical results based on qualitative (relative abundance) or quantitative (number/m^2) data. Analyses based on qualitative data were generally more variable and often unable to detect differences between metal-polluted and unpolluted sites.

To be fair, our appraisal of qualitative sampling employed in many RBPs neglects one major advantage of this approach. Because sample-processing times are greatly reduced using qualitative techniques, organisms can be collected from a larger and more diverse group of microhabitats. Sampling diverse habitats generally increases the total number of species collected compared to traditional quantitative techniques (e.g., 0.1 m^2 Surber sampler), which are often microhabitat-specific. Thus, species lists generated from qualitative sampling of diverse habitats will likely provide a more complete characterization of total species richness. Although quantitative techniques can be modified to sample different microhabitats, care must be taken to estimate relative habitat availability and to express the data accordingly.

22.4.2 Subsampling and Fixed-Count Sample Processing

The second major cost-saving measure in RBPs is the use of fixed-count sample processing (e.g., removal of 100, 200, or 300 individuals from a sample). Although fixed-count processing is standard in most RBPs, few studies have critically examined this procedure or determined the optimal number of individuals that should be removed from a sample (Barbour and Gerritsen 1996, Courtemanch 1996, Somers et al. 1998, Vinson and Hawkins 1996). Courtemanch (1996) argues that, because of the relationship between total abundance and species richness, fixed-count processing of samples can result in inconsistent and erroneous estimates of species richness. In addition, fixed-count processing is biased against rare taxa (although fixed counts can be supplemented by including large, rare taxa). Barbour and Gerritsen (1996) defend the use of fixed-count subsampling on the basis of significantly reduced costs and, more importantly, a greater ability to detect differences among sites compared to analyses using entire samples. Surprisingly, some studies have reported that removing a larger number of animals from samples does not necessarily improve the performance of RBP metrics. Using data collected from lakes, Somers et al. (1998) concluded that a two or three times increase in the number of organisms subsampled by fixed-count processing did not improve the ability of metrics to distinguish among locations. Analysis of more than 2000 benthic macroinvertebrate samples collected from the United States showed that, while fixed-count processing will significantly underestimate true species richness, this technique is quite robust with respect to distinguishing among locations (Vinson and Hawkins 1996). Furthermore, these authors conclude that fixed-count subsampling eliminates the need for using rarefaction techniques to estimate species richness when density varies greatly among locations.

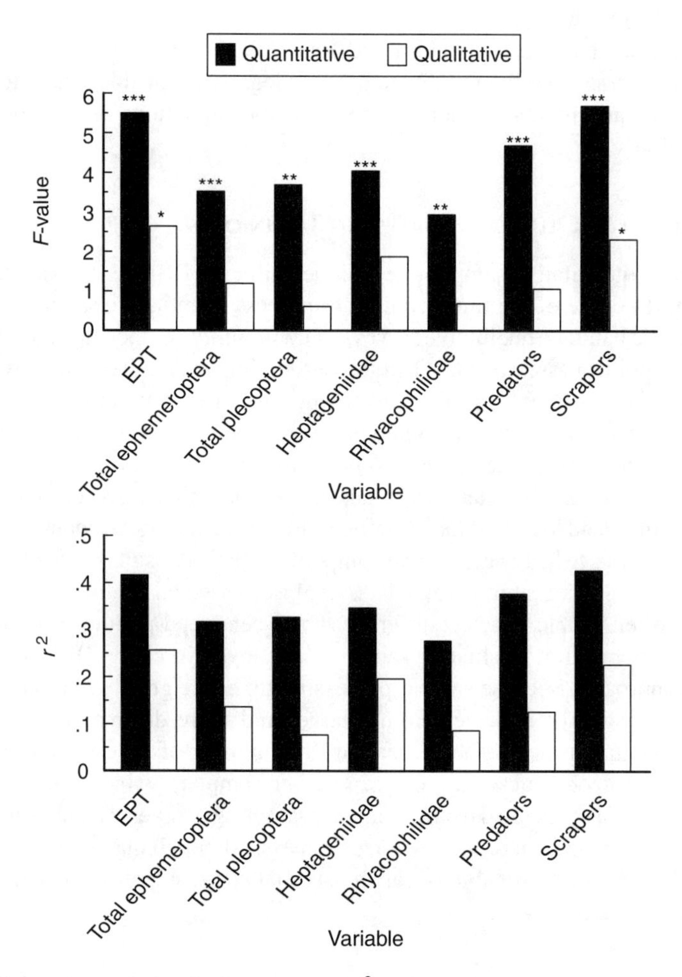

FIGURE 22.8 Comparison of quantitative (number/m^2) and qualitative (relative abundance) measures of macroinvertebrate community responses to metals in Rocky Mountain streams. Data were obtained from one-way ANOVA testing for differences among reference, moderately polluted, and highly polluted streams. All measures based on quantitative data were highly significant (*$P < .05$; **$P < .001$; ***$P < .0001$), whereas only two measures based on qualitative data (EPT and scrapers) were significant. In all instances, F-values and the amount of variation explained were much greater when based on quantitative measures. (From Clements, unpublished results.)

22.4.3 Pooling Samples

The third cost-saving measure common to many RBPs is the collection of a single, unreplicated sample from reference and polluted sites. The abandonment of replication has been criticized because it precludes estimating within-site variation and therefore limits statistical analyses (Resh et al. 1995). Although one could argue that since RBPs often integrate numerous metrics, each reflecting a unique component of ecological integrity, rigorous statistical analyses are less important. Indeed, summary metrics in RBPs are generally compared among sites without including estimates of variation. However, just like their constituent metrics, RBPs can vary among locations due to chance alone and therefore some analysis of variation would be useful.

From an experimental design perspective, the uneasiness that some ecotoxicologists feel about the abandonment of replication in RBPs may be irrelevant. Because samples collected from a single site are not true replicates, some argue that the use of inferential statistical analyses is not appropriate

(Hurlbert 1984). One practical solution to the lack of replication in RBPs is to collect data from many reference and polluted sites within a region (Clements and Kiffney 1995, Feldman and Connor 1992). Using this approach, sites are placed into categories (e.g., reference or impacted) and estimates of variation within and between categories are compared. This approach is the basis for the use of regional reference conditions described in Section 22.5. Because of the patchy distribution of organisms at any one location, it is recommended that collecting several pooled samples from a site is better than one large sample of equal area (Vinson and Hawkins 1996).

22.4.4 Relaxed Taxonomic Resolution

The appropriate level of taxonomic resolution is an important consideration in any biomonitoring study because of the difficulty and cost associated with identifying organisms to species. For many groups of organisms and in some regions, species-level identification is impossible because of the lack of sufficient taxonomic keys (e.g., many invertebrate groups in the tropics), difficulties with immature life stages (e.g., most aquatic insects), and large numbers of undescribed species (e.g., fungi, nematodes, and tropical beetles). Because of the difficulty in obtaining species-level identifications, some researchers have proposed abandoning traditional taxonomic approaches in favor of "recognizable taxonomic units" (RTUs) for assessing biological diversity. RTUs are taxa that are readily distinguished based on simple morphological characteristics and are generally developed by individuals who lack formal training in taxonomy. Oliver and Beattie (1993) reported that estimates of biodiversity of spiders, ants, and mosses based on RTUs were similar to those based on traditional taxonomic analysis (Figure 22.9). The correspondence for marine polychaetes was not as good, suggesting that applicability of RTUs for biomonitoring must be evaluated on a group-by-group basis. Although these nontaxonomic approaches can significantly reduce sample-processing costs, the lack of taxonomic information may hinder comparisons among studies.

Taxonomic resolution is a serious issue that deserves special consideration when employing RBPs. Large savings in sample-processing costs may be realized using relatively coarse (e.g., family level) taxonomic resolution (Lenat and Barbour 1994, Vanderklift et al. 1996). The major assumption when employing relaxed taxonomic resolution is similar to that of studies using species-level

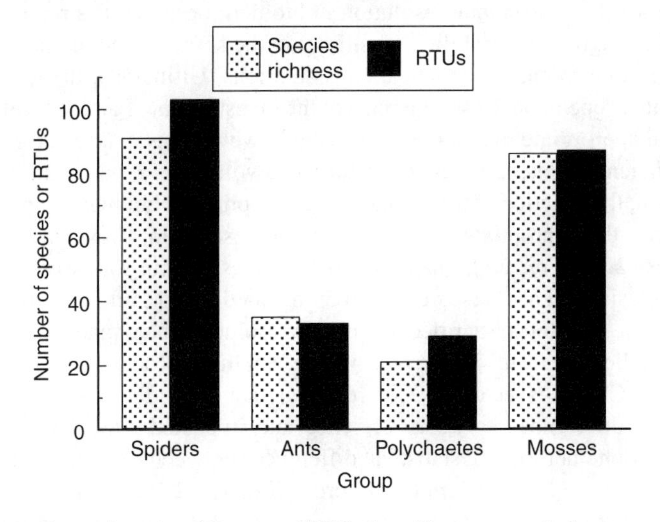

FIGURE 22.9 Comparison of species richness and RTUs for spiders, ants, polychaetes, and mosses. Measures of species richness were determined by taxonomic experts, whereas RTUs were determined by technicians with minimal training in taxonomy. Results show that for most groups, actual species richness and RTUs were similar. The major exception was for marine polychaetes, which were split into more groups by nonexperts. (Data from Table 1 in Oliver and Beattie (1993).)

identification, namely, that these taxonomic units respond predictably to environmental gradients (Olsgard et al. 1998). Several researchers have reported that relatively coarse levels of taxonomic resolution are sufficient to detect effects of pollution (Ferraro and Cole 1995, Olsgard et al. 1998, Vanderklift et al. 1996, Warwick 1993). For example, Bowman and Bailey (1997) concluded that patterns of community structure were similar when analyses were based on genus- or family-level identifications. Aggregate measures of phytoplankton community composition were actually more reliable indicators of eutrophication than species-level analyses in a whole-lake enrichment experiment (Cottingham and Carpenter 1998). Marchant et al. (1995) reported that analyses of benthic macroinvertebrate data collected over a large region were relatively robust to sampling techniques and taxonomic resolution. They showed that patterns of benthic communities measured using qualitative sampling techniques (presence/absence data) and family-level identification were similar to those using quantitative data and species-level identification. Feraro and Cole (1995) compared the ability of different indices to detect differences between polluted and unpolluted locations when analyses were conducted at the level of genus, family, order, and phylum. Results showed that the level of taxonomic resolution was relatively unimportant for detecting pollution. The most likely explanation for these results is that taxonomically related species often have similar ecological requirements and similar sensitivities to contaminants (Warwick 1988).

As previously noted, conservation biologists have also investigated the consequences of relaxed taxonomic resolution on their ability to estimate biological diversity. Williams and Gaston (1994) found that family-level richness was a highly significant predictor of species richness ($r^2 > .79$) for several groups, including ferns, butterflies, passerine birds, and bats. However, these researchers cautioned that the relationship between species richness and richness at higher levels of resolution could be influenced by the spatial scale of an investigation.

The appropriate level of taxonomic resolution will be determined by variation in sensitivity within groups, natural variation, and the spatial scale of the investigation (Box 22.2). When samples are

Box 22.2 The Relationship between Taxonomic Resolution, Sensitivity, and Natural Variation

The appropriate level of taxonomic resolution in biomonitoring studies represents a trade-off between natural background variability, sensitivity to the stressor, and, in the case of problematic groups such as chironomids, practical considerations. Ultimately, the level of taxonomic resolution may also depend on the spatial scale of the investigation. Family-level or higher identification may be appropriate over a regional scale; however, this coarse taxonomic resolution may not be sufficient to detect effects of disturbance within a single stream (Marchant et al. 1995). In addition, the practice of using qualitative sampling techniques typical of many RBPs may also influence the appropriate level of taxonomic resolution. Bowman and Bailey (1997) found that as taxa are aggregated, qualitative data are less useful for assessing differences in community composition. These researchers recommended that if trade-offs are necessary when employing RBPs, it is better to sacrifice taxonomic resolution than quantitative sampling.

Using data collected from 73 streams in the Southern Rocky Mountain ecoregion of Colorado (USA), Clements et al. (2000) reported that the effects of taxonomic aggregation on statistical differences between reference and impacted sites varied among groups. For the mayflies (Ephemeroptera), statistical differences between reference and metal-polluted sites were greatest at the level of family and order (Figure 22.10a). Although mayflies in the genus *Rhithrogena* sp. are sensitive to metals, high variability in abundance of *Rhithrogena* sp. among reference sites limited the ability to detect statistical differences. Aggregate taxonomic measures at the level of family and order were better indicators of pollution because most mayflies and almost all heptageniid mayflies in Rocky Mountain streams are sensitive to metals. In contrast to these results, total caddis fly abundance was a poor indicator of metal

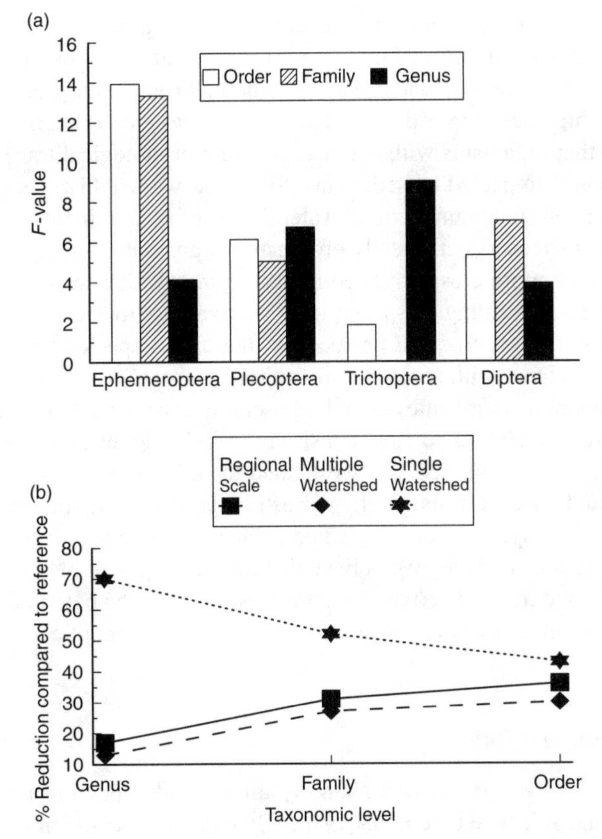

FIGURE 22.10 (a) Influence of taxonomic resolution on statistical differences between polluted and unpolluted sites based on the magnitude of *F*-values from one-way ANOVA. Separation of polluted and unpolluted sites was greatest at coarse levels of taxonomic resolution for some groups (e.g., Ephemeroptera), whereas differences were greatest at the level of genus for others (e.g., Trichoptera). (b) The relationship between sensitivity and taxonomic resolution across different spatial scales. Sensitivity was defined as the percent reduction compared to a reference site. Within a single watershed, responses at the level of genus were most sensitive (e.g., showed a greater reduction compared to reference streams). At multiple watershed and regional scales, sensitivity increased with taxonomic aggregation.

pollution. Unlike mayflies, the order Trichoptera includes taxa that are both highly tolerant (Brachycentridae and Hydropsychidae) and relatively sensitive (Rhyacophilidae) to heavy metal pollution. The relationship between taxonomic level and responses to stressors was also influenced by the spatial scale of the investigation. The percent reduction in abundance of mayflies was greatest at the level of genus in a single watershed study (Arkansas River, CO), but increased with taxonomic aggregation at larger spatial scales (Figure 22.10b).

---

collected over relatively large geographic areas, higher taxonomic aggregates (e.g., families, orders) may be necessary to characterize effects of stressors. Because relatively few species will occur at all sites across a large geographic region, it may be difficult to assess the effects of contamination using species-level data. In addition, abundances of individual species are more sensitive to natural environmental variability than aggregate indicators at coarse levels of taxonomic resolution (Warwick 1988). If all species within a group show similar responses to disturbance, then measures at coarse levels of taxonomic resolution will most likely be better indicators. Finally, the usefulness of genus and family-level abundance data for discerning effects of contamination will also be influenced by the severity of the stressor. Olsgard et al. (1998) reported better correlations between species' abundances and higher levels of taxonomic resolution at polluted sites compared to reference sites.

22.4.5 The Application of Species Traits in Biomonitoring

A recent development within the field of benthic ecology that has significant implications for bioassessments is the identification of species traits to characterize the functional composition of communities (Poff et al. 2006). The primary goal of this research is to relate life history, morphological, dispersal, feeding, and ecological characteristics of species to distinct environmental gradients. The species trait approach has several advantages over traditional measures of community composition in bioassessments. First, because species traits are associated with specific environmental gradients, we can establish predictive relationships between a priori selected traits and environmental stressors. For example, we would predict that organisms with external gills (a morphological trait) would be less common in agricultural streams impacted by sediments. Similarly, we would expect that grazing organisms would be less common in streams with elevated levels of contaminants in periphyton. Second, because environmental stressors act directly on functional attributes, changes in functional composition of communities are more closely related to underlying mechanisms. As described in Chapter 21, environmental factors operating on species traits act as a filter to determine membership in local species pools. Finally, from a practical perspective, the use of species traits may help resolve some of the problems associated with regional biomonitoring programs. Natural biogeographical variation in species composition complicates these large-scale assessments. In contrast, we expect that functional trait composition should be more consistent among regions and less influenced by biogeographical patterns. Doledec et al. (2006) investigated the effects of land-use changes on traditional structural indices and species traits. Both species composition and species traits responded to land use gradients; however, species traits explained greater variation and, more importantly, offered insights into underlying mechanisms. We believe that the development of ecologically meaningful and mechanistically based metrics derived from species traits has the potential to significantly improve our ability to predict effects of anthropogenic stressors on communities.

22.5 REGIONAL REFERENCE CONDITIONS

Ecologists have long recognized that patterns of vegetation vary naturally among landscapes and are influenced by regional climate, geology, and soil type (Clements 1916). Qualitative assessment of these patterns and compilation of ecoregion maps (Omernik 1987) improve our ability to define reference conditions. Much of the natural variability among reference sites can be reduced by restricting sites to a single ecoregion or subregion. Alternatively, variation among reference sites within a region can be explained using predictive models (Fausch et al. 1984).

The traditional approach in most biomonitoring studies is to compare communities at disturbed sites to a single reference site. In lotic surveys of point source discharges, upstream reference sites are often compared to downstream impacted and recovery sites within the same watershed. In assessments of sediment contamination, communities are collected from locations along a gradient of pollution, generally within the same watershed. As previously noted, replicates in these studies often consist of subsamples collected from a single location and are not considered true replicates (Hurlbert 1984). Within-site variance estimates are not useful for defining true reference conditions, and therefore, extrapolations to other locations are greatly limited. Multiple reference sites, even those within the same watershed, are more appropriate for assessing single point source discharges (Hughes 1985), especially in situations where locating ecologically similar watersheds is problematic. However, natural spatial variation also complicates assessment of point source discharges within the same habitat. For example, the river continuum concept predicts significant changes in the structure and function of streams from headwater sites to downstream sites (Vannote et al. 1980). The traditional upstream-reference versus downstream-impacted design employed in many stream surveys is confounded by natural changes in structure and function along a river continuum (Clements and Kiffney 1995). Community-level indicators that respond to both natural and anthropogenic variation will not be particularly useful for assessing disturbance unless these natural changes can be quantified.

Defining reference conditions and selecting appropriate reference sites are among the most difficult steps when designing biomonitoring studies. Unlike laboratory experiments where all variables except those of direct relevance to the experiment are controlled, field studies lack true controls and are often complicated by large amounts of natural variation. Reynoldson et al.

(1997) defined the reference condition as "the condition that is representative of a group of minimally disturbed sites organized by selected physical, chemical, and biological characteristics." Hughes (1995) reviewed the strengths and limitations of six approaches for determining reference conditions (regional reference sites, historical and paleoecological data, laboratory experiments, quantitative models, and best professional judgment) and concluded that multiple approaches are often required. Of these six approaches, the use of regional reference conditions holds the most promise. The regional reference approach involves selecting multiple sites within a single region to define expected conditions (Bailey et al. 1998). Establishment of regional reference conditions is a major improvement over traditional biomonitoring approaches that allows researchers to objectively characterize expected community composition and obtain legitimate estimates of natural variation.

22.6 INTEGRATED ASSESSMENTS OF BIOLOGICAL INTEGRITY

Physical and chemical measures of contaminant effects dominated the field of pollution assessment until the 1970s (Cairns and Pratt 1993). The historic emphasis on abiotic measures has gradually been replaced by an understanding that biological indicators of ecological integrity are equally important. Natural resource managers now realize that integrated assessments including chemical analyses, toxicity tests, and biological surveys are often necessary to discern impacts of contaminants (Figure 22.11). The sediment quality triad (Chapman 1986) is an example of an integrated approach that combines chemical measures of contaminants, toxicology, and field assessments of communities to characterize the degree of sediment contamination. The strength of the sediment quality triad lies in the weight of evidence approach and in its ability to discern direct toxicological effects from natural variation in habitat characteristics (Chapman 1996). For example, results that show altered community composition but no detectable levels of chemical contamination and no toxicological effects suggest that factors other than contaminants (e.g., substrate composition or habitat quality) are responsible for these differences. Conversely, results that show chemical contamination and toxic effects in the laboratory but no changes in community composition imply that the chemicals are not bioavailable in the field or that organisms have acclimated to these chemicals. Although the sediment quality triad was developed specifically to assess contamination in marine

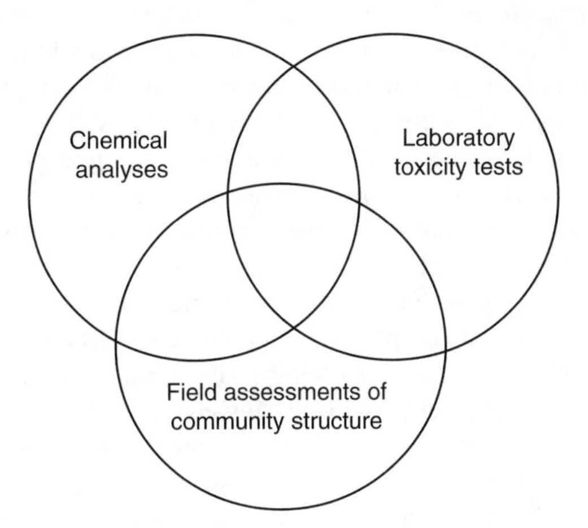

FIGURE 22.11 Integration of chemical analyses, laboratory toxicity tests, and field assessments of community structure can provide the strongest evidence for a causal relationship between the presence of chemical stressors and ecological impairment. (Modified from Chapman (1986).)

and freshwater ecosystems, the general integrated approach could be used in most biomonitoring programs. For example, toxicological effects of pesticides on surrogate species could be integrated with residue analysis and field assessments of community composition to estimate the subtle effects of pesticides on songbirds.

22.7 LIMITATIONS OF BIOMONITORING

Although integrated weight of evidence approaches such as the sediment quality triad can suggest a relationship between stressors and ecological responses, they do not demonstrate causation. Descriptive approaches such as biomonitoring studies provide support for hypotheses rather than direct tests of hypotheses. Results of biomonitoring studies are often equivocal because of the lack of adequate controls, nonrandom assignment of treatments, and lack of replication (Hurlbert 1984). Suter (1993) discusses the "ecological fallacy" of presuming that differences between polluted and unpolluted sites are a result of anthropogenic factors when alternative hypotheses (sensu Platt 1964) have not been tested experimentally. Several researchers have offered useful advice on how to strengthen causal relationships in descriptive studies (Beyers 1998, Hill 1965, Newman 2001, Suter 1993). The most often cited criteria for determining causation are derived from epidemiological studies of disease (Box 22.3), where cause-and-effect relationships are elusive (e.g., dioxin exposure and cancer) and identifying these associations has major societal implications (e.g., smoking and lung cancer).

Box 22.3 The Use of Causal Criteria in Community-Level Assessments

Hill's (1965) nine criteria and modifications of these guidelines have been employed in ecological risk assessment studies to strengthen causal relationships between stressors and ecological responses (Beyers 1998, Newman 2001, Suter 1993). However, most of these adaptations have been developed for population-level studies (see Box 13.2) and not directly applied to communities. We review Hill's nine criteria for determining causation within the context of assessing effects of stressors on communities as follows.

1. *Strength of the association*. The presence of a strong relationship between a stressor and alteration in community structure is one of the most important components for the formation of a logical argument of causation. The complete elimination of a sensitive species from a contaminated site or large shifts in the abundance of sensitive and tolerant species would be considered strong responses.
2. *Consistency of the association*. The responses of communities to the stressor should agree with those observed at other locations and by other researchers. The more diverse the situations, where consistent responses are observed, the stronger are the argument for causation. For example, is species richness of avian communities consistently reduced in areas sprayed with pesticides? Because the composition of communities will vary spatially and temporally, it may not be possible to satisfy this criterion by measuring effects on any single species. For example, responses of benthic communities to organic enrichment frequently result in increased abundance of certain groups; however, it is unlikely that we could predict the response of any particular species. Consistent community-level responses to contamination will necessarily incorporate multiple measures.
3. *Specificity*. Because the specificity of responses to contaminants often decreases at higher levels of biological organization, requiring that the observed response is

diagnostic of exposure will be problematic for most community- and ecosystem-level studies. Information on the relative sensitivity of dominant species to a particular chemical may allow researchers to predict specific community-level responses. However, most endpoints used in these studies (e.g., diversity or species richness) will likely show similar responses among contaminant classes.

4. *Temporality*. The requirement that exposure to the stressor must precede the responses is obvious for showing causation at any level of biological organization. However, demonstrating this temporal association is difficult when preexposure data are unavailable. Paleoecological studies of communities are especially appropriate for demonstrating temporal associations between stressors and community responses. Species composition of certain groups of organisms preserved in sediments can provide a long-term record of community change that could be associated with the onset of contamination.
5. *Plausibility*. A credible mechanistic explanation for the observed response of communities to a stressor strengthens the case for a causal relationship. However, this criterion may also be problematic in community-level studies. Identifying specific mechanistic explanations for changes in community-level endpoints may require an understanding of responses at lower levels of biological organization. For example, reduced species diversity could result from either the loss of sensitive species and/or increased dominance by tolerant species. An understanding of species-specific responses is necessary to provide a mechanistic explanation for reduced species diversity at polluted sites.
6. *Coherence*. Are the observed changes in community composition in agreement with our understanding of the stressor and the characteristics of the particular community? Often, identifying coherent community-level responses requires an understanding of the structure and function of reference and impacted communities. We generally expect that most toxic chemicals will have negative effects on species in a community. Therefore, increased abundance of a particular species at a contaminated site is difficult to explain unless we can attribute this response to indirect effects, such as the removal of a potential competitor or predator.
7. *Analogy*. If similar classes of stressors elicit similar community responses, then the case for a causal relationship is strengthened. For example, despite different modes of toxic action, responses of benthic communities to heavy metals and acidification are generally similar. Note that this criterion is somewhat contrary with the requirement that responses to contaminants should also be diagnostic.
8. *Ecological gradient*. Studies that show a gradient of responses to contamination (e.g., concentration–response relationship) provide stronger evidence for causation than all-or-none responses. Although relatively common in biological assessments, comparing a single contaminated site to a single reference site is a weak experimental design since differences between sites cannot be attributed directly to contamination. Because of differences in sensitivity among species, it is likely that many community-level variables will show a continuous distribution along contaminant gradients. The key challenges are to develop experimental designs that allow these gradients to be quantified and to separate contaminant effects from other sources of variation along the gradient. Where possible, study sites should be located along a known gradient of contamination and a suite of community variables should be measured at each site.
9. *Experimentation*. The last and perhaps the most important criterion to support a causal argument is direct experimental evidence. Changes in community composition resulting from experimental manipulation of stressors provide the most convincing evidence that the stressors are directly responsible for the observed

responses. Unfortunately, this is probably the most difficult type of evidence to obtain in community studies and, as a result, there are relatively few examples of whole community manipulations with chemical stressors. Conducting manipulative experiments on intact communities in the field is challenging and, depending on the nature of the stressor, often logistically impractical. Microcosm and mesocosm experiments are a reasonable alternative to field manipulations, but they are limited in terms of their spatial and temporal scale (Chapter 23).

22.7.1 Summary

Regardless of the strength and consistency of stressor–response relationships or the use of appropriate reference sites in a biomonitoring study, there is no substitute for experimentation to demonstrate causality. The transition from purely descriptive to manipulative approaches represents a major shift in scientific rigor and sophistication. It is the ability to test hypotheses with experiments that defines the maturity of a science (Popper 1972), whether it is physics, ecology, or ecotoxicology. In Chapter 23, we will review the development of experimental approaches in community ecology and show how these approaches can be employed to support causal arguments in ecotoxicology.

22.7.1.1 Summary of Foundation Concepts and Paradigms

- Biomonitoring is defined as the use of biological systems to assess the structural and functional integrity of aquatic and terrestrial ecosystems.
- Community-level biological monitoring is based on the assumption that composition and organization of communities reflect local environmental conditions and respond to anthropogenic alteration of those conditions.
- Identifying reference conditions and separating natural variation from contaminant-induced changes are major areas of research interest in community biomonitoring.
- The absence of a pollution-sensitive species from a contaminated site provides only weak support for the hypothesis that its absence is due to contamination. Similarly, the presence of pollution-tolerant species does not necessarily imply that a site is degraded.
- Because of differences in sensitivity among species, shifts in the relative abundance of tolerant and sensitive species at polluted sites should be reflected in the shape of species abundance curves.
- Despite broad support for the use of species richness to assess biological integrity, estimating the number of species for some groups is problematic.
- Species diversity indices condense large species abundance data sets into a single number and are generally based on dominance or derived from information theory.
- The dependence of the Shannon–Wiener diversity index on both species richness and evenness is considered a serious shortcoming by some.
- The most serious criticism of simple community-level endpoints such as species richness and diversity is the loss of information that occurs when details of community composition are aggregated in a single number.
- There is evidence that censoring rare species from data analysis in biomonitoring studies may underestimate effects of anthropogenic perturbations.
- Biotic indices have been developed to assess the state of a community based on abundance of sensitive and tolerant species.
- Because responses to contaminants are often species-specific, biotic indices based on sensitivity to one chemical stressor often fail to distinguish other types of perturbation.
- RBPs streamline biological assessments by (1) using qualitative sampling techniques, (2) using subsampling and fixed-count processing, (3) eliminating replication and

pooling samples collected from individual sites, and (4) relaxing the level of taxonomic resolution.

- Integrated weight-of-evidence approaches such as the sediment quality triad are often necessary to discern impacts of contaminants.
- Although biomonitoring studies can suggest a relationship between stressors and ecological responses, there is no substitute for experimentation to demonstrate causality.

REFERENCES

Alatalo, R.V., Problems in the measurements of evenness in ecology, *Oikos* 37, 199–204, 1981.

Bailey, R.C., Kennedy, M.G., Dervish, M.Z., and Taylor, R.M., Biological assessment of freshwater ecosystems using a reference condition approach: Comparing predicted and actual benthic invertebrate communities in Yukon streams, *Freshw. Biol.*, 39, 765–774, 1998.

Barbour, M.T. and Gerritsen, J., Subsampling of benthic samples: A defense of the fixed-count method, *J. N. Am. Benthol. Soc.*, 15, 386–391, 1996.

Beyers, D.W., Causal inference in environmental impact studies, *J. N. Am. Benthol. Soc.*, 17, 367–373, 1998.

Blair, R.B., Birds and butterflies along an urban gradient: Surrogate taxa for assessing biodiversity? *Ecol. Appl.*, 9, 164–170, 1999.

Bowman, M.F. and Bailey, R.C., Does taxonomic resolution affect the multivariate description of the structure of freshwater benthic macroinvertebrate communities?, *Can. J. Fish. Aquat. Sci.*, 54, 1802–1807, 1997.

Boyle, T.P., Smillie, G.M., Anderson, J.C., and Beeson, D.R., A sensitivity analysis of nine diversity and seven similarity indices, *Water Pollut. Con. Fed.*, 62, 749–762, 1990.

Bruns, D.A., Wiersma, G.B., and Minshall, G.W., Evaluation of community and ecosystem monitoring parameters at a high-elevation, Rocky Mountain study site, *Environ. Toxic. Chem.*, 11, 459–472, 1992.

Cairns, J., Jr., Indicator species versus the concept of community structure as an index of pollution, *Water Res. Bull.*, 10, 338–347, 1974.

Cairns, J., Jr. and Pratt, J.R., A history of biological monitoring using benthic macroinvertebrates, In *Freshwater Biomonitoring and Benthic Macroinvertebrates*, Rosenberg, D.M. and Resh, V.H. (eds.), Chapman & Hall, New York, 1993, pp. 10–27.

Cao, Y., Williams, D.D., and Williams, N.E., How important are rare species in aquatic community ecology and bioassessment?, *Limnol. Oceanogr.*, 43, 1403–1409, 1998.

Chapman, P.M., Sediment quality criteria from the sediment quality triad: An example, *Environ. Toxicol. Chem.*, 5, 957–964, 1986.

Chapman, P.M., A test of sediment effects concentrations: DDT and PCB in the Southern California Bight, *Environ. Toxic. Chem.*, 15, 1197–1198, 1996.

Chessman, B.C., Rapid assessment of rivers using macroinvertebrates: A procedure based on habitat-specific sampling, family level identification and a biotic index, *Aust. J. Ecol.*, 20, 122–129, 1995.

Chessman, B.C. and McEvoy, P.K., Towards diagnostic biotic indices for river macroinvertebrates, *Hydrobiologia*, 364, 169–182, 1998.

Clements, F.E., *Plant Succession*, Carnegie Institute of Washington, No. 242, Washington, D.C., 1916.

Clements, W.H., and Kiffney, P.M., The influence of elevation on benthic community responses to heavy metals in Rocky Mountain streams, *Can. J. Fish. Aquat. Sci.*, 52, 1966–1977, 1995.

Clements, W.H., Cherry, D.S., and Cairns, J., Jr., The impact of heavy metals on macroinvertebrate communities: A comparison of observational and experimental results, *Can. J. Fish. Aquat. Sci.*, 45, 2017–2025, 1988.

Clements, W.H., Cherry, D.S., and Van Hassel, J.H., Assessment of the impact of heavy metals on benthic communities at the Clinch River (Virginia): Evaluation of an index of community sensitivity, *Can. J. Fish. Aquat. Sci.*, 49, 1686–1694, 1992.

Cottingham, K.L. and Carpenter, S.R., Population, community, and ecosystem variates as ecological indicators: Phytoplankton responses to whole-lake enrichment, *Ecol. Appl.*, 8, 508–530, 1998.

Clements, W.H., Carlisle, D.M., Lazorchak, J.M., and Johnson, P.C., Heavy metals structure benthic communities in Colorado mountain streams, *Ecol. Appl.*, 10, 626–638, 2000.

Courtemanch, D.L., Commentary on the subsampling procedures used for rapid bioassessments, *J. N. Am. Benthol. Soc.*, 15, 381–385, 1996.

Davis, W.S., Biological assessment and criteria: Building on the past, In *Biological Assessment and Criteria: Tools for Water Resource Planning and Decision Making*, Davis, W.S. and Simon, T.P. (eds.), Lewis Publishers, Boca Raton, FL, 1995, pp. 15–29.

Doledec, S., Phillips, N., Scarsbrook, M., Riley, R.H., and Townsend, C.R., Comparison of structural and functional approaches to determining landuse effects on grassland stream invertebrate communities, *J. N. Am. Benthol. Soc.*, 25, 44–60, 2006.

Fausch, K.D., Karr, J.R., and Yant, P.R., Regional application of an index of biotic integrity based on stream fish communities, *Trans. Am. Fish. Soc.*, 113, 39–55, 1984.

Fausch, K.D., Lyons, J., Karr, J.R., and Angermeier, P.L., Fish communities as indicators of environmental degradation, *Am. Fish. Soc. Symp.*, 8, 123–144, 1990.

Feldman, R.S. and Connor, E.F., The relationship between pH and community structure of invertebrates in streams of the Shenandoah National Park, Virginia, U.S.A., *Freshw. Biol.*, 27, 261–276, 1992.

Ferraro, S.P. and Cole, F.A., Taxonomic level sufficient for assessing pollution impacts on the Southern California bight macrobenthos—Revisited, *Environ. Toxicol. Chem.*, 14, 1031–1040, 1995.

Fisher, R.A., Corbet, A.S., and Williams, C.B., The relation between number of species and the number of individuals in a random sample of an animal population, *J. An. Ecol.*, 12, 42–58, 1943.

Fore, L.S., Karr, J.R., and Wisseman, R.W., Assessing invertebrate responses to human activities: Evaluating alternative approaches, *J. N. Am. Benthol. Soc.*, 15, 212–231, 1996.

Gotelli, N.J. and Graves, G.R., *Null models in ecology*, Smithsonian Institution Press, Washington, D.C., 1996.

Green, R.H., *Sampling Design and Statistical Methods for Environmental Biologists*, John Wiley & Sons, New York, 1979.

Hannaford, M.J. and Resh, V.H., Variability in macroinvertebrate rapid-bioassessment surveys and habitat assessments in a northern California stream, *J. N. Am. Benthol. Soc.*, 14, 430–439, 1995.

Herricks, E.E. and Cairns, Jr., J., Biological monitoring. Part III—receiving system methodology based on community structure, *Water Res.*, 16, 141–153, 1982.

Hill, A.B., The environment and disease: Association or causation, *Proc. R. Soc. Med.*, 58, 295–300, 1965.

Hilsenhoff, W.L., An improved biotic index of organic pollution, *Great Lakes Ent.*, 20, 31–39, 1987.

Hilsenhoff, W.L., Rapid field assessment of organic pollution with a family-level biotic index, *J. N. Am. Benthol. Soc.*, 7, 65–68, 1988.

Hughes, R.M., Use of watershed characteristics to select control streams for estimating effects of metal mining wastes on extensively disturbed streams, *Environ. Manag.*, 9, 253–262, 1985.

Hughes, R.G., Theories and models of species abundance, *Am. Nat.*, 128, 879–899, 1986.

Hughes, R.M., Defining acceptable biological status by comparing with reference conditions, In *Biological Assessment and Criteria: Tools for Water Resource Planning and Decision Making*, Davis, W.S. and Simon, T.P. (eds.), Lewis Publ., Boca Raton, FL, 1995, pp. 31–47.

Hurlbert, S.H., The nonconcept of species diversity: A critique and alternative parameters, *Ecology*, 52, 577–586, 1971.

Hurlbert, S.H., Pseudoreplication and the design of ecological field experiments, *Ecol. Monogr.*, 54, 187–211, 1984.

Johnson, R.K., Wiederholm, T., and Rosenberg, D.M., Freshwater biomonitoring using individual organisms, populations, and species assemblages of benthic macroinvertebrates, In *Freshwater Biomonitoring and Benthic Macroinvertebrates*, Rosenberg, D.M. and Resh, V.H. (eds.), Chapman & Hall, New York, 1993, pp. 40–158.

Karr, J.R., Defining and assessing ecological integrity: Beyond water quality, *Environ. Toxicol. Chem.*, 12, 1521–1531, 1993.

Karr, J.R. and Dudley, D.R., Ecological perspective on water quality goals, *Environ. Manag.*, 5, 55–68, 1981.

Kolkwitz, R. and Marsson, M., Okologie der tierischen Saprobien, *Int. Rev. Gesaten.*, 2, 126–152, 1909.

Lenat, D.R., A biotic index for the southeastern United States: Derivation and a list of tolerance values, with criteria for assigning water quality ratings, *J. N. Am. Benthol. Soc.*, 12, 279–290, 1993.

Lenat, D.R. and Barbour, M.T., Using benthic macroinvertebrate community structure for rapid, cost-effective, water quality monitoring: Rapid bioassessment, In *Biological Monitoring of Aquatic Systems*, Loeb, S.L. and Spacie, A. (eds.), Lewis Publishers, Boca Raton, FL, 1994, pp. 187–215.

Ludwig, J.A. and Reynolds, J.F., *Statistical Ecology. A Primer on Methods and Computing*, John Wiley & Sons, New York, 1988.

MacArthur, R.H., On the relative abundance of species, *Am. Nat.*, 94, 25–36, 1960.

Magurran, A.E., *Ecological Diversity and Its Management*, Princeton University Press, Princeton, NJ, 1988.

Marchant, R., Barmuta, L.A., and Chessman, B.C., Influence of sample quantification and taxonomic resolution on the ordination of macroinvertebrate communities from running waters in Victoria, Australia, *Mar. Freshw. Res.*, 46, 501–506, 1995.

Matthews, R.A., Matthews, G.B., and Landis, W.G., Application of community level toxicity testing to environmental risk assessment. In *Risk Assessment. Logic and Measurement*, Newman, M.C. and Strojan, C.L. (eds.), CRC Press, Boca Raton, FL, 1998, pp. 225–253.

May, R.M., *Stability and Complexity in Model Ecosystems*, Princeton University Press, Princeton, NJ, 1973.

May, R.M., *Theoretical Ecology: Principles and Applications*, W.B. Saunders Co., Philadelphia, PA, 1976.

Motomura, I., On the statistical treatment of communities, *Zool. Mag. Tokyo*, 44, 379–383, 1932.

Mulvey, M. and Diamond, S.A., Genetic factors and tolerance acquisition in populations exposed to metals and metalloids. In *Ecotoxicology of Metals: Current Concepts and Applications*, Newman, M.C. and McIntosh, A.W. (eds.), Lewis Publishers, Chelsea, MI, 1991.

Newman, M.C., *Quantitative Methods in Aquatic Ecotoxicology*, CRC Publishers, Inc., Boca Raton, FL, 1995.

Newman, M.C., *Population Ecotoxicology*, John Wiley & Sons, Chichester, UK, 2001.

Niemi, G.J., Detenbeck, N.E., and Perry, J.A., Comparative analysis of variables to measure recovery rates in streams, *Environ. Toxicol. Chem.*, 12, 1541–1547, 1993.

Oliver, I., and Beattie, A.J., A possible method for the rapid assessment of biodiversity, *Conserv. Biol.*, 7, 562–568, 1993.

Olsgard, F., Somerfield, P.J., and Carr, M.R., Relationships between taxonomic resolution, macrobenthic community patterns and disturbance, *Mar. Eco. Prog. Ser.*, 172, 25–36, 1998.

Omernik, J.M., Ecoregions of the conterminous United States, *Ann. Assoc. Am. Geog.*, 77, 118–125, 1987.

Patrick, R., Diatom communities, In *The Structure and Function of Freshwater Microbial Communities*, Cairns, J., Jr. (ed.), Virginia Polytechnic Institute and State University, Blacksburg, VA, 1971, pp. 151–164.

Pielou, E.C., *An Introduction to Mathematical Ecology*, John Wiley & Sons, New York, 1969.

Pielou, E.C., *Ecological Diversity*, John Wiley & Sons, New York, 1975.

Plafkin, J.L., Barbour, M.T., Porter, K.D., Gross, S.K., and Hughes, R.M., *Rapid Bioassessment Protocols for Use in Streams and Rivers: Benthic Macroinvertebrates and Fish, EPA 440/4-89-001*, U.S. EPA, Washington, D.C., 1989.

Platt, J.R., Strong inference, *Science*, 146, 347–353, 1964.

Poff, N.L., Olden, J.D., Vieira, N.K.M., Finn, D.S., Simmons, M.P., and Kondratieff, B.C., Functional trait niches of North American lotic insects: Traits-based ecological applications in light of phylogenetic relationships, *J. N. Am. Benthol. Soc.*, 25, 730–755, 2006.

Popper, K.R., *The Logic of Scientific Discovery*, 3rd ed., Hutchinson, London, England, 1972.

Preston, F.W., The commonness and rarity of species, *Ecology*, 29, 254–283, 1948.

Qinghong, L., A model for species diversity monitoring at community level and its applications, *Environ. Monitor. Assess.*, 34, 271–287, 1995.

Rapport, D.J., Regier, H.A., and Hutchinson, T.C., Ecosystem behavior under stress, *Am. Nat.*, 125, 617–640, 1985.

Resh, V.H. and Jackson, J.K., Rapid assessment approaches to biomonitoring using benthic macroinvertebrates, In *Freshwater Biomonitoring and Benthic Macroinvertebrates*, Rosenberg, D.M. and Resh, V.H. (eds.), Chapman & Hall, New York, 1993, pp. 195–223.

Resh, V.H., Norris, R.H., and Barbour, M.T., Design and implementation of rapid assessment approaches for water resource monitoring using benthic macroinvertebrates, *Aust. J. Ecol.*, 20, 108–121, 1995.

Reynoldson, T.B., Norris, R.H., Resh, V.H., Day, K.E., and Rosenberg, D.M., The reference condition: A comparison of multimetric and multivariate approaches to assess water-quality impairment using benthic macroinvertebrates, *J. N. Am. Benthol. Soc.*, 16, 833–852, 1997.

Roback, S.S., Insects (Anthropoda:Insecta), In *Pollution Ecology of Freshwater Invertebrates*, Hart, C.W. Jr. and Fuller, S.L.H. (eds.), Academic Press, New York, 1974, pp. 313–376.

Simberloff, D.S., Properties of rarefaction diversity measures, *Am. Nat.*, 106, 414–415, 1972.

Simpson, E.H., Measurement of diversity, *Nature*, 163, 688, 1949.

Slooff, W., Benthic macroinvertebrates and water quality assessment: Some toxicological considerations, *Aquat. Toxicol.*, 4, 73–82, 1983.
Somers, K.M., Reid, R.A., and David, S.M., Rapid biological assessments: How many animals are enough? *J. N. Am. Benthol. Soc.*, 17, 348–358, 1998.
Suter, G.W., II, A critique of ecosystem health concepts and indexes, *Environ. Toxicol. Chem.*, 12, 1533–1539, 1993.
Tokeshi, M., Species abundance patterns and community structure, *Adv. Ecol. Res.*, 24, 112–186, 1993.
Vanderklift, M.A., Ward, T.J., and Jacoby, C.A., Effect of reducing taxonomic resolution on ordinations to detect pollution-induced gradients in macrobenthic infaunal assemblages, *Mar. Eco. Prog. Ser.*, 136, 137–145, 1996.
Vannote, R.L., Minshall, G.W., Cummins, K.W., Sedell, J.R., and Cushing, C.E., The river continuum concept, *Can. J. Fish. Aquat. Sci.*, 37, 130–137, 1980.
Vinson, M.R. and Hawkins, C.P., Effects of sampling area and subsampling procedure on comparisons of taxa richness among streams, *J. N. Am. Benthol. Soc.*, 15, 392–399, 1996.
Warwick, R.M., Environmental impact studies on marine communities: Pragmatic considerations, *Aust. J. Ecol.*, 18, 63–80, 1993.
Warwick, R.M., The level of taxonomic discrimination required to detect pollution effects on marine benthic communities, *Mar. Biol.*, 19, 259–268, 1988.
Washington, H.G., Diversity, biotic and similarity indices. A review with special relevance to aquatic ecosystems, *Water Research*, 18, 653–694, 1984.
Williams, P.H. and Gaston, K.J., Measuring more of biodiversity: Can higher-taxon richness predict wholesale species richness? *Biolog. Conserv.*, 67, 211–217, 1994.
Wilson, J.B., The cost of heavy-metal tolerance: An example, *Evolution*, 42, 408–413, 1988.

23 Experimental Approaches in Community Ecology and Ecotoxicology

Observational approaches may provide support for a causal relationship between stressors and community responses; however, descriptive studies alone cannot be used to show causation. While application of Koch's postulates, Hill's criteria, and other weight-of-evidence approaches such as Bayesian methods may strengthen arguments for causal relationships (Beyers 1998, Suter 1993), to many researchers controlled experimental manipulations remain the only way to rigorously demonstrate causation in scientific investigations. The relationship between descriptive and experimental approaches in ecotoxicology can be depicted as continua along two axes that reflect the degree of experimental control, replication, and ecological relevance (Figure 23.1). Experimental approaches, such as single species toxicity tests and microcosm experiments, provide rigorous control over confounding variables and are easily replicated, but lack ecological realism. Purely descriptive studies (e.g., routine biomonitoring) lack true replication and random assignment of treatments to experimental units. Because treatments are not assigned randomly, differences between reference and impacted sites in biomonitoring studies cannot be directly attributed to a particular stressor. Several alternative experimental designs have been proposed that address problems associated with the lack of replication and random assignment of treatments; however, Beyers (1998) argues that it is "fundamentally wrong to apply inferential statistics to pseudoreplicated data to show that an observed effect was caused by an impact." The widespread application of inferential statistics in published biomonitoring studies suggests that this opinion is not shared by many researchers or journal editors. As we will see, the use of inferential statistics is not an essential component of all experimental designs. In some instances, sustained manipulations at a large spatial or temporal scale may provide adequate evidence to demonstrate causation.

23.1 EXPERIMENTAL APPROACHES IN BASIC COMMUNITY ECOLOGY

> Anyone who has tried to perform a replicated experiment in community ecology knows that the replicates within a treatment have a perverse way of becoming different from each other, even when every effort is made to keep them identical.
>
> **(Wilson 1997)**

23.1.1 The Transition from Descriptive to Experimental Ecology

Observational approaches dominated the field of basic ecology during its early history, a period when ecology was primarily a concept-driven science instead of an experiment-driven science (Lubchenco and Real 1991). Descriptions of habitat requirements, feeding habits, and associations

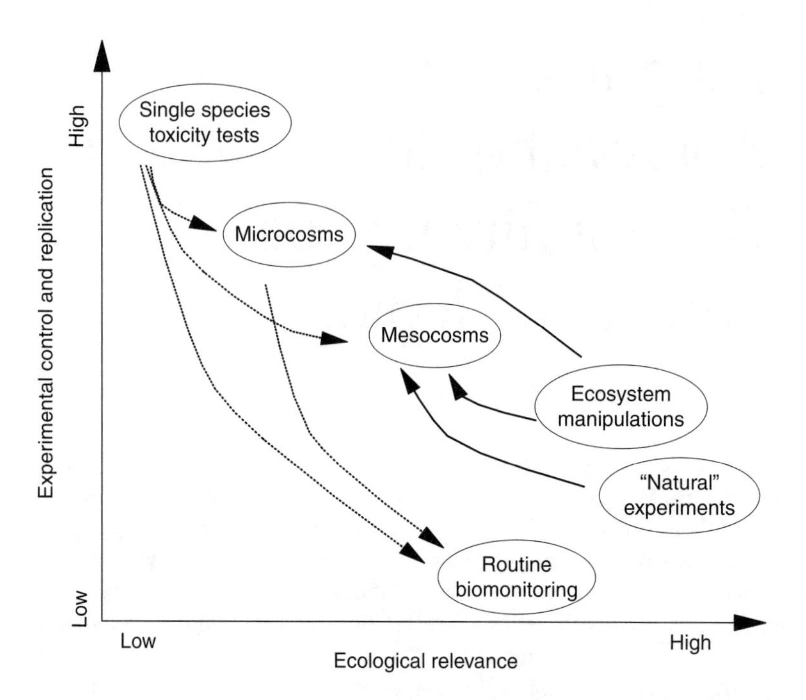

FIGURE 23.1 The relationship between ecological relevance, experimental control, and replication in ecotoxicological assessments is represented as continua along two axes. Small-scale laboratory and microcosm experiments lack ecological realism but are easily replicated and provide tight control over experimental variables. Experiments conducted at larger spatiotemporal scales (e.g., ecosystem manipulations, natural experiments) have greater ecological relevance but lack rigorous control and are difficult to replicate. A research program that integrates experimental approaches at different scales is optimal for determining causation. For example, the relevance of single species toxicity tests and microcosm experiments can be validated by conducting studies at larger spatial and temporal scales (represented by the dashed lines). The underlying mechanisms responsible for changes observed in unreplicated, large-scale experimental systems can be examined in microcosm and mesocosm studies (represented by the solid lines).

among populations formed the basis of most ecological research during this period. More recently, ecologists have recognized the importance of integrating purely descriptive and hypothesis-driven research by comparing patterns observed in natural communities to those predicted by theoretical studies (Werner 1998). Although this approach represented an important step in the transition of ecology to a more rigorous science, too often weak agreement between theory and observation was accepted as evidence for causal processes. The resulting harsh criticism of nonexperimental studies in ecology created a backlash against descriptive research that is still evident today. The acrimonious debate over the role of descriptive approaches is at least partially responsible for the rigor with which ecological experiments are conducted today. The transition from a purely descriptive to an experimental science is generally regarded as evidence of maturation in most fields of scientific inquiry, and ecology is no exception. The ability to test hypotheses with controlled experiments defines science and separates true science from pseudoscience (Popper 1972). Sciences that have progressed rapidly (e.g., physics, molecular biology, chemistry) have employed a particular form of inquiry that involves posing multiple hypotheses and testing these hypotheses with experiments (Newman 2001, Platt 1964).

In a survey of the three major ecology journals (*Ecology*, *The American Naturalist*, *The Journal of Animal Ecology*), Ives et al. (1996) reported a dramatic shift from laboratory studies to purely observational and descriptive studies that began in the 1960s (Figure 23.2). Although it was well established that an understanding of natural history was necessary to predict the distribution and abundance of organisms, ecologists realized the diminishing returns of purely descriptive studies

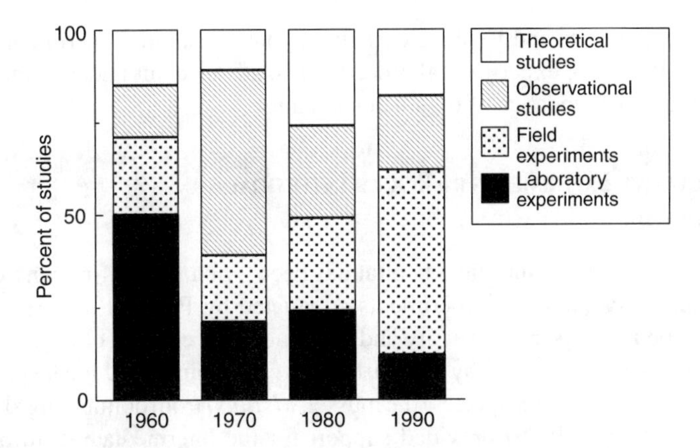

FIGURE 23.2 Changes in the approaches that ecologists employ to study populations and communities over a 40-year period. The results are based on the number of publications in each of four categories. Data were derived from a search of articles published in several leading ecological journals (*Ecology*, *The American Naturalist*, *The Journal of Animal Ecology*). (Modified from Figure 1 in Ives et al. (1996).)

and their inability to demonstrate causation. Thus, the late 1970s were characterized by another shift from descriptive and comparative approaches to field experiments (field cages and *in situ* manipulations). The role of experimental manipulations in the history of ecology is illustrated by the intense controversy over the importance of interspecific competition in regulating communities (Strong et al. 1984). Considerable research effort was devoted to showing that competition was a pervasive force in nature and that patterns of species abundances were a direct result of competition for limited resources. Comparisons of morphological characteristics and feeding habits of allopatric and sympatric populations supported the hypothesis that either competition or the "ghost of competition past" (Connell 1980) was a primary factor regulating communities. However, much of the corroborative evidence collected to support these hypotheses was based on observational studies. Comparative studies lacked the risky predictions required of experimental approaches and were virtually impossible to falsify (Popper 1972). Upon closer examination, results of these comparative studies were attacked as statistical artifacts (Connor and Simberloff 1979).

This transition from descriptive to experimental approaches in ecology was hampered by the tremendous natural variability of ecological systems and the difficulty in isolating specific components for investigation (Lubchenco and Real 1991). Natural variability adds uniqueness to ecological systems and limits our ability to generalize among systems. The interdependence and interactions among specific components in ecological systems, which are often of considerable interest to ecologists, makes it difficult to isolate effects of any single factor. Interestingly, similar concerns over complexity and natural variability contribute to the skepticism that many laboratory toxicologists have expressed for community and ecosystem studies.

Despite the logistical difficulties of conducting experiments on complex ecological systems, researchers began to realize that experimental manipulation was the most direct approach for showing causation and for resolving some of the more significant controversies in ecology. Although small-scale experiments investigating the importance of competition and predation have been conducted in the laboratory (Gause 1934, Park 1948), field manipulations were generally considered impractical and logistically difficult. All of this changed in the early 1960s. The pioneering experiments conducted by Connell (1961) investigating competition in the rocky intertidal zone are considered an important turning point in the history of ecology, providing the framework for field manipulations in a variety of other habitats. These conceptually simple, but elegant, experiments demonstrated that competition played an important role in structuring communities and that environmental factors can influence the outcome of species interactions. A critical period of self-evaluation followed as ecologists were introduced to the writings of Popper (1972) and Platt (1964), strong advocates of the need to falsify hypotheses and to test alternative hypotheses with experiments. Contemporary ecologists employ a variety of experimental procedures to advance our understanding of factors that limit the distribution and abundance of organisms in nature.

23.1.2 Manipulative Experiments in Rocky Intertidal Communities

Since the early 1960s, the rocky intertidal habitat has been a rich source for many of the significant hypotheses in community ecology. Experiments conducted by Paine (1966, 1969) illustrated the effects of predators on local species diversity and introduced the concept of keystone species. Paine (1969) showed that intense predation by the starfish *Pisaster* maintained local species diversity by preventing a competitively superior species (the mussel, *Mytilus*) from dominating all available space. Subsequent work by Sousa (1979) provided support for the intermediate disturbance hypothesis (see Chapter 25), which states that species diversity is influenced by competition and physical disturbance, and that greatest diversity is observed at intermediate levels of disturbance (Connell 1978). Disproportionate effects of a particular species or the notion that species diversity may be enhanced under moderate levels of disturbance are significant ecological concepts that have major implications for community ecotoxicology. The relationship between natural and anthropogenic disturbance will also be considered in Chapter 25.

It is no coincidence that several of the most significant contributions to the field of community ecology, namely the role of competition, the effects of predation on species diversity, the keystone species concept, and the intermediate disturbance hypothesis, were derived from experiments conducted in rocky intertidal habitats. The classic studies of Joseph Connell, Robert Paine, Paul Dayton, and Bruce Menge influenced a generation of ecologists and clearly demonstrated the effectiveness of field manipulations. Compared to other systems, rocky intertidal habitats are less complex and lend themselves to easy experimental manipulation. Removing competitors or excluding predators is relatively simple in these essentially two-dimensional systems, where most of the organisms are either sessile or very slow moving.

23.1.3 Manipulative Studies in More Complex Communities

Conducting manipulative experiments in more complex systems and at larger spatial scales has proven to be logistically challenging. However, there are several excellent examples where researchers have tested important principles of community ecology using large-scale field manipulations. The most striking example of a large-scale experiment designed to test specific theoretical predictions was Dan Simberloff's defaunation studies of mangrove islands in the Florida Keys (see Chapter 21). Simberloff and Wilson's (1969) demonstration of the dynamic equilibrium in number of species has important implications for conservation biology and restoration ecology. Interestingly, while these experiments were designed to test basic principles of island biogeography, removal of insects from the islands was accomplished by pesticide application. Thus, the results have direct relevance to community ecotoxicology from the perspective of studying recovery from chemical stressor.

A second set of large-scale experiments conducted in the 1960s involved direct measurement of ecosystem dynamics in a New Hampshire watershed. The box and arrow diagrams developed by ecologists in the 1950s and 1960s to describe energy flow and nutrient cycling were generally abstract and remained untested hypotheses. Manipulation of a watershed in the Hubbard Brook Experimental Forest provided an opportunity to test these models and to measure the response to deforestation (Likens et al. 1970). The researchers observed large export of nutrients and particulate materials in the deforested stream compared to a reference watershed.

In addition to testing theoretical predictions of ecosystem responses to perturbation, these early studies set the stage for a series of whole ecosystem manipulations that measured effects of chemical

TABLE 23.1
Comparison of the Strengths and Weaknesses of Different Types of Experiments in Community Ecology

Characteristic	Laboratory	Field	Natural Trajectory	Natural Snapshot
Regulation of independent variables	Highest	Medium to low	None	None
Site matching	Highest	Medium	Medium to low	Lowest
Ability to follow trajectory	Yes	Yes	Yes	No
Temporal scale	Lowest	Lowest	Highest	Highest
Spatial scale	Lowest	Low	Highest	Highest
Scope (range of manipulations)	Lowest	Medium to low	Medium to high	Highest
Realism	Low to none	High	Highest	Highest
Generality	None	Low	High	High

Source: After Diamond (1986).

stressors, including pesticides and acidification. These experiments also demonstrated that a powerful case can be made for causal relationships without true replication. Details of these experiments will be described in Section 23.4.1.

23.1.4 Types of Experiments in Basic Community Ecology

It is important to realize that all experimental approaches are not equal and that certain types of experimental systems may be more useful than others for investigating ecological responses to perturbations. Diamond (1986) distinguishes three types of experiments in ecological research: laboratory experiments, field experiments, and natural experiments (Table 23.1). He compares these experimental approaches in terms of control over independent variables, site matching (e.g., pretreatment similarity among experimental units), ability to follow a trajectory, spatiotemporal scale, scope, ecological realism, and generality. Laboratory experiments rank high in terms of control of independent variables and site matching, but are unrealistic because of their limited scope, spatiotemporal scale, ecological realism, and generality. Field experiments are conducted outdoors and often involve manipulation of natural communities, such as the removal or addition of a predator or competitor. Connell's studies in the rocky intertidal zone and Simberloff's defaunation studies in the Florida Keys are examples of field experiments. Although field experiments have played an important role in the development and testing of ecological theory, Diamond (1986) is critical of these approaches. Compared to laboratory experiments, field experiments are more realistic and offer a greater range of possible manipulations. However, field experiments have less control and may be confounded by pretreatment differences among experimental units. According to Diamond, field experiments are usually conducted at a small spatiotemporal scale and lack generality.

Natural experiments differ from field experiments in that the researcher does not directly manipulate the variables of interest, but selects sites where the perturbation is already present or will be present. Comparisons of species abundance, habitat preferences, and morphological characteristics in allopatric and sympatric populations are considered natural experiments. Probably the best example of a natural experiment is the comparison of beak sizes among allopatric and sympatric populations of Galapagos finches. Assuming that beak size is an appropriate surrogate for resource use, the greater separation of beak sizes on sympatric islands compared to allopatric islands is considered evidence for interspecific competition. Because researchers may investigate results of processes that occur over very large areas (island archipelagoes) and over evolutionary time periods, natural experiments have the greatest spatial and temporal scales. Diamond further distinguishes between natural snapshot experiments, in which a researcher compares sites that differ in a particular characteristic (e.g., presence or absence of a predator) and natural trajectory experiments, where a researcher makes comparisons before and after a perturbation.

It is important to note that Diamond's enthusiasm for natural experiments is not shared by all ecologists. Because treatment sites are not assigned by the investigator and because nothing is controlled or manipulated in natural experiments, differences between locations cannot be directly attributed to any particular cause. Lubchenco and Real (1991) consider these experiments a special case of observational studies and conclude that Diamond's "natural experiment" is a misnomer that masks the true contributions of comparative ecological studies.

23.2 EXPERIMENTAL APPROACHES IN COMMUNITY ECOTOXICOLOGY

Development of experimental techniques in basic ecology was partially motivated by the recognition that comparative approaches are insufficient for demonstrating causation and understanding mechanisms. Manipulative experiments gained popularity in the 1960s as ecologists realized that agreement between mathematical predictions and field observations did not necessarily demonstrate the truth of these predictions. Although this same realization provided some motivation for the development of experimental approaches in community ecotoxicology, other factors also played an important role. Some ecotoxicologists questioned the validity of using single species laboratory experiments to predict responses of more complex systems in the field (Cairns 1983). In addition, some ecotoxicologists realized that the relative influence of biotic and abiotic factors on responses of communities to contaminants could only be assessed using experiments.

Like ecology, the field of community ecotoxicology is currently undergoing a transition from purely descriptive, observational approaches to more rigorous experimental techniques. However, this transition has occurred much more slowly in ecotoxicology, as experiments investigating community and ecosystem responses to contaminants are still relatively rare. Laboratory experiments, such as standardized 96-h toxicity tests, have been the workhorse of the regulated community for many years (Cairns 1983). The historical focus on simple laboratory experiments using single species has at least partially impeded implementation of community-level experimental approaches. The continued emphasis on these "reductionist," lower-level techniques for predicting ecological consequences of contaminants has been criticized (Cairns 1983, 1986, Kimball and Levin 1985, Odum 1984) and is surprising given the widespread support for integrated assessments (Adams et al. 1992, Clements and Kiffney 1994, Joern and Hoagland 1996, Karr 1993). In addition, recent studies have shown that single species tests may not predict community-level responses to contaminants because of indirect effects and higher-order interactions (Clements et al. 1989, Gonzalez and Frost 1994, Pontasch et al. 1989, Schindler 1987). If communities are more than random associations of noninteracting species, it follows that experimental approaches are required to understand the effects of contaminants on these interactions.

Currently, there are no established protocols for investigating community responses to contaminants in experimental systems. Reviews of experimental approaches reveal an astonishing diversity of experimental conditions, communities, duration, spatiotemporal scale, experimental designs, and endpoints (Gearing 1989, Gillett 1989, Kennedy et al. 1995, Pontasch 1995, Shaw and Kennedy 1996). Most of these experimental studies have been conducted in aquatic systems (freshwater and marine). The limited number of studies conducted in terrestrial systems to investigate community responses to contaminants is considered a significant shortcoming in the field of ecotoxicology.

Ecotoxicologists have employed the same experimental approaches described in Table 23.1 to investigate the effects of contaminants on communities: laboratory experiments, field experiments, and natural experiments. Laboratory experiments using small-scale microcosms involve the exposure of natural or synthetic communities to specific chemicals. Larger experimental systems (mesocosms) are outdoors and generally have some interactions with the natural environment. Not surprisingly, field experiments (defined as the intentional addition of contaminants to natural systems) have received limited attention in ecotoxicology. However, this technique has become more common in the past few years. Researchers have also taken advantage of planned perturbations to assess the impacts of contaminants on communities. If data are collected before a particular chemical is released into the environment, the before–after control-impact (BACI) design (Stewart-Oaten et al. 1986) is a powerful quasiexperimental approach that can be employed to assess community responses. On the basis of their experiences following the *Exxon Valdez* oil spill, Wiens and Parker (1995) provide an excellent overview of quasiexperimental approaches for assessing the impacts of unplanned perturbations. They note that experimental designs that treat the level of contamination as a continuous variable are generally more precise and offer the greatest opportunity to detect nonlinear responses. Although relatively uncommon in community ecotoxicology, large-scale monitoring studies that compare communities with varying levels of perturbation are analogous to Diamond's (1986) natural experiments. Because treatments are not assigned randomly in comparative studies, these experimental designs also suffer from some of the same limitations as natural experiments.

23.3 MICROCOSMS AND MESOCOSMS

> While direct projection from the small laboratory microecosystem to open nature may not be entirely valid, there is evidence that the same basic trends that are seen in the laboratory are characteristic of succession on land and in large bodies of water.
>
> **(Odum 1969)**

> Most of the crucial questions in applied ecology are not open to attack by microcosms.
>
> **(Carpenter 1996)**

23.3.1 Background and Definitions

Because the application of microcosms and mesocosms to ecotoxicological research has been the subject of considerable controversy in recent years, it is important to place this research within the proper context. Model systems are effectively employed in a variety of fields, including engineering, architecture, and aviation. These scaled replicas are used to describe and evaluate performance of natural systems under a variety of experimental conditions. Similar to mathematical models, physical models make numerous simplifying assumptions to investigate the influence of specific variables. We contend that much of the criticism of model systems in ecotoxicological research is due to the failure of researchers to explicitly state these assumptions. To a certain extent, all experimental systems suffer from attempts to limit or control confounding variables (Drake et al. 1996). However, the strength of model systems lies in their ability to isolate key components and to investigate how these components respond to perturbation. Unlike field studies, microcosm and mesocosm experiments can provide clean tests of specific predictions of hypotheses (Daehler and Strong 1996). However, the degree of simplification necessary to obtain precise control often severs any connection to natural processes. This may or may not be a serious issue, depending on the specific goals of the study. If the primary objective of an experiment is to understand how a system works, then experiments should be as realistic as possible. However, if the primary objective is to obtain a mechanistic understanding of underlying processes, then realism may not be as significant (Peckarsky 1998). It is important to remember that microcosms and mesocosms do not attempt to duplicate all aspects of natural ecosystems. In fact, given our incomplete understanding of the structure and function of ecosystems, it is naive to think that we could reproduce the complexities of nature. We agree with Lawton (1996) that the best way to understand the operation of a complex ecological system is to construct a model and determine if it functions as expected. Despite criticism by some researchers

(Carpenter 1996), we feel that perturbations of model systems provide a powerful way to test basic and applied ecological hypotheses.

Recent reviews, essays, and special features have discussed the advantages and disadvantages of small-scale experiments in basic ecological research (Carpenter 1996, Daehler and Strong 1996, Resetarits and Bernado 1998, Schindler 1998). Ives et al. (1996) characterized complexity, time scale, and scientific impact of microcosm and mesocosm experiments relative to other approaches employed in basic ecology (e.g., observational studies, field manipulations, or theoretical studies). As expected, microcosm experiments generally included fewer species and were of shorter duration. However, there was relatively little difference in complexity and time scale between mesocosm experiments (field cages) and other approaches. The scientific impact of small-scale experiments was investigated by comparing the frequency of citations and prevalence in undergraduate ecology textbooks of microcosm and mesocosm experiments relative to other approaches. Ives et al. (1996) concluded that the type of study had a negligible role in determining scientific impact. In general, there were relatively few differences between small-scale experiments and other approaches employed in basic ecology.

Several chapters in the excellent book by Resetarits and Bernado (1998) address the issues of spatiotemporal scale and trade-offs between control and realism in ecological experiments. The consistent theme in this volume is the necessary link between small-scale experiments and well-planned observational studies. Resetarits and Fauth (1998) argue that the perceived trade-off between rigor and realism is partially a consequence of our lack of creativity in designing experiments. The importance of ecological realism in experimental design should be addressed in the same way scientists evaluate other research questions. That is, the criticism that model systems do not reflect processes in the natural world is simply a "hypothesis to be tested" (Resetarits and Fauth 1998).

Currently, the most significant challenge in microcosm and mesocosm research is to identify those key features that must be carefully reproduced in order to simulate structure and function of natural systems. How much simplification is possible in model systems before we lose the connection with the natural system we are attempting to simulate? In a comparison of microcosm, mesocosm, and whole ecosystem experiments, Schindler (1998) contends that small-scale studies may provide highly replicable but spurious results about community and ecosystem processes. Perez (1995) recommends the use of sensitivity analysis, a simulation technique that allows researchers to evaluate the relative importance of numerous variables, to identify critical aspects of model systems. Variables that significantly influence function of the model system must be reproduced carefully, whereas unimportant variables may receive less attention.

Although model systems are not typically included in ecological risk assessment or used for establishing chemical criteria, the value of microcosms and mesocosms to assess effects of contaminants on communities has been recognized for many years (see reviews by Gearing 1989, Gillett 1989, Graney et al. 1989). The emergence of model systems in ecotoxicological research represents an important transition from reductionist to holistic approaches (Odum 1984). Studies comparing results of microcosm and mesocosm experiments with mathematical models (Momo et al. 2006) and field data (Christensen et al. 2006) illustrate the likelihood of unexpected indirect effects and support a more holistic approach to ecological risk assessment. Although the distinction between microcosms and mesocosms is not always obvious in the literature, microcosms are generally smaller in size and commonly located indoors. Microcosms are defined as controlled laboratory systems that attempt to simulate a portion of the natural world. Odum (1984) defined mesocosms as "bounded and partially enclosed outdoor experimental setups." Because they are only partially enclosed, mesocosms generally have greater exchange with the natural environment. Despite these differences, one common feature of both microcosm and mesocosm experiments is that they can investigate the responses of numerous species simultaneously. Consequently, endpoints examined in microcosm and mesocosm experiments are not restricted to simple estimates of mortality and growth but generally include an array of structural and functional measures (e.g., community composition, species richness, or primary productivity).

A special series of articles published in *Ecology* entitled "Can we bottle nature?" (Daehler and Strong 1996) examined the role of microcosms in basic ecological research. Although the articles did not emphasize effects of contaminants, a general consensus that emerged was that small-scale experimental approaches should be used to solve problems in applied ecology. Most of the contributors agreed that, while microcosm experiments can provide very "clean" results with tight control of biotic and abiotic variables, microcosm research programs should be well integrated with field studies. Issues such as the simplicity of artificial communities and the lack of immigration and emigration can be addressed by comparing results of microcosm experiments with more traditional monitoring approaches conducted in the field. We agree with Carpenter (1996) that without the context of proper field studies, many microcosm experiments are "irrelevant and diversionary."

As noted above, microcosm experiments have played a major role in the development and testing of ecological theory (Drake et al. 1996). Many of the ideas proposed by early theoretical ecologists (e.g., the competitive exclusion principle) were tested in relatively simple experimental systems, and results provided insights for additional theoretical and empirical research. Unfortunately, microcosm and mesocosm research has not achieved a similar status in ecotoxicology. Although microcosms and mesocosms have been employed to assess impacts of contaminants on populations and communities, they have not played a major role in ecotoxicological research. Reviews of the major journals in aquatic and terrestrial toxicology reveal a surprisingly infrequent application of these tools. Notable exceptions include a few published symposia and special features that focused on microcosm and mesocosm experiments (*Environmental Toxicology and Chemistry*, 1992, 11; *Ecological Applications*, 1997, 7).

23.3.2 Design Considerations in Microcosm and Mesocosm Studies

A valid criticism of microcosm and mesocosm research is that the emphasis placed on increasing reproducibility and decreasing variability has come at the expense of ecological relevance to natural systems. Thus, one of the most important considerations when conducting microcosm or mesocosm research is to understand how biotic and abiotic conditions in model systems compare to the natural system. Surprisingly, few studies report information collected from the specific field sites represented by these experimental systems. In a review of aquatic microcosms, Gearing (1989) noted that only 9% of 339 published articles collected field data to verify that communities in microcosms were similar to those in natural systems. The most likely explanation for the failure to report ecological conditions is that many of these experiments were conducted simply to test the effects of a particular chemical. Relatively few microcosm or mesocosm experiments were designed to validate data from a specific field site. Nonetheless, information on the similarity or dissimilarity of the experimental systems and natural systems is necessary when evaluating the efficacy and ecological realism of microcosms.

23.3.2.1 Source of Organisms in Microcosm Experiments

The source of organisms is a major design issue when conducting microcosm and mesocosm experiments. One common approach is to add synthetic assemblages of organisms, generally obtained from laboratory cultures, to the experimental system. This technique ensures that replicates have similar initial community composition before the experimental units are assigned to treatments. In addition to providing a standardized technique for assessing effects of contaminants, variance is greatly reduced by controlling initial community composition. Freda Taub and others (Landis et al. 1997, Matthews et al. 1996, Taub 1989, 1997) have successfully employed this approach to investigate the effects of contaminants on microbial and planktonic assemblages. Taub's standardized aquatic microcosm (SAM) is now an American Society for Testing and Materials (ASTM) protocol (ASTM 1995), representing a major advance in the application of community-level endpoints in a regulatory framework. The same opportunities for comparisons among chemicals and among species that are cited as a major advantage of single species toxicity tests are also realized using a SAM. However, because of the synthetic composition of these communities, this standardized approach has been criticized because it lacks ecological relevance to natural systems (Perez 1995). As with most decisions in the development of model systems, trade-offs are often necessary between standardization and increased ecological realism.

The alternative methods for establishing organisms in microcosms and mesocosms are to add natural communities or to allow the system to colonize naturally. Both methods should result in communities that are initially similar to those in the natural system, thus improving ecological realism of the experiment. Samples of a known area or volume collected from the environment can be added to obtain realistic abundances of organisms. Perez et al. (1991) collected discrete samples of seawater and sediment cores containing indigenous organisms to investigate fate and effects of Kepone in microcosms. Experiments conducted with naturally derived microbial communities have investigated effects of herbicides and other chemicals on structural and functional endpoints (Niederlehner et al. 1990, Pratt and Barreiro 1998, Pratt et al. 1997). Colonized substrates obtained from reference systems are placed in replicate microcosms containing initially uncolonized "islands." Using principles derived from the theory of island biogeography (MacArthur and Wilson 1963), colonization rate of these islands over time is compared in control and contaminated microcosms (Cairns et al. 1980). Clements et al. (1989) developed a similar collection technique to expose natural communities of benthic macroinvertebrates to contaminants in stream microcosms. Substrate-filled trays were colonized in a natural stream and then transferred to replicate microcosms. The communities added to the streams were similar among replicates and, more importantly, similar to those in the natural system.

Natural colonization of microcosms and mesocosms is probably the best way to ensure that communities resemble natural systems. This approach is most appropriate in larger mesocosm experiments that have some exchange with the local environment. However, because initial densities are not controlled by the investigator, variability among replicates may be problematic. For example, Jenkins and Buikema (1998) showed that zooplankton communities established in 12 similar pond mesocosms were markedly different after 1 year of colonization. In addition to differences in structural characteristics among the ponds, secondary productivity and community-level respiration rates also varied. Wong et al. (2004) quantified spatial and temporal variation in the structure of stream benthic communities among control mesocosms. These researchers cautioned that variation in initial community composition and species sensitivity among control mesocosms must be considered when using mesocosm results for ecological risk assessment. Differences in structural and functional characteristics prior to the start of a mesocosm experiment will greatly complicate our ability to measure responses to contaminants. Unlike standard toxicity tests, initial abundances will not be known precisely; therefore, data cannot be expressed using conventional toxicological endpoints (e.g., percent mortality). Initial community composition can be compared to controls at the end of the experiment to obtain some estimate of variability; however, more commonly results are simply compared across treatments.

23.3.2.2 Spatiotemporal Scale of Microcosm and Mesocosm Experiments

The limited spatiotemporal scale of microcosms and mesocosms is considered one of their most serious weaknesses. Few studies have tested the hypothesis that experiments conducted at one scale are appropriate for predicting responses at a different scale. This question is central to the debate over the usefulness of model systems and clearly an important research need in ecotoxicology. Although increasing the size of a mesocosm may eliminate some potential artifacts, this does not make the study an ecosystem experiment (Schindler 1998). The relatively small spatial scale of microcosms greatly restricts the numbers and types of organisms that can be included. If larger or longer-lived organisms such as top predators are an essential component of the natural system (e.g., in systems regulated by top-down predators) or have a disproportionate influence on its structure (e.g., a keystone species), results of microcosm experiments that exclude these species may not be valid. However, because relatively few natural communities are controlled by top predators or keystone species, failure to include large, wide-ranging taxa in model systems may not be a serious issue. The fact remains that we do not know if the exclusion of certain species from microcosm experiments will influence results because of our poor understanding of these scaling issues.

A more serious issue related to the small spatial scale of microcosms and some mesocosms are container artifacts. Accumulation of biotic and abiotic materials on the container walls can complicate assessments of exposure, especially if contaminants are removed from the system either by bioaccumulation or adsorption. Periodic scraping of fouling material from the container walls is one solution to this problem. However, in a closed system this can result in pulses of organic enrichment unless the material is removed from the container. Because of surface area to volume relationships, container effects generally diminish with increased size of the microcosm.

Perez et al. (1991) provided one of the few detailed analyses of the effects of spatial scale on community responses to contaminants. Intact water column and benthic communities were exposed to Kepone in 9-, 35-, and 140-L containers. Results showed that fate and effects of Kepone on aquatic communities were size dependent. Phytoplankton density was actually greater in treated microcosms compared with controls due to reductions in abundance of grazing zooplankton; however, this effect was limited to small microcosms (Figure 23.3). Similarly, the concentration of Kepone in surface sediments and the potential exposure to benthic organisms increased with microcosm size due to greater mixing and bioturbation in larger microcosms. On the basis of these results, Perez et al. (1991) concluded that small microcosms would underestimate the effects of Kepone on aquatic

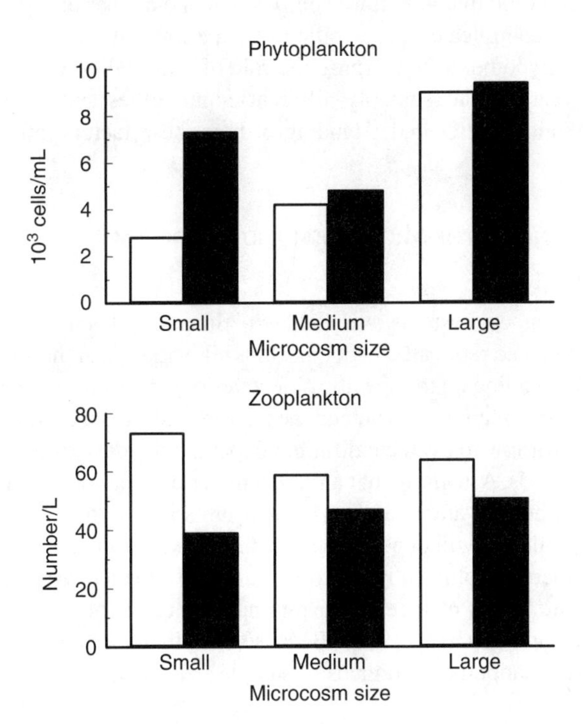

FIGURE 23.3 Response of phytoplankton and zooplankton communities to Kepone (solid bars) in small, medium, and large microcosms. Effects of Kepone on zooplankton abundance were greater in small microcosms. Reduced abundance of zooplankton and lower grazing pressure resulted in an increase in abundance of phytoplankton in small microcosms. (Modified from Perez et al. (1991).)

communities. The dependency of community responses on container size has obvious implications for ecological assessments using microcosms and mesocosms.

Finally, the relatively short temporal duration of most microcosm and mesocosm experiments limits the realism of these systems. The logistical difficulties of maintaining laboratory or large mesocosm experiments often prohibit long-term studies. More importantly, because environmental conditions in model systems deviate from natural systems over time, most experiments are conducted over relatively short-time periods (generally less than 6 months). In model systems where recruitment or immigration is absent, population abundances of most species will decrease and community composition may significantly deviate from the initial conditions. While comparisons across treatments partially alleviate this problem, separating these temporal changes in communities from those due to contaminants will complicate assessment of effects.

23.3.2.3 The Influence of Seasonal Variation on Community Responses

The time of year when microcosm or mesocosm experiments are conducted can influence the relative toxicity of contaminants and responses of communities. Because physical and chemical conditions that modify toxicity and bioavailability (e.g., temperature, pH, and dissolved organic carbon, [DOC]) may vary seasonally (Perez et al. 1991), it is important to document this information when conducting mesocosm studies. Experimental results of mesocosm studies will also be influenced by seasonal variation in community composition and contaminant bioavailability. Winner et al. (1990) used a mesocosm study to demonstrate that seasonal variation in sensitivity of planktonic communities to copper resulted from seasonal changes in DOC and community composition. Similarly, Le Jeune et al. (2006) attributed differences in effects of copper on spring and summer phytoplankton communities to seasonal variation in community composition and copper bioavailability. Although this temporal variation may complicate interpretation of experimental results, it also provides opportunities to test specific hypotheses concerning the role of seasonality. By conducting experiments at different times of year with presumably different communities and different physicochemical conditions, we can obtain a better understanding of how these factors influence responses in the field.

23.3.3 Statistical Analyses of Microcosm and Mesocosm Experiments

The major advantage of model systems over field experiments and ecosystem manipulations is the ability to randomly assign and replicate treatments, thus allowing researchers to analyze results using inferential statistics. Depending on the specific objectives of the study, a wide range of experimental designs have been employed in microcosm and mesocosm studies. An excellent overview of design considerations describing how to evaluate different experimental designs for community-level tests is provided by Smith (1995). Assuming that a finite number of experimental units are available, one of the first decisions is how to allocate experimental units among treatments and replicates. The necessary number of replicates will depend on the sampling variability, desired precision (e.g., how much change is considered ecologically relevant), and the selected α-value. Several algorithms are available to estimate power of an experiment and the necessary number of replicate samples based on these considerations (Green 1979). Because sampling variability and the number of replicates will differ among endpoints, estimates of sample size should be based on the most variable endpoint.

There has been considerable discussion in the literature concerning the relative merits of analysis of variance (ANOVA) and regression approaches for analyzing results of microcosm and mesocosm experiments (Liber et al. 1992). There is little difference in the statistical analyses used in ANOVA and regression designs. However, because the allocation of treatments and replicates among experimental units must occur prior to the start of the experiment, researchers must decide in advance which design to employ. Again, this decision will depend on the specific goals of the investigation. If the primary objective is to estimate a "safe" concentration of a particular chemical (e.g., the no observed effect concentration, NOEC), then an ANOVA approach might be most appropriate. The number of treatment levels will be determined after estimating the number of replicates required. For example, if only 12 experimental units are available and preliminary power calculations indicate that three replicates are necessary to detect significant differences, then four levels of treatment are possible. Unbalanced designs (unequal number of replicates in each treatment) are possible using ANOVA, but these are uncommon in community-level experiments (Smith 1995). More complex factorial designs are also useful in community experiments where researchers assess the relative importance of multiple stressors. For example, if we are interested in understanding the interaction of temperature and acidification, the same 12 mesocosms could be used in a 2×2 factorial design (three replicates each) with two levels of temperature (low, high) and two levels of acidification (control, acid dosed). In addition to estimating the relative importance of temperature and acidification (the main effects), this design allows us to test for potential synergistic or antagonistic interactions between these stressors. Although less common than traditional ANOVA, multivariate approaches are becoming increasingly popular for analyzing results of microcosm and mesocosm experiments (Clarke 1999, Landis et al. 1997, Matthews et al. 1996). Because the data generated from mesocosm experiments often involve multiple dependent variables, multivariate statistical techniques can be employed to obtain community-level NOECs (Wong et al. 2003).

If the goal of the experiment is to establish a relationship between the concentration of the chemical and community-level response, then regression analysis is more appropriate than ANOVA. In a regression approach, we are often less interested in a specific chemical concentration than in the slope of the concentration–response relationship. In the above example, each of the 12 experimental units could receive a different treatment (without replication) to establish this relationship. This approach would allow us to estimate the specific concentration that elicits a particular community response. For example, we may be interested in knowing the concentration that results in a 20% reduction in species richness. In addition, by comparing the slopes of the regression lines for several community-level endpoints, we could estimate their relative sensitivity to the particular chemical.

23.3.4 General Applications of Microcosms and Mesocosms

> Microcosms, in theory, should be one of the most valuable tools available to ecotoxicology.
>
> **(Gearing 1989)**

The original focus of most microcosm and mesocosm research was to predict transport and fate of contaminants under controlled conditions. Various processes involved in the movement of contaminants through biotic and abiotic compartments, including volatilization, microbial degradation, biotransformation, and food chain transfer, are readily quantified using microcosms and mesocosms. More recent applications of microcosms and mesocosms in community ecotoxicology emphasize assessment of ecological effects. Experimental systems have been used to support regulatory decisions regarding safe concentrations of pesticides (Giddings et al. 1996), establish concentration–response relationships in community-level experiments (Kiffney and Clements 1996a), validate single species toxicity tests (Pontasch et al. 1989), compare sensitivity of different endpoints (Carlisle and Clements 1999), validate field responses (Niederlehner et al. 1990), and evaluate interactions among multiple factors (Barreiro and Pratt 1994, Pratt and Barreiro 1998). Because community-level responses to contaminants may vary among locations, mesocosms also provide an efficient way to compare effects among different communities.

23.3.4.1 The Use of Mesocosms for Pesticide Registration

Mesocosm testing has been employed to measure effects of chemicals and estimate safe concentrations. Using an experimental design in which target concentrations bracket lowest observed effect concentrations (LOEC), researchers can determine if levels considered "safe," based on single species toxicity tests, are actually protective at the community and ecosystem level. Although this type of experimental design has been criticized, most studies conducted in pond mesocosms were designed to estimate ecological effects at a specific test concentration. The most controversial application of mesocosms in ecotoxicology was their use in a regulatory framework. The U.S. Environment Protection Agency (EPA)'s tiered approach for hazard assessment, the predecessor of contemporary ecological risk assessment, used a sequential series of tests to evaluate the risk of pesticides. Tier 4 tests, the most complex and ecologically relevant, involved field experiments that measured population, community, and ecosystem-level effects. A large number of studies published in the 1980s were designed to meet guidelines developed by the U.S. EPA for pesticide registration (Touart 1988, Touart and Maciorowski 1997). An excellent series of papers on the use of mesocosms for pesticide registration was published as a special issue of *Environmental Toxicology and Chemistry* (Volume 11, #1) in 1992. Although most of the studies examined fate and effects of pyrethroid insecticides (Fairchild et al. 1992, Heinis and Knuth 1992, Lozano et al. 1992, Webber et al. 1992), appropriate experimental designs were also discussed (Liber et al. 1992). A unifying theme for these studies, and indeed a primary motivation for conducting mesocosm research, is the opportunity to investigate direct and indirect effects simultaneously.

EPA's requirements for pesticide registration using mesocosm testing were rescinded in 1992. Not surprisingly, this decision created an outcry among ecotoxicologists who noted the paucity of ecological information in most risk assessments (Pratt et al. 1997). Institutions that had invested heavily in construction of mesocosm facilities in the United States were scrambling to identify other uses for these test systems. This decision, which was defended on grounds that the likelihood of false negative results based on single species tests did not justify the greater expense of multispecies experiments, was considered a major step backward by ecotoxicologists (Taub 1997).

The primary reasons for dropping mesocosm testing requirements were the problems obtaining reproducible results, variable data, and difficulties interpreting results. In addition, there was the belief that mesocosm experiments were not providing additional information beyond what was available based on single species laboratory tests. It is not surprising that data collected from mesocosm experiments were variable and complex. Indeed variability is a defining characteristic of most ecological systems and an understanding of this variability can greatly improve our ability to predict responses in nature. Simberloff (1980) characterizes ecologists' frustration with natural variability and their attempts to quantify ecological responses based on purely deterministic processes as "physics envy." He further states that, "What the physicist considers noise is music to the ears of the ecologist."

We feel that EPA's decision to abandon mesocosm testing represents a missed opportunity to increase our understanding of how natural systems respond to chemical stressors. Armed with an appreciation of natural variability of ecological systems and a greater commitment to more sophisticated data analysis procedures (e.g., multivariate techniques and nonlinear regression), a national mesocosm testing program could make a major contribution to the field of ecotoxicology. As long as regulatory agencies continue to rely on simplistic laboratory procedures for estimating field effects, ecological risk assessment will remain a reactive rather than a predictive science (Chapman 1995, Fairchild et al. 1992, Perez 1995).

23.3.4.2 Development of Concentration–Response Relationships

Another important application of microcosm and mesocosm research is to establish concentration–response relationships between contaminants and community-level endpoints (Figure 23.4).

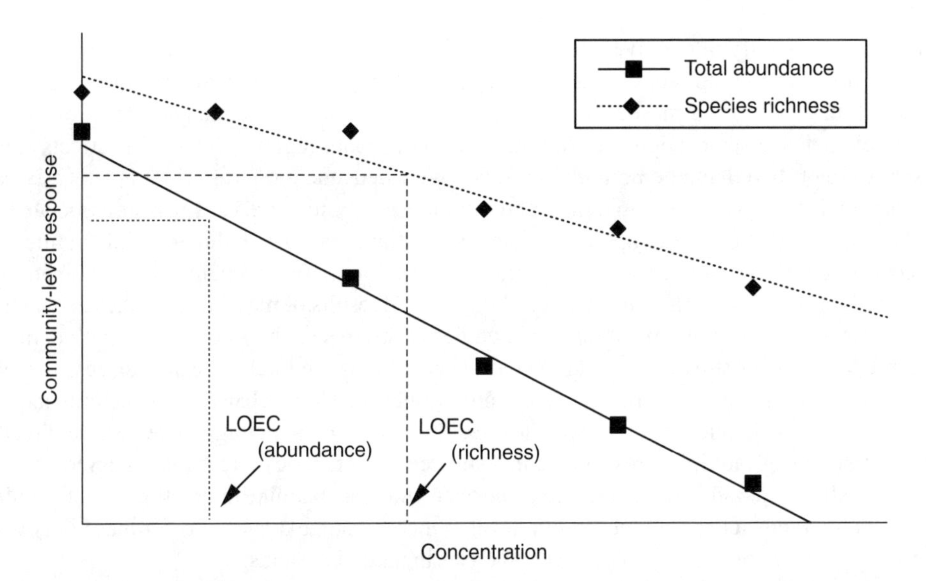

FIGURE 23.4 Hypothetical community-level responses to contaminants in microcosm or mesocosm experiments. The figure shows experimentally derived LOECs for total abundance and species richness. LOEC values were based on an estimated 20% reduction in treated systems compared to controls. In this example, species richness was less sensitive to the contaminant than total abundance.

If treatments are selected to represent a range of potential responses, researchers can estimate the level of impact expected to occur at a particular chemical concentration (e.g., the concentration that results in 20% reduction in species richness). Therefore, instead of extrapolating results of single species toxicity tests to community-level responses, the direct effect of a chemical on these responses could be quantified in a mesocosm experiment. Belanger et al. (2004) used mesocosm experiments to derive a community-level NOEC for benthic communities exposed to anionic surfactants. Wong et al. (2003) reported that community-level NOECs for anionic surfactants were similar to those developed for individual species. One significant advantage of mesocosm experiments is that NOECs and LOECs can be derived simultaneously for many species under environmentally realistic conditions in a single study. Mesocosm experiments can also be employed to compare the relative sensitivity of different community-level endpoints. Figure 23.4 shows that the estimated community-level LOEC is less for total abundance than for species richness. If these experimental results are correct, we would expect this particular chemical to have greater effects on abundance than species richness in the field. Because most microcosm and mesocosm experiments involve exposure of numerous species simultaneously, regression approaches can be used to estimate species-specific sensitivity to a particular contaminant. As described in Chapter 22 (Box 22.1), the slopes of concentration–response relationships for individual taxa provide an objective estimate of tolerance and can be used to develop biotic indices. These population and community responses observed in mesocosms could then be verified using routine field biomonitoring. Alternatively, relative sensitivity distributions derived from mesocosm studies could be used to link experimental results with biomonitoring data (Von der Ohr and Liess 2004).

23.3.4.3 Investigation of Stressor Interactions

Perhaps the most important contribution of microcosm and mesocosm research, which cannot be easily investigated in ecosystem manipulations or natural experiments, is the opportunity to measure interactions among stressors. Using a relatively simple factorial design, researchers can investigate effects of two different stressors simultaneously and estimate the potential interaction between stressors (Courtney and Clements 2000, Genter 1995, Genter et al. 1988). Genter (1995) used stream microcosms to quantify interactive effects of acidification and aluminum on periphyton communities and to measure the indirect effects of heavy metals on grazing by snails (Genter et al. 1988). Wiegner et al. (2003) used mesocosm experiments to examine the interactive effects of nutrients and trace elements (arsenic, copper, cadmium) on estuarine communities. Interactive effects of these stressors were observed, but community responses were dependent on trophic complexity. Because most communities exposed to contaminants are simultaneously subjected to stressors associated with global change (e.g., elevated temperature, increased ultraviolet (UV) radiation, and acidification), mesocosm experiments can be used to investigate interactions between chemical contamination and changing global conditions. Belzile et al. (2006) reviewed results of mesocosm studies investigating effects of UV on marine phytoplankton communities. These researchers concluded that interactions between UV and other stressors typical in coastal ecosystems are likely. We will discuss the use of microcosm and mesocosm experiments to quantify effects of global change on structural and functional responses to chemical stressors in Chapters 26 and 35. Conducting studies where direct and interactive effects of multiple stressors are investigated simultaneously requires a degree of control that is generally not possible in field studies. The opportunity to examine interactions among multiple stressors in microcosm experiments and to develop mechanistic explanations for these interactions will greatly improve our ability to predict responses in natural systems.

23.3.4.4 Influence of Environmental and Ecological Factors on Community Responses

One of the most consistent limitations of ecological data collected from field studies is the high amount of unexplained variability in natural communities. The same concentration of a particular chemical may have large effects on one community but negligible effects on another. Microcosm and mesocosm experiments can be designed to compare differences in responses among communities and to quantify the influence of environmental conditions on these responses. In addition, controlled experiments may elucidate mechanisms that show how environmental factors influence community responses. Simple factorial designs could be employed to compare the impacts of a stressor on communities collected during different seasons or obtained from different locations. Barreiro and Pratt (1994) used microcosms to demonstrate that effects of herbicides on periphyton communities were influenced by levels of nutrients and trophic status. Results showed that communities established under low nutrient conditions were more susceptible to chemical stress and required longer time to recover. Similar results were reported by Steinman et al. (1992) in which resilience of periphyton communities to chlorine stress increased with the rate of nutrient cycling. Mesocosm experiments were conducted to quantify effects of natural constituents in effluent-dominated streams on organism, population, and community responses to cadmium (Brooks et al. 2004, Stanley et al. 2005). Results of these experiments showed that Cd toxicity was overestimated by laboratory tests and generally supported application of the biotic ligand model (Di Toro et al. 2001) for establishing site-specific Cd criteria. Experiments conducted with protozoan communities examined the influence of community maturity on contaminant responses (Cairns et al. 1980). These studies showed that effects of copper on colonization rate were greater in immature communities compared to mature communities.

Microcosm and mesocosm experiments are the most effective way to evaluate the influence of community composition on stressor responses. Sallenave et al. (1994) reported that downstream transport of polychlorinatedbiphenyls (PCBs) was greater in experimental streams with grazers or shredders than in streams without these two functional groups. Kiffney and Clements (1996b) compared responses of benthic macroinvertebrate communities collected from low and high elevation streams to heavy metals in stream microcosms. Because low and high elevation communities were exposed to the same concentration of metals, the experiment provided an opportunity to estimate differences in sensitivity between locations. Results showed that headwater communities were more sensitive to heavy metals than communities from a low elevation stream (Figure 23.5). These differences in sensitivity between locations suggest that criterion values protective of low elevation

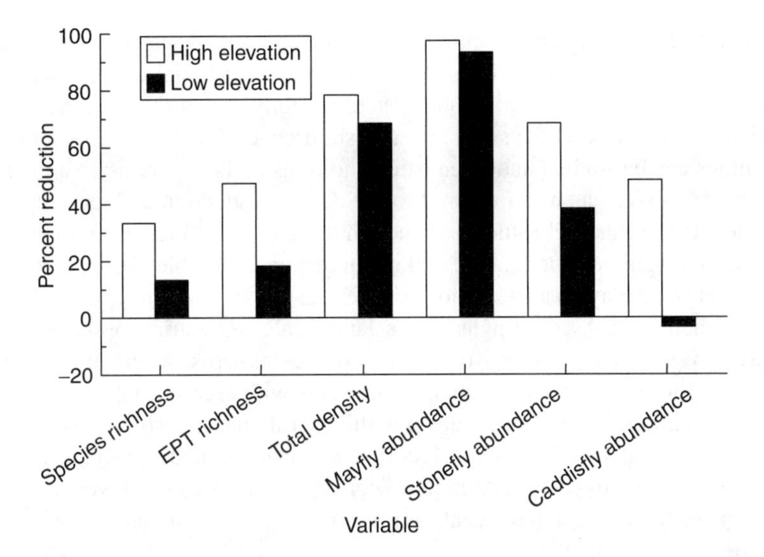

FIGURE 23.5 Comparison of the effects of contaminants on communities from different locations. The figure shows the responses of benthic macroinvertebrate communities to heavy metals in stream microcosms. Communities collected from low and high elevation sites were exposed to the same concentration of heavy metals. The responses were based on the percent reduction of benthic metrics in treated microcosms compared to control microcosms. For all metrics, the effects of metals were greater on the community from the high elevation site. (Modified from Figures 2 and 3 in Kiffney and Clements (1996b).)

communities may not be protective of those from high elevations (Kiffney and Clements 1996b). Interestingly, this pattern was reversed for diatom assemblages. Medley and Clements (1998) observed reduced effects of heavy metals on diatoms from headwater communities compared to those from lower elevations. Because headwater streams were naturally dominated by early successional species (*Achnanthes minutissima*), which are also tolerant of metals, communities showed little response to metals in experimental streams.

23.3.4.5 Species Interactions

Microcosms and mesocosms can also be employed to measure the effects of contaminants on species interactions such as competition or predation. For example, manipulation of predator density and contaminant concentration in a simple 2 × 2 factorial design allows researchers to determine if the susceptibility of prey species to predation is influenced by exposure to a chemical stressor (Clements 1999). Irfanullah and Moss (2005) used mesocosm experiments to quantify the interactive effects of pH and predation by *Chaoborus* larvae on lentic plankton communities. A more sophisticated experimental design was employed to determine the direct and indirect effects of predators, insecticides (malathion), and herbicides (Roundup) on amphibian assemblages in pond mesocosms (Relyea et al. 2005). The opportunity to quantify the significance of interactions between chemical stressors and susceptibility to predation is considered a major justification for the use of mesocosm experiments in ecotoxicology. Studies describing effects of contaminants on species interactions were reviewed in Chapter 21.

23.3.4.6 Applications in Terrestrial Systems

Microcosm and mesocosm research conducted at the community level has overwhelmingly focused on aquatic systems. Gillett's (1989) review of terrestrial microcosm and mesocosm experiments emphasized chemical fate and ecological effects of contaminants on populations. Relatively few studies cited in this review examined community-level responses. Nonetheless, there are important opportunities for microcosm and mesocosm experiments in terrestrial community ecotoxicology. In particular, soil microcosms offer a unique system to study structural and functional features of natural communities. Because of the small size and short generation times of many soil organisms, entire communities can be studied under realistic conditions in the laboratory. Verhoef (1996) compared ecosystem processes, such as carbon dioxide (CO_2) production and nutrient availability, in microcosms, mesocosms, and field studies. Although there were differences in functional attributes among systems, the magnitude and relative ranking of response variables were similar. These results suggest that spatial scale may be a less serious concern in soil microcosms.

The Ecotron facility in Ascot, England is a large-scale, terrestrial mesocosm system where researchers have investigated a variety of community processes across several trophic levels (Lawton 1996). The facility consists of 16 environmental chambers with precise control over light, temperature, humidity, and rainfall. Research conducted in this facility has investigated species interactions; the relationship between species diversity and ecosystem processes; and impact of CO_2 on population, community, and ecosystem dynamics. Although Ecotron has not been employed in ecotoxicological research, this type of facility would be ideal for investigating direct and indirect effects of chemical stressors on terrestrial communities.

Larger mesocosms have been employed to measure the effects of pesticides on terrestrial communities. Suttman and Barrett (1979) used a series of field enclosures to test Odum's (1969) hypothesis that effects of stress are greater on immature communities compared to mature communities. Field enclosures established in immature (monocultures of oats) and mature (late successional fields) systems were treated with the pesticide carbaryl, and responses of plant and arthropod communities were compared to those in control plots. Although results supported the hypothesis that insecticide effects were greater in the immature system, the period of recovery was greater in the mature community. More recently, Sheffield and Lochmiller (2001) used 0.1 ha (32 $\times$ 32 m) enclosures to examine the effects of the organophosphate insecticide diazinon on community structure and species interactions. Applications of 1.0 and 8.0 times the recommended field application rate of diazinon resulted in significant reproductive effects on small mammals, with considerable variation observed among species (Figure 23.6). Consumption of dead and dying insects was considered the most important route of exposure. Because of the significant reduction in arthropod density,

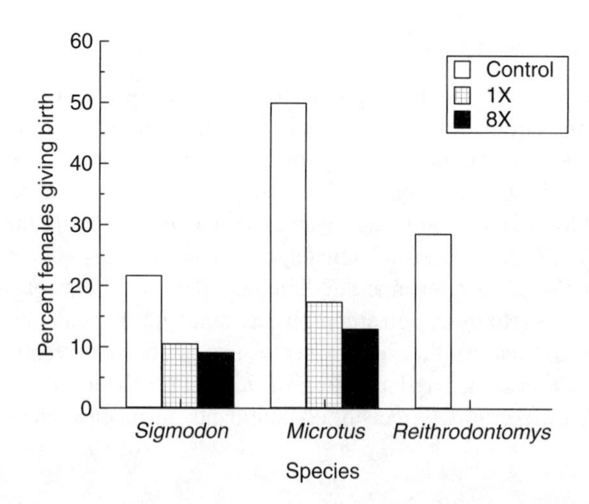

FIGURE 23.6 Effects of the insecticide diazinon at one time and eight times the recommended application rate on a small mammal community in outdoor enclosures. Results showed significant effects on reproduction at low levels of exposure and considerable variability among species. Because of the large reduction in insect abundance in the mesocosms, some of the effects on reproduction may have resulted from reduced prey availability. (Data from Table 4 in Sheffield, S.R. and Lochmiller, R.L., *Environ. Toxicol. Chem.*, 20, 284–296, 2001.)

it is possible that some of the observed reproductive effects resulted from lower prey abundance in treated plots. We agree with the authors that field studies using mesocosms that simultaneously investigate population and community-level responses are critical for evaluating indirect effects of pesticides.

23.3.5 Summary

Because microcosm and mesocosm experiments attempt to bridge the gap between single species toxicity tests and full-scale ecosystem studies, they receive criticism for being too simplistic and too complex. Ecologists consider model systems to be unrealistic simplifications of nature, whereas toxicologists are reluctant to consider community-level testing because results are complex and difficult to interpret. Despite their shortcomings, the use of microcosms and mesocosms have contributed significantly to our understanding of the effects of contaminants on communities. The degree of control over independent variables allows researchers to isolate specific components and investigate the mechanisms responsible for changes in community composition. Microcosm and mesocosm approaches represent a link between standardized, single species toxicity tests and more expensive, logistically difficult field experiments. However, questions of spatiotemporal scale remain largely unanswered and must be addressed if model systems are to play an important role in ecotoxicological research. As noted above, the significance of spatiotemporal scale in microcosm and mesocosm research is a hypothesis that remains to be tested. Just as researchers design studies to test the effects of a particular chemical, experiments should also be conducted to test responses at different spatiotemporal scales. When mesocosm experiments and field collections are conducted simultaneously, model systems can be used to support results of nonexperimental, descriptive studies and make a stronger argument for causal relationships.

23.4 WHOLE ECOSYSTEM MANIPULATIONS

> Accurate ecosystem management decisions cannot be made with confidence unless ecosystem scales are studied.
>
> **(Schindler 1998)**

Although microcosm and mesocosm studies have contributed to our understanding of community responses to contaminants and other forms of anthropogenic disturbance, critics argue that results of small-scale experiments reveal little about the natural world (Carpenter 1996). For example, because of the difficulty including top predators, results of mesocosm experiments should be viewed cautiously when predicting effects in systems where top-down effects are important. In addition, spatial and temporal scaling issues are a concern in most microcosm and mesocosm research. Container effects in small model systems may change environmental conditions, alter exposure regimes, and limit the duration of microcosm and mesocosm experiments.

One solution to the limited spatiotemporal scale and lack of ecological realism of model ecosystems is the direct application of contaminants in the field. Barrett (1968) treated a 0.4 ha (approximately 1.0 acre) fenced enclosure with the insecticide carbaryl and compared responses to those observed in a single control plot. Total biomass of arthropods was reduced by 95% in the treated plot, and patterns of recovery varied among taxa. The most significant response in the small mammal community was a dramatic shift in relative abundance of two species resulting from differential effects of carbaryl on reproduction. Barrett (1968) concluded that while direct toxicity of carbaryl was greatly reduced within a few days, long-term effects on structure and function persisted. Effects of disturbance on a forest ecosystem at a large spatial scale were investigated by Woodwell (1970). This classic experiment examined chronic effects of gamma radiation on structure and function of oak-pine forests at the Brookhaven National Laboratory, New York. Results showed a distinct alteration in community structure that diminished with distance from the radiation source. One of the first large-scale manipulations conducted in a riparian ecosystem examined the effects of clear cutting and herbicide applications on nutrient budgets in the Hubbard Brook Experimental Forest, New Hampshire (Likens et al. 1970). Results showed that disturbed watersheds exported large amounts of particulate matter and inorganic material.

These early experiments revealed the usefulness of whole ecosystem manipulations for assessing effects of contaminants on terrestrial communities. More importantly, they demonstrated that community responses to anthropogenic stressors were predictable and similar to natural disturbances. For example, patterns observed in response to chronic radiation were remarkably consistent with those observed following exposure of plant communities to salt spray, fire, and other natural disturbances (Woodwell 1970). The similarity of responses to natural and anthropogenic stressors illustrated in these early studies has been a consistent theme in subsequent whole ecosystem manipulations (Rapport et al. 1985) and will be further developed in Chapter 25.

23.4.1 Examples of Ecosystem Manipulations: Aquatic Communities

With their emphasis on ecological theory and principles of recovery, these early experiments set the stage for more focused studies of ecosystem-level responses to contaminants. Although numerous ecosystem-level manipulations have been conducted since the early 1970s (see review by Perry and Troelstrup 1988), two research programs deserve special attention because of their significant contributions to our understanding of how natural communities respond to chemical stressors. First, David Schindler's experiments conducted in the Experimental Lakes Area (ELA) (Ontario, Canada) measured structural and functional responses of lakes to a variety of anthropogenic stressors, including nutrients, acidification, and heavy metals (Schindler 1988). Subsequent whole lake manipulations conducted by researchers in other parts of North America verified the importance of this experimental approach. Next, Bruce Wallace's team at the University of Georgia has conducted a long-term study of watershed responses at Coweeta Hydrologic Laboratory (North Carolina, USA). Although these experiments were primarily limited to insecticides, results highlighted the importance of measuring direct and indirect effects of contaminants on ecological processes.

23.4.1.1 Experimental Lakes Area

The ELA consists of 46 natural, relatively undisturbed lakes located in northwestern Ontario. The lakes have been designated specifically for ecosystem-level research and have been used to investigate the effects of anthropogenic stressors on biotic and abiotic characteristics. The initial motivation for these manipulations was to increase fish productivity (Schindler 1988), but early experiments at the ELA also clarified important misconceptions about the causes of eutrophication in lentic ecosystems. Previously, many researchers believed that carbon was the primary limiting factor in lakes, and that reducing input of nutrients would have little beneficial effects. The striking results of phosphorus addition experiments in Lake 227, visually displayed on the cover of *Science* in 1974 (Schindler 1974), demonstrated unequivocally that phosphorus was a major cause of eutrophication.

One of the more insightful observations from the ELA studies was that, although these experiments were initiated with a set of explicit and testable hypotheses, researchers were consistently met with surprises (Schindler 1988). This statement is both a testimony to the importance of ecosystem manipulations and an admission of our relatively poor understanding of ecosystem processes. Many of these surprises were a result of indirect effects of contaminants on species interactions. Schindler states that, "in every aquatic experiment which we have done, the whole ecosystem response has involved complicated interactions between a number of species in the biotic community." The most striking example of this statement, with obvious relevance to community ecotoxicology, is from whole lake acidification experiments. Effects of acidification on lentic communities resulted from a complex interaction of direct toxicity, reproductive failure, increased parasitism, and starvation due to loss of prey species (Schindler 1987).

Another significant finding from the long history of ecosystem manipulations at ELA was the relative insensitivity of functional measures (e.g., decomposition, nutrient cycling, and primary productivity) compared to structural measures (e.g., species richness and community composition). Despite an initial emphasis on ecosystem processes, most studies found that functional measures were slower to respond and generally responded only to high levels of stress compared to structural measures (Schindler 1987). The general insensitivity of functional measures has been a consistent observation in ecosystem experiments (Howarth 1991), and has important implications for the selection of endpoints in contaminant research. Because of the insensitivity of functional measures, Schindler (1988) suggests that future studies should emphasize taxonomy and community ecology, possibly at the expense of more "fashionable" measures such as ecosystem metabolism and nutrient cycling.

23.4.1.2 Coweeta Hydrologic Laboratory

Experiments conducted by Bruce Wallace and colleagues at Coweeta Hydrologic Laboratory investigated effects of the pesticide methoxychlor on benthic communities in small headwater streams. Interestingly, the initial motivation for these experiments was not to assess effects of pesticides but rather to determine the functional role of benthic macroinvertebrates. The application of methoxychlor was simply the most direct method for eliminating large numbers of macroinvertebrates from the stream. Catastrophic macroinvertebrate drift, approximately 1000 times greater than pretreatment levels, occurred immediately following application of methoxychlor (Cuffney et al. 1984, Wallace et al. 1982). The resulting alterations in benthic community composition included a dramatic reduction in abundance of aquatic insects, especially shredders, and subsequent replacement by noninsects (oligochaetes).

Although documenting changes in community composition and differences in sensitivity among macroinvertebrate groups was important, the most significant contribution of Wallace's experiments was the establishment of a relationship between structural and functional characteristics of headwater streams. In contrast to results reported from ELA experiments, Wallace and colleagues found that functional measures were relatively sensitive to chemical stress. Application of methoxychlor resulted in significant alteration in detritus dynamics in the treated stream (Figure 23.7). The rate of leaf decomposition and the dry mass of suspended particulate organic matter (POM) was significantly lower in treated streams compared to controls. These alterations were directly attributable to loss of shredders, as there was relatively little influence of pesticide treatment on microbial communities (Wallace et al. 1982). More importantly, these results suggest that indirect effects of pesticides on organic matter processing and export of particulate material may exceed direct toxic effects (Wallace et al. 1989).

Because these manipulations were conducted over a relatively long time period, the findings also have important implications for the study of recovery from chemical stressors. Analysis of data collected several years after pesticide application showed that abundance data were not sufficient to evaluate recovery (Whiles and Wallace 1992). Total abundance of benthic macroinvertebrates was generally similar between treated and control streams within the first year following pesticide application. However, differences in ecosystem processes and taxonomic composition persisted for several years after treatment. Factors that influenced the rate of recovery in systems subjected to anthropogenic disturbance are considered in Chapter 25.

23.4.1.3 Summary

The ELA and Coweeta Hydrologic Laboratory are unique sites that were specifically established for manipulative, ecosystem-level research. Because of the expense and logistical difficulties associated

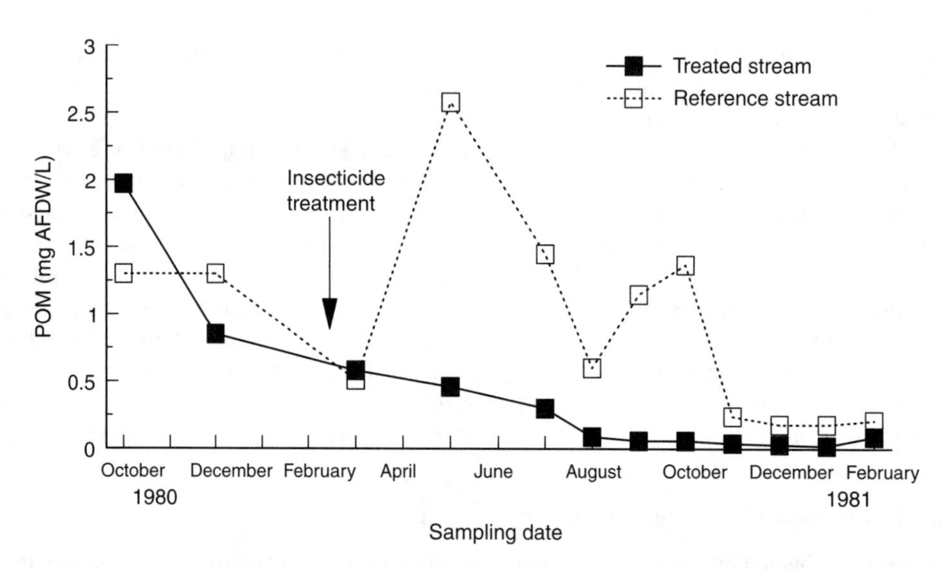

FIGURE 23.7 Export of particulate organic material (POM) in reference and treated streams in Coweeta Hydrologic Laboratory (North Carolina, USA). The treated stream was dosed with the insecticide methoxychlor in February 1980. The reduction in export of POM in the treated stream was hypothesized to result from the elimination of shredders, organisms that feed on coarse leaf detritus and convert this material to smaller particles. (Modified from Figure 1 in Wallace et al. (1982).)

with conducting these manipulations, it is unlikely that other large areas will be set aside exclusively for the purpose of assessing ecosystem responses to anthropogenic stressors. Thus, an important question is the relevance of these studies to understanding responses of other ecosystems and to other stressors. The answer to this question is quite encouraging. Indeed, the general patterns reported in Schindler's whole lake manipulations at ELA and Wallace's pesticide experiments at Coweeta are consistent with responses observed in numerous ecosystem studies, both descriptive and experimental. The similarity of responses among stressors and ecosystem types provides support for the "ecosystem distress syndrome" proposed by Rapport et al. (1985) and described in Chapter 25.

23.4.2 Examples of Ecosystem Manipulations: Avian and Mammalian Communities

Large-scale, experimental assessments of chemical effects on birds and mammals at the community level are uncommon in ecotoxicology. Like most applied research in wildlife biology, the primary emphasis in terrestrial ecotoxicology is at the level of populations. However, numerous studies have investigated impacts of other anthropogenic disturbances, particularly those related to forestry practices and other land use changes, on bird and mammal communities. Assuming that community-level responses to these disturbances are analogous to chemical stressors, results of large-scale experiments investigating effects of land use changes and other manipulations may provide some insight into how bird and mammal communities would be affected by chemicals. Chambers et al. (1999) measured community-level effects of silvicultural treatments on bird communities in the Pacific Northwest. This study is especially noteworthy because of the large spatial scale (treatment stands ranged from 5.5 to 17.8 ha) and because of the level of replication (n = 7–11). Results showed that total bird abundance declined along a disturbance gradient; however, species richness appeared to increase in treatments with intermediate levels of disturbance (Figure 23.8). As expected, these differences resulted from species-specific responses to silviculture treatments. Abundance of habitat generalists increased, whereas species with restricted geographical ranges decreased in response to disturbance.

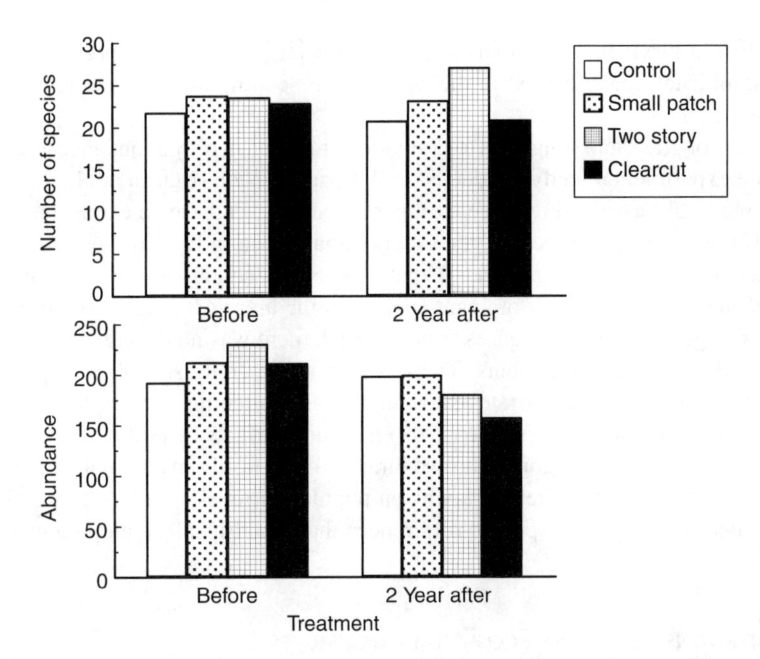

FIGURE 23.8 Community-level effects of disturbance on bird communities in the Pacific Northwest. The figure compares species richness and abundance across four levels of silviculture treatments. Total bird abundance declined along a disturbance gradient; however, species richness increased in treatments with intermediate levels of disturbance (two-story cut). (Data from Table 3 in Chambers, C.L., et al., *Ecol. Appl.*, 9, 171–185, 1999.)

A large-scale "natural" experiment compared responses of bird communities in boreal forests to harvesting and fire treatments over a 28-year period (Hobson and Schieck 1999). In addition to the large spatiotemporal scale, this study is especially relevant to our discussion of experimental approaches because of the unique 2×3 factorial design (2 disturbance types, 3 time periods following disturbance) used to detect treatment effects and recovery times. Researchers observed an increase in bird abundance 14 and 28 years after disturbance; however, patterns of recovery differed between disturbance types, primarily because of differences in community composition immediately after treatment. Although bird communities slightly converged after 14 years, differences in community composition persisted 28 years following disturbance. These results suggest that responses of bird communities to disturbance may persist for relatively long time periods and patterns of recovery may be disturbance-specific.

In addition to studies of effects of land use changes, a few large-scale field experiments have measured the effects of contaminants on bird and mammal communities. Schauber et al. (1997) tested the hypothesis that differences in diet and vegetation influenced susceptibility of small mammals (deer mice, voles) to organophosphorus pesticides. Using 3×2 factorial design (pesticide level $\times$ vegetation structure), organisms were exposed to pesticides in 24 relatively large (0.2 ha) enclosures. Results showed that variation in vegetation structure and timing of rainfall can affect susceptibility of small mammals to pesticides. In contrast to expectations, differences in diet between the insectivorous deer mice and herbivorous voles had little influence on toxicity of insecticides (Schauber et al. 1997). A similar large-scale experiment investigated the direct and indirect effects of organophosphate pesticides on growth and survival of passerines (Brewer's Sparrow, Sage Thrasher) (Howe et al. 1996). Application of malathion to a 520-ha treatment area significantly reduced abundance of insects, the primary prey of birds. Although this study focused on individual and population-level responses, the results are relevant to community ecotoxicology because of the emphasis on indirect effects. Despite a significant reduction in prey abundance, there were only moderate effects on nestling growth and survival. The authors speculated the large reduction in prey abundance was not biologically significant because food in the shrub-steppe community is superabundant during the breeding season.

The resilience of grassland songbirds to dramatic reductions in prey abundance was also observed in a large-scale experimental study conducted in Alberta, Canada (Martin et al. 2000). Study plots (56 ha) were randomly assigned to three treatments (control, carbamate exposure, and pyrethroid exposure). Despite a 90% reduction in grasshopper abundance in treated plots, there were no significant effects on nest success, number fledged, or body weight of chestnut-collared longspur nestlings (*Calcarius ornatus*), the dominant species in the area. Although birds in the pyrethroid-exposed plots foraged at greater distances from the nest, there was no difference in biomass of prey delivered to nestlings among treatments. These results are in contrast to those reported by Martin et al. (1998) in which depredation rates were higher and hatching success lower on pyrethroid-treated plots. Finally, Patnode and White (1991) measured effects of pesticides on productivity of several songbird species (mockingbirds, brown thrashers, and northern cardinals) in a Georgia pecan grove. Although the focus of the research was on population-level effects (e.g., survival and nestling growth), there were species-specific differences that could result in alterations in community structure.

23.4.3 Limitations of Whole Ecosystem Experiments

In their review of whole ecosystem manipulations, Perry and Troelstrup (1988) discuss several limitations of these experiments. In particular, the difficulty replicating treatments, high costs, and limited types of contaminants that may be investigated are important considerations. On the surface, the lack of replication may appear to be a major shortcoming of whole ecosystem experiments. Indeed, control, randomization, and replication are generally considered the major components of a legitimate experiment. Carpenter (1989) estimated that approximately 10 replicate lakes would be necessary to detect effects of contaminants on primary production because of high natural variability in these systems. It is unlikely that any research program can afford the luxury of this level of replication. Even in situations such as the ELA where a large number of lakes are available for manipulation, it is difficult to locate true replicates (Schindler 1998). Consequently, some researchers argue that sustained, long-term manipulations using unreplicated paired ecosystems are the best approach for assessing ecosystem responses (Carpenter 1989, Schindler 1998). Carpenter et al. (1998) make a strong case for evaluating "alternative explanations" in ecosystem experiments instead of the traditional emphasis on testing null hypotheses. Researchers should identify an explanation that is most plausible based on data from the manipulation and other relevant information. Carpenter et al. (1998) also argue that imposing different treatments on different ecosystems may be more informative than "wasting" precious resources on replicates for testing null hypotheses. This idea is the basis for a revolutionary approach advocated by some researchers who feel that ecologists have become too preoccupied with statistical significance at the expense of gaining mechanistic understanding of ecological processes (Box 23.1).

The cost of ecosystem manipulations will limit their widespread use in ecotoxicology. However, the expense may be justified in some instances because well-designed experiments generate extensive data on responses at different levels of organization. Ecosystem experiments often involve multiple investigators and promote cost-effective, interdisciplinary research (Perry and Troelstrup 1988). Interactions among investigators resulting from this collaboration may compensate for the greater expense of ecosystem experiments.

Finally, ecosystem experiments are limited by the types of manipulations that may be performed in natural systems. For example, experimental introduction of highly persistent compounds, such as PCBs and dioxins, would not (and should not) be allowed in most natural systems. Integration of smaller scale studies (microcosms) with ecosystem experiments and taking advantage of unexpected environmental perturbations (Wiens and Parker 1995) will be essential to understand effects of these persistent, highly toxic compounds.

Box 23.1 An Alternative Approach to Traditional Hypothesis Testing

> The statistical null hypothesis testing paradigm has become so catholic and ritualized as to seemingly impede clear thinking and alternative analysis approaches.
>
> **(Anderson et al. 2001)**

Statistical approaches in which null hypotheses are compared to alternatives are widely used in ecological and ecotoxicological research. Finding a statistically significant difference between treatment groups often improves the likelihood of publishing results, thus tempting researchers to employ iterative data mining and "fishing trips" to locate *P*-values (Anderson et al. 2001) (see also Box 10.2). Because researchers often confuse statistical significance with underlying processes of interest, data analysis has become synonymous with finding statistically significant differences. These exploratory approaches have recently been criticized because of their inherent subjectivity and reliance on post hoc techniques. In particular, model selection procedures, such as stepwise multiple regression, which identify "best" models based on maximizing R^2 values, have a high probability of identifying spurious results. Their criticism goes beyond the well-known problems of distinguishing statistical significance from biological significance and correcting for experiment-wise error rates. Anderson et al. (2001) argue that while chasing *P*-values, researchers often lose sight of the critical thinking processes that should precede any data analysis. Rejecting weak or sterile null hypotheses that researchers know are false (e.g., there is no difference in growth between exposed and unexposed groups) is not wrong, but arbitrary and uninformative (Burhnam and Anderson 2001). These approaches do little to advance science and often neglect the more important issue of estimating the magnitude of effects (Anderson et al. 2000).

Recognizing that we construct models to separate important processes from underlying noise and that we never know which model is best (e.g., closest to truth), objective approaches are necessary to distinguish among competing alternatives. The proposed solution to the unquestioning reliance on hypothesis testing is application of an information–theoretic approach as the basis for making inferences in scientific investigations (Burnham and Anderson 1998). The information–theoretic approach is an extension of classical likelihood methods that emphasizes a priori thinking and provides a formal ranking of statistical models. The approach uses Kullback–Leibler (K–L) information (Kullback and Leibler 1951) as a measure of the distance between a model and reality, and then ranks a set of competing models from best to worst using the likelihood of each model. Formally, K–L distance between conceptual truth and a model is given as $I(f, g)$, which is defined as the information that is lost when model g is used to estimate truth f. A significant breakthrough in the development of the information–theoretic approach occurred when Akaike found a formal relationship between K–L distance and maximum likelihood (Akaike 1992). Akaike's Information Criterion (AIC) can be used to estimate the expected value of K–L and provides a relative measure of the proximity of the model to the best model.

Although the focus of the K–L information approach is primarily on model selection, the issues addressed are relevant to all inferential methods. At the very least, researchers are reminded of the importance of a priori analyses and the need to distinguish between results derived from iterative processes of data mining and those obtained by an objective attempt to separate noise from underlying structure.

Despite their limitations, whole-ecosystem manipulations have revealed unique responses to anthropogenic disturbances that could not have been measured by microcosm and mesocosm studies. Although it is unlikely that whole ecosystem manipulations will be employed on a routine basis, large-scale experiments are the most direct method for demonstrating causation in natural systems. Kimball and Levin (1985) argue for establishment of research programs that integrate microcosm experiments and whole ecosystem manipulations to predict effects of chemicals. Because certain ecological processes are scale dependent, large-scale studies may be the only way to characterize responses to stressors. Finally, ecotoxicologists must become more creative in designing and implementing large-scale experimental studies. Taking advantage of planned (e.g., the intentional application of pesticides to control insect outbreaks) or unplanned (e.g., the *Exxon Valdez* oil spill) manipulations could be used to measure stressor effects at the scale of whole ecosystems.

23.5 WHAT IS THE APPROPRIATE EXPERIMENTAL APPROACH FOR COMMUNITY ECOTOXICOLOGY?

23.5.1 QUESTIONS OF SPATIOTEMPORAL SCALE

Perhaps the most serious criticism of most experimental approaches employed in community ecotoxicology is the limited spatiotemporal scale of these investigations. Carpenter (1996) argues that the statistical advantages and high degree of control of microcosm and mesocosm experiments do not compensate for their lack of ecological realism. Ironically, this has been the basis for criticism of laboratory toxicity tests for almost 25 years (Cairns 1983), where the underlying assumption is that results of single species toxicity tests can be extrapolated to more complex ecological systems. Some researchers are highly skeptical of extrapolation across spatial and temporal scales (Schindler 1998), and these same concerns should apply to more complex ecotoxicological experiments. The small size and short duration typical of most microcosm experiments will limit our ability to study some potentially important processes. Conducting experiments at different spatial and temporal scales and across different levels of biological organization will at least partially address these concerns. If the response to a particular stressor is scale dependent, then conducting experiments at different spatial scales may allow quantification of this effect (Perez et al. 1991). Conducting experiments at different scales may also reveal mechanistic explanations for observed responses to contaminants. For example, a mesocosm experiment could show that abundance of a grazing invertebrate increased after treatment with a particular chemical. Experiments conducted at a smaller spatial scale (e.g., microcosms) would be necessary to show if this unexpected response resulted from increased abundance of primary producers, reduced competition with other grazers, or release from predation by a higher trophic level. Single species toxicity tests could be used to document differences in sensitivity among these potentially interacting species.

23.5.2 INTEGRATING DESCRIPTIVE AND EXPERIMENTAL APPROACHES

There are important limitations of all experimental approaches employed in community ecotoxicology, regardless of the spatial or temporal scale of these investigations (Diamond 1986). Indirect effects of contaminants, stressor interactions, and potential artifacts introduced by the experimental system will complicate interpretation of experimental studies. Even complex factorial designs that investigate stressor interactions are limited in the number of variables that can be manipulated. Because the goal of many experiments is to demonstrate the importance of a single factor (e.g., the effects of a specific chemical or the abundance of a particular predator), the connection between the experiment and the natural system is often lost. Furthermore, while a well-designed experiment with sufficient power can demonstrate the statistical significance of a single factor, the importance of this factor relative to other unmanipulated variables remains unknown without supporting comparative data.

There is legitimate concern that the harsh criticisms directed at descriptive ecology in the 1970s may have resulted in premature abandonment of useful comparative approaches. Consequently, ecologists are often unable to address problems at relevant spatial scales where experimental manipulations are impractical (Power et al. 1998). The perception that experimentation and observation are opposing methodologies underlies a fundamental misconception about the importance of descriptive studies in ecological research. There is much to be learned by comparing patterns observed in nature to those predicted from theory, and a successful research program should combine theory, observations, and experiments. Werner (1998) describes the advantages of a research program that integrates experimental techniques with theoretical and comparative approaches for understanding basic ecological patterns. He makes a strong case for the importance of comparative components in a research program and argues that experiments lacking an obvious connection to observed patterns in nature may be irrelevant. Similar arguments can be made for research programs in ecotoxicology.

23.6 SUMMARY

Experimental studies to evaluate the effects of stressors on communities may be conducted at a variety of spatial and temporal scales. The most effective experimental approach in community ecotoxicology will be determined by the specific objectives of the research, cost, and logistical considerations. Dogmatic statements regarding the superiority of one experimental approach over another disregard the obvious fact that researchers have different goals in mind when designing and conducting experiments. If a researcher is primarily interested in studying interactions among stressors or quantifying the effects of abiotic variables on community responses to contaminants, a factorial experimental design is probably necessary. It is unlikely that a factorial experimental design will be practical at a large spatial scale (e.g., an entire ecosystem); therefore, microcosms or mesocosms are most appropriate. Similarly, microcosm and mesocosm experiments are required when investigating the toxicity of chemicals that cannot be intentionally released into the natural environment. Although small-scale laboratory experiments and mesocosm studies provide the greatest degree of control over independent and confounding variables, they lack realism and have limited temporal and spatial scales. If researchers are interested in comparing the consequences of long-term (e.g., greater than 1 year) exposure to a chemical or following the trajectory of a community response, a natural experiment is most appropriate. Although natural experiments have a high degree of ecological realism and offer greater opportunity to generalize to other systems, they often sacrifice control and replication. Finally, whole ecosystem manipulations are especially useful in situations where researchers wish to measure functional responses (e.g., primary productivity or nutrient cycling) of entire systems or where large, highly mobile species are believed to play an important role in community dynamics.

Ecological and ecotoxicological experiments are conducted for a variety of reasons. Most commonly, researchers are interested in establishing relationships among biotic and abiotic variables or measuring the effects of a particular stressor on ecologically important endpoints. Experiments may be conducted to test ecological and ecotoxicological theory, or simply to satisfy scientific curiosity. Regardless of whether experiments are conducted to test model predictions or to determine the relative importance of hypothesized causal factors, the key issues are generality and extrapolation. Ecotoxicological experiments should not be conducted without an appreciation of natural history or in isolation from underlying ecological theory. Experiments that lack grounding in natural history and theory may provide inconsistent, incomprehensible, and misleading results. Power et al. (1998) advocate a nested experimental, observational, and modeling approach designed to address three basic questions: (1) "What would happen if . . . ?," (2) "Does this new system work the same way?," and (3) "Is there quantitative agreement with predictions?" (Figure 23.9). We advocate a similar integration of these three approaches for community ecotoxicology. Observations of broad spatial patterns and a quantitative analysis of these patterns should precede design and implementation of

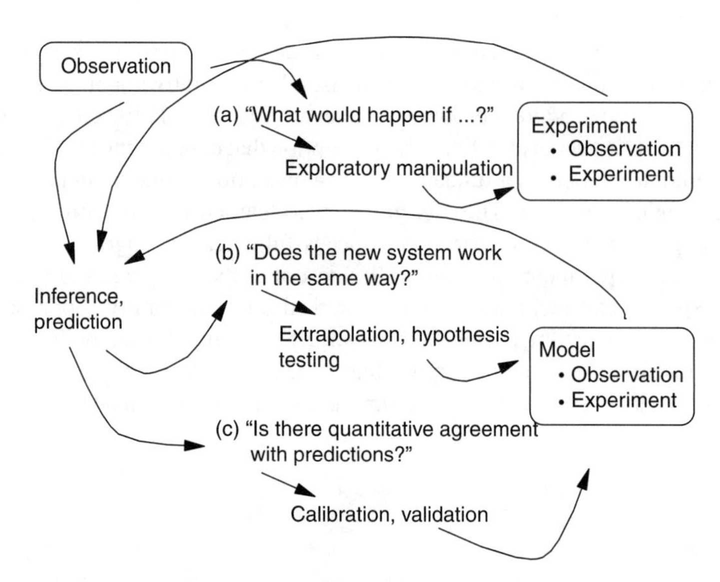

FIGURE 23.9 Integrating descriptive, experimental, and modeling approaches in an ecological research program. Initial observations in nature stimulate research questions that can be addressed by exploratory manipulations. Both observational and experimental studies allow researchers to make inferences and predictions about the system, which can be formalized into a conceptual model. The model is validated by comparing model predictions with observations in nature. (Modified from Figure 6.1 in Power et al. (1998).)

experiments. While experiments in ecotoxicology necessarily focus on a few factors (e.g., the concentration of a contaminant), understanding the importance of these factors relative to other variables is significantly enhanced by a research program that includes a descriptive component.

23.6.1 Summary of Foundation Concepts and Paradigms

- Although descriptive approaches provide support for a relationship between stressors and community responses, experimental studies are often necessary to demonstrate causation.
- Certain types of experimental systems may be more useful than other types for investigating ecological responses to anthropogenic perturbations.
- In addition to recognizing that comparative approaches are often insufficient for understanding mechanisms, some ecotoxicologists questioned the validity of using single species laboratory experiments to predict responses of more complex systems in the field.
- Ecotoxicologists have employed laboratory experiments, field experiments, and natural experiments to assess effects of contaminants on communities.
- Because microcosm and mesocosm experiments attempt to bridge the gap between single species toxicity tests and full-scale ecosystem studies, they receive criticism for being too simplistic and too complex.
- A valid criticism of microcosm and mesocosm research is that the emphasis placed on increasing reproducibility and decreasing variability has come at the expense of ecological relevance.
- The most significant challenge in microcosm and mesocosm research is to identify those key features that must be carefully reproduced in order to simulate structure and function of natural systems.
- Much of the criticism of model systems in ecotoxicological research is due to the failure of researchers to explicitly state the numerous simplifying assumptions required when using these systems.

- The major advantage of model systems over field experiments and ecosystem manipulations is the ability to randomly assign and replicate treatments, thus allowing researchers to analyze results using inferential statistics.
- Microcosm and mesocosm experiments are the most effective way to evaluate the influence of community composition on stressor responses.
- Although results of terrestrial studies have provided important insight into contaminant effects, microcosm and mesocosm research conducted at the community level has overwhelmingly focused on aquatic systems.
- Microcosm and mesocosm studies have contributed to our understanding of community responses to contaminants; however, because of spatial and temporal scaling issues, critics argue that results of these experiments reveal little about the natural world.
- One solution to the limited spatiotemporal scale and lack of ecological realism of model ecosystems is the direct application of contaminants in the field.
- Studies conducted at the ELA (Ontario, Canada) and Coweeta Hydrologic Laboratory (North Carolina, USA) have made significant contributions to our understanding of how natural communities respond to chemical stressors.
- Large-scale experimental assessments of chemical effects on birds and mammals at the community level are uncommon in ecotoxicology.
- Sustained long-term manipulations using unreplicated paired ecosystems may be a useful alternative to true replication in whole ecosystem experiments.
- Because of the difficulty replicating treatments, high costs, and limited types of contaminants that may be investigated, it is unlikely that whole ecosystem manipulations will be employed on a routine basis.
- One proposed solution to the unquestioning reliance on hypothesis testing is application of an information–theoretic approach, an extension of classical likelihood methods that emphasizes a priori thinking and provides a formal ranking of statistical models.
- The most effective experimental approach in community ecotoxicology will be determined by the specific objectives of the research, cost, and logistical considerations.

REFERENCES

Adams, S.M., Crumby, W.D., Greeley, M.S, Jr., Ryon, M.G., and Schilling, E.M., Relationships between physiological and fish population responses in a contaminated stream, *Environ. Toxicol. Chem.*, 11, 1549–1557, 1992.

Akaike, H., Information theory and the extension of the maximum likelihood principle, In *Breakthroughs in Statistics*, Kotz, S. and Johnson, N.L. (eds.), Springer-Verlag, London, 1992, pp. 610–624.

Anderson, D.R., Burnham, K.P., Gould, W.R., and Cherry, S., Concerns about finding effects that are actually spurious, *Biometrics*, 29, 311–316, 2001.

Anderson, D.R., Burnham, K.P., and Thompson, W.L., Null hypothesis testing: Problems, prevalence, and an alternative, *J. Wildl. Manag.*, 64, 912–923, 2000.

ASTM, E 1366-91: ASTM standard practice for standardized aquatic microcosm: Fresh water, In *1995 Annual Book of ASTM Standards, Volume 11.05*, American Society for Testing and Materials, Philadelphia, PA, 1995, pp. 1048–1082.

Barreiro, R. and Pratt, J.R., Interaction of toxicants and communities: The role of nutrients, *Environ. Toxicol. Chem.*, 13, 361–368, 1994.

Barrett, G.W., The effects of an acute insecticide stress on a semi-enclosed grassland ecosystem, *Ecology*, 49, 1019–1035, 1968.

Belanger, S.E., Lee, D.M., Bowling, J.W., and Leblanc, E.M., Responses of periphyton and invertebrates to a tetradecyl-pentadecyl sulfate mixture in stream mesocosms, *Environ. Toxicol. Chem.*, 23, 2202–2213, 2004.

Belzile, C., Demers, S., Ferreyra, G.A., Schloss, I., Nozais, C., Lacoste, K., Mostajir, B., et al., UV effects on marine planktonic food webs: A synthesis of results from mesocosm studies, *Photochem. Photobiol.*, 82, 850–856, 2006.

Beyers, D.W., Causal inference in environmental impact studies, *J. N. Am. Benthol. Soc.*, 17, 367–373, 1998.

Brooks, B.W., Stanley, J.K., White, J.C., Turner, P.K., Wu, K.B., and La Point, T.W., Laboratory and field responses to cadmium: An experimental study in effluent-dominated stream mesocosms, *Environ. Toxicol. Chem.*, 23, 1057–1064, 2004.

Burnham, K.P. and Anderson, D.R., *Model Selection and Inference: A Practical Information-Theoretic Approach*, Springer-Verlag, New York, 1998.

Burnham, K.P. and Anderson, D.R., Kullback–Leibler information as a basis for strong inference in ecological studies, *Wildl. Res.*, 28, 111–119, 2001.

Cairns, J., Jr., Are single species toxicity tests alone adequate for estimating environmental hazard? *Hydrobiologia*, 100, 47–57, 1983.

Cairns, J., Jr., The myth of the most sensitive species, *Bioscience*, 36, 670–672, 1986.

Cairns, J., Jr., Hart, K.M., and Henebry, M.S., The effects of a sublethal dose of copper sulfate on the colonization rate of freshwater protozoan communities, *Am. Mid. Nat.*, 104, 93–101, 1980.

Carlisle, D.M. and Clements, W.H., Sensitivity and variability of metrics used in biological assessments of running waters, *Environ. Toxicol. Chem.*, 18, 285–291, 1999.

Carpenter, S.R., Replication and treatment strength in whole-lake experiments, *Ecology*, 70, 453–463, 1989.

Carpenter, S.R., Microcosm experiments have limited relevance for community and ecosystem ecology, *Ecology*, 77, 677–680, 1996.

Carpenter, S.R., Cole, J.T., Essington, T.E., Hodgson, J.R., Houser, J.N., Kitchell, J.F., and Pace, M.L., Evaluating alternative explanations in ecosystem experiments, *Ecosystems*, 1, 335–344, 1998.

Chambers, C.L., McComb, W.C., and Tappeiner J.C., II, Breeding bird responses to three silvicultural treatments in the Oregon Coast Range, *Ecol. Appl.*, 9, 171–185, 1999.

Chapman, P.M., Extrapolating laboratory toxicity results to the field, *Environ. Toxicol. Chem.*, 14, 927–930, 1995.

Christensen, M.R., Graham, M.D., Vinebrooke, R.D., Findlay, D.L., Paterson, M.J., and Turner, M.A., Multiple anthropogenic stressors cause ecological surprises in boreal lakes, *Global Change Biology*, 12, 2316–2322, 2006.

Clarke, K.R., Nonmetric multivariate analysis in community-level ecotoxicology, *Environ. Toxicol. Chem.*, 18, 118–127, 1999.

Clements, W.H., Metal tolerance and predator-prey interactions in benthic macroinvertebrate stream communities, *Ecol. Appl.*, 9, 1073–1084, 1999.

Clements, W.H., Cherry, D.S., and Cairns, J., Jr., The influence of copper exposure on predator-prey interactions in aquatic insect communities, *Freshw. Biol.*, 21, 483–488, 1989.

Clements, W.H. and Kiffney, P.M., Assessing contaminant effects at higher levels of biological organization—Editorial, *Environ. Toxicol. Chem.*, 13, 357–359, 1994.

Connell, J.H., The influence of interspecific competition and other factors on the distribution of the barnacle *Chthamalus stellatus*, *Ecology*, 42, 710–723, 1961.

Connell, J.H., Diversity in tropical rain forests and coral reefs, *Science*, 199, 1302–1310, 1978.

Connell, J.H., Diversity and the coevolution of competitors, or the ghost of competition past, *Oikos*, 35, 131–138, 1980.

Connor, E.F. and Simberloff, D.S., The assembly of species communities: Chance of competition? *Ecology*, 60, 1132–1140, 1979.

Courtney, L.A. and Clements, W.H., Sensitivity to acidic pH in benthic invertebrate assemblages with different histories of exposure to metals, *J. N. Am. Benthol. Soc.*, 19, 112–127, 2000.

Cuffney, T.F., Wallace, J.B., and Webster, J.R., Pesticide manipulation of a headwater stream: Invertebrate responses and their significance for ecosystem processes, *Freshw. Invert. Biol.*, 3, 153–171, 1984.

Daehler, C.C. and Strong, D.R., Can you bottle nature? The roles of microcosms in ecological research, *Ecology*, 77, 663–664, 1996.

Diamond, J., Overview: Laboratory experiments, field experiments, and natural experiments, In *Community Ecology*, Diamond, J. and Case, T.J. (eds.), Harper & Row, New York, 1986, pp. 3–22.

Di Toro, D.M., Allen, H.E., Bergman, H.L., Meyer, J.S., Paquin, P.R., and Santore, R.C., Biotic ligand model of the acute toxicity of metals. 1. Technical basis, *Environ. Toxicol. Chem.*, 20, 2383–2396, 2001.

Drake, J.A., Huxel, G.R., and Hewitt, C.L., Microcosms as models for generating and testing community theory, *Ecology*, 77, 670–677, 1996.

Fairchild, J.F., La Point, T.W., Zajicek, J.L., Nelson, J.K., Dwyer, F.J., and Lovely, P.A., Population-, community- and ecosystem-level responses of aquatic mesocosms to pulsed doses of a pyrethroid insecticide, *Environ. Toxicol. Chem.*, 11, 115–129, 1992.

Gause, G.F., *The Struggle for Existence*, Williams and Wilkins, Baltimore, 1934.

Gearing, J.N., The role of aquatic microcosms in ecotoxicologic research as illustrated by large marine systems, In *Ecotoxicology: Problems and Approaches*, Levin, S.A., Harwell, M.A., Kelly, J.R., and Kimball, K.D. (eds.), Springer-Verlag, Inc., New York, 1989, pp. 409–470.

Genter, R.B., Benthic algal populations respond to aluminum, acid, and aluminum-acid mixtures in artificial streams, *Hydrobiologia*, 306, 7–19, 1995.

Genter, R.B., Colwell, F.S., Pratt, J.R., Cherry, D.S., and Cairns, J., Jr., Changes in epilithic communities due to individual and combined treatments of zinc and snail grazing in stream mesocosms, *Toxicol. Ind. Health*, 4, 185–201, 1988.

Giddings, J.M., Biever, R.C., Annunziato, M.F., and Hosmer, A.J., Effects of diazinon on large outdoor pond microcosms, *Environ. Toxicol. Chem.*, 15, 618–629, 1996.

Gillett, J.W., The role of terrestrial microcosms and mesocosms in ecotoxicologic research, In *Ecotoxicology: Problems and Approaches*, Levin, S.A., Harwell, M.A., Kelly, J.R., and Kimball, K.D. (eds.), Springer-Verlag, New York, 1989, pp. 41–67.

Gonzalez, M.J. and Frost, T.M., Comparisons of laboratory bioassays and a whole-lake experiment: Rotifer responses to experimental acidification, *Ecol. Appl.*, 4, 69–80, 1994.

Graney, R.L., Giesy, J.P., Jr., and DiToro, D., Mesocosm experimental design strategies: Advantages and disadvantages in ecological risk assessment, In *Using Mesocosms to Assess the Aquatic Ecological Risk of Pesticides: Theory and Practice*, Voshell, J.R., Jr. (ed.), Entomological Society of America, Lanham, MD, 1989, pp. 74–88.

Green, R.H., *Sampling Design and Statistical Methods for Environmental Biologists*, John Wiley & Sons, New York, 1979.

Heinis, L.J. and Knuth, M.L., The mixing, distribution and persistence of esfenvalerate within littoral enclosures, *Environ. Toxicol. Chem.*, 11, 11–25, 1992.

Hobson, K.A. and Schieck, J., Changes in bird communities in boreal mixedwood forest: Harvest and wildfire effects over 30 years, *Ecol. Appl.*, 9, 849–863, 1999.

Howarth, R.W., Comparative responses of aquatic ecosystems to toxic chemical stress, In *Comparative Analyses of Ecosystems: Patterns, Mechanisms, and Theories*, Cole, J., Lovett, G., and Findlay, S. (eds.), Springer-Verlag, New York, 1991, pp. 169–195.

Howe, F.P., Knight, R.L., McEwen, L.C., and George, T.L., Direct and indirect effects of insecticide applications on growth and survival of nestling passerines, *Ecol. Appl.*, 6, 1314–1324, 1996.

Irfanullah, H.M. and Moss, B., Effects of pH and predation by *Chaoborus* larvae on the plankton of a shallow and acidic forest lake, *Freshw. Biol.*, 50, 1913–1926, 2005.

Ives, A.R., Foufopoulos, J., Klopfer, E.D., Klug, J.L., and Palmer, T.M., Bottle or big-scale studies: How do we do ecology? *Ecology* 77, 681–685, 1996.

Jenkins, D.G. and Buikema, A.L., Jr., Do similar communities develop in similar sites? A test with zooplankton structure and function, *Ecol. Monogr.*, 68, 421–443, 1998.

Joern, A. and Hoagland, K.D., In defense of whole-community bioassays for risk assessment, *Environ. Toxicol. Chem.*, 15, 407–409, 1996.

Karr, J.R., Defining and assessing ecological integrity: Beyond water quality, *Environ. Toxicol. Chem.*, 12, 1521–1531, 1993.

Kennedy, J.H., Johnson, Z.B., Wise, P.D., and Johnson, P.C., Model aquatic ecosystems in ecotoxicology research: Considerations of design, implementation, and analysis, In *Handbook of Ecotoxicology*, Hoffman, D.J., Rattner, B.A., Burton, G.A., Jr., and Cairns, J., Jr. (eds.), CRC Press, Boca Raton, FL, 1995, pp. 117–162.

Kiffney, P.M. and Clements, W.H., Size-dependent response of macroinvertebrates to metals in experimental streams, *Environ. Toxicol. Chem.*, 15, 1352–1356, 1996a.

Kiffney, P.M. and Clements, W.H., Effects of metals on stream macroinvertebrate assemblages from different altitudes, *Ecol. Appl.*, 6, 472–481, 1996b.

Kimball, K.D. and Levin, S.A., Limitations of laboratory bioassays: The need for ecosystem-level testing, *Bioscience*, 35, 165–171, 1985.

Kullback, S. and Leibler, R.A., On information and sufficiency, *Ann. Math. Stat.*, 22, 79–86, 1951.

Landis, W.G., Matthews, R.A., and Matthews, G.B., Design and analysis of multispecies toxicity tests for pesticide registration, *Ecol. Appl.*, 7, 1111–1116, 1997.

Lawton, J.H., The Ecotron facility at Silwood Park: The value of "big bottle" experiments, *Ecology*, 77, 665–669, 1996.

Le Jeune, A.H., Charpin, M., Deluchat, V., and Briand, J.F., Effect of copper sulfate treatment on natural phytoplanktonic communities, *Aquat. Toxicol.*, 80, 267–280, 2006.

Liber, K., Kaushik, N.K., Solomon, K.R., and Carey, J.H., Experimental designs for aquatic mesocosm studies: A comparison of the "ANOVA" and "regression" design for assessing the impact of tetrachlorophenol on zooplankton populations in limnocorrals, *Environ. Toxicol. Chem.*, 11, 61–77, 1992.

Likens, G.E., Bormann, F.H., Johnson, N.M., Fisher, D.W., and Pierce, R.S., Effects of forest cutting and herbicide treatment on nutrient budgets in the Hubbard Brook Watershed-ecosystem, *Ecol. Monogr.*, 40, 23–47, 1970.

Lozano, S.J., O'Halloran, S.L., Sargent, K.W., and Brazner, J.C., Effects of esfenvalerate on aquatic organisms in littoral enclosures, *Environ. Toxicol. Chem.*, 11, 35–47, 1992.

Lubchenco, J. and Real, L.A., Manipulative experiments as tests of ecological theory, In *Foundations of Ecology*, Real, L.A. and Brown, J.H. (eds.), University of Chicago Press, Chicago, IL, 1991, pp. 715–733.

MacArthur, R.H. and Wilson, E.O., An equilibrium theory of insular zoogeography, *Evolution*, 17, 373–387, 1963.

Martin, P.A., Johnson, D.L., Forsyth, D.J., and Hill, B.D., Indirect effects of the pyrethroid insecticide deltamethrin on reproductive success of chestnut-collared longspurs, *Ecotoxicology*, 7, 89–97, 1998.

Martin, P.A., Johnson, D.L., Forsyth, D.J., and Hill, B.D., Effects of two grasshopper control insecticides on food resources and reproductive success of two species of grasshopper songbirds, *Environ. Toxicol. Chem.*, 19, 2987–2996, 2000.

Matthews, R.A., Landis, W.G., and Matthews, G.B., The community conditioning hypothesis and its application to environmental toxicology, *Environ. Toxicol. Chem.*, 15, 597–603, 1996.

Medley, C.N. and Clements, W.H., Responses of diatom communities to heavy metals in streams: The influence of longitudinal variation, *Ecol. Appl.*, 8, 631–644, 1998.

Momo, F., Ferrero, E., Eory, M., Esusy, M., Iribarren, J., Ferreyra, G., Schloss, I., Mostajir, B., and Demers, S., The whole is more than the sum of its parts: Modeling community-level effects of UVR in marine ecosystems, *Photochem. Photobiol.*, 82, 903–908, 2006.

Newman, M.C., *Population Ecotoxicology*, John Wiley & Sons, Chichester, UK, 2001.

Niederlehner, B.R., Pontasch, K.W., Pratt, J.R., and Cairns, J., Jr., Field evaluation of predictions of environmental effects from a multispecies-microcosm toxicity test, *Arch. Environ. Contam. Toxicol.*, 19, 62–71, 1990.

Odum, E.P., The strategy of ecosystem development, *Ecology*, 164, 262–270, 1969.

Odum, E.P., The mesocosm, *Bioscience*, 34, 558–562, 1984.

Paine, R.T., Food web complexity and species diversity, *Am. Nat.*, 100, 65–75, 1966.

Paine, R.T., A note on trophic complexity and community stability, *Am. Nat.*, 103, 91–93, 1969.

Park, T., Experimental studies of interspecies competition. I. Competition between poplations of the flour beetles, *Tribolium confusum* Duvall and *Tribolium castaneum* Herbst, *Ecol. Monog.*, 18, 267–307, 1948.

Patnode, K.A. and White, D.H., Effects of pesticides on songbird productivity in conjunction with pecan cultivation in southern Georgia: A multiple-exposure experimental design, *Environ. Toxicol. Chem.*, 10, 1479–1486, 1991.

Peckarsky, B.L., The dual role of experiments in complex and dynamic natural systems, In *Experimental Ecology: Issues and Perspectives*, Resetarits, W.J., Jr. and Bernardo, J. (eds.), Oxford University Press, Inc., New York, 1998, pp. 311–324.

Perez, K.T., Role and significance of scale to ecotoxicology, In *Ecological Toxicity Testing: Scale, Complexity, and Relevance*, Cairns, J., Jr. and Niederlehner, B.R. (eds.), CRC Press, Boca Raton, FL, 1995, pp. 49–72.

Perez, K.T., Morrison, G.E., Davey, E.W., Lackie, N.F., Soper, A.E., Blasco, R.J., Winslow, D.L., Johnson, R.L., Murphy, P.G., and Heltshe, J.F., Influence of size on fate and ecological effects of kepone in physical models, *Ecol. Appl.*, 1, 237–248, 1991.

Perry, J.A. and Troelstrup, N.H., Jr., Whole ecosystem manipulation: A productive avenue for test system research? *Environ. Toxicol. Chem.*, 7, 941–951, 1988.

Platt, J.R., Strong inference, *Science*, 146, 347–353, 1964.

Pontasch, K.W., The use of stream microcosms in multispecies testing, In *Ecological Toxicity Testing: Scale, Complexity, and Relevance*, Cairns, J., Jr. and Niederlehner, B.R. (eds.), CRC Press, Boca Raton, FL, 1995, pp. 169–191.

Pontasch, K.W., Niederlehner, B.R., and Cairns, J., Jr., Comparisons of single-species, microcosm and field responses to a complex effluent, *Environ. Toxicol. Chem.*, 8, 521–532, 1989.

Popper, K.R., *The Logic of Scientific Discovery*, 3rd ed., Hutchinson, London, England, 1972.

Power, M.E., Dietrich, W.E., and Sullivan, K.O., Experimentation, observation, and inference in river and watershed investigations, In *Experimental Ecology: Issues and Perspectives*, Resetarits, W.J., Jr. and Bernardo, J. (eds.), Oxford University Press, New York, 1998,

Pratt, J.R. and Barreiro, R., Influence of trophic status on the toxic effects of a herbicide: A microcosm study, *Arch. Environ. Contam. Toxicol.*, 35, 404–411, 1998.

Pratt, J.R., Melendez, A.E., Barreiro, R., and Bowers, N.J., Predicting the ecological effects of herbicides, *Ecol. Appl.*, 7, 1117–1124, 1997.

Rapport, D.J., Regier, H.A., and Hutchinson, T.C., Ecosystem behavior under stress., *Am. Nat.*, 125, 617–640, 1985.

Resetarits, W.J., Jr. and Bernardo, J. (eds.), *Experimental Ecology: Issues and Perspectives*, Oxford University Press, New York, 1998.

Resetarits, W.J., Jr. and Fauth, J.E., From cattle tanks to Carolina bays: The utility of model systems for understanding natural communities, In *Experimental Ecology: Issues and Perspectives*, Resetarits, W.H., Jr. and Bernardo, J. (eds.), Oxford University Press, New York, 1998, pp. 133–151.

Relyea, R.A., Schoeppner, N.M., and Hoverman, J.T., Pesticides and amphibians: The importance of community context, *Ecol. Appl.*, 15, 1125–1134, 2005.

Sallenave, R.M., Day, K.E., and Kreutzweiser, D.P., The role of grazers and shredders in the retention and downstream transport of a PCB in lotic environments, *Environ. Toxicol. Chem.*, 13, 1843–1847, 1994.

Schauber, E.M., Edge, W.D., and Wolff, J.O., Insecticide effects on small mammals: Influence of vegetation structure and diet, *Ecol. Appl.*, 7, 143–157, 1997.

Schindler, D.W., Eutrophication and recovery in experimental lakes: Implications for management, *Science*, 184, 897–899, 1974.

Schindler, D.W., Detecting ecosystem responses to anthropogenic stress, *Can J. Fish. Aquat. Sci.*, Suppl., 6–25, 1987.

Schindler, D.W., Effects of acid rain on freshwater ecosystems, *Science*, 239, 149–157, 1988.

Schindler, D.W., Replication versus realism: The need for ecosystem-scale experiments, *Ecosystems*, 1, 323–334, 1998.

Shaw, J.L. and Kennedy, J.H., The use of aquatic field mesocosm studies in risk assessment, *Environ. Toxic. Chem.*, 15, 605–607, 1996.

Sheffield, S.R. and Lochmiller, R.L., Effects of field exposure to diazinon on small mammals inhabitating a semienclosed prairie grassland ecosystem. I. Ecological and reproductive effects, *Environ. Toxicol. Chem.*, 20, 284–296, 2001.

Simberloff, D.S., A succession of paradigms in ecology: essentialism to materialism and probabilism, *Synthese*, 43: 3–39, 1980.

Simberloff, D.S. and Wilson, E.O., Experimental zoogeography of islands: The colonization of empty islands, *Ecology*, 50, 278–296, 1969.

Smith, E.P., Design and analysis of multispecies experiments, In *Ecological Toxicity Testing: Scale, Complexity, and Relevance*, Cairns, J., Jr. and Niederlehner, B.R. (eds.), CRC Press, Boca Raton, FL, 1995, pp. 73–95.

Sousa, W.P., Disturbance in marine intertidal boulder fields: The nonequilibrium maintenance of species diversity, *Ecology*, 60, 1225–1239, 1979.

Stanley, J.K., Brooks, B.W., and La Point, T.W., A comparison of chronic cadmium effects on *Hyalella azteca* in effluent-dominated stream mesocosms to similar laboratory exposures in effluent and reconstituted hard water, *Environ. Toxicol. Chem.*, 24, 902–908, 2005.

Steinman, A.D., Mulholland, P.J., Palumbo, A., DeAngelis, D., and Flum, T., Lotic ecosystem response to a chlorine disturbance, *Ecol. Appl.*, 2, 341–355, 1992.

Stewart-Oaten, A., Murdoch, W.W., and Parker, K.R., Environmental impact assessment: "Pseudoreplication" in time? *Ecology*, 67, 929–940, 1986.

Strong, D.R., Simberloff, D., Abele, L.G., and Thistle, A.B. (eds.), *Ecological Communities: Conceptual Issues and Evidence*, Princeton University Press, Princeton, NJ, 1984.

Suter, G.W., II, *Ecological Risk Assessment*, Lewis, Chelsea, MI, 1993.

Suttman, C.E. and Barrett, G.W., Effects of sevin on arthropods in an agricultural and an old field plant community, *Ecology*, 60, 628–641, 1979.

Taub, F.B., Standardized Aquatic Microcosms: Tools for assessing the ecological effects of chemicals and genetically engineered microorganisms, *Environ. Sci. Technol.*, 23, 1064–1066, 1989.

Taub, F.B., Are ecological studies relevant to pesticide registration decisions? *Ecol. Appl.*, 7, 1083–1085, 1997.

Touart, L.W., *Aquatic Mescoosm Tests to Support Pesticide Registrations, EPA 540-09-88-035*, U.S. Environmental Protection Agency, Washington, D.C., 1988.

Touart, L.W. and Maciorowski, A.F., Information needs for pesticide registration in the United States, *Ecol. Appl.*, 7, 1086–1093, 1997.

Verhoef, H.A., The role of soil microcosms in the study of ecosystem processes, *Ecology*, 77, 685–690, 1996.

Von der ohr, P.C. and Liess, M., Relative sensitivity distribution of aquatic invertebrates to organic and metal compounds, *Environ. Toxicol. Chem.*, 23, 150–156, 2004.

Wallace, J.B., Lugthart, G.J., Cuffney, T.F., and Schurr, G.A., The impact of repeated insecticidal treatments on drift and benthos of a headwater stream, *Hydrobiologia*, 179, 135–147, 1989.

Wallace, J.B., Webster, J.R., and Cuffney, T.F., Stream detritus dynamics: Regulation by invertebrate consumers, *Oecologia*, 53, 197–200, 1982.

Webber, E.C., Deutsch, W.G., Bayne, D.R., and Seesock, W.C., Ecosystem-level testing of a synthetic pyrethroid insecticide in aquatic mesocosms, *Environ. Toxicol. Chem.*, 11, 87–105, 1992.

Wiens, J.A. and Parker, K.R., Analyzing the effects of accidental environmental impacts: Approaches and assumptions, *Ecol. Appl.*, 5, 1069–1083, 1995.

Werner, E.E., Ecological experiments and a research program in community ecology, In *Experimental Ecology: Issues and Perspectives*, Resetarits, W.J., Jr. and Bernardo, J. (eds.), Oxford University Press, New York, 1998, pp. 3–26.

Whiles, M.R. and Wallace, J.B., First-year benthic recovery of a headwater stream following a 3-year insecticide-induced disturbance, *Freshw. Biol.*, 28, 81–91, 1992.

Wiegner, T.N., Seitzinger, S.P., Breitburg, D.L., and Sanders, J.G., The effects of multiple stressors on the balance between autotrophic and heterotrophic processes in an estuarine system, *Estuaries*, 26, 352–364, 2003.

Wilson, D.S., Biological communities as functionally organized units, *Ecol. Appl.*, 78, 2018–2024, 1997.

Winner, R.W., Owen, H.A., and Moore, M.V., Seasonal variability in the sensitivity of freshwater lentic communities to a chronic copper stress, *Aquat. Toxicol.*, 17, 75–92, 1990.

Wong, D.C.L., Maltby, L., Whittle, D., Warren, P., and Dorn, P.B., Spatial and temporal variability in the structure of invertebrate assemblages in control stream mesocosms, *Water Res.*, 38, 128–138, 2004.

Wong, D.C.L., Whittle, D., Maltby, L., and Warren, P., Multivariate analyses of invertebrate community responses to a C_{12-15}AE-3S anionic surfactant in stream mesocosms, *Aquat. Toxicol.*, 62, 105–117, 2003.

Woodwell, G.M., Effects of pollution on the structure and physiology of ecosystems, *Science*, 168, 429–433, 1970.

24 Application of Multimetric and Multivariate Approaches in Community Ecotoxicology

The most distinct and beautiful statement of any truth must take at last the mathematical form.

(Henry David Thoreau, in Walls 1999)

24.1 INTRODUCTION

Methods to assess the effects of contaminants and other anthropogenic stressors on communities range from computationally simple indices such as species richness to complex, computer-dependent algorithms such as multivariate analyses. The simplest community indices use species presence/absence or abundance data to show how individuals in the community are distributed among species. The advantages of these indices are their intuitive meaning and their ability to reduce complex data to a single number. Only slightly more involved but retaining more information, species abundance curves described in Chapter 22 characterize the distribution of individuals among the species by fitting abundance data to specified distributions. Estimated distributional parameters from species abundance models provide a parsimonious description of the community. Slightly more involved composite measures require additional knowledge about community qualities (e.g., the trophic status of a species) to produce indices developed specifically to gauge diminished community integrity due to anthropogenic stressors. Currently, the most popular of these composite indices is Karr's (1981) index of biological integrity (IBI). These composite indices require more ecological knowledge of the community than measures of species richness or species abundance models but have the advantage of being focused primarily on human effects on communities or species assemblages. More convenient, but perhaps applying less ecology than warranted, distributions of individual species effect metrics (e.g., distributions of 96-h LC50 values) are used to predict "safe concentrations" that presumably protect all but a specified, low percentage of the species making up the community. Even more computationally intense methods, such as multivariate analyses, aim to reduce the number of data dimensions to an interpretable low number, and to quantify similarities or differences among sampling units. These last methods tend to generate interpretive parsimony at the expense of methodological simplicity and straightforward terminology; therefore, considerable caution is needed to avoid errors during their application. However, the value of these methods in identifying clear explanations from complex data sets makes worthwhile any effort spent wading through obtuse computer manuals or dealing with the associated jargon.

Jargon, not argument, is your best ally in keeping him from the Church.

(Lewis 1942)

24.1.1 Comparison of Multimetric and Multivariate Approaches

Multimetric and multivariate approaches are applied to community data with the intent of rendering the associated complex array of information to a more parsimonious form. Because ecological assessments of biological integrity generally require analysis of numerous biotic and abiotic variables, sophisticated statistical approaches are often necessary to examine the complex relationships between species assemblages and multiple environmental factors. Multivariate approaches reduce complex, multidimensional data to two or three dimensions, thus allowing researchers to identify key environmental variables responsible for patterns of species abundance. In contrast, multimetric indices integrate a diverse suite of measures, often across several levels of biological organization, to assess biological integrity. It is appropriate to consider these two approaches together because the community data necessary to calculate a multimetric index or to conduct multivariate analyses are often the same (e.g., abundance, richness, and composition).

In their comparison of multivariate and multimetric approaches, Reynoldson et al. (1997) concluded that multivariate approaches provided greater accuracy and precision for assessing reference conditions in streams. Terlizzi et al. (2005) showed that univariate measures of molluscan community structure (species richness) showed little response to contamination whereas multivariate analyses identified significant differences between reference and polluted sites. Thomas and Hall (2006) compared the ability of individual metrics, multivariate approaches, and multimetric indices to identify impairment in periphyton, macroinvertebrate, and fish communities. Although some individual metrics were associated with large-scale habitat gradients, multivariate approaches were most useful for identifying spatial and temporal differences in each community. In a comprehensive analysis of community indices and multivariate approaches, Kilgour et al. (2004) compared the relative sensitivity of seven benthic community metrics and three multivariate indices to contamination associated with mines, pulp and paper mills, and urbanization. Multivariate approaches identified significant differences associated with each of the perturbations and greater effect sizes compared to the community metrics. Although the examples described above seem to highlight the greater discriminatory ability of multivariate approaches, the usefulness of univariate and multivariate techniques for distinguishing between reference and contaminated sites will likely vary with the spatial scale of an investigation (Quintino et al. 2006). Despite their growing popularity in Canada and Europe, multivariate approaches have received considerably less attention in the United States (Resh et al. 1995). Multivariate analyses have been criticized because of their inherent statistical complexity and because results are often difficult to interpret (Fore et al. 1996, Gerritsen 1995). The complex graphical representations of multivariate results are often of limited value to non-ecologists and managers. Although strict reliance on complex statistical algorithms may obscure important biological results, we believe that multivariate approaches are an essential set of tools for biological assessments of water quality. Because community–environment relationships are inherently multidimensional, approaches such as multivariate analyses that consider interactions among predictor variables and their effects on multiple response variables are necessary. New approaches, such as the application of principal response curves (Pardal et al. 2004), quantify multivariate community responses to contaminants in ways that are more accessible to managers and policymakers. We agree with the recommendations of Reynoldson et al. (1997) that multivariate and multimetric approaches are complementary and should be used in conjunction. For example, the variables used in multivariate analyses such as principal components could include species richness, abundance of sensitive groups, or other measures typically included in a multimetric index. Griffith et al. (2003) used this approach in their evaluation of the relationship between macroinvertebrate assemblages and environmental gradients. Multivariate statistical analysis (redundancy analysis) using metrics derived from an index of biotic integrity provided complementary results to canonical correspondence analysis based on macroinvertebrate abundance. Alternatively, a multimetric index similar to Karr's IBI could be developed using results of multivariate analyses. Loading coefficients from canonical discriminant analyses, principal component analyses (PCA), and other multivariate procedures identify variables that are most important for separation of groups (generally locations, sampling stations). Variables shown to be responsible for separation of reference and impacted stations could be combined in a multimetric index. Integration of multivariate and multimetric approaches may be necessary to detect perturbations when relatively weak relationships between stressors and community structure exist (Chenery and Mudge 2005). Finally, we note that our enthusiasm for multimetric and multivariate approaches in community-level bioassessment is not shared by all researchers. Weiss and Reice (2005) remind us that neither of these approaches provides causal linkages between stressors and community-level responses. These researchers advocate an alternative approach in which effects of stressors on individual taxa with known species-level tolerances are employed to develop an overall assessment of community-level impact.

24.2 MULTIMETRIC INDICES

> A principal objective of the 1972 Federal Water Pollution Control Act and its 1977 and 1987 amendments is to restore and maintain the biological integrity of the nation's waters.
>
> **(Miller et al. 1988)**

One of the most significant advances in the field of biological assessments over the past 20 years was the development and application of multimetric approaches for measuring ecological integrity. Because no single measure of impairment will respond to all classes of contaminants, and because some individual metrics may show unexpected changes (e.g., increased species richness at polluted sites), multimetric indices are an effective tool for measuring effects of stressors (Fausch et al. 1990, Karr 1981, Kerans and Karr 1994, Plafkin et al. 1989). The individual metrics in a multimetric index reflect different characteristics of life history, community structure, and functional organization. In general, as the number of metrics increases (up to some reasonable number), the ability to separate contaminant effects from natural variation increases (Karr 1993) (Figure 24.1). In addition, because

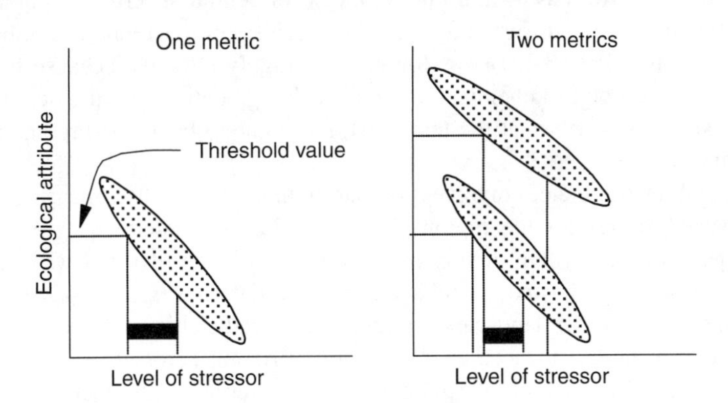

FIGURE 24.1 Hypothetical relationships between stressor levels and ecological attributes characterized using one or two metrics. The threshold value of the ecological attribute is defined as the response that is considered to be biologically significant. For example, a researcher may conclude that a 20% reduction in abundance of a sensitive species is a biologically significant response. The responses of the individual metrics are represented as clouds of points and the level of the stressor known to affect the ecological attribute is represented by the black bar. Note that addition of a second metric provides a more refined measure of the stressor level that causes a biologically significant response. (Modified from Figure 1 in Karr (1993).)

individual metrics respond differently to different classes of contaminants, multimetric approaches are useful for assessing a diverse suite of stressors and measuring impacts in systems receiving multiple stressors. The individual metrics included in a multimetric index may vary among perturbations, but should reflect important structural and functional characteristics of the system. In general, deviation of individual metrics from expected values at reference sites is estimated and a final value that includes the sum of all individual metrics is calculated.

Karr's (1981) IBI is the most widely used multimetric index for assessing the health of aquatic communities. The IBI was developed in response to the federally legislated mandate to "restore and maintain the chemical, physical, and biological integrity" of U.S. waters (Clean Water Act 1977, PL 95-217, also 1987 PL 100-4). Originally employed in Midwestern streams in the United States, the IBI is based on 12 attributes of fish assemblages in three general categories: species richness and composition, trophic composition, and fish abundance and condition. The individual metrics are assigned scores (1, 3, 5) based on their similarity to expected values in undisturbed or least impacted streams. Expected values for the individual metrics are obtained by sampling a large number of known reference sites in a region. Alternatively, expected values can be derived from surveys of reference and impacted sites and using the "best" values from these samples (Simon and Lyons 1995). Because expected values for species richness and total abundance vary with stream size, these metrics must be adjusted to reflect watershed area and other regional conditions. The scores of the 12 metrics are summed to yield a total IBI score for a site (which ranges from 12 to 60), with larger values indicating healthy fish assemblages. The IBI is sensitive to a diverse array of physical and chemical stressors, including industrial and municipal effluents, agricultural inputs, habitat loss, and introduction of exotic species.

The IBI works especially well for characterizing fish communities because environmental requirements and historic distributions of this group are well known. This greatly facilitates establishment of expected values for individual metrics. The structural and functional metrics included in the IBI are biologically relevant, and each individual metric responds to known gradients of degradation (Fausch et al. 1990). The general approach outlined in the IBI has been modified for other ecosystems (e.g., lakes and estuaries) and applied to other taxonomic groups (e.g., benthic macroinvertebrates and diatoms). Although the specific metrics vary among these applications, comparison of measured values to expected values and integration of a suite of metrics into a single index are consistent among approaches. A multimetric index for benthic macroinvertebrate communities was used to distinguish polluted from reference sites in rivers of the Tennessee Valley (Kerans and Karr 1994). The benthic IBI (B-IBI) was found to be highly effective because benthic macroinvertebrates generally respond to chemical and physical degradation in a predictable fashion. The IBI now enjoys such popularity that the term, IBI, has come to be applied to any new composite or multimetric index.

Calculating multimetric indices involves comparing individual metrics measured at an impacted site to the expected values for the region (Figure 24.2a). As described above, because some metrics (e.g., species richness) are greatly influenced by stream order and watershed area, these expected values must be adjusted to reflect natural variation (Figure 24.2b). Assuming that community responses to other landscape variables are predictable, a logical extension of this approach is to create models to account for natural variation across broad geographical areas. Bailey et al. (1998) found that simple geographic characteristics (distance from source, catchment area, elevation) and year sampled accounted for greater than 50% of the variation among reference sites. The performance of several bioassessment metrics was significantly improved when a predictive model that included this geographic variation was employed to identify impacted sites. The conventional approach of comparing metric values at impacted sites with expected values at reference sites has now advanced to the point where we can characterize habitat variation within subregions using more sophisticated multivariate statistics (Figure 24.2c). The application of multivariate techniques for assessing reference conditions is described below.

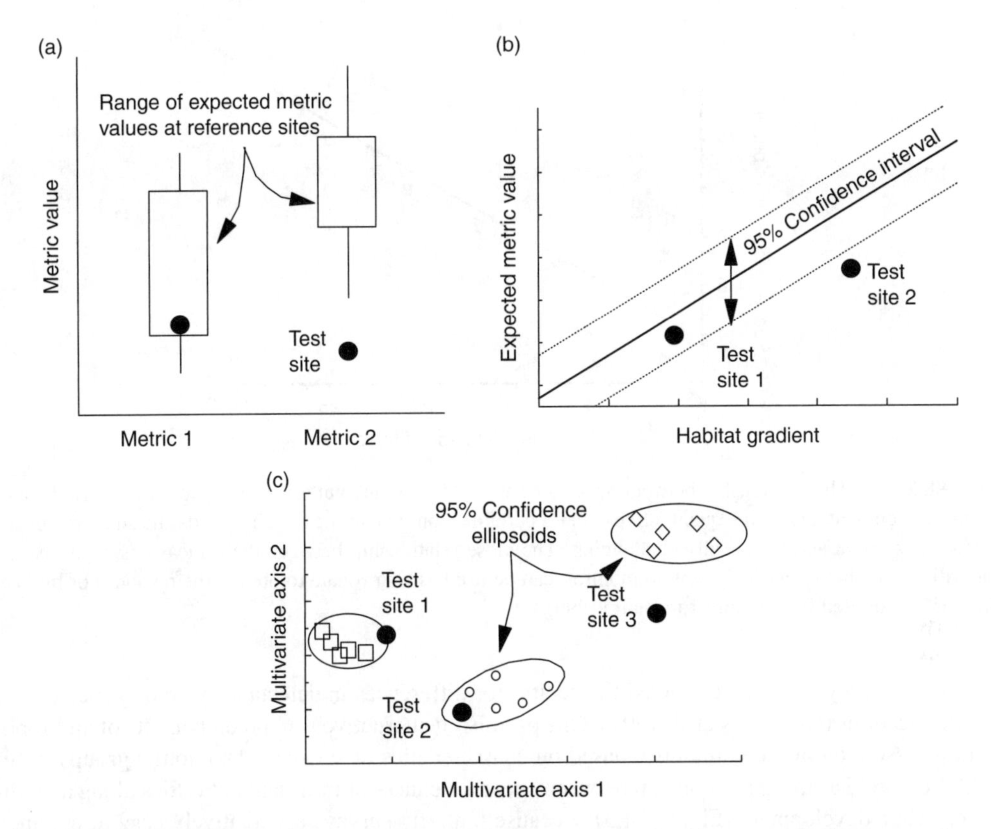

FIGURE 24.2 Multimetric and multivariate approaches for comparing test sites to expected values at reference sites. (a) Two metric values at a test site (indicated by solid circles) are compared to expected values. Values are within the expected range for metric 1, but below the range of expected values for metric 2. (b) Metric values are adjusted to reflect expected changes in habitat characteristics along a gradient. Although the metric value at test site 2 is greater than at test site 1, it is less than the expected value and would indicate impact. (c) Multivariate analysis of expected metric values based on regional differences in habitat characteristics. Test sites 1 and 2 are within the expected values whereas test site 3 falls outside the 95% confidence ellipsoid.

24.2.1 Multimetric Approaches for Terrestrial Communities

Although multimetric indices such as the IBI have been limited primarily to aquatic ecosystems, the general approach could be modified for terrestrial communities. Because of their sensitivity and rapid response to environmental stressors, terrestrial arthropods would be especially useful for assessing biological integrity (Kremen et al. 1993). Nelson and Epstein (1998) investigated the responses of lepidopterans to habitat modifications and concluded that butterfly communities integrate important structural and functional characteristics of terrestrial ecosystems. Kremen (1992) evaluated the indicator properties of butterfly communities and reported that this group was quite responsive to anthropogenic disturbance. Bird communities also offer opportunities for development of integrated measures of ecological integrity. The abundance, distribution, and habitat requirements of birds are generally well known, especially in North America. National monitoring programs, such as the Christmas Bird Counts conducted by the Audubon Society and Breeding Bird Surveys, have provided spatially extensive, long-term data on bird assemblages. Finally, responses of bird populations to some environmental stressors, especially pesticides and habitat alterations, have been well documented. However, given the logistical difficulties of sampling bird communities, developing a suite of ecologically relevant indicators for this group will be a challenge. In

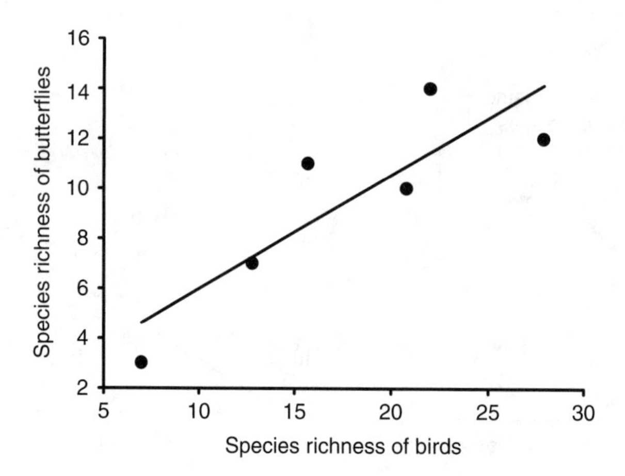

FIGURE 24.3 The relationship between species richness of birds and butterflies at 6 sites along a gradient of urban development. Obtaining quantitative data for certain taxonomic groups, such as birds and small mammals is often expensive and logistically challenging. The close relationship between these measures suggests that butterflies, which are relatively easy to monitor, can be used as a surrogate to predict the response of birds to stressors. (Modified from Figure 1 in Blair (1999).)

particular, surveys must be corrected to account for differences in detectability among species and among locations (Chambers et al. 1999). One promising alternative is to predict effects of anthropogenic stressors on bird communities based on characteristics of surrogate taxonomic groups. Blair (1999) reported a strong relationship between species richness of birds and butterflies along a gradient of urban development (Figure 24.3). Because butterfly surveys are relatively easy to conduct, Blair suggested that species richness of butterflies could be used as a surrogate for monitoring bird communities.

24.2.2 Limitations of Multimetric Approaches

One major advantage of multimetric approaches is that they integrate several ecologically relevant responses into a single measure, a characteristic that appeals to many water resource managers. However, some researchers are skeptical of multimetric indices and argue that a better approach is to assess an array of ecosystem responses, which provide a direct linkage between cause and effect (Suter 1993). Detailed critiques of multimetric indices as well as a discussion of their limitations have been published previously (Simon and Lyons 1995, Fausch et al. 1990, Suter 1993). Only a summary of the major limitations will be presented here.

First, multimetric indices are data intensive. Regardless of the specific system or taxonomic group, development and application of multimetric approaches require a thorough understanding of the ecology and habitat requirements of species as well as their tolerances for environmental stressors. For some taxonomic groups and in some systems, these data will not be available. Second, most multimetric approaches cannot be employed to identify specific causes of environmental impacts. This criticism reflects two mutually exclusive goals of many biological monitoring programs. While chemical-specific, diagnostic indicators may allow researchers to identify a single source of perturbation, more general measures such as the IBI are required to characterize the integrity of systems receiving multiple stressors. It is possible that the responses of individual metrics in a multimetric index could offer some insight into the specific source of contamination. For example, a multimetric index for benthic macroinvertebrates might include metrics for abundance and species richness of mayflies, stoneflies, and caddis-flies. All three groups are generally sensitive to organic enrichment; however, many caddis-flies and some stoneflies are tolerant of heavy metals (Clements et al. 1988, Clements and Kiffney 1995). Analysis of the responses of component metrics may allow researchers to quantify the relative importance of individual stressors in systems affected by multiple perturbations. Third, multimetric indices may not respond to some types of perturbation because changes in one metric may be offset by changes in another metric. Again, the obvious solution to this problem is to report not only the integrated scores but also the responses of component metrics. Finally, multimetric indices based on attributes of community composition will be less effective in areas with low species richness or naturally impoverished assemblages. Fausch et al. (1990) note that the low species richness of fish assemblages in western coldwater streams requires that many of the community-level metrics be replaced by life history and population-level responses.

24.3 MULTIVARIATE APPROACHES

Multivariate data sets are broadly defined here as those in which more than two dependent or independent variables are collected for each sampling unit. These variables typically include community characteristics (e.g., species abundances) that change or might be influenced together in complex ways. A wide range of multivariate statistical methods has been used to analyze these types of data. In contrast to the methods described to this point, multivariate analyses are not based on ecological concepts but are statistical constructs that reduce complex data sets to potentially meaningful patterns involving a few variables. Some, such as ordination methods, combine species abundance information for many sites or sampling units into functions that capture a portion of the total variance in the data. A small number of uncorrelated, linear combinations of the species abundances might be identified. Ecotoxicological meaning can be assigned to the positions of sampling units (e.g., sites) along these linear functions. Alternatively, the researcher may simply use the results to describe trends among sampling units. Other methods, such as cluster analysis, separate samples into groups in hopes of identifying some ecological or toxicological pattern that may emerge to explain the groupings. Another type of analysis might be applied to species abundance data to identify which qualities weigh most heavily in discriminating among known groups. Regardless of the applied method, the overarching idea is that multivariate analysis of the measured variables can reveal hidden or unmeasured qualities.

As with most parametric analyses, transformation of species abundance data is often advisable before applying a multivariate method. Transformation might be done to reduce the influence of one variable relative to others in the linear combinations of variables. One variable might have a much wider range of values and, in the absence of transformation, would have a disproportionately heavy influence on variance. In such a case, each variable (e.g., species' abundances at all sampling sites) may be standardized to a mean of 0 and standard deviation of 1. If a skewed distribution was to occur with the species abundance distributions, some transformation such as the square root or another power of abundance might be employed prior to standardization and multivariate analysis. This is often necessary when a few species are very abundant at some sites.

24.3.1 SIMILARITY INDICES

Although generally not included in treatment of multivariate analyses, similarity indices also reduce complex, multispecies data for the purpose of comparing communities among locations or over time. Similarity indices quantify the correspondence between two communities based on either presence–absence or abundance data. These indices are especially useful for comparing communities from regional reference sites to impacted sites. Alternatively, similarity indices are appropriate in studies of well-defined pollution gradients, where similarity to reference conditions is expected to increase with distance from a pollution source. The simplest and most frequently used similarity index based on presence–absence data is the Jaccard Index:

$$J = j/(a + b - j), \tag{24.1}$$

where a = the number of species in community a, b = the number of species in community b, and j = the number of species common to both sites.

Because the Jaccard Index does not account for differences in abundance between locations, rare species and abundant species are weighted equally. Thus, it is likely that the Jaccard Index will be relatively insensitive to low or moderate levels of contamination. More sophisticated similarity indices, such as the Morisita–Horn measure, compare the relative abundance of taxa between two communities. The Morisita–Horn Index is given as

$$\mathrm{MH} = 2\sum(an_i \times bn_i)/(\mathrm{d}a + \mathrm{d}b)aN \times bN, \tag{24.2}$$

where an_i = the number of individuals of the ith species at site a, bn_i = the number of individuals of the ith species at site b, aN = the total number of individuals at site a, and bN = the total number of individuals at site b. The terms da and db in the Equation 24.2 are calculated as

$$\mathrm{d}a = \sum an_i^2/aN^2, \quad \mathrm{d}b = \sum bn_i^2/bN^2.$$

The Morisita–Horn measure of similarity is favored by some researchers because it is relatively insensitive to sample size and species richness (Magurran 1988, Wolda 1981).

Dissimilarity among locations or between time points can also be used to evaluate responses to environmental stressors. Philippi et al. (1998) quantified spatial and temporal responses to perturbations by comparing the pairwise dissimilarity between sites with the average dissimilarity among replicate samples. These researchers noted that measures of dissimilarity (or similarity) can be employed to evaluate changes in community composition during recovery (Figure 24.4). If remediation was effective, the relative *dissimilarity* between reference and impacted sites would be expected to decrease over time.

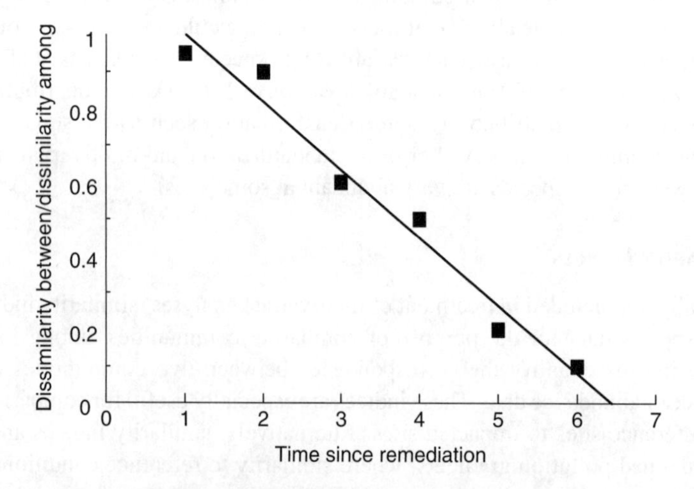

FIGURE 24.4 Hypothetical changes in community similarity between reference and impacted sites as a function of time since remediation was initiated. The relationship shows that the index of dissimilarity (expressed as the ratio of dissimilarity between sites to the average dissimilarity among sites) is reduced over time as a result of remediation.

While similarity indices provide a simple way to compare community composition, there are potential problems with these measures. Boyle et al. (1990) evaluated the ability of similarity indices to discriminate effects of simulated perturbations based on initial community structure, sensitivity to community change, stability in response to reduced richness and abundance, and consistency. These researchers concluded that some similarity indices were misleading because results were strongly influenced by initial community composition and the nature of the perturbation. Although similarity indices are useful when comparing communities from two locations, more sophisticated techniques are necessary to compare multiple sites. Cluster analysis, a logical extension of similarity indices, is applicable for comparing communities from several locations or for comparing the similarity of a single site with a group of sites. Cluster analysis employs a variety of similarity measures based on either presence–absence or abundance data. These data are often expressed using a dendrogram, with the most similar sites combined into a single cluster. Additional sites are included based on their similarity to the existing clusters. Several different clustering algorithms have been developed, and relatively simple software packages are available for most analyses. Details of the different clustering techniques and the justification for deciding how different sites and clusters should be joined have been published (Gauch 1982). These methods will be described below.

24.3.2 Ordination

Ordination is a process in which a large set of variables is reduced to a few variables with the intent of enhancing conceptual parsimony and tractability. With ordination analysis of community abundance data, the measured variables (e.g., abundance of each species for each sampling unit) are used to identify hidden patterns or unmeasured factors explaining the data structure. Mathematical constructs are sought to help interpret correlations among variables. There are five steps to ordination analysis, regardless of the specific method applied (Comrey 1973). (1) The relevant data are generated and selected for analysis. As noted above, the data might require transformation prior to use. (2) The correlation matrix for the variables is calculated. (3) Factors (mathematical functions) are extracted. (4) The factors might be rotated to enhance interpretation. (5) The factors are then interpreted. Ideally, plots of the sampling unit positions along the first few mathematical constructs reveal explanatory, or at least consistent, themes.

As an example, linear functions can be defined such as

$$\text{Function } 1 = b_1X_1 + b_2X_2 + b_3X_3 + b_4X_4 + \cdots, \tag{24.3}$$

where X_i = the normalized ln(abundance + 1) for each species sampled at the site. A first function is constructed that incorporates as much of the variance in the data as possible, and the process is repeated for additional functions with the remaining variance. Residual correlations after extraction of the first factor are used to produce a second, uncorrelated function that explains as much of the remaining variance as possible. The process is repeated to produce a series of functions. Ideally, most of the variance will be explained in the first few functions. A score for each sampling unit can be calculated for placement along each function. Plots for all sampling units using the formulated functions as axes should reveal an interpretable pattern. In this process, a matrix of many species abundances is reduced to a few sampling unit positions on a two- or three-dimensional plot. For example, the entire species abundance data set for a site might be reduced to one point in a two- or three-dimensional plot. The X, Y, and perhaps, Z dimensions are constructs that can be given physical meaning such as the influences of soil type (Function 1), heavy metal contamination (Function 2), and agricultural activity (Function 3) (Figure 24.5). Insight from additional information on soils, agricultural history, and soil metal concentrations might be used to interpret the distribution of the sampled plant communities along these three functions. The magnitude and signs of the b values (loading coefficients) in the linear functions are used to identify an underlying theme for each axis.

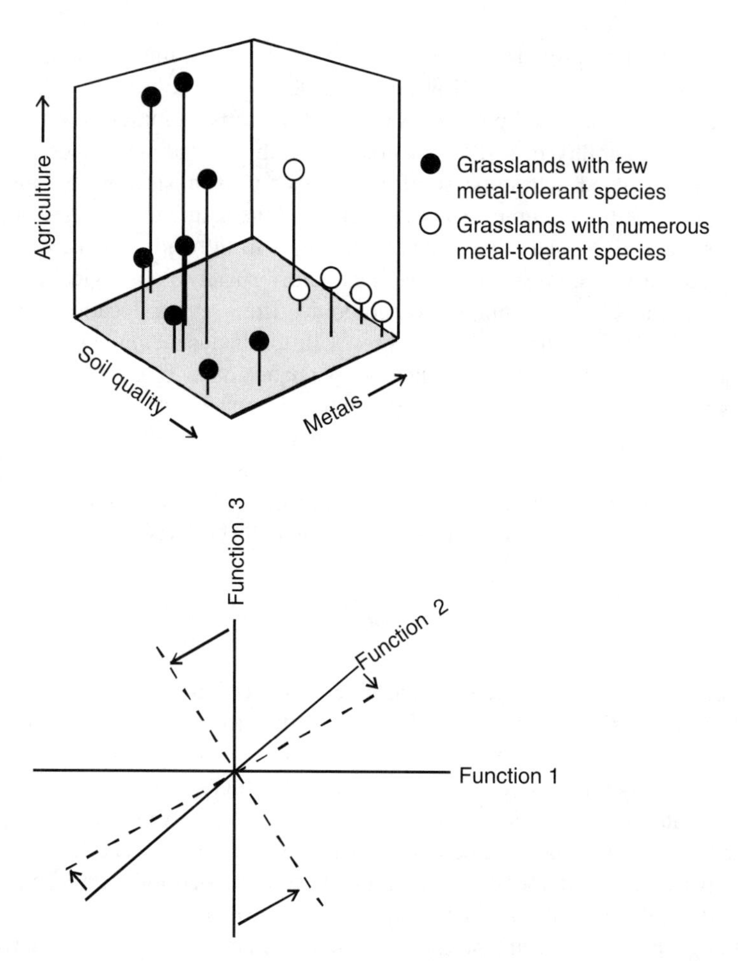

FIGURE 24.5 A hypothetical ordination analysis of plant communities relative to heavy metal contamination (top panel). Abundances of species are quantified at five sites near abandoned mines and another eight reference sites. Soil qualities and the history of agricultural use of the sites are also noted as potential confounding factors. After data transformation, ordination analysis results in three orthogonal, linear functions that are assigned interpretations of the influence of soil quality, soil metal concentrations, and agricultural history. The five mine sites clearly cluster away from the reference sites. There is a gradient of communities relative to soil quality and agricultural history. Ordination axes can be rotated to enhance interpretation using orthogonal and oblique methods (bottom panel).

These loadings represent the extent to which the variables are related to the hypothetical factor. For most factor extraction methods, these loadings may be thought of as correlations between the variables and the function (Comrey 1973). For example, very high loadings in Function 2 for species known to be tolerant to toxic metals and low or negative loadings for metal-sensitive species would suggest the influence of metal exposure on community composition. For Function 3, high loadings for species known to flourish in active agricultural areas might suggest the impact of active agriculture on community structure. The final result at this stage for ordination analyses would be to construct a table with rows of variables and associated loadings for each relevant factor (i.e., a table of unrotated factor loadings).

Several types of ordination methods exist (Boxes 24.1 and 24.2). PCA was the first, and remains the most popular method (Sparks 2000, Sparks et al. 1999). Using PCA, linear combinations of the original variables are extracted that sequentially account for the residual variance in a series of orthogonal (uncorrelated) components. The first component contains the most variance; the second

Box 24.1 Pollution's Signature on the Diversity of Estuarine Benthic Communities

To assess the influence of pollution on estuarine benthos, Diaz (1989) plotted species diversity on principal component axes generated from physical and chemical data for several James River (Virginia, USA) locations. Admittedly, one might object to this example because ordination is not being used directly to summarize community data. However, the study is a good illustration of applying two multivariate methods to interpret pollution effects on communities. The direct application of ordination to species abundance data will be described in Box 24.2 after illustrating key aspects of ordination analysis with this example.

The challenge faced by Diaz (1989) was to assess the influence of pollution on benthic communities relative to several other confounding variables. Stations were sampled at 5 nautical mile intervals from the fall line to within 10 miles of the river's mouth. Factors potentially influencing the benthic communities were measured, including sediment qualities, site-specific point discharges, and general water quality characteristics. Prior to ordination analysis, sites at salinity extremes were omitted to eliminate this obvious factor with a strong influence on community diversity.

Ordination analysis of physical and chemical data from James River sites was done after normalizing data with the formula

$$Z_{ij} = \frac{X_{ij} - M_j}{SD_j}, \tag{24.4}$$

where Z_{ij} = the standardized score of a datum for the jth variable of the ith site, X_{ij} = the datum for the jth variable for the ith site, and M_j and SD_j = the mean and standard deviation of the data for the jth variable, respectively. The normalized data were analyzed by principal components methods with no mention of any rotation of axes. Whether or not a rotation procedure would have produced more parsimonious principal components remains ambiguous.

Table 24.1 summarizes the PCA results. The percentage of total variance accounted for by each of the first five principal components is provided at the top of the table. Loadings (eigenvectors) for each chemical or physical factor are given for each principal component with

TABLE 24.1
Loadings (Eigenvectors) for Five Principal Components Derived by Diaz (1989) for James River Physical and Chemical Data

	Principal Component				
	1	2	3	4	5
Percentage of total variance (%)	36	22	15	12	8
Discharge biochemical oxygen demand	**0.33**	**0.37**	0.00	0.20	0.05
Discharge chemical oxygen demand	0.24	**0.46**	−0.07	0.20	0.16
Discharge coliform bacteria	**0.31**	−0.09	0.23	0.04	**−0.58**
Discharge total suspended solids	0.23	−0.19	−0.02	0.10	**0.73**
Ammonia concentration in water	0.13	0.02	−0.13	**−0.70**	0.03
Nitrite/nitrate concentration in water	−0.14	**0.49**	−0.26	−0.20	−0.03
Phosphate in water	**0.32**	**−0.30**	0.16	−0.10	−0.07
Suspended solids in water	−0.23	**0.33**	0.25	**−0.37**	0.00
Biochemical oxygen demand in water	**0.39**	0.14	−0.02	**−0.32**	0.00
Number of discharges	**0.32**	**0.31**	−0.01	0.28	−0.17
Percentage silt and clay	−0.19	−0.11	**−0.60**	0.13	−0.10
Cross-sectional area	**−0.36**	0.18	0.15	0.20	−0.02

Note: Boldface Indicates a variable with high loading.

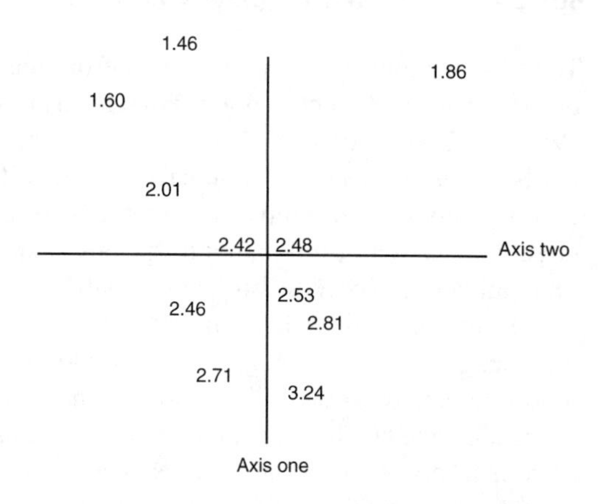

FIGURE 24.6 Ordination analysis (PCA) of physical and chemical qualities at sites along the James River (Virginia). Axes One and Two were interpreted as municipal waste discharge and industrial waste discharge, respectively. Numbers at each river site position on the plot are species diversities (H'). (Modified from Figure 7 of Diaz (1989).)

large eigenvectors in boldface. The large eigenvectors for specific variables in the first, fourth, and fifth principal components suggested to Diaz (1989) that these principal components reflected municipal waste discharges. Those variables with large eigenvectors in the second principal component suggested industrial discharges. The third principal component seemed to be related to physical characteristics of sediments.

The first two principal components were used as axes for plotting species diversity at the different sampling sites (Figure 24.6). Assuming the correct interpretation of the first principal component, an increase in municipal waste discharge was clearly associated with a decrease in species diversity (H'). The authors concluded from the plot that, "the greater the pollution load the lower the species diversity."

Box 24.2 Pesticide Spraying Changes Mesocosm Communities

Kedwards et al. (1999a,b) used ordination to study the impact of the pyrethroid pesticides, cypermethrin and lambda-cyhalothrin, on benthic communities established in 30-m^3 artificial ponds. Treatment involved duplicate mesocosms that were sprayed every 2 weeks for a total of four sprayings per mesocosm. Preapplication data were collected 5 weeks before the first spraying and sampling continued for 14 weeks after the final spraying occurred.

Redundancy analysis, an ordination technique, was applied to the results from cypermethrin-sprayed mesocosms (Figure 24.7). The two axes used in this figure accounted for 54% and 14% of the total variance in the data. Immediately after spraying began, the community in the treated mesocosms diverged from that of the controls, and each successive spraying moved the treated community further away. Several months after the last spraying, the communities remained quite divergent.

The authors interpreted the first two axes as being the influence of cypermethrin spraying (axis one) and the temporal changes in species abundances (axis two). The lines describing temporal changes in the reference mesocosms moved up and down along the second axis, but remained constant in its position relative to the first axis. The communities in the sprayed mesocosms changed with time and with spraying treatment. Spraying shifted community composition further to the right along the first axis, reflecting an increase in abundance of Chironomidae, Planorbidae, Hirudinea, and Lymnaeidea, and a decrease in Gammaridae and Asellidae.

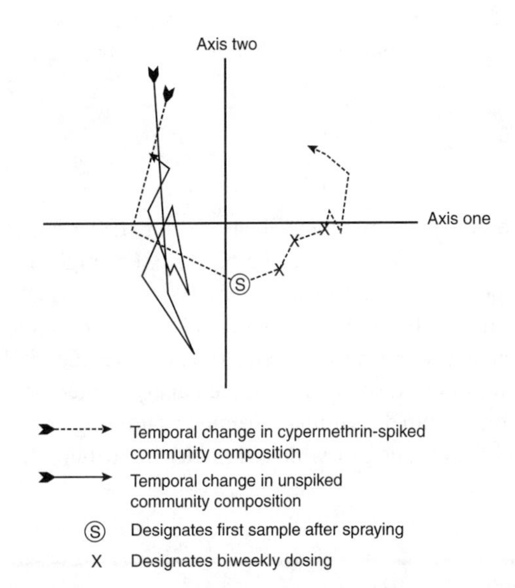

FIGURE 24.7 Ordination results for benthic invertebrate community composition for reference and cypermethrin-sprayed mesocosms. Community composition shifted abruptly along axis one at the sampling after spraying (denoted as S on diagram). Axis one and two were interpreted as the effect of spraying and the effect of time on community composition, respectively. (Modified from Figure 2 in Kedwards et al. (1999b).)

contains the most of the residual variance, and so forth. Ideally, the first few principal components account for most of the variance and the loadings allow sensible interpretations of these components. If this is not the case, some rotation method might be required.

Another general ordination method, factor analysis, is similar to PCA in that the variables are used to produce linear functions. Instead of being called principal components, these linear functions of the data are called factors. A factor is an unobservable variable that has attributes of a subset of the observed variables. In contrast to PCA in which components are calculated directly as linear functions of the observed variables, the observed variables in factor analysis are envisioned as linear functions of the factors (unobserved variables) plus random error (Sparks et al. 1999).

Numerous other ordination methods are available for applications with specific needs. Ordination can be done with discrete data using correspondence analysis or detrended correspondence analysis (Sparks et al. 1999). Discrete data might consist of presence/absence information or categorized species abundances such as rare, uncommon, common, abundant, or dominant. Although most multivariate ordination approaches employ traditional measures of community composition (e.g., abundance, presence/absence of species), other metrics may be necessary for groups where taxonomic issues limit our ability to identify species. Cao et al. (2006) used multivariate ordination to assess how bacterial community composition, as determined by phospholipid fatty acid and terminal restriction fragment length polymorphism analyses, responded to a mixture of contaminants. Nonmetric ordination methods exist (see Sparks 2000 for details) and have been used successfully to describe insect communities exposed to NEEM products (Kreutzweiser et al. 2000), Norwegian oilfield macrofauna (Clarke 1999), and benthic macroinvertebrates of the River Tees (Crane et al. 2002).

Methods for extracting functions aim to produce easily interpretable patterns. The mathematical functions or axes that are initially generated are uncorrelated or perpendicular. To enhance interpretation of these functions, some methods will rotate the axes at this stage of analysis based on some particular set of rules or criteria. Axes remain uncorrelated with orthogonal rotations but become correlated with oblique rotations. Many rotation methods are available for ordination; however, there is no formal statistical approach for determining which is best, and selection is usually based on user preferences. Among the most widely used rotation methods, the Kaiser Varimax produces orthogonal functions with as few variables with intermediate loadings as possible (Kaiser 1958, 1959, see also Comrey 1973). The concept is that a function with a few variables with very high or very low loadings will be more easily interpretable or parsimonious than one with many variables with intermediate loadings.

24.3.3 Discriminant and Cluster Analysis

Some multivariate methods, such as cluster and canonical discriminant analysis, explore differences or distances between sampling units. Groups for which differences are being assessed might be defined by the researcher (e.g., communities from polluted vs. clean sites), by design (e.g., treatment levels of copper added to a series of microcosms), or by statistical methods (e.g., community groupings identified by cluster analysis). Discriminant analysis aims to develop quantitative rules for separating groups or classes of sampling units. Similar to PCA, some discriminant analysis methods generate functions (canonical variates) that produce maximum discrimination among sampling units. Loading coefficients associated with the different variables suggest which variables contribute the most to the differences among sampling units (Box 24.3).

Box 24.3 Copper-Exposed Communities: What Separates Treatment Groups?

A series of triplicate 17-m^3 freshwater microcosms were spiked at 5 copper levels in an effort to define techniques for determining differences among toxicant-treated communities (Shaw and Manning 1996). *In situ* bioassays and species abundance data were collected, but only canonical discriminant analysis of macroinvertebrate species abundance data are presented here. Canonical variables, linear combinations of species abundance data that best distinguished among treatments, were produced for a series of times during the trial. Analysis for one sampling date during the spiking period (August 31, 1 month after spiking began and 19 days after the last spiking) is provided in Figure 24.8. The results show clear separation among treatments based on community composition. Surprisingly, species richness was not affected by copper spiking. However, abundances of annelids, crustaceans, mayflies, and chironomids did change. The mayfly *Caenis* was primarily responsible for separation among spiked treatments along the first canonical axis. (Importantly, *Caenis* bioassays in the spiked microcosms were also among the most useful for measuring effects of copper.) Orthocladiinae,

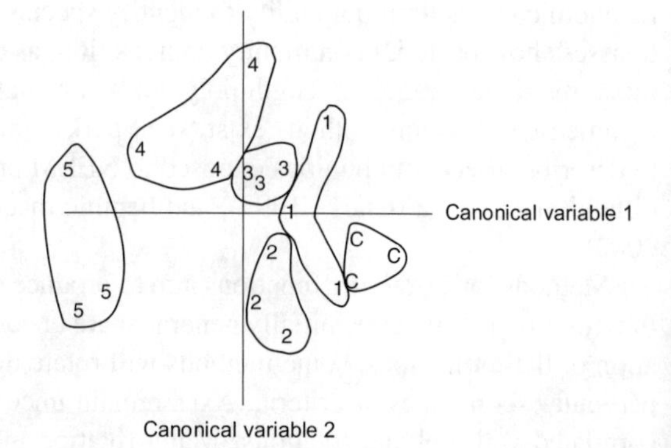

FIGURE 24.8 Separation of macroinvertebrate communities of microcosms receiving different copper treatments (spiked amounts being ranked as control < 1 < 2 < 3 < 4 < 5). Results are those obtained for canonical discriminant analysis of species abundance data for the August 31 sampling. The three observations plotted for each treatment are those for the triplicate microcosms. (Modified from Figure 8 of Shaw and Manning (1996).)

Chironominae, and Hydrozetes were also important. Only four taxa were needed to separate groups along copper treatments, suggesting that these species are useful indicators of metal pollution.

Cluster analysis also distinguishes among sampling units using multivariate data sets. As discussed in detail by Ludwig and Reynolds (1988) and Matthews et al. (1998), diverse metrics of resemblance or distance are applied to sampling units. Sampling units may be grouped in a hierarchical or nonhierarchical manner using a variety of algorithms. Hierarchical schemes produce tree-like structures (dendrograms) with branching points along groupings suggesting the degree of distinction or similarity among the groups on the various branches. Nonhierarchical methods simply place sampling units into groupings. Sparks et al. (1999) give the example of K-means clustering in which the number of groups is defined prior to analysis and the sampling units are sorted optimally into these groups. Using this method, differences are quantified as the square of the Euclidean distance (Matthews et al. 1998) and sampling units are distributed among the groups to produce maximum group separation.

Cluster analysis has many applications in community ecotoxicology. For example, Matthews et al. (1996) used nonmetric clustering (Matthews et al. 1995) to study microcosm community structural changes after turbine fuel exposure. The clustering methods revealed that differences among treated microcosms persisted for long periods of time, leading the authors to propose the community-conditioning hypothesis described in Chapter 25. In a field setting, Dauer et al. (1992) used cluster analysis to group benthic communities according to the influence of several physical and water quality characteristics (Box 24.4).

Box 24.4 Cluster Analysis Identifies Benthic Communities Affected by Anoxia

Physical and chemical qualities within estuaries greatly influence the composition of benthic communities. Dauer et al. (1992) explored Lower Chesapeake Bay (USA) benthic communities in an attempt to quantify the influence of such factors on community structure. Emphasis was placed on identifying communities modified by episodes of anoxia. Benthic species are subjected to anoxia when water produced during seasonal stratification is moved onto nearby shallows by wind-driven seiches. The extent and effect of anoxia are of concern because of potential exacerbation by increased nutrient influx from human activities.

Twenty-one samples were taken along the Lower Chesapeake Bay and in several tributaries. Water quality data, including oxygen concentrations, were available for interpreting benthic species abundance information. Site selection intentionally included those along salinity gradients, those with different sediment types, and those that experienced episodic anoxia. Cluster analysis was done using logarithm-transformed species abundance data and the Bray-Curtis similarity coefficient.

Cluster analysis identified groupings that were easily interpreted based on salinity, sediment type, and dissolved oxygen concentration (Figure 24.9). For explanatory convenience, six clusters are identified in Figure 24.9. There was a clear clustering of sites relative to salinity: freshwater (Cluster 6), transitional (Cluster 5), mesohaline (Cluster 4), and polyhaline (Clusters 2 and 3) sites. Within the polyhaline grouping, the communities split again into those associated with sandy (Cluster 2) and muddy (Cluster 3) substrates. Sites experiencing anoxia (four sites in Cluster 1) were set apart from the other sites (17 sites in Clusters 2 through 6) at a relatively high level (e.g., similarity of approximately 0.9). Relative to the other communities sampled, those experiencing periodic anoxia had lower species diversity, lower biomass, and

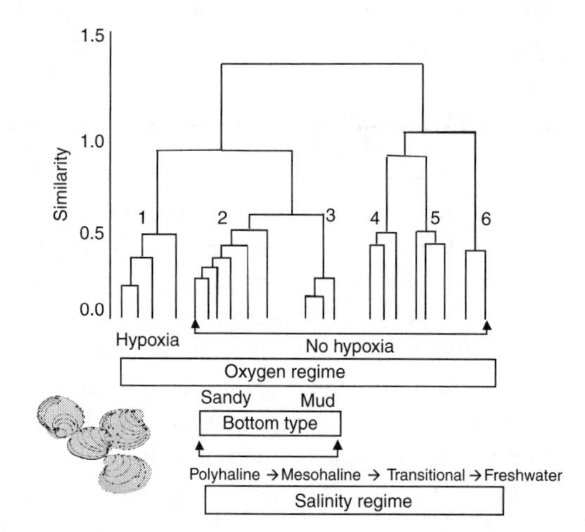

FIGURE 24.9 Clustering of 21 benthic macroinvertebrate communities based on Bray-Curtis similarity coefficient. The clustering was interpreted by applying knowledge of salinity, oxygen, and sediment conditions. (Modified from Figure 2 of Dauer et al. (1992).)

less biomass deeper than 5 cm in the sediments. Results also showed that dominant species tended to be opportunistic, with equilibrium species being less common than in the other communities.

24.3.4 Application of Multivariate Methods to Laboratory Data

With minor exceptions, most of the multivariate methods described to this point draw from species enumerations in order to describe community-level responses. However, other multivariate methods use results of single species toxicity tests to predict effects on communities. Box 24.5 describes an example that uses laboratory toxicity data for sediment and water to make predictions about community status.

Box 24.5 A Risk Ranking Model Based on Estuarine Fish Communities

The Maryland Department of Natural Resources (USA) developed a composite index (risk ranking) for Chesapeake Bay tributaries (Hartwell 1997, also see Hartwell et al. 1997) using laboratory toxicity tests of water and sediments from sites of interest. The intent was to initially "quantify the toxicological risk to populations due to the presence of toxic contamination ..." using ambient toxicity data. (See Newman (1998, 2001) for discussion of the problems in predicting population consequences based on these types of severity judgments.)

Four estuaries were selected to estimate a fish community-based IBI, fish species diversity, and this new ranking model. The ranking model employed water and sediment test results to quantify region status. On several dates, water samples from each site were collected for ambient toxicity tests, including sheepshead minnow (*Cyprinodon variegatus*) growth and survival, grass shrimp (*Palaemonetes pugio*) growth and survival, and copepod (*Eurytemora affinis*) reproduction and survival. Similarly, sediment toxicity tests were done including those quantifying sheepshead minnow embryo-larval survival and teratogencity, amphipod (*Leptocheirus plumulosus*) reburial, growth, and survival, and polychaete (*Streblospio benedicti*) survival and growth.

This ranking system of risk was influenced by a high hazard score for a particular measure and the uncertainty associated with producing a score for a region. The level of uncertainty influenced the score and the measured level of hazard. The severity of effect (mortality = 3, impaired reproduction = 2, impaired growth = 1), degree of response, variability in testing, site consistency, and the number of endpoints were components of the risk ranking model. The degree of response was the proportional difference from the control. The variability was expressed as the coefficient of variation (CV) for a particular metric for each set of laboratory replicates and each sample site during a particular sampling period. The last part of the ranking model involved consistency, or the level of agreement among assays for a site. Consistency was quantified as the cube root of the difference between half of the number of tests ($N/2$) and the number of statistically nonsignificant responses at each site (X):

$$\text{Consistency} = \sqrt[3]{N/2 - X}. \tag{24.5}$$

The consistency is then divided by the number of endpoints measured for a site. The site score is estimated with the following equation:

$$\text{Location score} = \frac{\left[\sum (\text{severity} \times \text{response} \times \text{CV})\right] + \text{consistency}}{\sqrt{N}}. \tag{24.6}$$

Scores were calculated for the four sites based on tests of water alone, sediment alone, or water and sediment combined. Pearson correlation coefficients were calculated for these scores versus a fish-based IBI, a benthic species diversity index, and a resident fish diversity index. There were no significant correlations ($\alpha = 0.05$) between the risk scores (water, sediment, or water sediment combined) and the IBI scores or species diversity based on all resident fish species. Similarly, no significant correlation was noted between water testing-based risk scores and bottom fish species diversity. However, there were significant correlations between bottom fish species diversity and the sediment test-based risk score ($P = .0092$) and the combined test risk score ($P = .0018$). These results suggest that scores for this risk index are related to bottom fish diversity. Notionally, the relationship involved responses to site-associated toxicant exposures.

The methods described to this point have involved data collected from potentially impacted sites in an attempt to document community changes. However, species sensitivity distribution (SSD) methods use mostly laboratory data to predict potential community changes on exposure to stressors. The approach extends the common use of one laboratory measure of effect, such as the 96-h LC50, to predict impact to an exposed community. Conventional prediction from one species can be made more credible by making predictions of effect based on information from the most sensitive test species. The SSD method modifies these laboratory-based approaches by using all available laboratory data to make predictions of effect concentrations for the ecological community. Its greatest advantage is that it uses all of the readily available information to predict community consequences. Its convenience and efficient use of single species data have led to a very rapid increase in its use (Newman 1995).

To apply the SSD method, effect concentrations such as acute LC50 or no-observed effect concentration (NOEC) values are collected for all relevant species. The effect concentration observations are ordered from the smallest to the largest value (e.g., smallest to largest 96-h LC50 values). The ordered values are then given a rank using one of several conventional methods. Currently popular is $i/(n+1)$, where i = the ith ranked observation and n = the total number of observations. A slightly better, but less commonly applied, approximation of rank for ordered observations is $(i - 0.5)/n$. At this point, the data set consists of a series of observations (e.g., 96-h LC50 observations and their corresponding ranks). A log normal model is often assumed and the probit transformation of each rank is taken. Another model and transformation can be used if there is evidence that the log normal

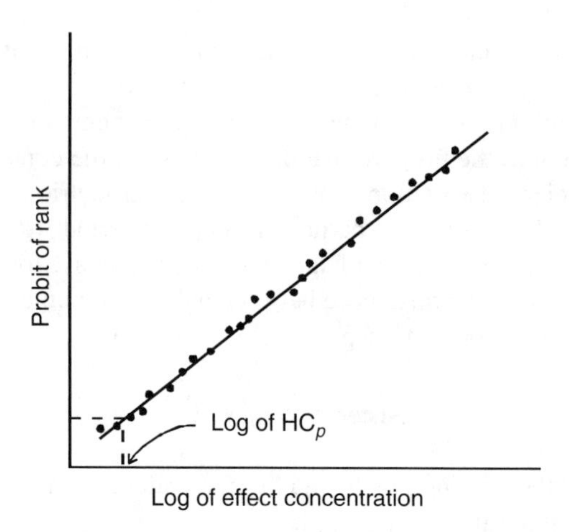

FIGURE 24.10 Log normal model for estimating the HC*p* using the SSD method. Transformations are easily done on effect concentrations and effect proportion in order to linearize species sensitivity data. The log of the effect concentration is plotted against the probit of the effect proportion for the log normal model assumed here.

model is inappropriate. Newman et al. (2000, 2001) indicate that the general assumption of a log normal model is often not appropriate. Regardless, a log normal model will be assumed here to illustrate the SSD method. A plot of logarithm of effect concentration versus probit of the rank is made, producing a straight line (Figure 24.10) if the log normal assumption is appropriate. A regression model is then used to estimate the concentration "protecting" all but a specified percentage ($p\%$) of species in the community. This concentration is often called the hazard concentration or HC_p.

Although the SSD approach enjoys increasingly widespread application (Posthuma et al. 2001), it does involve several unresolved shortcomings or ambiguities (Newman 2001, Newman et al. 2000). First, EC50, LC50, NOEC, lowest-observed effect concentration (LOEC), and maximum allowable toxicant concentration (MATC) effects metrics are used to generate models but they have significant deficiencies as predictors of population persistence in natural communities. Any HC_p derived using these effects metrics will consequently have deficiencies as a predictor of community consequences. Second, the selection of a specified p implies that some loss of species is acceptable for any community because of species redundancy. As will be described in Chapter 25, the extent to which this redundancy hypothesis can be validly applied is still hotly debated. Therefore, any predictions based on the redundancy hypothesis must be viewed as nonconservative predictions at this time. Third, application of the SSD method requires thorough knowledge of the dominant and keystone species, and the importance of species interactions. It has been our experience that this knowledge is often not available in studies applying the SSD method. Fourth, *in situ* exposure is more complex and species-dependent rather than reflected in the laboratory exposures done in toxicity testing. Fifth, there is a bias toward lethality information, although nonlethal effects can result in species disappearance from a community. Finally, the assumption of a specific model, such as the log normal model, is often made without careful scrutiny (Jagoe and Newman 1997, Newman et al. 2000, 2001).

24.3.5 Taxonomic Aggregation in Multivariate Analyses

Our previous discussion in Chapter 22 concerning how taxonomic aggregation and the exclusion of rare taxa influence our ability to distinguish reference and contaminated sites is also relevant to multivariate analyses. Ordination approaches are typically based on responses of individual species to environmental gradients. In fact, the argument frequently used to support these techniques is that multivariate approaches allow researchers to quantify the response of an entire community. However, depending on the degree of interspecific variation in sensitivity, there is likely some degree of redundancy in the responses of individual taxa to contamination. From a practical perspective, the complex taxonomy of some groups severely limits species-level identification. Caruso and Migliorini (2006) showed that multivariate analyses based on either genus- or family-level identification could detect effects of heavy metals on soil invertebrate communities. While these results are certainly encouraging, it is important to note that the loss of information associated with taxonomic aggregation may vary among groups. For example, Hirst (2006) showed that family-level identification was sufficient to identify multivariate patterns in marine invertebrates, but taxonomic aggregation of macroalgae resulted in a significant reduction in information.

24.4 SUMMARY

In summary, a diverse array of analytical approaches allows for the description of toxicant effects on communities. Some, such as species diversity indices, reduce abundance data to a single number while others, such as the IBI, apply considerable ecological knowledge to generate ad hoc measures of community integrity. Others, like the SSD approach, attempt to use available laboratory data to produce gross predictions of possible community-level effects. Finally, multivariate procedures are devoid of ecological theory and simply identify correlations or associations within a data set. All of these approaches can be extremely useful for detecting community differences or changes if applied insightfully.

24.4.1 SUMMARY OF FOUNDATION CONCEPTS AND PARADIGMS

- Methods to assess the effects of contaminants on communities range from computationally simple indices such as species richness to complex, computer-dependent algorithms such as multivariate analyses.
- The simplest community indices use species presence/absence or abundance data to show how individuals in the community are distributed among species.
- Computationally intense methods, such as multivariate analyses, aim to reduce the number of data dimensions to an interpretable low number, and to quantify similarities or differences among sampling units.
- One of the most significant advances in the field of biological assessments over the past 20 years was the development and application of multimetric approaches for measuring ecological integrity.
- The individual metrics in a multimetric index reflect different characteristics of life history, community structure, and functional organization that are integrated into a single measure.
- Karr's (1981) IBI is the most widely used multimetric index for assessing the health of aquatic communities.
- Similarity indices reduce complex, multispecies data and quantify correspondence between two communities based on either presence–absence or abundance.
- In contrast to multimetric indices, multivariate analyses are not based on ecological concepts but are statistical constructs that reduce complex data sets to illustrate potentially meaningful patterns involving a few variables.
- Multivariate data sets are broadly defined as those in which more than two dependent or independent variables are collected for each sampling unit.
- Ordination is a process in which a large set of variables is reduced to a few variables with the intent of enhancing conceptual parsimony and tractability.
- In PCA, linear combinations of the original variables are extracted to sequentially account for the residual variance in a series of orthogonal (uncorrelated) components.
- Nonmetric ordination methods have been used successfully to describe macroinvertebrate responses to a variety of contaminants.

- Some multivariate methods, such as cluster and canonical discriminant analysis, explore differences or distances between sampling units.
- Discriminant analysis develops quantitative rules for separating groups or classes of sampling units either defined by the researcher (e.g., communities from polluted vs. clean sites), by experimental design (e.g., treatment levels of copper added to a series of microcosms), or by statistical methods (e.g., community groupings identified by cluster analysis).
- Cluster analysis distinguishes among sampling units using multivariate data sets grouped in a hierarchical or nonhierarchical manner using a variety of algorithms.
- Despite their growing popularity, multivariate approaches have been criticized because of their inherent statistical complexity and because results are often difficult to interpret.
- Although strict reliance on complex statistical algorithms may obscure important biological results, multivariate approaches are an essential set of tools for assessments of water quality.
- Multivariate and multimetric approaches are complementary and should be used in conjunction. Variables used in multivariate analyses could include species richness, abundance of sensitive groups, or other measures typically included in a multimetric index. Alternatively, a multimetric index similar to Karr's IBI could be developed using results of multivariate analyses.

REFERENCES

Bailey, R.C., Kennedy, M.G., Dervish, M.Z., and Taylor, R.M., Biological assessment of freshwater ecosystems using a reference condition approach: Comparing predicted and actual benthic invertebrate communities in Yukon streams, *Freshw. Biol.*, 39, 765–774, 1998.

Blair, R.B., Birds and butterflies along an urban gradient: Surrogate taxa for assessing biodiversity? *Ecol. Appl.*, 9, 164–170, 1999.

Boyle, T.P., Smillie, G.M., Anderson, J.C., and Beeson, D.R., A sensitivity analysis of nine diversity and seven similarity indices, *Water Poll. Con. Fed.*, 62, 749–762, 1990.

Cao, Y., Cherr, G.N., Cordova-Kreylos, A.L., Fan, T.W.M., Green, P.G., Higashi, R.M., Lamontagne, M.G., et al., Relationships between sediment microbial communities and pollutants in two California salt marshes, *Microb. Ecol.*, 52, 619–633, 2006.

Caruso, T., and Migliorini, M., Micro-arthropod communities under human disturbance: Is taxonomic aggregation a valuable tool for detecting multivariate change? Evidence from Mediterranean soil oribatid coenoses, *Acta Oecol.-Int. J. Ecol.*, 30, 46–53, 2006.

Chambers, C.L., McComb, W.C., and Tappeiner, J.C., I., Breeding bird responses to three silvicultural treatments in the Oregon Coast Range, *Ecol. Appl.*, 9, 171–185, 1999.

Chenery, A.M. and Mudge, S.M., Detecting anthropogenic stress in an ecosystem: 3. Mesoscales variability and biotic indices, *Environ. Foren.*, 6, 371–384, 2005.

Clarke, K.R., Nonmetric multivariate analysis in community-level ecotoxicology, *Environ. Toxicol. Chem.*, 18, 118–127, 1999.

Clements, W.H. and Kiffney, P.M., The influence of elevation on benthic community responses to heavy metals in Rocky Mountain streams, *Can. J. Fish. Aquat. Sci.*, 52, 1966–1977, 1995.

Clements, W.H., Cherry, D.S., and Cairns, J., Jr., The impact of heavy metals on macroinvertebrate communities: A comparison of observational and experimental results, *Can. J. Fish. Aquat. Sci.*, 45, 2017–2025, 1988.

Comrey, A.L., *A Course in Factor Analysis*, Academic Press, New York, 1973.

Crane, M., Sorokin, N., Wheeler, J., Grosso, A., Whitehouse, P., and Morritt, D., European approaches to coastal and estuarine risk assessment, In *Coastal and Estuarine Risk Assessment*, Newman, M.C., Roberts, M.H. Jr., and Hale, R.C. (eds.), CRC Press, Boca Raton, FL, 2002, pp. 15–39.

Dauer, D.M., Rodi, A.J., Jr., and Ranasinghe, J.A., Effects of low dissolved oxygen events on the macrobenthos of the Lower Chesapeake Bay, *Estuaries*, 15, 384–391, 1992.

Diaz, R.J., Pollution and tidal benthic communities of the James River Estuary, Virginia, *Hydrobiologia*, 180, 195–211, 1989.

Fausch, K.D., Lyons, J., Karr, J.R., and Angermeier, P.L., Fish communities as indicators of environmental degradation, *Am. Fish. Soc. Symp.*, 8, 123–144, 1990.

Fore, L.S., Karr, J.R., and Wisseman, R.W., Assessing invertebrate responses to human activities: Evaluating alternative approaches, *J. N. Am. Benthol. Soc.*, 15, 212–231, 1996.

Gauch, H.G., Jr., *Multivariate Analysis in Community Ecology*, Cambridge University Press, Cambridge, England, 1982.

Gerritsen, J., Additive biological indices for resource management, *J. N. Am. Benthol. Soc.*, 14, 451–457, 1995.

Griffith, M.B., Husby, P., Hall, R.K., Kaufmann, P.R., and Hill, B.H., Analysis of macroinvertebrate assemblages in relation to environmental gradients among lotic habitats of California's Central Valley, *Environ. Monitor. Assess.*, 82, 281–309, 2003.

Hartwell, S.I., Demonstration of a toxicological risk ranking method to correlate measures of ambient toxicity and fish community diversity, *Environ. Toxicol. Chem.*, 16, 361–371, 1997.

Hartwell, S.I., Dawson, C.E., Durell, E.Q., Alden, R.W., Adolphson, P.C., Wright, D.A., Coehlo, G.M., Magee, J.A., Ailstock, S., and Norman, M., Correlation of measures of ambient toxicity and fish community diversity in Chesapeake Bay, USA, tributaries—urbanizing watersheds, *Environ. Toxicol. Chem.*, 16, 2556–2567, 1997.

Hirst, A.J., Influence of taxonomic resolution on multivariate analyses of arthropod and macroalgal reef assemblages, *Mar. Ecol. Prog. Ser.*, 324, 83–93, 2006.

Jagoe, R.H. and Newman, M.C., Bootstrap estimation of community NOEC values, *Ecotoxicology*, 6, 293–306, 1997.

Kaiser, H.F., The Varimax criterion for analytic rotation in factor analysis, *Psychometrika*, 31, 313–323, 1958.

Kaiser, H.F., Computer program for Varimax rotation in factor analysis, *Ed. Psych. Meas.*, 19, 413–420, 1959.

Karr, J.R., Assessment of biological integrity using fish communities, *Fisheries*, 6, 21–27, 1981.

Karr, J.R., Defining and assessing ecological integrity: Beyond water quality, *Environ. Toxicol. Chem.*, 12, 1521–1531, 1993.

Kedwards, T.J., Maund, S.J., and Chapman, P.F., Community level analysis of ecotoxicological field studies: I. Biological monitoring, *Environ. Toxicol. Chem.*, 18, 149–157, 1999a.

Kedwards, T.J., Maund, S.J., and Chapman, P.F., Community level analysis of ecotoxicological field studies: II. Replicated-design studies, *Environ. Toxicol. Chem.*, 18, 158–166, 1999b.

Kerans, B.L. and Karr, J.R., A benthic index of biotic integrity (B-IBI) for rivers of the Tennessee Valley, *Ecol. Appl.*, 4, 768–785, 1994.

Kilgour, B.W., Somers, K.M., and Barton, D.R., A comparison of the sensitivity of stream benthic community indices to effects associated with mines, pulp and paper mills, and urbanization, *Environ. Toxicol. Chem.*, 23, 212–221, 2004.

Kremen, C., Assessing the indicator properties of species assemblages for natural areas monitoring, *Ecol. Appl.*, 2, 203–217, 1992.

Kremen, C., Colwell, R.K., Erwin, T.L., Murphy, D.D., Noss, R.F., and Sanjayan, M.A., Terrestrial arthropod assemblages: Their use in conservation and planning, *Conserv. Biol.*, 7, 796–808, 1993.

Kreutzweiser, D.P., Capell, S.S., and Scarr, T.A., Community-level responses by stream insects to NEEM products containing azadirachtin, *Environ. Toxicol. Chem.*, 19, 855–861, 2000.

Lewis, C.S., *The Screwtape Letters*, HarperCollins Publishers, Inc., New York, 1942.

Ludwig, J.A., and Reynolds, J.F., *Statistical Ecology. A Primer on Methods and Computing*, John Wiley & Sons, New York, 1988.

Magurran, A.E., *Ecological Diversity and Its Management*, Princeton University Press, Princeton, NJ, 1988.

Matthews, G.M., Matthew, R., and Landis, W., Nonmetric conceptual clustering in ecology and ecotoxicology, *AI Appl.*, 9, 41–48, 1995.

Matthew, R.A., Matthew, G.M., and Landis, W.G., Application of community level toxicity testing to environmental risk assessment, In *Risk Assessment. Logic and Measurement*, Newman, M.C. and Strojan, C.L. (eds.), CRC Press, Boca Raton, FL, 1998, pp. 225–253.

Matthews, R.A., Landis, W.G., and Matthews, G.B., The community conditioning hypothesis and its application to environmental toxicology, *Environ. Toxicol. Chem.*, 15, 597–603, 1996.

Miller, D.I., Leonard, M., Hughes, R.M., Karr, J.R., Moyle, P.B., Schrader, L.H., Thompson, B.A., et al., Regional applications of an index of biotic integrity for use in water resource management, *Fisheries*, 13, 12–20, 1988.

Nelson, S.M., and Epstein, M.E., Butterflies (Lepidoptera: Papilionoidea and Hesperioidea) of Roxborough State Park, Colorado, USA: Baseline inventory, community attribute, and monitoring plan, *Environ. Manag.*, 22, 287–295, 1998.

Newman, M.C., *Quantitative Methods in Aquatic Ecotoxicology*, CRC Press, Boca Raton, FL, 1995.

Newman, M.C., *Fundamentals of Ecotoxicology*, CRC Press, Boca Raton, FL, 1998.

Newman, M.C., *Population Ecotoxicology*, John Wiley & Sons, Chichester, UK, 2001.

Newman, M.C., Ownby, D.R., Mezin, L.C.A., Powell, D.C., Christensen, T.R.L., Lerberg, S.B., and Anderson, B.A., Applying species sensitivity distributions in ecological risk assessment: Assumptions of distribution type and sufficient numbers of species, *Environ. Toxicol. Chem.*, 19, 508–515, 2000.

Newman, M.C., Ownby, D.R., MJzin, L.C.A., Powell, D.C., Christensen, T.R.L., Lerberg, S.B., Anderson, B.A., and Padma, T.V., Species sensitivity distributions in ecological risk assessment: Distributional assumptions, alternate bootstrap techniques, and estimation of adequate number of species, In *Species Sensitivity Distributions in Ecotoxicology*, Posthuma, L., Sutter, G.W., II, and Traas, T.P. (eds.), CRC Press, Boca Raton, FL, 2001, pp. 119–132.

Pardal, M.A., Cardoso, P.G., Sousa, J.P., Marques, J.C., and Raffaelli, D., Assessing environmental quality: A novel approach, *Mar. Ecol. Prog. Ser.*, 267, 1–8, 2004.

Philippi, T.E., Dixon, P.M., and Taylor, B.E., Detecting trends in species composition, *Ecol. Appl.*, 8, 300–308, 1998.

Plafkin, J.L., Barbour, M.T., Porter, K.D., Gross, S.K., and Hughes, R.M., *Rapid Bioassessment Protocols for Use in Streams and Rivers: Benthic Macroinvertebrates and Fish. EPA 440/4-89-001*, Washington, D.C., U.S. EPA, 1989.

Posthuma, L., Suter, G.W., II, and Traas, T.P., *Species Sensitivity Distributions in Ecotoxicology*, CRC Press, Boca Raton, FL, 2001.

Quintino, V., Elliott, M., and Rodrigues, A.M., The derivation, performance and role of univariate and multivariate indicators of benthic change: Case studies at differing spatial scales, *J. Exp. Mar. Biol. Ecol.*, 330, 368–382, 2006.

Resh, V.H., Norris, R.H., and Barbour, M.T., Design and implementation of rapid assessment approaches for water resource monitoring using benthic macroinvertebrates, *Aust. J. Ecol.*, 20, 108–121, 1995.

Reynoldson, T.B., Norris, R.H., Resh, V.H., Day, K.E., and Rosenberg, D.M., The reference condition: A comparison of multimetric and multivariate approaches to assess water-quality impairment using benthic macroinvertebrates, *J. N. Am. Benthol. Soc.*, 16, 833–852, 1997.

Shaw, J.L. and Manning, J.P., Evaluating macroinvertebrate population and community level effects in outdoor microcosms: Use of in situ bioassays and multivariate analysis, *Environ. Toxicol. Chem.*, 15, 608–617, 1996.

Simon, T.P. and Lyons, L., Application of the index of biotic integrity to evaluate water resource integrity in freshwater ecosystems, In *Biological assessment and criteria: Tools for water resource planning and decision making*, Davis, W.S. and Simon, T.P. (eds.), Lewis Publishers, Boca Raton, FL, 1995, pp. 245–262.

Sparks, T.H., *Statistics in Ecotoxicology*, John Wiley & Sons, Chichester, UK, 2000.

Sparks, T.H., Scott, W.A., and Clarke, R.T., Traditional multivariate techniques: Potential for use in ecotoxicology, *Environ. Toxicol. Chem.*, 18, 128–137, 1999.

Suter, G.W., II, A critique of ecosystem health concepts and indexes, *Environ. Toxicol. Chem.*, 12, 1533–1539, 1993.

Terlizzi, A., Scuderi, D., Fraschetti, S., and Anderson, M.J., Quantifying effects of pollution on biodiversity: A case study of highly diverse molluscan assemblages in the Mediterranean, *Mar. Biol.*, 148, 293–305, 2005.

Thomas, J.F., and Hall, T.J., A comparison of three methods of evaluating aquatic community impairment in streams, *J. Freshw. Ecol.*, 21, 53–63, 2006.

Walls, L.D., *Material Faith: Henry David Thoreau on Science*, Houghton Mifflin Co., Boston, MA, 1999.

Weiss, J.M. and Reice, S.R., The aggregation of impacts: Using species-specific effects to infer community-level disturbances, *Ecol. Appl.*, 15, 599–617, 2005.

Wolda, H., Similarity indices, sample size, and diversity, *Oecologia*, 50, 296–302, 1981.

25 Disturbance Ecology and the Responses of Communities to Contaminants

It is one of those refreshing simplifications that natural systems, despite their diversity, respond to stress in very similar ways.

(Rapport et al. 1998)

25.1 THE IMPORTANCE OF DISTURBANCE IN STRUCTURING COMMUNITIES

In this chapter, we will compare the ways in which communities respond to natural and anthropogenic disturbances. We suggest that many of the characteristics that determine resistance and resilience of communities to natural disturbance may also influence responses to chemical stressors. For the purposes of this discussion, disturbance is defined as any relatively discrete event that disrupts ecosystem, community, or population structure and changes resources, substrate availability, or the physical environment (White and Pickett 1985). Key features that determine the impact of disturbance on communities are the magnitude (e.g., how far the disturbance is outside the range of natural variability), frequency, and duration. Some ecologists define disturbance as any event that results in the removal of organisms and creates space. Indeed, some ecology textbooks (e.g., Begon et al. 1990) combine discussion of disturbance and predation in the same chapter because they ultimately have similar effects on communities: the removal of organisms from a community. The impact of a predator on a competitively superior species will have a qualitatively similar influence on community structure as the creation of space by physical disturbance. However, most community ecologists limit the definition of disturbance to include only events that are outside the range of natural variability. In other words, the predictability or novelty of a disturbance event greatly influences community responses and recovery times. Predictability of disturbance is largely influenced by the frequency of occurrence, but also varies among ecosystems and disturbance types (Table 25.1). Johnston and Keough (2005) conducted one of the few field experiments that compared the relative importance of frequency and intensity of contaminant exposure on communities. Interestingly, the influence of disturbance frequency and intensity varied among locations and was largely determined by recovery rates of competitively superior species.

Ecologists have long recognized the importance of natural disturbance in structuring communities (Connell 1978), and many consider disturbance a central organizing principle in community ecology (Peterson 1975, Sousa 1979, White and Pickett 1985). In particular, the biotic and abiotic factors that influence recovery from disturbance have received considerable attention. A large body of theoretical and empirical evidence supports the idea that most communities are subjected to natural disturbance and that disturbance regimes influence community structure and life history characteristics of

TABLE 25.1
Frequency and Predictability of Natural Disturbance Events in Ecosystems

Ecosystem	Disturbance Type	Frequency (Years)	Predictability
Forests	Fire	1/40–200	Moderate
	Windstorms	1/10–25	None
	Insect defoliation	Rare	None
Chaparral	Fire	1/15–25	High
Grasslands	Fire	1/5–10	Moderate
Deserts	Frost	1/50–200	None
Rivers	Floods	0–15	None
	Drought	0–2	Moderate to high
Lakes	Freezing	0–1	High
Intertidal zone	Log damage	Annual	Low

Source: Modified from Reice (1994).

component species. Most of this research has focused on physical perturbations (e.g., hurricanes, floods, volcanoes), whereas relatively few studies have employed basic ecological principles to describe responses to anthropogenic stressors. Just as variability and predictability determine the response of communities to natural disturbance, they also figure prominently in understanding the effects of anthropogenic disturbance (Rapport et al. 1985). The goal of this chapter is to describe ways in which ecotoxicologists can use this rich history of research in basic disturbance ecology to understand community responses to contaminants.

25.1.1 Disturbance and Equilibrium Communities

Much of the historical focus in disturbance ecology is closely aligned with the Clementsian paradigm of community succession and the "balance of nature" (Clements 1936). The equilibrium model of community structure asserts that overall community composition is relatively stable and that communities will return to equilibrium conditions if given sufficient time following a disturbance. The equilibrium model also assumes that species interactions, most notably competition, are the most important factors structuring the community. The idea that communities will return to predisturbance condition following perturbations implicitly assumes the existence of equilibrium conditions. The equilibrium model is in stark contrast to the idea that community structure is determined largely by stochastic processes, such as random colonization and highly variable environmental factors (Table 25.2). Proponents of the nonequilibrium theory assert that community composition is constantly changing over time and that natural systems are often recovering from the most recent disturbance (Reice 1994, Wiens 1984). Communities only give the illusion of stability if the frequency of disturbance is relatively low.

The debate over equilibrium and nonequilibrium determinants of community structure has important implications for the study of recovery from anthropogenic disturbance. If communities are determined largely by stochastic processes and therefore are constantly changing, then defining recovery as a return to predisturbance conditions will be difficult. In contrast, if communities are characterized by equilibrium conditions, then predictable recovery trajectories can be identified. Long-term investigations of predisturbance conditions may help define the range of natural variation in nonequilibrium communities. However, if communities show the degree of temporal variation expected on the basis of nonequilibrium models, it will possible to detect only the most severe disturbances.

TABLE 25.2
Characteristics of Equilibrium and Nonequilibrium Communities

	Equilibrium Communities	Non-Equilibrium Communities
Biotic interactions	Strong, especially competition	Weak
Number of species	Many	Few
Abiotic factors	Less important	Major importance
Community regulation	Density dependent	Density independent
Overall structure	Deterministic	Stochastic

Source: From Wiens, J.A., In *Ecological Communities: Conceptual Issues and the Evidence*, Strong, D.R., Simberloff, D., Abele, L.G., and Thistle, A.B. (eds.), Princeton University Press, Princeton, NJ, 1984, pp. 439–457.

25.1.2 Resistance and Resilience Stability

Ecologists recognize two different types of community stability when quantifying community responses to disturbance. Resistance stability refers to the ability of a community to maintain equilibrium conditions following a disturbance. Resistance can be quantified by measuring the magnitude of the response of a community compared to predisturbance conditions. If two communities are subjected to the same disturbance, the community that shows the least amount of change compared to predisturbance conditions has greater resistance. Resilience stability refers to the rate at which a community will return to predisturbance conditions. If two communities are exposed to the same disturbance, the community that recovers faster is considered to have greater resilience. Because resistance and resilience are fundamental properties of all ecological systems, some ecologists have proposed that they could be employed as indicators of ecological health (Box 25.1).

Box 25.1 Resistance and Resilience as "Fitness Tests" of Ecosystem Health

Measures of species richness, diversity, and ecosystem processes are routinely employed in biological monitoring to assess effects of anthropogenic stressors. The ability of a community to withstand and recover from natural disturbance is also recognized as a fundamental characteristic of ecological integrity. If exposure to contaminants or other anthropogenic stressors influences resilience or resistance of a community, responses to natural disturbance may be used as endpoints in ecological assessments. Whitford et al. (1999) measured resistance and resilience of a grassland community to a natural disturbance (drought) along a stress gradient induced by livestock grazing. Both resistance and resilience were compromised by grazing, suggesting that natural disturbance will have a greater and longer lasting effect on communities also subjected to anthropogenic disturbance. Whitford et al. (1999) proposed using measures of resistance and resilience as early warning "fitness tests" of ecosystem health. The strength of this approach is that it measures something that really matters (ability to withstand or recover from disturbance) and can be applied across different types of communities. Assuming that effects of natural disturbance in reference and impacted communities can be quantified, this approach provides a unique opportunity for comparisons among communities.

Resistance and resilience to disturbance are not necessarily correlated. Features that determine tolerance of a community to a stressor (resistance) do not always influence how quickly the community will recover (resilience). For example, a climax forest may show high resistance to outbreaks of an herbivorous pest (e.g., gypsy moths); however, resilience will be very low because of the time required for this community to return to predisturbance conditions. In contrast, grassland communities subjected to this same stressor may recover very quickly. Stream ecosystems are notoriously resilient and often recover very quickly from disturbance (Yount and Niemi 1990); however, most streams have low resistance and are relatively sensitive to many types of disturbance. Finally, coral reefs are an excellent example of an ecosystem with both low resistance and low resilience. Relatively few studies have simultaneously quantified resistance and resilience in communities and attempted to identify underlying mechanisms. Vieira et al. (2004) used a before–after control-impact (BACI) experimental design to determine effects of a large-scale wildfire disturbance on stream ecosystems. The magnitude of the initial response and the length of time necessary for communities to recover were related to species traits that conveyed resistance (e.g., body shape, mode of attachment to the substrate) and resilience (e.g., dispersal ability, resource use). Identifying the species-specific traits that confer tolerance and/or increase rates of recovery from contaminant exposure will greatly improve our ability to predict effects of anthropogenic disturbances.

While the above definitions of resilience and resistance stability are useful for classifying the diverse ways that communities may respond to either natural or anthropogenic disturbance, they are relatively simplistic concepts and their interpretation is context dependent. Although we can develop some general guidelines for predicting the magnitude of a response or the rate of recovery, it is unlikely that the specific details will be consistent across all types of perturbations. Therefore it is quite likely that underlying mechanisms responsible for conferring resistance and resilience of communities will be influenced by the nature and timing of the disturbance.

25.1.3 Pulse and Press Disturbances

In addition to understanding factors that influence susceptibility and recovery trajectories of communities following disturbance, ecologists also distinguish between two different types of perturbations. Pulse disturbances (Bender et al. 1984) are defined as instantaneous alterations in the abundance of species within a community (Figure 25.1). Factors that influence the recovery of a community as it returns to equilibrium are of particular interest in the study of pulse disturbances. The crown fire that occurred in Yellowstone National Park (YNP) (USA) in 1989 is an example of a large-scale pulse disturbance. Studies of the lodgepole forest communities in Yellowstone have

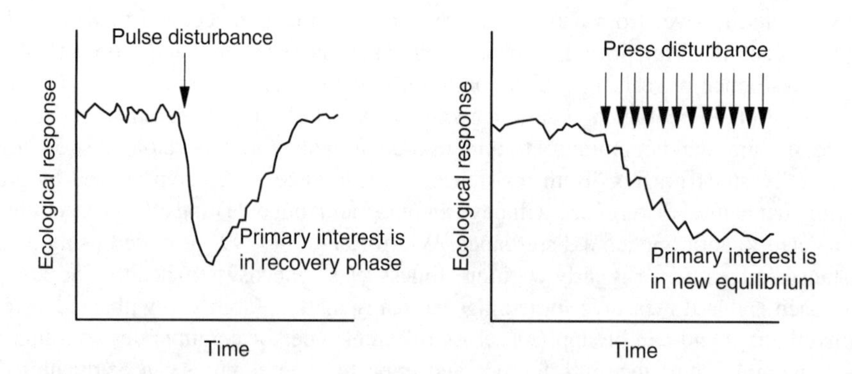

FIGURE 25.1 Comparison of pulse and press disturbances showing ecological responses of communities. Pulse disturbances result in instantaneous alterations of community structure and function. The primary research questions following pulse disturbances focus on processes that influence rate of recovery. Press disturbances are sustained alterations in ecological responses that may result in establishment of a new community. Following press disturbances ecologists are particularly interested in understanding characteristics of this new equilibrium.

focused primarily on identifying biotic and abiotic factors that influence the time required for this system to return to predisturbance conditions.

Press disturbances cause sustained alterations in abundance of species, often resulting in the elimination of some taxa and establishment of a new community. Here, ecologists are particularly interested in understanding community characteristics and factors that control this new equilibrium. Increased temperature associated with global climate change is an example of a press disturbance. Because communities affected by press disturbances are expected to establish new equilibria, investigators often focus on understanding characteristics of this altered community.

While the original theoretical treatment of pulse and press disturbances was developed to improve our quantitative understanding of species interactions (Bender et al. 1984), these concepts are also relevant to our discussion of how communities respond to contaminants. An ecotoxicological example of a pulse disturbance would be a chemical spill that temporarily reduced densities of certain species. Differences in sensitivity to the chemical among species may determine community composition immediately following the spill. However, assuming that the chemical was quickly degraded and there were no persistent effects, colonization ability of displaced species would be the primary factor influencing the rate of recovery. Recovery from this pulse disturbance may be rapid if an adequate supply of colonists is available to the system. In contrast to pulse disturbances, a press disturbance is continuous and the community is generally not expected to return to its original condition until the stressor is eliminated. An ecotoxicological example of a press disturbance would be the continuous input of toxic material into a system, such as acid deposition from coal-fired power plants. Here, differences in sensitivity among species will be the primary factor influencing community composition. If recovery is defined as a return to predisturbance conditions, it is unlikely that recovery will be observed until levels of the toxic materials are reduced. In the case of highly persistent contaminants (e.g., PCBs associated with lake sediments), recovery may not be observed even after the source has been eliminated.

The definitions used to distinguish between pulse and press disturbances have been criticized because they combine cause (e.g., disturbance) with effect (e.g., the response of the community) and assume a relatively simplistic response to perturbation (Glasby and Underwood 1996). For example, a pulse disturbance such as a chemical spill may have a lasting effect on community structure and function. Similarly, communities subjected to press disturbances could quickly return to equilibrium conditions if populations are able to acclimate or adapt to stressors. Glasby and Underwood (1996) refine these definitions and distinguish between discrete and protracted press and pulse perturbations (Table 25.3). They also suggest sampling procedures and experiments that allow investigators to identify these different categories of disturbance.

TABLE 25.3
Proposed Classification of Perturbations by Cause (Type of Disturbance) and Community Response

Classification	Type of Disturbance	Community Response
Discrete pulse	Short term	Short term
Protracted pulse	Short term	Continued
Protracted press	Continuous	Continued
Discrete press	Continuous	Short term

Source: From Glasby, T.M. and Underwood, A.J., *Environ. Monitor. Assess.*, 42, 241–252, 1996.

25.2 COMMUNITY STABILITY AND SPECIES DIVERSITY

One of the more impassioned debates in the field of community ecology has been over the positive relationship between species diversity and resistance/resilience stability (May 1973, Elton 1958). Darwin (1872) first proposed this intuitively pleasing idea and speculated that species-rich communities should be more stable than communities with few species. Complex food webs are assumed to allow communities to better tolerate disturbance because of greater functional redundancy among pathways of energy flow and nutrient cycling. According to this hypothesis, a species that was eliminated owing to disturbance would simply be replaced by a different species that performs a similar ecological functional. The hypothesis that greater species diversity results in greater stability also has significant implications for the study of anthropogenic disturbance. If complex systems are more stable, we would expect that the chronic effects of contaminants would be less pervasive in species-rich communities compared to depauperate communities.

In their synthesis of the relationship between diversity and ecological resilience, Peterson et al. (1998) describe four models of species richness and stability currently in the literature. The simplest model (the species richness-diversity model) proposes that the addition of species to a community increases the number of ecological functions, thereby increasing stability (Figure 25.2a). The model assumes that stability continues to increase as new species are added, and makes no allowances for saturation of ecological function. In contrast, the rivet model assumes that there is a limit to the number of functions in a community and that as new species are added functions begin to overlap (Figure 25.2b). Because of this functional redundancy in diverse communities, a few species can be removed with relatively little influence on stability. However, like removing rivets from the wing of an airplane, as more species are lost from a community, a critical threshold is eventually reached and stability will decrease rapidly. The idiosyncratic model (Figure 25.2c) proposes that the relationship

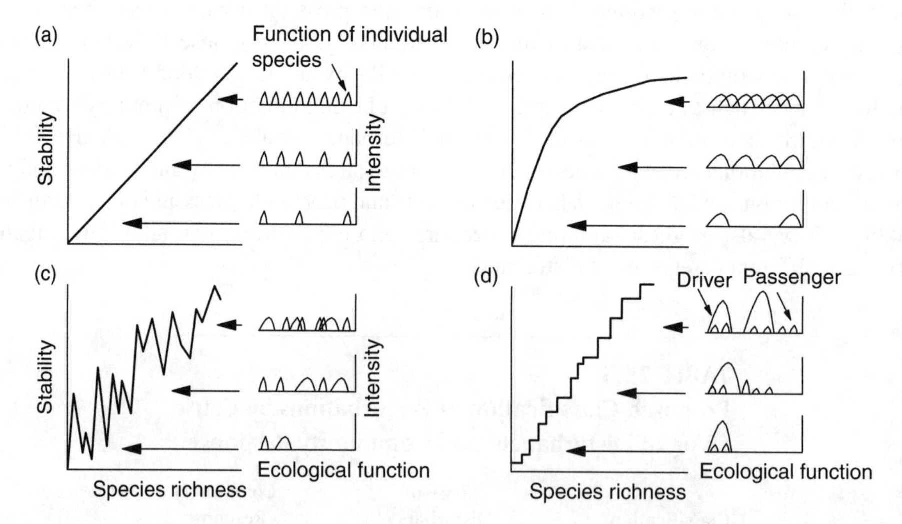

FIGURE 25.2 Four models showing the relationship between species richness and functional stability in communities. (a) The species diversity model assumes that stability decreases linearly as species are removed from the community. (b) The rivet model assumes that functional redundancy protects communities from loss of species, but that stability decreases rapidly once species are reduced to a critical threshold level. (c) The idiosyncratic model proposes that the effect of removing species is dependent on species interactions. (d) The drivers and passengers model assumes that the influence of species richness on stability depends on which species are removed from the community. Loss of driver species or keystone species have a greater impact on functional stability of a community than loss of passenger species. (Modified from Figures 1 through 4 in Peterson et al. (1998).)

between species richness and stability is highly variable and that the consequences of adding new species are dependent on species interactions. Addition of some species will stabilize ecological function whereas the addition of others will have relatively little influence on community stability. Finally, the drivers and passengers model (Figure 25.2d) assumes that the influence of species richness on stability depends on which particular species is added to the community. Driver species, including "ecological engineers" and other keystone species, have a greater impact on functional stability of a community than passenger species.

All four models described above assume a positive relationship between stability and diversity. However, despite its intellectual appeal, the relationship between diversity and stability is not straightforward, and relatively few experimental studies have provided strong support for this hypothesis. In fact, theoretical treatment of the diversity–stability relationship has suggested that complex communities are actually *less* stable than simple communities (May 1973). Microcosm experiments conducted with protists support these models and show that addition of more trophic levels resulted in reduced stability (Lawler and Morin 1993). One potential explanation for these conflicting results is that different researchers have used different measures to define stability. Peterson (1975) reported different relationships between diversity and stability depending on whether one measured stability at the species level (variation of individual populations) or at the community level (variation in community composition). In contrast to the theoretical studies of diversity–stability relationships, the most influential empirical studies have used temporal variation in productivity or biomass as a measure of stability (Doak et al. 1998). In a long-term experimental study of grassland plots Tilman (1996) reported that increased biodiversity stabilized community and ecosystem processes but not population-level processes (Figure 25.3). Variability of community biomass decreased (i.e., stability increased) as more species were added to the community, whereas variability of individual populations increased (although this relationship was relatively weak). These results may help resolve the long-standing debate over the diversity–stability relationship. It appears that increased diversity does stabilize community biomass and productivity as predicted by Elton (1958), but decreases population stability, consistent with May's (1973) mathematical models. The underlying mechanism responsible for these differences appears to be interspecific competition (Tilman 1996).

Some researchers have argued that the relationship between diversity and stability reported in the literature is an inevitable outcome of averaging the fluctuations of individual species' abundances (Doak et al. 1998). The premise for this argument is that community-level properties such as total

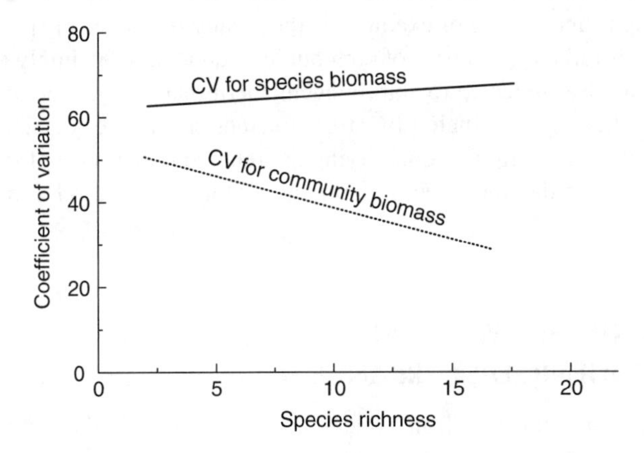

FIGURE 25.3 Proposed resolution of the diversity–stability debate. The figure shows a relationship between species richness and two measures of stability in plant communities. Population and community stability was characterized by measuring the coefficient of variation ($CV = (100 \times SD)/M$) for species and community biomass. As more species are added to the community, population stability decreases (the CV for species biomass increases), whereas community stability increases (the CV for community biomass decreases). (Modified from Figures 7 and 9 in Tilman (1996).)

biomass will be less variable as a greater number of species are included simply because of this averaging effect. This same statistical phenomenon is observed for other measures of community composition. For example, total abundance is generally less variable than abundance of individual species, especially for rare species. A practical aspect of this statistical averaging effect is that aggregate measures of community composition are often less variable and therefore more useful for assessing impacts of stressors than abundance of individual species (Clements et al. 2000). From an ecological perspective, the relative importance of this statistical relationship must be quantified in order to understand the role of species interactions in structuring communities. Previously, the diversity–stability relationship was assumed to be exclusively a result of species interactions. However, this statistical averaging effect associated with aggregate measures occurs regardless of the importance of competition or predation in a community (Doak et al. 1998).

Much of the experimental research investigating the relationship between diversity and stability has involved establishing a diversity gradient in which individual species are excluded from some treatments. While many of these experiments have shown a positive relationship between diversity and stability, it is uncertain if similar patterns occur in systems where diversity varies along natural gradients. Sankaran and McNaughton (1999) report results of a study of savannah grasslands in which plant communities along a natural disturbance gradient were exposed to experimental perturbations, including fires and grazing. These researchers observed that the relationship between diversity and resistance stability was dependent on the specific measure of stability being considered. Resistance to species turnover, measured as the proportion of species in both pre- and post-disturbance plots, increased with species diversity. This result is consistent with the hypothesis that stability is positively associated with diversity. In contrast, resistance to compositional change, measured as change in the relative contribution of different species before and after disturbance, decreased with species diversity. Because community composition is a reflection of numerous extrinsic factors, including disturbance regime and site history, it may be a more important determinant of stability than the actual number of species in a community. Sankaran and McNaughton's (1999) results demonstrate that the relationship between diversity and stability is largely influenced by these extrinsic factors and that species-rich communities may not necessarily be better at "coping" with disturbance.

The diversity–stability debate has serious implications for understanding how communities respond to anthropogenic stressors. Measures of stability based on aggregate properties, such as total abundance or biomass, appear to be related to the number of species in a community. The degree to which other measures of stability, such as community resistance and resilience, are influenced by this statistical relationship is uncertain. For example, is the greater resilience of species-rich communities to anthropogenic disturbances a result of community redundancy or simply a statistical artifact? Alternatively, communities subjected to anthropogenic perturbations may be resistant to additional disturbance because they are dominated by stress-tolerant species. Understanding the causes of the diversity–stability relationship and quantifying the relative importance of these statistical averaging effects requires that theoretical and empirical ecologists agree on common definitions of stability.

25.3 RELATIONSHIP BETWEEN NATURAL AND ANTHROPOGENIC DISTURBANCE

A unifying feature that has emerged from research on disturbance is the remarkable resilience of some communities to a wide range of natural disturbances. The characteristics that account for rapid recovery of communities following disturbance are diverse, but most often relate to the availability of colonists. One fundamental question from an ecotoxicological perspective is how can research on responses to natural disturbance be employed to predict recovery from anthropogenic disturbance. In particular, can we expect to see similar patterns of resistance and resilience to chemical stressors as to physical disturbances? Comparisons of natural and anthropogenic disturbance will

TABLE 25.4
Effects of Natural (Blowdown) and Anthropogenic (*N* Addition; Soil Warming) Disturbances in a Second Growth Forest

Process	Blowdown	*N* Addition	Soil Warming
Mineralization	+15.9	+138	+50
Methane uptake	−2.4	−36	+20
Soil respiration	+6.2	0	+76

Note: The table shows percentage changes of ecosystem processes.

Source: From Foster, D.R., et al., *Bioscience*, 47, 437–445, 1997.

allow researchers to answer these questions and improve their ability to predict responses to future disturbances.

Unfortunately, relatively few studies have compared responses of communities to both natural and anthropogenic disturbances. Foster et al. (1997) conducted several large-scale experiments designed to investigate the impacts of physical restructuring (a blowdown induced by a hurricane), nitrogen additions, and soil warming in a second-growth forest. Results of this study showed that despite obvious effect of the blowdown on forest structure, there was little change in ecosystem processes (Table 25.4). Because species in this forest were adapted to frequent disturbance associated with hurricanes, recovery was observed soon after the blowdown. In contrast, *N* addition and soil warming had a much greater impact on ecosystem processes but little influence on community composition. These researchers contend that because species in this community were not adapted to these novel stressors, little evidence of recovery was observed.

A long-term program of field monitoring and experiments conducted in Antarctica, "one of the most extreme physical environments in the world" compared the impacts of natural and anthropogenic disturbance on marine benthic communities (Lenihan and Oliver 1995). Anthropogenic disturbance included chemical contamination in sediments around McMurdo Station (primarily hydrocarbons, heavy metals, and PCBs), whereas natural disturbance included anchor ice formation and scour. Results showed remarkable similarity between anthropogenic and natural disturbances. Communities in contaminated sites and physically disturbed sites were dominated by the same assemblages of polychaete worms, species with highly opportunistic life history strategies. Despite the similarity in responses, these researchers suggested that recovery from chemical contamination would require considerably more time because of the slow degradation of these persistent contaminants in sediments.

25.3.1 The Ecosystem Distress Syndrome

Although there is some empirical support for the hypothesis that effects of contaminants vary among communities (Howarth 1991, Kiffney and Clements 1996, Medley and Clements 1998, Poff and Ward 1990), there have been few attempts to identify specific factors responsible for this variation. Fragility may be an inherent property of some communities, regardless of the history of disturbance (Nilsson and Grelsson 1995). Resistance and resilience to anthropogenic disturbances may vary among different communities or among similar communities in different locations. This variation greatly complicates our ability to predict community responses and recovery times. If some communities are inherently more fragile than others, identifying characteristics that increase sensitivity and the mechanisms responsible for ecosystem recovery are important areas of research.

Rapport et al. (1985) suggested that communities in unstable environments may be "preadapted" to moderate levels of anthropogenic stress. Howarth (1991) speculated that ecosystems with fewer opportunistic species, lower diversity, and closed element cycles would be sensitive to contaminants. In an experimental investigation of resistance and resilience, Steinman et al. (1992) reported that initial community structure was relatively unimportant in determining community responses to chlorine. In this study community biomass, which was regulated by grazing herbivores, determined resistance to chlorine exposure. These results are consistent with experiments showing that trophic status of a community influences resistance and resilience (Lozano and Pratt 1994).

Rapport et al. (1985) evaluated the responses of several communities to different types of disturbance and developed an "ecosystem distress syndrome." They argue that community responses to disturbance are analogous to the generalized adaptation syndrome that occurs when individual organisms are subjected to environmental stress (Seyle 1973) (see Section 9.1.1 and Box 9.1 in Chapter 9). Because the perturbations considered in their analysis included a range of natural and anthropogenic stressors (physical restructuring, overharvesting, pollution, exotic species, extreme natural events), the results may be used to compare responses across disturbance types and among communities (Table 25.5). Because it is not feasible to measure every potential indicator in all ecosystems, identifying general responses to disturbance across a diverse array of ecosystems and disturbance types is essential. Furthermore, identifying similarities between natural and anthropogenic disturbances will allow ecotoxicologists to benefit from the long history of research on natural disturbance to better understand how communities respond to chemical stressors.

25.3.2 The Intermediate Disturbance Hypothesis

Communities subjected to moderate levels of disturbance may have greater species richness or diversity compared to communities existing under benign conditions. The intermediate disturbance

TABLE 25.5
Characteristic Responses of the Ecosystem Distress Syndrome

Disturbance Type	Nutrient Pool	Primary Productivity	Species Diversity	Size Distribution	System Retrogression
Harvesting renewable resources					
Aquatic	*	*	−	−	+
Terrestrial	−	−	−	−	+
Pollutant discharges					
Aquatic	+	+	−	−	+
Terrestrial	−	−	−	−	+
Physical restructuring					
Aquatic	*	*	−	−	+
Terrestrial	−	−	−	−	+
Introduced species					
Aquatic	*	*	*	−	+
Terrestrial	*	*	*	*	+
Extreme natural events					
Aquatic	*	*	−	−	+
Terrestrial	−	−	−	−	+

Note: The table shows the expected response of each indicator as increasing (+), decreasing (−), or unknown (*).
Source: From Rapport, D.J., et al., *Am. Nat.*, 125, 617–640, 1985.

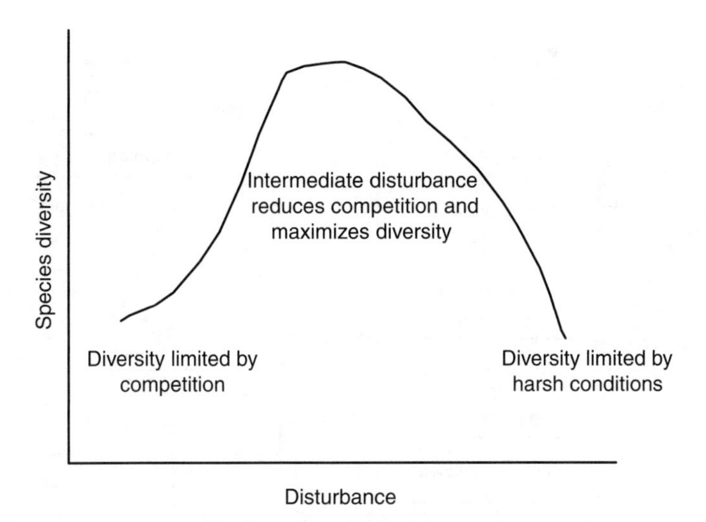

FIGURE 25.4 According to the IDH (Connell 1978) species diversity is maximized under conditions of intermediate levels of disturbance. Species diversity is low in stable, highly predictable communities because a small number of species dominate resources and are capable of excluding subordinate species. Species diversity increases with moderate levels of disturbance because the ability of dominant groups to exclude subordinates decreases. Species diversity is also low under extreme levels of disturbance because relatively few species are able to persist under these harsh environmental conditions.

hypothesis (IDH) was initially proposed by Connell (1978) to explain higher levels of species diversity observed in rocky intertidal habitats subjected to moderate levels of physical disturbance. The mechanism suggested to account for this somewhat counterintuitive observation was that moderate levels of disturbance reduced competition for limited resources and allowed more species to coexist. Diversity is low under benign conditions because a small number of species dominate resources and are capable of excluding subordinate species. Diversity is also low under extreme levels of disturbance because relatively few species are able to persist. Thus, according to predictions of the IDH we would expect the greatest species diversity under moderate levels of perturbation (Figure 25.4).

There is general support for the IDH in the literature, and natural communities in a variety of habitats seem to fit predictions of the IDH fairly well. According to this hypothesis, the rich biological diversity observed in tropical rainforests and coral reefs is maintained by a combination of high productivity, habitat complexity, and disturbance from hurricanes. Sousa (1979) conducted a series of experiments to test the IDH in marine intertidal communities associated with boulders. Because small boulders are more likely to be disturbed by waves, Sousa used boulder size as an index of the probability of disturbance. He initially demonstrated that the greatest number of species was found on intermediate-sized boulders, a finding consistent with predictions of the IDH. He then anchored the small boulders to prevent disturbance and observed an increase in the number of species. These results demonstrated that substrate stability was more important than size in determining species richness.

The IDH is now widely embraced by many ecologists, and examples of the positive effects of moderate disturbance on species diversity have been reported in many different systems. However, there are examples where the IDH was not supported, most notably in freshwater streams where rapid recolonization swamps the effects of disturbance. For example, Death and Winterbourn (1995) reported that species richness in New Zealand streams increased with habitat stability but showed no relationship with disturbance. Similar results were reported by Reice (1985) following experimental manipulation of cobble substrate designed to simulate flood disturbance. Although the importance of natural disturbance in structuring many communities was recognized, Reice concluded that the

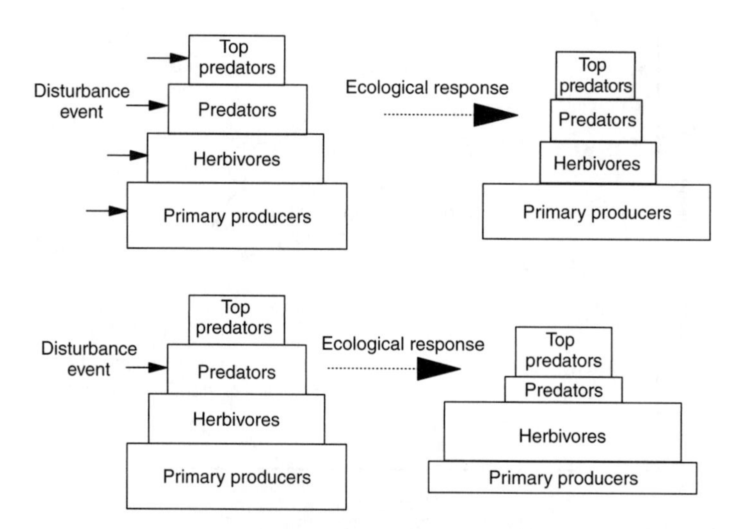

FIGURE 25.5 Conceptual model showing the effects of disturbance in multiple trophic-level systems. In the upper panel, disturbance to each of the trophic levels (represented by the solid arrows) results in a proportional reduction in biomass of each group. In the lower panel predators are disproportionately impacted by the disturbance, resulting in a cascading effect on lower trophic levels. (Modified from Figure 1 in Wootton (1998).)

IDH did not apply to streams. Failure to account for the effects of disturbance on multiple trophic levels may also limit the predictive ability of the IDH. Natural communities consist of several potentially interacting trophic levels, and disturbance to multitrophic communities may show very different results than disturbance to a single trophic level (Figure 25.5). Wootton (1998) developed a mathematical model to determine if predictions of the IDH were applicable to multiple trophic levels. Results of these analyses helped explain why the IDH successfully predicted patterns in some communities but not in others. Clearly, any application of the IDH to anthropogenic disturbances must consider systems with more than one trophic level.

Similar to research on disturbance in general, most tests of the IDH have focused on natural, physical perturbations in systems where space is the primary limiting resource. It is uncertain if predictions of this model can be applied to toxicological stressors. Rohr et al. (2006) hypothesized that contaminant-induced mortality is analogous to effects of a keystone predator that feeds selectively on competitively superior species. If low to moderate levels of contaminants have a disproportionate effect on competitive dominants, it is possible that species diversity could increase. Johnston and Keough (2005) reported that copper reduced abundance of large, dominant tunicates (Ascidiacea), thereby increasing recruitment of other competitively inferior species. Are there other examples where exposure to intermediate levels of toxic stressors prevents competitively superior species from dominating resources and reducing species diversity? Because species richness and diversity are common indicators of perturbation in biological assessments, the IDH has important practical implications that are relevant to community ecotoxicology. For example, if species diversity is enhanced under low levels of contaminant exposure as predicted by the IDH, then it may be difficult to detect subtle impacts on communities.

25.3.3 SUBSIDY–STRESS GRADIENTS

The theoretical treatment of subsidy–stress gradients by Odum et al. (1979) offers some insight into the responses of communities to different types of chemical stressors. According to this model, certain types of disturbances, such as the input of nutrients or organic material, may enhance or subsidize a community. However, when levels of these materials exceed a critical threshold, the system becomes stressed resulting in a unimodal response. In contrast to patterns observed for inputs

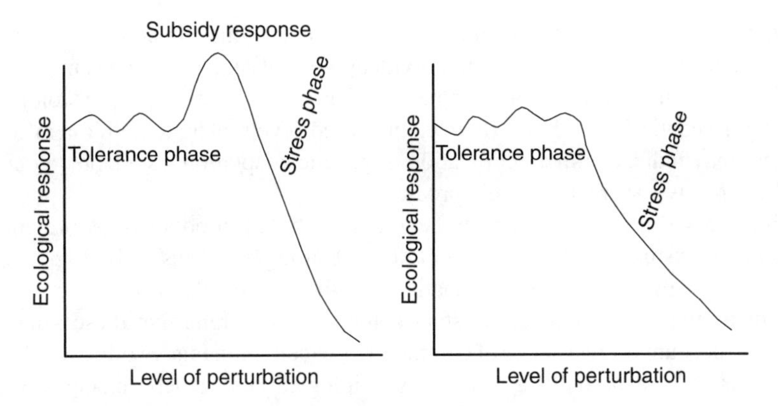

FIGURE 25.6 Odum's model of subsidy–stress gradients. The model predicts that certain types of stressors, such as the input of nutrients or organic material, may subsidize a community. When levels of these materials exceed some threshold of tolerance, the system becomes stressed resulting in a unimodal response to the stressor. In contrast, the addition of toxic materials generally does not subsidize ecological function and therefore results in a tolerance phase followed by a stress phase. (Modified from Figure 1 in Odum (1979).)

of usable resources, the input of toxicants into a system generally does not subsidize a community (Figure 25.6). In fact, very small amounts of toxic chemicals may have a similar effect on communities as large amounts of usable (e.g., subsidizing) materials. The shape of the perturbation–response curve for toxicant input or the location of the peak in the subsidy–stress gradient is dependent on numerous factors and varies greatly among communities. In addition, because of the hierarchical arrangement of natural systems, inputs of nutrients and organic matter may subsidize one level of organization (e.g., increase species diversity and productivity) but have a negative impact on some individual species. A good example to illustrate this point is the eutrophication observed in aquatic ecosystems resulting from the input of nutrients. In general, low input of nutrients into an oligotrophic system will stimulate primary and secondary productivity and may increase species diversity. However, these changes are likely to be accompanied by alterations in community structure, as nutrient-sensitive species are replaced by nutrient-tolerant species. The use of subsidy–stress models (Odum et al. 1979) for predicting responses to anthropogenic disturbances requires a thorough understanding of natural temporal changes in community composition. The initial increase in productivity and species diversity following the input of nutrients into an oligotrophic lake is often followed by a slow decline as the system adjusts to these novel conditions.

In summary, the input of either toxic chemicals or subsidizing materials can alter community composition because of differential sensitivity among species. The subsidy–stress model predicts that small inputs of usable materials in a system will increase primary productivity and may increase species diversity (Odum et al. 1979). In contrast, the input of toxic materials in a system will generally not increase productivity. It is unlikely that low concentrations of toxic materials will increase species diversity unless these chemicals remove competitively superior species or alter the outcome of species interactions, as predicted by the IDH (Section 25.4.2).

25.4 CONTEMPORARY HYPOTHESES TO EXPLAIN COMMUNITY RESPONSES TO ANTHROPOGENIC DISTURBANCE

Populations chronically exposed to contaminants often exhibit increased tolerance relative to naive populations (Chapter 18). Two general explanations are proposed to account for this observation: physiological acclimation and genetic adaptation. Physiological responses may include reduced contaminant uptake or increased production of detoxifying enzymes. In contrast, genetic adaptation results from higher survival rate of tolerant individuals and subsequent changes in gene frequencies. The distinction between acclimation and adaptation is somewhat arbitrary, as physiological processes may also have a genetic basis. For example, increased levels of metallothionein in response to metal exposure may indicate either acclimation or genetic adaptation, as adapted populations have developed the capacity for greater protein production.

Although increased tolerance has often been demonstrated in populations previously exposed to contaminants, few studies have examined tolerance at higher levels of biological organization. As noted above, the most common explanations for increased tolerance at the population level include acclimation and selection for resistant genotypes. We argue that these same intraspecific mechanisms also account for resistance of communities to contaminants. In other words, community-level tolerance is at least partially a result of physiological and genetic changes of populations. However, because communities consist of large numbers of interacting species, it is likely that other mechanisms, unique to these systems, will contribute to tolerance. For example, increased tolerance at the community level may result from replacement of sensitive species by tolerant species. This shift, termed "interspecific selection" (Blanck and Wangberg 1988), is a common response in contaminated systems and a consistent indicator of anthropogenic disturbance. Interspecific selection is also a likely explanation for pollution-induced community tolerance (PICT), a new ecotoxicological approach for demonstrating causation in community assessments.

25.4.1 Pollution-Induced Community Tolerance

Increased resistance of a population to a contaminant may indicate selection pressure and provide strong evidence that the population has been affected (Luoma 1977). Similarly, increased tolerance at the community level may also indicate ecologically important effects. PICT has been proposed as an ecotoxicological tool to assess the effects of contaminants on communities (Blanck 2002, Blanck and Wangberg 1988). PICT is tested by collecting intact communities from polluted and reference sites and exposing them to contaminants under controlled conditions. Increased community tolerance resulting from the elimination of sensitive species is considered strong evidence that community restructuring was caused by the pollutant. Proponents of the PICT argue that, while differences in traditional measures (abundance, richness, diversity) between communities from reference and polluted sites can be attributed to many factors, increased tolerance observed in communities is less sensitive to natural variation and most likely a result of contaminant exposure (Blanck and Dahl 1996). Furthermore, because acquisition of community tolerance is generally not influenced by environmental conditions, locating identical reference and polluted sites for comparison is less critical (Millward and Grant 2000). Because the restructuring of communities and the replacement of sensitive species by tolerant species are commonly observed at contaminated field sites, PICT holds tremendous potential as a monitoring tool in ecotoxicology that allows researchers to identify underlying causal relationships (Grant 2002).

The use of PICT to assess impacts of contaminants at the level of communities is based on three assumptions: (1) sensitivity to contaminants varies among species; (2) contaminants will restructure communities, with sensitive species being replaced by tolerant species; and (3) differences in tolerance among communities can be detected using short-term experiments (Gustavson and Wangberg 1995). The first two assumptions are relatively straightforward and easy to verify with field sampling. The third assumption is more problematic and significantly constrains application of PICT as an assessment tool. While tolerance at the population level can be assessed using a variety of species, logistical considerations will limit the types of communities where tolerance can be investigated experimentally. Although some researchers have speculated that the PICT approach can be applied to larger organisms by measuring biomarkers of exposure and effects in different communities (Knopper and Siciliano 2002), most PICT experiments have been conducted using small organisms with relatively fast life cycles (Table 25.6).

TABLE 25.6
Examples of Experimental Tests of the PICT Hypothesis Showing the Types of Stressors, Endpoints, and Diversity of Communities Examined

Community	Stressors	Endpoints	Reference
Soil microbes	Zn	Metabolic diversity	Davis et al. (2004)
Marine periphyton	Arsenate	Photosynthesis, biomass, species composition	Blanck and Wangberg (1988)
	TBT	Photosynthesis	Blanck and Dahl (1996)
Lentic phytoplankton	Arsenate, Cu	Photosynthesis, biomass, community composition	Wangberg (1995)
Lentic periphyton	Cu, atrazine	Photosynthesis	Gustavson and Wangberg (1995)
Marine phytoplankton	TBT	Primary production	Petersen and Gustavson (1998)
Freshwater protozoans	Zn	Primary production, biomass, species richness	Niederlehner and Cairns (1992)
Lotic microalgae	Cd, Zn	Biomass, carbohydrates, community composition	Ivorra et al. (2000)
Estuarine nematodes	Sediment Cu	Survival time	Millward and Grant (2000)
Benthic macroinvertebrates	Cd, Cu, Zn	Community composition, richness, susceptibility to predation	Clements (1999)

The PICT hypothesis was originally developed for marine periphyton, but has now been tested in several different communities. Protozoan communities developed under low levels of zinc stress were more tolerant of zinc than naive (e.g., unexposed) communities (Niederlehner and Cairns 1992). Relative resistance to zinc in acclimated communities increased by greater than three times compared to unacclimated communities. Schwab et al. (1992) reported that periphyton communities in experimental streams rapidly increased their tolerance to surfactants. Metal tolerance of nematodes collected from sediments along a contamination gradient increased with concentrations of copper in the environment (Millward and Grant 1995). Finally, benthic macroinvertebrate communities collected from a site with moderate levels of heavy metals were significantly more tolerant to subsequent cadmium, copper, and zinc exposure than those collected from pristine sites (Clements 1999, Courtney and Clements 2000, Kashian et al. 2007).

Studies testing the PICT hypothesis have also examined a variety of endpoints. As noted above, increased tolerance in communities may result from either population-level responses (acclimation or adaptation) or interspecific selection. For example, tolerance of nematode communities from a Cu-polluted estuary resulted from increased abundance of tolerant species, evolution of Cu tolerance, and exclusion of sensitive species (Millward and Grant 1995). Because of taxonomic challenges, PICT experiments conducted using soil microbial communities have quantified metabolic diversity based on substrate utilization profiles (Davis et al. 2004). Endpoints examined in PICT studies should be selected to allow investigators to distinguish between population and community-level mechanisms. Greater tolerance of populations can be evaluated by comparing responses of individual species collected from reference and polluted sites. Greater tolerance at the community level can be evaluated by measuring effects on structural and functional endpoints. An important consideration when selecting endpoints in PICT studies is the potential for functional redundancy in the restructured communities. Dahl and Blanck (1996) reported that some functional endpoints were inadequate for validating the PICT hypothesis because sensitive species were replaced by tolerant species with a similar functional role.

Although there has been widespread support for the PICT hypothesis in the literature, several issues must be resolved before the approach becomes a useful ecotoxicological tool. A number of attempts to demonstrate PICT in the field have not been successful, most likely because some populations fail to develop tolerance at polluted sites (Grant 2002). PICT is most likely to be observed in communities that show a large amount of variation in sensitivity among species. Nystrom et al. (2000) reported difficulty demonstrating PICT in algal communities exposed to atrazine because of the narrow distribution of tolerances among species. Development of tolerance in phytoplankton communities was reported to be size specific (Petersen and Gustavson 1998). Although microplankton showed tolerance to tributyltin (TBT), other size fractions of the community showed relatively little response. Finally, Ivorra et al. (2000) reported that the influence of exposure history on tolerance of periphyton is complicated by maturity of the community. Immature communities from a reference site were more sensitive to metals than those from a polluted site, supporting the PICT hypothesis; however, there was no difference in the responses of mature periphyton communities between the two sites.

One potential advantage of using PICT as an assessment tool is the opportunity to isolate effects of individual stressors in systems impacted by multiple stressors (Wangberg 1995). If we assume no interactions among stressors and that tolerance to one chemical does not influence tolerance to another, PICT could be used to quantify effects of a specific chemical. However, previous research has shown that co-tolerance may occur in some communities, especially when modes of action and detoxification mechanisms are similar (Blanck and Wangberg 1991). For example, Gustavson and Wangberg (1995) reported that communities exposed to copper also showed increased tolerance to zinc. In contrast, Wangberg (1995) observed that exposure to copper reduced tolerance for arsenate. These results indicate that some caution is necessary when using PICT to identify effects of specific chemicals in environments where multiple contaminants are present.

25.5 BIOTIC AND ABIOTIC FACTORS THAT INFLUENCE COMMUNITY RECOVERY

In addition to studying how communities respond to disturbance, ecologists are frequently interested in understanding how communities recover from disturbance. The definition of recovery, the characteristics of communities that influence rate of recovery, and the influence of disturbance type on recovery have been topics of considerable discussion in community ecology. From an applied perspective, predicting the rate of recovery from disturbance is at least as important as understanding the initial responses. If we assume that recovery is a non-stochastic process, then information on biotic and abiotic factors that influence rate of recovery may allow us to predict how long it will require communities to reach predisturbance conditions. More importantly, the study of recovery from natural disturbance may allow researchers to understand and predict how communities recover from anthropogenic disturbance (Box 25.2). For example, a study of lizard and spider populations in the Bahamas showed that the risk of extinction from hurricanes was related to population size only when disturbance was moderate (Spiller et al. 1998). Following a catastrophic disturbance large population size did not protect populations from extinction. Recovery of these assemblages was more related to fecundity and dispersal ability. Other research has demonstrated that species initially colonizing disturbed habitats are characterized by small body size and short life cycles. If these generalizations also apply to anthropogenic disturbances, we predict that disturbed communities would initially be dominated by relatively small species with short life cycles and high reproductive output and that recovery would be greatly influenced by the dispersal ability of the species.

Recovery from natural or anthropogenic disturbance is determined by a complex suite of factors related to the characteristics of the community, severity of the disturbance, and physical features of the disturbed habitat. Because disturbance is an integral part of the evolutionary history of many organisms, recovery from natural disturbance may be quite rapid. Communities dominated by opportunistic species capable of rapid colonization will generally recover quickly. Species that initially colonize disturbed habitats are often trophic generalists, capable of exploiting a wide range of resources.

Box 25.2 Recovery of Communities from Large-Scale Disturbances

Three large-scale disturbances that occurred over the past several decades have provided ecologists with unprecedented opportunities to examine recovery and test various hypotheses concerning biotic and abiotic factors that influence resistance and resilience. Two of these disturbances were natural (the eruption of Mt. St. Helens and the crown fires at YNP), whereas a third (the *Exxon Valdez* oil spill) was anthropogenic, providing an opportunity to examine recovery from different types of disturbances.

The eruption of Mt. St. Helens in May 1980 and the associated blowdown, mud flows, avalanches, and ash deposits affected over 700 km^2 in southwestern Washington (USA). The fires in YNP during the summer of 1988 were larger than any in the previous 200–300 years. A total of 2500 km^2 of the park burned, creating a complex mosaic of disturbed and undisturbed habitats. Finally, the breakup of the *Exxon Valdez* in March 1989 spilled approximately 41×10^6 L of crude oil in northeastern Prince William Sound (Alaska, USA) and oiled an estimated 800 km of shoreline. By any account, each of these disturbance events was large scale, novel, and had a major impact on the surrounding communities. Ecologists rushed to these sites to validate predictions of theoretical and empirical models derived from nearly a century of studying community succession. While some of the original predictions were well supported by field studies, others were not. For example, recovery of plant and animal communities on Mt. St. Helens occurred through a bewildering array of mechanisms, many of which involved the persistence of "biological legacies" (e.g., living and dead habitat structure that remained following the blast). In the Yellowstone fires, geographic location and the proximity of new colonists were more important for predicting recovery than burn severity and patch size (Turner et al. 1997). Finally, despite the dramatic impact of the *Exxon Valdez* oil spill on bird populations, which caused mortality of hundreds of thousands of birds, seabird communities in Prince William Sound showed unexpected resilience (Wiens et al. 1996). The lessons learned from intensive study of these large-scale disturbances have forced ecologists to reevaluate many of their models of community perturbation and recovery (Franklin and MacMahon 2000).

Magnitude (spatial extent) and novelty of disturbance will also influence recovery times. Thus, communities will require considerably more time to recover from severe, novel disturbances that have a large spatial extent (e.g., a large oil spill) compared to small scale, predictable perturbations.

The timing of a disturbance with respect to critical life stages for organisms will also influence the rate of recovery. For example, juvenile and immature life stages are generally more sensitive to disturbance than adults. Consequently, a disturbance that occurs when these immature life stages are present will have a disproportionately greater impact on a community. Other phenological considerations, such as the seasonal availability of seeds or other life stages that are critical for dispersal, also influence rates of recovery. Experiments conducted with salt marsh plants showed that differences in recovery rates among species were primarily determined by the season when the disturbance occurred (Allison 1995).

Specific features of the disturbed habitat, such as environmental heterogeneity and proximity to sources of colonists, must be considered when assessing potential for recovery. Communities in patchy environments that contain refugia and are located near undisturbed habitats will generally recover faster than communities in isolated, homogenous environments. Finally, rates of recovery will also be influenced by the potential interplay between these different features. For example, the effects of size of the disturbed area on recovery will depend on the colonization ability of nearby species. Recovery from a small-scale disturbance may require a significant amount of time if the dispersal ability of local species is limited.

Understanding the spatial and temporal dynamics of community recovery following anthropogenic disturbance is critical to the field of restoration ecology. Fundamental issues regarding the definition of recovery and the specific indicators of recovery must be considered before generalizations are possible. For example, is the return to an equilibrium number of species sufficient to demonstrate recovery or should the actual composition of the community matter? Has recovery occurred if the composition of a community returns to predisturbance conditions but certain ecosystem processes (e.g., decomposition, nutrient cycling) remain altered? Although our ability to predict contaminant effects on communities has increased greatly, there has been considerably less effort devoted to measuring recovery from anthropogenic disturbance. The vast majority of field investigations of disturbance focus on short-term effects and often fail to monitor recovery. A report prepared by an intergovernmental task force on biological monitoring in the United States concluded that despite the large amount of effort devoted to improving environmental quality, our understanding of the effectiveness of remediation programs is hampered by the failure to assess recovery (Hart 1994). In other words, restoration and remediation programs are often assumed to be successful; however, rigorous verification of this assumption is lacking because of the paucity of funding available to monitor communities during the recovery phase. As a consequence, much of the available data on recovery of ecosystems from contaminants is anecdotal, and relatively few studies have documented recovery using adequate experimental designs (Yount and Niemi 1990).

Failure to consider recovery may provide misleading information on the magnitude of a particular disturbance. For example, consider two different communities exposed to the same disturbance (Figure 25.7). Effects on one community are initially greater (e.g., a greater shift from predisturbance conditions), but the community eventually returns to predisturbance conditions. In contrast, the second community shows less impact but recovery is not observed. Samples collected soon after the disturbance would show that the effects on community A were greater than those on community B. However, samples collected later in the recovery trajectory would show just the opposite results. Thus, sampling programs designed to measure recovery must be conducted at the appropriate temporal scale to quantify the relative impacts of disturbance.

25.5.1 Cross-Community Comparisons of Recovery

Comparative studies of stream communities have shown that these systems often recover quite rapidly from both natural and anthropogenic disturbances. Because natural variation of streams is

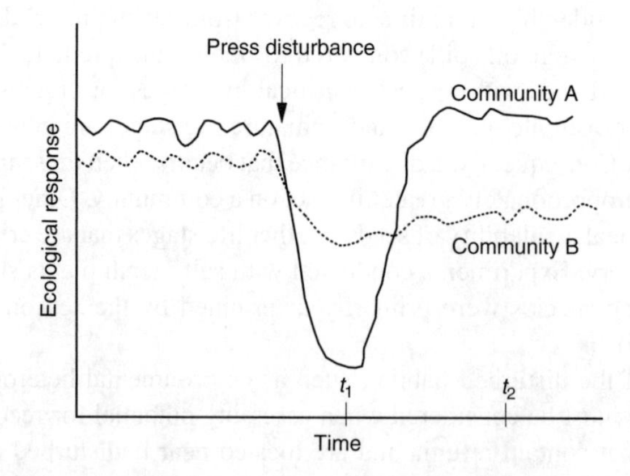

FIGURE 25.7 The importance of considering recovery trajectories when assessing responses to press disturbances is illustrated by comparing two communities subjected to the same stressor. Samples collected soon after the disturbance (t_1) would show that effects on community A were greater. However, because of the more rapid recovery observed in community A, samples collected later in time (t_2) would show greater effects on community B. (Modified from Figure 1 in Niemi et al. (1990).)

generally high, many species have flexible life history characteristics and are adapted to fluctuating conditions. These same characteristics may help explain the rapid recovery from anthropogenic disturbances. Niemi et al. (1990) reviewed over 150 case studies of stream communities in which some aspect of recovery was monitored. Because the study examined many types of disturbance (floods, drought, application of biocides, timber harvesting, mining, and toxic spills), the review provides an excellent opportunity to compare responses to natural and anthropogenic stressors. The most striking generalization from these studies was the rapid recovery observed in many lotic ecosystems. For macroinvertebrates, over 90% of the studies reported recovery of density, biomass, and richness within 1–2 years after disturbance. Although recovery for fish communities generally required more time, the majority of studies showed recovery within 2 years. In general, longer recovery times were reported for physical disturbance than for chemical disturbance. The most important exception to this pattern was for persistent organic chemicals that remained in systems for longer periods of time. The rapid recovery of lotic systems was determined by (1) life history characteristics of species that allowed for rapid recolonization; (2) the proximity of upstream and downstream undisturbed sites to provide a source of colonists; (3) the high flushing rate of lotic systems; and (4) the general adaptations of many stream organisms to natural disturbance.

These cross-community comparisons of recovery are especially useful for developing a broad understanding of responses to anthropogenic disturbances. However, these approaches provide relatively little insight into the mechanisms responsible for observed changes (Fisher and Grimm 1991). To test the hypothesis that communities resistant to one disturbance will also be resistant to a different disturbance requires that all variables except the disturbance be controlled by the investigator. Similarly, to test the hypothesis that one community is more resistant to a specific disturbance than another community requires that both systems be subjected to the same disturbance under the same conditions. Fisher and Grimm (1991) argue that, while cross community comparisons are useful for generating hypotheses, responses to disturbance should initially be investigated in similar systems.

25.5.2 Importance of Long-Term Studies for Documenting Recovery

Continuing research conducted at Mt. St. Helens and YNP has demonstrated the value of long-term studies for characterizing the range of natural variability in communities. Long-term data are especially important for developing a general model of recovery for communities dominated by long-lived species. In addition, conclusions based on short-term studies of recovery can often be misleading. For example, benthic macroinvertebrate communities in burned watersheds in YNP were progressing rapidly to predisturbance conditions within 1–2 years following the fires. However, these same communities showed an abrupt downturn in subsequent years (Minshall et al. 1997). Community inertia, defined as the tendency for species to remain dominant under unfavorable conditions, can mask recovery from anthropogenic disturbances for many years (Milchunas and Lauenroth 1995). Finally, because most communities are subjected to multiple stressors, long-term studies of recovery are essential for separating cumulative impacts from the responses to a specific perturbation. Long-term biomonitoring of systems following remediation can provide strong evidence that a specific stressor was responsible for observed changes in a community (Box 25.3).

25.5.3 Community-Level Indicators of Recovery

Currently, there is little agreement among ecologists as to the precise definition or the appropriate measures of recovery. This lack of objective indicators will hamper our ability to determine if a system has recovered from disturbance. For example, if 90% of the species that were affected by an oil spill return to predisturbance conditions, does this mean that the system has recovered? Does recovery require that community composition and relative abundance be exactly the same as before

Box 25.3 Long-Term Recovery of a Metal-Polluted Stream in Colorado

The upper Arkansas River basin is located in the Southern Rocky Mountain ecoregion in central Colorado (USA). Mining operations have had a major impact on this stream since the late 1800s when gold was discovered in California Gulch (CG). Concentrations of heavy metals (cadmium, copper, zinc) are greatly elevated in the Arkansas River and often exceed acutely toxic levels. Between 1989 and 1999, heavy metal concentrations and benthic macroinvertebrate community structure were examined seasonally (spring and fall) from stations located upstream and downstream from Leadville Mine Drainage Tunnel (LMDT) and CG, the primary sources of heavy metals in the system. In 1992, a large-scale restoration project was initiated to reduce metal concentrations in the river. Because data were collected before and after remediation, these long-term data provide an opportunity to examine community responses to improvements in water quality.

Heavy metal levels in the Arkansas River varied temporally (seasonally and annually) and spatially (Figure 25.8). The highest concentrations were observed during periods of spring runoff and downstream from CG (station AR3). Zinc concentrations upstream from CG (station EF5) prior to remediation were generally between 200 and 600 μg/L. After 1992, these levels decreased to less than 100 μg/L. In contrast, remediation of CG has resulted in relatively little change in metal concentrations downstream.

Heptageniid mayflies, organisms known to be highly sensitive to heavy metals (Clements 1994, 1999, Kiffney and Clements 1994), quickly responded to improvements in water quality. The density of heptageniids increased significantly following remediation of LMDT. In contrast, abundance of these metal-sensitive organisms has shown relatively little change downstream from CG where metal levels remained elevated. Results of this long-term "natural experiment" demonstrate the resilience of the Arkansas River and provide strong support for the hypothesis that elevated metal levels were responsible for observed changes in benthic communities.

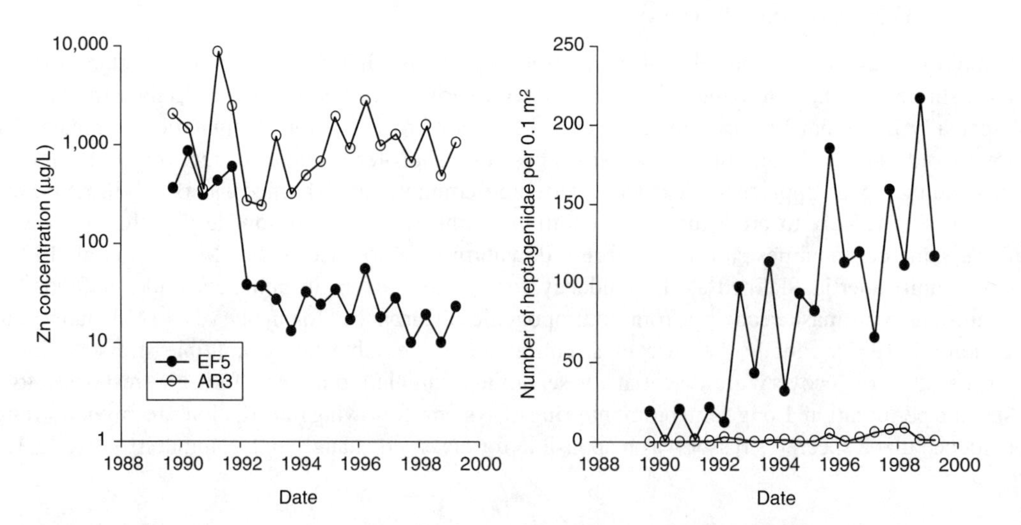

FIGURE 25.8 Results of a long term "natural" experiment showing changes in Zn concentration and responses of heptageniid mayflies (Ephemeroptera: Heptageniidae) at two stations in the Arkansas River, Colorado (USA). Reduction in Zn concentration at station EF5 was associated with an immediate increase in abundance of these metal-sensitive organisms. In contrast, there was little evidence of recovery at station AR3 where Zn levels remained elevated. These results provide evidence that the lower abundance of Heptageniidae at station EF5 before 1992 was a direct result of elevated Zn concentration.

the disturbance? If functional characteristics have returned to predisturbance conditions but community composition is different, has the system recovered? Our definition of recovery could also vary with community type. Characteristics of recovery for communities primarily regulated by stochastic factors will be quite different than for equilibrium communities. Objectively defining recovery is also complicated by the fact that different components of communities may recover at different rates. In the classic series of experiments conducted by Wallace and colleagues in Coweeta Hydrological Forest, recovery of trophic structure following additions of insecticides to a small stream occurred within 2 years. In contrast, taxonomic recovery required much longer (Wallace 1990). Disturbed communities generally show greater temporal variability than undisturbed communities, a finding that is consistent with theoretical predictions and some empirical investigations (Odum et al. 1979). Because disturbance-induced variability may be a reflection of reduced community stability (Lamberti et al. 1991), there is the intriguing possibility that temporal variability could be used as an indicator of recovery.

Although return to equilibrium conditions is an intuitively appealing definition of recovery, we know from long-term biogeographic studies that considerable turnover in species composition occurs in communities, even in the absence of disturbance. Simberloff and Wilson's (1969, 1970) study of recolonization of mangrove islands following insecticide fumigation demonstrated that the number of species rapidly returned to predisturbance values; however, the composition of these communities was often quite different, reflecting a high degree of species turnover. Similarly, research on metal pollution in the Arkansas River described in Box 25.3 has shown that the total number of species is similar between impacted and unimpacted sites. However, community composition is quite different because metal-tolerant species have replaced sensitive species (Figure 25.9). Similar results were reported for an eastern U.S. stream disturbed by flooding (Palmer et al. 1995). Despite rapid recovery in total abundance after a flood, composition of the recovering community remained distinct throughout the study period. These results demonstrate that before ecologists can determine if recovery has occurred in a community it will be necessary to identify appropriate endpoints.

Because of weaknesses associated with using any specific indicator of recovery, it is probably best to use a suite of biological measures to demonstrate a return to predisturbance conditions. Furthermore, if sensitivity of these indicators changes over time (e.g., Landis et al. 2000, Matthews et al. 1996), it is possible that no single variable will be useful throughout a study. Investigations that

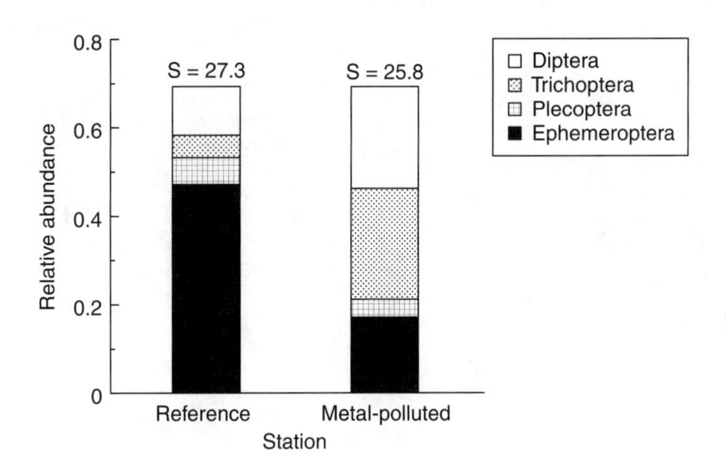

FIGURE 25.9 Comparison of species richness (S) and community composition of benthic macroinvertebrates at reference and metal-polluted sites in the Arkansas River, Colorado. Despite similar species richness at these two sites, the relative abundance of major macroinvertebrate groups was quite different. The reference site was dominated by mayflies (Ephemeroptera) whereas the metal-polluted site was dominated by caddis flies (Trichoptera) and dipterans.

focus on only a single indicator of recovery, such as abundance of economically important, rare, or charismatic species, often provide an incomplete picture of recovery. For example, studies of birds following the *Exxon Valdez* oil spill were generally limited to a few high-profile species that suffered significant mortality. The emphasis on individual species may miss important aspects of recovery that can only be assessed at the community level (Wiens et al. 1996). The same criteria used to select indicators of ecological effects in monitoring studies (Chapter 22) can also be used to select indicators of recovery. Because many disturbed systems are inherently variable, the most important characteristic of any indicator of recovery is a relatively large signal-to-noise ratio (defined as the ratio of indicator response to the sampling variability). Assuming that recovery is operationally defined as "not significantly different from predisturbance conditions," indicators that are highly variable or relatively insensitive can provide misleading information or lead to premature conclusions. Finally, because we are generally interested in knowing that both patterns and processes have returned to predisturbance conditions, indicators of recovery should include structural and functional measures.

Multivariate analysis of communities that considers spatial and temporal changes in composition is a powerful tool for assessing recovery from disturbance. Multivariate analyses provide a graphical representation of separation and overlap of communities based on linear combinations of a large number of variables (e.g., abundances of species). By conducting analyses at different time periods following disturbance, this approach could be used to test the hypothesis that a disturbed community has become more similar to either a reference community or to predisturbance conditions (Figure 25.10). A key strength of this approach is that it evaluates recovery based on the entire community, not just a few members. In addition, the relative importance of individual species in distinguishing the disturbed community from the predisturbed community can be evaluated. The use of multivariate approaches for assessing responses and recovery of communities after disturbance was described in Chapter 24.

The lack of predisturbance data imposes the greatest limitation on our ability to define recovery. Most commonly, disturbed sites are compared to reference sites that are presumed to represent predisturbance conditions. Although the selection of appropriate reference sites is a reasonable alternative to the lack of predisturbance data, this approach is also problematic. The same weaknesses

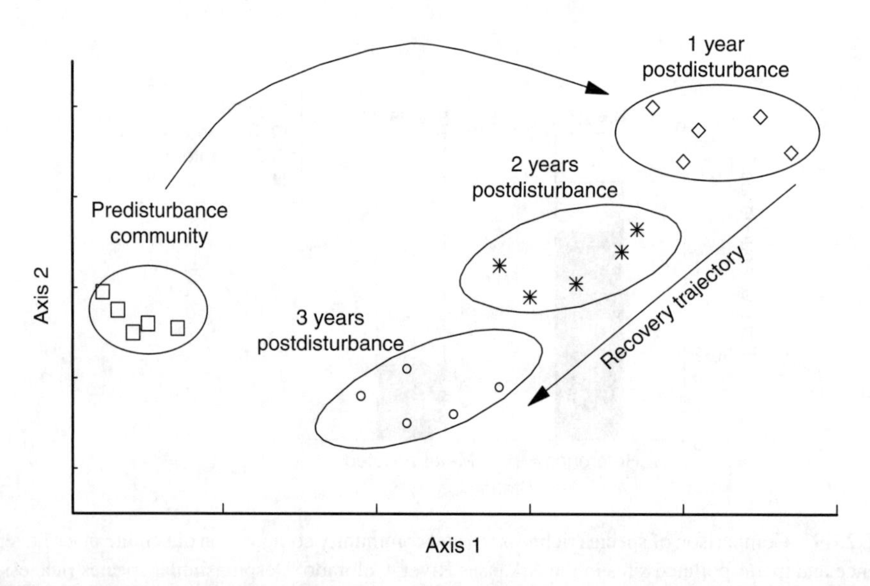

FIGURE 25.10 Multivariate analysis showing temporal changes in community composition 1, 2, and 3 years following disturbance. This hypothetical analysis is based on abundance of the dominant taxa sampled in each time period. The figure shows that the communities become more similar to predisturbance conditions over time.

associated with demonstrating causal relationships between stressors and responses in descriptive studies also apply to the study of recovery. For example, observed changes in community composition following improvements in water quality at a contaminated site could be a result of remediation. However, because other factors could be responsible for these changes, predisturbance data are necessary to support a causal relationship between restoration and community responses.

Most studies of recovery have focused on temporal changes in species richness, abundance, and community composition after a disturbance. However, some disturbances produce gradients where recovery can be observed over a large spatial scale. The best examples of spatial gradients of recovery are streams receiving point source discharges or other anthropogenic stressors that decrease in severity downstream. The most common experimental design in many stream biomonitoring studies compares upstream reference sites with downstream impacted sites along a gradient of contamination. Ignoring for a moment the concerns about pseudoreplication or the confounding influence of natural longitudinal variation (Chapter 22), the opportunity to substitute space for time to assess recovery is an attractive alternative. For example, communities along a gradient of reduced disturbance may provide some insight into the temporal patterns of recovery. This approach was used to evaluate fish communities along a disturbance gradient downstream from a hydroelectric dam (Kingsolving and Bain 1993). The design limitations associated with upstream versus downstream comparisons were addressed in this study by using a reference stream and by selecting a specific indicator (species richness of fluvial specialists) that was expected to respond to the disturbance gradient (Figure 25.11). By comparing this indicator with one considered insensitive to the disturbance gradient (species richness of fluvial generalists), the authors make a strong case that observed spatial patterns were a result of recovery.

25.5.4 Community Characteristics that Influence Rate of Recovery

Understanding factors that determine resilience may enable researchers to estimate the amount of time necessary for communities to return to predisturbance conditions. In this section, we describe some of the community-level characteristics that influence recovery from natural and anthropogenic disturbance. Steinman and McIntire (1990) reviewed biological factors that influence recovery rates of stream periphyton communities. Characteristics such as community maturity, life history features of dominant species, and frequency of disturbance were especially important predictors of recovery. Kaufman (1982) reported that communities obtained from high stress environments were more tolerant than those from stable environments. Recovery times were also influenced by community complexity as younger, simpler communities showed greater resilience than older communities. In contrast to these findings, Ivorra et al. (2000) reported that immature biofilm communities were more sensitive to heavy metals than mature communities. They attributed this difference to reduced metal penetration in thicker, mature biofilm communities. Finally, the striking resilience of seabird communities following the *Exxon Valdez* oil spill (Box 25.2) was attributed to recolonization over a large spatial scale, indicating the importance of regional factors when evaluating recovery from anthropogenic disturbances (Wiens et al. 1996).

Some researchers have questioned the suitability of traditional models used to explain recovery from natural and anthropogenic disturbance. These models often assume that communities will return to a predictable equilibrium condition following exposure to a stressor. Landis et al. (2000) state that "the search for the recovery of an ecological structure is meaningless in terms of the ecological system." The basis for their argument is that natural communities retain a long-term record of events that occurred in the past. This intriguing concept, called the community-conditioning hypothesis, has been proposed to account for the persistence of toxicant effects long after a contaminant has degraded (Landis et al. 1996, 2000, Matthews et al. 1996). Just like genetic structure reflects the unique history of a population over evolutionary time, communities are a reflection of their unique history and etiology. Events that occurred in the past are difficult to erase and can potentially influence

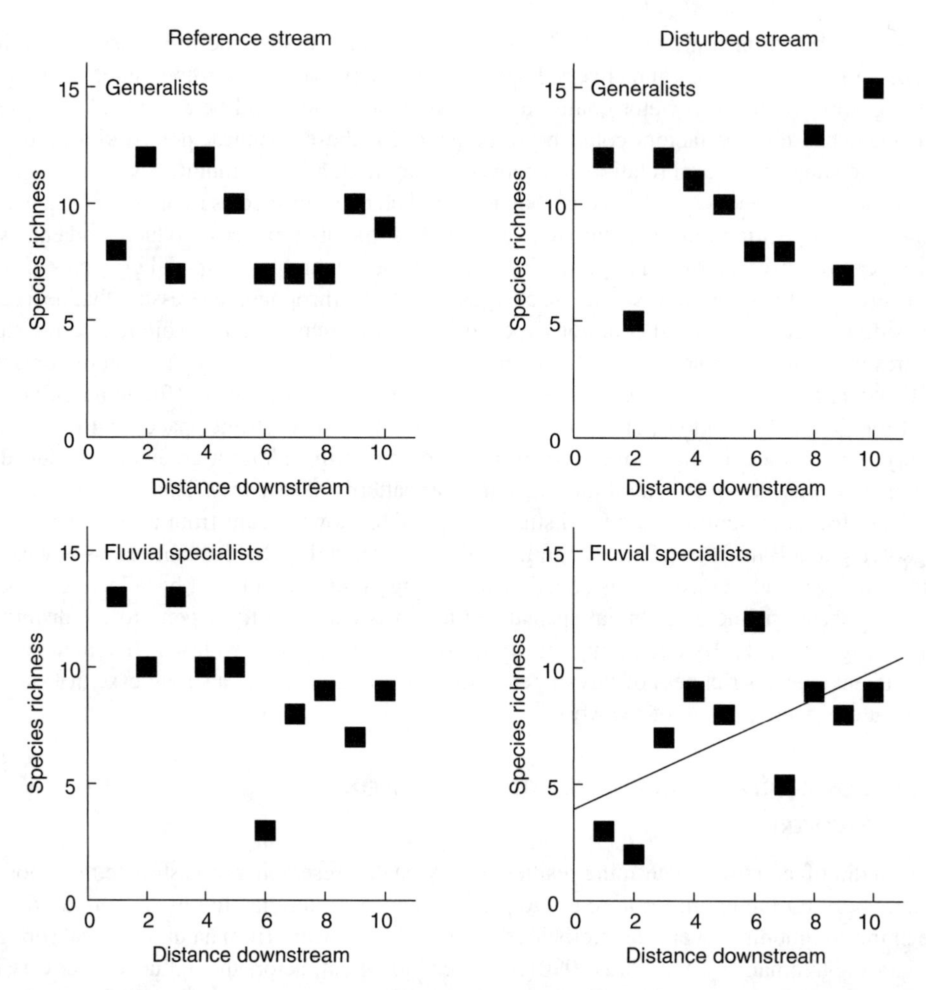

FIGURE 25.11 An alternative approach to evaluate recovery of fish communities along a disturbance gradient. Fluvial specialists are expected to be negatively affected by flow alterations and therefore should increase with distance downstream from impoundments. In contrast, fluvial generalists are not expected to be affected by flow alterations. By comparing responses of sensitive and insensitive indicators to this gradient in both reference and disturbed streams, the authors make a strong case that observed spatial patterns are a result of recovery. (Modified from Figure 6 in Kingsolving and Bain (1993).)

structural characteristics for long periods of time. Note that the community-conditioning hypothesis also provides a unified explanation for the PICT hypothesis described earlier. Previous exposure to a stressor is simply a special case of the community-conditioning hypothesis, and increased tolerance is a result of this unique historical event.

Support for the community-conditioning hypothesis is based on results of a series of standardized microcosm experiments and therefore the relevance of these findings to natural communities is open to debate (Box 25.4). However, field studies conducted by Jenkins and Buikema (1998) support the basic idea that communities established under very similar environmental conditions often differ in terms of structure. Our failure to recognize the stochastic nature of communities may be a result of our poor understanding of dispersal or, as suggested by Landis and colleagues (2000), our failure to measure relevant community variables. Nonmetric clustering and other multivariate

Box 25.4 The Community-Conditioning Hypothesis

The community-conditioning hypothesis was proposed by Matthews and Landis (Landis et al. 1996, Matthews et al. 1996) to describe the historical aspects of community structure. The hypothesis was initially proposed to explain results of a series of microcosm experiments that were inconsistent with predictions of traditional disturbance-recovery models. In these experiments, standardized communities were exposed to a nonpersistent chemical stressor (water soluble fraction of jet fuels). Because these chemicals degrade within about 10 days, differences observed several weeks after initial exposure are probably a result of persistent direct or indirect effects (Landis et al. 1996). Some of the population-level characteristics (e.g., abundance of the cladoceran, *Daphnia*) showed gradual recovery over time as predicted by equilibrium models of disturbance. However, differences in some community-level variables persisted long after the chemicals had degraded and were attributed to the unique history of each microcosm.

The assumptions of the community-conditioning hypothesis are that

(1) No two communities will ever be alike because each community is a function of its unique history and etiology.
(2) All events that influence the structure and function of a community remain a part of that community.
(3) Because historical information may be stored in a variety of compartments and layers, no single indicator will reflect the response of the entire community.
(4) Although information about previous events in a community may be difficult to extract, these events may continue to influence communities and alter future responses.
(5) Almost every environmental event has lasting effects on the community.

The community-conditioning hypothesis has important implications for how ecotoxicologists study disturbance and recovery (Matthews et al. 1996). According to this model, the effects of a disturbance event are long-lasting and communities are unlikely to return to predisturbance conditions. More importantly, if individual communities are unique, our definition of reference conditions in both field and experimental studies will require revision. Note that the individualistic nature of communities is in stark contrast to Rapport's ecosystem distress syndrome (Rapport et al. 1985), which argues that responses to disturbance are predictable and consistent among communities and stressors. The general validity of the community-conditioning hypothesis remains to be tested in other systems and will ultimately require field verification. However, if persistent historical events play a role in community organization, this will greatly complicate our attempts to define recovery using traditional equilibrium methods.

techniques advocated by these researchers for assessing contaminant effects in microcosms may also be applicable to the study of persistent ecological effects in the field.

25.6 INFLUENCE OF ENVIRONMENTAL VARIABILITY ON RESISTANCE AND RESILIENCE

Although the precise mechanism by which environmental variability influences community composition is a topic of considerable debate in ecology, the positive relationship between physical heterogeneity and species richness is well supported in the literature. Natural disturbance is likely to play an important role in maintaining environmental heterogeneity and increasing species richness. Natural and anthropogenic disturbance may also operate as a filter (sensu Southwood 1977) that constrains the types of species present in an area and ultimately determines life history characteristics, community composition, and trophic structure (Poff 1997) (see Figure 21.1). Local and regional habitat conditions create selective forces that determine life history characteristics and other traits of resident species. Taxa within a regional species pool must possess traits that allow them to pass through this hierarchical set of filters to be present within the local habitat. Using this model, the presence of anthropogenic stressors, which may occur at several different levels, is simply another filter that determines local species composition. The absence of a species is a result of its failure to pass through one of these filters, which is determined by functionally important species traits. Thus, the most important first step for predicting community composition is to identify functionally significant species traits at each level in this hierarchy. Recognizing the importance of these filters within the context of environmental heterogeneity offers an unprecedented opportunity to improve our ability to predict effects of anthropogenic disturbance on communities.

Natural disturbance is a highly selective filter that also operates at any level of this hierarchy. The influence of natural disturbance on life history characteristics, patterns of species diversity, and ecosystem processes has been examined by ecologists for many years (Matthaei et al. 1996, Poff and Ward 1990, Southwood 1977). Variable systems characterized by high levels of natural disturbance are often dominated by trophic generalists and other opportunistic species capable of rapid dispersal and high rates of reproduction. Thus, we expect that communities from these variable systems would show greater resistance and resilience to subsequent disturbances. Studies conducted in streams support the hypothesis that life history characteristics of species occupying highly variable habitats allow these organisms to tolerate natural disturbance. Matthaei et al. (1996) demonstrated that invertebrates from a more variable reach of a subalpine stream recovered from experimental disturbance faster than organisms from a site with less variability.

Despite an appreciation for the importance of environmental variability and natural disturbance on community composition, there is little information concerning how natural variability influences responses to contaminants. We propose that the same genetic, physiological, and life history characteristics that allow species to tolerate highly variable and physically stressful environments will also confer resistance and resilience to anthropogenic disturbances. In other words, frequency and intensity of natural disturbance in a community may influence its ability to withstand additional perturbations. For example, effects of an oil spill in a marine ecosystem will be much greater on a stable coral reef community than in a highly variable estuarine community. Similarly, communities in stable environments would probably require more time to recover from anthropogenic disturbance than those in variable environments (Figure 25.12). These predictions have important implications for the study of resistance and resilience among ecosystem types and across latitudinal gradients.

A review of studies conducted in aquatic systems suggests that natural variability may influence responses to contaminants. In general, communities from variable environments that were subjected to high levels of natural disturbance were more tolerant of anthropogenic stressors and recovered faster than communities from stable habitats (Poff and Ward 1990). Experiments conducted in microcosms also demonstrate that responses to chemical stressors vary among locations and that natural heterogeneity may influence these responses. Kiffney and Clements (1996) reported that effects of metals were significantly greater on communities from more stable, high elevation streams compared with those from more variable, low elevation streams. These results were supported by field studies, which also showed that impacts of metals were often greater on smaller headwater streams compared to large rivers. Experiments conducted by Angeler et al. (2006) demonstrated that wetland communities, which are often subjected to natural fluctuating conditions, were preadapted to disturbance and therefore relatively resistant to contaminants.

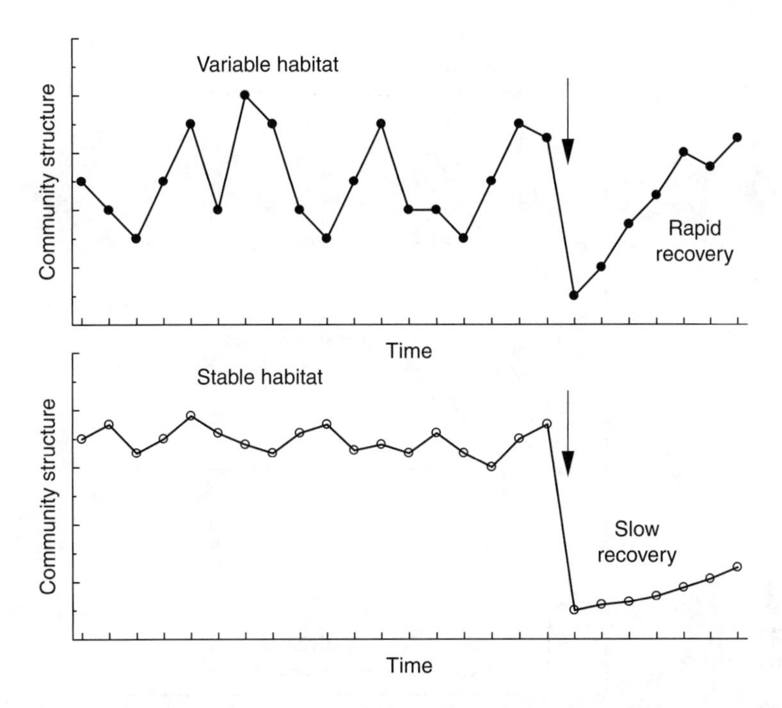

FIGURE 25.12 Conceptual model showing recovery in stable and naturally variable communities. The model assumes that because of genetic, physiological, and life history characteristics, populations in naturally variable habitats are preadapted to disturbance. Therefore, these populations are expected to recover from natural or anthropogenic disturbance faster than populations from stable habitats.

25.7 QUANTIFYING THE EFFECTS OF COMPOUND PERTURBATIONS

Like the historical emphasis on single stressors in aquatic toxicology, ecologists have generally limited their investigations of disturbance to single events. Because of the focus on individual stressors, we lack the ability to predict how communities will respond to multiple perturbations and have implicitly assumed that responses to multiple stressors are additive. Because anthropogenic stressors are often superimposed on natural disturbance regimes, multiple perturbations are probably common in nature. Paine et al. (1998) examined responses of communities subjected to a variety of single and multiple perturbations, including El Nino events, storms, exotic species, wildfires, deforestation, and hypoxia. Their theoretical model of ecological responses to multiple perturbations has important implications for the study of contaminants and other anthropogenic stressors. Although most communities can recover from natural disturbance (Figure 25.13a), a sequential perturbation, especially one that occurs during the recovery phase, may reset the community to an alternative stable state (Figure 25.13b). Finally, the addition of natural or anthropogenic disturbance to a community already subjected to a chronic stressor may alter patterns of recovery. Paine et al. (1998) suggest that disturbance in chronically stressed systems may impede recovery and force the system to a new stable state (Figure 25.13c). One very likely scenario for multiple perturbations involves systems currently subjected to widespread global perturbations such as climate change, ultraviolet-B (UV-B) radiation, and acidification (Chapter 26).

25.7.1 Sensitivity of Communities to Novel Stressors

Although there is evidence to support the hypothesis that previously exposed populations will be tolerant to contaminants, there is a need to understand mechanisms responsible for increased tolerance.

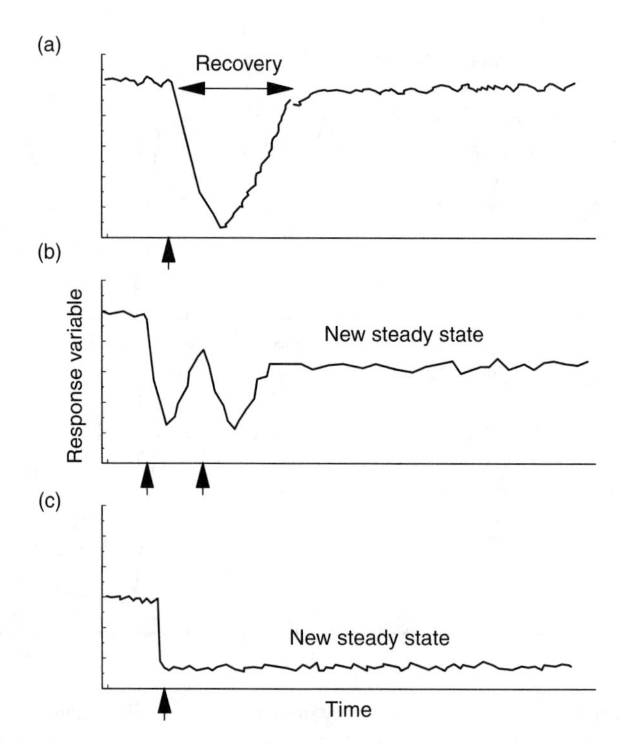

FIGURE 25.13 Compound perturbations in ecological systems. (a) A single disturbance is followed by gradual recovery to prepredisturbance conditions. (b) A single disturbance is followed by a second disturbance before recovery has occurred. The system is reset to a new steady state. (c) A single disturbance is superimposed on a system already altered by perturbations. The system is reset to a new steady state. (Modified from Figure 1 in Paine (1998).)

We argue that both physiological acclimation and adaptation to contaminants may have specific costs. For example, while induction of metal-binding proteins such as metallothionein increases tolerance to subsequent exposure, it also utilizes energy normally available for other metabolic processes (e.g., growth, reproduction). High metabolic costs associated with sequestering metals and/or producing metal-binding proteins decreases productivity in metal-tolerant plants (Wilson 1988). Springtails (Collembola) exposed to copper and lead exhibited reduced growth rates owing to increased molting necessary to eliminate metals (Hollaway et al. 1990). Similarly, genetic changes associated with exposure to contaminants may also have costs to populations. Reduced genetic heterozygosity has been reported in populations exposed to contaminants (Heagler et al. 1993, Mulvey and Diamond 1991) and may be caused by population bottlenecks. Furthermore, as metal-sensitive genotypes are eliminated from a population, the reduced genetic diversity may increase the susceptibility of this population to other stressors.

Acclimating or adapting to one set of environmental stressors may also increase an organism's sensitivity to novel stressors (Antonovics et al. 1971, Wilson 1988). Although most of the research on cost of tolerance has focused on populations, communities from chemically stressed environments may also be at greater risk from novel stressors. Support for this hypothesis is provided by results of experimental studies in which communities from metal-polluted and metal-unpolluted sites were exposed to acidification and UV radiation (Box 25.5). In all instances, communities from the metal-polluted sites were more sensitive to novel stressors (Clements 1999, Courtney and Clements 2000, Kiffney et al. 1997). An understanding of the potential effects of novel stressors on chronically disturbed systems is essential for predicting ecological responses. This understanding can only be achieved by integrating long-term monitoring with laboratory and field experiments.

Box 25.5 The Effects of Novel Stressors on Chronically Impacted Communities

Microcosm experiments have shown that organisms collected from the Arkansas River, a metal-polluted stream in central Colorado, were more tolerant of metals than organisms collected from an unpolluted stream (Figure 25.14). The mechanisms for increased tolerance most likely included physiological and genetic changes in these populations (Clements 1999, Courtney and Clements 2000). To test the hypothesis that these metal-tolerant organisms are more susceptible to novel stressors, a series of experiments was conducted in which benthic macroinvertebrate communities from the Arkansas River and a reference stream were exposed to acidification, UV-B radiation, and predation. Despite greater tolerance to metals, organisms from the Arkansas River were generally more susceptible to each of the novel stressors. Exposure to acidic pH decreased mayfly survival at all three sites; however, organisms from the most polluted station were especially sensitive. Similar results were observed for benthic communities exposed to UV-B radiation. Mayflies from a reference stream were relatively tolerant of UV-B radiation, whereas organisms from the Arkansas River were more sensitive. Finally, results of microcosm and field experiments showed that mayflies from the Arkansas River were more susceptible to stonefly predation than organisms from a reference stream.

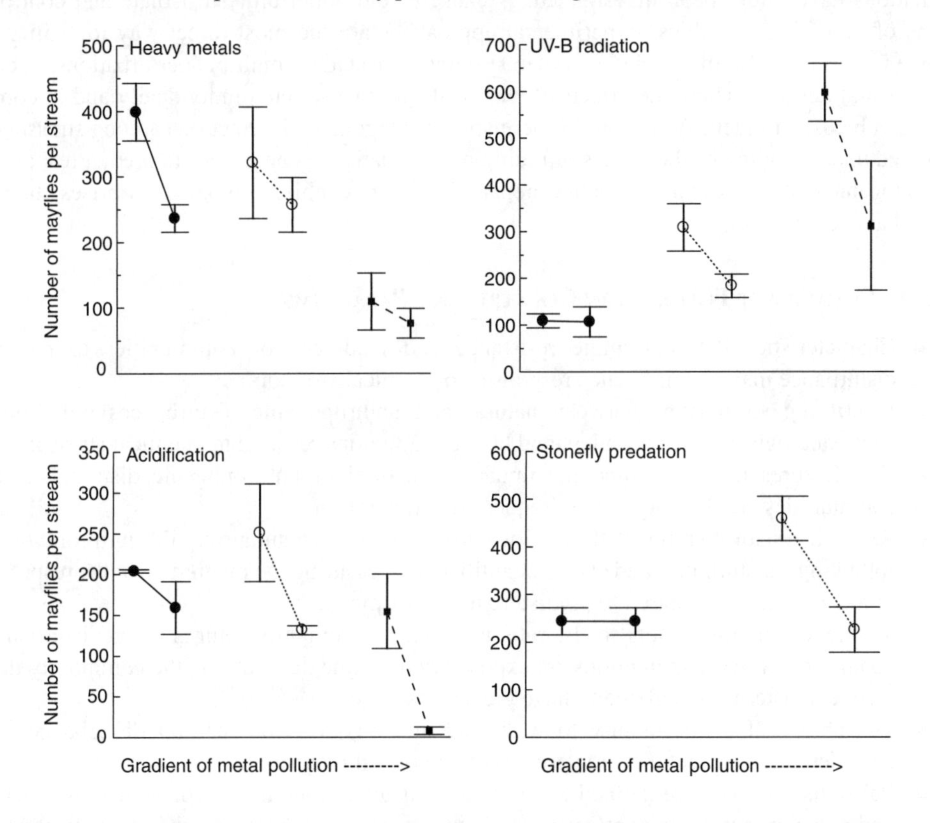

FIGURE 25.14 Results of microcosm experiments showing differences in sensitivity of mayflies collected from reference and metal-impacted sites to heavy metals, acidification, UV-B radiation, and stonefly predation. Macroinvertebrate communities for these experiments were collected from field sites along a gradient of metal pollution. (Solid circle = reference site; open circle = Moderately contaminated; closed square = heavily contaminated) Each pair of points connected by a line shows abundance of mayflies in control microcosms and treated microcosms. (Data from Clements, W.H., *Ecol. Appl.*, 9, 1073–1084, 1999; Courtney, L.A. and Clements, W.H., *J. N. Am. Benthol. Soc.*, 19, 112–127, 2000; and Peter M. Kiffney, Northwest Fisheries Science Center, Seattle, WA, unpublished.)

Results of these experiments demonstrate that these three communities responded with varying degrees of sensitivity to physical, chemical, and biological stressors, and that differences were related to differences in exposure history and community structure. Chronic metal pollution in the Arkansas River has influenced community structure and resulted in a community that is tolerant to metals but more sensitive to novel stressors. These results support the hypothesis that the physiological and genetic mechanisms responsible for increased tolerance to heavy metals may have a cost.

25.8 SUMMARY

In summary, although there has been significant improvement in our ability to quantify the effects of single chemical stressors on communities over the past 20 years, we generally have a poor understanding of the interactions among stressors and how these stressors interact with other physical and biological disturbances. Laboratory research has focused on quantifying interactions among chemicals using single species toxicity tests; however, the effects of stressor interactions on field populations have rarely been investigated. Because of the opportunity to isolate and control the effects of individual variables, experimental approaches are the most direct way to estimate the relative importance of individual stressors in systems impacted by multiple perturbations. Factorial experimental designs, where the effects of chemicals are investigated individually and in combination, can be used to identify potential synergistic or antagonistic interactions among stressors. An improved understanding of the effects of multiple perturbations is necessary to predict responses to anthropogenic disturbances and to better prepare for the inevitable "ecological surprises" that may occur (Paine et al. 1998).

25.8.1 SUMMARY OF FOUNDATION CONCEPTS AND PARADIGMS

- Characteristics that determine resistance and resilience of communities to natural disturbance may also influence responses to chemical stressors.
- Identifying similarities between natural and anthropogenic disturbances will allow ecotoxicologists to better understand how communities respond to chemical stressors.
- Key features that determine the impact of natural or anthropogenic disturbance on communities are the magnitude, frequency, and duration.
- Resistance stability refers to the ability of a community to maintain equilibrium conditions following a disturbance and can be quantified by measuring the magnitude of the response of a community compared to predisturbance conditions.
- Resilience stability refers to the rate at which a community returns to predisturbance conditions. If two communities are exposed to the same disturbance, the community that recovers faster is considered to have greater resilience.
- The ability of a community to withstand and recover from natural disturbance are considered fundamental characteristics of ecological integrity.
- Pulse disturbances are defined as instantaneous alterations in the abundance of species within a community. In contrast, press disturbances cause sustained alterations in abundance of species, often resulting in the elimination of some taxa and establishment of a new community.
- Definitions used to distinguish pulse and press disturbances have been criticized because they combine cause (e.g., disturbance) with effect (e.g., the response of the community).
- The positive relationship between species diversity and resistance or resilience to anthropogenic stressors has serious implications for understanding how communities respond to anthropogenic stressors.

- A unifying feature that has emerged from basic research on disturbance is the remarkable resilience of some communities to a wide range of natural perturbations.
- From an ecotoxicological perspective, a fundamental question is whether research on responses to natural disturbance can be employed to predict recovery from anthropogenic disturbance.
- Although there is empirical support for the hypothesis that effects of contaminants vary among communities, there have been few attempts to identify specific factors responsible for this variation.
- Variation in resistance and resilience among different communities greatly complicates our ability to predict community responses and ecological recovery.
- Identifying characteristics that increase community sensitivity to contaminants and the mechanisms responsible for recovery are important areas of research.
- Rapport et al. (1985) evaluated the responses of several communities to different types of disturbance and developed an "ecosystem distress syndrome."
- The IDH predicts that communities subjected to moderate levels of disturbance will have greater species richness or diversity compared to communities existing under benign conditions.
- PICT resulting from elimination of sensitive species and the replacement by tolerant species is considered strong evidence that community restructuring was caused by the pollutant.
- The use of PICT to assess impacts of contaminants is based on the assumptions that (1) sensitivity to contaminants varies among species; (2) contaminants will restructure communities, with sensitive species being replaced by tolerant species; and (3) differences in tolerance among communities can be detected using short-term experiments.
- Recovery from natural or anthropogenic disturbance is determined by a complex suite of factors related to the characteristics of the community, severity of the disturbance, and physical features of the disturbed habitat.
- The study of recovery from natural disturbance may allow researchers to understand and predict how communities recover from anthropogenic disturbance.
- The rapid recovery of lotic systems reported is determined by (1) life history characteristics of species that allowed for rapid recolonization; (2) the proximity of upstream and downstream undisturbed sites to provide a source of colonists; (3) the high flushing rate of lotic systems; and (4) the general adaptations of many stream organisms to natural disturbance.
- Because of limitations associated with using any specific indicator of recovery, it is probably best to use a suite of biological measures to demonstrate a return to predisturbance conditions.
- Understanding factors that determine resilience may enable researchers to estimate the amount of time necessary for communities to return to predisturbance conditions.
- The community-conditioning hypothesis (Landis et al. 1996) has been proposed to account for the persistence of toxicant effects long after a contaminant has degraded.
- Naturally disturbed systems are often dominated by trophic generalists and other opportunistic species. Consequently, we expect that communities from these systems would show greater resistance and resilience to anthropogenic disturbances.
- Because of the traditional focus on individual stressors, we lack the ability to predict how communities will respond to multiple perturbations and have implicitly assumed that responses to multiple stressors are additive.
- Acclimating or adapting to one set of environmental stressors may also increase an organism's sensitivity to novel stressors.

REFERENCES

Allison, S.K., Recovery from small-scale anthropogenic disturbances by northern California salt marsh plant assemblages, *Ecol. Appl.*, 5, 693–702, 1995.

Angeler, D.G., Sanchez, B., Garcia, G., and Moreno, J.M., Community ecotoxicology: Invertebrate emergence from fire Trol 934 contaminated vernal pool and salt marsh sediments under contrasting photoperiod and temperature regimes, *Aquat. Toxicol.*, 78, 167–175, 2006.

Antonovics, J., Bradshaw, A.D., and Turner, R.G., Heavy metal tolerance in plants, *Adv. Ecol. Res.*, 7, 1–85, 1971.

Begon, M., Harper, J.L., and Townsend, C.R., *Ecology: Individuals, Populations and Communities*, Blackwell Scientific Publications, Cambridge, MA, 1990.

Bender, E.A., Case, T.J., and Gilpin, M.E., Perturbation experiments in community ecology: Theory and practice, *Ecology*, 65, 1–13, 1984.

Blanck, H., A critical review of procedures and approaches used for assessing pollution-induced community tolerance (PICT) in biotic communities, *Human Ecol. Risk Assess.*, 8, 1003–1034, 2002.

Blanck, H. and Dahl, B., Pollution-induced community tolerance (PICT) in marine periphyton in a gradient of tri-n-butyltin (TBT) contamination, *Aquat. Toxicol.*, 35, 59–77, 1996.

Blanck, H. and Wangberg, S., Induced community tolerance in marine periphyton established under arsenate stress, *Can. J. Fish. Aquat. Sci.*, 45, 1816–1819, 1988.

Blanck, H. and Wangberg, S.A., Pattern of co-tolerance in marine periphyton communities established under arsenate stress, *Aquat. Toxicol.*, 21, 1–14, 1991.

Clements, F.E., Nature and structure of the climax, *J. Ecol.*, 24, 252–284, 1936.

Clements, W.H., Benthic invertebrate community responses to heavy metals in the Upper Arkansas River Basin, Colorado, *J. N. Am. Benthol. Soc.*, 13, 30–44, 1994.

Clements, W.H., Metal tolerance and predator-prey interactions in benthic macroinvertebrate stream communities, *Ecol. Appl.*, 9, 1073–1084, 1999.

Clements, W.H., Carlisle, D.M., Lazorchak, J.M., and Johnson, P.C., Heavy metals structure benthic communities in Colorado mountain streams, *Ecol. Appl.*, 10, 626–638, 2000.

Connell, J.H., Diversity in tropical rain forests and coral reefs, *Science*, 199, 1302–1310, 1978.

Courtney, L.A. and Clements, W.H., Sensitivity to acidic pH in benthic invertebrate assemblages with different histories of exposure to metals, *J. N. Am. Benthol. Soc.*, 19, 112–127, 2000.

Dahl, B. and Blanck, H., Pollution-induced community tolerance (PICT) in periphyton communities established under tri-n-butyltin (TBT) stress in marine microcosms, *Aquat. Toxicol.*, 34, 305–325, 1996.

Darwin, C., *The Origin of Species, Sixth London Edition*, Thompson and Thomas, Chicago, IL, 1872.

Davis, M.R.H., Zhao, F.J., and McGrath, S.P., Pollution-induced community tolerance of soil microbes in response to a zinc gradient, *Environ. Toxicol. Chem.*, 23, 2665–2672, 2004.

Death, R.G. and Winterbourn, M.J., Diversity patterns in stream benthic invertebrate communities: The influence of habitat stability, *Ecology*, 76, 1446–1460, 1995.

Doak, D.F., Bigger, D., Harding, E.K., Marvier, M.A., O'Malley, R.E., and Thomson, D., The statistical inevitability of stability–diversity relationships in community ecology, *Am. Nat.*, 151, 264–276, 1998.

Elton, C.S., *The Ecology of Invasions by Animals and Plants*, Methuen, London, England, 1958.

Fisher, S.G. and Grimm, N.B., Streams and disturbance: Are cross-ecosystem comparisons useful? In *Comparative Analyses of Ecosystems: Patterns, Mechanisms, and Theories*, Cole, J., Lovett, G., and Findlay, S. (eds.), Springer-Verlag, New York, 1991, pp 196–221.

Foster, D.R., Aber, J.D., Melillo, J.M., Bowden, R.D., and Bazzaz, F.A., Forest response to disturbance and anthropogenic stress, *Bioscience*, 47, 437–445, 1997.

Franklin, J.F. and MacMahon, J.A., Messages from a mountain, *Science*, 288, 1183–1185, 2000.

Glasby, T.M. and Underwood, A.J., Sampling to differentiate between pulse and press perturbations, *Environ. Monitor. Assess.*, 42, 241–252, 1996.

Grant, A., Pollution-tolerant species and communities: Intriguing toys or invaluable monitoring tools? *Human Ecol. Risk Assess.*, 8, 955–970, 2002.

Gustavson, K. and Wangberg, S.A., Tolerance induction and succession in microalgae communities exposed to copper and atrazine, *Aquat. Toxicol.*, 32, 283–302, 1995.

Hart, D.D., Building a stronger partnership between ecological research and biological monitoring, *J. N. Am. Benthol. Soc.*, 13, 110–116, 1994.

Heagler, M.G., Newman, M.C., Mulvey, M., and Dixon, P.M., Allozyme genotype in mosquitofish, *Gambusia holbrooki*, during mercury exposure: Temporal stability, concentration effects and field verification, *Environ. Toxicol. Chem.*, 12, 385–395, 1993.

Holloway, G.J., Sibly, R.M., and Povey, S.R., Evolution in toxin-stressed environments, *Funct. Ecol.*, 4, 289–294, 1990.

Howarth, R.W., Comparative responses of aquatic ecosystems to toxic chemical stress, In *Comparative Analyses of Ecosystems: Patterns, Mechanisms, and Theories*, Cole, J., Lovett, G., and Findlay, S. (eds.), Springer-Verlag, New York, 1991, pp. 169–195.

Ivorra, N., Bremer, S., Guasch, H., Kraak, M.H.S., and Admiraal, W., Differences in the sensitivity of benthic microalgae to Zn and Cd regarding biofilm development and exposure history, *Environ. Toxicol. Chem.*, 19, 1332–1339, 2000.

Jenkins, D.G. and Buikema, A.L., Jr., Do similar communities develop in similar sites? A test with zooplankton structure and function, *Ecol. Monogr.*, 68, 421–443, 1998.

Johnston, E.L. and Keough, M.J., Reduction of pollution impacts through the control of toxicant release rate must be site- and season-specific, *J. Exp. Mar. Biol. Ecol.*, 320, 9–33, 2005.

Kashian, D.R., Zuellig, R.E., Mitchell, K.A., and Clements, W.H., The cost of tolerance: Sensitivity of stream benthic communities to UV-B and metals, *Ecol. Appl.*, 17, 365–375, 2007.

Kaufman, L.H., Stream aufwuchs accumulation: Disturbance frequency and stress resistance and resilience, *Oecologia*, 52, 57–63, 1982.

Kiffney, P.M. and Clements, W.H., Effects of heavy metals on a Rocky Mountain stream macroinvertebrate assemblage in experimental streams, *J. N. Am. Benthol. Soc.*, 13, 511–523, 1994.

Kiffney, P.M. and Clements, W.H., Effects of metals on stream macroinvertebrate assemblages from different altitudes, *Ecol. Appl.*, 6, 472–481, 1996.

Kiffney, P.M., Clements, W.H., and Cady, T.A., Influence of ultraviolet radiation on the colonization dynamics of a Rocky Mountain stream benthic community, *J. N. Am. Benthol. Soc.*, 16, 520–530, 1997.

Kinsolving, A.D. and Bain, M.B., Fish assemblage recovery along a riverine disturbance gradient, *Ecol. Appl.*, 3, 531–544, 1993.

Knopper, L.D. and Siciliano, S.D., A hypothetical application, of the pollution-induced community tolerance concept in megafaunal communities found at contaminated sites, *Human Ecol. Risk Assess.*, 8, 1057–1066, 2002.

Lamberti, G.A., Gregory, S.V., Ashkenas, L.R., Wildman, R.C., and Moore, K.M.S., Stream ecosystem recovery following a catostrophic debris flow, *Can. J. Fish. Aquat. Sci.*, 48, 196–208, 1991.

Landis, W.G., Markiewicz, A.J., Matthews, R.A., and Matthews, G.B., A test of the community conditioning hypothesis: Persistence of effects in model ecological structures dosed with the jet fuel JP-8, *Environ. Toxicol. Chem.*, 19, 327–336, 2000.

Landis, W.G., Matthews, R.A., and Matthews, G.B., The layered and historical nature of ecological systems and the risk assessment of pesticides, *Environ. Toxicol. Chem.*, 15, 432–440, 1996.

Lawler, S.P. and Morin, P.J., Food web architecture and population dynamics in laboratory microcosms of protists, *Am. Nat.*, 141, 675–686, 1993.

Lenihan, H.S. and Oliver, J.S., Anthropogenic and natural disturbance to marine benthic communities in Antarctica, *Ecol. Appl.*, 5, 311–326, 1995.

Lozano, R. and Pratt, J.R., Interaction of toxicants and communities: The role of nutrients, *Environ. Toxicol. Chem.*, 13, 361–368, 1994.

Luoma, S.N., Detection of trace contaminant effects in aquatic ecosystems, *J. Fish. Res. Bd. Can.*, 34, 436–439, 1977.

Matthaei, C.D., Uehlinger, U., Meyer, E.I., and Frutiger, A., Recolonization by benthic invertebrates after experimental disturbance in a Swiss prealpine river, *Freshw. Biol.*, 35, 233–248, 1996.

Matthews, R.A., Landis, W.G., and Matthews, G.B., The community conditioning hypothesis and its application to environmental toxicology, *Environ. Toxicol. Chem.*, 15, 597–603, 1996.

May, R.M., *Stability and Complexity in Model Ecosystems*, Princeton University Press, Princeton, NJ, 1973.

Medley, C.N. and Clements, W.H., Responses of diatom communities to heavy metals in streams: The influence of longitudinal variation, *Ecol. Appl.*, 8, 631–644, 1998.

Milchunas, D.G. and Lauenroth, W.K., Inertia in plant community structure: State changes after cessation of nutrient enrichment stress, *Ecol. Appl.*, 5, 452–458, 1995.

Millward, R.N. and Grant, A., Assessing the impact of copper on nematode communities from a chronically metal-enriched estuary using pollution-induced community tolerance, *Mar. Pollut. Bull.*, 30, 701–706, 1995.

Millward, R.N. and Grant, A., Pollution-induced tolerance to copper of nematode communities in the severely contaminated Restronguet Creek and adjacent estuaries, Cornwall, United Kingdom, *Environ. Toxicol. Chem.*, 19, 454–461, 2000.

Minshall, G.W., Robinson, C.T., and Lawrence, D.E., Postfire responses of lotic ecosystems in Yellowstone National Park, USA, *Can. J. Fish. Aquat. Sci.*, 54, 2509–2525, 1997.

Mulvey, M. and Diamond, S.A., Genetic factors and tolerance aquisition in populations exposed to metals and metalloids, In *Ectoxicology of Metals: Current Concepts and Applications*, Newman, M.C. and McIntosh, A.W. (eds.), Lewis Publishers, Chelsea, MI, 1991.

Niederlehner, B.R. and Cairns, J., Jr., Community response to cumulative toxic impact: Effects of acclimation on zinc tolerance of aufwuchs, *Can. J. Fish. Aquat. Sci.*, 49, 2155–2163, 1992.

Niemi, G.J., Devore, P., Detenbeck, N., Taylor, D., Lima, A., Yount, J.D., and Naiman, R.J., Overview of case-studies on recovery of aquatic systems from disturbance, *Environ. Manage.*, 14, 571–587, 1990.

Nilsson, C. and Grelsson, G., The fragility of ecosystems: A review, *J. Appl. Ecol.*, 32, 677–692, 1995.

Nystrom, B., Paulsson, M., Almgren, K., and Blanck, H., Evaluation of the capacity for development of atrazine tolerance in periphyton from a Swedish freshwater site as determined by inhibition of photosynthesis and sulfolipid synthesis, *Environ. Toxicol. Chem.*, 19, 1324–1331, 2000.

Odum, E.P., Finn, J.T., and Franz, E.H., Perturbation theory and the subsidy-stress gradient, *Bioscience*, 29, 349–352, 1979.

Paine, R.T., Tegner, M.J., and Johnson, E.A., Compounded perturbations yield ecological surprises, *Ecosystems*, 1, 535–545, 1998.

Palmer, M.A., Arensburger, P., Botts, P.S., Hakenkamp, C.C., and Reid, J.W., Disturbance and the community structure of stream invertebrates: Patch-specific effects and the role of refugia, *Freshw. Biol*, 34, 343–356, 1995.

Petersen, S. and Gustavson, K., Toxic effects of tri-butyl-tin (TBT) on autotrophic pico-, nano-, and microplankton by a size fractionated pollution-induced community tolerance (SF-PICT) concept, *Aquat. Toxicol.*, 40, 253–264, 1998.

Peterson, C.H., Stability of species and of community for the benthos of two lagoons, *Ecology*, 56, 958–965, 1975.

Peterson, G., Allen, C.R., and Holling, C.S., Ecological resilience, biodiversity, and scale, *Ecosystems*, 1, 6–18.

Poff, N.L., Landscape filters and species traits: Towards mechanistic understanding and prediction in stream ecology, *J. N. Am. Benthol. Soc.*, 16, 391–409, 1997.

Poff, N.L. and Ward, J.V., Physical habitat template of lotic systems: Recovery in the context of historical pattern of spatiotemporal heterogeneity, *Environ. Manage.*, 14, 629–645, 1990.

Rapport, D.J., Regier, H.A., and Hutchinson, T.C., Ecosystem behavior under stress, *Am. Nat.*, 125, 617–640, 1985.

Rapport, D.J., Whitford, W.G., and Hilden, M., Common patterns of ecosystem breakdown under stress, *Environ. Monitor. Assess.*, 51, 171–178, 1998.

Reice, S.R., Experimental disturbance and the maintenance of species diversity in a stream community, *Oecologia*, 67, 90–97, 1985.

Reice, S.R., Nonequilibrium determinants of biological community structure, *Am. Sci.*, 82, 1994.

Rohr, J.R., Kerby, J.L., and Sih, A., Community ecology as a framework for predicting contaminant effects, *Trends Ecol. Evol.*, 21, 606–613, 2006.

Sankaran, M. and McNaughton, S.J., Determinants of biodiversity regulate compositional stability of communities, *Nature*, 401, 691–693, 1999.

Schwab, B.S., Maruscik, D.A., Ventullo, R.M., and Palmisano, A.C., Adaptation of periphytic communities to a quaternary ammonia surfactant, *Environ. Toxicol. Chem.*, 11, 1169–1177, 1992.

Selye, H., The evolution of the stress concept, *Am. Sci.*, 61, 692–699, 1973.

Simberloff, D.S. and Wilson, E.O., Experimental zoogeography of islands: The colonization of empty islands, *Ecology*, 50, 278–296, 1969.

Simberloff, D.S. and Wilson, E.O., Experimental zoogeography of islands: A two-year record of colonization, *Ecology*, 51, 934–937, 1970.

Sousa, W.P., Disturbance in marine intertidal boulder fields: The nonequilibrium maintenance of species diversity, *Ecology*, 60, 1225–1239, 1979.

Southwood, T.R.E., Habitat, the templet for ecological strategies?, *J. Anim. Ecol.*, 46, 337–367, 1977.

Spiller, D.A., Losos, J.B., and Schoener, T.W., Impact of a catostrophic hurricane on island populations, *Science*, 281, 695–697, 1998.

Steinman, A.D. and McIntire, C.D., Recovery of lotic periphyton communities after disturbance, *Environ. Manage.*, 14, 589–604, 1990.

Steinman, A.D., Mulholland, P.J., Palumbo, A., DeAngelis, D., and Flum, T., Lotic ecosystem response to a chlorine disturbance, *Ecol. Appl.*, 2, 341–355, 1992.

Tilman, D., Biodiversity: Population versus ecosystem stability, *Ecology*, 77, 350–363, 1996.

Turner, M.G., Romme, W.H., Gardner, R.H., and Hargrove, W.W., Effects of fire size and pattern on early succession in Yellowstone National Park, *Ecol. Monogr.*, 67, 411–433, 1997.

Vieira, N.K.M., Clements, W.H., Guevara, L.S., and Jacobs, B.F., Resistance and resilience of stream insect communities to repeated hydrologic disturbances after a wildfire, *Freshw. Biol.*, 49, 1243–1259, 2004.

Wallace, J.B., Recovery of lotic macroinvertebtate communities from disturbance, *Environ. Manage.*, 14, 605–620, 1990.

Wangberg, S.A., Effects of arsenate and copper on the algal communities in polluted lakes in the northern parts of Sweden assayed by PICT (Pollution-Induced Community Tolerance), *Hydrobiologia*, 306, 109–124, 1995.

White, P.S. and Pickett, S.T.A. (eds.), *The Ecology of Natural Disturbance and Patch Dynamics*, Academic Press, New York, 1985.

Whitford, W.G., Rapport, D.J., and deSoyze, A.G., Using resistance and resilience measurements for "fitness" tests in ecosystem health, *J. Environ. Manage.*, 57, 21–29, 1999.

Wiens, J.A., On understanding a non-equilibrium world: Myth and reality in community patterns and processes, In *Ecological Communities: Conceptual Issues and the Evidence*, Strong, D.R., Simberloff, D., Abele, L.G., and Thistle, A.B. (eds.), Princeton University Press, Princeton, NJ, 1984, pp. 439–457.

Wiens, J.A., Crist, T.O., Day, R.H., Murphy, S.M., and Hayward, G.D., Effects of the *Exxon Valdez* oil spill on marine bird communities in Prince William Sound, *Ecol. Appl.*, 6, 828–841, 1996.

Wilson, J.B., The cost of heavy-metal tolerance: An example, *Evolution*, 42, 408–413, 1988.

Wootton, J.T., Effects of disturbance on species diversity: A multitrophic perspective, *Am. Nat.*, 152, 803–825, 1998.

Yount, J.D. and Niemi, G.J., Recovery of lotic ecosystems from disturbance—a narrative review of case studies, *Environ. Manage.*, 14, 547–569, 1990.

26 Community Responses to Global and Atmospheric Stressors

26.1 INTRODUCTION

Several decades ago, our concerns about atmospheric pollutants were primarily limited to those associated with the combustion of fossil fuels (e.g., SO_2, NO_x, H^+). While fossil fuel combustion is still regarded as a major source for many atmospheric pollutants, our contemporary definition of atmospheric pollutants has broadened considerably. We now recognize that many stressors are globally distributed (Table 26.1) and that the temporal scale of effects ranges from days to centuries. Today we have serious concerns not only about the direct effects of atmospheric pollutants such as CH_4 and CO_2, but also their indirect effects on global climate. In addition, persistent organic pollutants (POPs), once regarded as a local problem primarily associated with industrial and agricultural discharges, are now globally distributed and occur in very remote environments such as the Canadian arctic. We now understand that not only is ozone (O_3) a serious atmospheric stressor for many plants, but that *loss* of stratospheric ozone owing to release of chlorofluorocarbons (CFCs) has significantly increased levels of ultraviolet radiation (UVR) striking the earth's surface over the past 20 years.

Effects of atmospheric stressors on communities are likely to be complex, interactive, and difficult to predict. The regional and global distribution of atmospheric pollutants presents unique challenges for study design and interpretation. Long-range transport of some atmospheric pollutants (e.g., SO_2, NO_x) and geographic variation in exposure to other stressors (e.g., UV-B radiation) complicate assessment of effects. Researchers are often forced to extrapolate results of relatively small-scale and short duration studies to much larger spatiotemporal scales. Some communities, particularly those characterized by long-lived species (e.g., forests), will respond very slowly to atmospheric stressors. Because long-term data from these systems are often unavailable, simply demonstrating that forest health has declined is challenging. Attempting to associate forest decline to a particular atmospheric stressor such as acidification is a daunting task. Finally, differences in sensitivity to global atmospheric stressors among communities further complicate our ability to make predictions about ecological effects. For example, Rusek (1993) observed that alpine communities were more sensitive to acidic deposition than either subalpine or forest communities. If the greater sensitivity of alpine communities to disturbance is a general phenomenon, changes in community structure and function observed in these habitats may provide an early warning of stress.

As with each class of anthropogenic disturbance we have considered in this book, community responses to global atmospheric stressors are a result of both direct and indirect effects. Significant changes in community composition will likely occur as a result of species-specific differences in exposure and sensitivity to atmospheric stressors. However, some researchers speculate that indirect effects of global warming, acidification, and UVR on species interactions will be greater than direct effects (Field et al. 1992). For example, increased susceptibility to disease or parasites may be a more likely cause of forest decline than direct effects of acidification.

Our discussion of atmospheric and global pollutants in this chapter will be limited to three stressors: CO_2 and associated global warming, acidic deposition, and UV-B radiation owing to stratospheric ozone depletion. Although we recognize the importance of other globally distributed

TABLE 26.1
Spatial and Temporal Scale, Sources, and Primary Concerns of Major Atmospheric Pollutants

Category	Pollutant	Temporal Scale	Spatial Scale	Primary Anthropogenic Source	Primary Concerns
Carbon	CH_4	Years	Globe	Agriculture, fossil fuels	Global warming
	CO_2	Decades	Globe	Fossil fuels, deforestation	Global warming, direct effects on plants
	CO	Days	Hemisphere	Fossil fuels	Toxic effects on plants
Nitrogen	NO_2	Days	Region	Fossil fuels	Nutrient enrichment
	HNO_3	Hours/Days	Region	Fossil fuels	Acidification, nutrient enrichment
	NH_3	Hours/Days	Region	Agriculture	Nutrient enrichment
Sulfur	SO_2	Days	Region	Fossil fuels	Acidification
Chlorinated compounds	CFCs	Century	Globe	Refrigerants, aerosols	Ozone depletion
	POPs	Decades	Globe	Agriculture, industry	Bioaccumulation
Miscellaneous	O_3	Days	Region	Photochemical reactions with fossil fuels	Toxic effects on plants
	Hg	Years	Globe	Industrial	Bioaccumulation

Source: Modified from Taylor et al. (1994).

pollutants, particularly POPs, very few studies have examined community-level responses to these contaminants. In contrast, effects of elevated CO_2, UV-B radiation, and acidification have received considerable attention in the literature and are known to have significant effects on terrestrial and aquatic communities. Furthermore, there is recent evidence that exposure of communities to any one of these global atmospheric stressors is likely to influence responses to other contaminants.

26.2 CO_2 AND CLIMATE CHANGE

The causes and consequences of global climate change and the specific role of CO_2 are among the most contentious environmental issues today. However, the connection between the atmosphere and biological processes, and the occurrence of a natural greenhouse effect are indisputable facts. The chemical composition of the atmosphere is largely determined by biological processes, and life on earth would probably not exist without the natural greenhouse phenomenon. The key controversies about global climate change relate to: (1) separating these natural changes from those related to anthropogenic stressors; and (2) predicting the ecological consequences of these changes. Although our understanding of the ecological effects of climate change is relatively poor, preliminary data suggest that effects on aquatic and terrestrial communities will be significant. At the very least, we expect that sustained alterations in global climate will have far reaching consequences for the distribution of plants and animals. Understanding the details of these alterations and how they may influence susceptibility to other stressors are among the greatest challenges in community ecology and ecotoxicology today.

Evidence from a variety of sources indicates that global temperatures have increased by approximately 0.5–1.0°C over the past century. The most comprehensive analyses of the relationship between climate change and greenhouse gases have been provided by the Intergovernmental Panel on Climate Change (IPCC). The IPCC, which was created in 1988 by the United Nations Environmental Program and the World Meteorological Organization, provides independent analyses of evidence derived from peer-reviewed sources to develop a scientific consensus on climate change. According to the most recent (2007) IPCC report, 11 of the past 12 years rank among the warmest years in the 150-year long instrumental record (www.ipcc-wgl.ucar.edu). These increased temperatures observed over the past 50 years are closely associated with an unprecedented increase in anthropogenic emissions of atmospheric CO_2 and other greenhouse gases. Although correlation between increased temperature and greenhouse gases strongly implicates CO_2 as a culprit, global temperatures are highly variable and have fluctuated greatly over the past several thousand years. Thus, one of the most significant challenges to understanding the effects of humans on global climate is to separate natural variation from that owing to anthropogenic emissions of CO_2. Understanding the effects of global warming is further complicated by the large spatial and temporal scales over which predicted changes will occur. Because global climate varies relatively little during a human lifetime (and even less during the tenure of most political leaders), society's willingness to act on this issue is limited. The difficulty obtaining empirical data and the necessary reliance on relatively coarse General Circulation Models (GCMs) to predict climate change is unsettling to many scientists. However, it is important to note that much of the debate within the scientific community is over the details of climate change (e.g., how much of the observed increase is owing to greenhouse gases; what is the role of carbon sinks in ameliorating increased CO_2 from anthropogenic sources; what are the most likely ecological effects). Despite uncertainty over these details, the majority of scientists today believe that global warming is real and a direct consequence of human activity. The portrayal of this debate in the media, as a sign of uncertainty or significant disagreement within the scientific community over the causes of global climate change, is both incorrect and dangerous. If even the most conservative estimates of increased temperatures are correct, global warming will undoubtedly be the most significant environmental issue faced by humanity during this century.

26.2.1 Facts and Evidence

The hypothesized relationship between global climate change and greenhouse gases is not a new idea. In the late 1800s, the Swedish chemist, Arrhenius, proposed that increased levels of CO_2 in the atmosphere could influence global temperatures. Short- and long-term records indicate that levels of CO_2 have increased dramatically and are currently the highest in human history. Ice core data reflecting CO_2 concentrations for 400,000 years prior to the industrial revolution showed that levels in the atmosphere remained relatively constant, fluctuating between 180 and 280 μL/L. More recent data from the Vostok ice core in Antarctica show that levels of CO_2 remained less than 300 μL/L until approximately 100 years before present (Figure 26.1a), followed by a steady increase. Finally, direct measurements obtained from the Mauna Loa Observatory indicate that CO_2 concentrations are now approximately 100 μL/L higher than historic levels and have steadily increased over the past 50 years (Figure 26.1b). This rate of increase is approximately 10–100 times faster than at any period before the industrial revolution. The Mauna Loa data also show a strong seasonal signal in CO_2, reflecting variation in photosynthesis and respiration in the northern hemisphere.

There is little doubt that the increased levels of atmospheric CO_2 over the past 50 years are a direct result of anthropogenic emissions. There is also convincing evidence that global temperatures have increased by approximately 0.6°C over the past century. The more challenging task and indeed the issue that generates the greatest controversy are attributing increased temperature to anthropogenic emissions of CO_2. The strongest evidence of a relationship between CO_2 and climate change is derived from paleoclimatic and geochemical data. Crowley and Berner (2001) report variation in CO_2 (estimated using several geochemical proxies) with global temperature and continental glaciation over the past 600 million years. They report good agreement between CO_2 and glaciation, indicating that CO_2 has played a major role in shaping the earth's climate. Data from marine systems also show a significant increase in global temperatures. Despite the fact that oceans cover greater than

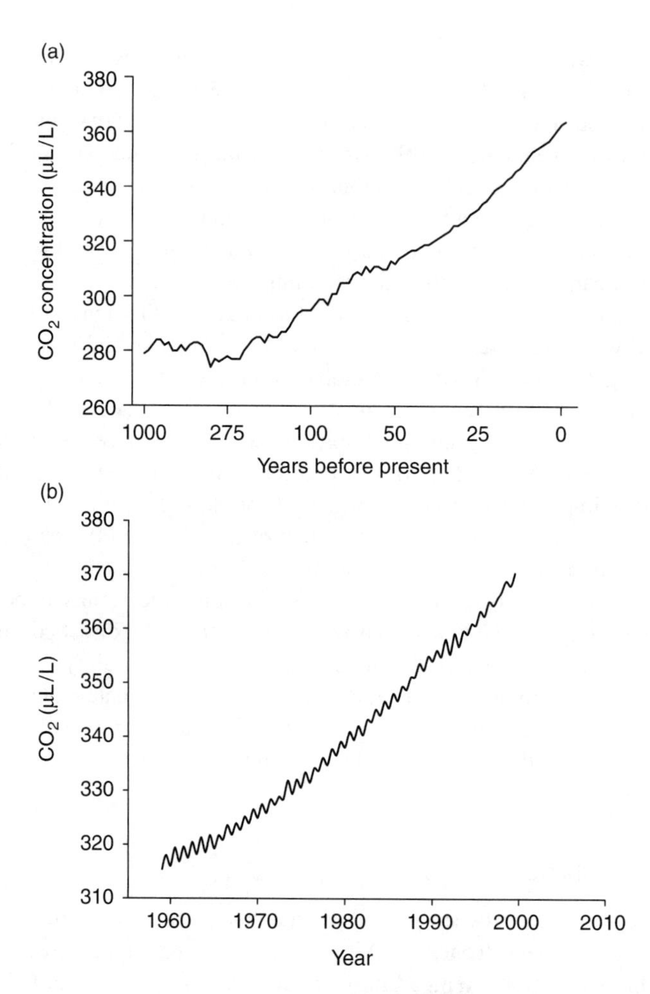

FIGURE 26.1 (a) Long-term changes in CO_2 concentrations based on Vostok ice core data (Boden et al. 1994). (b) Increased levels of atmospheric CO_2 collected from the Mauna Loa Observatory over the past 50 years (Keeling and Whorf 1998). Seasonal variation in CO_2 concentrations reflects seasonal changes in respiration and primary productivity in the northern hemisphere.

70% of the earth's surface, most models of climate change are based on atmospheric or near surface temperatures. Barnett et al. (2001) report that large-scale increases in heat content have also been observed in the world's oceans. The strong agreement between their model predictions and observed changes in ocean heat content supports the hypothesis that temperature increases are a direct result of anthropogenic emissions of greenhouse gases.

Not surprisingly, there is considerable uncertainty in estimates of future global warming derived from these models. The IPCC modified its projections of global warming over the next century, with the predicted upper limit of warming increasing from the 1995 estimate of 3.5–4.0°C. However, there is much greater certainty expressed in the recent IPCC report that humans are responsible for this warming (>90% likelihood that global warming is anthropogenic). The IPCC concluded that temperatures recorded in the Northern Hemisphere during the last half of the twentieth century were likely the highest in at least the past 1300 years. Some of the uncertainty concerning the range of potential increases in global temperatures involves the complex role of global carbon sinks (see Section 26.2.2). The influence of natural factors such as volcanic release of aerosols and variation in solar activity must also be considered relative to anthropogenic emissions of CO_2. For example, using data obtained from marine and lake sediments, tree rings, and glaciers, Overpeck et al. (1997)

TABLE 26.2
Carbon Pools in the Major Reservoirs on Earth

Carbon Pool	Quantity (Pg)
Atmosphere	720
Oceans	38,400
Terrestrial biosphere (living and dead biomass)	2,000
Aquatic biosphere	1–2
Fossil fuels	4,130

Source: From Falkowski, P., et al., *Science*, 290, 291–296, 2000.

report that changes in arctic temperatures resulted from a combination of natural and anthropogenic factors. Initiation of warming in the mid-nineteenth century most likely resulted from increased solar irradiance and decreased volcanic activity. However, most of the warming during the twentieth century was owing to greenhouse gases. When both natural and anthropogenic factors were considered, Stott et al. (2000) found good agreement between model simulations and observed temperature patterns from 1860 to present. More importantly, their results show that warming trends are expected to continue at a rate similar to that of recent decades.

26.2.2 Carbon Cycles and Sinks

> Although natural sinks can potentially slow the rate of increase in atmospheric CO_2, there is no natural savior waiting to assimilate all the anthropogenic CO_2 in the coming century.
>
> **(Falkowski et al. 2000)**

In order to estimate future changes in global temperature, we need to understand the sensitivity of climate to changes in CO_2. Levels of CO_2 in the atmosphere are determined by human activity and interactions with global carbon sinks (Table 26.2). Predicting the effects of increased CO_2 on global climate will require a better understanding of the size and spatial distribution of these sinks. The relatively constant glacial–interglacial concentrations of atmospheric CO_2 over the past 400,000 years suggests a strong feedback between the atmosphere and marine and terrestrial carbon sinks (Falkowski et al. 2000). Over the past 20 years, only about half of the CO_2 released from fossil fuel combustion has remained in the atmosphere. The remaining CO_2 has been sequestered by oceans and terrestrial ecosystems that on average have removed between 4 and 5 Pg C/year during the 1990s.[1] Although a large amount of inorganic carbon is stored in sediments, the major regulators of atmospheric CO_2 are oceans and forests. Biological processes in marine ecosystems (e.g., photosynthesis) remove significant amounts of CO_2 from the atmosphere and export carbon to deep ocean reservoirs. However, oceanic carbonate systems are primarily responsible for determining atmospheric CO_2 levels and maintaining equilibrium between the atmosphere and surface water (Falkowski et al. 2000). Finally, carbon storage in terrestrial ecosystems, especially forests, contributes significantly to the global flux of carbon. Although the total amount of carbon stored in terrestrial systems is relatively large, turnover is much slower than in marine ecosystems.

Most studies of global carbon cycles have considered marine and terrestrial systems separately, thus limiting the opportunity to develop a comprehensive model of carbon flux. Using conceptually similar models for terrestrial and marine primary producers, Field et al. (1998) estimated global net primary production (NPP) of 105 Pg year.[1] The contribution of marine and terrestrial components

[1] (1 Pg = 10^{15} g).

to global NPP was roughly equal (ocean = 48.5 Pg; terrestrial = 56.4 Pg), with a distinct latitudinal pattern. Spatial and temporal variation in NPP result from the limiting influences of light, nutrients, temperature, and water. Although marine ecosystems are a large sink for global carbon, the vast majority of the open ocean is relatively unproductive. An analysis of CO_2 balance in freshwater and marine ecosystems indicates that unproductive systems such as the open ocean tend to be heterotrophic, with a disproportionately higher rate of respiration than photosynthesis (Duarte and Agusti 1998). Unproductive aquatic ecosystems are generally sources of CO_2, whereas productive systems act as CO_2 sinks. The findings of Duarte and Agusti (1998) also illustrate that, despite low productivity of the open ocean, there is a balance between production and consumption on a global scale. While 80% of the open ocean is heterotrophic and a net carbon source, this excess carbon can be balanced by relatively high production of the remaining 20%.

Large-scale spatial patterns greatly complicate analysis of global carbon sinks. A latitudinal gradient of 3–4 ppm of CO_2 from the northern to the southern hemisphere has been attributed to greater CO_2 emissions from population centers in the North. Recently, scientists have also identified a temporal component to global carbon flux. Accumulation rates of CO_2 in the atmosphere have varied considerably over the past two decades, despite relatively little change in emissions from fossil fuels. This variation is most likely a result of changes in the flux of CO_2 from the atmosphere to marine and terrestrial sinks (Bousquet et al. 2000). Recognizing that atmospheric CO_2 levels are controlled by marine and terrestrial processes, some researchers have speculated that ecosystems can be managed to maximize CO_2 sequestration. In particular, adding nutrients to the oceans to stimulate primary productivity, reducing the rate of deforestation, and changing forestry management practices to increase NPP are being seriously considered as ways to mitigate anthropogenic CO_2 emissions (Dixon et al. 1994, Falkowski et al. 2000). Much of the discussion concerning ways to increase sequestration of carbon has focused on forests, especially low latitude tropical systems. The world's forests account for a large fraction of aboveground and belowground terrestrial carbon (Table 26.3). Changes in forest area and other carbon sinks, and flux of carbon from forests to the atmosphere vary greatly with latitude. Although tropical forests occupy approximately 13% of the total land surface, they account for about 40% of the world's plant carbon. On an annual basis, these systems naturally remove approximately 3% of the carbon from the atmosphere. Because of the importance of tropical ecosystems in sequestering carbon, the rapid rate of tropical deforestation has a significant impact on global carbon cycles, resulting in a relatively large (1.1–2.0 Pg C/year) net flux of carbon to the atmosphere.

Despite the obvious attraction of managing biological and biogeochemical systems to increase carbon storage and ameliorate effects of anthropogenic emissions, we must acknowledge that marine

TABLE 26.3
Carbon Pools and Flux in Forest Ecosystems of the World

Latitudinal Belt	Change in Forest Area (10^6 ha/year)	Carbon Pools in Terrestrial Vegetation and Soils (Pg)	Carbon Flux to (−) and from (+) the Atmosphere (Pg/year)
High (Russia, Canada, Alaska)	−0.7	559	+0.48
Mid (Continental USA, Europe, China, Australia)	+0.7	159	+0.26
Low (Asia, Africa, Americas)	−15.4	428	−1.65

Source: From Dixon, R.K., et al., *Science*, 263, 185–190, 1994.

and terrestrial ecosystems have a finite capacity to sequester carbon. In addition, it is likely that increased levels of atmospheric CO_2 and global temperature will directly influence the global carbon cycle. In a warmer, CO_2-enriched world, transport of carbon from the surface to deep oceans will be reduced, terrestrial plants will become less of a carbon sink, and increased microbial respiration may counteract effects of greater NPP (Falkowski et al. 2000). Most ecologists would agree that slowing the rate of tropical deforestation will have positive benefits aside from increased carbon storage. However, remediation strategies designed to increase sequestration of atmospheric carbon, especially at the large spatial and temporal scale necessary to influence global cycles, will likely have unpredictable effects on other biological and biogeochemical processes. Because of this uncertainty, we should not consider manipulation of global carbon cycles as an alternative to the more politically and socioeconomically challenging task of reducing global emissions of CO_2.

26.2.3 The Mismatch between Climate Models and Ecological Studies

Most ecological studies are carried out in areas roughly the size of a tennis court, while the resolution of most climate models is approximately the size of the state of Colorado.

(Root and Schneider 1993)

Much of the difficulty predicting the ecological consequences of global climate change on communities results from our inability to link large-scale climate models to smaller scale ecological studies. Currently we lack regional projections of climate change that can be applied to local ecosystems. General circulation models (GCMs) have allowed scientists to predict potential increases in global temperatures associated with elevated CO_2 and to quantify interactions among atmospheric, oceanic, and terrestrial compartments. However, the coarse spatial scale of GCMs (generally >500 km^2) is much larger than most ecological investigations. One proposed solution to this mismatch is to integrate regional models of climate change within GCMs (Hauer et al. 1997), thus allowing researchers to resolve the complexities of regional variation in climate, topography, vegetation, and hydrology. In addition, if we are to make any progress in understanding the ecological consequences of global climate change, interdisciplinary studies that integrate physiology, population biology, community ecology, and climatology are necessary. Clark et al. (2001) predicted climate change effects on trout populations in the southern Appalachians (USA) by integrating individual-based models with a geographic information system (GIS). Although the focus of this investigation was on life history characteristics (growth, spawning, feeding, mortality), the study demonstrates a unique approach for predicting regional population changes based on individual responses to climate. Root and Schneider (1993) show how large-scale climatic factors can be used to predict distribution of wintering North American birds. They describe a mechanism based on physiological constraints to explain the strong association between winter temperatures and geographical distributions. These types of studies represent an important step in resolving the mismatch between global climate models and ecological investigations.

Another way to link spatially extensive analyses of climate with ecological studies is to develop regional models to forecast changes in vegetation under various scenarios of climate change. Regional models have been used to predict the responses of grassland, forest, and tundra ecosystems to changes in climate (Pacala and Hurtt 1993). Most model projections for the northern hemisphere show a generally northward expansion of plant communities as a result of increased temperature. Under a scenario of doubled CO_2 levels, Lassiter et al. (2000) predicted northerly retraction and expansion of different mixed forests in the mid-Atlantic region of North America. These results demonstrate the potential for significant range shifts of dominant plant communities in response to moderate warming. More dramatic effects are expected in extreme northern and southern latitudes where climate change is predicted to be greatest. Because the boundary between boreal and tundra ecosystems is abrupt and closely associated with climate, the response of boreal ecosystems to global climate change has received considerable attention. Using a model to predict effects of transient changes in climate, Starfield and Chapin (1996) report that a 3°C increase in temperature would result in the transition of tundra to boreal forest within 150 years.

26.2.4 Paleoecological Studies of CO_2 and Climate Change

A significant challenge in the study of global climate change is to distinguish natural variation in climate from the variation associated with anthropogenic emissions of greenhouse gases. Because of the difficulty conducting manipulations at spatial scales compatible with GCMs, integrating models of environmental change with paleoecological records can improve our understanding of how climate influences communities. Today, interdisciplinary teams of atmospheric scientists, geologists, and paleoecologists integrate evidence from diverse sources to support the link between CO_2 concentrations and increased global temperatures (Table 26.4). Tree ring analyses provide high resolution of annual variation in climate over relatively short time periods (10^2–10^3 years), whereas pollen grains, ice cores and marine sediments yield much longer records (10^5–10^7 years). Recent studies have given atmospheric scientists a much better understanding of the correlation between atmospheric CO_2 levels and global temperature. Paleoecologists have contributed to this understanding by reconstructing relationships between global climate and prehistoric communities.

Modern plant species have persisted over the past 2.5 million years in the face of extensive changes in climate. Climate warming at the start of the Holocene was relatively rapid and provides a reasonable model for predicting changes associated with anthropogenic impacts. Climatic changes since the last glacial period have had profound effects on plant and animal communities in North America. Adaptations to climate change and extensive range expansion (e.g., migrations) have characterized plant responses over this period. For example, records based on pollen grain analyses showed that many forest tree species migrated northward at rates of 100–1000 m/year during the period of post-Pleistocene warming. Because of interspecific differences in tolerance to climate change and

TABLE 26.4
Paleoecological and Other Techniques Employed to Reconstruct Global Changes in Greenhouse Gases and Climate

Method	Information Obtained	Resolution	Typical Time Range
Tree rings	Temperature; rainfall; wildfires	Annual	500–700 years
Pollen grains	Changes in community composition related to temperature and precipitation	50 years	Present to several million years
Geomorphology	Extent of glaciers and ice sheets; sea level changes	Variable	Glaciation to 2.9 billion years
Ice cores	CO_2 concentration; volume of continental ice; snow accumulation rates	Seasonal to decades	Present to 440,000 years
Corals	Sea surface temperatures; precipitation cycles	Months	400 years
Marine sediments	Temperature; salinity; ice volume; atmospheric CO_2	Thousands of years to centuries	180 million years

Source: From Stokstad, E., *Science*, 292, 658–359, 2001.

migration rates, this northward movement generally occurred on a species-by-species basis and not at the level of assemblage. These results suggest that predicting future community structure may require an autecological focus (Harrison 1993).

Although the ability of some organisms to adapt to changing climate and disperse over relatively long distances during postglacial periods is encouraging, the unprecedented rate of climate change expected over the next century makes extrapolation from paleoecological records tenuous. Future climates may lie outside the range of historical records, and therefore caution is required when using paleoecological data to predict ecological effects. Because rapid climate change will most likely preclude the ability of plants to adapt, it is generally believed that range extension and retraction will be a common response. However, migration may not provide an alternative in the face of rapid climate change. On the basis of current climate projections for the next century, plants would be required to migrate 300–500 km/century, a rate significantly greater than previously reported for many tree species (Davis and Shaw 2001). For example, spruce trees, known to have a rapid rate of dispersal, have expanded their range about 200 km/century over the past 9000 years. Some model projections of forest succession in a changing climate are inconsistent with known rates of range expansion and illustrate our poor understanding of this process. Forest succession models predict that temperature increases associated with a twofold increase in CO_2 would force the boreal zone in central Sweden 1000 km northward within 150–200 years (Prentice et al. 1991). On the basis of paleoecological records, it is unlikely that species are capable of this unprecedented rate of range expansion. In addition, land use changes and habitat fragmentation represent significant impediments to range extension and gene flow, thus increasing the likelihood that many species will go extinct (Davis and Shaw 2001).

26.2.5 Effects of Climate Change on Terrestrial Vegetation

Unlike many of the anthropogenic stressors considered in our examination of community ecotoxicology, significant research on effects of CO_2 has focused on terrestrial ecosystems. For example, most of the chapters in the book, *Biotic Interactions and Global Change*, by Karieva et al. (1993) examine effects on terrestrial communities. Community-level responses to elevated levels of atmospheric CO_2 include direct effects associated with alterations in primary productivity and indirect effects attributed to changes in global climate, especially temperature and precipitation. If CO_2 limits primary productivity (Bazzaz 1990), we would expect to see alterations in community composition as a direct result of species-specific responses to elevated CO_2. Faster growing species or those that employ C_3 photosynthetic pathways will likely be favored by increased levels of CO_2. In addition, differential responses of C_3 and C_4 plants to CO_2 enrichment may modify competitive relationships. Finally, these changes in plant community composition will likely have significant impacts on grazers and other herbivores. For example, plants that respond to elevated CO_2 generally have lower nutrient content, thus requiring herbivores to consume more food (Vitousek 1994).

Small-scale experiments have been conducted to measure responses of plant communities to both elevated concentrations of CO_2 and increased temperature. To manipulate temperature, researchers have employed a variety of approaches, including plastic enclosures, snow fences, heating cables, and overhead heaters. Robinson et al. (1998) used polythene tents to investigate the response of an arctic plant community to warming. Results showed that a 3.5°C increase in air temperature increased total plant cover over a season. However, this response was not consistent between years, suggesting that short-term responses to warming may be poor predictors of longer-term impacts.

As with communities located at higher latitudes, we expect greater effects of global warming on communities at higher elevations because of relatively short growing seasons. Harte and Shaw (1995) used overhead heaters suspended above 30 m^2 plots to simulate effect of warming on composition of a montane plant community. Results of these experiments showed that aboveground biomass of forbs decreased and biomass of shrubs (primarily sagebrush) increased in response to warmer soil

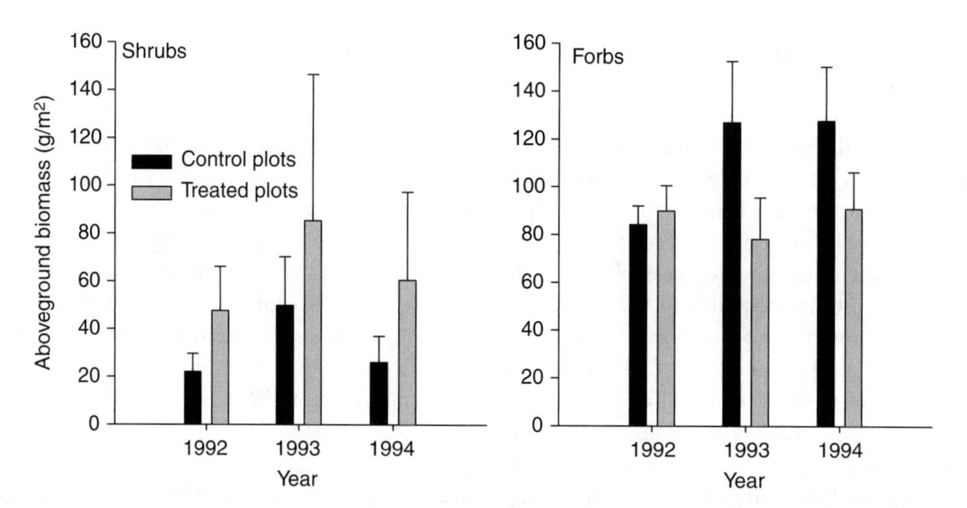

FIGURE 26.2 Results of a climate warming experiment showing shifts in dominance of montane plant communities in the Rocky Mountains. The figure shows changes in aboveground biomass of shrubs and forbs following experimental manipulation of soil temperature using overhead radiators. Increased temperature in these treatments corresponded to a concentration of CO_2 approximately two times greater than preindustrial levels. (Data from Table 2 in Harte and Shaw (1995).)

temperatures and lower soil moisture (Figure 26.2). The response of forbs to warming was species-specific, and differences were attributed to effects on soil resource availability (de Valpine and Harte 2001). Although the warming-induced shift from forbs to drought-tolerant sagebrush reported by Harte and Shaw (1995) is consistent with our expectations, reanalysis of these data using a different statistical model casts some doubt on the findings. Price and Waser (2000) suggest that differences in sagebrush biomass between control and heated plots reported by Harte and Shaw (1995) were attributable to pretreatment differences. These researchers observed no effect of warming in their study, and argued that soil desiccation and reduced microbial activity in treated plots offset the influence of earlier snowmelt. The contradictory findings of these two investigations highlight the difficulty of conducting field experiments and the need for long-term studies to assess community responses to climate change.

Most experimental investigations of the effects of climate change on terrestrial plant communities have focused on relatively short-term and direct effects on dominant species. As described in Chapter 21, the outcome and importance species interactions are often influenced by perturbation. Because ecological complexity increases with spatial and temporal scale, whereas the number of experiments conducted at relatively large spatial or temporal scales is quite limited, there is the tendency to underestimate the importance of these interactive effects (Walther 2007). Results of a large-scale field experiment conducted in a California grassland community demonstrate the importance of spatiotemporal scale and the necessity of considering indirect effects on food web structure (Suttle et al. 2007). Initial increases in biomass of nitrogen-fixing forbs observed in the first 2 years of the experiment were reversed as annual grasses increased in treated plots. These shifts in community composition had dramatic consequences for biomass and diversity of higher trophic levels. The important point is that evaluating findings after 2–3 years (the average duration of many field experiments) would have provided very different results when compared to conditions after 5 years.

Several characteristics influence responses of plant communities to global warming, including previous exposure to climatic extremes, species richness, functional composition, and successional stage (Grime et al. 2000). Consequently, we expect that different plant communities will respond to climate change in very different ways. This hypothesis was tested by comparing responses of a mature, stable grassland community to those of an immature, early successional community (Grime et al. 2000). Soil temperatures in treated plots were increased by 3°C using heating cables placed at the soil surface. Results showed that early successional communities composed of fast-growing or short-lived species were more sensitive to warming than mature communities. Because landscape alterations that maintain early successional communities are becoming increasingly common, these authors speculate that climate change may have disproportionate effects on these previously disturbed communities.

26.2.6 Ecological Responses to CO_2 Enrichment

Although the effects of increased CO_2 on global climate change have received considerable attention, relatively few studies have investigated the direct response of plant communities to CO_2 enrichment. Elevated levels of CO_2 are likely to have profound effects on plant community composition as well as a significant influence on belowground processes. Owensby et al. (1993) investigated effects of CO_2 enrichment (2 $\times$ ambient levels) on species composition, biomass production, and leaf area in a tall grass prairie ecosystem. These authors note that because rangelands account for 47% of the earth's land area, responses of these ecosystems to elevated CO_2 have important implications for global carbon budgets. In contrast to expectations, elevated levels of CO_2 increased production of C_4 grass species but not C_3 species. The enhanced productivity of C_4 species was related to greater water-use efficiency. There was little indication of a shift in competitive relationships between C_3 and C_4 species.

Tree species that employ the C_3 photosynthetic pathway are carbon-limited and are expected to increase productivity in response to enhanced CO_2. Increased NPP in forests dominated by C_3 trees may therefore reduce the amount of CO_2 from anthropogenic sources. While some studies have shown that productivity of seedlings is increased under an enriched CO_2 regime, analysis of tree rings shows relatively little relationship between growth rate and atmospheric CO_2 (DeLucia et al. 1999). To reconcile the differences between results of growth chamber experiments and these paleoecological investigations, research conducted at larger spatial scales is necessary. DeLucia et al. (1999) investigated responses of loblolly pines to CO_2 enrichment (+200 μL) in 30 m diameter experimental plots (Figure 26.3a). Results showed that growth rate was approximately 26% greater after two years of exposure to elevated CO_2 (Figure 26.3b). In contrast to model simulations that predict only a 9% increase in NPP in response to doubling CO_2, DeLucia et al. observed that ecosystem NPP increased by 25% in enriched plots relative to controls. If applied globally, this increase in NPP could sequester about 50% of the total anthropogenic carbon expected to be released by 2050; however, these researchers speculate that this may represent the upper limit of forest carbon uptake.

Because the amount of carbon stored in soil organic matter is 2–3 times greater than in terrestrial vegetation, changes in soil processes can significantly influence global carbon cycles and sequestration. Elevated CO_2 is expected to control belowground processes in terrestrial ecosystems by influencing NPP, soil respiration, decomposition, and nitrogen mineralization. The relationship between CO_2 enrichment and belowground processes was investigated using experimental plots in the loblolly pine forest described above (Allen et al. 2000). Although litterfall mass and fine root biomass increased in treated plots, there was no influence of enriched CO_2 on litterfall C:N ratios, nutrient cycling, microbial biomass, or nitrogen mineralization. These results are consistent with other studies of belowground processes and indicate that elevated CO_2 may accelerate the input of organic matter to carbon pools in soils. Changes in soil organic matter and carbon pools are likely to influence belowground communities and food chains. Experiments conducted in terrestrial microcosms showed increased abundance and changes in community composition of fungal-feeding arthropods (Collembola) in response to CO_2 enrichment (Jones et al. 1998). The authors concluded that these structural changes were a result of alterations in the fungal community, which responded to increased dissolved organic carbon (DOC) in soil.

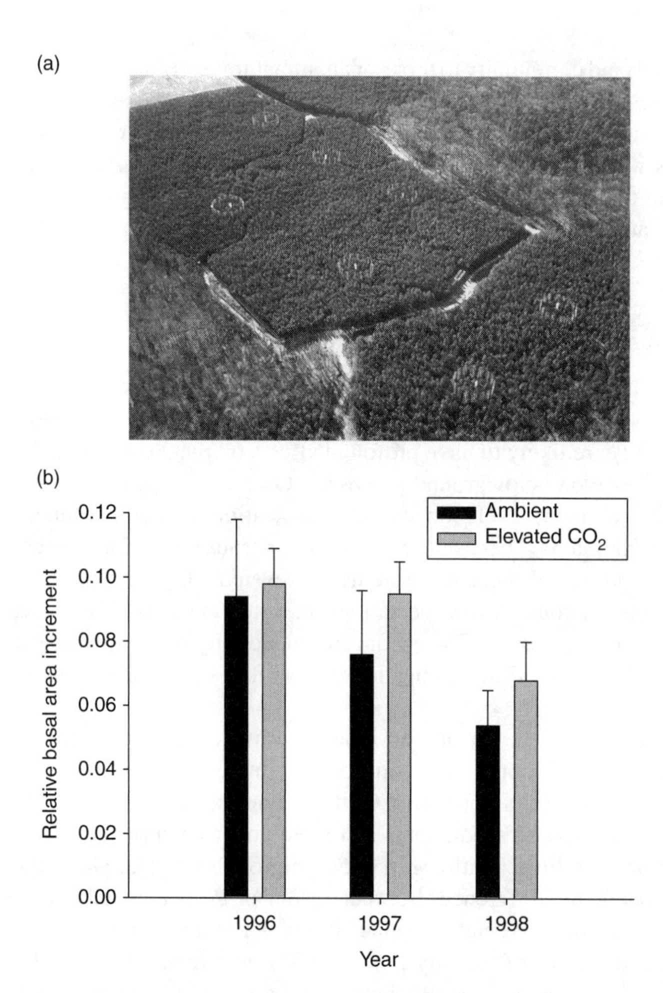

FIGURE 26.3 (a) Experimental plots used for the free air CO_2 enrichment (FACE) studies of the effects of elevated CO_2 on growth of loblolly pines in North Carolina, USA. Each ring in the photograph is 30 m in diameter and contains about 100 trees. (From Figure 3 in Allen, A.S., et al., *Ecol. Appl.*, 10, 437–448, 2000. Reproduced by permission of the Ecological Society of America.) (b) Relative basal area increment for loblolly pines growing in ambient and elevated CO_2 conditions. (Data from Table 1 in DeLucia et al. (1999).)

26.2.7 Effects of Climate Change on Terrestrial Animal Communities

In addition to the direct effects of increased temperature on animals, changes in the distribution and abundance of plants will likely have significant impacts on animal communities. Alterations in climate may modify terrestrial food webs in systems regulated by top-down or bottom-up control (Box 26.1). Assuming other environmental factors are favorable, the most consistent responses of species limited by temperature will be a northward (or southward in the southern hemisphere) range expansion. The geographic distributions of many animal species are strongly correlated with vegetation, and some species are obligate associates of a particular vegetation type. Thus, while many animals are expected to migrate in response to changes in climate, their dependence on slower dispersing plants could limit these range shifts and result in extinctions (Root 1993).

Much of the research on effects of climate change on birds and mammals has been conducted at the population level (Larson 1994). Sophisticated modeling approaches that couple large-scale estimates of species' distributions with regional climate have proven very useful for describing

Box 26.1 The Influence of Global Climate Change on Interactions between Wolves and Moose

As noted above, it is generally assumed that most effects of climate change on animals are a secondary result of changes in distribution and abundance of plants. However, in systems where top predators exert control over community structure and function (see Chapter 27), animals may actually regulate responses of plants to climate. Long-term (40 years) records of predator–prey interactions between wolves and moose on Isle Royale, USA have demonstrated top-down control in this system (McLaren and Peterson 1994). Wolves regulate moose density and moose control abundance of balsam fir, their primary winter forage. Recent analyses have shown that variation in global climate also plays an important role in these interactions. Annual variation in snow depth associated with the North Atlantic Oscillation influences the foraging behavior and efficiency of wolves. During years with heavy snowfall, wolves tend to hunt in larger packs and their predation rate on moose is increased. Thus, densities of moose are lower during years of heavy snowfall and growth of balsam fir is greater owing to reduced herbivory. In contrast, predation is reduced during years of low snowfall, moose populations are larger, and growth of balsam fir is limited by grazing. Results of this study demonstrate the unique influence of climate on top-down regulation of plant production. Assuming that winter snowpack in this region will be reduced owing to climate warming, results of these long-term studies suggest that moose populations will increase and growth of balsam fir will be reduced.

current conditions. For example, Sillett et al. (2000) report that regional variation in climate affects survival and reproductive success of migratory songbirds. However, because predictions of range shifts based on bioclimatic models are quite variable, model calibration using backward predictions ("hindcasting") or "space-for-time" substitutions is essential to improve forecasts of future distributions (Araujo and Rahbek 2006). The timing of reproduction in many passerines corresponds with peak abundance in local food supply. Thomas et al. (2001) show that the earlier leaf flush and the associated pulse of food expected under climate warming will result in a mismatch between peak food abundance and nestling demand. In a comprehensive analysis of 148 land bird species, Root (1988) identified six major environmental factors (minimum January temperature, length of frost-free period, humidity, precipitation, elevation, and vegetation) that limited the distribution of North American land birds (Figure 26.4). With the exception of elevation, all of these variables are expected to change in response to global warming.

McDonald and Brown (1992) provide one of the more insightful approaches for investigating effects of climate change on small mammal communities. By integrating the theory of island biogeography with data on distribution and abundance of mammals inhabiting isolated mountain ranges, these researchers develop a quantitative model to predict the number and identity of species expected to go extinct as a result of global warming. Assuming a relatively conservative (on the basis of recent estimates) 3°C temperature increase, McDonald and Brown first estimated the amount of boreal habitat that will be lost on 19 isolated mountain ranges in the Great Basin (USA). Next, they estimated the response of 14 boreal mammal species based on the proportion of lost habitat. Their results were striking. Under a 3°C increase in temperature, greater than 50% of the boreal mammal species on individual mountain ranges would go extinct locally and an additional three species would go extinct throughout the region. Although this analysis makes several simplifying assumptions, McDonald and Brown have provided a useful framework for predicting the probability of extinction based on model projections of vegetation change and present geographic distributions. Boggs and Murphy (1997) used a similar approach to predict effects of climate change on butterfly communities in this same region. Their analysis showed that the butterfly community would

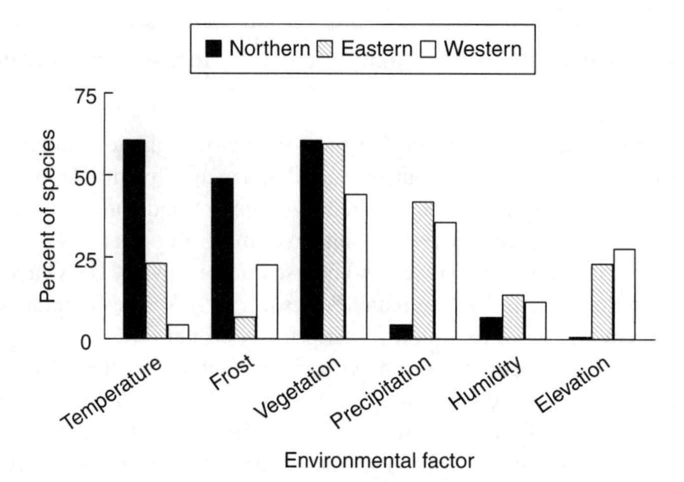

FIGURE 26.4 The influence of environmental factors on the distribution of birds. The figure shows the percent of bird species' northern, eastern, and western boundaries that are associated with six environmental variables. Note that five of these factors (temperature, frost, vegetation, precipitation, and humidity) will likely be affected by climate change. (Modified from Figure 4 in Root (1993).)

experience a 23% reduction in number of species, with the greatest effects on less mobile species. These analyses support the hypothesis that montane communities are at high risk of extinction from global warming and associated habitat loss.

26.2.8 Effects of Climate Change on Freshwater Communities

> The effects of climate change on freshwaters have been largely disregarded in major global change programs.
>
> **(Schindler et al. 1990)**

Although much of the basic research in aquatic ecology is relevant to global climate change, relatively few studies have considered how freshwater communities will respond to climate change (Carpenter et al. 1992). Complex changes in lakes and streams in response to global climate are expected as a result of alterations in thermal regime and hydrologic characteristics. Many aquatic organisms are adapted to a relatively narrow range of temperature. In particular, coldwater, stenothermal species (e.g., salmonids in high elevation lakes and streams) are likely to be impacted by increased water temperatures. The longitudinal distribution of net-spinning caddis-flies (Trichoptera) in Rocky Mountain (USA) streams provides a good example of the close association between elevation (and presumably water temperature) and community composition (Figure 26.5). Predictable changes in abundance of dominant species are observed from headwater streams to larger, warmer rivers. With increased water temperatures associated with climate change, the distribution of temperature-sensitive species will likely shift to higher elevations. These shifts are likely to result in the extirpation of many coldwater species currently restricted to alpine habitats.

Increased temperature may also modify species interactions in aquatic ecosystems, thus indirectly altering community composition. Laboratory experiments have shown that brook trout (an introduced species in many western streams) have a competitive advantage over native cutthroat trout at higher temperatures (DeStato and Rahel 1994). Thus, we would expect a greater rate of extirpation of some native species under conditions of increased water temperatures.

A long-term study of boreal lakes and streams in northwestern Ontario (Canada) provides some insight into potential physicochemical and ecological modifications associated with climate change

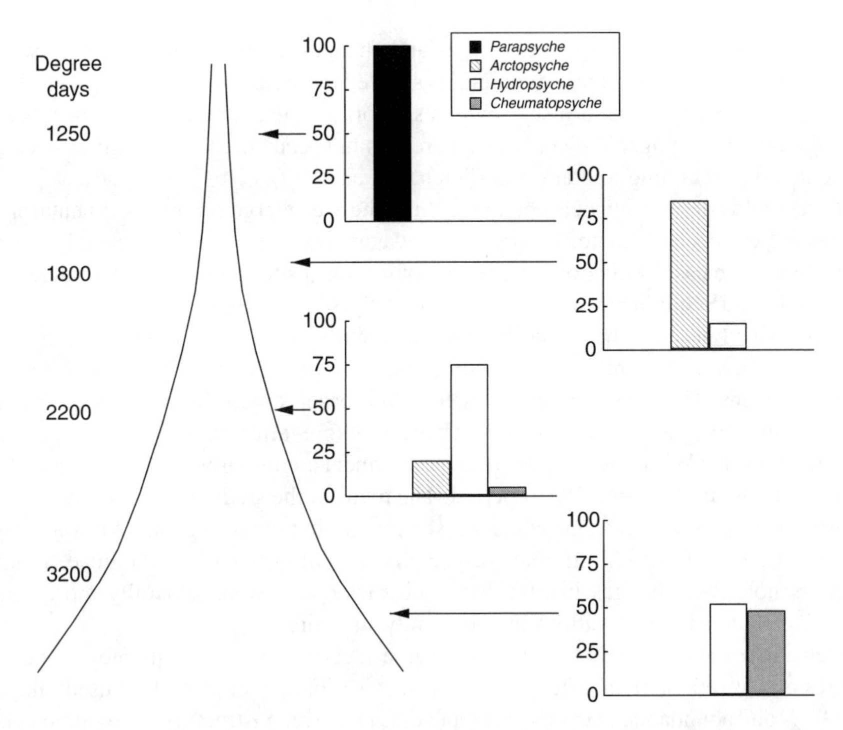

FIGURE 26.5 The longitudinal distribution of net-spinning caddis-flies (Trichoptera: Hydropsychidae) along an elevation and temperature gradient in Rocky Mountain streams of southern British Columbia. The figure shows the relative abundance of four dominant genera at four locations from headwaters to downstream reaches. It is expected that increased water temperature will shift the distribution of some species to higher elevations and may result in the extirpation of stenothermal taxa such as *Parapsyche elsis*. (Modified from Figure 3 in Hauer et al. (1997).)

(Schindler et al. 1996). Between 1970 and 1990, researchers at the Experimental Lakes Area (ELA) observed a gradual increase in air temperature (approximately 1.6°C) and a decrease in precipitation (approximately 200 mm). While it is uncertain if these changes are a direct result of global climate change, they provide an excellent opportunity to document the influence of climate on hydrologic, biogeochemical, and ecological characteristics of freshwater systems. Some permanent first-order streams became ephemeral and stream discharge and export of base cations were significantly reduced as a result of lower precipitation. Physicochemical changes in lakes included increased surface water temperature, increased water clarity and light penetration, and a deeper thermocline. Complex changes in biomass and diversity of phytoplankton were also associated with these physicochemical alterations. However, the most striking ecological response was the complete loss of habitat for lake trout and other stenothermal species as a result of lower dissolved oxygen levels and a deeper thermocline. Because the magnitude and duration of climatic changes observed at ELA were less than predicted by relatively conservative GCMs, these results show that modest alterations in temperature and precipitation can have significant consequences for freshwater ecosystems (Schindler et al. 1996). Finally, an increase in water clarity and reduced levels of dissolved organic matter (DOM) may cause significant interactions between climate change, UV-B exposure, and acidification in aquatic ecosystems (Section 26.5).

Alterations in functional characteristics of aquatic ecosystems may result from direct physiological effects of increased temperature as well as changes in trophic structure. Using microcosm experiments containing bacteria, algae, and diatoms, Petchey et al. (1999) showed significant extinction of species (30–40%) and altered ecosystem function in response to warming. Frequency of extinction varied among trophic levels, with greatest impacts on herbivores and top predators.

Temperature-dependent physiological responses and changes in community structure were related to alterations in function, as warmer communities showed increased rates of primary production and decomposition. These experiments also provided support for the hypothesis that impacts of climate change (and other anthropogenic disturbances) are less in species-rich communities because of the greater likelihood of retaining tolerant taxa (Petchey et al. 1999).

As in terrestrial systems, logistical challenges limit the use of large-scale experimental approaches for investigating effects of climate change on freshwater communities. Consequently, most studies in aquatic systems are based either on long-term monitoring at sites where known changes in climate have been recorded (Schindler et al. 1990), or relatively small-scale microcosm experiments such as the one described above. The study by Hogg and Williams (1996) is unique because it measured responses of a stream benthic community to increased temperature in a relatively large-scale experimental system. These researchers divided a first-order stream longitudinally and increased water temperature on one side by 2–3.5°C. Using a before–after control-impact (BACI) design (Chapter 23), Hogg and Williams characterized pretreatment community composition and then measured responses to warming over a 2-year period. The focus of the study was primarily on life history characteristics (insect emergence, growth, size, sex ratios), but the results also showed a significant reduction in total density of chironomids, especially the coldwater Orthocladiinae (Figure 26.6). These authors note that changes in life history characteristics were generally more sensitive to increased temperatures than alterations in community structure.

Although most of our discussion has focused on the negative impacts of climate change, increased temperatures can have beneficial effects on some species. Finney et al. (2000) used lake sediment records of δ^{15}N and abundance of cladocerans and diatoms to reconstruct sockeye salmon population densities over a 300-year period. Because salmon migrating from the North Pacific to freshwater systems have a strong marine-derived isotopic signature (e.g., high δ^{15}N relative to terrestrial sources), stable N isotopes in lake sediments can be used to track changes in salmon-derived nitrogen. Finney et al. (2000) show good agreement between salmon-derived N and abundance of higher trophic levels, indicating the importance of salmon carcasses to productivity in these otherwise oligotrophic systems. More importantly, they report a positive relationship between salmon abundance and documented changes in sea surface temperature from 1750 to about 1850 (Figure 26.7). Colder than average temperatures generally resulted in below-average salmon abundance. This relationship breaks down over the last few decades, primarily as a result of over harvesting salmon populations. The study

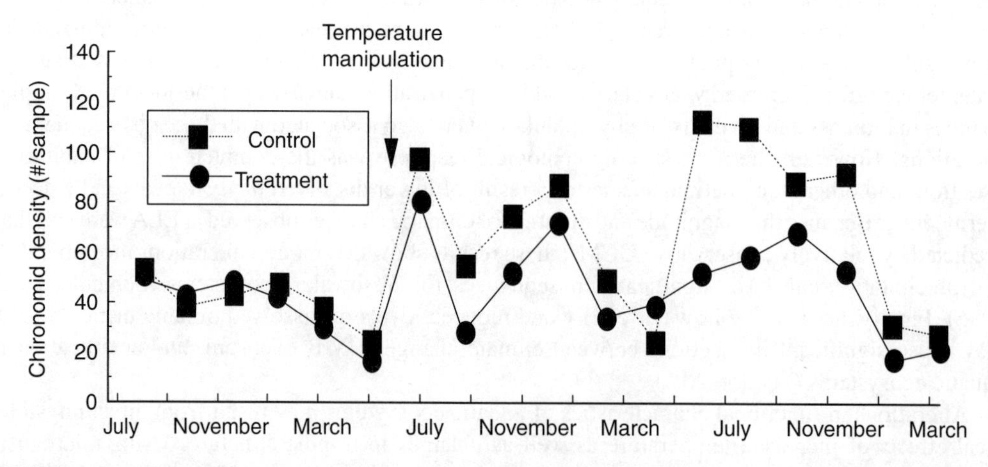

FIGURE 26.6 Response of chironomids to a 2–3.5°C increase in water temperature. Data were collected 1 year before and two years after temperature treatments in a first-order stream. (Modified from Figure 6 in Hogg and Williams (1996).)

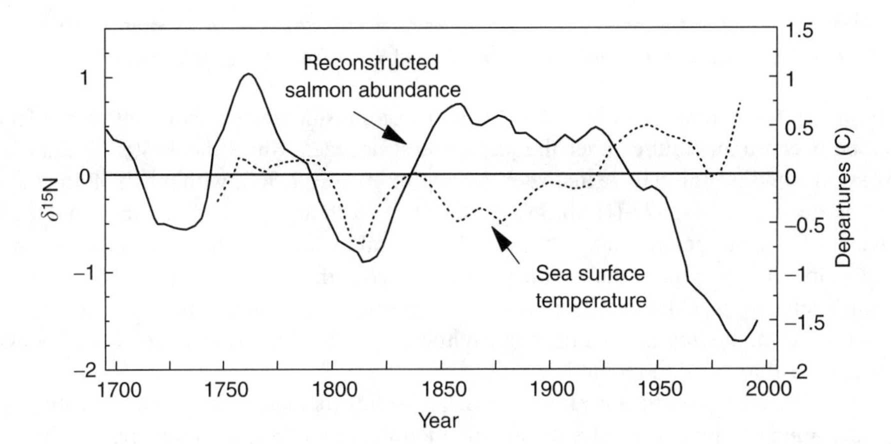

FIGURE 26.7 The relationship between salmon abundance and sea surface temperatures. Stable isotope data derived from sediment records were used as an indicator of salmon abundance over the 300-year period. Sea surface temperatures are based on tree ring analyses. There was a positive relationship between salmon abundance and sea surface temperature from about 1700 to 1850. The poor relationship between temperature and salmon abundance in the past few decades is likely a result of commercial harvesting. (Modified from Figure 4 in Finney et al. (2000).)

demonstrates the strength of integrating sensitive analytical approaches such as stable isotopes with paleoecological studies to characterize ecological effects of climate change.

Because of the potential interactions among temperature, biogeochemical processes, and hydrologic characteristics, predicting community responses based only on modified thermal regimes may provide misleading results (Clark et al. 2001). For example, changes in the magnitude or timing of spring runoff in snowmelt-dominated watersheds (Hauer et al. 1997) may alter biogeochemical cycles and have direct impacts on stream communities. Interactions between riparian vegetation and watershed processes will likely be altered as a result of modified flow regimes. In their review of global change and aquatic ecosystems, Carpenter et al. (1992) distinguish between transitional changes and perturbational changes. Transitional changes occur over relatively long time periods (10–100 years) and result from alterations in landscapes, hydrological and geomorphological features, and community persistence. In contrast, perturbational changes occur over relatively short time periods (1–10 years) and are associated with floods, droughts, and temperature extremes. Because hydrologic characteristics, biogeochemical processes, and thermal regimes play a prominent role in determining the distribution and abundance of aquatic organisms, both perturbational- and transitional-scale changes must be considered when predicting impacts of climate change on freshwater communities.

26.2.9 Effects of Climate Change on Marine Communities

Elevated ocean temperatures are expected to have significant and diverse impacts on most marine communities (Box 26.2). Effects ranging from increased incidence of disease (Harvell et al. 1999) to reduced ecosystem productivity (Roemmich and McGowan 1995) have been associated with warming of ocean waters. Mass mortalities of seagrasses, corals, urchins, and abalone have been attributed to elevated ocean temperatures, resulting in dramatic changes in community composition (Harvell et al. 1999). Warmer ocean temperatures may directly influence organisms or make them more susceptible to other stressors. For example, elimination of grazing sea urchins from many Caribbean reefs, which was attributed to an unidentified pathogen, shifted the reefs from a coral-dominated community to an algae-dominated community. In addition to these direct effects, alterations in ocean surface temperatures may have indirect effects on marine food webs.

Box 26.2 The Southern California Bight: A Test Case of Global Warming?

The Southern California Bight has been the focus of considerable research into the effects of increased ocean temperatures over the past several decades. Since the 1940s, ocean surface temperatures in this region have increased by approximately 1.5°C, with much of this increase occurring during 1976–1977 (Holbrook et al. 1997). Although it is uncertain if this increase is a result of global climate change or part of a natural cycle, the effects are widespread. The altered temperature regime of the region has been linked to dampened upwelling events, reduced nutrient levels and productivity, loss of species diversity, and alterations in community composition. In short, the Southern California Bight may provide important lessons on how marine ecosystems will respond to global warming.

One of the more significant effects associated with increased temperature in this region is the alteration of upwelling currents and the modification of marine food chains. Upwelling currents off the Pacific Coast of North and South America supply inorganic nutrients from colder, deeper waters to nutrient-poor surface waters. Marine primary producers and the complex food chains they support are highly dependent on this supply of nutrients. Warming trends over the past several decades have increased vertical stratification in areas around the Southern California Bight and reduced the supply of nutrients from upwelling. Roemmich and McGowan (1995) report that over a 43-year period (1951–1993) zooplankton biomass declined by 80% in this region. They attribute this dramatic response to a decrease in inorganic nutrients caused by dampened upwelling. Because of the importance of zooplankton in marine food chains and carbon cycling, this trend could have devastating consequences for coastal marine ecosystems.

Changes in the composition of benthic communities and reef fishes in the Southern California Bight were also associated with increased ocean surface temperatures. Barry et al. (1995) reported an increase in abundance of benthic invertebrates with a more southern distribution and a decrease in northern species from the 1930s to 1994. Holbrook et al. (1997) observed a 15–25% decrease in total species richness at two sites off Los Angeles, CA in the year immediately after a 1°C increase in annual seawater temperature. In addition, the proportion of the community consisting of northern species gradually declined and the fraction of southern species increased over a 20-year period. These observations are consistent with predictions that species will shift their geographic distribution in response to increased temperatures. Changes in community composition were accompanied by large reductions in abundance of reef fishes, especially northern species that declined by 88%. These dramatic reductions in abundance are not predicted from current models of climate change (Holbrook et al. 1997) and may represent the same long-term trend of lower productivity observed for zooplankton abundance (Roemmich and McGowan 1995). These data underscore the importance of developing a better mechanistic understanding of the relationship between climate change, ecosystem productivity, and community dynamics (Holbrook et al. 1997).

Sanford (1999) speculates that warmer temperatures will have a significant impact on predator–prey interactions that regulate communities in the rocky intertidal zone of southern California. Because the starfish *Pisaster ochraceus*, a keystone predator in this community, is highly sensitive to temperature, community structure of the rocky intertidal zone could be altered by temperature-induced changes in feeding rates.

Because of their narrow range of temperature tolerance, coral reefs and associated communities are likely to suffer significant damage as a result of moderate global warming (Smith and Buddemeier 1992). The response of coral reefs to global climate change will depend on numerous factors, including the rate of temperature increase, the ability of reef systems to tolerate and adapt to warmer temperatures, and their geographic location. Elevated temperatures during the 1998 El Nino event are a suspected cause of the massive die-offs of corals observed in the Caribbean and other locations. Temperatures exceeding 30°C triggered reef-building hermatypic corals to expel their zooxanthellae (the symbiotic algae living in corals), a phenomenon known as coral bleaching. During 1998, 46% of the reefs in the Indian Ocean were severely damaged by elevated surface temperatures and 16% of the reefs globally experienced bleaching. In addition to the impacts of elevated temperature, there are also concerns about the direct influence of elevated CO_2 on the process of calcification and reef formation. Although coral reef communities are highly sensitive to contaminants and many other types of anthropogenic stressors, the widespread devastation following the 1998 El Nino event indicates that global climate change is probably the most serious threat.

The distribution of fishes and other mobile species in marine ecosystems is likely to respond to climate change, with most groups shifting toward the poles. As in terrestrial ecosystems, the extent to which species distributions shift to the poles will likely depend on life history characteristics. Perry et al. (2005) analyzed the distribution of demersal (bottom-living) fish communities in the North Sea from 1977 to 2001, a period that corresponded to a 1.05°C increase in water temperature. Centers of distribution shifted for 15 of 36 species, and most of these shifts (13 species) were northward (mean $= 172.3 \pm 98.8$ km). Species that shifted distributions were significantly smaller and matured at a younger age compared to nonshifting species. Global analyses of climate effects on marine fishes have generally been limited to studies on movement and recruitment of individual commercially important species. Research conducted by Worm et al. (2005) is unique because it examined species richness patterns of multiple trophic levels. Their analysis of global patterns of diversity of oceanic predators (tuna, billfish) over the past 50 years showed significant relationships with temperature and dissolved oxygen. Large (10–50%) declines in diversity resulting from increased fishing pressure and climate were observed in all oceans.

26.2.10 Conclusions

In the final chapter of *Biotic Interactions and Global Change*, Kingsolver et al. (1993) outline a research agenda comprised of eight specific goals designed to help ecologists "understand and forecast the consequences of global environmental change for (1) biological diversity, (2) community integrity, and (3) ecosystem services." Because these research goals are relevant not only to global change but also to our broader understanding of how communities respond to stressors, it is appropriate to review these eight recommendations (Table 26.5). We feel that the most critical research need for assessing ecological consequences of global change is to identify specific community responses to expected spatial and temporal variation in climate. This will occur only through better integration of basic ecological principles into experimental, monitoring, and modeling studies that focus on climate change. Relating individual responses to fitness, population abundance, and community structure will allow researchers to identify sensitive and ecologically important indicators of climate change. Research that matches the coarse spatial scale of GCMs with the smaller scale of most ecological investigations is essential for predicting regional responses to climate change. Because effects of climate change will vary among locations and among community types, a rigorous approach for assessing community susceptibility will improve our ability to predict these responses. This approach should distinguish between the direct effects of climate change and the indirect effects associated with alterations in species interactions. Finally, ecologists are only beginning to understand the influence of multiple anthropogenic stressors on species assemblages. Because community composition and ecosystem function will be quite different in a warmer climate, these relationships will likely change. Thus, predicting effects of contaminants on communities will require a better understanding of interactions between global climate change and other anthropogenic stressors.

TABLE 26.5
Proposed Research Agenda for Understanding and Predicting the Consequences of Global Climate Change on Communities

1. Relate temporal and spatial patterns of global change to likely biotic responses.
2. Relate stress responses of individual organisms to changes in fitness, population abundance, species distribution, and interactions.
3. Identify the critical rates of environmental change that determine ecological and evolutionary outcomes.
4. Understand the reassortment of ecological communities as a source of environmental change.
5. Identify sensitive and reliable indicators for current and future ecological research.
6. Understand factors that determine changes in the location and nature of ecological transition zones and species margins.
7. Develop standards and criteria for simplification of complex models.
8. Identify the ecological variables that contribute to important changes in regional or global climate, disturbance regimes, and patterns of habitat fragmentation.

Source: From Kingsolver, J.G., et al., In *Biotic Interactions and Global Change*, Kareiva, P.M., Kingsolver, J.G., and Huey, R.B. (eds.), Sinauer Associates, Inc., Sunderland, MA, 1993, pp. 480–486.

26.3 STRATOSPHERIC OZONE DEPLETION

Decreasing levels of stratospheric ozone (O_3) have been observed in polar and mid-latitude regions for about two decades (Madronich 1992). There is now conclusive evidence that reduced levels of ozone is a direct result of anthropogenic activities, particularly the release of CFCs. Although production and release of CFCs occurred primarily in the northern hemisphere, the greatest reductions in ozone levels have been reported in Antarctica. In September 2000, measurement of the ozone depletion-area using NASA's Total Ozone Mapping Spectrometer showed that the highly publicized "ozone hole" over Antarctica was the largest ever observed, covering approximately 28 million km^2. Other locations in the southern hemisphere, especially southern Australia and New Zealand, have also reported greatly reduced levels of ozone.

Because ozone limits the penetration of UVR through the earth's atmosphere, lower levels of stratospheric ozone are associated with increased levels of UVR. Recent studies conducted in New Zealand reveal that current levels of ozone during summer are approximately 10–15% less than in the 1970s, resulting in a significant increase in UVR (McKenzie et al. 1999). Although production and release of CFCs have decreased as a result of international agreements (e.g., the 1987 Montreal Protocol on Substances that Deplete the Ozone Layer), global CFC levels will likely remain elevated because of atmospheric persistence. Therefore, it is anticipated that levels of stratospheric ozone will continue to decline over the next several decades (Smith et al. 1992). In addition, there is concern that potential interactions between ozone depletion and global warming may significantly delay the return of stratospheric ozone to preindustrial levels (Shindell et al. 1998).

Increased UVR as a result of ozone depletion is a significant environmental hazard and is expected to have negative effects on humans and other organisms. For example, a 1% reduction in ozone is estimated to result in a 3% increase in certain forms of skin cancer in humans. Indeed, there is speculation that the high incidence of skin cancer in New Zealand and Australia is partially a result of elevated levels of UVR. Of the three categories of UVR, UV-B (280–320 nm) is most closely associated with loss of ozone. Ozone depletion has relatively little effect on UV-A (320–400 nm), UV-C (190–280 nm), or photosynthetically active radiation (PAR; 400–700 nm).

As noted above, ozone depletion has most frequently been reported over Antarctica where ambient doses of UV-B have increased approximately 140% per decade (Madronich 1992); however, a similar pattern has been observed in other regions (Kerr and McElroy 1993). Increases of 10–20% UV-B per decade have been reported in temperate regions of the northern and southern hemispheres, and this trend is expected to continue. Spectral measurements of UV-B in Toronto, Canada showed that the intensity of radiation at 300 nm has increased by 35% in winter and 7% in summer since 1989 (Kerr and McElroy 1993). In addition to these latitudinal differences, other factors such as elevation will increase exposure to UVR. For example, Kinzie et al. (1998) attributed a significant reduction in photosynthesis in a high elevation (3980 m) lake to UV-B. Thus, despite the focus on Antarctic communities, effects of UV-B radiation are widespread and also likely to impact mid-latitude regions.

Although ozone depletion and associated increases in UV-B are a recent phenomenon, the presence of UV-B has played an important role in the evolution of life on earth. Because of low levels of ozone, UV-B readily penetrated earth's primitive atmosphere and restricted organisms to aquatic habitats for most of their evolutionary history. Migration to terrestrial habitats occurred only after sufficient levels of ozone had accumulated in the atmosphere (Cloud 1968, Fisher 1965). As a consequence of the long-term exposure to UV-B, many organisms evolved protective mechanisms to reduce UV-B effects. For example, the presence of photoprotective pigments, natural sunscreens, and various DNA repair mechanisms allow organisms to survive in habitats saturated with UV-B radiation. Although it is likely that many organisms show some tolerance to UV-B, there is considerable variation among taxa. Differences in the ability of organisms to tolerate UV-B may account for the patterns of community structure observed in some habitats, particularly alpine areas with naturally high levels of exposure.

All wavelengths of UVR are potentially harmful to organisms, but UV-B radiation is of particular concern because of the dramatic increase associated with ozone depletion. Studies conducted with a variety of plants and animals have shown effects of UV-B at all levels of biological organization (Figure 26.8). At the molecular level, photochemical damage resulting from adsorption of specific wavelengths by macromolecules (e.g., DNA, RNA) and inactivation of photosystem II in plants are typical responses to UV-B exposure (Vincent and Roy 1993). Mutagenic effects resulting from DNA damage and reduced photosynthesis owing to alteration of light and dark reactions occur at the cellular level. These molecular and cellular alterations affect individual growth rates, community structure, and ecosystem function. Environmentally relevant, background levels of UV-B radiation are lethal to some taxa, and dramatic changes in primary production, community composition, and

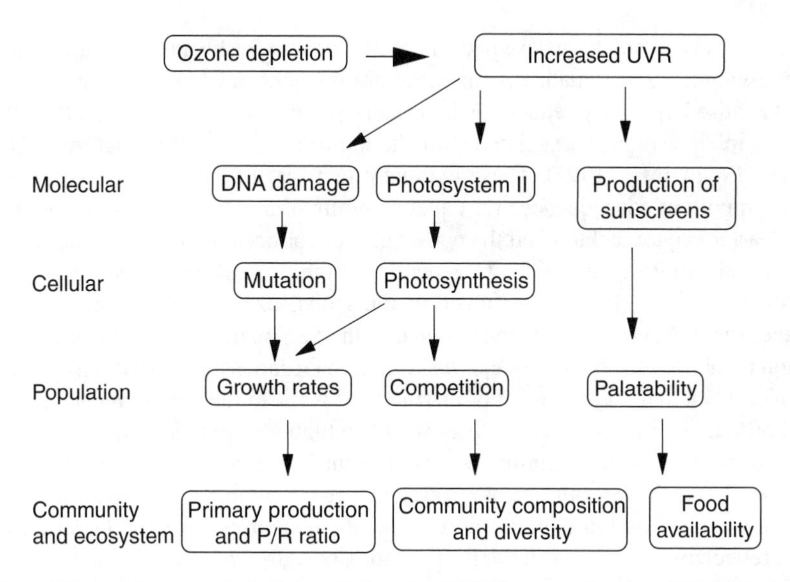

FIGURE 26.8 Effects of ozone depletion and increased UVR across levels of biological organization. (Modified from Figure 6 in Vincent and Roy (1993).)

trophic structure have been reported following UV-B exposure. The direct and indirect effects of UV-B radiation on communities are diverse and will be the primary focus of this section.

26.3.1 Methodological Approaches for Manipulating UVR

A variety of approaches, including UV cutoff filters, UV lamps, and more sophisticated solar simulators have been employed in field and laboratory studies to manipulate levels of UVR. UV cutoff filters can either transmit or remove UV-A and UV-B, whereas UV lamps can enhance exposure to different wavelengths of UVR. For logistical reasons, laboratory and microcosm experiments tend to rely on lamps to increase UV-B exposure, whereas field experiments tend to rely on filters to remove natural levels of UV-B. Both approaches have limitations, and there is some concern that differences between field and laboratory studies may be an artifact of different experimental techniques. To expose planktonic organisms to high levels of UV-B in the field, researchers will restrict organisms to shallow, high UV-B habitats. Because this experimental approach limits vertical migration, relevance to natural populations is questionable. Artificial lamps are commonly used in laboratory experiments (Bothwell et al. 1994, Kiffney et al. 1997a) and in some field studies (Rader and Belish 1997a) to enhance exposure to UV-B. Although this approach allows establishment of dose–response relationships, there are concerns about the unnatural spectral properties of UV lamps (Kelly et al. 2001).

26.3.2 The Effects of UVR on Marine and Freshwater Plankton

> The alarmist predictions of immediate and large scale impairment of primary production in response to ozone depletion seem to us to be greatly exaggerated . . . .
>
> **(Vincent and Roy 1993)**

> One of the most important caveats to working with the impact of UV-B radiation on freshwater ecosystems is that complex rather than simple responses are likely to be the rule.
>
> **(Williamson 1995)**

Negative effects of UV-B radiation have been measured in both freshwater and marine environments. While some researchers feel aquatic communities are resilient to UV-B and that declines in ozone are unlikely to cause large-scale reductions in primary productivity (Vincent and Roy 1993), others suggest that complex responses associated with alterations in community structure and aquatic food webs are likely (Williamson 1995). These indirect effects are often subtle and difficult to predict, but may have important consequences for aquatic communities. In particular, the effects of UV-B radiation on marine phytoplankton and the consequences for oceanic food webs have received considerable attention (Smith et al. 1992). In general, marine primary producers (algae and diatoms) are highly sensitive to UV-B. Because much of the global ozone depletion has taken place over Antarctica, there are concerns that enhanced UV-B in this region may have serious consequences for phytoplankton inhabiting the photic zone of the Southern Ocean. More importantly, because Antarctic phytoplankton is a major component of marine food webs in the region, reduced phytoplankton biomass and production may have cascading effects on higher trophic levels.

Despite mounting evidence from the laboratory that UV-B radiation negatively affects phytoplankton, extrapolating these results to natural systems is challenging. Owing to the unnatural spectral properties of UV lamps used in laboratory experiments, responses observed under artificial conditions may not reflect responses in the field. Furthermore, because organisms cultured in the laboratory may lack the protective pigments and repair mechanisms found in natural populations, effects of UVR may be exaggerated in laboratory experiments (Mostajir et al. 1999). Thus, comprehensive field experiments are essential for understanding the direct and indirect effects of enhanced UV-B. Using combinations of filters to remove different wavelengths of UVR, field experiments have shown that exposure to naturally occurring levels of UVR significantly affected primary production and community composition (Kinzie et al. 1998, Mostajir et al. 1999, Smith et al. 1992). In an extensive survey of phytoplankton communities in Antarctic waters, Smith et al. (1992) related ozone depletion and UV-B levels to phytoplankton production. Measurements were taken inside and outside the ozone depletion zone during a 6-week cruise. UV-B was detected at depths exceeding 60 m, and depth of penetration was greater inside the ozone hole than outside. Ozone-related UV-B inhibition of photosynthesis was observed at depths of 25 m, and primary production was 6–12% lower inside the ozone hole than outside. These results correspond to a 2–4% reduction in primary productivity and an estimated loss of 7×10^{12} g of carbon per year over the entire Antarctic marginal ice zone.

26.3.2.1 Direct and Indirect Effects of UV-B Radiation

In a provocative review of the role of UV-B radiation in freshwater ecosystems, Williamson (1995) posed four hypotheses to describe the direct and indirect effects of UV-B on planktonic communities (Table 26.6). These hypotheses are especially relevant to our discussion of community ecotoxicology because they emphasize species interactions and trophic ecology. Because most research on UV-B effects has focused on molecular, cellular, and physiological responses, Williamson's (1995) description of the potential ecological effects is quite illuminating.

The solar ambush hypothesis proposes that aquatic organisms unable to detect and respond to UV-B may be "ambushed" by differential wavelength changes in total solar radiation. These wavelength-specific changes occur as a result of differences in elevation, light attenuation, cloud cover, and other factors. According to this hypothesis, sessile or relatively immobile organisms are at high risk because of their inability to respond behaviorally to increased UV-B.

The solar cascade hypothesis highlights the effects of UV-B radiation on trophic interactions in lakes. The influence of top-down and bottom-up trophic regulation in aquatic systems is well established, and the ecotoxicological implications of these interactions are discussed in Chapter 27. According to the solar cascade hypothesis, differential effects of UV-B among trophic levels may

TABLE 26.6
Four Hypotheses Describing Potential Direct and Indirect Effects of Increased UV-B Radiation on Planktonic Communities

Hypothesis	Description	Ecological concept
Solar ambush	Wavelength-selective changes in solar radiation result in differential abilities of organisms to detect and respond to increased UV-B.	Differential tolerance among species will result in variation in community structure
Solar cascade	Differential effects of UV-B across trophic levels will have cascading influences on energy flow.	Trophic cascades; top-down versus bottom-up effects
Acid transparency	Effects of UV-B radiation will be greater in anthropogenically acidified lakes than in naturally acidic lakes.	Differential responses among communities; stressor interactions
Solar bottleneck	Small zooplankton in clear lakes will experience a "bottleneck" as a result of intense UV-B in upper surface water and predation pressure from below.	Predator–prey interactions; importance of vertical migrations in lakes

Source: From Williamson, C.E., *Limnol. Oceanogr.*, 40, 386–392, 1995.

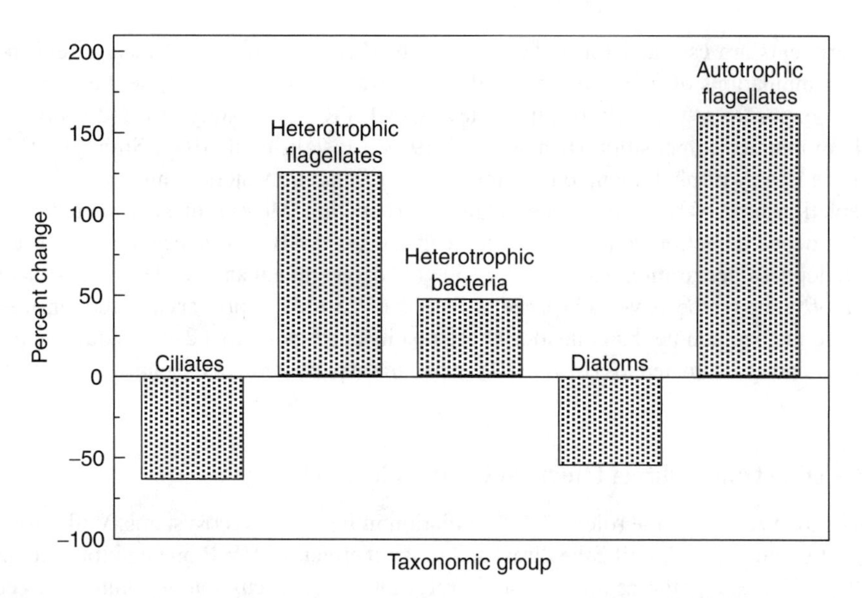

FIGURE 26.9 Experimental test of the solar cascade hypothesis in microbial and plankton communities. The figure compares the percent change in carbon biomass between natural and enhanced UV-B treatments for ciliates, flagellates, bacteria, and diatoms in mesocosms. The decrease in biomass of predatory ciliates resulted in an increase in biomass of flagellates and bacteria, thus channeling more energy into the microbial food web. (Data from Table 2 in Mostajir et al. (1999).)

have cascading effects on energy flow and community structure. For example, if grazing herbivores are more sensitive to UV-B than phytoplankton, primary production will increase in grazer-limited lakes. The solar cascade hypothesis has been tested in a mesocosm experiment where microbial and planktonic communities were exposed to natural and enhanced levels of UV-B radiation (Figure 26.9). Results showed considerable variation in responses to UV-B among trophic levels, and that elimination of predatory ciliates caused an increase in abundance of their prey (bacteria, heterotrophic flagellates, and small phytoplankton). The direct effects of UV-B on ciliates reduced the transfer of energy to higher trophic levels and channeled carbon into the microbial food web.

The acid transparency hypothesis describes the potential interactions between UV-B and acid deposition in freshwater lakes. Because of lower levels of humic materials, DOC, and other light attenuating substances, UV-B penetration and ecologically significant effects are expected to be greater in acidified lakes. It is possible that alterations in community composition and trophic structure observed in lakes receiving acidic deposition are at least partially a result of greater UV-B exposure.

From an ecological perspective, the most intriguing hypothesis advanced by Williamson (1995) describes the influence of UV-B on predator–prey interactions between large and small zooplankton species in lakes (Figure 26.10). The solar bottleneck hypothesis proposes that small zooplankton may experience a bottleneck near the surface in clear oligotrophic lakes because of intense UV-B radiation from above and predation pressure by large zooplankton from below. Although small zooplankton could avoid predation and UV-B exposure during most of the year, intense solar radiation in summer months would eliminate this refuge.

26.3.3 Responses of Benthic Communities

Most research investigating responses of aquatic ecosystems to UVR has focused on marine and freshwater plankton. The likely explanation for the emphasis on planktonic communities, especially in marine systems, was that rapid attenuation of UVR was expected to limit exposure to benthic

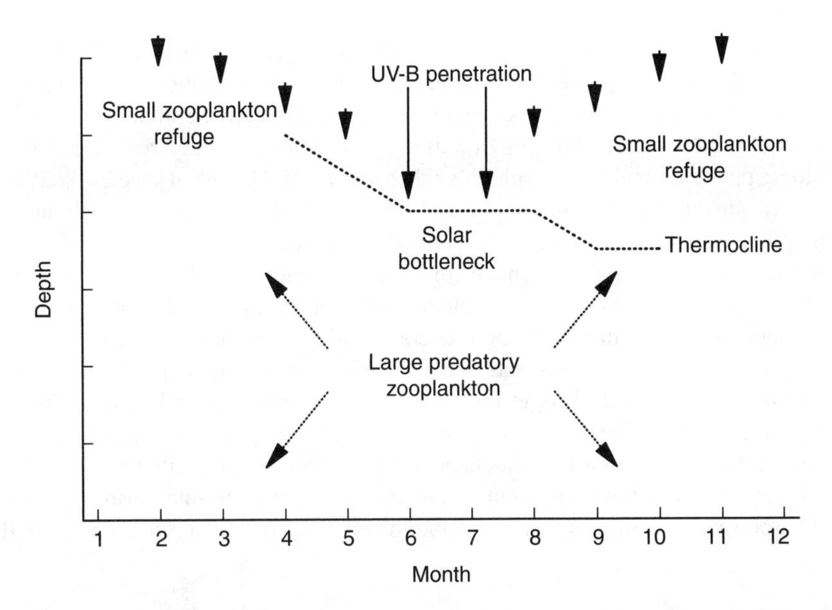

FIGURE 26.10 The solar bottleneck hypothesis. Seasonal changes in the depth of the thermocline and penetration of UV-B radiation create a bottleneck for small zooplankton in clear, oligotrophic lakes. Small zooplankton have a refuge from predation in shallow waters during most months of the year. However, deeper penetration of UV-B in summer forces these organisms to deeper water where they are exposed to large predatory zooplankton. (Modified from Figure 1 in Williamson (1995).)

communities. However, investigators have recently documented significant alterations in periphyton and benthic macroinvertebrate communities during exposure to UVR. Not only can UVR reach benthic communities in shallow aquatic environments, UVR may actually penetrate sediments and influence microbial photosynthetic communities within these habitats (Garcia-Pichel and Bebout 1996). Unlike planktonic organisms, periphyton communities and some benthic taxa are sessile and unable to avoid exposure to UV-B. Even relatively mobile macroinvertebrate taxa may be exposed to intense UV-B while grazing on periphyton in shallow, clear lakes, and streams.

The pioneering experiments of Bothwell et al. (1994) highlighted the complex ecosystem responses to UVR and demonstrated the importance of accounting for indirect effects. In contrast to expectations based on studies of freshwater phytoplankton, accrual of periphyton was actually enhanced in experimental streams receiving UVR. This apparent paradox was explained by lower abundance of grazing chironomids in UVR-treated streams, which allowed algal communities to flourish under conditions of reduced herbivory. Interestingly, the complex trophic response observed in these experiments was similar to that reported for streams exposed to the insecticide, malathion (Bothwell et al. 1994). It is likely that similar trophic cascades may occur in freshwater systems where herbivores are more sensitive to UVR than primary producers.

26.3.4 Responses of Terrestrial Plant Communities

Responses of plants to UV-B include reduced growth and biomass, reduced photosynthesis, and increased concentrations of UV-B absorbing compounds, particularly flavonoids (Caldwell et al. 1995, van de Staaij et al. 1995a,b). Because most research on plant responses to UV-B has been conducted in the laboratory and focused on individual species, the direct application of these studies to natural communities is uncertain (Gehrke 1999). Significant interspecific variation in sensitivity to UV-B has been observed among plant species, suggesting that changes in composition of communities are likely. In a series of field experiments conducted in northern Swedish Lapland,

Gehrke (1999) measured responses of two species of bryophytes to enhanced UV-B radiation over a 3-year period. Bryophytes possess a number of unique characteristics that increase their susceptibility to UV-B, including thin leaves that often lack a protective cuticle, reduced photorepair mechanisms and rootless shoots that cannot buffer against aboveground stress (Gehrke 1999). Differences in these physiological and morphological characteristics between species likely contributed to differences in sensitivity. The authors speculated that UV-B suppressed growth and altered the outcome of competition between sensitive and tolerant species.

Exposure of plants to UV-B may have important consequences for associated belowground communities (e.g., bacteria, mycorrhizal fungi, microarthropods). Because belowground communities play an important role in mineralization, decomposition, and plant nutrition, UV-B may have ecosystem-level effects. Klironomos and Allen (1995) exposed sugar maple (*Acer saccharum*) seedlings to various UVR treatments to measure effects on belowground communities. Although there were no direct effects of UVR on seedlings, large changes in the structure and function of fungal and bacterial communities associated with the roots were attributed to UV-B exposure. Changes in fungal and bacterial populations significantly affected soil microarthropod communities and shifted carbon flow from a mycorrhizal-dominated system to a saprobe/pathogen-dominated system.

26.3.5 Biotic and Abiotic Factors That Influence UV-B Effects on Communities

26.3.5.1 Dissolved Organic Materials

Dissolved organic materials (DOM) consist of a complex mixture of organic molecules that influence a variety of processes in aquatic ecosystems. Levels of dissolved organic and particulate materials play an important role in moderating effects of UV-B radiation in aquatic ecosystems. In fact, some researchers suggest that changes in DOM may have a greater influence on UVR exposure in aquatic ecosystems than impacts resulting from increased global levels of UVR (Schindler et al. 1996). The generally low levels of DOM typical of most marine ecosystems, especially the open ocean, allow UV-B to penetrate deep into the photic zone. In contrast, the much higher levels of UV-B-absorbing materials found in many freshwater ecosystems significantly reduce UV-B penetration in all but the most oligotrophic lakes (Williamson 1995). Because of rapid attenuation of light, UV-B effects in freshwater systems are generally considered to be less than those in marine systems. Nonetheless, some freshwater communities, especially those inhabiting shallow, clear lakes and streams where damaging levels of UV-B may penetrate several meters are at significant risk (Bothwell et al. 1994, Mostajir et al. 1999, Williamson 1995). In field experiments conducted in a Rocky Mountain stream, Kiffney et al. (1997b) observed effects of UV-B on benthic communities only after a seasonal decrease in natural levels of DOM.

Interactions between UVR and DOM in aquatic ecosystems are complex. The ability of DOM to attenuate light varies with source (allochthonous versus autochthonous) and chemical composition. For example, DOM derived from terrestrial sources absorb more UVR than autochthonous materials (McKnight et al. 1997). Exposure to UVR can also degrade DOM through a process known as photobleaching (De Haan 1993). Thus, long-term exposure of DOM to UV-B may result in reduced light attenuation and actually enhance UV-B effects in aquatic communities. Because the amount of dissolved and particulate material in some freshwater systems varies with productivity, the trophic status of lakes and streams may also influence light penetration and UV-B effects. Williamson et al. (1994) compared responses of zooplankton communities to UV-B radiation in an oligotrophic and a eutrophic lake in Pennsylvania (USA). Experiments showed significant mortality owing to UV-B exposure in the oligotrophic lake, whereas communities in the eutrophic lake were protected. Because DOM may have direct effects on microbial productivity and community composition, separating the chemical effects of DOM from the light attenuating influences on UV-B penetration is difficult. Kelly

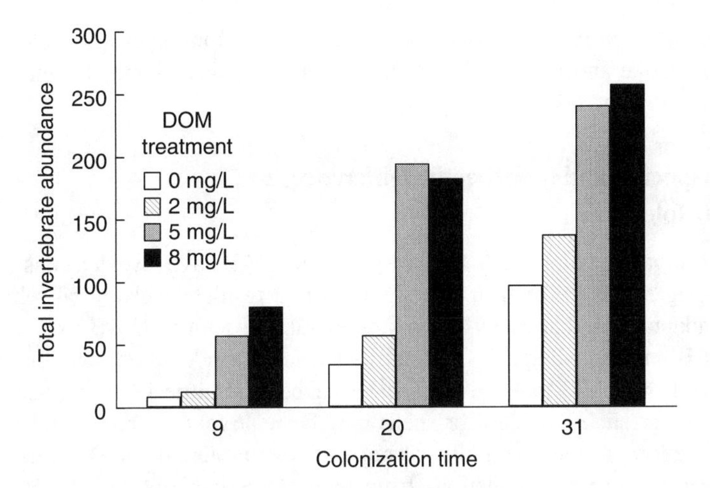

FIGURE 26.11 The influence of DOM on attenuation of UVR and colonization by benthic macroinvertebrates in outdoor experimental streams. Filters containing varying concentrations of DOM were suspended above replicate experimental streams to reduce levels of UVR. Note the sharp decrease in macroinvertebrate colonization between 2 and 5 mg/L DOM. (Modified from Figure 5 in Kelly et al. (2001).)

et al. (2001) used filters containing several concentrations of DOM (0, 2, 5, 8 mg/L) suspended above experimental streams to isolate effects of DOM and UVR on benthic communities. Results showed highly significant increases in abundance of macroinvertebrates with increased DOM concentration and UVR attenuation (Figure 26.11). A sharp threshold in response of benthic macroinvertebrate grazers was observed between 2 and 5 mg/L, indicating that streams with low levels of DOM (<5 mg/L) may be particularly sensitive to UVR.

26.3.5.2 Location

Because of spatial variation in UV exposure and species' sensitivity (Hill et al. 1997), the ecological effects of UV-B radiation on aquatic ecosystems are likely to differ among geographic regions. Levels of UV-B reaching the earth are affected by a variety of large-scale factors, including cloud cover (Lubin and Jensen 1995), snow depth, elevation (Caldwell et al. 1980), and latitude. Seasonal changes in day length can also significantly alter exposure to UV-B radiation. Exposure to UV-B in polar regions is especially elevated during the summer owing to extended periods of sunshine. Because the intensity of UV-B radiation increases with elevation, communities located at higher altitudes are exposed to greater levels of UV-B than communities at lower elevations. For example, Caldwell et al. (1980) reported that the biologically effective UVR dose was 25% higher at 3300 m than at 1500 m (40°N latitude). On a smaller spatial scale, Rader and Belish (1997b) showed that effects of ambient and enhanced UV-B on periphyton communities varied among locations, and that effects were greater on communities inhabiting open reaches than shaded reaches. Santos et al. (1997) showed that depth influenced responses of marine diatom communities to UV-B, and concluded that accurate predictions about UV-B effects cannot be made without accounting for interactions with other environmental factors.

Factors such as elevation and DOM may interact to influence exposure of aquatic communities to UV-B. Because of shallow depth and naturally low levels of DOM, aquatic communities in many alpine habitats are subjected to intense levels of UV-B radiation during summer months (Kiffney et al. 1997b). In addition, because solar intensity increases significantly with elevation, communities located at higher altitudes are at particular risk from UV-B. Vinebrooke and Leavitt (1996) measured effects of UVR on periphyton in an ultraoligotrophic lake in Banff National Park (Canada). Colonization of periphyton was significantly suppressed under ambient UVR, primarily owing to effects on the diatom *Achnanthes minutissima*. Because *A. minutissima* dominates lower elevation lakes with high DOM, Vinebrooke and Leavitt (1996) suggest that this species may be an indicator of UVR effects in western montane lakes.

26.3.5.3 Interspecific and Intraspecific Differences in UV-B Tolerance

Experiments investigating community-level responses to UV-B have revealed considerable variation in tolerance among species. Much of this variation is a result of behavioral, physiological, and morphological adaptations that allow organisms to survive natural UV-B exposure. Variation in tolerance to UV-B among fish species was attributed to an unidentified component in skin tissue that protected fish from UV-B-induced sunburn (Fabacher and Little 1995). Organisms that possess a protective covering, such as a case or shell, may be relatively tolerant to UVR exposure (Hill et al. 1997). It is expected that the ability of populations to tolerate UV-B is related to their level of natural exposure, and that populations from high UV-B habitats have greater UV tolerance. For example, organisms in oligotrophic, high elevation lakes and streams produce UV-absorbing compounds and pigments that act as natural sunscreens. Leech and Williamson (2000) define UV-B tolerance as the sum of photoprotection and photorepair processes that vary with microhabitat and location. Sommaruga and Garcia-Pichel (1999) reported that levels of UV-absorbing compounds in planktonic organisms decreased with depth. Similarly, Gleason and Wellington (1995) observed that survivorship of coral larvae exposed to UV-B decreased with depth of collection, most likely owing to lower amounts of UV-B-protective amino acids. Zellmer (1995) compared UV-B tolerance of transparent and heavily melanized *Daphnia* collected from lakes. Significant mortality and lower reproduction were observed in transparent populations, whereas the pigmented populations showed no effects of UV-B exposure. Finally, plants also produce photoprotective compounds in response to UV-B (van de Staaij 1995a), and populations exposed to naturally high levels of UVR are generally more efficient at UV-B absorption. However, in contrast to expectations, van de Staaij et al. (1995b) found no significant difference in UV-B tolerance between highland populations (where natural levels of UV-B were greater) and lowland populations of a perennial herb.

Since small phytoplankton species ($<2.0\ \mu m$) often comprise a significant portion of total biomass in plankton communities, understanding size-dependent variation in sensitivity to UV-B is important for predicting ecological effects. Because the protective efficiency of natural sunscreens to phytoplankton decreases with cell size (Laurion and Vincent 1998), we expect that small cells may be more sensitive to UV-B than large cells. Despite the intuitive appeal of the size-dependency hypothesis, it is obvious that other factors will influence tolerance of planktonic communities to UVR. Laurian and Vincent (1998) reported that small phytoplankton were relatively tolerant to UV-B exposure and that interspecific differences were more important than size for predicting effects on phytoplankton. The relationship between organism size and UVR tolerance is further complicated by the presence of visual predators, which often select larger zooplankton species and can modify patterns of vertical migration in the water column. In addition, lake transparency influences tolerance of zooplankton to UV-B and diel migration patterns. Leech and Williamson (2000) developed a conceptual model relating zooplankton body size, lake transparency, and the presence of fish predators to UVR tolerance (Figure 26.12). The model predicts that body size will be inversely related to tolerance in high UVR lakes with fish predators. Large zooplankton in lakes with fish predators will spend most of the daylight hours in deeper water to avoid predation, thus reducing UVR exposure and the development of tolerance. In contrast, small zooplankton will occupy the upper surface waters in these lakes where they are exposed to intense UVR, thus increasing tolerance. In clear, fishless lakes, the model predicts that zooplankton should be relatively tolerant of UVR and that body size is unimportant. Finally, because of the limited exposure to UVR, zooplankton inhabiting low UVR lakes should show little tolerance. In a series of short-term field experiments to test these predictions, Leech and Williamson (2000) reported that taxonomic variation was more important

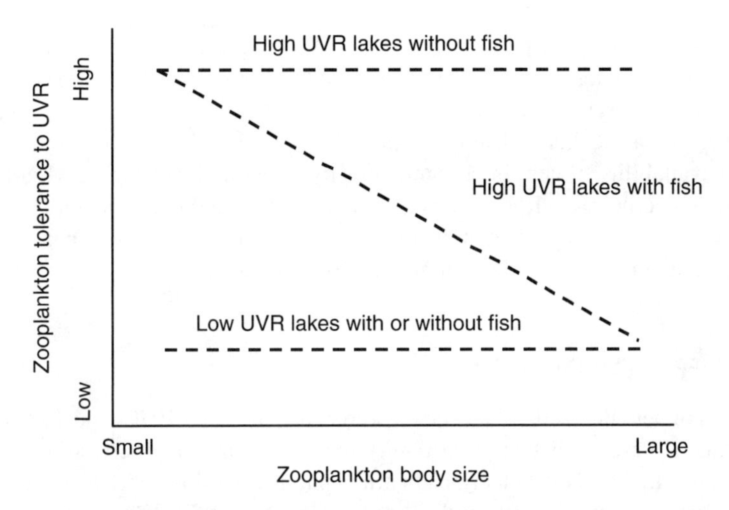

FIGURE 26.12 Conceptual model relating zooplankton body size, lake transparency, and the presence of fish predators to UVR tolerance. In high UVR lakes with fish predators, UVR tolerance will be inversely related to body size. To avoid predation, small zooplankton will generally occupy the lake surface where they are exposed to higher levels of UVR and are expected to develop enhanced tolerance. In contrast, larger zooplankton that occupy deeper sections of the lake receive less natural UVR and are expected to show little tolerance. There is no relationship between body size and tolerance in high UVR lakes without fish because zooplankton of all sizes can migrate freely between the surface and deeper water. These organisms are expected to show greater tolerance than individuals from low UVR lakes because of their greater natural exposure. (Modified from Figure 2 in Leech and Williamson (2000).)

than body size or lake transparency for predicting UVR tolerance. They speculated that results may change during longer-term exposure and that UVR may interact with other stressors to influence zooplankton community structure. Potential interactions between UVR and other stressors will be discussed below.

Some aquatic organisms may avoid exposure to UVR by migrating to deeper water or moving to shaded habitats during periods of intense solar radiation. These behavioral responses are greatly limited in sessile organisms or in species that spend significant amounts of time in shallow water. For example, filter-feeding blackfly larvae are often found attached to the substrate in clear, shallow streams and lake outlets. Donahue and Schindler (1998) observed significant emigration when organisms were exposed to UVR in experimental streams. Some grazing organisms, such as mayflies, chironomids, and caddis-flies, are obligate inhabitants of the upper surface of cobble substrate and likely subjected to intense UVR. Kiffney et al. (1997a) measured the drift response of benthic communities to UVR and observed greatest effects on grazing organisms associated with the substrate surface.

26.3.5.4 Interactions with Other Stressors

In addition to the direct effects of UVR on aquatic communities, synergistic interactions between UVR and other anthropogenic stressors may enhance effects and complicate our ability to predict changes observed in the field (Blaustein et al. 1995). The increased toxicity of polycyclic aromatic hydrocarbons (PAHs) under UV-B exposure is well documented (Oris and Giesy 1986). Not only are organisms more sensitive to certain PAHs when exposed to UV light but also this increased toxicity persists following exposure (Ankley et al. 1994) and may be transferred maternally (Hall and Oris 1991). Unfortunately, because most of the research on interactions between UV-B and PAHs has been conducted in the laboratory, our understanding of community-level phototoxic effects is limited. Long et al. (1995) reported a synergistic interaction between pH and UV-B in amphibians. Mortality of *Rana pipens* was unaffected by low pH or high-UV-B exposure; however, significantly greater mortality was observed when organisms were exposed to both stressors. Although the mechanism responsible for this synergistic interaction was not specified, the results are significant because of concerns over global declines of amphibians (Blaustein and Wake 1990). Finally, UVR may increase the bioavailability of certain contaminants by affecting levels of DOM and by disrupting ligand–contaminant complexes. It is well established that the quality and quantity of DOM greatly influences bioavailability of many contaminants. Exposure to UVR degrades DOM (De Haan 1993), thus increasing light penetration and contaminant bioavailability.

26.4 ACID DEPOSITION

Interest in effects of acid deposition has waned somewhat since the 1970s and 1980s when literally thousands of papers were published describing effects on aquatic and terrestrial ecosystems. Part of the explanation is that modest reductions in atmospheric emissions of SO_2 and particulates from power plants and smelters have significantly improved local conditions. In addition, there are greater concerns for contemporary global stressors such as CO_2 and UV-B radiation. However, acid deposition has significant negative effects on aquatic and terrestrial communities, and remains a global environmental problem (Galloway 1995). More importantly, there is a growing awareness that acidic deposition may interact with other atmospheric stressors to affect communities. These interactions will be described in Section 26.5.

26.4.1 Descriptive Studies of Acid Deposition Effects in Aquatic Communities

Acid deposition from anthropogenic sources is a well established cause of acidification of lakes and streams. In an extensive survey of aquatic ecosystems in the United States, Baker et al. (1991) concluded that atmospheric deposition was responsible for acidification of 75% of 1180 acidic lakes and 47% of 4670 acidic streams examined. The spatial distribution of these atmospheric pollutants is determined by a complex interplay of geographic and climatic factors. In particular, patterns of prevailing winds and proximity to industrialized and urban sources greatly influence concentration and deposition. The effects of acidification on aquatic communities have been studied extensively, especially in the northeastern United States, Canada, and Europe (Heard et al. 1997, Herrmann 1993, Juggins et al. 1996, Lancaster et al. 1996, Rosemond et al. 1992, Somers and Harvey 1984, VanSickle et al. 1996). A variety of approaches have been employed, including large-scale biomonitoring, whole ecosystem manipulations, and paleolimnological studies (Box 26.3). The most convincing evidence for a relationship between acidification and community structure has been obtained from long-term, experimental studies (e.g., Schindler's long-term acidification studies in the Experimental Lakes Area described in Chapter 23). However, well-designed biomonitoring studies conducted at large spatial scales and/or over relatively long time periods have also contributed to our understanding of the ecological effects of acid deposition on aquatic communities. By taking advantage of established gradients in acid deposition downwind from known sources (e.g., smelters and coal-fired power plants), it is possible to quantify effects of acidification at a large spatial scale. Somers and Harvey (1984) described the fish communities from 50 lakes located at varying distances (up to 65 km) downwind from a large point source of SO_2 emissions in Ontario, Canada. Species richness was significantly correlated with distance from the source. Feldman and Connor (1992) used a replicated "natural experiment" to quantify the relationship between pH and benthic community structure in Shenandoah National Park (Virginia). Six sites (three acidic and three circumneutral) were selected from 60 candidate streams on the basis of physical, chemical, and habitat conditions. Differences between acidic and nonacidic streams were generally greatest for mayflies. Because their unique experimental design avoids the problem of pseudoreplication (Hurlbert 1984), Feldman and

Box 26.3 The Role of Paleolimnological Studies

Additional support for a causal link between acid deposition and acidification of surface waters has been provided by paleolimnological studies. The preservation of certain groups of organisms in lake sediments provides a historical record of community change and, in some instances, can compensate for the lack of long-term data. By dating the strata in sediment cores, the approximate time period when changes in community composition occurred can be identified and related to a hypothesized environmental change (e.g., increased acid deposition during the industrial revolution). Materials commonly used in paleolimnological studies include pollen grains, chironomid head capsules, and diatoms. Most of the paleolimnological research in lentic ecosystems has focused on diatoms, which are well preserved in lake sediments and known to be excellent indicators of water quality. Changes in abundance of pH-sensitive and pH-tolerant taxa over time can indicate the onset of acidification or recovery after reductions of acidic deposition. Comparisons of temporal changes in community composition across broad spatial scales can help identify factors that influence watershed sensitivity to acidification.

Juggins et al. (1996) analyzed long-term temporal patterns (1850-present) in diatom communities to develop a model of diatom-inferred pH values for 11 lakes in the United Kingdom. Results showed that all lakes were acidified over the past 150 years, but that the timing of acidification varied among lakes. Hall and Smol (1996) compared present-day and preindustrial diatom communities in 54 southern Ontario lakes. An initial relationship between diatom community composition, lake acidity, and total phosphorus (an indicator of shoreline development) was established by sampling present-day diatoms in surface sediments. These data were employed to develop a predictive model for water quality based on diatom communities, which was used to estimate long-term changes in pH and total phosphorus. Hall and Smol (1996) conclude that most acidification has occurred during the postindustrial period, probably as a result of atmospheric deposition. Despite increased shoreline development in the lakes, diatom-inferred phosphorus levels actually declined in some lakes. These researchers speculated that lake or watershed acidification may have reduced loading or increased loss of phosphorus.

Most paleolimnological studies of acidification have been limited to relatively short time periods (<150 years) and have focused on anthropogenic sources of lake acidity. However, it is important to note that natural changes in lake acidity will influence communities and may occur over much longer time periods. By characterizing pollen grains, diatoms, and chemical composition of sediment cores, Ford (1990) presents a 10,000-year history of pH changes in two northeastern U.S. lakes. Results showed that a unique combination of geological features (e.g., thin tills derived from base-poor gneisses and schists) predisposed one of these lakes (Cone Pond, New Hampshire) to become acidic over 6,500 years ago. These studies highlight the usefulness of paleolimnological data for complementing biomonitoring results and for placing community assessments within the proper historical context.

Connor (1992) suggest that their conclusions are broadly applicable to the larger population of acidified streams in the region. In addition to spatially extensive surveys, long-term monitoring data can be used to support hypothesized relationships between pH and community composition. Hall and Ide (1987) related temporal changes in benthic communities collected from two Canadian streams in 1937 and 1985 to increased acidification.

Biomonitoring studies conducted at relatively large spatial scales (e.g., multiple watersheds) have reported consistent effects on benthic communities (Feldman and Connor 1992, Rosemond et al. 1992). In general, acid-sensitive mayflies and stoneflies are replaced by acid-tolerant dipterans and caddisflies at pH values less than 5.5. Interestingly, this is about the same pH level where researchers conducting whole ecosystem manipulations observe significant effects on community structure (see below). It appears that many aquatic organisms are able to tolerate pH excursions to approximately 5.5, but greater levels of acidification generate all-or-none responses. Herrmann et al. (1993) reviewed effects of acidification on streams and discussed mechanisms of toxicity to invertebrates. Benthic macroinvertebrate community responses to acidification are quite similar to those observed in circumneutral streams polluted by heavy metals (Clements et al. 2000). These results suggest that similar mechanisms may influence toxic effects of acidification and heavy metals on benthic communities (Courtney and Clements 1998). Alternatively, toxic effects on macroinvertebrates observed in acidified streams may be a result of increased metal leaching rather than direct effects of low pH. This possibility highlights one of the key limitations of biomonitoring studies of acidification. Because acidification and metal leaching often co-occur, quantifying the relative importance of these stressors is difficult.

Shifts in phytoplankton and zooplankton communities in response to low pH are well documented. In particular, the reduced species richness of zooplankton has been proposed as an indicator of lake acidification. However, in a comprehensive review of zooplankton responses to acidification, Brett (1989) noted that despite sensitivity of many species to acidification, relationships between community composition and pH in spatially extensive surveys are often weak. The most likely explanation for the poor explanatory power of pH is that most surveys have focused exclusively on abiotic factors and ignored species interactions. It is well established that fish predation and other trophic interactions have a powerful influence on zooplankton biomass, diversity, and size structure. Because elimination of fish is a common occurrence in acidified lakes, it is not surprising that relationships between community composition and abiotic factors are often weak (Brett 1989). Not only can predation and competition confound relationships between pH and community composition, but acidification may alter the outcome of species interactions (Locke 1992). Separating the direct toxic effects of acidification from changes in trophic structure will require a better understanding of these interactions.

Predicting effects of acid deposition on aquatic communities is also complicated by the ability of some taxa to acclimate or adapt to low pH. Although reduced species richness and shifts in community composition from acid-sensitive to acid-tolerant species are typical responses in acidic lakes and streams, researchers studying naturally acidic environments have reported different results. Winterbourn and Collier (1987) found no relationship between low pH (pH 3.5–5.5) and benthic community structure in naturally acidic streams in New Zealand. In addition to improving our understanding of toxic mechanisms associated with acidification, insights into the process of acclimation and adaptation to low pH are essential for predicting community responses.

26.4.2 Episodic Acidification

Assessments of acidification in lotic ecosystems have focused primarily on chronic effects that result from relatively long-term changes in pH. However, acute exposure to the combined effects of low pH and metals occurs in some streams during periods of high discharge (Figure 26.13). This phenomenon, known as episodic acidification, is a short-term decrease in acid neutralizing capacity (ANC) of a watershed. Episodic acidification is common in the northeastern United States and is generally associated with snowmelt and rainfall events. During periods of high discharge, base cations are diluted and materials from upper, more acidic soil layers are flushed into the surrounding watershed. The result is a short-term pulse of solutes, especially H^+ and Al^*, and a rapid decrease in pH. These episodic events have practical implications for monitoring effects of acidification on streams. Because of the relationship between stream discharge and ANC, assessments of chemical conditions during base flow may provide a relatively poor indication of water quality.

The ecological effects of episodic acidification on aquatic organisms relative to long-term, chronic exposures are not well known. In a series of *in situ* bioassays conducted as part of a large-scale assessment of episodic acidification in the northeastern United States, Van Sickle et al. (1996) concluded

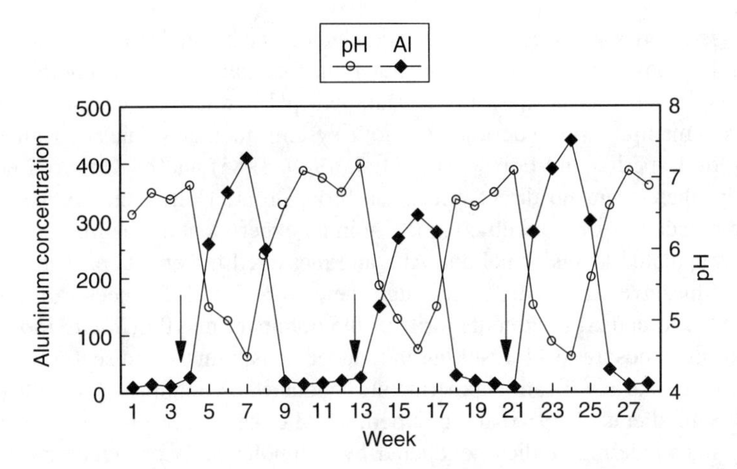

FIGURE 26.13 Hypothetical relationship between aluminum concentration and pH during three periods of episodic acidification (indicated by arrows). Decreases in pH as a result of high discharge from rainfall or snowmelt result in rapid increases in Al* concentrations. Because of the dynamic relationship between pH, Al*, and stream discharge, snapshots of chemical conditions at any point in time are probably inadequate for characterizing ecological integrity.

that exposure to Al* accounted for most (86–98%) of the observed mortality in native fish species. In a companion study, Baker et al. (1996) showed that episodic acidification significantly influenced fish movement, mortality, and community composition. Responses to episodic acidification were similar to those observed in chronically acidified streams, suggesting that short-term acute exposure is an equally important stressor to aquatic organisms.

26.4.3 Experimental Studies of Acid Deposition Effects in Aquatic Communities

Spatially extensive surveys of lakes and streams can provide important insight into the relationship between pH and community structure. However, because of the confounding influences of other physical, chemical, and biological factors, these studies only provide a potential explanation for alterations in aquatic communities. Experimental studies have been used extensively to demonstrate effects of acidification on lake and stream communities. In laboratory microcosms, Niederlehner et al. (1990) observed that functional measures (e.g., nutrient retention, community respiration) were relatively insensitive to acidification compared to taxonomic measures. The insensitivity of functional measures was also observed by Burton et al. (1985) who reported that differences in decomposition between treated (pH = 4.0) and control (pH = 7.0–7.4) streams did not occur until 6 months after exposure.

Because acidification of aquatic ecosystems frequently results in increased concentrations of metals, particularly Al, experimental studies are an excellent way to separate the relative importance of these two stressors. Using experimental stream channels, Allard and Moreau (1987) reported that addition of Al (160–520 μg/L) did not increase the effects of acidification (pH = 4.0) on benthic communities. Genter (1995) used a 2×2 factorial design to measure interactive effects of acidification and Al on benthic algal communities in experimental streams. In contrast to results of Allard and Moreau (1987), this study found that aluminum was more toxic than acidification for several taxa. Similar results were reported by Havens and DeCosta (1987) who assessed effects of low pH (5.0) in combination with Al (100–200 μg/L) on zooplankton communities. Results showed little effects of acidification alone; however, the combination of low pH and increased Al* significantly impacted most major groups.

Whole-ecosystem manipulations have been conducted in lakes and streams to assess the impacts of acidification on community structure and function. Compared to other chemical and physical stressors such as organochlorines and UV-B radiation, pH is relatively easy to manipulate in the field. Examples of manipulative experiments in lotic systems include studies by Hall and colleagues at Hubbard Brook Experimental Forest (Hall et al. 1980, 1985) and by Ormerod et al. (1987) in Wales. Although there were no direct effects on brook trout, Hall et al. (1980) reported lower diversity and a reduction of food web complexity in an experimentally acidified stream. Ormerod et al. (1987) manipulated levels of pH and Al*, and measured the effects on fish and invertebrate survival and benthic invertebrate drift. Although there were few differences in macroinvertebrate survival between acid and Al treatments, drift of the dominant mayfly (*Baetis rhodani*) increased dramatically during exposure to Al, resulting in reduced density in treated sections.

Details of whole-lake acidification experiments conducted by Schindler and colleagues (Schindler 1987, 1988; Schindler et al. 1985) at the Experimental Lakes Area were described in Chapter 23. Biomonitoring and modeling studies conducted by Schindler and colleagues corroborated these experimental findings and showed that significant changes in community composition of phytoplankton, zooplankton, and fish occurred at pH values less than 6.0. Greater levels of acidification (pH approximately 5.0) caused species-specific responses ranging from acute toxicity of H^+ to indirect effects on food chains. Alterations in community composition were generally more sensitive indicators than functional endpoints (e.g., primary production, decomposition, nutrient cycling) (Schindler 1987). In particular, changes in abundance of small, rapidly reproducing species were the earliest indicators of stress. Although it is likely that these changes in community composition will alter ecosystem processes, functional endpoints alone appear to be inadequate for detecting early signs of acidification stress.

26.4.4 Recovery of Aquatic Ecosystems from Acidification

Reduced emissions of SO_2 and particulates as a result of federal regulations is expected to have a positive influence on physical, chemical, and biological characteristics in aquatic ecosystems. However, our understanding of recovery of lakes and streams from the deleterious effects of acidification is relatively poor. Changes in aquatic communities over time following the removal of a stressor should reflect conditions expected in naturally recovering systems. However, the rate of recovery from acidic deposition will also be influenced by local climate and geology of the watershed. More importantly, studies of recovery from experimental acidification may not adequately predict recovery from atmospheric deposition. Experimental manipulation of pH in lakes and streams generally involves direct acidification of the waterbody, with no treatment of the surrounding watershed. It is likely that recovery rates will be much slower in systems subjected to atmospheric deposition owing to loss of buffering capacity from the watershed (Likens et al. 1996).

A comprehensive assessment of recovery of lentic systems was conducted near Sudbury, Ontario where local emissions of SO_2 and particulates from smelters have been reduced by approximately 90% since the 1970s. Mallory et al. (1998) reported chemical characteristics from a 12-year survey of 161 lakes in the region. Most lakes showed significant signs of recovery between 1983 and 1995 owing to reductions in sulfur inputs. Physicochemical changes included increased pH and ANC, and decreased base cations, metals, and SO_4. However, long-term temporal patterns in recovery were greatly influenced by annual precipitation and lake type. Recovery was greatest during drought years, but conditions deteriorated during wet years, especially in glacial headwaters and lakes located on highly sensitive bedrock.

Despite improvements in water quality as a result of local reductions in atmospheric emissions, acid deposition still affects many aquatic ecosystems. In addition, after several decades of acid deposition many watersheds have lost their natural buffering capacity and are expected to recover only over geological time scales (Likens et al. 1996). Experimental treatment of streams with lime

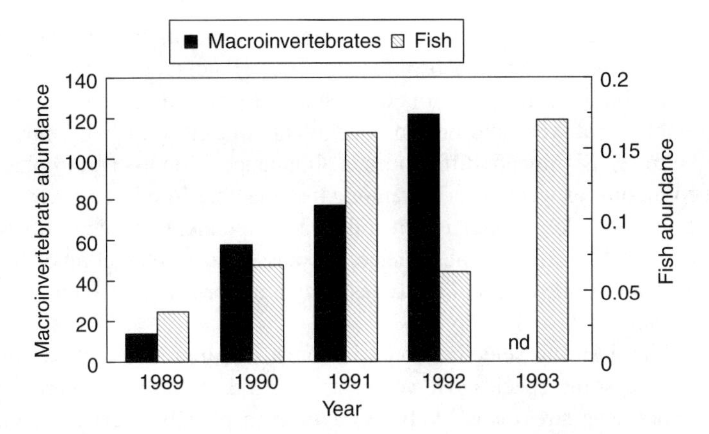

FIGURE 26.14 Improvements in macroinvertebrate density and fish abundance (number per m^2) in an acidified stream (Dogway Fork, West Virginia, USA) following treatment with limestone. (Macroinvertebrate data from Table 4 in Clayton and Menendez (1996). Fish data from Table 5 in Menendez et al. (1996).) nd = no data.

has been proposed to ameliorate effects of acidic deposition in systems not expected to recover naturally. Lime treatments in Dogway Fork, a second-order stream in West Virginia, significantly increased pH and ANC, and reduced Al* concentrations (Clayton and Menendez 1996, Menendez et al. 1996). Pretreatment pH values were generally less than 5.0, whereas a pH of greater than 6.0 was maintained following lime treatments. These improvements in water quality resulted in rapid recovery of fish and macroinvertebrate communities over a 4-year period (Figure 26.14). Similar results were reported by Simmons et al. (1996) in which brook trout density increased significantly following liming treatment of a Massachusetts stream degraded by acid precipitation.

26.4.5 Effects of Acid Deposition on Forest Communities

> Discussion of current forest declines and the possible role of acidic deposition in these declines has been characterized by considerable unsupported claims and speculation in the popular press and even semiscientific publications.
>
> **(Pitelka and Raynal 1989)**

Few examples of the effects of acidic deposition have received as much attention as the observed correlation between acidification and forest decline. Massive declines in forest health have been reported in Europe and North America over the past several decades, and research efforts initiated in the 1980s to understand the role of acidic deposition have been unprecedented (Pitelka and Raynal 1989). Although forest decline occurs as a result of numerous natural and anthropogenic factors, the simultaneous loss of several acid-sensitive species across broad geographical regions suggests a global cause. In particular, populations of Norway Spruce (*Picea abies*) in the Bavarian Alps and red spruce (*Picea rubens*) in eastern North America and Canada have been significantly impacted. Ollinger et al. (1993) measured broad spatial patterns of major ions associated with precipitation across the northeastern United States. Highly significant increases in SO_4^{2-}, NO_3, and H^+ from East to West were attributed to industrial activities. These authors noted that broad spatial patterns of these atmospheric pollutants were closely correlated with declines in forest health.

The proposed mechanism most likely responsible for the increase in tree mortality is nutrient deficiency resulting from nutrient leaching from soils. The loss of nutrients from forest soils is a typical response to acidification, and a variety of physical, chemical, and biological processes influence leaching. While most research has focused on the role of soil biota (bacteria, fungi, nematodes, microarthropods), the relative importance of different components within the soil community is poorly understood. In a series of microcosm experiments, Heneghan and Bolger (1996) measured various functional attributes of microarthropod assemblages (primarily Acari and Collembola) isolated from soil samples that were exposed to constituents of acid rain (e.g., nitric acid, sulphuric acid, ammonium nitrate). Significant differences in abundance, biomass and trophic composition of microarthropod communities were observed among treatments. More importantly, loss of nutrients, exchangeable cations, and CO_2 production in soils were dependent on microarthropod community composition. These results show that abundance, biomass, and trophic composition of microarthropod communities, independent of soil microbes, play a major role in decomposition and nutrient cycling in forest soils.

Acid deposition is a strong selective force in forest communities and has likely resulted in increased resistance of some species. However, as with other classes of contaminants we have considered, resistance to a stressor is likely to have some physiological or ecological cost. For example, we expect that resistant populations or species may be at a competitive disadvantage in relatively undisturbed habitats. In situations where pollutant exposure is episodic, there may be a temporal component to patterns of resistance and sensitivity. Taylor et al. (1994) present a theoretical model showing the relationship between selection coefficients and allocation of carbon resources for growth, maintenance and defense in polluted and pristine environments (Figure 26.15). In polluted environments, resistant genotypes have a competitive advantage over sensitive genotypes and can allocate greater carbon resources for growth, maintenance, and defense. However, because of the predicted cost of tolerance, these resistant genotypes have a competitive disadvantage in unpolluted environments.

Some researchers are skeptical of the relationship between acidic deposition and forest health. Pitelka and Raynal (1989) argue that because forest decline occurs for numerous reasons, quantifying the relative influence of acidification is not possible without long-term investigations. Despite the difficulties isolating specific causes of decline, there is strong evidence that the combined effects of atmospheric stressors impact forest communities. In a comprehensive review of air pollution effects on forests, Taylor et al. (1994) conclude that current levels of atmospheric pollutants are having negative effects on forest ecosystems worldwide. The consequences of changes in community

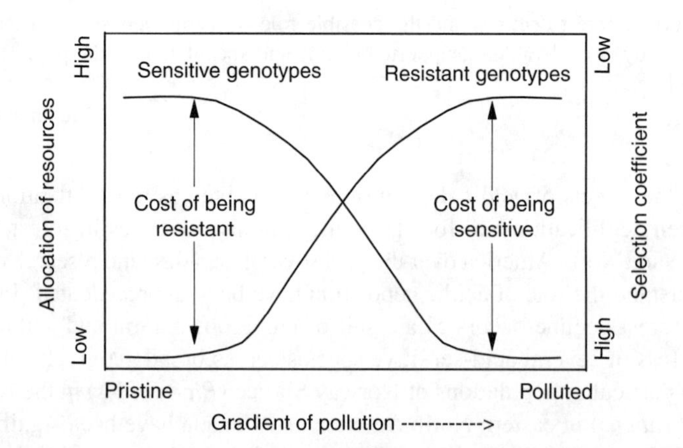

FIGURE 26.15 The hypothesized influence of air pollution on allocation of carbon resources and selection coefficients in trees. The figure characterizes the trade-offs between tolerance to stress and allocation of resources along a pollution gradient. In pristine environments, sensitive genotypes have a selective advantage because they can allocate more resources to growth, maintenance, and defense. In contrast, resistant genotypes have an advantage over sensitive genotypes in polluted environments. The differences in performance of these genotypes in pristine and polluted environments are defined as the cost of resistance or the cost of sensitivity. (Modified from Figure 6 in Taylor, G.E., Jr., et al., *Ecol. Appl.*, 4, 662–689, 1994. Reproduced by permission of the Ecological Society of America.)

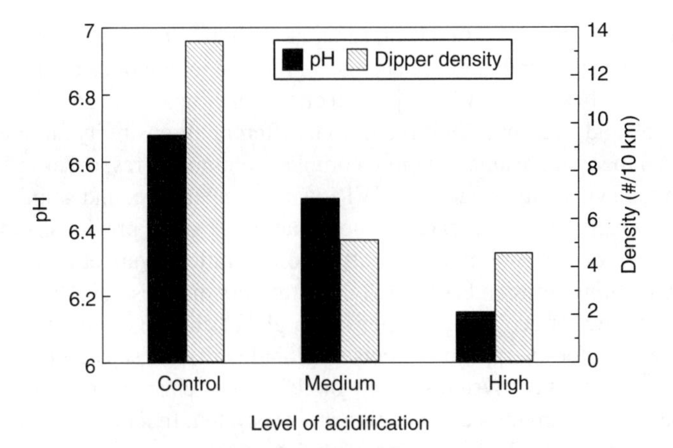

FIGURE 26.16 The relationship between pH and abundance of dippers (*Cinclus cinclus*), an obligate riverine predator, in Scotland. Data collected from 27 sites were placed in three categories based on the frequency that pH exceeded a critical threshold. Because dippers are dependent on aquatic prey, standardized counts of dipper abundance may be used to assess the chemical and biological status of streams. (Data from Table 3 in Logie et al. (1996).)

composition on long-term health of forests and the animal communities they support are a significant cause for concern.

26.4.6 Indirect Effects of Acidification on Terrestrial Wildlife

Because most terrestrial research on acid deposition has focused on forest communities, our understanding of responses of terrestrial wildlife, particularly birds and mammals, is somewhat limited. Although acidification is assumed to have relatively little direct effects on terrestrial wildlife (Schreiber and Newman 1988), indirect effects owing to loss of habitat and prey resources have been reported. Simplification of forest canopy resulting from acidic deposition will reduce habitat complexity and is likely to impact bird communities and other wildlife species. Indirect effects of acid deposition on a variety of species that consume fish and aquatic invertebrates have been observed (Schreiber and Newman 1988). Reduced abundance of feral mink (*Mustela vison*) populations in Norway has been attributed to declines of fish in acidified aquatic ecosystems (Bevanger and Albu 1986). The dominant prey species of dippers (*Cinclus cinclus*) in Scottish streams include mayflies, caddisflies, and stoneflies, organisms known to be sensitive to acidification. Because of the dependence of dippers on macroinvertebrates, reductions in prey density owing to acidification are likely to impact these bird populations. Buckton et al. (1998) reported that dippers were generally absent from acidified streams in Wales owing to reduced density of macroinvertebrates. They also note that despite modest improvements in water quality between 1984 and 1995, dipper populations and macroinvertebrates showed no indication of recovery. In a related study, Logie et al. (1996) developed a regression model based on census data of dipper abundance to assess the degree of acidification in streams (Figure 26.16). Results showed that dippers were generally absent or less abundant in watersheds with pH less than 6.5.

26.5 INTERACTIONS AMONG GLOBAL ATMOSPHERIC STRESSORS

Increased levels of CO_2, ozone depletion, and atmospheric deposition of H^+, SO_2, and NO_x occur simultaneously across large spatiotemporal scales. Thus, it is difficult to study the effects of these global atmospheric stressors in isolation or to separate their effects from other disturbances such as contaminants, exotic species, and land use changes. Potential interactions between atmospheric and local stressors will complicate our ability to predict effects of climate change, UV-B, and acidification on communities. Rae and Vincent (1998) report that differential sensitivity among microbial organisms to increased temperature resulted in more complex community responses to UVR. Przeslawski et al. (2005) reported synergistic effects of UVR, water temperature, and salinity stress on marine rocky intertidal organisms. These researchers note that while some previous studies have reported relatively weak effects of UV-B on marine benthic communities (Wahl et al. 2004), failure to consider potential interactions between UVR and other stressors may result in dramatic underestimates of effects. In their review of ecosystem responses to global change, Field et al. (1992) developed a conceptual framework for interpreting effects of global stressors, separately and in combination, across a variety of terrestrial ecosystems. Their model relates changes in resource availability and utilization to impacts on community composition and ecosystem function. They note the importance of including assessments of land use changes when interpreting responses to global anthropogenic stressors. The most obvious interaction between global climate change and landscape alteration is the rapid rate of deforestation occurring in tropical ecosystems. Because tropical forests are a major carbon sink, deforestation will likely influence global carbon cycles and atmospheric CO_2 levels. Complex interactions between global warming, acidification, and UV-B will also influence microbial production and decomposition. In terrestrial ecosystems, warmer soil temperatures will increase the rate of microbial mineralization and result in greater release of CO_2. However, these increases may be offset in systems subjected to acidification because of the negative effects of low pH on decomposition. Similar interactions among global atmospheric stressors are likely to occur in marine ecosystems. For example, the oceans, also considered a major carbon sink, may be responsible for buffering the planet from increased global temperatures owing to anthropogenic emissions of CO_2. The deleterious effects of UV-B radiation on phytoplankton and NPP in marine ecosystems could therefore influence carbon storage.

Increased levels of atmospheric CO_2 and associated climate change could influence responses of plant communities to air pollutants, but predicting the direction of these responses is difficult. Increased temperature is likely to increase bioavailability and uptake of many pollutants. Taylor et al. (1994) suggest that effects of atmospheric pollutants on plants will be reduced in a CO_2-enriched world because greater availability of carbon will provide more resources for growth, maintenance, and defense. In addition, increased levels of CO_2 will reduce stomatal transfer and subsequent uptake of airborne toxicants. In contrast, elevated levels of CO_2 may result in greater surface area of leaves, thus providing a larger surface for deposition of pollutants (Bazzaz 1990). Similar complex interactions between climate change and contaminant exposure are likely to occur in aquatic ecosystems. Warmer water temperatures increase uptake rates and bioavailability of many chemicals in lakes and streams. In addition, reduced precipitation and changes in hydrologic characteristics could influence chemical concentrations. For example, decreased precipitation and discharge in northwestern Ontario lakes resulted in dramatically lower renewal rates and increases in concentrations of many chemical solutes (Schindler et al. 1990). We expect that similar changes in hydrologic characteristics of other lakes and streams could affect concentrations of contaminants.

Interactions among global atmospheric stressors may indirectly affect responses of aquatic communities to chemical stressors. Because of the well-established influence of DOM on contaminant bioavailability and toxicity in aquatic ecosystems, alterations of DOM could increase or decrease exposure to other stressors. Figure 26.17 shows a conceptual model that describes the predicted influence of acidification, UVR, and climate change on the quality (contaminant binding affinity) and quantity of DOM. The model could apply to any class of contaminants where bioavailability and toxicity are influenced by DOM. Acidification and UV-B radiation both decrease the quality and quantity of DOM in aquatic ecosystems (De Haan 1993, Morris and Hargreaves 1997, Williamson et al. 1999). Because DOM also reduce light penetration, lower levels will result in greater exposure to UV-B and possibly greater photodegradation of DOM. The direct effects of climate change on

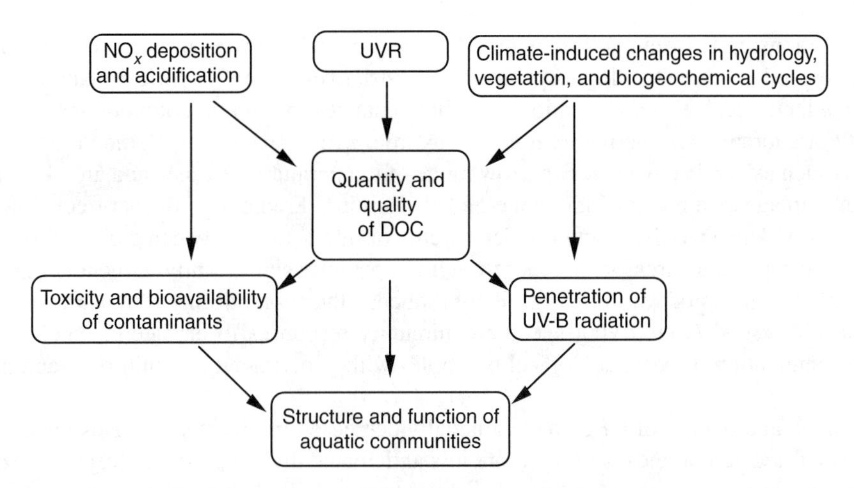

FIGURE 26.17 Conceptual model showing the hypothesized influence of climate change, UVR, and acidification on aquatic ecosystems. Each of these stressors will directly affect quantity and quality of DOC, which influences toxicity and bioavailability of contaminants and determines penetration of UV-B radiation.

DOM will likely be complex and influenced by alterations in riparian vegetation, hydrologic characteristics, and biogeochemical cycles. Studies conducted in subarctic and Canadian Shield lakes showed that declines in DOM resulting from climate change increased exposure to UVR (Pienitz and Vincent 2000, Schindler et al. 1996). Hader et al. (2003) described some of the potential interactions among climate change, UV-B, and DOM in aquatic communities and concluded that complex responses of individual species will likely be more common than ecosystem-level alterations. In addition to the direct effects of reduced DOM and greater UV-B penetration in aquatic ecosystems, contaminant bioavailability will likely increase in systems with lower concentrations of DOM. West et al. (2003) reported that interactions between UVR and DOC increased bioavailability and toxicity of Cu to green algae. Photooxidation of natural DOM during exposure to artificial sunlight significantly decreased Cu complexation and increased cupric ion ($\{Cu^{2+}\}$), the most bioavailable Cu species (Brooks et al. 2007). These examples illustrate the need to consider not only direct effects of global atmospheric stressors on communities but also their interactions with and influences on other anthropogenic disturbances. Developing a mechanistic understanding of these interactions will be a major research topic over the next several decades.

26.6 SUMMARY

The regional and global distribution of atmospheric stressors such as SO_2, NO_x, and CO_2 present unique challenges. Because of long-range transport and geographic variation in exposure, predicting effects of these stressors on communities is difficult. Integrating results of relatively small-scale experiments and regional models to predict effects of globally distributed atmospheric stressors on communities has also proven quite challenging. Research that matches the coarse spatial scale of GCMs with the smaller scale of most ecological investigations is essential for predicting regional responses. Reconstructing relationships between global atmospheric stressors and prehistoric communities using paleoecological investigations has significantly contributed to our understanding of potential effects.

The causes and consequences of global climate change and the specific role of CO_2 are highly contentious environmental issues, primarily because of the potential political and socioeconomic implications. It is important to note that most of the debate within the scientific community is over the details of climate change and the relative uncertainty associated with GCMs used to predict the magnitude of warming. Relatively few contemporary scientists question the existence of a causal relationship between anthropogenic CO_2 and global climate change. Considerable effort has been devoted to understanding responses of terrestrial vegetation to climate change, but aquatic ecosystems have received less attention. Complex changes in aquatic communities are expected as a result of alterations in thermal regime and hydrologic characteristics. Some groups of aquatic organisms such as corals have a very narrow range of temperature tolerance and are likely to suffer significant damage as a result of moderate global warming. Predicting effects of contaminants on communities will also require a better understanding of interactions between global climate change and other anthropogenic stressors. Because water temperature significantly influences bioavailability and toxicity, effects of warming will likely influence the susceptibility of aquatic communities to chemical stressors. Quantifying specific community responses to climate change will require thoughtful integration of basic ecological principles with experimental, monitoring, and modeling studies.

Production and release of CFCs have gradually decreased in recent years. However, CFCs are highly recalcitrant compounds, and therefore it is anticipated that levels of stratospheric ozone will continue to decline over the next several decades. Increased UVR as a result of ozone depletion is a significant environmental hazard, with greatest effects observed on freshwater and marine phytoplankton. Because much of the global ozone depletion has occurred over Antarctica and because Antarctic phytoplankton are a major component of marine food webs, reduced phytoplankton biomass and production may have cascading effects on higher trophic levels. Although terrestrial plants are relatively tolerant to UV-B radiation, exposure may have important consequences for associated belowground communities. A variety of local and regional characteristics influence exposure and effects of UV-B, including latitude, elevation, DOMs, life history, and the presence of other stressors.

Spatially extensive surveys, paleolimnological studies, mesocosm experiments, and whole ecosystem manipulations have quantified effects of acid deposition on aquatic and terrestrial communities. The effects of acid deposition on aquatic communities are strongly influenced by local geology. Unlike many of the other chemical stressors we have considered in this section, recovery of communities from the effects of acidic deposition may require a significant period of time, particularly if base cations and natural buffering capacity are lost from a watershed.

Although researchers have generally considered effects of climate change, UV-B radiation, and acidic deposition in isolation, the potential for interaction among these global atmospheric stressors complicates our ability to predict effects on communities. Research at appropriate spatial and temporal scales is necessary to evaluate potential interactions among global atmospheric stressors and how they influence responses to chemical stressors.

26.6.1 Summary of Foundation Concepts and Paradigms

- Global atmospheric stressors are widely distributed physical and chemical perturbations that impact the structure and function of ecosystems.
- Effects of global atmospheric stressors on communities are likely to be complex, interactive, and difficult to predict.
- Long-range transport of some atmospheric pollutants (e.g., SO_2, NO_x) and geographic variation in exposure to other stressors (e.g., UV-B radiation) complicate assessment of effects.
- Although our understanding of the specific ecological consequences of global climate is relatively limited, the connection between the atmosphere, biological processes and the occurrence of a natural greenhouse effect are indisputable facts.
- Evidence from a variety of sources indicates that global temperatures have increased by approximately 0.5–1.0°C over the past century and that this increase is closely associated with an unprecedented increase in anthropogenic emissions of atmospheric CO_2 and other greenhouse gases.

- Direct measurements obtained from the Mauna Loa Observatory indicate that CO_2 concentrations are now approximately 100 μL/L higher than historic levels and these levels have steadily increased over the past 50 years at 10–100 times faster than at any period prior to the industrial revolution.
- Remediation strategies designed to increase sequestration of atmospheric carbon, especially at the large spatial and temporal scale necessary to influence global cycles, will likely have unpredictable effects on other biological and biogeochemical processes.
- Much of the difficulty predicting the ecological consequences of global climate change on communities results from our inability to link large-scale climate models to smaller scale ecological studies.
- Because of the difficulty in conducting manipulations at spatial scales compatible with general circulation models (GCMs), integrating model predictions with paleoecological records can improve our understanding of how climate influences communities.
- Several characteristics influence responses of plant communities to global warming, including previous exposure to climatic extremes, species richness, functional composition, and successional stage.
- Because the amount of carbon stored in soil organic matter is 2–3 times greater than in terrestrial vegetation, changes in soil processes can significantly influence global carbon cycles and sequestration.
- Complex changes in lakes and streams in response to global climate are expected as a result of alterations in thermal regime and hydrologic characteristics; however, relatively few studies have considered how freshwater communities will respond to climate change (Carpenter et al. 1992).
- Elevated ocean temperatures are expected to have diverse impacts on marine communities, ranging from increased incidence of disease to reduced ecosystem productivity.
- Because of their narrow range of temperature tolerance, coral reefs and associated communities are likely to suffer significant damage as a result of moderate global warming.
- The most critical research need for assessing ecological consequences of global change is to identify specific community responses to expected spatial and temporal variation in climate.
- There is now conclusive evidence that reduced levels of ozone are a direct result of anthropogenic activities, particularly the release of CFCs.
- Although production and release of CFCs have decreased as a result of international agreements, CFCs are very persistent compounds and therefore stratospheric ozone will continue to decline over the next several decades.
- Increased UVR as a result of ozone depletion is a significant environmental hazard and is expected to have negative effects on humans and other organisms.
- Studies conducted with a variety of plants and animals have shown effects of UV-B at all levels of biological organization.
- Because phytoplankton are a major component of marine food webs in Antarctica and other areas where global ozone depletion has been most severe, reduced phytoplankton biomass may have cascading effects on higher trophic levels.
- While most research investigating responses of aquatic communities to UVR has focused on marine and freshwater plankton, investigators have documented significant alterations in benthic communities exposed to UVR.
- Responses of plants to UV-B include reduced growth and biomass, reduced photosynthesis, and alterations in belowground communities.
- Interspecific variation in behavioral, physiological, and morphological adaptations to natural UV-B exposure may account for differences in UV-B tolerance observed in experiments.

- In addition to the direct effects of UVR on aquatic communities, synergistic interactions between UVR and other anthropogenic stressors may enhance effects and complicate our ability to predict changes observed in the field.
- Acid deposition has significant negative effects on aquatic and terrestrial communities and remains a global environmental problem.
- Biomonitoring studies conducted at relatively large spatial scales (e.g., multiple watersheds) have reported consistent effects of acidification on aquatic communities.
- Assessments of acidification in lotic ecosystems have focused primarily on chronic effects resulting from relatively long-term changes in pH. However, acute exposure to the combined effects of low pH and metals from episodic acidification occurs in some streams during periods of high discharge.
- Reduced emissions of SO_2 and particulates as a result of federal regulations is expected to have a positive influence on physical, chemical, and biological characteristics in aquatic ecosystems. However, our understanding of recovery of lakes and streams from the deleterious effects of acidification is relatively poor.
- Increased levels of CO_2, ozone depletion, and atmospheric deposition of H^+, SO_2, and NO_x occur simultaneously across large spatiotemporal scales, and therefore it is difficult to study the effects of these global atmospheric stressors in isolation or to separate their effects from other disturbances such as contaminants, exotic species, and land use changes.

REFERENCES

Allard, M. and Moreau, G., Effects of experimental acidification on a lotic macroinvertebrate community, *Hydrobiologia*, 144, 37–49, 1987.

Allen, A.S., Andrews, J.A., Finzi, A.C., Matamala, R., Richter, D.D., and Schlesinger, W.H., Effects of free-air CO_2 enrichment (FACE) on belowground processes in a *Pinus taeda* forest, *Ecol. Appl.*, 10, 437–448, 2000.

Ankley, G.T., Collyard, S.A., Monson, P.D., and Kosian, P.A., Influence of ultraviolet light on the toxicity of sediments contaminated with polycyclic aromatic hydrocarbons, *Environ. Toxicol. Chem.*, 13, 1791–1796, 1994.

Araujo, M.B. and Rahbek, C., How does climate change affect biodiversity? *Science*, 313, 1396–1397, 2006.

Baker, L.A., Herlihy, A.T., Kaufmann, P.R., and Eilers, J.M., Acidic lakes and streams in the United States: The role of acidic deposition, *Science*, 252, 1151–1154, 1991.

Baker, J.P., VanSickle, J., Gagen, C.J., DeWalle, D.R., Sharpe, W.E., Carline, R.F., Baldigo, B.P., et al., Episodic acidification of small streams in the northeastern United States: Effects on fish populations, *Ecol. Appl.*, 6, 422–437, 1996.

Barnett, T.P., Pierce, D.W., and Schnur, E., Detection of anthropogenic climate change in the world's oceans, *Science*, 292, 270–274, 2001.

Barry, J.P., Baxter, C.H., Sagarin, R.D., and Gilman, S.E., Climate-related long-term faunal changes in a California rocky intertidal community, *Science*, 267, 672–675, 1995.

Bazzaz, F.A., The response of natural ecosystems to the rising global CO_2 levels, *Ann. Rev. Ecol. Syst.*, 21, 167–196, 1990.

Bevanger, K. and Albu, O., Decrease in a Norwegian feral mink (*Mustels vison*) population—a response to acid precipitation? *Biol. Cons.*, 38, 75–78, 1986.

Blaustein, A.R., Edmond, B., Kiesecker, J.M., Beatty, J.J., and Hokit, D.J., Ambient ultraviolet radiation causes mortality in salamander eggs, *Ecol. Appl.*, 5, 740–743, 1995.

Blaustein, A.R. and Wake, D.B., Declining amphibian populations: A global phenomenon? *Trends Ecol. Evolut.*, 5, 203–204, 1990.

Boggs, C.L. and Murphy, D.D., Community composition in mountain ecosystems: Climatic determinants of montane butterfly distributions, *Glob. Ecol. Biogeor. Let.*, 6, 39–48, 1997.

Bothwell, M.L., Sherbot, D.M.J., and Pollock, C.M., Ecosystem response to solar ultraviolet-B radiation: Influence of trophic level interactions, *Science*, 265, 97–100, 1994.

Bousquet, P., Peylin, P., Ciais, P., Le Quere, C., Friedlingstein, P., and Tans, P.P., Regional changes in carbon dioxide fluxes of land and oceans since 1980, *Science*, 290, 1342–1346, 2000.

Brett, M.T., Zooplankton communities and acidification processes (a review), *Water Air Soil Poll.*, 44, 387–414, 1989.

Brooks, M.L., McKnight, D.M., and Clements, W.H., Photochemical control of copper complexation by dissolved organic matter in Rocky Mountain Streams, Colorado, *Limnol. Oceanogr.*, 52, 766–779, 2007.

Buckton, S.T., Brewin, P.A., Lewis, A., Stevens, P., and Ormerod, S.J., The distribution of dippers, *Cinclus cinclus* (L.), in the acid-sensitive region of Wales, 1984–95, *Freshw. Biol.*, 39, 387–396, 1998.

Burton, T.M., Stanford, R.M., and Allan, J.W., Acidification effects on stream biota and organic matter processing, *Can. J. Fish. Aquat. Sci.*, 42, 669–675, 1985.

Caldwell, M.M., Robberecht, R., and Billings, W.D., A steep latitudinal gradient of solar untraviolet-B radiation in the arctic-alpine life zone, *Ecology*, 61, 600–611, 1980.

Caldwell, M.M., Teramura, A.H., Tevini, M., Bornman, J.F., Bjorn, L.O., and Kulandaivelu, G., Effects of increased solar ultraviolet radiation on terrestrial plants, *Ambio*, 14, 166–173, 1995.

Carpenter, S.R., Fisher, S.G., Grimm, N.B., and Kitchell, J.F., Global change and freshwater ecosystems, *Ann. Rev. Ecol. Syst.*, 23, 119–139, 1992.

Clark, M.E., Rose, K.A., Levine, D.A., and Hargrove, W.W., Predicting climate change effects on Appalachian trout: Combining GIS and individual-based modeling, *Ecol. Appl.*, 11, 161–178, 2001.

Clayton, J.L. and Menendez, R., Macroinvertebrate responses to mitigative liming of Dogway Fork, West Virginia, *Res. Ecol.*, 4, 234–246, 1996.

Clements, W.H., Carlisle, D.M., Lazorchak, J.M., and Johnson, P.C., Heavy metals structure benthic communities in Colorado mountain streams, *Ecol. Appl.*, 10, 626–638, 2000.

Cloud, P.E., Atmospheric and hydrospheric evolution of primitive earth, *Science*, 160, 729–736, 1968.

Courtney, L.A. and Clements, W.H., Effects of acidic pH on benthic macroinvertebrate communities in stream microcosms, *Hydrobiologia*, 379, 135–145, 1998.

Crowley, T.J. and Berner, R.A., CO_2 and climate change, *Science*, 292, 870–872, 2001.

Davis, M.B. and Shaw, R.G., Range shifts and adaptive responses to quaternary climate change, *Science*, 292, 673–679, 2001.

De Haan, H., Solar UV-light penetration and photodegradation of humic substances in peaty lake water. *Limnol. Oceangr.,* 38, 1072–1076, 1993.

de Valpine, P. and Harte, J., Plants responses to experimental warming in a montane meadow, *Ecology*, 82, 637–648, 2001.

DeLucia, E.H., Hamilton, J.G., Naidu, S.L., Thomas, R.B., Andrews, J.A., Finzi, A., Levine, M., et al., Net primary production of a forest ecosystem with experimental CO_2 enrichment, *Science*, 284, 1177–1179, 1999.

DeStato, J., III and Rahel, F.J., Influence of water temperature on interactions between juvenile Colorado River cutthroat trout and brook trout in a laboratory streams, *Trans. Am. Fish. Soc.*, 123, 289–297, 1994.

Dixon, R.K., Brown, S., Houghton, R.A., Solomon, A.M., Trexler, M.C., and Wisniewski, J., Carbon pools and flux of global forest ecosystems, *Science*, 263, 185–190, 1994.

Donahue, W.F. and Schindler, D.W., Diel emigration and colonization responses of blackfly larvae (Diptera: Simuliidae) to ultraviolet radiation, *Freshw. Biol.*, 40, 357–365, 1998.

Duarte, C.M. and Agusti, S., The CO_2 balance of unproductive aquatic ecosystems, *Science*, 281, 234–236, 1998.

Fabacher, D.L. and Little, E.E., Skin component may protect fishes from ultraviolet-B radiation, *Environ. Sci. Poll. Res.*, 2, 30–32, 1995.

Falkowski, P., Scholes, R.J., Boyle, E., Canadell, J., Canfield, D., Elser, J., Gruber, N., et al., The global carbon cycle: A test of our knowledge of earth as a system, *Science*, 290, 291–296, 2000.

Feldman, R.S. and Connor, E.F., The relationship between pH and community structure of invertebrates in streams of the Shenandoah National Park, Virginia, U.S.A., *Freshw. Biol.*, 27, 261–276, 1992.

Field, C.B., Behrenfeld, M.J., Randerson, J.T., and Falkowski, P., Primary production of the biosphere: Intergrating terrestrial and oceanic components, *Science*, 281, 237–240, 1998.

Field, C.B., Chapin, F.S., III, Matson, P.A., and Mooney, H.A., Responses of terrestrial ecosystems to the changing atmosphere: A resource-based approach, *Ann. Rev. Ecol. Syst.*, 23, 201–235, 1992.

Finney, B.P., Gregory-Eaves, I., Sweetman, J., Douglas, M.S.V., and Smol, J.P., Impacts of climatic change and fishing on Pacific salmon abundance over the past 300 years, *Science*, 290, 795–799, 2000.

Fisher, A.G., Fossils, early life, and atmospheric history, *Proc. Natl. Acad. Sci. USA*, 53, 1205–1215, 1965.

Ford, M.S., A 10,000-year history of natural ecosystem acidification, *Ecol. Monogr.*, 60, 57–89, 1990.

Galloway, J.N., Acid deposition: Perspectives in time and space, *Water Air Soil Pollut.*, 85, 15–24, 1995.

Garcia-Pichel, F. and Bebout, B.M., Penetration of ultraviolet radiation into shallow water sediments: High exposure for photosynthetic communities, *Mar. Eco. Prog. Ser.*, 131, 257–262, 1996.

Gehrke, C., Impacts of enhanced ultraviolet-B radiation on mosses in a subarctic heath ecosystem, *Ecology*, 80, 1844–1851, 1999.

Genter, R.B., Benthic algal populations respond to aluminum, acid, and aluminum-acid mixtures in artificial streams, *Hydrobiologia*, 306, 7–19, 1995.

Gleason, D.F. and Wellington, G.M., Variation in UVB sensitivity of planula larvae of the coral *Agaricia agaricites* along a depth gradient, *Mar. Biol.*, 123, 693–703, 1995.

Grime, J.P., Brown, V.K., Thompson, K., Masters, G.J., Hillier, S.H., Clarke, I.P., Askew, A.P., Corker, D., and Kielty, J.P., The response of two contrasting limestone grasslands to simulated climate change, *Science*, 289, 762–765, 2000.

Hader, D.P., Kumar, H.D., Smith, R.C., and Worrest, R.C., Aquatic ecosystems: Effects of solar ultraviolet radiation and interactions with other climatic change factors, *Photochem. Photobiol. Sci.*, 2, 39–50, 2003.

Hall, R.J. and Ide, F.P., Evidence of acidification effects on stream insect communities in central Ontario between 1937 and 1985, *Can. J. Fish. Aquat. Sci.*, 44, 1652–1657, 1987.

Hall, A.T. and Oris, J.T., Anthracene reduces reproductive potential and is maternally transferred during long-term exposure in fathead minnows, *Aquat. Toxicol.*, 19, 249–264, 1991.

Hall, R.I. and Smol, J.P., Paleolimnological assessment of long-term water-quality changes in south-central Ontario lakes affected by cottage development and acidification, *Can. J. Fish. Aquat. Sci.*, 53, 1–17, 1996.

Hall, R.J., Driscoll, C.T., Likens, G.E., and Pratt, J.M., Physical, chemical, and biological consequences of episodic aluminum additions to a stream, *Limnol. Oceanogr.*, 30, 212–220, 1985.

Hall, R.J., Likens, G.E., Fiance, S.B., and Hendrey, G.R., Experimental Acidification of a stream in the Hubbard Brook Experimental Forest, New Hampshire, *Ecology*, 61, 976–989, 1980.

Harrison, S., Species diversity, spatial scale, and global change, In *Biotic Interactions and Global Change*, Kareiva, P.M., Kingsolver, J.G., and Huey, R.B. (eds.), Sinauer Associates, Inc., Sunderland, MA, 1993, pp. 388–401.

Harte, J. and Shaw, R., Shifting dominance within a montane vegetation community: Results of a climate-warming experiment, *Science*, 267, 876–880, 1995.

Harvell, C.D., Kim, K., Burkholder, J.M., Colwell, R., Epstein, P.R., Grimes, D.J., Hofmann, E.E., et al., Emerging marine diseases—climate links and anthropogenic factors, *Science*, 285, 1505–1510, 1999.

Hauer, F.R., Baron, J.S., Campbell, D.H., Fausch, K.D., Hostetler, S.W., Leavesley, G.H., Leavitt, P.R., McKnight, D.M., and Stanford, J.A., Assessment of climate change and freshwater ecosystems of the Rocky Mountians, USA and Canada, *Hydrol. Proc.*, 11, 903–924, 1997.

Havens, K.E. and DeCosta, J., The role of aluminum contamination in determining phytoplankton and zooplankton responses to acidification, *Water Air Soil Pollut.*, 33, 277–293, 1987.

Heard, R.M., Sharpe, W.E., Carline, R.F., and Kimmel, W.G., Episodic acidification and changes in fish diversity in Pennsylvania headwater streams, *Trans. Amer. Fish. Soc.*, 126, 977–984, 1997.

Heneghan, L. and Bolger, T., Effects of components of "acid rain" on the contribution of soil microarthropods to ecosystem function, *J. Appl. Ecol.*, 33, 1329–1344, 1996.

Herrmann, J., Degerman, E., Gerhardt, A., Johansson, C., Lingdell, P., and Muniz, I.P., Acid-stress effects on stream biology, *Ambio*, 22, 298–307, 1993.

Hill, W.R., Dimick, S.M., McNamara, A.E., and Branson, C.A., No effects of ambient UV radiation detected in periphyton and grazers, *Limnol. Oceanogr.*, 42, 769–774, 1997.

Hogg, I.D. and Williams, D.D., Response of stream invertebrates to a global-warming thermal regime: An ecosystem-level manipulation, *Ecology*, 77, 395–407, 1996.

Holbrook, S.J., Schmitt, R.J., and Stephens, J.S., Jr., Changes in an assemblage of temperate reef fishes associated with a climate shift, *Ecol. Appl.*, 7, 1299–1310, 1997.

Hurlbert, S.H., Pseudoreplication and the design of ecological field experiments, *Ecol. Monogr.*, 54, 187–211, 1984.

Jones, T.H., Thompson, L.J., Lawton, J.H., Bezemer, T.M., Bardgett, R.D., Blackburn, T.M., Bruce, K.D., et al., Impacts of rising atmospheric carbon dioxide on model terrestrial ecosystems, *Science*, 280, 441–442, 1998.

Juggins, S., Flower, R.J., and Battarbee, R.W., Paleolimnological evidence for recent chemical and biological changes in UK acid waters monitoring network sites, *Freshw. Biol.*, 36, 203–219, 1996.

Kareiva, P.M., Kingsolver, J.G., and Huey, R.B., *Biotic Interactions and Global Change*, Sinauer Associates, Inc., Sunderland, MA, 1993.

Keeling, C.D. and Whorf, T.P., *Atmospheric CO_2 Concentrations Mauna Loa Observatory, Hawaii, 1958–1997*, Carbon Dioxide Information Analysis Center, Oak Ridge National Laboratory, Oak Ridge, Tennessee, 1998.

Kelly, D.J., Clare, J.J., and Bothwell, M.L., Attenuation of solar ultraviolet radiation by dissolved organic matter alters benthic colonization patterns in streams, *J. N. Amer. Benthol. Soc.*, 20, 96–108, 2001.

Kerr, J.B. and McElroy, C.T., Evidence for large upward trends of ultraviolet-B radiation linked to ozone depletion, *Science*, 1032–1034, 1993.

Kiffney, P.M., Little, E.E., and Clements, W.H., Influence of ultraviolet-B radiation on the drift response of stream invertebrates, *Freshw. Biol.*, 37, 485–492, 1997a.

Kiffney, P.M., Clements, W.H., and Cady, T.A., Influence of ultraviolet radiation on the colonization dynamics of a Rocky Mountain stream benthic community, *J. N. Amer. Benthol. Soc.*, 16, 520–530, 1997b.

Kingsolver, J.G., Huey, R.B., and Karieva, P.M., An agenda for population and community research on global change, In *Biotic Interactions and Global Change*, Kareiva, P.M., Kingsolver, J.G., and Huey, R.B., (eds.), Sinauer Associates, Inc., Sunderland, MA, 1993, pp. 480–486.

Kinzie, R.A., Banaszak, A.T., and Lesser, M.P., Effects of ultraviolet radiation on primary productivity in a high altitude tropical lake, *Hydrobiologia*, 385, 23–32, 1998.

Klironomos, J.N. and Allen, M.F., UV-B-mediated changes on below-ground communities associated with the roots of *Acer saccharum*, *Funct. Ecol.*, 9, 923–930, 1995.

Lancaster, J., Real, M., Juggins, S., Monteith, D.T., Flower, R.J., and Beaumont, W.R.C., Monitoring temporal changes in the biology of acid waters, *Freshw. Biol.*, 36, 179–201, 1996.

Larson, D.L., Potential effects of anthropogenic greenhouse gases on avian habitats and populations in the northern Great Plains, *Amer. Midl. Nat.*, 131, 330–346, 1994.

Lassiter, R.R., Box, E.I., Wiegert, R.G., Johnston, J.M., Bergengren, J., and Suarez, L.A., Vulnerability of ecosystems of the mid-Atlantic regions, USA, to climate change, *Environ. Toxicol. Chem.*, 19, 1153–1160, 2000.

Laurion, I. and Vincent, W.F., Cell size versus taxonomic composition as determinants of UV-sensitivity in natural phytoplankton communities, *Limnol. Oceanogr.*, 43, 1774–1779, 1998.

Leech, D.M. and Williamson, C.E., Is tolerance to UV radiation in zooplankton related to body size, taxon, or lake transparency? *Ecolog. Appl.*, 10, 1530–1540, 2000.

Likens, G.E., Driscoll, C.T., and Buso, D.C., Long-term effects of acid rain: Response and recovery of a forest ecosystem, *Science*, 272, 244–246, 1996.

Locke, A., Factors influencing community structure along stress gradients: Zooplankton responses to acidification, *Ecology*, 73, 903–906, 1992.

Logie, J.W., Bryant, D.M., Howell, D.L., and Vickery, J.A., Biological significance of UK critical load exceedance estimates for flowing waters: Assessments of dipper *Cinclus cinclus* populations in Scotland, *J. Appl. Ecol.*, 33, 1065–1076, 1996.

Long, L.E., Saylor, L.S., and Soule, M.E., A pH/UV-B synergism in amphibians, *Conserv. Biol.*, 9, 1301–1303, 1995.

Lubin, D. and Jensen, E.H., Effects of clouds and stratospheric ozone depletion on ultraviolet radiation trends, *Nature*, 377, 710–713, 1995.

Madronich, S., Implications of recent total atmospheric total ozone measurements for biologically active untraviolet radiation reaching the earth's surface, *Geophys. Res. Let.*, 19, 37–40, 1992.

Mallory, M.L., McNicol, D.K., Cluis, D.A., and Laberge, C., Chemical trends and status of small lakes near Sudbury, Ontario, 1983–1995: Evidence of continued chemical recovery, *Can. J. Fish. Aquat. Sci.*, 55, 63–75, 1998.

McDonald, K.A. and Brown, J.H., Using montane mammals to model extinctions due to global change, *Conserv. Biol.*, 6, 409–415, 1992.

McKenzie, R., Connor, B., and Bodeker, G., Increased summertime UV radiation in New Zealand in response to ozone loss, *Science*, 285, 1709–1711, 1999.

McKnight, D.M., Harnish, R., Wershaw, R.L., Baron, J.S., and Schiff, S., Chemical characteristics of particulate, colloidal, and dissolved organic material in Loch Vale Watershed, Rocky Mountain National Park, *Biogeochemistry*, 36, 99–124, 1997.

McLaren, B.E., and R.O., Peterson, Wolves, Moose, and tree rings on Isle Royale, *Science*, 266, 1555–1558, 1994.

Menendez, R., Clayton, J.L., and Zurbuch, P.E., Chemical and fishery responses to mitigative liming of an acidic stream, Dogway Fork, West Virginia, *Res. Ecol.*, 4, 220–233, 1996.

Morris, D.P. and Hargreaves, B.R., The role of photochemical degradation of dissolved organic carbon in regulating the UV transparency of three lakes on the Pocono Plateau, *Limnol. Oceanogr.*, 42, 239–249, 1997.

Mostajirm, B., Demers, S., deMora, S., Belzile, C., Chanut, J.P., Gosselin, M., Roy, S., et al., Experimental test of the effect of ultraviolet-B radiation in a planktonic community, *Limnol. Oceanogr.*, 44, 586–596, 1999.

Niederlehner, B.R., Pontasch, K.W., Pratt, J.R., and Cairns, J., Jr., Field evaluation of predictions of environmental effects from a multispecies-microcosm toxicity test, *Arch. Environ. Contam. Toxicol.*, 19, 62–71, 1990.

Ollinger, S.V., Aber, J.D., Lovett, G.M., Millham, S.E., Lathrop, R.G., and Ellis, J.E., A spatial model of atmospheric deposition for the northeastern U.S., *Ecol. Appl.*, 3, 459–472, 1993.

Oris, J.T. and Giesy, J.P., Jr., Photoinduced toxicity of anthracene to juvenile bluegill sunfish (*Lepomis macrochirus* rafinesque): Photoperiod effects and predictive hazard evaluation, *Environ. Toxicol. Chem.*, 5, 761–768, 1986.

Ormerod, S.J., Boole, P., McCahon, C.P., Weatherley, N.S., Pascoe, D., and Edwards, R.W., Short-term experimental acidification of a Welsh stream: Comparing the biological effects of hydrogen ions and aluminum, *Freshw. Biol.*, 17, 341–356, 1987.

Overpeck, J., Hughen, K., Hardy, D., Bradley, R., Case, R., Douglas, M., Finney, B., et al., Arctic environmental change of the last four centuries, *Science*, 278, 1251–1256, 1997.

Owensby, C.E., Coyne, P.I., Ham, J.M., Auen, L.M., and Knapp, A.K., Biomass production in a tallgrass prairie ecosystem exposed to ambient and elevated CO_2, *Ecol. Appl.*, 3, 644–653, 1993.

Pacala, S.W. and Hurtt, G.C., Terrestrial vegetation and climate change: Integrating models and experiments, In *Biotic Interactions and Global Change*, Karieva, P.M., Kingsolver, J.G., and Huey, R.B. (eds.), Sinauer Associates, Inc., Sunderland, MA, 1993, pp. 57–74.

Perry, A.L., Low, P.J., Ellis, J.R., and Reynolds, J.D., Climate change and distribution shifts in marine fishes, *Science*, 308, 1912–1915, 2005.

Petchey, O.L., McPhearson, P.T., Casey, T.M., and Morin, P.J., Environmental warming alters food-web structure and ecosystem function, *Nature*, 402, 69–72, 1999.

Pienitz, R. and Vincent, W.F., Effect of climate change relative to ozone depletion on UV exposure in subarctic lakes, *Nature*, 404, 484–487, 2000.

Pitelka, L.F. and Raynal, D.J., Forest decline and acidic deposition, *Ecology*, 70, 2–10, 1989.

Prentice, I.C., Sykes, M.T., and Cramer, W., The possible dynamic response of northern forests to global warming, *Glob. Ecol. Biogeogr. Let.*, 1, 129–135, 1991.

Price, M.V. and Waser, N.M., Responses of subalpine meadow vegetation to four years of experimental warming, *Ecol. Appl.*, 10, 811–823, 2000.

Przeslawski, R., Davis, A.R., and Benkendorff, K., Synergistic effects associated with climate change and the development of rocky shore molluscs, *Global Change Biology*, 11, 515–522, 2005.

Rader, R.B. and Belish, T.A., Short-term effects of ambient and enhanced UV-B on moss (*Fontinalis neomexicana*) in a mountain stream, *J. Freshw. Ecol.*, 12, 395–403, 1997a.

Rader, R.B. and Belish, T.A., Effects of ambient and enhanced UV-B radiation on periphyton in a mountain stream, *J. Freshw. Ecol.*, 12, 615–628, 1997b.

Rae, R. and Vincent, W.F., Effects of temperature and ultraviolet radiation on microbial foodweb structure: Potential responses to global change, *Freshw. Biol.*, 40, 747–758, 1998.

Robinson, C.H., Wookey, P.A., Lee, J.A., Callaghan, T.V., and Press, M.C., Plant community responses to simulated environmental change at a high arctic polar semi-desert, *Ecology*, 79, 856–866, 1998.

Roemmich, D. and McGowan, J., Climatic warming and the decline of zooplankton in the California current, *Science*, 267, 1324–1326, 1995.

Root, T.L., Environmental factors associated with avian distributional boundaries, *J. Biogeogr.*, 15, 489–505, 1988.

Root, T.L., Effects of global climate change on North American birds and their communities, In *Biotic Interactions and Global Change*, Kareiva, P.M., Kingsolver, J.G., and Huey, R.B. (eds.), Sinauer Associates, Inc., Sunderland, MA, 1993, pp. 280–292.

Root, T.L. and Schneider, S.H., Can large-scale climatic models be linked with multiscale ecological studies? *Conserv. Biol.*, 7, 256–270, 1993.

Rosemond, A.D., Reice, S.R., Elwood, J.W., and Mulholland, P.J., The effects of stream acidity on benthic invertebrate communities in the south-eastern United States, *Freshw. Biol.*, 27, 193–209, 1992.

Rusek, J., Air-pollution-mediated changes in alpine ecosystems and ecotones, *Ecol. Appl.*, 3, 409–416, 1993.

Sanford, E., Regulation of keystone predation by small changes in ocean temperature, *Science*, 283, 1999.

Santas, R., Lianou, C., and Danielidis, D., UVB radiation and depth interaction during primary succession of marine diatom assemblages of Greece, *Limnol. Oceanogr.*, 42, 986–991, 1997.

Schindler, D.W., Detecting ecosystem responses to anthropogenic stress, *Can J. Fish. Aquat. Sci.*, (Suppl.), 6–25, 1987.

Schindler, D.W., Effects of acid rain on freshwater ecosystems, *Science*, 239, 149–157, 1988.

Schindler, D.W., Bayley, S.E., Parker, B.R., Beaty, K.G., Cruiksjank, D.R., Fee, E.J., Schindler, E.U., and Stainton, M.P., The effects of climate warming on the properties of boreal lakes and streams in the Experimental Lakes Area, Northwestern Ontario, *Limnol. Oceanogr.*, 41, 1004–1017, 1996.

Schindler, D.W., Beaty, K.G., Fee, E.J., Cruikshank, D.R., DeBruyn, E.R., Findlay, D.L., Linsey, G.A., et al., Effects of climatic warming on lakes of the Central Boreal Forest, *Science*, 250, 967–970, 1990.

Schindler, D.W., Mills, K.H., Malley, D.F., Findlay, D.L., Shearer, J.A., Davies, I.J., Turner, M.A., Linsey, G.A., and Cruikshank, D.R., Long-term ecosystem stress: The effects of years of experimental acidification on a small lake, *Science*, 228, 1395–1401, 1985.

Schreiber, R.K. and Newman, J.R., Acid precipitation effects on forest habitats: Implications for wildlife, *Cons. Biol.*, 2, 249–259, 1988.

Shindell, D.T., Rind, D., and Lonergan, P., Increased polar stratospheric losses and delayed eventual recovery owing to increasing greenhouse gas concentrations, *Nature*, 392, 589–592, 1998.

Sillett, T.S., Holmes, R.T., and Sherry, T.W., Impacts of a global climate cycle on population dynamics of a migratory songbird, *Science*, 288, 2040–2042, 2000.

Simmons, K.R., Cieslewicz, P.G., and Zajicek, K., Limestone treatment of Whetstone Brook, Massachusetts. 2. Changes in the brown trout (*Salmo trutta*) and brook trout (*Salvelinus fontinalis*) fishery, *Res. Ecol.*, 4, 273–283, 1996.

Smith, R.C., Prezelin, B.B., Baker, K.S., Bidigare, R.R., Boucher, N.P., Coley, T., Karentz, D., MacIntyre, S., Matlick, H.A., Menzies, D., Ondrusek, M., Wan, Z., and Waters, K.J., Ozone depletion: Ultraviolet radiation and phytoplankton biology in Antarctic waters, *Science*, 255, 1992.

Smith, S.V. and Buddemeier, R.W., Global change and coral reef exosystems, *Annu. Rev. Ecol. Syst.*, 23, 89–118, 1992.

Somers, K.M. and Harvey, H.H., Alteration of fish communities in lakes stressed by acid deposition and heavy metals near Wawa, Ontario, *Can. J. Fish. Aquat. Sci.*, 41, 20–29, 1984.

Sommaruga, R. and Garcia-Pichel, F., UV-absorbing mycosporine-like compounds in planktonic and benthic organisms from a high-mountain lake, *Arch. für Hydrobiol.*, 144, 255–269, 1999.

Starfield, A.M. and Chapin, III, F.S., Model of transient changes in arctic and boreal vegetation in response to climate and land use change, *Ecol. Appl.*, 6, 842–864, 1996.

Stokstad, E., Myriad ways to reconstruct past climates, *Science*, 292, 658–359, 2001.

Stott, P.A., Tett, S.F.B., Jones, G.S., Allen, M.R., Mitchell, J.F.B., and Jenkins, G.J., External control of 20th century temperature by natural and anthropogenic forcings, *Science*, 290, 2133–2137, 2000.

Suttle, K.B., Thomsen, M.A., and Power, M.E., Species interactions reverse grassland responses to changing climate, *Science*, 315, 640–642, 2007.

Taylor, G.E., Jr., Johnson, D.W., and Anderson, C.P., Air pollution and forest ecosystems: A regional to global perspective, *Ecol. Appl.*, 4, 662–689, 1994.

Thomas, D.W., Blondel, J., Perret, P., Lambrechts, M.M., and Speakman, J.R., Energetic and fitness costs of mismatching resource supply and demand in seasonally breeding birds, *Science*, 291, 2598–2600, 2001.

van de Staaij, J.W.M., Ernst, W.H.O., Hakvoort, H.W.J., and Rozema, J., Ultraviolet-B (280–320 nm) absorbing pigments in the leaves of *Silene vulgaris*: Their role in UV-B tolerance, *J. Plant Physiol.*, 147, 75–80, 1995a.

van de Staaij, J.W.M., Huijsmans, R., Ernst, W.H.O., and Rozema, J., The effect of elevated UV-B (280–320 nm) radiation levels on *Silene vulgaris*: A comparison between a highland and a lowland population, *Environ. Pollut.*, 90, 357–362, 1995b.

VanSickle, J., Baker, J.P., Simonin, H.A., Baldigo, B.P., Kretser, W.A., and Sharpe, W.E., Episodic acidification of small streams in the northeastern United States: Fish mortality in field bioassays, *Ecol. Appl.*, 6, 408–421, 1996.

Vincent, W.F. and Roy, S., Solar ultraviolet-B radiation and aquatic primary productivity: Damage, protection, and recovery, *Environ. Rev.*, 1, 1–12, 1993.

Vinebrooke, R.D. and Leavitt, P.R., Effects of ultraviolet radiation on periphyton in an alpine lake, *Limnol. Oceanogr.*, 41, 1035–1040, 1996.

Vitousek, P.M., Beyond global warming: Ecology and global change, *Ecology*, 75, 1861–1876, 1994.

Wahl, M., Molis, M., Davis, A., Dobretsov, S., Durr, S.T., Johansson, J., Kinley, J., et al., UV effects that come and go: A global comparison of marine benthic community level impacts, *Global Change Biology*, 10, 1962–1972, 2004.

Walther, G.R., Tackling ecological complexity in climate impact research, *Science*, 315, 606–607, 2007.

West, L.J.A., Li, K., Greenberg, B.M., Mierle, G., and Smith, R.E.H., Combined effects of copper and ultraviolet radiation on a microscopic green alga in natural soft lake waters of varying dissolved organic carbon content, *Aquat. Toxicol.*, 64, 39–52, 2003.

Williamson, C.E., What role does UV-B radiation play in freshwater ecosystems? *Limnol. Oceanogr.*, 40, 386–392, 1995.

Williamson, C.E., Hagreaves, B.R., Orr, P.S., and Lovera, P.A., Does UV play a role in changes in predation and zooplankton community structure in acidified lakes? *Limnol. Oceanogr.*, 44, 774–783, 1999.

Williamson, G.E., Zagarese, H.E., Schulze, P.C., Hargreaves, B.R., and Seva, J., The impact of short-term exposure to UV-B radiation on zooplankton communities in north temperate lakes, *J. Plank. Res.*, 16, 205–218, 1994.

Winterbourn, M.J. and Collier, K.J., Distribution of benthic invertebrates in acid, brown water streams in the South Island of New Zealand, *Hydrobiologia*, 155, 277–286, 1987.

Worm, B., Sandow, M., Oschlies, A., Lotze, H.K., and Myers, R.A., Global patterns of predator diversity in the open oceans, *Science*, 309, 1365–1369, 2005.

Zellmer, I.D., UV-B-tolerance of alpine and arctic Daphnia, *Hydrobiologia*, 307, 153–159, 1995.

27 Effects of Contaminants on Trophic Structure and Food Webs

The empirical patterns are widespread and abundantly documented, but instead of an agreed explanation there is only a list of possibilities to be explored.

(May 1981)

There has been little synthesis of the relative roles of different ecological forces in determining population change and community structure. Rather, there is a collection of idiosyncratic systems, with their associated protagonists, in which opposing views on the importance of particular factors are debated.

(Hunter and Price 1992)

27.1 INTRODUCTION

An understanding of trophic interactions and food web structure is critical to the study of basic ecology and ecotoxicology. Early in the history of ecology, feeding relationships were recognized as a fundamental characteristic that defined communities. Trophic interactions provide the fundamental linkages among species that determine the structure of terrestrial and aquatic communities. For some ecologists, the study of food webs and trophodynamics is the central, unifying theme in ecology (Fretwell 1987). Because energy is a common currency required by all living organisms, the study of bioenergetics of individuals, populations, communities, and ecosystems allows researchers to integrate their findings across several levels of biological organization (Carlisle 2000).

Despite the importance of food webs and trophic interactions in basic ecology, ecotoxicologists have not incorporated significant components of basic food web theory into investigations of contaminant effects. This reluctance is ironic because the concern about food chain transport of contaminants in wildlife populations was at least partially responsible for much of the environmental legislation in the early 1960s. Reports of biomagnification of organochlorine pesticides and the subsequent effects on birds of prey (Carson 1962) eventually resulted in the ban of organochlorine pesticides.

One important exception to the general neglect of basic food web theory in ecotoxicology is the application of models to predict contaminant fate. Contaminant transport models used in ecotoxicology are analogous to energy flow models derived from the ecological literature. The application of these models for understanding fate and transport of contaminants in ecosystems will be described in Chapter 34. Quantifying the movement of contaminants through an ecosystem is only one of several potential applications of food web theory. An understanding of the ecological factors that determine energy flow in communities, such as food chain length, interaction strength, and connectedness, are also necessary to quantify contaminant fate and effects. For example, studies have shown that trophic structure and food chain length regulate contaminant concentrations in top predators (Bentzen et al. 1996, Rasmussen et al. 1990, Stemberger and Chen 1998, Wong et al. 1997). It is likely that other ecological processes, either directly or indirectly related to trophic structure, will play a role in determining contaminant transport. Recent refinements in transport models have been primarily limited to quantifying the role of physicochemical characteristics that modify contaminant bioavailability. Further improvement of these models will require that ecotoxicologists develop a better understanding of the ecological factors that influence contaminant fate and transport.

Another potential contribution of basic feed web theory to ecotoxicology is the measurement of food web structure and function as indicators of contaminant effects. Although the relationship between trophic structure and natural disturbance has been recognized for many years (Odum 1969, 1985), there have been few attempts to determine how food webs respond to contaminants (Carlisle 2000). There is some evidence that food chains are shorter in systems subjected to frequent disturbance, but the mechanistic explanation for this observation has not been determined. Using food web structure and function as indicators of contaminant effects is appropriate for several reasons. Bioenergetic approaches at the level of individuals and populations have a long history in toxicology. Growth is a common end point in many toxicological investigations that integrates numerous physiological characteristics. Energetic cost of contaminant exposure may be interpreted within the context of growth and metabolism. For example, recent studies have combined measurements of metabolism, food consumption, and growth into an individual-based bioenergetic model to assess effects of organochlorines (Beyers et al. 1999a,b). Similar approaches could be used to measure the effects of contaminants on flow of energy through a community. Finally, because energy is a common currency in all biological systems, understanding ecological effects of contaminant exposure on communities may help establish mechanistic linkages to lower (individuals, populations) and higher (ecosystems) levels of biological organization.

27.2 BASIC PRINCIPLES OF FOOD WEB ECOLOGY

27.2.1 Historical Perspective of Food Web Ecology

The strength of trophic interactions and the relationship between energy flow and community structure have been topics of considerable interest to ecologists for many years. Charles Elton's (1927) studies of feeding relationships in a tundra community and his representation of trophic levels as an energy pyramid (Figure 27.1a) focused the attention of ecologists on the importance of food as a "burning question in animal society." Subsequent representations of Elton's trophic pyramids included biomass and numbers of individuals as the fundamental components.

Ecologists soon recognized that this simple depiction of energy flow treated all species within a trophic level equally and, more importantly, did not account for microbial processes. In addition, there was no attempt to quantify the movement of energy among trophic levels. Raymond Lindeman's (1942) classic paper introduced the "trophic-dynamic" aspect of natural systems and revolutionized the study of food webs. On the basis of an extensive analysis of Cedar Bog Lake (MN), this work formalized the concept of energy flow through ecosystems and influenced a generation of systems ecologists. The study of ecology shifted from habitat associations and species lists to a more quantitative analysis of trophic relationships and energy flow. Lindeman also recognized the inherent inefficiency of energy flow in ecological systems, setting the stage for a contentious debate concerning the importance of biotic and abiotic factors that limit the number of trophic levels in communities.

Lindemen's box and arrow diagrams depicting energy flow and cycling of materials through a community were further refined by Eugene P. Odum (1968) (Figure 27.1b), widely regarded as the father of systems ecology. The emergence of ecosystems ecology in the 1950s also highlighted philosophical differences between holistic and reductionist approaches. While some ecologists felt that understanding complex systems required sophisticated and quantitative analysis of all interacting components, others felt that characteristics of ecosystems transcended those of individual components and could only be investigated by considering emergent properties. Unfortunately, these philosophical differences between proponents of holism and reductionism still persist in ecology and ecotoxicology today (Section 1.2 in Chapter 1 and Box 20.1 in Chapter 20).

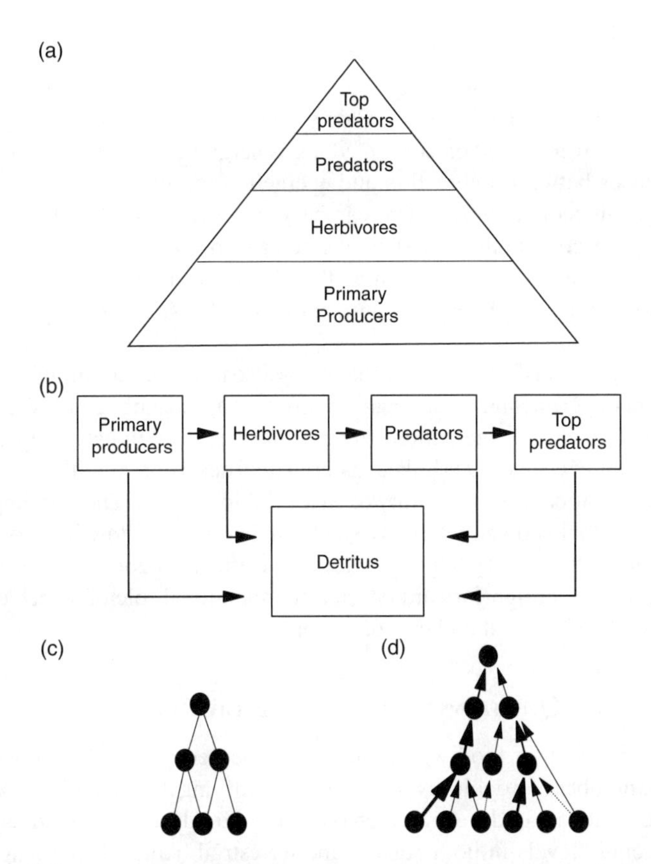

FIGURE 27.1 Four different representations of trophic structure and food chains in the ecological literature. (a) Eltonian trophic pyramid showing biomass at each trophic level. (b) Box and arrow diagram showing energy flow through a community. (c) Descriptive food chain showing potential interactions among species. (d) Energetic or interaction food chain showing energy flux or strength of interactions (represented by thickness of the lines) between dominant species in a community.

27.2.2 Descriptive, Interactive, and Energetic Food Webs

Food webs depicted in the contemporary ecological literature fall into three general categories: descriptive, interactive, and energetic (Figure 27.1c,d). Descriptive food webs are probably the most common and are produced by simply characterizing feeding habits of dominant species. Descriptive food webs are analogous to the use of presence–absence data in community monitoring because they provide no information on the relative importance of linkages among species. In contrast, interaction and energetic food webs quantify the importance of linkages among species and energy flow. Interaction food webs are constructed by manipulating the abundance of either predators or prey and measuring responses. Interaction food webs have a long history in experimental ecology and have been employed to identify keystone species (e.g., Paine 1980). The best examples of interaction food webs are from marine rocky intertidal habitats where experimental manipulation is simplified because of the low mobility of species and the essentially two-dimensional nature of the habitat. Energetic-based food webs are constructed by quantifying energy flow between species. This is generally accomplished by characterizing feeding habits and measuring secondary production of dominant species in a community (Benke and Wallace 1997). Either interaction or energetic food webs would be appropriate for assessing contaminant effects; however, it is important to recognize that both approaches are data intensive and require a significant amount of effort to develop.

Because the strength of species interactions are not necessarily related to the amount of energy flow between trophic levels, bioenergetic and interaction approaches can yield different results.

For example, relatively little energy flows between kelp and sea urchins in marine ecosystems; however, as described in the following section, removal of sea urchins may have a large impact on kelp populations and associated species. Paine (1980) showed very different patterns resulted when marine food webs were based on connectedness, energy flow, or interaction strength. Because of potential differences between interaction and energetic food webs, these approaches may have different applications in ecotoxicology. If researchers are interested in modeling the movement of contaminants through a community, an energetic food web may be most appropriate. However, if the purpose of an investigation is to examine the direct and indirect effects of contaminants on community structure, it may also be very important to know the strength of species interactions and construct an interaction food web.

The strength of interactions within a food chain may also influence community stability; however, because of the lack of experimental studies and the different approaches employed by theoretical and empirical ecologists to measure interaction strength, the relationship between stability and energy flow is uncertain. de Ruiter et al. (1995) linked material flow descriptions with measures of interaction strength to quantify the influences on stability of terrestrial food webs. Their findings were consistent with previous research that showed relatively small rates of energy flow in a community can have large effects on community stability. Thus, predicting the effects of contaminant-induced alterations on energy flow will not be straightforward because the functional role of a species in a community may not be directly related to its abundance or biomass.

27.2.3 Contemporary Questions in Food Web Ecology

Most contemporary research in food web ecology has focused on two key topics: (1) identifying factors that limit the number of trophic levels; and (2) quantifying the strength of species interactions. One consistent observation in food web research is that most food webs are relatively short, averaging between 3 and 5 trophic levels in both aquatic and terrestrial habitats. The length of food chains and the number of trophic levels is assumed to be limited by the inefficient transfer of energy. Ecological systems conform to laws of thermodynamics, and the loss of energy from prey resources to consumers limits the number of possible trophic levels. On the basis of this argument, we would expect shorter food webs in unproductive systems where resources are limited. We also know that top predators tend to occur in low numbers and are sparsely distributed compared to herbivores and other secondary consumers. In an insightful essay on this topic, Colinvaux (1978) argued that the rarity of large, fierce predators (e.g., tigers, great white sharks) in many ecosystems resulted from the inefficiency of energy transfer from lower trophic levels.

Despite the intuitive appeal and broad theoretical support, few studies of food chains in natural communities have found consistent relationships between productivity and food chain length. Primary productivity may vary by orders of magnitude among communities, but the number of trophic levels remains remarkably consistent. Food chains are not necessarily shorter in unproductive environments (e.g., arctic tundra) compared with productive environments (e.g., tropical rainforests). Ricklefs (1990) estimated the average number of trophic levels based on net primary production, ecological efficiency, and energy available to predators for a variety of communities (Table 27.1). In contrast to theoretical predictions, there was no consistent relationship between net primary productivity and the estimated number of trophic levels.

Hairston, Smith, and Slobodkin's (HSS) (1960) revolutionary paper offered an alternative explanation for the relationship between productivity and food web structure. According to the HSS model, species interactions (competition, predation) within and between trophic levels determined the structure of food webs. In a three trophic level system typical of many terrestrial communities, abundance of herbivores was controlled by predators, thus allowing primary producers to compete for resources. Support for this model in terrestrial food webs has been widespread, and predator control of herbivores is proposed as an explanation for the dominance of green plants in most terrestrial ecosystems. A general extension of this argument to other communities suggests that plants are controlled by

TABLE 27.1
Estimated Number of Trophic Levels Based on Primary Production, Energy Flux to Consumers, and Ecological Efficiencies

Community	Net Primary Production (kcal/m^2/year)	Number of Trophic Levels
Open ocean	500	7.1
Coastal marine	8000	5.1
Temperate grassland	2000	4.3
Tropical forest	8000	3.2

Source: Modified from Table 11.5 in Ricklefs (1990).

resources (nutrients, light, and space) in systems with an odd number of trophic levels and controlled by herbivores in systems with an even number of trophic levels. In an alternative synthesis of the relationship between energy flow and trophic structure, Hairston and Hairston (1993) observed that the mean number of trophic levels in pelagic (i.e., open water) systems (3.6) is significantly greater than in terrestrial systems (2.6). On the basis of the relative importance of competition among primary producers in pelagic and terrestrial systems, they provide a compelling argument for the hypothesis that trophic structure determines food web energetics instead of visa versa.

The hypothetical relationship between food chain length and community stability is also somewhat tenuous. Briand and Cohen (1987) reported that the average food chain length in food webs from constant and fluctuating environments was 3.60 and 3.66, respectively. Interestingly, these researchers reported that the complexity and dimensionality of a habitat had a greater influence on food chain length than community stability. In general, two-dimensional habitats (e.g., stream bottoms, rocky intertidal zones) had shorter food chains than three-dimensional habitats (e.g., coral reefs, open ocean). Thus far, an adequate mechanistic explanation for this relationship has not been provided. However, results are consistent with the observation that more complex habitats have a greater number of species.

Experimental manipulations of food webs provide the most direct tests of the relationship between trophic structure, productivity, and disturbance. Experiments conducted by Power and colleagues (Power 1990, Wootton et al. 1996) extended the HSS model to aquatic ecosystems and demonstrated the role of disturbance in regulating trophic structure. As predicted by HSS, primary producers were limited by resources (nutrients, space, and light) in communities with an odd number of trophic levels, whereas communities with an even number of trophic levels were regulated by herbivores. Disturbance also played a prominent role by controlling abundance of grazers and shifting energy to predatory fish. These results indicate the need to advance from a single species perspective to a community perspective when assessing the effects of disturbance (Wootton et al. 1996). More importantly, these results demonstrate that disturbance may alter trophic structure and energy flow in food webs by removing key species.

Food chain length and the number of trophic levels of a community may also influence resistance and resilience stability. Mathematical models predict that communities with longer food chains will experience extreme population fluctuations, resulting in a greater probability of extinction of top predators. This hypothesis has important implications for the study of systems subjected to anthropogenic disturbance. For example, we expect that effects of contaminants would be greater in communities with greater trophic complexity and longer food chains.

Food web interactions involving otters and sea urchins in kelp beds of western Alaska provide some insight into how disturbance can dramatically alter trophic structure (Estes et al. 1998). The role of otters as a keystone species in marine kelp beds is well established. Otter predation on sea urchins, major consumers of early growth stages of kelp, maintains the structure of kelp forests.

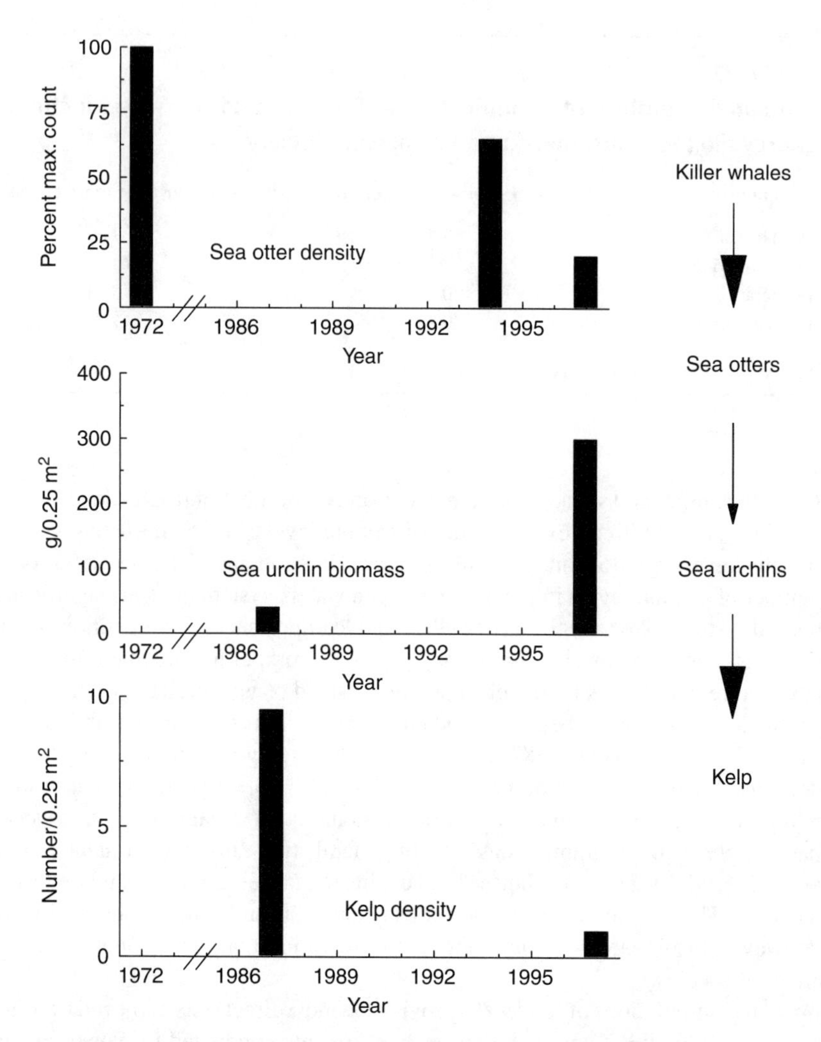

FIGURE 27.2 Effects of predation by killer whales on trophic structure of nearshore marine ecosystems in western Alaska. The figure depicts changes in otter abundance, sea urchin biomass, and effects on kelp density following increased predation by killer whales. (Modified from Figure 1 in Estes et al. (1998).)

Recovery of otter populations following protection from overhunting resulted in recovery of kelp forests along the Pacific Northwest coast. However, a dramatic decline of sea otters over large areas in western Alaska in the 1990s caused increased abundance of urchins and a corresponding decline in kelp abundance (Figure 27.2). Surprisingly, increased sea otter mortality was attributed to predation by killer whales, which shifted their foraging to coastal areas following reductions in their preferred prey: seals and sea lions. Estes et al. (1998) speculated that reduced abundance of seals and sea lions resulted from unexplained declines of forage fish stocks. Thus, addition of a top predator (killer whales) to coastal Alaska converted this three trophic level system to a four level system. This spectacular example illustrates the connectance and interdependence of multiple trophic links and the interactions between oceanic and nearshore communities. More importantly, this study demonstrates the difficulty predicting indirect effects of reduced prey abundance in natural communities. It is unlikely that researchers could have anticipated that declines in fish forage stocks in the oceanic environment would cause a collapse of coastal kelp beds. Similar "ecological surprises" (sensu Paine et al. 1998) are likely to occur in systems where important predator or prey species are eliminated as a result of contaminants.

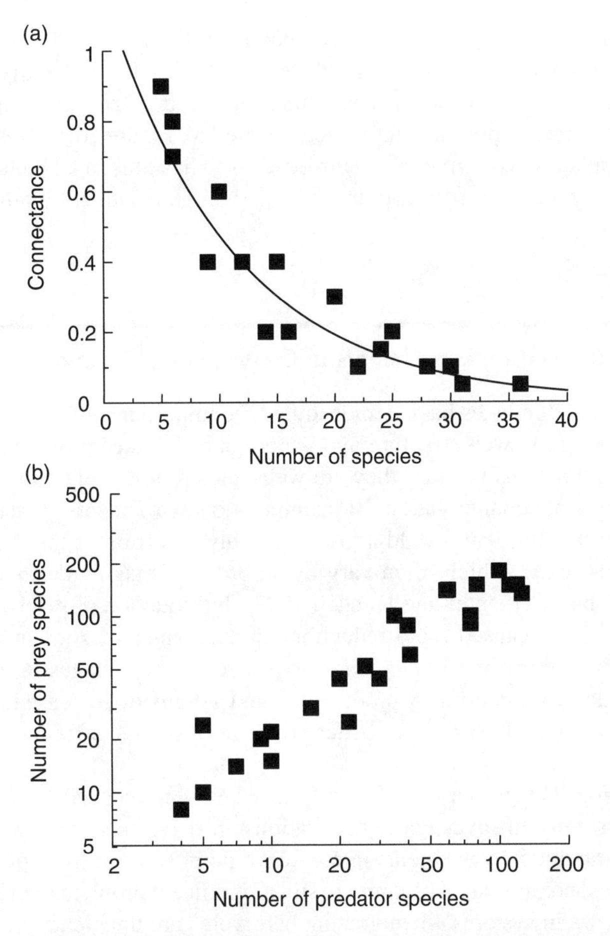

FIGURE 27.3 (a) Hypothetical relationship between connectance (number of interactions/number of possible interactions) and the total number of species in a food web (upper panel). (b) Hypothetical relationship between number of predator species and number of prey species (lower panel).

The other major generalizations regarding the structure of food webs are the relatively constant number of species interactions and the ratio of predators to prey. Food web connectance, defined as the observed number of trophic interactions divided by the total number of possible interactions, generally decreases with species richness (Pimm 1982) (Figure 27.3a). As a result, each species tends to average about two trophic interactions, regardless of the number of species in the community. Similarly, the ratio of predator species to prey species in a community is relatively constant (generally between two and three prey species per predator species), regardless of the total number of species in the community (Jeffries and Lawton 1985) (Figure 27.3b). Assuming that these theoretical predictions are consistent among communities, connectance and the ratio of predators to prey may prove to be useful endpoints for assessing effects of stressors on food web structure.

27.2.4 Trophic Cascades

The trophic cascade hypothesis (Carpenter and Kitchell 1993) predicts that each trophic level in a community is influenced by trophic levels directly above (e.g., consumers) and directly below (e.g., resources). According to this hypothesis, nutrients determine the potential productivity of a system, but deviations from this potential are owing to food web structure. Thus, two conditions define a trophic cascade: (1) top-down control of community structure by predators; and (2) strong indirect effects of two or more links away from the top predator (Frank et al. 2005). For example, increased abundance of piscivorous fish in a lake can reduce abundance of zooplanktivorous fish, allowing abundance of zooplankton to increase. The resulting increased grazing pressure by zooplankton is predicted to reduce biomass of phytoplankton (see Chapter 20, Figure 20.1). Researchers conducting large-scale biomanipulation experiments in eutrophic lakes have taken advantage of these relationships and attempted to control primary productivity and eliminate algal blooms by introducing top predators (Box 27.1).

Box 27.1 Biomanipulation Experiments to Control Eutrophication

Experiments conducted in lakes have demonstrated the importance of trophic linkages and the relationship between food web structure and water quality. Lakes provide an ideal habitat to examine trophic interactions because they are well-defined, relatively closed systems and are amenable to experimental manipulation. Biomanipulation experiments were initially motivated by the observation that nutrients could account for only a portion of the variation in primary productivity among lakes, which often vary by an order of magnitude in systems with similar levels of nutrients (Carpenter and Kitchell 1993). Introduction of piscivorous fish to Peter Lake, Wisconsin (USA) caused rapid reductions in abundance of zooplanktivorous fish and an increase in herbivore (primarily *Daphnia*) body size. These changes in food web structure resulted in a 37% decrease in primary productivity and a dramatic increase in light penetration. Interestingly, herbivore body size was a better predictor of trophic effects on productivity than abundance.

The observation that primary productivity in lakes is influenced by food web structure provided an opportunity to investigate the relationship between trophic structure and water quality. Despite dramatic improvements in control of point source inputs of nutrients over the past several decades, noxious algal blooms are still a significant problem in many lakes. Cultural eutrophication occurs in systems when grazing herbivores are unable to control abundance of phytoplankton, especially blue-green algae. If introduction of piscivorous fish can reduce predation on herbivores by limiting abundance of zooplanktivorous fish, then grazing pressure on noxious algae is expected to increase. This idea was the impetus for a large-scale biomanipulation experiment conducted in Lake Mendota (WI) during the late 1980s (Kitchell 1992). As was expected, increased stocking of northern pike and walleye in Lake Mendota caused increased abundance of large, grazing zooplankton. However, because of a combination of unexpected events, including unusual weather patterns, greater runoff, and greater fishing pressure, the results of this experiment were mixed. Primary productivity did not respond throughout the experiment as predicted, suggesting that food web interactions were not the sole determinant of primary productivity in Lake Mendota. However, results of this study and others conducted by Kitchell and colleagues demonstrate that predation played a major role in structuring lower trophic levels in lakes.

These experiments highlight the close connection between trophic interactions and energy flow in lentic ecosystems. It is important that ecotoxicologists recognize the significance of these interactions when characterizing food chain transport of contaminants in lake communities. Simple models of contaminant transport generally do not consider direct effects on trophic structure or potential feedback between adjacent trophic levels. In addition, food web manipulations conducted in lakes have generally not included a littoral or benthic component. Because sediments are a major sink for contaminants in most lentic systems, a complete understanding of how trophic structure will influence contaminant transport requires that processes involving sediments should also be considered.

Although there has been strong support for the trophic cascade hypothesis in lakes, the generality of this hypothesis and the relative importance of top-down (predator control) and bottom-up (nutrient driven) effects in other systems have been subjects of considerable debate. An understanding of the relative importance of top-down versus bottom-up regulation is necessary to predict the consequences of anthropogenic nutrient inputs into ecosystems and has important management implications. For example, protecting top predators may be more important than nutrient control in systems regulated by top-down processes (Halpern et al. 2006). Because much of the research documenting the importance of top-down effects has been conducted in systems with relatively simple food webs and low diversity, the significance of trophic cascades in complex and species-rich communities remains uncertain (Frank et al. 2005). The removal of top predators from marine continental shelf ecosystems has provided the best opportunity to test the generality of the trophic cascade hypothesis at relatively large spatiotemporal scales. Stock assessments of commercial fisheries over the past 50 years have shown significant reductions in biomass and size of top predators such as tuna and billfish, but relatively minor effects on trophic structure (Sibert et al. 2006). In contrast, removal of cod from the northwest Atlantic Ocean resulted in dramatic effects on lower trophic levels and nutrient concentrations that were consistent with the trophic cascade hypothesis. Halpern et al. (2006) reported strong top-down control by top predators in 16 kelp forests located around the Channel Islands, California. Despite strong spatial gradients in chlorophyll a among sites, top-down control accounted for 7–10 times greater variability in abundance of lower and mid-level trophic levels than primary productivity. These researchers noted that removal of top predators may convert ecosystems from top-down to bottom-up control, making these systems more sensitive to nutrient enrichment.

Although relatively strong support for the trophic cascade hypothesis has been obtained for some aquatic ecosystems, few studies have documented top-down effects in terrestrial environments. Strong (1992) argues that trophic cascades in lakes are an exception and generally restricted to species-poor habitats. He suggests that terrestrial systems and more diverse aquatic communities are more frequently characterized by "trophic trickles" rather than cascades. Because predator control is weaker and more diffuse in these species-rich communities, the effects of trophic interactions are buffered. More importantly, unlike aquatic systems where manipulative studies are common, the lack of experimental research in terrestrial habitats limits our ability to identify trophic cascades (Strong 1992). In one of the few experimental studies conducted with terrestrial communities to characterize trophic cascades, Salminen et al. (2002) constructed food webs in laboratory microcosms consisting of three trophic levels (soil microbes, microbivorous-detritivorous worms, and predatory mites). Results showed strong top-down effects of predatory mites on trophic structure and that lead contamination in soil disrupted these interactions. Because some of the responses were an unexpected outcome of indirect effects of lead, these investigators urged caution when using traditional food web models to quantify contaminant effects. Croll et al. (2005) took advantage of a large-scale natural experiment to investigate the effects of top predators on plant biomass and community structure in the Aleutian archipelago (Alaska). The introduction of arctic foxes to some islands greatly reduced abundance of seabirds, resulting in a two order of magnitude decline in guano. Elimination of marine-derived nutrient subsidies to these islands had dramatic effects on plant biomass and community composition.

An important exception to the general absence of trophic cascades in terrestrial ecosystems is the interaction between moose and wolves on Isle Royale reported in Chapter 26 (McLaren and Peterson 1994). Results of long-term monitoring of wolves and moose have described a tightly coupled predator–prey system. Periods of low wolf and high moose numbers are correlated with intense grazing pressure on balsam fir, the primary forage of moose. These results are especially significant because they provide strong support for top-down control in a nonaquatic, three trophic level system. However, it is important to note that because spatial boundaries are well defined and trophic complexity is low, Isle Royale may represent a relatively unique situation.

Quantifying the relative importance of consumer versus resource control in communities will require a more sophisticated understanding of population dynamics, species interactions, and the abiotic environment. Resource enrichment experiments conducted in a terrestrial, detritus-based food chain showed strong bottom-up limitation of top predators (Chen and Wise 1999). Conversely, Stein et al. (1995) reported that food webs in temperate reservoirs were regulated by complex weblike interactions rather than chainlike trophic cascades. The lack of a zooplankton response to introduced piscivorous fish (northern pike) and reduced abundance of planktivores were explained by poor food quality for these grazers. Brett and Goldman (1996) conducted a meta-analysis of 54 different experiments to test the generality of the trophic cascade hypothesis. Meta-analysis is a powerful statistical approach for analyzing patterns and central tendencies of large datasets derived from multiple investigations. Results of this analysis provided strong support for the trophic cascade hypothesis. However, a subsequent analysis of 11 mesocosm experiments showed no relationship between nutrient enrichment and the number of trophic levels (Brett and Goldman 1997). Another meta-analysis of 47 mesocosm experiments and 20 time-series studies conducted in marine habitats demonstrated the importance of nitrogen enrichment and predation on pelagic food webs (Micheli 1999). As expected, based on research conducted in freshwater systems, nutrient enrichment increased primary production and addition of planktivorous fish reduced zooplankton abundance. However, unlike patterns observed in lakes and streams, consumer–resource interactions did not cascade through other trophic levels because of the weak interactions between grazers and phytoplankton. As a result, it is unlikely that biomanipulation of marine food chains would have the same effects on algal productivity as those observed in lakes (Micheli 1999). Finally, the presence of trophic cascades may also influence the recovery of some aquatic ecosystems from anthropogenic disturbance. Long-term (18 years) records of trophic structure in a hypereutrophic lake following reductions in total phosphorus and organic matter showed that cascading influences of fish predators on zooplankton grazing had much greater influence on recovery than changes in nutrient input (Jeppesen et al. 1998).

27.2.5 Limitations of Food Web Studies

Significant progress has been made in the development of food webs and the quantification of energy flow among trophic levels since the publication of Elton's energy pyramids in 1927. Because transport of contaminants in a community is often intimately associated with the flow of energy, a better understanding of trophic interactions will improve our ability to predict contaminant fate. However, as with any general ecological model, it is important to recognize the limitations and simplifying assumptions of food webs. Although grouping organisms into broad trophic categories has facilitated the development of mathematical models for estimating energy flow, this representation of food webs is greatly oversimplified. In addition, most studies of food webs either ignore or minimize the importance of omnivory, which may be the dominant mode of feeding for many species. Relatively few consumers feed exclusively on resources from one trophic level. Many consumers are opportunistic generalists that feed on the most abundant, available, or energetically profitable food resources. Thus, pollution-induced alterations in prey communities may shift feeding habits of predators to tolerant prey species with little impact on bioenergetics (Clements and Rees 1997).

Traditional representations of food webs often ignore the role of detritus, which is a major contributor of energy to many aquatic and terrestrial food chains. Experiments conducted by Wallace et al. (1997) showed reduced biomass of most functional feeding groups when allochthonous detritus was excluded from a headwater stream. In addition, most characterizations of food webs are limited to a single habitat, and often fail to consider energy flow between adjacent habitats. Experiments conducted by Nakano et al. (1999) demonstrated the importance of terrestrial arthropods to trophic structure of a small stream and the linkages between terrestrial and aquatic habitats. Exclusion of terrestrial arthropods shifted feeding habits of predatory fish to aquatic prey and caused significant changes in energy flow and trophic structure.

Food web studies are also limited by the general lack of information on interaction strength among species. Knowing that a particular trophic interaction occurs in a community does not provide any indication of the strength of this interaction. Thus, some assessment of interaction strength,

Reference stream

Metal-polluted stream

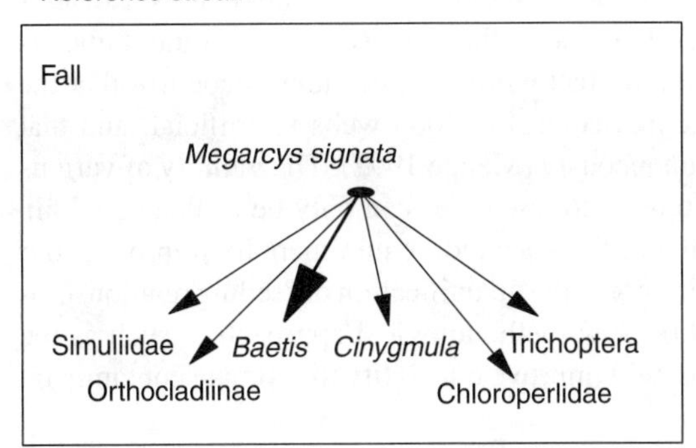

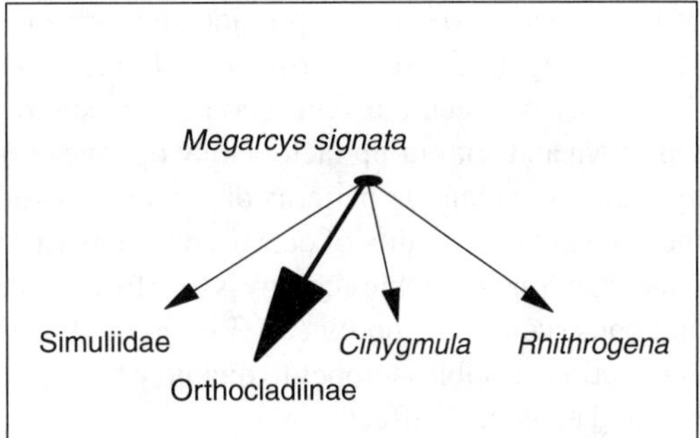

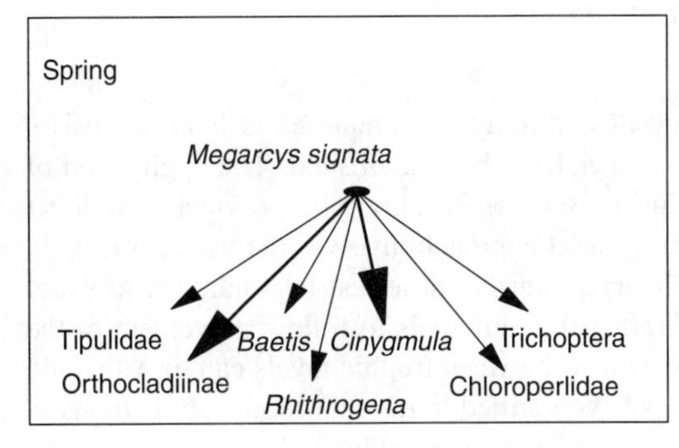

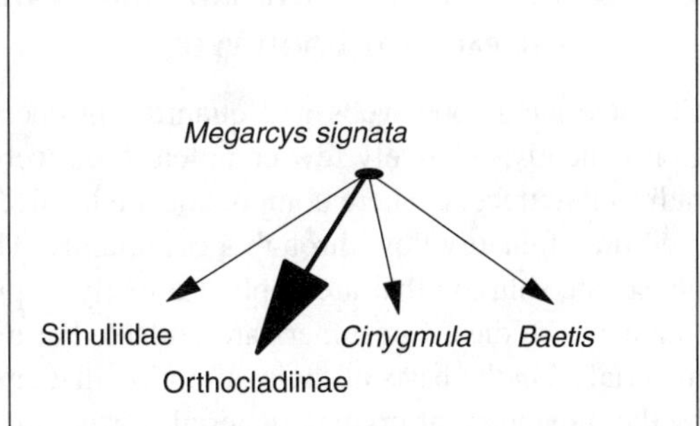

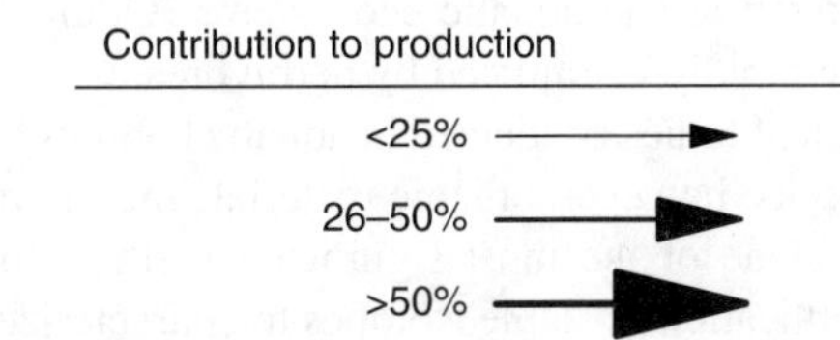

FIGURE 27.4 Seasonal changes in energy flow in a reference and metal-polluted stream. The figure shows the relative contribution of each prey species to production of a stonefly predator (*Megarcys signata*). (Modified from Figure 4 in Carlisle (2000).)

either by manipulative experiments or by analysis of energy flow, is necessary to understand the importance of trophic interactions. In addition, information on seasonal and ontogenetic shifts in trophic interactions should be considered when constructing food webs. Carlisle (2000) developed production-based bioenergetic food webs for reference and metal-polluted streams in the Rocky Mountains, USA (Figure 27.4). The pathways of major energy flow and relative contribution of prey species to predator biomass differed greatly between streams. There was also significantly more seasonal variation in food web structure in reference streams compared with polluted streams.

Finally, the number of species in natural food webs is usually much greater than the number considered in most published food web studies. Relatively few studies have provided complete food web analyses of all species within a community. If the majority of trophic interactions in a community are relatively weak, it may not be necessary to quantify the importance of all species. However, very different food web characteristics may emerge when all species in a community are included (Polis 1991).

Despite these limitations, analysis of food webs is a productive area of basic ecological research with important applications to ecotoxicology. Changes in the abundance of consumers or their resources often result in strong cascading effects across trophic levels. To predict fate and effects of contaminants, ecotoxicologists must develop a better understanding of these interactions. Reviews of basic food web ecology provide support for the hypothesis that energy flow in communities is regulated by a diverse assortment of biotic and abiotic factors. Ecologists now recognize that the dichotomy between consumer control versus resource control of food webs is artificial, and that top-down and bottom-up factors may operate simultaneously (Menge 1992). The validity of various hypotheses explaining patterns of food web structure in communities can only be evaluated within the context of the quality of data used to construct them. Thus, a major requirement for improving our understanding of trophic ecology is to infuse greater rigor in the quantification of feeding relationships and construction of food webs (Begon et al. 1990). New methodological approaches, such as the application of stable isotopes to quantify food webs, will improve our ability to estimate contaminant fate and ecological effects.

27.2.6 Use of Radioactive and Stable Isotopes to Characterize Food Webs

Characterizing food webs and quantifying energy flow through communities is labor-intensive. Consequently, relatively few complete food web studies have been published. The application of radioactive tracers, where compounds are labeled using isotopes (usually ^{32}P), provides an indirect measure of energy flow through a community. The general approach involves the use of tracers in phosphate solution that are applied directly to primary producers in terrestrial studies or to water in aquatic studies. Consumers are collected at different time intervals to follow movement of the materials. On the basis of the assumption that energy flow between trophic levels can be estimated by the movement of organic material, tracer studies have verified traditional food web methods in terrestrial and aquatic ecosystems (Odum 1968). Results of tracer studies indicate that materials are rapidly assimilated by herbivores, whereas uptake by predators and decomposers requires more time. Studies conducted in aquatic habitats showed that most of the energy is dissipated within a few weeks; however, organic materials incorporated into bottom sediments may persist for many years.

One of the most significant methodological developments in the study of food webs is the application of stable isotopes to characterize feeding habits and quantify energy flow. Over the past 20 years, studies have shown a strong relationship between stable isotope ratios of consumers and those of their diet. Stable isotope ratios, particularly $^{13}C/^{12}C$ and $^{15}N/^{14}N$, are determined by a variety of geochemical, meteorological, and biological characteristics that vary among habitats and trophic groups. Thus, isotopic analyses of organisms can provide a unique signature that is representative of their habitat and feeding habits. By comparing stable isotope ratios of predators and prey across different communities, it is possible to obtain time-integrated estimates of energy flow, trophic position, and carbon sources for major consumers (Figure 27.5). Stable isotope studies have been used to assess effects of disturbances on food web structure and energy flow and to investigate the food web consequences of introduced species on native species (Vander Zanden et al. 1999). The application of these approaches to the study of contaminant transport is described in Chapter 34.

27.3 EFFECTS OF CONTAMINANTS ON FOOD CHAINS AND FOOD WEB STRUCTURE

Although there has been significant progress in the development and testing of food web theory over the past 20 years, investigators generally have not considered contaminant-induced alterations in food web structure as endpoints in ecotoxicological investigations. The limited application of basic food web ecology to ecotoxicological research is partially a result of the logistical difficulties and uncertainty associated with constructing food webs. New technical approaches, such as stable isotope analyses and bioenergetic modeling, will likely increase the integration of food web theory into ecotoxicology.

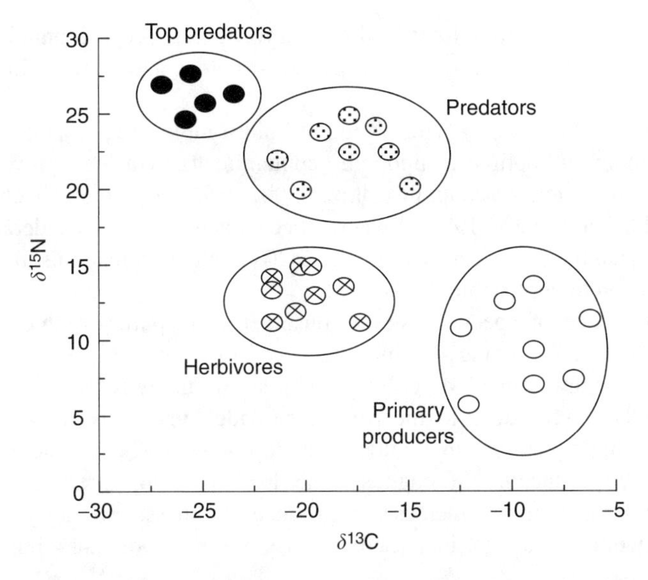

FIGURE 27.5 Example of the use of stable isotopes to characterize food webs. Variation in stable isotopes of N reflects trophic levels, whereas variation in stable isotopes of C reflects food sources.

General patterns of food web structure described in the basic ecological literature provide some insight into how communities may respond to anthropogenic stress. The number of trophic levels, the connectance of food webs, and the strength of interactions are likely to be affected by exposure to contaminants. For example, one of the most consistent observations at contaminated sites is reduced species richness and simplification of community structure. Loss of species will probably be accompanied by a reduced number of trophic levels and a reduced degree of connectance. Contaminated sites also tend to be characterized by a disproportionate loss of larger, longer-lived species and a switch to smaller, more opportunistic taxa. The loss of keystone and other important predators from these systems will likely have cascading effects on lower trophic levels. Finally, because the stability of food chains is often influenced by external forces, factors such as disturbance may decouple food chains and weaken trophic cascades.

27.3.1 Interspecific Differences in Contaminant Sensitivity

Interspecific differences in sensitivity to contaminants are well established, and the effects of contaminant exposure on food webs will depend on these differences. Distinguishing the direct effects of toxic chemicals on abundance of organisms from the indirect effects resulting from changes in abundance of predators or prey is of critical importance for understanding how food webs respond to anthropogenic stressors. Contaminant-induced elimination of a sensitive, but ecologically important, species from a food web would have significant consequences for trophic levels above and below. Currently, there is little evidence indicating that any one trophic level will be more or less sensitive to contaminants than another. Alteration in the structure of food webs has been proposed as an indicator of anthropogenic disturbance (Pimm et al. 1991). For example, *Daphnia* and other large zooplankton play an important role in controlling phytoplankton in lakes. Because these organisms may be more sensitive to some xenobiotics than phytoplankton, contaminant-induced reductions in abundance or feeding rates may have important consequences for net production. Bengtsson et al. (2004) observed significant reductions in grazing rates of *Daphnia* exposed to DDE and subsequent increases in primary producer biomass as a result of reduced top-down control by herbivores. Riddell et al. (2005) reported that exposure to sublethal levels of cadmium (0.5 μg/L) significantly reduced capture efficiency of brook trout feeding on invertebrate prey. Contaminant-induced alterations in predation will likely influence the trophic structure and energy flow through aquatic food webs.

Because of the potential for biomagnification of lipid soluble contaminants, we expect that top predators will be more susceptible to lipophilic contaminants than their prey. For example, the extremely high sensitivity and susceptibility of mink (*Mustela vison*) to organochlorines, especially PCBs, is well established (Peterle 1991). Because these organisms are considerably more sensitive to organochlorines than their prey (primarily fish), PCBs could eliminate this top predator resulting in significant effects on lower trophic levels.

The relative influence of species loss on consumer or resource trophic levels will depend on a number of factors, including the importance of top-down vs. bottom-up regulation. In many ways, contaminant-induced mortality is similar to the effects of a selective predator (Carman et al. 1997). According to the trophic cascade hypothesis, removal of a top predator in a three trophic level system regulated by top-down forces would result in decreased biomass of primary producers. In contrast, the loss of a top predator in a four trophic level system would result in an increase in producer biomass. Similarly, elimination of an important grazer would result in increased primary producer biomass but reduced predator biomass.

Mathematical modeling techniques can improve our ability to predict risks of chemical stressors to multiple trophic levels and provide important opportunities to generalize among stressors, locations, and seasons. However, the effectiveness of modeling approaches to predict risks across trophic levels requires that results are validated using field data collected from natural or experimental systems. Hanratty and Stay (1994) used data from a littoral enclosure experiment on the effects of the insecticide chlorpyrifos to validate a bioenergetic effects model. The Littoral Ecosystem Risk Assessment Model (LERAM) links single-species toxicity data to a bioenergetic model of trophic structure to predict community responses to chemical stressors. Changes in biomass are estimated from LC50 values, and LERAM simulates population growth by modeling changes in energy and biomass to each trophic level. Model predictions for most populations across several trophic levels were generally within 2 times the results observed from field experiments (Hanratty and Stay 1994).

27.3.2 Indirect Effects of Contaminant Exposure on Feeding Habits

Exposure to contaminants may have both direct and indirect effects on feeding relationships and trophic interactions. Altered prey behavior following exposure to pesticides may increase vulnerability to predation, thus increasing predator exposure to contaminants. Schauber et al. (1997) reported that deer mice (*Peromyscus maniculatus*) consumed more insects immediately following application of insecticides, suggesting that these organisms opportunistically selected dead or dying prey. Selective predation and the ability of predators to switch to contaminated prey will greatly complicate our ability to predict effects of contaminants on trophic interactions.

Elimination or reduction of prey resources will cause predators to shift to alternative prey species and may have important energetic consequences for predators. Experimental studies investigating effects of large-scale insecticide sprays have provided an unprecedented opportunity to evaluate responses of bird predators to reductions in prey abundance. These studies also demonstrate the important linkage between evolutionary ecology and ecotoxicology. Because many avian species are adapted to take advantage of seasonal increases in insect abundance, the application of pesticides often coincides with critical life history periods. The greatest exposure and potential for loss of prey resources occurs when adults are caring for their young. Surprisingly, despite large reductions in prey abundance following application of pesticides, some studies have shown relatively little indirect impacts on insectivorous birds (Adams et al. 1994, Howe et al. 1996, Powell 1984). These results suggest that either prey resources are superabundant in these systems or that predators are able to switch to alternative prey.

The opportunistic feeding habits of many predators may allow them to compensate for contaminant-induced reductions in abundance of preferred prey. Clements and Rees (1997) examined the effects of heavy metals on prey abundance and feeding habits of brown trout (*Salmo trutta*). Prey communities at an unpolluted station were dominated by metal-sensitive mayflies (Ephemeroptera) and black flies (Diptera: Simuliidae), whereas those at a polluted site were dominated by metal-tolerant chironomids (Diptera: Chironomidae) and caddisflies (Trichoptera). Differences in prey community composition were reflected in the feeding habits of brown trout, which consumed primarily chironomids and caddisflies at the metal-polluted station. Despite these alterations in prey communities, the mean biomass of prey consumed by brown trout was actually greater at the polluted site. These results are consistent with the hypothesis that a predator's feeding habits are flexible and can shift to take advantage of locally abundant prey resources. Similar results were reported by Wipfli and Merritt (1994) for invertebrate predators. Experimental treatment of two streams with the black fly larvicide B.t.i. (*Bacillus thuringiensis* var. *israelensis*) resulted in a switch to alternative, less preferred prey species. The ability to utilize alternative prey varied among species, as effects of prey reduction were greater for specialized predators than for generalized predators.

27.3.3 Alterations in Energy Flow and Trophic Structure

Energy flow and trophic structure of communities may be altered by exposure to contaminants if important functional groups are eliminated. Experimental and descriptive studies conducted in Rocky Mountain streams have shown that grazing mayflies are highly sensitive to heavy metals and usually eliminated from polluted streams (Clements et al. 2000). These organisms are generally replaced by metal-resistant groups, resulting in a shift in energy flow and greater utilization of detritus by other consumers. Carlisle (2000) used stable isotopes to characterize the food webs of stream communities impacted by heavy metals. As expected, isotopic analysis of food web structure showed a greater reliance on detritus in the metal-polluted stream and a greater utilization of periphyton in the unpolluted stream. These findings are in agreement with Odum's (1985) predictions that stressed ecosystems tend to be more detritus-based compared with unstressed ecosystems.

Exposure of meiofaunal communities to diesel fuel in sediment microcosms resulted in significantly increased algal biomass owing to lower grazing pressure by hydrocarbon-sensitive copepods (Carman et al. 1997). These researchers observed a similar pattern of reduced grazing pressure and increased algal biomass in a field study. Although stimulation of algae by hydrocarbons has been reported previously, this was the first study to demonstrate the role of grazers. Clearly, consideration of multiple trophic levels is necessary to understand mechanisms by which natural communities respond to contaminants.

A long-term series of studies conducted by Wallace and colleagues in the Coweeta Experimental Forest demonstrated significant alterations in food chains and energy flow after experimental introductions of the larvicide, methoxychlor (Wallace et al. 1982, 1987, 1989). Catastrophic drift following methoxychlor treatments resulted in a 90% reduction in total abundance and biomass of stream invertebrates. Changes in abundance of dominant prey taxa also caused shifts in feeding habits of predators. More importantly, the elimination of shredders (organisms that consume leaf litter) reduced leaf decomposition rates and the amount of particulate organic material transferred downstream.

Alterations in trophic structure observed in contaminated systems may depend on specific characteristics of the food web. In simple, two trophic level systems (producers and herbivores), toxic substances may modify trophic structure by reducing abundance of important grazers, resulting in an increase in primary producers (Webber et al. 1992). In a three trophic level system exposed to the insecticide, esfenvalerate, Fairchild et al. (1992) reported increased biomass of primary producers resulted from lower predation by bluegill on grazing zooplankton. Whole ecosystem acidification studies conducted by Schindler et al. (1985) have shown that elimination of mysids (*Mysis relicta*) and fathead minnows (*Pimephales promelas*), major prey organisms in the lake's pelagic food web, had significant effects on lake trout condition. Interestingly, abundance of lake trout and changes in ecosystem processes (primary productivity, decomposition, and nutrient cycling) were less sensitive to acidification than food web alterations.

Because species interactions in natural communities are often subtle, experimental manipulation of both contaminants and consumers may be necessary to understand consumer-resource dynamics. In experimental rockpools, Koivisto et al. (1997) manipulated levels of Cd, grazers, and predators to measure interactions between chemical stressors and trophic structure. Results showed that Cd directly reduced phytoplankton and zooplankton abundance, but did not alter trophic interactions. In contrast, reduction of herbivores by predators resulted in increased phytoplankton productivity, demonstrating top-down control of this system. Experiments conducted in salt marshes tested the interactive effects of climate-induced drought and snail grazing on plant biomass and trophic structure (Silliman et al. 2005). These researchers concluded that synergistic interactions between physicochemical stressors and trophic dynamics amplified top-down effects and that were likely responsible for massive die-offs of salt marshes in the southeastern United States.

Although most research on contaminant-induced alterations in trophic structure has been limited to aquatic systems, a few investigators have examined food web responses in soil communities (Parker et al. 1984, Parmelee et al. 1993, 1997). Organisms inhabiting soil communities control important ecosystem processes such as decomposition and mineralization. Because soil microfaunal communities are often naturally diverse and consist of large numbers of organisms (10^4–10^6 per m^2), effects on multiple species at several trophic levels can be investigated at ecologically realistic spatial and temporal scales (Parmelee et al. 1993). When integrated with microbial assays and measures of functional characteristics (soil respiration, decomposition), soil microcosm experiments provide a relatively complete assessment of contaminant effects.

A major challenge of working with soil fauna, especially meiofauna, is the difficulty identifying certain taxonomic groups. This limitation can be partially resolved by categorizing organisms based on guilds or functional feeding groups (e.g., fungivore, bacterivore, herbivore, and omnivore-predator) instead of relying on traditional taxonomic measures. Parmelee et al. (1993) observed considerable variation in sensitivity among trophic groups to contaminants, and noted that reduced abundance of predators resulted in increased abundance of herbivores (Figure 27.6). Similar results were reported in desert soil communities treated with insecticides (Parker et al. 1984). The large amount of variation among feeding groups demonstrates the importance of understanding trophic structure when assessing ecological impacts on soil communities.

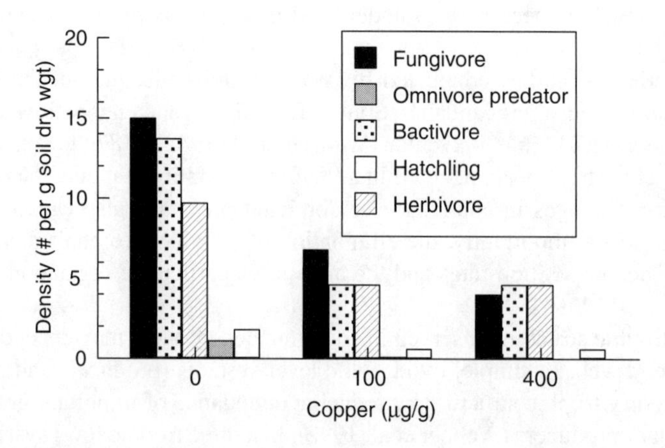

FIGURE 27.6 Responses of nematode functional groups to copper in soil microcosms. (Modified from Figure 1 in Parmelee et al. (1997).)

Finally, it should be noted that complete removal of species is not necessary to disrupt the structure and function of food webs. Significantly reduced abundance of predators or prey could influence lower or higher trophic levels. As noted above, many vertebrate predators are capable of switching to alternative prey when abundance of preferred prey drops below a particular threshold. Because contaminants influence either the ability of predators to capture prey or the ability of prey to escape predation (Chapter 21), species-specific differences in relative sensitivity will ultimately influence how food webs are affected by these stressors.

27.4 SUMMARY

In summary, a significant amount of empirical and theoretical research has been devoted to understanding the structure and function of food webs in basic ecology. The historical application of food web theory in ecotoxicology has been limited primarily to predicting the flux of contaminants through food chains and estimating potential exposure to top predators. More recently, ecotoxicologists have developed a better understanding of how ecological characteristics (e.g., primary production, food chain length, and trophic complexity) influence transport of contaminants through food chains. However, there has been relatively little effort devoted to predicting the direct effects of contaminants on food chain structure or using food web characteristics as endpoints in ecotoxicological assessments. We see this as an important area for future research in community ecotoxicology. If food is the "burning question" in animal communities as suggested by Elton (1927), the field of ecotoxicology would benefit from a greater understanding of how contaminants directly and indirectly influence food chain structure.

27.4.1 Summary of Foundation Concepts and Paradigms

- An understanding of food webs is critical to the study of ecotoxicology because trophic interactions provide the fundamental linkages among species that determine the structure of terrestrial and aquatic communities.
- Contaminant transport models used in ecotoxicology are analogous to energy flow models derived from the basic ecological literature.
- An understanding of the ecological factors that determine energy flow in communities, such as food chain length, interaction strength, and connectedness, are also necessary to quantify contaminant fate and effects.
- Food webs depicted in the contemporary ecological literature fall into three general categories: descriptive, interactive, and energetic.
- Contemporary research in food web ecology has focused on two key topics: (1) identifying factors that limit the number of trophic levels; and (2) quantifying the strength of species interactions.
- The strength of interactions within a food chain may influence community stability; however, because of the lack of experimental studies and the different approaches employed by theoretical and empirical ecologists to measure interaction strength, the relationship between stability and energy flow is uncertain.
- Experimental manipulations of food webs provide the most direct tests of the relationships among trophic structure, productivity, and disturbance.
- The trophic cascade hypothesis predicts that each trophic level in a community is influenced by trophic levels directly above (e.g., consumers) and directly below (e.g., resources).
- Although there has been strong support for the trophic cascade hypothesis in lakes, the generality of this hypothesis and the relative importance of "top-down" and "bottom-up" effects in terrestrial systems have been subjects of considerable debate.

- Grouping organisms into broad trophic categories has facilitated the development of mathematical models for estimating energy flow; however, it is important to realize that this representation of food webs is greatly oversimplified.
- One of the most significant methodological developments in the study of food webs is the application of stable isotopes, particularly $^{13}C/^{12}C$ and $^{15}N/^{14}N$, to characterize feeding habits and quantify energy flow.
- Distinguishing the direct effects of toxic chemicals on abundance of organisms from the indirect effects resulting from changes in abundance of predators or prey is of critical importance for understanding how food webs respond to anthropogenic stressors.
- Because species interactions in natural communities are often subtle, experimental manipulation of both contaminants and consumers may be necessary to understand consumer-resource dynamics.
- Bioenergetic food web approaches are an important step in linking observed toxicological effects on population growth and production to ecologically significant responses at higher levels of organization.

REFERENCES

Adams, J.S., Knight, R.L., McEwen, L.C., and George, T.L., Survival and growth of nestling Vesper Sparrows exposed to experimental food reductions, *Condor*, 96, 739–748, 1994.

Begon, M., Harper, J.L., and Townsend, C.R., *Ecology: Individuals, Populations and Communities*, Blackwell Scientific Publications, Cambridge, MA, 1990.

Bengtsson, G., Hansson, L.A., and Montenegro, K., Reduced grazing rates in *Daphnia pulex* caused by contaminants: Implications for trophic cascades, *Environ. Toxicol. Chem.*, 23, 2641–2648, 2004.

Benke, A.C. and Wallace, J.B., Trophic basis of production among riverine caddisflies: Implications for food web analysis, *Ecology*, 78, 1132–1145, 1997.

Bentzen, E., Lean, D.R.S., Taylor, W.D., and Mackay, D., Role of food web structure on lipid and bioaccumulation of organic contaminants by lake trout (*Salvelinus namycush*), *Can. J. Fish. Aquat. Sci.*, 53, 2397–2407, 1996.

Beyers, D.W., Rice, J.A., and Clements, W.H., Evaluating biological significance of chemical exposure to fish using a bioenergetics-based stressor-response model, *Can. J. Fish. Aquat. Sci.*, 56, 823–829, 1999a.

Beyers, D.W., Rice, J.A., Clements, W.H., and Henry C.J., Estimating physiological cost of chemical exposure: Integrating energetics and stress to quantify toxic effects in fish, *Can. J. Fish. Aquat. Sci.*, 56, 814–822, 1999b.

Brett, M.T. and Goldman, C.R., A meta-analysis of the freshwater trophic cascade, *Proc. Natl. Acad. Sci. USA*, 93, 7723–7726, 1996.

Brett, M.T. and Goldman, C.R., Consumer versus resource control in freshwater pelagic food webs, *Science*, 275, 384–386, 1997.

Briand, F. and Cohen, J.E., Environmental correlates of food chain length, *Science*, 238, 956–960, 1987.

Carlisle, D.M., Bioenergetic food webs as a means of linking toxicological effects across scales of ecological organization, *J. Aquat. Ecosys. Stress Recov.*, 7, 155–165, 2000.

Carman, K.R., Fleeger, J.W., and Pomarico, S.M., Response of a benthic food web to hydrocarbon contamination, *Limnol. Oceanogr.*, 42, 561–571, 1997.

Carpenter, S.R. and Kitchell, J.F., *The Trophic Cascade in Lakes*, Cambridge University Press, Cambridge, UK, 1993.

Carson, R., *Silent Spring*, Houghton Mifflin, Boston, MA, 1962.

Chen, B. and Wise, D.H., Bottom-up limitation of predaceous arthropods in a detritus based terrestrial food web, *Ecology*, 80, 761–772, 1999.

Clements, W.H., Carlisle, D.M., Lazorchak, J.M., and Johnson, P.C., Heavy metals structure benthic communities in Colorado mountain streams, *Ecol. Appl.*, 10, 626–638, 2000.

Clements, W.H. and Rees, D.E., Effects of heavy metals on prey abundance, feeding habits, and metal uptake of brown trout in the Arkansas River, Colorado, *Trans. Am. Fish. Soc.*, 126, 774–785, 1997.

Colinvaux, P., *Why Big Fierce Animals are Rare. An Ecologists Perspective*, University Press, Princeton, NJ, 1978.

Croll, D.A., Maron, J.L., Estes, J.A., Danner, E.M., and Byrd, G.V., Introduced predators transform subarctic islands from grassland to tundra, *Science*, 307, 1959–1961, 2005.

de Ruiter, P.C., Neutel, A., and Moore, J.C., Energetics, patterns of interaction strengths, and stability in real ecosystems, *Science*, 269, 1257–1260, 1995.

Elton, C., *Animal Ecology*, Macmillan, New York, 1927.

Estes, J.A., Tinker, M.T., Williams, T.M., and Doak, D.F., Killer whale predation on sea otters linking oceanic and nearshore ecosystems, *Science*, 282, 473–476, 1998.

Fairchild, J.F., La Point, T.W., Zajicek, J.L., Nelson, M.K., Dwyer, F.J., and Lovely, P.A., Population-, community- and ecosystem-level responses of aquatic mesocosms to pulsed doses of a pyrethroid insecticide, *Environ. Toxicol. Chem.*, 11, 115–129, 1992.

Frank, K.T., Petrie, B., Choi, J.S., and Leggett, W.C., Trophic cascades in a formerly cod-dominated ecosystem, *Science*, 308, 1621–1623, 2005.

Fretwell, S.D., Food chain dynamics: The central theory of ecology? *Oikos*, 50, 291–301, 1987.

Hairston, N.G., Jr. and Hairston, N.G., Sr., Cause-effect relationships in energy flow, trophic structure, and interspecific interactions, *Am. Nat.*, 142, 379–411, 1993.

Hairston, N.G., Sr., Smith, F.E., and Slobodkin, L.B., Community structure, population control, and interspecific competition, *Am. Nat.*, 94, 421–425, 1960.

Halpern, B.S., Cottenie, K., and Broitman, B.R., Strong top-down control in southern California kelp forest ecosystems, *Science*, 312, 1230–1232, 2006.

Hanratty, M.P. and Stay, F.S., Field-evaluation of the littoral ecosystem risk assessment models predictions of the effects of chlorpyrifos, *J. Appl. Ecol.*, 31, 439–453, 1994.

Howe, F.P., Knight, R.L., McEwen, L.C., and George, T.L., Direct and indirect effects of insecticide applications on growth and survival of nestling passerines, *Ecol. Appl.*, 6, 1314–1324, 1996.

Hunter, M.D. and Price, P.W., Playing chutes and ladders: Heterogeneity and the relative roles of bottom-up and top-down forces in natural communities, *Ecology*, 73, 724–732, 1992.

Jeffries, M.J. and Lawton, J.H., Predator-prey ratios in communities of freshwater invertebrates: The role of enemy free space, *Freshw. Biol.*, 15, 105–112, 1985.

Jeppesen, E., Sondergaard, M., Jensen, J.P., Mortensen, E., Hansen, A.M., and Jorgensen, T., Cascading trophic interactions from fish to bacteria and nutrients after reduced sewage loading: An 18-year study of a shallow hypertrophic lake, *Ecosystems*, 1, 250–267, 1998.

Kitchell, J.F., *Food Web Management—A Case Study of Lake Mendota*, Springer-Verlag, New York, 1992.

Koivisto, S., Arner, M., and Kautsky, N., Does cadmium pollution change trophic interactions in rockpool food webs? *Environ. Toxicol. Chem.*, 16, 1330–1336, 1997.

Lindeman, R.L., The trophic-dynamic aspect of ecology, *Ecology*, 23, 399–418, 1942.

May, R.M., Patterns in multi-species communities, In *Theoretical Ecology*, May, R.M. (ed.), Blackwell Scientific Publications, Boston, 1981, pp. 197–227.

McLaren, B.E. and Peterson, R.O., Wolves, moose, and tree rings on Isle Royale, *Science*, 266, 1555–1558, 1994.

Menge, B.A., Community regulation: Under what conditions are bottom-up factors important on rocky shores? *Ecology*, 73, 755–765, 1992.

Micheli, F., Eutrophication, fisheries, and consumer-resource dynamics in marine pelagic ecosystems, *Science*, 285, 1396–1398, 1999.

Nakano, S., Miyasaka, H., and Kuhara, N., Terrestrial-aquatic linkages: Riparian arthropod inputs alter trophic cascades in a stream food web, *Ecology*, 80, 2435–2441, 1999.

Odum, E.P., Energy flow in ecosystems: A historical review, *Am. Zool.*, 8, 11–18, 1968.

Odum, E.P., The strategy of ecosystem development, *Ecology*, 164, 262–270, 1969.

Odum, E.P., Trends expected in stressed ecosystems, *BioScience*, 35, 419–422, 1985.

Paine, R.T., Food webs: Linkage, interaction strength and community infrastructure, *J. Anim. Ecol.*, 49, 667–685, 1980.

Paine, R.T., Tegner M.J., and Johnson E.A., Compounded perturbations yield ecological surprises, *Ecosystems*, 1, 535–545, 1998.

Parker, L.W., Santos, P.F., Phillips, J., and Whitford, W.G., Carbon and nitrogen dynamics during the decomposition of litter and roots of a Chihuahuan desert annual, *Lepidium lasiocarpum*, *Ecol. Monogr.*, 54, 339–360, 1984.

Parmelee, R.W., Wentsel, R.S., Phillips, C.T., Simini, M., and Checkai, R.T., Soil microcosm for testing effects of chemical pollutants on soil fauna communities and trophic structure, *Environ. Toxicol. Chem.*, 12, 1477–1486, 1993.

Parmelee, R.W., Phillips, C.T., Checkai, R.T., and Bohlen, P.J., Determining the effects of pollutants on soil faunal communities and trophic structure using a refined microcosm system, *Environ. Toxicol. Chem.*, 16, 1212–1217, 1997.

Peterle, T.J., *Wildlife Toxicology*, Van Nostrand Reinhold, New York, 1991.

Pimm, S.L., *Food Webs*, Chapman and Hall, London, 1982.

Pimm, S.L., Lawton, J.H., and Cohen, J.E., Food web patterns and their consequences, *Nature*, 350, 669–674, 1991.

Polis, G.A., Complex trophic interactions in deserts: An empirical critique of food-web theory, *Am. Nat.*, 138, 123–155, 1991.

Powell, G.V.N., Reproduction by an altricial songbird, the red-winged blackbird, in fields treated with the organophosphate insecticide fenthion, *J. Appl. Ecol.*, 21, 83–95, 1984.

Power, M.E., Effects of fish in river food webs, *Science*, 250, 811–814, 1990.

Rasmussen, J.B., Rowan, D.J., Lean, D.R.S., and Carey, J.H., Food chain structure in Ontario lakes determines PCB levels in lake trout (*Salvelinus namaycush*) and other pelagic fish, *Can. J. Fish. Aquat. Sci.*, 47, 2030–2038, 1990.

Ricklefs, R.E., *Ecology*, 3rd ed., W.H. Freeman and Company, New York, 1990.

Riddell, D.J., Culp, J.M., and Baird, D.J., Behavioral responses to sublethal cadmium exposure within an experimental aquatic food web, *Environ. Toxicol. Chem.*, 24, 431–441, 2005.

Salminen, J., Korkama, T., and Strommer, R., Interaction modification among decomposers impairs ecosystem processes in lead-polluted soil, *Environ. Toxicol. Chem.*, 21, 2301–2309, 2002.

Schauber, E.M., Edge, W.D., and Wolff, J.O., Insecticide effects on small mammals: Influence of vegetation structure and diet, *Ecol. Appl.*, 7, 143–157, 1997.

Schindler, D.W., Mills, K.H., Malley, D.F., Findlay, D.L., Shearer, J.A., Davies, I.J., Turner, M.A., Linsey, G.A., and Cruikshank, D.R., Long-term ecosystem stress: The effects of years of experimental acidification on a small lake, *Science*, 228, 1395–1401, 1985.

Sibert, J., Hampton, J., Kleiber, P., and Maunder, M., Biomass, size, and trophic status of top predators in the Pacific Ocean, *Science*, 314, 1773–1776, 2006.

Silliman, B.R., van de Koppel, J., Bertness, M.D., Stanton, L.E., and Mendelssohn, I.A., Drought, snails, and large-scale die-off of southern U.S. salt marshes, *Science*, 310, 1803–1806, 2005.

Stein, R.A., DeVries, D.R., and Dettmers, J.M., Food-web regulation by a planktivore: Exploring the generality of the trophic cascade hypothesis, *Can. J. Fish. Aquat. Sci.*, 52, 2518–2526, 1995.

Stemberger, R.S. and Chen, C.Y., Fish tissue metals and zooplankton assemblages of northeastern lakes, *Can. J. Fish. Aquat. Sci.*, 55, 339–352, 1998.

Strong, D.R., Are trophic cascades all wet? Differentiation and donor-control in speciose ecosystems, *Ecology*, 73, 747–754, 1992.

Vander–Zanden, M.J., Casselman, J.M., and Rasmussen, J.B., Stable isotope evidence for the food web consequences of species invasions in lakes, *Nature*, 401, 464–467, 1999.

Wallace, J.B., Cuffney, T.F., Lay, C.C., and Vogel, D., The influence of an ecosystem-level manipulation on prey consumption by a lotic dragonfly, *Can. J. Zool.*, 65, 35–40, 1987.

Wallace, J.B., Eggert, S.L., Meyer, J.L., and Webster, J.R., Multiple trophic levels of a forested stream linked to terrestrial litter inputs, *Science*, 277, 102–104, 1997.

Wallace, J.B., Lugthart, G.J., Cuffney, T.F., and Schurr, G.A., The impact of repeated insecticidal treatments on drift and benthos of a headwater stream, *Hydrobiologia*, 179, 135–147, 1989.

Wallace, J.B., Webster, J.R., and Cuffney, T.F., Stream detritus dynamics: Regulation by invertebrate consumers, *Oecologia*, 53, 197–200, 1982.

Webber, E.C., Deutsch, W.G., Bayne, D.R., and Seesock, W.C., Ecosystem-level testing of a synthetic pyrethroid insecticide in aquatic mesocosms, *Environ. Toxicol. Chem.*, 11, 87–105, 1992.

Wipfli, M.S. and Merritt, R.W., Disturbance to a stream food web by a bacterial larvicide specific to black flies: Feeding responses of predatory macroinvertebrates, *Freshw. Biol.*, 32, 91–103, 1994.

Wong, A.H.K., McQueen, D.J., Williams, D.D., and Demers, E., Transfer of mercury from benthic invertebrates to fishes in lakes with contrasting fish community structures, *Can. J. Fish. Aquat. Sci.*, 54, 1320–1330, 1997.

Wootton, J.T., Parker, M.S., and Power, M.E., Effects of disturbance on river food webs, *Science*, 273, 1558–1561, 1996.

28 Conclusions

28.1 GENERAL

The study of community ecology is primarily concerned with understanding how biotic and abiotic factors influence patterns of distribution, abundance, and species diversity. Deriving mechanistic explanations for patterns observed in nature, including the log normal distribution of abundance, species–area relationships, and the spatial changes that occur across latitudinal or elevational gradients, has occupied the attention of community ecologists for several decades. Our treatment of community ecotoxicology has attempted to explain how communities will respond to one major abiotic factor: contaminants. Direct effects of contaminants may result in reduction or elimination of local populations and are generally easier to interpret than indirect effects. An autecological understanding of life history characteristics and species-specific sensitivity to contaminants may be sufficient for predicting direct effects on populations. In contrast, indirect effects of contaminants are often subtle and difficult to predict without conducting manipulative experiments. Our limited understanding of the indirect effects of contaminants is surprising, especially given the prominent role that research on species interactions (e.g., competition, predation, and mutualism) has played in basic community ecology. If ecotoxicologists accept the notion that species interactions are important in regulating natural communities, then focused research should be directed at understanding the influence of contaminants on these interactions.

28.2 SOME PARTICULARLY KEY CONCEPTS

28.2.1 Improvements in Experimental Techniques

One of the greatest challenges in community ecotoxicology is separating contaminant-induced changes in species diversity and composition from variation owing to natural factors. While observational studies can provide critical support for hypotheses concerning relationships between contaminants and community responses, some researchers feel that experiments are the only way to demonstrate causation. The lack of true replication and random assignment of treatments to experimental units limits the use of inferential statistics in most observational studies. The recent emphasis on experimental approaches for assessing the effects of contaminants is seen as a major development in community ecotoxicology. If the ability to test hypotheses with critical experiments truly defines the maturity of a science (Popper 1972), we suggest that experimental approaches should play an increasingly important role in ecotoxicology.

Our enthusiasm for experimental approaches is somewhat tempered by recognition of the important tradeoffs between replication and ecological realism. Some ecologists argue that the degree of control afforded by small-scale experiments does not compensate for the lack of realism (Carpenter 1996). We feel that the importance of spatiotemporal considerations for predicting how communities will respond to contaminants should be treated like any other scientific hypothesis (Resetarits and Fauth 1998). Thus, an emerging area of research in microcosm and mesocosm testing is to identify those key ecological processes that must be accurately reproduced in order to have an adequate representation of nature.

Another major research goal for community ecotoxicology should be to determine the context dependency of community responses to contaminants. Because communities from different locations will vary naturally in diversity, abundance, and history of disturbance, effects of stressors on these communities are also likely to differ. Experimental approaches may be the only way to investigate context-dependent responses to contaminants. Relatively simple experimental designs allow researchers to manipulate several variables simultaneously and investigate the importance of direct and indirect effects. The U.S. Environmental Protection Agency (EPA)'s decision to abandon mesocosm testing for pesticide registration is seen by many as a missed opportunity to improve our understanding of context-dependent responses and indirect effects (Pratt et al. 1997, Taub 1997).

Finally, while improvements in experimental approaches have strengthened our ability to demonstrate causation, ecotoxicological experiments should not be conducted without an appreciation of natural history or in isolation from underlying ecological theory. We do not consider descriptive and experimental approaches as opposing ends of a continuum, but rather advocate a research program in which well-designed experiments are integrated with observational and theoretical approaches. When small-scale experiments are linked with observational studies, ecotoxicologists will gain a more realistic understanding of how communities respond to contaminants and other stressors.

28.2.2 Use of Multimetric and Multivariate Approaches to Assess Community-Level Responses

Community-level data used to assess effects of contaminants range from simple lists that reflect the presence or absence of species to more sophisticated compilations that include abundance, trophic structure, life history characteristics, and measures of species-specific sensitivity. Because the occurrence of an indicator species is influenced by numerous factors other than contaminants, presence–absence data alone are insufficient for assessing all but the most severe forms of pollution. Although there has been significant progress in biomonitoring research since the development of the Saprobien system of classification, most community-level assessments rely on the assumption that species vary in their sensitivity to a particular stressor and that community responses will reflect this variation. Multimetric and multivariate approaches are particularly useful for community-level studies because they reduce the typically complex, multidimensional data to readily interpretable patterns. Unfortunately, the complex and often unwieldy statistical algorithms of many multivariate approaches are considered major obstacles to their widespread application. The new generation of software packages designed to perform multivariate analyses has increased the use of these approaches; however, the widespread availability of "point-and-click" software does not eliminate the obligation of users to fully understand the output. Multimetric approaches are computationally simple, but are data-intensive and often require a comprehensive understanding of ecology and natural history. These approaches have been especially effective for assessing impacts of contaminants in aquatic systems. The development of multimetric indices for other taxonomic groups, particularly for those in terrestrial communities, is seen as an important research need.

Although they were developed independently, multimetric and multivariate approaches are complementary and can be used together to assess biological integrity (Reynoldson 1997). For example, output from multivariate analyses could be used to identify sensitive metrics in a multimetric index. Conversely, multivariate analyses could be conducted using traditional metrics from a multimetric index. By selecting metrics that respond to different classes of stressors, results of multivariate analyses may be useful for identifying specific stressors in systems receiving multiple perturbations.

28.2.3 Disturbance Ecology and Community Ecotoxicology

One of the most significant contributions of basic ecology to ecotoxicology is the application of disturbance theory in the study of community responses to contaminants. Disturbance is considered a major regulator of community structure and has been the subject of intense debate for several decades. Assuming that responses to natural and anthropogenic disturbance are somewhat analogous, theoretical and empirical studies of resistance and resilience may help ecotoxicologists predict effects of contaminants. Rapport's Ecosystem Distress Syndrome (Rapport et al. 1985) provides an important framework for understanding how communities respond to and recover from natural and anthropogenic disturbances.

Basic research in disturbance ecology may also help explain the significant variation in responses to contaminants observed among communities. If diversity in some communities is enhanced under moderate levels of contaminant stress, as predicted by the intermediate disturbance hypothesis, we would not expect that concentration–response relationships between contaminants and species diversity to be linear. In addition, communities subjected to natural disturbance may be preadapted to anthropogenic stressors, thus reducing their sensitivity to contaminants. The ability of communities to tolerate contaminants forms the basis of the pollution-induced community tolerance (PICT) hypothesis (Blanck and Wangberg 1988). In contrast to community composition, which can vary significantly owing to natural factors, increased community tolerance for a particular contaminant is considered to be a direct result of exposure.

It is likely that the same biological factors that determine the rate of recovery from natural disturbance will also determine how quickly communities recover from exposure to contaminants. Thus, an understanding of colonization abilities of dominant species, proximity of colonists, and life history characteristics may help ecotoxicologists predict recovery times following remediation. Because of natural changes in community composition and species turnover over time, there remains considerable uncertainty in our ability to identify specific endpoints of recovery. For example, we know that some measures of recovery, such as number of species, can quickly return to predisturbance conditions despite persistent differences in community composition. Similarly, because of redundancy in many communities, functional measures may recover faster than structural measures. Our understanding of recovery is further complicated by the uncritical acceptance of equilibrium theories and our failure to recognize the role of historic events. The community-conditioning hypothesis (Landis et al. 1996, Matthews et al. 1996) acknowledges that communities are a reflection of their unique history. Because this history may include exposure to contaminants and other anthropogenic disturbances, traditional models of recovery based on equilibrium conditions may not apply.

Finally, throughout this section we have attempted to make a strong case for the importance of long-term research. The National Science Foundation's Long Term Ecological Research (LTER) programs have contributed significantly to our understanding of basic ecology. Unfortunately, there are relatively few examples where long-term studies have been conducted to assess recovery from contaminants. Bruce Wallace's research on responses of headwater streams to pesticides (Wallace et al. 1982) and David Schindler's long-term research on acidification in lakes (Schindler 1988) described in Chapter 23 are two of the more prominent examples. We suggest that a national program monitoring responses to anthropogenic disturbance, analogous to NSF's LTER program would greatly enhance our understanding of biotic and abiotic factors that determine recovery.

28.2.4 An Improved Understanding of Trophic Interactions

Few topics in basic community ecology are as relevant to understanding ecotoxicological effects of contaminants as food webs and trophic interactions. Quantitative approaches used by ecologists to measure energy flow have been modified to estimate potential transport of contaminants. The primary focus of this effort has been on measuring the concentrations of contaminants in organisms and attempting to quantify contaminant transport among biotic and abiotic compartments. What is generally lacking from many ecotoxicological investigations has been a critical understanding of the ecological factors that influence contaminant transport. In addition to information on physiochemical properties of contaminants (e.g., molecular structure, lipophilicity), ecotoxicologists now realize that predicting contaminant transport also requires an understanding of ecological characteristics (e.g., feeding habits, food chain length and complexity, and habitat use). Our ability to quantify the importance of ecological factors on contaminant transport has been greatly improved by the application of stable isotopes analyses. Few advances in the study of food webs have had as great an impact on our understanding of feeding relationships. Time-integrated estimates of energy flow, trophic position, and carbon sources can be obtained by comparing the unique isotopic signatures of consumers and resources.

Despite the broad interest in quantifying contaminant transport in ecotoxicological investigations, relatively few studies have considered the effects of contaminants on trophic structure or have used food web characteristics as endpoints in assessments of ecological integrity. We feel that understanding the effects of contaminants on food web length, complexity, and trophic structure is a significant research need in ecotoxicology. The ecological effects of contaminants on trophic structure and the transport of contaminants to higher trophic levels is dependent on the number of trophic levels and whether the system is regulated by top-down or bottom-up factors. These ecological factors have important applications for the management of sport fisheries. Because contaminant transfer is greatly influenced by food chain length and other aspects of trophic structure (Rasmussen et al. 1990), size-selective stocking and other fisheries management programs may influence contaminant levels in game species (Jackson 1997).

The most important research limitation in food web ecotoxicology has been the inability to relate concentrations of contaminants measured in different trophic levels with biologically important effects. Bioenergetic approaches may provide the conceptual framework to quantify biologically significant responses associated with contaminant uptake. Because energy is a common currency that unifies all biological systems (Carlisle 2000), studying the effects of contaminants on energy flow provides an opportunity to integrate responses across levels of organization. Integrating contaminant transport models with bioenergetic models will allow researchers to link exposure with ecologically significant effects.

28.2.5 Interactions between Contaminants and Global Atmospheric Stressors

Although generally not included in discussions of ecotoxicology, global climate change, increased UV-B radiation, and acidic deposition represent three of the most serious threats to ecological communities. Assessing the direct effects of these stressors is complicated by their large geographic extent, which requires extrapolation across broad spatial and temporal scales. Evidence that global climate change, increased UV-B radiation, and acidification are directly related to anthropogenic emissions has been obtained from a variety of sources. However, the direct and indirect effects on aquatic and terrestrial communities are largely uncertain. The coarse spatial scale of most general circulation models (GCMs) limits our ability to accurately predict regional responses to climate change. For example, although global declines of amphibians have been related to increased levels of UV-B (Blaustein and Wake 1990) and the worldwide degradation of forest health has been attributed to acidification (Ollinger et al. 1993), there is tremendous uncertainty in these relationships.

Because of the pervasive and widespread distribution of global atmospheric stressors, ecotoxicologists cannot continue to study the effects of contaminants in isolation. Interactions between atmospheric and local stressors complicate our ability to predict effects of climate change, UV-B, and acidification on communities. It is quite likely that increased temperatures will have a significant influence on physiochemical characteristics of contaminants and also influence community responses to contaminants. In addition, the structure of communities and their susceptibility to contaminants will most likely change in a warmer climate. Synergistic interactions between UV-B radiation and polycyclic aromatic hydrocarbons (Oris and Giesy 1986) and between acidification and heavy metals (Genter 1995) are well documented. To predict effects of contaminants on communities simultaneously subjected to increased temperature, greater UV-B radiation, and/or increased acidification will require a better appreciation for the impacts of multiple stressors. Currently, our ability to predict interactive effects of multiple stressors on natural systems is greatly limited. It is likely that interactions among stressors will be common in natural systems, and therefore, community ecotoxicologists should anticipate the "ecological surprises" resulting from these interactions (Paine et al. 1998).

28.3 SUMMARY

We hope that our treatment of community ecotoxicology has convinced the reader of the importance of understanding how contaminants may affect distribution, abundance, and species diversity. Within the context of the hierarchical arrangement of living systems, communities are intermediate between populations and ecosystems. Although the responses of individual species to contaminants will influence patterns of diversity and abundance in nature, community responses often transcend those observed in populations. Recognition of the emergent properties of communities and higher levels of biological organization remains a significant point of contention between proponents of reductionism and holism (Odum 1984). The emergence of ecotoxicology as a distinct discipline within the field of toxicology was at least partially a result of criticism of traditional reductionist approaches such as laboratory toxicity tests (Cairns 1983, 1986). Until ecotoxicologists develop a better appreciation for the importance of species interactions and indirect effects of contaminants, extrapolation of laboratory results based on responses of single species will remain tenuous.

Predicting the indirect effects of contaminants on species interactions and trophic structure has been a major theme in our discussion of community ecotoxicology. Experience has shown that population surveys of charismatic or economically important species often provide an incomplete picture of how communities will respond to or recover from anthropogenic disturbance (Wiens 1996). Similarly, functional characteristics such as primary productivity, nutrient cycling, and detritus processing, the endpoints typically included in ecosystem-level studies, are often less sensitive to anthropogenic stressors (Schindler 1987). Because the endpoints evaluated in community-level assessments are generally sensitive, ecologically significant, and socially relevant, communities are an appropriate focus for ecotoxicological investigations.

28.3.1 SUMMARY OF FOUNDATION CONCEPTS AND PARADIGMS

- The study of community ecology is primarily concerned with understanding how biotic and abiotic factors influence patterns of distribution, abundance, and species diversity.
- Direct effects of contaminants may result in reduction or elimination of local populations and are generally easier to interpret than indirect effects. In contrast, indirect effects of contaminants are often subtle and difficult to predict without conducting manipulative experiments.
- One of the greatest challenges in community ecotoxicology is separating contaminant-induced changes in species diversity and community composition from variation owing to natural factors.
- The recent emphasis on experimental approaches such as microcosms and mesocosms for assessing the effects of contaminants is seen as a major development in community ecotoxicology.
- An emerging area of research in microcosm and mesocosm testing is to identify key ecological processes that must be accurately reproduced in order to have an adequate representation of nature.
- Because communities from different locations will vary naturally in diversity, abundance, and history of disturbance, effects of stressors on these communities are also likely to differ.

- Although the development of more ecologically realistic experimental approaches has strengthened our ability to demonstrate causation, these experiments should not be conducted without an appreciation of natural history or in isolation from underlying ecological theory.
- Multimetric and multivariate approaches are particularly useful for community-level studies because they reduce the typically complex, multidimensional data to readily interpretable patterns.
- Assuming that responses to natural and anthropogenic disturbance are somewhat analogous, theoretical and empirical studies of resistance and resilience may help ecotoxicologists predict effects of contaminants.
- An understanding of colonization abilities of dominant species, proximity of colonists, and life history characteristics may help ecotoxicologists predict recovery times following remediation.
- Despite the broad interest in quantifying contaminant transport in ecotoxicological investigations, few studies have considered the effects of contaminants on trophic structure or have used food web characteristics as endpoints in assessments of ecological integrity.
- Bioenergetic approaches provide the conceptual framework to relate concentrations of contaminants measured in different trophic levels with biologically important effects.
- Because of the pervasive and widespread distribution of global atmospheric stressors such as CO_2, UV-B radiation, and acidification, ecotoxicologists cannot continue to study the effects of contaminants in isolation.
- Interactions between atmospheric and local stressors complicate our ability to predict effects of climate change, UV-B, and acidification on communities.

REFERENCES

Blanck, H. and Wangberg, S., Induced community tolerance in marine periphyton established under arsenate stress, *Can. J. Fish. Aquat. Sci.*, 45, 1816–1819, 1988.

Blaustein, A.R. and Wake, D.B., Declining amphibian populations: A global phenomenon? *Trends Ecol. Evol.*, 5, 203–204, 1990.

Cairns, J., Jr., The myth of the most sensitive species, *BioScience*, 36, 670–672, 1986.

Cairns, J., Jr., Are single species toxicity tests alone adequate for estimating environmental hazard? *Hydrobiologia*, 100, 47–57, 1983.

Carlisle, D.M., Bioenergetic food webs as a means of linking toxicological effects across scales of ecological organization, *J. Aquat. Ecosys. Stress Recov.*, 7, 155–165, 2000.

Carpenter, S.R., Microcosm experiments have limited relevance for community and ecosystem ecology, *Ecology*, 77, 677–680, 1996.

Genter, R.B., Benthic algal populations respond to aluminum, acid, and aluminum-acid mixtures in artificial streams, *Hydrobiologia*, 306, 7–19, 1995.

Jackson, L.J., Piscivores, predation, and PCBs in Lake Ontario's pelagic food web, *Ecol. Appl.*, 7, 991–1001, 1997.

Landis, W.G., Matthews, R.A., and Matthews, G.B., The layered and historical nature of ecological systems and the risk assessment of pesticides, *Environ. Toxicol. Chem.*, 15, 432–440, 1996.

Matthews, R.A., Landis, W.G., and Matthews, G.B., The community conditioning hypothesis and its application to environmental toxicology, *Environ. Toxicol. Chem.*, 15, 597–603, 1996.

Odum, E.P., The mesocosm, *Bioscience*, 34, 558–562, 1984.

Ollinger, S.V., Aber, J.D., Lovett, G.M., Millham, S.E., Lathrop, R.G., and Ellis, J.E., A spatial model of atmospheric deposition for the northeastern U.S., *Ecol. Appl.*, 3, 459–472, 1993.

Oris, J.T. and Giesy, J.P., Jr., Photoinduced toxicity of anthracene to juvenile bluegill sunfish (*Lepomis macrochirus* rafinesque): Photoperiod effects and predictive hazard evaluation, *Environ. Toxicol. Chem.*, 5, 761–768, 1986.

Paine, R.T., Tegner, M.J., and Johnson, E.A., Compounded perturbations yield ecological surprises, *Ecosystems*, 1, 535–545, 1998.

Popper, K.R., *The Logic of Scientific Discovery*, 3rd ed., Hutchinson, London, England, 1972.

Pratt, J.R., Melendez, A.E., Barreiro, R., and Bowers, N.J., Predicting the ecological effects of herbicides, *Ecol. Appl.*, 7, 1117–1124, 1997.

Rapport, D.J., Regier, H.A., and Hutchinson, T.C., Ecosystem behavior under stress, *Am. Nat.*, 125, 617–640, 1985.

Rasmussen, J.B., Rowan, D.J., Lean, D.R.S., and Carey, J.H., Food chain structure in Ontario lakes determines PCB levels in lake trout (*Salvelinus namaycush*) and other pelagic fish, *Can. J. Fish. Aquat. Sci.*, 47, 2030–2038, 1990.

Resetarits, W.J., Jr. and Fauth, J.E., From cattle tanks to Carolina bays: The utility of model systems for understanding natural communities, In *Experimental Ecology: Issues and Perspectives*, Resetarits, W.H., Jr. and Bernardo, J. (eds.), Oxford University Press, Inc., New York, 1998, pp. 133–151.

Reynoldson, T.B., Norris, R.H., Resh, V.H., Day, K.E., and Rosenberg, D.M., The reference condition: A comparison of multimetric and multivariate approaches to assess water-quality impairment using benthic macroinvertebrates, *J. N. Am. Benthol. Soc.*, 16, 833–852, 1997.

Schindler, D.W., Detecting ecosystem responses to anthropogenic stress, *Can. J. Fish. Aquat. Sci.*, (Suppl.), 6–25, 1987.

Schindler, D.W., Experimental studies of chemical stressors on whole lake ecosystems, *Verh. Internat. Verein. Limnol.*, 23, 11–41, 1988.

Taub, F.B., Are ecological studies relevant to pesticide registration decisions? *Ecol. Appl.*, 7, 1083–1085, 1997.

Wallace, J.B., Webster, J.R., and Cuffney, T.F., Stream detritus dynamics: Regulation by invertebrate consumers, *Oecologia*, 53, 197–200, 1982.

Washington, H.G., Diversity, biotic and similarity indices. A review with special relevance to aquatic ecosystems, *Water Res.*, 18, 653–94, 1984.

Wiens, J.A., Crist, T.O., Day, R.H., Murphy, S.M., and Hayward, G.D., Effects of the *Exxon Valdez* oil spill on marine bird communities in Prince William Sound, *Ecol. Appl.*, 6, 828–841, 1996.

[illegible] 1994.

[illegible] Pergamon [illegible] 1998.

[illegible]

[illegible]

[illegible]

[illegible]

[illegible]

[illegible]

[illegible]

[illegible]

[illegible]

[illegible]

Part V

Ecosystem Ecotoxicology

> In amnesiac revery it is also easy to overlook the services that ecosystems provide humanity. They enrich the soil and create the very air we breathe. Without these amenities, the remaining tenure of the human race would be nasty and brief.
>
> **(E.O. Wilson 1999)**

Ecosystems represent the highest and final level of biological organization that we will consider in our treatment of ecotoxicology. It is appropriate that we conclude with a discussion of ecosystems, which have been considered by some ecologists to be the fundamental units of nature (Tansley 1935). The critical defining feature of ecosystems that is unique from other levels of biological organization we have considered is the inclusion of abiotic variables. Ecosystem ecotoxicology is necessarily a multidisciplinary science, and the ecosystem processes that respond to contaminants go beyond those of populations and communities. Because these processes are often scale dependent (Carpenter and Turner 1998), effects of contaminants on ecosystem function also vary across spatiotemporal scales. Ecosystem ecologists have made tremendous progress developing biogeochemical models of nutrient dynamics, and these models can be readily adapted to predict contaminant movement within and between ecosystems. Quantifying effects of contaminants on ecosystem processes and demonstrating causal relationships between stressors and responses is challenging. As a consequence, ecosystem responses are not routinely measured in ecological risk assessments. However, characterization of ecological integrity based exclusively on structural measurements has provided a somewhat incomplete picture of how ecosystems respond to anthropogenic perturbations (Gessner and Chauvet 2002). Furthermore, the unprecedented rate of species extinction occurring at a global scale (Wilson 1999) requires that ecologists and ecotoxicologists develop a better appreciation of the relationship between community patterns and ecosystem processes. Finally, many ecosystem processes are intimately connected to ecosystem goods and services that are essential for the welfare of humanity. The goal of this section is to demonstrate how contaminants and other anthropogenic stressors affect these critical ecosystem processes and related services.

REFERENCES

Carpenter, S.R. and Turner, M.G., At last a journal devoted to ecosystem science, *Ecosystems*, 1, 1–5, 1998.
Gessner, M.O. and Chauvet, E., A case for using litter breakdown to assess functional stream integrity, *Ecol. Appl.*, 12, 498–510, 2002.
Tansley, A.G., The use and abuse of vegetational concepts and terms, *Ecology*, 16, 284–307, 1935.
Wilson, E.O., *The Diversity of Life*, W.W. Norton & Company, New York, 1999.

29 Introduction to Ecosystem Ecology and Ecotoxicology

29.1 BACKGROUND AND DEFINITIONS

> Ecosystems behave in ways that are very different from the systems described by other sciences.
>
> **(Ulanowicz 1997)**

> Ecosystems can be seen more powerfully as sequences of events rather than as things in a place. These events are transformations of matter and energy that occur as the ecosystem does its work. Ecosystems are process-oriented and more easily seen as temporally rather than spatially ordered.
>
> **(Allen and Hoekstra 1992)**

Although the term ecosystem is broadly recognized by the general public and appears frequently in the nonscientific literature, ecologists and ecotoxicologists still struggle with a precise definition. Recognition that groups of plants form predictable associations across broad geographic regions was a significant breakthrough in the history of ecology (Clements 1916), and early plant ecologists devoted considerable effort to understanding the mechanisms responsible for these patterns. Perhaps because of the tremendous influence of Frederic Clements on the field of ecology, the contentious debates regarding holistic and reductionist interpretations of natural systems continued well into the 1930s. These debates figured prominently in the establishment and maturation of the emerging field of ecosystem ecology. Rejecting the Clementsian superorganism perspective that growth, development, and senescence of a community was analogous to that of individual organisms, the term *ecosystem* was first introduced by Arthur Tansley in 1935 when he appropriately recognized the difficulty of studying biotic and abiotic components of natural systems in isolation.

> Though the organisms may claim our primary interest, when we are trying to think fundamentally we cannot separate them from their special environment, with which they form one physical system.
>
> **(Tansley 1935)**

Thus, one distinguishing feature of ecosystem ecology, which was recognized early in its history, was the necessity of considering integrated physical, chemical, and biological processes. Ecosystem ecologists are not simply recognizing the influences of the physical environment but are considering organisms and the abiotic environment as part of a single system. This holistic perspective is fundamentally different than how lower levels of organization have been treated in ecology. Likens (1992) defined an ecosystem as a "spatially explicit unit of the earth that includes all of the organisms along with all components of the abiotic environment within its boundaries." One can see by this broad definition that while the spatial extent of an ecosystem remains somewhat vague, the emphasis is on including organisms and the environment. We will also see that because of the focus on movement of energy and abiotic materials (e.g., C, N, P), ecosystem ecology integrates the fields of chemistry, physics, and biology and is, therefore, necessarily a multidisciplinary science.

29.1.1 The Spatial Boundaries of Ecosystems

Because of the loosely defined spatial and temporal boundaries, some ecologists have argued that ecosystems lack the logical interconnectedness typical of other levels of biological organization (Reiners 1986). Clearly, the spatial boundaries of an ecosystem often extend beyond those of its component populations and communities. These broad spatial and temporal boundaries of ecosystems are necessary because they provide ecologists with the flexibility to match questions with appropriate scales. For example, to quantify the mass balance of nitrogen or phosphorus in a lake ecosystem, it is necessary to include materials contributed from the surrounding watershed. Similarly, to quantify the transport of organochlorines or other persistent organic pollutants through an aquatic food web, assessment of atmospheric sources may be required. Although flexibility in defining the spatial and temporal scale of an ecosystem is necessary, the classic studies of ecosystem dynamics have been conducted in systems with well-defined boundaries such as watersheds and lakes. Thus, ecologists recognize the necessity of including inputs of materials from outside sources, but in practice ecosystem boundaries are more precisely defined.

While Tansley considered ecosystems "*the basic units of nature on the face of the earth*," there remains some debate in the literature over whether ecosystems actually exist or are simply an artifact of our inability to adequately describe nature (Goldstein 1999). Contemporary ecologists still question whether the ecosystem is a physical construct, as defined by Tansley, or more like a theoretical concept that serves to organize our thoughts and ideas. Early definitions attempted to place specific boundaries on ecosystems, lakes being the most obvious example. However, we now recognize that ecosystems are connected to and influenced by features outside these traditional borders. Allen and Hoekstra (1992) note that it is unworkable to consider an ecosystem simply as a place on a landscape. Thus the question becomes, is ecosystem science simply the study of processes (as opposed to patterns)? We can readily discuss properties of ecosystems (e.g., trophic structure), but recognize that it may not be possible or prudent to enclose ecosystems in arbitrary boundaries.

A relatively broad delineation of ecosystem boundaries will also influence the scope and coverage of ecosystem ecotoxicology considered in the following sections. In our previous discussion of food web ecotoxicology, we described the structure of food webs and how contaminants may influence linkages among trophic levels. Analyses of connectance, trophic linkages, and food chain length provide important insights into community organization and help explain variation in contaminant levels among consumers. In the following sections, we will emphasize factors that affect contaminant transport in ecosystems and the potential effects of contaminants on bioenergetics, nutrient cycling, and other ecosystem processes.

29.1.2 Contrast of Energy Flow and Materials Cycling

Although the flow of energy and the transport of materials through an ecosystem are generally treated separately in most ecosystem assessments, these processes are so intimately linked that it is often more practical to consider them simultaneously. For example, the flow of energy is closely associated with the transfer of carbon through photosynthesis and respiration. One important distinction between the movement of energy and abiotic materials through ecosystems concerns the second law of thermodynamics, which essentially states that some energy is dissipated as heat with each energy transformation. It is well established that energy flow through biological systems is a highly inefficient, one-way process, with approximately 10% of energy transferred from one trophic level to the next (Slobodkin 1961). This inefficiency greatly limits the number of trophic levels in an ecosystem and accounts for the rarity of large predators (Colinvaux 1978). In contrast, abiotic materials such as nutrients and carbon are cycled through ecosystems, and the amount of these materials increases with trophic level (Figure 29.1). These differences between energy flow and materials cycling are at least partially responsible for the process of biomagnification in top predators observed for many organic chemicals. Although the amount of energy decreases,

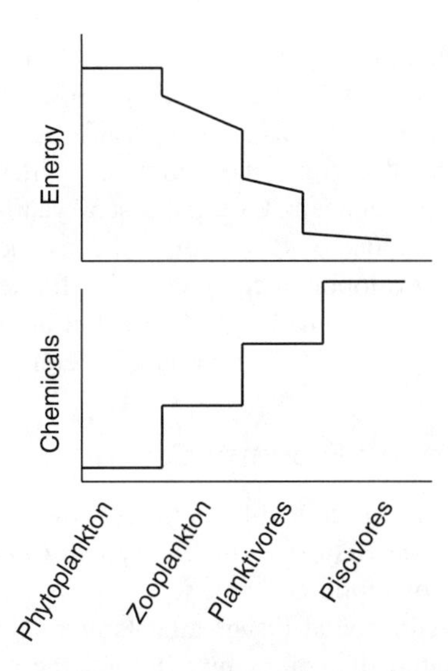

FIGURE 29.1 Hypothetical changes in energy and chemicals in an aquatic food web. Because energy is dissipated as heat as it is transferred through a food chain, it decreases with trophic level. In contrast, many chemicals, including toxic and bioaccumulative substances, cycle through an ecosystem and may increase with trophic level. (Modified from Stiling (1999).)

many abiotic materials, including contaminants, tend to increase in concentration with trophic level.

29.1.3 Community Structure, Ecosystem Function and Stability

The precise characterization of ecosystem properties has important implications for how we define ecosystem resistance and resilience. Previous studies have reported that structural characteristics, such as abundance or the number of species, are generally more sensitive than ecosystem processes, such as energy flow or nutrient cycling (Schindler 1987). Consider the example of acidified lakes, which have been studied extensively in ecosystem ecology. If we define resistance based on alterations in primary productivity of an acidified lake, we may conclude the ecosystem was relatively stable. However, if we assessed stability based on loss of species or changes in community composition, responses known to be considerably more sensitive, we may conclude that the system had low stability. The important point is that populations and communities may appear to behave quite differently when they are considered in isolation from ecosystems. The simplification of ecosystems to component parts has also contributed to the controversy over the relationship between stability and diversity described in previous sections. Attempts to define the stability of ecosystem processes based on the diversity of its components (e.g., number of species) have met with mixed success.

29.2 ECOSYSTEM ECOLOGY AND ECOTOXICOLOGY: A HISTORICAL CONTEXT

Compared to the study of population and community ecology, an ecosystem perspective is relatively new in the history of ecology. There is also considerable variation among ecologists in their precise descriptions of ecosystems, which have been compared to individual organisms and precisely engineered (though relatively inefficient) machines. Ecosystems have been described as static or dynamic, as open or closed, and as predictable or stochastic collections of unrelated, noninteracting species

(Ulanowicz 1997). For some contemporary ecologists, the field of ecology is predominantly a study of the movement of energy and materials through ecosystems. Others consider the movement of materials to be an outcome of the interactions among organisms and with the abiotic environment. These different characterizations reflect some uncertainty in the literature with respect to the ecosystem as an object of study or simply a concept. Over the past 50 years, the predominant perspective of an ecosystem has evolved from the idea of spatiotemporal constancy to coupled dynamics in space in time. Despite this evolution, developing a comprehensive framework to address spatiotemporal issues in ecosystem ecology remains a challenge (O'Neill et al. 1986), and how we describe an ecosystem is often influenced by personal bias or point of reference.

29.2.1 Early Development of the Ecosystem Concept

As noted above, Tansley (1871–1955) coined the term ecosystem and was the first to publish the concept in a technical paper. In the *History of the Ecosystem Concept in Ecology*, Golley (1993) argued that Tansley's inclusion of biotic and abiotic processes in the definition of an ecosystem was an attempt to resolve the conceptual disagreements among plant ecologists concerning the hierarchical versus organismic nature of a community. In 1942, the ecosystem concept was formalized by Raymond Lindeman into the "trophic dynamic aspect," widely recognized as one of the most significant contributions in the early history of ecology (Lindeman 1942). The most striking aspect of this original work was Lindeman's attempts to quantify seasonal dynamics of vegetation and animal production in a small lake (Cedar Bog Lake, Minnesota) and to characterize an ecosystem based on energy flow. He also organized different groups of species into categories based on their feeding habits or trophic level (e.g., browsers, plankton predators, benthic predators). More importantly, he highlighted the interactions between biotic and abiotic components of the ecosystem. Important concepts such as the substitution of units of energy (calories) for biomass, estimates of production based on turnover, and calculation of ecological efficiencies anticipated questions that would figure prominently in contemporary ecosystem research. However, the most significant contribution of the work was the recognition that energy, or more specifically calories, was the most appropriate currency by which to characterize ecosystems. Ironically, Lindeman's original manuscript was rejected by *Ecology*, primarily because of its overly theoretical nature. The paper was accepted only after strong appeal from Lindeman's Ph.D. advisor, the famous Yale limnologist G.E. Hutchinson, and published after Lindeman's death in 1942.

While Lindeman's classic paper introduced the trophic dynamic concept and formalized the study of ecosystem ecology, it was the publication of Eugene P. Odum's (1953) classic text *Fundamentals of Ecology* a decade later that placed ecosystem studies in the mainstream of ecological research. This textbook greatly influenced a generation of ecologists during a critical period of development and allowed the ecosystem concept to finally emerge as a legitimate topic of ecological research. Ecology was gradually attempting to move from a predominantly descriptive science concerned primarily with natural history to a more mechanistic-based science that sought to achieve the status of chemistry and physics. Interpretation of ecological processes using laws of thermodynamics appealed to many ecologists. This work also initiated a series of disputes among ecologists regarding the usefulness of mathematical models for quantifying ecosystem dynamics. Ecosystem ecologists were criticized for reducing the complexity of ecosystems to fewer and fewer components, and for simplifying interactions among these components using strictly deterministic models. Golley (1993) notes that much of the ecosystem research conducted during this period was little more than "machine theory applied to nature." The ecosystem as a machine concept and the application of large-scale ecosystem models, referred to as "brute force reductionism" (Allen and Starr 1982) figured prominently in the early history of ecosystem research. Although there is some dispute that the complex box-and-arrow models of system ecologists represent testable hypotheses (Golley 1993), they at least provided ecologists with mechanistic explanations for patterns observed in nature. Providing insight into mechanisms, which has long been considered the holy grail of ecological research (Ulanowicz 1997), is likely to improve the ability of ecosystem ecologists to address applied issues. Although Odum's textbook preceded the environmental movement by over a decade, it appealed to a growing number of ecologists concerned with human impacts on natural systems. At a time when humans were only beginning to understand the potential effects of their actions on the environment, this book stands out as one of the first to emphasize the importance of including anthropogenic activities in any assessment of ecosystem structure and function.

29.2.2 Quantification of Energy Flow through Ecosystems

The flow of energy described in the conceptual diagrams of Elton and Lindeman was quantified in the early 1960s. These initial analyses confirmed theoretical predictions showing the relative inefficiency of energy transfers from primary producers to herbivores and predators. Golley's (1960) classic study of energy dynamics conducted in an old field with a relatively simple food chain from plants to herbivores (mice) and predators (weasels) (kcal/ha/year) showed that only a small fraction of the energy in primary producers results in predator production (Figure 29.2). About 50% of the sunlight striking the field is of the wavelength that can be used by plants, and only about 1% of this is converted to Net Primary Production (NPP). Fisher and Likens (1973) quantified all organic material input and output to develop an energy budget for Bear Brook, a small second-order stream in the northeastern United States. Over 99% of the energy input to the stream was allochthonous, indicating that Bear Brook was a strongly heterotrophic system.

Ecosystem-level studies by Golley (1960), Fisher and Likens (1973), and others demonstrated that the movement of energy through an ecosystem could be quantified; however, the food chains in these initial studies were relatively simple. Quantifying energetics of more complex systems proved to be a daunting task. One significant event during this period facilitated the development of new techniques to quantify energy and materials flow in ecosystems. Funding provided by the Atomic Energy Commission (AEC) allowed researchers to study the distribution of radioactive materials in biotic

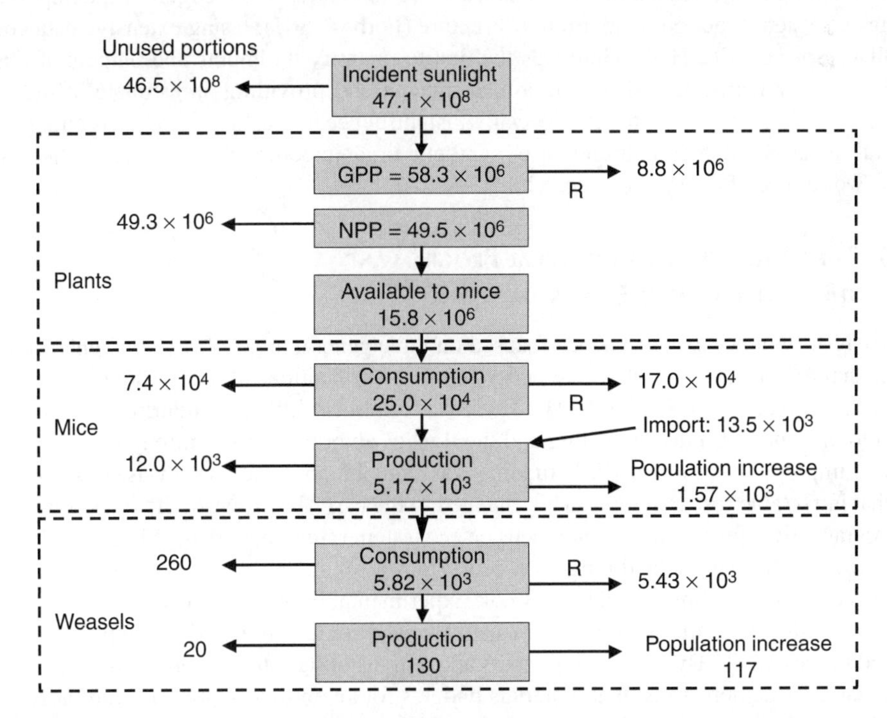

FIGURE 29.2 Energy flow in an old field ecosystem showing the relative amounts of energy (kcal/ha/year) from incident sunlight to top predators. (Modified from Golley (1960).)

and abiotic compartments following intentional releases associated with tests of explosive devices. It is no coincidence that several prominent centers for ecosystem studies in the United States, including Oak Ridge National Laboratory and Savannah River Ecology Laboratory, were associated with nuclear testing facilities and involved the emerging area of radiation ecology. Recognition that radioactive materials moved between biotic and abiotic compartments and accumulated in food chains was a significant discovery that linked basic and applied ecological research. A readily available source of funding from the AEC certainly facilitated this association (Golley 1993). Experimental techniques such as the addition of radioactive tracers improved the ability of ecologists to quantify the movement of energy and materials through an ecosystem. By labeling primary producers with a radioactive isotope, most commonly phosphorus-32 (^{32}P), ecologists can trace the movement of energy through a foodweb. Whittaker (1961) pioneered this technique in aquatic ecology and used microcosms to measure movement of ^{32}P through an aquatic food web. Similar experiments were conducted by Ball and Hooper (1963) in a Michigan trout stream. Tracer experiments became a mainstay for the emerging field of radioecology that allowed ecosystem ecologists to estimate the rate of movement of materials and energy through the system. The large number of studies that followed reflected the growing perspective that energy is the universal currency in ecosystems and that an understanding of energy flow was critical to the study of ecosystem ecology. This development proved to be especially significant to the study of ecosystem ecotoxicology because many of the transport and fate models used to quantify the movement of radioactive materials were eventually adopted and modified for the study of contaminants.

Quantifying the movement of energy and materials through an ecosystem generally required a mass budget approach in which inputs and outputs were measured. Thus, lakes and streams became appropriate models for the study of ecosystems because, unlike terrestrial ecosystems, the boundaries were well defined. As we will see, recognizing the exchanges that take place between ecosystems, such as the movement of materials from terrestrial to aquatic habitats, has become an important area of ecosystem research (Ulanowicz 1997). In fact, despite precedence for the term ecosystem being attributed to Tansley, earlier writings of Stephen Forbes, an American limnologist, also highlighted the role of abiotic processes and interactions within communities and recognized the importance of studying ecosystem function in addition to structure (Forbes 1887). Using extensive data collected from Silver Springs, FL, H.T. Odum was the first to quantify the inputs and outputs of materials through an ecosystem, thus calculating a mass budget and providing an estimate of metabolism (Odum 1957). This approach proved especially insightful because estimates of mass budgets, either of natural materials such as nutrients or of synthetic organic compounds such as pesticides, have been the workhorse of ecosystem research.

29.2.3 The International Biological Program and the Maturation of Ecosystem Science

The early history of ecosystem science focused on three general areas of research: characterization of the structure and function of whole ecosystems, quantification of energy flow, and estimation of ecosystem productivity (Golley 1993). There was relatively little effort during this initial period devoted to the study of nutrient cycling and the flow of abiotic materials through ecosystems. Perhaps more importantly, relatively little funding was available to pursue what was considered to be a somewhat intractable research topic. This changed in the early 1960s when the International Biological Program (IBP) provided a unique focus on ecosystem research and, more importantly, funding opportunities for large-scale and long-term ecosystem-level studies. The pioneering investigations into biogeochemical cycling at Hubbard Brook Experimental Forest, New Hampshire demonstrated that ecosystem-level questions were both manageable and could address critical applied issues (Bormann and Likens 1967). By quantifying inputs and outputs of various nutrients, cations, and anions, these researchers demonstrated that materials budgets for an entire ecosystem could be developed. More importantly, they expanded the traditional boundaries of stream ecosystems to include the surrounding upland areas and pioneered the field of watershed research. From an applied ecotoxicological perspective, the creation of a watershed budget for Hubbard Brook also provided some of the first concrete evidence of the effects of acid rain on ecosystems in the United States (Likens et al. 1996).

29.3 CHALLENGES TO THE STUDY OF WHOLE SYSTEMS

> The answer for ecosystems lies neither in the elegant simplicity of classical physics nor in the fascination for detail of natural history.
>
> **(Holling and Allen 2002)**

Ecologists who believed whole ecosystems were the most appropriate scale of their investigations soon realized they faced several significant challenges. An ecosystem perspective would require estimates of biomass and production of all resident species—clearly an impossible task. If ecologists were to study ecosystems in their entirety, a system-level approach was necessary. Various solutions to this dilemma were offered, including limiting analyses to the few dominant species and assuming that related species performed similar functions. The second alternative, the approach used by most contemporary ecologists, was to assign species to functional groups and characterize energy and materials flow through these groups. This approach required numerous simplifying assumptions, and many ecologists were critical of the loss of information that occurred when aggregating feeding habits of different species. In addition to minimizing species-specific differences, categorizing organisms into functional feeding groups ignored seasonal and ontogenetic variation. Suter (1993) lists several additional impediments to ecosystem-level assessments, including greater costs, lack of standardization, lack of consensus over relevant endpoints, ecosystem complexity, high variation, and relative insensitivity. Because much of ecosystem ecology remains purely descriptive, there has been criticism that the hypothetico-deductive approach advocated by many philosophers of science (Popper 1972) has been neglected. These practical and conceptual impediments partially explain why ecotoxicologists have not pursued a more rigorous program of research in ecosystem assessments. Clearly, the relevant question for many of these issues is how much detail can we ignore and still have an adequate representation of overall ecosystem function. However, downplaying the importance of species in favor of characterizing ecosystems based entirely on processes has received harsh criticism, particularly in the field of conservation biology (Goldstein 1999).

Finally, our ability to understand effects of contaminants on ecosystems is both facilitated and impeded by their self-organizing and cybernetic characteristics (O'Neill et al. 1986). The perspective that ecosystems are controlled by stabilizing negative feedback relationships is relatively widespread in ecology. Indeed, the ability of some ecosystems to quickly return to predisturbance conditions following perturbation implies some degree of organization and homeostasis. This is encouraging and suggests that, despite inherent complexity, ecosystems are legitimate objects of study and that patterns and processes are tractable. However, the resilience and resistance of ecosystem processes to disturbance may hamper our ability to quantify these responses.

29.3.1 TEMPORAL SCALE

In addition to questions regarding the appropriate boundaries and spatial scale, the relevant time scale required to adequately quantify ecosystem-level processes requires careful consideration. According to hierarchy theory, responses at higher levels of organization occur slower than those at lower levels (Figure 29.3). For example, bioaccumulation of contaminants and resulting physiological and behavioral alterations in organisms may occur rapidly following the discharge of toxic materials to an ecosystem. However, it may require considerably longer time before we observe discernible effects on ecosystem processes. The temporal scale of ecosystem responses is an important consideration

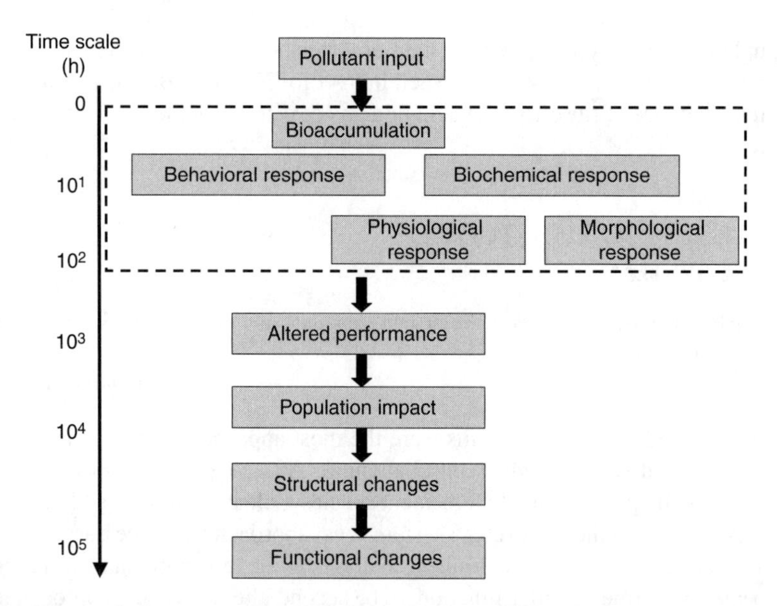

FIGURE 29.3 Time scale for responses to chemical pollutants at different levels of biological organization. Effects of chemicals on physiological and biochemical endpoints are expected to occur within hours to days, whereas community- and ecosystem-level responses may require months to years. (Modified from Sheehan (1984).)

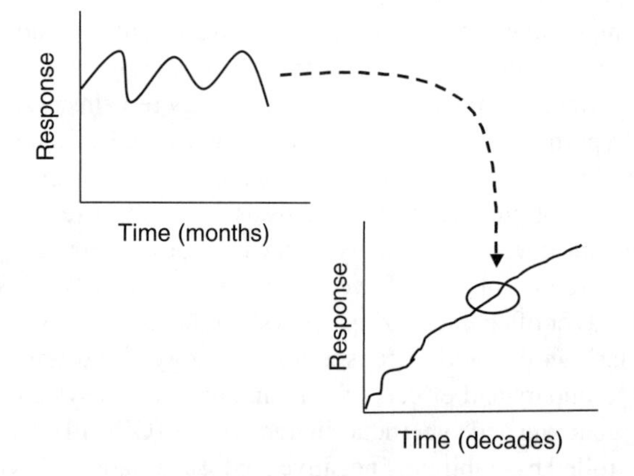

FIGURE 29.4 Importance of temporal scale in ecosystem assessments. Response trajectories to ecosystem perturbations measured over very short time scales (e.g., months), the typical duration of many ecological investigations, will likely be quite different from those measured over longer time scales (e.g., decades).

in assessing impacts of anthropogenic perturbations. Ecosystem responses to global climate change at one temporal scale (e.g., hundreds of years) may show very different responses over shorter time periods (Figure 29.4). A sampling regime that is too short will not capture the patterns occurring over longer time periods. It is therefore critical that the time scale of ecosystem responses and the methodological approaches and sampling frequency designed to assess these responses match the expected time scale of the perturbation. In a review of over 800 experimental and field studies, Tilman (1989) reported that over 75% were limited to 1–2 years. Although a 2-year duration may be adequate to characterize some ecosystem processes, these short-term ecosystem studies will likely miss novel events, such as droughts or other natural disturbances, which are important features of ecosystems.

29.4 THE ROLE OF ECOSYSTEM THEORY

29.4.1 SUCCESSION THEORY AND THE STRATEGY OF ECOSYSTEM DEVELOPMENT

Ecologists have long recognized that ecosystems change over time and that these changes are often orderly and predictable. Characterizing long-term temporal changes in ecosystem structure and function is problematic, and therefore researchers studying succession have relied heavily on space for time substitutions. For example, by comparing vegetation patterns along newly formed sand dunes of Lake Michigan, ecologists could visualize the transition from pioneer to climax species along a defined spatial gradient (Cowles 1899). Because ecological succession is often initiated following natural or anthropogenic disturbance, it is appropriate to consider temporal changes within the context of ecosystem perturbation and recovery.

Odum's (1969) classic paper "The strategy of ecosystem development" recognized the parallels between developmental biology of organisms and succession of ecosystems. He stated that ... *the "strategy" of succession as a short-term process is basically the same as the "strategy" of long-term evolutionary development of the biosphere.* The overall strategy of ecosystem development according to Odum was to achieve as large and diverse an organic structure as possible within the constraints of available energy and materials. Due to the misgivings of many ecologists concerning the organismal or superorganismal properties of ecosystems, Odum's use of the word "strategy" is somewhat unfortunate. However, this paper is especially significant because it was one of the first to relate ecosystem processes such as succession and stability to anthropogenic disturbance. Ecosystem developmental changes, such as shifts in the ratio of primary production (P) to respiration (R) and changes in species diversity, are considered functional indices of ecosystem maturity. Because of the solid theoretical underpinnings and the generality of these responses, they may represent useful measures of ecosystem responses to contaminants (Table 29.1).

Ecologists generally employ two very different methodologies to study ecosystem processes. The first approach employs the traditional hypothetical-deductive method and relies on a combination of induction, observation, and experimentation. This approach tends to be more site specific, and the results often pertain to a specific set of questions in a particular ecosystem. Much of applied ecosystem ecology, in which researchers attempt to identify the causes of specific alterations in ecological processes, uses this form of inquiry. In the second approach, ecologists develop theoretical principles and mathematical models to draw inferences about processes that can be generalized across different ecosystems. Sagoff (2003) discusses several conceptual obstacles faced by researchers using these theoretical or "top-down" approaches. Because these obstacles pertain to how we define ecosystems and how we isolate cause and effect relationships, a brief discussion is warranted here.

Perhaps the most serious challenge to theoretical ecosystem ecology is defining the class of objects that constitute an ecosystem. Sagoff (2003) argues that most definitions are either over- or under-inclusive. The broad definition cited above would include such diverse systems as the bacterial assemblages living in a cow's intestines as well as all of Lake Superior. A second challenge is the remarkably diverse ways in which ecosystem ecologists have attempted to explain processes occurring in nature. Ecologists have borrowed heavily from information systems, chaos theory, statistical mechanics, thermodynamics, cybernetic systems, and hierarchy theory (to name a few) to develop ecosystem models (Sagoff 2003). Finally, the use of mathematical models to address applied issues in environmental biology may represent the greatest challenge to theoretical ecosystem ecology. Although some researchers are pessimistic about our ability to integrate ecosystem models in applied ecology (Sarkar 1996), it is likely that the following decades will see numerous opportunities to test model predictions in altered ecosystems. Deviation in the behavior of these systems from model predictions may be used as a measure of the level of perturbation.

We recognize that it will not be possible or practical to study the response of all ecosystem components to anthropogenic stressors. To measure effects of soil acidification on decomposition rate,

TABLE 29.1
Examples of Population, Community, and Ecosystem Attributes of Developing and Mature Systems

Attribute	Developing Stages	Mature Stages
Life history characteristics		
Niche specialization	Broad	Narrow
Organism size	Small	Large
Life cycles	Short; simple	Long, complex
Community structure		
Total organic matter pool	Small	Large
Inorganic nutrients	Extrabiotic	Intrabiotic
Species diversity and evenness	Low	High
Spatial heterogeneity	Low	High
Ecosystem attributes		
Gross production:respiration (P/R)	$>$ or <1.0	~1.0
Gross production:biomass (P/B)	High	Low
Net community production	High	Low
Food chains	Linear	Web-like
Mineral cycles	Open	Closed
Nutrient exchange	Rapid	Slow
Role of detritus	Less important	Important

We suggest that many of these same attributes may be employed to characterize the level of anthropogenic disturbance in ecosystems.

Source: Modified from Odum (1969).

ecosystem ecologists generally will not attempt to characterize structural composition of microbial communities but simply rely on surrogate responses such as litter decay or respiration rates. Therefore, in addition to deciding which groups of objects are appropriately classified as ecosystems, we must also decide which processes or component parts are important and which can be omitted from our characterizations.

In summary, because ecosystems are complex, dynamic, spatially variable, and often controlled by interacting physical, chemical, biological, and socioeconomic factors (Gosz 1999), improving our basic understanding of ecosystem processes and identifying solutions to applied problems will not be easy. As noted in Chapter 20, mechanistic explanations for processes occurring at higher levels of biological organization are often lacking or difficult to identify. While simplification of these processes may be necessary to facilitate progress, we must acknowledge ecosystem complexity and incorporate complexity into our investigations (Pace and Groffman 1998).

29.4.2 Hierarchy Theory and the Holistic Perspective of Ecosystems

The idea that ecosystems can be viewed as parts within parts has served as a focal point for the debate between holistic and reductionist interpretations of ecological phenomena since Tansley first introduced the ecosystem concept. Fundamentally, the question becomes, can we learn more by studying ecosystem-level processes such as the flow of energy and the movement of materials or should we devote more effort to studying the behavior of component parts (e.g., populations and communities)? Central to this debate is the idea that emergent properties of an ecosystem cannot be understood by studying the behavior of component parts. It is unlikely that we would make much headway in understanding how energy flows or materials are cycled through an ecosystem by detailed analyses of populations and communities. The analysis of these processes requires an ecosystem-level perspective. Similarly, although we can readily measure concentrations of persistent organic chemicals in various compartments of an ecosystem, a complete assessment of fate and transport requires a holistic ecosystem perspective. We agree with Carney (1987) that there is a critical need for better integration among population/community ecology and ecosystem-level ecology. This integration will provide a much broader context for population and community ecologists to interpret demographic patterns and species interactions. Integrated studies across these levels of organization will provide ecosystem ecologists with a better understanding of mechanisms responsible for energy flow, material fluxes, and nutrient cycling.

29.5 RECENT DEVELOPMENTS IN ECOSYSTEM SCIENCE

In 1998, six decades after Tansley first introduced the term, ecosystem, a new journal dedicated entirely to the study of ecosystem science was introduced. The debut of *Ecosystems* represented an important turning point in the history of ecosystem science. Questions within the field regarding the definition of an ecosystem and specific spatiotemporal boundaries may be resolved now that ecosystem ecologists have a forum for discussion. Among the recent accomplishments within the field of ecosystem science, Carpenter and Turner (1998) list several examples in applied ecosystem ecology and ecotoxicology, including a better understanding of fate and transport of contaminants (Table 29.2). Most notable among the contributions of ecosystem science to applied issues is the improved understanding of the effects of acid deposition on ecosystem processes. Applied issues will continue to play a prominent role in ecosystem science as ecologists struggle with the dual challenges of basic research and environmental problem solving. The emphasis on applied issues in ecosystem science is obvious from a casual review of the first few volumes of this new journal, which reveals an impressive assortment of papers devoted to topics such as effects of contaminants on ecosystem processes, effects of nutrient deposition and eutrophication, and responses to climate change.

Another important development in ecosystem science is the recognition that humans play an increasingly important role in controlling the flux of materials through aquatic and terrestrial ecosystems. Although the inclusion of humans as components of ecosystems was initially proposed by Odum (1953), ecologists have traditionally conducted research in areas presumed to be free from human influences. While there remains some debate as to whether human-dominated landscapes can be defined as ecosystems, most ecologists recognize that there are relatively few pristine ecosystems and that any analysis of the structure and function of an ecosystem must account for anthropogenic inputs. Even relatively remote areas located within protected habitats or wilderness areas receive

TABLE 29.2
Significant Accomplishments in Basic and Applied Ecosystem Science

- Understanding the movement of energy and materials through freshwater and marine ecosystems
- Assessment of feedbacks between plants and animals and the abiotic environment
- Understanding the causes and consequences of eutrophication
- Quantification of the fate and transport of contaminants through ecosystems
- Understanding abiotic controls of production and the relationship with climate change
- Understanding the importance of belowground processes
- Recognizing the scale dependence of ecosystem processes

Source: Modified from Carpenter and Turner (1998).

inputs from atmospheric deposition. For example, Baron et al. (2000) reported increased nitrogen deposition in Rocky Mountain National Park, Colorado and associated changes in forest and watershed processes. Finally, because humans also are an important component of the history of ecosystems, processes measured in these systems may reflect previous disturbance events. Similar to the community conditioning hypothesis described in Chapter 25, our assessments of ecosystems must account for ecological legacies that may continue to influence processes (Carpenter and Turner 1998).

29.5.1 General Methodological Approaches

The general methods employed by ecosystem ecologists and ecotoxicologists to study ecosystem processes are similar to those used by population and community ecologists (Figure 29.5). A combination of modeling, comparative, long-term monitoring, and experimental approaches have been used to address a variety of basic and applied research questions (Pace and Groffman 1998). Experimental approaches often involve pulsing a system and measuring the associated signal as the pulse passes through the system (Allen and Hoekstra 1992). Some of the classic experiments in ecosystem ecology involved the addition of tracers or other materials that are then followed over time to different compartments of the ecosystem. Measurements of energy and material transformations within a system provide information on fate, whereas measurements of inputs and outputs provide estimates of mass balance and ecosystem processing. Although some combination of modeling, comparative, long-term monitoring, and experimental approaches will be required to address many applied questions, it is unlikely that these four approaches will contribute equally to the advancement of ecosystem science. Borrowing from a metaphor developed by Karr (1993) to describe issues in water quality management, it seems reasonable to represent each approach as the adjustable legs of a four-legged table placed on uneven ground. In this example, the relative length of each leg represents the

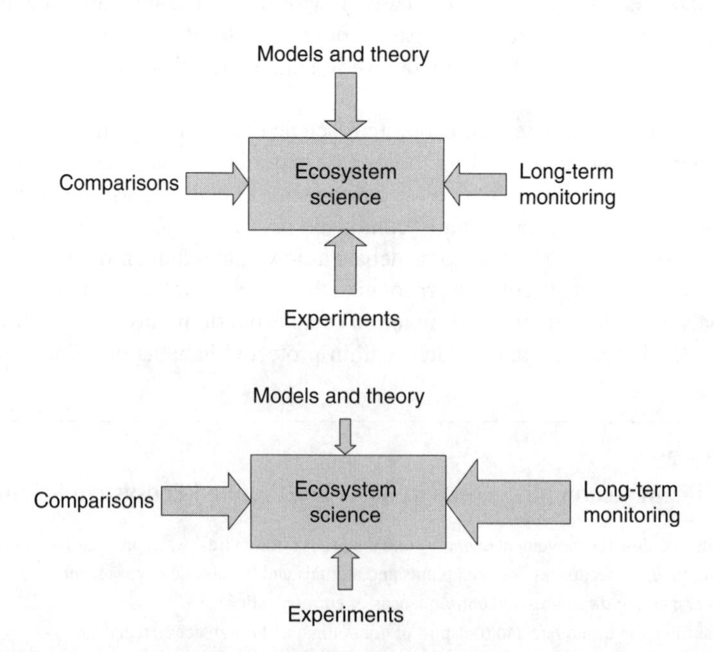

FIGURE 29.5 Descriptive, experimental, and theoretical approaches used to assess ecosystem processes to perturbations. The relative size of the arrows indicates the importance of each approach. The lower panel is an example common in many ecotoxicological investigations where long-term monitoring and comparative studies were more important than experimental or theoretical approaches. (Modified from Pace and Groffman (1998).)

significance of the approach for addressing a specific question. For example, because experimental introduction of persistent toxic chemicals into whole ecosystems is neither practical nor ethical, many of the questions related to effects of toxic chemicals on ecosystem processes will be addressed primarily using a combination of comparative and long-term studies. Experimental assessments of the fate of persistent chemicals in ecosystems will likely be limited to small-scale microcosm or mesocosm approaches. In contrast, modeling and long-term paleoecological approaches may be more appropriate for predicting ecosystem responses to global climate change.

29.5.2 The Importance of Multidisciplinary Research in Ecosystem Ecology and Ecotoxicology

It has become clear that basic and applied ecosystem research requires a multidisciplinary approach, making it difficult to conduct ecosystem-level studies as an individual investigator. A review of papers published in the new journal *Ecosystems* reveals relatively few single-investigator publications. However, institutional barriers have impeded development of multidisciplinary approaches and often stymied collaboration between researchers from different fields. Ecosystem-level research was also hampered because of the difficulty in fitting abiotic aspects of ecosystem ecology into traditional biology departments where emphasis remained on individual research (Golley 1993). The overwhelming complexity of ecosystems requires a multidisciplinary approach, but this same complexity segregates investigators into more traditional and specialized areas of research (Maciorowski 1988). The existence of disciplines within a field of study is often necessary to understand this complexity; however, some of the most innovative developments in science result from cooperation among researchers in different disciplines. These interactions produce new research questions and require participants to view complex issues from very different perspectives. Naiman (1999) provides a unique perspective on the benefits and pitfalls of conducting interdisciplinary research as well as several strategies for overcoming these obstacles. Opportunities for interdisciplinary research may be found within the broad field of ecosystem ecology, with distinctly different disciplines in the physical and chemical sciences, and across the range of spatiotemporal scales that engage ecologists (Pickett et al. 1999).

29.5.3 Strong Inference versus Adaptive Inference: Strategies for Understanding Ecosystem Dynamics

In the early 1970s, many ecologists were strongly influenced by the writings of philosophers of science such as Popper, Kuhn, and Platt. Popper's (1972) emphasis on falsification and Platt's (1964) process of using strong inference and multiple working hypotheses in scientific investigations proved to be highly successful in physics and molecular biology and resonated well with population and community ecologists. Strong aversion to committing type I errors (e.g., incorrectly rejecting a null hypothesis) greatly influenced experimental designs and generated criticism of unreplicated ecosystem-level experiments (Carpenter 1989). The process of systematically culling unlikely alternative hypotheses and deriving truth was perceived as the most rigorous application of Platt's "strong inference." However, because of the complexity and dynamic nature of ecosystems, single explanations are often insufficient, and unambiguous tests of hypotheses are unlikely to explain the patterns observed in nature. Holling and Allen (2002) argue that strictly guarding against type I errors may inhibit the creativity necessary to unravel these complexities. They propose a course of inquiry called "adaptive inference" in which researchers shift concerns between type I and type II errors (Figure 29.6). Attempts to minimize type II errors dominate the process initially, thereby allowing researchers uninhibited speculation about cause and effect relationships. These initial propositions (or "bold conjectures" as described by Popper) are themselves untestable, but capable of generating testable hypotheses. The process gradually shifts from propositions to hypotheses and eventually to models as researchers become more concerned about type I errors.

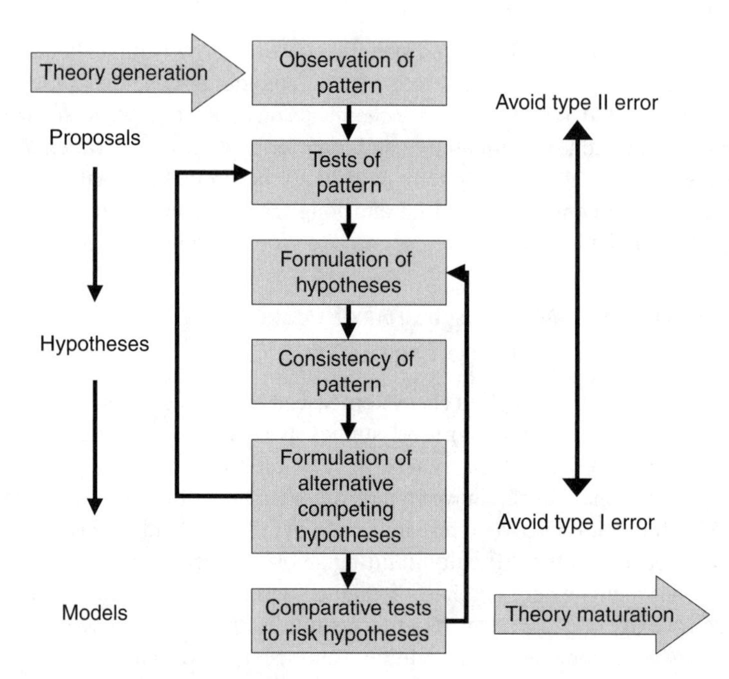

FIGURE 29.6 Adaptive inference in ecosystem research. Adaptive inference is a process in which concerns over type II errors (i.e., incorrectly rejecting an alternative hypothesis) are less important early in the investigation, thereby allowing unrestrained speculation about patterns. The emphasis shifts over time to concerns over type I errors (i.e., incorrectly rejecting a null hypothesis) to sort out false hypotheses. (Modified from Holling and Allen (2002).)

29.6 ECOSYTEM ECOTOXICOLOGY

One primary focus of the field of ecology is to describe patterns and processes in the natural world. Effects of contaminants on these patterns, such as differences in abundance or species richness upstream and downstream from a toxic discharge, was the subject of community ecotoxicology described in the previous chapters. Ecosystem ecologists are generally concerned with processes, such as the rate of energy flow through a food web or the rate of nutrient cycling in a forest. Following the recommendations of Ulanowicz (1997), these exchanges can be summarized in four simple categories: (1) imports from the external environment; (2) internal transfers between compartments (or species); (3) exports of usable materials of energy; and (4) dissipation of energy or conversion of materials to its base form. Therefore, in its simplest form, ecosystem ecology attempts to quantify energy and materials exchanges within these four categories. If the goal of ecosystem ecology is to understand and quantify the flow of energy and materials through a system, our treatment of ecosystem ecotoxicology will therefore be limited to the influence of contaminants and other anthropogenic stressors on these processes. From a more practical perspective, a fundamental question is whether alterations in the rate of energy and materials exchanges can serve as sensitive indicators of anthropogenic perturbation.

Although ecosystems are widely regarded as fundamental units of ecology, effects of contaminants on ecosystem processes have not received significant attention in the ecotoxicological literature and are rarely considered within a regulatory framework. Despite the stated goal of maintaining both structural and functional integrity of ecological systems, biological assessments in support of the U.S. Clean Water Act emphasize population and community-level analyses. Other federal programs within the United States, such as the Department of Interior Natural Resource Damage Assessment program, rely almost exclusively on assessing responses at lower levels of organization (e.g., individuals and populations). Sheehan (1984) reviewed the effects of physical and chemical stressors on several ecosystem processes, including detritus processing, ecosystem metabolism, and food web regulation. In light of the complexity of ecosystems and uncertainty in defining their spatiotemporal boundaries, the focus on populations and communities in most ecotoxicological research is understandable. However, an ecosystem perspective is essential for predicting and understanding impacts of chemicals and other stressors.

29.7 LINKS FROM COMMUNITY TO ECOSYSTEM ECOTOXICOLOGY

> Neither the process-functional approach nor the population–community approach can form a complete theoretical foundation for ecosystem analysis.
>
> **(O'Neill et al. 1986)**

29.7.1 ECOSYSTEMS WITHIN THE HIERARCHICAL CONTEXT

> Ecology has many general theories but few have been fruitful in producing new insights.
>
> **(O'Neill et al. 1986)**

Most ecologists readily acknowledge the existence of levels of organization within biological systems and therefore accept the simple application of hierarchy theory to ecosystem ecology. The application of hierarchy theory to the study of ecosystems may help resolve many of the issues inherent in the study of complex systems. By decomposing complex systems into component parts, we can more easily study mechanisms responsible for higher level, emergent processes. However, it is important to recognize that, while this approach may be sufficient to study some ecological processes, such as the influence of species interactions on energy flow, other processes will require a more holistic analysis of emergent properties. Although hierarchy theory appears to be an appropriate model for the study of ecosystems, significant progress has not been realized because of an overly simplistic view of hierarchies and ecosystems (O'Neill et al. 1986). Hierarchy theory goes well beyond the familiar level of organization concept described in many ecology textbooks. The levels within a hierarchy are defined on the basis of differences in rate structures rather than an arbitrary assignment to levels of organization.

Within ecological systems, hierarchies are often considered synonymous with levels of biological organization. Within this hierarchy, ecosystems are located between communities and landscapes. The critical distinction between communities and ecosystems is not simply a matter of size but the relative emphasis on biotic and abiotic components. Population and community ecologists are concerned with the interactions among individuals and species within an area. The abiotic environment is certainly recognized as being important and may influence these interactions, but it is largely peripheral to the focus of the investigation. To the ecosystem ecologist, the organisms and surrounding abiotic environment are integrated and constitute the unit of study. It is difficult to study the discrete components of an ecosystem in isolation. On the basis of hierarchy theory, we would expect that ecosystems are composed of the next lower level (communities) and are controlled by processes at the next higher level (landscapes). Because we have defined ecosystems based on processing and flow of materials and energy, the hierarchical analysis is not especially satisfactory because it assumes that these processes can be understood by studying the populations and communities that comprise an ecosystem. O'Neill et al. (1986) warn against this "naive reductionism" and note that, while ecosystems are indeed comprised of populations and communities, these lower levels of organization are not necessarily the most appropriate for studying ecosystem processes. Furthermore, most analyses of populations and communities either do not consider abiotic factors, a critical component of ecosystems, or treat the abiotic environment separately.

To place the ecosystem perspective in its proper context within a hierarchy of biological organization, consider the following example. Assume that a brown trout is feeding on aquatic insects in a small mountain stream. A population ecologist would be especially interested in the consequences of this predation event on demographic characteristics of either predators or prey. For example, does the removal of prey species from a population contribute significantly to overall prey mortality? Is survivorship or reproduction of the brown trout improved because of consumption of these prey species? In contrast, a community ecologist would be more interested in the nature of the species interactions and the implications for other species. For example, does the removal of individuals of a particular prey species influence abundance and distribution of other species in the community? Finally, the ecosystem ecologist is likely to consider this interaction strictly from a bioenergetics perspective and focus on the transfer of calories (or other units of energy) between these two trophic levels. Ecosystem ecologists are generally not concerned with the demographic consequences or the strength of species interactions, but simply consider individual species as energy transformers.

O'Neill et al. (1986) contrast the population–community and process–functional approaches to the study of ecosystems and discuss the consequences of viewing ecosystems from these very different perspectives. The false dichotomy between structural and process-functional approaches to the study of ecosystem science is similar to the historic controversy between the individualistic and holistic viewpoints of Clements and Gleason on the nature of the community. Both perspectives were narrowly focused on a set of observations and readily dismissed observations that were inconsistent with these predisposed ideas. Although many ecologists believe otherwise, the relationship between the structure and process of an ecosystem is not always obvious and in some situations may not exist. This is at least partially owing to the fact that community and ecosystem ecologists use different theories, vocabularies, and methodologies to study nature (Carney 1987). The redundancy of functional processes and the fact that different assemblages of species are capable of performing the same function further complicates this relationship. The long list of ecosystem processes considered critical to the assessment of ecosystem function further complicates the relationship between pattern and process. Goldstein laments that

> The list of possible descriptors of ecosystem processes and functions is limited neither by the myriad of definitions of ecosystems nor by organismal considerations. Only the imagination of ecologists and the shortcomings of language place a ceiling on the alleged number of ecosystem properties.
>
> **(Goldstein 1999)**

Only recently, with the renewed interest in understanding how species diversity may influence ecosystem function, have ecologists begun to clarify these relationships (Tilman et al. 1997). Despite this progress, it is unlikely that we will ever be able to characterize the functioning of an ecosystem based entirely on community structure and composition. Except in those rare instances where a keystone species dominates a community and exerts strong control over ecosystem processes will connections between pattern and process be clearly obvious. Splitting the traditional levels of organization and treating population–community and process-functional approaches as dual hierarchies resolves many of the problems that ecologists have encountered when attempting to establish connections between structural and functional components (O'Neill et al. 1986). According to this model, the population–community hierarchy is regulated by organisms and species interactions (e.g., competition, predation) whereas the process-functional hierarchy is regulated by mass balance and principles of thermodynamics. More importantly, it is not necessary (or possible) to reduce one dimension of the hierarchy to the other. As an example, Figure 29.7 contrasts the structural and functional perspectives for the study of energy flow through a stream ecosystem. The structural perspective may consider questions related to energy flow, but most of the focus is on the importance of the interactions among the groups. A functional perspective is less focused on effects of predators on

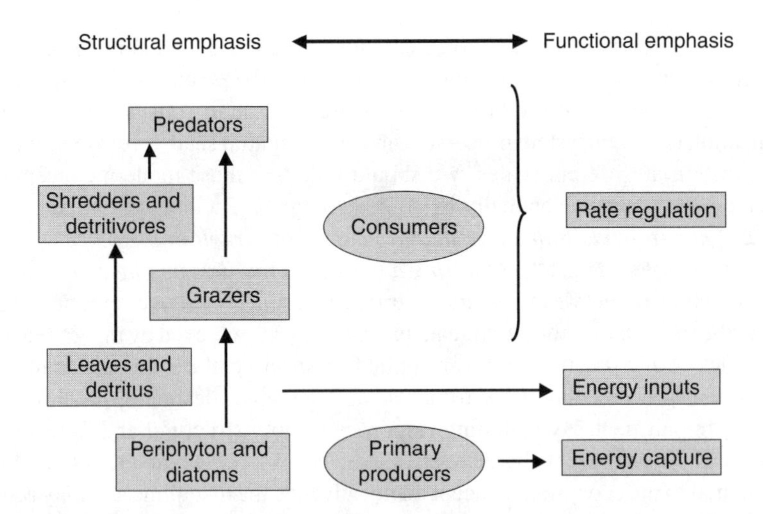

FIGURE 29.7 Contrasting structural and functional emphasis in an aquatic food chain. The structural hierarchy emphasizes population and community interactions whereas the functional hierarchy emphasizes ecosystem processes and exchanges with the abiotic environment. (Modified from O'Neill et al. (1986).)

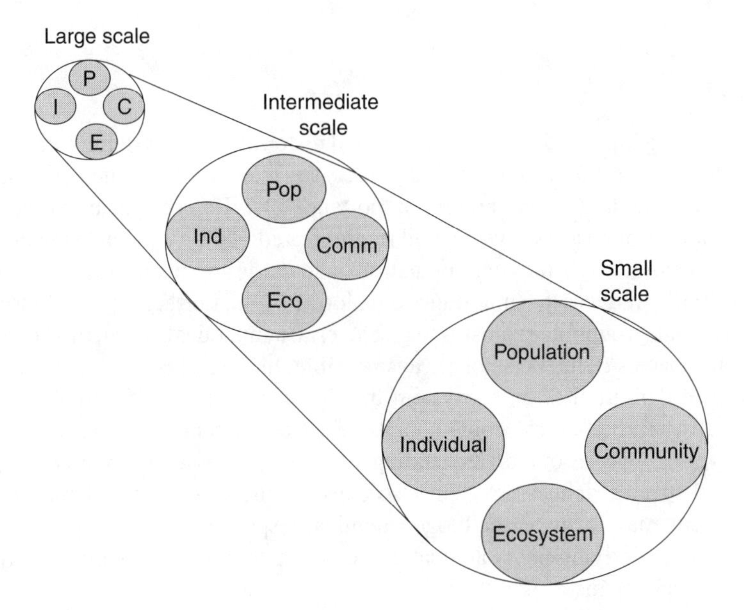

FIGURE 29.8 An alternative depiction of individual, population, community, and ecosystem responses across different spatial scales. Each level of organization is represented on each layer. Comparisons among levels of the hierarchy can be made within a single spatial scale or between different spatial scales. This representation allows investigators to consider how properties at one level of organization (e.g., ecosystem processes) and at one spatial scale level are influenced by other levels of organization at different scales. (Modified from Allen and Hoekstra (1992).)

prey populations and places more emphasis on the rate and regulation of energy inputs and capture to the system.

Allen and Hoekstra (1992) provide a very different conceptual arrangement of the hierarchy of biological organization. They suggest that separating an ecosystem to its component parts does not reveal populations and communities, and use the analogy of a layer cake to illustrate connections among levels of biological organization across different spatial scales (Figure 29.8). In their depiction, each level of organization is represented on each layer of the cake and comparisons among levels of the hierarchy can be made within a single spatial scale or between different spatial scales. This representation is intuitively appealing because it also allows us to consider how properties at one level of organization (e.g., ecosystem processes) and at one spatial scale level are influenced by other levels of organization at different scales. We should note that these modern conceptualizations of ecosystems have been viewed skeptically by some ecologists. Ricklefs (1990) suggests that these new approaches "*may provide little more than a new set of jargon in the place of old conceptual frustrations*" and cautions against "*legitimizing a mysticism that has no place in science.*"

Regardless of which model we choose to illustrate where and how ecosystems fit within a hierarchical framework, the fact remains that significant progress can be achieved by integrated investigations of ecosystem structure and function. From an applied perspective, the somewhat tenuous connection between pattern and process may work to our advantage when designing monitoring systems to assess ecological impairment. By including responses of both structural and functional endpoints, we are clearly measuring different characteristics of an ecosystem. Arguments over which approach is more fundamental to the ecosystem concept do not advance the discipline and may actually impede progress. While recent developments in ecosystem science have significantly improved our ability to assess effects of contaminants on mass flow through ecosystem compartments, the perception remains that structural and functional approaches are clearly at odds (Cairns and Pratt 1986).

29.8 SUMMARY

In summary, measuring input and output of energy and materials in ecosystems with no consideration of the species responsible for regulating these processes misses a fundamental component of the system. Some understanding of the relative importance of different species to the movement of energy and materials through an ecosystem will improve predictability and help understand variation among ecosystems. Similarly, studying populations and communities in isolation from the abiotic environment arbitrarily reduces the importance of abiotic factors. Furthermore, because of taxonomic difficulties with certain groups of organisms (e.g., bacteria, nematodes), a structural perspective based on abundance and species richness is not always possible. In the following chapters, we will adhere to the conventional definition of an ecosystem that includes both biotic and abiotic components. Although our focus will be on how contaminants affect ecosystem processes such as energy flow and materials cycling, where our understanding of these processes is improved by information on abundance and species richness we will also consider these structural elements. Although the ecosystem ecologists may focus on the biogeochemical responses of an ecosystem to stressors, an appreciation for the underlying physiological processes of individual organisms may enhance our understanding of mechanisms.

In assessing the potential impacts of contaminants and other anthropogenic stressors on ecosystems, we will distinguish between responses to localized disturbances, such as the discharge of persistent chemicals to a lake, from the more pervasive effects of global stressors, such as increasing atmospheric CO_2 levels. Although both disturbances may similarly influence ecosystem processes, the consequences and interpretations of responses are quite different. In the former example, ecosystems are reservoirs of these chemicals and the transformations within compartments are of primary concern. In the second example, ecosystems are not only directly affected by increasing CO_2 levels, but may also directly influence global atmospheric concentrations by sequestration in various compartments. Pace and Groffman (1998) characterize effects of the localized and widespread stressors as footprints and fingerprints. Ecological footprints reflect the direct effects of localized stressors and are a consequence of human appropriation of natural resources or discharge of toxic substances. In contrast, the ecological fingerprints associated with global and atmospheric stressors are likely to be subtle and more difficult to detect. In the following chapters, we will first describe important ecosystem-level processes and then discuss how these processes are likely to be affected by ecological footprints and fingerprints.

29.8.1 Summary of Foundation Concepts and Paradigms

- The goal of ecosystem ecology is to understand and quantify the flow of energy and materials through a system. Ecosystem ecotoxicology will assess the influence of contaminants and other anthropogenic stressors on these processes.
- One distinguishing feature of ecosystem ecology, which was recognized early in its history, was the necessity of considering integrated physical, chemical, and biological processes.
- Although the flow of energy and the transport of materials through an ecosystem are generally treated separately in most ecosystem assessments, these processes are so intimately linked it is often more practical to consider them simultaneously.
- The publication of Eugene P. Odum's (1953) classic text, *Fundamentals of Ecology*, placed ecosystem studies in the mainstream of ecological research and allowed the ecosystem concept to emerge as a legitimate topic.
- Ecosystem-level studies conducted in systems with relatively simple food webs demonstrated that the movement of energy through an ecosystem could be quantified; however, quantifying energetics of more complex systems proved to be a daunting task.
- Experimental techniques such as the addition of radioactive tracers (e.g., ^{32}P) improved our ability to quantify the movement of energy and materials through ecosystems.
- Because quantifying the movement of energy and materials through an ecosystem generally required a mass budget approach in which inputs and outputs were measured, lakes and streams became appropriate models because the boundaries were well defined.
- Our ability to understand effects of contaminants on ecosystems is both facilitated and impeded by the fact that ecosystem processes are controlled by negative feedback relationships.
- A better understanding of fate and transformation of contaminants is considered a major accomplishment in the field of applied ecosystem science.
- Some combination of modeling, comparative, long-term monitoring and experimental approaches will be required to address applied questions in ecosystem ecotoxicology.
- Despite the stated regulatory goal of maintaining both structural and functional integrity of ecological systems, biological assessments in support of the Clean Water Act rarely assess effects of contaminants on ecosystem processes.
- The critical distinction between communities and ecosystems is not simply a matter of size but the relative emphasis on biotic and abiotic components. While population and community ecologists are concerned primarily with the interactions among individuals and species within an area, ecosystem ecologists consider these interactions within the context of the surrounding abiotic environment.

REFERENCES

Allen, T.F.H. and Hoekstra, T.W., *Toward a Unified Ecology*, Columbia University Press, New York, 1992.

Allen, T.F.H. and Starr, T.B., *Hierarchy: Perspectives for Ecological Complexity*, University of Chicago Press, Chicago, 1982.

Ball, R.C. and Hooper, F.F., Translocation of phosphorous in a trout stream ecosystem, In *Radioecology*, Schultz, V. and Klement, W. (eds.), Reinhold, New York, 1963, pp. 217–228.

Baron, J.S., Rueth, H.M., Wolfe, A.M., Nydick, K.R., Allstott, E.J., Minear, J.T., and Moraska, B., Ecosystem responses to nitrogen deposition in the Colorado Front Range, *Ecosystems*, 3, 352–368, 2000.

Bormann, F.H. and Likens, G.E., Nutrient cycling, *Science*, 155, 424–429, 1967.
Cairns, J., Jr. and Pratt, J.R., On the relation between structural and functional analyses of ecosystems, *Environ. Toxicol. Chem.*, 5, 785–786, 1986.
Carney, H.J., On competition and the integration of population, community, and ecosystem studies, *Funct. Ecol.*, 3, 637–641, 1987.
Carpenter, S.R., Replication and treatment strength in whole-lake experiments, *Ecology*, 70, 453–463, 1989.
Carpenter, S.R. and Turner, M.G., At last a journal devoted to ecosystem science, *Ecosystems*, 1, 1–5, 1998.
Clements, F.E., *Plant Succession: Analysis of the Development of Vegetation*, Carnegie of Washington Publication, No. 242, Washington, D.C., 1916.
Colinvaux, P., *Why Big Fierce Animals are Rare. An Ecologists Perspective*, University Press, Princeton, NJ, 1978.
Cowles, H.C., The ecological relations of the vegetation on the sand dunes of Lake Michigan, *Bot. Gaz.*, 27, 95–117, 167–202, 281–308, 361–391, 1899.
Fisher, S.G. and Likens, G.E., Energy flow in Bear Brook, New Hampshire: An integrative approach to stream ecosystem metabolism, *Ecol. Monogr.*, 43, 421–439, 1973.
Forbes, S.A., The lake as a microcosm, III. *Nat. Hist. Surv. Bull.*, 15, 537–550, 1887.
Goldstein, P.Z., Functional ecosystems and biodiversity buzzwords, *Con. Bio.*, 13, 247–255, 1999.
Golley, F.B., Energy dynamics of a food chain of an old field community, *Ecol. Monogr.*, 30, 187–206, 1960.
Golley, F.B., *A History of the Ecosystem Concept in Ecology*, Yale University Press, New Haven, 1993.
Gosz, J.R., Ecology challenged? Who? Why? Where is this headed?, *Ecosystems*, 2, 475–478, 1999.
Holling, C.S. and Allen, C.R., Adaptive inference for distinguishing credible from incredible patterns in nature, *Ecosystems*, 5, 319–328, 2002.
Karr, J.R., Defining and assessing ecological integrity: Beyond water quality, *Env. Toxicol. Chem.*, 12, 1521–1531, 1993.
Likens, G.E., *The Ecosystem Approach: Its Use and Abuse*, Oldendorf/Luhe, Germany, Ecology Institute, 1992.
Likens, G.E., Driscoll, C.T., and Buso, D.C., Long-term effects of acid rain: Response and recovery of a forest ecosystem, *Science*, 272, 244–246, 1996.
Lindeman, R.L., The trophic-dynamic aspect of ecology, *Ecology*, 23, 399–418, 1942.
Maciorowski, A.F., Populations and communities: Linking toxicology and ecology in a new synthesis, *Env. Toxicol. Chem.*, 7, 677–678, 1988.
Naiman, R.J., A perspective on interdisciplinary science, *Ecosystems*, 2, 292–295, 1999.
O'Neill, R.V., DeAngelis, D.L., Waide, J.B., and Allen, T.F.H., *A Hierarchical Concept of Ecosystems*, Princeton University Press, Princeton, NJ, 1986.
Odum, E.P., *Fundamentals of Ecology*, Saunders, Philadelphia, 1953.
Odum, E.P., The strategy of ecosystem development, *Science*, 164, 262–270, 1969.
Odum, H.T., Trophic structure and productivity of Silver Springs, Florida, *Ecol. Monogr.*, 27, 55–112, 1957.
Pace, M.L. and Groffman, P.M., Successes, limitations, and frontiers in ecosystem science: Reflections on the seventh Cary Conference, *Ecosystems*, 1, 137–142, 1998.
Pickett, S.T.A., Burch, W.R., and Grove, J.M., Interdisciplinary research: Maintaining the constructive impulse in a culture of criticism, *Ecosystems*, 2, 302–307, 1999.
Platt, J.R., Strong inference, *Science*, 146, 347–353, 1964.
Popper, K.R., *The Logic of Scientific Discovery*, 3rd ed., Hutchinson, London, England, 1972.
Reiners, W.A., Complementary models for ecosystems, *Am. Nat.*, 127, 59–73, 1986.
Ricklefs, R.E., *Ecology*, 3rd ed., W.H. Freeman and Company, New York, 1990.
Sagoff, M., The plaza and the pendulum: Two concepts of ecological science, *Biol. Phil.*, 18, 529–552, 2003.
Sarkar, S., Ecological theory and anuran declines, *BioScience*, 46, 199–207, 1996.
Schindler, D.W., Detecting ecosystem responses to anthropogenic stress, *Can. J. Fish. Aquat. Sci.* 44 (Suppl. 1), 6–25, 1987.
Sheehan, P.J., Functional changes in the ecosystem, In *Effects of Pollutants at the Ecosystem Level,* Sheehan, P.J., Miller, D.R., Butler, G.C., and Bourdeau, P. (eds.), John Wiley & Sons, Chichester, 1984, pp. 101–145.
Slobodkin, L.B., *Growth and Regulation of Animal*, Holt, Rinehart and Winston, New York, 1961.
Stiling, P., *Ecology: Theories and Application*, 3rd ed., Prentice Hall, Upper Saddle River, NJ, 1999.
Suter, G.W., II, *Ecological Risk Assessment*, Lewis Publishers, Chelsea, MI, 1993.
Tansley, A.G., The use and abuse of vegetational concepts and terms, *Ecology*, 16, 284–307, 1935.

Tilman, D., Ecological experimentation: Strengths and conceptual problems, In *Long-Term Studies in Ecology*, Likens, G.E. (ed.), Springer-Verlag, New York, 1989, pp. 136–157.

Tilman, D., Knops, J., Wedin, D., Reich, P., Ritchie, M., and Siemann, E., The influence of functional diversity and composition on ecosystem processes, *Science*, 277, 1300–1302, 1997.

Ulanowicz, R.E., *Ecology, The Ascendant Perspective*, Columbia University Press, New York, 1997.

Whittaker, R.H., Experiments with radiophosphorus tracer in aquarium microcosms, *Ecol. Monogr.*, 31, 157–188, 1961.

30 Overview of Ecosystem Processes

A major stumbling block in the study of ecosystems is their bewildering complexity.

(O'Neill and Waide 1981)

30.1 INTRODUCTION

The perspective offered by O'Neill and Waide (1981) in the above quote illustrates one of the more significant challenges faced by ecotoxicologists when attempting to understand the potential impacts of anthropogenic stressors on ecosystem processes. This perspective may also partially explain the relative infrequency with which ecosystem processes are measured in biological assessments. Because ecosystem "surprises" (sensu Paine et al. 1998) may result from focusing on isolated components, one potential solution to this "bewildering complexity" is to develop a comprehensive understanding of emergent ecosystem properties (O'Neill and Waide 1981). For example, we can readily quantify the contributions of decomposers to nutrient cycling or the influence of predators on energy flow; however, it is unlikely that we can predict ecosystem consequences based exclusively on abundance or biomass estimates of these functional groups. As described in the previous chapters, the idea that behavior of a complex system often cannot be understood solely by analysis of its components is a major thesis of hierarchy theory. The order that emerges from complex systems and the constraints placed on the range of potential interactions in these systems are fundamental differences between randomly assembled populations and a stable ecosystem. The functional redundancy of ecosystems that results from species replacement is a good example of our inability to predict ecosystem responses based on understanding of components.

In one of the earlier theoretical treatments of ecosystem ecotoxicology, O'Neill and Waide (1981) provide several recommendations for research programs in this emerging field:

1. Focus on functionally intact systems that reflect ecosystem-level properties.
2. Focus on integrative properties that reflect interactions among physical, chemical, and biological properties.
3. Treat the ecosystem as a biogeochemical system that focuses on movement of energy and materials.
4. Rate processes (e.g., mineralization, decomposition, and nitrification) may be better indicators of contaminant effects than the amount of materials or energy stored in ecosystem pools.

Thus, before we can understand how ecosystems respond to contaminants and other anthropogenic perturbations, it is necessary to develop an appreciation for the complex ecosystem processes that are most likely to be affected by physical and chemical stressors. In the previous chapter, we characterized ecosystems in terms of energy flow and materials cycling. Much of our discussion of how ecosystems respond to stressors will focus on these processes. Although general ecology textbooks and much of the ecological literature treat energy flow and materials cycling through an ecosystem separately, it is important to realize that these processes are intimately related. Patterns of primary and secondary production in ecosystems are often limited by the amount of available nutrients. Biogeochemical processes, the size of nutrient pools, and the rate of materials cycling in an ecosystem can, in turn, be regulated by primary productivity. Finally, while our focus in this section will be on characterizing functional attributes of ecosystems, recent findings that demonstrate strong links between species richness, diversity, and ecosystem processes require that we also consider structural features.

30.2 BIOENERGETICS AND ENERGY FLOW THROUGH ECOSYTEMS

In addition to viewing ecosystems within a hierarchical context, contemporary ecologists routinely characterize ecosystems based on bioenergetic and biogeochemical processes. Captured solar radiation stored in chemical bonds by autotrophic organisms is made available to heterotrophs. As described in the previous chapter, the perception that ecosystems are energy-transforming systems emerged relatively early in the history of ecology. Elton's (1927) depiction of a tundra food web and his recognition that a large number of herbivores are necessary to support a smaller number of predators preceded Tansley's definition of ecosystem by a decade. Elton's description of this relatively simple food web also made ecologists aware of the difficulties associated with accurately characterizing ecosystem energetics. Although the use of calories or other units of energy as the currency to integrate Elton's trophic levels did not occur for several decades, these early investigations helped to formalize contemporary perspectives of ecosystem dynamics. Elton's (1927) food web became Lindeman's (1942) food cycle that was eventually formalized as a universal energy model by Odum (1968) that also included a material-cycling component.

30.2.1 PHOTOSYNTHESIS AND PRIMARY PRODUCTION

Flux of energy through an ecosystem is determined by the rate at which plants assimilate energy by photosynthesis, the transfer of this energy to herbivores and other consumers, and the efficiency of these conversions. Because contaminants and other stressors can affect any of these processes, energy flux through an ecosystem is an important indicator in ecosystem-level assessments. Photosynthesis in plants is the conversion of light energy and raw materials (carbon dioxide and water) to carbohydrates and oxygen:

$$6CO_2 + 6H_2O + \text{Light energy} \rightarrow C_6H_{12}O_2 + 6O_2 \quad (30.1)$$

Although this stoichiometrically balanced chemical reaction to describe photosynthesis is correct, it is not especially satisfying from an ecological perspective and should be expanded to include both elemental and energy components to reflect biomass accrual in the following way (Sterner and Elser 2002):

$$\text{Inorganic carbon} + \text{Nutrients} + \text{Light energy} \rightarrow \text{Biomass} + \text{Heat} \quad (30.2)$$

The energy necessary for the conversion of CO_2 to a reduced state in carbohydrates is provided by visible light and the total amount of energy fixed by plants is referred to as gross primary production (GPP). Plants require a portion of this fixed energy for their own metabolic needs (e.g., respiration) and the difference between GPP and these metabolic costs is called net primary production (NPP). NPP is defined as the total amount of energy available to the plant for growth and reproduction after accounting for respiration:

$$\text{NPP} = \text{GPP} - \text{Respiration} \quad (30.3)$$

30.2.1.1 Methods for Measuring Net Primary Production

Methods for measuring NPP in terrestrial and aquatic ecosystems are diverse, but typically focus on assessing changes in biomass, CO_2, or O_2. The most direct method for estimating NPP in terrestrial ecosystems is the harvest method, which generally involves measuring the increase in plant standing crop or biomass (B) over a growing season.

$$\Delta B = B_2 - B_1 \tag{30.4}$$

where B_2 is the biomass at time 2 and B_1 is the biomass at time 1.

Note that primary production of an ecosystem is a functional measure of the instantaneous rate of biomass generation, generally expressed as dry weight of plant material (or carbon) per unit area per unit time (g/m^2/year). In contrast, biomass is a structural measure of the amount of plant material present at one particular point in time. A more energetically appropriate measure may be obtained by converting dry weight of plant material to calories. Estimates of both NPP and biomass have been used as endpoints in assessing stressor impacts on ecosystem energetics.

Other approaches for estimating primary production involve measuring gas exchange (e.g., uptake of CO_2 or release of O_2) and the use of radioactive carbon isotopes, ^{14}C. Although harvest methods provide the most direct measure of NPP in terrestrial ecosystems and have been employed to estimate production of larger marine plants (macrophytes, kelp), they are less common in aquatic ecosystems because of small size and rapid turnover of primary producers. Three approaches have been employed in aquatic ecosystems to estimate primary productivity: light and dark bottles oxygen techniques, radioisotopes such as ^{14}C, and *in situ* diel approaches. The traditional approach for aquatic systems uses light and dark bottles containing water with ambient phytoplankton populations. This approach compares changes in dissolved oxygen concentration ($\Delta[O_2]$) in bottles held in the light with changes in the dark. Because $\Delta[O_2]$ in the light is a result of both GPP and respiration whereas change in the dark bottle is a result of respiration only, GPP can be estimated by the difference between these measures:

$$\text{GPP} = \Delta[O_2]\text{light} - \Delta[O_2]\text{dark} \tag{30.5}$$

A similar approach has been used to estimate metabolism in stream ecosystems in which cobble substrate collected from the streambed is placed in light and dark chambers. This approach provides an estimate of whole community metabolism because the cobble substrate typically includes both autotrophic and heterotrophic organisms (e.g., algae, bacteria, fungi, and invertebrates).

The carbon-14 (^{14}C) technique provides a considerably more sensitive estimate of primary productivity, which may be necessary in oligotrophic systems where GPP is very low. The ^{14}C technique is also preferred by some ecologists because it allows researchers to explicitly follow carbon flow through an ecosystem (Howarth and Michales 2000). Clear bottles with water and ambient phytoplankton are incubated with a tracer amount of ^{14}C—labeled dissolved inorganic carbon. The accumulation of carbon in organic matter relative to the dissolved inorganic fraction provides a measure of primary production.

Note that the light and dark bottle technique and the ^{14}C incubation technique may be compromised by container artifacts. Isolation of primary producers from natural systems by placement in bottles may result in depleted nutrient concentrations, decreased turbulence and mixing, and growth of organisms on the sides of the container. *In situ* approaches that measure diel changes in O_2 or CO_2 eliminate these bottle effects and provide ecosystem-level estimates of GPP and respiration. Changes in CO_2 or O_2 during daylight are a result of GPP, whole ecosystem respiration, and exchange with the atmosphere. Thus, whole ecosystem GPP and respiration can be estimated by measuring changes in O_2 or CO_2 during the daylight and at night, and correcting for atmospheric exchange. A variation of this approach is used in streams, where whole ecosystem metabolism is determined by comparing O_2

or CO_2 concentrations at upstream and downstream locations and measuring the travel time between stations.

30.2.1.2 Factors Limiting Primary Productivity

Numerous abiotic factors limit primary productivity in both terrestrial and aquatic ecosystems; however, light, temperature, nutrients, and moisture (in terrestrial habitats) are generally considered the most important limiting factors. Because plants differ in the efficiency with which they capture and convert incident sunlight, an understanding of factors that limit the efficiency of GPP is necessary to understand how contaminants may influence these processes. In general, phytoplankton communities have very low efficiencies ($<1\%$), whereas higher values are observed in forests (2–3.5%) (Cooper 1975). Most of the energy fixed by plants, approximately 50–70%, is used for plant respiration.

Sufficient light levels are necessary for primary production; however, intense light can saturate pigments and inhibit photosynthesis. Similarly, the rate of photosynthesis generally increases with temperature, up to some optimal value, and then declines. Because respiration also increases with temperature, optimal temperatures for NPP and photosynthesis will likely differ, complicating our ability to predict the precise relationship between temperature and NPP. Finally, the amount of available nutrients, especially nitrogen (N) and phosphorus (P), limit primary production in many ecosystems. Primary production of aquatic ecosystems is particularly sensitive to nutrient limitation, as evidenced by studies showing that even slightly increased levels of nutrients can significantly increase algal productivity (Ryther and Dunstan 1971). Although most of the research on nutrient limitation has focused on N and P, some ecosystems may be limited by other materials. Studies conducted using water collected from the Sargasso Sea, a highly oligotrophic ecosystem, showed that enrichment with N and P had relatively little effect on phytoplankton (Menzel and Ryther 1961). Primary productivity in much of the open ocean is limited by iron, which has stimulated interest in the use of iron to fertilize the oceans as a measure to increase sequestration of anthropogenic CO_2. Not surprisingly, many of the most comprehensive studies demonstrating effects of nutrients on productivity have been conducted in lakes where the association between primary productivity and abiotic factors has been documented experimentally (Schindler 1974).

Because light is rapidly attenuated in aquatic ecosystems, the amount of light available to primary producers decreases as a function of depth according to the following equation:

$$\mathrm{d}I/\mathrm{d}z = -kI \tag{30.6}$$

where I = amount of solar radiation, z = depth, and k = extinction coefficient. The extinction coefficient varies among ecosystems, from about 0.02 in pure water to 0.10 in open seawater. The amount of light at 10 m depth in open seawater is about 50% lower than at the surface. Because of greater amounts of light absorbing materials, values of k in lakes and other productive ecosystems are considerably greater. In deep rivers, lakes, and marine ecosystems, the reduction in light limits the depth at which many plants can occur to a narrow band called the euphotic zone, and is defined as the area near the surface where photosynthesis is greater than respiration.

30.2.1.3 Interactions Among Limiting Factors

In addition to the direct effects of these limiting factors, combined and interactive effects of light levels, nutrients, and other abiotic factors can affect primary production in aquatic ecosystems. In a large-scale comparison across several ecoregions in North America, Bott et al. (1985) concluded that the combined effects of photosynthetically active radiation (PAR), chlorophyll a, and water temperature accounted for $>70\%$ of the variation in community metabolism among streams. Similarly, Fleituch (1999) reported that benthic community metabolism along a river continuum was primarily influenced by physical factors, including solar radiation, riparian canopy, water temperature, and conductivity. It is well established that enrichment of aquatic ecosystems caused by excessive nutrients often stimulates primary production and causes excessive plant growth, including blooms of potentially toxic blue-green algae. Because these dense populations of algae limit light penetration, dramatic shifts in the structure and function of major primary producers may occur. Attached macrophytes, which are dependent on sufficient light levels penetrating from the surface, are often replaced by phytoplankton communities that are capable of remaining near the surface.

Because of its association with global climate change, ecologists have recently given special attention to the influence of CO_2 and other abiotic factors on primary productivity. Researchers hypothesize that if elevated CO_2 increases primary productivity, some of the excess anthropogenic carbon released from burning fossil fuels and land use changes may be sequestered into plant biomass. This response remains uncertain because of the potential for other factors (e.g., nutrients, light, temperature) to limit primary production in terrestrial and aquatic ecosystems. Melillo et al. (1993) used a terrestrial ecosystem model (TEM) to predict the effects of climate change and elevated CO_2 on NPP. Spatially referenced information on climate, soils, vegetation, water availability, and elevation were used to predict current NPP values for a wide variety of ecosystems. Model predictions of current NPP were very close to values based on field measurements. The model was then run to simulate responses of NPP to a doubling of CO_2 and associated changes in temperature, precipitation, and cloud cover as predicted by general circulation models (Figure 30.1). Overall global NPP increased by approximately 23%, but there was considerable variation among ecosystems. This geographic variation reflects not only how different ecosystems will respond to climate change but also the underlying mechanisms. For example, moist temperate ecosystems responded primarily to elevated temperature and increased nitrogen cycling whereas dry temperate ecosystems responded primarily to elevated levels of CO_2.

30.2.1.4 Global Patterns of Productivity

Productivity is not evenly distributed among regions of the world, and comparisons of NPP and biomass estimated by Whittaker and Likens (1973) for some of the world's major biomes reveal

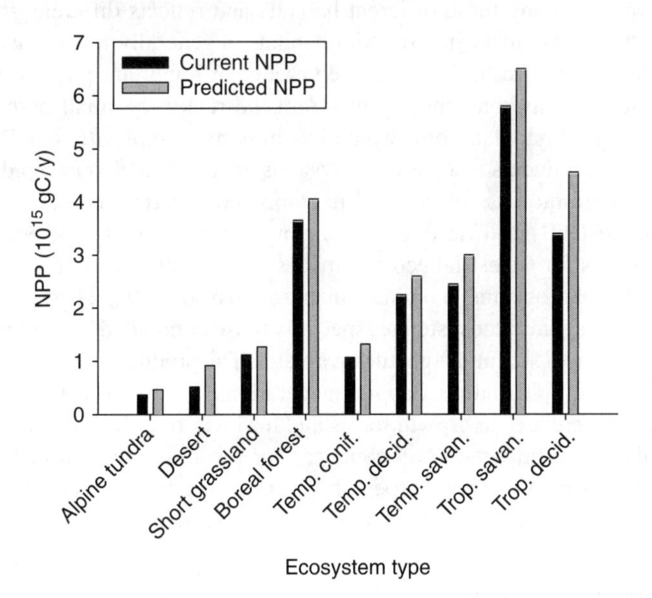

FIGURE 30.1 Results from a TEM used to predict the effects of a 2× increase in CO_2 and associated changes in temperature, precipitation, and cloud cover on NPP of different terrestrial ecosystems. (Data from Table 2 in Melillo et al. (1993).)

TABLE 30.1
Estimates of Net Primary Productivity and Biomass in the Earth's Major Biomes

	NPP ($g/m^2/y^2$)	Biomass (kg/m^2)
Terrestrial ecosystems		
Tropical forest	1800	42
Temperate forest	1250	32
Boreal forest	800	20
Temperate grassland	500	1.5
Alpine and tundra	140	0.6
Desert scrub	70	0.7
Aquatic ecosystems		
Algal beds and reefs	2000	2
Estuaries	1800	1
Lakes and streams	500	0.02
Continental shelf	360	0.01
Open ocean	125	0.003

Source: Data from Whittaker and Likens (1973).

several interesting patterns (Table 30.1). Although sunlight is necessary for primary production, it is evident from Table 30.1 that other factors contribute to global patterns. If adequate moisture or nutrients are not available, as in arid ecosystems or the open ocean, NPP will be low regardless of the levels of sunlight. In forest ecosystems, a general decrease in NPP is seen as we move to colder and more arid climates. The combination of sufficient sunlight, warm temperature, and abundant moisture results in very high productivity for tropical forests. Despite the generally low productivity of open ocean ecosystems, estuaries, algal beds, and coral reefs are among the most productive aquatic habitats.

Biomass also varies among these different habitats and reflects different growth forms of the major primary producers. Biomass in terrestrial habitats is generally much greater than in aquatic ecosystems, and this large terrestrial biomass represents an important pool of global carbon. The lower biomass in aquatic environments results from the relatively small body size of dominant primary producers (e.g., phytoplankton), which has important implications for trophic dynamics. Because small primary producers in aquatic ecosystems are capable of very rapid turnover, they can support a relatively large biomass of consumers compared to terrestrial ecosystems. The ratio of productivity to biomass (P:B) also varies greatly among different biomes and ecosystems. As shown in Table 30.1, P:B ratios for terrestrial ecosystems, especially forests, are relatively low, reflecting the large amount of nonphotosynthetic biomass in these ecosystems (e.g., bark, trunk, and branches). In contrast, P:B ratios in aquatic ecosystems, especially those dominated by phytoplankton, are much higher because of their small size and rapid turnover rates. The production values of lentic and marine phytoplankton reflect multiple and overlapping generations, but biomass is measured at a specific point in time. These differences in growth forms and turnover rates between terrestrial and aquatic ecosystems may also have important consequences for responses to anthropogenic disturbances. Because of their rapid growth rates, we expect that primary producers in aquatic ecosystems would respond more rapidly to contaminants than in terrestrial ecosystems.

30.2.2 Secondary Production

Secondary production is defined as the rate of productivity of consumers such as herbivores and predators that obtain their energy from plant or animal biomass. Consumers such as bacteria and

TABLE 30.2
Measures and Definitions of Ecosystem Energetics and Efficiencies

Measure	Definition
Consumption (C)	Total amount of energy consumed
Egestion (E)	Total amount of energy lost to egestion
Assimilation (C–E)	Total amount of energy available for production and respiration
Production (C–A)	Total amount of energy available for growth and reproduction
Assimilation efficiency (A/C × 100%)	Portion of consumed food that is assimilated
Net production efficiency (P/A × 100%)	Portion of assimilated food that is converted to new biomass
Gross production efficiency (P/C × 100%)	Portion of consumed food that is converted to new biomass
Trophic level efficiency (A_n/A_{n-1} × 100%)	Efficiency of transfer of assimilated energy between two trophic levels n, a consumer, and $n - 1$, the resource

fungi, organisms that obtain energy from decomposing plant and animal material, should also be included in measures of secondary production. Secondary productivity is similar to primary productivity in that we must distinguish between the portion of energy for growth and reproduction (and thus available to higher trophic levels) and the portion associated with maintenance costs of the consumer (Table 30.2). As noted above, the amount of energy available to consumers is ultimately determined by NPP and the efficiency with which fixed energy is converted to biomass. Similar to the bioenergetic approaches described for populations, ecosystem ecologists have identified several processes that limit efficiency of secondary production. Only a portion of the biomass consumed by herbivores or predators is actually assimilated. Because food quality for predators is generally greater than herbivores (i.e., proteins vs. recalcitrant cellulose and lignin), assimilation efficiency, which is defined as the fraction of consumed biomass that is assimilated (e.g., available for growth, reproduction, respiration, and maintenance), is generally greater in predators. In addition to the recalcitrant materials in plant tissue, herbivores must also contend with a diverse assortment of defensive chemicals produced by plants, which also limits consumption. Interestingly, coevolutionary responses to these natural defensive chemicals may also explain the well-developed detoxification systems in herbivores, which coincidentally provide protection against some xenobiotics. In contrast to terrestrial herbivores, assimilation efficiency is relatively high for zooplankton and other herbivores feeding on unprotected phytoplankton or algae.

30.2.2.1 Ecological Efficiencies

Only a small fraction of the assimilated energy in consumers is available for growth and reproduction; the remaining is necessary for maintenance and respiration. Because metabolic costs are generally greater in homeotherms than in poikilotherms, net production efficiency, defined as the amount of assimilated food available for new biomass, is generally lower in "warm-blooded" organisms (1–2%) than in "cold-blooded" organisms (5–10%). Gross production efficiency, defined as the amount of consumed food available for biomass, is a function of both assimilation and production efficiencies. Finally, trophic level efficiency (also called Lindeman's efficiency) is the efficiency of energy transfer between two trophic levels. Although trophic level efficiency averages around 10%, there is considerable variation among ecosystems (Pauly and Christensen 1995). The important point is not the universality of the figure but the relative inefficiency of ecological systems. The inefficiency of energy transfer also limits food chain length and the number of trophic levels in an ecosystem (Table 30.3). Ricklefs (1990) estimated the average length of food chains based on NPP, ecological efficiency, and energy flux of top predators for several different ecosystems using

TABLE 30.3
Comparison of Average NPP, Predator Ingestion Rates, Ecological Efficiencies, and Number of Trophic Levels in Marine and Terrestrial Ecosystems

Community Type	NPP ($kcal/m^2/y^2$)	Predator Ingestion ($kcal/m^2/y^2$)	Ecological Efficiency (%)	Number of Trophic Levels
Open ocean	500	0.1	25	7.1
Coastal marine	8000	10.0	20	5.1
Temperate grassland	2000	1.0	10	4.3
Tropical forest	8000	10.0	5	3.2

Source: Data from Ricklefs (1990).

the following equations:

$$E(n) = \text{NPP Eff}^{n-1} \tag{30.7}$$

$$n = 1 + \frac{\log[E(n)] - \log(\text{NPP})}{\log(\text{Eff})} \tag{30.8}$$

where n = number of trophic levels, $E(n)$ = energy available to a predator at a trophic level n, and Eff = geometric mean of the ecological efficiencies of transfer between each level. Results showed that the number of trophic levels was more closely related to ecological efficiency than overall NPP.

30.2.2.2 Techniques for Estimating Secondary Production

Estimates of secondary production for some species can be derived from measures of feeding rates, assimilation efficiencies, and respiration in the laboratory (Fitzpatrick 1973) or under controlled conditions (West 1968). However, determining secondary production in natural populations is more challenging and generally requires estimates of consumption, growth, and reproduction. Sophisticated bioenergetics models have been developed for some aquatic species such as large-mouth bass (Kitchell 1983). These individual-based models generally use laboratory-derived estimates of consumption, respiration and elimination, and then solve for growth.

$$\text{Consumption} = \text{Respiration} + \text{Wastes} + \text{Growth} \tag{30.9}$$

Several practical issues complicate our ability to estimate whole ecosystem production using these individual-based models. While estimates of secondary production for individual species, especially those for which we have a thorough understanding of natural history (Jordan et al. 1971, Kilgore and Armitage 1978), have been developed, integrating this information to derive secondary production estimates for whole ecosystems or even major components of ecosystems is challenging. Wiens (1973) estimated secondary production of grassland bird communities, and Chew and Chew (1970) examined energy relationships of dominant mammals in a desert shrub community. Perhaps the best examples have been developed in aquatic ecosystems where researchers have derived community-level estimates of secondary production for major taxonomic or functional feeding groups (Benke and Wallace 1980, 1997, Carlisle 2000, Fisher and Gray 1983). Secondary production (P) in benthic macroinvertebrates is obtained from estimates of biomass and growth rate using the following simple relationship:

$$P = \Sigma B_i \times g_i \tag{30.10}$$

where B_i and g_i are biomass and growth rates of the ith species.

While the methodology for estimating secondary production in aquatic ecosystems is well established, these are labor-intensive efforts. Estimating biomass for macroinvertebrates is relatively straightforward; however, the intensive sampling frequency necessary to determine growth rates of macroinvertebrates often limits application of this technique. Because of these methodological challenges, with few exceptions, secondary production has not received significant attention in the ecotoxicological literature. An example of one such exception, Carlisle (2000) constructed food webs based on quantitative analyses of macroinvertebrate secondary production for six different streams along a gradient of heavy metal pollution.

Difficulties quantifying the role of detritus and the imprecise assignment of organisms to different trophic groups are also impediments to studies of secondary production. Early attempts to quantify the relationship between NPP and secondary production should be evaluated cautiously because of the failure to appreciate the dominant role of decomposers and microbial production. The opinion of O'Neill et al. (1986) that "the trophic level concept is most useful as a heuristic device and tends to obscure, rather than illuminate, organizational principles of ecosystems" is likely shared by many ecosystem ecologists. The use of stable isotopes, described in Section 34.4.4, is one potential solution to this problem; however, relatively few studies have employed this technique in ecosystem-level studies of secondary production.

30.2.3 The Relationship between Primary and Secondary Production

Numerous studies have reported a direct quantitative relationship between primary productivity and secondary productivity or biomass of consumers (Coe et al. 1976, Cyr and Pace 1993, McNaughton et al. 1989). A major emphasis of the International Biological Program described in Chapter 29 was to understand the biological basis of productivity and to quantify relationships between primary and secondary production. Much of this research focused on understanding the underlying mechanisms and consequences of interactions between plants and consumers. Some of the strongest evidence to support the relationship between NPP and secondary production has been obtained from experimental introductions of nutrients to whole ecosystems. The predictable increases in both primary and secondary production illustrate the need to consider energy flow and nutrient cycling together when investigating ecosystem energetics. Intuitively, we would expect that herbivore biomass or production would increase with NPP; however, the nature of this relationship will likely vary among ecosystems and herbivore types. For example, because grassland herbivores consume a larger portion of NPP than forest herbivores (Whittaker 1975), the relationship between NPP and herbivore abundance in forest ecosystems is relatively weak (Figure 30.2). Concentrations of structural compounds, such as lignins and other recalcitrant materials that limit herbivory in terrestrial ecosystems, are generally lower in aquatic primary producers. Consequently, grazers in many aquatic ecosystems consume a large fraction of available biomass (30%–40%) and the relationship between primary and secondary production in these systems is generally much stronger. Because a greater fraction of NPP is removed in aquatic ecosystems, we also predict that predators would play a more important role in energy flow here than in terrestrial ecosystems (Cebrian and Lartigue 2004). These expectations are supported by studies showing the relative importance of top-down effects in aquatic ecosystems compared to terrestrial ecosystems (Strong 1992). Predator control over lower trophic levels, termed trophic cascades, has been frequently observed in aquatic ecosystems but only rarely in terrestrial ecosystems. Similarly, bottom-up control of herbivores and other consumers by nutrients and primary producers is quite common in many lentic ecosystems and is the mechanism responsible

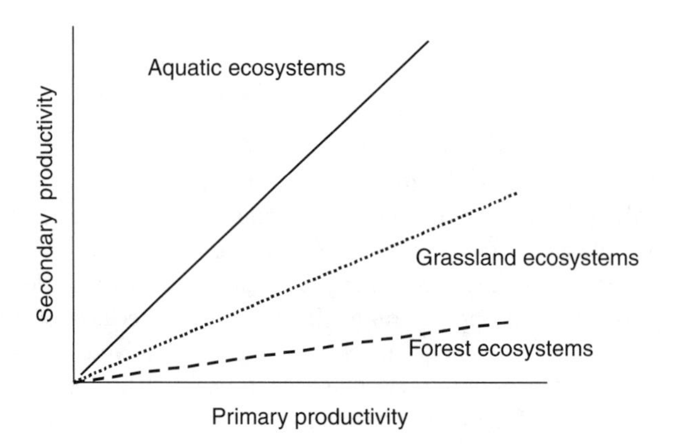

FIGURE 30.2 Hypothetical relationship between net primary productivity (NPP) and secondary productivity in aquatic and terrestrial ecosystems. Stronger relationships are expected in aquatic ecosystems because grazers consume a larger portion of plant biomass compared to terrestrial ecosystems.

for cultural eutrophication. Understanding the relationship between NPP and secondary production in ecosystems is important for predicting potential contaminant effects. It is possible that some of the variation in this relationship may account for differences in contaminant transfer rates among ecosystems. These issues will be explored in Chapter 34.

30.2.4 The River Continuum Concept

The movement of materials and energy in ecosystems has been investigated using a variety of descriptive, theoretical, and empirical approaches. Attempts to develop comprehensive explanatory models that connect physical, chemical, and biological processes have been especially successful in aquatic ecosystems. Vannote's classic paper "The river continuum concept" (Vannote et al. 1980) recognized that patterns and processes in streams change predictably from headwaters to the mouth. In addition to linking geomorphologic characteristics of a watershed to biological processes, this paper elucidated mechanisms responsible for the downstream transport, utilization, and storage of energy and materials. The major tenets of the river continuum concept (RCC) can be summarized by considering longitudinal changes in the sources of energy and materials from upstream to downstream (Figure 30.3). The relative importance of allochthonous and autochthonous sources of energy shift from upstream to downstream, resulting in changes in the ratio of NPP to respiration and structural alterations in the composition of stream communities. Shaded headwater streams are generally heterotrophic ($P/R<1$) because the dense riparian canopy in these systems limits primary productivity and contributes significant amounts of allochthonous materials. Further downstream, as the canopy opens, shading and the relative input of organic materials from riparian areas is reduced, and the stream becomes autotrophic ($P/R>1$). Finally, large rivers may return to heterotrophic conditions ($P/R<1$) because of increased depth and greater light attenuation.

Longitudinal changes in the abundance and composition of macroinvertebrate functional feeding groups (Cummins 1973) along the river continuum are hypothesized to reflect the relative importance of allochthonous and autochthonous inputs. The abundance of organisms that utilize coarse particulate organic material (CPOM) (e.g., leaf litter) is greatest in headwater streams and decreases downstream. Grazers, organisms that consume attached algae and periphyton, are more important in mid-order streams where light levels are highest. Finally, organisms that collect fine particulate organic material (FPOM), including collector-gatherers and collector-filterers, dominate larger rivers.

Tests of the predictions of the RCC in different geographic regions have provided good support for the major tenets in North America (Bott et al. 1985, Minshall et al. 1983) and Europe (Fleituch 1999). Minshall et al. (1983) measured benthic organic matter, community metabolism, decomposition,

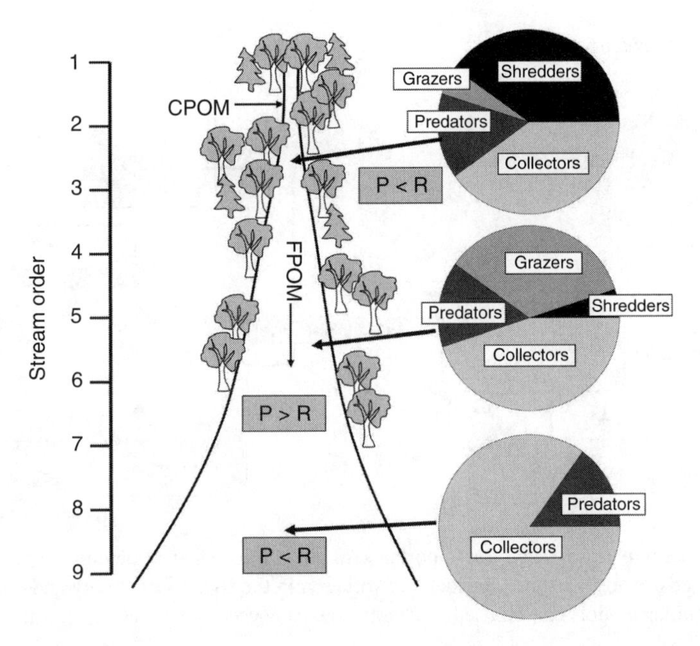

FIGURE 30.3 Major predictions of the river continuum concept showing changes in ecosystem energetics and abundance of major functional feeding groups along a longitudinal stream gradient. (Modified from Figure 1 in Vannote et al. (1980).)

and functional feeding group composition along longitudinal gradients in streams from four distinct geographic areas in North America. Although regional and local variation was observed, changes in structure and function from headwaters to downstream sites were consistent with predictions of the RCC. The RCC provided a unified theory describing structural and functional organization in stream ecosystems that clearly illustrated the important connections between upstream and downstream processes. Because of difficulty defining the spatial extent of stream ecosystems, streams of different size within the same drainage had previously been treated as completely different systems. By visualizing streams as a continuum of processes along a longitudinal gradient, ecologists recognized that ecosystem function occurring in small headwater streams or large rivers could be described using similar models (Figure 30.4). Despite differences in community composition and the relative importance of allochthonous and autochthonous inputs, similar processes operate in both headwater and mid-order streams. The RCC provided a conceptual framework for testing hypotheses about factors that regulate stream ecosystem dynamics and radically modified the way that ecologists visualized these systems.

30.3 NUTRIENT CYCLING AND MATERIALS FLOW THROUGH ECOSYSTEMS

In Chapter 29, we defined ecosystem ecology as the study of the movement of energy and materials through biotic and abiotic compartments of ecosystems. Figure 30.5 is a simple model illustrating the connection between biotic and abiotic compartments and the processes that influence the movement of materials through these compartments. Nutrients and other materials are assimilated from soil or water by autotrophic organisms (plants and autotrophic bacteria), passed on to consumers, and released back to abiotic compartments. The amount and availability of nutrients are among the most important factors that limit primary productivity. In addition to limiting growth rates of primary producers and heterotrophic microbes, nutrient availability also influences decomposition rates. The rate of nutrient cycling may also influence ecosystem resistance and the rate of recovery from natural

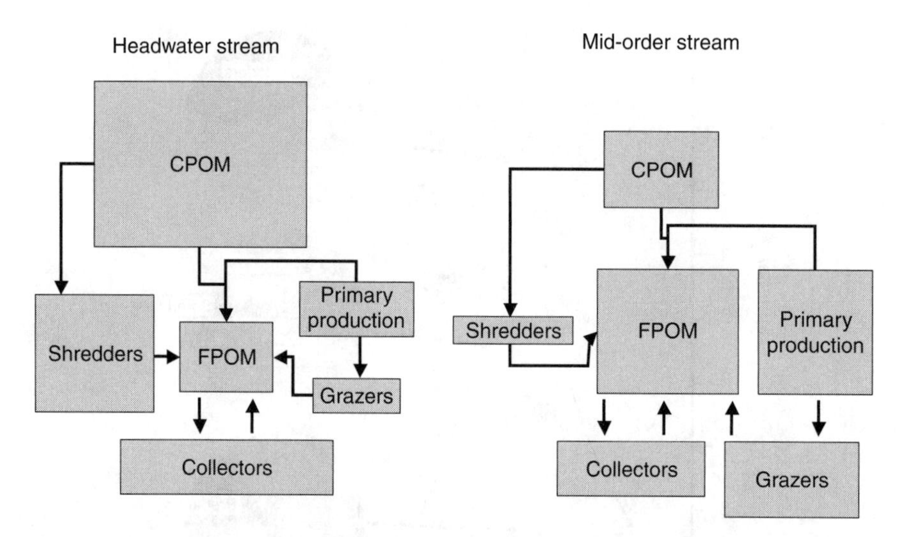

FIGURE 30.4 Energy pathways and the importance of allochthonous and autochthonous materials in headwater and mid-order streams. Despite considerable variation in the sources of energy and dominant functional feeding groups, similar models can be used to characterize ecosystem dynamics. (Modified from Figure 1 in Minshall et al. (1983).)

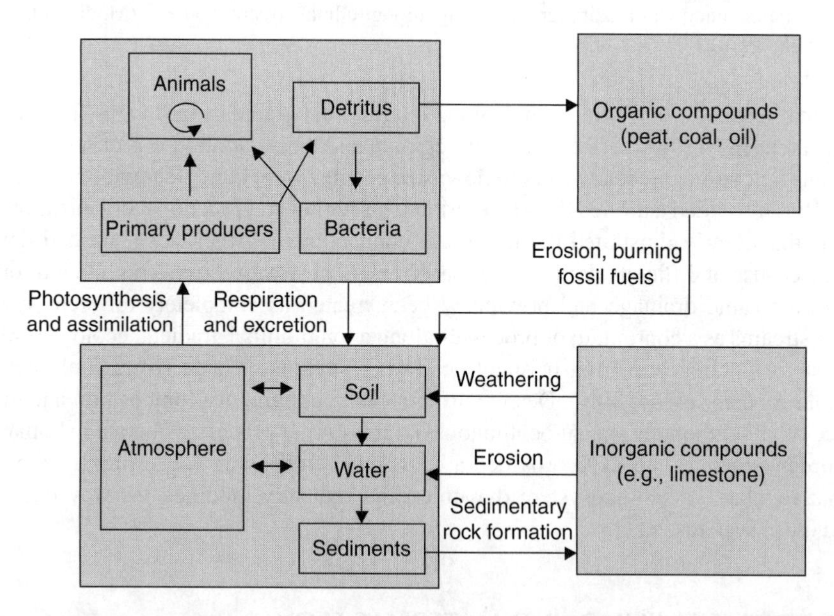

FIGURE 30.5 Simple model showing the connection between biotic and abiotic compartments and the dominant processes that regulate movement of organic and inorganic materials among compartments. (Modified from Figure 12.3 in Ricklefs (1990).)

and anthropogenic disturbances (DeAngelis et al. 1989). More importantly, an understanding of nutrient and material cycles is essential for predicting ecosystem consequences of increased anthropogenic inputs of certain materials, especially carbon, nitrogen, sulfur, and phosphorus. Predicting the consequences of altered biogeochemical cycles also requires that we consider the ecosystem as a unit instead of focusing only on component parts (O'Neill and Waide 1981). Because the behavior and pathway of many toxic chemicals follow those of natural elements in ecosystems, consideration of biogeochemical processes is essential for understanding fate and transport of contaminants.

Finally, we will see that many toxic chemicals have direct effects on ecosystems because they alter biogeochemical processes.

30.3.1 Energy Flow and Biogeochemical Cycles

The primary difference between movement of energy and nutrients through an ecosystem is that nutrients are retained and constantly cycled between biotic and abiotic components by a variety of meteorological, geological, and biological processes. In other words, the earth is considered an open system with respect to energy but a closed system with respect to materials. Energy fixed by primary producers is constantly supplied from an outside source and flows through the system. In contrast, nutrients such as C, N, and P are assimilated, transformed, and released back to the ecosystem, often in a very different form, where they can be used again. Meteorological processes include precipitation, snowmelt, and atmospheric deposition. The major geological process is weathering of materials from soils and underlying geological formations. Biological processes are analogous to those discussed for energy flow and include transfers and transformations that occur within food chains. If input of materials exceeds output, these unused materials may also accumulate in nutrient pools, and their rate of movement between different pools is called the flux rate. If a nutrient pool size is relatively constant, we can calculate the length of time an average molecule resides in this pool. Residence time is calculated as the pool volume divided by the outflow of materials and is a good measure of the accessibility of materials to organisms. For example, the atmosphere is a relatively active pool for oxygen, with a residence time of about 20,000 years. In contrast, the atmosphere is a storage pool for nitrogen, with a residence time of about 20 million years, reflecting the limited accessibility of atmospheric N to organisms.

At a local level, if we assume that nutrient concentrations within a compartment are at equilibrium (e.g., uptake is approximately equal to export), then measurement of uptake or loss provides an estimate of turnover time within a compartment (Figure 30.6). Turnover time of organic materials and nutrients varies among ecosystems and is closely related to climate and temperature. Table 30.4 shows the accumulation and mean turnover times for organic material and nitrogen in several different forest types located in different climatic zones. Nitrogen and organic matter accumulations are greatest in temperate coniferous forests, reflecting low rates of decomposition and efficient use of nutrients. The low rates of decomposition observed in cold boreal forests result in a larger fraction of organic material in soils relative to trees. Turnover time, the average amount of time that a molecule remains in soil before it is assimilated by plants, increases in colder climates and is longer in coniferous forests because foliage is not replaced each year.

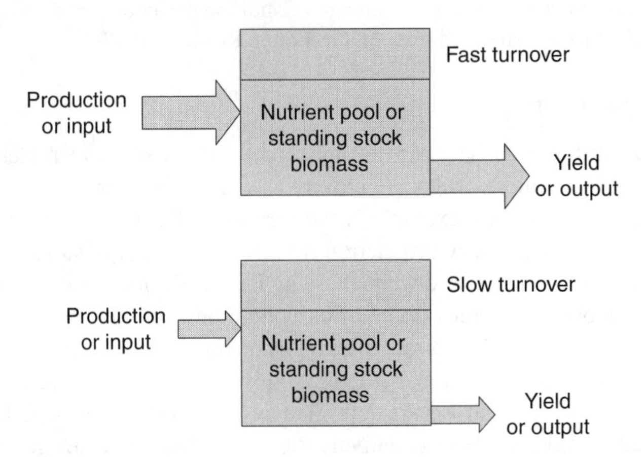

FIGURE 30.6 The relationship between nutrient input, output, and turnover time. (Modified from Figure 25.1 in Krebs (1994).)

TABLE 30.4
Accumulation and Turnover Times of Organic Matter and Nitrogen in Different Forest Types from Different Biomes

Forest Type	Organic Matter (kg/ha)	Turnover Time (y)	Nitrogen (kg/ha)	Turnover Time (y)
Boreal coniferous	226,000	353	3,250	230.0
Boreal deciduous	491,000	26	3,780	27.0
Temperate coniferous	618,000	17	7,300	17.9
Temperate deciduous	389,000	4	5,619	5.5

Source: Data from Cole and Rapp (1981).

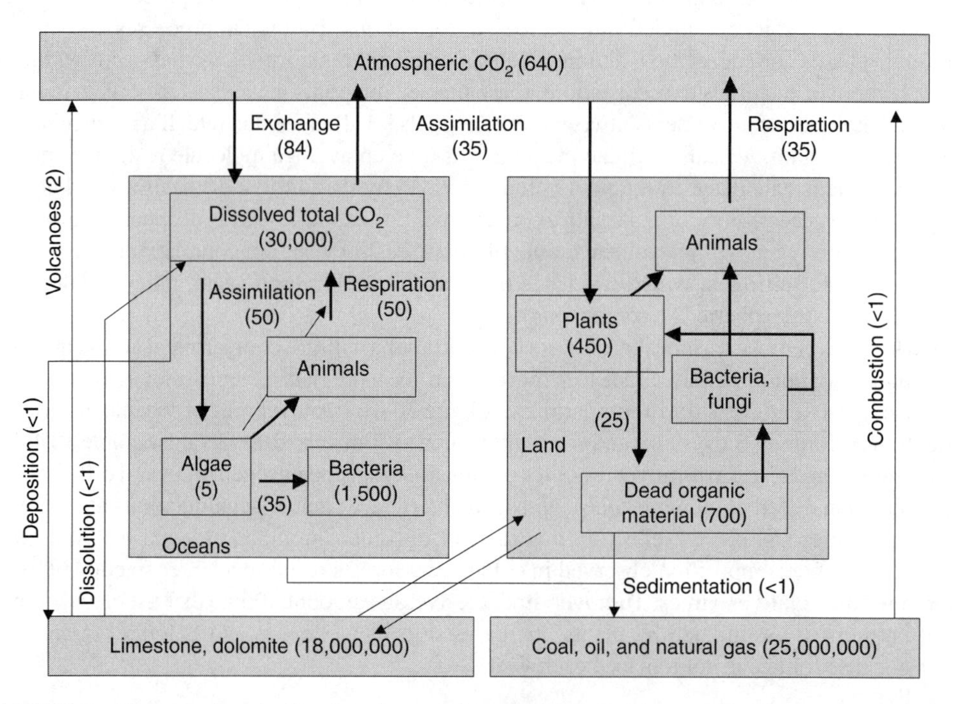

FIGURE 30.7 The carbon cycle. Numbers in parentheses indicate the amount of C in each compartment and moving between compartments. (Modified from Figure 12.6 in Ricklefs (1990).)

30.3.1.1 The Carbon Cycle

Despite fundamental differences between the movement of energy and materials, nutrients and other elements in biogeochemical cycles are closely associated with primary production and often follow the flow of energy. The best example is the carbon cycle, which is intimately connected to ecosystem metabolism and secondary production (Figure 30.7). Energy stored in carbohydrates by primary producers is ultimately released when these high energy compounds are oxidized to CO_2 by consumers and decomposers. The movement of carbon among compartments integrates biological processes such as assimilation and respiration with physical processes such as atmospheric-oceanic exchange, dissolution, and sedimentation. Although significant amounts of dissolved CO_2 are present in the oceans, the largest pools of carbon are deposited in sediments (limestone, dolomite) and stored as fossil fuels. Carbon dioxide occurs in a relatively low concentration in the atmosphere. Autotrophic organisms (primarily plants) assimilate CO_2 and incorporate it into organic matter by photosynthesis, and a portion of the CO_2 is returned to the atmosphere by respiration.

These physical, chemical, and biological processes are closely integrated in aquatic ecosystems through the carbonate–bicarbonate system. The CO_2 dissolved in water forms carbonic acid, which readily disassociates to bicarbonate and carbonate ions in the following reactions:

$$CO_2 + H_2O \leftrightarrow H_2CO_3 \tag{30.11}$$

$$H_2CO_3 \leftrightarrow H^+ + HCO_3^- \tag{30.12}$$

$$HCO_3^- \leftrightarrow H^+ + CO_3^- \tag{30.13}$$

Because these reactions are dependent on pH, the amount of calcium (which equilibrates with the bicarbonate and carbonate ions), and metabolism, they illustrate the close connection between biotic and abiotic components as well as the relationship between carbon flux and energy flow. For example, removal of CO_2 by photosynthesis or addition of CO_2 by respiration will drive these reactions to the left or right, respectively, thus influencing pH. Because the concentrations of H^+ and CO_3^- in aquatic ecosystems significantly influence the bioavailability of some contaminants, especially heavy metals, alterations in the carbonate–bicarbonate system have important toxicological implications.

30.3.1.2 Nitrogen, Phosphorus, and Sulfur Cycles

Phosphorus is a major limiting nutrient in aquatic ecosystems and primarily responsible for eutrophication of many lakes and streams. The P cycle is relatively simple because the atmosphere plays a relatively small role. Consequently, transport of P in ecosystems is primarily sedimentary and at a local scale. The major source of P to ecosystems is from underlying rocks (Figure 30.8), and loss from soils is usually balanced by releases of inorganic P from weathering. Plants assimilate phosphorus as phosphate (PO_4^{3-}), and availability and rate of uptake are dependent on pH. Herbivores

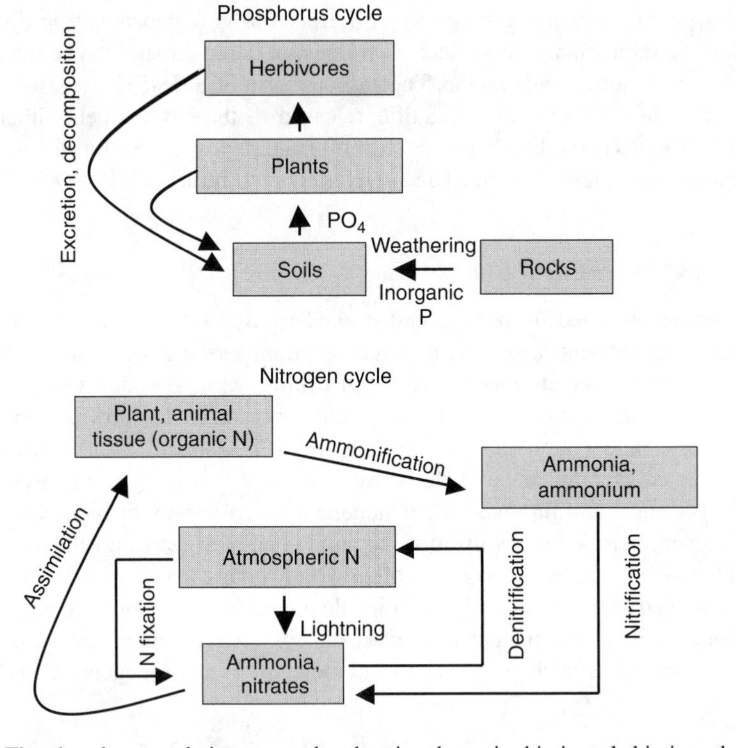

FIGURE 30.8 The phosphorus and nitrogen cycles showing the major biotic and abiotic pathways.

obtain all of their required P from consumption of plants, and P is returned to soil by excretion and decomposition. Whittaker's (1961) tracer study quantified the movement of ^{32}P in an aquatic microcosm from primary producers (phytoplankton, periphyton) to zooplankton. This study is an excellent demonstration of the intimate relationship between nutrient dynamics and energy flow through an ecosystem. Similar to the situation for many contaminants, this study also demonstrated that the ultimate fate of ^{32}P was sediments, which was shown to be a major reservoir of P in aquatic ecosystems.

The N cycle is considerably more complex because it involves a major atmospheric component and because N can exist in numerous oxidation states. Five basic processes drive the N cycle: nitrogen fixation, nitrification, assimilation, ammonification, and denitrification (Figure 30.8). The vast majority of N occurs in the atmosphere as molecular N_2, a form that is unavailable to plants. Nitrogen-fixing bacteria in soil (*Rhizobium*) and cyanobacteria (blue-green algae) in aquatic environments convert atmospheric N_2 to ammonia. Nitrification is the process by which ammonia (NH_3) is oxidized first to nitrites (NO_2^-) and then to nitrates (NO_3^-) by several different groups of bacteria, including *Nitrosomonas* and *Nitrobacter*. Plants assimilate N primarily as nitrate, and N is released back to soils through decomposition and ammonification. Under anoxic conditions, another group of denitrifying bacteria (*Pseudomonas*) reduces nitrates and releases inorganic N back to the atmosphere. On a global scale, nitrogen fixation is balanced by denitrification. Anthropogenic release of N to the biosphere has doubled the global rate of N fixation (Vitousek et al. 1997). These increases in global emissions of N have significant consequences for aquatic ecosystems, including eutrophication, toxic algal blooms (Burkholder and Glasgow 1997), and the formation of anoxic conditions in the Gulf of Mexico (Rabalais et al. 1998).

Compared with other biogeochemical cycles such as C, N, and P, anthropogenic alteration of the global sulfur (S) cycle by combustion of fossil fuels has been extreme. Approximately 60% of the global S emissions are anthropogenic, resulting in widespread acidification of terrestrial and aquatic ecosystems (Likens et al. 1996). The influence of acidification on ecosystem processes will be described in Section 35.2. The S cycle is also relatively complex because S can exist in several oxidation states, and conversion among these different forms is dependent on different types of bacteria. Sulfur in a sedimentary phase such as organic matter and rocks can be released by natural processes such as weathering and erosion. The gaseous form of S (H_2S) is released from volcanoes and decomposition of organic material. Sulfur released to the environment either by natural or anthropogenic processes is oxidized to sulfate (SO_4^{2-}) and deposited. Sulfur dioxide released from the combustion of fossil fuels is oxidized and converted to sulfuric acid (H_2SO_4).

30.3.2 Nutrient Spiraling in Streams

Unlike the situation observed in mature, undisturbed forests where most nutrients are generally retained, a fraction of nutrients and other materials are transported downstream in lotic ecosystems either in dissolved or particulate forms. Instead of cycling as observed in terrestrial systems, the movement of nutrients in lotic ecosystems is generally represented as a downstream spiral (Elwood et al. 1983, Webster and Patten 1979). Spiraling length is defined as the average distance that a molecule travels as it completes a cycle between organic and inorganic phases. The length of the spiral reflects uptake and turnover and is dependent on a variety of biotic and abiotic factors including the rate of microbial mineralization, stream temperature, stream velocity, the shape of the stream channel, and the number of snags and other woody debris that reduce downstream transport. Uptake length, defined as the distance that a molecule travels before sorption to particulate matter or uptake by organisms, is an important characteristic of ecosystem function. Uptake length essentially measures nutrient uptake efficiency and is a useful indicator of anthropogenic disturbance.

$$S_w = F_w / wU \tag{30.14}$$

where S_w = uptake length, F_w = downstream nutrient flux in water, U = uptake rate of nutrients from water, and w = average stream width. Uptake length generally increases with stream discharge and decreases with temperature, the amount of riparian vegetation, and biomass of detritus and algae.

In relatively small streams, attached algae, fungi, bacteria, and periphyton are responsible for most of the uptake of nutrients, which generally follows Michaelis–Menten kinetics. Some of these materials may be released back to the water column, but a fraction enters benthic food chains through grazing organisms. Nutrients and materials continue to spiral downstream in larger rivers, but as stream size increases nutrient dynamics in these systems more closely resemble patterns observed in lentic ecosystems. The same processes that determine retention and transport of nutrients have also been hypothesized to influence fate of contaminants in streams (Stewart and Hill 1993). It is well established that periphyton and attached algae are important sinks for contaminants in lotic ecosystems, and highly persistent chemicals may spiral downstream as they move between biotic and abiotic compartments.

30.3.3 Nutrient Budgets in Streams

The relative importance of allochthonous inputs to stream energy budgets has been well established. However, the contribution of these terrestrial inputs to nutrient dynamics has received considerably less attention. Whole ecosystem nutrient budgets have been calculated for a few relatively small watersheds. In one of the most comprehensive studies, Triska et al. (1984) measured inputs from litterfall, subsurface flow, and nitrogen fixation to develop an annual nitrogen budget for a small stream in the H.J. Andrews Experimental Forest, Oregon, USA. Most of the total annual nitrogen input (15.25 g/m^2) was from subsurface flow (11.06 g/m^2), with biological inputs contributing an additional 4.19 g/m^2. Direct and indirect biologically derived inputs from litterfall, throughfall, lateral movement, groundwater, and nitrogen-fixation accounted for over 90% of the total nitrogen to the stream. Total input of nitrogen was 34% greater than output, indicating that the stream was not operating at a steady state. The difference between input and output was primarily a result of storage of nitrogen as particulate organic matter.

30.3.3.1 Case Study: Hubbard Brook Watershed

Constructing nutrient budgets for whole ecosystems requires that we identify and measure the processes that control inputs and outputs. Researchers at Hubbard Brook Experimental Forest, a deciduous forest located in the White Mountains of New Hampshire, USA, have developed mass budgets for a variety of nutrients (Likens et al. 1970). Because the Hubbard Brook watershed is underlain by relatively impermeable metamorphic bedrock, inputs and outputs could be quantified by measuring stream discharge, precipitation, and concentrations of materials in precipitation and stream water. Because terrestrial losses of nutrients were eventually released to streams, measures of atmospheric input from precipitation and fluvial output from streams could be used to construct mass budgets for the entire watershed (Table 30.5). With the exceptions of NH_4^+ and NO_3^-, output of materials exceeded input, reflecting the weathering and leaching from soils and underlying rock. One of the most significant findings of the research was that flux of nutrients was relatively small compared to the pools of materials. This result demonstrates that, in stable watersheds such as Hubbard Brook, the majority of nutrients are usually retained and recycled. The export of sulfur, derived primarily from atmospheric sources, was an important exception to this pattern. The significance of this result will be discussed in Section 35.2.

30.3.3.2 Nutrient Injection Studies

Experimental injection of nutrients and tracers is the most direct method to examine retention and transport of nutrients in streams. The general approach is to add a small amount of a radioactive

TABLE 30.5
Mean Annual (1963–1968) Nutrient Budgets in the Hubbard Brook Experimental Forest, New Hampshire, USA (kg/ha/year)

	NH_4^+	NO_3^-	SO_4^{2-}	K^+	Ca^{2+}	Mg^{2+}	Na^+
Input	2.7	16.3	38.3	1.1	2.6	0.7	1.5
Output	0.4	8.7	48.6	1.7	11.8	2.9	6.9
Net change	2.3	7.6	−10.3	−0.6	−9.2	−2.2	−5.4

Source: Data from Likens et al. (1970).

or stable isotope at one point, and measure concentrations at several points downstream. Triska et al. (1989) reported that 29% of the N injected into a third-order forested stream in California was retained, while the remaining portion was transported downstream. Despite uptake by autotrophs, which was greatest during the day, nitrate concentrations increased downstream, indicating that the stream reach was a source of dissolved N to benthic communities. Decreased respiration and tissue C:N ratios downstream of the injection point indicated a biological response to N enrichment.

Although the preferred method to measure nutrient dynamics and uptake length in streams is to use tracers (e.g., ^{32}P, ^{15}N) that maintain ambient concentrations, because of expense and logistical issues, short-term additions that increase ambient nutrient concentrations are becoming increasingly common (Davis and Minshall 1999, Hall et al. 2002). Although this approach may be useful for comparing different streams, Mulholland et al. (2002) reported that uptake length was overestimated by short-term nutrient addition experiments compared to tracer additions. Because the degree of overestimation was related to the level of nutrient addition, these authors concluded that nutrient additions should be as low as possible, while maintaining the ability to accurately measure concentrations at several locations downstream.

30.3.4 Transport of Materials and Energy among Ecosystems

Our attempts to place spatial boundaries on ecosystems are compromised by our recognition that materials and energy readily move between ecosystems. Aquatic ecologists have long recognized the important connections between upland riparian and stream ecosystems. The significance of allochthonous detritus to streams from surrounding upland areas was first described by Hynes (1970) and figured prominently in the development of the RCC (Vannote et al. 1980). Streams serve to connect processes occurring in upland terrestrial systems to lakes and oceans (Fisher et al. 1998). However, streams are not simply conduits for nutrients and other chemicals but significantly alter the quality and quantity of transported materials through input, storage, and instream biogeochemical processes. The boundaries of stream ecosystems were previously considered to extend only a short distance into the riparian zone. However, stream ecologists now recognize the important linkages between streams and upland ecosystems (Fausch et al. 2002). Rather than visualizing stream ecosystems as longitudinal corridors, recent studies emphasize vertical and lateral connections between the stream and surrounding landscape. Fisher et al. (1998) extended the nutrient spiraling concept to include processes that occur outside of the stream. They suggested that the nested, concentric arrangement of subsystems within stream ecosystems (e.g., surface water, hyporheic zone, and riparian zone) is analogous to a telescope, where the length of the telescoping cylinders reflects the processing length of materials. The processing length is the linear distance required to transform an amount of material in transport. Thus, because processing length increases with disturbance, we would expect that impacted systems would have lower rates of materials cycling.

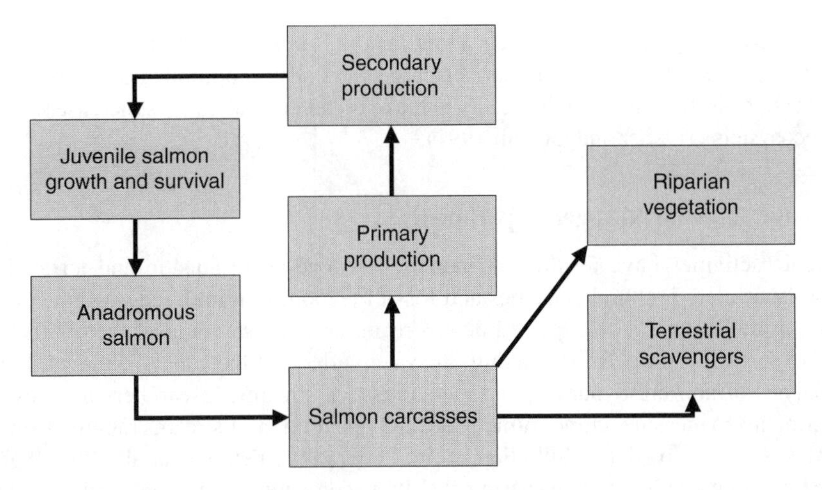

FIGURE 30.9 The influence of anadromous salmon on production in aquatic and terrestrial ecosystems.

Although we have traditionally considered the transport of nutrients and other materials in streams as a one-way process, in some instances nutrients exported downstream may be returned to headwaters. Marine-derived nutrients from anadromous Pacific salmon contribute significant amounts of organic matter and N when the salmon return to their natal streams to spawn (Figure 30.9). Because many of these streams are naturally oligotrophic and have a heavy canopy that limits primary productivity, these nutrient subsidies can be very important to ecosystem productivity. Studies using stable isotopic tracers of ^{15}N and ^{15}C have quantified the amount of marine-derived N and C delivered to streams and adjacent riparian habitats. Because salmon are enriched with heavier isotopes of N and C, comparisons of primary producers, consumers, riparian vegetation, and wildlife in streams with and without spawning salmon have revealed the importance of these subsidies. In addition to stimulation of primary producers and bottom-up effects on higher trophic levels (Bilby et al. 1996, Wipfli et al. 1998), nutrients and organic material from salmon carcasses increase productivity and diversity of riparian vegetation, and provide up to 25% of the N to riparian plants and 30–90% of the N to the diet of terrestrial scavengers (Naiman et al. 2002). Increased primary and secondary productivity associated with salmon carcasses translated to greater growth rates and survival of juveniles inhabiting these streams, providing a positive feedback for returning salmon (Bilby et al. 1996). Similarly, declines in the abundance and biomass of spawning salmon that return to their natal streams have important consequences for the function of both aquatic and adjacent terrestrial ecosystems. In addition to transporting marine-derived nutrients to headwater streams, anadromous fish also transfer contaminants from the oceans to relatively pristine streams (Ewald et al. 1998). Although there is little information on the effects of these marine-derived pollutants on headwater communities, levels of contaminants in resident (nonmigratory) species may increase significantly (Ewald et al. 1998).

30.3.5 Cross-Ecosystem Comparisons

The vast majority of studies investigating processes that control rates of material cycling have been conducted in single ecosystems or in a single type of ecosystem. The primary factors that influence the rates of material cycling and energy flow (e.g., temperature, precipitation, vegetation, underlying geology, and disturbance regime) vary significantly among ecosystems and geographic locations. Comparisons of nutrient cycles among different ecosystems (e.g., tropical vs. temperate forests; arid vs. humid grasslands; headwater streams vs. large rivers), at different altitudes, and among different geomorphological units improve our understanding of these regulating factors and provide insights into underlying mechanisms. Comparative studies across ecosystems may also provide the best opportunity to develop generalized models about material transfer and energy flow (Essington and Carpenter 2000). Cross-ecosystem comparisons are essentially an optimization problem, where the suitability of controls decreases but the ability to make broad generalizations increases among highly dissimilar ecosystems (Fisher and Grimm 1991).

30.3.5.1 Lotic Intersite Nitrogen Experiment

Anthropogenic activities have resulted in increased N loading to aquatic and terrestrial ecosystems primarily from agricultural activities and fossil fuel combustion. Because some of the excess N deposited in headwaters is transported downstream, an understanding of factors that control N uptake and export is essential for predicting ecosystem effects at the watershed level. In a comprehensive analysis of nutrient dynamics and metabolism in streams, researchers have used nutrient tracer experiments to measure ammonium uptake and retention in 11 streams ranging from the North Slope of Alaska to Puerto Rico (Mulholland et al. 2001, 2002, Peterson et al. 2001, Webster et al. 2003). The Lotic Intersite Nitrogen Experiment (LINX) compared stream metabolism and N dynamics in tropical, arid, temperate, and tundra streams. The goal of this large-scale comparative study was to relate inter-biome variability in stream metabolism and nutrient uptake to physical, chemical, and biological characteristics. Stream metabolism (i.e., autotrophic primary production, and autotrophic and heterotrophic respiration) was measured using the upstream–downstream diurnal dissolved oxygen technique. To measure ammonium uptake, $^{15}NH_4$ was injected in each stream and samples were collected at downstream sites to determine uptake length. Stream metabolism (GPP) was closely related to PAR (400–700 nm) and P concentration (Mulholland et al. 2001). Comparison of results across different biomes indicated that ammonium uptake length varied by approximately two orders of magnitude (14–1350 m) and increased with stream discharge (Webster et al. 2003). In shallow headwater streams with a higher surface-to-volume ratio, most of the uptake and removal processes occurred through assimilation by benthic autotrophic and heterotrophic organisms and by sorption to sediments (Peterson et al. 2001). Because a large fraction of N inputs to headwater streams was retained, especially during periods of high productivity, these systems regulated downstream transport to lakes, rivers, and estuaries and thereby may reduce eutrophication.

30.3.5.2 Comparison of Lakes and Streams

Comparative studies of lakes and streams provide an opportunity to assess factors that regulate movement of materials in two different types of ecosystems that vary widely in their major hydrologic characteristics. The unidirectional flow of water is the defining physical property of stream ecosystems. In contrast, lakes generally have low flushing rates, and therefore movement of materials is dominated by vertical exchanges between epilimnetic, metalimnetic, and hypolimnetic zones. The relative importance of allochthonous and autochthonous input is also quite different in lakes and streams. Because of these differences, approaches used by ecologists to quantify nutrient cycling in lakes and streams are quite different. Lake ecologists typically quantify the influence of nutrients on primary production, whereas stream ecologists are more concerned with uptake length. In lakes, uptake of nutrients occurs primarily by phytoplankton and bacteria, whereas periphyton and attached algae play more important roles in lotic ecosystems. Grazing by zooplankton and subsequent return of nutrients to the water column via excretion and mortality can occur rapidly in lakes. The rate at which nutrients are deposited to and released from sediments is dependent on lake size, the volume of throughflow, and seasonal changes in water temperature.

Because of differences in the movement of materials in lakes and streams, and the different methodological approaches, a common currency is necessary to compare factors that regulate nutrient cycling in lentic and lotic ecosystems. In both lakes and streams, cycling of nutrients is a function of uptake and export of dissolved and particulate materials (Figure 30.10). Essington and Carpenter (2000) developed a conceptual model to compare the recycling ratio, defined as the number of times

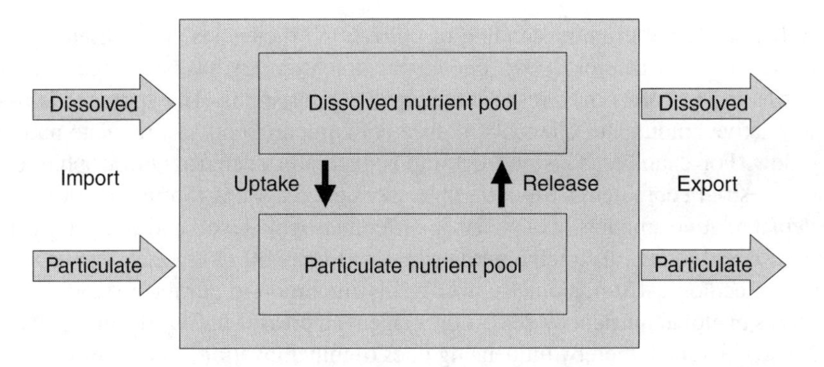

FIGURE 30.10 Conceptual model showing the import and export of dissolved and particulate nutrients in aquatic ecosystems. (Modified from Figure 1 in Essington and Carpenter (2000).)

that a nutrient molecule is used before it is exported from the system, in lakes and streams. The recycling ratio is a dimensionless number calculated as the ratio of uptake rate (U, mass × time^{-1}) to export rate (E, mass × time^{-1}). According to this model, mass-specific export rates of nutrients in particulate form (primarily to sediments) will be greater in lakes, whereas export rates of dissolved materials will be greater in streams. Nutrient cycling in streams is controlled primarily by the association of nutrients with particles and downstream transport. In lakes, nutrient cycling is controlled by physical processes that reduce the rate of sedimentation and biological processes that increase remineralization. Essington and Carpenter's (2000) model also predicts that the effects of consumers (grazers in streams; zooplankton in lakes) will be fundamentally different in lakes and streams. Although the model was developed to compare nutrient cycling in lakes and streams, it can be used to quantify the movement of contaminants through these systems. We would expect that movement of contaminants in streams would be dominated by downstream export in the dissolved phase, whereas movement in lakes would be dominated by sedimentation.

30.3.5.3 Comparisons of Aquatic and Terrestrial Ecosystems

Relatively few studies have compared ecosystem processes across terrestrial and aquatic ecosystems. In a comprehensive analysis of >800 aquatic and terrestrial systems, Cebrian and Largitue (2004) examined factors that controlled herbivory and decomposition rates. Although NPP varied greatly within aquatic and terrestrial ecosystems, there was surprisingly little variation between these ecosystem types when analyzed across all studies. Nutritional quality of primary producers was an important predictor of the proportion of NPP consumed by herbivores in both aquatic and terrestrial ecosystems, indicating the extent of top-down regulation of producer biomass. In contrast, while the total consumption by herbivores (g C/m^2/year) was correlated with NPP, it was unrelated to nutritional quality.

30.3.6 Ecological Stoichiometry

Energy has been considered the universal currency for studying ecosystem processes for several decades. Although the transfer of energy through aquatic and terrestrial foodwebs can reveal much about how ecosystems operate, our understanding of this process remains somewhat incomplete. Ecological stoichiometry improves on this single currency approach by using ratios of certain elements (C, N, and P) to characterize how composition of organisms and their prey affect nutrient cycling, production, and energy flow. Ecological stoichiometry is defined as the balance of multiple chemical substances in ecological interactions and processes (Sterner and Elser 2002). By comparing elemental ratios in abiotic compartments, primary producers, and consumers, limiting elements can be identified and a better understanding of utilization efficiencies and nutrient cycling can be obtained (Anderson and Boersma 2004). Ecological stoichiometry has been used to examine linkages between N and C cycles in terrestrial and aquatic ecosystems. The approach is based on the concept that relative amounts of elements in different compartments can regulate nutrient cycling and energy flow. For example, C:N and C:P ratios generally decrease with trophic level in both aquatic and terrestrial ecosystems. Simple trophochemical diagrams (Sterner and Elser 2002) may be used to depict relative amounts of elements in different trophic levels and compare spatial (among ecosystem types) and temporal stoichiometric patterns (Figure 30.11).

The study of ecological stoichiometry is critically important to our understanding of anthropogenic alterations in global nutrient cycles. Anthropogenic increases in CO_2, N, or P will significantly modify C:N and C:P ratios, thereby influencing rates of mineralization, energy flow, and decomposition. Analyses of global increases in these materials indicate that N and P are disproportionately enriched relative to C, thus potentially disrupting natural stoichiometric ratios of these elements (Table 30.6). Because N and P enrichment will likely influence nutrient use efficiency, models describing CO_2 effects on primary productivity cannot be considered in isolation but should also account for anthropogenic increases in other nutrients (Sterner and Elser 2002). Similarly, elevated

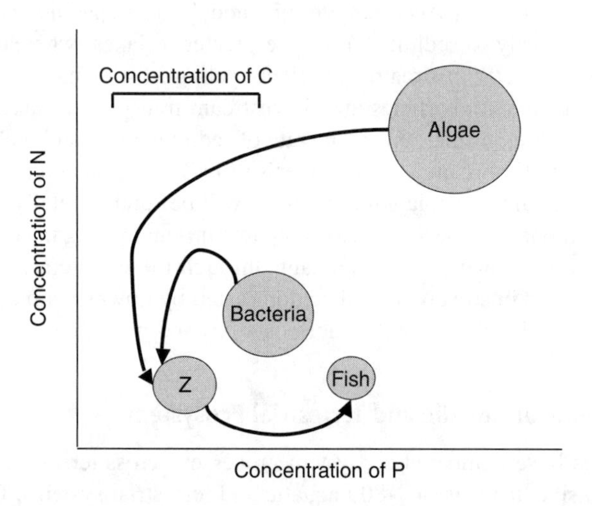

FIGURE 30.11 Simple trophochemical diagram used in ecological stoichiometry to depict relative amounts of elements in different trophic levels and to compare spatial (among ecosystem types) and temporal stoichiometric patterns. (Modified from Figure 8.1 in Sterner and Elser (2002).)

TABLE 30.6
Estimates of Natural and Anthropogenic Releases of Carbon, Nitrogen, and Phosphorus to Global Biogeochemical Cycles (Billions of Metric Tons per Year)

Source	C	N	P
Natural	61	0.13	0.003
Anthropogenic	8	0.14	0.012
Percent increase from anthropogenic sources	13	108	400

Source: From Falkowski, P., et al., *Science*, 290, 291–296, 2000.

CO_2 will likely produce leaf litter with greater C:N ratios, resulting in lower rates of mineralization and decomposition.

At a more local level, enrichment of certain elements resulting from anthropogenic releases may alter processes occurring in ecosystems. On the basis of an analysis of major ecosystem compartments (detritus, producers, and consumers), Dodds et al. (2004) measured C:N ratios and N-specific uptake rates in streams located in several biomes. The C:N ratios increased from consumers to primary producers to detritus. These researchers also reported a significant inverse relationship between C:N ratios and N-specific uptake. These findings have important implications for N export and eutrophication of downstream systems. Although undisturbed headwater streams are highly retentive of N, anthropogenic deposition of N to these watersheds will decrease C:N ratios and increase N export downstream (Dodds et al. 2004).

30.4 DECOMPOSITION AND ORGANIC MATTER PROCESSING

Decomposition is the process by which dead organic matter (detritus) is broken down to its component parts. Decomposition is a critical characteristic of ecosystem function and has been shown to vary among ecosystems and with levels of disturbance. Decomposition rates are determined by a complex interplay of physical, chemical, and biological processes. Factors such as chemical composition and nutritional quality of detritus, physicochemical characteristics of the habitat, and the abundances of decomposers influence decomposition rates. Decomposing organic material may be consumed by detritivores or further processed by bacteria and fungi, with CO_2 and inorganic materials (e.g., NH_4) as the end products. Organic material may be fragmented into smaller sizes by physical processes and invertebrates, thereby increasing surface area of particles. The net result is the conversion of particulate organic materials to dissolved constituents and the transfer of carbon to decomposers and detritivores. In terrestrial ecosystems, the most important macroinvertebrate decomposers include nematodes, insects, isopods, crustaceans, and oligochaetes. In aquatic ecosystems, CPOM is processed by shredders (primarily aquatic insects) and converted to FPOM.

30.4.1 ALLOCHTHONOUS AND AUTOCHTHONOUS MATERIALS

Secondary production of some aquatic ecosystems, particularly canopied headwater streams, may be heavily subsidized by inputs of allochthonous materials. In contrast to autochthonous energy derived from macrophytes, periphyton, phytoplankton, and other primary producers, allochthonous energy sources consist of organic materials and detritus derived from outside sources. Detritus-based food webs are common in many marine and lotic ecosystems. The relative contribution of allochthonous and autochthonous materials to secondary production is dependent on the dimensions and shape of the stream or lake and will be described in Section 31.2.3.

Although the elemental composition of leaf litter varies significantly among leaf species, much of the nutritional quality provided to detritivores is derived from bacteria and fungi that rapidly colonize leaf surfaces during decomposition (Suberkropp and Klug 1976). High nutritional quality of detritus is expected to increase growth rates and metabolic activity of detritivores and decomposers, thereby increasing the rate of decomposition. In general, nutritional quality of detritus, expressed as concentrations of N and P, is greater in aquatic ecosystems than in terrestrial systems. The rate of detrital production is also highly correlated with NPP in both aquatic and terrestrial ecosystems (Cebrian and Lartigue 2004). Therefore, factors discussed earlier that influence NPP such as light and nutrients also play a role in determining decomposition rates and the size of detritus pools.

Because the fraction of NPP consumed by herbivores is greater in aquatic than in terrestrial ecosystems (Cyr and Pace 1993), the amount of carbon channeled to detritus is generally greater in terrestrial systems. Nonetheless, detrital pathways dominate the flux of carbon and nutrients in most ecosystems, with greater than 50% of NPP going to decomposers (Cebrian and Lartigue 2004). The fate of this material and the relative importance of invertebrate detritivores and microbes have received considerable attention in the ecological literature. Studies conducted in aquatic and terrestrial ecosystems have quantified decomposition rates, usually expressed as rates of leaf litter decay. Various approaches have been employed to assess the relative importance of physical processes, macroinvertebrates, and microbial processes in litter decomposition. Computer simulations were used to calculate an organic matter budget for a second-order stream at Coweeta Hydrologic Laboratory, North Carolina (Webster 1983). Results showed that macroinvertebrate shredders were responsible for approximately 27% of the downstream transport of particulate organic material. Elimination of macroinvertebrate shredders using the pesticide methoxychlor (discussed in Section 23.4.1.2) from a stream in this same watershed resulted in significant changes in dynamics of detritus (Wallace et al. 1982).

The most compelling evidence demonstrating the significance of detritus to a stream food web comes from experimental studies that increased nutrient concentrations or eliminated terrestrial inputs. Elwood et al. (1981) showed that nutrient enrichment of a small heterotrophic stream increased decomposition rates, resulting in greater abundance of consumers and higher trophic levels. Wallace et al. (1997) used an overhead canopy to eliminate leaf litter inputs to a small headwater stream for 3 years. Benthic organic matter was lower in the treated stream, resulting in significant reductions in abundance and biomass of most major taxa and declines in overall secondary production. Predator production also declined in streams where leaf litter was eliminated, demonstrating strong bottom-up effects and important linkages between detritus and higher trophic levels.

30.4.2 Methods for Assessing Organic Matter Dynamics and Decomposition

Although the complex interactions that occur within ecosystems are grossly oversimplified in many food web diagrams, measuring the outcome of decomposition is relatively straightforward. Changes in mass (dry weight) or nutrient content of organic material over time is generally used as an indicator of decomposition rates. Long term assessments of organic matter dynamics can be obtained in terrestrial habitats by measuring changes in total soil organic carbon pools using combustion techniques. However, these methods do not provide an estimate of carbon availability, which is often independent of the amount of carbon stored in pools (Robertson and Paul 2000). Fractionation of organic carbon into humic acids, fulvic acids, and other constituents may provide some insight into the availability of carbon to microbial decomposers (McKnight 2001).

To measure decomposition rates, organic material such as leaf litter is placed in mesh bags in the field and then collected over a period of time (usually a geometric sequence of dates) to evaluate mass loss and nutrient changes. Leaf or litter materials are usually collected from representative and relevant species in the ecosystem. The decomposition rate constant, k, is estimated from the following equation:

$$X_t = X_0 \times e^{-kt} \tag{30.15}$$

where X_t is the mass remaining at time t and X_0 is the initial mass. The (first-order) decomposition rate constant (k) is the slope of the least squares regression of proportion mass remaining versus time. The magnitude of k reflects the rate of decomposition and varies greatly among different leaf species and different ecosystems (Figure 30.12). Mesh size of litter bags is often manipulated to exclude or include certain macroinvertebrate groups and to discern the relative importance of macroinvertebrates and microbes in decomposition. Mass loss may be related to the abundance or biomass of detritivores that colonized the bags during the exposure period. Because small mesh of litter bags may impede aeration or water flow, or create unnatural conditions, results of this technique should be interpreted cautiously. Leaf litter decomposition studies in terrestrial ecosystems are generally conducted over

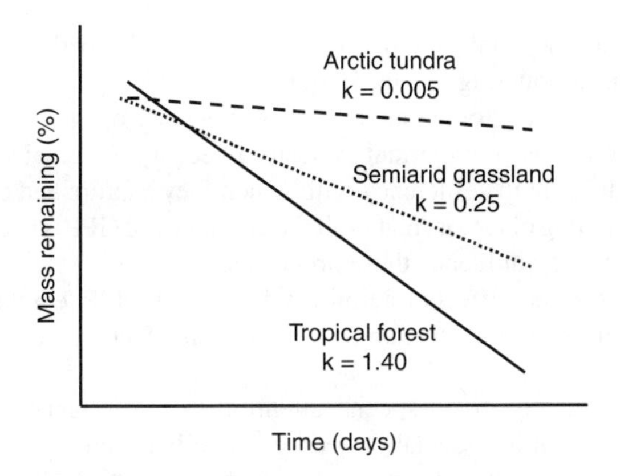

FIGURE 30.12 Hypothetical relationship between percent mass remaining and time in three ecosystems with varying decomposition rates. k = the decomposition rate constant.

longer time periods (1–3 years) than in aquatic ecosystems (3–6 months). Because contaminants will affect both microbial processes and abundance of macroinvertebrate decomposers, values of k are expected to decrease in stressed ecosystems.

30.5 SUMMARY

Flow of energy, cycling of nutrients, and decomposition of litter are fundamental processes in all ecosystems. A basic understanding of these processes and how they vary spatially and temporally is necessary to predict how these systems may respond to anthropogenic stressors (Howarth, 1991). Because the components of ecosystems (e.g., populations and communities) may respond quite differently to anthropogenic disturbances, an ecosystem perspective is necessary to limit the likelihood of ecological surprises described by Paine et al. (1998). Because pathways of energy flow and biogeochemical cycles in ecosystems are intimately coupled with the movement of contaminants, basic models developed by ecosystem ecologists may also help explain fate of xenobiotic chemicals. Each of the processes we have discussed in this chapter is likely to be affected by physical or chemical stressors to some degree. The relative sensitivity and variability of these responses compared with more traditional population- and community-level endpoints will likely vary among stressors and among ecosystem types. In the following two chapters we will consider descriptive and experimental approaches that have examined effects of stressors on these ecosystem processes.

30.5.1 SUMMARY OF FOUNDATION CONCEPTS AND PARADIGMS

- The perspective that ecosystems are highly complex may explain the relative infrequency with which ecosystem processes are measured in biological assessments.
- Before we can understand how ecosystems respond to contaminants and other anthropogenic perturbations, it is necessary to develop an appreciation for the ecosystem processes that are most likely to be affected.
- In addition to viewing ecosystems within a hierarchical context, ecologists routinely characterize ecosystems based on bioenergetic and biogeochemical processes.
- Flux of energy through an ecosystem is determined by the rate at which plants assimilate energy by photosynthesis, the transfer of this energy to herbivores and other consumers, and the efficiency of these conversions.
- The energy necessary for the conversion of CO_2 to a reduced state in carbohydrates is provided by visible light, and the total amount of energy fixed by plants is referred to as GPP.

- NPP is defined as the total amount of energy available to the plant for growth and reproduction after accounting for respiration.
- Because contaminants and other stressors can affect any of these processes, energy flux through an ecosystem is an important indicator in ecosystem-level assessments.
- Because plants differ in the efficiency with which they capture and convert incident sunlight, an understanding of factors that limit the efficiency of GPP is necessary to understand how contaminants may influence these processes.
- In addition to the direct effects of limiting factors on NPP, combined and interactive effects of light, nutrients, and other abiotic factors can affect primary production in aquatic ecosystems.
- Ecologists have recently given special attention to the influence of CO_2 on primary productivity because of its association with global climate change.
- Secondary production is defined as the rate of productivity of consumers such as herbivores and predators that obtain their energy from plant or animal biomass.
- The inefficiency of energy transfer limits food chain length and the number of trophic levels in an ecosystem.
- While the methodology for estimating secondary production in aquatic ecosystems is well established, most techniques are highly labor-intensive and consequently secondary production has not received significant attention in the ecotoxicological literature.
- Difficulties quantifying the role of detritus and the imprecise assignment of organisms to different trophic groups are significant impediments to the quantification of secondary production.
- Intuitively, we would expect that herbivore biomass or production would increase with NPP; however, the nature of this relationship varies among ecosystems and herbivore types.
- Because a greater fraction of NPP is removed in aquatic ecosystems, it is likely that predators would play a more important role in energy flow here than in terrestrial ecosystems.
- Variation in the relationship between NPP and secondary production may account for differences in contaminant transfer rates among ecosystems.
- The amount and availability of nutrients are among the most important factors that limit primary productivity and decomposition rates.
- Because the behavior of many toxic chemicals is similar to that of natural elements in ecosystems, consideration of biogeochemical processes is essential for understanding fate and transport of contaminants.
- The earth is considered an open system with respect to energy but a closed system with respect to materials such as C, N, and P.
- Despite fundamental differences between the movement of energy and materials, nutrients and other elements in biogeochemical cycles are closely associated with primary production and often follow the flow of energy.
- Unlike the situation observed in mature, undisturbed forests where most nutrients are generally retained, a significant fraction of nutrients and other materials are transported downstream in lotic ecosystems, a process known as nutrient spiraling.
- Experimental injection of nutrients and tracers is the most direct method to examine retention and spiraling of nutrients in streams.
- Rather than visualizing lotic ecosystems as longitudinal conduits for transporting nutrients and other chemicals downstream, ecologists now recognize the important linkages between streams and upland ecosystems.
- Although we have traditionally considered the transport of nutrients and other materials in streams as a one-way process, in some instances materials exported downstream may be returned to headwaters.

- Comparative studies across ecosystems such as the LINX project may provide the best opportunity to develop generalized models about material transfer and energy flow.
- Ecological stoichiometry, defined as the balance of multiple chemical substances in ecological interactions and processes (Sterner and Elser 2002), uses ratios of certain elements (C, N, and P) to characterize how composition of organisms and their prey affect nutrient cycling, production, and energy flow.
- Decomposition, defined as the process by which dead organic matter (detritus) is broken down to its component parts, is a fundamental process of ecosystems that varies significantly with levels of disturbance.
- Because the fraction of NPP consumed by herbivores is greater in aquatic than terrestrial ecosystems, the amount of carbon channeled to detritus is generally greater in terrestrial systems.
- Changes in mass (dry weight) or nutrient content of organic material over time is generally used as an indicator of decomposition rates.

REFERENCES

Anderson, T.R. and Boersma, M.R.D., Stoichiometry: Linking elements to biochemicals, *Ecology*, 85, 1193–1202, 2004.

Benke, A.C. and Wallace, J.B., Trophic basis of production among net-spinning caddisflies in a southern Appalachian stream, *Ecology*, 61, 108–118, 1980.

Benke, A.C. and Wallace, J.B., Trophic basis of production among riverine caddisflies: Implications for food web analysis, *Ecology*, 78, 1132–1145, 1997.

Bilby, R.E., Fransen, B.R., and Bisson, P.A., Incorporation of nitrogen and carbon from spawning coho salmon into the trophic system of small streams: Evidence from stable isotopes, *Can. J. Fish. Aquat. Sci.*, 53, 164–173, 1996.

Bott, T.L., Brock, J.T., Dunn, C.S., Naiman, R.J., Ovink, R.W., and Petersen, R.C., Benthic community metabolism in 4 temperate stream systems—An inter-biome comparison and evaluation of the river continuum concept, *Hydrobiologia*, 123, 3–45, 1985.

Burkholder, J.M. and Glasgow, H.B., *Pfiesteria piscida* and other *Pfiesteria*-like dinoflagellates: Behavior, impacts, and environmental controls, *Limnol. Oceanogr.*, 42, 1052–1075, 1997.

Carlisle, D.M., Bioenergetic food webs as a means of linking toxicological effects across scales of ecological organization, *J. Aquat. Ecosys. Stress Recov.*, 7, 155–165, 2000.

Cebrian, J. and Lartigue, J., Patterns of herbivory and decomposition in aquatic and terrestrial ecosystems, *Ecol. Monogr.*, 74, 237–259, 2004.

Chew, R.M. and Chew, A.E., Energy relationships of the mammals of a desert shrub (*Larrea tridentata*) community, *Ecol. Monogr.*, 40, 1–21, 1970.

Coe, M.J., Cumming, D.H., and Phillipson, J., Biomass and production of large African herbivores in relation to rainfall and primary production, *Oecologia*, 22, 341–354, 1976.

Cole, D.W. and Rapp, M., Elemental cycling in forest ecosystems, In *Dynamic Properties of Forest Ecosystems*, Reichle, D.E. (ed.), Cambridge University Press, Cambridge, UK, 1981.

Cooper, J.P. (ed.), *Photosynthesis and Productivity in Different Environments*, Cambridge University Press, London, 1975.

Cummins, K.W., Trophic relations of aquatic insects, *Ann. Rev. Entomol.*, 18, 183–206, 1973.

Cyr, H. and Pace, M.L., Magnitude and patterns of herbivory in aquatic and terrestrial ecosystems, *Nature*, 361, 148–150, 1993.

Davis, J.C. and Minshall, G.W., Nitrogen and phosphorus uptake in two Idaho (USA) headwater streams, *Oecologia*, 119, 247–255, 1999.

DeAngelis, D.L., Bartell, S.M., and Brenkert, A.L., Effects of nutrient cycling and food-chain length on resilience, *Nature*, 134, 778–805, 1989.

Dodds, W.K., Marti, E., Tank, J.L., Pontius, J., Hamilton, S.K., Grimm, N.B., Bowden, W.B., et al., Carbon and nitrogen stoichiometry and nitrogen cycling rates in streams, *Oecologia*, 140, 458–467, 2004.

Elton, C., *Animal Ecology*, Macmillan, New York, 1927.

Elwood, J.W., Newbold, J.D., O'Neill, R.V., and van Winkle, W., Resource spiralling: An operational paradigm for analyzing lotic ecosystems, In *Dynamics of Lotic Ecosystems*, Fontaine, T.D. and Bartell, S.M. (eds.), Ann Arbor Science Publishers, Ann Arbor, MI, 1983, pp. 3–28.

Elwood, J.W., Newbold, J.D., Trimble, A.F., and Stark, R.W., The limiting role of phosphorus in a woodland stream ecosystem: Effects of P enrichment on leaf decomposition and primary producers, *Ecology*, 62, 146–158, 1981.

Essington, T.E. and Carpenter, S.R., Nutrient cycling in lakes and streams: Insights from a comparative analysis, *Ecosystems*, 3, 131–143, 2000.

Ewald, G., Larsson, P., Linge, H., Okla, L., and Szarzi, N., Biotransport of organic pollutants to an inland Alaska lake by migrating sockeye salmon (*Oncorhynchus nerka*), *Arctic*, 51, 40–47, 1998.

Falkowski, P., Scholes, R.J., Boyle, E., Canadell, J., Canfield, D., Elser, J., Gruber, N., et al., The global carbon cycle: A test of our knowledge of earth as a system, *Science*, 290, 291–296, 2000.

Fausch, K.D., Torgersen, C.E., Baxter, C.V., and Li, H.W., Landscapes to riverscapes: Bridging the gap between research and conservation of stream fishes, *Bioscience*, 52, 483–498, 2002.

Fisher, S.G. and Gray, L.J., Secondary production and organic matter processing by collector macroinvetebrates in a desert stream, *Ecology*, 64, 1217–1224, 1983.

Fisher, S.G. and Grimm, N.B., Streams and disturbance: Are cross-ecosystem comparisons useful?, In *Comparative Analyses of Ecosystems: Patterns, Mechanisms, and Theories*, Cole, J.J., Lovett, G., and Findlay, S. (eds.), Springer-Verlag, New York, 1991, pp. 196–221.

Fisher, S.G., Grimm, N.B., Marti, E., Holmes, R.M., and Jones, J.B., Material spiraling in stream corridors: A telescoping ecosystem model, *Ecosystems*, 1, 19–34, 1998.

Fitzpatrick, L.C., Energy allocation in the Allegheny Mountain salamander, *Desmognathus ochrophaeus*, *Ecol. Monogr.*, 43, 43–58, 1973.

Fleituch, T., Responses of benthic community metabolism to abiotic factors in a mountain river in southern Poland, *Hydrobiologia*, 380, 27–41, 1999.

Hall, R.O., Bernhardt, E.S., and Likens, G.E., Relating nutrient uptake with transient storage in forested mountain streams, *Limnol. Oceanogr.*, 47, 255–265, 2002.

Howarth, R.W., Comparative responses of aquatic ecosystems to toxic chemical stress, In *Comparative Analyses of Ecosystems: Patterns, Mechanisms, and Theories*, Cole, J., Lovett, G., and Findlay, S. (eds.), Springer-Verlag, New York, 1991, pp. 169–195.

Howarth, R.W. and Michales, A.F., The measurement of primary production in aquatic ecosystems, In *Methods in Ecosystem Science*, Sala, O.E., Jackson, R.B., Mooney, H.A., and Howarth, R.A. (eds.), Springer, New York, 2000, pp. 72–85.

Hynes, H.B.N., *The Ecology of Running Waters*, University of Toronto Press, Toronto, 1970.

Jordan, P.A., Botkin, D.B., and Wolfe, M.L., Biomass dynamics in a moose population, *Ecology*, 52, 147–152, 1971.

Kilgore, D.L. and Armitage, K.B., Energetics of yellow-bellied marmot populations, *Ecology*, 59, 78–88, 1978.

Kitchell, T.B., Energetics, In *Fish Biomechanics*, Webb, P.W. and Weih, D. (eds.), Praeger, New York, 1983, pp. 312–338.

Krebs, C.J., *Ecology: The Experimental Analysis of Distribution and Abundance*, 4th ed., Harper Collins College Publisher, New York, 1994.

Likens, G.E., Bormann, F.H., Johnson, N.M., Fisher, D.W., and Pierce, R.S., Effects of forest cutting and herbicide treatment on nutrient budgets in Hubbard Brook Watershed-Ecosystem, *Ecol. Monogr.*, 40, 23–47, 1970.

Likens, G.E., Driscoll, C.T., and Buso, D.C., Long-term effects of acid rain: Response and recovery of a forest ecosystem, *Science*, 272, 244–246, 1996.

Lindeman, R.L., The trophic-dynamic aspect of ecology, *Ecology*, 23, 399–418, 1942.

McKnight, D.M., Boyer, E.W., Westerhoff, P.K., Doran, P.T., Kulbe, T., and Andersen, D.T., Spectrofluorometric characterization of dissolved organic matter for indication of precursor organic material and aromaticity, *Limnol. Oceanogr.*, 46, 38–48, 2001.

McNaughton, S.J., Oesterheld, M., Frank, D.A., and Williams, K.J., Ecosystem-level patterns of primary productivity and herbivory in terrestrial habitats, *Nature*, 341, 142–144, 1989.

Melillo, J.M., McGuire, A.D., Kicklighter, D.W., Moore, B., Vorosmarty, C.J., and Schloss, A.L., Global climate change and terrestrial net primary production, *Nature*, 363, 234–240, 1993.

Menzel, D.W. and Ryther, J.H., Nutrients limiting the production of phytoplankton in the Sargasso Sea, with special reference to iron, *Deep Sea Res.*, 7, 276–281, 1961.

Minshall, G.W., Petersen, R.C., Cummins, K.W., Bott, T.L., Sedell, J.R., Cushing, C.E., and Vannote, R.L., Interbiome comparison of stream ecosystem dynamics, *Ecol. Monogr.*, 53, 1–25, 1983.

Mulholland, P.J., Fellows, C.S., Tank, J.L., Grimm, N.B., Webster, J.R., Hamilton, S.K., Marti, E., et al., Inter-biome comparison of factors controlling stream metabolism, *Freshw. Biol.*, 46, 1503–1517, 2001.

Mulholland, P.J., Tank, J.L., Webster, J.R., Bowden, W.B., Dodds, W.K., Gregory, S.V., Grimm, N.B., et al., Can uptake length in streams be determined by nutrient addition experiments? Results from an interbiome comparison study, *J. N. Am. Benthol. Soc.*, 21, 544–560, 2002.

Naiman, R.J., Bilby, R.E., Schindler, D.E., and Helfield, J.M., Pacific salmon, nutrients, and the dynamics of freshwater and riparian ecosystems, *Ecosystems*, 5, 399–417, 2002.

O'Neill, R.V., Perspectives on economics and ecology, *Ecol. Appl.*, 6, 1031–1033, 1996.

O'Neill, R.V., DeAngelis, D.L., Waide, J.B., and Allen, T.F.H., *A Hierarchical Concept of Ecosystems*, Princeton University Press, Princeton, NJ, 1986.

O'Neill, R.V. and Waide, J.B., Ecosystem theory and the unexpected: Implications for environmental toxicology, In *Management of Toxic Substances in our Ecosystems*, Cornaby, B.W. (ed.), Ann Arbor Science Publ., Inc., Ann Arbor, MI, 1981.

Odum, E.P., Energy flow in ecosystems: A historical review, *Am. Zool.*, 8, 11–18, 1968.

Paine, R.T., Tegner, M.J., and Johnson, E.A., Compounded perturbations yield ecological surprises, *Ecosystems*, 1, 535–545, 1998.

Pauly, D. and Christensen, V., Primary productivity required to sustain global fisheries, *Nature*, 374, 255–257, 1995.

Peterson, B.J., Wollheim, W.M., Mulholland, P.J., Webster, J.R., Meyer, J.L., Tank, J.L., Marti, E., et al., Control of nitrogen export from watersheds by headwater streams, *Science*, 292, 86–90, 2001.

Rabalais, N.N., Turner, R.E., Wiseman, W.J., and Dortch, Q., Consequences of the 1993 Mississippi River flood in the Gulf of Mexico, *Regul. Rivers Res. Manag.*, 14, 161–177, 1998.

Ricklefs, R.E., *Ecology*, 3rd ed., W.H. Freeman & Company, New York, 1990.

Robertson, G.P. and Paul, E.A., Decomposition and soil organic dynamics, In *Methods in Ecosystem Science*, Sala, O.E., Jackson, R.B., Mooney, H.A., and Howarth, R.A. (eds.), Springer, New York, 2000, pp. 104–116.

Ryther, J.H. and Dunstan, W.M., Nitrogen, phosphorus, and eutrophication in the coastal marine environment, *Science*, 171, 1008–1013, 1971.

Schindler, D.W., Eutrophication and recovery in experimental lakes: Implications for management, *Science*, 184, 897–899, 1974.

Sterner, R.W. and Elser, J.J., *Ecological Stoichiometry: The Biology of Elements from Molecules to the Biosphere*, Princeton University Press, Princeton, NJ, 2002.

Stewart A.J. and Hill, W.R., Grazers, periphyton and toxicant movement in streams, *Environ. Toxicol. Chem.*, 12, 955–957, 1993.

Strong, D.R., Are trophic cascades all wet? Differentiation and donor-control in speciose ecosystems, *Ecology*, 73, 747–754, 1992.

Suberkropp, K. and Klug, M.J., Fungi and bacteria associated with leaves during processing in a woodland stream, *Ecology*, 57, 707–719, 1976.

Triska, F.J., Kennedy, V.C., Avanzino, R.J., Zellweger, G.W., and Bencala, K.E., Retention and transport of nutrients in a third-order stream: Channel processes, *Ecology*, 70, 1877–1892, 1989.

Triska, F.J., Sedell, J.R., Cromack, K., Jr., Gregory, S.V., and McCorison, F.M., Nitrogen budget for a small coniferous forest stream, *Ecol. Monogr.*, 54, 119–140, 1984.

Vannote, R.L., Minshall, G.W., Cummins, K.W., Sedell, J.R., and Cushing, C.E., The river continuum concept, *Can. J. Fish. Aquat. Sci.*, 37, 130–137, 1980.

Vitousek, P.M., Aber, J.D., Howarth, R.W., Likens, G.E., Matson, P.A., Schindler, D.W., Schlesinger, W.H., and Tilman, D.G., Human alteration of the global nitrogen cycle: Sources and consequences, *Ecol. Appl.*, 7, 737–750, 1997.

Wallace, J.B., Eggert, S.L., Meyer, J.L., and Webster, J.R., Multiple trophic levels of a forested stream linked to terrestrial litter inputs, *Science*, 277, 102–104, 1997.

Wallace, J.B., Webster, J.R., and Cuffney, T.F., Stream detritus dynamics: Regulation by invertebrate consumers, *Oecologia*, 53, 197–200, 1982.

Webster, J.R., The role of benthic macroinvertebrates in detritus dynamics of streams: A computer simulation, *Ecol. Monogr.*, 53, 383–404, 1983.

Webster, J.R., Mulholland, P.J., Tank, J.L., Valett, H.M., Dodds, W.K., Peterson, B.J., Bowden, W.B., et al., Factors affecting ammonium uptake in streams—An inter-biome perspective, *Freshw. Biol.*, 48, 1329–1352, 2003.

Webster, J.R. and Patten, B.C., Effects of watershed perturbation on stream potassium and calcium dynamics, *Ecol. Monogr.*, 49, 51–72, 1979.

West, G.C., Bioenergetics of captive willow ptarmigan under natural conditions, *Ecology*, 49, 1035–1045, 1968.

Whittaker, R.H., Experiments with radiophosphorus tracer in aquarium microcosms, *Ecol. Monogr.*, 31, 157–188, 1961.

Whittaker, R.H., *Communities and Ecosystems*, 2nd ed., Macmillan, New York, 1975.

Whittaker, R.H. and Likens, G.E., Primary production: The biosphere and man, *Science*, 1, 357–369, 1973.

Wiens, J.A., Pattern, and process in grassland bird communities, *Ecol. Monogr.*, 43, 237–270, 1973.

Wipfli, M.S., Hudson, J., and Caouette, J., Influence of salmon carcasses on stream productivity: Response of biofilm and benthic macroinvertebrates in southeastern Alaska, USA, *Can. J. Fish. Aquat. Sci.*, 55, 1503–1511, 1998.

31 Descriptive Approaches for Assessing Ecosystem Responses to Contaminants

31.1 INTRODUCTION

Now that we have an appreciation of the important processes that characterize ecosystems and the general approaches used to quantify these processes, we will turn our attention to the primary objective of this section. As with community-level assessments, ecotoxicologists interested in ecosystem responses to anthropogenic stressors employ descriptive, quasi-experimental, and experimental approaches. In the following two chapters, we will explore the use of these observational and experimental studies to link changes in primary and secondary production, nutrient cycling, and decomposition to contaminants and other anthropogenic stressors. In a separate chapter, we will consider effects of globally distributed and atmospheric stressors (e.g., acidification, NO_x deposition, elevated CO_2, and UV radiation) on these ecosystem processes.

Investigations of ecosystem processes may be conducted across a range of spatial and temporal scales. Functional measures such as community metabolism or nutrient transport can be measured in isolated soil microbial systems or in whole forests or watersheds. However, as we move up the hierarchy of biological organization from individuals → populations → communities → ecosystems, we generally increase the spatial and temporal scales of our investigations. Because many experimental studies of ecosystem processes are often limited in spatial and temporal scale, descriptive approaches can provide very compelling and ecologically realistic results. As discussed in Chapter 23 for communities, the typical trade-off is that observational or correlative investigations only provide a catalog of potential causal explanations. A more powerful case for causation in descriptive studies can be established by the application of strong inference (Platt 1964), other formal inferential methods such as stressor identification (Suter et al. 2002), or Bayesian inferential techniques.

The initial definition of ecological integrity proposed by Karr (1991) included both structural and functional measures, and most ecologists would agree that assessing effects of anthropogenic stressors on ecosystems requires adequate characterization of both patterns and processes. The efficacy of using functional measures to assess ecosystem responses to contaminants has received limited attention. As a consequence, development of functional criteria as indicators of ecological integrity has lagged behind more traditional approaches based on community structure (Bunn and Davies 2000, Gessner and Chauvet 2002, Hill et al. 1997). Kersting (1994) provides an excellent review of literature on the use of functional endpoints in freshwater field tests for hazard assessment of chemicals. Some assessments of ecological integrity measure patterns of community composition as a surrogate for ecosystem processes (Bunn and Davies 2000); however, patterns and processes are not necessarily related in some instances, especially in systems where disturbance is relatively weak. For example, Bunn and Davies (2000) measured stream metabolism at seven sites in southwestern Australia and related ecosystem processes to community structure. Although changes in gross primary production (GPP) and respiration were related to water quality, there was no relationship between water quality and macroinvertebrate community structure.

The characterization of ecological integrity based exclusively on structural measures is inconsistent with how most ecologists view ecosystems (Gessner and Chauvet 2002). We believe that restricting our analyses to mainly structural measures has provided a somewhat incomplete picture of how ecosystems respond to and recover from anthropogenic disturbances. Issues such as relative sensitivity, response variability, and functional redundancy have been considered when comparing the usefulness of structural and functional measures (Howarth 1991, Schindler 1987, 1988). Despite concern that changes in some ecosystem processes occur only after compositional changes and are therefore less useful, alterations in material cycling and energy flow are such fundamental properties of ecosystems that they should be included in ecological assessments. Leland and Carter (1985) argue that some functional processes in ecosystems are easier to quantify than relationships between abundance and environmental variables. These functional measures also integrate general characteristics of diverse communities, thus facilitating comparisons among different ecosystems. Given the recent interest in making comparisons across relatively broad geographic regions, functional measures may be more useful than structural measures because they are not dependent on specific taxa that are often restricted to a specific region. Finally, because causal mechanisms that control ecosystem processes are generally well understood, restoration strategies may be more obvious when based on functional measures (Bunn and Davies 2000).

Although papers that report relative sensitivity of community metrics to contaminants are common in the literature (Carlisle and Clements 1999, Kilgour et al. 2004), surprisingly few studies have compared responses across levels of biological organization (Adams et al. 2002, Bendell-Young et al. 2000, Cottingham and Carpenter 1998, Niemi et al. 1993, Sheehan 1984, Sheehan and Knight 1985). There is also the perception that quantifying ecosystem responses is logistically challenging compared to structural measures (Crossey and La Point 1988), an idea that has not been rigorously examined in the literature. Thus, we believe that it is premature to conclude that ecosystem processes are less sensitive or less reliable indicators of stress. In fact, some studies have reported that changes in ecosystem processes may occur in the absence of alterations in community structure (Bunn and Davies 2000). Niemi et al. (1993) reported that functional measures such as GPP were more sensitive indicators of recovery than structural measures. Similarly, Clements (2004) observed that community respiration was generally more sensitive to heavy metals than common structural measures such as abundance and species richness. Because alterations in community structure are not necessarily related to ecosystem processes, we view these as complementary measures for assessing ecological integrity. More important, simultaneous assessment of pattern and process can provide insight into the mechanistic linkages between stressors and responses. Studies by Wallace and colleagues (Wallace et al. 1996) provide some of the best examples demonstrating how contaminant-induced alterations in structural characteristics (e.g., elimination of macroinvertebrate shredders) directly influence ecosystem processes (e.g., litter decomposition and export). There is also some evidence that functional measures may be more directly related to specific types of stressors (Gessner and Chauvet 2002).

Although many different processes could be used to assess ecosystem integrity, we will focus in this section on three functional measures: ecosystem metabolism (respiration, primary and secondary production), litter decomposition, and nutrient cycling. As described in Chapter 30, a significant amount of background information characterizing these processes is available, although the level of development varies among ecosystem types. For example, lake ecologists have historically relied on functional measures, especially primary production, whereas lotic ecologists have tended to rely on structural measures (Gessner and Chauvet 2002). These differences have resulted in divergent approaches used to assess effects of contaminants in aquatic ecosystems. Similarly, studies of biogeochemical processes in terrestrial habitats, especially in agricultural systems, focus primarily on factors that increase primary production, whereas aquatic ecologists have been more concerned with understanding factors that limit production as a way to control eutrophication (Grimm et al. 2003). The methodological approaches used to assess effects of contaminants on ecosystem metabolism are consequently different in aquatic and terrestrial ecosystems.

31.2 DESCRIPTIVE APPROACHES IN AQUATIC ECOSYSTEMS

31.2.1 Ecosystem Metabolism and Primary Production

Energy flow and metabolism are fundamental properties of ecosystems that are also closely related to the transport of contaminants. Many of the same physical, chemical, and biological processes that influence the flow of energy between biotic and abiotic compartments also regulate the fate of chemicals. Primary production in aquatic ecosystems is particularly sensitive to many anthropogenic stressors. The effects of nutrient subsidies and input of organic materials on productivity have received considerable attention in streams, lakes, and marine ecosystems. In their description of subsidy-stress gradients, Odum et al. (1979) contrast the ecosystem-level effects of "utilizable" inputs such as nutrients and organic materials with toxic materials (Chapter 25). This analysis contrasts the role of nutrients such as N and P as both regulators of ecosystem production as well as stressors when threshold levels are exceeded.

Input of nutrients associated with agricultural, domestic, industrial, and atmospheric sources are widely regarded as major stressors of aquatic ecosystems (National Research Council 1992). Whole ecosystem nutrient budgets calculated for several ecosystems reveal that inputs often exceed outputs, resulting in large amounts of nutrients being stored in a watershed (Bennett et al. 1999, Jowarski et al. 1992, Lowrance et al. 1985). Bennett et al. (1999) used a mass-balance approach to estimate P storage based on inputs and outputs in the Lake Mendota (Wisconsin, USA) watershed. They reported that approximately 50% of the P entering the watershed was retained and could be readily mobilized by climatic, geologic, or hydrologic events. These increased nutrient levels in aquatic ecosystems are often associated with toxic algal blooms, increased plant growth, oxygen depletion, fish kills, and major shifts in community composition. Land-based inputs of nutrients also increase eutrophication and have negative effects on primary production of macrophytes in coastal areas. Using data compiled from an extensive literature survey, Valiela and Cole (2002) reported a strong inverse relationship between N loading and primary production of seagrass meadows in coastal marine areas. The percent of seagrass cover lost reached 100% as N loading approached 100 Kg N/ha/y. These effects resulted from reduced light supply associated with increased phytoplankton production. The damaging effects of N enrichment were significantly reduced in areas protected by salt marshes and mangroves.

As described in Chapter 30, availability of N and P can directly regulate primary production and biomass accrual in aquatic ecosystems (Biggs 2000). However, the direct effects of nutrients on primary production are complex and may be mediated by other factors such as hydrologic characteristics and abundance of grazers. Riseng et al. (2004) used covariance structure analysis (CVA) to examine effects of hydrologic regime and nutrients in 97 midwestern U.S. streams. Increased nutrients in streams with high hydrologic variability resulted in greater algal abundance because grazers were reduced. In contrast, in more stable streams where grazers were abundant, algal production was limited and the net effect was an increase in herbivore production. Because hydrologic characteristics of a watershed are dependent on watershed physiography and climate, these factors may ultimately control responses to nutrient additions (Riseng et al. 2004).

Alterations in primary productivity and respiration have been measured in response to chemical stressors other than nutrients in aquatic ecosystems. Crossey and La Point (1988) could not detect differences in GPP between metal-impacted and reference sites, but when data were normalized to algal biomass (as chlorophyll a), GPP was higher at the reference site. Hill et al. (1997) measured variance and sensitivity of several functional measures in the Eagle River, a Colorado (USA) stream impacted by metals. Results showed that measures of community metabolism (GPP, NPP, and respiration) were lower at stations located downstream from heavy metal inputs (Figure 31.1). These functional measures were correlated with mortality of *Ceriodaphnia dubia*, and inhibition concentrations (IC50 values) for respiration were comparable to LC50 values derived from these more traditional toxicological approaches. These results suggest that functional measures were about

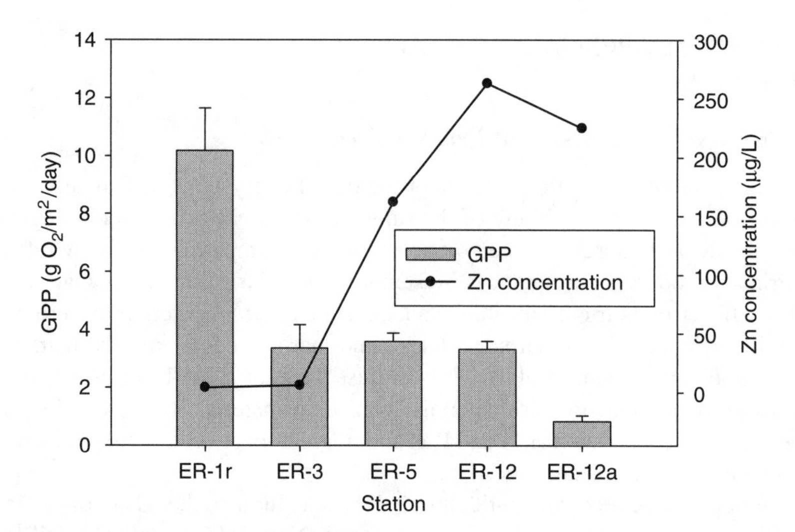

FIGURE 31.1 Effects of Zn on community respiration measured at different stations in the Eagle River, Colorado, United States. (Data from Tables 2 and 3 in Hill et al. (1997).)

as sensitive as acute toxicity for quantifying effects of heavy metals. Clements (2004) compared structural (species richness and abundance of metal-sensitive mayflies) and functional (community respiration) responses of benthic macroinvertebrate communities to a mixture of Cd and Zn in stream microcosms (Figure 31.2). Both structural and functional measures were significantly related to metal concentration, but effects on community respiration were generally greater than effects on species richness or abundance of metal-sensitive mayflies.

Unlike studies conducted with diatoms and attached algae, research investigating effects of contaminants on emergent macrophytes has shown that primary production and photosynthesis of these groups are relatively insensitive. Bendell-Young et al. (2000) compared the response of several structural and functional endpoints (mutagenic responses, morphological deformities, mortality, community structure, and plant productivity) measured in wetlands receiving oil sands effluents. Photosynthetic rates of cattails (*Typha latifolia* L.), measured as CO_2 uptake, were actually greater in wetlands receiving processed water from oil sands, a response that contradicted expectations. These researchers concluded that structural changes in benthic communities and blood chemistry of fish were more sensitive indicators of stress than functional measures. Photosynthetic rate of salt marsh plants (*Spartina alterniflora*) was measured at reference and contaminated sites in the southeastern United States (Wall et al. 2001). Although significant negative effects on benthic detritivores were observed at a site heavily contaminated by mercury and PCBs, photosynthesis of *Spartina* was not affected.

31.2.2 Secondary Production

In addition to direct effects on primary producers, contaminants and other stressors may alter the amount and rate of energy flow to higher trophic levels. The utilization of available energy in an ecosystem is thus an important measure of ecological integrity. Perhaps the most common functional response related to energetics measured in aquatic ecosystems is the abundance of different functional feeding groups (Rawer-Jost et al. 2000, Wallace et al. 1996). In part, the utility of functional feeding groups as a metric in ecological assessments is based on the assumption that specialist feeders such as scrapers and shredders are more sensitive to contaminants than generalist feeders such as collector-gatherers and filterers (Barbour et al. 1996). Although data on functional feeding groups are generally presented as abundance or density per unit area, and therefore not strictly a functional measure,

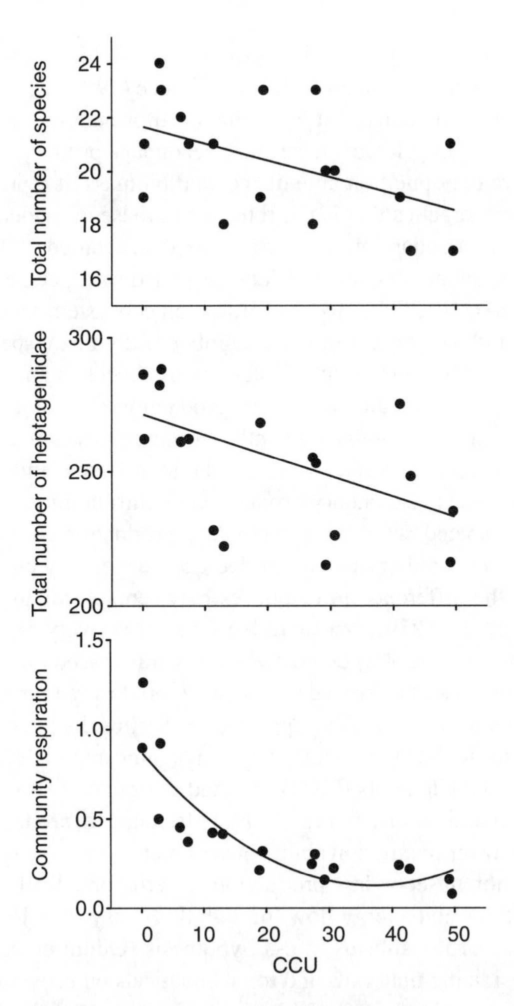

FIGURE 31.2 Relationship between structural (total number of species; abundance of metal-sensitive heptageniid mayflies) and functional (community respiration) endpoints and heavy metal (Cd and Zn) concentration in stream microcosms. Heavy metal concentration was expressed as the cumulative criterion unit (CCU), defined as the ratio of the measured metal concentration to the hardness adjusted chronic criterion values for Cd and Zn. (Data from Clements (2004).)

the assumption is that composition of different feeding groups reflects important ecosystem processes. For example, abundance of grazers, organisms that feed directly on periphyton and algae, is related to primary productivity in streams. Similarly, abundance of shredders, organisms that process leaf litter, regulates downstream transport of coarse particulate organic material (Wallace et al. 1982).

Secondary production, which we have defined as the production of heterotrophic organisms, has been used to document effects of several stressors in aquatic ecosystems, including hydrologic modification (Raddum and Fjellheim 1993), pesticides (Whiles and Wallace 1995), urbanization (Shieh et al. 2002), and heavy metals (Carlisle and Clements 2003). Because secondary production integrates individual growth rates and population dynamics, it captures in a single measure several important aspects of energy flow through ecosystems. Although integration of these measures across levels of biological organization is a laudable goal in ecosystem ecotoxicology, measures of secondary production are rarely included in biological assessments. Even measuring secondary production of individual species is highly labor intensive because it requires sampling populations with sufficient frequency to quantify individual growth rates, mortality, immigration, and emigration. The logistical challenges associated with measuring secondary production will likely deter some researchers from using this endpoint in ecological assessments. Indeed, France (1996) argues that because secondary production is dependent on abundance, little additional information is gained by including these more labor-intensive approaches. However, because secondary production is a composite of individual mortality, growth rate, population abundance, and biomass, it represents a potential holistic indicator of ecosystem bioenergetics that is not reflected in these individual measures.

Because most studies of secondary production are based on detailed analysis of individual species or groups of related species, our understanding of energetics from the perspective of entire ecosystems is somewhat limited (Shieh et al. 2002). To be useful as an ecosystem-level indicator, a measure of secondary production should include a significant number of dominant species in an ecosystem and should also be combined with data on trophic interactions. Sheehan and Knight (1985) compared patterns of community composition and secondary production at several sites along a gradient of metal contamination. Chironomids dominated benthic communities at metal-polluted sites, a finding commonly reported in the literature. However, despite large shifts in community composition among sites, relatively little difference in secondary production of chironomids was observed.

Another challenge associated with using secondary production as an indicator of ecosystem integrity is that production may either increase or decrease, depending on the nature of the stressor. The theoretical basis for the difference in responses between subsidizing and toxic stressors was first described by Odum et al. (1979), but there have been relatively few empirical studies documenting this pattern. Shieh et al. (2002) estimated energy flow based on secondary production and trophic interactions at polluted and reference sites in a Colorado stream receiving urban discharges. Secondary production, which was primarily supported by detritus, increased by more than two times at the most impacted site due to the input of nutrients and organic materials (Figure 31.3a). In contrast to these patterns, Carlisle and Clements (2003) reported a decline in secondary production along a gradient of heavy metal contamination (Figure 31.3b). Differences in production among these streams were primarily a result of lower population abundance of metal-sensitive species, especially grazing mayflies. The large reduction in secondary production of herbivores likely had important cascading effects on trophic interactions and energy flow through this ecosystem. Results of these studies are consistent with predictions of the subsidy-stress hypothesis (Odum et al. 1979) and illustrate the contrasting effects of subsidizing materials and toxic chemicals on ecosystem processes.

Reduced secondary production of zooplankton in lake ecosystems may result from increased mortality, lower growth rates, and/or shifts in size composition of dominant species. Hanazato (2001) reviewed effects of pesticides on zooplankton across levels of organization, from individuals to ecosystem-level responses. A general trend observed in lakes receiving pesticides was a reduction in mean body size of zooplankton as a result of differential sensitivity among species. Hanazato (2001) speculated that one potential ecosystem-level consequence of altered size distributions was a reduction in the amount of energy transferred from primary producers to higher trophic levels. This reduced transfer efficiency was associated with a variety of anthropogenic stressors, including heavy metals, acidification, and nutrient enrichment.

31.2.3 Decomposition

Litter decomposition is a fundamental ecological process that has been studied extensively, especially in lotic ecosystems (Chapter 30). It integrates responses of a variety of biota, from bacteria and fungi to shredding macroinvertebrates (Niyogi et al. 2001, 2003). There is a large database available reporting decomposition rates of leaves and quantifying biotic and abiotic factors that influence litter decay under a variety of environmental conditions. Gessner and Chauvet (2002) provided an excellent and comprehensive review of numerous studies that used litter breakdown to quantify effects of physical and chemical stressors in streams. They make a compelling argument for the use of decomposition as an ecosystem indicator and provide specific criteria for assessing ecological integrity. Breakdown rate coefficients (k), measured by regressing remaining mass of

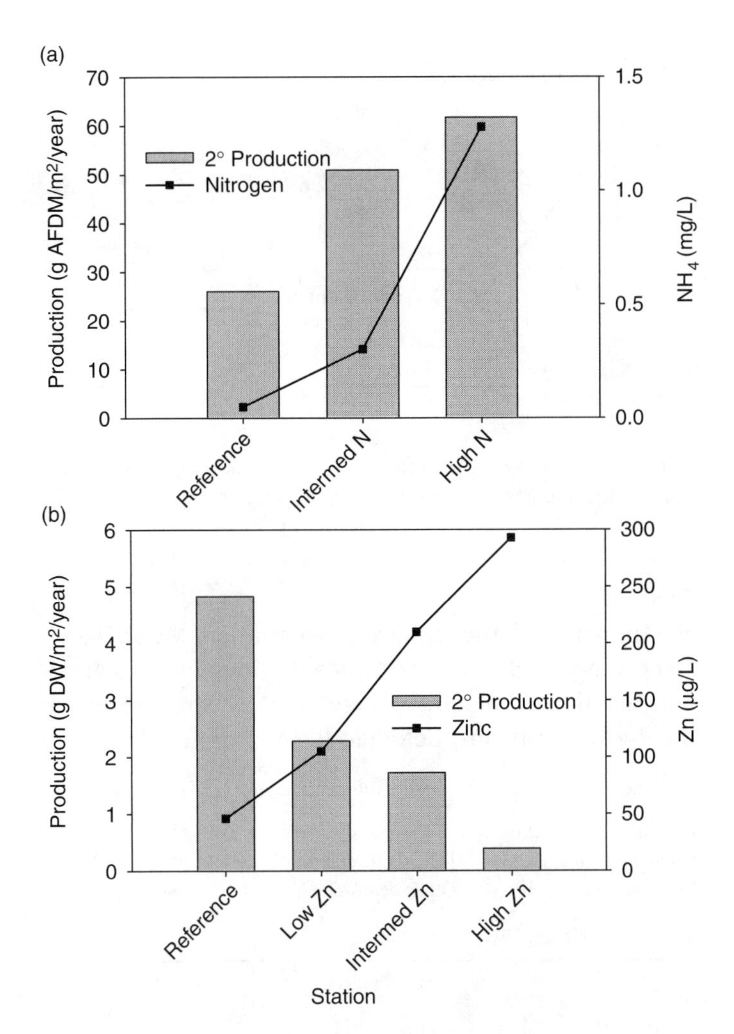

FIGURE 31.3 Contrasting effects of nutrients (a) and heavy metals (b) on invertebrate secondary production. (Effects of nutrients from Tables 1 and 4 in Shieh et al. (2002). Effects of Zn from Tables 1 and 2 in Carlisle and Clements (2003).)

litter against time, are generally reduced in disturbed ecosystems. Effects of contaminants on litter decay may result either from alterations in microbial processes or reduced abundance of macroinvertebrate shredders (Figure 31.4). It is also necessary to distinguish effects of contaminants on biological processes, such as the elimination of shredders or reduced microbial activity, from effects due to physical characteristics of the system. Methodological approaches that exclude or include different groups of organisms can be used to separate the relative importance of these processes, thus allowing ecologists to isolate underlying mechanisms. Because of the diversity of approaches used to quantify litter decay and the large number of environmental factors that influence k, development of standardized techniques for assessing effects of contaminants is essential.

The most comprehensive applications of leaf litter methodologies to investigate effects of contaminants have been conducted in metal-polluted and acidified streams. Schulthesis and Hendricks (1999) and Schulthesis et al. (1999) measured macroinvertebrate community composition and leaf decomposition at sites upstream and downstream from an abandoned pyrite mine in southwestern Virginia (USA). Shredder abundance was greater and decomposition rates were 1.4–2.7 times faster at the reference site compared Cu-polluted sites. Remediation activities initiated during the

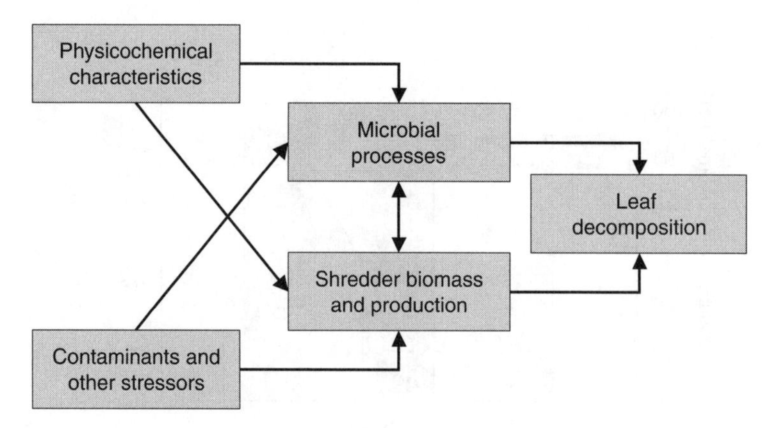

FIGURE 31.4 Conceptual model showing the effects of contaminants and physicochemical characteristics on microbial processes, shredder biomass, and leaf litter decomposition.

TABLE 31.1
The Influence of Aqueous Zn Concentration, Metal Oxide Deposition, and Nutrient Concentrations on Structural and Functional Endpoints Measured at 27 Stream Sites in the Rocky Mountains, Colorado (USA)

Variable	Independent Variables	R^2	*P*-Value
Leaf breakdown rate (*k*)	[Zn], oxide deposition	.72	.0001
Shredder biomass	[Zn], oxide deposition	.64	.0001
Microbial respiration	Oxide deposition, nutrients	.54	.0001

Data from Table 2 in Niyogi et al. (2001).

study period allowed these researchers to compare recovery of structural and functional responses. Although community composition and abundance of shredders increased following improvements in water quality, the rate of leaf processing did not increase as expected, suggesting some residual effects of Cu on microbial processes. In contrast to these studies, Nelson (2000) reported little effects of Zn contamination on decomposition rates of aspen (*Populus tremuloides*) in a Colorado Rocky Mountain stream, despite significant changes in community composition. These researchers speculated that the lack of a response in their study resulted from the relative insensitivity of microbes, especially fungi, to the moderate levels of Zn contamination. Alternatively, because microbial activity is limited by cold temperatures in Rocky Mountain streams, leaf processing may be more dependent on invertebrate shredders (Niyogi et al. 2001), which were unaffected by Zn in this study (Nelson 2000).

Comparative studies in streams across a gradient of heavy metal pollution provide the best opportunity to quantify effects of stressors relative to other factors that regulate leaf decomposition. Niyogi et al. (2001) measured decomposition rates at 27 stream sites (8 reference and 19 metal-polluted) in the Rocky Mountains of Colorado, USA. In addition to its broad spatial scale, this study is unique because researchers quantified the relative influence of several stressors associated with mining pollution, including acidification, elevated Zn concentration, and metal oxide deposition. Litter decay coefficients (*k*) and shredder biomass decreased with increasing aqueous Zn concentration and deposition of metal oxides (Table 31.1). In contrast, microbial respiration was more influenced by metal oxide deposition and nutrients. Because decay coefficients were more closely related to shredder biomass than microbial respiration, results of this study suggest that macroinvertebrates were more important than microbial processes in regulating leaf litter processing in streams (Niyogi et al. 2001). Carlisle and Clements (2005) related leaf decomposition to shredder secondary production and microbial respiration in reference and metal-polluted streams in Colorado, USA. Because leaf decomposition was measured as a function of shredder secondary production instead of shredder biomass, this study provided a unique opportunity to quantify effects of stressors on energy flow through an allochthonous food web. Results showed that shredders disproportionately contributed to leaf litter decay, and that species-specific differences in sensitivity to metals among shredders helped explain differences among streams.

Stream acidification by atmospheric deposition or other sources can have direct effects on litter decomposition (Dangles et al. 2004, Griffith and Perry 1993, Tuchman 1993, Webster and Benfield 1986). Griffith and Perry (1993) attributed slower processing rate of litter in acidic streams to lower biomass of shredders. In contrast, differences in community composition of shredders were primarily responsible for differences in processing rates between neutral and more alkaline streams. Tuchman (1993) reported that declines in invertebrate shredders in acidified lakes were correlated with decreased litter breakdown rates. As with studies of metal-polluted streams, the most convincing evidence demonstrating a relationship between acidification and leaf decomposition has been obtained from spatially extensive surveys. Dangles et al. (2004) measured microbial respiration, litter decay, and shredder abundance and composition in 25 streams along a gradient of acidification in the Vosges Mountains, France. Breakdown rates varied 20-fold between acidified and neutral streams, with alkalinity and aluminum concentration explaining 88% of the variation. Reduced leaf decomposition in acidified streams was related to lower abundance and biomass of the amphipod, *Gammarus fossarum*, a functionally important and acid-sensitive species. The greater breakdown rate observed in coarse mesh bags (5.0 mm), which allowed shredder colonization, compared to fine mesh bags (0.3 mm), which excluded shredders, supported the hypothesis that microbial processes were relatively unimportant in this investigation (Dangles and Guerold 2001).

Although pesticides and other organic contaminants are likely to have significant effects on litter decomposition by altering microbial processes and shredder communities, these stressors have received considerably less attention than heavy metals and acidification (Gessner and Chauvet 2002). Delorenzo et al. (2001) provided a comprehensive review of the effects of pesticides on microbial processes related to decomposition. The best examples showing the effects of organic chemicals on shredder biomass and subsequent alterations in leaf processing involve long-term experimental studies (Whiles and Wallace 1992, 1995) and stream mesocosm experiments (Stout and Cooper 1983), which will be described in Chapter 32. Swift et al. (1988) examined effects of dimilin, an insect growth regulator used for control of gypsy moths, on litter decomposition. Although laboratory bioassays with shredders showed significant mortality when shredders were fed dimilin-treated leaves, decomposition rates of treated leaves in the field were actually greater than controls. The faster processing rate of treated leaves was attributed to the potential carbon source that dimilin provided for bacteria.

Most investigations of leaf litter decomposition report that decay coefficients are reduced in contaminated streams. However, stressors that subsidize an ecosystem (e.g., nutrients or organic materials) may have the opposite effect. Niyogi et al. (2003) measured breakdown of tussock grass (*Chionocloa rigida*) in 12 New Zealand streams along a gradient of agricultural development. Nutrients (N and P), the predominant stressors in this system, increased along this gradient and stimulated microbial respiration, invertebrate abundance, and the rate of litter decomposition. In contrast, the macroinvertebrate community index (MCI), a biotic index of organic pollution, showed increased stress along this same gradient. Similar findings were reported by Pascoal et al. (2001) for a stream in Portugal receiving elevated nutrients. Despite reductions in abundance of shredders at polluted sites, leaf breakdown rates were greater. These results serve to illustrate the importance of understanding mechanistic linkages among stressors, microbial processes, and macroinvertebrate community composition when using leaf decomposition to assess ecological integrity.

In addition to the direct effects of contaminants on the rate of litter decay, the concentrations of toxic chemicals may increase in decomposing of plant material, thus providing a direct link to detritus-based food chains. Windham et al. (2004) measured reduced decomposition of marsh grass (*Spartina alterniflora*) in a metal-contaminated marsh as compared to a reference site. The 10–100 times increase in metal concentrations in decomposing litter was attributed to adsorption and microbial processes.

31.2.4 Nutrient Cycling

The majority of studies investigating effects of contaminants on nutrient cycling in aquatic ecosystems have focused on nitrification, denitrification, and other processes associated with N flux (Kemp and Dodds 2002, Kemp et al. 1990, Royer et al. 2004). Most of this research has been conducted within the context of understanding effects of nutrient enrichment, especially N and P, on freshwater and estuarine ecosystems. Eutrophication, caused by the release of excess nutrients, is regarded as the major threat to freshwater and coastal ecosystems in the United States (U.S. EPA 1990). Greater than 50% of the impaired lake area and river reaches in the United States result from excess nutrients. Most of this impairment is associated with nonpoint source inputs from agricultural and urban activities (Carpenter et al. 1998), although atmospheric deposition is considered an important source of N to some areas. In particular, N inputs from the upper Midwest to the Gulf of Mexico have increased dramatically in the past several decades, and excess nutrients have had severe effects on water quality and community composition. A significant portion of the N from nonpoint sources is retained in aquatic ecosystems by biological processes such as microbial uptake as well as lateral exchange with the hyporheic zone. However, despite N retention in some aquatic ecosystems, a large amount of excess N is transported downstream. Royer et al. (2004) measured denitrification in headwater stream sediments located in agricultural areas. Because denitrification rates were low in these streams, there was relatively little influence on instream concentrations and therefore most of the NO_3–N was transported downstream. These researchers concluded that previous estimates of denitrification rates may have overestimated N loss to the sediments.

Because rates of nitrification and denitrification in aquatic ecosystems are dependent on concentrations of ammonium (NH_4) and nitrate (NO_3), these processes are likely to increase in areas receiving anthropogenic inputs. Kemp and Dodds (2002) measured effects of anthropogenic N on rates of nitrification and denitrification in pristine and agriculturally influenced watersheds. Whole stream nitrification and denitrification rates were greater at agriculturally influenced sites, most likely due to greater input of N. Despite greater denitrification, the large amount of anthropogenic N exceeded the natural retentive ability of the stream and a significant amount was transported downstream.

As described in Chapter 30, retention of nutrients and organic materials is dependent on a number of physical, chemical, and biological characteristics. Headwater streams often represent the largest portion of the linear dimension of a watershed and are closely connected to surrounding riparian and terrestrial ecosystems. These systems are generally considered to be highly retentive of nutrients (Peterson et al. 2001). A similar situation exists in coastal marine ecosystems. Meta-analysis of data collected in coastal areas demonstrated that denitrification by fringing wetlands (e.g., salt marshes, mangroves) serves to intercept excess nutrients and protect seagrass meadows from anthropogenic N (Valiela and Coe 2002). Physical disturbances such as removal of vegetation and reduced habitat complexity may decouple denitrification and nitrification processes in aquatic ecosystems, thereby exacerbating the effects of nutrient enrichment (Kemp and Dodds 2002).

31.3 TERRESTRIAL ECOSYSTEMS

31.3.1 Respiration and Soil Microbial Processes

Functional measures of ecosystem processes have also been used to characterize the impacts of contaminants in terrestrial ecosystems. In particular, effects of contaminants on microbial and soil ecosystem processes, especially soil respiration, have been examined in considerable detail (Aceves et al. 1999, Dai et al. 2004, Megharaj et al. 1998, 2000, Zimakowska-Gnoinska et al. 2000). There is also evidence suggesting that ecosystem processes in soils are significantly more sensitive to contaminants than the plant communities they support (see review by Giller et al. 1998). The emphasis on soil processes in terrestrial ecosystems is at least partially a result of taxonomic difficulties associated with characterizing microbial communities. Consequently, most soil ecologists tend to focus on functional measures instead of structural characteristics such as biodiversity and species composition. Dai et al. (2004) reported a strong inverse relationship between heavy metal concentrations in soils and respiration rates. An increase in the accumulation of organic C, total N, and the C:N biomass ratio was attributed either to reduced microbial activity in soils or to changes in microbial community composition. Similar results were reported by Edvantoro et al. (2003) where microbial respiration was greatly reduced in soils contaminated by DDT and arsenic. Interestingly, although microbial biomass was significantly lower in polluted soils, bacterial populations measured using plate counts showed little difference between reference and contaminated sites.

Shifts in community structure, such as an increase in biomass of fungal populations and a decrease in bacterial populations, can also result in changes in soil ecosystem function. Megharaj et al. (1998) reported that changes in community composition of soil microalgae were associated with large (>90%) reductions in microbial activity (dehydrogenase, nitrate reductase) at field sites contaminated by pentachlorophenol (PCP). Megharaj et al. (2000) measured changes in soil microbial function at sites contaminated by DDT and its metabolites. Because sensitive bacteria were replaced by DDT-tolerant microorganisms, total microbial biomass was found to be less sensitive than dehydrogenase activity, a measure of total microbial activity that correlates with respiration (Brookes 1995). Changes in respiration can result in carbon accumulation in ecosystems and therefore specific respiration rate, measured as the ratio of CO_2 production to biomass C, may be a better indicator of stress than either measure alone (Brookes 1995). For example, Megharaj et al. (2000) reported that specific activity, defined as the ratio of dehydrogenase activity to microbial biomass C, decreased with DDT concentration (Figure 31.5). Aceves et al. (1999) also determined that biomass-specific respiration rate was a better indicator of heavy metal contamination in soils than either microbial biomass or respiration.

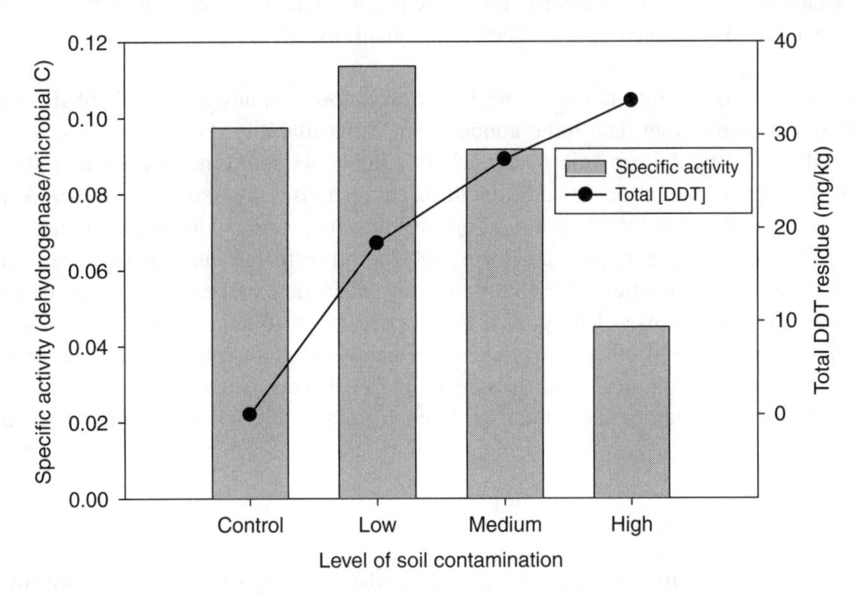

FIGURE 31.5 Relationship between DDT concentration in soil and specific microbial activity. Specific activity was calculated as the ratio of dehydrogenase activity (mg 2,3,5, triphenyltetrazoluim formazan/kg) to soil microbial biomass (mg/kg). (Data from Tables 2 and 3 in Megharaj et al. (2000).)

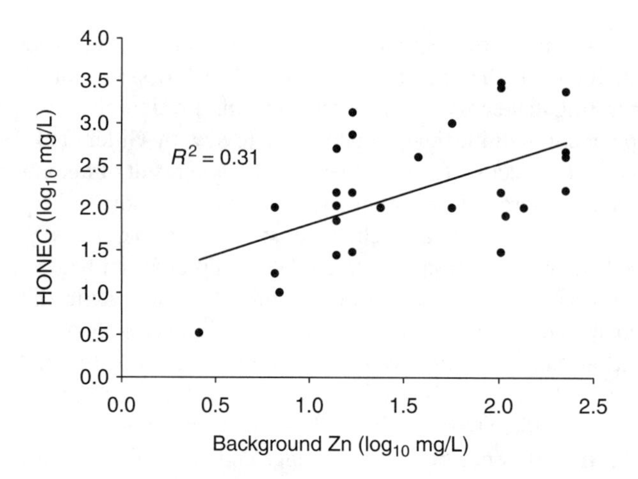

FIGURE 31.6 Relationship between background Zn concentration and the highest observed no effect concentration (HONEC) for a variety of soil microbial processes. (Data from Table 1 in McLaughlin and Smolders (2001).)

Toxic effects of heavy metals on respiration rate has been reported for terrestrial ecosystems (Dai et al. 2004, Laskowski et al. 1994, Niklinska et al. 1998); however, unlike our understanding of aquatic ecosystems, we lack a clear understanding of biotic and abiotic factors that influence toxicity. For example, toxic metal concentrations that affect respiration can vary by more than 100 times (Giller et al. 1998). Biological factors such as acclimation or adaptation of soil microorganisms to contaminants may also explain some of the variation observed in these studies. McLaughlin and Smolders (2001) used literature values to examine the influence of background Zn concentration on the responses of soil microbial processes (e.g., respiration, nitrification, and ammonification) to Zn. Background Zn concentrations in soil accounted for a significant amount of the variation in sensitivity of several processes to Zn (Figure 31.6). Although considerable unexplained variation resulted from methodological differences among studies and from comparing different endpoints, the results clearly indicate that background concentration affected sensitivity to Zn. It is uncertain if this increased tolerance to Zn affects the sensitivity of soil processes to other classes of contaminants.

As in aquatic ecosystems, much of the variation among studies is a result of differences in physicochemical factors that determine contaminant bioavailability. For example, metal toxicity in soils is influenced by soil organic matter, clay content, pH, and other factors that regulate the amount of free metals in solution. In addition, different responses are likely to be observed between short-term experiments conducted in the laboratory and long-term monitoring studies conducted in the field. The difficulty extrapolating from the laboratory to the field greatly complicates our ability to define safe concentrations of contaminants necessary to protect soil processes. Giller et al. (1998) advocate the use of long-term field experiments to explain effects of heavy metals on microbial respiration and other soil processes. Because litter respiration has been proposed as a potential endpoint in ecological risk assessment of forests ecosystems (Niklinska et al. 1998), a better understanding of factors that influence effects of metals and other stressors on respiration is necessary.

31.3.2 Litter Decomposition

Litter decomposition is an important process in terrestrial ecosystems that is closely related to primary productivity, energy flow, and nutrient cycling. Large portions (>90%) of the net production in temperate forests may be deposited on the forest floor as leaf litter. Thus, reduced rates of decomposition

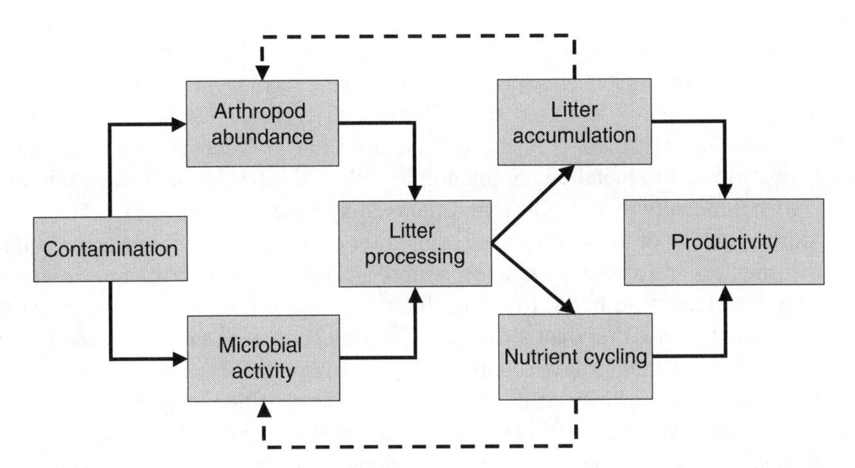

FIGURE 31.7 Conceptual model showing the potential mechanisms of contaminant effects on leaf litter processing, nutrient cycling, and productivity. Dashed lines indicate potential feedback between certain processes.

in polluted terrestrial ecosystems may result in the accumulation of leaf litter and reduced supply of nutrients through microbial mineralization (Figure 31.7). These changes in nutrient supply can affect primary productivity and subsequent litter production. Because litter decomposition is relatively easy to measure and sensitive to a variety of stressors, it is a useful endpoint for ecological risk assessment in terrestrial ecosystems. Litter-bag studies measuring decomposition rates are frequently included in assessments of contaminant effects (De Jong 1998, Johnson and Hale 2004, McEnroe and Helmisaari 2001, Strojan 1978). Slower rates of decomposition are often observed at contaminated sites, resulting in accumulation of organic material (Freedman and Hutchinson 1980, Strojan 1978). Reduced litter decomposition can also affect nutrient cycling and growth of vegetation, thereby reducing soil organic content and increasing contaminant bioavailability (Derome and Nieminen 1998, Johnson and Hale 2004). Because contaminants may affect processing of detritus and accumulation of organic material in soils, the ability of an ecosystem to assimilate contaminants may be reduced because of lower organic content and subsequently greater contaminant bioavailability (Derome and Nieminen 1998). Alterations in litter decomposition resulting from contaminant deposition may also affect soil processes and subsequent movement of contaminants. For example, changes in heavy metal concentrations in forest soils were attributed to altered decomposition and depletion of organic matter at sites affected by a nearby aluminum industry (Egli et al. 1999). Derome and Nieminen (1998) reported much greater flux of heavy metals through soils at disturbed sites adjacent to a smelter as compared to distant sites. The increased flux of metals was associated with reduced interception of precipitation by the disturbed canopy and a subsequent greater movement of water through denuded soils.

Approaches employed in terrestrial ecosystems to measure litter decomposition are similar to those in streams and generally involve measurement of weight loss of litter placed at reference and contaminated field sites. Because decomposition rates are generally lower in terrestrial systems, litter bags must be deployed for longer periods of time (e.g., 12–24 months) to obtain reliable decay coefficients (k). Terrestrial insects and other macroinvertebrate decomposers are likely to play a key role in regulating rates of decomposition. However, unlike research conducted in aquatic ecosystems, most terrestrial studies employ relatively fine mesh litter bags designed to exclude invertebrate decomposers. As a consequence, the relative importance of invertebrate decomposers, microbial processes, and physical processes in regulating litter decomposition at contaminated sites is uncertain in terrestrial ecosystems.

Much of the research employing litter bags to examine decomposition rates has focused on soils contaminated with heavy metals (Breymeyer et al. 1997, Coughtrey et al. 1979, Johnson and

Hale 2004, McEnroe and Helmisaari 2001, Post and Beeby 1996). Accumulation of litter in a forest near a lead-zinc-cadmium smelter was attributed to reduced decomposition associated with heavy metal contamination (Coughtrey et al. 1979). McEnroe and Helmisaari (2001) measured decomposition of pine needles along a gradient of metal contamination in soil. Decreased mass loss and reduced C:N ratios were attributed to metal contamination in soils and litter. Metal concentrations in litter, which increased significantly over the 30-month study, may have contributed to reduced decomposition rates. Although many of the studies measuring litter decomposition have focused on localized sources such as smelters and abandoned mines, nonpoint sources may alter litter decay rates. Metals deposited along roadsides were found to inhibit litter decomposition, but this effect resulted from contamination of plant litter rather than differences in metal content of soils (Post and Beeby 1996).

Although a majority of studies have reported a negative relationship between contaminant levels and rates of decomposition in terrestrial ecosystems, decomposition may be enhanced in some polluted habitats. Breymeyer et al. (1997) measured decomposition rates at 15 pine forest sites along a gradient of metal contamination. Positive correlations between litter decomposition and metal concentrations were attributed to fertilization effects in the nutrient-poor soils typical of pine forest ecosystems. Similar results were reported by Post and Beeby (1996) where increased microbial respiration along urban roadways was attributed to elevated concentrations of resources in the form of hydrocarbons. Kauppi et al. (1992) attributed an observed increase in forest production in Europe between 1960 and 1990 to the fertilization effects of some pollutants, which apparently outweighed adverse effects in these systems.

31.3.2.1 Mechanisms of Terrestrial Litter Decomposition

Despite the large number of studies over the past several decades that have examined the relationship between contaminant levels in soils and reduced decomposition rates, the underlying mechanisms are not well understood (Coughtrey et al. 1979, Johnson and Hale 2004, Kohler et al. 1995). Because contaminants may affect microbial processes in the soil, litter palatability, and/or invertebrate abundance, additional studies that quantify the relative importance of these processes are necessary. Freedman and Hutchinson (1980) conducted a comprehensive analysis of the effects of smelter emissions on leaf litter decomposition and the underlying mechanisms responsible for litter accumulation at field sites (Figure 31.8). Reduced abundance of soil arthropods and lower microbial activity (measured as soil respiration and acid phosphatase activity) resulted in lower rates of litter decomposition at contaminated sites compared with reference sites. The lower rates of litter processing resulted in a significant increase in litter accumulation at sites adjacent to the smelter. Interactions between soil arthropods and microbial communities may also influence leaf litter processing. Kohler et al. (1995) reported that direct stimulation of microbial activity by arthropods in soils enhanced litter decomposition.

Although it is generally recognized that microbial processes controlling decomposition are particularly sensitive to contaminants (Derome and Lindroos 1998), few studies have quantified the relative importance of contaminants in soils versus those deposited directly on litter. Contaminants deposited on the surface of leaves before litterfall could have direct effects on both macroinvertebrate decomposers and microbial processes. Distinguishing the relative importance of these mechanisms has important implications for remediation of contaminated sites. Clean-up of polluted soils will likely have little effect on litter decomposition rates if microbial processes are inhibited by contaminants deposited on leaf surfaces. Breymeyer et al. (1997) reported that atmospheric deposition of metals to litter was more important in controlling decomposition than metal concentrations in soils. Reciprocal transplant studies, in which decomposition of clean and contaminated litter is measured at reference and impacted sites, is an effective approach for distinguishing the effects of contaminants in soils from those deposited on leaf litter. Johnson and Hale (2004) measured metal accumulation and decomposition of white birch leaves at metal-polluted sites in Canada. Decomposition rates were reduced at a metal-polluted site, but concentrations of metals in litter, which accumulated over time, did not influence dry mass loss.

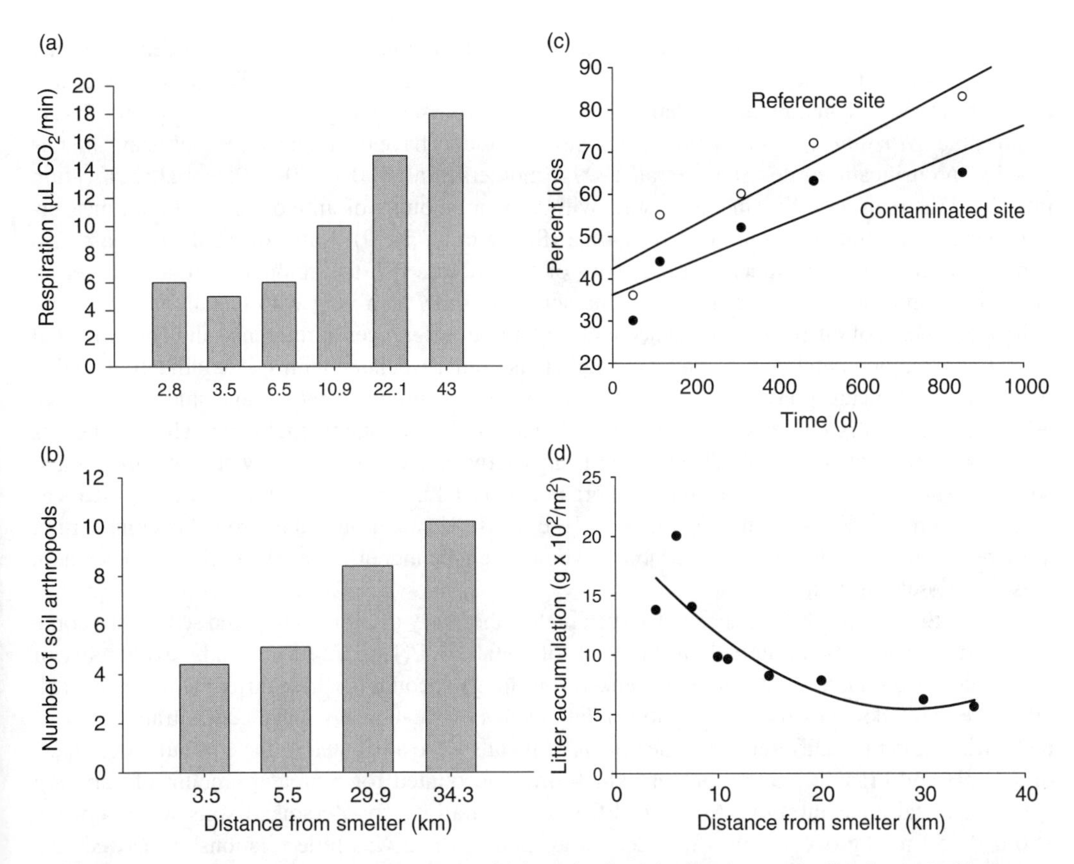

FIGURE 31.8 Effects of heavy metals on microbial respiration (a), abundance of soil arthropods (b), decomposition rate of birch leaves at reference (open symbols) and impacted (closed symbols) sites (c), and litter accumulation (d) in forest soils in the vicinity of a nickel-copper smelter at Sudbury, Ontario. (Data from Figure 1 and Tables 2, 5, and 8 in Freedman and Hutchinson (1980).)

In addition to studying decomposition of leaves, some researchers have used standardized materials, such as cellulose strips buried in soils, to quantify effects of contaminants on decomposition (De Jong 1998, Post and Beeby 1996). Because these materials degrade rapidly, decomposition studies can be completed in a much shorter period of time when compared to natural litter. As with any attempt to standardize approaches used in ecological assessments, the advantages of greater control must be weighed against the loss of ecological realism. Post and Beeby (1996) concluded that natural litter bags were more appropriate for measuring effects of contaminants on activity of microbial decomposers because of the relatively unrealistic resources provided by cellulose strips. De Jong (1998) conducted a series of experiments designed to assess the effects of pesticides on litter decomposition. Because of considerable variability and susceptibility of decomposition studies to external confounding factors, this researcher concluded that "the chance of developing a standardized field trial with litterbags is deemed too slim." If litter decomposition is to become a widely used endpoint in risk assessments of terrestrial ecosystems, a better appreciation for the underlying mechanisms and complexities that control this process is essential.

31.3.3 Nutrient Cycling

The majority of studies investigating the influence of contaminants on nutrient dynamics in terrestrial ecosystems have focused on nitrification, which is defined as the conversion of ammonium to nitrite (NO_2) and nitrate (NO_3). Nitrification is a critical component of the nitrogen cycle that ultimately determines the availability of soil N to plants and other organisms. The sensitivity of this two-step process to contaminants is dependent on the relative sensitivity of the two groups of bacteria responsible (*Nitrosomonas* and *Nitrobacter*). Several studies have examined effects of heavy metals in soils on nitrification rate (Sauve et al. 1999, Smolders et al. 2001, 2003). Potential nitrification rate (PNR) is measured by amending soils with known amounts of ammonium and then measuring changes in nitrate concentration over time (Sauve et al. 1999). Rates of nitrification are also dependent on physicochemical characteristics of soil such as pH, the amount of organic material, and soil composition. Because these soil characteristics may also vary with metal contamination, field assessments of nitrification must account for natural differences at reference and contaminated sites. For example, metal deposition from smelters is often associated with soil acidification, which would increase metal bioavailability. In contrast, soils contaminated by sewage sludge may have elevated levels of organic material, which would likely reduce contaminant bioavailability. There is also the potential for positive feedback relationships between the bioavailability of contaminants and soil properties. For example, the amount of organic material in soils is in part regulated by nitrification and microbial decomposition (Figure 31.7). Because nitrification and microbial decomposition are sensitive to contaminants, bioavailability would likely be increased in contaminated soils where these processes are inhibited.

Although reduced PNR has been observed in the laboratory using metal-spiked sediments, some field studies have shown relatively little effects of metals on PNR. Differences in bioavailability of Zn in field-contaminated and laboratory-spiked soils likely account for these differences. Sauve et al. (1999) reported both positive and negative relationships between heavy metal concentration in soils and nitrification rate. Differences in the effects of metals were attributed to the confounding effects of soil pH and organic matter content, which often accounted for a greater amount of variation than soil metal concentration. Although EC50 concentrations for Zn-spiked soils were between 150 and 350 mg/kg dry weight (Smolders et al. 2001), there was little relationship between Zn concentration and nitrification in field-contaminated soils. These researchers questioned the utility of PNR in assessments of contaminated soils because of high background variability and inconsistent responses in uncontaminated ecosystems.

Smolders et al. (2003) compared PNR in laboratory-spiked and natural sediments contaminated with Zn. Nitrification rate in the field increased significantly with soil pH, but Zn had little effect at concentrations shown to be toxic in the laboratory. These differences between natural and laboratory-spiked sediments were attributed to either microbial acclimation to metals or lower metal bioavailability in the field. Rusk et al. (2004) reported that adaptation of soil microbes to metals significantly reduced effects on nitrification. Sauve et al. (1999) questioned the utility of nitrification as an indicator of contamination because of its sensitivity to other factors such as pH and organic carbon. Because of potential differences in effects of metals and other contaminants in field-collected and laboratory-spiked soils, studies that validate laboratory responses are necessary. Furthermore, because soil pH and organic matter may affect nitrification rates and contaminant bioavailability, experimental approaches are necessary to determine if the relationship between nitrification and soil characteristics is a result of direct or indirect effects.

31.3.4 An Integration of Terrestrial and Aquatic Processes

Biogeochemists generally recognize that a holistic understanding of the factors that regulate transport and transformation of nutrients in ecosystems requires integration of aquatic and terrestrial processes. Increased alteration of local, regional, and global cycles of C, N, and P requires that ecologists develop a broader perspective of factors that influence nutrient movements between ecosystems. Regardless of the ecosystem, the primary focus should be on factors that limit primary productivity, affect nutrient retention, and alter rates of nutrient transformation. However, theories describing biogeochemical processes in aquatic and terrestrial ecosystems have largely been developed in isolation, with relatively fewer attempts to integrate these disciplines (Grimm et al. 2003). As a result, very different approaches have been employed in aquatic and terrestrial ecosystems to quantify effects of natural and anthropogenic stressors on nutrient processing. Grimm et al. (2003) discuss underlying physical, chemical, and biological characteristics of aquatic and terrestrial ecosystems that led to the application of disparate approaches used in these systems. An important distinction between terrestrial and aquatic ecosystems is that N often limits productivity in terrestrial ecosystems whereas P is frequently the primary limiting nutrient in aquatic ecosystems. Although this generalization is somewhat overstated, it is at least partially responsible for the different research questions addressed in aquatic and terrestrial systems (Grimm et al. 2003). Research in terrestrial ecosystems, particularly in agricultural systems, has largely focused on understanding factors to increase plant productivity. In contrast, much of the emphasis in aquatic systems has been devoted to understanding factors that limit productivity and to assessing the negative effects of eutrophication. Not surprisingly, many of the differences in key structural features (geomorphology, hydrology, nutrient pools, lifespan, and size of dominant primary producers) and flux of elements between aquatic and terrestrial ecosystems are closely linked to the quantity and movement of water. A better understanding of the effects of contaminants on nutrient dynamics in terrestrial and aquatic ecosystems will require development of methodological approaches that transcend boundaries and allow direct comparison of processes between systems.

31.4 SUMMARY

Descriptive studies have greatly improved our understanding of contaminant effects on ecosystem processes such as production, decomposition, and nutrient cycling. The potential for conducting surveys of contaminant effects on ecosystem processes over a relatively large spatiotemporal scale is a major strength of these comparative approaches. While attempts to predict ecosystem-level responses to contaminants based on changes in community composition have been reasonably successful, we feel that direct measurement of ecosystem processes is considerably more informative. We will explore the ecotoxicological implications of the relationship between structural and functional characteristics in Chapter 33. One of the more significant developments in ecosystem ecology is the recognition that traditional boundaries between certain ecosystems are often arbitrary (Fausch et al. 2002). Although these boundaries have historically defined the spatial extent of individual ecosystems, the exchange of contaminants and other materials between ecosystems requires a more holistic perspective. We consider the integration of methodological approaches for aquatic and terrestrial ecosystems, which have been largely developed in isolation (Grimm et al. 2003), to be a major research need in ecosystem ecotoxicology.

Because the ecosystem processes described in this chapter are strongly interrelated and influenced by natural biotic and abiotic factors (Figure 31.7), quantifying the direct effects of contaminants is challenging. A more comprehensive understanding of underlying mechanisms would facilitate our ability to quantify the importance of natural changes in ecosystem processes relative to contaminant-induced alterations. In the following chapter, we will examine the application of experimental approaches such as the use of microcosms, mesocosms, and whole ecosystem manipulations to assess effects of contaminants on ecosystem processes.

31.4.1 Summary of Foundation Concepts and Paradigms

- Ecotoxicologists interested in ecosystem responses to anthropogenic stressors employ descriptive, quasi-experimental, and experimental approaches.
- Because experimental studies of ecosystem processes are often limited by spatiotemporal scale, descriptive approaches can provide compelling and ecologically realistic results.

- Restricting our analyses to mainly structural measures has provided a somewhat incomplete picture of how ecosystems respond to and recover from anthropogenic disturbances.
- Lake ecologists have historically relied on functional measures, especially primary production, whereas lotic ecologists have tended to focus on structural measures.
- Primary production in aquatic ecosystems is particularly sensitive to anthropogenic stressors.
- Input of nutrients associated with agricultural, domestic, industrial, and atmospheric sources are widely regarded as major stressors of aquatic ecosystems.
- Direct effects of nutrients on primary production are complex and may be mediated by other factors such as hydrologic characteristics and abundance of grazers.
- The most common functional response related to energetics measured in aquatic ecosystems is the abundance of different functional feeding groups.
- Secondary production integrates estimates of individual growth rates and population dynamics and therefore captures in a single measure several important aspects of energy flow through ecosystems.
- One significant challenge associated with using secondary production as an indicator of ecosystem integrity is that production may increase or decrease, depending on the nature of the stressor.
- Effects of contaminants on litter decay may result either from alterations in microbial processes or reduced abundance of macroinvertebrate shredders.
- The most convincing evidence demonstrating a relationship between acidification and leaf decomposition has been obtained from spatially extensive surveys.
- Most investigations of leaf litter decomposition report that decay coefficients are reduced in contaminated streams; however, stressors that subsidize an ecosystem (e.g., nutrients or organic materials) may have the opposite effect.
- Most studies investigating effects of contaminants on nutrient dynamics in aquatic ecosystems have focused on nitrification, denitrification, and other processes associated with N flux.
- Effects of contaminants on microbial and soil ecosystem processes, especially soil respiration, have been examined in considerable detail.
- There is evidence suggesting that ecosystem processes in soils are more sensitive to contaminants than the plant communities they support.
- Contaminant-induced shifts in community structure can also result in changes in soil ecosystem function.
- Litter decomposition is an important process in terrestrial ecosystems that is closely related to primary productivity, energy flow, and nutrient cycling.
- Because decomposition rates are generally lower in terrestrial systems, litter bags must be deployed for longer periods of time (e.g., 12–24 months) to obtain reliable decay coefficients (k).
- Despite the large number of studies that have examined the relationship between contaminant levels in soils and decomposition rates, the underlying mechanisms are not well understood.
- The majority of studies investigating the influence of contaminants on nutrient dynamics in terrestrial ecosystems have focused on nitrification.
- Because nitrification and microbial decomposition are sensitive to contaminants and influence the amount of organic materials in soils, contaminant bioavailability would likely be increased in ecosystems where these processes are inhibited.
- A holistic understanding of factors that regulate transport and transformation of nutrients in ecosystems requires integration of aquatic and terrestrial processes.
- Theories describing biogeochemical processes in aquatic and terrestrial ecosystems have largely been developed in isolation.

REFERENCES

Aceves, M.B., Grace, C., Ansorena, J., Dendooven, L., and Brookes, P.C., Soil microbial biomass and organic C in a gradient of zinc concentrations in soils around a mine spoil tip, *Soil Biol. Biochem.*, 31, 867–876, 1999.

Adams, S.M., Hill, W.R., Peterson, M.J., Ryon, M.G., Smith, J.G., and Stewart, A.J., Assessing recovery in a stream ecosystem: Applying multiple chemical and biological endpoints, *Ecol. Appl.*, 12, 1510–1527, 2002.

Barbour, M.T., Gerritsen, J., Griffith, G.E., Frydenborg, R., McCarron, E., White, J.S., and Bastian, M.L., A framework for biological criteria for Florida streams using benthic macroinvertebrates, *J. N. Am. Benthol. Soc.*, 15, 185–211, 1996.

Bendell-Young, L.I., Bennett, K.E., Crowe, A., Kennedy, C.J., Kermode, A.R., Moore, M.M., Plant, A.L., and Wood, A., Ecological characteristics of wetlands receiving an industrial effluent, *Ecol. Appl.*, 10, 310–322, 2000.

Bennett, E.M., Reed-Andersen, T., Houser, J.N., Gabriel, J.R., and Carpenter, S.R., A phosphorus budget for the Lake Mendota Watershed, *Ecosystems*, 2, 69–75, 1999.

Biggs, B.J.F., Eutrophication of streams and rivers: Dissolved nutrient-chlorophyll relationships for benthic algae, *J. N. Am. Benthol. Soc.*, 19, 17–31, 2000.

Breymeyer, A., Degorski, M., and Reed, D., Decomposition of pine-litter organic matter and chemical properties of upper soil layers: Transect studies, *Environ. Pollut.*, 98, 361–367, 1997.

Brookes, P.C., The use of microbial parameters in monitoring soil pollution by heavy-metals, *Biol. Fertil. Soils*, 19, 269–279, 1995.

Bunn, S.E. and Davies, P.M., Biological processes in running waters and their implications for the assessment of ecological integrity, *Hydrobiologia*, 422, 61–70, 2000.

Carlisle, D.M. and Clements, W.H., Sensitivity and variability of metrics used in biological assessments of running waters, *Environ. Toxicol. Chem.*, 18, 285–291, 1999.

Carlisle, D.M. and Clements, W.H., Growth and secondary production of aquatic insects along a gradient of Zn contamination in Rocky Mountain streams, *J. N. Am. Benthol. Soc.*, 22, 582–597, 2003.

Carlisle, D.M. and Clements, W.H., Leaf litter breakdown, microbial respiration and shredder production in metal-polluted streams, *Freshw. Biol.*, 50, 380–390, 2005.

Carpenter, S.R., Caraco, N.F., Correll, D.L., Howarth, R.W., Sharpley, A.N., and Smith, V.H., Nonpoint pollution of surface waters with phosphorus and nitrogen, *Ecol. Appl.*, 8, 559–568, 1998.

Clements, W.H., Small-scale experiments support causal relationships between metal contamination and macroinvertebrate community responses, *Ecol. Appl.*, 14, 954–967, 2004.

Cottingham, K.L. and Carpenter, S.R., Population, community, and ecosystem variates as ecological indicators: Phytoplankton responses to whole-lake enrichment, *Ecol. Appl.*, 8, 508–530, 1998.

Coughtrey, P.J., Jones, C.H., Martin, M.H., and Shales, S.W., Litter accumulation in woodlands contaminated by Pb, Zn, Cd and Cu, *Oecologia*, 39, 51–60, 1979.

Crossey, M.J. and La Point, T.W., A comparison of periphyton community structural and functional responses to heavy metals, *Hydrobiologia*, 162, 109–121, 1988.

Dai, J., Becquer, T., Rouiller, J.H., Reversat, G., Bernhard-Reversat, F., and Lavelle, P., Influence of heavy metals on C and N mineralisation and microbial biomass in Zn-, Pb-, Cu-, and Cd-contaminated soils, *Appl. Soil Ecol.*, 25, 99–109, 2004.

Dangles, O. and Guerold, F., Linking shredders and leaf litter processing: Insights from an acidic stream study, *Intern. Rev. Hydrobiol.*, 86, 395–406, 2001.

Dangles, O., Gessner, M.O., Guerold, F., and Chauvet, E., Impacts of stream acidification on litter breakdown: Implications for assessing ecosystem functioning, *J. Appl. Ecol.*, 41, 365–378, 2004.

De Jong, F.M.W., Development of a field bioassay for the side effects of pesticides on decomposition, *Ecotoxicol. Environ. Saf.*, 40, 103–114, 1998.

Delorenzo, M.E., Scott, G.I., and Ross, P.E., Toxicity of pesticides to aquatic microorganisms: A review, *Environ. Toxicol. Chem.*, 20, 84–98, 2001.

Derome, J. and Lindroos, A.J., Effects of heavy metal contamination on macronutrient availability and acidification parameters in forest soil in the vicinity of the Harjavalta Cu-Ni smelter, SW Finland, *Environ. Pollut.*, 99, 225–232, 1998.

Derome, J. and Nieminen, T., Metal and macronutrient fluxes in heavy-metal polluted Scots pine ecosystems in SW Finland, *Environ. Pollut.*, 103, 219–228, 1998.

Edvantoro, B.B., Naidu, R., Megharaj, M., and Singleton, I., Changes in microbial properties associated with long-term arsenic and DDT contaminated soils at disused cattle dip sites, *Ecotoxicol. Environ. Saf.*, 55, 344–351, 2003.

Egli, M., Fitze, P., and Oswald, M., Changes in heavy metal contents in an acidic forest soil affected by depletion of soil organic matter within the time span 1969–93, *Environ. Pollut.*, 105, 367–379, 1999.

Fausch, K.D., Torgersen, C.E., Baxter, C.V., and Li, H.W., Landscapes to riverscapes: Bridging the gap between research and conservation of stream fishes, *BioScience*, 52, 483–498, 2002.

France, R.L., Biomass and production of amphipods in low alkalinity lakes affected by acid precipitation, *Environ. Pollut.*, 94, 189–193, 1996.

Freedman, B. and Hutchinson, T.C., Effects of smelter pollutants on forest leaf litter decomposition near a nickel-copper smelter at Sudbury, Ontario, *Can. J. Bot.*, 58, 1722–1736, 1980.

Gessner, M.O. and Chauvet, E., A case for using litter breakdown to assess functional stream integrity, *Ecol. Appl.*, 12, 498–510, 2002.

Giller, K.E., Witter, E., and McGrath, S.P., Toxicity of heavy metals to microorganisms and microbial processes in agricultural soils: A review, *Soil Biol. Biochem.*, 30, 1389–1414, 1998.

Griffith, M.B. and Perry, S.A., Colonization and processing of leaf-litter by macroinvertebrate shredders in streams of contrasting pH, *Freshw. Biol.*, 30, 93–103, 1993.

Grimm, N.B., Gergel, S.E., Mcdowell, W.H., Boyer, E.W., Dent, C.L., Groffman, P., Hart, S.C., et al., Merging aquatic and terrestrial perspectives of nutrient biogeochemistry, *Oecologia*, 137, 485–501, 2003.

Hanazato, T., Pesticide effects on freshwater zooplankton: An ecological perspective, *Environ. Pollut.*, 112, 1–10, 2001.

Hill, B.H., Lazorchak, J.M., McCormick, F.H., and Willingham, W.T., The effects of elevated metals on benthic community metabolism in a Rocky Mountain stream, *Environ. Pollut.*, 95, 183–190, 1997.

Howarth, R.W., Comparative responses of aquatic ecosystems to toxic chemical stress, In *Comparative Analyses of Ecosystems: Patterns, Mechanisms, and Theories*, Cole, J., Lovett, G., and Findlay, S. (eds.), Springer-Verlag, New York, 1991, pp. 169–195.

Jaworski, N.A., Groffman, P.M., Keller, A.A., and Prager, J.C., A watershed nitrogen and phosphorus balance—The Upper Potomac River Basin, *Estuaries*, 15, 83–95, 1992.

Johnson, D. and Hale, B., White birch (*Betula papyrifera* Marshall) foliar litter decomposition in relation to trace metal atmospheric inputs at metal-contaminated and uncontaminated sites near Sudbury, Ontario and Rouyn-Noranda, Quebec, Canada, *Environ. Pollut.*, 127, 65–72, 2004.

Karr, J.R., Biological integrity: Along-neglected aspect of water resource management, *Ecol. Appl.*, 1, 66–84, 1991.

Kauppi, P.E., Mielikainen, K., and Kuusela, K., Biomass and carbon budget of European forests, 1971 to 1990, *Science*, 256, 70–74, 1992.

Kemp, M.J. and Dodds, W.K., Comparisons of nitrification and denitrification in prairie and agriculturally influenced streams, *Ecol. Appl.*, 12, 998–1009, 2002.

Kemp, W.M., Sampou, P., Caffrey, J., Mayer, M., Henriksen, K., and Boynton, W.R., Ammonium recycling versus denitrification in Chesapeake Bay sediments, *Limnol. Oceanogr.*, 35, 1545–1563, 1990.

Kersting, K., Functional endpoints in field testing, In *Freshwater Field Tests for Hazard Assessment of Chemicals*, Hill, I.R., Heimbach, F., Leeuwangh, P., and Matthiessen, P. (eds.), CRC Press, Boca Raton, FL, 1994, pp. 57–81.

Kilgour, B.W., Somers, K.M., and Barton, D.R., A comparison of the sensitivity of stream benthic community indices to effects associated with mines, pulp and paper mills, and urbanization, *Environ. Toxicol. Chem.*, 23, 212–221, 2004.

Kohler, H.R., Wein, C., Reiss, S., Storch, V., and Alberti, G., Impact of heavy metals on mass and energy flux within the decomposition process in deciduous forests, *Ecotoxicology*, 4, 114–137, 1995.

Laskowski, R., Maryanski, M., and Niklinska, M., Effect of heavy-metals and mineral nutrients on forest litter respiration rate, *Environ. Pollut.*, 84, 97–102, 1994.

Leland, H.V. and Carter, J.L., Effects of copper on production of periphyton, nitrogen fixation and processing of leaf litter in a Sierra Nevada, California, stream, *Freshw. Biol.*, 15, 155–173, 1985.

Lowrance, R.R., Leonard, R.A., Asmussen, L.E., and Todd, R.L., Nutrient budgets for agricultural watersheds in the southeastern coastal plain, *Ecology*, 66, 287–296, 1985.

McEnroe, N.A. and Helmisaari, H.S., Decomposition of coniferous forest litter along a heavy metal pollution gradient, South-West Finland, *Environ. Pollut.*, 113, 11–18, 2001.

McLaughlin, M.J. and Smolders, E., Background zinc concentrations in soil affect the zinc sensitivity of soil microbial processes—A rationale for a metalloregion approach to risk assessments, *Environ. Toxicol. Chem.*, 20, 2639–2643, 2001.

Megharaj, M., Kantachote, D., Singleton, I., and Naidu, R., Effects of long-term contamination of DDT on soil microflora with special reference to soil algae and algal transformation of DDT, *Environ. Pollut.*, 109, 35–42, 2000.

Megharaj, M., Singleton, I., and McClure, N.C., Effect of pentachlorophenol pollution towards microalgae and microbial activities in soil from a former timber processing facility, *Bull. Environ. Contam. Toxicol.*, 61, 108–115, 1998.

National Research Council, *Restoration of Aquatic Ecosystems: Science, Technology, and Public Policy*, National Academy Press, Washington, D.C., 1992.

Nelson, S.M., Leaf pack breakdown and macroinvertebrate colonization: Bioassessment tools for a high-altitude regulated system? *Environ. Pollut.*, 110, 321–329, 2000.

Niemi, G.J., Detenbeck, N.E., and Perry, J.A., Comparative analysis of variables to measure recovery rates in streams, *Environ. Toxicol. Chem.*, 12, 1541–1547, 1993.

Nieminen, T.M., Derome, J., and Helmisaari, H.S., Interactions between precipitation and Scots pine canopies along a heavy-metal pollution gradient, *Environ. Pollut.*, 106, 129–137, 1999.

Niklinska, M., Laskowski, R., and Maryanski, M., Effect of heavy metals and storage time on two types of forest litter: Basal respiration rate and exchangeable metals, *Ecotoxicol. Environ. Saf.*, 41, 8–18, 1998.

Niyogi, D.K., Lewis, W.M., and McKnight, D.M., Litter breakdown in mountain streams affected by mine drainage: Biotic mediation of abiotic controls, *Ecol. Appl.*, 11, 506–516, 2001.

Niyogi, D.K., Simon, K.S., and Townsend, C.R., Breakdown of tussock grass in streams along a gradient of agricultural development in New Zealand, *Freshw. Biol.*, 48, 1698–1708, 2003.

Odum, E.P., Finn, J.T., and Franz, E.H., Perturbation theory and the subsidy-stress gradient, *BioScience*, 29, 349–352, 1979.

Pascoal, C., Cassio, F., and Gomes, P., Leaf breakdown rates: A measure of water quality? *Intl. Rev. Hydrobiol.*, 86, 407–416, 2001.

Peterson, B.J., Wollheim, W.M., Mulholland, P.J., Webster, J.R., Meyer, J.L., Tank, J.L., Marti, E., et al., Control of nitrogen export from watersheds by headwater streams, *Science*, 292, 86–90, 2001.

Platt, J.R., Strong inference, *Science*, 146, 347–353, 1964.

Post, R.D. and Beeby, A.N., Activity of the microbial decomposer community in metal-contaminated roadside soils, *J. Appl. Ecol.*, 33, 703–709, 1996.

Raddum, G.G. and Fjellheim, A., Life-cycle and production of *Baetis rhodani* in a regulated river in Western Norway. Comparison of pre-regulation and post-regulation conditions, *Reg. Rivers Res. Manag.*, 8, 49–61, 1993.

Rawer-Jost, C., Bohmer, J., Blank, J., and Rahmann, H., Macroinvertebrate functional feeding group methods in ecological assessment, *Hydrobiologia*, 422, 225–232, 2000.

Riseng, C.M., Wiley, M.J., and Stevenson, R.J., Hydrologic disturbance and nutrient effects on benthic community, structure in Midwestern US streams: A covariance structure analysis, *J. N. Am. Benthol. Soc.*, 23, 309–326, 2004.

Royer, T.V., Tank, J.L., and David, M.B., Transport and fate of nitrate in headwater agricultural streams in Illinois, *J. Environ. Qual.*, 33, 1296–1304, 2004.

Rusk, J.A., Hamon, R.E., Stevens, D.P., and McLaughlin, M.J., Adaptation of soil biological nitrification to heavy metals, *Environ. Sci. Technol.*, 38, 3092–3097, 2004.

Sauve, S., Dumestre, A., Mcbride, M., Gillett, J.W., Berthelin, J., and Hendershot, W., Nitrification potential in field-collected soils contaminated with Pb or Cu, *Appl. Soil Ecol.*, 12, 29–39, 1999.

Schindler, D.W., Detecting ecosystem responses to anthropogenic stress. *Can. J. Fish. Aquat. Sci.*, 44 (Suppl. 1), 6–25, 1987.

Schindler, D.W., Experimental studies of chemical stressors on whole lake ecosystems, *Verh. Internat. Verein. Limnol.*, 23, 11–41, 1988.

Schulthesis, A.S. and Hendricks, A.C., The role of copper accumulations on leaves in the inhibition of leaf decomposition in a mountain stream, *J. Freshw. Ecol.*, 14, 31–40, 1999.

Schulthesis, A.S., Sanchez, M., and Hendricks, A.C., Structural and functional responses of stream insects to copper pollution, *Hydrobiologia*, 346, 85–93, 1999.

Sheehan, P.J., Functional changes in the ecosystem, In *Effects of Pollutants at the Ecostsyem Level*, Sheehan, P.J., Miller, D.R., Butler, G.C., and Bourdeau, P. (eds.), John Wiley and Sons, Chichester, UK, 1984, pp. 101–145.

Sheehan, P.J. and Knight, A.W., A multilevel approach to the assessment of ecotoxicological effects in a heavy metal polluted stream, *Verh. Intern. Verein. Limnol.*, 22, 2364–2370, 1985.

Shieh, S.H., Ward, J.V., and Kondratieff, B.C., Energy flow through macroinvertebrates in a polluted plains stream, *J. N. Am. Benthol. Soc.*, 21, 660–675, 2002.

Smolders, E., Brans, K., Coppens, F., and Merckx, R., Potential nitrification rate as a tool for screening toxicity in metal-contaminated soils, *Environ. Toxicol. Chem.*, 20, 2469–2474, 2001.

Smolders, E., McGrath, S.P., Lombi, E., Karman, C.C., Bernhard, R., Cools, D., Van den Brande, K., van Os, B., and Walrave, N., Comparison of toxicity of zinc for soil microbial processes between laboratory-contaminated and polluted field soils, *Environ. Toxicol. Chem.*, 22, 2592–2598, 2003.

Stout, R.J. and Cooper, W.E., Effect of p-cresol on leaf decomposition and invertebrate colonization in experimental outdoor streams, *Can. J. Fish. Aquat. Sci.*, 40, 1647–1657, 1983.

Strojan, C.L., Forest leaf litter decomposition in vicinity of a zinc smelter, *Oecologia*, 32, 203–212, 1978.

Suter, G.W., Norton, S.B., and Cormier, S.M., A methodology for inferring the causes of observed impairments in aquatic ecosystems, *Environ. Toxicol. Chem.*, 21, 1101–1111, 2002.

Swift, M.C., Smucker, R.A., and Cummins, K.W., Effects of dimlin on freshwater litter decomposition, *Environ. Toxicol. Chem.*, 7, 161–166, 1988.

Tuchman, N.C., Relative importance of microbes versus macroinvertebrate shredders in the process of leaf decay in lakes of differing pH, *Can. J. Fish. Aquat. Sci.*, 50, 2707–2712, 1993.

U.S. Environmental Protection Agency, National water quality inventory, 1988 Report to Congress, Office of Water, U.S. Government Printing Office, Washington, D.C., 1990.

Valiela, I. and Cole, M.L., Comparative evidence that salt marshes and mangroves may protect seagrass meadows from land-derived nitrogen loads, *Ecosystems*, 5, 92–102, 2002.

Wall, V.D., Alberts, J.J., Moore, D.J., Newell, S.Y., Pattanayek, M., and Pennings, S.C., The effect of mercury and PCBs on organisms from lower trophic levels of a georgia salt marsh, *Arch. Environ. Contam. Toxicol.*, 40, 10–17, 2001.

Wallace, J.B., Grubaugh, J.W., and Whiles, M.R., Biotic indices and stream ecosystem processes: Results from an experimental study, *Ecol. Appl.*, 6, 140–151, 1996.

Wallace, J.B., Webster, J.R., and Cuffney, T.F., Stream detritus dynamics: Regulation by invertebrate consumers, *Oecologia*, 53, 197–200, 1982.

Webster, J.R. and Benfield, E.F., Vascular plant breakdown in fresh-water ecosystems, *Ann. Rev. Ecol. Syst.*, 17, 567–594, 1986.

Whiles, M.R. and Wallace, J.B., First-year benthic recovery of a headwater stream following a 3-year insecticide-induced disturbance, *Freshw. Biol.*, 28, 81–91, 1992.

Whiles, M.R. and Wallace, J.B., Macroinvertebrate production in a headwater stream during recovery from anthropogenic disturbance and hydrologic extremes, *Can. J. Fish. Aquat. Sci.*, 52, 2402–2422, 1995.

Windham, L., Weis, J.S., and Weis, P., Metal dynamics of plant litter of *Spartina alterniflora* and *Phragmites australis* in metal-contaminated salt marshes. Part 1: Patterns of decomposition and metal uptake, *Environ. Toxicol. Chem.*, 23, 1520–1528, 2004.

Zimakowska-Gnoinska, D., Bech, J., and Tobias, F.J., Assessment of the heavy metal pollution effects on the soil respiration in the Baix Llobregat (Catalonia, NE Spain), *Environ. Monit. Assess.*, 61, 301–313, 2000.

32 The Use of Microcosms, Mesocosms, and Field Experiments to Assess Ecosystem Responses to Contaminants and Other Stressors

When factors are chosen for investigation, it is not because we anticipate that laws of nature can be expressed with any particular simplicity in terms of these variables, but because they are variables that can be controlled or measured with comparative ease.

(Fisher 1960)

32.1 INTRODUCTION

Results of field surveys and other descriptive approaches have provided a solid foundation by which to evaluate the effects of contaminants on ecosystem processes. These studies have shown that certain functional characteristics of ecosystems, especially productivity, nutrient flux, and decomposition, are quite sensitive to anthropogenic disturbance. However, as we noted in the previous chapter, descriptive studies are limited because of the inability to demonstrate cause-and-effect relationships and because of difficulties identifying underlying mechanisms responsible for changes in these ecosystem processes. Complex interactions and indirect effects of chemicals are likely to be the rule rather than the exception in many ecosystems. In addition, community inertia, defined as the tendency for communities to persist under unfavorable conditions following disturbance (Milchunas and Lauenroth 1995), complicates evaluation of ecosystem responses to perturbation. Isolating causal mechanisms is particularly important in ecosystem studies because these processes are often complex and controlled by an assortment of direct and indirect effects. For example, litter decomposition in aquatic and terrestrial ecosystems is regulated by microbial processes and activity of invertebrates. Because effects of contaminants on decomposition rate are dependent on the relative sensitivity of microbial and macroinvertebrate communities, experimental approaches that isolate these different mechanisms are necessary to predict effects. This is an ideal application of microcosm and mesocosm experiments, which are often designed to manipulate single or multiple environmental variables, providing an opportunity to isolate specific factors and identify underlying mechanisms. It is the ability to isolate and manipulate individual factors that makes application of microcosm and mesocosm experiments particularly powerful in ecotoxicological research.

In this chapter we turn our attention to experimental approaches that have been employed to demonstrate effects of contaminants and other stressors on ecosystem processes. We will examine the use of both small-scale approaches such as microcosms, as well as larger, more ecologically realistic and field-based approaches such as mesocosms and whole ecosystem manipulations. Because of the limited spatiotemporal scale, measuring responses of ecosystem processes to contaminants in microcosms presents significant challenges. The duration of microcosm and mesocosm experiments is of critical importance when assessing effects of contaminants. For example, a common phenomenon in soil microcosms is the natural reduction in microbial biomass or activity over time. As a consequence, effects of stressors on soil microbial processes become more difficult to quantify as experiments progress. Although microcosms are designed to simulate specific portions of a natural ecosystem, the most valuable experiments investigating effects of contaminants on ecosystem processes have been conducted in larger systems.

32.2 MICROCOSM AND MESOCOSM EXPERIMENTS

> There is an increasing belief amongst risk assessors that model ecosystems do not possess ecological advantages that were originally assumed, and that an instrumentalist approach to the prediction of toxic effects in ecosystems will yield the most cost-effective results.
>
> **(Crane 1997)**

> Current approaches to ecological risk assessment of chemicals are ecologically naive and fail to include current knowledge about effects of stressors on ecological communities.
>
> **(Pratt et al. 1997)**

The genesis of microcosm and mesocosm research emerged from uncertainties regarding the usefulness of single species approaches for predicting effects of contaminants in nature (Cairns 1986). While questions about ecological relevance of laboratory toxicity tests persist, microcosm and mesocosm experiments are now routinely employed in ecotoxicological research and to test ecological principles (Fraser and Keddy 1997). The use of controlled experimental systems in aquatic ecology has clearly increased over time (Figure 32.1), and these systems have been used to address both basic and applied ecological questions. For example, if we consider contaminants as simply another form of anthropogenic disturbance, model ecosystems can be used to characterize ecological resistance and resilience. However, in his critique of model ecosystems used in ecotoxicological research, Crane (1997) argues that research priorities should shift from understanding these ecological complexities to questions regarding repeatability, precision, and the relationship between experimental

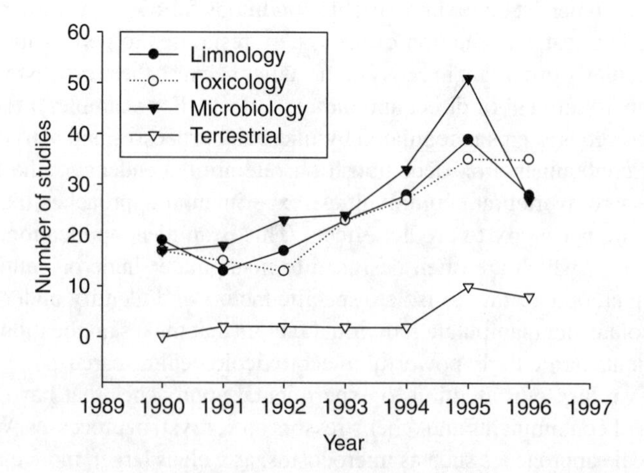

FIGURE 32.1 The number of studies published between 1990 and 1996 that included the words microcosm or mesocosm in the title or abstract. (Data from Table 1 in Fraser and Keddy (1997).)

and natural systems. He notes the "dangers of allowing model ecosystem studies to be driven by ecological theory" and argues that model ecosystems are most appropriate as a tool to provide environmentally realistic exposure conditions. Because microcosm and mesocosm experiments often provide complex responses across levels of biological organization, interpreting results can be challenging. Recall that problems with data interpretation were provided as a primary justification for the decision by the U.S. EPA to drop mesocosm testing for pesticide registration (Touart 1988, Touart and Maciorowski 1997, Chapter 23). Stay et al. (1988) argued that a lack of correspondence between population, community, and ecosystem-level responses observed in their experiments indicated that measurements at one hierarchical level may not be useful for predicting effects at other levels. Sorting out these complex responses and relating alterations in community structure to ecological processes should be a priority for microcosm and mesocosm research. We also believe that a critical area of research is to determine the extent to which these experimental systems reflect natural conditions. One of the challenges associated with the use of microcosms and mesocosms is the change in functional measures over time, independent of the effects of contaminants. These changes, which are often a result of container artifacts, compromise our ability to make comparisons among treatments. Williams et al. (2002) compared structural characteristics of microcosms and natural ponds, and recommended refinements to the design of model systems to improve their ecological realism. Unfortunately, few studies have made this comparison based on functional measures or ecosystem processes. Kurtz et al. (1998) measured reproducibility and stability of structural and functional processes in estuarine microcosms. Both structural (relative abundance and density of sulfate reducing bacteria) and functional (CO_2 assimilation, sulfate reduction) measures in microcosms were similar to conditions in natural sediments after 7 days. However, the authors cautioned against longer term experiments without modifying the system. Suderman and Thistle (2003) examined changes in structural and functional measures in sediment microcosms derived from a shallow estuary. Chlorophyll *a*, primary production and most measures of meiofauna community composition remained relatively stable over the 3-month period.

Another potential criticism of microcosm and mesocosm experiments is their relatively limited temporal scale. Because long-term mesocosm experiments (e.g., >1 year) are rare, our understanding of prolonged exposure to stressors is incomplete. This is an especially important issue when considering ecosystem processes that often show delayed responses compared to structural alterations. Bokn et al. (2003) exposed rocky intertidal communities to long-term nutrient enrichment in marine mesocosms. Despite large inputs of N and P (maximum target concentrations were 32 and 2.0 μM, respectively) and significant increases in periphyton biomass, there were essentially no effects on NPP, GPP, or respiration. These unexpected results were attributed to competition among macroalgal species, grazing by herbivores, and physical disturbance. Although this study was conducted for 2.5 years, a relatively long time period when compared with most mesocosm studies, this was not a sufficient amount of time for opportunistic algal species to become established and respond to nutrient enrichment (Bokn et al. 2003).

In addition to comparing processes in microcosms and mesocosms with those in natural ecosystems and assessing changes in controls over time, additional research is necessary to optimize experimental designs and to evaluate the statistical power of these systems (Kennedy et al. 1999). In Chapter 23, we discussed strengths and weaknesses of different experimental designs (e.g., ANOVA versus regression; assignment of replicates to experimental units) for community-level assessments. Consideration of statistical power is especially critical for assessments of ecosystem processes because variability of these measures is often greater than for structural measures. Kraufvelin (1998) estimated the number of replicates necessary to detect significant differences for 50 different variables derived from land-based, brackish water mesocosms. Although calculations were based on population and community-level variables, the results have important implications for mesocosm experiments designed to assess functional endpoints. Relatively few of the structural variables examined had coefficients of variation (CV) less than 20%. Using an endpoint with a modest CV of approximately 30%, 24 replicates were necessary to detect a statistically significant difference

($\alpha = 0.05$) of 25% between control and treatment mesocosms. Kraufvelin (1998) also noted large differences in the amount of variation among response variables. Assuming that ecosystem processes will show a similar or greater level of variation, statistical power will obviously be an important consideration when selecting functional endpoints.

32.2.1 Microcosms and Mesocosms in Aquatic Research

Microcosms and mesocosms have been employed extensively to assess the effects of contaminants on processes in aquatic ecosystems. Although the majority of these investigations have focused on changes in primary production, respiration and other aspects of ecosystem metabolism, endpoints related to nutrient processing and decomposition rates have also been considered. Most experiments conducted in stream microcosms and mesocosms have focused on the response of periphyton. Because of their small size, rapid rate of development, and diverse taxonomic composition, periphyton are sensitive indicators of water quality in natural and experimental streams (Lowe et al. 1996). Changes in the structure and function of epilithic assemblages exposed to contaminants can occur very rapidly. Colwell et al. (1989) attributed increased respiration in outdoor stream microcosms treated with Zn to the establishment of Zn-tolerant bacteria and algae. We should also remember that structural characteristics of ecosystems may directly or indirectly influence ecological processes. For example, physicochemical characteristics in macrophyte-dominated systems are often controlled by biological processes that are directly related to ecosystem metabolism (Brock et al. 1993). Kersting and van den Brink (1997) describe a dissolved oxygen–pH–alkalinity–conductivity syndrome, in which each of these variables is expected to respond to toxic substances in parallel. These interrelated responses may result in feedback between contaminants and ecosystem processes. For example, it is likely that alterations in community metabolism resulting from exposure to contaminants may affect pH and thereby modify contaminant bioavailability. To improve our understanding of the complex responses frequently observed in microcosm and mesocosm experiments, sampling protocols should be designed to quantify relationships between these physicochemical and biological variables.

32.2.1.1 Separating Direct and Indirect Effects

One of the important applications of microcosm and mesocosm research has been to separate the direct effects of contaminants from the indirect or secondary effects on ecosystem processes. Selective application of contaminants that have specific effects on one group of organisms but relatively limited effects on another group is a useful approach for quantifying these direct and indirect effects (Pratt et al. 1997, Slijkerman et al. 2004). Pearson and Crossland (1996) measured photosynthesis and respiration in outdoor experimental streams exposed to the herbicide atrazine and the insecticide lindane. Atrazine had a direct inhibitory effect on photosynthesis at 100 μg/L. In contrast, photosynthesis increased in lindane-treated streams due to the elimination of grazing invertebrates. Because the direct toxicological effect on invertebrates is limited, atrazine has been used to identify bottom-up responses in model systems (Pratt et al. 1997) (Figure 32.2). In this example, reduced food availability to higher trophic levels would be considered an indirect effect of herbicide exposure. Conversely, exposure of model systems to insecticides can have a direct effect on invertebrate grazers, resulting in increased algal biomass and production. Predicting effects of contaminants on intermediate trophic levels where elimination of one group may impact both lower and higher trophic levels presents special challenges. Boyle et al. (1996) quantified indirect effects of the insecticide diflubenzuron, a chitin inhibitor, on ecosystem processes in 0.1 ha mesocosms. Significant declines in abundance of grazing insects and zooplankton following diflubenzuron treatment resulted in increased chlorophyll *a* biomass and GPP (a top-down response due to reduced grazing) and reduced biomass of juvenile bluegill (a bottom-up response due to reduced food supply). Brock et al. (1993) reported similar increases in periphyton and phytoplankton when

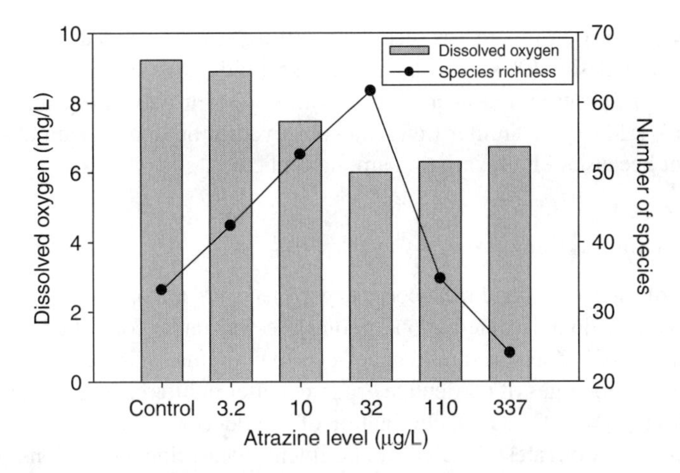

FIGURE 32.2 Changes in structural (species richness) and functional (dissolved oxygen concentration) variables in microcosms treated with the herbicide atrazine. (Data from Table 1 in Pratt et al. (1997).)

mesocosms were treated with the insecticide chlorpyrifos, which reduced abundance of grazing invertebrates.

The relationships between structural and functional components of model ecosystems are often complex and may be dependent on contaminant concentration. Slijkerman et al. (2004) observed that at intermediate concentrations of the fungicide carbendazim (17 μg/L), structural changes were observed but there were no corresponding effects on ecosystem function. Functional impairment occurred at higher exposure concentrations (219 μg/L), indicating that functional redundancy could not compensate for changes in community structure. Similar results were reported by Carman et al. (1995) for meiofauna exposed to PAHs in sediment microcosms. Despite significant changes in meiofaunal community composition in high PAH treatments, there were no effects on bacterial or microalgal activity.

Elimination of invertebrates by chlorpyrifos in experimental ditches had modest effects on ecosystem metabolism by decreasing respiration and increasing oxygen concentration; however, effects on community structure and decomposition rates were much more dramatic (Kersting and van den Brink 1997). Exposure of macroinvertebrates to chlorpyrifos reduced abundance of shredders and resulted in decreased litter decomposition (Cuppen et al. 1995). These researchers also speculated that elimination of grazing invertebrates by insecticides may enhance effects of eutrophication by reducing top-down control of primary producers. Detenbeck et al. (1996) measured biomass, GPP, and respiration in mesocosms treated with the herbicide atrazine in wetland mesocosms. GPP was reduced at the lowest exposure level (15 μg/L), but respiration was either reduced (25 μg/L) or enhanced (75 μg/L). An increase in ammonium, dissolved N, and dissolved P in treated mesocosms was attributed to reduced nutrient uptake by periphyton. Bester et al. (1995) observed significant reductions in primary production at low levels of atrazine exposure (0.12 μg/L) in marine mesocosms. An increase in concentrations of dissolved organic N and P in treated microcosms was attributed to release from damaged cell walls.

Mesocosm experiments also allow investigators to compare responses across levels of biological organization, thereby providing opportunities to examine underlying mechanisms and relate structural changes to functional alterations. As noted in Chapter 31, reduced decomposition rate observed in contaminated streams may result from either lower abundance of macroinvertebrate shredders or changes in microbial activity. Stream mesocosm experiments have been used to assess the relative importance of these two explanations. Newman et al. (1987) measured litter processing rates, shredder abundance, and microbial colonization in outdoor experimental streams dosed with chlorine. Although no effects were measured at intermediate concentrations (64 μg/L total residual chlorine), lower rates of decomposition in streams receiving 230 μg/L were attributed primarily to reduced abundance of amphipod shredders. The bacterial insecticide *Bacillus thuringiensis* significantly increased microbial respiration and decreased decomposition in laboratory microcosms (Kreutzweiser et al. 1996). Although a similar trend was observed in outdoor stream channels, this trend was not significant because of high variation among replicates.

32.2.1.2 Stressor Interactions

The key strengths of microcosm and mesocosm experiments are the opportunity to assess effects of chemical mixtures and to quantify interactions among stressors under controlled experimental conditions. Cuppen et al. (2002) observed significant effects of a mixture of insecticides (chlorpyrifos and lindane) on decomposition rates of particulate organic matter in litterbags. Despite rapid dissipation of both insecticides ($t_{1/2} = 9$–22 days), elimination of shredders and reduced microbial activity resulted in lower decomposition rates. Results of experiments measuring interactions between nutrients and agricultural contaminants (e.g., herbicides, insecticides, sediments) are especially enlightening because these stressors frequently co-occur. More importantly, bioavailability of contaminants may vary depending on the nutrient status and the amount of organic material in an ecosystem. For example, sorption of contaminants is likely to be greater in more productive ecosystems. Barreiro and Pratt (1994) used a factorial experimental design to measure the interactive effects of nutrient enrichment and the herbicide diquat on primary productivity in microcosms. Although structural variables responded to nutrients, there was no effect of diquat on algal biovolume, chlorophyll *a*, or protein levels. In contrast, GPP was significantly reduced in treated microcosms. These researchers also reported that recovery was greater in systems with higher nutrient levels, most likely due to faster contaminant dissipation (Pratt and Barreiro 1998). The influence of nutrient concentration on community resistance and resilience was also reported by Steinman et al. (1992), indicating that functional responses to chemical perturbations are often context-dependent.

Comparatively few studies have examined effects of contaminants on N cycling and flux in aquatic microcosms. Petersen et al. (2004) compared the effects of two antifouling biocides (zinc pyrithione, ZPT; and copper pyrithione, CPT) on nitrification and denitrification processes in sediments. Flux of nitrate from sediment increased significantly after additions of ZPT and CPT (Figure 32.3). This increase was a result of increased nitrification ($NH_4 \rightarrow NO_2 \rightarrow NO_3$) and/or a decrease in denitrification ($NO_3 \rightarrow N_2$). The greater sensitivity of nitrification observed in this experiment was

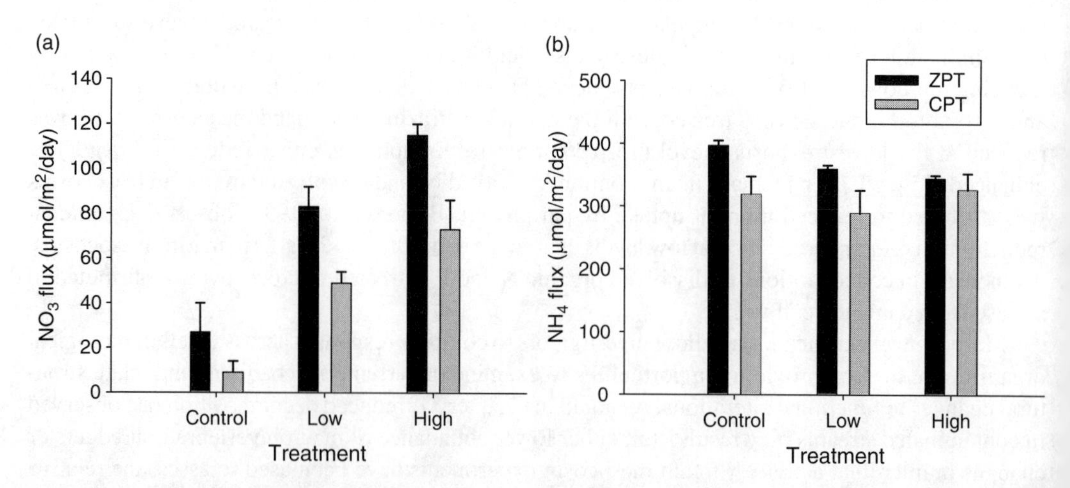

FIGURE 32.3 Flux of NO_3 and NH_4 in microcosms exposed to zinc pyrithione (ZPT) and copper pyrithione (CPT). Low and high treatments in the ZPT and CPT experiments were: 1.0 and 10.0 nmol ZPT/g and 0.1 and 1.0 nmol CPT/g, respectively. (Data from Table 1 in Petersen et al. (2004).)

likely a result of greater functional redundancy of denitrification processes (Petersen et al. 2004). Nitrification is a process performed by a limited number of bacteria, whereas denitrification is a general process performed by many species.

Microcosm and mesocosm experiments can also be used to compare functional responses of communities derived from different sources, thereby providing an opportunity to understand how intrinsic features of an ecosystem may influence susceptibility to contaminants. Stay et al. (1988) reported that effects of fluorene on respiration and rates of recovery differed among communities depending on the source of these organisms. Fate of the insecticide chlorpyrifos in microcosms and its effects on community metabolism, decomposition, and nutrient cycling was influenced by the presence of macrophytes (Kersting and van den Brink 1997). Balczon and Pratt (1994) compared effects of Cu on littoral and open water communities. Effects of Cu on oxygen production and respiration were reduced in microcosms with an established littoral zone, most likely because of greater adsorption and complexation by macrophytes and sediments. Although these results showing variable responses in different ecosystems complicate our ability to make broad generalizations, understanding the underlying mechanisms responsible for this variation may ultimately improve our predictive ability.

Interactions between biotic and abiotic factors may also influence the response of primary producers to contaminants. Steinman et al. (1992) observed that the physical structure and integrity of periphyton mats influenced resistance and resilience of carbon fixation rates (a measure of primary productivity) to chlorine exposure. Hill et al. (2000) measured bioaccumulation of Cd by periphyton and subsequent effects on photosynthesis. Effects of Cd on photosynthesis were regulated by periphyton biomass, with greater effects observed in treatments with less biomass. Although there were differences in community composition among biomass treatments, reduced effects in high biomass treatments were attributed to contaminant dilution and lower Cd bioavailability.

32.2.1.3 Ecosystem Recovery

Although the short duration of many microcosm and mesocosm experiments precludes assessment of recovery, some researchers have used these experimental systems to evaluate improvements in ecosystem processes when contaminants are reduced or eliminated. Oviatt et al. (1984) measured recovery of benthic respiration and nutrient flux for 21 months in mesocosms containing sediments collected along a pollution gradient. Within 5 months, water quality characteristics (nutrients, chlorophyll *a*, and dissolved oxygen) and net system production were similar among treatments, indicating that recovery may occur rapidly after pollutants are eliminated. Rapid recovery (4 weeks) of photosynthesis following exposure of marsh plants to crude oil was also reported by Pezeshki and Deluane (1993). Similarly, periphyton productivity in outdoor experimental stream channels dosed with the herbicide, hexazinone, was reduced by 80%, but recovered within 24 h following treatment (Schneider et al. 1995). The estimated LC50 of hexazinone for periphyton production (3.6 μg/L) was reported to be less than published values based on single species tests, demonstrating the greater sensitivity of this functional measure.

32.2.1.4 Comparisons of Ecosystem Structure and Function

The majority of published microcosm and mesocosm experiments measure either structural or functional characteristics. Because of concerns over sensitivity, variability, and the rate of response of some functional indicators, we suggest that a practical application of these experimental systems is to compare the efficacy of structural and functional endpoints. Questions such as the number of replicates required to detect statistical differences between reference and treated microcosms and the rate at which structural and functional variables respond to chemical stressors are of particular importance. Rigorous control over exposure conditions and the ability to manipulate several variables simultaneously in microcosm and mesocosm experiments provide a unique opportunity to compar effects of stressors on structural and functional characteristics (Culp et al. 2003).

The conventional wisdom is that, because of functional redundancy and greater variability o functional measures, changes in community composition are likely to occur before alterations in eco system processes are observed (Schindler 1987, Schindler et al. 1985). However, like many example of conventional wisdom, there are exceptions to these generalizations in the literature. Some studie have reported that functional measures are equally sensitive or even more sensitive than measure of abundance, biomass, or community composition. Functional measures (periphyton productivity were considerably more sensitive than structural measures (periphyton biomass; macroinvertebrat abundance and drift) to the herbicide hexazinone in outdoor stream mesocosms (Schneider et al 1995). Concentration–response relationships between copper and several functional endpoints wer established by Hedtke (1984) in laboratory microcosms. GPP and respiration were reduced at 9.3 μg Cu/L, but changes in community composition were observed only at higher concentrations of Cu (30 μg/L), suggesting that ecosystem processes were more sensitive than structure in these experi ments. Clements (2004) reported that EC_{10} values for heavy metals based on community respiratio and abundance of metal-sensitive species were similar. Jorgensen et al. (2000) calculated no effec concentrations (NECs) for a variety of structural and functional measures in large pelagic meso cosms exposed to anionic surfactants (linear alkylbenzene sulfonates). Biomass (as chlorophyll *a* and biovolume of the dominant taxonomic groups were affected only at the highest concentration tested, whereas phytosynthetic activity was the most sensitive parameter for phytoplankton. Afte 4.5-day exposure, NECs for photosynthetic activity were similar to values for structural characterist ics (abundance of protozoans, crustaceans, and diatoms). Detenbeck et al. (1996) reported that gros productivity of periphyton was significantly reduced in microcosms exposed to 15 μg/L of atrazine a concentration that significantly reduced survival of *Daphnia* but had no effect on other respons variables measured (biomass of cattails; growth of tadpoles and fathead minnows). Fairchild et al (1987) compared community composition, nutrient dynamics, leaf decomposition, and primary pro duction in experimental streams exposed to clean and contaminated (triphenyl phosphate) sediments Sediment exposures altered patterns of macroinvertebrate drift and increased nutrient retention, bu had no effects on leaf decomposition.

Some stream microcosm experiments have been conducted specifically to validate results o laboratory toxicity tests and provide an opportunity to compare ecosystem functional measure with more traditional toxicological endpoints. Exposure of stream mesocosms to relatively hig levels of Cd (143 μg/L) resulted in reduced abundance of grazing snails and increased periphyto biomass, but had no effects on gross or net primary productivity (Brooks et al. 2004). Concurren single species toxicity tests with *Ceriodaphnia dubia* and *Pimephales promelas* showed that surviva was significantly reduced at this concentration. The lack of a response at lower Cd concentration (15 μg/L) was attributed to high concentrations of dissolved organic materials in these effluent dominated streams, which likely reduced metal bioavailability. Richardson and Kiffney (2000 compared structural and functional measures in outdoor experimental streams dosed with mixture of metals. Significant concentration–response relationships were developed for several measure related to mayfly abundance and drift, but no effects of metals on algal biomass or bacterial respiratio were observed. These researchers recommended that regulatory agencies should include estimate of mayfly abundance and richness as indicators of metal impacts in streams.

Balczon and Pratt (1994) derived maximum allowable toxicant concentrations (MATCs) fo littoral and aquatic microbial microcosms exposed to Cu. The MATCs were generally greater fo process (photosynthesis, respiration) as compared to measures of community composition (specie richness, chlorophyll *a* biomass), indicating greater sensitivity of structural responses. Similar res ults were reported by Melendez et al. (1993) in which microbial communities were exposed to th herbicide diquat. Changes in productivity and respiration were observed only at the two highest con centrations (10 and 30 mg/L), and these ecosystem-level responses recovered after 2 weeks exposure In contrast, the MATC for protozoan species richness and bacterial cell density was 0.32 mg/L, an these responses showed little evidence of recovery. Barreiro and Pratt (1994) observed that gross community productivity of periphyton was considerably more sensitive than chlorophyll *a* to diquat. The lowest observable effect concentration (LOEL) for P/R in planktonic communities exposed to fluorene, a polycyclic aromatic hydrocarbon (0.12 mg/L), was comparable to chronic toxicity values based on single species tests with cladocerans, chironomids, and bluegill (Stay et al. 1988). However, the magnitude of change in ecosystem processes did not reflect the near complete elimination of most zooplankton at concentrations exceeding 2 mg/L. Exposure of microbial communities derived from natural sediments to a fungicide, herbicide, or insecticide reduced microbial biomass but had no significant effects on respiration or denitrification (Widenfalk et al. 2004). These differences among experiments suggest that not only is the relative sensitivity of structural and functional measures contaminant-specific, it may also vary with level of contamination and characteristics of the exposure system.

In a comprehensive analysis of structural and functional responses of outdoor aquatic mesocosms to the insecticide diflubenzuron, Boyle et al. (1996) observed relatively little effects on GPP, but a significant increase in chlorophyll *a*, and reduced abundance and species richness of secondary consumers (zooplankton, insects, and bluegill) in treated mesocosms (Figure 32.4). Although community metabolism and decomposition rates were affected in microcosms treated with chlorpyrifos, these processes were generally less sensitive and occurred only after changes in structural measures, suggesting functional redundancy of these systems (Brock et al. 1993). Cuppen et al. (2002) reported that no observable effects concentrations (NOECs) for decomposition rate of *Populus* leaves and abundance of several macroinvertebrate shredders were similar. Interestingly, the structural and

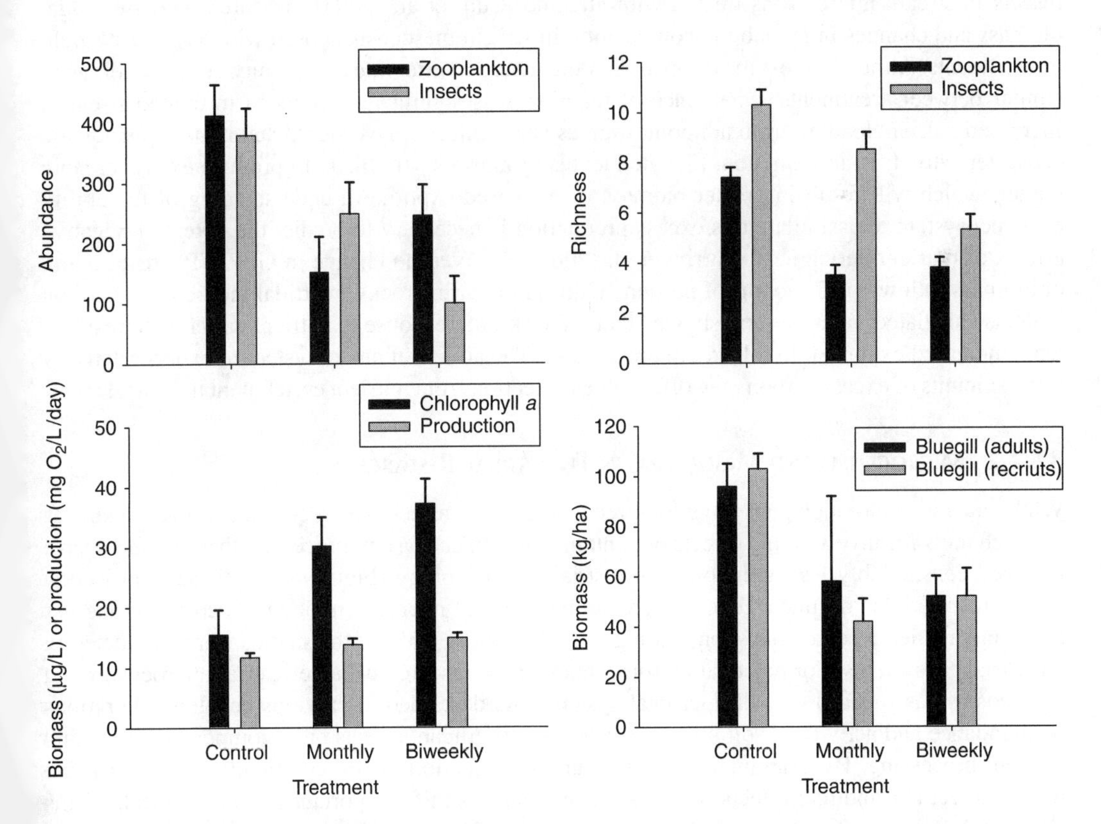

FIGURE 32.4 Effects of the insecticide, diflubenzuron, on structural (abundance, biomass, richness) and functional (primary production) measures in lentic mesocosms. Increased chlorophyll *a* biomass in mesocosms treated monthly and biweekly compared with controls was attributed to reduced grazing pressure. (Data from Boyle et al. (1996).)

functional NOECs derived from this microcosm experiment were considerably less than the LC50 value derived from standard toxicity tests using known sensitive organisms. Because of the complex and often unpredictable relationship between structural and functional measures observed in some studies, we suggest that an appropriate strategy will be to include endpoints reflecting both pattern and process when designing microcosm and mesocosm experiments. We agree with Brock et al. (1993) that an understanding of contaminant effects on ecosystem function cannot be fully appreciated without an understanding of community structure.

32.2.1.5 Effects of Contaminants on Other Functional Measures

Although we traditionally consider changes in community composition to be a structural measure, some researchers consider alteration in the abundance of groups that play an important functional role (e.g., abundance of shredders in streams; abundance of grazing zooplankton in lakes) to be intimately related to ecosystem processes and therefore an appropriate surrogate functional measure (Gruessner and Watzin 1996, Wallace et al. 1996). Field (Wallace et al. 1982) and stream microcosm experiments (Carlisle and Clements 1999) have assessed the effects of contaminants on functional feeding group composition. The export or loss of materials from an ecosystem is an important functional process that has received relatively little attention in the ecotoxicological literature. Similarly, emergence of adult insects represents a net transfer of energy from aquatic to terrestrial habitats and therefore could be considered a functional response. Gruessner and Watzin (1996) reported increased emergence of insects in stream microcosms treated with atrazine. Culp et al. (2003) measured increased algal biomass and changes in taxonomic composition in stream mesocosms dosed with 5% or 10% pulp mill effluents. Although most measures of benthic macroinvertebrate community composition were similar between treatments, emergence of mayflies was significantly reduced in treated streams. Increased nutrient loading from nonpoint sources is expected to have significant impacts on aquatic ecosystem structure and function. Elevated levels of nutrients are likely to produce excess organic matter, which will result in greater biomass or increased export. An understanding of the ability of an ecosystem to assimilate this excess production is necessary to predict the potential negative effects of nutrient enrichment. Barron et al. (2003) observed no change in GPP, NPP, respiration, or biomass following 27 months of nutrient addition in marine rocky intertidal mesocosms. Carbon budgets calculated in this system showed that the lack of a response to nutrient enrichment resulted from increased export of dissolved organic carbon. The ability of an ecosystem to export relatively large amounts of excess carbon may offer some protection from nutrient enrichment in coastal areas.

32.2.2 Microcosms and Mesocosms in Terrestrial Research

While aquatic ecotoxicologists have long recognized the value of microcosms and mesocosms as research tools for investigating effects of contaminants on ecosystem processes, these systems have received considerably less attention in terrestrial ecotoxicology (Figure 32.1). Fraser and Keddy (1997) reported that despite a general increase in the use of microcosms and mesocosms to address basic and applied research questions during the mid-1990s, <5% of the studies were conducted in terrestrial ecosystems. For practical reasons, much of the research using terrestrial microcosms and mesocosms has focused on soil microbial systems. As described in previous chapters, alterations in abundance and activity of soil microbes can have significant effects on decomposition rates and nutrient processing. By examining both structure and function of soil communities, it is possible to link direct and indirect effects of contaminants, and identify important regulating mechanisms (Bogomolov et al. 1996). Although the vast majority (>95%) of soil respiration in terrestrial ecosystems is a result of microbial activity, nematodes, arthropods, annelids, and other organisms contribute significantly to decomposition. Experiments have been conducted to determine the relative contributions of microbes and invertebrates to detrital food webs. Salminen et al. (2001) measured the effects of heavy metals and detritivores (enchytraeid oligochaetes) on respiration in soil microcosms. Invertebrate detritivores were eliminated at the highest Zn concentrations (>2500 mg/kg) and effects of Zn on microbial respiration were dependent on detritivore density. Clear effects of Zn were only observed in treatments with the greatest density of detritivores. These researchers reported some evidence of functional redundancy, but noted that elimination of species that play an important role in regulating soil microbial processes will have disproportionate impacts on ecosystems.

Unique properties of the soil environment may complicate our ability to assess bioavailability and contaminant effects. For example, in aquatic ecosystems, the assumption that contaminants are evenly distributed within the water column is generally valid. Although most experiments conducted with soil microcosms attempt to achieve a relatively homogeneous distribution of contaminants, chemicals in natural soils are often patchily distributed. In addition, small-scale spatial variation in the physicochemical characteristics of soils may alter chemical bioavailability (Salminen and Sulkava 1997). Some of the characteristics of soils that modify chemical bioavailability, such as particle size and amount of organic material, are analogous to properties of aquatic sediments. We will see that experimental designs that account for soil type and modify soil characteristics are a common feature of many terrestrial microcosm experiments. In the following sections, we will review some of the experiments conducted to assess the effects of heavy metals, organics, and other stressors on soil processes.

32.2.2.1 Heavy Metals

Effects of heavy metals on ecosystem processes have been measured in soil microcosms containing both natural and synthetic assemblages of microbes. In addition to measuring bacterial, fungal, nematode, and arthropod abundance and biomass, typical functional endpoints reported in these studies include soil respiration, nitrification, and N mineralization. Although some experimental studies have directly measured leaf litter decay rates (Cotrufo et al. 1995, Kohler et al. 1995), most of the research has focused on underlying microbial processes that regulate decomposition. Bogomolov et al. (1996) measured a suite of structural and functional characteristics in microcosms exposed to Cu. Increased pools of dissolved organic N and ammonium, reduced soil respiration, and reduced litter decay were observed in Cu-treated microcosms. Soil respiration was the most sensitive process examined, with effects observed at 50 mg Cu/L. These changes in ecosystem processes were the result of direct toxic effects on structural measures (reduced microbial biomass and abundance of nematodes).

As in aquatic ecosystems, one major advantage of microcosm and mesocosm experiments is the ability to manipulate several independent variables or site characteristics to quantify factors that determine contaminant effects and bioavailability. Khan and Scullion (2000) examined effects of heavy metals on microbial biomass, respiration, and mineralization in soils with varying clay and organic content. Metal bioavailability and effects were generally greater in sandy loams as compared to soils with higher organic content. Ammonification and nitrification were found to be more sensitive to Cd in calcareous soils than noncalcareous soils (Dusek 1995). Nitrate accumulated in Cd-treated calcareous soils primarily as a result of the greater sensitivity of nitrite oxidizers. Niklinska et al. (1998) established concentration–response relationships between heavy metals and respiration in litter collected from beech-pine and oak forests. Despite differences in physical and chemical characteristics of the two litter types, storage time was more important in controlling effects of metals than litter type. Decreases in litter respiration rates with storage time were most likely a result of rapid reduction in the amount of easily degraded material and/or increased respiration rate immediately following litter collection (Niklinska et al. 1998). For litter respiration to be a useful indicator of ecosystem responses to contaminants, a better understanding of the effects of storage time is required. On the basis of the estimated EC50 values for respiration (Figure 32.5), Niklinska et al. (1998) reported the following range of toxicity: $Cu > Zn \geq Cd \gg Pb$. It is interesting to note

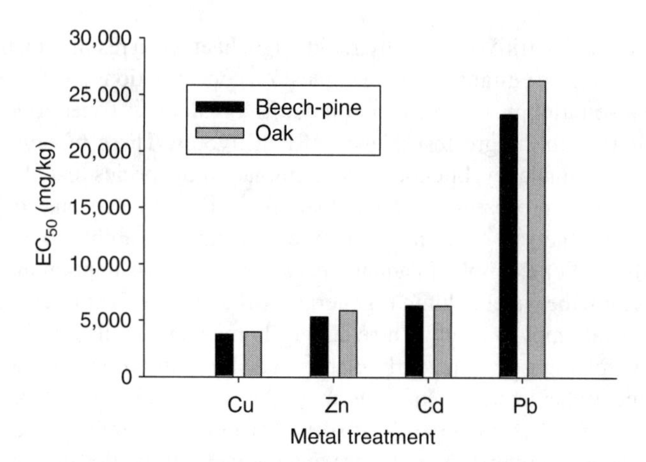

FIGURE 32.5 Average EC50 values for soil respiration rates measured in microcosms exposed to heavy metals. Forest litter was collected from beech-pine and oak-hornbeam forests. (Data from Niklinska et al. (1998).)

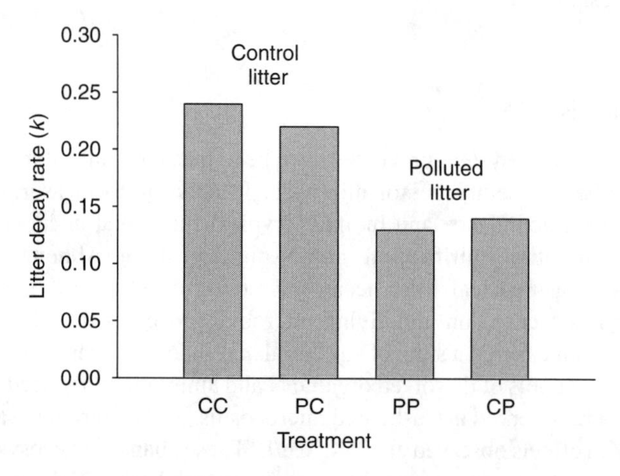

FIGURE 32.6 Litter decay rates of oak leaves (*Quercus ilex*) collected from control and metal polluted sites after 2 months exposure to clean and metal contaminated soil. CC = soil and litter from control site; PC = polluted soil and clean litter; PP polluted soil and polluted litter; CP = clean soil and polluted litter. (Data from Table 3 in Cotrufo et al. (1995).)

that this order of toxicity is quite different from most aquatic studies in which Cd is significantly more toxic to primary producers than either Cu or Zn.

Information concerning the relative toxicity of heavy metals in soil versus leaf litter is necessary for remediation of metal contaminated sites. In microcosm experiments Cotrufo et al. (1995) compared soil respiration, microbial and fungal biomass, and litter decomposition rates of oak leaves collected from clean and metal-polluted sites. Lower respiration rates and reduced fungal abundance were observed in litter contaminated by metals. Decomposition rates were significantly reduced for metal-polluted litter, regardless of the soil source (Figure 32.6). These data indicate that heavy metals in litter were directly responsible for reduced decomposition rates. The addition of organic material through natural decomposition processes can reduce the effects of heavy metals in terrestrial ecosystems. Boon et al. (1998) observed that the combined effects of low pH and Cu contamination on soil ecosystem processes were significantly reduced in microcosms planted with Cu-tolerant grass. The positive influence of Cu-tolerant plants on soil processes, which provided organic material and reduced metal bioavailability, has important implications for restoring habitats impacted by heavy metals.

32.2.2.2 Organic Contaminants and Other Stressors

Microcosms and mesocosms have been used to examine the effects of organic contaminants in terrestrial ecosystems, with considerable effort devoted to assessing responses of soil communities. Some papers have adopted a comparative approach and examined responses to a large number of organic chemicals on a few ecosystem-level endpoints (Pell et al. 1998). Others have examined effects of a single chemical or class of chemicals on several ecosystem processes. An excellent series of papers describing the use of terrestrial microcosms in ecological risk assessment was published in the journal *Ecotoxicology* in 2004 (Volume 13, Issue 4). This series of publications was the result of a joint effort by university, private, and governmental partners to develop a standardized method for conducting microcosm experiments in terrestrial ecosystems. The terrestrial model ecosystems (TME) experiments were conducted with intact soil cores collected from several different field sites. Initial experiments focused on ecosystem responses to carbendazim, a fungicide used extensively for agricultural applications in Europe. Experiments conducted at sites in the United Kingdom, Germany, Portugal, and the Netherlands examined effects of carbendazim on nutrient cycling and organic matter processing. In general, nutrient dynamics were not affected by contaminant exposure during the 16-week experiment, a result that was also supported by field experiments conducted simultaneously (Van Gestel et al. 2004). In contrast to these results, Burrows and Edwards (2004) reported significant effects of carbendazim on nutrient dynamics, soil dehydrogenase activity, and several structural measures in soil microcosms treated with similar carbendazim concentrations.

Alteration in decomposition rate of organic matter is likely to affect nutrient dynamics and is therefore considered an integrative functional endpoint in microcosm tests. Because decomposition and nutrient dynamics in soils are closely coupled and regulated by microbial processes, structural changes in microbial communities are likely to have important effects on ecosystem function. The fungicide dithianon had relatively little influence on decomposition rates but significantly inhibited microbial activity in soil microcosms (Liebich et al. 2003). These alterations in functional processes corresponded to changes in microbial community composition and fungal biomass. Forster et al. (2004) used cellulose paper as a standardized material to measure decomposition and assess invertebrate feeding activity. A significant concentration–response relationship between carbendazim concentration and decomposition rate was observed, with treated microcosms having 40%–80% lower decomposition and showing a significant reduction in invertebrate feeding activity. Similar LC50 values were calculated for both microcosm and field experiments (7.1–9.5 kg/Ha), lending additional support for the application of TMEs to assess contaminant effects on soil processes.

Soil respiration, measured as evolution of CO_2, is also a sensitive endpoint in microcosm experiments. Salminen et al. (2002) compared the effects of several organic chemicals and heavy metals on microbial respiration rate in soil microcosms. This functional measure was correlated with microbial biomass and estimates of microarthropod and nematode abundance. Soil respiration was reduced in all contaminant treatments, and effects increased with chemical concentration. Changes in respiration were accompanied by decreases in microbial biomass and abundance of soil organisms resulting from direct toxicity in most treatments. Because some of these responses were not observed until late in the study, Salminen et al. (2002) recommend that soil microcosm experiments should be conducted for a sufficient period of time to avoid false conclusions regarding responses.

Few investigators have conducted comparative studies of organic chemicals under different physicochemical conditions in soil microcosm experiments. Chen and colleagues (Chen and Edwards 2001, Chen et al. 2001) employed soil microcosms to examine the effects of fungicides on several ecosystem processes. Soil properties, especially soil texture, regulated the effects of fungicides on soil microbial activity and nutrient dynamics (Chen and Edwards 2001). Soil microcosms amended with either alfalfa leaves or wheat straw, materials with very different C:N ratios (alfalfa leaves,

C:N = 8.9; wheat straw C:N = 84), showed variable responses to the fungicides benomyl, captan, and chlorothalonil (Chen et al. 2001). At recommended application rates, each of the fungicides reduced soil respiration rates by 30%–50% in unamended soils; however, effects in soils amended with alfalfa leaves or wheat straw varied among the three fungicides. Martinez-Toledo et al. (1996) measured the effects of the herbicide simazine in soil microcosms with different physicochemical characteristics. Simazine had relatively little effect on most structural and functional characteristics of soil microflora. However, abundance of nitrifying bacteria was significantly reduced at application rates normally used in agriculture, with greatest effects in soils with low organic content. These researchers speculated that alterations in abundance of nitrifying bacteria are likely to have long-term consequences for nutrient dynamics in soils and may disrupt the balance among nitrogen fixation, denitrification, and nitrification.

Riparian habitats are located at the interface between terrestrial and aquatic ecosystems, and often function as buffers to protect lakes and streams from anthropogenic disturbances. For example, processes that occur in riparian soils, such as denitrification and microbial mineralization, can reduce the effects of N enrichment in riparian ecosystems. The size of riparian buffers required to protect stream ecosystems from negative effects of agricultural runoff, urban discharges, and other stressors has received considerable attention in the literature (Harding et al. 1998, Stewart et al. 2001, Wang et al. 1997). Ettema et al. (1999) compared the effect of N addition to soil microcosms placed in two zones along a riparian corridor. Although microbial biomass and respiration were not affected by N treatments in either zone, N added to microcosms located in the near-stream zone was effectively removed by denitrification. In contrast, addition of N to an upslope zone did not stimulate denitrification. These results suggest that denitrification can provide some protection against increased N addition, but that responses vary among soil types and are relatively ephemeral (Ettema et al. 1999).

Although top-down models of food web structure are frequently used to characterize aquatic ecosystems, it is uncertain if these models are appropriate for soil communities. Many soil communities are highly dependent on processes occurring in decomposer food webs and are therefore likely to be controlled by bottom-up processes. Salminen and Sulkava (1997) measured microbial biomass, abundance of soil invertebrates, and nutrient dynamics in soil microcosms treated with sodium pentachlorophenol (PCP), a wood preservative known to be toxic to many organisms. Reduced microbial biomass resulted in the accumulation of nutrients in treated microcosms (Figure 32.7). Because of the strong influence of PCP on the microbial food web and the similar sensitivity of predators and prey to PCP, no evidence for top-down control was observed. On the basis of these results, it appears that

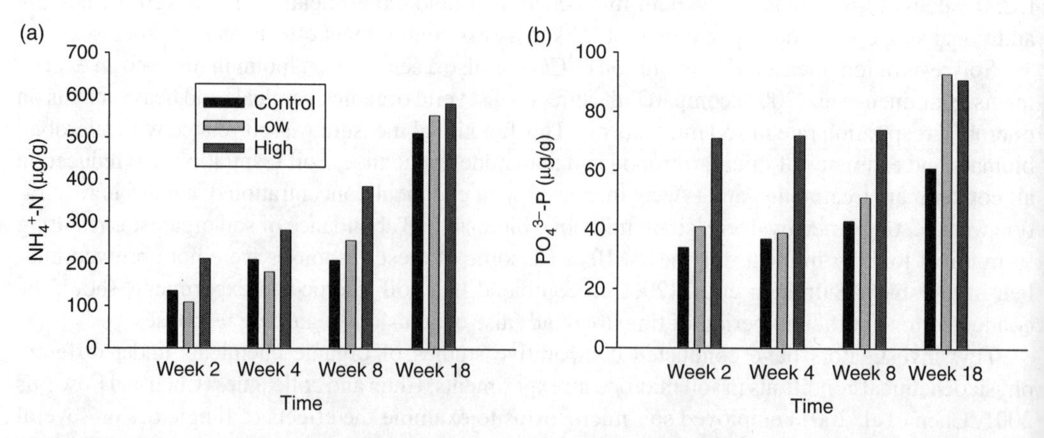

FIGURE 32.7 Concentrations of NH_4 and PO_4 in soil microcosms exposed to sodium pentachlorophenate (PCP) over an 18-week period. Low and high treatments correspond to 50 and 500 mg PCP/kg, respectively. (Data from Table 3 in Salminen and Sulkava (1997).)

soil decomposer food webs are regulated by bottom-up processes and therefore traditional top-down models employed in many aquatic systems are unsuitable for measuring effects of contaminants in soils (Salminen and Sulkava 1997).

32.3 WHOLE ECOSYSTEM EXPERIMENTS

> Ecosystem experiments are the most direct method available for improving predictions of environmental response to management or inadvertent perturbation.
>
> **(Carpenter et al. 1995)**

Whole ecosystem experiments have often been employed to assess the effects of contaminants on functional processes. In fact, ecosystem-scale experiments designed to test the effectiveness of new fertilizers, pesticides, and other crop treatments for agricultural applications played a prominent role in the history of experimental design and statistical analyses. However, unlike argoecosystem experiments that often have well-defined management objectives (e.g., increase crop production) and can provide results in relatively short periods of time, conducting experiments in natural ecosystems is considerably more challenging (Schindler 1990). Tests of Odum's (1985) hypotheses concerning how ecosystem processes should be affected by stress are generally not well supported in the literature (Rapport et al. 1985, 1998; Schindler 1990), highlighting the difficulty of predicting how ecosystems will respond to perturbations.

Although the distinction between large mesocosm and whole ecosystem experiments is often blurred, especially in terrestrial and marine systems, we will use the conventional definition of Odum (1984) and consider mesocosms as partially enclosed experimental systems. Thus, whole ecosystem experiments involve the planned application of contaminants or other stressors directly to a system that is large enough to contain all of the important physical, chemical, and biological processes of interest (Carpenter et al. 1998). In Chapter 23, we discussed the use of whole ecosystem experiments for examining effects of contaminants on community structure and composition. Here we will limit our discussion primarily to effects of stressors on ecosystem processes that regulate the movement of materials and energy. As with experiments designed to assess changes in community structure, there has been considerable discussion in the ecological literature concerning the trade-offs between replication and ecological realism. High variability and the large number of replicates necessary to obtain sufficient statistical power in ecosystem experiments complicate the use of hypothesis testing approaches (Carpenter 1989). In fact, because most ecosystem experiments do not involve true replication, traditional approaches that require hypothesis testing for assessing statistical significance are often inappropriate. Instead of relying on hypothesis tests that often involve insufficient statistical power (Carpenter 1989), approaches that evaluate alternative explanations may be more useful. Instead of using additional ecosystems as replicates, investigators should employ strong inference (Platt 1964), and experiments should be designed to evaluate multiple alternative explanations (Carpenter et al. 1998). For example, comparing temporal trends in several ecosystems that differed in important physical, chemical, or biological characteristics subjected to the same stressor will allow researchers to evaluate alternative models. Note that this approach differs in an important way from simple descriptive or comparative studies because the systems are actually manipulated. A variety of statistical techniques, including intervention analyses, multivariate autoregressive models, dynamic linear models, and repeated measures, have been employed to analyze changes in unreplicated ecosystem experiments.

32.3.1 AQUATIC ECOSYSTEMS

Experimental manipulation of ecosystems was originally employed in aquatic habitats when Chancey Juday, a limnologist at the University of Wisconsin, conducted one of the first whole lake nutrient enrichment experiments (Juday and Schloemer 1938). The tradition of whole ecosystem experiments continued at the University of Wisconsin when Arthur D. Hasler, a graduate student of Juday, added lime to a brown water lake (Hasler et al. 1951). The primary objective of these early experiments was to improve fisheries, but subsequent experiments were designed to examine effects of anthropogenic stressors. Future students of Hasler's, recognizing the usefulness of this experimental approach, went on to establish two of the best-known whole ecosystem experimental facilities in North America: The Hubbard Brook Ecosystem study area in New Hampshire and the Experimental Lakes Area in Ontario, Canada.

As a consequence of these early experiments, a large amount of information concerning ecosystem responses of lakes and streams to anthropogenic stressors has been obtained (Carpenter 1989, Likens 1992, Likens et al. 1970, Schindler 1988, Wallace et al. 1986). The application of valid experimental controls, analogous to experimental approaches used in agriculture, was a major development in ecosystem sciences and allowed researchers to rigorously apply the scientific method to test hypotheses regarding how ecosystems responded to perturbations (Likens 1985). Many of these experiments have emphasized functional responses to anthropogenic perturbations and some have provided a quantitative assessment of responses across levels of biological organization (Cottingham and Carpenter 1998). The classic set of whole lake experiments conducted by Schindler and colleagues at the Experimental Lakes Area (ELA) in Canada represented an important turning point in the history of ecosystem research.

Some of the findings of this research seemed to question the usefulness of functional measures for understanding how ecosystems respond to anthropogenic disturbances (Carpenter et al. 1995, Howarth 1991, Odum 1985, Schindler 1987, 1990). In a review of several years of experiments investigating whole-lake responses to eutrophication, acidification, and heavy metals, Schindler (1987) concluded that measures of ecosystem processes, including primary production, nutrient cycling, and respiration were relatively insensitive to perturbations and therefore "poor indicators of early stress." Table 32.1 summarizes major findings from ecosystem-level experiments conducted at the ELA and illustrates Schindler's perspectives on the potential limitations of ecosystem processes as early warning indicators of perturbation. This somewhat pessimistic evaluation of the usefulness of ecosystem processes for assessing stressor effects has led some researchers to abandon functional endpoints altogether and rely entirely on structural measures. Issues such as functional redundancy, feedback mechanisms, lower sensitivity, and high variability of ecosystem process variables were discussed in Chapter 23 and are routinely cited in the literature as justification for focusing on

TABLE 32.1
General Responses of Aquatic Ecosystems and Mesocosms in the Experimental Lakes Area to Chemical Perturbations

1. Phytoplankton biomass and production are limited primarily by the input of phosphorus and the rate of water renewal, regardless of pollutant stress.
2. The rate of decomposition is governed by primary production and is not affected by pollutants.
3. Nutrient cycling is not affected by input of toxicants.
4. Abundance of species with short life cycles can respond quickly to perturbations and maintain ecosystem function. Abundance of these species may serve as early indicators of ecosystem stress.
5. Diversity indices are less useful than changes in community composition for detecting stress in ecosystems.
6. Species with short life cycles and poor powers of dispersal are generally most sensitive to perturbation.
7. The most sensitive indicators of stress in most aquatic ecosystems include demographic changes in short-lived species and changes in community structure.
8. Short-term toxicity tests and measures of ecosystem-level processes are not sufficiently sensitive to be used as early indicators of ecosystem stress.

From Table 3 in Schindler (1987).

lower levels of biological organization. However, we feel that abandonment of process variables as ecological endpoints in ecosystem experiments may be premature for some situations. For example, Schindler (1987) notes that production is a very sensitive indicator of perturbation in some terrestrial ecosystems. In contrast to the generalization that ecosystem variables are less sensitive, Cottingham and Carpenter (1998) reported that primary production was a better indicator of nutrient enrichment than abundance of individual species. A comprehensive characterization of how ecosystems respond to chemical perturbations requires that both patterns (structural measures) and processes (functional measures) be considered. We agree with Cottingham and Carpenter (1998) that rather than asking the question about which endpoints are better, a more appropriate approach is to develop general guidelines for understanding why some indicators work better in some systems or for some classes of stressors. It is quite likely that a list of sensitive indicators for toxic chemicals would be quite different from a list of sensitive indicators for nutrient enrichment, climate change, or physical disturbance.

Two sets of experiments conducted in stream ecosystems deserve special attention because they were designed specifically to assess the relationship between structural and functional measures. Details of experiments conducted by Bruce Wallace and colleagues at Coweeta Hydrologic Laboratory (North Carolina, USA) were discussed in Chapter 23. Briefly, experimental introduction of the pesticide, methoxychlor, into a small headwater stream significantly reduced leaf litter decomposition (Figure 32.8) and downstream export of particulate organic material (Cuffney et al. 1984, Wallace et al. 1982). Results showed significant variation among leaf species and that litter decomposition was a sensitive indicator of contaminant effects. These alterations in detritus dynamics were not associated with changes in microbial activity and were shown to be a direct result of mortality to macroinvertebrate shredders and associated reductions in litter fragmentation. Higher production to biomass ratios (P/B) in the treated stream resulted from a shift in community structure to smaller organisms with shorter generation times and faster turnover.

A second set of whole ecosystem experiments conducted in the Kuparuk River, an Alaskan (USA) tundra stream, was designed to assess the effects of P additions on community structure and function (Peterson et al. 1993). Similar to the whole lake experiments conducted at the ELA, this set of experiments relied on long-term assessments of unreplicated treatments. The experiments are especially significant because of the long duration (16 years) of the manipulations and the large diversity of response variables measured (Slavik et al. 2004). Summarizing the first 4 years of these

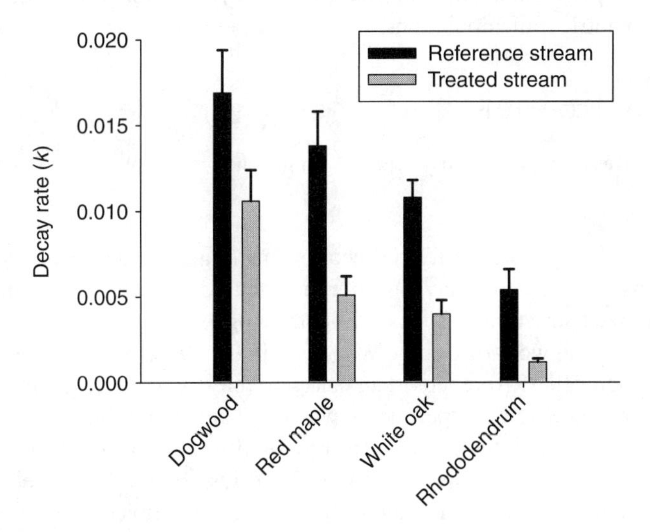

FIGURE 32.8 Exponential breakdown rates of leaves (mean ± 95% confidence limit) in reference streams and streams treated with the pesticide methoxychlor at Coweeta Hydrologic Laboratory. (Data from Table 2 in Wallace et al. (1982).)

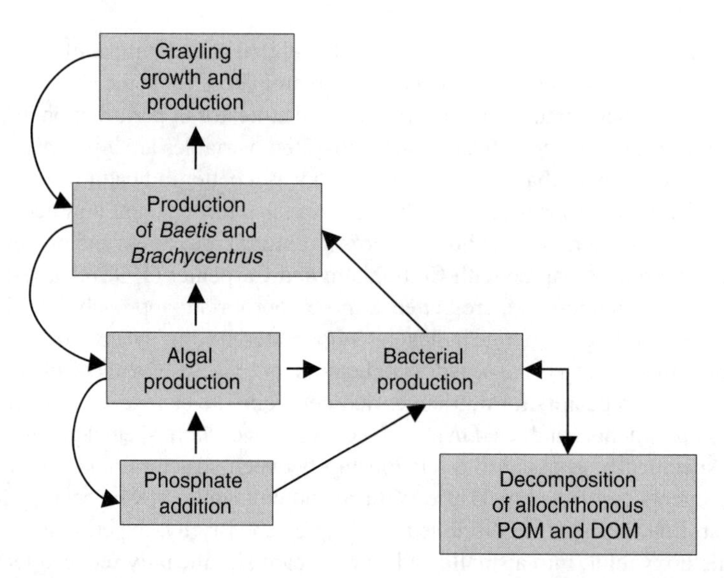

FIGURE 32.9 Observed and hypothesized top-down and bottom-up responses in an Arctic stream food web following experimental addition of phosphorus. (Modified from Figure 15 in Peterson et al. (1993).)

experiments, Peterson et al. (1993) reported increased algal biomass and productivity in fertilized reaches, but little change in diatom community composition. Decomposition rate of *Carex* litter was also unaffected by P addition. Although total secondary productivity of insects was also unaffected by nutrient addition, there was considerable variation among species, with some groups increasing and others decreasing in the treated reach. Interestingly, Peterson et al. (1993) found evidence for strong bottom-up effects of P additions on benthic food webs in the first 2 years of the experiment, followed by strong top-down feedback in later years (Figure 32.9). A follow-up paper published after 16 years of P enrichment reported a dramatic increase in abundance of the bryophyte *Hygrohypnum*, an aquatic moss that replaced epilithic diatoms, resulting in significant changes in the benthic habitat and a four times increase in NH_4^+ uptake (Slavik et al. 2004). The authors concluded that even relatively long-term experimental manipulations (e.g., 4–8 years) were not adequate to predict these striking responses to nutrient manipulations.

32.3.2 Terrestrial Ecosystems

> Changes in many different types of ecosystems are similar and predictable.
>
> **(Woodwell 1970)**

Experiments conducted in natural terrestrial ecosystems (e.g., non-agroecosystems) to assess the effects of anthropogenic stressors have been conducted for at least 50 years. Many of these experiments emphasized structural responses to anthropogenic disturbance and were discussed in Chapter 23. The pioneering work of George Woodwell in the 1960s at Brookhaven National Laboratory, New York (Woodwell 1970) exposed an oak-pine forest to chronic gamma radiation (^{137}Cs). Most of the focus of Woodwell's experiments was on changes in vegetation structure and community composition. However, the mechanism proposed to account for these structural changes invoked alterations in the relationship between GPP and respiration. Woodwell (1970) speculated that larger plants such as trees are at their physiological limit in terms of the amount of surface area available for respiration. Consequently, smaller plants are favored in disturbed environments because they can withstand more damage to respiratory surfaces. Woodwell's findings regarding responses to anthropogenic disturbance were quite consistent with Odum's "strategy of ecosystem development"

(Odum 1969) and stimulated considerable interest among terrestrial ecologists to seek broader generalizations. Enthusiasm for identifying consistent responses to disturbance persists (Rapport et al. 1998), although many terrestrial ecologists have been frustrated by the variability in responses among ecosystems. A general consensus has emerged that greater understanding of terrestrial ecosystem responses to disturbance can be achieved through controlled, long-term experiments.

Field experiments in terrestrial ecosystems have been designed to assess both direct toxicological effects of contaminants and the indirect effects associated with the loss of food resources. Barrett (1968) measured effects of the insecticide carbaryl on structural and functional characteristics in a grassland ecosystem. Exposure to carbaryl had significant direct and indirect effects on insect abundance, litter decomposition, and small mammal production. Not surprisingly, there was no effect of the insecticide on NPP or plant community composition. However, biomass and abundance of arthropods were reduced by >95% and litter decomposition was significantly lower in treated plots. Reduced litter decomposition was attributed to the loss of microarthropods.

Field experiments conducted to assess effects of heavy metals on ecosystem processes are relatively uncommon in the ecotoxicological literature. Korthals et al. (1996) examined long-term effects (10-year exposure) of copper and low pH on microbial activity, respiration, and the abundance of bacteria and nematodes in an agroecosystem (Figure 32.10). In addition to the relatively long duration of the study, this field experiment is significant because it examined interactions between two stressors. The combination of high Cu and low pH reduced bacterial growth and abundance of nematodes, but had little effect on other structural or functional endpoints. Significant reductions in the abundance of herbivorous nematodes were attributed to changes in primary production and loss of food resources. Laskowski et al. (1994) tested the hypothesis that mineral nutrients such as Ca, Mg, and K would reduce the effects of heavy metals on forest litter respiration. Exposure of microbial communities to Cu, Cd, Pb, and Zn reduced respiration rate as predicted, but mineral nutrients either increased toxicity or had no effects on respiration rates.

Many of the recent experiments that emphasized functional processes in ecosystems examined responses to nutrient additions, especially N. Inputs of N from agricultural, industrial, and domestic sources to terrestrial ecosystems have been recognized as a serious environmental problem for several decades. Deposition of N and the process of N saturation were considered potential threats to forest ecosystems in Europe in the mid-1980s (Nihlgard 1985). Unlike toxic chemicals in which ecosystem responses are often immediate and typically result in removal of sensitive species, subsidizing

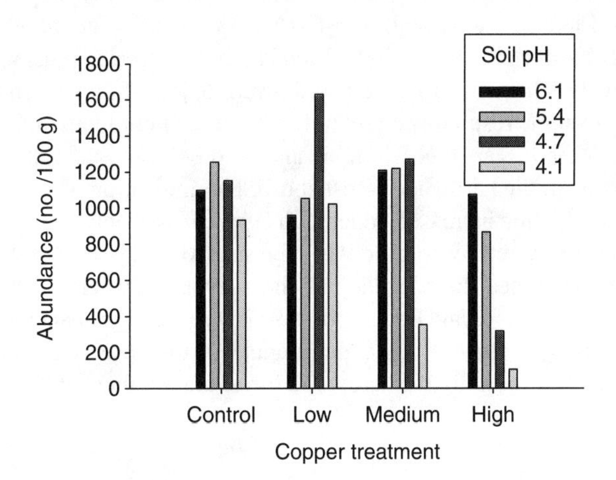

FIGURE 32.10 Mean abundance of the plant-feeding nematode *Pratylenchus* in an agroecosystem after 10 years of exposure to copper and pH. Low, medium and high Cu treatments refer to 250, 500, and 750 kg Cu per Ha, respectively. Reduced abundance of these herbivores was attributed to reduced primary production and the loss of food resources. (Data from Table 2 in Korthals et al. (1996).)

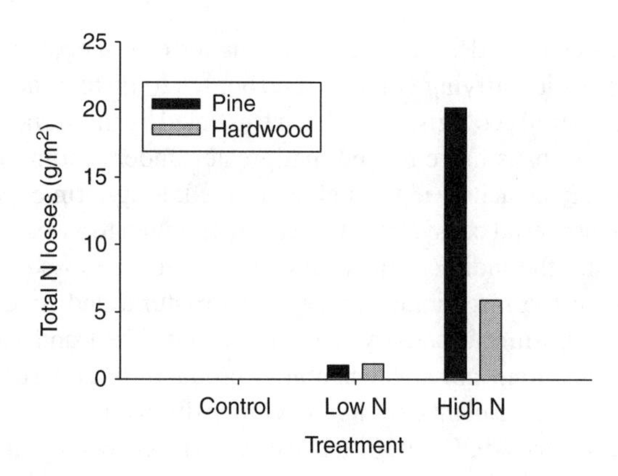

FIGURE 32.11 Total N losses following 9 years of nitrogen addition (1988–1996) to experimental plots on a pine plantation and hardwood forest at the Harvard Forest Long-Term Ecological Research site. (Data from Table 3 in Magill et al. (2000).)

stressors such as N deposition can initially have positive effects. This "paradox of enrichment" (Rosenzweig 1971) greatly complicates assessments of N deposition effects on ecosystems. For example, while productivity of forest ecosystems is often related to N availability, long-term chronic exposure to N deposition is detrimental and export of excess N may damage adjacent watersheds. N saturation of an ecosystem occurs when N output approximates or exceeds input (Argen and Bosatta 1988). Although N saturation is an extreme and relatively unusual phenomenon, many disturbed forest ecosystems either "leak" nutrients or have experienced damages because of N overloads. The effects of N deposition on forest ecosystems were examined at Hubbard Brook Experimental Forest, New Hampshire (Christ et al. 1995). These researchers reported that treated plots retained >95% of the N added, and that effects of excess N on soil processes such as ammonification varied with soil horizon. In contrast to expectations, excess N was not nitrified and appeared to be limited by some other factor(s).

Although ecologists have long speculated about the potential ecosystem-level threats of N deposition, there is relatively little information concerning how different forests will respond to N saturation. The Chronic Nitrogen Amendment experiment conducted at the Harvard Forest Long-Term Ecological Research site (Massachusetts, USA) compared responses of a pine plantation and native deciduous broad-leaved (hardwood) forest to N enrichment (as ammonium nitrate, NH_4NO_3) over a 9-year period (Magill et al. 1997, 2000). Total N inputs in treated plots were increased 6–19× above background levels. These two forest types showed major differences in responses to N addition (Figure 32.11), and some responses could not be discerned until late in the experiment. Nitrate losses were observed after 1 year of N additions and woody biomass production decreased in the pine plantation. In contrast, the hardwood forest showed no indication of nitrate loss until year 8 of treatment, and wood production increased under high N levels. Not only do these results demonstrate variability in responses of two ecosystems to the same stressor, they support the contention of other researchers that have questioned the adequacy of short-term ecosystem experiments for assessing effects of N deposition (Milchunas and Lauenroth 1995). Despite significant responses to treatments, one of the most interesting results of these experiments was the observation that forest ecosystems were highly retentive of N, especially the hardwood forest that retained 96% of the N added (Magill et al. 2000).

32.4 SUMMARY

Microcosm experiments, mesocosm experiments, and ecosystem-level manipulations are powerful approaches for quantifying direct and indirect effects of contaminants on ecosystem processes. Losses of structural components of an ecosystem, such as reduced abundance of important functional feeding groups, often have significant consequences for movement of materials and energy. Conversely, alterations in some ecosystem processes, such as lower primary production, can result in structural changes in communities dependent on these resources. Experimental approaches are often necessary to characterize the complex relationships between patterns and processes. Despite the strength of large scale manipulations for this purpose, clearly it will not be possible to conduct experiments for all ecosystem and stressor combinations. One area where whole ecosystem experiments have clearly lagged is in marine ecosystems. Some of the pioneering experiments in mesocosm research were initially conducted in marine habitats to investigate effects of oil spills (Chapter 23). However, whole-ecosystem experiments in marine environments offer unique challenges. Because of the expense and logistical difficulties associated with conducting ecosystem-level manipulations, as well as the limited number of sites where these experiments can be conducted, a more efficient strategy is necessary that will allow investigators to make generalizations from the current body of literature (Schindler 1988). Attempts to generalize among ecosystem responses using conceptual models such as the ecosystem distress syndrome (Chapter 25) will be useful for this purpose. However, large-scale experimental manipulation of some classes of stressors, including elevated CO_2, UV-B radiation and atmospheric N, present significant logistical problems and will be examined in a later chapter. We believe that a more efficient strategy for generalizing among ecosystem and stressor types is to integrate what we already know about ecosystem responses to perturbations with a well-designed program of observational and small-scale experimental approaches to identify critical questions that can only be addressed using large-scale experiments. This strategy will allow researchers to focus limited resources on those issues that are central to understanding how ecosystems respond to anthropogenic perturbations.

32.4.1 Summary of Foundation Concepts and Paradigms

- Isolating causal mechanisms is particularly important in ecosystem-level studies because ecosystem processes are often complex and controlled by direct and indirect effects.
- Because of the limited spatiotemporal scale, measuring responses of ecosystem processes to contaminants in microcosms presents significant challenges.
- The use of controlled experimental systems in aquatic ecology has clearly increased over time.
- One of the more significant challenges associated with the use of microcosms and mesocosms is the change in functional measures over time, independent of the effects of contaminants.
- Duration of experiments is especially important when considering ecosystem processes that often show delayed responses compared with structural alterations.
- Most experiments conducted in stream microcosms and mesocosms have focused on functional responses of periphyton.
- Mesocosm experiments also allow investigators to compare responses across levels of biological organization, thereby providing opportunities to examine underlying mechanisms and correlate structural and functional alterations.
- Microcosm and mesocosm experiments can also be used to compare functional responses of communities derived from different sources, thereby providing an opportunity to understand how intrinsic features of an ecosystem may influence susceptibility to contaminants.
- Because of questions concerning the sensitivity and variability of ecosystem processes, one important practical application of mesocosm experiments is to compare the efficacy of structural and functional endpoints.
- While aquatic ecotoxicologists have long recognized the value of microcosms and mesocosms as research tools for investigating effects of contaminants on ecosystem

processes, these systems have received considerably less attention in terrestrial ecotoxicology.
- Typical functional endpoints reported in soil microcosm experiments include soil respiration, nitrification, and N mineralization.
- Because decomposition and nutrient dynamics in soils are regulated by microbial processes, structural changes in microbial communities are likely to have important effects on ecosystem function.
- Alterations in abundance of nitrifying bacteria in soils are likely to have long-term consequences for nutrient dynamics.
- Many soil communities are highly dependent on processes occurring in decomposer food webs and are therefore more likely to be controlled by bottom-up processes than top-down processes.
- Whole ecosystem experiments involve the planned application of contaminants or other stressors directly to a system that is large enough to contain all of the important physical, chemical, and biological processes of interest.
- A comprehensive characterization of how ecosystems respond to chemical perturbations requires that both patterns (structural measures) and processes (functional measures) be considered.
- It is quite likely that a list of sensitive indicators for toxic chemicals would be quite different from a list of sensitive indicators for nutrient enrichment, climate change, or physical disturbance.
- Improved understanding of terrestrial ecosystem responses to anthropogenic disturbance can be achieved through controlled, long-term experiments.
- Unlike toxic chemicals in which ecosystem responses are often immediate and typically result in removal of sensitive species, subsidizing stressors such as N deposition can initially have positive effects on ecosystem processes.
- Alterations in structural components of an ecosystem, such as reduced abundance of important functional feeding groups, often have significant consequences for movement of materials and energy.
- Alterations in ecosystem processes, such as reduced primary production, can result in structural changes in communities dependent on these resources.

REFERENCES

Argen, G.I. and Bosatta, E., Nitrogen saturation of terrestrial ecosystems, *Environ. Pollut.*, 54, 185–197, 1988.

Balczon, J.M. and Pratt, J.R., A comparison of the responses of 2 microcosm designs to a toxic input of copper, *Hydrobiologia*, 281, 101–114, 1994.

Barreiro, R. and Pratt, J.R., Interaction of toxicants and communities: The role of nutrients, *Environ. Toxicol. Chem.*, 13, 361–368, 1994.

Barrett, G.W., The effects of an acute insecticide stress on a semi-enclosed grassland ecosystem, *Ecology*, 49, 1019–1035, 1968.

Barron, C., Marba, N., Duarte, C.M., Pedersen, M.F., Lindblad, C., Kersting, K., Moy, F., and Bokn, T., High organic carbon export precludes eutrophication responses in experimental rocky shore communities, *Ecosystems*, 6, 144–153, 2003.

Bester, K., Huhnerfuss, H., Brockmann, U., and Rick, H.J., Biological effects of triazine herbicide contamination on marine phytoplankton, *Arch. Environ. Contam. Toxicol.*, 29, 277–283, 1995.

Bogomolov, D.M., Chen, S.K., Parmelee, R.W., Subler, S., and Edwards, C.A., An ecosystem approach to soil toxicity testing: A study of copper contamination in laboratory soil microcosms, *Appl. Soil Ecol.*, 4, 95–105, 1996.

Bokn, T.L., Duarte, C.M., Pedersen, M.F., Marba, N., Moy, F.E., Barron, C., Bjerkeng, B., et al., The response of experimental rocky shore communities to nutrient additions, *Ecosystems*, 6, 577–594, 2003.

Boon, G.T., Bouwman, L.A., Bloem, J., and Romkens, P.F.A.M., Effects of a copper-tolerant grass (*Agrostis capillaris*) on the ecosystem of a copper-contaminated arable soil, *Environ. Toxicol. Chem.*, 17, 1964–1971, 1998.

Boyle, T.P., Fairchild, J.F., Robinson-Wilson, E.F., Haverland, P.S., and Lebo, J.A., Ecological restructuring in experimental aquatic mesocosms due to the application of diflubenzuron, *Environ. Toxicol. Chem.*, 15, 1806–1814, 1996.

Brock, T.C.M., Vet, J.J.R.M., Kerkhofs, M.J.J., Lijzen, J., Vanzuilekom, W.J., and Gijlstra, R., Fate and effects of the insecticide Dursban(R) 4e in indoor *Elodea*-dominated and macrophyte-free fresh-water model-ecosystems: III. Aspects of ecosystem functioning, *Arch. Environ. Contam. Toxicol.*, 25, 160–169, 1993.

Brooks, B.W., Stanley, J.K., White, J.C., Turner, P.K., Wu, K.B., and La Point, T.W., Laboratory and field responses to cadmium: An experimental study in effluent-dominated stream mesocosms, *Environ. Toxicol. Chem.*, 23, 1057–1064, 2004.

Burrows, L.A. and Edwards, C.A., The use of integrated soil microcosms to assess the impact of carbendazim on soil ecosystems, *Ecotoxicology*, 13, 143–161, 2004.

Cairns, J., Jr., The myth of the most sensitive species, *BioScience*, 36, 670–672, 1986.

Carlisle, D.M. and Clements, W.H., Sensitivity and variability of metrics used in biological assessments of running waters, *Environ. Toxicol. Chem.*, 18, 285–291, 1999.

Carman, K.R., Fleeger, J.W., Means, J.C., Pomarico, S.M., and McMillin, D.J., Experimental investigation of the effects of polynuclear aromatic hydrocarbons on an estuarine sediment food web, *Mar. Environ. Res.*, 40, 289–318, 1995.

Carpenter, S.R., Replication and treatment strength in whole-lake experiments, *Ecology*, 70, 453–463, 1989.

Carpenter, S.R., Chisholm, S.W., Krebs, C.J., Schindler, D.W., and Wright, R.F., Ecosystem experiments, *Science*, 269, 324–327, 1995.

Carpenter, S.R., Cole, J.T., Essington, T.E., Hodgson, J.R., Houser, J.N., Kitchell, J.F., and Pace, M.L., Evaluating alternative explanations in ecosystem experiments, *Ecosystems*, 1, 335–344, 1998.

Chen, S.K. and Edwards, C.A., A microcosm approach to assess the effects of fungicides on soil ecological processes and plant growth: Comparisons of two soil types, *Soil Biol. Biochem.*, 33, 1981–1991, 2001.

Chen, S.K., Edwards, C.A., and Subler, S., Effects of the fungicides benomyl, captan and chlorothalonil on soil microbial activity and nitrogen dynamics in laboratory incubations, *Soil Biol. Biochem.*, 33, 1971–1980, 2001.

Christ, M., Zhang, Y.M., Likens, G.E., and Driscoll, C.T., Nitrogen-retention capacity of a northern hardwood forest soil under ammonium-sulfate additions, *Ecol. Appl.*, 5, 802–812, 1995.

Clements, W.H., Small-scale experiments support causal relationships between metal contamination and macroinvertebrate community responses, *Ecol. Appl.*, 14, 954–967, 2004.

Colwell, F.S., Hornor, S.G., and Cherry, D.S., Evidence of structural and functional adaptation in epilithon exposed to zinc, *Hydrobiologia*, 171, 79–90, 1989.

Cotrufo, M.F., Desanto, A.V., Alfani, A., Bartoli, G., and Decristofaro, A., Effects of urban heavy-metal pollution on organic-matter decomposition in *Quercus ilex* L Woods, *Environ. Pollut.*, 89, 81–87, 1995 .

Cottingham, K.L. and Carpenter, S.R., Population, community, and ecosystem variates as ecological indicators: Phytoplankton responses to whole-lake enrichment, *Ecol. Appl.*, 8, 508–530, 1998.

Crane, M., Research needs for predictive multispecies tests in aquatic toxicology, *Hydrobiologia*, 346, 149–155, 1997.

Cuffney, T.F., Wallace, J.B., and Webster, J.R., Pesticide manipulation of a headwater stream: Invertebrate responses and their significance for ecosystem processes, *Freshw. Invert. Biol.*, 3, 153–171, 1984.

Culp, J.M., Cash, K.J., Glozier, N.E., and Brua, R.B., Effects of pulp mill effluent on benthic assemblages in mesocosms along the Saint John River, Canada, *Environ. Toxicol. Chem.*, 22, 2916–2925, 2003.

Cuppen, J.G.M., Crum, S.J.H., van den Heuvel, H.H., Smidt, R.A., and van den Brink, P.J., Effects of a mixture of two insecticides in freshwater microcosms: I. Fate of chlorpyrifos and lindane and responses of macroinvertebrates, *Ecotoxicology*, 11, 165–180, 2002.

Cuppen, J.G.M., Gylstra, R., Vanbeusekom, S., Budde, B.J., and Brock, T.C.M., Effects of nutrient loading and insecticide application on the ecology of *Elodea*-dominated fresh-water microcosms. 3. Responses of macroinvertebrate detritivores, breakdown of plant litter, and final conclusions, *Arch. für Hydrobiol.*, 134, 157–177, 1995.

Detenbeck, N.E., Hermanutz, R., Allen, K., and Swift, M.C., Fate and effects of the herbicide atrazine in flow-through wetland mesocosms, *Environ. Toxicol. Chem.*, 15, 937–946, 1996.

Dusek, L., The effect of cadmium on the activity of nitrifying populations in two different grassland soils, *Plant Soil*, 177, 43–53, 1995.

Ettema, C.H., Lowrance, R., and Coleman, D.C., Riparian soil response to surface nitrogen input: Temporal changes in denitrification, labile and microbial C and N pools, and bacterial and fungal respiration, *Soil Biol. Biochem.*, 31, 1609–1624, 1999.

Fairchild, J.F., Boyle, T., English, W.R., and Rabeni, C., Effects of sediment and contaminated sediment on structural and functional components of experimental stream ecosystems, *Water Air Soil Pollut.*, 36, 271–293, 1987.

Fisher, R.A., *The Design of Experiments*, 7th ed., Oliver and Boyd, Edinburgh, 1960.

Forster, B., van Gestel, C.A.M., Koolhaas, J.E., Nentwig, G., Rodrigues, J.M.L., Sousa, J.P., Jones, S.E., and Knacker, T., Ring-testing and field-validation of a terrestrial model ecosystem (TME)—An instrument for testing potentially harmful substances: Effects of carbendazim on organic matter breakdown and soil fauna feeding activity, *Ecotoxicology*, 13, 129–141 , 2004.

Fraser, L.H. and Keddy, P., The rate of experimental microcosms in ecological research, *Trends Ecol. Evol.*, 12, 478–481, 1997.

Gruessner, B. and Watzin, M.C., Response of aquatic communities from a Vermont stream to environmentally realistic atrazine exposure in laboratory microcosms, *Environ. Toxicol. Chem.*, 15, 410–419, 1996.

Harding, J.S., Benfield, E.F., Bolstad, P.V., Helfman, G.S., and Jones, E.B.D., Stream biodiversity: The ghost of land use past, *Proc. Natl. Acad. Sci. USA*, 95, 14843–14847, 1998.

Hasler, A.D., Brynildson, O.M., and Helm, W.T., Improving conditions for fish in brown-water bog lakes by alkalization, *J. Wildl. Manage.*, 15, 347–352, 1951.

Hedtke, S.F., Structure and function of copper-stressed aquatic microcosms, *Aquat. Toxicol.*, 5, 227–244, 1984.

Hill, W.R., Bednarek, A.T., and Larsen, I.L., Cadmium sorption and toxicity in autotrophic biofilms, *Can. J. Fish. Aquat. Sci.*, 57, 530–537, 2000.

Howarth, R.W., Comparative responses of aquatic ecosystems to toxic chemical stress, In *Comparative Analyses of Ecosystems: Patterns, Mechanisms, and Theories*, Cole, J., Lovett, G., and Findlay, S. (eds.), Springer-Verlag, New York, 1991, pp. 169–195.

Jorgensen, E. and Christoffersen, K., Short-term effects of linear alkylbenzene sulfonate on freshwater plankton studied under field conditions, *Environ. Toxicol. Chem.*, 19, 904–911, 2000.

Juday, C. and Schloemer, C.L., Effect of fertilizers on plankton production and on fish growth in a Wisconsin lake, *Prog. Fish Cult.*, 40, 24–27, 1938.

Kennedy, J.H., Ammann, L.P., Waller, W.T., Warren, J.E., Hosmer, A.J., Cairns, S.H., Johnson, P.C., and Graney, R.L., Using statistical power to optimize sensitivity of analysis of variance designs for microcosms and mesocosms, *Environ. Toxicol. Chem.*, 18, 113–117, 1999.

Kersting, K. and van den Brink, P.J., Effects of the insecticide Dursban(R)4e (active ingredient chlorpyrifos) in outdoor experimental ditches: Responses of ecosystem metabolism, *Environ. Toxicol. Chem.*, 16, 251–259, 1997.

Khan, M. and Scullion, J., Effect of soil on microbial responses to metal contamination, *Environ. Pollut.*, 110, 115–125, 2000.

Kohler, H.R., Wein, C., Reiss, S., Storch, V., and Alberti, G., Impact of heavy metals on mass and energy flux within the decomposition process in deciduous forests, *Ecotoxicology*, 4, 114–137, 1995.

Korthals, G.W., Alexiev, A.D., Lexmond, T.M., Kammenga, J.E., and Bongers, T., Long-term effects of copper and pH on the nematode community in an agroecosystem, *Environ. Toxicol. Chem.*, 15, 979–985, 1996.

Kraufvelin, P., Model ecosystem replicability challenged by the "soft" reality of a hard bottom mesocosm, *J. Exp. Mar. Biol. Ecol.*, 222, 247–267, 1998.

Kreutzweiser, D.P., Gringorten, J.L., Thomas, D.R., and Butcher, J.T., Functional effects of the bacterial insecticide *Bacillus thuringiensis* var Kurstaki on aquatic microbial communities, *Ecotoxicol. Environ. Saf.*, 33, 271–280, 1996.

Kurtz, J.C., Devereux, R., Barkay, T., and Jonas, R.B., Evaluation of sediment slurry microcosms for modeling microbial communities in estuarine sediments, *Environ. Toxicol. Chem.*, 17, 1274–1281, 1998.

Laskowski, R., Maryanski, M., and Niklinska, M., Effect of heavy-metals and mineral nutrients on forest litter respiration rate, *Environ. Pollut.*, 84, 97–102, 1994.

Liebich, J., Schaffer, A., and Burauel, P., Structural and functional approach to studying pesticide side-effects on specific soil functions, *Environ. Toxicol. Chem.*, 22, 784–790, 2003.

Likens, G.E., An experimental approach for the study of ecosystems—The 5th Tansley Lecture, *J. Ecol.*, 73, 381–396, 1985.

Likens, G.E., *The Ecosystem Approach: Its Use and Abuse*, Oldendorf/Luhe, Germany, Ecology Institute, 1992.

Likens, G.E., Bormann, F.H., Johnson, N.M., Fisher, D.W., and Pierce, R.S., Effects of forest cutting and herbicide treatment on nutrient budgets in Hubbard Brook Watershed-Ecosystem, *Ecol. Monogr.*, 40, 23–47, 1970.

Lowe, R.L., Guckert, J.B., Belanger, S.E., Davidson, D.H., and Johnson, D.W., An evaluation of periphyton community structure and function on tile and cobble substrata in experimental stream mesocosms, *Hydrobiologia*, 328, 135–146, 1996.

Magill, A.H., Aber, J.D., Berntson, G.M., McDowell, W.H., Nadelhoffer, K.J., Melillo, J.M., and Steudler, P., Long-term nitrogen additions and nitrogen saturation in two temperate forests, *Ecosystems*, 3, 238–253, 2000.

Magill, A.H., Aber, J.D., Hendricks, J.J., Bowden, R.D., Melillo, J.M., and Steudler, P.A., Biogeochemical response of forest ecosystems to simulated chronic nitrogen deposition, *Ecol. Appl.*, 7, 402–415, 1997.

Martinez-Toledo, M.V., Salmeron, V., Rodelas, B., Pozo, C., and Gonzalezlopez, J., Studies on the effects of the herbicide Simazine on microflora of four agricultural soils, *Environ. Toxicol. Chem.*, 15, 1115–1118, 1996.

Melendez, A.L., Kepner, R.L., Balczon, J.M., and Pratt, J.R., Effects of diquat on fresh-water microbial communities, *Arch. Environ. Contam. Toxicol.*, 25, 95–101, 1993.

Milchunas, D.G. and Lauenroth, W.K., Inertia in plant community structure: State changes after cessation of nutrient enrichment stress, *Ecol. Appl.*, 5, 452–458, 1995.

Newman, R.M., Perry, J.A., Tam, E., and Crawford, R.L., Effects of chronic chlorine exposure on litter processing in outdoor experimental streams, *Freshw. Biol.*, 18, 415–428, 1987.

Nihlgard, B., The ammonium hypothesis—An additional explanation for forest dieback in Europe, *Ambio*, 14, 2–8, 1985.

Niklinska, M., Laskowski, R., and Maryanski, M., Effect of heavy metals and storage time on two types of forest litter: Basal respiration rate and exchangeable metals, *Ecotoxicol. Environ. Saf.*, 41, 8–18, 1998.

Odum, E.P., The strategy of ecosystem development, *Science*, 164, 262–270, 1969.

Odum, E.P., The mesocosm, *BioScience*, 34, 558–562, 1984.

Odum, E.P., Trends expected in stressed ecosystems, *BioScience*, 35, 419–422, 1985.

Oviatt, C.A., Pilson, M.E.Q., Nixon, S.W., Frithsen, J.B., Rudnick, D.T., Kelly, J.R., Grassle, J.F., and Grassle, J.P., Recovery of a polluted estuarine system: A mesocosm experiment, *Mar. Ecol. Prog. Ser.*, 16, 203–217, 1984.

Pearson, N. and Crossland, N.O., Measurement of community photosynthesis and respiration in outdoor artificial streams, *Chemosphere*, 32, 913–919, 1996.

Pell, M., Stenberg, B., and Torstensson, L., Potential denitrification and nitrification tests for evaluation of pesticide effects in soil, *Ambio*, 27, 24–28, 1998.

Petersen, D.G., Dahllof, I., and Nielsen, L.P., Effects of zinc pyrithione and copper pyrithione on microbial community function and structure in sediments, *Environ. Toxicol. Chem.*, 23, 921–928, 2004.

Peterson, B.J., Deegan, L., Helfrich, J., Hobbie, J.E., Hullar, M., Moller, B., Ford, T.E., et al., Biological responses of a tundra river to fertilization, *Ecology*, 74, 653–672, 1993.

Pezeshki, S.R. and Delaune, R.D., Effect of crude-oil on gas-exchange functions of *Juncus roemerianus* and *Spartina alterniflora*, *Water Air Soil Pollut.*, 68, 461–468, 1993.

Platt, J.R., Strong inference, *Science*, 146, 347–353, 1964.

Pratt, J.R. and Barreiro, R., Influence of trophic status on the toxic effects of a herbicide: A microcosm study, *Arch. Environ. Contam. Toxicol.*, 35, 404–411, 1998.

Pratt, J.R., Melendez, A.E., Barreiro, R., and Bowers, N.J., Predicting the ecological effects of herbicides, *Ecol. Appl.*, 7, 1117–1124, 1997.

Rapport, D.J., Regier, H.A., and Hutchinson, T.C., Ecosystem behavior under stress, *Am. Nat.*, 125, 617–640, 1985.

Rapport, D.J., Whitford, W.G., and Hilden, M., Common patterns of ecosystem breakdown under stress, *Environ. Monit. Assess.*, 51, 171–178, 1998.

Richardson, J.S. and Kiffney, P.M., Responses of a macroinvertebrate community from a pristine, Southern British Columbia, Canada, stream to metals in experimental mesocosms, *Environ. Toxicol. Chem.*, 19, 736–743, 2000.

Rosenzweig, M.L., Paradox of enrichment: Destabilization of exploitation ecosystems in ecological time, *Science*, 171, 385–387, 1971.

Salminen, J., Anh, B.T., and van Gestel, C.A.M., Indirect effects of zinc on soil microbes via a keystone enchytraeid species, *Environ. Toxicol. Chem.*, 20, 1167–1174, 2001.

Salminen, J., Liiri, M., and Haimi, J., Responses of microbial activity and decomposer organisms to contamination in microcosms containing coniferous forest soil, *Ecotoxicol. Environ. Saf.*, 53, 93–103, 2002.

Salminen, J.E. and Sulkava, P.O., Decomposer communities in contaminated soil: Is altered community regulation a proper tool in ecological risk assessment of toxicants? *Environ. Pollut.*, 97, 45–53, 1997.

Schindler, D.W., Detecting ecosystem responses to anthropogenic stress. *Can. J. Fish. Aquat. Sci.*, 44(Suppl. 1), 6–25, 1987.

Schindler, D.W., Experimental studies of chemical stressors on whole lake ecosystems, *Verh. Internat. Verein. Limnol.*, 23, 11–41, 1988.

Schindler, D.W., Experimental perturbations of whole lakes as tests of hypotheses concerning ecosystem structure and function, *Oikos*, 57, 25–41, 1990.

Schindler, D.W., Mills, K.H., Malley, D.H., Findlay, D.L., Shearer, J.A., Davies, I.J., Turner, M.A., Linsey, G.A., and Cruikshank, D.R., Long-term ecosystem stress: The effects of years of experimental acidification on a small lake, *Science*, 228, 1395–1401, 1985.

Schneider, J., Morin, A., and Pick, F.R., The response of biota in experimental stream channels to a 24-hour exposure to the herbicide Velpar L, *Environ. Toxicol. Chem.*, 14, 1607–1613, 1995.

Slavik, K., Peterson, B.J., Deegan, L.A., Bowden, W.B., Hershey, A.E., and Hobbie, J.E., Long-term responses of the Kuparuk River ecosystem to phosphorus fertilization, *Ecology*, 85, 939–954 , 2004.

Slijkerman, D.M.E., Baird, D.J., Conrad, A., Jak, R.G., and van Straalen, N.M., Assessing structural and functional plankton responses to carbendazim toxicity, *Environ. Toxicol. Chem.*, 23, 455–462, 2004.

Stay, F.S., Katko, A., Rohm, C.M., Fix, M.A., and Larsen, D.P., Effects of fluorene on microcosms developed from 4 natural communities, *Environ. Toxicol. Chem.*, 7, 635–644, 1988.

Steinman, A.D., Mulholland, P.J., Palumbo, A.V., DeAngelis, D.L., and Flum, T.E., Lotic ecosystem response to a chlorine disturbance, *Ecol. Appl.*, 2, 341–355, 1992.

Stewart, J.S., Wang, L.Z., Lyons, J., Horwatich, J.A., and Bannerman, R., Influences of watershed, riparian-corridor, and reach-scale characteristics on aquatic biota in agricultural watersheds, *J. Am. Water Res. Assoc.*, 37, 1475–1487, 2001.

Suderman, K. and Thistle, D., A microcosm system for the study of pollution effects in shallow, sandy, subtidal communities, *Environ. Toxicol. Chem.*, 22, 1093–1099, 2003.

Touart, L.W., Aquatic mesocosm tests to support pesticide registrations, EPA 540-09-88-035, U.S. Environmental Protection Agency, Washington, D.C., 1988.

Touart, L.W. and Maciorowski, A.F., Information needs for pesticide registration in the United States, *Ecol. Appl.*, 7, 1086–1093, 1997.

van Gestel, C.A.M., Koolhaas, J.E., Schallnass, H.J., Rodrigues, J.M.L., and Jones, S.E., Ring-testing and field-validation of a terrestrial model ecosystem (TME)—An instrument for testing potentially of harmful substances: Effects of carbendazim on nutrient cycling, *Ecotoxicology*, 13, 119–128, 2004.

Wallace, J.B., Grubaugh, J.W., and Whiles, M.R., Biotic indices and stream ecosystem processes: Results from an experimental study, *Ecol. Appl.*, 6, 140–151, 1996.

Wallace, J.B., Ross, D.H., and Meyer, J.L., Seston and dissolved organic carbon dynamics in a southern Appalachian stream, *Ecology*, 63, 824–838, 1982.

Wallace, J.B., Vogel, D.S., and Cuffney, T.F., Recovery of a headwater stream from an insecticide-induced community disturbance, *J. N. Am. Benthol. Soc.*, 5, 115–126, 1986.

Wang, L.Z., Lyons, J., Kanehl, P., and Gatti, R., Influences of watershed land use on habitat quality and biotic integrity in Wisconsin streams, *Fisheries*, 22, 6–12, 1997.

Widenfalk, A., Svensson, J.M., and Goedkoop, W., Effects of the pesticides captan, deltamethrin, isoproturon, and pirimicarb on the microbial community of a freshwater sediment, *Environ. Toxicol. Chem.*, 23, 1920–1927, 2004.

Williams, P., Whitfield, M., Biggs, J., Fox, G., Nicolet, P., Shillabeer, N., Sherratt, T., Heneghan, P., Jepson, P., and Maund, S., How realistic are outdoor microcosms? A comparison of the biota of microcosms and natural ponds, *Environ. Toxicol. Chem.*, 21, 143–150, 2002.

Woodwell, G.M., Effects of pollution on the structure and physiology of ecosystems, *Science*, 168, 429–433, 1970.

33 Patterns and Processes: The Relationship between Species Diversity and Ecosystem Function

The economies of the Earth would grind to a halt without the services of ecological life-support systems ....

(Costanza et al. 1997)

The biodiversity–ecosystem function linkage appears to be another concept for which enthusiasm outweighs supportive evidence.

(Schwartz et al. 2000)

33.1 INTRODUCTION

A fundamental characteristic of most biological systems is their remarkable diversity. Our accounting of global diversity should not be restricted to the large number of species inhabiting the biosphere, estimated to be between 10 and 100 million (Wilson 1999), but should also include the genetic variation residing within each species as well as the functional diversity of processes for which they are responsible. The rapid loss of genetic, species, and functional diversity resulting from habitat destruction, exotic species, climate change, overharvesting, chemical stressors, and other sources of anthropogenic disturbance is a significant environmental concern with global consequences. Arguments for the protection of biological diversity have traditionally been based on moral or aesthetic perspectives. However, researchers and policymakers are becoming increasingly aware that species also provide ecological goods and services that are essential for human welfare. In this chapter, we describe theoretical and empirical evidence that supports the hypothesis that species diversity controls ecosystem function, describe the limitations in our understanding of this relationship, and discuss the implications for ecotoxicology. Because of the controversy surrounding the significance of the diversity–ecosystem function relationship and its practical importance in managing ecosystems, the topic has received considerable attention in the ecological literature. The recent comprehensive review published by Hooper et al. (2005) is especially noteworthy because coauthors included both proponents and critics of the diversity–ecosystem function relationship. This review, which characterizes major points of agreement and uncertainty, represents a broad consensus within the scientific community (Table 33.1).

Because reduced genetic, species, and functional diversity resulting from contaminants has important consequences for the services provided by ecosystems, we believe that the diversity–ecosystem function relationship has significant implications for ecotoxicology. We first review the observational and experimental studies that support the theoretical relationship between

TABLE 33.1
Major Points of Agreement and Remaining Uncertainty within the Ecological Community Regarding the Relationship between Species Diversity and Ecosystem Function

Points of certainty:

1. Species and functional diversities influence ecosystem function.
2. Anthropogenic alteration of ecosystem function and related services has been well documented.
3. The relationship between species diversity and ecosystem processes is context-dependent and varies among ecosystem properties and types.
4. The relative insensitivity of some ecosystem processes to loss of species or changes in composition results from the redundancy of some species, the fact that some species contribute relatively little to ecosystem function, and the dominant influence of abiotic environmental factors.
5. As spatial and temporal variability increases, more species are required to maintain the stability of ecosystem processes and services.

Points of high confidence:

1. Although some combinations of species are complementary and can increase ecosystem function, environmental factors can influence the importance of complementarity.
2. The effects of exotic species are determined by community composition and are generally lower in communities with high species richness.
3. Variation in the sensitivity and susceptibility of species to anthropogenic stressors within a community can provide stability to ecosystem processes.

Points of uncertainty and the need for future research:

1. Additional research is necessary to understand the mechanisms responsible for the relationship between species diversity, functional diversity, and ecosystem function.
2. Because most studies have focused on the relationship between diversity of primary producers and ecosystem function, effects of diversity across trophic levels are poorly understood.
3. Although there is broad theoretical support for the relationship between species diversity and stability, long-term field experiments are necessary to determine the importance of this relationship in natural ecosystems.
4. Because ecosystem function simultaneously influences and responds to biological diversity, understanding the feedback between these variables is critical.
5. Because the focus of diversity–ecosystem function research has been in terrestrial ecosystems and to a limited extent in freshwater ecosystems, little is known about this relationship in marine ecosystems.

Source: From Table 1 in Hooper et al. 2005.

species diversity and ecosystem processes. We then discuss the practical implications of this relationship and argue that, in addition to the direct effects of pollutants on energy flow and biogeochemical cycles described in previous chapters, contaminant-induced change in diversity can negatively impact ecosystem processes and services. We also review the evidence supporting the hypothesis that ecosystems with lower biological diversity have lower resistance and resilience to natural and anthropogenic stressors compared to species-rich ecosystems. Finally, we demonstrate that recent studies investigating the concept of ecological thresholds have important implications for understanding the diversity–ecosystem function relationship. We suggest that quantifying the level of perturbation (or species loss) where ecosystem function is significantly impaired will improve our ability to predict effects of anthropogenic perturbations.

Many of the important ecosystem processes we have discussed in the preceding chapters, including primary productivity, nutrient dynamics, decomposition, and energy flow respond directly to anthropogenic perturbations. We also know that changes in abundance of keystone species and other ecologically important taxa as a result of physical and chemical stressors affect ecosystem function at local and global scales (Chapin et al. 1997). In this chapter, we consider the implications of reduced species richness and diversity on ecosystem processes. The fundamental argument supporting this relationship is quite simple. From a probabilistic perspective, greater species richness increases the likelihood that functionally important species will be present in an ecosystem. Elimination of species in ecosystems with lower species diversity increases the likelihood that critical ecosystem processes will be affected. As described previously (Chapter 25), greater species diversity provides functional redundancy and increases the resistance and resilience of ecosystems to anthropogenic perturbations. For example, Frost et al. (1999) attributed the functional redundancy and increased resilience of acidified lakes to high zooplankton diversity.

Although our understanding of the mechanistic linkages between community structure and ecosystem function remains limited, it is becoming increasingly apparent that alterations in ecosystem processes as a result of species loss have important implications for ecosystem services provided to humans. Species loss within functionally related assemblages, such as pollinators and flowering plants, may impact ecosystem services at very large spatial scales (Biesmeijer et al. 2006). The unprecedented rate of species extinction occurring at a global scale requires that ecologists and ecotoxicologists develop a better appreciation of the relationship between patterns and processes. Despite the recent interest in this relationship within the basic ecological literature, surprisingly few studies have examined the consequences of contaminant-induced species loss on ecosystem function and services. In addition to measuring the direct effects of chemical stressors on ecosystems, we believe that it is important that ecotoxicologists recognize the indirect effects on ecosystem processes owing to loss of species. The goal of this chapter is to provide an ecotoxicological perspective on the critical relationship between community patterns and ecosystem processes.

33.2 SPECIES DIVERSITY AND ECOSYSTEM FUNCTION

Although biologists have hypothesized about the relationship between species diversity and ecosystem function for over a century, the topic remains controversial in the contemporary ecological literature (Grime 1997, Hooper and Vitousek 1997, Huston 1997). There is general agreement that high species diversity provides benefits to ecosystems beyond simple aesthetics. In a review of evidence supporting the diversity–ecosystem function relationship, Chapin et al. (1998) concluded that high species diversity maximizes resource acquisition across trophic levels, reduces the risk associated with stochastic changes in environmental conditions, and protects communities from exposure to pathogens or exotic species. Because diversity's relationship with ecosystem goods and services has important socioeconomic implications, this debate has also generated significant attention among policymakers.

A key issue in the diversity–ecosystem function debate is that many of the ecosystem processes that have been linked directly to species diversity, such as the primary productivity of tropical rainforests, are clearly influenced by other environmental factors in addition to the number of species (Figure 33.1). Second, while most research has investigated the influence of diversity on ecosystem function, ecosystem processes such as primary productivity also regulate species diversity. The influence of ecosystem processes on species diversity may be very complex. For example, Chase and Leibold (2002) reported that the shape of the relationship between productivity and species diversity was scale dependent. At a local scale, the relationship was hump shaped, with diversity increasing up to a certain level of productivity and then declining at higher levels. In contrast, species diversity increased linearly with productivity at the regional scale. As a consequence of these complex, scale-dependent relationships, an understanding of potential feedbacks between diversity and ecosystem processes is critical. The relationships between species diversity, ecosystem function, and ecosystem services also must be interpreted within the context of changing global climate. The most contentious debates within the ecological community pertain to the mechanisms by which species diversity controls ecosystem function. It is possible that the positive relationship is simply a sampling artifact, by which greater species richness increases ecosystem function by increasing

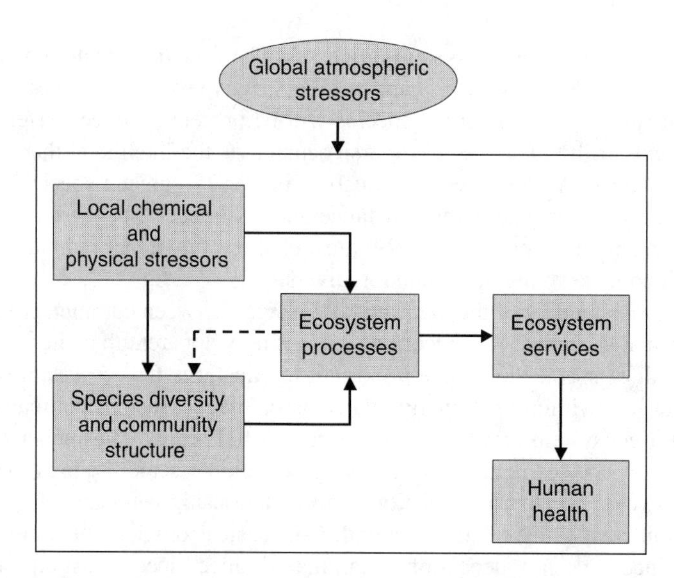

FIGURE 33.1 The influence of anthropogenic stressors and species diversity on ecosystem processes and services. Because ecosystem processes also influence species diversity, an understanding of the potential feedback is necessary for characterizing these relationships. All of these interactions must be interpreted within the context of global atmospheric stressors such as climate change, *N* deposition, and increased UV-B radiation.

the likelihood that functionally important species are present. Alternatively, increased ecosystem function at higher species richness could be a result of positive interactions among species, described as complementarity or facilitation effects. Thus, while many researchers accept the existence of a relationship between species diversity and ecosystem function, the mechanisms responsible for this relationship remain a significant source of controversy.

33.2.1 Experimental Support for the Species Diversity–Ecosystem Function Relationship

Large-scale field experiments conducted in grasslands by Tilman and colleagues have contributed significantly to our understanding of the relationship between diversity and plant productivity (Tilman et al. 1997). By adding a known number of species (0–32) or functional groups (0–5) to large (169 m^2) grassland plots, these researchers found that both species diversity and functional diversity influenced plant productivity (Figure 33.2). When results were analyzed based on functional composition, the relationships were stronger, suggesting that composition of the community was more important than the number of species. Similarly, Hooper and Vitousek (1997) reported that the composition of plant functional groups was more closely related to ecosystem processes than functional group richness. These results demonstrate that the different functional roles of species may be more important predictors of ecosystem integrity than the actual identity of those species.

A large-scale experimental test of the relationship between grassland plant diversity and productivity was conducted at eight European field sites (Hector et al. 1999). Five levels of species richness were established at each site across a broad geographic region (Germany, Portugal, Switzerland, Greece, Ireland, Sweden, and two sites in the United Kingdom). Productivity (measured as aboveground biomass) varied among locations, but the overall pattern at all sites was greater productivity with higher species richness. The mechanisms proposed to account for this pattern included positive mutualistic interactions among species and niche complementarity, whereby variation among species resulted in more complete utilization of resources. Although distinguishing between these alternative explanations will not be simple, the results demonstrate that loss of species

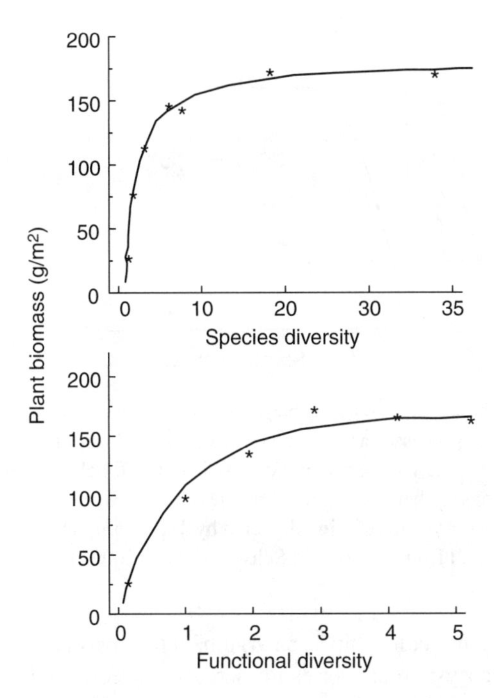

FIGURE 33.2 The influence of species diversity and functional diversity on productivity (measured as above-ground biomass) in grassland plots. Species diversity and functional diversity were manipulated by adding a known number of species or functional groups to experimental plots. (Modified from Figure 1 in Tilman et al. (1997).)

and the alteration in community composition will significantly alter ecosystem processes. Consistent with predictions of the drivers and passengers model described in Chapter 25 (Figure 25.2d), the loss of some functionally important species will have greater impacts on ecosystem function than the loss of other species. Taylor et al. (2006) reported that removal of a single detritivorous fish species from a species-rich tropical river had large effects on carbon flow and ecosystem metabolism. These results were contrary to the theoretical prediction that high species diversity at lower trophic levels provides insurance against changes in ecosystem function. If one of the key goals of basic ecology is to identify these functionally important species, we believe that one of the challenges in ecotoxicology is to predict the consequences of their local extinction owing to the presence of chemical stressors.

33.2.2 Functional Redundancy and Species Saturation in Ecosystems

The positive influence of species richness on ecosystem function reported in many studies has attained greater significance as conservation biologists have used this relationship to argue for species protection. The accelerating loss of biodiversity has intensified efforts to clarify the diversity–productivity relationship and to identify mechanistic explanations. From a species conservation perspective, the shape of the relationship between richness and ecosystem processes may be at least as important as the actual existence of this relationship. A linear relationship between ecosystem processes and richness implies that all species in a community are important and contribute to ecosystem function (Figure 33.3). However, if the relationship is curvilinear and ecosystem processes can be supported by a relatively small number of species, then ecosystems could potentially lose a significant number of species without affecting function. Schwartz et al. (2000) reviewed observational, experimental, and theoretical studies and found relatively little support for the linear dependence of ecosystem processes on species richness. These researchers recommended caution when using the

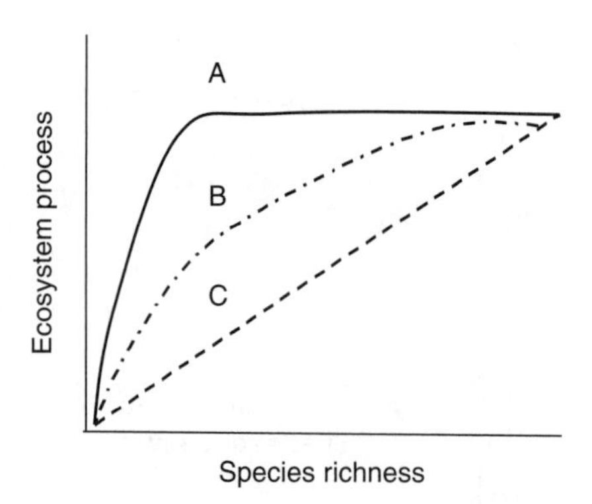

FIGURE 33.3 Three hypothetical relationships between species richness and ecosystem processes. Type A is an example of where ecosystem processes are saturated at a relatively low number of species. This response also shows an abrupt threshold response when species richness is reduced below a certain critical number. Type B shows an intermediate relationship between species richness and ecosystem function. The linear relationship between species richness and ecosystem function depicted by Type C implies that all species contribute equally to ecosystem function. (Modified from Figure 1 in Schwartz et al. (2000).)

diversity–ecosystem function relationship as an argument to support species conservation. Although a saturating response of ecosystem processes to increasing species richness is the most commonly observed pattern (Hooper et al. 2005), in a global analysis of marine ecosystems Worm et al. (2006) found no evidence of functional redundancy and reported a linear relationship between richness and ecosystem processes.

33.2.3 Increased Stability in Species-Rich Ecosystems

In many respects, the relationship between species richness and ecosystem function is closely related to the diversity–stability relationship described in Chapter 25. Indeed, one of the responses frequently cited to support the existence of a positive diversity–ecosystem function relationship is greater stability in species-rich ecosystems. This relationship is supported by mathematical models that predict that, if species abundances vary randomly or are negatively correlated, ecosystem processes will be more stable in diverse communities than in species-poor communities. This statistical averaging phenomenon, which has been termed the "portfolio" effect, provides a type of insurance for ecosystems where species have varying sensitivities to environmental conditions. Despite its broad theoretical support and intuitive appeal, there have been few long-term experiments testing the relationship between species diversity and ecosystem stability in nature. We agree with Hooper et al. (2005) that linking results of long-term experiments with theoretical and mathematical models will improve our understanding of the role that biological diversity plays in stabilizing ecosystem function.

33.2.4 Criticisms of the Diversity–Ecosystem Function Relationship

Critics of the diversity–ecosystem function relationship argue that ecosystem properties are not a direct consequence of species richness or diversity per se, but simply an outcome of the functional composition of dominant species. Experimental studies conducted in grasslands, greenhouses, and growth chambers that controlled for potential confounding variables have demonstrated a strong positive relationship between species diversity and plant productivity. However, large-scale comparative field studies showed that this relationship was not consistent, suggesting that factors other than species diversity determined ecosystem processes (Chapin et al. 1997). Wardle et al. (1997) took advantage of natural variation in community composition of an island archipelago to examine the relationship between island size and ecosystem processes. In contrast to predictions based on many studies, ecosystem processes were inversely related to species diversity.

Another legitimate criticism of studies reporting a relationship between community structure and ecosystem processes is that biological diversity within a community is frequently reduced to a single number (e.g., species richness). However, there are other important community characteristics that are equally likely to respond to anthropogenic stressors and influence ecosystem processes. For example, in addition to reduced species richness and alterations in community composition, one of the most consistent responses to many chemical stressors is increased dominance of pollution-tolerant species. Dangles and Malmqvist (2004) reported that the relationship between species richness and leaf breakdown rates in 36 European streams was determined by dominance of invertebrate shredders. Detrital processing rates were higher and showed an asymptotically increasing relationship with species richness in streams with high dominance, indicating considerable functional redundancy. In contrast, the relationship between species richness and detrital processing in streams with an even distribution of individuals was linear, indicating that all shredder species were important and contributed to ecosystem function.

33.2.5 Mechanisms Responsible for the Species Diversity–Ecosystem Function Relationship

> Sacrificing those aspects of ecosystems that are difficult or impossible to reconstruct, such as diversity, simply because we are not yet certain about the extent and mechanisms by which they affect ecosystem properties, will restrict future management options even further.
>
> **(Hooper et al. 2005)**

If we assume that different species in a community have different functional roles and that the functions performed by individual species are limited, it follows that alterations in community composition resulting from anthropogenic disturbance will affect ecosystem processes. However, identifying the specific mechanistic explanations for the diversity–ecosystem function relationship and characterizing its form (e.g., linear vs. curvilinear) have been challenging. Some ecologists suggest that this relationship is inconsistent among communities because the relative contributions of individual species to ecosystem function are context dependent and vary with environmental conditions (Cardinale et al. 2000). Others argue that the observed pattern is a sampling artifact resulting from inappropriate experimental designs and hidden treatment effects (Grime 1997, Huston 1997). An important research challenge will be to distinguish among these alternatives and to identify the specific mechanisms responsible for the diversity–ecosystem function relationship. Fox (2006) developed a framework to partition effects of species loss on ecosystem function. Effects were partitioned into those resulting from random loss of species, nonrandom loss of species, and changes in functioning of remaining species.

Much of the debate about the relationship between biodiversity and ecosystem function centers on the hypothesis that ecosystem integrity is dependent on the number of species and that loss of species will affect critical ecosystem services. In addition, there is an obvious inconsistency between the hypothesis that all species in an ecosystem are important and the alternative that ecosystems with a large number of species have significant functional redundancy (Figure 33.3). Chapin et al. (1997) argue that this issue can be resolved by considering functional traits of species instead of simple measures of species richness and diversity. Species richness is predicted to influence ecosystem function in several fundamental ways. First, ecosystems with a large number of species have a greater probability of containing taxa with important functional roles. Second, ecosystems with more species will likely use available resources more efficiently, resulting in greater productivity. Finally, a large number of species provide functional redundancy in an ecosystem and a buffer against species loss owing to anthropogenic disturbance. Yachi and Loreau (1999) developed a stochastic dynamic model to test this "insurance hypothesis" and concluded that greater species richness had both a buffering effect on temporal variance and a performance-enhancing effect on productivity.

As described above, ecosystem processes are more likely to respond to the functional diversity of a community rather than the total number of species. Heemsbergen et al. (2004) measured the effects of species richness and functional dissimilarity (defined as the range of species traits that determine their functional role) on soil processes (leaf litter mass loss, nitrification, and respiration). Although the number of species had relatively little impact on soil processes, leaf litter decomposition and soil respiration significantly increased with functional dissimilarity. Finally, while the focus of the biodiversity–ecosystem function debate has been primarily on the role of species diversity, we should remember that genetic diversity within populations may also influence ecosystem processes. Crutsinger et al. (2006) reported that genotypic diversity in a population of plant species increased aboveground net primary productivity and species richness of arthropod herbivores and predators. This increase in consumer species richness was a result of both greater resource productivity and greater diversity of these resources.

33.3 THE RELATIONSHIP BETWEEN ECOSYSTEM FUNCTION AND ECOSYSTEM SERVICES

> Although a number of uncertainties remain, the importance of ecosystem services to human welfare requires that we adopt the prudent strategy of preserving biodiversity in order to safeguard ecosystem processes vital to society.
>
> **(Naeem et al. 1999)**

The practical significance of understanding the relationship between community patterns and ecosystem processes is best illustrated by considering the services provided by ecosystems. We know that natural ecosystems supply irreplaceable benefits to society, and that many of these benefits are critical for the health and survival of humanity. Some ecosystem services, such as removal of nutrients and other wastes, soil stabilization, pollination, and regulation of climate and atmospheric gasses, contribute directly to human welfare. The ecosystem service most familiar to ecotoxicologists is the biotic and abiotic attenuation of contaminants, often referred to as the assimilative capacity of an ecosystem. The purifying function of ecosystems has been widely reported in the literature (Havens and James 2005, Ng et al. 2006, Richardson and Qian 1999), but only recently have researchers considered specific management practices that facilitate assimilative capacity (Vorenhout et al. 2000). Although researchers and policymakers have long recognized the qualitative importance of ecosystem services, collaboration between ecologists and economists has improved our ability to estimate their total economic value. Costanza et al. (1997) estimated that the global economic value of 17 ecosystem services across a range of aquatic and terrestrial biomes was US$ 16–54 trillion (average = US$ 33 trillion) per year, which was approximately 1.8 times the global gross domestic product (Table 33.2). These researchers also note that as ecosystems providing services become increasingly stressed, it is likely that their economic value will significantly increase.

Disruption of ecosystem services is a result of alterations in ecosystem processes that are linked either directly or indirectly to physical, chemical and biological stressors (Figure 33.1). These linkages all occur within the context of global climate change, which operates at a much larger spatiotemporal scale. The dependence of ecosystem processes on community characteristics described in this chapter provides additional justification for the protection of biological diversity. Identifying quantitative relationships among community patterns, processes, and ecosystem services should be a research priority in ecotoxicology. Many ecologists now recognize that research conducted exclusively in undisturbed ecosystems provides an important but somewhat biased perspective of ecosystem processes (Palmer et al. 2004). Although the inclusion of humans and associated anthropogenic

TABLE 33.2
Ecosystem Function, Services, and Annual Economic Global Value

Ecosystem Function	Ecosystem Services	Examples	Value (US$ 10^9)
Regulation of atmospheric chemical composition	Gas regulation	CO_2/O_2 balance; O_3 for UV-B protection	1,341
Regulation of global temperature	Climate regulation	Greenhouse gas regulation	684
Damping ecosystem response to environmental fluctuations	Disturbance regulation	Storm protection, flood control, and other response to environmental variability	1,779
Regulation of hydrological flows	Water regulation	Provision of water for agricultural and industrial processes	1,115
Storage and retention of water	Water supply	Provision of water by watersheds, reservoirs, and aquifers	1,692
Soil retention	Erosion control	Prevention of soil loss by wind and runoff	576
Soil formation processes	Soil formation	Geological weathering and accumulation of organic material	53
Nutrient storage, cycling and processing	Nutrient cycling	N fixation; N and P cycling	17,075
Retention of nutrients and immobilization of toxic chemicals	Waste treatment	Pollution control and detoxification	2,277
Movement of floral gametes	Pollination	Providing pollinators for plant reproduction	117
Trophodynamic regulation of populations	Biological control	Keystone predator control of prey species; herbivore control by top predators	417
Habitat for resident and transient populations	Refugia	Nurseries and habitat for migratory and commercially important species	124
Portion of GPP used as food	Food production	Production of fish, game, fruits, nuts, and crops	1,386
Portion of GPP used as raw materials	Raw materials	Production of lumber, fuel, or fodder	721
Sources of unique biological materials and products	Genetic resources	Medicine, products for materials science, and genes for resistance to plant pathogens	79
Opportunities for recreational activities	Recreation	Ecotourism, sport fishing, and other outdoor activities	815
Opportunities for noncommercial uses	Cultural	Aesthetic, artistic, educational, and spiritual value	3,015

Source: From Table 2 in Costanza et al. (1997).

disturbances into the study of basic ecosystem processes is controversial, we believe this step is fundamental to understanding the complex relationship between ecosystems and the services they provide. Because there will likely be variation in the sensitivity among ecosystem processes to chemical stressors, quantifying stressor–response relationships should be a research priority. The lack of a consensus on which ecosystem services are critical and therefore should be protected also impedes our ability to make policy decisions based on the diversity–ecosystem function relationship (Schwartz et al. 2000). A critical step will be to prioritize the importance of ecosystem services and to determine which are irreplaceable and which can be maintained with technological advances (Palmer et al. 2004).

33.4 FUTURE RESEARCH DIRECTIONS AND IMPLICATIONS OF THE DIVERSITY–ECOSYSTEM FUNCTION RELATIONSHIP FOR ECOTOXICOLOGY

33.4.1 EFFECTS OF RANDOM AND NONRANDOM SPECIES LOSS ON ECOSYSTEM PROCESSES

Because most experimental investigations of the diversity–productivity relationship have focused on terrestrial primary producers, the widespread generality of these patterns in other ecosystems is uncertain. In addition, most studies investigating this relationship have assumed that elimination of species is a random process. However, the susceptibility of a species to anthropogenic disturbance in natural systems will be influenced by a wide range of life history features, including mobility, longevity, reproductive rates, and body size (Bunker et al. 2005, Raffaelli 2004, Solan et al. 2004). For example, specialized species are likely to be more sensitive to stressors than generalized species that rely on a broader range of resources. Our understanding of the diversity–ecosystem function relationship is also limited because species removals are generally restricted to a single trophic level. We can be confident that the relationship between species diversity and ecosystem processes is considerably more complex in natural systems with multiple trophic levels than what is predicted based on single-trophic models. The oft-cited metaphor that keystone species represent a critical supporting stone in an arch of subordinate species has recently been modified to account for the dynamic nature of food webs (de Ruiter et al. 2005). The potentially complex functional interactions among trophic groups require that ecologists adopt a multitrophic approach to predict ecosystem responses to changes in species diversity. For example, loss of species occupying higher trophic levels will likely have very different consequences for energy flow and other ecosystem processes compared to the loss of primary producers and herbivores. Furthermore, because species richness generally decreases at higher trophic levels and because species at higher trophic levels are often more susceptible to anthropogenic disturbances, an understanding of food web structure is necessary to predict the consequences of local species extinctions on ecosystem function (Petchey et al. 2004).

In addition to the cascading effects through food webs, species occupying higher trophic levels may also have direct effects on ecosystem processes. Ngai and Srivastava (2006) reported that consumption of detritivores by damselfly predators reduced the export of *N* and increased *N* cycling. In systems regulated by top–down trophic interactions, we would expect that removal of species at higher trophic levels will have greater effects (Downing and Leibold 2002). However, the relationship between species richness and ecosystem processes is context dependent and will be influenced by many environmental factors (de Ruiter et al. 2005). For example, major changes in community composition of Tuesday Lake (Michigan, USA) resulting from removal of three planktivorous fish species and addition of one piscivorous fish species had remarkably little effect on trophic dynamics (Jonsson et al. 2005). Naeem et al. (2000) also reported that increased producer or decomposer diversity could not account for greater algal production observed in freshwater microcosms. Duffy et al. (2001) observed that species composition of grazers in marine seagrass beds strongly influenced productivity and was more important than species richness. These results indicate that studies focusing on a single trophic level may underestimate ecosystem effects of anthropogenic disturbance on biodiversity.

Failure to consider the consequences of ordered versus random species losses may cause researchers to underestimate the effects of species extinction on ecosystem function (Zavaleta and Hulvey 2004). Assuming that the loss of species from ecosystems will likely be nonrandom, understanding factors that influence the susceptibility of species to local extinction will improve our ability to predict ecosystem consequences. Solan et al. (2004) compared effects of species loss on ecosystem processes in marine sediments under random and nonrandom species extinction models. Removal of abundant, large, and highly mobile marine invertebrates had much greater effect on ecosystem

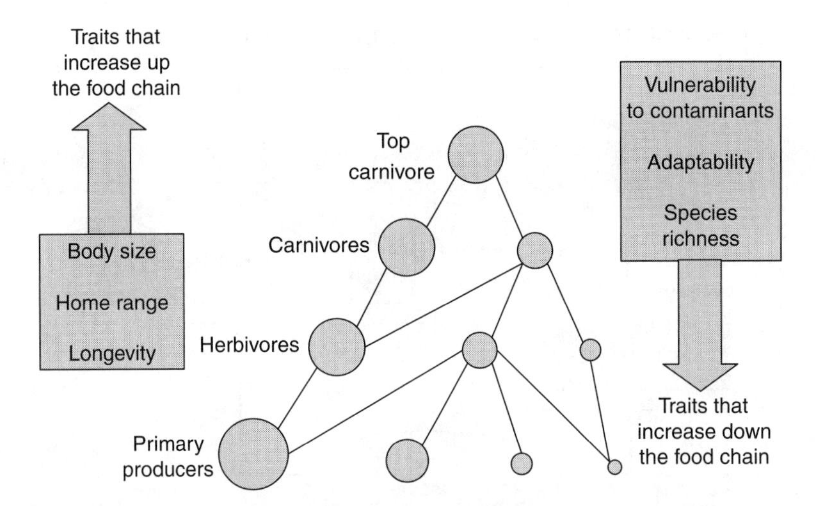

FIGURE 33.4 The effect of species loss on ecosystem function depends on life history traits (body size, susceptibility to stress), trophic level, and stressor type. Large individuals at higher trophic levels are predicted to be more susceptible to habitat loss whereas small individuals at lower trophic levels are predicted to be more sensitive to contaminants. Predicting the effects of species removal on ecosystem function is challenging because some species traits increase with trophic level whereas others decrease. (Modified from Raffaelli (2004).)

processes than removal of smaller, less abundant species. Bunker et al. (2005) performed model simulations based on random and nonrandom extinction scenarios and showed that effects of species extinction on carbon storage were strongly influenced by which species were removed first. Results of these studies also suggested that the consequences of species loss on ecosystem processes were determined by the particular stressors responsible. Raffaelli (2004) provides a conceptual model to show how life history traits that determine vulnerability to anthropogenic stressors vary among trophic levels (Figure 33.4). Because the effects of physical (e.g., habitat loss) and chemical stressors vary among trophic levels, this model could be used to predict which species are most likely to be eliminated from an ecosystem and the potential effects on ecosystem processes.

Contaminants often reduce abundance of sensitive species and alter evenness of communities, but may not result in the complete elimination of a species from an ecosystem. Consequently, relationships between community structure and ecosystem processes should account for changes in relative abundance of important species. The relative contribution of a species to ecosystem processes is a function of both its relative abundance and functional importance. Balvanera et al. (2005) developed a model analogous to the dominance–diversity curves described in Chapter 22 to quantify the relative contribution of a species to ecosystem processes. In addition to expanding our understanding of diversity–ecosystem function relationships beyond simple measures of species richness, this model allows researchers to quantify the relative contributions of each species to ecosystem processes.

33.4.2 The Need to Consider Belowground Processes

Most research investigating the relationship between species diversity and ecosystem function in terrestrial ecosystems has focused on aboveground processes. Because of the intimate connection between plants and soil microbial communities, some researchers have argued that plant diversity will directly influence microbial communities in soils and thereby affect belowground ecosystem processes such as decomposition and nutrient cycling (Heemsbergen et al. 2004, Zak et al. 2003). In terrestrial ecosystems, complex positive and negative feedbacks between aboveground plant communities and belowground microbial communities ultimately determine the nature of the

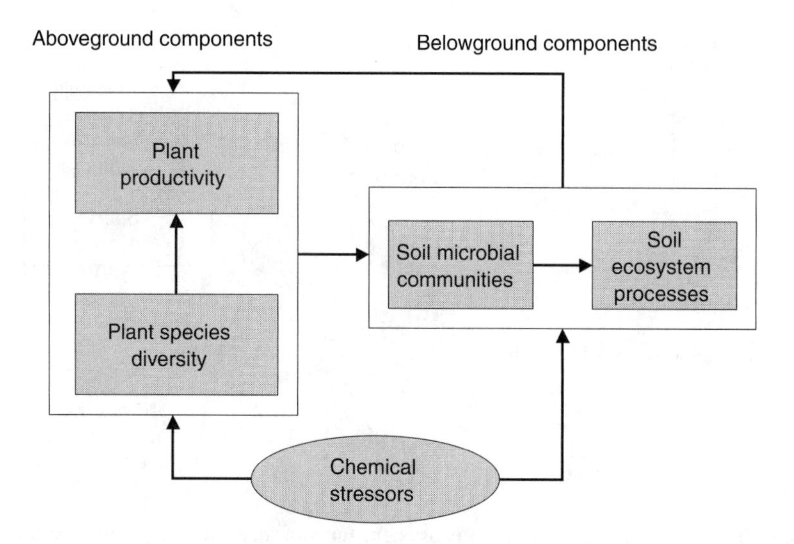

FIGURE 33.5 The influence of chemical stressors on the relationship between above- and belowground components in soil ecosystems. Complex positive and negative feedback determines the relationship between species diversity and ecosystem processes.

relationship between biodiversity and ecosystem function (Figure 33.5). For example, if increased plant diversity affects resource availability and community composition of soil communities, we would expect changes in rates of *C* and *N* cycling. Zak et al. (2003) reported that increased plant diversity altered community composition and increased soil microbial biomass, resulting in greater rates of *N* mineralization. It is important to note that these linkages between above- and belowground communities are not unidirectional. Alterations in microbial community composition and biogeochemical cycles within soil communities resulting from greater plant diversity will also affect aboveground productivity; however, these relationships may be either positive or negative (Wardle et al. 2004). These linkages between above- and belowground communities and the processes they control also have important implications for distinguishing between direct and indirect effects of chemical stressors in terrestrial ecosystems. We agree with Wardle et al. (2004) that a combined aboveground–belowground approach may be necessary to understand the effects of global change and other anthropogenic stressors on the relationship between biodiversity and ecosystem function.

33.4.3 The Influence of Scale on the Relationship between Diversity and Ecosystem Processes

Most research supporting the relationship between species diversity and ecosystem function has been conducted at relatively small to moderate spatial scales. Because species loss is a regional and global phenomenon, an understanding of the consequences of reduced biological diversity at larger spatiotemporal scales is essential. Schroter et al. (2005) quantified the vulnerability of ecosystem services to climate and land use changes across broad geographic regions in Europe. Significant species loss and increased vulnerability were observed in all regions, but effects were greatest in Mediterranean and mountainous areas. Lotze et al. (2006) reconstructed historical baselines (300–1000 years bp) of species richness and ecosystem processes for 12 temperate estuaries and coastal ecosystems in Europe, North America, and Australia. Patterns were remarkably consistent among these diverse geographic regions, showing gradual declines immediately after human settlement followed by accelerating degradation during the past 150–300 years. These long-term, global declines in abundance, diversity, and ecosystem function were attributed to several causes, including overexploitation, habitat degradation, pollution, and disturbance.

Worm et al. (2006) conducted one of the most spatially extensive investigations of the relationship between biodiversity and ecosystem processes in marine ecosystems. Meta-analysis of 32 controlled experiments showed consistent positive relationships between biodiversity (both genetic and species diversity) and primary productivity, secondary productivity, and stability. Results of these experiments scaled up regionally and globally, also demonstrating a positive relationship between biodiversity and recovery potential, stability and water quality. Worm et al. (2006) conclude that reduced biodiversity in marine ecosystems has significantly impaired stability and productivity of the world's oceans. More importantly, at the current rate of diversity loss their model projects a complete collapse of all marine fisheries by the year 2048.

33.4.4 How Will the Structure–Function Relationship Be Influenced by Global Change?

Descriptive and observational studies conducted to quantify the influence of species diversity on ecosystem processes have generally not considered the effects of climate change and other global atmospheric stressors. If the nature of the relationship between diversity and ecosystem function is context dependent as some researchers have suggested (de Ruiter 2005, Wardle 2004), then consideration of how elevated CO_2, UV-B radiation, NO_x, and other global stressors will influence this relationship is necessary (Figure 33.1). For example, ecosystem processes associated with *C* sequestration will be directly influenced by elevated CO_2 and NO_x and indirectly influenced by effects of these stressors on species diversity (Aber et al. 2001). Similarly, UV-B radiation simultaneously eliminates sensitive phytoplankton species and reduces primary productivity in marine ecosystems (Day and Neale 2002). Separating the relative importance of these direct and indirect effects will require that researchers move away from relatively small-scale, closed experimental systems to larger and more ecologically realistic outdoor systems.

33.4.5 Biodiversity–Ecosystem Function in Aquatic Ecosystems

A key limitation to our understanding of the diversity–ecosystem function relationship is the lack of research conducted in aquatic ecosystems. Because diversity in aquatic ecosystems is particularly susceptible to anthropogenic disturbance and declining rapidly, we believe that quantifying diversity–ecosystem function relationships in lentic, lotic, and marine aquatic ecosystems should be a research priority. Covich et al. (2004) reviewed the role of biodiversity in the functioning of freshwater and marine benthic communities and describe some of the unique features of aquatic ecosystems that affect this relationship. Approximately one half of the 32 diversity–ecosystem function relationships were significantly positive, and most of the remaining relationships either showed no effect or were positive but not statistically significant. Variability in the magnitude and direction of the diversity–ecosystem function relationship among aquatic ecosystems was partially attributed to spatiotemporal variation and the dependency of these relationships on environmental conditions (Covich et al. 2004).

33.5 ECOLOGICAL THRESHOLDS AND THE DIVERSITY–ECOSYSTEM FUNCTION RELATIONSHIP

One of the fundamental issues that must be resolved before we can fully understand the influence of species diversity and richness on ecosystem function is the shape of these relationships. As explained above, linear and curvilinear increases in ecosystem processes as a function of species diversity have very different implications. A linear relationship between ecosystem function and species diversity implies that all species contribute equally to ecosystem processes. In contrast, a decrease in ecosystem function at some threshold of species richness (Figure 33.3) implies inherent redundancy and that ecosystem processes are saturated until species richness is reduced to a critical level. Characterizing the nature of the relationship between diversity and ecosystem function and identifying the existence and location of any threshold would be useful for predicting ecosystem responses to species loss. The existence of ecological thresholds and the statistical techniques used to identify their location along stressor gradients have recently received attention in the literature. We believe that quantifying the relationship between stressors and ecosystem responses and identifying potential ecological thresholds is fundamental to understanding the influence of species richness and has important implications for ecosystem ecotoxicology.

33.5.1 Theoretical and Empirical Support for Ecological Thresholds

Theoretical and empirical studies suggest that some ecosystems show abrupt, nonlinear changes in structure and function in response to perturbations (e.g., Connell and Sousa 1983, Estes and Duggins 1995, May 1977). While gradients of nutrient or toxic chemical concentrations may be gradual, responses to these changes can occur rapidly and without warning (Figure 33.6). Catastrophic shifts to alternative stable states have been reported in a variety of ecosystems including lakes, coral reefs, deserts, and oceans (Scheffer et al. 2001). Shifts to alternative stable states can be triggered by natural disturbance, such as fire or flooding, or anthropogenic factors such as climate change, nutrient accumulation, exotic species, and toxic chemicals. The loss of natural resistance caused by long-term exposure to a chronic stressor may increase the likelihood that an ecosystem will shift to an alternative stable state. Although most ecosystems recover from natural disturbance through successional processes, human-induced disturbances are often unique and may move ecological

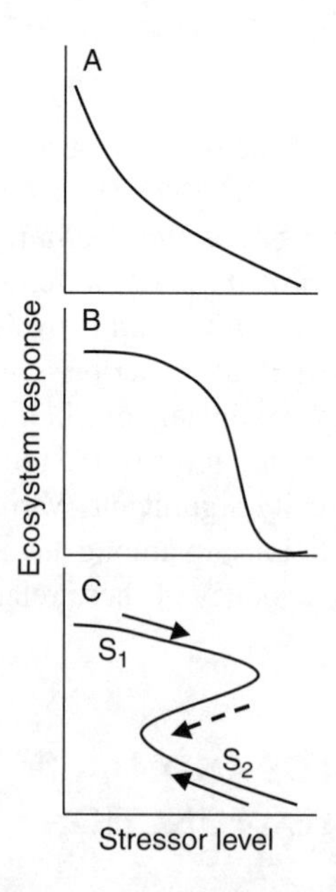

FIGURE 33.6 Hypothetical responses of ecosystems to stressors. A. Gradual response to increased stressor levels where a distinct threshold does not occur. B. Threshold response to increased stressor level. C. Threshold response to increased stressor level showing two alternative stable states, S_1 and S_2. Note that for ecosystem C to recover and return to the initial stable state, stressor levels must be reduced below those causing the initial shift. (Modified from Figure 1 in Scheffer et al. (2001).)

systems to novel, alternative states (Holling 1973). If ecosystems are chronically stressed owing to natural or anthropogenic disturbances, these alternative states may remain stable even when stressors are removed (Paine et al. 1998, Scheffer et al. 2001).

Thresholds, or ecological discontinuities, represent significant changes in an ecological state variable as a consequence of continuous changes in an independent (stressor) variable (Muradian 2001). The point at which rapid change initially occurs defines the threshold. Near this point, small changes in stressor intensity may produce large effects on response variables. Given this, it is possible to interpret ecological thresholds within the context of theoretical relationships between biodiversity and resistance stability. Both the diversity–stability hypothesis (Johnson et al. 1996) and the rivet hypothesis (Walker 1995) predict that ecosystem-level functions should decline with declining biodiversity. The diversity–stability hypothesis predicts that the decline should be smooth and linear. In contrast, the rivet hypothesis predicts that the decline will be nonlinear, showing a steep threshold at a critical level of diversity loss.

Unfortunately, there is an inherent arbitrariness to the above definition of ecological threshold because of uncertainty in determining if the magnitude of change in the response variable at the threshold is ecologically relevant. Nevertheless, numerous statistical models have been developed to detect these thresholds. For example, consider a nonlinear relationship between a response variable and a stressor variable. If this function shows a dramatic change in slope at some point along a stressor gradient, then a threshold point has been defined. However, this does not necessarily imply that the system has been shifted to an alternative stable state. One important challenge in the evaluation of data used to estimate these functions is to distinguish true thresholds from background variation in a response variable. It is also important to identify changes in slope of the function owing to external or internal factors that affect the response variable but act independently of the stressor. Because of the uncertainties associated with quantifying a precise ecological threshold, some researchers have suggested that we identify a range of stressor values where threshold responses may occur instead of a specific stressor level (Figure 33.7) (Muradian 2001).

The ecological threshold concept is closely related to other ecosystem properties such as resistance and resilience that we have discussed previously; however, there remains confusion over the use of these terms in the ecological literature describing thresholds. We have previously defined resistance stability as the ability of a community (or an ecosystem) to maintain equilibrium conditions following a disturbance. In contrast, we defined resilience stability as the ability of the system to return to predisturbance conditions following a disturbance. One way to distinguish between these concepts is to note that resistance is estimated from the relationship between an ecosystem response variable and stressor level. In contrast, resilience is measured as the relationship between an ecosystem response variable and the length of time since stressor removal (Figure 33.8). Just as the catastrophic response to a stressor may shift an ecosystem to an alternative stable state, the recovery

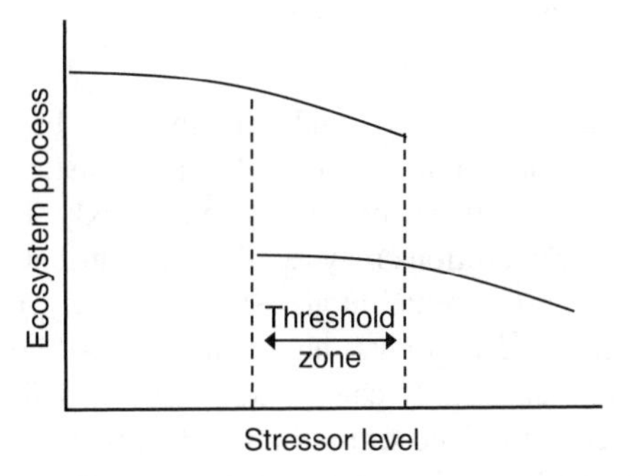

FIGURE 33.7 A hypothetical relationship between stressors and ecosystem processes. Because of inherent uncertainty in quantifying the precise location of an ecological threshold, a threshold zone is identified where the system is most likely to move to an alternative stable state. (Modified from Figure 1 in Muradian (2001).)

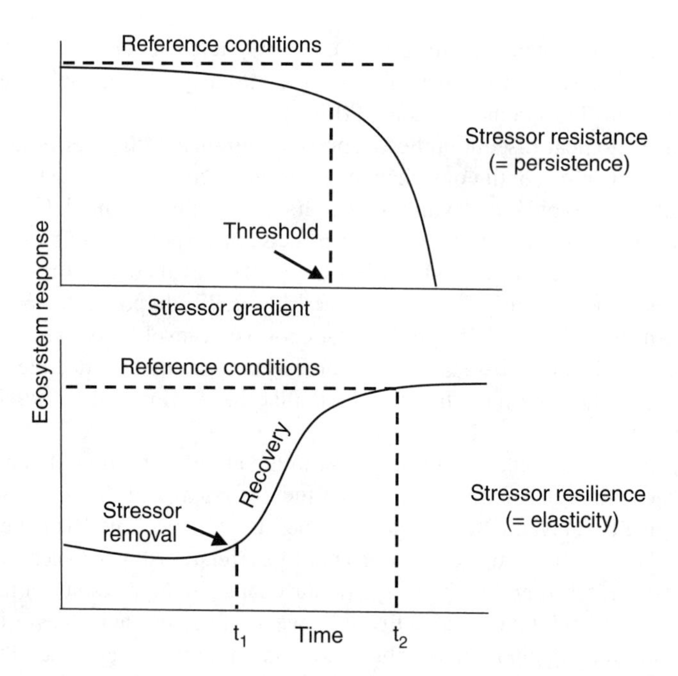

FIGURE 33.8 Resistance, resilience, and ecological thresholds. Resistance (ecosystem persistence) is estimated from the relationship between an ecosystem response and stressor level. The threshold is the point where we see an abrupt change in the ecosystem response. Resilience (elasticity) is measured as the relationship between an ecosystem response and the length of time since stressor removal. The threshold is the point where the ecosystem response returns to background conditions.

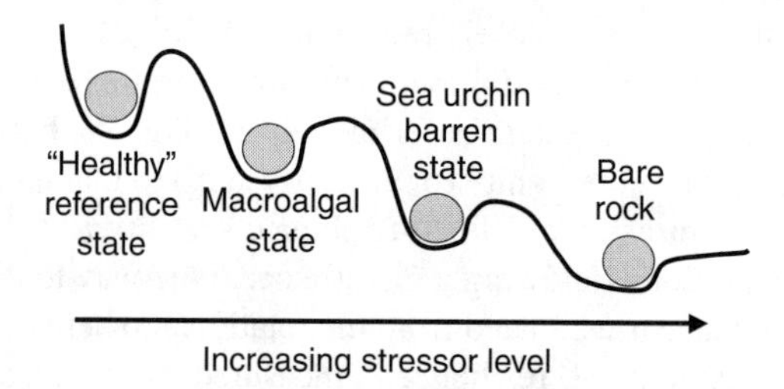

FIGURE 33.9 Ecosystem shifts between alternative steady states in coral reefs as a result of anthropogenic disturbance. The figure depicts a decline in ecosystem characteristics as a result of increasing stressor level. Returning to a previous steady state requires that stressors be reduced below levels that initially triggered the transition. (Modified from Figure 2 in Bellwood et al. (2004).)

after a stressor is removed may require that conditions improve beyond where the initial switch occurred (Scheffer et al. 2001). For example, many lakes are resistant to increases in nutrient concentrations up to a critical threshold, after which the system quickly shifts from clear to turbid water conditions. Restoring clear water conditions may require that nutrient levels be reduced well below the initial threshold concentration (Carpenter et al. 1999). Similarly, recovery of coral reef ecosystems following exposure to elevated nutrients, temperature, or siltation may require that stressor levels be reduced below those that initially caused shifts to alternative stable states (Figure 33.9) (Bellwood et al. 2004). The need to reduce stressor levels below those that initially caused the shift to an alternative stable state clearly has important implications for the restoration and recovery of damaged ecosystems.

33.5.2 Ecological Thresholds in Streams

Despite evidence that some lotic ecosystems may show abrupt responses to disturbance, few studies have examined threshold responses in streams. This is surprising given that streams are highly susceptible to anthropogenic disturbance, but also may recover rapidly when stressors are removed (Clements 2004, Lake 2000, Yount and Niemi 1990). Lotic organisms may respond to these disturbances such that community structure is continually maintained in a nonequilibrium state (Reice 1994). Given this momentum of community change, lotic communities may quickly shift between alternate states in response to natural or anthropogenic disturbance (Strange et al. 1993). Anthropogenic disturbances (i.e., contamination, habitat restructuring, exotic species, or climate change) can dramatically alter lotic communities (Yount and Niemi 1990) and hold the greatest potential for eliciting threshold responses. For instance, thresholds have been observed in fish communities in urban streams where 10–20% of the watershed consisted of impervious surfaces (Paul and Meyer 2001). Similarly, Wang et al. (1997) found that nonlinear declines in fish communities occurred once >20% of a watershed was urbanized, or after 50% of a watershed was converted to agriculture. Others have found that catastrophic wildfires, and the flooding and sediment scouring that follow, can result in relatively permanent shifts in community composition (Minshall et al. 1997, 2001; Vieira et al. 2004).

33.6 SUMMARY

The relationship between species diversity and ecosystem function is well established in the ecological literature. Although uncertainties regarding specific mechanisms and appropriate experimental designs remain, the fundamental relationship and its practical importance for protecting biological diversity are well supported. We agree with Hooper et al. (2005) that we should not restrict management options for protecting important ecosystem processes and services simply because of debate over these details. One of the greatest challenges in quantifying the potential effects of species loss on ecosystem function is to identify the important species traits that are most likely to alter processes. Because species loss from ecosystems will likely be nonrandom, models that estimate consequences based on removal of species from a single trophic level are unrealistic. Chapin et al. (1997) argue that species traits that modify resource dynamics, trophic structure and disturbance regimes have the greatest potential to affect ecosystem processes. Because species at higher trophic levels are more susceptible to extinction, understanding the relationship between top predators and ecosystem function should be a research priority. From an ecotoxicological perspective, we believe that alterations in trophic structure are most likely to impact contaminant transfer through ecosystems. For example, introduction of exotic fish species into food chains can significantly increase the transfer of contaminants to higher trophic levels (Johnston et al. 2003, Kidd et al. 1995).

33.6.1 Summary of Foundation Concepts and Paradigms

- Reduced genetic, species, and functional diversity resulting from contaminants and other stressors has important consequences for the services provided by ecosystems.
- Although many ecologists acknowledge the relationship between species diversity and ecosystem function, the mechanisms responsible for this relationship remain a significant source of controversy.
- We should not restrict management options for protecting important ecosystem processes and services simply because of debate over details of the underlying mechanisms.
- Many of the ecosystem processes that have been linked directly to species diversity are influenced by other environmental factors in addition to the number of species.

- Large-scale field experiments conducted in grasslands have contributed significantly to our understanding of the relationship between diversity and plant productivity.
- From a species conservation perspective, the shape of the diversity–ecosystem function relationship (e.g., linear, curvilinear) may be at least as important as the actual existence of this relationship.
- There is an obvious inconsistency between the hypothesis that all species in an ecosystem are important and the alternative that ecosystems with a large number of species have inherent functional redundancy.
- Failure to consider the consequences of ordered versus random species losses may cause researchers to underestimate the effects of species extinction on ecosystem function.
- Because species loss is a regional and global phenomenon, an understanding of the consequences of reduced biological diversity at larger spatiotemporal scales is necessary.
- Ecological thresholds represent significant changes in an ecological state variable as a consequence of continuous changes in an independent (stressor) variable.
- Quantifying the existence and location of ecological thresholds is fundamental to understanding the influence of species richness on ecosystem function.
- Ecosystem recovery after a stressor is removed may require that conditions improve beyond where the initial state transition occurred.
- Natural ecosystems supply irreplaceable benefits to society, some of which are essential for human welfare.

REFERENCES

Aber, J., Neilson, R.P., McNulty, S., Lenihan, J.M., Bachelet, D., and Drapek, R.J., Forest processes and global environmental change: Predicting the effects of individual and multiple stressors, *BioScience*, 51, 735–751, 2001.

Balvanera, P., Kremen, C., and Martinez-Ramos, M., Applying community structure analysis to ecosystem function: Examples from pollination and carbon storage, *Ecol. Appl.*, 15, 360–375, 2005.

Bellwood, D.R., Hughes, T.P., Folke, C., and Nystrom, M., Confronting the coral reef crisis, *Nature*, 429, 827–833, 2004.

Biesmeijer, J.C., Roberts, S.P.M., Reemer, M., Ohlemuller, R., Edwards, M., Peeters, T., Schaffers, A.P., et al., Parallel declines in pollinators and insect-pollinated plants in Britain and the Netherlands, *Science*, 313, 351–354, 2006.

Bunker, D.E., Declerck, F., Bradford, J.C., Colwell, R.K., Perfecto, I., Phillips, O.L., Sankaran, M., and Naeem, S., Species loss and aboveground carbon storage in a tropical forest, *Science*, 310, 1029–1031, 2005.

Cardinale, B.J., Nelson, K., and Palmer, M.A., Linking species diversity to the functioning of ecosystems: On the importance of environmental context, *Oikos*, 91, 175–183, 2000.

Carpenter, S.R., Ludwig, D., and Brock, W.A., Management of eutrophication for lakes subject to potentially irreversible change, *Ecol. Appl.*, 9, 751–771, 1999.

Chapin, F.S., Sala, O.E., Burke, I.C., Grime, J.P., Hooper, D.U., Lauenroth, W.K., Lombard, A., et al., Ecosystem consequences of changing biodiversity—Experimental evidence and a research agenda for the future, *BioScience*, 48, 45–52, 1998.

Chapin, F.S., Walker, B.H., Hobbs, R.J., Hooper, D.U., Lawton, J.H., Sala, O.E., and Tilman, D., Biotic control over the functioning of ecosystems, *Science*, 277, 500–504, 1997.

Chase, J.M. and Leibold, M.A., Spatial scale dictates the productivity-biodiversity relationship, *Nature*, 416, 427–430, 2002.

Clements, W.H., Small-scale experiments support causal relationships between metal contamination and macroinvertebrate community responses, *Ecol. Appl.*, 14, 954–967, 2004.

Connell, J.H. and Sousa, W.P., On the evidence needed to judge ecological stability or persistence, *Am. Nat.*, 121, 789–824, 1983.

Costanza, R., Darge, R., Degroot, R., Farber, S., Grasso, M., Hannon, B., Limburg, K., et al., The value of the world's ecosystem services and natural capital, *Nature*, 387, 253–260, 1997.

Covich, A.P., Austen, M.C., Barlocher, F., Chauvet, E., Cardinale, B.J., Biles, C.L., Inchausti, P., et al., The role of biodiversity in the functioning of freshwater and marine benthic ecosystems, *BioScience*, 54, 767–775, 2004.

Crutsinger, G.M., Collins, M.D., Fordyce, J.A., Gompert, Z., Nice, C.C., and Sanders, N.J., Plant genotypic diversity predicts community structure and governs an ecosystem process, *Science*, 313, 966–968, 2006.

Dangles, O. and Malmqvist, B., Species richness-decomposition relationships depend on species dominance, *Ecol. Lett.*, 7, 395–402, 2004.

Day, T.A. and Neale, P.J., Effects of UV-B radiation on terrestrial and aquatic primary producers, *Ann. Rev. Ecol. System.*, 33, 371–396, 2002.

de Ruiter, P.C., Wolters, V., Moore, J.C., and Winemiller, K.O., Food web ecology: Playing jenga and beyond, *Science*, 309, 68, 2005.

Downing, A.L. and Leibold, M.A., Ecosystem consequences of species richness and composition in pond food webs, *Nature*, 416, 837–841, 2002.

Duffy, J.E., Macdonald, K.S., Rhode, J.M., and Parker, J.D., Grazer diversity, functional redundancy, and productivity in seagrass beds: An experimental test, *Ecology*, 82, 2417–2434, 2001.

Estes, J.A. and Duggins, D.O., Sea otters and kelp forests in Alaska—generality and variation in a community ecological paradigm, *Ecol. Monogr.*, 65, 75–100, 1995.

Fox, J.W., Using the price equation to partition the effects of biodiversity loss on ecosystem function, *Ecology*, 87, 2687–2696, 2006.

Frost, T.M., Montz, P.K., Kratz, T.K., Badillo, T., Brezonik, P.L., Gonzalez, M.J., Rada, R.G., et al., Multiple stresses from a single agent: Diverse responses to the experimental acidification of Little Rock Lake, Wisconsin, *Limnol. Oceanogr.*, 44, 784–794, 1999.

Grime, J.P., Biodiversity and ecosystem function: The debate deepens, *Science*, 277, 1260–1261, 1997.

Havens, K.E. and James, R.T., The phosphorus mass balance of Lake Okeechobee, Florida: Implications for eutrophication management, *Lake Reserv. Manag.*, 21, 139–148, 2005.

Hector, A., Schmid, B., Beierkuhnlein, C., Caldeira, M.C., Diemer, M., Dimitrakopoulos, P.G., Finn, J.A., et al., Plant diversity and productivity experiments in European grasslands, *Science*, 286, 1123–1127, 1999.

Heemsbergen, D.A., Berg, M.P., Loreau, M., van Haj, J.R., Faber, J.H., and Verhoef, H.A., Biodiversity effects on soil processes explained by interspecific functional dissimilarity, *Science*, 306, 1019–1020, 2004.

Holling, C.S., Resilience and stability of ecological systems, *Ann. Rev. Ecol. System*, 4, 1–24, 1973.

Hooper, D.U., Chapin, F.S., Ewel, J.J., Hector, A., Inchausti, P., Lavorel, S., Lawton, J.H., et al., Effects of biodiversity on ecosystem functioning: A consensus of current knowledge, *Ecol. Monogr.*, 75, 3–35, 2005.

Hooper, D.U. and Vitousek, P.M., The effects of plant composition and diversity on ecosystem processes, *Science*, 277, 1302–1305, 1997.

Huston, M., Hidden treatments in ecological experiments: Re-evaluating the ecosystem function of biodiversity, *Oecologia*, 110, 449–460, 1997.

Johnson, K.H., Vogt, K.A., Clark, H.J., Schmitz, O.J., and Vogt, D.J., Biodiversity and the productivity and stability of ecosystems, *Trends Ecol. Evol.*, 11, 372–377, 1996.

Johnston, T.A., Leggett, W.C., Bodaly, R.A., and Swanson, H.K., Temporal changes in mercury bioaccumulation by predatory fishes of boreal lakes following the invasion of an exotic forage fish, *Environ. Toxicol. Chem.*, 22, 2057–2062, 2003.

Jonsson, T., Cohen, J.E., and Carpenter, S.R., Food webs, body size, and species abundance in ecological community description, *Adv. Ecol. Res.*, 36, 1–84, 2005.

Kidd, K.A., Schindler, D.W., Muir, D.C.G., Lockhart, W.L., and Hesslein, R.H., High concentrations of toxaphene in fishes from a subartic lake, *Science*, 269, 240–242, 1995.

Lake, P.S., Disturbance, patchiness, and diversity in streams, *J. N. Am. Benthol. Soc.*, 19, 573–592, 2000.

Lotze, H.K., Lenihan, H.S., Bourque, B.J., Bradbury, R.H., Cooke, R.G., Kay, M.C., Kidwell, S.M., et al., Depletion, degradation, and recovery potential of estuaries and coastal seas, *Science*, 312, 1806–1809, 2006.

May, R.M., Thresholds and breakpoints in ecosystems with a multiplicity of stable states, *Nature*, 269, 471–477, 1977.

Minshall, G.W., Robinson, C.T., and Lawrence, D.E., Postfire responses of lotic ecosystems in Yellowstone National Park, USA, *Can. J. Fish. Aquat. Sci.*, 54, 2509–2525, 1997.

Minshall, G.W., Royer, T.V., and Robinson, C.T., Response of the Cache Creek macroinvertebrates during the first 10 years following disturbance by the 1988 Yellowstone wildfires, *Can. J. Fish. Aquat. Sci.*, 58, 1077–1088, 2001.

Muradian, R., Ecological thresholds: A survey, *Ecol. Econ.*, 38, 7–24, 2001.

Naeem, S., Chapin III, F.S., Costanza, R., Ehrlich, P.R., Golley, F.B., Hooper, D.U., Lawton, J.H., et al., Biodiversity and ecosystem functioning: Maintaining natural life support processes, *Issues Ecol.*, 4, 2–12, 1999.

Naeem, S., Hahn, D.R., and Schuurman, G., Producer-decomposer co-dependency influences biodiversity effects, *Nature*, 403, 762–764, 2000.

Ng, A.W.M., Perera, B.J.C., and Tran, D.H., Improvement of river water quality through a seasonal effluent discharge program (SEDP), *Water, Air Soil Poll.*, 176, 113–137, 2006.

Ngai, J.T. and Srivastava, D.S., Predators accelerate nutrient cycling in a bromeliad ecosystem, *Science*, 314, 963, 2006.

Paine, R.T., Tegner, M.J., and Johnson, E.A., Compounded perturbations yield ecological surprises, *Ecosystems*, 1, 535–545, 1998.

Palmer, M., Bernhardt, E., Chornesky, E., Collins, S., Dobson, A., Duke, C., Gold, B., et al., Ecology for a crowded planet, *Science*, 304, 1251–1252, 2004.

Paul, M.J. and Meyer, J.L., Streams in the urban landscape, *Ann. Rev. Ecol. System*, 32, 333–365, 2001.

Petchey, O.L., Downing, A.L., Mittelbach, G.G., Persson, L., Steiner, C.F., Warren, P.H., and Woodward, G., Species loss and the structure and functioning of multitrophic aquatic systems, *Oikos*, 104, 467–478, 2004.

Raffaelli, D., How extinction patterns affect ecosystems, *Science*, 306, 1141–1142, 2004.

Reice, S.R., Nonequilibrium determinants of biological community structure, *Am. Sci.*, 82, 1994.

Richardson, C.J. and Qian, S.S., Long-term phosphorus assimilative capacity in freshwater wetlands: A new paradigm for sustaining ecosystem structure and function, *Environ. Sci. Technol.*, 33, 1545–1551, 1999.

Scheffer, M., Carpenter, S., Foley, J.A., Folke, C., and Walker, B., Catastrophic shifts in ecosystems, *Nature*, 413, 591–596, 2001.

Schroter, D., Cramer, W., Leemans, R., Prentice, I.C., Araujo, M.B., Arnell, N.W., Bondeau, A., et al., Ecosystem service supply and vulnerability to global change in Europe, *Science*, 310, 1333–1337, 2005.

Schwartz, M.W., Brigham, C.A., Hoeksema, J.D., Lyons, K.G., Mills, M.H., and van Mantgem, P.J., Linking biodiversity to ecosystem function: Implications for conservation ecology, *Oecologia*, 122, 297–305, 2000.

Solan, M., Cardinale, B.J., Downing, A.L., Engelhardt, K.A.M., Ruesink, J.L., and Srivastava, D.S., Extinction and ecosystem function in the marine benthos, *Science*, 306, 1177–1180, 2004.

Strange, E.M., Moyle, P.B., and Foin, T.C., Interactions between stochastic and deterministic processes in a stream fish community assembly, *Environ. Biol. Fish.*, 36, 1–15, 1993.

Taylor, B.W., Flecker, A.S., and Hall, R.O., Loss of a harvested fish species disrupts carbon flow in a diverse tropical river, *Science*, 313, 833–836, 2006.

Tilman, D., Knops, J., Wedin, D., Reich, P., Ritchie, M., and Siemann, E., The influence of functional diversity and composition on ecosystem processes, *Science*, 277, 1300–1302, 1997.

Vieira, N.K.M., Clements, W.H., Guevara, L.S., and Jacobs, B.F., Resistance and resilience of stream insect communities to repeated hydrologic disturbances after a wildfire, *Freshw. Biol.*, 49, 1243–1259, 2004.

Vorenhout, M., van Straalen, N.M., and Eijsackers, H.J.P., Assessment of the purifying function of ecosystems, *Environ. Toxicol. Chem.*, 19, 2161–2163, 2000.

Walker, B., Conserving biological diversity through ecosystem resilience, *Conserv. Biol.*, 9, 747–752, 1995.

Wang, L.Z., Lyons, J., Kanehl, P., and Gatti, R., Influences of watershed land use on habitat quality and biotic integrity in Wisconsin streams, *Fisheries*, 22, 6–12, 1997.

Wardle, D.A., Bardgett, R.D., Klironomos, J.N., Setala, H., van der Putten, W.H., and Wall, D.H., Ecological linkages between aboveground and belowground biota, *Science*, 304, 1629–1633, 2004.

Wardle, D.A., Zackrisson, O., Hornberg, G., and Gallet, C., The influence of island area on ecosystem properties, *Science*, 277, 1296–1299, 1997.

Wilson, E.O., *The Diversity of Life*, W. W. Norton and Company, New York, 1999.

Worm, B., Barbier, E.B., Beaumont, N., Duffy, J.E., Folke, C., Halpern, B.S., Jackson, J.B.C., et al., Impacts of biodiversity loss on ocean ecosystem services, *Science*, 314, 787–790, 2006.

Yachi, S. and Loreau, M., Biodiversity and ecosystem productivity in a fluctuating environment: The insurance hypothesis, *Proc. Natl. Acad. Sci.*, 96, 1463–1468, 1999.

Yount, J.D. and Niemi, G.J., Recovery of lotic ecosystems from disturbance—a narrative review of case studies, *Environ. Manag.*, 14, 547–569, 1990.

Zak, D.R., Holmes, W.E., White, D.C., Peacock, A.D., and Tilman, D., Plant diversity, soil microbial communities, and ecosystem function: Are there any links? *Ecology*, 84, 2042–2050, 2003.

Zavaleta, E.S. and Hulvey, K.B., Realistic species losses disproportionately reduce grassland resistance to biological invaders, *Science*, 306, 1175–1177, 2004.

34 Fate and Transport of Contaminants in Ecosystems

34.1 INTRODUCTION

Food web investigations have a relatively long history in ecotoxicological research. Rachel Carson's *Silent Spring* (1962) placed bald eagles and other birds of prey at the top of Elton's trophic pyramid and introduced the lay public to the important, but often misunderstood, concept of biomagnification. Since the publication of Carson's influential book, literally hundreds of studies have reported concentrations of contaminants across trophic levels and attempted to relate trophic position to biomagnification. The goal of this chapter is not to provide a comprehensive review of these studies, which have been adequately described in several recent publications (Barber 2003, Borgå et al. 2004, Fisher and Wang 1998, Iannuzzi et al. 1996, Zaranko et al. 1997). Instead, the primary goal of this section is to characterize the ecological factors that influence transport of contaminants through ecosystems. Because of the difficulty developing reliable food web models, researchers are keenly aware that predicting food chain transport requires more than an understanding of the physicochemical properties of contaminants. Quantification of feeding habits of organisms, especially those with mixed diets or that show ontogenetic changes, is often challenging. The structure of food webs and the dynamics of energy and contaminant flow also vary greatly among locations. Consequently, predictive models have become increasingly sophisticated as investigators attempt to quantify the influence of ecological factors, such as feeding habits, food chain length, and habitat characteristics, on contaminant transport and biomagnification. The inclusion of these ecological factors into transport models represents a major improvement in our understanding of how contaminants are distributed in ecosystems. However, knowing the concentration of contaminants in a particular species or trophic level tells very little about the consequences of exposure. The next logical step in the refinement of food web models is to relate predicted tissue concentrations to ecologically significant effects (Cain et al. 2004, Toll et al. 2005).

34.2 BIOCONCENTRATION, BIOACCUMULATION, BIOMAGNIFICATION, AND FOOD CHAIN TRANSFER

The traditional application of food web ecology to ecotoxicological research has been to quantify uptake and transport of contaminants between biotic and abiotic compartments. Inconsistent usage of terms such as bioconcentration, bioaccumulation, and biomagnification has caused some confusion in the literature, especially in aquatic communities (Dallinger et al. 1987). Here, we define bioconcentration as the uptake of contaminants directly from water. Thus, bioconcentration factors (BCFs) are calculated as the ratio of chemical concentration in the organism to the concentration in water. Bioaccumulation is defined as the uptake of chemicals from either biotic (food) or abiotic (sediment) compartments, and bioaccumulation factors (BAFs) are calculated as the ratio of the concentration in organisms to the concentration in these compartments. Biomagnification refers specifically to the increase in contaminant concentration with trophic level (often after adjusting for lipid content of the organism). If biomagnification occurs, we would expect that lipid-based concentrations of lipophilic contaminants should increase with trophic level. Although the highest levels of contaminants such as polychlorinated biphenyls (PCBs) and other lipophilic chemicals are frequently measured in top predators, biomagnification is a complex phenomenon influenced by many physicochemical, physiological, and ecological factors (Moriarty et al. 1984, Moriarty and Walker 1987). In addition to feeding habits, factors such as metabolism, growth rates, and habitat preferences of predators and prey may regulate contaminant transfer to higher trophic levels.

Bioaccumulation and bioconcentration of chemical substances are widely recognized as useful indicators of biological effects. BCFs and BAFs have been employed to predict hazard of hydrophobic organic chemicals to aquatic organisms. Persistent organic compounds with relatively large BCF or BAF values are generally considered to be of greater environmental concern than less recalcitrant materials. The application of these concepts to predict effects of other compounds, especially metals and other inorganic substances, is problematic. Physicochemical differences between hydrophobic organic chemicals and heavy metals limit the applicability of the BCF/BAF approach for heavy metals. Furthermore, many aquatic organisms are capable of regulating internal metal concentrations, especially essential metals such as Cu and Zn, through a variety of physiological processes. McGeer et al. (2003) observed extreme variability in BCF/BAF values for several metals and an inverse relationship between BCF/BAF and exposure concentrations. Assuming that high values should be indicative of greater hazard, the observed inverse relationship between BCF/BAF values and exposure concentration is inconsistent with known toxicological data. These results indicate that application of BCF and BAF values to assess hazard is inappropriate for metals (McGeer et al. 2003) and possibly other classes of contaminants.

Criticism of the use of BCFs and BAFs in hazard assessment highlights a more fundamental issue concerning the significance of contaminant bioaccumulation. Although observing elevated levels of a contaminant in organisms is a reasonable indicator of exposure, few studies have attempted to quantify the ecological effects of bioaccumulation. This is a particularly important issue for heavy metals and other classes of contaminants that are regulated. What is often lacking is a fundamental understanding of the mechanisms associated with bioaccumulation and a direct link to biological effects. Studies conducted by Cain et al. (2004) and Buchwalter and Luoma (2005) have provided important insight into the mechanisms of metal bioaccumulation in invertebrates and attempted to explain differential sensitivity among species based on these mechanisms. These researchers related interspecific variation in morphological characteristics of aquatic insects to heavy metal uptake and sensitivity. Cain et al. (2004) quantified interspecific variation in subcellular distributions of heavy metals between metal-sensitive and detoxified compartments in aquatic insects. These differences were then related to observed distributions of sensitive and tolerant invertebrate species in the field. Longitudinal distributions of most species were explained by partitioning of metals between metal-sensitive and detoxified fractions. These two studies represent important steps in improving our understanding of the relationship between metal bioaccumulation and ecological effects. They also demonstrate that important insights can be achieved by linking mechanistic-based studies of physiology and toxicology to ecological investigations conducted at higher levels of biological organization.

34.2.1 Lipids Influence the Patterns of Contaminant Distribution among Trophic Levels

The positive relationship between the concentration of lipophilic chemicals and trophic level is a consistent pattern reported in the literature. However, the precise mechanistic explanation for this phenomenon is not well understood. The high concentration of contaminants often observed in upper trophic levels may simply be explained by the greater levels of lipids in these organisms. Kiriluk et al. (1995) reported a significant positive relationship between lipid content and trophic position in a pelagic food web. Similar results were reported by Rasmussen et al. (1990) for lake trout.

The observation that organisms representing higher trophic levels often have greater levels of lipids complicates assessments of biomagnification and requires that lipid content be considered. If lipids increase with trophic level, the greater concentration of hydrophobic contaminants observed in top predators reported by many studies may simply be a result of equilibrium partitioning. One alternative is to measure lipid content in different compartments and then simply express all contaminant concentrations on a lipid basis. Using this approach, our definition of biomagnification is restricted only to those instances where lipid-based concentrations increase with trophic level. However, if the concentration of a chemical does not vary in direct proportion with lipids, this approach can provide biased results (Hebert and Keenleyside 1995). Various statistical approaches, such as analysis of covariance (ANCOVA), have been employed to estimate the influence of lipid content and food chain length on organochlorine concentrations in fish (Bentzen et al. 1996). Kidd et al. (1998) observed a strong positive relationship between food chain length and organochlorine concentration after accounting for lipid content in fish from subarctic lakes. The strength of the relationship between contaminant concentration and trophic position will also be influenced by lipophilicity of the chemicals (Figure 34.1). In general, more lipophilic chemicals show stronger relationships between concentration and trophic level (Kiriluk et al. 1995).

Physicochemical characteristics, such as that reflected by the octanol–water partition coefficient (K_{ow}), greatly influence uptake and transport of contaminants through food webs. There is considerable evidence that the molecular configuration of PCBs, particularly the number and arrangement of chlorine molecules, significantly influences uptake (Oliver and Niemi 1988). Trowbridge and Swackhamer (2002) observed preferential biomagnification of dioxin-like PCB congeners in a Lake Michigan food web. Because of preferential uptake, the ratio of these highly toxic PCBs to total PCBs increased with trophic level. Because of this relationship, ecological risk assessments based on food web models using total PCBs may underestimate potential effects on higher trophic levels. Russell et al. (1999) examined the roles of chemical partitioning and ecological factors in determining transfer of organic contaminants in the Detroit River. Biomagnification of high-K_{ow} organic chemicals ($\log_{10} K_{ow} > 6.3$) was observed in this food web, but simple equilibrium partitioning between lipids and water explained patterns for low-K_{ow} chemicals ($\log_{10} K_{ow} < 5.5$). Principal component analysis (PCA) based on chemical concentrations in organisms showed greater similarity to the observed diets of these organisms than assigned trophic positions. Similar results were reported by Kucklick et al. (1996) for a pelagic food web in Lake Baikal. BAFs, defined as the ratio of lipid-corrected PCB concentrations in predators to those in prey, increased with $\log_{10} K_{ow}$ for both predatory zooplankton and fish.

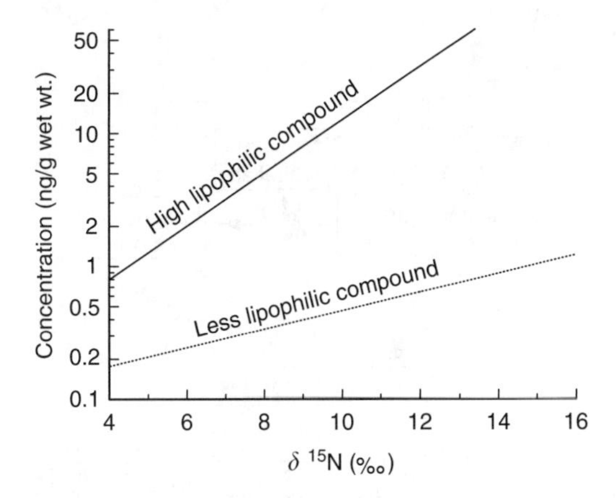

FIGURE 34.1 Hypothetical relationship between trophic position (as indicated by stable isotope δ^{15}N value) and organochlorine concentration in fish for highly lipophilic and less lipophilic compounds. It is expected that highly lipophilic compounds will have a greater potential for biomagnification than less lipophilic chemicals.

34.2.2 Relative Importance of Diet and Water in Aquatic Ecosystems

Much of the debate regarding the significance of food chain transfer of contaminants in aquatic systems focuses on the relative importance of food and water pathways. For many lipophilic organic contaminants, especially PCBs and other organochlorines, accumulation from food is generally considered the primary route of exposure. Although sophisticated models have been developed to predict bioconcentration from water, models that ignore aqueous exposure can provide reasonably accurate estimates of contaminant levels in fish (Jackson and Schindler 1996). In contrast, attempts to predict chemical concentrations in predators based only on physiological features of organisms and physicochemical characteristics of contaminants are fraught with uncertainty (Owens et al. 1994, Russell et al. 1999). Failure to account for food chain transport will significantly underestimate concentrations of organochlorines and other lipophilic chemicals (Zaranko et al. 1997). Indeed, contemporary models describing fate and transport of highly lipophilic contaminants generally include a food chain component and account for input from sediment (Figure 34.2). Comparative studies of different food webs have been conducted to quantify the relative importance of trophic transfer and passive uptake. Wallberg et al. (2001) compared uptake and food chain transfer of a PCB (2,2′,4,4′,6,6′-hexachlorobiphenyl) in an autotrophic food web consisting of algae and bacteria and a heterotrophic food web consisting of bacteria, flagellates, and ciliates. Results showed that trophic transfer was the dominant pathway in the heterotrophic food web, resulting in significantly elevated concentrations in higher trophic levels. Russell et al. (1999) employed multivariate analyses to investigate the relationship between trophic level and organochlorine concentrations in a Detroit River food chain. Lipid-based concentrations of organochlorines increased with trophic level, supporting the hypothesis that these chemicals biomagnified through the food chain. In addition to an increase in concentration with trophic level, PCA showed that the specific constituents of organochlorines varied among trophic groups. Morrison et al. (1997) developed and field validated a model to predict transfer of PCBs in a pelagic food chain. Results showed that 95% of the observed concentrations in invertebrates and fish were within a factor of two times the predicted concentrations. The close agreement between measured and predicted concentrations suggests that the model ultimately may be useful for assessing effects of PCBs on aquatic organisms. Mathematical models

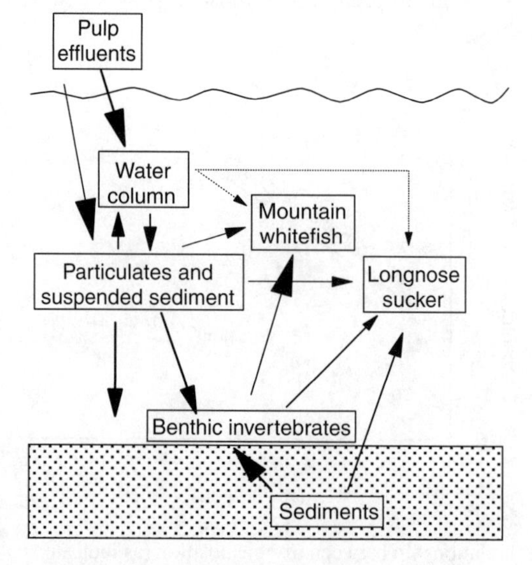

FIGURE 34.2 Food chain model showing transport of contaminants in an aquatic ecosystem. The size of the arrows indicates the relative importance of each pathway. (Modified from Figure 6 in Owens et al. (1994).)

developed by Thomann (1981) that quantify the relative importance of exposure from diet and water are discussed in Section 34.3.2.

Unlike the situation for PCBs and many other lipophilic organic contaminants, the relative importance of aqueous and dietary exposure to heavy metals is uncertain. Most of the evidence derived from laboratory studies indicates that uptake from water is a more important route of exposure than food, particularly for fish.[1] However, some investigators have suggested that dietary uptake may also contribute significantly to total body burdens of heavy metals (see review by Dallinger et al. 1987). For example, Hatakeyama and Yasuno (1987) reported that 90% of cadmium (Cd) accumulation in the guppy, *Poecilia reticulata*, was derived from feeding on contaminated chironomids. Similarly, Dallinger and Kautzky (1985) demonstrated that rainbow trout accumulated metals primarily through the diet, particularly when levels in the water were low. Munger and Hare (1997) measured the relative importance of diet and water as sources of Cd uptake for the predatory insect *Chaoborus* in a laboratory food chain. They reported no significant difference in organisms exposed to Cd in food alone versus Cd in food and water, indicating that uptake from water was relatively unimportant.

Although food chain transfer of most metals is probably a less serious issue than for lipophilic organic contaminants, dietary exposure should not be ignored when assessing ecological risk of heavy metals (Hansen et al. 2004). Dietary exposure to heavy metals is especially contentious because water quality criteria are based exclusively on aqueous exposure and assume no effects from dietary uptake. Because concentrations of metals in certain biotic and abiotic compartments may be very high, relatively inefficient transfer of metals through food chains can result in harmful levels. For example, periphyton and attached algae in streams concentrate metals and other contaminants several orders of magnitude above aqueous levels. Organisms grazing these materials, such as mayflies and other benthic macroinvertebrates, are exposed to significantly elevated concentrations. Irving et al. (2003) compared effects of aqueous and dietary cadmium on grazing mayflies. Organisms were very tolerant of aqueous exposure (96-h median lethal concentration = 1611 μg/L), whereas exposure to Cd through the diet significantly inhibited feeding and reduced mayfly growth. Several researchers have reported that despite low transfer efficiencies for some metals, dietary exposure may have negative effects on upper trophic levels (Farag et al. 1998, Woodward et al. 1994, Woodward et al. 1995). This point was demonstrated convincingly in a series of laboratory experiments in which rainbow trout were fed benthic invertebrates collected from a metal-contaminated stream (Woodward et al. 1994). Fish consuming metal-contaminated prey showed reduced growth and greater mortality as compared to fish feeding on organisms collected from an unpolluted stream.

At least part of the controversy surrounding the relative importance of aqueous versus dietary exposure to metals involves differences in experimental designs used to expose organisms. Some studies using artificial diets have reported relatively minor effects (Mount et al. 1994), whereas those using field-collected organisms have observed increased mortality and reduced growth (Woodward et al. 1994). Although natural diets collected from reference and contaminated sites are more ecologically realistic, differences in prey composition between locations confound interpretation of growth effects because of potential differences in nutritional quality. An alternative experimental design that addresses this problem is to expose prey species to contaminated media (e.g., periphyton or sediments) collected from field sites and then feed these prey to fish predators. Hansen et al. (2004) used this experimental design to assess the effects of dietary exposure to metals on the growth of rainbow trout. Fish were fed freshwater oligochaetes that had been exposed to reference and metal-contaminated sediments collected from the Clark Fork River (Montana, USA), a stream receiving metals from historic mining and mineral processing facilities. Significant reductions in growth of fish feeding on metal-contaminated prey were attributed to elevated levels of arsenic in tissues. This is one of the first studies to demonstrate a relationship between contaminated sediments and effects on

[1] The notable exception is mercury that, as methylmercury, has a dominant food-linked transfer among species.

fish through dietary exposure of metals. It is important to note that, from a management perspective, concerns over differences in prey nutritional quality between reference and metal-contaminated sites may be relatively unimportant. While differences in community composition of prey may confound our understanding of mechanisms of toxicity of dietary exposure, effects on fish are ultimately a result of heavy metals, either through direct dietary exposure or because of metal-induced alterations in prey nutritional quality.

34.2.3 Energy Flow and Contaminant Transport

Quantitative approaches developed to measure energy flow in ecosystems can also be employed to estimate the movement of contaminants across trophic levels and between biotic and abiotic compartments. Odum's (1968) box and arrow diagrams showing energy and material flow among trophic levels are the predecessors of contemporary contaminant transport models. Although ecotoxicologists have done a reasonable job quantifying contaminant concentrations in biotic and abiotic compartments, validation of transport models requires accurate estimates of transfer rates between trophic levels. Because these estimates are typically obtained from laboratory studies, there is some uncertainty concerning their relevance to conditions in the field. Jackson and Schindler (1996) used a long-term monitoring program to estimate transfer efficiencies of PCBs from prey fishes to salmonids in Lake Michigan. Despite significant temporal changes in concentrations of PCBs in prey, transfer efficiencies remained relatively constant over the 15-year study. These findings demonstrate that temporal changes in PCB levels in top predators are determined primarily by concentrations in prey species. Thus, the steady decline in PCB levels in Lake Michigan salmonids over the past 20 years (Stow et al. 1995) is likely a direct result of both reduced inputs and lower PCB concentrations in prey species.

Alterations in food web structure resulting from anthropogenic perturbations have important implications for energy flow and trophic dynamics in aquatic ecosystems. Some of the most comprehensive examples demonstrating the cascading influences of contaminants on predator populations and energy flow are from estuaries subjected to hypoxia (Buzzelli et al. 2002, Peterson et al. 2000). Loss of oysters and other benthic suspension feeders reduces the capacity of estuarine ecosystems to regulate phytoplankton, making these systems more susceptible to nutrient enrichment. Baird et al. (2004) used network analysis to quantify the movement of energy through the Neuse River Estuary (North Carolina, USA), a eutrophic system receiving high levels of N from agricultural, industrial, and urban sources. By taking advantage of annual variation in the level of hypoxia over two consecutive summers (1997 and 1998), researchers demonstrated that impairments in water quality cascaded through several trophic levels and diverted energy from consumers to microbial pathways. These researchers also speculated that reduced transfer of energy to higher trophic levels increased the susceptibility of the Neuse River estuary to other stressors.

34.3 MODELING CONTAMINANT MOVEMENT IN FOOD WEBS

In the past several decades, there has been significant progress in the development of food web models to predict contaminant concentrations in aquatic organisms and transport among compartments. The goal of these models is often to estimate concentrations in organisms at different trophic levels based on measured concentrations in abiotic compartments such as water or sediments. Alternatively, researchers often use food web models to predict events outside the range of existing empirical data. The relatively simple equilibrium partitioning models based on physicochemical characteristics of organic contaminants (e.g., K_{OW}) have been replaced by more sophisticated compartmental, kinetic, bioenergetic, and physiological models (Landrum et al. 1992). Much of this research has focused on improving our understanding of factors that contribute to variation among species. In their simplest forms, these steady-state models predict that the concentration of contaminants in organisms is a function of uptake from water and food minus loss due to depuration, growth dilution, metabolism, and excretion. Recognition of the importance of dietary contributions to total body burdens and the incorporation of biological factors such as lipid content, reproduction, body size, age, sex, life cycle, habitat use, feeding ecology, and trophic position into these models represent major improvements in their predictive capability. However, as with the development of any mathematical model, these improvements have a cost. Incorporating these additional parameters increases the complexity of food web models, thereby reducing their generality and increasing uncertainty of predictions (Borgå et al. 2004). Researchers also recognize that because of the large number of species and potential feeding interactions in most ecosystems, predicting contaminant concentrations in all species is not practical. Consequently, it is often necessary to select representative taxa from different functional groups when constructing contaminant transport models (Arnot and Gobas 2004). Finally, comparison of model results with empirical data is a critical step in this process and is required to give food web models the necessary environmental realism.

34.3.1 Kinetic Food Web Models

Food web models developed by Thomann et al. (1992) and Gobas et al. (1993) have been widely employed to predict the bioaccumulation and transport of hydrophobic organic compounds (HOCs) in aquatic ecosystems. The models are similar in the use of lipid-normalized contaminant levels in organisms and expressing sediment contaminant concentrations based on organic carbon levels. There are important differences between the models in the treatment of contaminant dynamics in the benthic and planktonic compartments that may result in different estimates of bioaccumulation. Using empirical data collected from Lake Ontario, Burkhard (1998) compared the ability of both models to predict BAFs of HOCs in phytoplankton, zooplankton, macroinvertebrates, and fish. BAFs were generally similar for most groups; however, BAFs for compounds with $\log_{10} K_{ow}$ values >8.0 diverged significantly. Although the Thomann model had greater predictive ability for phytoplankton, zooplankton, and benthic invertebrates, predicted BAFs had lower uncertainty in the Gobas model (Burkhard 1998).

Although kinetic food web models have been validated using data from several freshwater ecosystems, especially Lake Ontario, these approaches have received considerably less attention in other ecosystems. Borgå et al. (2004) conducted an extensive review of biological factors that determined uptake and food chain transfer of HOCs in Arctic marine food webs. They note that Arctic ecosystems offer unique advantages for the study of trophic transfer of contaminants because of their remote location and distance from point sources, relatively simple but long food chains, and high dependence on lipid levels in most organisms. The relative importance of various biological factors varied among HOCs and among different species, but diet and trophic levels were the most important biological factors for seabirds and marine mammals.

Parameters included in most food web models are based on point estimates of organism body weight, lipid content, ingestion rate, metabolism, growth, and other physiological characteristics that determine bioaccumulation. However, it is generally recognized that there is considerable variability in estimates of these exposure factors, even at specific locations. Iannuzzi et al. (1996) conducted a comprehensive literature review to develop probabilistic distributions for factors that determine contaminant exposure and uptake. Mechanistic food web models developed by Thomann et al. (1992) and Gobas et al. (1993) were applied to a relatively simple estuarine food web that included polychaetes, benthic forage fish, blue crabs, and stripped bass. Exposure factors were represented by one of four distributional forms (uniform, triangular, beta, or truncated normal) to derive a probabilistic food web model. Estimated concentrations of five PCB congers were within an order of magnitude of measured concentrations, suggesting this probabilistic approach is appropriate for screening level risk assessment (Iannuzzi et al. 1996).

Compounds that may be rapidly metabolized by aquatic organisms, such as polycyclic aromatic hydrocarbons (PAHs), pose significant challenges to the development of food web models. Iannuzzi et al. (1996) argue that because metabolites are generally more toxic than parent compounds and because metabolites are often detected in specific target organs, food web models developed for these and other compounds are not very effective. Nonetheless, PAHs are widely distributed in aquatic ecosystems and pose significant risks to many aquatic organisms, especially higher trophic levels. Thus, some understanding of the potential transfer of these contaminants among trophic levels is critical for developing ecological risk assessments. Using a similar framework employed for PCBs, Thomann and Komlos (1999) developed a steady-state food web model for PAHs and applied this model using data from a small creek in Alabama (USA). Biota-sediment accumulation factors (BSAF), defined as the ratio of the lipid-normalized concentration of PAHs in the organism to the organic-carbon normalized concentration in the sediment, were calculated for PAHs over a range of K_{ow} values. Measured concentrations of PAHs in crayfish and fish were considerably less than in sediments, indicating significant loss due to metabolism of the parent compounds. Model components to account for this loss of PAHs included organism weight, lipid content, growth rate, respiration rate, food assimilation efficiency, and food ingestion rate.

Sensitivity analysis of the model showed that metabolism in fish had a large effect on bioaccumulation of PAHs with $\log_{10} K_{ow} > 4.5$. In contrast, relatively low metabolism of the crayfish resulted in much higher BSAF values. The analysis also showed that relative contributions of food and water varied with K_{ow} values for the unsubstituted PAHs. Water was the predominant route of exposure for PAHs with $\log_{10} K_{ow}$ values between 4 and 6, and food was the predominant route at lower and higher values.

Arnot and Gobas (2004) described an innovative bioaccumulation model that represented significant improvement in the original kinetic model developed by Gobas et al. (1993). These new elements included: (1) a new model to predict contaminant partitioning; (2) a new model to predict contaminant levels in algae and phytoplankton; (3) improved estimates of gill ventilation rates based on allometric relationships; and (4) a mechanistic model to predict gastrointestinal magnification. Improvements in the model were evaluated using empirical data collected for 64 chemicals in 35 species from three different ecosystems. The modifications in the original model significantly reduced model bias and improved predictions for each ecosystem. Arnot and Gobas (2004) note that further improvements in the model will be challenging because of the large amount of variation among individuals within a population.

34.3.2 Models for Discrete Trophic Levels

Trophic exchange of contaminants can be defined with a simple model that includes contaminant concentration, biomass in the trophic level of interest, biomass consumed from the lower trophic level, contaminant bioavailability, and the fraction of contaminant excreted daily (Ramade 1987) by organisms in the trophic level of interest. To begin developing such a model, the BAF is defined as the ratio of the contaminant concentration (C) at trophic level $n+1$ and the concentration in the next lowest trophic level, n:

$$\mathrm{BAF} = \mathrm{BAF}_{n,n+1} = \frac{C_{n+1}}{C_n}. \tag{34.1}$$

Rearranging this equation, the concentrations in the two trophic levels can be defined,

$$C_{n+1} = \mathrm{BAF}_{(n,n+1)} C_n. \tag{34.2}$$

The BAF for transfer $n \rightarrow n+1$ can be described in more detail by inclusion of the weight of organisms in Level $n+1$ (b_n), the weight of level n organisms consumed (a_n), the fraction of contaminant absorbed from ingested food (f_n), and the fraction of accumulated contaminant that is excreted daily (k_n):

$$\mathrm{BAF}_{n,n+1} = \frac{a_n f_n}{b_n k_n}. \tag{34.3}$$

Substituting this more detailed version of $\mathrm{BAF}_{n,n+1}$ into the relationship between C_n and C_{n+1} given above, the following model is generated:

$$C_{n+1} = \left[\frac{a_n f_n}{b_n k_n}\right] C_n. \tag{34.4}$$

This model can be easily expanded to predict the concentration at Level C_{n+2} by adding the explicit form of $\mathrm{BAF}_{n+1,n+2}$ into this model.

$$C_{n+2} = \left[\frac{a_{n+1} f_{n+1}}{b_{n+1} k_{n+1}}\right]\left[\frac{a_n f_n}{b_n k_n}\right] C_n. \tag{34.5}$$

Generalizing this approach, one could theoretically predict the concentration in any trophic level (r) knowing the contaminant concentration at the lowest level (C_0) and the variables a_i, f_i, b_i, and k_i for each trophic level:

$$C_r = \left[\prod_{i=1}^{r} \frac{a_i f_i}{b_i k_i}\right] C_0. \tag{34.6}$$

Close inspection of this model reveals a general lack of realism as well as its conceptual parsimony. Considerable information is needed to parameterize this model, but more importantly the trophic sequence is based on overly simplified exchanges. Species only feed on those prey in the next lower trophic level and are only consumed by species at the next highest trophic level. This might be adequate in some situations, but it is inadequate for modeling many food webs.

Thomann (1981) expanded this steady-state approach by including organism growth rate and uptake of contaminants from water. Organism growth was incorporated because any increase in body mass has an apparent dilution effect on contaminant concentration. Inclusion of uptake from water allowed comparison of the relative importance of food and water sources. A food chain transfer number (f) serves this purpose.

$$f = \frac{\alpha C}{k} + G. \tag{34.7}$$

In this equation, α = the chemical absorption efficiency (f_i in the simple BAF model above), C = the specific consumption (weight-specific consumption rate in units of mass of prey/(mass of predator $\times$ day), k = excretion rate (k_i in the BAF model above), and G = net organism growth rate. Thomann (1981) generalized that significant food chain transfer was indicated if $f > 1$, but uptake of contaminants from water was more important than food sources if $f < 1$. Applying this rule to PCBs, ^{239}Pu and ^{137}Cs data, he concluded that PCB and radiocesium concentrations in top predators were predominantly determined by food sources but accumulated plutonium came primarily from water sources. Thomann (1981) also added explicit details to this steady-state model for predicting water $\rightarrow$ phytoplankton, phytoplankton $\rightarrow$ zooplankton, zooplankton $\rightarrow$ small fish, and small fish $\rightarrow$ large fish transfers of contaminants. Later (Thomann 1989), this approach was focused on predicting transfer of organic chemicals in food chains by relating relevant model parameters to K_{ow}. Trophic transfer in simple aquatic systems was predicted to be insignificant if $\log_{10} K_{ow} < 5$. Food chain transfer was important for top predators in aquatic systems if $\log_{10} K_{ow}$ was between 5 and 7.

34.3.3 Models Incorporating Omnivory

A major shortcoming of the approaches described above is the assumption that no species feeds on more than one trophic level. Although unrealistic in many cases, this assumption allows a level of accuracy in predicting trophic transfer of some contaminants. After noting that such an approach was insufficient to define trophic transfer in a pelagic food web, Cabana and Rasmussen (1994) expanded trophic models to include "omnivory." Here, omnivory means that a species is feeding on food items coming from several trophic levels. Although the approach is similar to that described above, matrix formulation accommodates the increased number of trophic exchanges. In this approach, fractions of the total amount of the ith level's diet coming from specific trophic levels (j) are designated ρ_{ij}. Obviously, all ρ_{ij} fractions sum to 1 in order to include the entire diet of level i. The total ration to the ith level (C_i) is defined as follows:

$$C_i = \sum_{1}^{j} \rho_{ij} C_j. \tag{34.8}$$

The fractions of the ith level's diet coming from the different sources (j levels) can be placed into a matrix with the subdiagonal reflecting the fractions for the simple Level 1 → Level 2, Level 2 → Level 3, Level 3 → Level 4, and so forth transfers. The fractions entered below the subdiagonal are those for the transfers not accommodated in Thomann's model (e.g., Level 1 → Level 3 and Level 2 → Level 4 transfers). The following relationship describes a vector of the total rations for all trophic levels i in a trophic scheme with four levels:

$$C_i = \begin{bmatrix} C_1 \\ C_2 \\ C_3 \\ C_4 \end{bmatrix} \begin{bmatrix} 0 & 0 & 0 & 0 \\ \rho_{21} & 0 & 0 & 0 \\ \rho_{31} & \rho_{32} & 0 & 0 \\ 0 & \rho_{42} & \rho_{43} & 0 \end{bmatrix}. \tag{34.9}$$

Such a matrix was called an omnivory matrix by Cabana and Rasmussen (1994). The food chain model reduces to the simple one described by Thomann if fractions for all matrix elements are 0 except those in the subdiagonal. There are more complex exchanges in the omnivory matrix illustrated above because neither ρ_{31} nor ρ_{42} is equal to 0.

Using matrix notation and omitting accumulation for all sources except food, Cabana and Rasmussen (1994) redefined Thomann's steady-state model as the following:

$$\mathbf{B} = \boldsymbol{\alpha}\mathbf{C}[(\mathbf{K} + \mathbf{G})\mathbf{I}]^{-1}, \tag{34.10}$$

where, for the different trophic levels, $\mathbf{B}$ = a vector of BAFs, $\boldsymbol{\alpha}$ = a vector of assimilation (chemical absorption) efficiencies, $\mathbf{C}$ = a vector of rations, $\mathbf{K}$ = a vector of excretion rates, $\mathbf{G}$ = a vector of growth rates, and $\mathbf{I}$ = the identity matrix. They expanded this formulation to include exchanges other than those depicted in the matrix subdiagonal (e.g., ρ_{42} and ρ_{43}) in the example above. The following matrix-formulated model predicts a vector of contaminant concentrations (ν) expected for the i trophic levels in a food web incorporating omnivory:

$$\nu(\rho\nu I)^{-1} = \alpha C[(K + G)I]^{-1}. \tag{34.11}$$

In this model, ρ is the omnivory-adjusted mean dietary concentration for each trophic level.

A major challenge to applying this approach is to obtain estimates of elements in the omnivory matrix. Some estimates of the trophic position must be obtained that includes the possibility that species are feeding at various lower levels. In Section 34.4.4, a technique will be described that can be applied to these estimates.

34.3.4 The Influence of Life History, Habitat Associations, and Prey Tolerance on Contaminant Transport

Species-specific feeding habits, habitat associations, and tolerance of prey will greatly influence food chain transfer and levels of contaminants in top predators. Gewurtz et al. (2000) reported significant variation in PAH and PCB concentrations among benthic macroinvertebrate taxa collected from Lake Erie, USA (Figure 34.3). The highest concentrations of both classes of compounds were observed in the mayfly *Hexagenia*, organisms that inhabit and consume highly contaminated sediments and detritus. Because of the importance of *Hexagenia* in the diet of both aquatic and terrestrial predators, and because abundance of these organisms is increasing as a result of improvements in water quality (primarily reduced anoxia), it is likely that greater PAH and PCB exposure to the Lake Erie food web will occur (Gewurtz et al. 2000). Differences in organochlorine concentrations among waterfowl species from the Great Lakes were directly related to consumption of zebra mussels (*Dreissena polymorpha*), an introduced species that has dramatically altered food chains in this region (Mazak et al. 1997). Variation in contaminant concentrations within populations were also explained by the proportion of zebra mussels in the diet. Similarly, differences in feeding habits between populations of small mammals also accounted for large variation in Hg bioaccumulation (Figure 34.4). Higher levels of contamination in prey and greater transfer efficiency resulted in a 20 times higher concentrations of Hg in insectivorous mammals (shorttail shrew) compared to omnivorous mammals (white-footed mouse) (Talmage and Walton 1993). Finally, several investigations have reported that concentrations of contaminants in aquatic systems are often higher in small prey organisms compared to larger individuals (van Hattum et al. 1991, Kiffney and Clements 1993). This phenomenon may be partially explained by the greater surface area to volume ratio of small individuals. Regardless of the explanation, predators that select smaller prey species, such as juveniles and early life stages, may be at greater risk from contaminant exposure (Farag et al. 1998).

Habitat associations of prey species will contribute to variation in contaminant levels among predators. Contaminated habitats are typically characterized by reduced species diversity and a shift in community composition from sensitive to tolerant species. Prey species directly associated with

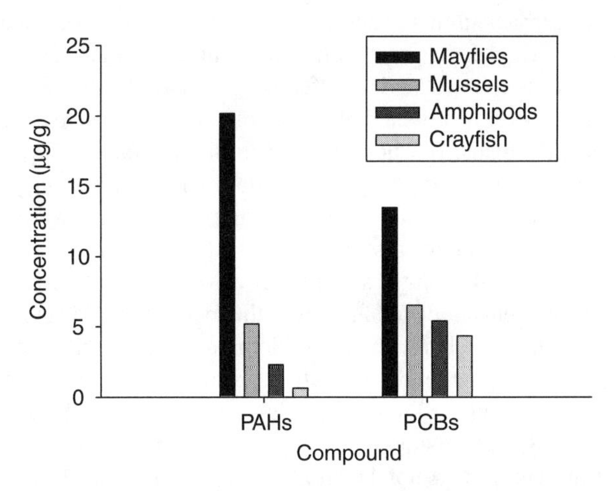

FIGURE 34.3 Concentrations (μg/g, lipid basis) of total PAHs and total PCBs measured in benthic macroinvertebrates collected from Lake Erie, USA. (Data from Table 1 in Gewurtz et al. (2000).)

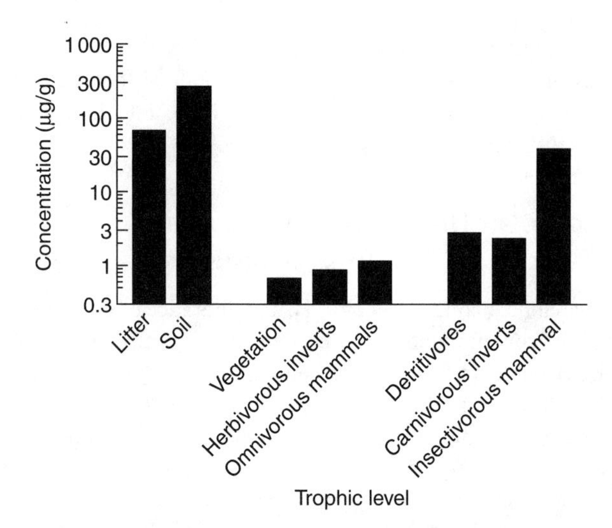

FIGURE 34.4 Hg concentrations (µg/g wet weight) in soil, litter, vegetation, invertebrates, and small mammals (kidney tissue) collected from a terrestrial field site at Oak Ridge National Laboratory. Dietary differences and variation in transfer coefficients were hypothesized to account for the differences in Hg levels between omnivorous and insectivorous mammals. (Data from Figure 1 in Talmage and Walton (1993).)

the most contaminated compartments in these systems (e.g., sediments, periphyton) are likely to have significantly elevated levels of chemicals. Several investigators have shown that feeding habits of predators at impacted sites may be modified to include these tolerant and highly contaminated prey species (Clements and Livingston 1983, Jeffree and Williams 1980, Livingston 1984). For example, Jeffree and Williams (1980) reported that fish switched from pollution-sensitive to pollution-tolerant prey in streams polluted by mining effluents. As described above, these shifts in feeding habits are likely to influence contaminant levels in top predators.

Pollution-tolerant species employ a variety of mechanisms to detoxify contaminants, including increased excretion, storage, and compartmentalization. The specific method of detoxification employed by prey species in polluted environments may influence bioavailability and food chain transfer. In particular, organisms that store or compartmentalize contaminants may pose a significant risk to predators. This phenomenon, called the "food chain effect," has been reported for species inhabiting metal-polluted environments (Dallinger et al. 1987). In a laboratory study, fish fed Cd-contaminated mussels accumulated approximately two times higher metal levels than fish fed Cd-contaminated chironomids, despite greater metal concentrations in the chironomids (Langevoord et al. 1995). These differences were related to differences in detoxification mechanisms between the two species. Wallace et al. (1998) showed that metal-tolerant oligochaetes accumulated four times more Cd than nonresistant organisms when exposed in the laboratory. However, because of differences in regulatory mechanisms employed by resistant and nonresistant prey (storage in metal rich granules vs. metallothionein), metals in nonresistant oligochaetes were more bioavailable to predators. Cain et al. (2006) noted that rates of Cd uptake by caddisflies were similar in organisms collected from reference and metal-polluted streams. However, a larger fraction of Cd was associated with metallothionein-like proteins in caddisflies from the metal-polluted stream.

Finally, variation in life history characteristics of dominant prey species may control contaminant uptake and transfer to higher trophic levels. Elevated concentrations of persistent organic pollutants in alpine and subalpine lakes compared to montane lakes have been attributed to greater deposition by snowfall (Blais et al. 1998). Life history characteristics of dominant prey species in these systems play an important role in determining uptake and transport of organochlorines. Blais et al. (2003) measured levels of persistent organic contaminants in amphipods from lakes along a 1300 m elevation gradient in Alberta, Canada. Concentrations of semivolatile organic compounds in *Gammarus lacustris* increased with elevation. Most of the variation in contaminant accumulation was explained by the slower growth rates and higher lipid content of amphipods from alpine lakes. Because amphipods are an important component of the food web in these lakes, it is likely that top predators will also be exposed to higher levels of these persistent contaminants.

34.3.5 Transport from Aquatic to Terrestrial Communities

While the majority of studies investigating food chain transport of contaminants have focused on invertebrates and fish, a few researchers have attempted to quantify movement from aquatic systems to avian and mammalian predators. Export of contaminants from aquatic to terrestrial ecosystems can be significant in some situations, posing risks to terrestrial predators. By integrating estimates of secondary production with measures of Cd concentration in emerging insects, Currie et al. (1997) calculated that 1.3–3.9 g Cd was exported annually by aquatic insects (dipterans, dragonflies, and mayflies) from Cd-treated Lake 382 in the Experimental Lakes Area, Ontario. Fairchild et al. (1992) estimated that as much as 2% of 2,3,7,8-tetrachlorodibenzofuran (TCDF) in sediments are exported annually by emerging insects, posing a significant risk to terrestrial predators (primarily birds and bats). Froese et al. (1998) measured transport of PCBs from emerging aquatic insects to tree swallows in Saginaw Bay, Michigan. Relative concentrations of PCB congeners were markedly different between sediments, benthic invertebrates, and swallows, possibly reflecting metabolic differences among trophic levels. This relationship between contaminant concentrations in sediments and levels in terrestrial predators is often complex and will be influenced by trophic relationships and life history characteristics of emerging aquatic insects. Maul et al. (2006) reported that biomagnification of PCBs in nestling tree swallows was dependent on feeding habits of adults birds, which were quite variable. These researchers cautioned that risk assessments based exclusively on a single component (e.g., contaminant concentrations in emerging insects) that do not consider life history characteristics of prey species and feeding habits of predators may be biased. Muir et al. (1988) measured PCBs and other organochlorines in a marine food chain consisting of arctic cod (*Boreogadus saida*), ringed seals (*Phoca hispida*), and polar bears (*Ursus maritimus*). In addition to increased concentrations with trophic level, major differences in the constituents of PCBs and chlordane-related compounds were observed among species. Elevated levels of organochlorines in bald eagles collected from Lake Superior were attributed to consumption of highly contaminated gulls (Kozie and Anderson 1991), which feed predominately on fish. Finally, food chain transport and biomagnification of PCBs have likely contributed to the decline of otter *(Lutra lutra)* populations in western Europe (Leonards et al. 1997). In addition to significant biomagnification of PCBs, results of multivariate analyses showed changes in the distribution of PCB congeners among trophic levels and enrichment of the most toxic constituents in otters (Figure 34.5).

34.3.6 Food Chain Transfer of Contaminants from Sediments

Because sediments are an important sink for contaminants in aquatic ecosystems, models of contaminant transport should include a sediment compartment. Concentrations of contaminants in sediments are often several orders of magnitude greater than in overlying water, and benthic organisms associated with sediments may influence the transport of these contaminants. In addition to their role in food chain transport of contaminants to higher trophic levels, the activities and movements of benthic organisms may indirectly affect bioconcentration and bioaccumulation. For example, Reynoldson (1987) reported that 0.2–7.4 g/m^2/year PCBs are ingested by oligochaete worms in contaminated sections of the Detroit River. Similarly, Evans et al. (1991) estimated that 30% of the PCBs deposited annually in Lake Michigan sediments are recycled by amphipods. Vertical migration of the invertebrate planktivore, *Mysis relicta*, transports sediment contaminants back to the water column where they are available to higher trophic levels (Bentzen et al. 1996). Bioturbation, defined as the

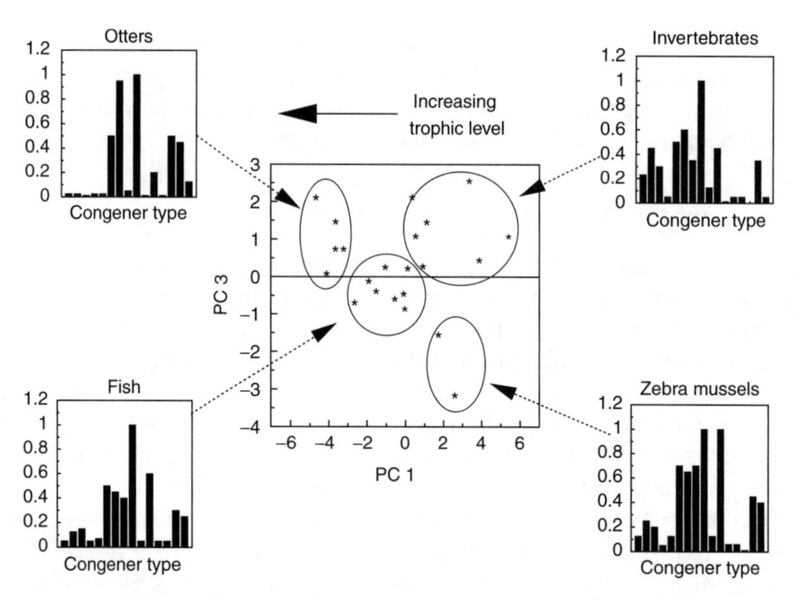

FIGURE 34.5 Results of PCA showing the relationship between trophic level and patterns of PCB constituents in an aquatic food web. (Modified from Figure 4 in Leonards et al. (1997).)

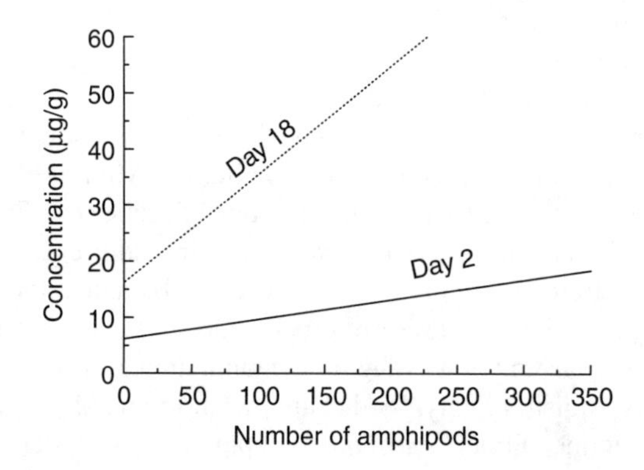

FIGURE 34.6 The influence of bioturbation by benthic invertebrates on concentration of fluoranthene in filter-feeding mussels. The figure shows results after 2- and 18-day exposure. (Modified from Figure 4 in Ciarelli et al. (1999).)

reworking of sediments resulting from various activities of benthic organisms, releases sediment contaminants into overlying water where they are bioconcentrated by other organisms and transferred to higher trophic levels. Ciarelli et al. (1999) observed that activities of amphipods in sediments resulted in significant transfer of PAHs to filter-feeding mussels (Figure 34.6). Finally, consumption of contaminated sediments, either directly or incidentally, can result in elevated concentrations in predators. DiPinto and Coull (1997) estimated the transfer of PCBs in a simple benthic food chain (sediments → copepods → fish). Approximately 33% of the PCB Aroclor 1254 accumulated by copepods was transferred to fish. Interestingly, PCB levels in predators foraging on clean prey in contaminated sediments were five times greater than those in fish feeding on contaminated prey in clean sediments. These results suggest that incidental ingestion of sediments is a significant route of exposure in benthic-feeding fish.

Comparative studies of food webs in different ecosystems provide an opportunity to evaluate the relative importance of sediment and aqueous exposure to contaminants. Morrison et al. (2002)

compared transport and fate of PCBs in the eastern and western basins of Lake Erie, areas that differ in important limnological and geomorphological characteristics related to sediment–water interactions. Compared to the deeper eastern basin, the western basin of Lake Erie is relatively shallow, highly productive, and subjected to high winds that result in sediment resuspension. Concentrations of PCBs in organisms were much higher in the western basin, and PCBs in water contributed significantly to these body burdens compared to organisms from the eastern basin. These differences also have important implications for the responses of organisms to hypothetical decreases in PCBs in water and sediment. In the eastern basin, fugacity of PCBs in sediment was much greater than fugacity in water, indicating that organisms accumulate most of their PCBs from sediment. Thus, remediation efforts to reduce PCB levels in sediments would likely be successful. In contrast, because organisms from western Lake Erie receive significant amounts of PCBs from water, remediation efforts should focus on reducing levels of dissolved PCBs.

34.3.7 Biological Pumps and Contaminant Transfer in Ecosystems

Persistent organic pollutants such as PCBs, HCB, and dichlorodiphenyltrichloroethane (DDT), as well as Hg are widely distributed by the atmosphere and oceans. The global distribution of these contaminants is indicated by their elevated levels in food webs of remote arctic and subarctic ecosystems. Transport of persistent pollutants in remote marine ecosystems is facilitated by migratory salmon that accumulate contaminants from the ocean and deliver them to their native lakes when they return to spawn. These migrating organisms act as biological pumps, delivering contaminants upstream where they may accumulate in aquatic food webs. Krummel et al. (2003) observed a highly significant relationship ($r^2 \geq .9$) between the density of spawning sockeye salmon (*Oncorhynchus nerka*) and PCB concentrations in lake sediments. Concentrations of PCBs in lakes with spawning salmon were approximately six times greater than in lakes without fish, and the pattern of PCB congeners in lake sediments was very similar to those in fish. Persistent pollutants that are pumped upstream may be accumulated in arctic food webs of receiving systems. Ewald et al. (1998) reported elevated levels of PCBs and DDT in arctic grayling (*Thymallus arcticus*) collected from lakes with returning migratory salmon. Similar transport of marine-derived contaminants has been reported in arctic seabirds. Blais et al. (2005) collected sediments from ponds at the base of cliffs along a gradient of petrel (*Fulmarus glacialis*) use in the Canadian Arctic. Concentrations of Hg, DDT, and HCB were 10–60 times greater in sediments collected from ponds with high petrel use as a result of inputs from guano. This research indicates that in some instances biological transport can have a much greater influence on levels of organic contaminants in arctic and subarctic ecosystems than atmospheric deposition.

34.4 ECOLOGICAL INFLUENCES ON FOOD CHAIN TRANSPORT OF CONTAMINANTS

Most studies that describe uptake and food chain transport of contaminants usually do not focus on the ecology of these systems, but simply report tissue concentrations in biotic and abiotic compartments. More recently, researchers have recognized that ecological characteristics of communities influence contaminant transfer and the concentrations in upper trophic levels. Because food web interactions strongly influence energy flow and biogeochemical cycling, understanding the relative importance of consumer versus resource control is important for predicting chemical transport. For example, the concentration of lipophilic contaminants in top predators will be influenced by food web interactions and the relative strength of top-down versus bottom-up controls. The development of new techniques to quantify feeding preferences, such as stable isotope analyses, allows investigators to better characterize relationships between trophic level and contaminant concentrations.

In addition, the larger spatial scale of many contemporary food web studies provides an opportunity to investigate how landscape features influence food chain transport of chemicals. Quantifying the relative importance of ecological factors on contaminant transport is greatly improved by making comparisons across communities. For example, studying contaminant levels in systems that lack point source discharges allows investigators to isolate the relative importance of ecological and habitat features. The best examples of this research have been conducted in remote systems where atmospheric deposition is the primary source of contamination (Berglund et al. 1997, Kidd et al. 1995, Kidd et al. 1998, Larsson et al. 1992, Rasmussen et al. 1990). Better integration of ecological and landscape concepts into kinetic and bioenergetic models will allow for a more comprehensive understanding of contaminant transport in communities.

34.4.1 Food Chain Length and Complexity

Understanding the relative importance of ecological factors such as food chain length, primary and secondary productivity, and linkage strength will help explain the large amount of variability in contaminant concentrations often observed in predators collected from different ecosystems. The early work by Rasmussen et al. (1990) stimulated a significant amount of interest in the relationship between food web structure and contaminant transport. These investigators classified lakes into three types based on the presence of invertebrate planktivores (*Mysis*) and pelagic forage fish. Trout collected from lakes with long food chains (i.e., more trophic levels) generally had higher PCB levels than fish from lakes with simple food chains (Figure 34.7). Similar results were reported by Kidd et al. (1995) in which elevated levels of toxaphene in fish collected from a subarctic lake were attributed to an "exceptionally long" food chain.

The influence of food chain length on contaminant levels in top predators may have important implications for systems where food webs are altered by exotic species. Introduced species that lengthen food chains may increase levels of persistent chemicals in top predators (Cabana et al. 1994, Cabana and Rasmussen 1994, Kidd et al. 1995), especially if these species link contaminated benthic habitats to pelagic consumers. However, results of studies attempting to demonstrate enhanced food chain transport in ecosystems where exotic species have invaded are mixed. Rainbow smelt (*Osmerus mordax*) have recently invaded many freshwater ecosystems of North America. Because rainbow

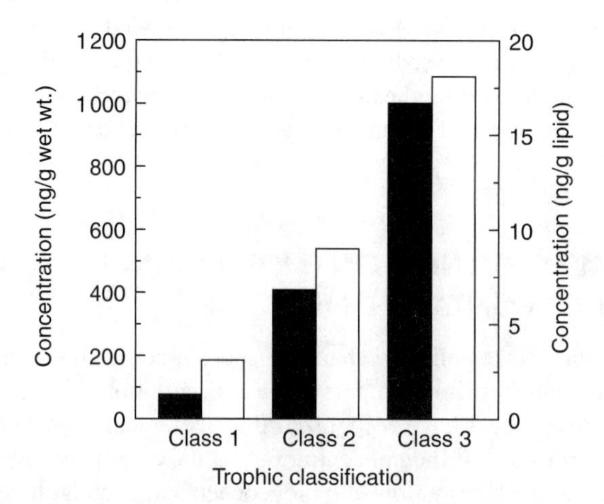

FIGURE 34.7 Influence of trophic structure on concentrations of PCBs in lake trout from central Ontario lakes. Data are shown as total PCBs (solid bars) and after correcting for lipid content (open bars). Class 1 lakes with short food chains lack *Mysis* and pelagic forage fish. Class 2 lakes with intermediate length food chains lack *Mysis* but have pelagic forage fish. Class 3 lakes with long food chains have both *Mysis* and pelagic forage fish. (Data from Table 1 in Rasmussen et al. (1990).)

smelt are generally more piscivorous than native forage fish, top predators in these systems may be exposed to higher levels of contaminants. Johnston et al. (2003) reported that the decline in Hg levels of top predators over time was less in smelt-invaded lakes than in reference lakes; however, these differences were not statistically significant. Similarly, Swanson et al. (2003) found that despite the elevated trophic position of rainbow smelt relative to other forage fish, there was little evidence of increased bioaccumulation of Hg in top predators. These researchers concluded that some predictions based on food web theory and contaminant transfer among trophic levels may not be applicable to boreal lakes and that contaminant levels may not be appropriate measures of trophic position in these systems.

The complexity of food webs and the presence of key species may also influence concentrations of contaminants in top predators. Wong et al. (1997) attributed high rates of Hg transport from benthic communities to fish in an Ontario lake to the presence of piscivorus fish. These top predators reduced abundance of benthic-feeding fish, resulting in greater biomass of macroinvertebrates. Presence of the invertebrate planktivore, *Mysis relicta*, was found to regulate food chain transport of organic contaminants (DDT, PCBs) in pelagic food webs (Bentzen et al. 1996). Stemberger and Chen (1998) observed a negative relationship between metal levels in fish tissue and the number of trophic links. They speculated that complex food webs may contain a large number of lateral or horizontal links that do not terminate in top predators, resulting in reduced metal transfer within the community.

Although most studies relating levels of contaminants to food chain length have been conducted in aquatic habitats, recent evidence suggests that trophic complexity will also influence bioaccumulation and biomagnification in terrestrial species. Differences in food chain structure may explain why concentrations of organochlorines and other lipophilic contaminants are often higher in aquatic mammalian predators (e.g., mink, otters) compared to terrestrial predators. Bremle et al. (1997) speculated that the shorter food chains typical of terrestrial systems may account for lower levels of PCBs in pine marten (a forest-dwelling mustelid) compared to those in aquatic predators. Similarly, the elevated levels of organochlorines and Hg measured in bald eagles from the Aleutian Archipelago supports the hypothesis that food chain length influences bioaccumulation and biomagnification. Levels of contaminants were greater and reproductive success was lower in eagles that consumed fish-eating seabirds compared to eagles that fed directly on fish (Anthony et al. 1999).

34.4.2 Primary Productivity and Trophic Status

Productivity and trophic status of ecosystems can greatly influence the fate and transport of contaminants through food webs. Several researchers have shown that levels of organochlorines in phytoplankton and the potential transport of these chemicals to higher trophic levels are largely determined by primary productivity (Hanten et al. 1998, Larsson et al. 1992, Taylor et al. 1991). Because the flow of some chemicals is closely related to carbon flux, turnover rates and productivity can affect the transfer of pollutants (Wallberg and Andersson 2000). In general, productive lakes have higher rates of sedimentation and greater biomass dilution, resulting in lower contaminant transfer. However, depending on the composition of these dissolved or particulate fractions, movement of contaminants from sediments back to the water column can be increased. Wallberg and Andersson (2000) compared transfer of carbon and PCBs through a microbial food web during a rainy season and a dry season. Net *C* flux was approximately three times greater during the more productive rainy season, corresponding to a three times increase in PCBs concentrations in plankton. Larsson et al. (1992) reported significant variation in concentrations of PCBs and DDE in predatory fish from 61 Scandinavian lakes, despite similar inputs of pollutants. In general, levels of persistent chemicals decreased with lake productivity and concentration of humic substances. These researchers speculated that lower levels of chemicals in more productive lakes resulted from higher growth rates of fish (and corresponding growth dilution) and faster turnover of phytoplankton.

Experimental enrichment provides an opportunity to assess effects of productivity on contaminant transfer under controlled conditions. Currie et al. (1998) added N and P to littoral enclosures to examine effects of nutrient enrichment on contaminant transfer and uptake by organisms. Increased productivity was associated with higher Cd concentrations in the water column and greater Cd accumulation by zooplankton and chironomids. Greater uptake in the pelagic and benthic components of enriched enclosures was attributed to exposure to Cd-enhanced particulate materials (Currie et al. 1998). Using a 2 × 2 factorial design, Ridal et al. (2001) manipulated nutrients and planktivorous fish in large lake enclosures to examine effects on accumulation and transfer of organochlorine pesticides. Removal of large grazing zooplankton (*Daphnia*) by planktivorous fish in nutrient-enriched enclosures resulted in significantly greater biomass of small plankton. Concentrations of pesticides in zooplankton and fish were reduced in nutrient-treated enclosures because greater amounts of contaminants were sorbed by phytoplankton. The observation that concentrations of organochlorines were greater in low-nutrient mesocosms was consistent with results of studies conducted in oligotrophic lakes (Ridal et al. 2001).

Although studies of lentic systems have generally shown a negative relationship between productivity and contaminant levels in top predators, food chain transfer of contaminants in streams may be quite different. Streams differ from lakes and large rivers in several important ways, including major physical structuring forces (flow vs. thermal stratification), sources of energy (allochthonous vs. autochthonous), major primary producers (periphyton vs. phytoplankton), and factors that control primary productivity (light vs. nutrients). As a consequence, the ecological factors that regulate contaminant transport in lotic and lentic communities may be quite different. In contrast to results observed in lakes, Berglund et al. (1997) reported that levels of organochlorines in brown trout from streams increased with primary productivity. Differences between lotic and lentic systems were attributed to spiraling of pollutants, a shift from heterotrophic to autotrophic production, and the greater influence of watershed area on streams. Hill et al. (2000) reported that Cd sorption by periphyton increased with biomass. This finding has important implications for the downstream transport of contaminants in lotic ecosystems, especially in shallow streams where a relatively large portion of water is in contact with periphyton.

The relationship between phytoplankton biomass and levels of contaminants in higher trophic levels has important implications for the biomanipulation experiments described in Chapter 27 (Box 27.1). If the introduction of piscivorous fish to lakes results in lower abundance of planktivores and greater abundance of zooplankton as predicted by the trophic cascade hypothesis (Carpenter and Kitchell 1993), we may expect organochlorine concentrations in top predators to increase (Taylor et al. 1991). Other potential conflicts exist between managing fisheries for maximum sustainability and controlling PCB levels in sport fishes (Figure 34.8). The sport fisheries in the Great Lakes is an approximately $10 billion per year industry. Thus, understanding the relationship between PCB levels in top predators and sport fishery management has major socioeconomic implications. Because of size-selective predation, stocking programs for salmonids and other predators may influence size structure and growth rates of prey (Jackson 1997). Lower stocking rates of salmonids would most likely reduce predation pressure and result in older, more contaminated prey species. Although higher stocking rates would result in less contaminated prey, increased predation pressure would increase the probability of a prey population crash. Jackson (1997) developed an age-structured model for Lake Ontario that considered the trade-offs between managing the Great Lakes for a sustainable fishery and the potential problems associated with elevated PCBs in top predators. Results of this model showed that small changes in stocking levels had significant effects on PCB concentrations in predators and the probability of a crash in prey communities.

Trophic status of lakes can control the biogeochemical cycling of contaminants within an ecosystem and the potential export of these materials to other ecosystems. Jeremiason et al. (1999) used a paired whole-lake experiment to document the effects of nutrient enrichment on mass budget of PCBs in two lakes receiving the same atmospheric source. Greater productivity in the nutrient-enriched ecosystem resulted in higher settling rates of particle-associated PCBs, presumably removing some dissolved PCBs from the water column. Because the dissolved concentrations

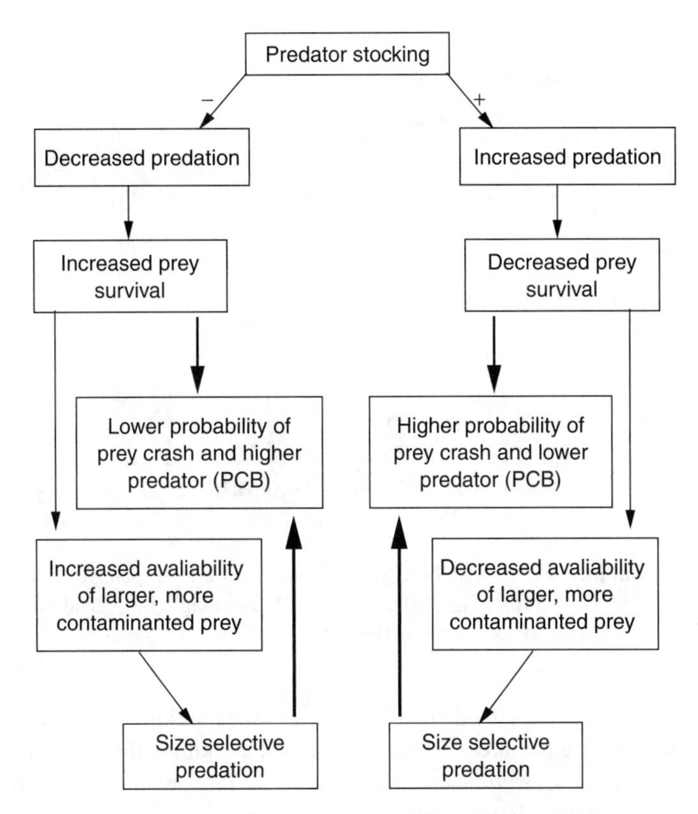

FIGURE 34.8 Trade-off between managing a sustainable Great Lakes salmon fishery and maintaining PCB concentrations below the consumption advisory. (Modified from Figure 1 in Jackson, L.J., *Ecol. Appl.*, 7, 991–1001, 1997. Reproduced by permission of the Ecological Society of America.)

were actually similar between control and nutrient-enriched lakes, these researchers concluded that the *net* volatilization was reduced in the eutrophic system.

34.4.3 Landscape Characteristics

Large scale, comparative studies have documented the influence of landscape features such as watershed area, land use, and hydrologic characteristics on food web structure and contaminant transport. In one of the first comprehensive investigations of landscape influences on the distribution of organochlorines, Munn and Gruber (1997) reported that land use determined DDT and PCB concentrations in fish collected from a 34,000 km^2 study area in Washington and Idaho, USA. The concentrations of these persistent contaminants in predators will also be influenced by local hydrologic characteristics and trophic dynamics within a watershed. Macdonald et al. (1993) reported greater bioavailability of PCBs in shallow lakes compared to deep lakes, and speculated that food web processes were more important determinants of contaminant transport in these larger systems. Similarly, Hanten et al. (1998) observed that watershed and hydrological characteristics explained a significant amount of variation in Hg concentrations in fish from 46 Connecticut (USA) lakes. In contrast, Paterson et al. (1998) found no relationship between lake size and levels of PCBs in zooplankton and fish. However, because organic carbon content decreased with lake size, levels of PCBs in sediment expressed on an organic carbon basis were greater in larger lakes. Chen et al. (2000) examined food webs in 20 lakes across a gradient of metal contamination. Landscape-level characteristics (elevation, lake area,

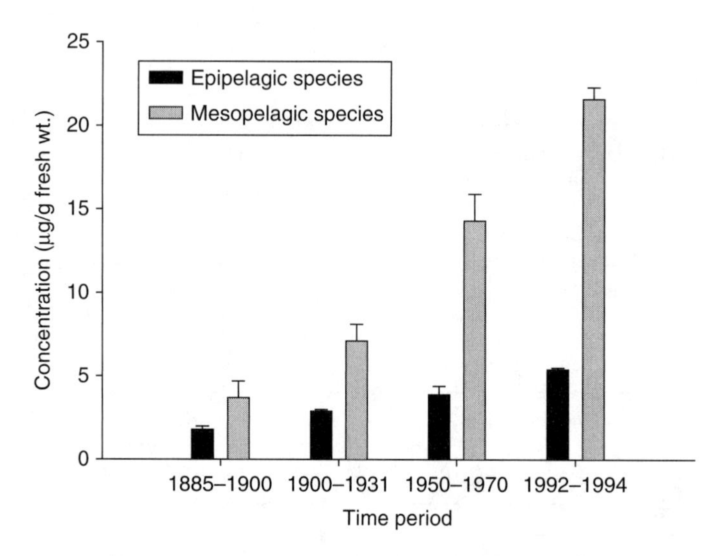

FIGURE 34.9 Long term increase in mean (+ SE) Hg concentration in epipelagic (Cory's shearwater) and mesopelagic (Bulwer's petrel) seabirds in the North Atlantic. Mesopelagic birds feed on more contaminated prey. (Data from Table 1 in Monteiro and Furness (1997).)

watershed area, and land use) explained significant amounts of variation in aqueous metal concentrations, which were then used to predict sources of metal contamination in higher trophic levels. Results showed that water was a significant source of Cd and Zn contamination in fish, whereas diet was a more important source of Hg contamination.

Land use within a watershed may interact with trophic characteristics to influence contaminant transport. Berglund et al. (1997) reported that the percent of land in agriculture was positively associated with higher levels of nutrients and higher concentrations of organochlorines in top predators. Evers et al. (1998) attributed variation in Hg concentrations measured in the common loon (*Gavia immer*) to large-scale geographic patterns of anthropogenic deposition. However, variation within a region was explained primarily by geochemical variables and lake morphology. Long-term (100 year) trends in Hg contamination were documented by comparing concentrations in bird communities collected from the North Atlantic (Monteiro and Furness 1997). Using museum specimens from the late 1800s, these researchers showed that concentrations of mercury in mesopelagic birds (those feeding on mesopelagic fish) were generally higher and increased more over the 100-year period compared to concentrations in epipelagic birds (Figure 34.9). Although increased Hg over time in epipelagic birds was consistent with increases observed in global Hg concentrations, levels in mesopelagic birds were considerably greater. Differences between these two groups were attributed to the greater production of methylmercury in low oxygen, mesopelagic seawater.

A spatially extensive survey of PCB concentrations in herring gull eggs collected from the Great Lakes showed that levels have declined significantly between 1974 and 1996 (Hebert 1998). However, concentrations appear to have stabilized through the 1990s (Figure 34.10). Interestingly, much of the annual variation in PCB concentrations resulted from the effects of winter severity on gull feeding and migration behavior. Previous studies demonstrated that winter severity influenced PCB concentrations in gulls by altering the proportion of fish in the diet (Hebert et al. 1997). In a subsequent study, Hebert (1998) speculated that extreme winters forced gulls to migrate farther south where they fed in more contaminated locations. This study illustrates the importance of accounting for behavioral characteristics of animals when assessing long-term patterns in contaminant accumulation.

In addition to the direct effects on food webs and contaminant transport, certain landscape characteristics can also influence factors that regulate contaminant bioavailability. Variation among watersheds in pH, water hardness and other factors known to influence metal bioaccumulation is

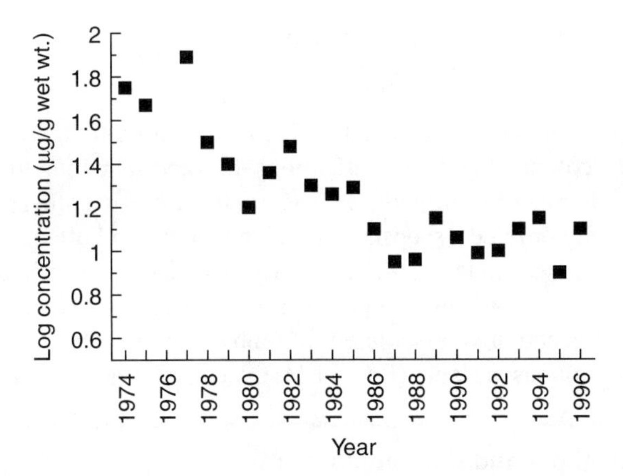

FIGURE 34.10 Temporal patterns in PCB concentrations ($\log_{10}$) in herring gull eggs collected from Double Island, Lake Huron, USA. The data show that PCB levels have declined significantly since 1974, but that concentrations have stabilized around 10 µg/g wet weight. Some of the annual variation resulted from the effects of winter severity on migration patterns. (Data from Table 4 in Hebert (1998).)

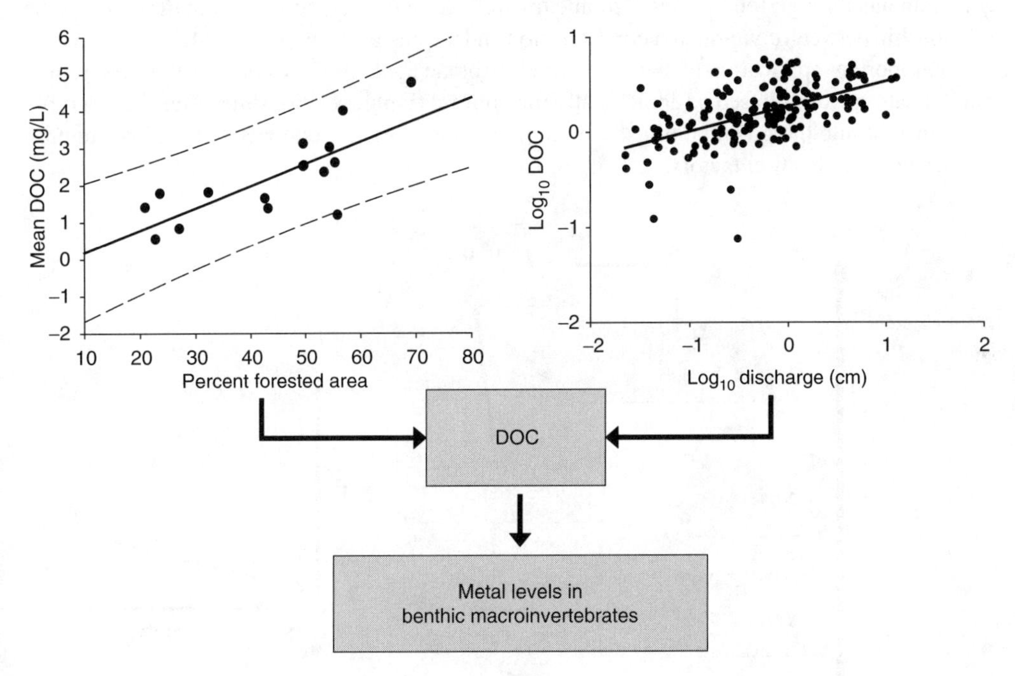

FIGURE 34.11 The influence of landscape characteristics on dissolved organic carbon (DOC) and metal bioavailability in benthic macroinvertebrates collected from Rocky Mountain streams, USA. The concentration of DOC, an important determinant of metal bioavailability in aquatic organisms, is influenced by stream discharge and the amount of vegetation in a watershed. (Data showing the relationship between forested area and DOC from Prusha and Clements (2004).)

influenced by geological features (Quinlan et al. 2003, Xie et al. 2005, Young et al. 2005). Prusha and Clements (2004) measured metal concentrations and landscape characteristics in 16 watersheds located in central Colorado (USA). Results showed that metal concentrations in caddisflies (Trichoptera) were inversely related to concentrations of dissolved organic carbon, which in turn were controlled by stream discharge and the amount of terrestrial vegetation within a watershed (Figure 34.11).

34.4.4 Application of Stable Isotopes to Study Contaminant Fate and Effects

The same problems and limitations associated with characterizing food webs for basic ecological studies also complicate ecotoxicological investigations. In particular, obtaining reliable estimates of biomagnification of contaminants is difficult because trophic levels are often poorly defined. Dietary analyses of consumers only provide a snapshot of feeding habits, and often omit important seasonal and ontogenetic changes. Indeed much of the variability associated with estimating biomagnification of different compounds results from uncertainty of assigning organisms to trophic levels. The use of stable isotopes improves quantitative assessment of food chain transfer of contaminants by treating trophic position as a continuous variable (Box 34.1). Instead of simply characterizing a predator as

---

Box 34.1 Stable Isotopes and Contaminant Transport

Food web structure influences the distribution and partitioning of lipophilic chemicals and may account for significant variation in contaminant concentrations within ecosystems (Rasmussen et al. 1990). By linking stable isotope analyses with models of contaminant transport, it is possible to examine the relationship between trophic ecology of a system and movement of contaminants through food chains. Trophic magnification factors, derived from the slope of the relationship between contaminant concentration and trophic position (Figure 34.12) may represent an important application of stable isotopes (Broman et al. 1992). Chemical-specific trophic magnification factors based on ecologically meaningful trophic relationships offer considerably more insight into the potential transfer of contaminants through food webs than bioaccumulation or biomagnification factors.

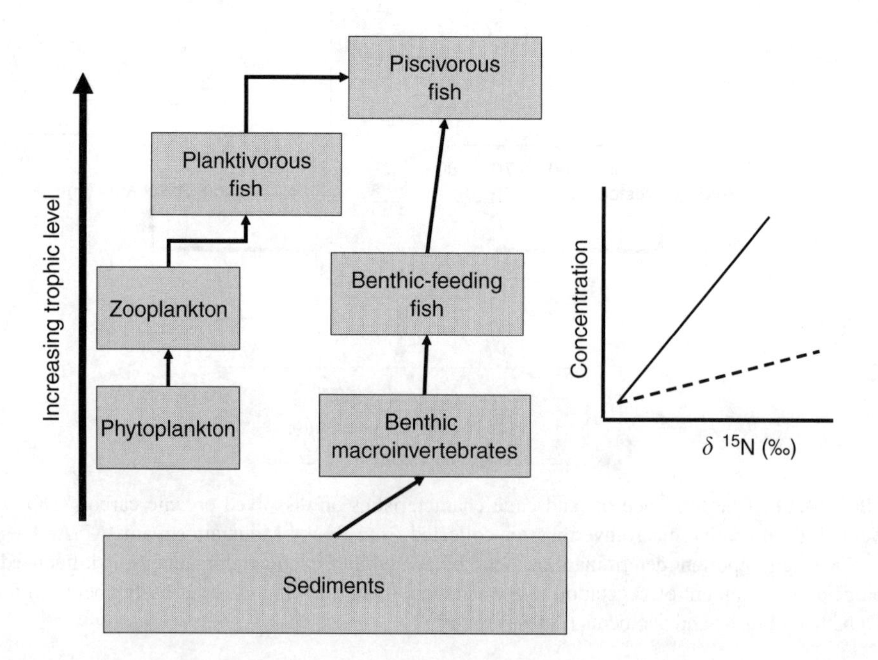

FIGURE 34.12 Hypothetical relationship between trophic level, as determined by stable isotopes, and the concentration of contaminants in pelagic (solid line) and benthic (dashed line) food webs. The slope of the relationship between trophic level and contaminant concentration is defined as the trophic magnification factor. In this example, greater trophic magnification is observed in the pelagic food web compared to the benthic food web.

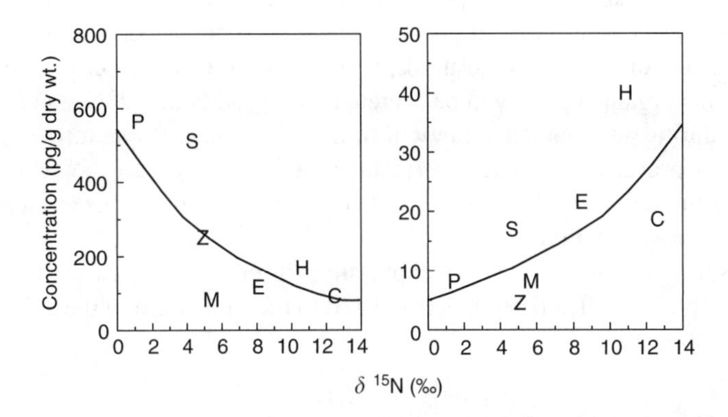

FIGURE 34.13 Relationship between trophic position (as indicated by $\delta\ ^{15}N$ values) and total concentration of 2378-substituted PCDD/Fs (left panel) and toxic content of the 2378-substituted PCDD/Fs (right panel). P = phytoplankton, S = seston, Z = zooplankton, M = mussels, E = eider ducks, H = herring, C = cod. (Modified from Figure 3 in Broman et al. (1992).)

Broman et al. (1992) used stable isotopes of C and N to characterize pelagic and littoral food chains and to estimate biomagnification of polychlorinated dibenzo-p-dioxins (PCDDs). Levels of $\delta\ ^{15}N$ in pelagic and littoral food chains increased from phytoplankton → seston → grazers → top predators, reflecting known trophic differences among these groups. Results showed that the total concentration of PCDDs and PCDFs decreased, whereas concentrations of the more toxic constituents increased with trophic level (Figure 34.13). Comparison of the slopes of the relationship between contaminant concentration and trophic level (as determined by $\delta\ ^{15}N$) for the different PCDD isomers provide an estimate of the biomagnification potential.

Kiriluk et al. (1995) employed a similar approach to assess biomagnification of DDE, mirex, and PCBs in a pelagic food web. Stable isotope signatures for C and N characterized this food web and defined lake trout as the top predator in the system. Results showed that $\delta\ ^{15}N$ increased with trophic level and was significantly correlated with contaminant concentrations within the major groups. Similar to results of Broman et al. (1992), these researchers concluded that the slope of the relationship between $\delta\ ^{15}N$ and organochlorine concentration (b) was an indication of biomagnification potential. This study also demonstrated the role of omnivory in aquatic food chains and showed that omnivory may be important in top predators. Considerable variation in $\delta\ ^{15}N$ was measured among individual lake trout, reflecting the opportunistic feeding habits of these fish.

Quantification of food web structure using stable isotopes may also help clarify relationships between long-term changes in food webs and contaminant concentrations. Gradual declines in levels of persistent organochlorines and other contaminants in fish from the Great Lakes have been observed for the past 20 years (Stow et al. 1995). It is uncertain how much of this decline reflects reduced levels of contamination versus long-term alterations in food web structure. Stable isotope analyses may allow researchers to quantify the relative importance of reduced concentrations of organochlorines and changes in trophic structure (Kiriluk et al. 1995).

---

a secondary or tertiary consumer, stable isotope analyses provide a quantitative and time-integrated measure of trophic position. Because $\delta\ ^{15}N$ values are enriched with trophic position, the relative degree of omnivory in a predator can also be quantified.

Great advantage has been taken of the fact that stable, naturally occurring nitrogen isotopes are excreted at different rates by organisms. The heavy isotope (^{15}N) is not excreted as readily as the lighter ^{14}N by any species, regardless of its specific excretory biochemistry and physiology. Differential isotopic excretion leads to differential isotopic retention. A consumer will ingest a source of nitrogen slightly enriched with ^{15}N relative to atmospheric N, the ultimate source to all trophic levels. During the lifetime of that organism, ^{14}N will be excreted more readily than ^{15}N, and the enrichment of ^{15}N relative to ^{14}N during an organism's life will further shift this isotopic ratio in favor of ^{15}N. The consequence of this process is a gradual enrichment of ^{15}N relative to ^{14}N with each trophic exchange. Quantifying the degree of ^{15}N enrichment relative to ^{14}N produces an index of trophic status that readily incorporates omnivory.

The $\delta\,^{15}N$ metric is the most common way of expressing ^{15}N enrichment relative to ^{14}N. The ratio of ^{15}N and ^{14}N in a sample (e.g., fish tissue), is compared to the same ratio in the atmosphere using the following equation:

$$\delta\,^{15}\mathrm{N} = 1000\left[\frac{(^{15}\mathrm{N}_{\mathrm{tissue}})/^{14}\mathrm{N}_{\mathrm{tissue}}}{(^{15}\mathrm{N}_{\mathrm{air}})/(^{14}\mathrm{N}_{\mathrm{air}})} - 1\right]. \tag{34.12}$$

The units for $\delta\,^{15}N$ are parts per thousand (‰ or per mill). The average $\delta\,^{15}N$ increase with each trophic exchange is 3.4‰ (Cabana and Rasmussen 1994, Minawaga and Wada 1984), but it can vary from 1.3‰ to 5.3‰ (Minawaga and Wada 1984). A species feeding at several trophic levels will have a $\delta\,^{15}N$ value lower than would be predicted if omnivory was not occurring. Cabana and Rasmussen (1994) related $\delta\,^{15}N$ to ρ_{ij} values (the fraction of the *i*th's level diet coming from trophic level *j*) with lake trout (*Salvelinus namaycush*) as the apex predator using matrix algebra: $\delta\,^{15}N - 3.4 = \rho\delta\,^{15}N$. Assuming that the average increase in $\delta\,^{15}N$ is 3.4‰ per trophic exchange in these lakes, the realized $\delta\,^{15}N$ for a lake trout feeding at different trophic levels is estimated by accounting for the fraction of the diet consumed from the different levels.

Numerous studies have related $\delta\,^{15}N$ quantitatively to contaminant concentration in various species occupying different positions in a food web (Broman et al. 1992, Cabana et al. 1994, Hesslein et al. 1991, Kidd et al. 1995, Rasmussen et al. 1990, Rolff et al. 1993). Instead of relating changes in $\delta\,^{15}N$ to ρ values, statistical models are constructed relating $\delta\,^{15}N$ to contaminant concentration. These statistical models can be linear (e.g., linear fit of Figure 3 of Cabana and Rasmussen (1994)), but application of an exponential model is more common:

$$\text{Concentration} = e^{a+b\delta\,^{15}\mathrm{N}}. \tag{34.13}$$

This model can be fit to data directly with nonlinear regression methods or with linear regression methods after transformation of concentrations to a log scale:

$$\ln \text{Concentration} = a + b\delta\,^{15}\mathrm{N}. \tag{34.14}$$

The *a* and *b* in the above model are the intercept and slope derived by linear regression. (See Newman (1993) about the importance of correcting the backtransformation bias that appears while converting results of linearized exponential models back to the original exponential form.) As a good example of applying linear regression after concentration transformation, Kidd et al. (1995) related total toxaphene concentration to $\delta\,^{15}N$ in the pristine Laberge Lake (Canada) with the following model, $\log_{10}$ of total toxaphene (ng/g wet wt.) $= 0.23(\delta\,^{15}N$ per mill$) - 0.33$. Relative to the same fish species in lakes nearer to toxaphene sources, trout (*S. namaycush*), burbot (*Lota lota*), and lake whitefish (*Coregonus clupeaformis*) had very high toxaphene tissue concentrations due to their more piscivorus habits in Laberge Lake. The $\delta\,^{15}N$ and gut content analysis of these species from several lakes supported this notion. The longer than normal food chain was the only reason for the relatively elevated toxaphene concentrations in Laberge Lake.

The term *b* is described by some authors as the trophic magnification factor (Broman et al. 1992, Rolff et al. 1993). If $b > 0$, the transfer of a contaminant is more efficient than the trophic transfer of biomass and biomagnification of the contaminant will occur. If $b < 0$, the transfer of a contaminant is less efficient than the trophic transfer of biomass and contaminant concentrations will decrease with an increase in trophic position ($\delta\ ^{15}N$). Finally, if $b = 0$, the transfer of a contaminant through trophic levels is the same as that for biomass and there will be no discernible change in concentration with increase in $\delta\ ^{15}N$. The a in the model is related to the amount of contaminant available at the base of the food chain (Rolff et al. 1993). However, there will be a bias in this estimate if the linearizing transformation was used to fit the data to the model and no bias correction was made (Newman 1993).

34.4.5 The Development and Application of Bioenergetic Food Webs in Ecotoxicology

Qualitative food web models can illustrate potential links among consumers, but are limited because the flow of carbon and energy are not quantified. Conversely, quantitative population models allow researchers to estimate the flow of energy, but data are rarely available for all dominant species in an ecosystem. There are relatively few examples in the literature where researchers have integrated results of quantitative food web models with empirical estimates of energy flow to predict contaminant transport. Carlisle (2000) has proposed a bioenergetic approach for assessing impacts of contaminants on communities that combines population measures of secondary production with energetic food webs. Just as growth integrates numerous physiological processes in individual organisms, secondary production integrates important population processes and is a useful endpoint in ecotoxicological investigations. This approach assumes that exposure to contaminants will modify population energetics directly by effects on growth, mortality, production efficiency, and assimilation efficiency. Indirect effects include alterations in feeding behavior, quality/quantity of resources, and susceptibility to predation (Figure 34.14). Using data collected from Rocky Mountain streams impacted by heavy metals, Carlisle (2000) measured community-level production in a guild of aquatic insects

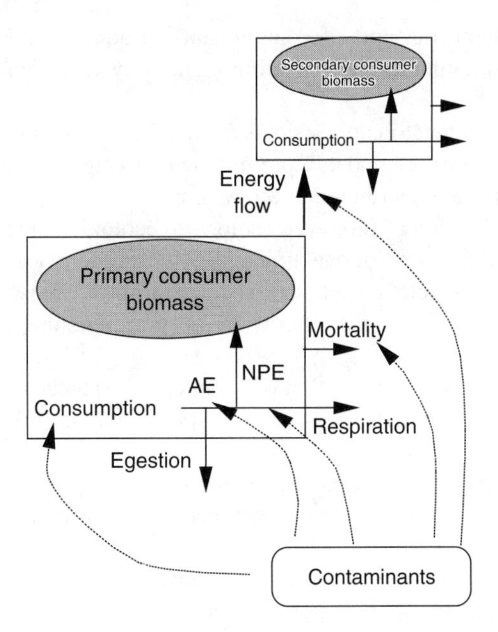

FIGURE 34.14 Model describing the potential effects of contaminants on bioenergetics. Solid lines show the flow of energy within and between trophic levels. Dashed lines show the aspects of production that are most likely to be affected by contaminants. AE = assimilation efficiency; NPE = net production efficiency. (Modified from Figure 3 in Carlisle (2000).)

and linked these estimates using energetic food webs. Results showed that metals significantly reduced secondary production of several species, primarily through reduced abundance, and that these changes had cascading effects on predators. This bioenergetic food web approach is an important step in linking observed toxicological effects on population growth and production to ecologically significant responses at higher levels of organization.

34.5 SUMMARY

The transport of contaminants within ecosystems is determined by a complex assortment of biotic and abiotic factors operating at different spatiotemporal scales. Contemporary research in contaminant transport has gone well beyond the conventional focus on simple physicochemical factors that determine uptake and depuration. In an attempt to improve our mechanistic understanding of factors that influence contaminant transport, these more sophisticated models include both autecological and synecological components. Recognition that models used to quantify energy and material flow among ecosystem compartments can be modified to characterize contaminant movement is a significant development. Furthermore, application of stable isotopes allows investigators to quantify relationships between trophic level and contaminant concentrations. A better understanding of large-scale ecological factors that influence contaminant transport, such as food chain length, ecosystem productivity, and landscape features, will also improve our ability to predict concentrations of chemicals in higher trophic levels. Comparative studies of remote ecosystems that are located away from point sources of contamination allow researchers to isolate the relative importance of these ecological factors. Finally, the traditional perspective that contaminants only move downstream has been challenged by studies showing that migrating organisms act as biological pumps, delivering contaminants upstream where they accumulate in aquatic food webs.

34.5.1 SUMMARY OF FOUNDATION CONCEPTS AND PARADIGMS

- Predicting food chain transport of contaminants in aquatic and terrestrial ecosystems requires more than a simple understanding of the physicochemical properties of these materials.
- The inclusion of ecological factors such as feeding habits, food chain length, and habitat characteristics into transport models represents a major improvement in our understanding of how contaminants are distributed in ecosystems.
- The traditional application of food web ecology to ecotoxicological research has been to quantify uptake and transport of contaminants between biotic and abiotic compartments.
- Inconsistent usage of terms such as bioconcentration, bioaccumulation, and biomagnification has caused some confusion in the literature, especially in aquatic ecosystems.
- Bioaccumulation and bioconcentration of chemical substances are widely recognized as useful indicators of contaminant exposure and *potentially* useful measures of biological effects.
- Although observing elevated levels of a contaminant in organisms is a reasonable indicator of exposure, few studies have attempted to quantify the ecological effects of bioaccumulation.
- The strength of relationships between contaminant concentration and trophic position will be influenced by lipophilicity of the chemicals.
- Much of the debate regarding the significance of food chain transfer of contaminants in aquatic systems focuses on the relative importance of food and water pathways.
- For many lipophilic organic contaminants, especially PCBs and other organochlorines, accumulation from food is generally considered the primary route of exposure.

- Unlike the situation for PCBs and many other lipophilic organic contaminants, the relative importance of aqueous and dietary exposure to heavy metals is uncertain.
- Because concentrations of metals in certain biotic and abiotic compartments may be very high, relatively inefficient transfer of metals through food chains can result in harmful levels.
- Quantitative approaches developed to measure energy flow in ecosystems can be employed to estimate the movement of contaminants across trophic levels and between biotic and abiotic compartments.
- Species-specific feeding habits, habitat associations, and tolerance of prey will greatly influence food chain transfer and levels of contaminants in top predators.
- The specific method of detoxification and contaminant storage employed by prey species in polluted environments may influence bioavailability and food chain transfer.
- Incorporation of biological factors (e.g., lipid content, age, life cycle, habitat use, trophic position) into contaminant transport models represents a major improvement in their predictive capability.
- Compounds that are rapidly metabolized by aquatic organisms (e.g., PAHs) pose significant challenges to the development of food web models.
- Export of contaminants from aquatic to terrestrial ecosystems can be significant in some situations, posing risks to terrestrial predators.
- Concentrations of contaminants in sediments are often several orders of magnitude greater than in overlying water and therefore transport models should include a sediment compartment.
- Comparative studies of food webs in different ecosystems provide an opportunity to evaluate the relative importance of sediment and aqueous exposure to contaminants.
- Transport of persistent pollutants in remote marine ecosystems is facilitated by migratory salmon that accumulate contaminants from the ocean and deliver them to their native lakes when they return to spawn.
- Exotic species introduced into aquatic ecosystems that lengthen food chains may increase levels of persistent chemicals in top predators.
- Productivity and trophic status of ecosystems can greatly influence the fate and transport of contaminants through food webs.
- Large scale, comparative studies have documented the influence of landscape features such as watershed area, land use, and hydrologic characteristics on food web structure and contaminant transport.
- By linking stable isotope analyses (e.g., $\delta\ ^{15}N$) with models of contaminant transport, it is possible to quantify the relationship between trophic ecology and movement of contaminants through food chains.
- Trophic magnification factors, derived from the slope of the relationship between contaminant concentration and trophic position, represent an important application of stable isotopes.
- Development of bioenergetic food webs is an important step in linking observed toxicological effects on population growth and production to ecologically significant responses at higher levels of organization.

REFERENCES

Anthony, R.G., Miles, A.K., Estes, J.A., and Isaacs, F.B., Productuvity, diets, and environmental contaminants in nesting bald eagles from the Aleutian Archipelago, *Environ. Toxicol. Chem.*, 18, 2054–2062, 1999.

Arnot, J.A., and Gobas, F.A.P.C., A food web bioaccumulation model for organic chemicals in aquatic ecosystems, *Environ. Toxicol. Chem.*, 23, 2343–2355, 2004.

Baird, D., Christian, R.R., Peterson, C.H., and Johnson, G.A., Consequences of hypoxia on estuarine ecosystem function: Energy diversion from consumers to microbes, *Ecol. Appl.*, 14, 805–822, 2004.

Barber, M.C., A review and comparison of models for predicting dynamic chemical bioconcentration in fish, *Environ. Toxicol. Chem.*, 22, 1963–1992, 2003.

Bentzen, E., Lean, D.R.S., Taylor, W.D., and Mackay, D., Role of food web structure on lipid and bioaccumulation of organic contaminants by lake trout (*Salvelinus namycush*), *Can. J. Fish. Aquat. Sci.*, 53, 2397–2407, 1996.

Berglund, O., Larsson, P., Bronmark, C., Greenberg, L., Eklov, A., and Okla, L., Factors influencing organochlorine uptake in age-0 brown trout (*Salmo trutta*) in lotic environments, *Can. J. Fish. Aquat. Sci.*, 54, 2767–2774, 1997.

Blais, J.M., Kimpe, L.E., McMahon, D., Keatley, B.E., Mallory, M.L., Douglas, M.S.V., and Smol, J.P., Arctic seabirds transport marine-derived contaminants, *Science*, 309, 445, 2005.

Blais, J.M., Schindler, D.W., Muir, D.C.G., Kimpe, L.E., Donald, D.B., and Rosenberg, B., Accumulation of persistent organochlorine compounds in mountains of western Canada, *Nature*, 395, 585–588, 1998.

Blais, J.M., Wilhelm, F., Kidd, K.A., Muir, D.C.G., Donald, D.B., and Schindler, D.W., Concentrations of organochlorine pesticides and polychlorinated biphenyls in amphipods (*Gammarus lacustris*) along an elevation gradient in mountain lakes of western Canada, *Environ. Toxicol. Chem.*, 22, 2605–2613, 2003.

Borgå, K., Fisk, A.T., Hoekstra, P.F., and Muir, D.C.G., Biological and chemical factors of importance in the bioaccumulation and trophic transfer of persistent organochlorine contaminants in arctic marine food webs, *Environ. Toxicol. Chem.*, 23, 2367–2385, 2004.

Bremle, G., Larsson, P., and Helldin, J.O., Polychlorinated biphenyls in a terrestrial predator, the pine marten (*Martes martes* L.), *Environ. Toxicol. Chem.*, 16, 1779–1784, 1997.

Broman, D., Naf, C., Rolff, C., Zebuhr, Y., Fry, B., and Hobbie, J., Using ratios of stable nitrogen isotopes to estimate bioaccumulation and flux of polychlorinated dibenzo-p-dioxins (PCDDs) and dibenzofurans (PCDFs)in two food chains from the northern Baltic, *Environ. Toxicol. Chem.*, 11, 331–345, 1992.

Buchwalter, D.B. and Luoma, S.N., Differences in dissolved cadmium and zinc uptake among stream insects: Mechanistic explanations, *Environ. Sci. Technol.*, 39, 498–504, 2005.

Burkhard, L.P., Comparison of two models for predicting bioaccumulation of hydrophobic organic chemicals in a Great Lakes food web, *Environ. Toxicol. Chem.*, 17, 383–393, 1998.

Buzzelli, C.P., Luettich, R.A., Powers, S.P., Peterson, C.H., McNinch, J.E., Pinckney, J.L., and Paerl, H.W., Estimating the spatial extent of bottom water hypoxia and habitat degradation in a shallow estuary, *Mar. Ecol. Prog. Ser.*, 230, 103–112, 2002.

Cabana, G. and Rasmussen, J.B., Modeling food chain structure and contaminant bioaccumulation using stable nitrogen isotopes, *Nature*, 372, 255–257, 1994.

Cabana, G., Tremblay, A., Kalff, J., and Rasmussen, J.B., Pelagic food chain structure in Ontario lakes: A determinant of mercury levels in lake trout (*Salvelinus namacush*), *Can. J. Fish. Aquat. Sci.*, 51, 381–389, 1994.

Cain, D.J., Buchwalter, D.B., and Luoma, S.N., Influence of metal exposure history on the bioaccumulation and subcellular distribution of aqueous cadmium in the insect *Hydropsyche californica, Environ. Toxicol. Chem.*, 25, 1042–1049, 2006.

Cain, D.J., Luoma, S.N., and Wallace, W.G., Linking metal bioaccumulation of aquatic insects to their distribution patterns in a mining-impacted river, *Environ. Toxicol. Chem.*, 23, 1463–1473, 2004.

Carlisle, D.M., Bioenergetic food webs as a means of linking toxicological effects across scales of ecological organization, *J. Aquat. Ecosys. Stress Recov.*, 7, 155–165, 2000.

Carpenter, S.R. and Kitchell, J.F., *The Trophic Cascade in Lakes*, Cambridge University Press, New York, 1993.

Carson, R., *Silent Spring*, Houghton Mifflin, Boston, MA, 1962.

Chen, C.Y., Stemberger, R.S., Klaue, B., Blum, J.D., Pickhardt, P.C., and Folt, C.L., Accumulation of heavy metals in food web components across a gradient of lakes, *Limnol. Oceanogr.*, 45, 1525–1536, 2000.

Ciarelli, S., van Straalen, N.M., Klap, V.A., and van Wezel, A.P., Effects of sediment bioturbation by the estuarine amphipod *Corophium volutator* on fluoranthene resuspension and transfer into the mussel (*Mytilus edulis*), *Environ. Toxicol. Chem.*, 18, 318–328, 1999.

Clements, W.H. and Livingston, R.J., Overlap and pollution-induced variability in the feeding habits of filefish (Pisces: Monacanthidae) from Apalachee Bay, Florida, *Copeia*, 2, 331–338, 1983.

Currie, R.S., Fairchild, W.L., and Muir, D.C.G., Remobilization and export of cadmium from lake sediments by emerging insects, *Environ. Toxicol. Chem.*, 16, 2333–2338, 1997.

Currie, R.S., Muir, D.C.G., Fairchild, W.L., Holoka, M.H., and Hecky, R.E., Influence of nutrient additions on cadmium bioaccumulation by aquatic invertebrates in littoral enclosures, *Environ. Toxicol. Chem.*, 17, 2435–2443, 1998.

Dallinger, R. and Kautzky, H., The importance of contaminated food for the uptake of heavy metals by rainbow trout (*Salmo gairdneri*): A field study, *Oecologia*, 67, 82–89, 1985.

Dallinger, R., Prosi, F., Segner, H., and Back, H., Contaminated food and uptake of heavy metals by fish: A review and a proposal for further research, *Oecologia*, 73, 91–98, 1987.

DiPinto, L.M. and Coull, B.C., Trophic transfer of sediment-associated polychlorinated biphenyls from meiobenthos to bottom-feeding fish, *Environ. Toxicol. Chem.*, 16, 2568–2575, 1997.

Evans, M.S., Noguchi, G.E., and Rice, C.P., The biomagnification of polychlorinated biphenyls, toxaphene and DDT compounds in a Lake Michigan offshore food web, *Arch. Environ. Contam. Toxicol.*, 20, 87–93, 1991.

Evers, D.C., Kaplan, J.D., Meyer, M.W., Reaman, P.S., Braselton, W.E., Major, A., Burgess, N., and Scheuhammer, A.M., Geographic trend in mercury measured in common loon feathers and blood, *Environ. Toxicol. Chem.*, 17, 173–183, 1998.

Ewald, G., Larsson, P., Linge, H., Okla, L., and Szarzi, N., Biotransport of organic pollutants to an inland Alaska lake by migrating sockeye salmon (*Oncorhynchus nerka*), *Arctic*, 51, 40–47, 1998.

Fairchild, W.L., Muir, D.C.G., Currie, R.S., and Yarechewski, A.L., Emerging insects as a biotic pathway for movement of 2,3,7,8-tetrachlorodibenzofuran from lake sediments, *Environ. Toxicol. Chem.*, 11, 867–872, 1992.

Farag, A.M., Woodward, D.F., Goldstein, J.N., Brumbaugh, W., and Meyer, J.S., Concentrations of metals associated with mining waste in sediments, biofilm, benthic macroinvertebrates and fish from the Coeur d'Alene River Basin, Idaho, *Arch. Environ. Contam. Toxicol.*, 34, 119–127, 1998.

Fisher, N.S. and Wang, W.X., Trophic transfer of silver to marine herbivores: A review of recent studies, *Environ. Toxicol. Chem.*, 17, 562–571, 1998.

Froese, K.L., Verbrugge, D.A., Ankley, G.T., Niemi, G.J., Larsen, C.P., and Giesy, J.P., Bioaccumulation of polychlorinated biphenyls from sediments to aquatic insects and tree swallow eggs and nestlings in Saginaw Bay, Michigan, USA, *Environ. Toxicol. Chem.*, 17, 484–492, 1998.

Gewurtz, S.B., Lazar, R., and Haffner, G.D., Comparison of polycyclic aromatic hydrocarbon and polychlorinated biphenyl dynamics in benthic invertebrates of Lake Erie, USA, *Environ. Toxicol. Chem.*, 19, 2943–2950, 2000.

Gobas, F.A.P.C., McCorquodale, J.R., and Haffner, G.D., Intestinal absorption and biomagnification of organochlorines, *Environ. Toxicol. Chem.*, 12, 567–576, 1993.

Hansen, J.A., Lipton, J., Welsh, P.G., Cacela, D., and MacConnell, B., Reduced growth of rainbow trout (*Oncorhynchus mykiss*) fed a live invertebrate diet pre-exposed to metal-contaminated sediments, *Environ. Toxicol. Chem.*, 23, 1902–1911, 2004.

Hanten, R.P., Neumann, R.M., and Ward, S.M., Relationships between concentrations of mercury in largemouth bass and physical and chemical characteristics of Connecticut lakes, *Trans. Amer. Fish. Soc.*, 127, 807–818, 1998.

Hatakeyama, S. and Yasuno, M., Chronic effects of Cd on the reproduction of the guppy (*Poecilia reticulata*) through Cd-accumulated midge larvae (*Chrionomus yoshimatsui*), *Ecotoxicol. Environ. Saf.*, 14, 191–207, 1987.

Hebert, C.E., Winter severity affects migration and contaminant accumulation in northern Great Lakes Herring Gulls., *Ecol. Appl.*, 8, 669–679, 1998.

Hebert, C.E., Shutt, J.L., and Norstrom, R.J., Dietary changes cause temporal fluctuations in polychlorinated biphenyl levels in Herring Gull eggs from Lake Ontario, *Environ. Sci. Technol.*, 31, 1012–1017, 1997.

Herbert, C.E. and Keenleyside, K.A., To normalize or not to normalize—Fat is the question, *Environ. Toxicol. Chem.*, 14, 801–807, 1995.

Hesslein, R.H., Capel, M.J., Fox, D.E., and Hallard, K.A., Stable isotopes of sulfur, carbon, and nitrogen as indicators of trophic level and fish migration in Lower MacKenzie River Basin, Canada, *Can. J. Fish. Aquat. Sci.*, 48, 2258–2265, 1991.

Hill, W.R., Bednarek, A.T., and Larsen, I.L., Cadmium sorption and toxicity in autotrophic biofilms, *Can. J. Fish. Aquat. Sci.*, 57, 530–537, 2000.

Iannuzzi, T.J., Harrington, N.W., Shear, N.M., Curry, C.L., Carlson-Lynch, H., Henning, M.H., Su, S.H., and Rabbe, D.E., Distributions of key exposure factors controlling the uptake of xenobiotic chemicals in an estuarine food web, *Environ. Toxicol. Chem.*, 15, 1979–1992, 1996.

Irving, E.C., Baird, D.J., and Culp, J.M., Ecotoxicological responses of the mayfly *Baetis tricaudatus* to dietary and waterborne cadmium: Implications for toxicity testing, *Environ. Toxicol. Chem.*, 22, 1058–1064, 2003.

Jackson, L.J. and Schindler, D.E., Field estimates of net trophic transfer of PCBs from prey fishes to Lake Michigan salmonids, *Environ. Sci. Technol.*, 30, 1861–1865, 1996.

Jackson, L.J., Piscivores, predation, and PCBs in Lake Ontario's pelagic food web, *Ecol. Appl.*, 7, 991–1001, 1997.

Jeffree, R.A. and Williams, N.J., Mining pollution and the diet of the purple-striped gudgeon *Mogurnda mogurnda* Richardson (Eleotridae) in the Finniss River, Northern Territory, Australia, *Ecol. Monogr.*, 50, 457–485, 1980.

Jeremiason, J.D., Eisenreich, S.J., Paterson, M.J., Beaty, K.G., Hecky, R., and Elser, J.J., Biogeochemical cycling of PCBs in lakes of variable trophic status: A paired-lake experiment, *Limnol. Oceanogr.*, 44, 889–902, 1999.

Johnston, T.A., Leggett, W.C., Bodaly, R.A., and Swanson, H.K., Temporal changes in mercury bioaccumulation by predatory fishes of boreal lakes following the invasion of an exotic forage fish, *Environ. Toxicol. Chem.*, 22, 2057–2062, 2003.

Kidd, K.A., Schindler, D.W., Hesslein, R.H., and Muir, D.C.G., Effects of trophic position and lipid on organochlorine concentrations in fishes from subarctic lakes in Yukon Territory, *Can. J. Fish. Aquat. Sci.*, 55, 869–881, 1998.

Kidd, K.A., Schindler, D.W., Muir, D.C.G., Lockhart, W.L., and Hesslein, R.H., High concentrations of toxaphene in fishes from a subartic lake, *Science*, 269, 240–242, 1995.

Kiffney, P.M. and Clements, W.H., Bioaccumulation of heavy metals by benthic invertebrates at the Arkansas River, Colorado, U.S.A., *Environ. Toxicol. Chem.*, 12, 1507–1517, 1993.

Kiriluk, R.M., Servos, M.R., Whittle, D.M., Cabana, G., and Rasmussen, J.B., Using ratios of stable nitrogen and carbon isotopes to characterize the biomagnification of DDE, mirex, and PCB in a Lake Ontario pelagic food web, *Can. J. Fish. Aquat. Sci.*, 52, 2660–2674, 1995.

Kozie, K.D. and Anderson, R.K., Productivity, diet, and environmental contaminants in bald eagles nesting near the Wisconsin shoreline of Lake Superior, *Arch. Environ. Contam. Toxicol.*, 20, 41–48, 1991.

Krummel, E.M., Macdonald, R.W., Kimpe, L.E., Gregory-Eaves, I., Demers, M.J., Smol, J.P., Finney, B., and Blais, J.M., Delivery of pollutants by spawning salmon—Fish dump toxic industrial compounds in Alaskan lakes on their return from the ocean, *Nature*, 425, 255–256, 2003.

Kucklick, J.R., Harvey, H.R., Ostrom, P.H., Ostrom, N.E., and Baker, J.E., Organochlorine dynamics in the pelagic food web of Lake Baikal, *Environ. Toxicol. Chem.*, 15, 1388–1400, 1996.

Landrum, P.F., Lee, H., II, and Lydy, M.J., Toxicokinetics in aquatic systems: Model comparisons and use in hazard assessment, *Environ. Toxicol. Chem.*, 11, 1709–1725, 1992.

Langevoord, M., Kraak, M.H.S., Kraal, M.H., and Davids, C., Importance of prey choice for Cd uptake by carp (*Cyprinus carpio*) fingerlings, *J. N. Amer. Benthol. Soc.*, 14, 423–429, 1995.

Larsson, P., Collvin, L., and Meyer, G., Lake productivity and water chemistry as governors of the uptake of persistent pollutants in fish, *Environ. Sci. Technol.*, 26, 346–352, 1992.

Leonards, P.E.G., Zierikzee, Y., Brinkman, U.A.Th., Cofino, W.P., van Straalen, N.M., and van Hattum, B., The selective dietary accumulation of planar polychlorinated biphenyls in the otter (*Lutra lutra*), *Environ. Toxicol. Chem.*, 16, 1807–1815, 1997.

Livingston, R.J., Trophic response of fishes to habitat variability in coastal seagrass systems, *Ecology*, 65, 1258–1275, 1984.

Macdonald, C.R., Metcalfe, C.D., Balch, G.C., and Metcalfe, T.L., Distribution of PCB congeners in seven lake systems: Interactions between sediment and food-web transport, *Environ. Toxicol. Chem.*, 12, 1991–2003, 1993.

Maul, J.D., Belden, J.B., Schwab, B.A., Whiles, M.R., Spears, B., Farris, J.L., and Lydy, M.J., Bioaccumulation and trophic transfer of polychlorinated biphenyls by aquatic and terrestrial insects to tree swallows (*Tachycineta bicolor*), *Environ. Toxicol. Chem.*, 25, 1017–1025, 2006.

Mazak, E.J., MacIsaac, H.J., Servos, M.R., and Hesslein, R., Influence of feeding habits on organochlorine contaminant accumulation in waterfowl on the Great Lakes, *Ecol. Appl.*, 7, 1133–1143, 1997.

McGeer, J.C., Brix, K.V., Skeaff, J.M., Deforest, D.K., Brigham, S.I., Adams, W.J., and Green, A., Inverse relationship between bioconcentration factor and exposure concentration for metals: Implications for hazard assessment of metals in the aquatic environment, *Environ. Toxicol. Chem.*, 22, 1017–1037, 2003.

Minawaga, M. and Wada, E., Stepwise enrichment of ^{15}N along food chains: Further evidence and the relation between δ ^{15}N and animal age, *Geochim. Cosmochim. Acta.*, 48, 1135–1140, 1984.

Monteiro, L.R., and Furness, R.W., Accelerated increase in mercury contamination in north Atlantic mesopelagic food chains as indicated by time series of seabird feathers, *Environ. Toxicol. Chem.*, 16, 2489–2493, 1997.

Moriarty, F. and Walker, C.H., Bioaccumulation in food chains—a rational approach, *Ecotox. Environ. Saf.*, 13, 208–215, 1987.

Moriarty, F., Hanson, H.M., and Freestone, P., Limitations of body burden as an index of environmental contamination: Heavy metals in fish *Cottus gobio* L. from the River Ecclesbourne, Derbyshire, *Environ. Pollut.*, 34, 297–320, 1984.

Morrison, H.A., Gobas, F.A.P.C., Lazar, R., Whittle, D.M., and Haffner, G.D., Development and verification of a benthic/pelagic food web bioaccumulation model for PCB congeners in western Lake Erie, *Environ. Sci. Technol.*, 31, 3267–3273, 1997.

Morrison, H.A., Whittle, D.M., and Haffner, G.D., A comparison of the transport and fate of polychlorinated biphenyl congeners in three Great Lakes food webs, *Environ. Toxicol. Chem.*, 21, 683–692, 2002.

Mount, D.R., Barth, A.K., Garrison, T.D., Barten, K.A., and Hockett, J.R., Dietary and waterborne exposure of rainbow trout (*Oncorhynchus mykiss*) to copper, cadmium, lead and zinc using a live diet, *Environ. Toxicol. Chem.*, 13, 2031–2041, 1994.

Muir, D.C.G., Norstrom, R., and Simon, M., Organochlorine contaminants in arctic marine food chains: Accumulation of specific polychlorinated biphenyls and chlordane-related compounds, *Environ. Sci. Technol.*, 22, 1071–1979, 1988.

Munger, C. and Hare, L., Relative importance of water and food as cadmium sources to an aquatic insect (*Chaoborus punctipennis*): Implications for predicting Cd bioaccumulation in nature, *Environ. Sci. Technol.*, 31, 891–895, 1997.

Munn, M.D. and Gruber, S.J., The relationship between land use and organochlorine compounds in streambed sediment and fish in the Central Columbia Plateau, Washington and Idaho, USA, *Environ. Toxicol. Chem.*, 16, 1877–1887, 1997.

Newman, M.C., Regression analysis of log-transformed data: Statistical bias and its correction, *Environ. Toxicol. Chem.*, 12, 1129–1133, 1993.

Odum, E.P., Energy flow in ecosystems: A historical review, *Amer. Zool.*, 8, 11–18, 1968.

Oliver, B.G. and Niemi, A.J., Trophodynamic analysis of polychlorinated biphenyl congeners and other chlorinated hydrocarbons in the Lake Ontario ecosystem, *Environ. Sci. Technol.* 22, 338–397, 1988.

Owens, J.W., Swanson, S.M., and Birkolz, D.A., Bioaccumulation of 2,3,7,8-tetrachlorodibenzo-p-dioxin, 2,3,7,8-tetrachlorodibenzofuran and extractable organic chlorine at a bleached-kraft mill site in a northern Canadian river system, *Environ. Toxicol. Chem.*, 13, 343–354, 1994.

Paterson, M.J., Muir, D.C.G., Rosenberg, B., Fee, E.J., Anema, C., and Franzin, W., Does lake size affect concentrations of atmospherically derived polychlorinated biphenyls in water, sediment, zooplankton, and fish? *Can. J. Fish. Aquat. Sci.*, 55, 544–553, 1998.

Peterson, C.H., Summerson, H.C., Thomson, E., Lenihan, H.S., Grabowski, J., Manning, L., Micheli, F., and Johnson, G., Synthesis of linkages between benthic and fish communities as a key to protecting essential fish habitat, *Bull. Mar. Sci.*, 66, 759–774, 2000.

Prusha, B.A. and Clements, W.H., Landscape attributes, dissolved organic C, and metal bioaccumulation in aquatic macroinvertebrates (Arkansas River Basin, Colorado), *J. N. Amer. Benthol. Soc.*, 23, 327–339, 2004.

Quinlan, R., Paterson, A.M., Hall, R.I., Dillon, P.J., Wilkinson, A.N., Cumming, B.F., Douglas, M.S.V., and Smol, J.P., A landscape approach to examining spatial patterns of limnological variables and long-term environmental change in a southern Canadian Lake District, *Freshw. Biol.*, 48, 1676–1697, 2003.

Ramade, F., *Ecotoxicology*, John Wiley & Sons, New York, 1987.

Rasmussen, J.B., Rowan, D.J., Lean, D.R.S., and Carey, J.H., Food chain structure in Ontario lakes determines PCB levels in lake trout (*Salvelinus namaycush*) and other pelagic fish, *Can. J. Fish. Aquat. Sci.*, 47, 2030–2038, 1990.

Reynoldson, T.B., Interactions between sediment contaminants and benthic organisms, *Hydrobiologia*, 149, 53–66, 1987.

Ridal, J.J., Mazumder, A., and Lean, D.R.S., Effects of nutrient loading and planktivory on the accumulation of organochlorine pesticides in aquatic food chains, *Environ. Toxicol. Chem.*, 20, 1312–1319, 2001.

Rolff, C., Broman, D., Näf, C., and Zebühr, Y., Potential biomagnification of PCDD/Fs—New possibilities for quantitative assessment using stable isotope trophic position, *Chemosphere*, 27, 461–468, 1993.

Russell, R.W., Gobas, F.A.P.C., and Haffner, D.G., Role of chemical and ecological factors in trophic transfer of organic chemicals in aquatic food webs, *Environ. Toxicol. Chem.*, 18, 1250–1257, 1999.

Stemberger, R.S. and Chen, C.Y., Fish tissue metals and zooplankton assemblages of northeastern lakes, *Can. J. Fish. Aquat. Sci.*, 55, 339–352, 1998.

Stow, C.A., Carpenter, S.R., Eby, L.A., Amrhein, J.F., and Hesselberg, R.J., Evidence that PCBs are approaching stable concentrations in Lake Michigan fishes, *Ecol. Appl.*, 5, 248–260, 1995.

Swanson, H.K., Johnston, T.A., Leggett, W.C., Bodaly, R.A., Doucett, R.R., and Cunjak, R.A., Trophic positions and mercury bioaccumulation in rainbow smelt (*Osmerus mordax*) and native forage fishes in northwestern Ontario lakes, *Ecosystems*, 6, 289–299, 2003.

Talmage, S.S. and Walton, B.T., Food chain transfer and potential renal toxicity of mercury to small mammals at a contaminated terrestrial field site, *Ecotoxicology*, 2, 243–256, 1993.

Taylor, W.D., Carey, J.H., Lean, D.R.S., and McQueen, D.J., Organochlorine concentrations in the plankton of lakes in southern Ontario and their relationship to plankton biomass, *Can. J. Fish. Aquat. Sci.*, 48, 1960–1966, 1991.

Thomann, R.V., Equilibrium model of fate of microcontaminants in diverse aquatic food chains, *Can. J. Fish. Aquat. Sci.*, 38, 280–296, 1981.

Thomann, R.V., Bioaccumulation model of organic chemical distribution in aquatic food chains, *Environ. Sci. Technol.*, 23, 699–707, 1989.

Thomann, R.V. and Komlos, J., Model of biota-sediment accumulation factor for polycyclic aromatic hydrocarbons, *Environ. Toxicol. Chem.*, 18, 1060–1068, 1999.

Thomann, R.V., Connolly, J.P., and Parkerton, T.F., An equilibrium model of organic chemical accumulation in aquatic food webs with sediment interaction, *Environ. Toxicol. Chem.*, 11, 615–629, 1992.

Toll, J.E., Tear, L.M., DeForest, D.K., Brix, K.V., and Adams, W.J., Setting site-specific water-quality standards by using tissue residue criteria and bioaccumulation data. Part 1. Methodology, *Environ. Toxicol. Chem.*, 24, 224–230, 2005.

Trowbridge, A.G. and Swackhamer, D.L., Preferential biomagnification of aryl hydrocarbon hydroxylase-inducing polychlorinated biphenyl congeners in the Lake Michigan, USA, lower food web, *Environ. Toxicol. Chem.*, 21, 334–341, 2002.

van Hattum, B., Timmermans, K.R., and Govers, H.A., Abiotic and biotic factors influencing in situ trace metal levels in macroinvertebrates in freshwater ecosystems, *Environ. Toxicol. Chem.*, 10, 275–292, 1991.

Wallace, W.G., Lopez, G.R., and Levinton, J.S., Cadmium resistance in an oligochaete and its effect on cadmium trophic transfer to an omnivorous shrimp, *Mar. Ecol. Prog. Ser.*, 172, 225–237, 1998.

Wallberg, P. and Andersson, A., Transfer of carbon and a polychlorinated biphenyl through the pelagic microbial food web in a coastal ecosystem, *Environ. Toxicol. Chem.*, 19, 827–835, 2000.

Wallberg, P., Jonsson, P.R., and Andersson, A., Trophic transfer and passive uptake of a polychlorinated biphenyl in experimental marine microbial communities, *Environ. Toxicol. Chem.*, 20, 2158–2164, 2001.

Wong, A.H.K., McQueen, D.J., Williams, D.D., and Demers, E., Transfer of mercury from benthic invertebrates to fishes in lakes with contrasting fish community structures, *Can. J. Fish. Aquat. Sci.*, 54, 1320–1330, 1997.

Woodward, D.F., Brumbaugh, W.G., DeLonay, A.J., Little, E.E., and Smith, C.E., Effects on rainbow trout of a metals-contaminated diet of benthic invertebrates from the Clark Fork River, Montana, *Trans. Amer. Fish. Soc.*, 123, 51–62, 1994.

Woodward, D.F., Farag, A.M., Bergman, H.L., Delonay, A.J., Little, E.E., Smith, C.E., and Barrows, F.T., Metals-contaminated benthic invertebrates in the Clark Fork River, Montana: Effects on age-0 brown trout and rainbow trout, *Can. J. Fish. Aquat. Sci.*, 52, 1994–2004, 1995.

Xie, X.D., Norra, S., Berner, Z., and Stuben, D., A GIS-supported multivariate statistical analysis of relationships among stream water chemistry, geology and land use in Baden-Wurttemberg, Germany, *Water Air Soil Pollut.*, 167, 39–57, 2005.

Young, R.G., Quarterman, A.J., Eyles, R.F., Smith, R.A., and Bowden, W.B., Water quality and thermal regime of the Motueka River: influences of land cover, geology and position in the catchment, *New Zeal. J. Mar. Freshw. Res.*, 39, 803–825, 2005.

Zaranko, D.T., Griffiths, R.W., and Kaushik, N.K., Biomagnification of polychlorinated biphenyls through a riverine food web, *Environ. Toxicol. Chem.*, 16, 1463–1471, 1997.

35 Effects of Global Atmospheric Stressors on Ecosystem Processes

35.1 INTRODUCTION

Global atmospheric stressors create widely distributed physical and chemical perturbations that impact the structure and function of ecosystems. In addition to their ubiquitous distribution and diffuse sources, global atmospheric stressors are especially problematic because they are not restricted by geopolitical boundaries. In Chapter 26, we described the effects of global atmospheric stressors on abundance, species diversity, community composition, and other structural characteristics. Here we will focus on descriptive and experimental studies that assess effects of increased CO_2, N deposition, acidification, and ultraviolet radiation (UVR) on the function of aquatic and terrestrial ecosystems.

35.2 NITROGEN DEPOSITION AND ACIDIFICATION

In Chapter 30, we described biogeochemical cycles and the important processes that control movement of elements in aquatic and terrestrial ecosystems. Significant increases in the global reservoirs of C, N, and S as a result of combustion of fossil fuels and agricultural/land use changes disrupt these natural cycles and have contributed to a variety of local and global environmental concerns. Global emissions of biologically reactive N compounds (e.g., NH_3, NH_4, HNO_3, and NO_3) have increased from about 15 teragrams (Tg) in 1860 to more than 165 Tg in 2000 (Galloway et al. 2003). Although effects of increased N deposition have not attracted the same attention from scientists and the public as other global atmospheric stressors such as chlorofluorocarbons (CFCs) and CO_2, N poses serious threats to ecosystem processes. Potential negative effects of excess N on forest ecosystems were first described by Nihlgard (1985). Biologically reactive N compounds that accumulate in the atmosphere are rapidly deposited on the earth's surface where they can affect net primary productivity (NPP) (Aber et al. 1995), disrupt N dynamics in soils (Gundersen et al. 1998), and contribute to eutrophication (Rabalais et al. 2002), acidification (Vitousek 1994), and subsequent loss of biological diversity (Stevens et al. 2004). In addition, N_2O is a potent greenhouse gas that contributes to global climate change. Because the rates of production of reactive N in the biosphere greatly exceed rates of removal by denitrification, biologically active N rapidly accumulates in the environment. Predicting effects of N deposition on aquatic and terrestrial ecosystems is complicated by variation in regional climate, hydrologic characteristics, vegetation type, and other sources of anthropogenic disturbance (Aber et al. 2003). Assessing effects of N on ecosystems is also complicated because deposition often co-occurs with other stressors, such as heavy metals (Gawel et al. 1996). Finally, input and output of N are not necessarily coupled in all ecosystems, and leaching of nitrate will depend on nutrient status (Gundersen et al. 1998).

35.2.1 THE NITROGEN CASCADE

Accumulation, transfer, and denitrification of biologically reactive N compounds through the biosphere and the changes that result have been termed the nitrogen cascade (Table 35.1) (Galloway et al. 2003). Forests and grassland ecosystems, especially the soil components, are major

TABLE 35.1
Factors That Influence the Nitrogen Cascade in Aquatic and Terrestrial Ecosystems

Ecosystem	Accumulation Potential	Transfer Potential	Denitrification Potential	Biological Effects
Grasslands and forests	High	Moderate (high in some places)	Low	Biodiversity; NPP; mortality; groundwater
Freshwater	Low; higher in sediments	Very high	Moderate to high	Biodiversity; altered community structure; eutrophication
Coastal marine	Low to moderate; higher in sediments	Moderate	High	Biodiversity; altered community structure; algal blooms

Source: Modified from Table 1 in Galloway et al. (2003).

reservoirs for N. Because output of N in undisturbed forest and grassland ecosystems is generally quite low, residence time can be many years. Forest ecosystems are often N limited, and therefore N is cycled internally with little export to surface water, groundwater, or the atmosphere. As excess N deposition increases, other environmental factors will limit NPP and unused N leaches below the rooting zone, a process known as nitrogen saturation (Aber et al. 1995). Land use changes in forests and grasslands also have the potential to significantly alter internal N cycling and increase the export of N to aquatic ecosystems and the atmosphere.

Fertilizers and runoff associated with agricultural and urban areas are the primary contributors of N to aquatic ecosystems and have been considered in previous chapters. However, nitrogen oxides (NO_x) from fossil fuel combustion account for about 25% of the reactive N in the environment. Although there is relatively little storage of N in overlying water, sediments represent a significant reservoir of N in aquatic ecosystems. Enrichment of aquatic ecosystems by N deposition can result in eutrophication, anoxia, and reduced biodiversity. Although natural aquatic ecosystems are highly retentive of N, this capacity for internal processing can be exceeded, especially in disturbed habitats, and downstream transport of N can contribute to eutrophication of coastal areas. Transport of N from rivers to coastal areas is generally regarded as one of the most serious threats to marine ecosystems (Rabalais et al. 2002). Accumulation potential of N in estuaries is relatively low; however, as with freshwater ecosystems coastal marine sediments may represent a significant N reservoir. Because of large amounts of organic material and low concentrations of dissolved oxygen in sediments, coastal marine ecosystems also have the greatest potential for conversion of reactive N to N_2 by denitrification, a process that is often enhanced by excess reactive N. In fact, denitrification in rivers and estuaries greatly reduces the amount of N transported from terrestrial to coastal and offshore areas (Galloway et al. 2003).

35.2.2 Effects of N Deposition and Acidification in Aquatic Ecosystems

Although most of the earlier studies of atmospheric deposition focused on ecosystem effects of sulfur (S), attention in North America and Europe has shifted to concerns about effects of atmospheric N deposition. Atmospheric deposition of S and N from the mid-1980s to the mid-1990s show contrasting temporal patterns in North America (Sirois et al. 2001). While declines of SO_4 in precipitation were observed, concentrations of NO_3 and NH_4 generally remained constant or increased. These changes corresponded to decreased emissions of SO_4 and increased emissions of NO_x over this same period. As previously N-limited forests became saturated, NO_3 is released to watersheds causing eutrophication, acidification and reductions in acid-neutralizing capacity (ANC; defined as the capacity of the watershed to withstand strong acid inputs based on the difference between total cations and total anions). In fact, one of the most consistent indicators of N saturation in ecosystems is an increase in concentrations of NO_3 in stream water. Detecting these trends in streams requires access to long-term data that are often unavailable for many watersheds. Much of the research that describes biogeochemical responses of streams to N deposition has been conducted in Europe and the northeastern United States. In particular, experimental studies and long-term monitoring of SO_4 and NO_3 in stream water at Hubbard Brook Experimental Forest documented acidification effects on watersheds and significant decreases in base cation concentrations (Likens et al. 1996).

Assuming that ecosystems can only tolerate a certain level of acidification before important processes become disrupted, defining the threshold point at which the capacity of an ecosystem to withstand additional inputs is exceeded is an important exercise. The concept of critical soil acidification load for a watershed is analogous to assimilative capacity in ecotoxicology. The idea assumes that once the net neutralizing capacity of an ecosystem is reached, additional atmospheric deposition will result in soil acidification. Critical soil acidification load is determined primarily by a balance between the weathering of base cations and leaching associated with deposition of SO_4 and NO_x. Moayeri et al. (2001) developed a model to calculate critical soil acidification loads for a watershed in Ontario, Canada. Model results showed that soil acidification would occur faster in a harvested watershed compared to an old-growth watershed.

Even relatively remote areas located away from sources of N can experience effects of N deposition. In general, water quality in alpine and subalpine lakes of the Rocky Mountains is relatively pristine, with low background concentrations of NO_3 (Williams and Tonnessen 2000). A survey of 44 high-elevation lakes (>3000 m a.s.l.) located on both sides of the continental divide in Colorado indicated that higher NO_3 concentration and lower ANC of eastern lakes corresponded with greater atmospheric N deposition (Baron et al. 2000). Long-term changes in community composition and productivity of diatoms, as revealed by paleolimnological records of lake sediments, were consistent with increased eutrophication. These increases in N deposition and shifts in diatom flora corresponded with increases in urban, agricultural, and industrial development on the Front Range of Colorado. Williams and Tonnessen (2000) used long-term monitoring data and synoptic surveys of 91 high-elevation lakes in the central Rocky Mountains to establish critical loads for inorganic N deposition. Episodic acidification of sensitive headwater catchments in remote Wilderness Areas has resulted from increased wet deposition of N.

The concept of N saturation, originally developed to describe export of N in forest ecosystems, may also apply to watersheds. Similar to the acidification effects associated with S emissions, sensitivity of watersheds to N deposition will be influenced by underlying geology, hydrologic characteristics, soil type, and vegetation. Alpine and high-elevation watersheds, especially those located above treeline, may be especially susceptible to N deposition because of their relatively nonreactive bedrock, short growing season, and limited vegetation (Fenn et al. 1998). Williams et al. (1996a) reported that N deposition in the catchments of the Colorado Front Range was similar to that in other well-studied northeastern locations, including Hubbard Brook (New Hampshire) and Acadia National Park (Maine). A shift in nutrient dynamics from N-limited to N-saturated conditions was also observed in these high-elevation watersheds as a result of increased anthropogenic N deposition. Williams et al. (1996a) suggested that N saturation in high-elevation catchments may serve as an early warning of disruption in N cycling.

Ecosystem responses of oligotrophic lakes to N deposition will also depend on nutrient conditions and the history of NO_3 availability. Nydick et al. (2004) measured effects of NO_3 enrichment in two alpine lakes with very different background levels of N. Enrichment significantly increased photosynthetic rate and chlorophyll *a* in a low N lake, but had no effects on a high N lake. Both NO_3 and PO_4 additions were necessary to increase productivity in the high N lake. Nydick et al. (2004) also reported that despite relatively little effect on benthic algal biomass, epilithon, surface sediment, and subsurface sediment accounted for 57–92% of the NO_3 uptake, indicating the importance of benthic processes in these lakes.

Ecosystem-level effects of SO_2 and NO_x deposition on acidification of aquatic ecosystems has been studied extensively in Europe and North America (Hornung and Reynolds 1995, Likens et al. 1996, Schindler 1988). Much of the research conducted in the Experimental Lakes Area (ELA) focused on ecosystem processes, especially primary productivity (Schindler 1987, Schindler et al. 1985). Results of these and other studies of acidification showed reduced rates of primary and secondary production, decomposition, and nutrient cycling. Although the whole lake experimental studies conducted at Little Rock Lake in northern Wisconsin focused on community responses to acidification (Gonzalez and Frost 1994), Frost et al. (1999) speculated that loss of sensitive species and shifts in community composition would diminish the ability of acidified lakes to maintain system function.

Effects of acidification on organic matter processing have been examined experimentally in natural and artificial streams. Burton et al. (1985) measured effects of acidification on decomposition of white birch and sugar maple in experimental stream channels (Figure 35.1). Reduced decomposition rates in acidified stream channels were attributed to lower density of macroinvertebrates, particularly shredder caddisflies and detritivorous isopods. It is interesting to note that significant effects of acidification were not observed until relatively late in the study (>80 days), demonstrating the importance of long-term experiments. However, results of long-term experiments do not necessarily demonstrate significant ecological effects. Smock and Gazzera (1996)

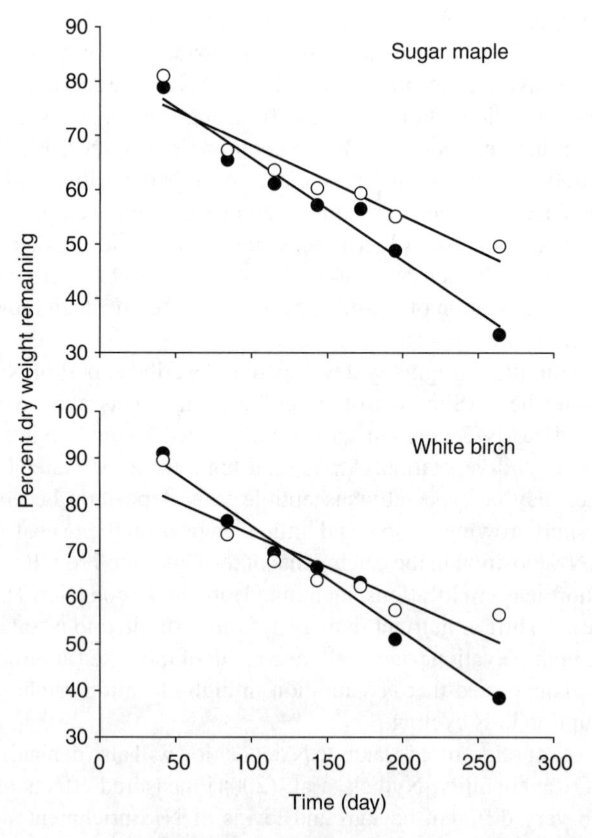

FIGURE 35.1 Decomposition rate (as percent dry weight remaining) of sugar maple and white birch in reference (closed symbols) and acidified (open symbols) stream channels. (Data from Table 2 in Burton et al. (1985).)

introduced H_2SO_4 to a low gradient, blackwater stream in Virginia (USA). Monthly additions over a 1-year period reduced benthic microbial respiration, but had no effect on processing rates of red maple leaves or abundance of macroinvertebrates. These results suggest that ecological effects of acidification on ecosystem processes will likely vary with location. Blackwater streams with naturally high levels of tannins and organic acids that typify this region are relatively insensitive to acidification.

35.2.3 Effects of N Deposition and Acidification in Terrestrial Ecosystems

Although negative effects of NO_x deposition on lakes and streams have been frequently observed, changes in terrestrial ecosystems, especially forests, have received the most attention. Deposition rates of N to forest ecosystems range from 2 kg N/ha/year in remote reference areas to 40 kg N/ha/year in forests downwind of industrial sources (Aber et al. 1989). A large-scale survey of 68 grassland ecosystems across Great Britain showed a strong negative relationship between species richness and N deposition (Stevens et al. 2004). Although this paper focused on changes at the community level, it served to illustrate the widespread nature of this problem. At the current rate of N deposition in central Europe (17 kg N/ha/year), these researchers estimated that species richness was reduced by approximately 23% compared to grassland ecosystems receiving the lowest levels. Similar spatial gradients in N deposition and ecological effects are also evident in the northeastern United States. For example, relatively high rates of deposition have been measured in southern New York and Pennsylvania (12 kg N/ha/year), whereas low rates of deposition are measured in eastern Maine (<4 kg N/ha/year) (Aber et al. 2003). Although it is well documented that alpine and other high-elevation ecosystems are at greater risk from N pollution because of higher rates of deposition and increased sensitivity, other landscape factors that influence N deposition are not well understood. Direct measurement of wet and dry deposition rates in forest ecosystems is difficult. Weathers et al. (2000) developed a model to predict the influence of several landscape factors on N deposition in montane ecosystems. Using concentrations of Pb in forest floor soils as an index of N deposition, these investigators quantified effects of forest edges, elevation, aspect, and vegetation type on N deposition in montane forests.

Fenn et al. (1998) published a comprehensive review of factors that predispose ecosystems to N saturation and the general ecosystem responses to N deposition. Because of the intimate linkages between forests and surrounding watersheds, research describing transport of N in terrestrial ecosystems has important implications for N transport to lakes and streams. For example, export of base cations, increased acidification, and elevated levels of nitrate and aluminum in streams are likely to be associated with N deposition in forests. Aber et al. (1989) provided a formal definition of the term N saturation and developed a hypothesized time course describing responses of forest ecosystems to N deposition (Figure 35.2). At a critical stage in this sequence of events, forests will likely become net sources of N rather than sinks. Aber et al. (1989) also described the limited capacity of some forest ecosystems to assimilate excess N and the potential interactions of N deposition with other atmospheric stressors such as ozone, sulfate, and heavy metals.

35.2.3.1 The NITREX Project

One of the most comprehensive and spatially extensive experimental assessments of N deposition was conducted in several coniferous forests in northeastern Europe. The NITREX (nitrogen saturation experiments) project involved experimental additions of N to sites along a gradient of N pollution to examine changes in structural and functional characteristics (Emmett et al. 1998). Input rates of N ranged from 13 to 59 kg/ha/year across sites. Similar to the lotic intersite nitrogen experiment (LINX) experiments described in Chapter 30, an important objective of the NITREX project was to compare responses to N addition among sites. Experimental treatments involved both enhanced

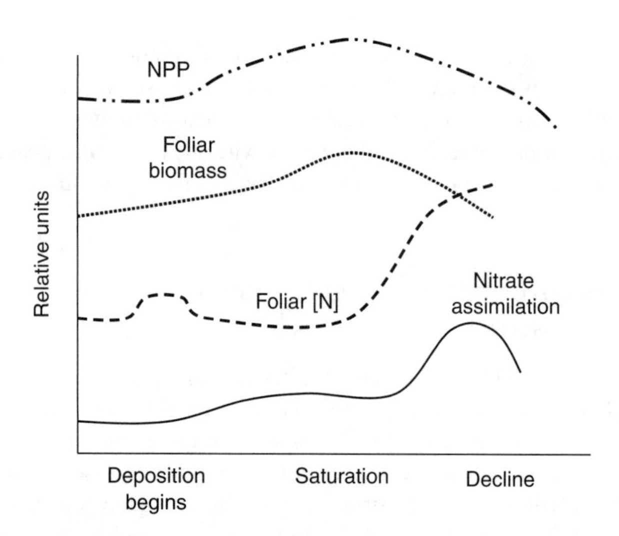

FIGURE 35.2 Predicted time course for changes in NPP, biomass, foliar N concentration, and nitrate (NO_3) assimilation in a forest ecosystem in response to chronic deposition of N. (Modified from Figure 1 in Aber et al. (1989).)

N input at sites with naturally low atmospheric deposition and reduced N input (using exclusion roofs and an ion exchange system) at sites with high deposition. Researchers predicted that N addition in N-limited forests would have relatively little effect on leaching because of the high capacity for N retention in these systems. Most of the observed responses to N manipulations were consistent with expectations, with reductions in N status and NO_3 leaching occurring at sites where N deposition was reduced and increases occurring at sites where N deposition was enhanced (Gundersen et al. 1998). Shifts in N status and cycling rates following treatments generally supported the N saturation hypothesis (Aber et al. 1989).

The most consistent responses to enhanced N deposition were changes in water quality, particularly increased NO_3. A strong relationship between N leaching and N status suggested that the distinction between N-limited and N-saturated forests could be quantitatively demonstrated (Gundersen et al. 1998). Nitrate leaching was observed within the first year of the experiments, whereas biological responses were often delayed. The strong link between N deposition and acidification was also demonstrated in the NITREX project. N additions caused a decrease in ANC, whereas experimental reductions in N caused an increase in ANC (Emmett et al. 1998). Ratios of carbon to nitrogen (C:N) were also good predictors of the onset of NO_3 leaching (Figure 35.3) because nitrification rates are stimulated as C:N declines, resulting in a decrease in the retention efficiency of N (Emmett et al. 1998). These results suggest a simple threshold response of N export. At C:N greater than 24, only a small proportion of nitrate leaches (approximately 10%); however, as C:N decreases, a rapid increase in the proportion of NO_3 leached was observed.

Comparison of N dynamics at N-limited and N-saturated sites showed that microbial cycling of C and N was characterized by low NH_4 transformation and respiration rates at N-limited sites. Surprisingly, despite a wide range of variation in N deposition rates, N transformation showed relatively little variation. Gross mineralization and immobilization rates of NH_4 were highly correlated with respiration rates across ecosystems (Figure 35.4), indicating the important linkage between C and N dynamics and showing that simple measurements of CO_2 in soils could potentially serve as a measure of N cycling (Tietema 1998).

The potential damaging effects of atmospheric deposition on forest productivity were clearly illustrated by the NITREX project. Experimental reduction of N and sulfur inputs to a highly N-saturated site resulted in a 50% increase in tree growth (Emmett et al. 1998). However, in general, ecosystem responses to experimental manipulation of N were relatively modest and required longer periods

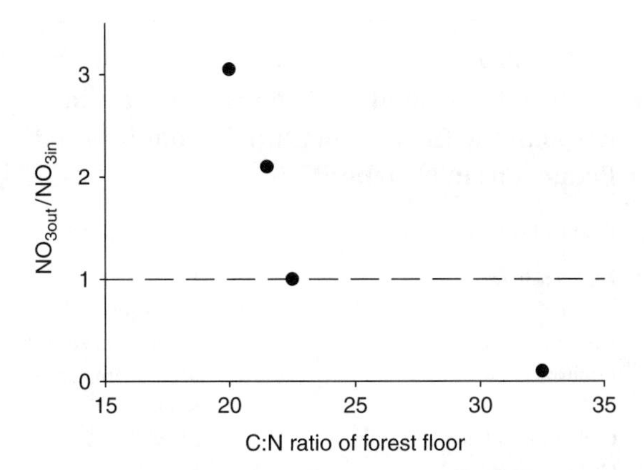

FIGURE 35.3 Relationship between N leaching and carbon:nitrogen (C:N) ratios in the forest floor. Data are from experimental results of the NITREX project in northwestern Europe. (Modified from Figure 5 in Emmett et al. (1998).)

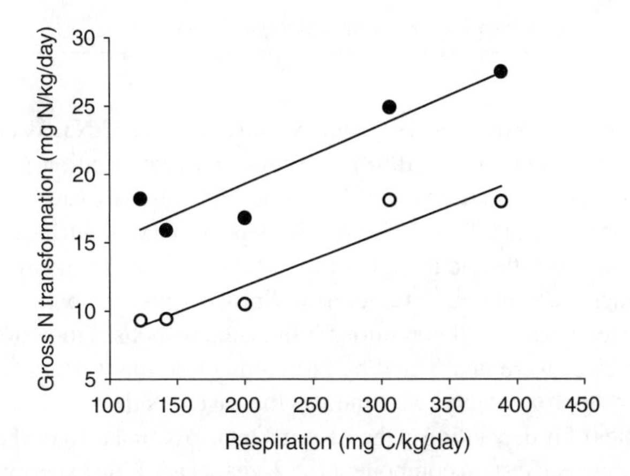

FIGURE 35.4 Relationship between gross N transformation rates and respiration in forests soils from the NITREX experiment. Solid circles = gross NH_4 mineralization rate; open circles = gross NH_4 immobilization rate. (Data from Tables 2 and 3 in Tietema (1998).)

of time. Boxman et al. (1998) commented that it was "remarkable that no ecosystem components have responded with increasing vitality to the high N levels in the initially N limited, oligotrophic forests." Mass loss in litter bags also increased along the gradient of N status, but effects of N manipulation were not significant. Responses of soil fauna to N treatments were generally less than initial differences among sites along the N gradient. The modest biological responses to N treatments may have resulted from the relatively short duration of these experiments. Gundersen et al. (1998) provided estimates of the amount of time required for several chemical and biological responses to N treatment (Table 35.2). In general, decomposition rates and changes in NPP were relatively slow processes compared with NO_3 leaching. These results again highlight the importance of conducting long-term manipulations for assessing ecosystem effects of N deposition.

35.2.3.2 Variation in Responses to N Deposition among Ecosystems

Responses of forest ecosystems to N deposition are often variable, and factors that control differences in N retention and export are not completely understood. As described above, C:N ratios in soil are

TABLE 35.2
Predicted Timing of Selected Forest Ecosystem Responses to Changes in Chronic Additions and Reductions in N Deposition

Pool or Process	Response
NO_3 leaching	Fast
Net mineralization	Intermediate
Decomposition	Slow intermediate
Denitrification	Slow intermediate
NPP	Slow
C:N of forest floor	Slow
C:N of mineral soil	Very slow

Fast = 1 year; intermediate = 2–4 years; slow $\geq$ 5 years.

Note: Results are based on experimental manipulations of nitrogen from the NITREX project.

Source: From Table 8 in Gundersen et al. (1998).

linked to the capacity of forest ecosystems to retain N, and decreased C:N may occur under conditions of chronic N deposition (Emmett et al. 1998). Surveys of wet deposition in old-growth stands of Engelmann Spruce showed elevated levels of NO_3 and NH_4 on the eastern side of the continental divide in Colorado (Baron et al. 2000). Greater N deposition was attributed to agricultural and atmospheric sources and was reflected in higher rates of mineralization and nitrification. Subsequent N fertilization experiments conducted in Engelmann Spruce forests on both sides of the continental divide showed that differences in soil conditions influenced responses to N treatments (Rueth et al. 2003). Mineralization rates were unaffected by N treatment on the western side of the continental divide but increased by approximately two times on the eastern side.

Experimental studies of N deposition at the ELA in Ontario (Canada) have been conducted to contrast responses of different ecosystem components. A 2-year N addition experiment (40 kg N/ha/year) was conducted in a boreal forest to compare effects in N-limited "forest islands" with naturally N-saturated lichen outcrops (Lamontagne and Schiff 1999). Responses to N treatments differed between habitat types. In contrast to forests, N-saturated lichen outcrops were highly sensitive to N deposition. After 2 years of treatment, lichen outcrops no longer retained additional N inputs, whereas the proportion of N retained in treated and reference forest–islands was similar (Figure 35.5). These data highlight the role that relatively small habitat patches in a landscape play in controlling ecosystem processes.

Differences in abundance of dominant species will also complicate our ability to predict ecosystem responses to N deposition in forests. For example, species-specific differences in litter quality and rates of decomposition influence N cycling and availability. Surveys conducted along a gradient of atmospheric N deposition in the northeastern United States showed that responses differed between tree species (Lovett and Rueth 1999). Rates of mineralization and nitrification of soils were significantly related to N deposition in maple plots but not in beech plots. Thus, while increased rates of mineralization and nitrification are typical responses to N deposition in forest ecosystems, the species composition of these forests should be considered when developing predictive models. Similarly, changes in community composition of forest ecosystems as a result of natural or anthropogenic disturbance will affect responses to N deposition. Lovett and Rueth (1999) also demonstrated that although descriptive studies do not allow researchers to directly infer causality, comparison of ecosystem responses to chronic N deposition along a gradient can be a useful alternative to experimental treatments.

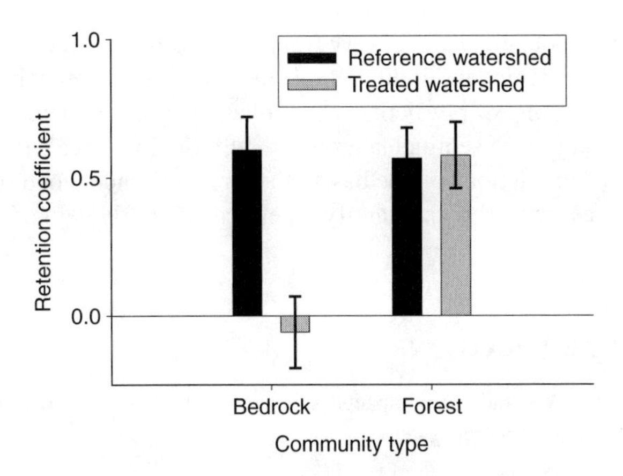

FIGURE 35.5 Responses of lichen bedrock communities and forest communities after 2 years of N additions in the ELA, Ontario, Canada. The figure shows retention coefficients ([Total N Input – Total N Output]/Total N Input). (Data are from Table 4 in Lamontagne and Schiff (1999).)

35.2.4 Ecosystem Recovery from N Deposition

Mitigating the effects of N deposition on forest ecosystems will be challenging because of the widespread distribution of sources and tremendous variation in ecological effects. However, potential for recovery is relatively high if N deposition is significantly reduced. Experiments conducted in The Netherlands and Germany showed rapid improvements in water quality following reductions in N deposition to a watershed (Emmett et al. 1998). Increased tree growth was also observed at two of the three sites where N deposition was reduced (Boxman et al. 1998). Soils are a major sink for excess N in forests, and recent evidence suggests that microbial assimilation of NO_3 may be an important regulator of N retention. Because major reductions in N emissions are unlikely for some areas, management options should also include strategies to enhance the incorporation of N into soils (Fenn et al. 1998).

Recovery of watersheds from the long-term effects of acidification may require many years if base cations are significantly depleted. Reductions in atmospheric deposition of sulfate as a result of the Clean Air Act have been associated with significant improvements in stream water chemistry. However, export of base cations, especially Ca and Mg, from acidified watersheds will likely delay recovery (Likens et al. 1996). Although the loss of base cations has been attributed primarily to prolonged exposure to acid rain and a decline in Ca in precipitation, ecological factors also influence Ca export. Hamburg et al. (2003) measured levels of Ca in the forest floor, abundance of snails (organisms that require Ca for growth), and Ca export in stream water from hardwood forests of various ages. Results showed that Ca concentration in snails, litterfall, and the forest floor and export of Ca in stream water increased with forest age. Calcium mobilization in young stands (4.6–6.0 g Ca/m/year) was much greater than in old stands (0.4 g Ca/m/year), indicating that forest aging significantly influenced Ca dynamics.

Recent amendments to the 1990 Clean Air Act in the United States are expected to have significant effects on air quality and water chemistry across large broad geographic regions. Assessing these changes will require integrated studies of physicochemical and biological responses over large regional areas and for relatively long periods of time. Because long-term monitoring data at a regional level are generally lacking, researchers are often required to use trends from site-specific results to infer regional patterns. Stoddard et al. (1998) analyzed trends in water chemistry from 44 Adirondack and New England (USA) lakes that were sampled from 1982 to 1994. Long-term trends in measures of acidic deposition (SO_4 and NO_3 concentrations in stream water) and watershed responses to acidification (ANC and export of base cations) differed between subregions. In particular, ANC increased over time in New England lakes but decreased in Adirondack Lakes. These results were not expected and indicate that potential for recovery from acid deposition would be considerably less in the Adirondacks. The most significant finding of this research is that even with long-term data and a solid mechanistic understanding of physicochemical relationships, predicting regional trends based on well-defined subpopulations of sentinel lakes is difficult (Stoddard et al. 1998). The challenges of making accurate regional predictions on the basis of a very well-understood phenomenon highlight the potential difficulties associated with quantifying effects of poorly defined stressors such as CO_2 and UV-B radiation.

35.3 ULTRAVIOLET RADIATION

> ... to really demonstrate UV-B radiation impacts at the ecosystem level requires establishing a chain of cause and effect from molecule to ecosystem.
>
> **(Bassman 2004)**

> Ecosystem-level experiments are the only method of detecting UV influences on the myriad of competitive and trophic interactions present in nature.
>
> **(Flint et al. 2003)**

35.3.1 Aquatic Ecosystems

Effects of UVR on ecosystem processes have been studied extensively, especially in pelagic marine and lentic systems. As a result of this comprehensive research effort, there exists sufficient information concerning effects of UVR to develop reasonably detailed ecological risk assessments for certain groups of organisms and processes (Hansen et al. 2003). Several excellent reviews on the effects of UVR have been published recently. Day and Neale (2002) have provided the most comprehensive treatment of UV-B effects in aquatic and terrestrial ecosystems, with an emphasis on primary producers. One of the most consistent observations in studies of UV-B effects on marine and freshwater phytoplankton is reduced primary production (Kinzie et al. 1998, Mostajir et al. 1999, Neale et al. 1998, Smith et al. 1992, Williamson 1995). UV-B causes damage to Photosystem II, and effects of UVR on primary production and other ecosystem processes can be extensive. Gala and Giesy (1991) estimated that exposure to UV-B reduced primary production of a natural assemblage of phytoplankton from Lake Michigan by 25%. Exposure to simulated UV-B reduced photosynthesis by 40% in Georgian Bay (Furgal and Smith 1997). Remarkably, even short-term exposure to surface radiation (e.g., 30 min) can be sufficient to inhibit photosynthesis (Marwood et al. 2000).

Because of proximity to the ozone depletion zone and the intense exposure to UVR during the early austral spring (October–November), considerable research effort has focused on phytoplankton in Antarctic waters, where a 50% reduction in ozone has been documented. Smith et al. (1992) conducted transects in the Antarctic marginal ice zone and reported a 6–12% reduction in primary production associated with ozone depletion during a 6-week cruise. Assuming that this reduction is representative of the entire area and integrating results over the austral spring, Smith et al. (1992) concluded that this change corresponded to an approximately 2% reduction in annual production of the Southern Ocean. Similar experiments conducted in the Weddell Sea showed that productivity of marine phytoplankton decreased as a cumulative function of UV exposure, indicating little evidence of photorepair (Neale et al. 1998). These researchers also documented considerable variation in sensitivity of primary production among sites, and attributed this variation to UV exposure before sampling.

35.3.1.1 Methodological Considerations

With respect to experimental methodology, most aquatic investigations involved the removal of different wavelengths of UVR using various filters, although some have employed experimental enhancement of UVR using lamps. In contrast, studies conducted in terrestrial ecosystems have more commonly employed lamps to enhance UVR. Because of the significantly elevated levels of UV-B in the Southern Hemisphere, researchers have relied primarily on UV exclusion methodology. In contrast, experimental designs in the Northern Hemisphere where UV increases are less dramatic have more commonly employed UV enhancement (Flint et al. 2003). These are important methodological distinctions with significant implications for how results are interpreted. UVR exclusion experiments document effects of current levels of UVR, whereas enhancement experiments attempt to estimate effects of predicted increases. Advantages and disadvantages of the different experimental techniques used to enhance or reduce UVR were described in Chapter 26. In particular, there has been considerable discussion of the artificial wavelength spectra produced by UV lamps. Unrealistic combinations of UV-A, UV-B, and photosynthetically active radiation (PAR) may artificially enhance effects of UV-B (Caldwell and Flint 1997). These problems may be partially addressed by either calculating biological spectral weighting functions or by using modulated lamp systems that measure incoming UV-B and adjust lamp outputs accordingly. Comparisons of ecosystem processes under ambient and UVR-excluded treatments provide important information on effects of current levels of UVR. However, it may be difficult to extrapolate these results to conditions of enhanced UVR because it requires that we make predictions beyond those used in the experiments (Behrenfeld et al. 1995). More importantly, because of the significant influence of PAR on photosynthesis, it is essential that filtering materials that exclude and transmit UV-B allow the same amount of PAR (Flint et al. 2003). Because of these issues, some combination of UV exclusion and enhancement may be necessary to reliably estimate effects of current and increased UVR under conditions of ozone depletion.

The duration of experiments designed to assess UVR effects on primary production is also an important consideration. In contrast to research conducted in terrestrial systems, most investigations of UVR effects in aquatic ecosystems have been limited to relatively short-term experiments. Watkins et al. (2001) measured effects of UVR on epilithic metabolism, pigment concentrations, nutrients, and community composition in a boreal lake over a 4-month period. Although chlorophyll *a* was not affected, photosynthetic rates were increased by 37–46% and shifts in community composition were observed when UVR was eliminated. Most of the observed response was a result of exposure to UV-A. These results strongly support the hypothesis that current levels of UVR penetrating clear-water lakes have detrimental effects on primary productivity. Because the experiments were conducted in summer and fall, investigators were able to document seasonal responses to declining UVR. Although differences among treatments were negligible in fall as a result of lower incident UVR, differences in taxonomic composition persisted.

35.3.1.2 Factors that Influence UV-B Exposure and Effects in Aquatic Ecosystems

Responses of marine and freshwater phytoplankton to UVR are complicated by numerous environmental factors, and quantification of effects on ecosystem processes is often challenging (Marwood et al. 2000). In addition to the elevated UV-B levels in the Southern Ocean, increases in UV-B radiation occur at higher elevations (increasing by approximately 20% for each 1000 m), placing alpine and subalpine ecosystems at considerable risk (Blumthaler and Ambach 1990, Sommaruga 2001). Alpine lakes and streams are also more susceptible to UV-B because they often have naturally low concentrations of light-attenuating dissolved organic material (DOM), which protect communities from exposure (Vinebrooke and Leavitt 1998). UV-B exposure will also be elevated at low latitudes and in tropical ecosystems because of the naturally thin layer of ozone and direct angle of exposure for most of the year (Kinzie et al. 1998). Thus, tropical ecosystems located at higher elevations would be expected to receive significant UV-B exposure. Kinzie et al. (1998) measured effects of UV-B on photosynthesis of benthic and planktonic communities in a tropical alpine lake. Net oxygen production of phytoplankton was actually lower in microcosms exposed to UV-B than in the dark. Effects of UV-B were greater on phytoplankton than on benthic algal mats, which were likely protected by UVR-absorbing amino acids.

Habitat features, behavioral characteristics, and morphological adaptations of organisms will influence exposure and sensitivity to UVR. Microcosm experiments conducted with artificial light to enhance UV-B showed no effects on phytoplankton, zooplankton, periphyton, or macroinvertebrates (De Lange et al. 1999). Some species-specific responses were observed, but overall ecosystem characteristics were unaffected. The lack of a response in this system was attributed to UV-B attenuation resulting from high concentrations of dissolved organic carbon (DOC), which protected organisms from exposure. Interestingly, bioassays conducted with *Daphnia pulex* showed higher growth of organisms fed seston from the control microcosms than organisms consuming seston from UV-B-treated microcosms. These results suggest the intriguing possibility that energy transfer from phytoplankton to zooplankton could be affected by UV-B.

Because penetration of UVR through the water column is dependent on water clarity, factors that influence turbidity, trophic status, and levels of DOM will potentially influence ecosystem responses. Anthropogenic changes in water clarity, such as those resulting from acidification, climate change, or exotic species will also influence UVR exposure. Invasion of exotic filter-feeding zebra and quagga mussels (*Dreissena* spp.), which remove phytoplankton from the water column, has significantly increased water clarity and UVR penetration in lakes. Experiments conducted in Lake Erie (USA) showed that UVR inhibited primary production, but that effects were mediated by N availability (Hiriart et al. 2002). There are also concerns over the sustainability of planktonic food webs in Lake Erie as a result of removal of phytoplankton by dreissenid mussels. The stability of these food webs will be further compromised if UVR affects phytoplankton production. Allen and Smith (2002) observed that UVR significantly inhibited phosphate uptake capacity in plankton, which may result in a potential negative feedback by diminishing P availability in this system.

Vertical profiles of primary production showed that relative effects of UV-A and UV-B on photosynthesis vary with depth and water clarity (Palffy and Voros 2003). As expected, greatest effects on photosynthesis were observed near the surface, but these effects were primarily attributable to UV-A (Figure 35.6). A multiple regression model showed that UVR and vertical light attenuation accounted for 90% of the variation in photoinhibition. Effects of UV-B on ecosystem processes may be ephemeral and change as a result of alterations in community composition and maturity. Santos et al. (1997) measured successional changes in tropical marine diatoms exposed to varying UVR treatments in the field. During initial stages of colonization, primary production was reduced by >40% when exposed to a full solar spectrum of UV-A + UV-B + PAR. These changes corresponded to differences in composition of diatoms among treatments. Effects of UVR treatments were reduced over time, suggesting that diatoms were most sensitive during the initial stages of succession.

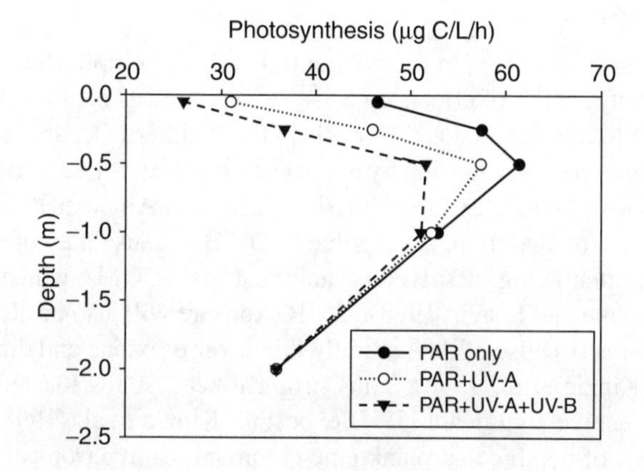

FIGURE 35.6 Vertical profile of phytoplankton primary production in the western basin of Lake Balaton (Central Europe) measured on July 19, 1999. (Data from Table 1 in Palffy and Voros (2003).)

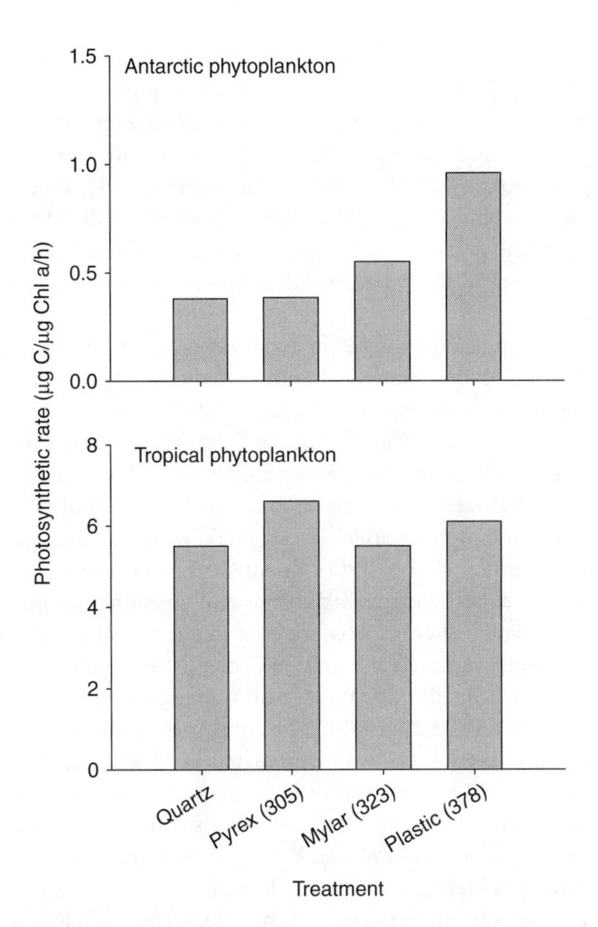

FIGURE 35.7 Photosynthetic rates (μg C/μg Chl a/h) of Antarctic and tropical phytoplankton when solar radiation is filtered using quartz, pyrex, mylar, and plastic film. Numbers in parentheses are the wavelengths (in nanometers) corresponding to 50% transmission for each treatment. (Data from Figure 9 in Helbling et al. (1992).)

As described previously, documented losses of ozone and subsequent increases in UVR have been greatest in Antarctic marine ecosystems. Because phytoplankton in these southern oceans have historically been exposed to relatively low levels of UVR, it is likely that they are especially sensitive to anthropogenic increases. Helbling et al. (1992) compared effects of UVR on photosynthesis of tropical and Antarctic phytoplankton populations. Results showed relatively little effects of eliminating UVR on tropical phytoplankton but dramatic effects on Antarctic organisms (Figure 35.7). These researchers also noted that most of these effects were a result of reducing UV-A, whereas UV-B had relatively minor effects on photosynthesis. In contrast to these findings, Banaszak and Neale (2001) observed that photosynthesis of phytoplankton from a shallow estuarine environment was more strongly inhibited by UV-B than UV-A. Biological weighting functions (BWFs) that quantified effects of different wavelengths on photosynthesis were similar to those derived for Antarctic systems and showed relatively little seasonal variation, despite considerable variation in physicochemical characteristics in this ecosystem (Banaszak and Neale 2001).

35.3.1.3 Comparing Direct and Indirect Effects of UVR on Ecosystem Processes

Because UVR affects both primary producers and consumers, responses to UVR manipulations observed in field experiments are often a combination of direct and indirect effects. Organisms representing different trophic levels will likely show differential sensitivity to UVR. Despite considerable speculation that indirect effects of UVR will be important, relatively few studies have documented these food web responses. Research by Bothwell et al. (1994) was one of the first studies to quantify the importance indirect effects on benthic communities. Although accrual rates of algae were initially inhibited by UVR, changes in abundance of algal consumers (chironomids) mediated these responses. McNamara and Hill (2000) measured effects of UV-B on photosynthesis and food resources available to grazers in experimental streams. These researchers observed a dose–response relationship between UV-B irradiance and photosynthesis in both short- (4-h) and long-term (13-day) experiments.

Direct or indirect UVR-induced changes in food webs can have important consequences for aquatic ecosystems. If UV-B inhibits growth and nutrient uptake of primary producers or alters size and species composition, the quality of food resources for grazers may be affected (Hessen et al. 1997). Tank et al. (2003) measured direct and indirect effects of UVR in four montane lakes of varying water transparency in Jasper National Park, Alberta (Canada). Results of mesocosm experiments using filters showed that UVR altered trophic structure and function of benthic communities, but direct and indirect effects were highly variable among lakes. In contrast to expectations, exposure to UVR generally did not reduce the quality or quantity of food resources to invertebrates. UVR exposure decreased species richness and resulted in lower photosynthetic pigments in organisms from two clear lakes, but other factors such as nutrient concentration and grazers were more important than UVR in structuring communities (Tank and Schindler 2004). Indirect effects of UVR on food webs in a British Columbia (Canada) stream varied among locations, but were generally weak compared to direct effects (Kelly et al. 2003). This study also failed to show effects of UVR on food quality. Vinebrooke and Leavitt (1999) manipulated UVR and density of macroinvertebrates in an oligotrophic alpine lake to test the relative importance of direct and indirect effects of UVR. Responses of primary producers and consumers varied by species and habitat, with greatest effects observed on epilithic standing crop. These researchers speculated that direct effects of UVR would be more important in extreme environments, such as alpine lakes or other stressed ecosystems, where abiotic factors regulate ecosystem processes. The hypothesis that UVR will have greater effects in stressed ecosystems has important implications for understanding potential interactions between UVR and contaminants and will be considered in Section 35.5.3.

Most studies investigating effects of UVR in marine and lentic ecosystems have focused on inhibition of photosynthesis. However, a more comprehensive understanding of the potential ecosystem-level effects of UVR requires that other processes be considered. Mesocosm experiments using natural assemblages of marine phytoplankton showed that exposure to enhanced UV-B significantly affected N transport rates (Mousseau et al. 2000). Research conducted by Behrenfeld et al. (1995) also documented effects of UV-B on N uptake in natural plankton assemblages collected from mid-latitudes of the North Pacific Ocean. Results showed that exclusion of UV-B increased uptake of ammonium and nitrate compared to ambient levels, whereas enhancement of UV-B reduced uptake. These researchers also established dose–response relationships between N uptake and UV-B dose. Results of these analyses showed that rates of N uptake were more sensitive to UV-B than C fixation, suggesting that assessment of effects based exclusively on photosynthesis may underestimate total UV-B damage to ecosystems (Behrenfeld et al. 1995).

35.3.1.4 Effects of UV-B on Ecosystem Processes in Benthic Habitats

Effects of UVR in benthic communities have received considerably less attention than planktonic communities, presumably because these organisms should be protected by overlying water and because UVR does not penetrate into sedimentary habitats. However, benthic communities occupying clear, shallow water environments are likely to be exposed to intense levels of UVR. In addition, Garcia-Pichel and Bebout (1996) reported that UVR penetrated a range of sediment types, with relatively low attenuation in sandy quartz sediments where effects on photosynthetic organisms are likely to be significant. These predictions are consistent with results of experiments measuring UVR effects on benthic algal and meiofaunal communities. Odmark et al. (1998) exposed microbenthic communities collected from sandy sediments to several UVR treatments. After 3 weeks of exposure to natural UV-B, carbon fixation rates were significantly reduced as compared to the no UV-B treatments. These researchers speculated that UV-B would have greater effects on communities inhabiting sandy sediments compared to sediments with a high silt and clay content. Roux et al. (2002) observed reduced photosynthesis in microphytobenthic communities (primarily small diatoms) from an intertidal mudflat exposed to UV-B; however, these effects were limited to periods of high solar irradiance. Finally, unlike some planktonic organisms that are able to avoid UVR in the photic zone, behavioral avoidance in some benthic habitats is limited. Because benthic algae and diatoms can account for a significant portion of primary production in aquatic ecosystems, exposure to UVR could have serious consequences for energy flow.

35.3.2 Effects of UVR in Terrestrial Ecosystems

While the major focus of UVR research in aquatic ecosystems has been on primary production, research in terrestrial ecosystems has documented effects on other ecosystem processes, including litter decomposition and biogeochemical cycles (Newsham et al. 1997, Pancotto et al. 2003, Zepp et al. 1995). The consensus of these investigations is that terrestrial ecosystem processes are generally less sensitive to UVR than processes in aquatic ecosystems. Caldwell and Flint (1994) predicted that the occurrence of UVR effects on plants from most frequent to least frequent was the following: increased production of UV-absorbing compounds > reduced growth and morphological changes ≫ reduced photosynthesis.

35.3.2.1 Direct and Indirect Effects on Litter Decomposition and Primary Production

Effects of UVR on litter decomposition have been described as a result of both direct and indirect processes. Direct effects on decomposing litter are usually a result of inhibition of microbial, fungal, and other components of the soil community, which reduces decomposition rates. Because these effects may be offset by enhanced photodegradation, which enhances decomposition rate, predicting direct effects of UV-B on litter decomposition rates is complex. Indirect effects of UV-B occur during growth and senescence of plants and can result in changes in leaf chemistry (e.g., lignin content) or physical characteristics of leaves. One of the most consistent responses of plants to UV-B exposure is increased production of protective secondary plant metabolites, including phenolics and flavonoids. If these changes in leaf chemistry influence feeding habits of other trophic levels or alter plant–herbivore interactions, there exists the possibility that higher trophic levels will be indirectly affected by UV-B (Bassman 2004). In most instances these secondary plant compounds serve as deterrents to herbivory and therefore are likely to mediate trophic responses to UV-B radiation. Although aquatic ecologists routinely consider implications of cascading trophic interactions, these ideas have received less attention from terrestrial ecologists (Bassman 2004), perhaps because top-down control in terrestrial ecosystems is considered relatively unimportant (Strong 1992). Nonetheless, the often subtle direct effects of UV-B on terrestrial plants may be less important than the indirect effects on trophic interactions.

Because effects of UV-B exposure will likely vary among locations, comparisons of plant communities across sites is a valuable approach for understanding factors that determine ecosystem-level effects. Moody et al. (2001) measured direct effects on litter decomposition of *Betula pubescens* exposed to ambient and elevated UV-B at sites in Norway, Sweden, the Netherlands, and Greece. Although the fungal community was significantly affected by UV-B, differences in mass loss and chemical composition of litter between treatments were modest. Verhoef et al. (2000) also reported that litter decomposition and nutrient fluxes in a grassland ecosystem were not affected by UV-B; however, abundance of soil decomposers was significantly reduced in both UV-A and UV-B treatments. There is also likely to be a strong seasonal component to UVR effects that will vary among terrestrial ecosystems. For example, UVR exposure to leaf litter in deciduous forests is likely to be greatest in early spring when leaf canopies are absent and incident UV-B is high. Newsham et al. (1997) observed subtle and transient effects of enhanced UV-B on decomposition of oak (*Quercus robur*) leaf litter. Lower decomposition in UV-B treatments was associated with increases in C content of leaves and reduced fungal colonization. However, in a subsequent study of UV-B effects on decomposition, Newsham et al. (2001) reported that *Q. robur* saplings exposed to a 30% increase in UV-B (corresponding to an 18% reduction in ozone) for 2 years showed little change in chemical composition. These researchers concluded that recent increases in UV-B in the Northern Hemisphere are unlikely to have significant effects on organic matter pools, nutrient cycling, and decomposition through alterations in litter quality. Experiments conducted at high latitudes of the Southern Hemisphere where ozone depletion is greatest showed quite different results. Pancotto et al. (2003) employed a 2 × 2 factorial experimental design to assess both direct and indirect effects of UV-B on a native shrub community in Tierra del Fuego National Park (Argentina). Plants were grown under ambient or reduced UV-B and decomposition rate of litter produced by these plants was measured under ambient or reduced UV-B. Decomposition rate (mass loss) was significantly (14–34%) lower under ambient UV-B compared to reduced UV-B treatments. These direct effects were found to be more important in controlling decomposition rates than indirect effects on litter quality. Pancotto et al. (2003) speculated that changes in decomposition rates have important implications for other ecosystem-level processes, including nutrient mineralization and carbon storage, in high latitudes of the Southern Hemisphere.

Although effects of UV-B on primary production and nutrient cycling have been examined in terrestrial habitats (Klironomos and Allen 1995, Gehrke 1998, Shi et al. 2004), these processes have received considerably less attention compared to aquatic ecosystems. Assessing direct effects on primary productivity is complicated because UVR can either increase or decrease physiological processes that determine production. For example, growth of *Sphagnum* in a subarctic bog was significantly reduced by exposure to UV-B (Gehrke 1998). However, total production was not affected because photosynthesis was enhanced and dark respiration was reduced. Klironomos and Allen (1995) exposed sugar maple (*Acer saccharum*) seedlings to enhanced UV-B and measured shoot and root biomass. Despite significant shifts in belowground carbon flow and microarthropod abundance in UV treatments, shoot and root biomass was not affected. Plants inhabiting alpine ecosystems are naturally exposed to greater levels of UV-B and are therefore expected to possess repair mechanisms to reduce the damaging effects on photosynthesis. In field experiments, Shi et al. (2004) exposed alpine plants to enhanced UV-B radiation that simulated a 14% reduction in ozone depletion. Photosynthesis and respiration were either similar or increased slightly under moderate UV-B exposure. These researchers speculated that alpine plants are acclimated to UV-B and that photosynthetic processes are protected by morphological adaptations such as increased leaf thickness.

Meta-analysis offers a quantitative approach for integrating results of multiple studies to assess complex relationships among variables. This approach is especially appropriate for assessing terrestrial ecosystem responses to UV-B because effects are expected to be relatively subtle and often indirect. Searles et al. (2001) conducted meta-analysis of 62 papers that investigated effects of UV-B radiation on the concentration of UV-B-absorbing compounds, growth, morphological variables, and photosynthetic processes. With the exception of UV-B-absorbing compounds, most variables showed relatively minor response to UV-B treatments. These researchers concluded that indirect effects in the form of alterations in herbivory are likely to be the most significant responses of terrestrial ecosystems to elevated UV-B radiation.

Finally, our understanding of effects of UV-B on terrestrial ecosystems is seriously limited by the lack of long-term investigations (Aphalo 2003). In the meta-analysis of terrestrial studies described above (Searles et al. 2001), over 80% of the studies were conducted for less than 1 year. Long-term studies are even less common in aquatic ecosystems where manipulation of UVR is more problematic because of experimental artifacts. Although short-term experiments may help understand underlying mechanisms, they often provide very different results than those of longer duration. Experiments conducted in Tierra del Fuego represent one of the best examples of long-term UV-B studies (Robson et al. 2003). These researchers used filters to reduce ambient levels of UV-B in a peatland ecosystem for six field seasons. It is important to note that 6 years was an insufficient time period to detect subtle effects of UV-B for several of the responses measured.

35.4 INCREASED CO_2 AND GLOBAL CLIMATE CHANGE

35.4.1 AQUATIC ECOSYSTEMS

Despite widespread recognition of the potential ecological effects of global climate change associated with increased levels of atmospheric CO_2, research on ecosystem-level responses in aquatic systems has been lacking. Several excellent reviews describing predicted effects of climate change on distribution and extirpation of species have been published (Carpenter et al. 1992b, Clark et al. 2001, Firth and Fisher 1992, Grimm 1992, Lodge 2001, Meyer et al. 1999, Smith and Buddemeier 1992); however, relatively few studies have investigated effects on ecosystem processes. A recent report published by the Pew Center on Global Climate Change (Poff et al. 2002) summarized the current state of knowledge on effects of climate change on aquatic ecosystems, but contained very little information on changes in ecosystem function (Table 35.3). It is expected that increased surface water temperatures associated with global climate change will affect ecosystem productivity, materials transport, nutrient dynamics and decomposition; however, little data have been collected to support this hypothesis. Increased water temperature will likely increase rates of respiration and photosynthesis, and the relative magnitude of these increases will determine overall effects on ecosystem metabolism. A survey of factors related to lake productivity along a latitudinal gradient showed that primary production was directly related to water temperature (Brylinsky and Mann 1973). Long-term records (1970–1990) indicated that increased air temperature and reduced precipitation in northwestern Ontario were most likely responsible for reduced discharge, increased water temperature, greater light penetration, and reduced concentration of DOC in boreal lakes (Schindler et al. 1996). Because

TABLE 35.3
Major Conclusions of the Pew Center on Global Climate Change Report Regarding Aquatic Ecosystem Responses to Global Climate Change

- Aquatic and wetland ecosystems are very vulnerable to climate change.
- Increases in water temperature will cause a shift in the thermal suitability of aquatic habitats for resident species.
- Seasonal shifts in stream runoff will have significant negative effects on many aquatic ecosystems.
- Wetland loss in boreal regions of Alaska and Canada is likely to result in additional releases of CO_2 into the atmosphere.
- Coastal wetlands are particularly vulnerable to sea level rise associated with increasing global temperatures.
- Most specific ecosystem responses to climate change cannot be predicted because new combinations of native and nonnative species will interact in novel situations.
- Increased water temperatures and seasonally reduced streamflows will alter many ecosystem processes with potential direct societal costs.
- The manner in which humans adapt to a changing climate will greatly influence the future status of inland freshwater and coastal wetland ecosystems.

Source: Poff et al. (2002).

of cascading trophic-level interactions in many lake ecosystems, alterations of one trophic level will likely have consequences for both upper and lower trophic levels. Results of microcosm experiments with aquatic microbes showed that warming increased primary production and decomposition by both direct effects on temperature-dependent physiological processes and indirect effects on trophic structure. Finally, climate-induced alterations in the composition of riparian canopies may have significant effects on the quality and quantity of allochthonous detritus delivered to lakes and streams.

35.4.1.1 Linking Model Results with Monitoring Studies in Aquatic Ecosystems

Much of the research in lakes and streams documenting potential effects of climate change has been limited to hydrologic models that predict modifications in discharge resulting from altered precipitation patterns. Alterations in the flow regime of aquatic ecosystems are likely to be significant, especially in western U.S. watersheds where modest changes in precipitation are expected to result in dramatic reductions in stream runoff (Carpenter et al. 1992a). Long-term records of stream discharge are available for many watersheds (e.g., U.S. Geological Survey); therefore, predictive models that relate regional changes in climate to altered flow regimes within a watershed can be developed. Associating local weather patterns and stream discharge over the past several decades may also provide useful insights into potential trends associated with global climate change. However, extrapolation from global and regional models to local conditions may not be appropriate for some areas. For example, general circulation models (GCMs) for the Rocky Mountains predict that temperature should increase under a two times CO_2 scenario. However, long-term monitoring in this region showed a decline in mean annual temperature and an increase in precipitation (Williams et al. 1996b), demonstrating that climate in alpine areas may be controlled more by local conditions than by regional trends.

The oceans have long been recognized as an important sink for excess CO_2 released to the atmosphere. Recent evidence indicates that approximately 48% of the anthropogenic C released between 1800 and 1994 was sequestered by oceans and that, without this oceanic uptake, atmospheric CO_2 levels would be about 55 ppm greater than current levels (Sabine et al. 2004). These authors also suggest that the strength of the oceans as a sink for atmospheric CO_2 has diminished and that the fraction of CO_2 currently stored in the oceans is approximately one-third of the total potential storage. Relatively complex feedback mechanisms will determine the effects of increased CO_2 and global temperatures on oceanic ecosystems. Using results of long-term (1958–2002) surveys of marine phytoplankton in the Northeast Atlantic Ocean, Richardson and Schoeman (2004) associated increased sea surface temperatures with increased primary production in cooler areas and decreased production in warmer areas. Changes in primary production were also related to production of grazers, which propagated to other trophic levels. The effects of changes in primary production on global carbon flux are difficult to predict because other factors, especially nutrient availability, will determine CO_2 uptake. In one of the more ambitious attempts to understand the relationship among nutrients, primary production, and CO_2 flux, Coale et al. (2004) performed multiple iron injection experiments in large areas (15 km^2) of the Southern Ocean. Rates of photosynthesis increased from 0.29 to 6.9 mmol C/m^3/day and nitrate concentrations decreased by approximately 2 μM following treatment, indicating that iron may play an important role in controlling CO_2 uptake in this region.

Unlike research in marine ecosystems, the role of freshwater ecosystems as a source or sink for atmospheric carbon has received little attention. Community metabolism varies greatly among aquatic ecosystems, with the highest productivity observed in marshes (Figure 35.8). Although it is generally assumed that freshwater ecosystems export CO_2 to the atmosphere, the importance of aquatic biota as sources or sinks depend on overall ecosystem productivity (Duarte and Agusti 1998). Factors such as nutrient enrichment and trophic structure also regulate primary production and CO_2 flux in freshwater ecosystems. Whole ecosystem experiments conducted in Wisconsin (USA) showed

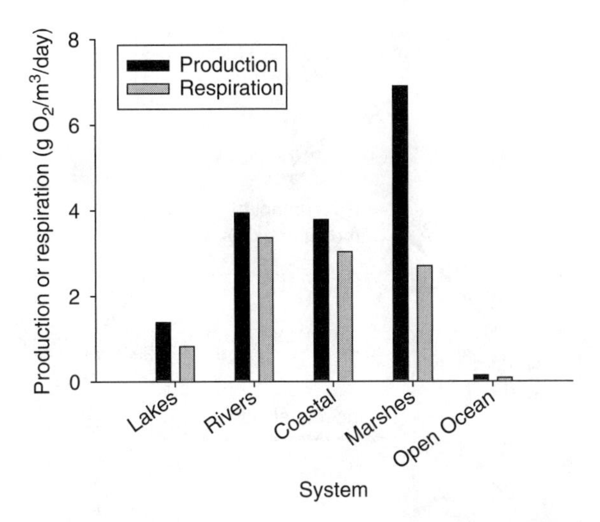

FIGURE 35.8 Median gross primary production and respiration of freshwater and marine ecosystems. Results were compiled from five decades of studies that reported O_2 evolution as a surrogate for carbon flux. (Data from Table 1 in Duarte and Agusti (1998).)

that shifts in top predators in experimentally enriched lakes regulated the flow of C and determined if a lake was a sink or source of CO_2 (Schindler et al. 1997).

The relative lack of information concerning potential effects of climate change on primary production, nutrient cycling, decomposition, and other processes in aquatic ecosystems is surprising and in sharp contrast to tremendous research efforts currently underway in terrestrial habitats. Considerably more research effort is necessary to understand responses of aquatic ecosystems to elevated CO_2 and associated climate change. Climate-induced changes in water temperature, hydrology, and physicochemical characteristics in aquatic ecosystems are likely to influence contaminant transport, bioavailability, uptake, and toxicity. Aquatic ecotoxicologists need to develop a better appreciation for how these processes will be affected in a warmer, CO_2-enriched world.

35.4.2 Terrestrial Ecosystems

In contrast to the relatively limited research on effects of climate change in freshwater ecosystems, studies conducted in terrestrial ecosystems have been extensive. These studies have considered several facets of the CO_2 problem (Figure 35.9): direct influences of CO_2 enrichment and indirect effects of increased temperature (Körner 2000) and shifts in terrestrial vegetation (Wolters et al. 2000). Much of this research has focused on assessing the role of forests, grasslands, and other ecosystems in C sequestration, a problem that requires a better understanding of the complex relationship between CO_2 and C storage and the role of soil nutrients, especially N, in regulating storage. Quantifying global C sequestration is a difficult problem because of variation among ecosystems and because of complex feedback processes. For example, boreal peatlands occupy only about 2% of the earth's surface but sequester about 33% of the global soil C. Increased soil temperature associated with climate change could result in positive feedback by increasing decomposition rate of peatlands thereby releasing large quantities of stored C to the atmosphere. Alternatively, warmer temperatures could cause negative feedback by enhancing productivity and C storage in these areas. Storage of C in terrestrial ecosystems will be largely dependent on the rate of turnover within C pools and the availability of N. Carbon allocated to pools with relatively fast turnover such as leaves and roots will likely result in little long-term storage. In contrast, C allocated to pools with slow turnover will result in a large increase in soil C (Allen et al. 2000). Finally, if elevated CO_2 significantly increases

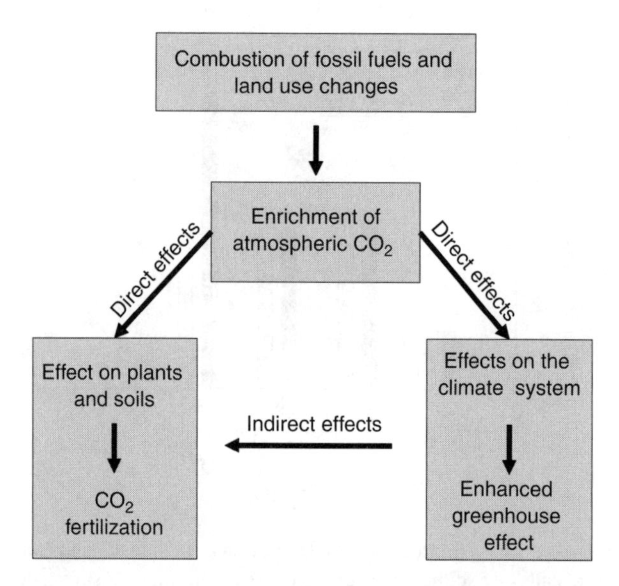

FIGURE 35.9 The two potential aspects of increased CO_2 on terrestrial ecosystems. Increased CO_2 directly stimulates photosynthesis and may alter NPP. Direct effects on the global climate system include changes in atmospheric temperatures and patterns of precipitation. These climatic changes will likely have indirect effects on terrestrial vegetation. (Modified from Figure 2 in Körner (2000).)

soil respiration or decreases decomposition by changing litter quality (e.g., altering C:N ratios), any excess C removed by stimulation of NPP may be returned to the atmosphere.

Three general approaches have been employed to predict how terrestrial ecosystems will respond to increased CO_2 and associated warmer temperatures: modeling, monitoring, and experimentation. In the section below, we will briefly discuss important findings using these three approaches as well as their strengths and limitations.

35.4.2.1 Simulation Models

Because of limitations associated with conducting experiments at appropriate spatial or temporal scales (Chapter 26), computer modeling has played a prominent role in predicting effects of climate change on terrestrial ecosystems. These models have attempted to quantify how various ecosystem processes, especially photosynthesis and production, will respond to global increases in CO_2. The most realistic simulation models integrate biogeographic analyses showing forest redistribution under various climate change scenarios with biogeochemistry models that simulate movement of C, nutrients, and water. Aber et al. (2001) reviewed contemporary models developed to predict these changes in forest ecosystems, with an emphasis on interactions among potential stressors. One of the most consistent findings of these models was increased NPP in response to elevated CO_2. However, there was considerable uncertainty regarding the sustainability of increased NPP because of limitations imposed by other factors, especially nutrients. Because increased NPP and C storage in forest ecosystems will largely be determined by N availability, greater anthropogenic N deposition has the potential to enhance C sequestration (Rastetter et al. 1997). Much of the spatial variation among ecosystem responses to CO_2 enrichment is partially a result of N cycling and the rates of movement between organic and inorganic pools. Simulation modeling predicts that short-term responses to increased CO_2 are quite different from long-term responses and are characterized by at least four different time scales (Rastetter et al. 1997): (1) instantaneous physiological responses include greater NPP because photosynthesis is stimulated by CO_2; (2) acclimation occurs over a period of several years as uptake of N from soils is increased; (3) over a time scale of decades, increased

C and N in vegetation results in greater litter production and accumulation of soil organic matter; (4) over a time period of centuries, the most important response is increased ecosystem-level N and the increased organic matter in vegetation and soils. These different temporal responses also influence the sequestration of C in vegetation and soils. Carbon is initially stored in vegetation as ratios of C:N increase. At intermediate (10–100 years) and longer time scales, C storage will be driven by movement of N from soil to vegetation and by increases in total ecosystem N, respectively (Rastetter et al. 1997).

Much of the simulation modeling conducted to assess ecosystem responses to increased CO_2 has addressed C sequestration in terrestrial habitats. As described above, approximately 50% of the CO_2 released into the environment over the past 20 years has remained in the atmosphere. The remaining proportion has been sequestered in either oceanic or terrestrial ecosystems. Significant annual variation in the amount of CO_2 in the atmosphere has also been observed during this period. However, because fossil fuel emissions show little annual variation, changes in atmospheric CO_2 most likely reflect annual variation in the CO_2 flux to oceans and continents (Bousquet et al. 2000). Improved understanding of C storage in terrestrial ecosystems has helped identify the terrestrial biosphere as a "missing sink" for anthropogenic CO_2 that previously could not be accounted for in global models. The importance of identifying this sink is obvious because it explains why CO_2 levels in the atmosphere have not increased as much as expected on the basis of known anthropogenic releases.

Despite significant progress, there remains considerable variation among the different models used to predict changes in C storage in the continental United States (Aber et al. 2001). As part of the Vegetation Ecosystem Modeling and Analysis Project (VEMAP), Schimel et al. (2000) integrated historical climate information with three different ecosystem models to predict spatial and temporal variation in C storage in the United States for the period 1895–1993. Results showed that C storage was evenly distributed among different biomes in the United States (100–150 kg/ha), but there was considerable annual variation (<0.1 Pg C efflux to >0.2 Pg C uptake). Because of this high annual variability, any 2- to 3-year period would be insufficient to assess potential C storage, thus highlighting the need for long-term assessments (Schimel et al. 2000). These researchers also note that effects of forest management and agricultural abandonment are similar to effects of climate and CO_2 on C storage. Again, there is considerable uncertainty with respect to the role of limiting factors such as N and water availability on C storage in terrestrial ecosystems. However, estimates of C balance by various models are relatively consistent and predict that the capacity of terrestrial ecosystems to store C over the next century will decline as a result of continued increases in CO_2 emissions (Prentice et al. 2000).

In addition to understanding effects of CO_2 enrichment on C sequestration and ecosystem processes, simulation models have predicted effects of increased temperature and precipitation on C balance. Dramatic changes in the distribution of vegetation types, including the extirpation of alpine habitats, spruce-fir forests, aspen-birch forests, and sagebrush in the United States are likely to occur as a result of temperature increases (Hansen and Dale 2001, Iverson and Prasad 2001, Shafer et al. 2001). These shifts in vegetation types are likely to have significant effects on C storage. Bachelet et al. (2001) used equilibrium and dynamic models to show that moderate increases in temperature resulted in increased vegetation density and C storage. Esser (1992) compared modeled responses of grasslands and coniferous forests to increased temperature (3.5°C) and a 10% increase in precipitation. Relative changes in NPP in response to warming were greatest in temperature-limited grasslands and coniferous forests at low latitudes. Because of greater NPP, annual storage will be more important in grasslands; however, because of their greater biomass and broader spatial distribution, disruptions to coniferous forests will have the greatest effects on global C cycles (Esser 1992).

Results of simulation models are most convincing when they are validated using monitoring or experimental results and applied to other ecosystems. These reality checks on model predictions are powerful tools for enhancing our understanding of complex responses to climate change and potential interactions with other stressors (see Section 35.5). Mechanistic models are especially appropriate for scaling between levels of organization (e.g., relating physiological responses of individual leaves to C storage in whole ecosystems). Williams et al. (1997) developed a fine-scale model for a temperate deciduous forest and then applied this model to predict gross primary production (GPP) for Alaskan tundra and several forest types in Oregon. Results showed excellent agreement between modeled and measured GPP in both systems ($r^2 = .76$ in Alaska; $r^2 = .97$ in Oregon). Luo et al. (2001) validated modeling results designed to predict effects of elevated CO_2 on GPP with measurements of photosynthesis and C flux at the Duke Free Air CO_2 Enrichment (FACE) experimental site. Again, modeled and measured values for photosynthesis and canopy C flux were in good agreement ($r^2 = 0.80$–0.85).

35.4.2.2 Monitoring Studies

> We cannot predict with certainty, through direct experimentation, what the responses of forests to global change will be because we cannot carry out the multisite, multifactorial experiments required for doing so.
>
> **(Aber et al. 2001)**

> The ecological effects of global climate change are too complex to be comprehensively addressed by experimental approaches, at least in the short term.
>
> **(Ringold and Groffman 1997)**

Studies monitoring changes in temperature and CO_2 concentrations from regional to global scales have provided useful information regarding effects of climate change on NPP and C sequestration. Paleoclimate data have been linked with geochemical and climate models to show that long-term variation in CO_2 strongly influences global temperatures (Crowley and Berner 2001). Because of the unprecedented rate of change in CO_2 and global temperatures in the past several decades, it is now possible to relate NPP to climatic changes over a much shorter time scale. Tree ring chronologies have correlated increased growth with increases in CO_2 levels observed over the past 100 years in several ecosystems (Graybill and Idso 1993, Nicolussi et al. 1995). Braswell et al. (1997) measured CO_2, temperature anomalies, and vegetation to examine the response of vegetation to temperature variability at a global scale. Correlations among vegetation growth, temperature, and CO_2 were lagged, suggesting that responses are regulated by biogeochemical feedbacks. Patterns observed at a global scale were found to be a composite of individualistic responses among biomes, and changes in the distribution of these systems could have important consequences for C sequestration (Braswell et al. 1997). NPP has increased by >6% globally in the period from 1982 to 1999 (Nemani et al. 2003), but the magnitude and mechanisms explaining these increases differ among regions. The greatest increase in NPP was observed in tropical ecosystems, with Amazonian rain forests accounting for 42% of the global increase. Increased NPP observed in mid-latitudes and high latitudes were attributed to climate, CO_2 enrichment, N fertilization, and forest regrowth, whereas increases in the tropics were primarily a result of increased CO_2 (Nemani et al. 2003). The disproportionate increase in NPP for tropical forests highlights the role that these systems play in the global C budget. The importance of terrestrial ecosystems, especially tropical forests, as a sink for excess C is well established (Phillips et al. 1998). Grace et al. (1995) measured CO_2 flux in an undisturbed tropical rain forest in Brazil and concluded that this system was a net sink for CO_2. Using data from permanent plots of >600,000 individual trees, Phillips et al. (1998) reported that biomass of tropical forests increased by 0.71 t/ha/year between 1975 and 1996. This increased biomass was likely a response to greater atmospheric CO_2 and nutrient deposition and may account for approximately 40% of the "missing" terrestrial C. Although the sequestration of C by tropical forests may reduce the impact of increasing CO_2 on global climate, there is an upper limit to the amount of C that tropical forests can remove from the atmosphere (Phillips et al. 1998).

While much attention has been placed on the role that forest ecosystems play in C storage, other terrestrial ecosystems are also capable of removing significant amounts of C from the atmosphere.

For example, wet tundra ecosystems retain large stocks of C that have been stored for long periods of time. These permafrost-dominated ecosystems are highly sensitive to global climate change and modest increases in soil temperatures could return significant amounts of C to the atmosphere. Oechel et al. (1995) compared ecosystem GPP and whole ecosystem respiration measured in 1971 with measurements 20 years later. Results showed dramatic changes in C storage during this period. Wet meadow ecosystems stored approximately 25 g C/m^2/year in 1971 but released 1.3 g C/m^2/year two decades later. Oechel et al. (1995) speculated that this shift from a strong C sink to a slight C source in tundra ecosystems was related to increased surface temperatures and drying in the upper soil layers.

35.4.2.3 Experimental Manipulations of CO_2 and Temperature

Manipulating concentrations of CO_2 or temperature to assess effects of climate change on processes in terrestrial ecosystems is challenging. In fact, because of the limited spatiotemporal scale and difficulty conducting multifactor manipulations, some researchers question the usefulness of these experimental approaches (Aber et al. 2001, Luo and Reynolds 1999, Ringold and Groffman 1997). Much of what we know about the direct physiological effect of CO_2 enrichment on forests is based on relatively small-scale experiments conducted with isolated seedlings or saplings. Our ability to link these responses of individual plants measured in a greenhouse to whole ecosystem processes is questionable. In addition, the relationship between effects of step increases in CO_2 versus gradual changes that occur in nature requires careful consideration (Luo and Reynolds 1999). However, as described in Chapters 23 and 32, we feel that experimental manipulations remain one of the most effective approaches for establishing cause-and-effect relationships between stressors and responses. Observed positive correlations between NPP and CO_2 based on monitoring studies cannot separate the effects of other factors such as increased N deposition or precipitation. Because of the time scales and complexity of potential interactions among climate warming, CO_2, and other related stressors, it is essential that modelers and experimental ecologists collaborate to study effects of climate change on ecosystem processes (Luo et al. 2004). Several large-scale experimental programs have been established to quantify direct effects of elevated CO_2 on NPP and other ecosystem processes in grasslands and forests. We will highlight findings of these programs in this section. Smaller-scale studies have been conducted using individual species to identify mechanisms and to investigate stressor interactions.

Experimental approaches to measure effects of increased CO_2 in terrestrial ecosystems include closed-controlled containers, open-top chambers, FACE, and natural experiments (Körner 2000). Hattenschwiler et al. (1997) compared production of trees grown adjacent to a naturally elevated CO_2 source (a freshwater spring) to production of trees grown at sites with ambient CO_2. Results showed that growth was 12% greater at the CO_2-enriched sites, but this difference did not persist with older trees. The decline in CO_2 stimulation for older trees has important implications for predicting C sequestration because many CO_2 enrichment experiments are conducted with seedlings or saplings. Using open-top chambers, Zak et al. (2000) measured effects of enhanced CO_2 on microbial community composition and N cycling. Despite increases in fine-root production, elevated CO_2 had no effect on microbial biomass or N cycling because the elevated C was not sufficient to affect the large natural pools of soil organic matter. After 2 years of exposure, enhanced CO_2 increased net CO_2 assimilation rate of aspen (*Populus tremuloides*) grown under low or high N levels, with little evidence of photosynthetic acclimation (Figure 35.10) (Curtis et al. 2000). However, after the third year, plants grown under elevated CO_2 in low N conditions showed reduced photosynthetic activity. Dukes and Hungate (2002) reviewed results of experiments conducted using open-top chambers to assess effects of CO_2 on decomposition in grassland ecosystems. The direct effects of elevated CO_2 on litter quality and decomposition were relatively modest. However, these researchers speculated that because decomposition rates vary

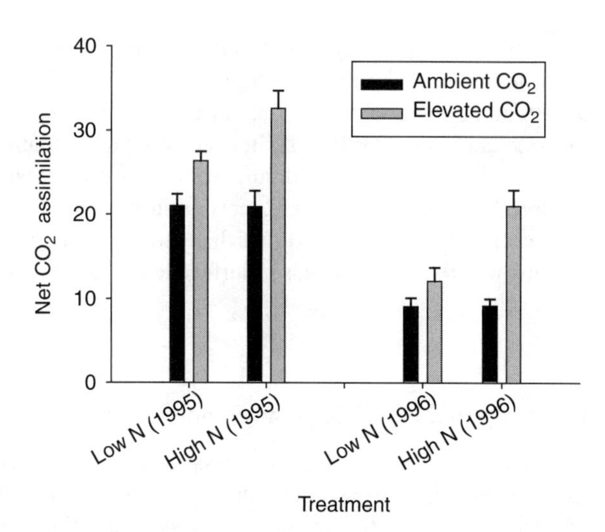

FIGURE 35.10 Effects of elevated CO_2 on net assimilation in aspen (*P. tremuloides*) grown under conditions of low and high N. (Data from Table 2 in Curtis et al. (2000).)

greatly among species, CO_2-induced changes in community composition may alter ecosystem-level decomposition rate.

Research findings from the FACE studies conducted in a loblolly pine (*Pinus taeda*) plantation in North Carolina, a sweetgum (*Liquidambar styraciflua*) monoculture in Tennessee, and a mixed deciduous forest in Wisconsin (USA) provide the most comprehensive experimental evidence showing a relationship between CO_2 and ecosystem processes in forests (DeLucia et al. 1999, Norby et al. 2002, Zak et al. 2003). Increases in NPP in response to experimentally elevated CO_2 at approximately 1.5 times ambient levels (537–560 μL/L) ranged from 21% in the sweetgum forest to 25% in the loblolly pine forest. Despite a significant increase in NPP, the long-term potential of temperate forests to remain a sink for excess C is limited. For example, most of the increased C sequestered in the sweetgum forests was allocated to leaves and roots, compartments with relatively fast turnover (Norby et al. 2002).

Because soil N pools and rates of mineralization are closely related to C flux, quantifying the effects of CO_2 on these belowground processes is of critical importance for predicting C storage in terrestrial ecosystems. Interactions between plants and soil microbial communities will largely determine C sequestration, but the precise role of N availability in this process is controversial (Luo et al. 2004). Experimental studies have consistently shown that elevated CO_2 enhances NPP, but the sustainability of these increases will largely be determined by belowground processes that influence N turnover and availability. Under increased levels of CO_2, N availability in soils may either decrease as a result of lower rates of litter decomposition or increase as a result of increased C substrate. If greater amounts of C enter the soil as a result of CO_2 enrichment, microbial uptake of nutrients may be enhanced and availability to plants may be diminished. CO_2 enrichment will result in increased demand for N to support greater plant productivity and the greater storage of N in long-lived plant biomass or soil organic matter pools (Luo et al. 2004). Finally, the potential long-term effects of N limitation may be offset if N is supplied from other sources, such as increased atmospheric deposition. Zak et al. (2003) addressed uncertainties associated with these belowground processes using results from the three FACE studies described above. Because researchers in each experiment used similar techniques, this study represents the first comprehensive evaluation of soil responses to elevated CO_2 in different forest types. Although elevated CO_2 increased plant and litter production in all experiments, there was no evidence of effects on N availability or transformation in soils. Zak et al. (2003) speculated that because of relatively large pools of organic matter and

N naturally present in soils, increases in plant and litter production associated with elevated CO_2 were not sufficient to affect N cycling.

Experiments to measure terrestrial ecosystem responses to warming have been conducted in a variety of habitats, but for logistical reasons most research has been restricted to low vegetation such as grasslands, alpine meadows, and peatlands (Price and Waser 2000, Shaw and Harte 2001, Shaw et al. 2002, Updegraff et al. 2001). Subalpine vegetation in the Rocky Mountains showed relatively little response to warming over a 4-year period (Price and Waser 2000). These researchers speculated that despite earlier snowmelt in treated plots, drying soils limited photosynthesis and microbial activity. Soil warming in a subalpine meadow had relatively little effect on litter decomposition, but indirect effects on litter quality were observed in heated treatments (Shaw and Harte 2001). Results of a decade-long soil warming experiment in a mid-latitude hardwood forest showed that warming enhanced CO_2 flux and N mineralization rates, but effects diminished over time (Melillo et al. 2002). Experiments conducted in northern peatlands compared responses of a sedge fen and a *Sphagnum* bog to warming (Updegraff et al. 2001). Ecosystem respiration was closely related to soil temperature but showed little difference between community types. These researchers concluded that the flux of CO_2 from peatlands will increase exponentially with climate warming.

Separating effects of CO_2 enrichment, warming, and other changes in climate on ecosystem processes is a difficult undertaking. For example, elevated CO_2 and increases in soil moisture will likely increase C storage in forest ecosystems, whereas lower soil moisture and greater respiration will likely reduce storage (Melillo et al. 2002). Experimental manipulation of both CO_2 and temperature has been conducted under controlled conditions in single species and mesocosm studies such as the Ecotron facility (Lawton 1996), but larger-scale experiments conducted at the ecosystem level are uncommon (Shaw et al. 2002, Wright 1998). Because responses to increased CO_2 will likely occur simultaneously with changes in temperature, precipitation, and other environmental factors, multifactorial manipulations are essential for predicting realistic responses. The CLIMEX (climate change experiment) project in Norway is one of the more ambitious ecosystem manipulations of CO_2 and temperature (van Breemen et al. 1998, Wright 1998). Researchers enclosed a forested headwater catchment in an 860 m^2 greenhouse and measured a variety of ecosystem responses. There were no changes in decomposition rates or photosynthesis of dominant species as a result of increased temperature or CO_2. However, experimentally elevated soil temperatures increased N mineralization and resulted in greater N export in the catchment. The possibility that climate change could exacerbate N pollution and acidification in these areas is a serious concern that deserves additional attention (Wright 1998). Shaw et al. (2002) measured individual and combined effects of CO_2 enrichment, warming, precipitation, and N deposition on NPP in a California grassland. In contrast to many greenhouse experiments and ecosystem studies, increased CO_2 did not stimulate NPP. More importantly, responses to individual factors were quite different from responses to combinations of factors. These results should send a strong message to researchers and funding agencies regarding the importance of multifactorial experiments for understanding how ecosystems will respond to global environmental changes.

Alterations in the composition of terrestrial communities under conditions of increased CO_2 levels and warmer temperatures were discussed in Chapter 26. These changes in the structure of vegetation are also likely to affect ecosystem processes by modifying terrestrial food webs. Changes in the chemical composition of plants exposed to elevated CO_2, including increases in nonstructural carbohydrates, secondary compounds, and C:N ratios, are among the most consistent responses observed in terrestrial studies (Körner 2000). Effects of CO_2-induced changes in food quality have important implications for food webs and energy flow. In a comparative study of three grasslands, Wilsey et al. (1997) measured effects of elevated CO_2 and simulated ungulate grazing on productivity of plants collected from Yellowstone National Park (USA), Flooding Pampa of Buenos Aires (Argentina), and the Serengeti Ecosystem (Tanzania). Increased productivity in response to CO_2 treatment was observed only in Yellowstone National Park. These results were expected because Yellowstone species were dominated by C3 plants that are more susceptible to increased

CO_2 (Wilsey et al. 1997). Lower N content was also observed in Yellowstone species, but there was no CO_2 treatment–grazer treatment interaction, indicating that herbivores did not influence how plants responded to CO_2.

With few exceptions (e.g., Hattenschwiler et al. 1997), most of the experimental evidence showing effects of enhanced CO_2 on terrestrial vegetation was based on step increases in CO_2 (e.g., 1.5–2 times). These step increases are necessary because responses to enhanced CO_2 are often subtle and because the time scale required to conduct experiments with gradual increases (e.g., 1–2% per year) is too long. However, step increases in CO_2 are quite different from what is occurring in nature and there has been some question as to their ecological relevance. A C sequestration model developed by Luo and Reynolds (1999) tested the hypothesis that responses of terrestrial vegetation to step increases in CO_2 can be extrapolated to the gradual increases occurring in nature. Their modeling results showed that responses of terrestrial vegetation to step increases in CO_2 were transient and differed markedly from responses to gradual increases. If step increases cannot be used to predict effects of CO_2 enrichment on C sequestration and other processes in natural ecosystems, new experimental designs and analytical approaches may be necessary. At the very least, experimental studies comparing ecosystem responses to gradual and step increases should be immediately undertaken.

In summary, the FACE and CLIMEX projects have contributed significantly to our understanding of how forests and grasslands in northern and mid-latitudes respond to CO_2 enrichment. Because of high NPP and greater C sequestration, it is likely that tropical ecosystems, especially tropical forests, will have a disproportionate influence on global C flux. Large-scale experiments similar to the design of CLIMEX and FACE should be conducted in tropical forests to obtain a better understanding of how these ecosystems will respond to enhanced CO_2.

35.5 INTERACTIONS AMONG GLOBAL ATMOSPHERIC STRESSORS

Because ecosystem responses to global environmental stressors are often complex and not a simple combination of responses to individual factors (Shaw et al. 2002), improved understanding of stressor interactions is critical. Synergistic interactions among climate change, ozone depletion, and acidification may increase exposure of aquatic and terrestrial ecosystems to UVR (Watkins et al. 2001). Alterations in physicochemical characteristics of aquatic ecosystems resulting from climate change, N deposition, or UV-B can affect contaminant transport, bioavailability, and toxicity. In this section, we will explore the potential interactions among global atmospheric stressors and discuss how these stressors may influence contaminant effects on ecosystem processes.

35.5.1 INTERACTIONS BETWEEN CO_2 AND N

We have previously described the role that N plays in determining responses of terrestrial ecosystems to CO_2 enrichment. Early stages of N deposition in forests, especially boreal forests where plant growth is N limited, may stimulate primary productivity, thereby increasing C storage in vegetation and soil (Lamontagne and Schiff 1999). However, the duration of this increase in production is likely to be ephemeral as forests become N saturated (Aber et al. 1989). N deposition can also affect decomposition rates and thereby regulate C sequestration in soils; however, because N can either stimulate or inhibit decomposition the overall effect on C storage is variable. For example, decomposition of lignin and cellulose, two primary constituents of plant litter, is dependent on different microbial pathways that vary in response to N availability (decomposition of cellulose is stimulated by N whereas decomposition of lignin is inhibited). Thus, the chemical composition of leaf litter will determine the response to N deposition in forests (Sinsabaugh et al. 2002). Waldrop et al. (2004) observed significant increases in C storage in an oak-dominated forest and significant loss of soil C

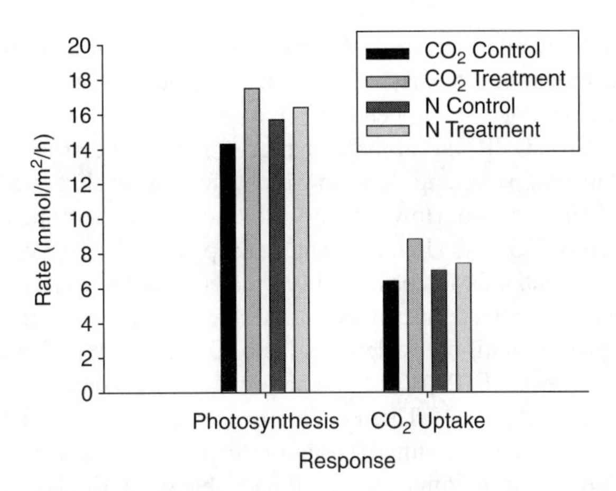

FIGURE 35.11 Response of boreal vegetation plots to either CO_2 or N enrichment (NH_4NO_3) after 2 years in experimental microcosms. (Data from Table 3 in Saarnio et al. (2003).)

in a sugar maple-dominated forest, demonstrating that interactions between N deposition and C will be ecosystem specific. Ultimately, experiments may be necessary to quantify potential interactions between CO_2 and N deposition. Saarnio et al. (2003) conducted microcosm experiments to compare effects of CO_2 and N addition in a boreal ecosystem. Using a miniature version of the FACE experimental system, plots were exposed to either increased NH_4NO_3 or increased CO_2. Although effects of N additions were relatively minor, CO_2 enrichment resulted in greater photosynthesis, respiration, and net CO_2 uptake (Figure 35.11).

35.5.2 Interactions between Global Climate Change and UVR

Interactions between climate warming and UVR are likely to be significant in some ecosystems and may have global consequences. For example, because the ocean is an important sink for global C, reduced primary production in marine ecosystems resulting from UV-B exposure may have important consequences for climate change. As described in Chapter 26, increases in water clarity resulting from acidification and climate-induced reductions in DOC increase penetration of UV-B radiation in aquatic ecosystems (Frost et al. 1999, Schindler et al. 1996, Williamson et al. 1999). Although organisms will likely be exposed to greater levels of UV-B in low DOC and/or acidic lakes and streams (Schindler et al. 1996, Watkins et al. 2001), few studies have examined effects on ecosystem processes. Changes in the composition and structure of stream riparian habitats as a result of climate change may alter light regimes and influence the exposure of aquatic organisms to UV-B. Kelly et al. (2003) manipulated UV-A and UV-B levels in streams with varying amounts of riparian vegetation. Chlorophyll *a* biomass and C accrual were lower under UVR exposure, but this response was only observed at a site with no canopy cover. Shifts in species composition in response to climate change and increased UVR in northern ecosystems will likely influence carbon flux through microbial food webs. Microcosm experiments conducted with natural lake and stream assemblages measured microbial responses to the combined effects of increased temperature and UVR (Rae and Vincent 1998). Although microbial food webs were relatively insensitive to short-term exposure, effects of UVR on bacterial production were dependent on temperature. On the basis of mesocosm experiments conducted in an alpine lake, Vinebrooke and Leavitt (1998) concluded that food webs in these systems were more sensitive to climate warming and associated changes in DOM and nutrients than to increased UV-B exposure.

Sommaruga (2001) noted the importance of distinguishing between short- and long-term changes when evaluating interactions between UVR and climate change. Decreased cover of snow and ice in alpine streams and lakes could increase UVR exposure in the short term; however, over a longer time period the establishment of terrestrial vegetation in alpine habitats may increase DOM export to watersheds, resulting in reduced UVR penetration. In general, long-term responses of ecosystems to the interactive effects of UVR and climate change will be difficult to predict because factors that determine the relationships among decomposition, nutrient cycling, and carbon storage on a regional scale are poorly understood. However, any modification in the storage of C has the potential to influence global levels of CO_2 and climate warming (Zepp et al. 2003). For example, on an annual basis, carbon storage in peatlands is considerably greater than the total amount of carbon fixed by terrestrial vegetation. The effects of converting these global carbon sinks to sources as a result of UVR may not be apparent until many decades later (Gehrke 1998). Comprehensive reviews of the potential interactive effects of UV-B and global climate change on biogeochemical cycles were published by Zepp et al. (1995, 2003). The most likely scenario involves UVR-induced alterations in storage and release of organic C, either through effects on decomposition or by alterations in primary production. The long-term influences of climate change on levels of DOM and exposure of algae to UV-B were described for several Canadian lakes using paleoecological techniques (Leavitt et al. 2003, Pienitz and Vincent 2000). By reconstructing UVR penetration, algal biomass, DOM, and surrounding vegetation cover over the past 10,000 years, these researchers showed that algal biomass was closely associated with long-term climatic changes in DOM and UVR penetration. Algal biomass, which was reduced by 10–25 times during the early stages of lake formation when DOM was low, also responded to climate-induced changes in DOM (Leavitt et al. 2003).

35.5.3 Interactions between Global Atmospheric Stressors and Contaminants

Interactions between global atmospheric stressors and other contaminants have been reported in several aquatic ecosystems. In general, reduced or altered streamflow patterns as a result of lower precipitation may concentrate chemicals in streams receiving pollutants (Carpenter et al. 1992a). Some contaminants may become more bioavailable under conditions of increased temperature or acidification. Frost et al. (1999) observed greater accumulation of methylmercury in the food web of an experimentally acidified lake. Increases of methylmercury in phytoplankton, zooplankton, and fish were attributed to changes in microbial activity and/or increases in the pool of Hg used as a substrate for microbial methylation. Tingey et al. (2001) used simulation modeling to predict effects of increased temperature and CO_2 in combination with tropospheric ozone (O_3) on production of Ponderosa pine. Results showed that increases in growth and production from CO_2 partially offset the negative phytotoxic effects of O_3.

Kashian et al. (2004) measured interactive effects of UV-B radiation and heavy metals in stream microcosms with and without DOM. UV-B did not affect community respiration, but increased drift was observed in treatments with enhanced UV-B. The combination of UV-B + Zn had a greater effect on community structure and ecosystem processes than Zn alone, suggesting that Zn may enhance UV-B effects. In a subsequent experiment, Kashian et al. (2007) compared the combined and interactive effects of UVR and heavy metals on metabolism of benthic communities derived from reference and metal-polluted streams. The combined effects of UV-B and metals on metabolism were greater then either stressor alone at both sites (Figure 35.12a). More importantly, the interactive effects of UV-B and metals varied between streams, suggesting that communities responded differently to these two stressors. These results were consistent with findings of a large field study in which UVR was excluded from 12 streams along a gradient of metal pollution (Figure 35.12b). Results showed that effects of UV-B removal on chlorophyll *a* differed among streams and was greatest in unpolluted streams compared to metal-polluted streams. Winch et al. (2002) proposed a potential mechanism to explain interactive effects of heavy metals and UV-B exposure. Exposure of water to UV-B for 10-day reduced concentrations of DOC by 20%, thus increasing toxicity and bioavailability of heavy metals (Pb, Cu, Zn, Co) to algae (Figure 35.13). Effects of UV-B on DOC and metal bioavailability

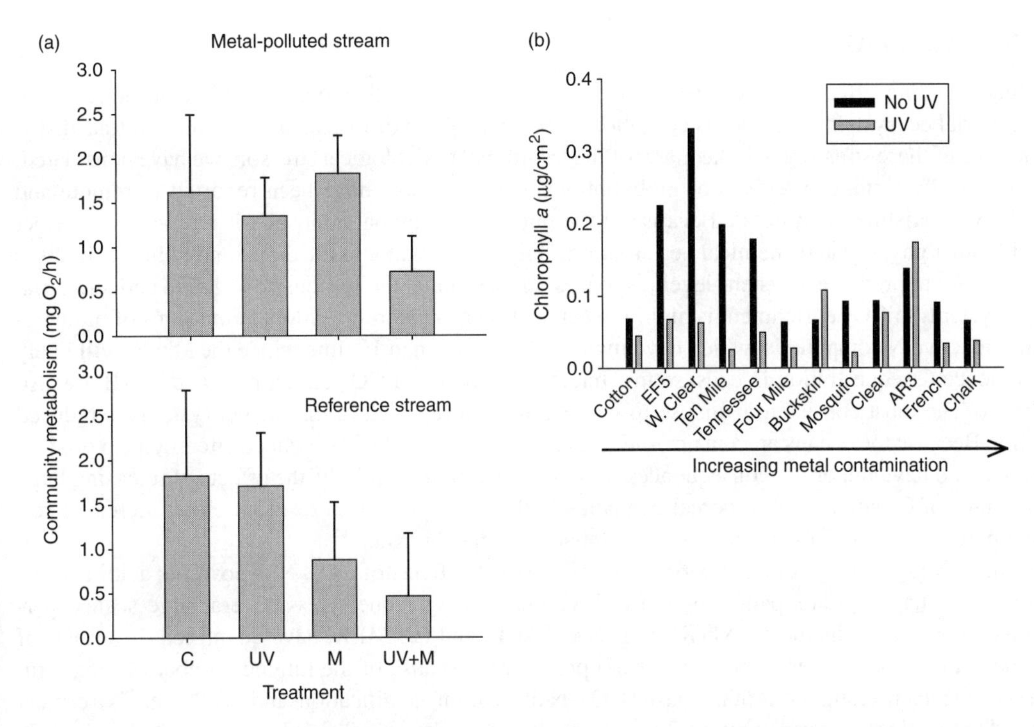

FIGURE 35.12 (a) Effects of heavy metals and UV-B on community metabolism of benthic macroinvertebrates collected from metal-polluted and reference streams in Colorado, USA. Modified from Kashian et al. (2007). (b) Effects of UV-B removal on chlorophyll *a* in 12 streams along a gradient of metal contamination in the Rocky Mountains, Colorado, USA). (Data from Zuellig (2006).)

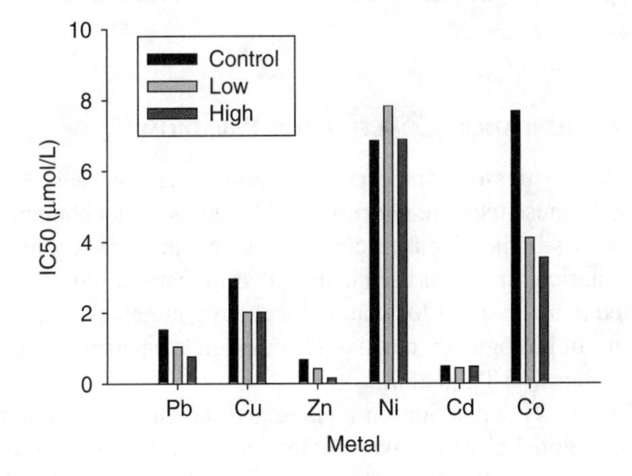

FIGURE 35.13 Effects of UVR on metal toxicity to green algae, expressed as the concentration resulting in 50% growth inhibition (IC50). Water collected from headwaters of the Raisin River (Ontario, Canada) was exposed for 0 (control), 5 (low), and 10 (high) days to UV-B lamps. DOC concentrations in low and high UV-B exposures were reduced by approximately 10% and 20%, respectively. (Data from Table 4 in Winch et al. (2002).)

were not observed in a relatively clear-water lake with lower natural concentrations of DOC. These researchers speculated that effects observed in the field under long-term exposure to natural UVR would be much greater and that photodegradation of DOC by UVR would significantly influence primary productivity in metal-contaminated ecosystems.

35.6 SUMMARY

Understanding effects of increased CO_2, N deposition, acidification, and UVR on aquatic and terrestrial ecosystem processes is complicated because of the diffuse sources and ubiquitous distributions of these stressors. Unlike many of the traditional toxicological stressors we have considered, ecologically significant effects of global atmospheric stressors have been reported in remote and relatively pristine ecosystems. Because of the potential for global atmospheric stressors to interact with other physical and chemical perturbations, an understanding of the direct and indirect effects is necessary to predict ecosystem-level responses. For example, the response of aquatic and terrestrial ecosystems to CO_2 enrichment is highly dependent on N availability. Global emissions of biologically reactive N compounds, which have increased by more than 10 times since the 1860s, will likely influence C sequestration in ecosystems. Interactions between CO_2 enrichment and UV-B are also likely. The most consistent response to UV-B radiation reported in marine ecosystems is reduced NPP. Because the oceans are a major sink for anthropogenic C, UV-B-induced alterations in production could have important consequences for C sequestration. Similarly, the effects of releasing large amounts of C sequestered in boreal peatlands to the atmosphere as a result of either increased soil temperature or UV-B exposure may also impact global C budgets.

Designing and executing experiments to investigate effects of CO_2, N deposition, acidification, and UVR at appropriate spatiotemporal scales is challenging. Nonetheless, several large-scale experimental programs, including NITREX, LINX, FACE, and CLIMEX, have quantified the effects of global atmospheric stressors on ecosystem processes. Because of the limited number of long-term experimental investigations of increased CO_2, N deposition, acidification, and UVR, there is a critical need to link these experimental results with monitoring and regional simulation models. Large-scale comparative studies conducted across broad spatial gradients have quantified variation in responses to global atmospheric stressors and identified factors that influence these responses. Finally, research into effects of global atmospheric stressors provides a unique opportunity to employ recent technological advances (e.g., geographic information system, GIS; remote sensing) to study impacts at landscape-level scales.

35.6.1 SUMMARY OF FOUNDATION CONCEPTS AND PARADIGMS

- Global atmospheric stressors are especially problematic because of their ubiquitous distribution and because they are not restricted by geopolitical boundaries.
- Significant increases in the global reservoirs of C, N, and S as a result of combustion of fossil fuels and agricultural/land use changes disrupt natural biogeochemical cycles and have contributed to a variety of local and global environmental concerns.
- Global emissions of biologically reactive N compounds have increased from about 15 Tg in 1860 to more than 165 Tg in 2000.
- Predicting effects of N deposition on aquatic and terrestrial ecosystems is complicated by variation in regional climate, hydrologic characteristics, vegetation type, and other sources of anthropogenic disturbance.
- Accumulation, transfer, and denitrification of biologically reactive N compounds through the biosphere and the changes that result have been termed the nitrogen cascade.
- As previously N-limited forests became saturated, NO_3 is released to watersheds causing eutrophication, acidification, and reductions in ANC.
- Critical soil acidification load is determined primarily by a balance between the weathering of base cations and leaching associated with deposition of SO_4 and NO_x.
- Even relatively remote areas located away from sources of N can experience effects of atmospheric N deposition.
- Sensitivity of watersheds to N deposition will be influenced by underlying geology, hydrologic characteristics, soil type, and vegetation.

- Ecosystem-level studies of SO_2 and NO_x deposition and associated acidification have consistently shown reduced rates of primary and secondary production, decomposition, and nutrient cycling.
- Because of the intimate linkages between forests and surrounding watersheds, research describing transport of N in terrestrial ecosystems has important implications for N export to lakes and streams.
- The most consistent responses to experimentally enhanced N deposition in forests were changes in water quality, particularly increased export of NO_3.
- Responses of forest ecosystems to N deposition are often variable, and factors that control differences in N retention and export are not completely understood.
- Species-specific differences in litter quality and rates of decomposition will complicate our ability to predict ecosystem responses to N deposition in forests.
- One of the most consistent observations in studies of UV-B effects on marine and freshwater phytoplankton is reduced primary production.
- UVR exclusion experiments document effects of current levels of UVR, whereas enhancement experiments attempt to estimate effects of predicted increases.
- Responses of marine and freshwater phytoplankton to UVR are complicated by numerous environmental factors, and quantification of effects on ecosystem processes is often challenging.
- Because penetration of UVR through the water column is dependent on water clarity, factors that influence turbidity, trophic status, and levels of DOM will potentially influence ecosystem responses.
- Because UVR affects both primary producers and consumers, responses to UVR manipulations are often a combination of direct and indirect effects.
- Benthic communities occupying clear, shallow water environments are likely to be exposed to intense levels of UVR.
- The general consensus is that terrestrial ecosystem processes are less sensitive to UVR than those in aquatic ecosystems.
- Direct negative effects of UV-B on decomposition in terrestrial communities may be offset by enhanced photodegradation, which enhances litter decay.
- Assessing direct effects of UVR on primary productivity is complicated because UVR can either increase or decrease physiological processes that determine production.
- One of the most consistent responses of plants to UV-B exposure is increased production of protective secondary plant metabolites (e.g., phenolics and flavonoids).
- Changes in leaf chemistry may influence feeding habits of consumers or alter plant–herbivore interactions.
- Our understanding of effects of UV-B on terrestrial ecosystems is seriously limited by the small number of long-term investigations.
- Despite widespread recognition of the potential ecological effects of global climate change, research on ecosystem-level responses in aquatic systems is limited.
- Climate-induced changes in water temperature, hydrology, and physicochemical characteristics of aquatic ecosystems are likely to influence contaminant transport, bioavailability, uptake, and toxicity.
- Studies conducted in terrestrial ecosystems have considered direct influences of CO_2 enrichment and indirect effects of increased temperature and shifts in terrestrial vegetation.
- Because of limitations associated with conducting experiments at appropriate spatial or temporal scales, computer modeling has played a prominent role in predicting effects of climate change on terrestrial ecosystems.
- The most realistic simulation models integrate biogeographic analyses showing forest redistribution under various climate change scenarios with biogeochemistry models that simulate movement of C, nutrients, and water.

- Simulation models have predicted effects of increased temperature and precipitation on C balance and dramatic changes in the distribution of vegetation types.
- Studies monitoring changes in temperature and CO_2 concentrations at regional and global scales have provided useful information regarding effects of climate change on NPP and C sequestration.
- Much of what we know about the direct physiological effect of CO_2 enrichment on forests is based on relatively small-scale experiments conducted with isolated seedlings or saplings.
- The effects of step increases in CO_2 versus gradual changes that are likely to occur in nature require critical evaluation.
- Experimental approaches to measure effects of increased CO_2 in terrestrial ecosystems include closed-controlled containers, open-top chambers, FACE, and natural experiments.
- Experimental studies have consistently shown that elevated CO_2 enhances NPP, but the sustainability of these increases will largely be determined by belowground processes that influence N turnover and availability.
- Experiments to measure terrestrial ecosystem responses to warming have been conducted in a variety of habitats, but most research has been restricted to low vegetation such as grasslands, alpine meadows, and peatlands.
- Distinguishing effects of CO_2 enrichment, warming, and other changes in climate on ecosystem processes is difficult.
- Changes in the composition of terrestrial communities under increased CO_2 conditions may affect ecosystem processes by modifying food webs.
- Alterations in physicochemical characteristics of aquatic ecosystems resulting from climate change, N deposition, or UV-B can affect contaminant transport, bioavailability, and toxicity.
- Because the ocean is an important sink for global C, reduced primary production in marine ecosystems resulting from UV-B exposure may have important consequences for climate change.

REFERENCES

Aber, J., Neilson, R.P., McNulty, S., Lenihan, J.M., Bachelet, D., and Drapek, R.J., Forest processes and global environmental change: Predicting the effects of individual and multiple stressors, *BioScience*, 51, 735–751, 2001.

Aber, J.D., Goodale, C.L., Ollinger, S.V., Smith, M.-L., Magill, A.H., Martin, M.E., Hallett, R.A., and Stoddard, J.L., Is nitrogen deposition altering the nitrogen status of northeastern forests? *BioScience*, 53, 375–389, 2003.

Aber, J.D., Magill, A., McNulty, S.G., Boone, R.D., Nadelhoffer, K.J., Downs, M., and Hallett R., Forest biogeochemistry and primary production altered by nitrogen saturation, *Water Air Soil Pollut.*, 85, 1665–1670, 1995.

Aber, J.D., Nadelhoffer, K.J., Steudler, P., and Melillo, J.M., Nitrogen saturation in northern forest ecosystems, *BioScience*, 39, 378–386, 1989.

Allen, A.S., Andrews, J.A., Finzi, A.C., Matamala, R., Richter, D.D., and Schlesinger, W.H., Effects of free-air CO_2 enrichment (FACE) on belowground processes in a *Pinus taeda* forest, *Ecol. Appl.*, 10, 437–448, 2000.

Allen, C.D. and Smith, R.E.H., The response of planktonic phosphate uptake and turnover to ultraviolet radiation in Lake Erie, *Can. J. Fish. Aquat. Sci.*, 59, 778–786, 2002.

Aphalo, P.J., Do current levels of UV-B radiation affect vegetation? The importance of long-term experiments, *New Phytol.*, 160, 273–276, 2003.

Bachelet, D., Neilson, R.P., Lenihan, J.M., and Drapek, R.J., Climate change effects on vegetation distribution and carbon budget in the United States, *Ecosystems*, 4, 164–185, 2001.

Banaszak, A.T. and Neale, P.J., Ultraviolet radiation sensitivity of photosynthesis in phytoplankton from an estuarine environment, *Limnol. Oceanogr.*, 46, 592–603, 2001.

Baron, J.S., Rueth, H.M., Wolfe, A.M., Nydick, K.R., Allstott, E.J., Minear, J.T., and Moraska, B., Ecosystem responses to nitrogen deposition in the Colorado Front Range, *Ecosystems*, 3, 352–368, 2000.

Bassman, J.H., Ecosystem consequences of enhanced solar ultraviolet radiation: Secondary plant metabolites as mediators of multiple trophic interactions in terrestrial plant communities, *Photochem. Photobiol.*, 79, 382–98, 2004.

Behrenfeld, M.J., Lean, D.R.S., and Lee, H., Ultraviolet-B radiation effects on inorganic nitrogen uptake by natural assemblages of oceanic plankton, *J. Phycol.*, 31, 25–36, 1995.

Blumthaler, M. and Ambach, W., Indication of increasing solar ultraviolet-B radiation flux in alpine regions, *Science*, 248, 206–208, 1990.

Bothwell, M.L., Sherbot, D.M.J., and Pollock, C.M., Ecosystem response to solar ultraviolet-B radiation: Influence of trophic level interactions, *Science*, 265, 97–100, 1994.

Bousquet, P., Peylin, P., Ciais, P., Le Quere, C., Friedlingstein, P., and Tans, P.P., Regional changes in carbon dioxide fluxes of land and oceans since 1980, *Science*, 290, 1342–1346, 2000.

Boxman, A.W., Blanck, K., Brandrud, T.E., Emmett, B.A., Gundersen, P., Hogervorst, R.F., Kjonaas, O.J., Persson, H., and Timmermann, V., Vegetation and soil biota response to experimentally-changed nitrogen inputs in coniferous forest ecosystems of the NITREX Project, *Forest Ecol. Manag.*, 101, 65–79, 1998.

Braswell, B.H., Schimel, D.S., Linder, E., and Moore, B.I., The response of global terrestrial ecosystems to interannual temperature variability, *Science*, 278, 870–873, 1997.

Brylinsky, M. and Mann, K.H, An analysis of factors governing productivity in lakes and reservoirs, *Limnol. Oceanogr.*, 18, 1–14, 1973.

Burton, T.M., Stanford, R.M., and Allan, J.W., Acidification effects on stream biota and organic matter processing, *Can. J. Fish. Aquat. Sci.*, 42, 669–675, 1985.

Caldwell, M.M. and Flint, S.D., Stratospheric ozone reduction, solar UV-B radiation and terrestrial ecosystems, *Climate Change*, 28, 375–394, 1994.

Caldwell, M.M. and Flint, S.D., Uses of biological spectra weighting functions and the need of scaling for the ozone reduction problem, *Plant Ecol.*, 128, 66–76, 1997.

Carpenter, S.R., Fisher, S.G., Grimm, N.B., and Kitchell, J.F., Global change and freshwater ecosystems, *Ann. Rev. Ecol. Syst.*, 23, 119–139, 1992a.

Carpenter, S.R., Frost, T.M., Kitchell, J.F., and Kratz, T.K., Species dynamics and global environmental change: A perspective from ecosystem experiments, In *Biotic Interactions and Global Change*, Karieva, P.M., Kinsolver, J.G., and Huey, R.B. (eds.), Sinauer Associates, Inc., Sunderland, MA, 1992b, pp. 267–279.

Clark, M.E., Rose, K.A., Levine, D.A., and Hargrove, W.W., Predicting climate change effects on Appalachian trout: Combining GIS and individual-based modeling, *Ecol. Appl.*, 11, 161–178, 2001.

Coale, K.H., Johnson, K.S., Chavez, F.P., Buesseler, K.O., Barber, R.T., Brzezinski, M.A., Cochlan, W.P., et al., Southern ocean iron enrichment experiment: Carbon cycling in high- and low-Si waters, *Science*, 304, 408–414, 2004.

Crowley, T.J. and Berner, R.A., CO_2 and climate change, *Science*, 292, 870–872, 2001.

Curtis, P.S., Vogel, C.S., Wang, X.Z., Pregitzer, K.S., Zak, D.R., Lussenhop, J., Kubiske, M., and Teeri, J.A., Gas exchange, leaf nitrogen, and growth efficiency of *Populus tremuloides* in a CO_2-enriched atmosphere, *Ecol. Appl.*, 10, 3–17, 2000.

Day, T.A. and Neale, P.J., Effects of UV-B radiation on terrestrial and aquatic primary producers, *Ann. Rev. Ecol. System.*, 33, 371–396, 2002.

De Lange, H.J., Verschoor, A.M., Gylstra, R., Cuppen, J.G.M., and Van Donk, E., Effects of artificial ultraviolet-B radiation on experimental aquatic microcosms, *Freshw. Biol.*, 42, 545–560, 1999.

DeLucia, E.H., Hamilton, J.G., Naidu, S.L., Thomas, R.B., Andrews, J.A., Finzi, A., Levine, M., et al., Net primary production of a forest ecosystem with experimental CO_2 enrichment, *Science*, 284, 1177–1179, 1999.

Duarte, C.M. and Agusti, S., The CO_2 balance of unproductive aquatic ecosystems, *Science*, 281, 234–236, 1998.

Dukes, J.S. and Hungate, B.A., Elevated carbon dioxide and litter decomposition in California annual grasslands: Which mechanisms matter? *Ecosystems*, 5, 171–183, 2002.

Emmett, B.A., Boxman, D., Bredemeier, M., Gundersen, P., Kjonaas, O.J., Moldan, F., Schleppi, P., Tietema, A., and Wright, R.F., Predicting the effects of atmospheric nitrogen deposition in conifer stands: Evidence from the NITREX ecosystem-scale experiments, *Ecosystems*, 1, 352–360, 1998.

Esser, G., Implications of climate change for production and decomposition in grasslands and coniferous forests, *Ecol. Appl.*, 2, 47–54, 1992.

Fenn, M.E., Poth, M.A., Aber, J.D., Baron, J.S., Bormann, B.T., Johnson, D.W., Lemly, A.D., McNulty, S.G., Ryan, D.F., and Stottlemyer, R., Nitrogen excess in North American ecosystems: Predisposing factors, ecosystem responses, and management strategies, *Ecol. Appl.*, 8, 706–733, 1998.

Firth, P. and Fisher, S.G. (eds.), *Global Climate Change and Freshwater Ecosystems,* Springer-Verlag, New York, 1992.

Flint, S.D., Ryel, R.J., and Caldwell, M.M., Ecosystem UV-B experiments in terrestrial communities: A review of recent findings and methodologies, *Agri. Forest Meteorol.*, 120, 177–189, 2003.

Frost, T.M., Montz, P.K., Kratz, T.K., Badillo, T., Brezonik, P.L., Gonzalez, M.J., Rada, R.G., et al., Multiple stresses from a single agent: Diverse responses to the experimental acidification of Little Rock Lake, Wisconsin, *Limnol. Oceanogr.*, 44, 784–794, 1999.

Furgal, J.A. and Smith, R.E.H., Ultraviolet radiation and photosynthesis by Georgian Bay phytoplankton of varying nutrient and photoadaptive status, *Can. J. Fish. Aquat. Sci.*, 54, 1659–1667, 1997.

Gala, W.R. and Giesy, J.P., Effects of ultraviolet-radiation on the primary production of natural phytoplankton assemblages in Lake-Michigan, *Ecotoxicol. Environ. Saf.*, 22, 345–361, 1991.

Galloway, J.N., Aber, J.D., Erisman, J.W., Seitzinger, S.P., Howarth, R.W., Cowing, E.B., and Cosby, B.J., The nitrogen cascade, *BioScience*, 53, 341–356, 2003.

Garcia-Pichel, F. and Bebout, B.M., Penetration of ultraviolet radiation into shallow water sediments: High exposure for photosynthetic communities, *Mar. Ecol. Prog. Ser.*, 131, 257–262, 1996.

Gawel, J.E., Ahner, B.A., Friedland, A.J., and Morel, F.M.M., Role for heavy metals in forest decline indicated by phytochelatin measurements, *Nature*, 381, 64–65, 1996.

Gehrke, C., Effects of enhanced UV-B radiation on production-related properties of a *Sphagnum fuscum* dominated subarctic bog, *Funct. Ecol.*, 12, 940–947, 1998.

Gonzalez, M.J. and Frost, T.M., Comparisons of laboratory bioassays and a whole-lake experiment: Rotifer responses to experimental acidification. *Ecol. Appl.*, 4, 69–80, 1994.

Grace, J., Lloyd, J., Mcintyre, J., Miranda, A.C., Meir, P., Miranda, H.S., Nobre, C., et al., Carbon-dioxide uptake by an undisturbed tropical rain-forest in southwest Amazonia, 1992 to 1993, *Science*, 270, 778–780, 1995.

Graybill, D.A. and Idso, S.B., Detecting the aerial fertilization effect of atmospheric CO_2 enrichment in tree-ring chronologies, *Global Biogeochem. Cycles*, 7, 81–95, 1993.

Grimm, N.B., Implications of climate change for stream communities, In *Biotic Interactions and Global Change*, Karieva, P.M., Kinsolver, J.G., and Huey, R.B. (eds.), Sinauer Associates, Inc., Sunderland, MA, 1992, pp. 293–314.

Gundersen, P., Emmett, B.A., Kjonaas, O.J., Koopmans, C.J., and Tietema, A., Impact of nitrogen deposition on nitrogen cycling in forests: A synthesis of NITREX data, *Forest Ecol. Manag.*, 101, 37–55, 1998.

Hamburg, S.P., Yanai, R.D., Arthur, M.A., Blum, J.D., and Siccama, T.G., Biotic control of calcium cycling in northern hardwood forests: Acid rain and aging forests, *Ecosystems*, 6, 399406, 2003.

Hansen, A. and Dale, V., Biodiversity in US forests under global climate change, *Ecosystems*, 4, 161–163, 2001.

Hansen, L., Hedtke, S.F., and Munns, W.R., Integrated human and ecological risk assessment: A case study of ultraviolet radiation effects on amphibians, coral, humans, and oceanic primary productivity, *Hum. Ecol. Risk Assess.*, 9, 359–377, 2003.

Hattenschwiler, S., Miglietta, F., Raschi, A., and Körner, C., Thirty years of in situ tree growth under elevated CO_2: A model for future forest responses? *Global Change Biol.*, 3, 463–471, 1997.

Helbling, E.W., Villafane, V., Ferrario, M., and Holmhansen, O., Impact of natural ultraviolet-radiation on rates of photosynthesis and on specific marine phytoplankton species, *Mar. Ecol. Prog. Ser.*, 80, 89–100, 1992.

Hessen, D.O., De Lange, H.J., and van Donk, E., UV-induced changes in phytoplankton cells and its effects on grazers, *Freshw. Biol.*, 38, 513–524, 1997.

Hiriart, V.P., Greenberg, B.M., Guildford, S.J., and Smith, R.E.H., Effects of ultraviolet radiation on rates and size distribution of primary production by Lake Erie phytoplankton, *Can. J. Fish. Aquat. Sci.*, 59, 317–328, 2002.

Hornung, M. and Reynolds, B., The effects of natural and anthropogenic environmental changes on ecosystem processes at the catchment scale, *Trends Ecol. Evol.*, 10, 443–449, 1995.

Iverson, L.R. and Prasad, A.M., Potential changes in tree species richness and forest community types following climate change, *Ecosystems*, 4, 186–199, 2001.

Kashian, D.R., Prusha, B.A., and Clements, W.H., Influence of total organic carbon and UV-B Radiation on zinc toxicity and bioaccumulation in aquatic communities, *Environ. Sci. Technol.*, 38, 6371–6376, 2004.

Kashian, D.R., Zuellig, R.E., Mitchell, K.A., and Clements, W.H., The cost of tolerance: Sensitivity of stream benthic communities to UV-B and metals, *Ecol. Appl.*, 17, 365–375.

Kelly, D.J., Bothwell, M.L., and Schindler, D.W., Effects of solar ultraviolet radiation on stream benthic communities: An intersite comparison, *Ecology*, 84, 2724–2740, 2003.

Kinzie, R.A., Banaszak, A.T., and Lesser, M.P., Effects of ultraviolet radiation on primary productivity in a high altitude tropical lake, *Hydrobiologia*, 385, 23–32, 1998.

Klironomos, J.N. and Allen, M.F., UV-B-mediated changes on below-ground communities associated with the roots of *Acer saccharum*, *Funct. Ecol.*, 9, 923–930, 1995.

Körner, C., Biosphere responses to CO_2 enrichment, *Ecol. Appl.*, 10, 1590–1619, 2000.

Lamontagne, S. and Schiff, S.L., The response of a heterogeneous upland boreal shield catchment to a short term NO_3 addition, *Ecosystems*, 2, 460–473, 1999.

Lawton, J.H., The Ecotron facility at Silwood Park: The value of "big bottle" experiments, *Ecology*, 77, 665–669, 1996.

Leavitt, P.R., Cumming, B.F., Smol, J.P., Reasoner, M., Pienitz, R., and Hodgson, D.A., Climatic control of ultraviolet radiation effects on lakes, *Limnol. Oceanogr.*, 48, 2062–2069, 2003.

Likens, G.E., Driscoll, C.T., and Buso, D.C., Long-term effects of acid rain: Response and recovery of a forest ecosystem, *Science*, 272, 244–246, 1996.

Lodge, D.M., Responses of lake biodiversity to global changes, In *Future Scenarios of Global Biodiversity*, Sala, O.E., Chapin, F.S., and Huber-Sannwald, E. (eds.), Springer-Verlag, New York, 2001, pp. 277–313.

Lovett, G.M. and Rueth, H., Soil nitrogen transformations in beech and maple stands along a nitrogen deposition gradient, *Ecol. Appl.*, 9, 1330–1344, 1999.

Luo, Y., Medlyn, B., Hui, D., Ellsworth, D., Reynolds, J., and Katul, G., Gross primary productivity in Duke Forest: Modeling synthesis of CO_2 experiment and eddy-flux data, *Ecol. Appl.*, 11, 239–252, 2001.

Luo, Y., Su, B., Currie, W.S., Dukes, J.S., Finzi, A., Hartwig, U., Hungate, B., et al., Progressive nitrogen limitation of ecosystem responses to rising atmospheric carbon dioxide, *BioScience*, 54, 731–739, 2004.

Luo, Y.Q. and Reynolds, J.F., Validity of extrapolating field CO_2 experiments to predict carbon sequestration in natural ecosystems, *Ecology*, 80, 1568–1583, 1999.

Marwood, C.A., Smith, R.E.H., Furgal, J.A., Charlton, M.N., Solomon, K.R., and Greenberg, B.M., Photoinhibition of natural phytoplankton assemblages in Lake Erie exposed to solar ultraviolet radiation, *Can. J. Fish. Aquat. Sci.*, 57, 371–379, 2000.

McNamara, A.E. and Hill, W.R., UV-B irradiance gradient affects photosynthesis and pigments but not food quality of periphyton, *Freshw. Biol.*, 43, 649–662, 2000.

Melillo, J.M., Steudler, P.A., Aber, J.D., Newkirk, K., Lux, H., Bowles, F.P., Catricala, C., Magill, A., Ahrens, T., and Morrisseau, S., Soil warming and carbon-cycle feedbacks to the climate system, *Science*, 298, 2173–2176, 2002.

Meyer, J.L., Sale, M.J., Mulholland, P.J., and Poff, N.L., Impacts of climate change on aquatic ecosystem functioning and health, *J. Amer. Water Res. Assoc.*, 35, 1373–1386, 1999.

Moayeri, M., Meng, F.R., Arp, P.A., and Foster, N.W., Evaluating critical soil acidification loads and excedances for a deciduous forest at the Turkey Lakes Watershed, *Ecosystems*, 4, 555–567, 2001.

Moody, S.A., Paul, N.D., Bjorn, L.O., Callaghan, T.V., Lee, J.A., Manetas, Y., Rozema, J., et al., The direct effects of UV-B radiation on *Betula pubescens* litter decomposing at four European field sites, *Plant Ecol.*, 154, 27–36, 2001.

Mostajir, B., Demers, S., deMora, S., Belzile, C., Chanut, J.P., Gosselin, M., Roy, S., et al., Experimental test of the effect of ultraviolet-B radiation in a planktonic community, *Limnol. Oceanogr.*, 44, 586–596, 1999.

Mousseau, L., Gosselin, M., Levasseur, M., Demers, S., Fauchot, J., Roy, S., Villegas, P.Z., and Mostajir, B., Effects of ultraviolet-B radiation on simultaneous carbon and nitrogen transport rates by estuarine phytoplankton during a week-long mesocosm study, *Mar. Ecol. Prog. Ser.*, 199, 69–81, 2000.

Neale, P.J., Cullen, J.J., and Davis, R.F., Inhibition of marine photosynthesis by ultraviolet radiation: Variable sensitivity of phytoplankton in the Weddell-Scotia confluence during the Austral spring, *Limnol. Oceanogr.*, 43, 433–448, 1998.

Nemani, R.R., Keeling, C.D., Hashimoto, H., Jolly, W.M., Piper, S.C., Tucker, C.J., Myneni, R.B., and Running, S.W., Climate-driven increases in global terrestrial net primary production from 1982 to 1999, *Science*, 300, 1560–1563, 2003.

Newsham, K.K., Mcleod, A.R., Roberts, J.D., Greenslade, P.D., and Emmett, B.A., Direct effects of elevated UV-B radiation on the decomposition of *Quercus robur* leaf litter, *Oikos*, 79, 592–602, 1997.

Newsham, K.K., Splatt, P., Coward, P.A., Greenslade, P.D., McLeod, A.R., and Anderson, J.M., Negligible influence of elevated UV-B radiation on leaf litter quality of *Quercus robur*, *Soil Biol. Biochem.*, 33, 659–665, 2001.

Nicolussi, K., Bortenschlager, S., and Korner, C., Increase in tree-ring width in sub-alpine *Pinus cembra* from the Central Alps that may be CO_2 related, *Trees Struct. Funct.*, 9, 181–189, 1995.

Nihlgard, B., The ammonium hypothesis—An additional explanation for forest dieback in Europe, *Ambio*, 14, 2–8, 1985.

Norby, R.J., Hanson, P.J., O'Neill, E.G., Tschaplinski, T.J., Weltzin, J.F., Hansen, R.A., Cheng, W.X., et al., Net primary productivity of a CO_2-enriched deciduous forest and the implications for carbon storage, *Ecol. Appl.*, 12, 1261–1266, 2002.

Nydick, K.R., Lafrancois, B.M., and Baron, J.S., NO_3 uptake in shallow, oligotrophic, mountain lakes: The influence of elevated NO_3 concentrations, *J. N. Amer. Benthol. Soc.*, 23, 397–15, 2004.

Odmark, S., Wulff, A., Wangberg, S.A., Nilsson, C., and Sundback, K., Effects of UVB radiation in a microbenthic community of a marine shallow-water sandy sediment, *Mar. Biol.*, 132, 335–345, 1998.

Oechel, W.C., Vourlitis, G.L., Hastings, S.J., and Bochkarev, S.A., Change in arctic CO_2 flux over two decades: Effects of climate change at Barrow, Alaska, *Ecol. Appl.*, 5, 846–855, 1995.

Palffy, K. and Voros, L., Effect of ultraviolet radiation on phytoplankton primary production in Lake Balaton, *Hydrobiologia*, 506, 289–295, 2003.

Pancotto, V.A., Sala, O.E., Cabello, M., Lopez, N.I., Robson, T.M., Ballare, C.L., Caldwell, M.M., and Scopel, A.L., Solar UV-B decreases decomposition in herbaceous plant litter in Tierra Del Fuego, Argentina: Potential role of an altered decomposer community, *Global Change Biol.*, 9, 1465–1474, 2003.

Phillips, O.L., Malhi, Y., Higuchi, N., Laurance, W.F., Nunez, P.V., Vasquez, R.M., Laurance, S.G., et al., Changes in the carbon balance of tropical forests: Evidence from long-term plots, *Science*, 282, 439–442, 1998.

Pienitz, R. and Vincent, W.F., Effect of climate change relative to ozone depletion on UV exposure in subarctic lakes, *Nature*, 404, 484–487, 2000.

Poff, N.L., Brinson, M.M., and Day, J.W., Jr., *Aquatic Ecosystems and Global Climate Change: Potential Impacts on Inland Freshwater and Coastal Wetland Ecosystems in the United States*, Pew Center on Global Climate Change, 2101 Wilson Boulevard, Arlington, VA 22201, 2002.

Prentice, I.C., Heimann, M., and Sitch, S., The carbon balance of the terrestrial biosphere: Ecosystem models and atmospheric observations, *Ecol. Appl.*, 10, 1553–1573, 2000.

Price, M.V. and Waser, N.M., Responses of subalpine meadow vegetation to four years of experimental warming, *Ecol. Appl.*, 10, 811–823, 2000.

Rabalais, N.N., Turner, R.E., and Scavia, D., Beyond science into policy: Gulf of Mexico hypoxia and the Mississippi River, *BioScience*, 52, 129–142, 2002.

Rae, R. and Vincent, W.F., Effects of temperature and ultraviolet radiation on microbial foodweb structure: Potential responses to global change, *Freshw. Biol.*, 40, 747–758, 1998.

Rastetter, E.B., Agren, G.I., and Shaver, G.R., Responses of N-limited ecosystems to increased CO_2: A balanced-nutrition, coupled-element-cycles model, *Ecol. Appl.*, 7, 444–460, 1997.

Richardson, A.J. and Schoeman, D.S., Climate impact on plankton ecosystems in the northeast Atlantic, *Science*, 305, 1609–1612, 2004.

Ringold, P.L. and Groffman, P.M., Inferential studies of climate change, *Ecol. Appl.*, 7, 751–752, 1997.

Robson, T.M., Pancotto, V.A., Flint, S.D., Ballare, C.L., Sala, O.E., Scopel, A.L., and Caldwell, M.M., Six years of solar UV-B manipulations affect growth of *Sphagnum* and vascular plants in a Tierra Del Fuego Peatland, *New Phytol.*, 160, 379–389, 2003.

Roux, R., Gosselin, M., Desrosiers, G., and Nozais, C., Effects of reduced UV radiation on a microbenthic community during a microcosm experiment, *Mar. Ecol. Prog. Ser.*, 225, 29–43, 2002.

Rueth, H.M., Baron, J.S., and Allstott, E.J., Responses of Engelmann spruce forests to nitrogen fertilization in the Colorado Rocky Mountains, *Ecol. Appl.*, 13, 664–673, 2003.

Saarnio, S., Jarvio, S., Saarinen, T., Vasander, H., and Silvola, J., Minor changes in vegetation and carbon gas balance in a boreal mire under a raised CO_2 or NH_4NO_3 supply, *Ecosystems*, 6, 46–60, 2003.

Sabine, C.L., Feely, R.A., Gruber, N., Key, R.M., Lee, K., Bullister, J.L., Wanninkhof, R., et al., The oceanic sink for anthropogenic CO_2, *Science*, 305, 367–371, 2004.

Santos, R., Lianou, C., and Danielidis, D., UVB radiation and depth interaction during primary succession of marine diatom assemblages of Greece, *Limnol. Oceanogr.*, 42, 986–991, 1997.

Schimel, D., Melillo, J., Tian, H., McGuire, A.D., Kicklighter, D., Kittel, T., Rosenbloom, N., et al., Contribution of increasing CO_2 and climate to carbon storage by ecosystems in the United States, *Science*, 287, 2004–2006, 2000.

Schindler, D.E., Carpenter, S.R., Cole, J.J., Kitchell, J.F., and Pace, M.L., Influence of food web structure on carbon exchange between lakes and the atmosphere, *Science*, 277, 248–251, 1997.

Schindler, D.W., Detecting ecosystem responses to anthropogenic stress. *Can. J. Fish. Aquat. Sci.*, 44 (Suppl.), 6–25, 1987.

Schindler, D.W., Effects of acid rain on freshwater ecosystems, *Science*, 239, 149–157, 1988.

Schindler, D.W., Bayley, S.E., Parker, B.R., Cruiksjank, D.R., Fee, E.J., Schindler, E.U., and Stainton, M.P., The effects of climate warming on the properties of boreal lakes and streams in the Experimental Lakes Area, *Limnol. Oceanogr.*, 41, 1004–1017, 1996.

Schindler, D.W., Mills, K.H., Malley, D.F., Findlay, D.L., Shearer, J.A., Davies, I.J., Turner, M.A., Linsey, G.A., and Cruikshank, D.R., Long-term ecosystem stress: The effects of years of experimental acidification on a small lake, *Science*, 228, 1395–1401, 1985.

Searles, P.S., Flint, S.D., and Caldwell, M.M., A meta analysis of plant field studies simulating stratospheric ozone depletion, *Oecologia*, 127, 1–10, 2001.

Shafer, S.L., Bartlein, P.J., and Thompson, R.S., Potential changes in the distributions of western North America tree and shrub taxa under future climate scenarios, *Ecosystems*, 4, 200–215, 2001.

Shaw, M.R. and Harte, J., Control of litter decomposition in a subalpine meadow-sagebrush steppe ecotone under climate change, *Ecol. Appl.*, 11, 1206–1223, 2001.

Shaw, M.R., Zavaleta, E.S., Chiariello, N.R., Cleland, E.E., Mooney, H.A., and Field, C.B., Grassland responses to global environmental changes suppressed by elevated CO_2, *Science*, 298, 1987–1990, 2002.

Shi, S.B., Zhu, W.Y., Li, H.M., Zhou, D.W., Han, F., Zhao, X.Q., and Tang, Y.H., Photosynthesis of *Saussurea superba* and *Gentiana straminea* is not reduced after long-term enhancement of UV-B radiation, *Environ. Expt. Botany*, 51, 75–83, 2004.

Sinsabaugh, R.L., Carreiro, M.M., and Repert, D.A., Allocation of extracellular enzymatic activity in relation to litter decomposition, N deposition, and mass loss, *Biogeochemistry*, 60, 1–24, 2002.

Sirois, A., Vet, R., and Mactavish, D., Atmospheric deposition to the Turkey Lakes watershed: Temporal variations and characteristics, *Ecosystems*, 4, 503–513, 2001.

Smith, R.C., Prezelin, B.B., Baker, K.S., Bidigare, R.R., Boucher, N.P., Coley, T., Karentz, D., et al., Ozone depletion: Ultraviolet radiation and phytoplankton biology in Antarctic waters, *Science*, 255, 1992.

Smith, S.V. and Buddemeier, R.W., Global change and coral reef ecosystems, *Ann. Rev. Ecol. System.*, 23, 89–118, 1992.

Smock, L.A. and Gazzera, S.B., Effects of experimental episodic acidification on a southeastern USA blackwater stream, *J. Freshw. Ecol.*, 11, 81–90, 1996.

Sommaruga, R., The role of solar UV radiation in the ecology of alpine lakes, *J. Photochem. Photobiol. B.*, 62, 35–42, 2001.

Stevens, C.J., Dise, N.B., Mountford, J.O., and Gowing, D.J., Impact of nitrogen deposition on the species richness of grasslands, *Science*, 303, 1876–1879, 2004.

Stoddard, J.L., Driscoll, C.T., Kahl, J.S., and Kellogg, J.P., Can site-specific trends be extrapolated to a region? An acidification example for the northeast, *Ecol. Appl.*, 8, 288–299, 1998.

Strong, D.R., Are trophic cascades all wet? Differentiation and donor-control in speciose ecosystems, *Ecology*, 73, 747–754, 1992.

Tank, S.E. and Schindler, D.W., The role of ultraviolet radiation in structuring epilithic algal communities in Rocky Mountain montane lakes: Evidence from pigments and taxonomy, *Can. J. Fish. Aquat. Sci.*, 61, 1461–1474, 2004.

Tank, S.E., Schindler, D.W., and Arts, M.T., Direct and indirect effects of UV radiation on benthic communities: Epilithic food quality and invertebrate growth in four montane lakes, *Oikos*, 103, 651–667, 2003.

Tietema, A., Microbial carbon and nitrogen dynamics in coniferous forest floor material collected along a European nitrogen deposition gradient, *Forest Ecol. Manag.*, 101, 29–36, 1998.

Tingey, D.T., Laurence, J.A., Weber, J.A., Greene, J., Hogsett, W.E., Brown, S., and Lee, E.H., Elevated CO_2 and temperature alter the response of *Pinus ponderosa* to ozone: A simulation analysis, *Ecol. Appl.*, 11, 1412–1424, 2001.

Updegraff, K., Bridgham, S.D., Pastor, J., Weishampel, P., and Harth, C., Response of CO_2 and CH_4 emissions from peatlands to warming and water table manipulation, *Ecol. Appl.*, 11, 311–326, 2001.

van Breemen, N., Jenkins, A., Wright, R.F., Beerling, D.J., Arp, W.J., Berendse, F., Beier, C., et al., Impacts of elevated carbon dioxide and temperature on a boreal forest ecosystem (CLIMEX Project), *Ecosystems*, 1, 345–351, 1998.

Verhoef, H.A., Verspagen, J.M.H., and Zoomer, H.R., Direct and indirect effects of ultraviolet-B radiation on soil biota, decomposition and nutrient fluxes in dune grassland soil systems, *Biol. Fertil. Soils*, 31, 366–371, 2000.

Vinebrooke, R.D. and Leavitt, P.R., Direct and indirect effects of allochthonous dissolved organic matter, inorganic nutrients, and ultraviolet radiation on an alpine littoral food web, *Limnol. Oceanogr.*, 43, 1065–1081, 1998.

Vinebrooke, R.D. and Leavitt, P.R., Differential responses of littoral communities to ultraviolet radiation in an alpine lake, *Ecology*, 80, 223–237, 1999.

Vitousek, P.M., Beyond global warming: Ecology and global change, *Ecology*, 75, 1861–1876, 1994.

Waldrop, M.P., Zak, D.R., and Sinsabaugh, R.L., Microbial community response to nitrogen deposition in northern forest ecosystems, *Soil Biol. Biochem.*, 36, 1443–1451, 2004.

Watkins, E.M., Schindler, D.W., Turner, M.A., and Findlay, D., Effects of solar ultraviolet radiation on epilithic metabolism, and nutrient and community composition in a clear-water boreal lake, *Can. J. Fish. Aquat. Sci.*, 58, 2059–2070, 2001.

Weathers, K.C., Lovett, G.M., Likens, G.E., and Lathrop, R., The effect of landscape features on deposition to Hunter Mountain, Catskill Mountains, New York, *Ecol. Appl.*, 10, 528–540, 2000.

Williams, M., Rastetter, E.B., Fernandes, D.N., Goulden, M.L., Shaver, G.R., and Johnson, L.C., Predicting gross primary productivity in terrestrial ecosystems, *Ecol. Appl.*, 7, 882–894, 1997.

Williams, M.W., Baron, J.S., Caine, N., Sommerfeld, R., and Sanford, R., Jr., Nitrogen saturation in the Rocky Mountains, *Environ. Sci. Technol.*, 30, 640–646, 1996a.

Williams, M.W., Losleben, M., Caine, N., and Greenland, D., Changes in climate and hydrochemical responses in a high-elevation catchment in the Rocky Mountains, USA, *Limnol. Oceanogr.*, 41, 939–946, 1996b.

Williams, M.W. and Tonnessen, K.A., Critical loads for inorganic nitrogen deposition in the Colorado Front Range, USA, *Ecol. Appl.*, 10, 1648–1665, 2000.

Williamson, C.E., What role does UV-B radiation play in freshwater ecosystems? *Limnol. Oceanogr.*, 40, 386–392, 1995.

Williamson, C.E., Hargreaves, B.R., Orr, P.S., and Lovera, P.A., Does UV play a role in changes in predation and zooplankton community structure in acidified lakes? *Limnol. Oceanogr.*, 44, 774–783, 1999.

Wilsey, B.J., Coleman, J.S., and McNaughton, S.J., Effects of elevated CO_2 and defoliation on grasses: A comparative ecosystem approach, *Ecol. Appl.*, 7, 844–853, 1997.

Winch, S., Ridal, J., and Lean, D., Increased metal bioavailability following alteration of freshwater dissolved organic carbon by ultraviolet B radiation exposure, *Environ. Toxicol.*, 17, 267–274, 2002.

Wolters, V., Silver, W.L., Bignell, D.E., Coleman, D.C., Lavelle, P., van der Putten, W.H., de Ruiter, P., et al., Effects of global changes on above- and belowground biodiversity in terrestrial ecosystems: Implications for ecosystem functioning, *BioScience*, 50, 1089–1098, 2000.

Wright, R.F., Effect of increased carbon dioxide and temperature on runoff chemistry at a forested catchment in southern Norway (CLIMEX Project), *Ecosystems*, 1, 216–225, 1998.

Zak, D.R., Holmes, W.E., Finzi, A.C., Norby, R.J., and Schlesinger, W.H., Soil nitrogen cycling under elevated CO_2: A synthesis of forest FACE experiments, *Ecol. Appl.*, 13, 1508–1514, 2003.

Zak, D.R., Pregitzer, K.S., Curtis, P.S., and Holmes, W.E., Atmospheric CO_2 and the composition and function of soil microbial communities, *Ecol. Appl.*, 10, 47–59, 2000.

Zepp, R.G., Callaghan, T., and Erickson, D., Effects of increased solar ultraviolet-radiation on biogeochemical cycles, *Ambio*, 24, 181–187, 1995.

Zepp, R.G., Callaghan, T.V., and Erickson, D.J., Interactive effects of ozone depletion and climate change on biogeochemical cycles, *Photochem. Photobiol. Sci.*, 2, 51–61, 2003.

Zuellig, R.E., The influence of ultraviolet-B radiation on benthic communities in Rocky Mountain streams with different metal exposure histories, Ph.D. Dissertation, Colorado State University, Fort Collins, CO, 2006.

Part VI

Ecotoxicology: A Comprehensive Treatment—Conclusion

36 Conclusion

36.1 OVERARCHING ISSUES

> ... science students accept theories on the authority of teacher and text, not because of evidence.
>
> **(Kuhn 1977)**

> Except in their occasional introductions, science textbooks do not describe the sorts of problems that the professional may be asked to solve and the variety of techniques available for their solution. Rather, these books exhibit concrete problem solutions that the profession has come to accept as paradigms ... [however, students] must, we say, learn to recognize and evaluate problems to which no unequivocal solution has been given; they must be supplied with an arsenal of techniques for approaching these future problems ....
>
> **(Kuhn 1977)**

So what is left to say? The twin goals of differentiation and integration (see Section 1.1) were attained in Chapters 1 through 35. Facts and paradigms[1] relevant at each level of the biological hierarchy were presented and then interconnected as much as presently possible. Relative to the conduct of normal science,[2] this will foster the "determination of significant fact, matching of facts with theory, and articulation of theory" (Kuhn 1970) at all relevant scales. Hopefully, an appropriate balance was achieved by including more ecology than usually found in ecotoxicology primers. The imbalance between autecotoxicological and synecotoxicological themes is generally recognized as an important shortcoming of ecotoxicology as currently practiced (e.g., Cairns 1984, 1989; Chapman 2002). A more congruent treatment was also attempted by including relevant human effects information rather than taking the contrived approach of "asking humans to step out of the picture" when discussing human influences on the biosphere. As justifying examples, discussion of multidrug resistance transporter proteins (Chapter 3), inflammation (Chapter 4), endocrine modifiers (Chapter 5), Seyle's General Adaptation Syndrome (Chapter 9), and human epidemiology metrics (Chapter 13) surely provided enriching detail to considerations of contaminant effects on nonhuman individuals and populations. So, what more can be said? The last important issues to be explored are ways of recognizing when innovation is needed and the best way of fostering change in this extremely important science.

Statements about possible changes to the science of ecotoxicology were made throughout this book. Yet, other than some general discussion in Chapter 1, no specific advice was given about how exactly one recognizes the need for and then contributes to healthy change. As expressed in Kuhn's quote above, the absence of this kind of guidance is a fundamental shortcoming of most textbooks. To avoid committing such a sin of omission, this chapter provides context and specific advice about recognizing and then contributing to necessary change. Effective change is particularly crucial in our applied science in which much harm can be done to the biosphere and our well-being if a failing

[1] Kuhn's concept of scientific paradigm is applied here because it is as useful as it is hackneyed. According to Kuhn (1970), paradigms are "universally recognized scientific achievements that for a time provide model problems and solutions to a community of practitioners." They are the best explanations or approaches currently available to the scientific community.

[2] Activities of scientists are divided into normal and innovative science (Kuhn 1970). When participating in innovative science, a scientist intends to directly challenge an existing paradigm, or to propose a novel one. In contrast, the practice of normal science involves filling in or refining important details surrounding an existing paradigm, or building ancillary concepts or connections that reinforce or enrich a current paradigm. Normal and innovative science are complementary activities essential to the health of any science.

paradigm is maintained too long. The emerging global warming paradigm is one example where irreparable harm at a global scale could occur if effective paradigm scrutiny and possible replacement were put off any longer. The approach here will be to focus on foibles impeding innovation, hoping that understanding impediments helps minimize their influence.

> For if we learn more about resistance to scientific discovery, we shall know more also about the sources of acceptance, just as we know more about health when we successfully study disease. By knowing more about both resistance and acceptance in scientific discovery, we may be able to reduce the former by a little bit and thereby increase the latter in the same measure.
>
> **(Barber 1961)**

General cognitive and social psychology concepts will be blended with those more directly concerned with sciences. General psychological concepts were included because understanding them is pivotal to understanding how ecotoxicology might be moved out of its scientific nonage. As put simply by Bourdieu (2004), "... the obstacles to the progress of science are fundamentally social." Footnotes will be applied liberally to improve continuity despite numerous digressions about unfamiliar materials.

36.1.1 Generating and Integrating Knowledge in the Hierarchical Science of Ecotoxicology

The ecotoxicologist's vocation is eminently important and enormous. In our opinion, this new science could emerge as one of the most important for addressing society's major challenges of the millennium. The working knowledge needed to make useful ecotoxicological predictions spans broad temporal, spatial, and informational content scales (Figure 36.1). Outstanding challenges are the full articulation of major issues, development of predictive tools for handling problems arising at every scale, and most critically, the integration of concepts and tools for every scale into

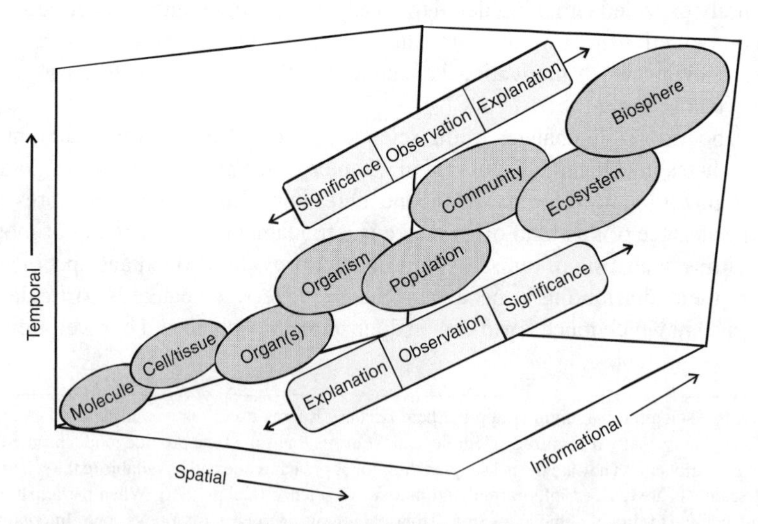

FIGURE 36.1 The temporal, spatial, and informational complexity scales relevant to ecotoxicology. In order for ecotoxicology to emerge as a self-consistent science, more integration of facts and paradigms is essential along all three scales. The explanation–observation-significance concatenation described in Chapter 1 is equally useful at any level from molecule to biosphere. Although applied most often from the lower (explanation) to higher (significance) levels, this concatenation is often useful in the opposite ("top-down") direction.

a congruent whole. A haphazard "feeling our way" (Platt 1964) approach is insufficient for this task so thoughtful discussion is needed to meet these challenges.

> Heterophilous[3] communications between dissimilar individuals may cause cognitive dissonance because an individual is exposed to messages that are inconsistent with existing beliefs, an uncomfortable psychological state.
>
> **(Rogers 1995)**

> Homophily can act as an invisible barrier to flow of innovations within a system. New ideas usually enter a system through higher status and more innovative members. A high degree of homophily means that these elite individuals interact mainly with each other, ...
>
> **(Rogers 1995)**

Fundamental social processes determine how readily a novel idea, vantage, or technique diffuses into any group, including a scientific community.[4] The only, albeit important, distinction between scientific and nonscientific communities is the rules by which beliefs acquire favored status (Chapter 1). Despite this distinction, scientists remain subject to rules governing acceptance of ideas and innovation in any social group. As stated by Barber (1961), the rational, open-minded tradition has a powerful influence in scientific communities yet it "works in conjunction with a number of other cultural and social elements, which sometimes reinforce it, sometimes give it limits." As reflected in the above quotes, an obstacle to accepting new ideas is often the barrier presented by homophily: scientists trained in a particular discipline encounter cognitive dissonance, resist, and then react against ideas from outside their immediate training or practice; for example, a systems ecologist's negative response to ideas of a mammalian toxicologist or vice versa. The discomfort invoked by dissonance prompts a person to seek association with those sharing similar ideas and to actively thwart, or minimally isolate, those with different ideas. Most likely, this is the true root of the distracting reductionism–holism debate criticized in Chapters 1 (Section 1.2) and 20 (Section 20.2.1), not any definitive superiority of one or the other as an investigative vantage. These dynamics are altered only when the discomfort of maintaining a failing paradigm or stance becomes harder to bear than that associated with confronting cognitive dissonance. Kuhn (1977) describes such a crisis in scientific communities as one of "pronounced professional insecurity" that forces resolution. Contrary to popular belief,[5] most scientists only abandon "the idol of certainty" (Popper 1959) under duress.

Described as an essential tension by Kuhn (1970, 1977), scientists resist the discomfort of change until that of bolstering a failing paradigm becomes harder to bear. It should come as no surprise that strong conformists are most often the opinion leaders of groups, including those of scientific communities, and innovators are the most actively censored members (Bourdieu 2004, Rogers 1995) *except during special occasions requiring change*.

It is our opinion that ecotoxicology is in a period of tension relative to the integration of core concepts from all pertinent scales of organization. Out of immediate necessity, the individual-based paradigms and approaches of mammalian toxicology were wisely adopted in our young science to address immediate problems. However, enough data and collective experience has accumulated to expose the discomforting inconsistencies among level-specific paradigms and approaches that, together, constitute a self-contradictory system. For example, most risk assessments reluctantly

[3] Homophily is the degree to which two interacting or communicating individuals are similar in relevant attributes. The opposite of homophily is heterophily (McPherson et al. 2001, Rogers 1995).

[4] A scientific community is a "group whose members are united by a common objective and culture" (Hagstrom 1965).

[5] Typifying the presumption of scientists being open-minded is Berkeley's quote, "He must surely be either very weak, or very little acquainted with the sciences, who shall reject a truth, that is capable of demonstration, for no other reason but because it is newly known and contrary to the prejudices of mankind" (Berkeley 1710). In reality, strong-willed and informed scientists often display these behaviors. Barber (1961) provides an insightful counterpoint to this conventional image.

ignore central ecological principles, relying on LC50 and NOEL data for effects to individuals. Recently, a pragmatic species sensitivity distribution approach combining individual LC50 or NOEC values has been proposed to estimate a concentration protective of ecosystems (see Posthuma et al. 2002 for details). Such tests tacitly ignore critical species interactions such as those described in Chapter 27 among killer whales, pinnipeds, otters, urchins, and kelp.[6] As explained by Popper (1959), such self-contradictory systems eventually fail because they are uninformative systems from which "... no statement can be singled out, either as incompatible or as derivable, since all are derivable." As evidenced in the primary literature and reflected in this textbook, a paradigm shift seems to have begun in which ecological paradigms and techniques are applied in an increasingly congruent and concerted manner with those emerging from mammalian toxicology. In times like this, the ideal opinion leader is more innovative than typical and innovators are given more credibility than otherwise warranted.[7] Ecotoxicology will shatter into a loose collection of interface disciplines (Odum 1996) in the absence of such leadership and the recognition of contributions made by innovators.

Sociology provides clues about how best to act as an individual practitioner in a science undergoing necessary change. The most useful is related to the above discussion of homophilic and heterophilic diffusion of innovations. Homophilic communication, the form of exchange that is most frequent and least likely to produce cognitive dissonance, is also the form least likely to contain novel information with which to solve an emerging problem. The more challenging and frustrating, heterophilic communication is more likely to produce rapid diffusion of crucial knowledge into one's evolving social group. This is the basis of the Strength of Weak Ties Theory, i.e., weak, heterophilic communication forms bridging links containing the most novel information with which to address challenges faced by a social group (Rogers 1995). What is the specific message to be taken from this by a nonleader practitioner of ecotoxicology today? Instead of seeing ecotoxicologists working at a different scale than your immediate group as erring competitors to be coped with, consider the possibility that your scientific activities would be most enhanced by exploring the vantage or ideas of these heterophilic ecotoxicologists. In doing this, strong emphasis should be on discovering consistency among levels and enriching interpretation at any particular level.

Heterophilic communication among ecotoxicologists working at different hierarchical levels is not enough to meet the challenge: the communication must also be thoughtful in order to reach sound judgments. For example, satisficing is a common danger in decision making by well-intended individuals. Satisficing is a flawed form of decision making that often emerges in groups of individuals with very different agendas and vantages (i.e., heterophilic groups). Instead of seeking the best possible decision, the group takes "a course of action that is 'good enough,' that meets a minimal set of requirements" of all parties (Janis and Mann 1977). The operating premise is that a "barely 'acceptable' course ... [is] better than the way things are now." Janis and Mann (1977) and similar books on decision making theory and practice provide concrete means of minimizing suboptimal heterophilic interactions like satisficing. Certainly, the influence of satisficing during heterophilic scientific deliberations can be reduced by adopting Chamberlin's multiple working hypothesis scheme (Chamberlin 1897) in combination with Bayesian abductive inference methods

---

[6] This is an example of the behavior called collusive lying, that is, "two parties, knowing full well that what they are saying or doing is false, collude in ignoring the falsity" (Bailey 1991) (see also Section 16.4). It is a technique of groups attempting to establish a useful paradigm in which they push away an inconsistent fact until the fledgling paradigm has been given sufficient clarity and detail to be assessed properly. Unfortunately, this understandable behavior also carries the risk of uncritical acceptance by the ecotoxicology community based on the stature of advocates with consequent long delay in eliminating such a compromised paradigm when its shortcomings are eventually revealed.

[7] It is equally important to understand that innovators are very poor opinion leaders when no change is needed (Rogers 1995). Their distracting exploration of unhelpful innovations can slow the accumulation of facts and ancillary concepts by normal science around useful, established paradigms.

(Howson and Urbach 1989, Josephson and Josephson 1996, Pearl 2000) that will be discussed at the end of this chapter.

36.1.2 Optimal Balance of Imitation, Innovation, and Inference

> Transmission [of ideas and innovations] withers on the vine when the present is taken as the only model. And innovation itself withers with it, scorn for the past being the greatest enemy of progress.
>
> **(Debray 2000)**

> Very often the successful scientist must simultaneously display the characteristics of the traditionalist and of the iconoclast.[8]
>
> **(Kuhn 1977)**

In the rest of this chapter, we try to sketch out the most efficient way of incorporating crucial innovations while preserving existing, valuable ecotoxicological theory and practice. The next two sections explore the relative virtues of resisting and embracing change to scientific knowledge.

36.1.2.1 The Virtues of Imitation

> Among the forces that support social rules there is the imperative of regularization ... of "falling into the line with the rule" ....
>
> **(Bourdieu 2004)**

> The dominant players [in a science] impose by their very existence, as a universal norm, the principles that they engage in their own practice. This is what is called into question by revolutionary innovation .... A major scientific innovation may destroy whole swathes of research and researchers as a side-effect, without being inspired by the slightest intention of doing damage .... It is not surprising that innovations are not well received, that they arouse formidable resistance ....
>
> **(Bourdieu 2004)**

> Invention of alternates is just what scientists seldom undertake except during the pre-paradigm stage of their science's development and at very special occasions during its subsequent evolution. So long as the tools a paradigm supplies continue to prove capable of solving the problems it defines, science moves fastest and penetrates most deeply through confident employment of those tools. The reason is clear. As in manufacture so in science ... retooling is an extravagance to be reserved for the occasion that demands it.
>
> **(Kuhn 1977)**

Change is resisted for obvious reasons, some high minded, and others not. Often a scientist's behavior and activities have committed them to a paradigm that is now questioned, resulting in their beliefs and hard work also being questioned—"A threat to theory is therefore a threat to the scientific life" (Kuhn 1977). To expect open and immediate acceptance from such a scientist is to expect the superhuman. Such a person will resist dismissing the value of their past work or giving up their hard-earned professional status. More importantly, if a certain level of resistance to change were not present in a scientific community, forward progress would be stymied by frequent detours or side trips to explore novel paradigms that eventually turned out to be dead ends. That is the point being made in the quote above from Kuhn (1977). It is inefficient to keep "retooling" a science

[8] Kuhn (1977) adds an important footnote to this statement, "Strictly speaking, it is the professional group rather than the individual scientist that must display both these characteristic simultaneously."

when there is no need. Furthermore, Loehle (1987) notes that insistence on testing and potentially rejecting a concept before sufficient related details have accumulated by normal science can lead to premature ("dogmatic") falsification of a perfectly sound paradigm. A current case in point is the species sensitivity distribution approach mentioned briefly above. There is real virtue to healthy resistance to change when faced with a novel explanation or solution. The key is avoiding pathological resistance.

36.1.2.2 The Wisdom of Insecurity

> Our innate social psychology is probably that bequeathed to us by our Pleistocene ancestors.
>
> **(Richerson and Boyd 2005)**

> We have analyzed this problem using several mathematical models of the evolution of imitation, and all of them tell the same story. Selection favors a heavy reliance on imitation whenever individual learning is error prone or costly, and environments are neither too variable nor too stable. When these conditions are satisfied, our models suggest that natural selection can favor individuals who pay almost no attention to their own experience, and are almost totally bound to what Francis Bacon called the "dead hand of custom."
>
> **(Richerson and Boyd 2005)**

> ... social influences exist that tend to form habits of thought leading to inadequate and erroneous beliefs.
>
> **(Dewey 1910)**

As evidenced by the above quotes, the strategy of imitation—uncritical acceptance of actions or beliefs of those around us—is prominent in human interactions. Evolution favored imitation in our Pleistocene ancestors as they attempted to survive in bands of hunter-gatherers. It remains intact in modern social groups (Richerson and Boyd 2005), including scientific communities. However, modern groups exist in a very different environment and have goals other then hunting and gathering: sometimes our evolved behaviors remain useful, but in others, they are maladaptive. One healthy modern manifestation is informational mimicry, a behavior in financial corporations in which a group mimics another particularly knowledgeable group, rather than attempting to formulate its own strategy (Vernimmen et al. 2005). A maladaptive example is faculty mobbing, an odd tendency of university faculty to mob[9] overachieving members "who stand out from the crowd," endangering the common good (Westhues 1998, 2005). These evolved strategies, which were optimized for Pleistocene hunter-gatherers social groups, arguably do not always result in optimal fitness of a modern group of scientists and, consequently, require some understanding and coping behavior. As already discussed, optimal fitness relative to a scientific community's goal involves a thoughtful balance of adherence to traditional explanation (mimicking) and openness to plausible alternatives. Imitation has clear advantages as long as a minimum number of competent innovators exist when change is required.

A group's ability to respond to change is placed at risk when too few innovators exist, either because of too active exclusion or because of passive neglect in teaching innovation skills. Clearly articulating to students the valued role of innovation and then actively teaching problem solving skills are two pedagogical necessities in a healthy modern science, especially a nascent one like ecotoxicology.

Fitting Gigerenzer's (2000) metaphor to the present topic, trying to change a scientific discipline can be like trying to move a cemetery. Given the social psychological backdrop described above, the

---

[9] "Workplace mobbing is 'a common and bloodless form of workplace mayhem' (Maguire 1999), usually carried out politely and without violence" (Westhues 2005).

challenge for leaders in our field is facilitating necessary change in the presence of natural resistance. An inappropriately conservative (or innovative) opinion leader will be swept aside after a period of ineffective confusion. The challenge for opinion leaders is discerning when and where innovation is truly needed. A major component of any solution is adopting a way of reliably discerning the relative plausibility of candidate explanations, and then effectively updating these plausibility estimates as new information emerges from normal science activities. An ideal system would allow the most effective identification and then communication of the need for change to the group's members. Such a system should be resistant to processes such as satisficing or groupthink (a suboptimal process prevalent in homophilic group decision making in which group concurrence is the objective, not the best decision Box 36.1) (Janis and Mann 1977). Unfortunately, the unstructured expert opinion systems on which we currently depend for making regulatory and many scientific judgments are prone to these kinds of decision making errors (Cooke 1991). The approach of melding Chamberlin's multiple working hypothesis schemes and Bayesian abductive inference methods described at the end of this chapter provides one possible means of doing this.

Box 36.1 Minimizing Groupthink

Groupthink, a pervasive problem in homophilic group decision making, has significantly impacted our recent history. A classic example is the flawed decision making that occurred in the Kennedy administration that led to the failed Bay of Pigs invasion. More recently, NASA groupthink contributed substantially to the 1986 Challenger space shuttle disaster. Groupthink will be examined closer here because of our pervasive dependence on the error-prone expert opinion approach for establishing much ecotoxicological consensus and assessing ecological risk. Although underexploited by natural scientists in their decision making, there are simple ways of reducing groupthink's influence during group activities.

The qualities of groupthink are described by Janis and Mann (1977). Importantly, groupthink occurs with "in-groups" (homophilic groups) in which concurrence is a highly valued characteristic of the group dynamics. Rationalizations are invoked to preserve and foster concurrence. In-groups experiencing groupthink often manifest eight characteristics:

1. Members tend to feel minimally vulnerable to mistakes and become overconfident in their abilities. Consequently, they take on more risk in decisions than warranted by facts.
2. There is a consorted effort to dismiss contrary facts or opinions, and to rationalize.
3. The group's "inherent morality" becomes a given during deliberations.
4. Foibles are exaggerated and strengths trivialized for rivals or those with contrasting opinions.
5. Direct pressure is applied to any member who questions the group's actions or stances.
6. Members who have doubts practice self-censoring in which they remain silent despite their misgivings.
7. That silence indicates concurrence is a shared assumption of the group.
8. Self-appointed "mindguards" emerge who aggressively act to protect the group from erring members or information inconsistent with emerging consensus.

Groupthink can erode the quality of a group's deliberations. Fortunately, there are means of reducing its influence (Table 36.1). Panels, committees, and less-formal scientific teams can benefit greatly from being aware of them. It is easy to imagine groupthink emerging during a panel's deliberations when applying Fox's or Miller's qualitative rules (see Section 13.2.1 and Box 13.2 in Chapter 13, and Box 22.3 in Chapter 22) or codified Environmental Protection

TABLE 36.1
Tools for Reducing Groupthink

1. The leader or empowering agency should objectively emphasize the need for impartiality at the onset of the decision-making process.
2. The leader or empowering agency should state the importance of expressing objections and concerns, making this an obligation of each group member.
3. One member of the group should be assigned the role of skeptic or challenger during each decision-making session.
4. Perhaps by dividing the group occasionally to conduct separate assessments, the group should periodically assess the feasibility of the group's current stance.
5. If a rival group with contrasting views can be identified, enough members of that group should be as engaged as possible in establishing possible alternate decisions.
6. After the consensus-building meetings have occurred, a "second chance upon reflection" meeting should be planned to air any concerns emerging after the group breaks up.
7. Significant experts not associated with the core group should be invited to engage with the group with the request to act as a "fresh pair of eyes."
8. Each group member should be asked to discuss the group's thoughts and progress with trusted peers and report back the results of these independent exchanges.
9. If possible, separate groups can be established that address the same problem or question. The decisions from different groups are then used to come to a final decision.

Source: Summarized from pages 399–400 of Janis and Mann (1977).

Agency (EPA) guidance (2000) to determine the cause or risk associated with a particular scenario.

36.1.2.3 Strongest Possible Inference: Bounding Opinion and Knowledge

It is therefore worth while to search out the bounds between opinion and knowledge.

(Locke 1690, reprinted 1959)

What current approach best balances conceptual stability and change? How can we define the bounds between mere opinion and sound knowledge? Clues can be found in many places. Some may be familiar while others may induce a level of discomfort, which was identified earlier as a characteristic of heterophilic exploration of concepts.

Previously, we explored the concept of strong inference as articulated in the classic *Science* paper by Platt (1964). The conventional Baconian scientific method is advocated by Platt with emphasis being placed on consistent application of such an inductive inference approach in a scientific discipline. He holds up as an exemplary example of a conditional logic tree the process that a chemist might employ to qualitatively determine the nature of a substance. A series of positive/negative tests are conducted in an exclusionary manner until only one possibility remains. Unfortunately, there are substantial shortcomings associated with advocating such an approach as the kingpin of scientific inquiry. Even assuming that the qualitative (positive/negative) nature of such a process is adequate for all tasks, the presence of Type I and II error rates restricts the value of such a simple approach, making it prone to error in the hands of the naïve practitioner (e.g., Box 10.2). Type I and II errors must be considered in order to make sensible decisions. Also, many ecotoxicological judgments of plausibility are based on quantitative information for which such a dichotomous approach is suboptimal or prone to logical error (see again Box 10.2). Some require the adaptive inference strategy described in Section 29.5.3.

An even more serious problem with using such a strictly Popperian falsification scheme arises from trying to incorporate the second major component of Platt's strong inference approach, the method of multiple hypotheses, which Platt (1964) claims to be "... the second great intellectual invention ... which is what is needed to round out the Baconian scheme." Chamberlin's (1897) multiple hypotheses approach attempts to reduce the bias toward any particular hypothesis(ses) during testing by requiring that all plausible hypotheses be given equal amounts of effort during testing. Unfortunately, adhering to this approach is extremely difficult and often becomes contrived in ecotoxicology. The complexity (high dimensionality) and high uncertainty of many ecotoxicological issues limits the value of any testing method that requires the classic dichotomous "accept/reject" context advocated by Popper and Platt, and institutionalized in Fisher's significance testing. While current null hypothesis-based testing remains invaluable, dogmatic rejection of any other logical approach of gauging plausibility of an explanation creates an impasse for the ecotoxicologist.

> As long as there is an institutionalized [null hypothesis testing] methodology that does not encourage researchers to specify their hypotheses, there is little incentive to think hard and develop theories from which such hypotheses could be derived.
>
> **(Gigerenzer 2000)**

Fortunately, an approach for reducing these difficulties exists for which the approach advocated by Platt (1964) is a special case. It is called the Strongest Possible Inference approach in this book only for the purposes of identifying it as a simple extension of Platt's Strong Inference and emphasizing the conditional nature of any ensuing judgments of scientific hypothesis/explanation plausibility. It is not novel, being a straightforward application of quantitative abductive inference. The strongest possible inferences available to ecotoxicologists can be made at this time with abductive inference as formalized in Bayesian inference formulations.[10] Associated Bayesian methods are pervasive and widely available, and many textbooks (e.g., Gelman et al. 1995, Neapolitan 2004, Pearl 2000, Woodworth 2004) and software [e.g., Netica® (Norsys Software)] facilitate their implementation.

Explanation of Strongest Possible Inference will begin by repeating Locke's premise that "our assent ought to be regulated by the grounds of probabilities" (Locke 1690). This seventeenth-century quote contains the essence of abductive inference and Bayesian statistical inference. Abductive inference is simply inference that favors the most probable explanation or hypothesis (Newman and Evans 2002). Josephson and Josephson (1996) use the following syllogism for abductive inference:

D is a data collection about a phenomenon.
H explains the data collection, D.
No alternate hypothesis (H_A) explains D as effectively as H does.
$\therefore H$ is probably true.

The key to applying abductive inference is quantifying "as effectively as" and "probably." Bayesian techniques permit quantification of abductive inference. The logic can be shown for using data to judge a single hypothesis:

D provides support for H if $P(H \mid D) > P(H)$.
D draws support away from H if $P(H \mid D) < P(H)$.
D provides neither undermining nor supportive information if $P(H \mid D) = P(H)$.

where $P(H)$ = the probability of the hypothesis being true before any consideration of the data, and $P(H \mid D)$ = the probability of the hypothesis being true given the data. The task becomes estimating the probabilities. Evidence is combined with a prior probability of H being true to produce a statement of probability given the evidence—a new probability of an explanation being true is established. If more evidence (D_{NEW}) was then collected during an inquiry, the newly established probability can be used as the new "prior probability"[11] and combined with D_{NEW} to calculate a new post

[10] See Boxes 10.2 and 13.3 as instances in which we have already applied Bayesian methods for this purpose.
[11] The probability is a "prior probability" relative to the collection of the new data.

probability reflecting the plausibility of the H, given D and D_{NEW}. Bayes's theorem (Equation 36.1) can be used to estimate $P(H \mid D)$ in this case

$$P(H \mid D) = \frac{P(H)P(D \mid H)}{P(D)}, \tag{36.1}$$

where $P(D \mid H)$ = the probability of getting the data given the hypothesis was true, and $P(D)$ = the probability of getting the data regardless of whether or not the hypothesis was true. The resulting $P(H \mid D)$ can become the new prior probability ($P(H_{NEW})$) with the collection of additional data

$$P(H \mid D_{NEW}) = \frac{P(H_{NEW})P(D_{NEW} \mid H)}{P(D_{NEW})}. \tag{36.2}$$

This process can be repeated with the addition of data until the associated probability is "good enough" to make an evidence-based judgment of hypothesis or explanation plausibility. Obviously, if more information becomes available, it can be modified again. This same process can be applied to judging any hypothesis against its negation ($\sim H$), a single alternate (H_A), several alternate hypotheses (e.g., H_{A1}, H_{A2}, H_{A3} ...). Equation 36.3 illustrates how the posterior odds for H versus H_A being true ($P(H \mid D)/P(H_A \mid D)$) can be calculated from the prior odds ($P(H)/P(H_A)$) and likelihood ratio ($P(D \mid H)/P(D \mid H_A)$):

$$\frac{P(H \mid D)}{P(H_A \mid D)} = \frac{P(H)}{P(H_A)} \frac{P(D \mid H)}{P(D \mid H_A)}. \tag{36.3}$$

Equation 36.4 estimates the probability of an hypothesis from its prior ($P(H)$), the prior of its negation ($P(\sim H)$), $P(D \mid H)$, and $P(D \mid \sim H)$:

$$P(H \mid D) = \frac{P(H)P(D \mid H)}{P(H)P(D \mid H) + P(\sim H)P(D \mid P(D \mid \sim H))}. \tag{36.4}$$

If one thinks carefully for a moment about Equation 36.4, it will become clear that $P(H \mid D)$ expresses the Positive Predictive Value (PPV) of a test, the usefulness of which was illustrated in Box 10.2 for conventional hypothesis testing. (See page 81 in Gigerenzer 2000 for more details.) Equation 36.5 is a generalization in which the probability of the ith hypothesis or explanation (H_i) of n hypotheses/explanations being true given information (D)

$$P(H_i \mid D) = \frac{P(D \mid H_i)P(H_i)}{\sum_{i=1}^{n} P(D \mid H_i)P(H_i)}. \tag{36.5}$$

So, the plausibility of an explanation or relative plausibilities of a set of alternate explanations can be judged using evidence-based probabilities. These estimates of belief warranted by evidence can be recalculated periodically, and associated explanations reinforced or discarded as evidence accumulates. Calculations can include the simple "accept/reject" context described by Popper (1959), or more complex contexts with higher uncertainty.

An excellent illustration of this approach can be found in the publications of Lane, Hutchinson, and coworkers (Cowell et al. 1991, Hutchinson et al. 1989, Lane 1989, Lane et al. 1987) although they focus on applications during medical diagnosis. Newman and Evans (2002) and Newman et al. (2007) discuss their direct application to ecological risk assessment. Gigerenzer (2000, 2002) provides many examples of applying these methods in risk assessment and communication.

For the reader who is put off by the above equations, it may be helpful to realize that fairly complex situations can be rendered to this framework by using natural frequencies instead of probability equations (Gigerenzer 2000, 2002). The reader is urged to review Box 13.3 in which Bayesian methods were applied with more intuitive diagrams.

Box 36.2 Fish Kills due to Toxic Dinoflagellate Blooms or Hypoxia?

> *P. piscicida* was implicated as the causative agent of 52 ± 7% of the major fish kills . . . on an annual basis in North Carolina estuaries and coastal waters.
>
> **(Burkholder et al. 1995)**

A series of large fish kills began in mid-Atlantic USA estuaries and coastal waters in the early 1990s. Regional resource managers and politicians asked scientists to determine the cause so a remedy could be found. Early in the process, the notionally toxic dinoflagellate *Pfiesteria piscicida* was identified as the cause by North Carolina researchers (e.g., Burkholder et al. 1992). This conclusion leaned heavily on statements like the one quoted above. The basis for this statement was 3 years of monitoring data in which high levels of *P. piscicida* were found at 8 of 15, 5 of 8, and 4 of 10 large fish kills. When counterarguments were presented that episodic low oxygen events could be the cause, a confrontation took place that "became mired in accusations of ethical misconduct, risk exaggeration, and legislative stonewalling" (Newman et al. 2007). The maladaptive features of heterophilic exchange manifested, leading to the resource managers being poorly served.

Newman et al. (2007) recently used this situation to illustrate how Bayesian methods could quantify the relative plausibility of two competing explanations. They began by reiterating the point of Stow (Stow 1999, Stow and Borsuk 2003) that the above quote demonstrates a common error. The above quote describes data with which *P*(*Pfiesteria* | Fish Kill) can be estimated, yet the direct conclusion was made about *P*(Fish Kill | *Pfiesteria*). These probabilities are different. Bayes's theorem (Equation 36.1) can be used to show how they are related,

$$P(\text{Fish Kill} \mid \textit{Pfiesteria}) = \frac{P(\text{Fish Kill})P(\textit{Pfiesteria} \mid \text{Fish Kill})}{P(\textit{Pfiesteria})},$$

where *P*(Fish Kill) = the probability of a fish kill, *P*(*Pfiesteria*) = the probability of finding high *Pfiesteria* levels at an estuarine or coastal location regardless of whether or not a fish kill occurred. Newman and Evans (2002) used published reports to derive estimates of *P*(*Pfiesteria*) (either .205 or .345 depending on the data applied) and *P*(Fish Kill) (approximately .081), and used these probabilities along with Burholder et al.'s *P*(*Pfiesteria* | Fish Kill) to estimate the *P*(Fish Kill | *Pfiesteria*). Depending on which data were used for *P*(*Pfiesteria*), the probability of getting a large fish kill given the presence of high *Pfiesteria* levels (*P*(Fish Kill | *Pfiesteria*)) was estimated to be between .122 and .205. The likelihood of getting a large fish kill if high levels of *Pfiesteria* were present (12–21%) was lower than the originally inferred 52%.

What about the relative likelihoods of large fish kills being associated with *Pfiesteria* versus hypoxia? Rearrangement of Equation 36.3 produces a likelihood ratio that answers this question:

$$\frac{P(H \mid D)}{P(H_A \mid D)} = \frac{P(H)}{P(H_A)} \frac{P(D \mid H)}{P(D \mid H_A)}$$

$$\frac{P(\textit{Pfiesteria} \mid \text{Fish Kill})}{P(\text{Low DO} \mid \text{Fish Kill})} = \frac{P(\textit{Pfiesteria})}{P(\text{Low DO})} \frac{P(\text{Fish Kill} \mid \textit{Pfiesteria})}{P(\text{Fish Kill} \mid \text{Low DO})}$$

$$\frac{P(\text{Fish Kill} \mid \textit{Pfiesteria})}{P(\text{Fish Kill} \mid \text{Low DO})} = \frac{P(\textit{Pfiesteria} \mid \text{Fish Kill})}{P(\text{Low DO} \mid \text{Fish Kill})} \frac{P(\text{Low DO})}{P(\textit{Pfiesteria})}$$

where *P*(Low DO | Fish Kill) = probability of low dissolved oxygen concentration if a large fish kill occurred, *P*(Low DO) = probability of low dissolved oxygen concentration at a location, and $P(\text{Fish Kill} \mid \text{Low DO})$ = the probability of a large fish kill given low dissolved oxygen concentration at a location. Newman et al. (2007) used published or North Carolina State agency records for the period and region studied in Burkholder et al. (1995) to obtain estimates of the probabilities, $P(\text{Low DO} \mid \text{Fish Kill}) = .220$ and $P(\text{Low DO}) = .095$. These probabilities were inserted into the equation for the likelihood ratio:

$$\frac{P(\text{Fish Kill} \mid \textit{Pfiesteria})}{P(\text{Fish Kill} \mid \text{Low DO})} = \frac{P(\textit{Pfiesteria} \mid \text{Fish Kill})P(\text{Low DO})}{P(\text{Low DO} \mid \text{Fish Kill})P(\textit{Pfiesteria})}$$

$$= \frac{(0.520)(0.095)}{(0.220)(0.345 \text{ or } 0.205)} = 0.651 \text{ or } 1.095.$$

A conditional, evidence-based inference from these likelihoods is that hypoxia is as, or slightly more, likely to have caused a large fish kill than *Pfiesteria*.

This kind of reasoning is more easily understood and communicated using natural frequencies (e.g., "42 out of 100") as argued persuasively by Gigerenzer (2000, 2002) and illustrated in Box 13.3. Figure 36.2 illustrates how the natural frequency approach can lead to these same inferences using the $P(\textit{Pfiesteria})$ estimate of 0.205. From that figure, the likelihood ratio can be calculated:

$$\frac{421 \text{ Cases of large Fish Kills with high } \textit{Pfiesteria} \text{ levels}}{1884 \text{ Cases of no large Fish Kills with high } \textit{Pfiesteria} \text{ levels}} = 0.22346$$

$$\frac{178 \text{ Cases of large Fish Kills with low dissolved oxygen concentrations}}{873 \text{ Cases of no large Fish Kills with low dissolved oxygen concentrations}} = 0.20389$$

$$\text{Likelihood ratio} = \frac{0.22346}{0.20389} = 1.096.$$

The advantage of Bayesian inference is the ability to quantitatively express the degree of belief warranted by evidence in a particular explanation. Any quantitative expression can be

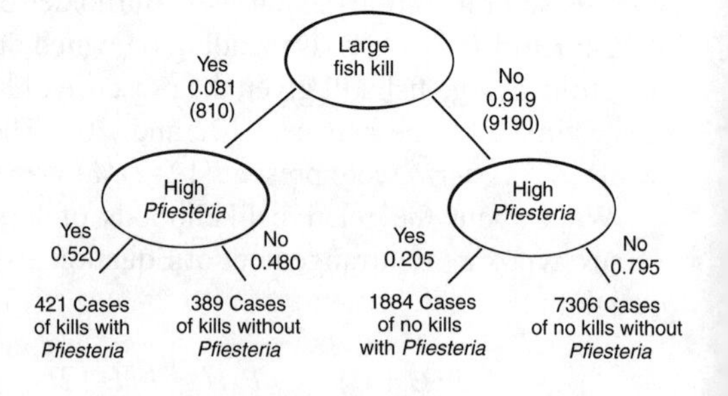

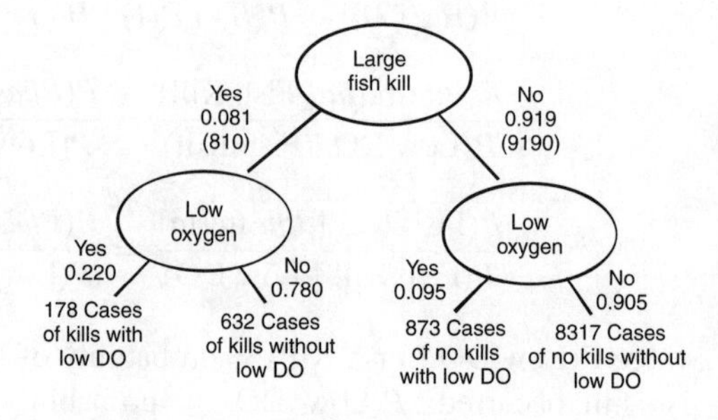

FIGURE 36.2 An example of applying Bayesian inference using natural frequencies. Probabilities are placed alongside arrows associated with different states. The numbers associated with each branch are based on 10,000 cases. As examples, 810 of 10,000 cases will involve a large fish kill or 421 of 10,000 cases will involve a large fish kill and the presence of high levels of *Pfiesteria*.

updated as new evidence is obtained. The equations above also suggest what information is most needed to permit belief-based action (i.e., funding and legislative decision making) to be effectively aligned with or withdrawn from a particular explanation. In this case, more accurate and precise information for the associated probabilities would be extremely helpful for both the hypoxia and *Pfiesteria* explanations for fish kills.

The one issue unresolved relative to the Strongest Inference Possible approach is establishing the probability at which confidence is sufficiently high to accept a choice or explanation. Just as we concluded in Chapter 10 relative to hypothesis and equivalence testing, there is no simple answer because the issue covers a broad array of situations. However, this difficulty does not force us to fall back on qualitative approaches such as Hill's rules: conditional answers can be provided.

Acceptable probabilities can be chosen depending on the stage of scientific inquiry or the seriousness of decision error consequences. This logic is similar to that employed by the EPA and other regulatory agencies in establishing thresholds for acceptable "excess mortality" (i.e., the 1 in 10^4 to 1 in 10^6 rule). To be of most utility, probabilities should be agreed on *a priori*. As an overly simplified illustration, a civil action in U.S. courts requires a preponderance of evidence (i.e., $P > .50$), but a criminal trial requires a level of belief "beyond a reasonable doubt" ($P > .70$–$.90$?), which experience has shown to be more difficult to pinpoint (Cohen 1977). Attempts have been made during expert elicitation exercises to define probabilities associated with expressions of warranted belief. The Kent chart provided by Cooke (Table 2.4) is one example:

Highly likely to near certainty	90–99%
Probable	60%
Even or about even chances	40–50%
Improbable	10%
Nearly impossible	1%

Cooke (1991) describes techniques to achieve best consensus about probabilities during expert opinion elicitations. Obviously, the same type of process is needed to choose between alternate explanations or decisions. How different do the associated probabilities have to be in order to decide to conditionally accept one and reject another?

36.2 SUMMARY: *SAPERE AUDE*[12]

Ecotoxicology's ambitious goals, immediate obligations to society, and unquestionable success in generating a rich information base have created the need for integration of information and explanations into a congruent whole. Drawing on the concepts sketched out in this chapter, information created via normal science has succeeded in producing enough cognitive dissonance that change in existing paradigms must occur. Suggestions about how an ecotoxicologist might recognize and facilitate effective change are provided and a Strongest Inference Possible approach advocated as the best means of judging the relative merits of hypotheses or explanations in situations ranging from high to very low certainty. The Strongest Inference Possible approach is simply an extension of Platt's Strong Inference approach. Platt himself extended the Baconian approach by simply insisting on consistent application of the exclusionary "scientific method" and incorporation of Chamberlin's multiple hypotheses approach. The Strongest Inference Possible approach only adds

[12] "Dare to Know," an intellectual challenge made famous by Immanual Kant.

to Platt's Strong Inference the integration of Bayesian inference methods that allow most effective inferences in situations varying in levels of uncertainty. The ecotoxicologist's task of judging, explaining, and integrating information for all relevant levels requires the most effective approach. We suggest that the Strongest Inference Possible approach is the most effective approach currently available.

36.2.1 Summary of Foundation Concepts and Paradigms

- Ways of fostering appropriate innovation in ecotoxicology were highlighted in this chapter. Effective change is crucial in this applied science in which delay can result in substantial harm to the biosphere and our well-being.
- Unnecessary stasis or change can impede the forward movement of any science.
- The major impediments to scientific progress are social; therefore, social solutions are required.
- Diffusion of innovation can be accelerated by understanding the value and shortcomings of heterophilic and homophilic interactions.
- On the basis of the Strength of Weakest Links theory, the most productive (and challenging) interactions for an ecotoxicologist working at a particular scale will be those with ecotoxicologists working at a different scale.
- During periods requiring change, opinion leaders should become more innovative and heterophilic interactions should increase. During periods in which unnecessary change would impede normal science, the opposite is most useful in a scientific community.
- Ecotoxicology is currently in a period of tension that requires more innovation, especially in connecting facts and concepts emerging at different temporal, spatial, and informational content scales (Figure 36.1).
- Platt's classic Strong Inference approach can be extended via Bayesian inference techniques to create a powerful tool for judging plausibility of explanations.

REFERENCES

Bailey, F.G., *The Prevalence of Deceit*, Cornell University Press, Ithaca, NY, 1991.

Barber, B., Resistance by scientists to scientific discovery, *Science*, 134, 596–602, 1961.

Berkeley, G., *Principles of Human Knowledge/Three Dialogues*, Penguin Books Ltd., London, UK, 1710.

Bourdieu, P., *Science of Science and Reflexivity* (translated by R. Nice), University of Chicago Press, Chicago, IL, 2004.

Burkholder, J.M., Glasgow, H.B., Jr., and Hobbs, C.W., Fish kills linked to a toxic ambush-predator dinoflagellate: Distribution and environmental conditions, *Mar. Biol. Prog. Series*, 124, 43–61, 1995.

Burkholder, J.M., Noga, E.J., Hobbs, C.H., Glasgow, H.B., Jr., and Smith, S.A., New "phantom" dinoflagellate is the causative agent of major fish kills, *Nature*, 358, 407–410, 1992.

Cairns, J., Jr., Are single species toxicity tests alone adequate for estimating environmental hazard? *Environ. Monit. Assess.*, 4, 259–273, 1984.

Cairns, J., Jr., Will the real ecotoxicologist please stand up, *Environ. Toxicol. Chem.*, 8, 843–844, 1997.

Chamberlin, T.C., The method of multiple working hypotheses, *J. Geol.*, 5, 837–848, 1897.

Chapman, P.M., Integrating toxicology and ecology: Putting the "eco" into ecotoxicology, *Mar. Pollut. Bull.*, 44, 7–15, 2002.

Cohen, L.J., *The Probable and the Provable*, Oxford University Press, Oxford, UK, 1977.

Cooke, R.M., *Experts in Uncertainty. Opinion and Subjective Probability in Science*, Oxford University Press, New York, 1991.

Cowell, R.G., Dawid, A.P., Hutchinson, T., and Spiegelhalter, D.J., A Bayesian expert system for the analysis of an adverse drug reaction, *Artif. Intell. Med.*, 3, 257–270, 1991.

Debray, R., *Transmitting Culture* (translated by E. Rauth), Columbia University Press, New York, 2000.

Dewey, J., *How We Think*, Reprinted in 2005 by Barnes & Nobles Books, New York, 1910.

EPA., *Stressor Identification Guidance Document*, EPA 822-B-00-025, December 2000, Office of Research and Development, Washington, D.C., 2000.

Gelman, A., Carlin, J.B., Stern, H.S., and Rubin, D.B., *Bayesian Data Analysis*, Chapman & Hall/CRC, Boca Raton, FL, 1995.

Gigerenzer, G. *Adaptive Thinking*, Oxford University Press, Oxford, UK, 2000.

Gigerenzer, G., *Calculated Risks*, Simon & Schuster, New York, 2002.

Hagstrom, W.O., *The Scientific Community*, Basic Books, New York, 1965.

Howson, C. and Urbach, P., *Scientific Reasoning. The Bayesian Approach*, Open Court, La Salle, IL, 1989.

Hutchinson, T.A. and Lane, D.A., Assesing methods for causality assessment of suspected adverse drug reactions, *J. Clin. Epidemiol.*, 42, 5–16, 1989.

Janis, I.L. and Mann, L., *Decision Making. A Psychological Analysis of Conflict, Choice, and Commitment*, The Free Press, New York, 1977.

Josephson, J.R. and Josephson, S.G., *Abductive Inference. Computation, Philosophy, Technology*, Cambridge University Press, Cambridge, UK, 1996.

Kuhn, T.S., *The Structure of Scientific Revolutions*, 2nd ed., The University of Chicago Press, Chicago, IL, 1970.

Kuhn, T.S., *The Essential Tension*, The University of Chicago Press, Chicago, IL, 1977.

Lane, D.A., Subjective probability and causality assessment, *Appl. Stochastic Models Data Analy.*, 5, 53–76, 1989.

Lane, D.A., Kramer, M.S., Hutchinson, T.A., Jones, J.K., and Naranjo, C., The causality assessment of adverse drug reactions using a Bayesian approach, *Pharmaceut. Med.*, 2, 265–283, 1987.

Locke, J., *An Essay Concerning Human Understanding*, Collated by A.C. Fraser and reprinted by Dover Publications, Inc., New York, 1690, reprinted 1959.

Loehle, C., Hypothesis testing in ecology: Psychological aspects and the importance of theory maturation, *Q. Rev. Biol.*, 31, 145–164, 1987.

McPherson, M., Smith-Lovin, L., and Cook, J.M., Birds of a feather: Homophily in social networks, *Annu. Rev. Sociol.*, 27, 415–444, 2001.

Neapolitan, R.E., *Learning Bayesian Networks*, Pearson Prentice Press, Upper Saddle River, NJ, 2004.

Newman, M.C. and Evans, D.A., Enhancing belief during causality assessments: Cognitive idols or Bayes's theorem? In *Coastal and Estuarine Risk Assessment*, Newman, M.C., Roberts, M.H., Jr., and Hale, R.C., (eds.), CRC Press, Boca Raton, FL, 2002, pp. 73–96.

Newman, M.C., Zhao, Y., and Carriger, J.F., Coastal and estuarine ecological risk assessment: The need for a more formal approach to stressor identification, *Hydrobiol.*, 577, 31–40, 2007.

Odum, E.P., Preface, In *Ecotoxicology. A Hierarchical Treatment*, Newman, M.C. and Jagoe, C.H. (eds.), CRC Press/Lewis Publishers, Boca Raton, FL, 1996.

Pearl, J., Causality, *Models, Reasoning and Inference*, Cambridge University Press, Cambridge, 2000.

Platt, J.R., Strong inference, *Science*, 146, 347–352, 1964.

Popper, K.R., *The Logic of Scientific Discovery*, Routledge, London, UK, 1959.

Posthuma, L., Suter, G.W., II, and Traas, T.P. (eds.), *Species Sensitivity Distributions in Ecotoxicology*, CRC Press/Lewis Publishers, Boca Raton, FL, 2002.

Richerson, P.J. and Boyd, R., *Not By Genes Alone*, University of Chicago Press, Chicago, IL, 2005.

Rogers, E.M., *Diffusion of Innovations*, 4th ed., The Free Press, New York, 1995.

Stow, C.A., Assessing the relationship between Pfiesteria and estuarine fish kills, *Ecosystems*, 2, 237–241, 1999.

Stow, C.A. and Borsuk, M.E., Enhancing casual assessment of estuarinc fish kills using graphical models, *Ecosystems*, 6, 11–19, 2003.

Vernimmen, P., Quiry, P., Le Fur, Y., Dallocchio, M., and Salvi, A., *Corporate Finance. Theory and Practice*, Wiley & Sons, New York, 2005.

Westhues, K., *Eliminating Professors. A Guide to the Dismissal Process*, Kempner Collegium Publications, Queenston, ON, 1998.

Westhues, K., *The Envy of Excellence. Administrative Mobbing of High-achieving Professors*, The Edwin Mellen Press, Lewiston, NY, 2005.

Woodworth, G.G., *Biostatistics. A Bayesian Introduction*, John Wiley & Sons, Hoboken, NJ, 2004.

Index

A

B

C

D

E

F

G

H

I

J

K

L

M

N

O

P

Q

R

S